SOLIDIFICATION
AND CASTING
OF
METALS

SOLIDIFICATION AND CASTING OF METALS

Proceedings of an international conference on solidification, organized jointly by the Sheffield Metallurgical and Engineering Association and the University of Sheffield, in association with The Metals Society. The meeting was held at the University of Sheffield on 18–21 July 1977.

Book 192
Published 1979 by
The Metals Society
1 Carlton House Terrace
London SW1Y 5DB

669·94SØL
24187
F 9 OCT 1979

ISBN 0 904357 16 3

Text set in 10/11 pt Times
Titles set in 28 pt Univers Light
and 20 pt Univers Bold

Printed and made in England by
J. W. Arrowsmith Ltd
London and Bristol

CONTENTS

SESSION 3: NUCLEATION AND GRAIN REFINEMENT

SESSION 4: SOLIDIFICATION AND QUALITY CONTROL—CONTINUOUS CASTING

SESSION 5: SOLIDIFICATION AND QUALITY CONTROL—MACROSEGREGATION, POROSITY, AND SIMULATION OF SOLIDIFICATION

SESSION 6: STRUCTURE AND PROPERTY RELATIONSHIPS AND WELDING

SESSION 7: NEW PROCESSES AND PRODUCTS

Preface

After an interval of 10 years since the last international conference on solidification, held in Brighton in 1967, it seemed appropriate that a second meeting should be held to assess the state of the art.

A number of research workers in the Sheffield area, both in industry and in the university, have been working on solidification for many years, and in view of their interests this meeting was jointly organized by the Sheffield Metallurgical and Engineering Association and Sheffield University in association with The Metals Society. Papers were openly invited from all workers in the field and the Technical Committee chose a cross section of those submitted for presentation at the conference. The programme, therefore, covered a wide range of topics and although we could not hope to cover the field completely, we believe that there was something of interest to all scientists and technologists working on solidification.

We are grateful to Dr H. Morrogh F.I.M., who organized the 1967 conference, for opening the 1977 conference and for chairing the first session.

B. B. Argent

Session 1

Growth morphology, solute distribution, and ripening

Chairman: H. Morrogh (Director, British Cast Iron Research Association)

Editor: J. Hunt (Department of Metallurgy and Science of Materials, University of Oxford)

Keynote Address: Cellular and primary dendrite spacings

J. D. Hunt

A crude order of magnitude analysis has been made to predict the relationship between primary dendrite spacings or cell spacings and growth rate and temperature gradient. It has been found that the derived relationship agrees reasonably well with the available experimental results.

The author is in the Department of Metallurgy and Science of Materials, University of Oxford

Many alloys and commercially pure materials freeze with a cellular or dendrite growth front, the rate of movement and direction of the front being determined by heat flow. Most of the detailed theoretical treatments of dendrite growth[1-7] have considered single isothermal (or isoconcentrate) dendrites growing into an originally infinite isothermal medium. These treatments cannot readily be applied to an array of dendrites or cells growing in a temperature gradient. In other more intuitive treatments[8,9] important aspects of the growth process have been neglected.

An attempt is made in this paper to extend the order of magnitude analysis of Burden and Hunt[10] to include cell and primary dendrite spacings. In the earlier work it was possible to describe reasonably satisfactorily the complex variation of tip temperature with velocity and temperature gradient of cellular or dendritic arrays.[11] The object of the present work is to put forward a simple expression which, it is hoped, adequately describes the growth process near the actual growth conditions rather than produce an accurate expression valid over a wide range of conditions.

As in the previous work it will be assumed that a dendrite or cell tip may be approximated by a smooth steady state shape even when dendrite arms have been formed. The analysis should thus be applicable to both cells and dendrites. The assumption is illustrated in Fig. 1. The problem to be solved or approximated is as follows: the diffusion equation (1) using coordinates moving with the tip should be solved for the steady state interface shape

$$D\nabla^2 C + V \frac{\partial C}{\partial x} = 0 \qquad (1)$$

where C is the composition in atom fraction, and V the velocity of the interface in the x direction (Fig. 1). The

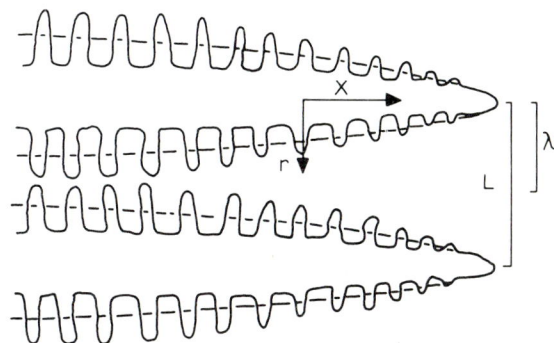

1 Schematic dendrites: the dotted line shows the smooth shape assumed in the present model; the figure also shows the definitions of x, r, λ and L

continuity of matter equation must be satisfied everywhere on the solid–liquid interface

$$V_n(C_I^L - C_I^S) + D \frac{\partial C}{\partial \bar{n}} = 0 \qquad (2)$$

where V_n is the velocity normal to the interface, the subscript I refers to the interface, the superscript L and S refer to liquid and solid, and $\partial C/\partial \bar{n}$ is the concentration gradient normal to the interface.

Similarly, a heat flow solution should be obtained satisfying the heat continuity equation. Finally, an equation which relates temperature composition and curvature must be satisfied on the solid–liquid interface. This equation may be written, neglecting kinetic undercooling,[10] as

$$\Delta T = \Delta T_D + \Delta T_\sigma \qquad (3)$$

3

where

$$\Delta T = T_0 - T_I$$

$$\Delta T_D = m(C_\infty - C_I)$$

$$\Delta T_\sigma = \theta(R_1^{-1} + R_2^{-1})$$

T_0 and C_∞ are the temperature and composition of the starting alloy at equilibrium on a flat solid–liquid interface; T_I, C_I are the values at a point on the growing interface; m is the liquidus slope (negative for $K < 1$). R_1 and R_2 are the principal radii of curvature of the solid. θ is the curvature undercooling constant[10] at constant liquid composition. Allowing for curvature in this way means that $(C_I^L - C_I^S)$ of equation (2) may be replaced (with negligible error) by the values obtained from the equilibrium phase diagram, i.e. $(C_I^L - C_I^S) = (1 - K)C_I^L$ where K is the equilibrium distribution coefficient.

The temperature of the interface T_I should be given by a solution to the heat flow equation taking into account heat flow as well as heat evolution. To simplify the problem, since heat flow is so rapid compared with solute flow, the temperature is assumed to vary only in the x direction and heat evolution is ignored.

Each cell or dendrite is assumed to have radial symmetry. The hexagonal prism (assuming an hexagonal array) is replaced by a cylinder containing the cell. Because lateral growth is slow in typical structures, well behind the tip the composition in the liquid is almost homogeneous in the r direction (the radial direction perpendicular to x, Fig. 1). It can be seen that if diffusion in the x direction is neglected the fraction liquid g_L varies with $\bar{C}$, the average liquid composition, as in the Scheil equation $\bar{C} = C_\infty(g_L)^{K-1}$. This composition $\bar{C} \simeq C_I$ must vary according to the undercooling equation (3). Since the average curvature undercooling well behind the tip is small from equation (3)

$$\Delta T = T_0 - T_I = m(C_\infty - C_I) \tag{4}$$

Thus, knowing the temperature distribution the interface shape can be calculated.

Diffusion in the x direction may be included, as was done by Bower, Brody and Flemings[12]. If it is assumed that $\bar{C}(x)$ is approximately equal to $C_I(x)$, considering Fig. 2 the volume of the element times the rate of change of composition must be equal to the difference

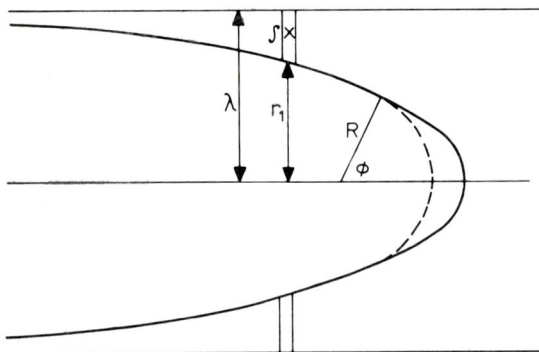

2   Schematic view of the hypothetical cell or dendrite growing with homogenous liquid in the *r* direction but some diffusion in the *x* direction: showing *R* and *ϕ*

in the material diffusing in and out of the element plus that rejected into the element due to freezing.

$$g_L \frac{\partial \bar{C}}{\partial t} = \frac{\partial[Dg_L(\partial \bar{C}/\partial x)]}{\partial x} + \bar{C}(K-1)\frac{\partial g_L}{\partial t} \tag{5}$$

where t is time. From equation (4)

$$\frac{dT_I}{dx} = G = m\frac{dC_I}{dx} = m\frac{\partial \bar{C}}{\partial x} \tag{6}$$

where G is the temperature gradient. Assuming that the temperature gradient is almost constant

$$\frac{\partial^2 \bar{C}}{\partial x^2} \to 0$$

and noting that at steady state

$$\frac{\partial g_L}{\partial x} = -\frac{1}{V}\frac{\partial g_L}{\partial t}$$

equation (5) becomes

$$g_L \frac{\partial \bar{C}}{\partial t} = -\frac{DG}{mV}\frac{\partial g_L}{\partial t} + \bar{C}(K-1)\frac{\partial g_L}{\partial t} \tag{7}$$

Integrating from $\bar{C} = C_T$ to C_I and $g_L = 1$ to g_L gives

$$g_L^{(K-1)} = \frac{C_I(1-K) + (DG/mV)}{C_T(1-K) + (DG/mV)} \tag{8}$$

Then relating g_L to r_1 and λ from Fig. 2

$$\pi r_1^2 = (1 - g_L)\pi\lambda^2$$

or

$$g_L = \left(1 - \frac{r_1^2}{\lambda^2}\right) \tag{9}$$

and writing

$$\frac{Gx}{m} = C_I - C_{x=0}$$

from equation (6) gives

$$\left(\frac{Gx}{m} + C_{x=0}\right)(1-K) + \frac{DG}{mV}$$

$$= \left[C_T(1-K) + \frac{DG}{mV}\right]\left(1 - \frac{r_1^2}{\lambda^2}\right)^{k-1} \tag{10}$$

This expression gives the hypothetical shape of a cell or dendrite growing with the liquid homogeneous in the r direction but where there is some diffusion in the x direction. The shape is shown schematically in Fig. 2. This shape describes reasonably accurately the shape of the cell or dendrite far from the tip in a real structure. It does not however describe the shape near the tip since the liquid is obviously not homogeneous in the r direction in this region. It is suggested that an approximation of the interface shape may be obtained by fitting part of a sphere to the derived shape, as is shown by the dotted line in Fig. 2. From equation (10)

$$\frac{dx}{dr_I} = \tan\phi = -\frac{m}{G}\left[C_T(1-K) + \frac{DG}{mV}\right]\frac{2r_1}{\lambda^2}\left(1 - \frac{r_1^2}{\lambda^2}\right)^{K-2} \tag{11}$$

Noting that near the tip $r_I^2/\lambda^2 \ll 1$ and that the tip radius $R = r_I/\sin\phi$ gives

$$R = -G\lambda^2 \Big/ \cos\phi\, 2m\left[C_T(1-K)+\frac{DG}{mV}\right] \quad (12)$$

letting $\phi = 45°$ gives

$$R = -G\lambda^2 \Big/ \sqrt{2}\left[mC_T(1-K)+\frac{DG}{V}\right] \quad (13)$$

The relationship is relatively insensitive to the particular value of ϕ assumed. A variation of only $\sqrt{2}$ is obtained if $\phi = 60°$ or $0°$.

It is now necessary to obtain an approximate diffusion solution to this interface shape. Proceeding as in the Burden and Hunt[10] analysis, the composition ahead of the dendrite tip is divided into two parts, $\bar{C}$ the composition in the liquid averaged across the interface and $\int C$ an additional radial part where $C = \bar{C} + \int C$. The average composition $\bar{C}$ decays exponentially away from the tip and may be obtained by solving

$$\frac{d^2\bar{C}}{dx^2}+\frac{V}{D}\frac{d\bar{C}}{dx}=0 \quad (14)$$

giving the composition gradient

$$\frac{d\bar{C}}{dx}=\frac{V}{D}(C_\infty - \bar{C}) \quad (15)$$

For a planar interface the average gradient is just sufficient to take away rejected solute. After breakdown an additional radial gradient $\partial\int C/\partial\bar{n}$ is necessary to take away solute from the tip. Using equation (3)

$$\Delta T = m(C_\infty - \bar{C}_T) - m\int C_T + \Delta T_\sigma \quad (16)$$

where the subscript T refers to the tip. Using equation (15)

$$\Delta T = \frac{mD}{V}\frac{d\bar{c}}{dx} - m\int C_T + \Delta T_\sigma \quad (17)$$

The problem is thus to determine approximate expressions for $\int C_T$ and the average gradient.

As was pointed out earlier, between the dendrites away from the tip there is sufficient time for the liquid to be almost homogeneous in the r direction dc_I/dx is approximately equal to $d\bar{c}/dx$ so that from equation (6)

$$G = m\frac{d\bar{C}}{dx} \quad (18)$$

The almost linear composition gradient must match the exponential variation in $\bar{C}$ ahead of the tip. It was suggested[10] that since the average gradient could not change rapidly with distance the average gradient immediately ahead of the tip would also be given by equation (7). The approximation is illustrated in Fig. 3. $\int C_T$ may be estimated using the approximation first proposed by Zener[13] for the growth of a sphere and used to describe dendrite like growth.[14] It is suggested that the radial gradient decays in a distance of about R (where R is the tip radius of curvature) or that

$$\left(\frac{\partial\int C}{\partial x}\right)_{x=x_T} = -\frac{\int C_T}{R} \quad (19)$$

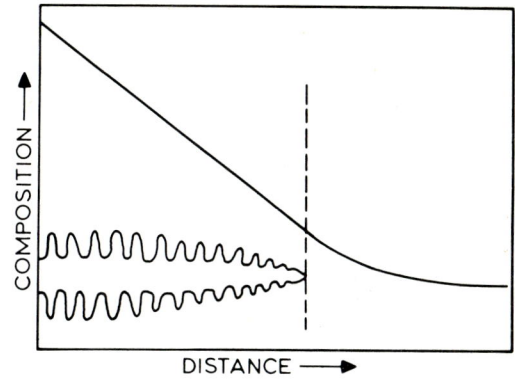

3 Showing the average liquid composition between the dendrites matching the exponential decay of the average composition ahead of the dendrites

By definition

$$\frac{\partial C}{\partial x}=\frac{d\bar{C}}{dx}+\frac{\partial\int C}{\partial x} \quad (20)$$

Substituting equations (18) and (19) and the continuity of matter equation

$$\left(\frac{\partial c}{\partial\chi}=\frac{V}{D}(1-K)C_T\right)$$

into equation (20) gives

$$\int C_T = \left(\frac{V}{D}(1-K)C_T+\frac{G}{m}\right)R \quad (21)$$

Thus equation (17) becomes

$$\Delta T = \frac{GD}{V}-\frac{mVR}{D}(1-K)C_T - GR + \Delta T_\sigma \quad (22)$$

Putting $\Delta T_\sigma = 2\theta/R$ and substituting λ for R using equation (13) gives

$$\Delta T = \frac{GD}{V}+\frac{G\lambda^2 V}{\sqrt{2}D}-\frac{2\theta\sqrt{2}}{G\lambda^2}\left[m(1-K)C_T+\frac{GD}{V}\right] \quad (23)$$

Substituting $m(1-K)C_\infty - (1-K)\Delta T_D = m(1-K)C_T$ from the definition of ΔT_D (equation 3) and noting that from equation (23) $\Delta T_D = GD/V + G\lambda^2 V/\sqrt{2}D$ gives

$$\Delta T = \frac{GD}{V}+\frac{G\lambda^2 V}{\sqrt{2}D}+\frac{2\theta(1-K)V}{D}$$
$$-\frac{2\theta\sqrt{2}}{G\lambda^2}\left[m(1-K)C_\infty+\frac{KGD}{V}\right] \quad (24)$$

Some other condition must now be specified. As in the previous analysis[10] it is assumed that the cells or dendrites grow at the minimum undercooling for a given velocity (this is equivalent to the maximum growth rate at a given temperature). This assumption is difficult to justify but there is no other obvious alternative. (The assumption has been shown to hold reasonably well with eutectic growth.[15,16]) Differentiating with respect to λ and equating to zero gives the final result

$$VG^2\lambda^4 = -4\theta D[m(1-K)C_\infty+KGDV^{-1}] \quad (25a)$$

Most measurements of spacing are taken from the centre of one cell or dendrite to the next if this is L

$$VG^2L^4 = -64\theta D[m(1-K)C_\infty + KGDV^{-1}]$$

$$L^4 = -64\theta D[m(1-K)C_\infty + KGDV^{-1}]G^{-2}V^{-1}$$

$$= 64\theta D\beta \qquad (25b)$$

where $-G^{-2}V^{-1}[m(1-K)C_\infty + KGDV^{-1}] = \beta$

It should be noted that at the critical condition when consistutional undercooling is just present on a planar interface the bracket $[m(1-K)C_\infty + KGDV^{-1}]$ is equal to zero. When the critical condition is exceeded by increasing V or C_∞ or decreasing G, $m(1-K)C_\infty$ rapidly becomes the dominant term giving for most of the growth range

$$VG^2L^4 = -64\theta Dm(1-K)C_\infty \qquad (26)$$

or $L \propto G^{-\frac{1}{2}}$, $L \propto V^{-\frac{1}{4}}$.

A typical plot of L against V is shown in Fig. 4. The expression for tip temperature is given as before by equation (18) of reference 10.

COMPARISON WITH EXPERIMENT
There has been considerable experimental work on the measurement of cell and primary dendrite spacings.[8,9,17-31] The work has often been presented as empirical plots of spacing against some growth parameter. Because of the experimental scatter and the very small ranges of change in velocity and temperature gradient other empirical relationships can and have been shown to be equally valid. Much of the experimental work considers either cell or primary dendrite spacings. It is suggested that the derivation should apply to both structures. A considerable amount of the work on primary dendrite spacings has been correlated with cooling rate[8,27,28] $(\partial T/\partial t)$ and with freezing rate[9,29,30] $(\partial g_s/\partial t)$. Velocities and temperature gradients have often not been measured or reported. In other work the results have been published[20,27] in combined forms of the gradient and velocity without the original data.

Qualitatively the derived expression agrees with the experimental findings. The spacing decreases with increasing temperature gradient and velocity and decreasing composition. As expected from the expression the spacing is more sensitive to changes in

temperature gradient than to changes in composition and velocity.[21,22]

A number of observations need to be considered before the experimental data are examined in detail. Sharp and Hellawell[21] have recently shown that the cell spacing is different if measured at various distances behind the interface, i.e. cells coarsen behind the interface. They found that the final spacing had a different dependence on velocity to that at the tips. The coarsening is most rapid in alloys with low solute concentrations and becomes negligible when there is enough solute to form dendrites.[24] Thus data for cells from specimens which have either been decanted or quenched will be more reliable than those from fully grown specimens. A number of early workers[17,18] on cell formation pointed out that the cell size varied with crystal orientation. The considerable scatter in much of the work may well be a result of variations in orientation. Recently it has been shown that convective fluid flow has an effect on spacing[31]. It seems probable that large discrepancies between specimens grown horizontally and vertically are a result of this phenomenon. Since convective flow is not considered in the present work results of specimens grown vertically upwards rejecting a slightly heavier solute (or slightly lighter solute grown downwards) would be expected to give more reliable data.

CELL GROWTH
Rutter and Chalmers[17] measured the spacing of cells on impure Sn. They decanted their specimens and only used grains of a specific orientation. The range of velocity was very small and the temperature gradient was not recorded. Their results are replotted in Fig. 5a as a plot of $\log L \rightarrow -\log V$. The filled line has a slope of 0·25 as expected by the present model.

Tiller and Rutter[18] carried out similar experiments on Pb 0·25 wt-% Sn. They selected grains and used a decanted interface. Their results are plotted on Fig. 6a as $\log L \rightarrow \log \beta$ where

$$\beta = -\frac{1}{G^2V}\left[m(1-K)C_\infty + \frac{kGD}{V}\right]$$

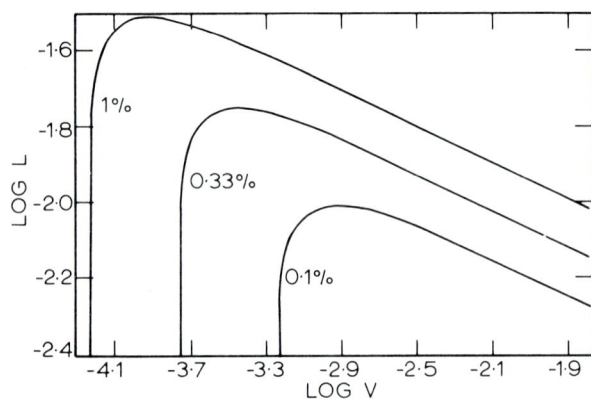

4   A typical plot of log L → log V for three different compositions at a constant temperature gradient

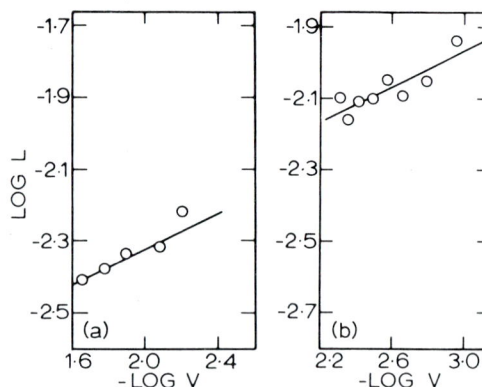

5   *a* A plot of log L → −log V from the results of Rutter and Chalmers[17] on Sn: the line shows the expected slope of 0·25; *b* a plot of log L → −log V from the results of Sharp[22] on Al 0·2 wt-% Cu: the line shows the expected slope (L is expressed in cm, V in cm/s)

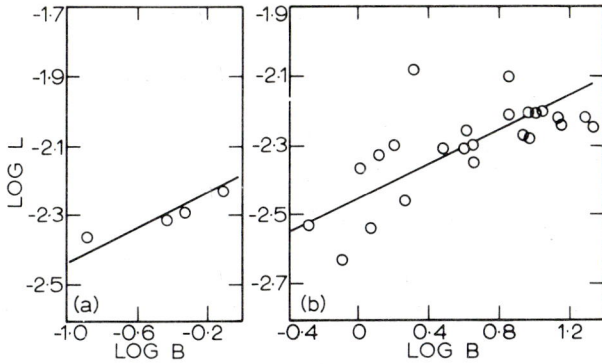

6 *a* A plot of log $L \rightarrow$ log β from the results of Tiller and Rutter[18] Pb 0·25 wt-% Sn: the line shows the expected slope; *b* A plot of log $L \rightarrow$ log β from the results of Plaskett and Winegard[19] on Sn–Bi alloys (*L* is expressed in cm, β in cm/s K^{-1})

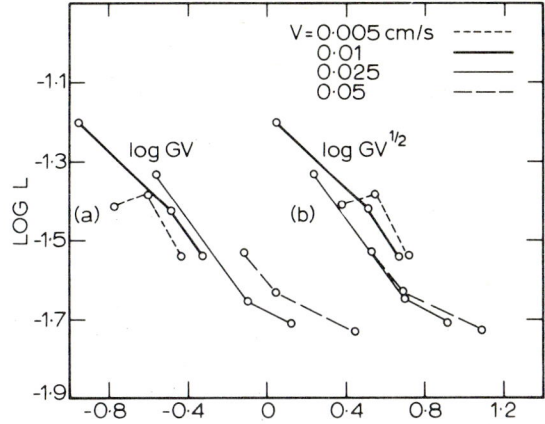

7 *a* A plot of log $L \rightarrow$ log GV; *b* A plot of log $L \rightarrow$ log $GV^{1/2}$ from the results of Young and Kirkwood[24] for Al 10·1 wt-% Cu; specimens grown at the same velocity have been joined by lines (*L* is expressed in cm, β in cm/s K^{-1})

assuming $m = 162$ K per atom fraction, $k = 0.64$, and $D = 10^{-5}$ cm^2/s. The expected slope is shown by the line; again there is reasonable agreement. Plaskett and Winegard[19] measured cell spacings in Sn–Pb, Sn–Bi, and Sn–Sb alloys using decanted specimens. The results for Sn–Bi are shown in Fig. 6*b* as a plot of log $L \rightarrow$ log β, assuming $m = 216$ K per atom fraction, $k = 0.2$, and $D = 10^{-5}$ cm^2/s. There is some agreement with the expected slope given by the line. There is much more scatter in this figure; the scatter, however, appears to be present in the original results, specimens grown under similar conditions have very different spacings. The scatter probably arises because of different orientations of different grains and appears to be present however the results are plotted. Similar results are obtained for Sn–Pb and Sn–Sb systems.

More recently Jesse and Giller[23] have investigated the cell spacing of Cd in Zn and Zn in Cd for a fixed temperature gradient and crystal orientation. From their Fig. 2 it appears that the slope of log $\lambda \rightarrow$ log V is about 0·41 and 0·44, respectively, for Cd in Zn and Zn in Cd. It is possible that the discrepancy is a result of the coarsening mentioned earlier, but it is more likely that it is a result of a different geometry. The cells considered in the analysis and those discussed earlier grow as a hexagonal array. Both Zn and Cd cells grow as plates or elongated cells in which the basal plane contains the cell boundary. The analysis can readily be modified for this geometry. Instead of equation (9), g_L is given by

$$g_L = \left(1 - \frac{r}{\lambda}\right) \quad (27)$$

Proceeding as before gives

$$R = \frac{G\lambda\sqrt{2}}{m[C_T(1-K)+(DG/mV)]} \quad (28)$$

and

$$VG^2\lambda^2 = -D\theta\left(m(1-K)C_\infty + \frac{KGD}{V}\right) \quad (29)$$

or $\lambda \propto V^{-\frac{1}{2}}$. There is less justification for this expression since r/λ may not be small enough to neglect in respect to one and the Zener approximation is less good for a cylinder. Nonetheless, it can be seen that a larger slope is to be expected. This effect may be part of the reason

for the variation in spacing with orientation, since some orientations tend to give platelike cells.

As was mentioned earlier Sharp and Hellawell[21] found that the final spacing of cells was different to that present at the tips during growth. They found that $L \propto V^{-1}$ for the final spacing but that at the tips the variation in spacing was within their experimental scatter for the small range of velocities examined. Plotting their results[22] as log $L \rightarrow$ log $1/V$, Fig. 5*b* shows that their results are in reasonable agreement with the present relationship.

DENDRITIC GROWTH

Perhaps the most extensive work on dendrites in which composition velocity and temperature gradient are reported is that of Young and Kirkwood[24] on Al–Cu alloys. These workers suggested that their results could be expressed as $\lambda \propto (GV)^{-\frac{1}{2}}$ instead of $\lambda \propto G^{-\frac{1}{2}}V^{-\frac{1}{4}}$ predicted in the present work. They examined the dependence of their results on velocity and although they found that the dependence was less than a half they felt that their results approximated $\lambda \propto (GV)^{-\frac{1}{2}}$. Because of the experimental scatter and the small differences between possible relationships it is very difficult to determine the precise dependence. Their

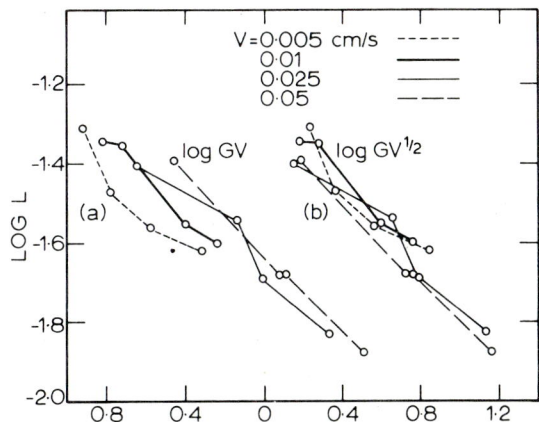

8 As for Fig. 7, but with alloy composition Al 4·4 wt-% Cu

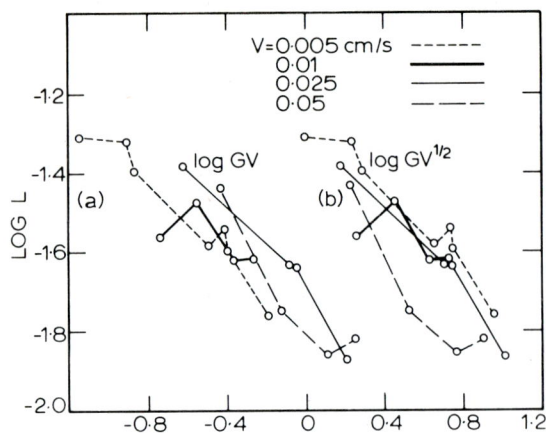

9   As for Figs 7 and 8, but with alloy composition Al 2·4 wt-% Cu

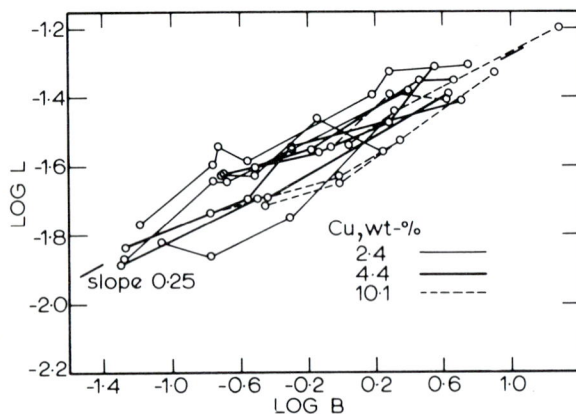

11   A plot of log $L \rightarrow$ log β for cell and dendrites from the results of Sharp[22] and Sharp and Hellawell[21] on the Al–Cu system; the filled line indicates a slope of 0·25, the dotted line is the line used to indicate the slope of 0·25 on Fig. 10 (L is expressed in cm and β in cm/s K^{-1})

results are replotted in Figs 7, 8, and 9. Figures 7a, 8a, and 9a show a plot of log $L \rightarrow$ log(GV) and Figs 7b, 8b, and 9b show a plot of log $L \rightarrow$ log $GV^{\frac{1}{2}}$ for the Al 10·1, 4·4, and 2·4 wt-% Cu alloys, respectively. Specimens grown at the same velocity have been joined by lines. A systematic displacement of high-velocity specimens to the left or right indicates that the wrong relationship has been used. On this criterion it appears there is a slightly better fit obtained for Figs 7b, 8b, and 9b, i.e. $\lambda \propto G^{-\frac{1}{2}} V^{-\frac{1}{4}}$. These results indicate how difficult it is to distinguish between $\lambda \propto (GV)^{-\frac{1}{2}}$ and $\lambda \propto G^{-\frac{1}{2}} V^{-\frac{1}{4}}$. Figure 10 shows a combined plot of log $L \rightarrow$ log β for the Al 10·1, 4·4, and 2·4 wt-% Cu alloys assuming $m = 647·4$ K per atom fraction, $k = 0·144$, $D = 3 \times 10^{-5}$ cm^2/s. Again specimens of the same composition grown at the same rate have been joined with lines. There appears to be no systematic shift with composition, and a good fit is obtained with the expected slope of 0·25. Figure 11 shows Sharp,[22] and Sharp and Hellawell's[21] results for the Al–Cu system. Three of the points do not fit at all well; apart from these the scatter is large but not inconsistent with the known experimental scatter due to orientation. It is of interest to note that the result for cells plotted in Fig. 5b have been included in this figure and that they fall near the line determined by the dendrite results. The filled line shows the expected slope.

Other work on dendrite spacings includes that of Suzuki et al.[26] Using vertical semi-steady state solidification they suggested that the primary spacing of two steels Fe–25 Cr–20 Ni and Fe–1 Cr–0·25 Mo gave a relationship of $\lambda \propto G^{-0·4} V^{-0·2}$ in good agreement with the present predictions. Their results can be replotted to give a good fit to a $\lambda \propto G^{-0·5} V^{-0·25}$ relationship.

The work of Bell and Winegard[25] on vertically grown Sn–Pb alloys is shown in Fig. 12 (assuming $m = 293$ K per atom fraction, $k = 0·087$, and $D = 10^{-5}$ cm^2/s). The points are reasonably well grouped about a line having the expected slope of 0·25. Points from Plaskett and Winegard[19] on cells have not been included on Fig. 12, because the two sets of results do not agree with one another. Specimens grown with similar velocity, gradient, and composition have very different spacings. The discrepancy is probably the result of the different fluid flow in the two sets of experiments. Vertical growth was used by Bell and Winegard whereas Plaskett and Winegard used horizontal growth.

A number of workers[8,27,28] have correlated primary dendrite spacings with cooling rate $(\partial T/\partial t)$ and found

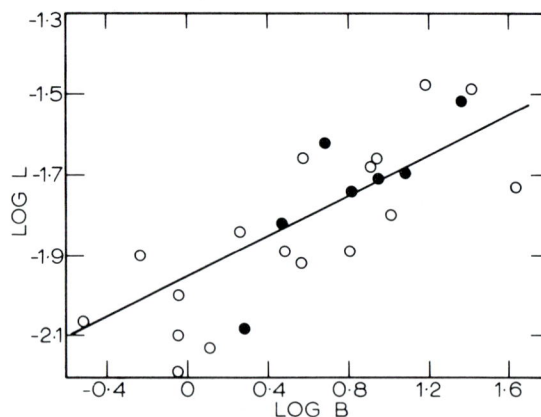

10   A combined plot of log $L \rightarrow$ log β for the specimens shown on Figs 7, 8, and 9; the expected slope of 0·25 is indicated (L is expressed in cm, and β in cm/s K^{-1})

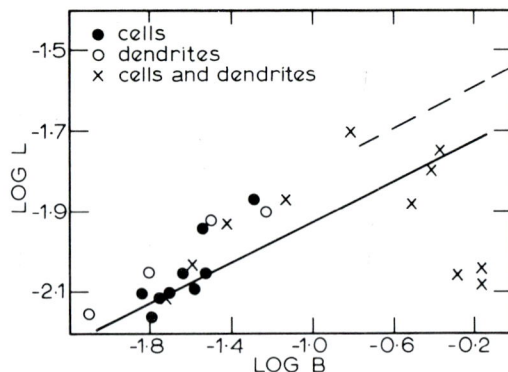

12   A plot of log $L \rightarrow$ log β for Sn–Pb alloys from the results of Bell and Winegard[25]; the filled circles are Sn 1·132 at-% Pb and the open circles Sn 0·286 at-% Pb (L is expressed in cm and β in cm/s K^{-1})

$\lambda \propto (\partial T/\partial t)^{-\frac{1}{2}}$. At steady state $(\partial T/\partial t) = GV$. As is shown by Young and Kirkwood's results it is very difficult to distinguish between $\lambda \propto (GV)^{-\frac{1}{2}}$ and $\lambda \propto G^{-\frac{1}{2}}V^{-\frac{1}{2}}$ so that these results are in qualitative agreement and may well be in agreement with the predictions made in the present work. Similarly, the prediction is in qualitative agreement[9,29,30] with the results which have been correlated with $(\partial g_s/\partial t)$.

Comparison of the analysis with experiment has been made by considering the slope of $\log L$ against $\log \beta$. The remaining surface energy constant $4\theta D$ in equation (26) calculated assuming $\sigma_{Sn} = 60$ erg cm^{-2}, $\sigma_{Al} = 121$ erg cm^{-2}, $\Delta S_{Sn} = 8\cdot 8 \times 10^6$ erg cm^{-3} K^{-1}, $\Delta S_{Al} = 1\cdot 16 \times 10^7$ erg cm^{-3} K^{-1}, $D = 10^{-5}$ and $D_{Al} = 3 \times 10^{-5}$ gives $4\theta D = 2\cdot 7 \times 10^{-10}$ K cm^3/s for Sn and 5×10^{-8} K cm^3/s for Al. These values should be compared with $4\theta D = 1\cdot 9 \times 10^{-10}$ K cm^3/s for Sn from Fig. 12 and $1\cdot 25 \times 10^{-9}$ K cm^3/s for Al from Fig. 10. The agreement of the Al is not good, but it should be noted that all the numerical errors made in the order of magnitude assumptions become grouped together with uncertainties in the surface energies and entropies of melting. It is not sufficiently wrong to invalidate (or validate) the minimum undercooling assumption. This must await a more accurate analysis.

CONCLUSION

An order of magnitude expression has been derived for the variation of cell and primary dendrite spacing with growth rate and temperature gradient; over most of the growth range it is predicted that $\lambda^4 G^2 VC_\infty^{-1} =$ constant. The expression has been shown to be consistent with the available experimental data. However, more detailed experimental work is necessary to make a definite comparison. The analysis has been shown previously to satisfactorily explain the variation in tip temperature with velocity and temperature gradient.

Neglecting convective fluid flow, perhaps the four most serious approximations or assumptions made in the analysis are

(i) the approximations used to relate λ to R
(ii) the approximation used for the average concentration gradient at the tip
(iii) the assumption that the radial concentration gradient decays in the order of R
(iv) the assumption that the structure grows at the minimum undercooling.

Any one of these could lead to large discrepancies. The first three can be validated or invalidated by more detailed analysis, the last must await a full perturbation analysis of the problem.

REFERENCES
1 G. P. IVANTZOV: *Dokl. Akad. Nauk SSSR*, 1947, **58**, 567
2 G. HORVAY AND J. W. CAHN: *Acta Metall.*, 1961, **9**, 695
3 G. F. BOLLING AND W. A. TILLER: *J. Appl. Phys.*, 1961, **32**, 2 587
4 D. E. TEMKIN: *Dokl. Akad. Nauk SSSR*, 1960, **132**, 1 307
5 G. R. KOTLER AND L. A. TARSHIS: *J. Cryst. Growth*, 1969, **5**, 90
6 R. TRIVEDI: *Acta Metall.*, 1970, **18**, 287
7 G. E. NASH AND M. E. GLICKSMAN: *Acta Metall.*, 1974, **22**, 1 283
8 T. OKAMOTO AND K. KISHITAKE: *J. Cryst. Growth*, 1975, **29**, 137
9 P. K. ROHATGI AND C. M. ADAMS: *Trans. AIME*, 1967, **239**, 1 737
10 M. H. BURDEN AND J. D. HUNT: *J. Cryst. Growth*, 1974, **22**, 109
11 M. H. BURDEN AND J. D. HUNT: *J. Cryst. Growth*, 1974, **22**, 99
12 T. F. BOWER *et al.*: *Trans AIME*, 1966, **236**, 624
13 C. ZENER: *Trans TIM–AIME*, 1946, **167**, 550
14 C. ZENER: *J. Appl. Phys.*, 1949, **20**, 950
15 R. M. JORDAN AND J. D. HUNT: *Metall. Trans.*, 1972, **3**, 1 385
16 M. TASSA AND J. D. HUNT: *J. Cryst. Growth*, 1976, **34**, 38
17 J. W. RUTTER AND B. CHALMERS: *Can. J. Phys.*, 1953, **31**, 15
18 W. A. TILLER AND J. W. RUTTER: *Can. J. Phys.*, 1956, **34**, 96
19 T. S. PLASKETT AND W. C. WINEGARD: *Can. J. Phys.*, 1960, **38**, 1 077
20 J. O. COULTHARD AND R. ELLIOTT: *J. Inst. Met.*, 1967, **95**, 21
21 R. M. SHARP AND A. HELLAWELL: *J. Cryst. Growth*, 1971, **11**, 77
22 R. M. SHARP: D.Phil. Thesis, Oxford University, 1971
23 R. E. JESSE AND H. F. J. I. GILLER: *J. Cryst. Growth*, 1970, **7**, 348
24 K. P. YOUNG AND D. H. KIRKWOOD: *Metall. Trans.*, 1975, **6A**, 197
25 J. A. E. BELL AND W. C. WINEGARD: *J. Inst. Met.*, 1963, **92**, 357
26 A. SUZUKI AND Y. NAGOAKA: *J. Jpn Inst. Met.*, 1969, **33**, 658
27 M. C. FLEMINGS *et al.*: *J. Iron Steel Inst.*, 1970, **208**, 371
28 T. OKAMOTO *et al.*: *J. Cryst. Growth*, 1975, **29**, 131
29 P. K. ROHATGI AND C. M. ADAMS: *Trans AIME*, 1967, **239**, 850
30 P. K. ROHATGI AND C. M. ADAMS: *Trans AIME*, 1967, **239**, 1 729
31 M. H. BURDEN AND J. HUNT: *Met. Sci.*, 1976, **10**, 156

Morphological instability at a grain-boundary groove

J. D. Ayers and R. J. Schaefer

A quantitative study was made of the development of morphological instabilities at grain-boundary grooves in high-purity succinonitrile, a transparent organic material which behaves like a metal during solidification. Succinonitrile was selected for this study because it is the only material with metal-like solidification behaviour for which the solid/liquid interfacial free energy, γ, is known with enough confidence to permit an unambiguous test of the theories of morphological instability. This study showed that a parametric fit of the data to the theory yields a γ of only two-thirds the correct value. This discrepancy between theory and experimental findings is probably attributable to the fact that the linearized theory does not adequately treat the small contact angle observed in the experiments.

The authors are with the Material Sciences Division, Naval Research Laboratory, Washington DC, USA

There have been numerous theoretical treatments of the development of crystallite morphology during solidification, most of which derive from the analysis of the morphological stability of a growing sphere carried out by Mullins and Sekerka.[1] In this study, interest was focused on two versions of the theory; one of these describes the breakdown of an initially smooth cylinder as developed by Hardy and Coriell,[2] while the other, proposed by Coriell and Sekerka,[3] describes the development of morphological instabilities at a grain-boundary groove. Of the various forms in which the theory has been developed, these two are probably the best suited to experimental verification. The first form was tested by Hardy and Coriell who made extensive growth-kinetics measurements on the periodic perturbations which develop on the surface of an initially smooth cylinder of ice growing in supercooled water. These measurements yielded a value for the solid/liquid interfacial free energy γ of about $25 \, \mathrm{MJ/m^2}$, in reasonably good agreement with the accurate value of $29 \cdot 1 \pm 0 \cdot 8 \, \mathrm{MJ/m^2}$ recently determined by Hardy.[4] As discussed by Sekerka,[5] other tests of morphological stability theory have been quite approximate in nature. The morphological development of growing single crystal cylinders has not been studied in any material other than ice, and the theoretical treatment of morphological stability at a grain-boundary groove has not previously been subjected to quantitative testing; however, morphological observations were reported by Schaefer and Glicksman[6] even before presentation of the theory by Coriell

and Sekerka.[3] This lack of an experimental test of the theory is partly attributable to the fact that such a test requires the value for the solid/liquid interfacial free energy to be known with a high degree of confidence. This requirement was satisfied when Schaefer et al.[7] made accurate measurements of γ for succinonitrile (also known as ethylene dicyanide). This study of interfacial free energy and the experimental study of dendritic growth in succinonitrile by Glicksman et al.[8] also generated purification techniques, and accurate values for the thermal conductivity and melting point. This information helped to minimize experimental uncertainties in the determination of thermal gradients.

EXPERIMENTAL

Succinonitrile of high purity was prepared in a unified purification–crystal-growth apparatus similar to that described previously.[7] Sample purity, assessed by the method described in Ref. 8, was found to be greater than 99·99%.

The crystal-growth chamber of the apparatus used in this study consisted of a glass cylinder 15 cm long with an i.d. of 1·3 cm. It contained an axial, fine bore (0·018 cm i.d., 0·030 cm o.d.) nickel cooling tube, through which chilled methyl alcohol was forced by pressurized gas. Crystal-growth experiments were conducted with the growth chamber immersed in a thermostatic bath which had a temperature stability of $\pm 0 \cdot 001 °C$ over 1 h. The growing crystals were observed

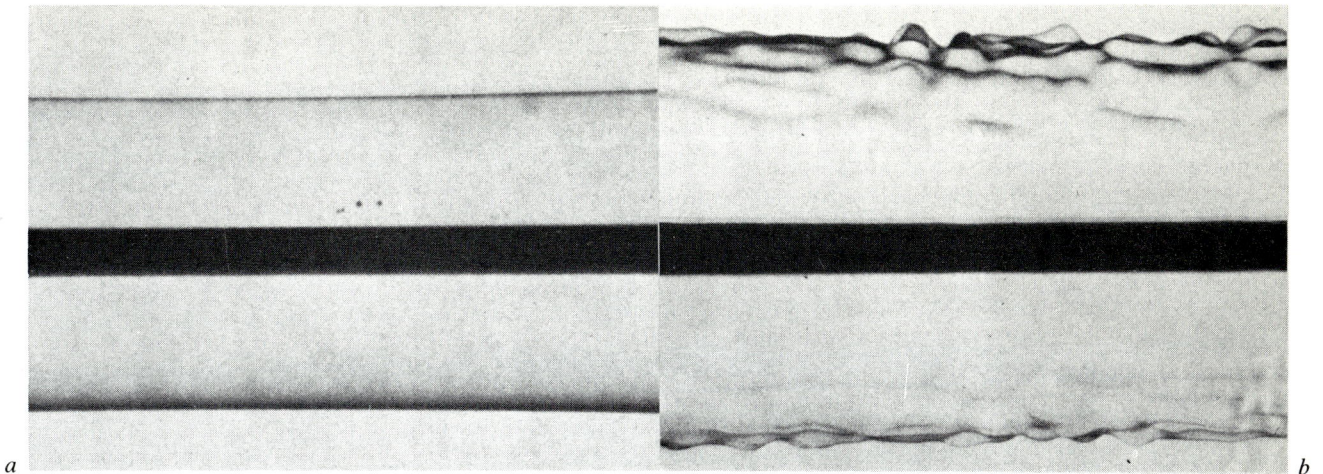

a cylinder immediately after thermostatic bath temperature was lowered to 0·2°C below melting point; *b* cylinder 400 s later

1   Development of morphological instabilities on initially smooth cylinder of succinonitrile with 0·03 cm diameter axial cooling tube   [×21]

and photographed with a long focal-length microscope through a double-glazed window in the thermostat.

Crystals of the purified succinonitrile were initially prepared by passing cooled methyl alcohol through the axial cooling tube while maintaining the thermostat at a temperature a few millidegrees above the melting point. It was possible to propagate single crystals along the tube by gradually increasing the rate of coolant flow. Bamboo-type crystals were formed by grain growth from a rapidly solidified polycrystalline cylinder. Crystals to be studied during growth were melted back to a small diameter (typically 0·1–0·15 cm) by elevating the thermostat temperature to one or two degrees above the melting point, while maintaining the coolant flow through the axial tube. Radial growth for the morphology studies was initiated by quickly cooling the thermostat (with the aid of a copper cooling coil) to a preselected temperature below the melting point. This change in temperature could be accomplished in about 1 min. The axial coolant flow was continued, at a reduced flowrate, until the thermostat was stabilized at the growth temperature. At these temperatures below the melting point, short auxiliary heaters wound on the ends of the growth chamber prevented growth of the crystals around the ends and along the inner surface of the chamber.

GROWTH OF A SMOOTH CYLINDER

One of the initial objectives of this investigation was to test, on a material closely analogous to metal, the theory which describes the time-dependent growth of perturbations on an initially smooth, cylindrical single crystal. This objective was not achieved, because, under all experimental conditions, the perturbations which developed were spatially aperiodic. This fact is evident in Fig. 1, which shows an initially smooth-surfaced crystal, and a different area of the same crystal after it had been allowed to grow for 400 s at a bath undercooling of 0·2°C. Perturbations such as those shown were not suitable for quantitative comparison with theory because the theory describes only periodic perturbations of the type which develop on ice cylinders.[2]

When a crystal such as that shown in Fig. 1 was maintained long enough below the melting point the larger bumps evolved into dendrites, as is evidenced in Fig. 2. Because these dendrites grow in ⟨100⟩ directions of the bcc crystals, the orientation of individual crystals could be conveniently determined. Even when crystals with a ⟨100⟩ direction nearly parallel to the central cooling tube were melted back to less than 0·1 cm dia. and then allowed to grow at bath undercoolings of less than 0·05°C, they developed aperiodic perturbations. There was a tendency for portions of the crystal with the surface nearly normal to a ⟨100⟩ direction to break down more rapidly than other portions, but this tendency was not strong enough to produce the 'n'-fold fluting which promotes periodic structures on ice crystals.

A likely reason for the irregular type of interface breakdown observed is that imperfections within the crystals are effective sites for the formation of small

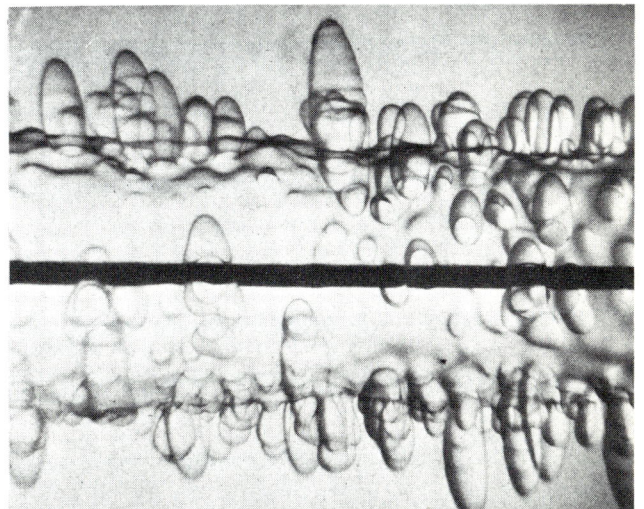

2   Larger bumps evident on crystal shown in Fig. 1*b* evolved with time into dendrites of type shown here: at later times well evolved side branches were evident   [× 10]

bumps. Barely visible defects such as low-angle grain boundaries were conspicuously effective sites for breakdown, and the crystals surely contained many other defects which could not be resolved by optical microscopy. In fact, repeated thermal cycling produced resolvable sub-boundaries where none were evident initially. Ice crystals of the type used by Hardy and Coriell[2] do not suffer from this problem because they are nearly free of dislocations.[9]

GROWTH NEAR A GRAIN BOUNDARY

It was possible to make good quantitative measurements on morphological instabilities at grain-boundary grooves, for they were quite regular in shape. Ten experimental runs were photographed with bath undercoolings ranging from 0·056 to 0·527°C. Figure 3, a photograph taken during a growth run at an undercooling of 0·33°C, shows how the shape evolves with time; note that the two maxima adjacent to the boundary increase in amplitude with time, and that the distance between them also increases. Using photographs like these from the ten experimental runs, time-dependent measurements were made of the cylinder diameter, the groove depth, the height of the maxima adjacent to the groove, and the distance between maxima. In addition, on certain photos, the *r* and *z* coordinates were measured at about 40 positions in order to permit a detailed comparison with the theory proposed by Coriell and Sekerka.[3]

The theory considers a grain-boundary groove at a solid/liquid interface, in a material having thermal conductivities k_S and k_L in the solid and liquid, respectively. The temperature gradients in the solid and liquid, beyond the perturbing influence of the grain-boundary groove, are G_S and G_L. The initial shape of the interface is assumed to have the form

$$w = \frac{s}{b}\exp(-b|x|) \qquad (1)$$

so that the slope of the interface at $x = 0$ is s, and the depth of the groove at this point is s/b. The morphological development of the grain-boundary groove is then calculated in terms of the parameters s and b of equation (1), and two additional parameters which, in the case $k_S = k_L$ (as is true for succinonitrile) are defined by

$$a^2 = \frac{|G_S + G_L|L_V}{2T_M\gamma} \qquad (2)$$

and

$$\tau = \frac{3^{\frac{3}{2}}}{2^{\frac{1}{2}}}\frac{L_V^{\frac{1}{2}}T_M^{\frac{1}{2}}\gamma^{\frac{1}{2}}}{k|G_S + G_L|^{\frac{3}{2}}} \qquad (3)$$

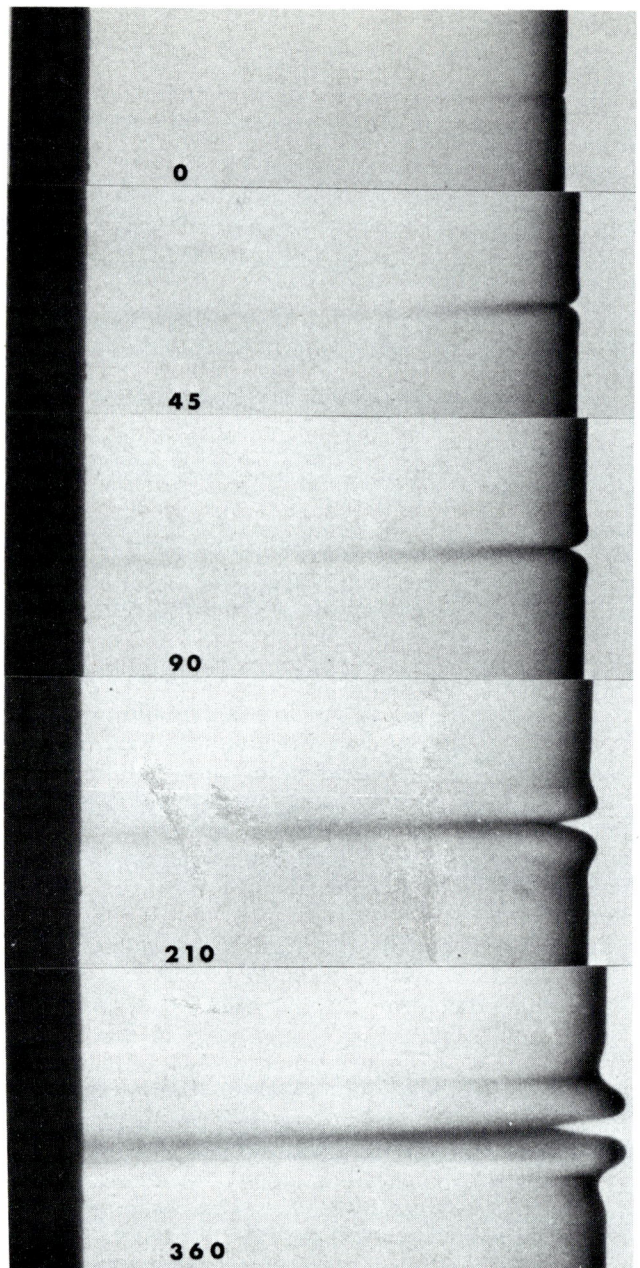

a growth time 0 s; *b* growth time 45 s; *c* growth time 90 s; *d* growth time 210 s; *e* growth time 360 s

3   Development with time of grain-boundary groove at bath undercooling of 0·33°C; 0·03 cm diameter tube is evident in these photos   [×31]

In several respects, the experiments did not exactly duplicate the conditions described in the theory proposed by Coriell and Sekerka. One of these differences lay in the temperature gradients G_S and G_L, which in the theory are assumed to be constants independent of time. In the experiments, it was necessary to start with a large value of G_S in order to make the initial groove depth small. Subsequently, G_S was allowed to drop to zero, thus producing the conditions favouring morphological instability. The effect of this transient behaviour of G_S could not have been large, however, because the characteristic time associated with the

decay of G_S was significantly smaller than the time interval over which the interface shapes were measured. The gradients also varied with time because the perturbed solid/liquid interface was a growing cylinder rather than a plane, but for this experiment the greatest change in the gradient caused by interface motion was only about 1%.

Another difference between the theory and experiment was the initial groove shape. It is assumed in the theory that the parameter *s* in equation (1) is ≪ 1, corresponding to the situation in which the dihedral angle at the grain-boundary groove is close to 180°, as when a very low-energy grain boundary intersects the solid/liquid interface. It was found, however, that low energy grain boundaries were relatively immobile, and that only high-energy boundaries migrated into the 'bamboo' geometry such that their shape could be accurately determined by profile views. The experiments were therefore carried out on grooves in which the dihedral angle at the bottom of the groove was 0° instead of being near 180°. It was not initially apparent to what extent this difference would affect the development of the interface shape at points not in the immediate vicinity of the grain boundary.

Finally, it is assumed in the theory that the slope of the interface is small at all times, whereas in the experiments the interface shapes were observed until the slopes became rather large—one objective of this experiment being to identify the approximate amplitudes at which non-linear effects start to invalidate a theory of the type proposed by Coriell and Sekerka.

In spite of these differences between the theoretical and experimental conditions, reasonable estimates can be made for the experimental values of the theoretical parameters a, b, s, and τ. Using the values of the latent heat, surface energy, thermal conductivity, and melting temperature as given in Refs. 7 and 8, and noting that $G_S = 0$ after the decay of the initial transient, it is found that $a = 285\ G_L^{\frac{1}{2}}$ and $\tau = 195\ G_L^{-\frac{3}{2}}$. The parameters s and b were evaluated by calculating the approximate temperature gradient in the solid when the chilled fluid was flowing through the central tube, and calculating the equilibrium groove depth[7] for this gradient. This procedure yielded a groove depth (s/b) of $\sim 10^{-3}$ cm.

The photographs of the groove confirm approximately this initial size, although this measurement was approaching the resolution limit of the optical system. The combination $s = 3$, $b = 3 \times 10^3$ gives an initial groove shape closely resembling the shape of an equilibrated groove with 0° contact angle. Because of the manner in which theory and experiment were compared, it would have made a negligible difference if the initial groove depth had been assumed to be different by a factor of two.

As mentioned above, the dimensions measured were the cylinder diameter, the groove depth, the height of the maxima, and the distance between the maxima, together with a complete profile in some cases. The measurements of the cylinder diameter were very useful in estimating the duration of the initial temperature transient. These measurements also provided a growth rate which made possible a check of the calculated gradients. As for the parameters which describe the shape of the groove, the groove depth cannot be measured with precision because it is difficult to locate the exact root of the groove. Moreover, because of the large difference between the experimental (0°) and theoretical ($\sim 180°$) values of the dihedral angle at the root of the groove, good agreement between the theoretical and experimental groove depth can hardly be expected. The heights of the maxima were also found to yield little significant information because the relative error was quite large until the shape was well evolved. The most pertinent single parameter was found to be the distance between the maxima. For a given value of the parameter a/b, this distance depends only on the normalized time t/τ, and the parameter a, both of which are known with considerable precision in

this experiment. Moreover, any reasonable value of b yields a value for a/b which is small in comparison with unity, in which case the exact value of a/b has an insignificant effect on the theoretical curves.

Since its inception, a major application of morphological stability theory has been for the determination of solid/liquid surface energies γ. Because the value of γ for succinonitrile is known with a high degree of confidence, a determination of its value from the present experiments constitutes a test of the validity of determining γ from this type of morphological stability measurement. In order to make such a determination of γ, note that the theory may be used to predict, for a given value of a/b, the normalized position ax_m of the interface maximum as a function of the normalized time t/τ. Assuming that γ is unknown, one can use the definitions of a and τ to write

$$a\gamma^{\frac{1}{2}}x_m = \frac{G_L L_V}{2T_M}x_m \tag{4}$$

and

$$\gamma^{\frac{1}{2}}\frac{t}{\tau} = \frac{2^{\frac{1}{2}}k|G_S + G_L|^{\frac{3}{2}}}{L_V^{\frac{1}{2}}T_M^{\frac{1}{2}}}t \tag{5}$$

All of the quantities on the right-hand sides of equations (4) and (5) are known or measured directly in the experiment, and one can therefore plot experimental values of $a\gamma^{\frac{1}{2}}x_m$ v. $\gamma^{\frac{1}{2}}t/\tau$, as shown in Fig. 4. It is found

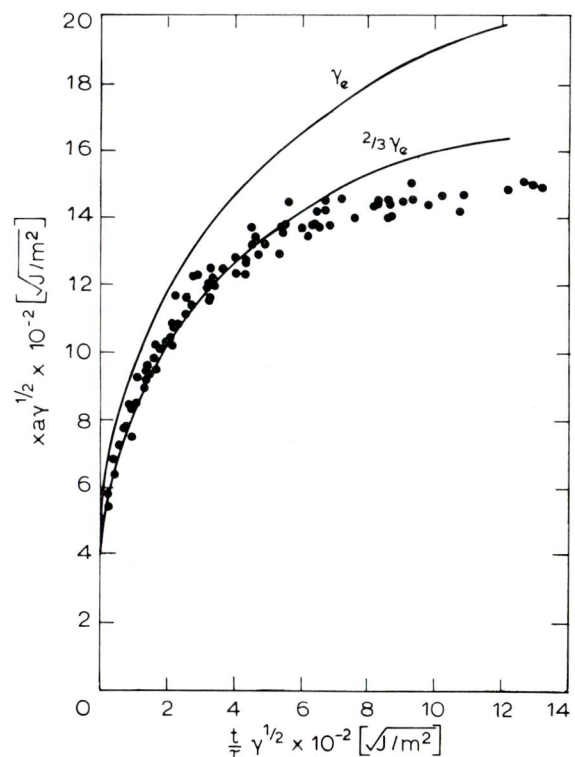

4 Plot in normalized dimensions of distance between grain-boundary groove maxima x as function of time t; coordinate normalizing constants a and τ are from theory[3]: data points represent measurements made on ten different experiments with undercoolings that ranged from 0·056 to 0·527°C; final groove picture shown in Fig. 3 falls at a normalized time of 1·32 × 10⁻³ (J/m²)¹/² in this figure

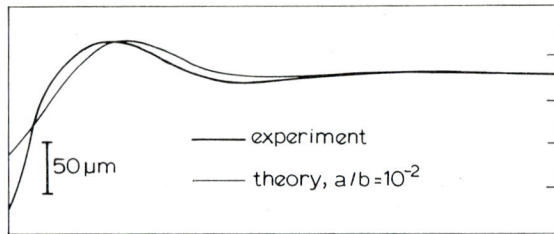

5 Comparison of theoretical and experimental solid/liquid interface profiles; experimental profile was obtained from Fig. 3e

from this master plot, which contains the data points from all ten runs (in which G_L varied from 0·204 to 2·10°C/cm), that the points lie on a single universal curve as would be expected from the theory. The curve predicted by the theory depends only on the value of γ in this representation. Two curves have been drawn, one corresponding to the γ_e value (8·94 MJ/m²) determined earlier,[7] and the other corresponding to two-thirds of γ_e. Good agreement at small times is obtained with the smaller γ value, which however lies well outside the range of uncertainty in the previous experiments. It is therefore seen that if the theory in its present form were used to derive γ values from moving solid/liquid interfaces containing high-angle grain boundaries, substantial errors would be expected.

While the theory does not describe the time evolution of the high-angle boundary groove shape accurately enough to permit a quantitative determination of γ, it does provide quite a good qualitative description of the groove profile, except in the near vicinity of the boundary. Figure 5 shows the type of differences observed when experimentally determined groove profiles were compared with theoretical profiles. In this figure, the two curves are normalized to have the same heights at the maxima. The difference in groove depths, which relate to the discrepancy between the experimental and theoretical contact angles, can be

seen quite clearly. The theoretical depth can, of course, be altered somewhat by varying the parameter a/b (in Fig. 5 a value of 0·01 was assumed), but this variation results in a less satisfactory fit over the remainder of the curves.

Perhaps the most important qualitative difference between theory and experiment is that the additional, more distant, maxima predicted by the theory were never observed experimentally. This difference may derive from a more rapid breakdown into dendrites of the primary maxima than would result with the lower theoretical slope in the vicinity of the groove. An alternative possibility is that a more complete nonlinear theory would not predict the secondary maxima. Only after additional theoretical work will it be possible to determine which explanation is correct.

ACKNOWLEDGMENT

The authors are grateful to S. R. Coriell for many useful discussions and for providing us with computed interface shapes for specific values of the experimental parameters.

REFERENCES

1 W. W. MULLINS AND R. F. SEKERKA: *J. Appl. Phys.*, 1963, **34**, 323
2 S. C. HARDY AND S. R. CORIELL: *J. Cryst. Growth*, 1968, **3–4**, 569
3 S. R. CORIELL AND R. F. SEKERKA: *ibid.*, 1973, **19**, 90
4 S. C. HARDY: *Philos. Mag.*, 1977, **35**, (2), 471
5 R. F. SEKERKA: 'Crystal growth: an introduction' (ed. P. Hartman), 403, 1973, Amsterdam, North-Holland Publishing Company
6 R. J. SCHAEFER AND M. E. GLICKSMAN: *Metall. Trans.*, 1970, **1**, 1 973
7 R. J. SCHAEFER et al.: *Philos. Mag.*, 1975, **32**, 725
8 M. E. GLICKSMAN et al.: *Metall. Trans.*, 1976, **7A**, 1 747
9 A. HIGASHI et al.: *J. Cryst. Growth*, 1968, **3–4**, 728

Dendrite arm spacings in hypoeutectic Pb–Sb alloys directionally solidified under steady and non-steady conditions

J. A. Spittle and D. M. Lloyd

Dendrite arm spacings have been measured in hypoeutectic Pb–Sb alloys directionally solidified under steady and non-steady conditions. When solidifying under steady conditions the temperature gradient in the liquid G_L was either $\sim 2°$ or $\sim 20°C/mm$. For steady growth conditions it was observed (a) that primary d_1 and secondary d_2 arm spacings decreased as G_L or the rate of advance of the primary tips R increased; (b) that with the low G_L, d_1, and d_2 spacings decreased as the initial alloy composition C_0 increased; and (c) with the high G_L, d_1, and d_2 spacings were independent of C_0. In the case of non-steady solidification d_1 and d_2 spacings decreased as the cooling rate $G_L R$ or C_0 increased. Logarithmic plots of arm spacings v. local solidification time t_s indicated, for the range of conditions investigated, that a reasonably linear relationship fits d_2 arm spacings but not d_1 spacings and for a given local solidification time, (a) the smallest d_1 spacings always occurred (except for short times) in alloys solidified under steady conditions with a high G_L; and (b) there was a tendency for d_2 spacings to also be smaller under the conditions given in (a).

J. A. Spittle is in the Department of Metallurgy and Materials Science, University College, Cardiff, and D. M. Lloyd is in the Department of Metallurgy, UMIST, Manchester

Many studies have been made of the 'as-solidified' microstructures of binary alloys in order to determine experimentally the interdependence of dendrite arm spacings and solidification parameters.[1-15] Data on arm spacings have been derived from alloys that have been cast,[1,2] melted, and then cooled *in situ* from the liquid state,[4,9] solidified from arc deposits,[8] and unidirectionally solidified under controlled[5,10,11,13,14] or uncontrolled[3,7,12,15] conditions. With the exception of the results obtained from controlled unidirectional growth studies, the remaining data relates to non-steady solidification. During unidirectional growth under steady conditions the significant controllable variables, G_L (the temperature gradient in the liquid ahead of the macroscopic solidification front) and R (the rate of advance of the primary dendrite tips) are maintained constant.

A number of factors have been correlated with arm spacings. However, in recent years, there has been a tendency to relate them either to the cooling rate[6,7,12,14,15] or, in the case of secondary (d_2) arm spacings, the local solidification time t_s.[13-16] The latter term is defined as

$$t_s = \frac{\Delta T_s}{\text{cooling rate}} \qquad (1)$$

where ΔT_s is the non-equilibrium temperature range of freezing. Under conditions of steady directional freezing, equation (1) can be rewritten as

$$t_s = \frac{\Delta T_s}{G_s R} \qquad (2)$$

where G_s is is the temperature gradient in the solid + liquid region parallel to the principal growth direction. The product $G_s R$ has the units of cooling rate. When calculating t_s, some investigators have used G_L instead of G_s.[14]

15

There is absolute agreement in the literature that, in columnar grains, the primary (d_1) arm spacings decrease as the cooling rate increases. Likewise, for both equiaxed and columnar grains the same inter-dependence of spacings and cooling rate has been noted for secondary arms. From studies carried out under conditions of steady directional solidification, it has usually been observed, for an alloy of given composition, that d_1 and d_2 spacings decrease as G_s, G_L, or R increase.

The above observations have led, over the past decade, to the establishment of relationships of the following general type

$$d_{(1 \text{ or } 2)} = \text{const.} \, (G_{(S \text{ or } L)}R)^{-n} \qquad (3)$$

where the exponent n has a value in the approximate range 0.25–0.5. It follows from equations (2) and (3) that if the alloy composition is constant

$$d_{(1 \text{ or } 2)} = \text{const.} \, (t_s)^n \qquad (4)$$

However, Sharp and Hellawell[10] found that d_1 arm spacing is almost independent of R, and Kotler et al.[11] reported that d_2 arm spacing is almost independent of G_L.

There has been considerable disagreement in the literature regarding the influence of the initial alloy composition (C_0) on arm spacings. Most of the results pertaining to d_2 arm spacings in hypoeutectic alloys indicate a decrease in spacing with increasing C_0. This complies with the trend that would be anticipated from equation (4), assuming that the equation is applicable if ΔT_s is a variable and that ΔT_s decreases with increasing solute content, i.e. $\Delta T_s = T_L - T_E$, where T_L and T_E are the equilibrium liquidus and eutectic temperatures, respectively. However, at least two studies have concluded that C_0 has hardly any measurable effect on d_2 spacing,[7,11] and one paper reports a direct relationship between d_2 arm spacing and C_0.[1] In addition, Dann et al.[13] have recently reported that d_2 arm spacings are proportional to the equilibrium liquidus–solidus temperature interval.

It has been claimed, in the majority of cases, that d_1 spacing increases as C_0 increases.[5,7,8,11,12,14,15] Badon-Clerc and Durand[12] also comment that this increase is significant only at high cooling rates. However, Sharp and Hellawell[10] found that C_0 has little effect on d_1 spacings and Dann et al.[13] reported that d_1 arm spacings, like d_2 arm spacings, are proportional to the equilibrium freezing range of the alloy.

The objective of this study was to examine further the dependence of arm spacings on solidification conditions, and in particular the significance of solute content, under steady and non-steady freezing conditions.

EXPERIMENTAL PROCEDURE

The investigation was carried out on hypoeutectic Pb–Sb alloys and the phase diagram for the system is shown in Fig. 1.

Steady growth

The influence of growth rate and alloy composition on arm spacings was examined at two different tempera-ture gradients, i.e. ~2° and 20°C/mm. For runs at the low temperature gradient, cylindrical specimens,

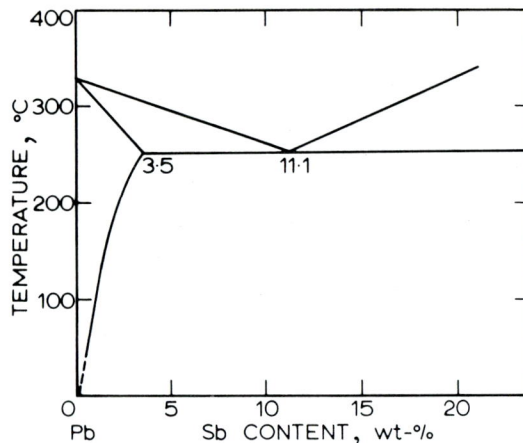

1 Pb–Sb equilibrium diagram: Pb-rich end

13 mm in diameter and 160 mm in length, contained in pyrex tubes, were lowered at known velocities through a stationary furnace, maintained at a fixed temperature, into a water-cooled copper tube. In runs at the high-temperature gradient, specimens, 7 mm in diameter and 160 mm in length, contained in silica tubes were lowered through a stationary furnace directly into water.

It was ascertained from trial runs that steady condi-tions existed over the length within which arm spacings would ultimately be determined, and also that the growth rate R corresponded to the lowering rate.

In each experimental run G_L was determined from a single thermocouple located at the specimen mid-height. On completion of solidification d_1, and where possible d_2, arm spacings were measured on transverse and longitudinal sections, respectively. The measure-ments were made at the mid-height position and each recorded spacing represented an average of about 40 measurements of individual nearest neighbour spacings.

Non-steady growth

Three hypoeutectic alloys, of different composition, were cast with constant superheat into a cylindrical graphite mould 25 mm in diameter and 150 mm in height, located in a furnace which was held at a temperature just above the liquidus of the alloy. The base of the mould was a copper block to which water cooling was applied immediately on completion of pouring. During directional freezing, temperatures were continuously monitored by four thermocouples, known distances apart, located at different heights in the alloy. The lower and upper thermocouples were 10 and 50 mm, respectively, from the copper base.

After each ingot had solidified a plot was obtained, from the temperature recordings, of the distance X that the macroscopic front had travelled from the base as a function of the square root of the time t. From the resulting linear relationship the proportionally constant K was determined:

$$X = K\sqrt{t} \qquad (5)$$

The ingot was then sectioned at the position of each of the thermocouple beads. The d_1 and d_2 arm spacings measured at these locations, as described above for steady growth, were plotted v. the product $G_L R$. R was determined at each position from the differential of

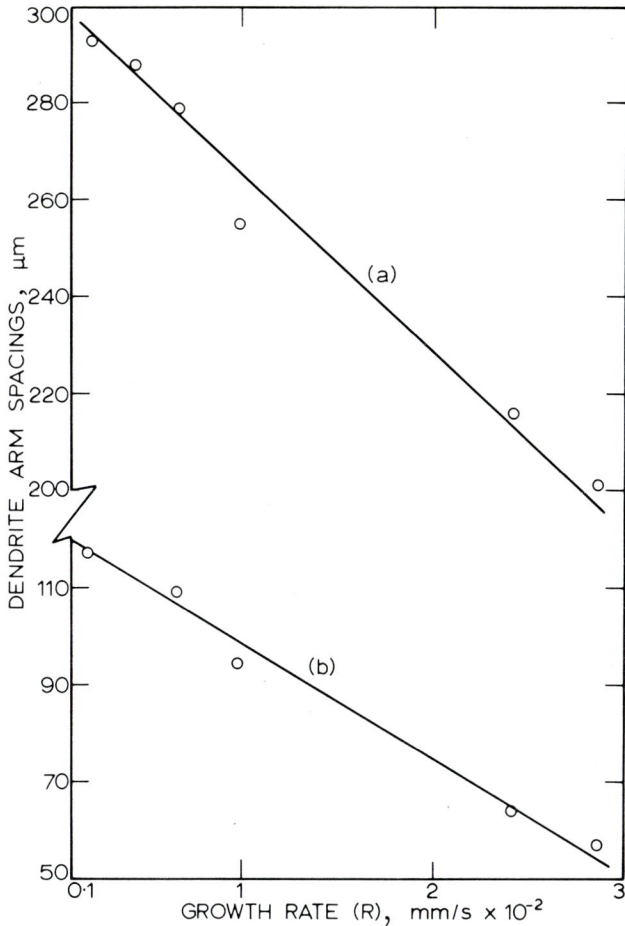

a primary (d_1) arm spacing; b secondary (d_2) arm spacing

2 Dendrite arm spacings as function of growth rate for Pb–5·2 Sb alloy and $G_L = 2°C/mm$

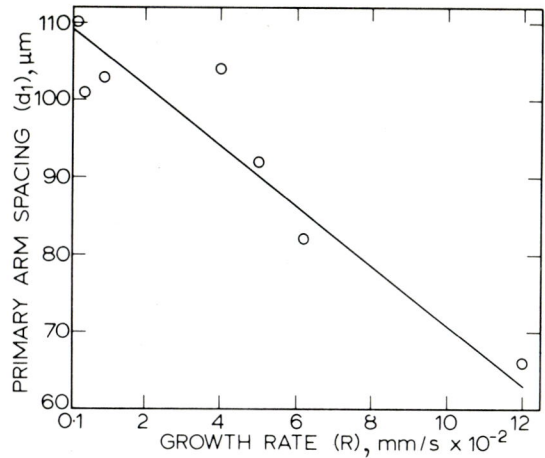

3 Primary (d_1) arm spacing as function of growth rate for Pb–5·7 Sb alloy and $G_L = 19·7°C/mm$

trated for a growth rate of 5×10^{-2} mm/s and $G_L = 19·7°C/mm$. The measurement of secondary arm spacings again became increasingly difficult as C_0 decreased, i.e. as the primary arm morphology became less branched (and finally rod like) and the volume fraction of the primary phase increased. Primary and secondary arm spacings can be seen to be independent of C_0 with the high G_L value.

equation (5) and G_L was calculated, from the thermocouple data, over a distance 5 mm ahead of the macroscopic solidification front. The alloys were analysed using atomic absorption spectrometry.

RESULTS
Steady growth

The dependence of d_1 and d_2 spacings on growth velocity with $G_L = 2°C/mm$ and $C_0 = 5·2$ wt-% Sb is shown in Fig. 2. Figure 3 shows the effect of growth rate on d_1 arm spacing with $G_L = 19·7°C/mm$ and $C_0 = 5·7$ wt-% Sb. At the lower end of the growth-rate range in Fig. 3 the growth morphology of the primary arms was unbranched rod or cruciform and although secondary branching became evident with increasing growth rate, the relatively large volume fraction of primary phase in the microstructure made accurate measurement of d_2 arm spacing difficult. Secondary arm spacings were therefore not recorded for this alloy as a function of growth rate.

Both d_1 and d_2 arm spacings decreased with composition for alloys solidified at $1·2 \times 10^{-2}$ mm/s with $G_L = 2·3°C/mm$, (see Fig. 4). The rate of change of d_2 arm spacing with solute content was reasonably constant but the rate of change of d_1 arm spacing decreased markedly as the alloy composition increased. In Fig. 5 the influence of solute content is again illus-

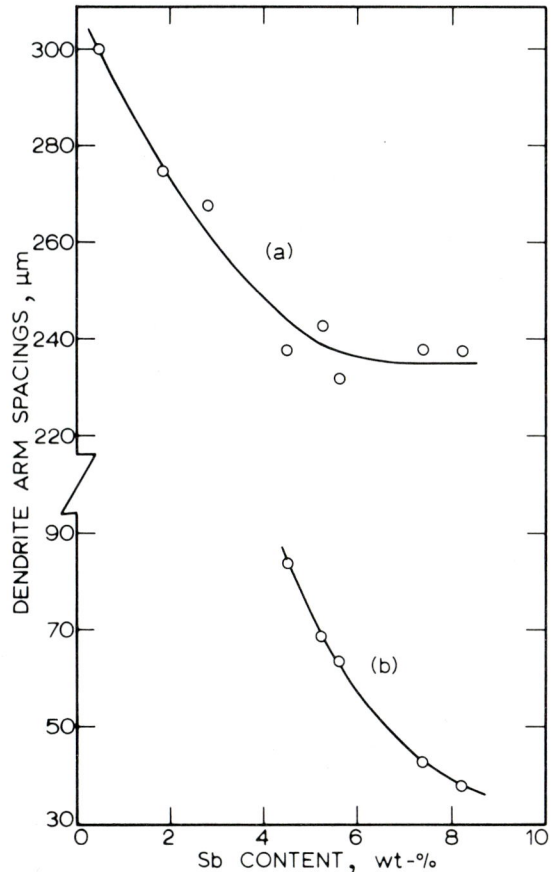

a primary (d_1) arm spacing; b secondary (d_2) arm spacing

4 Dendrite arm spacings as function of Sb content in Pb–Sb alloys solidified with $R = 1·2 \times 10^{-2}$ mm/s and $G_L = 2·3°C/mm$

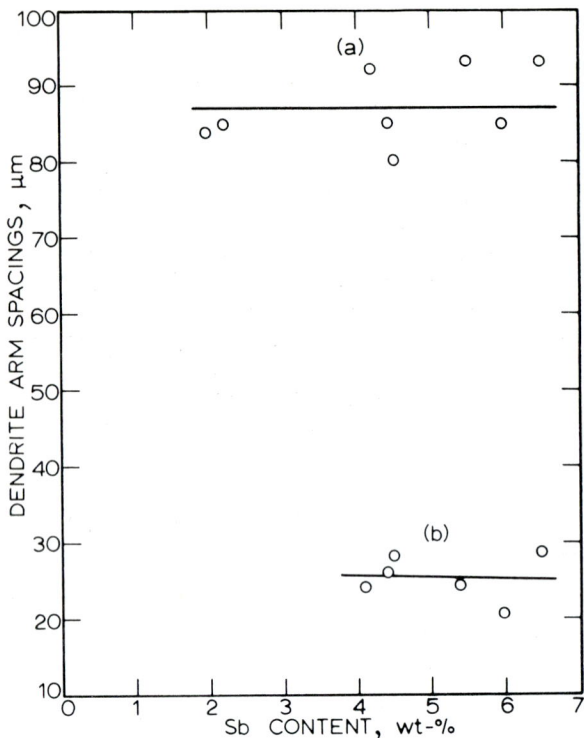

a primary (d_1) arm spacing; *b* secondary (d_2) arm spacing

5   Dendrite arm spacings as function of Sb content in Pb–Sb alloys solidified with $R = 5 \times 10^{-2}$ mm/s and $G_L = 19 \cdot 7°$C/mm

Observations at the low temperature gradient (Fig. 4) were also confirmed for hypoeutectic Pb–Sn alloys (Fig. 6) for the same growth rate and $G_L = 6 \cdot 6°$C/mm. The same comments apply, regarding the rate of change of the arm spacings with composition, as for Fig. 4.

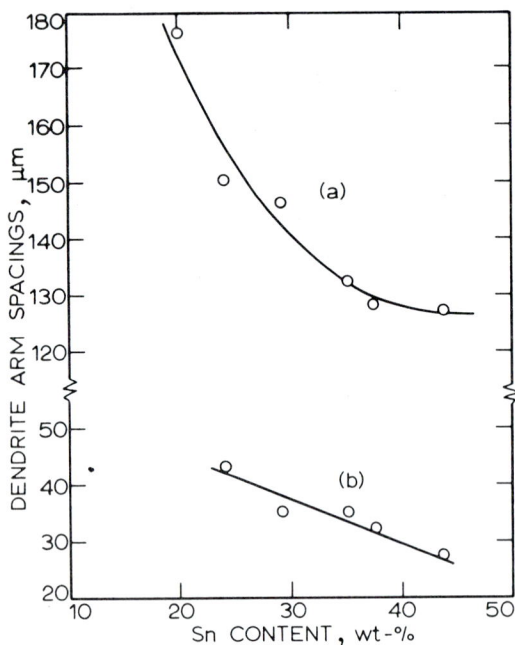

a primary (d_1) arm spacing; *b* secondary (d_2) arm spacing

6   Dendrite arm spacings as function of Sn content in Pb–Sb alloys solidified with $R = 1 \cdot 2 \times 10^{-2}$ mm/s and $G_L = 6 \cdot 6°$C/mm

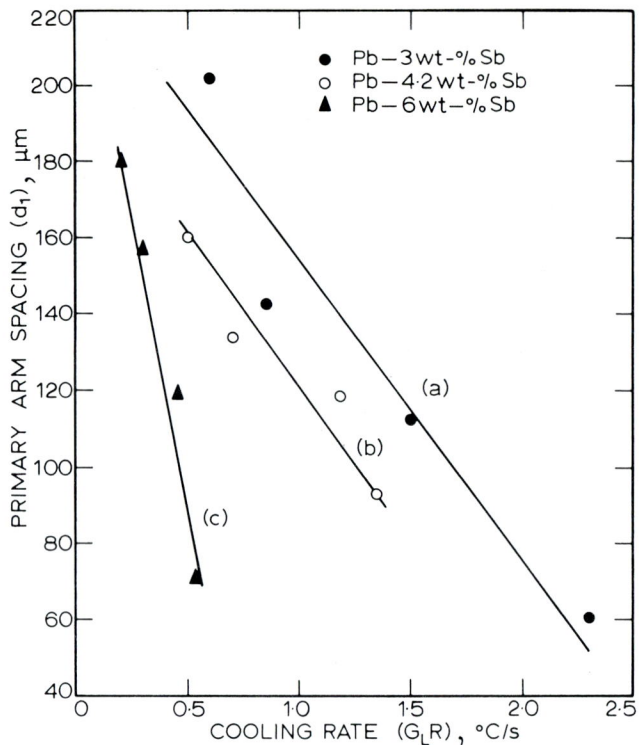

a 3·0 wt-% Sb; *b* 4·2 wt-% Sb; *c* 6·0 wt-% Sb

7   Primary (d_1) dendrite arm spacing as function of $G_L R$ for Pb–Sb alloys directionally solidified under non-steady conditions

Non-steady growth

The influence of composition and of the product $G_L R$ on d_1 and d_2 arm spacings is shown in Figs 7 and 8. It can again be seen, as for steady growth with low G_L values, that for a fixed value of the product $G_L R$, both d_1 and d_2 spacings decrease with increasing solute content. The temperature gradients calculated in these runs varied between 0·7 and 3·2°C/mm.

Significance of local solidification time

The results shown in Figs 2–5 and 7 and 8, together with additional data obtained from the present study, are shown in Fig. 9 in the form of logarithmic plots of d_1 and d_2 arm spacings *v.* local solidification time. The local solidification times were calculated assuming that the equilibrium liquidus and solidus lines are straight and that $\Delta T_s = T_L - T_E$. The latter assumption is substantiated by the appearance of eutectic in every specimen examined.

It is evident from Fig. 9 that the variation of $\log d_1$ arm spacing with $\log t_s$ does not conform to a linear relationship. In addition, for a given local solidification time, the d_1 arm spacings under steady growth conditions with $G_L \sim 20°$C/mm, are consistently lower, except at very short times, than those that were measured for steady growth with $G_L \sim 2°$C/mm or for the non-steady runs.

In the case of the secondary arms, it was assumed that a linear relationship existed and the value of n was determined to be 0·25. As for d_1 spacings, the d_2 spacings tend to be smaller for the same local solidification time, in the alloys solidified under steady growth conditions with the high gradient.

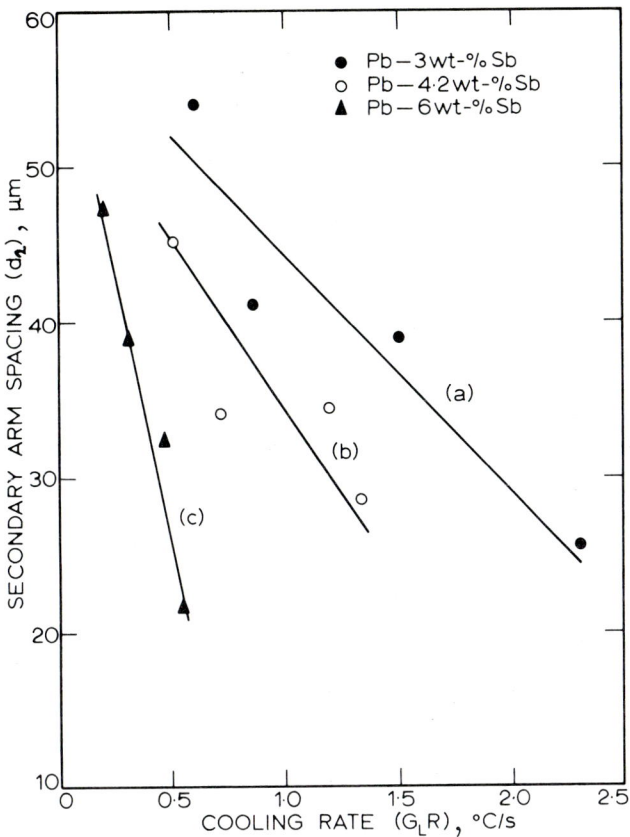

a 3·0 wt-% Sb; *b* 4·2 wt-% Sb; *c* 6·0 wt-% Sb

8 Secondary (d_2) arm spacing as function of $G_L R$ for Pb–Sb alloys directionally solidified under non-steady conditions

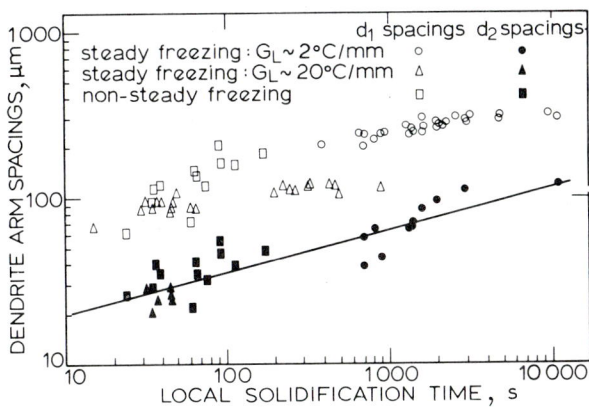

9 Primary (d_1) and secondary (d_2) dendrite arm spacings in Pb–Sb alloys as function of local solidification time t_s

DISCUSSION

The results of the present investigation of hypoeutectic Pb–Sb alloys are self consistent. The general observations, for a fixed alloy composition, that the tendency is for arm spacings to decrease as either G_L or R increase or, under non-steady conditions, that arm spacings decrease as the product $G_L R$ increases, confirm previous reports in the literature. However, the observation, for steady growth with low G_L values or for non-steady conditions, that both d_1 and d_2 arm spacings decrease as C_0 increases, has not previously been reported for binary alloys. Under steady growth conditions, when G_L was high, d_1 and d_2 spacings were found to be independent of composition. This finding, in respect of the former spacing, is in agreement with that in the work of Sharp and Hellawell.[10]

It is generally accepted that the principal factor determining the d_2 arm spacings in solidified alloys is coarsening,[13,14] which arises from remelting or more probably coalescence of arms during freezing.[14] The independent influences of G_L, R, and C_0 on d_2 arm spacings found in the present study, therefore conform reasonably to the trends anticipated from an empirical relationship of the type shown in equation (4). Nevertheless, as mentioned by other workers[14,15] it is still debatable whether or not d_2 spacing is governed solely by coarsening. The smaller spacing, for a given local solidification time, when G_L is high, may signify some dependence of d_2 arm spacing on the initial spacing before coarsening.

During solidification d_1 spacing either remains constant with respect to time (steady growth) or continuously readjusts in response to the changing conditions at the advancing macroscopic solidification front (non-steady growth). Since in the present investigation d_1 and d_2 spacings responded identically to changes in G_L, R, or C_0 it is tempting to postulate, as did Dann *et al.*,[13] that in controlled directional-growth experiments primary arm spacing is determined, to some degree or entirely, by spacing changes that take place before the establishment of steady growth. If this is so, the d_1 spacing presumably readjusts by the occlusion of individual primary arms by their neighbours. Termination of growth of individual arms will continue to occur throughout the period of solidification if directional freezing takes place under non-steady conditions. The latter statement is presumably true provided that the conditions will enable readjustment to take place. Sharp and Hellawell[18] have shown, by quenching during directional freezing, that d_1 spacing can remain constant even though the solidification environment may be changing drastically.

Under non-steady conditions it may be possible that for a given value of $G_L R$, d_1 spacing can be smaller than the spacing that would be observed under steady conditions for the same values of G_L and R. Non-steady directional freezing is invariably initiated by chilling which must inevitably induce supercooling, in the liquid adjacent to the chill, to an extent dependent on the initial alloy composition and melt temperature. The initial spacing of the primary arms will therefore presumably reflect the degree of undercooling in the chilled liquid and adjustment of this spacing, to that value that would be observed under steady conditions for the same G_L and R values may be prevented (particularly in the early stages of directional growth), by the extremely high cooling rates. This could account for the observation of Badon-Clerc and Durand[12] that composition markedly influenced spacings only at high cooling rates. The cooling rates, in the non-steady growth experiments in the present investigation, were considerably lower than in the study carried out by Badon-Clerc and Durand.

It is evident from Fig. 9 that d_1 arm spacings cannot be expressed, over a wide range of local solidification times, by empirical relationships of the type given in equation (4) and quoted in solidification texts.[17] In fact, there appears to be no logical basis for expecting that d_1 spacings will become exceedingly large with increasing t_s. Young and Kirkwood[14] also commented that equation (3) will break down for d_1 spacings with continuously decreasing G_L or R.

Sharp and Hellawell[10] demonstrated, for the steady growth of Al–Cu alloys, that d_1 spacing can remain constant over a wide range of values of the product RC_0. A d_1 spacing of $70 \pm 10 \ \mu m$ was quoted down to $3 \cdot 3 \ \mu m$ wt-%/s. An estimated cooling rate $G_L R$ for this RC_0 value is of the order of a few degrees per minute. The d_1 spacing they observed is considerably smaller than that found by Young and Kirkwood[14] in Al–Cu alloys for the same order of magnitude of the cooling rate. This is presumably because the cooling rates used in the work of Sharp and Hellawell correspond to higher gradients and lower growth rates than those employed by Young and Kirkwood. In the present study it was also found that for the same local solidification time, d_1 spacing was reduced as the temperature gradient became higher and the growth rate decreased (Fig. 9).

It is interesting to compare the results of a recent study[19] of arm spacings in low-carbon manganese steels with the present results. It was observed that d_2 (but not d_1) arm spacings could be described by equation (3). Except at high cooling rates, d_1 spacing decreased for a fixed cooling rate as the temperature gradient increased. Also, it was found that for a given cooling rate, d_2 spacings decreased with increasing carbon content, and that for a given G and R, d_1 spacings also decreased with carbon content.

Analytical attempts[8,15,17,20] to relate arm spacings to the solidification variables, from consideration of mass transport in the interdendritic regions, have been unsuccessful. Although a decrease in spacing with increasing cooling rate, or independently with increasing temperature gradient or growth rate, has been predicted there are conflicting opinions regarding the influence of alloy composition on arm spacing. Rohatgi and Adams[8] observed experimentally that d_1 spacing increased as C_0 increased. However, it is not implicit in their mathematical analysis that spacing should increase with C_0. A mathematical treatment by Flemings,[17] of solute redistribution in the presence of a temperature gradient, predicts that cell (or unbranched primary dendrite spacing) will be independent of composition and that the spacing will adjust to reduce constitutional supercooling in the intercellular (interdendritic) regions to a very low value. On the other hand, Okamoto and Kishitake[15] have determined theoretically that the primary spacings should be directly proportional to the square root of the solute content.

CONCLUSIONS

1 Under steady freezing conditions, primary d_1 and secondary d_2 spacings were observed to decrease as (a) the temperature gradient in the liquid G_L or the growth rate R of the macroscopic solidification front increased,

for a given alloy composition; or (b) as the initial alloy composition (C_0) increased, with G_L ($\sim 2°C/mm$) and R maintained constant. When G_L was $\sim 20°C/mm$, d_1 and d_2 spacings were independent of C_0.

2 Under non-steady freezing conditions, d_1 and d_2 spacings were observed to decrease with (a) increasing cooling rate ($G_L R$), for a given alloy composition; and (b) increasing solute content, for a given cooling rate.

3 Logarithmic plots of the arm spacings v. local solidification time t_s indicated that d_2 spacing reasonably conforms to an empirical relationship of the type $d_2 = \text{const.}(t_s)^{0.25}$. However, the logarithmic plot for the d_1 arm spacings was not linear. For the same local solidification time d_1 spacings, and to a lesser extent d_2 spacings, were the smallest for those alloys solidified under steady conditions with the high G_L.

ACKNOWLEDGMENTS
The authors would like to thank Professor H. K. Lloyd for the provision of laboratory facilities. One of the authors, D. M. Lloyd is also indebted to University College, Cardiff, for financial support.

REFERENCES

1 B. ALEXANDER AND F. RHINES: *Trans. AIME*, 1950, **188**, 1267
2 A. B. MICHAEL AND M. B. BEVER: *Trans. AIME*, 1954, **200**, 47
3 J. A. HORWATH AND L. F. MONDOLFO: *Acta Metall.*, 1962, **10**, 1037
4 R. E. SPEAR AND G. R. GARDNER: *Trans. AFS*, 1963, **71**, 209
5 J. A. E. BELL AND W. D. WINEGARD: *J. Inst. Met.*, 1964, **92**, 357
6 T. F. BOWER et al.: *Trans. AIME*, 1966, **236**, 624
7 I. IBARAKI et al.: *Mem. Inst. Sci. Ind. Res. Osaka University*, 1967, **24**, 107
8 P. K. ROHATGI AND C. M. ADAMS, JR: *Trans. AIME*, 1967, **239**, 1737
9 L. BACKERUD AND B. CHALMERS: *ibid.*, 1969, **245**, 309
10 R. M. SHARP AND A. HELLAWELL: *J. Cryst. Growth*, 1970, **6**, 253
11 G. R. KOTLER et al.: *Metall. Trans.*, 1972, **3**, 723
12 M. BADON-CLERC AND F. DURAND: *Mém. Sci. Rev. Métall.*, 1974, **LXXI**, (7–8), 451
13 P. C. DANN et al.: *J. Aust. Inst. Met.*, 1974, **19**, 140
14 K. P. YOUNG AND D. H. KIRKWOOD: *Metall. Trans.*, 1975, **6A**, 197
15 T. OKAMOTO AND K. KISHITAKE: *J. Cryst. Growth*, 1975, **29**, 137
16 B. P. BARDES AND M. C. FLEMINGS: *Trans. AFS*, 1966, **74**, 406
17 M. C. FLEMINGS: 'Solidification processing', 1974, New York, McGraw-Hill
18 R. M. SHARP AND A. HELLAWELL: *J. Cryst. Growth*, 1969, **5**, 155
19 T. EDVARDSSON et al.: *Met. Sci.*, 1976, **10**, 298
20 P. E. BROWN AND C. M. ADAMS JR: *Trans. AFS*, 1961, **69**, 879

Aberrations observed in the relationship of dendrite size-alloying elements for low-alloy steel

D. J. Hurtuk and A. A. Tzavaras

This paper covers some of the results of a rather extensive effort at the Republic Steel Corporation Research Centre in the last few years to combine laboratory and industrial experiments to improve the understanding of the solidification process in strand casting. An experimental strand cast evaluation programme has been developed to establish cause and effect relationships in the continuous casting of low-alloy high strength steel blooms. This was based on a detailed recording of all macrofeatures of a large number of deep etched cross-sections of blooms in the as-cast condition, such as columnar growth, centreline segregation, internal cracks, along with nearly every operating parameter. Computerized processing, on an experimental basis, of a large amount of such data from more than 100 casts revealed a direct relationship between the extent of columnar growth and the frequency and severity of casting defects in the blooms such as internal cracks (ghost lines) and centreline segregation. On the other hand, and in spite of the plethora of parameters affecting the strand cast process, a clear relationship was established between the carbon content of the steel and the extent of the columnar zone in these castings. To verify these results, an apparatus was designed in the laboratory and a large number of small steel ingots were cast. These castings were made under various, but controlled, thermal and chemical conditions. Results support the data from the strand cast evaluation programme and indicate that, contrary to current established thinking, increased carbon concentrations enhance the size of the solidification parameters, such as columnar growth and dendrite arm spacing in low-alloy steels up to a maximum for a certain carbon content. Beyond this point increased carbon concentrations have the opposite effect, as theory would indicate, and the size and spacing of the dendrites diminish at approximately the same rate. This behaviour may be unique for carbon since other elements, such as sulphur, do not show any such effect but comply with the accepted thinking regarding the effect of alloying elements on solidification structure parameters. The behaviour of carbon may be caused by the peritectic reaction that occurs for iron–carbon alloys.

D. J. Hurtuk is Research Engineer with the Republic Steel Corporation, Ohio, USA, and A. A. Tzavaras is Professor of Metallurgy at the Aristotelian University, Salonica, Greece

The mechanical and physical properties of cast steel products are significantly influenced by the solidification structure of the steel.[1-4] This influence is even more intense in continuous casting than in ingot casting because of the intensely anisotropic solidification structure usually found in continuously-cast steel. Yet the factors affecting the solidification structures, particularly regarding chemistry variations, and the degree of influence of these factors have not been fully defined. Therefore, it is highly desirable to develop an understanding of the factors influencing solidification structures and a means to control these relevant factors.

In the last few years an extensive effort has been underway at the Republic Steel Corporation Research Centre to combine industrial and laboratory experiments to improve the understanding of the solidification process in strand casting. This is being accomplished through a programme of collecting and analysing data from a continuous-casting facility and through · a series of controlled laboratory experiments involving the production and examination of unidirectionally solidified steel ingots. The results of these studies are then applied to production conditions to develop even better means of control for the parameters pertinent to the production of better quality strand-cast steel.[5]

This paper discusses the effect of some of the parameters involved in the development of solidification structures in continuously-cast steel, namely the effect of superheat, carbon content, and sulphur content. In particular, a unique effect of carbon content on solidification parameters that contradicts current established theory for a substantial range of carbon contents was observed and verified in both industrial and laboratory experiments. This information is of particular interest in controlling structurally related defects in continuously-cast products.

LITERATURE; THEORY

In steel castings, columnar growth usually begins near the chilled surface and proceeds inwards in a cellular-dendritic fashion as a result of heat flow into the heat sink, which can be predicted by equations for solidification of the nonlinear type.[6,7] During growth, the total number of primary dendrites decreases, the cross-section of those remaining increases, and the development of a preferred orientation occurs with the growth of crystals which are favourably oriented.[6] In typical low-alloy steel this takes place along the $\langle 100 \rangle$ direction.[8] Columnar growth eventually ceases when interference from equiaxed growth overrides the driving force for columnar growth or the driving force ceases to exist. It is thought that dendrite remelting and an increased solute content play an important role in reducing columnar growth.[6,8,9]

A basic problem in continuously-cast steel is the development of large columnar dendritic zones.[10] The effect of columnar growth on the mechanical properties and loss of ductility in steel has been investigated by Wieser.[11] Alberny *et al.*[12] have shown that centreline defects in continuous casting can be significantly reduced by controlling the columnar growth regions. Japanese investigators have attempted to control columnar growth in stainless-steel castings after recognizing the adverse effects of extended columnar growth on centreline casting defects.[13] Other research work on induction stirring and segregation has indicated that the control of columnar growth is crucial in producing good quality strand-cast products.[5,12,14]

EXPERIMENTAL APPROACH

The industrial experiments involved the development and implementation of a computerized strand-cast evaluation system for gathering and processing continuous-caster data from a production facility. An evaluated cast involves sampling three times per strand corresponding to the beginning, middle, and end of the cast. These cross-sectional samples are etched, as shown in Fig. 1, and the frequency and severity of various casting defects are recorded. This includes columnar growth, centre defects, rhomboidity, ghost lines, fronts, corner cracks, and inclusions. Data from over 100 evaluated casts were used in this investigation.

The laboratory experiments initially involved designing an apparatus for producing unidirectionally

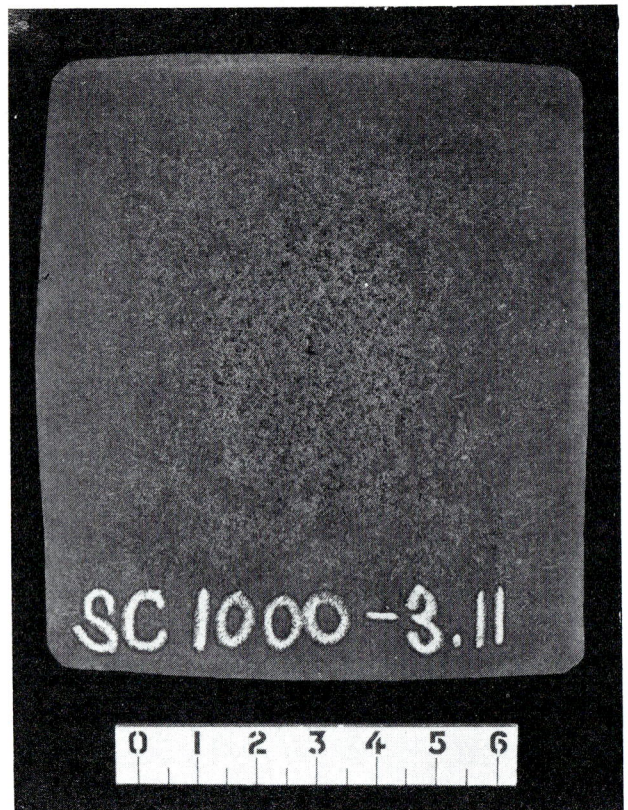

1   Macroetch of a cross-section of a continuously-cast billet in the as-cast condition showing solidification structure for a billet from the middle of a cast

solidified 5 lb ingots of steel on a water-cooled copper chill in a controlled atmosphere. Figure 2 is a schematic drawing of the apparatus. A graphite mould is lined with insulating paper and clamped on to a water-cooled copper chill. An insulating lid with an inlaid exothermic compound is positioned so that it can be closed once the ingot is teemed. Thus, a 5 lb heat is made under a vacuum in an induction furnace and tapped in an argon atmosphere into the mould, immediately after which the lid is closed and the exothermic ignites. Growth begins at the copper chill and proceeds upwards. Side growth is inhibited by the insulating lining, and the

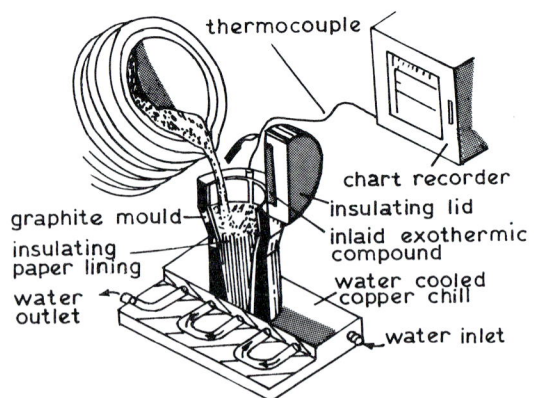

2   Schematic drawing of the apparatus for producing unidirectionally solidified 5 lb ingots of steel for investigating columnar growth under controlled laboratory conditions

exothermic compound and the insulating lid keep the riser above liquidus temperature until the ingot solidifies. A thermocouple is positioned in the riser portion of the mould to record the cooling curve on a chart recorder. Cooling curves at various distances from the chill were produced by placing seven thermocouples along the longitudinal axis of certain moulds for each steel at 10–15 mm intervals.

Experiments were made involving three types of steel—8620, 4140, and 5160. Ingots of 8620 type steel were made with about 20 different carbon contents ranging from 0·02% C–1·80% C and ingots of 4140 type steel were made with eight sulphur contents from 0·010% S–0·045% S. The superheat (casting temperature—liquidus temperature) was varied between 2°F (1°C) and as much as 140°F (78°C) for each steel. Multiple cooling curves were recorded for ingots cast with both 60°F (34°C) and 100°F (56°C). Ingots were sectioned and etched, and examples are shown in Fig. 3.

The measurement of various solidification parameters was obtained from the ingots and cooling curves. The length of the columnar dendritic zone was measured in each ingot produced. Metallographic specimens were taken from ingots cast with superheats of 20°F (11°C), 60°F (34°C), and 100°F (56°C); these samples were polished and etched to obtain secondary dendrite arm spacing measurements at various distances from the chill. The results are listed in Table 1 and typical micrographs are shown in Fig. 4. The multiple cooling curves were used to determine the local solidification time and solidus–liquidus temperature ranges for the corresponding ingots.

RESULTS AND DISCUSSION
Industrial experiments

Results from industrial experiments indicated that the frequency and severity of several structurally related casting defects are significantly influenced by columnar growth and reach high levels when the columnar growth exceeds 40 mm.

Defective centres may be particularly undesirable in strand-cast billets because of the low reduction applied to those billets before final use. It can be seen in Fig. 5

Table 1 Measured secondary dendrite arm spacings at various distances from the chill for experimental ingots of 8260 type steel with varying carbon content and superheat

Sample No.	Type grade	% C	Superheat, °F	Secondary dendrite arm spacing, μm Distance from chill			
				5 mm	10 mm	20 mm	30 mm
0·06–17	8620	0·06	18	—	88·9	—	—
–12			44	—	57·8	—	—
–5			117	—	69·8	—	—
0·16–13	8620	0·12	21	—	100·6	205·4	277·6
–1			104	59	72·4	137	191·7
0·18–9	8620	0·18	20	55·9	55·9	—	—
–7			52	61·4	66·7	81·2	—
–5			100	41·5	100	120·3	112
0·20–11	8620	0·20	16	—	65·3	—	—
–3			60	—	100·6	130·5	169·8
–5			108	—	54·1	99·4	96·9
0·22–13	8620	0·22	25	70·1	114·3	333	261
–16			66	69·6	67·9	200	170
–4			103	56·6	83·3	120·3	233
0·30–12	8620	0·26	20	78·5	112	138·9	250
–8			52	58·3	108·3	216·7	250
–15			100	71·7	75·6	120	158·7
0·57–13	8620	0·57	24	66·7	164	178·6	333
–7			63	92·3	108·3	103·6	86·7
–4			107	80·1	111	169·9	158·8
0·90–11	8620	0·90	33	—	100	—	—
–5			78	—	100	117·8	—
–4			104	—	100	111·1	166·5
1·2–15	8620	1·2	20	—	59·8	—	—
–1			55	—	96·9	132·3	—
–12			113	67·9	66·1	108·3	—
1·20–15	8620	1·2	30	65·6	96·9	—	—
–17			63	62·7	118·6	125	—
1·35–18	8620	1·35	34	63·7	57·7	—	—
–7			68	52·7	71·7	138·9	136·8
1·5–25	8620	1·5	10	65·3	79·7	—	—
–7			65	—	50	—	—
–4			99	85·7	—	112	—
1·8–24	8620	1·8	26	—	51·7	—	—
–4			61	62·9	65·2	80·3	—
2·0–13	8620	2·0	16	89·3	—	112	—
–8			57	—	75·1	97	94·4

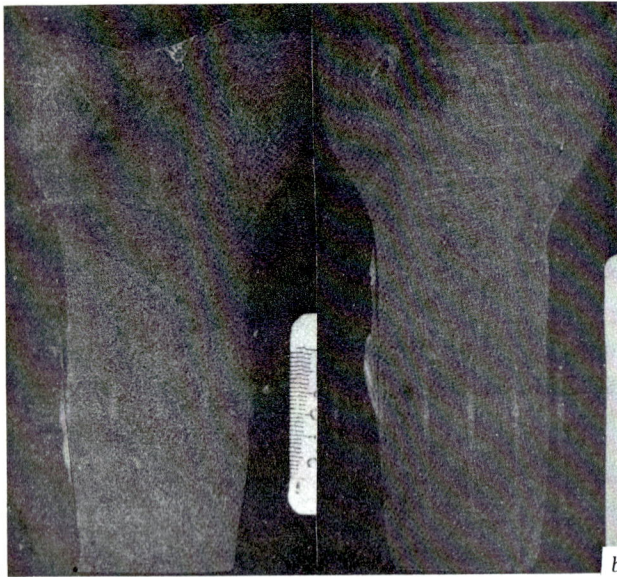

a an ingot cast with low superheat; *b* an ingot cast with high superheat

3 Etched cross-section of unidirectionally solidified 5 lb ingots of 8629 steel showing the difference in the columnar growth regions

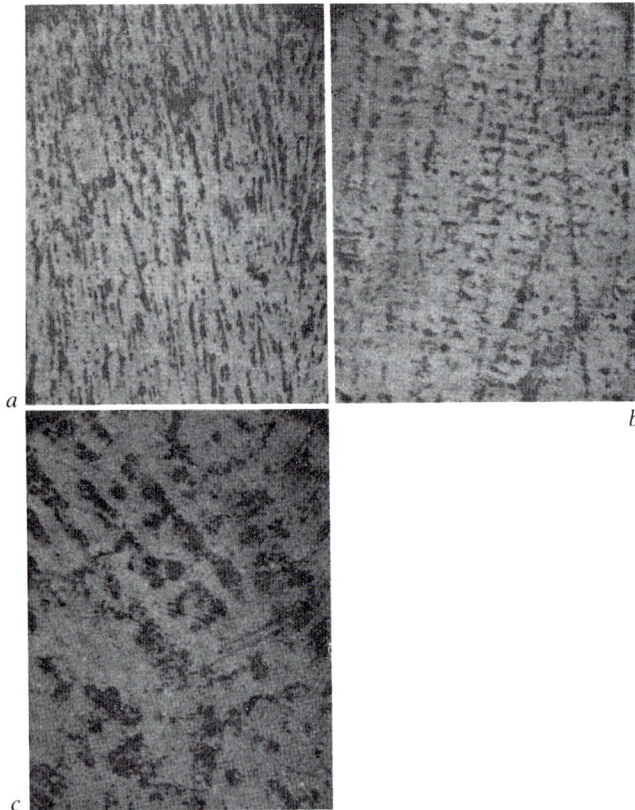

a 5 mm from chill (M516551DM-3, [×28];
b 10 mm from chill (M516551DM-4, [×28];
c 20 mm from chill (M516551DM-5, [×28])

4 Photomicrographs of an experimental ingot of 8620 type steel with 0·20% C at three distances from the chill

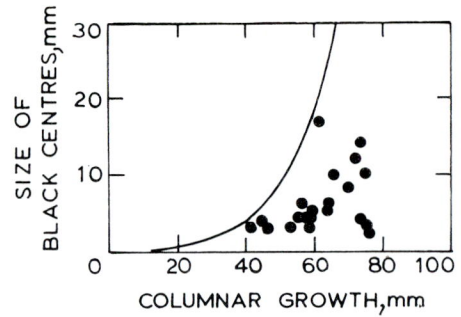

5 A plot based on 100 evaluated casts of centreline segregation against columnar growth indicating that when columnar growth is less than 40 mm centreline segregation is practically absent

that the frequency and severity of centreline segregation is related to the length of the columnar growth zone, especially when the columnar zone exceeds 40 mm. Similar results were found for centre shrinkage defects. The mechanism of formation for both of these centre defects are thought to be related. Centre segregation is formed by movement of segregated liquids, and centre shrinkage occurs when dendritic growth inhibits the motion of liquid metal along the centreline of the billet (bridging). The superheat of a cast varies by increasing rapidly at the beginning of the cast, reaching a plateau within the first few minutes, and then decreasing rapidly near the end of the cast. The columnar growth follows a similar behaviour. It has been found that centre defects occur most frequently at high casting temperatures. Thus, it is necessary to investigate the effects of superheat on solidification structures under controlled laboratory conditions.

During this investigation, certain grades of steel were found to be more susceptible to centreline defects than others. This led to a study of the effect of chemistry, particularly carbon content, on columnar growth. Figure 6 indicates the relationship found between carbon content and the length of the columnar zone in the eighth billet of casts made with normal casting temperatures (liquidus temperature $+50°F$ (28°C)). Despite the plethora of chemistry variations within the many grades of steel examined and the dynamic conditions on the caster, it is seen that carbon content tends to increase the length of columnar growth for the range of carbon contents examined. This supports the belief

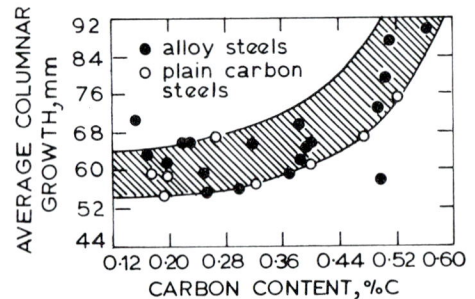

6 A plot of caster data of the average columnar growth against carbon content showing that columnar growth increases with increasing carbon contents for the range of 0·12% C–0·60% C

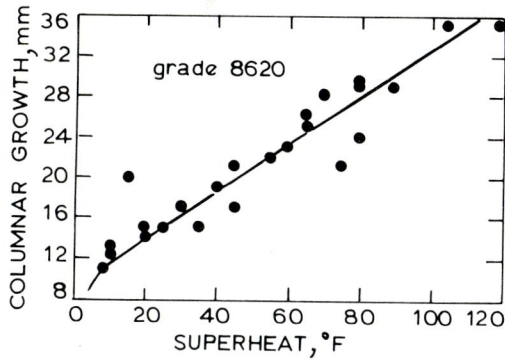

7 Plot of laboratory data of columnar growth against superheat for 8620 steel showing existence of nearly linear relationship

that superheat and carbon content are indeed two of the main variables affecting the solidification structures and must be studied under controlled laboratory conditions to obtain a meaningful correlation.

Laboratory experiments
The effect of superheat
The initial laboratory studies involved the investigation of the effects of superheat and carbon content on columnar growth for three grades of steel, 8620, 4140, and 5160, having average carbon contents of 0·22% C, 0·40% C, and 0·57% C, respectively. A total of 24 ingots of 8620 type steel were cast with superheats ranging from 8–120°F (4–67°C). Figure 7 shows the relationship between the length of the columnar zone and superheat to be very nearly linear. A computerized simple linear regression method was used to find the curve of best fit for these data. The program calculates the least-squares estimate of the intercept and slope. Through a variety of statistical tests, the relationship between columnar growth and superheat was found to be linear. Similar experiments were conducted for 4140 and 5160 type steels, and the relationships obtained for the three grades of steel examined are shown in Fig. 8. In each case, a linear relationship was found to exist between columnar growth and superheat.

The effect of carbon
In Fig. 8 it can be seen that increasing carbon contents increase the level of columnar growth for the range of carbon content examined. This essentially verifies the

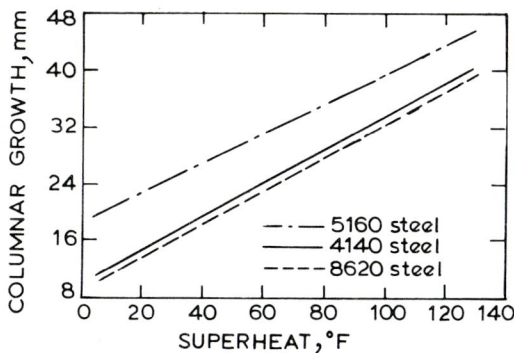

8 Plot of laboratory data of columnar growth v. superheat for the three grades analysed showing increased columnar growth for increased carbon content

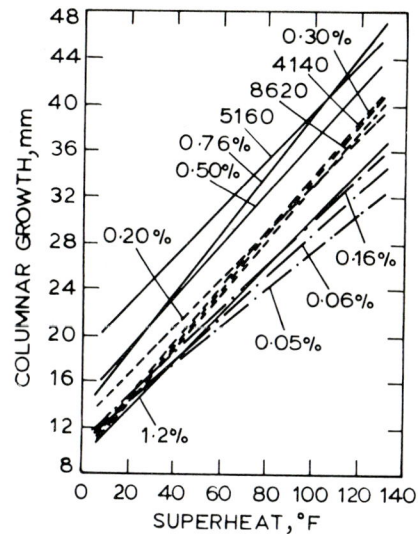

9 Plot of laboratory data of columnar growth against superheat showing the effect of increasing carbon content on columnar growth

trend shown in Fig. 6, established on the basis of data taken from evaluated strand-cast steel.

Since these results suggested an aberration from solidification theory,[6] a more thorough investigation of the effects of carbon on solidification parameters was conducted. An 8620 type steel was made with about 20 different carbon contents ranging from 0·02% C to 1·80% C and superheats from 2°F (1°C) to 140°F (78°C). Some of these results, along with the earlier results, are shown in Fig. 9. Increasing carbon contents appear to increase the length of columnar growth up to a maximum for a certain carbon content, after which carbon exerts the expected influence of decreasing columnar growth. In each steel examined a linear relationship between columnar growth and superheat was verified statistically.

Results from the complete carbon investigation can be found in Fig. 10 where the effects of carbon content on columnar growth for various superheats are indicated. The curves were fitted to the data by computer processing as third degree polynomials. It is clearly seen that, contrary to current theory,[6] carbon content initially increases the length of columnar growth through a maximum at about 0·60% C. Beyond this point, increasing carbon contents reduce the length of columnar growth as theory would indicate. In addition, the effect of increasing superheat on columnar growth is seen.

To verify the observed effect of carbon content on columnar growth, measurements of secondary dendrite arm spacings were obtained. Samples were taken from ingots cast with about 20°F (11°C), 60°F (34°C), and 100°F (56°C) superheat, and three measurements were taken at each of four distances from the chill (5, 10, 20, and 30 mm) when possible, and averaged. Complete results are listed in Table 1, and results from measurements at 10 mm from the chill are represented in Fig. 11. The curves were fitted by the computer processing as third degree polynomials. It can be seen that increasing carbon contents initially increase the secondary dendrite arm spacing to a maximum at about 0·60% C, after which the spacings are reduced. This

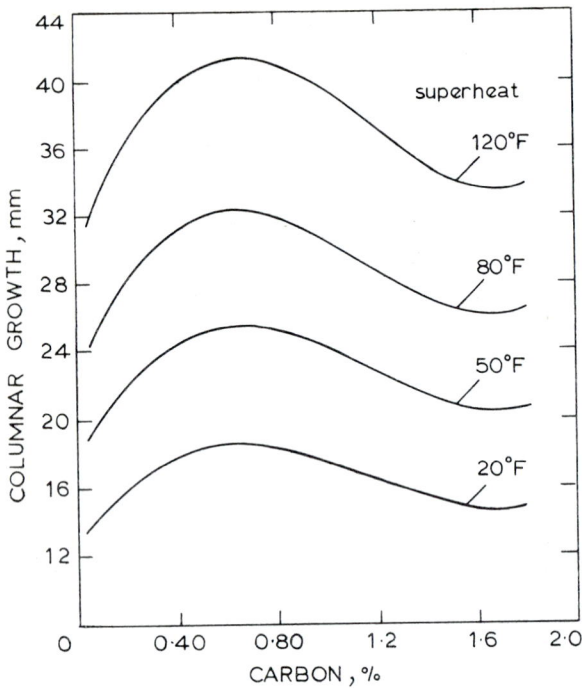

10   **Plot of laboratory data of columnar growth against carbon content for four superheats, showing that increasing carbon contents initially increase the length of columnar growth to a maximum at 0·60% C after which the opposite is true; this plot is based on data from over 400 experimental ingots**

corresponds closely to the effect of carbon content on columnar growth and apparently verifies this effect. These results agree with the findings of other investigators[8,15] where decreased columnar growth zones were accompanied by decreased secondary dendrite arm spacings. These refined structures are usually the result of more effective constitutional supercooling.[15] Figure 11 also shows the influence of increasing superheat on secondary dendrite arm spacing. As theory would predict, increasing superheat decreases the dendritic spacings[8] as a result of the higher thermal gradient.

The measurement of local solidification times was accomplished using the multiple cooling curves from ingots of 8620 type steel with carbon contents from 0·02% C to 1·80% C and cast with superheats of 60°F (34°C) and 100°F (56°C). For the production of these ingots, the moulds were equipped with seven thermocouples along the longitudinal axis of the mould at the following distances from the chill: 20, 35, 45, 60, 70, 85, and 95 mm. The results from these studies are depicted in Figs 12 and 13. Again, a computer program was used to find the curves of best fit, and a third degree polynomial was found to satisfy the data.

It is seen in these figures that increasing carbon contents initially increase the local solidification time to a maximum at about 0·40% C, after which increasing carbon causes a decrease in local solidification time. These results closely correspond to and apparently verify the columnar growth and secondary dendrite arm spacing data. The slight shift in the point of maximum is probably a result of the thermal diffusivity, *a*, change since

$$d = a t_f^n$$

and

$$a = \left(\frac{\lambda}{\rho C_p} \right)^{\frac{1}{2}}$$

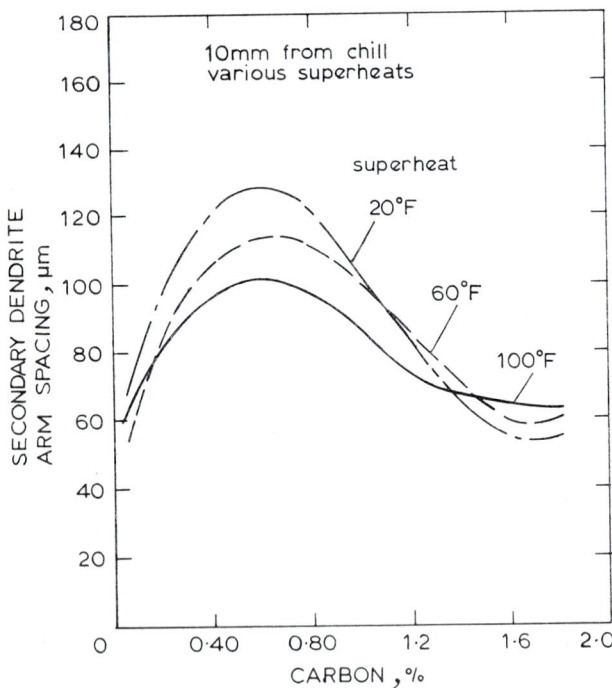

11   **Plot of laboratory data listed in Table 1 of secondary dendrite arm spacing against carbon content for three superheats, showing that increasing carbon contents initially increase the secondary spacing to a maximum at 0·60% after which the opposite is true; these curves correspond closely with those in Fig. 10**

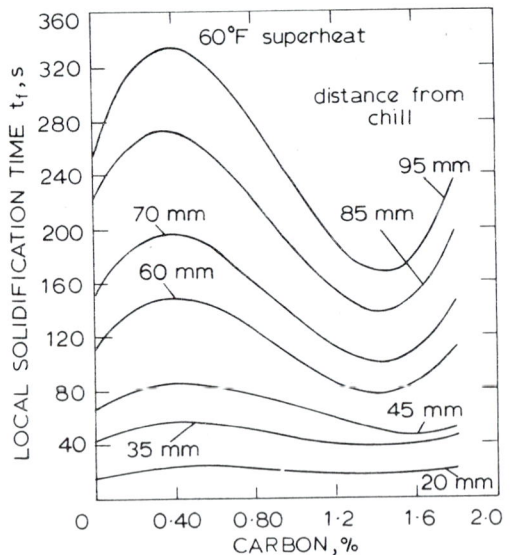

12   **Plot of laboratory data of local solidification time against carbon content at various distances from the chill for ingots cast with 60°F superheat showing that increasing carbon contents initially increase local solidification time to a maximum at 0·40% C, after which the opposite is true**

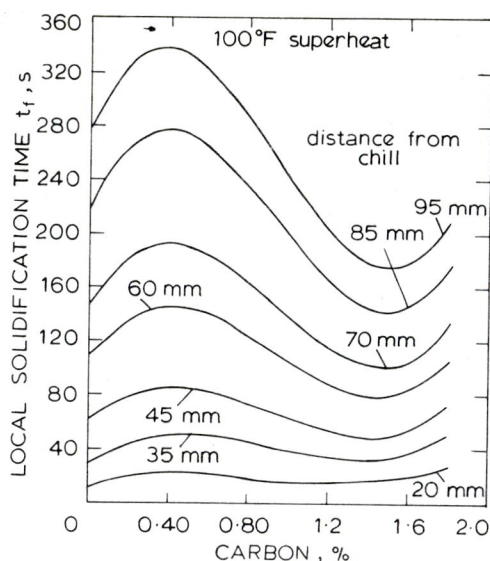

13 Plot of laboratory data of local solidification time against carbon content at various distances from the chill for ingots cast with 100°F superheat showing that increasing carbon contents initially increase local solidification time to a maximum at 0·40% C, after which the opposite is true

where t_f is the local solidification time, n is in the range of $\frac{1}{2}$ to $\frac{1}{3}$ for secondary dendrite arm spacings, λ is the thermal conductivity, ρ is the density, and C_p is the volumetric heat capacity.[8] The effect of longer local solidification times at increasing distances from the chill is also seen in Figs 12 and 13, as theory predicts.[6,8]

The liquidus and solidus temperatures for the steels examined were read from the cooling curves, and the results are listed in Table 2 and plotted as the solidus–liquidus temperature range against carbon content in Fig. 14. It is seen in this figure that the solidus–liquidus temperature range is affected by increasing carbon contents in much the same manner as the other solidification parameters investigated. The peak is once again observed at about 0·60% C, and the slopes on either side are approximately the same as in the other curves for columnar growth, secondary dendrite arm spacing, and local solidification time.

Thus, it has been found that increasing carbon contents enhance the size of solidification parameters to a maximum at about 0·60% C after which further increases of carbon have the opposite effect, as theory would indicate. This unique effect of carbon has been verified by the measurement of four solidification parameters including columnar growth, secondary dendrite arm spacing, local solidification times, and solidus–liquidus temperature ranges. The result that the solidus–liquidus temperature range increases with local solidification time, secondary dendrite arm spacing, and length of columnar growth conforms well with theory.[6,8,15] The aberration from current thinking, which has been based largely on work with substitutional solutes, involves the effect of carbon, an interstitial solute, on these parameters below 0·60% C. It is thought that this behaviour is caused by the peritectic reaction that occurs for iron–carbon alloys. The fact that the 0·60% C point is so important adds credence to this thinking. The complete mechanism is not yet

Table 2 Experimentally measured liquidus and solidus temperatures for laboratory ingots of 8620 type steel with varying carbon contents

% C	Liquidus temperature, °F	Solidus temperature, °F	Liquidus–solidus temperature range, °F
0·05	2 767	2 670	97
0·06	2 766	2 668	98
0·09	2 757	2 660	97
0·10	2 750	2 650	100
0·20	2 742	2 650	92
0·22	2 740	2 640	100
0·30	2 735	2 630	105
0·40	2 730	2 625	105
0·50	2 727	2 615	112
0·57	2 725	2 605	120
0·60	2 723	2 600	123
0·70	2 710	2 590	120
0·80	2 700	2 585	115
0·90	2 695	2 580	115
1·00	2 675	2 578	97
1·10	2 660	2 550	110
1·20	2 640	2 540	100
1·30	2 620	2 535	85
1·40	2 600	2 525	75
1·50	2 590	2 520	70
1·60	2 575	2 510	65
1·70	2 560	2 505	65
1·80	2 550	2 480	70

fully understood and additional research will be required to clarify this phenomenon so that it can be applied to actual production conditions to produce better quality ingot and continuously-cast products.

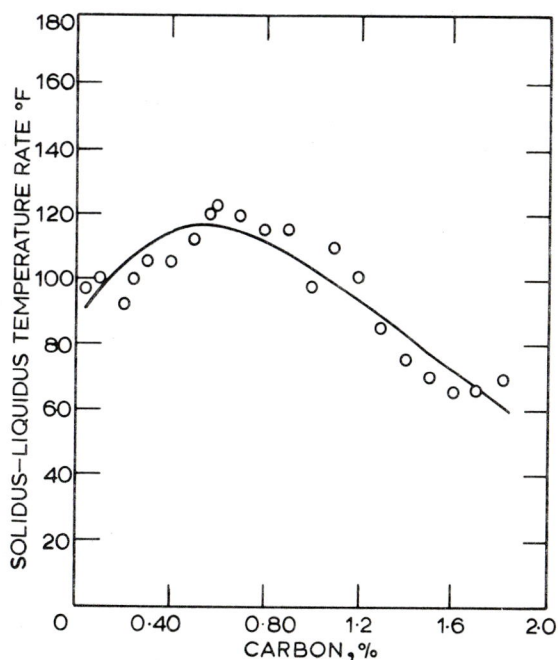

14 Plot of experimentally measured solidus–liquidus temperature ranges against carbon content, showing that increased carbon contents initially increase the temperature range to a maximum at 0·6% C, after which the opposite is true

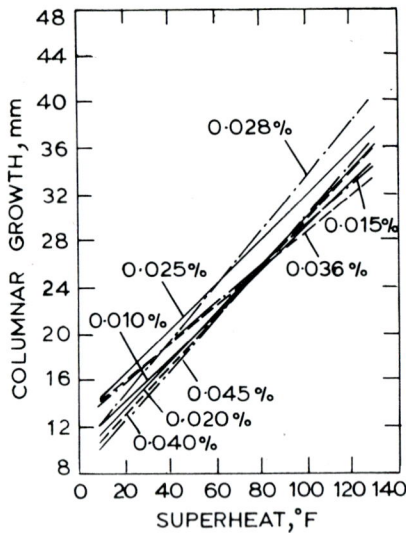

15 Plot of laboratory data of columnar growth against superheat showing the effect of increasing sulphur content on columnar growth

The effect of sulphur

Experiments were conducted to assess the influence of sulphur content on solidification structures by casting ingots of 4140 type steel with eight sulphur levels from 0·010% S–0·045% S and superheats ranging from 2°F (1°C) to 140°F (78°C). Results are depicted in Fig. 15. In each case, a linear relationship was found to exist between columnar growth and superheat for the range of superheats investigated. The effects of sulphur content on columnar growth appear to be slight with the expected tendency for increasing sulphur content to reduce, somewhat, the level of columnar growth. The sulphur experiment, then, in no way helped to clarify the unique effect of carbon content on solidification parameters.

CONCLUSIONS; OUTLOOK

It is well known that steel solidification structures exert a significant effect upon the properties of cast-steel products. This investigation has been concerned with the main factors affecting steel solidification structures, namely superheat and chemistry, particularly carbon content.

It has been found from industrial experiments that the frequency and severity of several structurally related continuous-casting defects are significantly influenced by the length of the columnar dendritic zone and reach high levels when the columnar growth exceeds 40 mm. These defects, particularly centre defects, were found to occur most frequently at high casting temperatures.

Laboratory experiments revealed that a nearly linear relationship exists between superheat and the length of columnar growth for the practical range of superheats investigated. Statistical processing of laboratory data verified that the relationship was linear.

Results also indicated that, contrary to current thinking, increased carbon concentrations enhance the length of columnar growth up to a maximum at approximately 0·60% C. This phenomenon was seen in the industrial experiments as well. Beyond this carbon point, further increases in carbon content reduce the length of columnar growth as theory would predict. These results were verified with the measurement of other solidification parameters including secondary dendrite arm spacing, local solidification time, and solidus–liquidus temperature range.

Sulphur was found to comply with accepted thinking and had only a slight effect on columnar structures. The mechanism for the unique effect of carbon on solidification parameters is thought to be related to the peritectic reaction in iron–carbon alloys. Since the phenomenon is not fully understood, further research must be performed to develop an understanding of the effect so that it can be applied to production conditions to improve the quality of both ingot and strand-cast steel.

REFERENCES

1 J. F. WALLACE: *J. Met.*, May 1963, 372
2 J. F. WALLACE *et al.*: *Trans. AFS*, **59**, 1951, 223
3 S. Z. URAN *et al.*: *Trans. AFS*, **68**, 1960, 347
4 N. L. CHURCH: 'Nucleation and Constitutional Supercooling Effects on Cast Steel Properties', Ph.D. thesis, Case Institute of Technology, 1965
5 A. A. TZAVARAS: 'Continuous casting', 197, 1973, New York, TMS–AIME
6 B. CHALMERS: 'Principles of solidification', 1964, John Wiley & Sons
7 J. T. BERRY AND D. R. DURHAM: *Trans. AFS*, 1974, **52**
8 M. C. FLEMINGS: 'Solidification processing', 1974 McGraw-Hill
9 K. A. JACKSON *et al.*: *Trans. Met. Soc. AIME*, 1966, **236**, 149
10 D. J. HURTUK AND A. A. TZAVARAS: *Trans. AFS*, 1975, **78**, 423,
11 P. WIESER: 'Inoculation and solidification control effect on cast steel properties', Ph.D. thesis, Case Institute of Technology, 1967
12 P. ALBERNY *et al.*: *Electric Furnace Proceedings*, 1973, **31**, 237, TMS–AIME, New York
13 Y. ITO AND Y. SUZUKI: US Patent 3952791, 1976
14 D. J. HURTUK AND A. A. TZAVARAS: Paper presented at the IMD Solidification Committee Meeting of the TMS–AIME Conference, Las Vegas, Nevada, Feb. 1976, to be published in *Metall. Trans. B.*
15 N. CHURCH *et al.*: *Trans. AFS*, 1966, **74**

Segregation and the determination of phase equilibria in multicomponent systems: Al–Cu–Mn

D. M. L. Bartholomew, M. Jezuit, B. Watts and A. Hellawell

Using aluminium-base alloys in the system Al–Cu–Mn, it is shown how directional solidification methods can be used to produce specimens, the examination of which provides complete information about liquidus and solidus data. Chemical and thermal analyses of some six directionally frozen alloy specimens yielded equilibrium data in good quantitative agreement with the published diagram, and demonstrate the general feasibility of the approach for rapid determination of multicomponent phase equilibria.

This work was carried out when all the authors were in the Department of Metallurgy and Science of Materials, University of Oxford. D. M. L. Bartholomew is now in the Admiralty Compass Observatory, Slough, M. Jezuit is in the Instytut Materialoznawska Politechniki, Warsaw, Poland. A. Hellawell and B. Watts are in the Materials Department, University of Wisconsin, Milwaukee, USA

In this paper, the authors are concerned with liquid–solid phase equilibria and segregation profiles which are to be expected during the solidification of multicomponent alloys (i.e. those with three or more components). Conventional methods for the determination of such systems involve combinations of thermal analysis and solidus detection, e.g. by microscopical methods, and also require sampling and analytical techniques to determine solid–liquid tie lines,[1-3] the last information being particularly important if accurate predictions are to be made about micro- or macrosegregation during solidification.

These experimental procedures are tedious and time consuming and the object of this work has been to demonstrate that it is possible to obtain the necessary data from relatively few directionally solidified specimens, and simultaneously provide complete segregation traces across a system. The principle of the method, first suggested by Tiller,[4] does not seem to have received the attention it merits, and depends on the near equilibrium freezing of alloy specimens which are then examined (*a*) microscopically; (*b*) thermally; and (*c*) chemically. Microscopical examination, coupled with microprobe analysis, identifies phases and probable reactions; from chemical analyses the solid–liquid distribution coefficients can be obtained, and from

thermal analyses the liquidus temperatures as a function of composition which, when combined with tie line information, yields the corresponding solidus.

The method is applicable to any number of components, but for experimental convenience as a demonstration exercise the system Al–Cu–Mn was chosen, the form of the phase diagram for aluminium-rich alloys being relatively well established[5] (Fig. 1), although other sources[6] suggest a similar diagram with the same reactions occurring at much higher manganese concentrations and at higher temperatures. As shown below, this work confirms the data in Fig. 1.

EXPERIMENTAL REQUIREMENTS AND PROCEDURES
Equilibrium segregation
In order to approximate to equilibrium conditions it is necessary to solidify an alloy in such a way that the liquid is fully mixed as closely to the solid/liquid interface as possible. From composition–distance profiles along specimens so frozen (for each solute) it is then possible to use the Scheil equation[7]:

$$C_x = k_0 C_0 (1-x)^{(k_0-1)}$$

where C_0 is the initial composition, k_0 the equilibrium solid–liquid distribution coefficient and C_x the solid

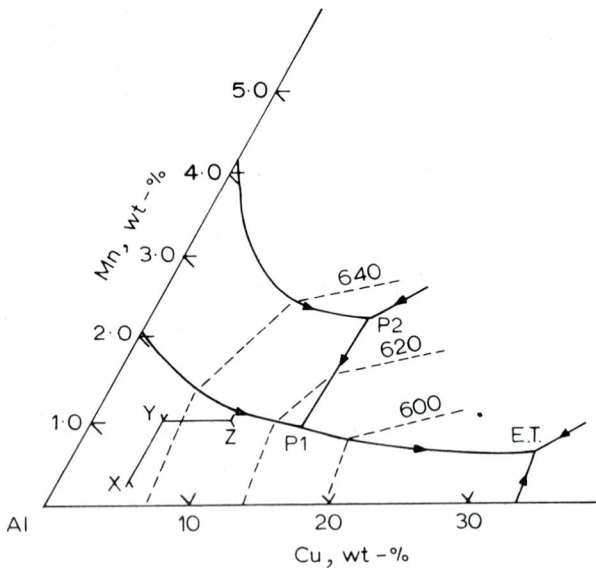

1   Liquidus for Al-base corner of Al–Cu–Mn diagram, after Day and Phillips[5]; P_1: 14·85 Cu, 0·90 Mn, and 616 K, P_2: 15·75 Cu, 2·2 Mn, and 625 K, E.T.: 32·5 Cu, 0·6 Mn and 547·5 K alloys used in this work were between X–Y–Z

composition for a fraction x solidified. From this equation k_0 can be obtained for each component at variable compositions along the alloy specimen, i.e. within the alloy system.

The conditions necessary to obtain almost uniform liquid composition can be achieved with efficient convective mixing at very slow rates of growth,[8] and in this work alloy specimen rods ~150 mm long and 4 mm dia. were withdrawn upwards from a resistance furnace in alumina tubes. The heavier solute elements, Cu and Mn, enhanced the density inversion and increased the convective mixing. A steep temperature gradient of ~10 K/mm was achieved by means of a water-cooled copper insert into the furnace[9] and the unstirred boundary layer was then some 100 μms thick. Estimates[10] of the effective distribution coefficient, k_e, then show that it is within 1% of k_0 at freezing rates of 0·1 to 0·2 μms (*see* Fig. 2); there is negligible solute accumulation at the interface.

Planar interface
The conditions for segregation given above involve a high ratio of temperature gradient G, freezing rate V,

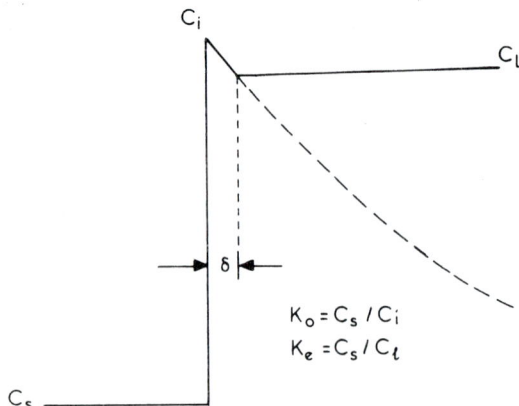

2   Definition of solid–liquid solute profiles, boundary layer, δ, and k_0 and k_e

of ~10^{-1} K μm^{-2} s, and using the simple expression for planar breakdown,[11] taking data, e.g. for the Al–Cu system, it can be shown that a planar growth front should be maintained for Al-base alloys having a solid composition as high as ~20 wt-% Cu. Thus, in these growth experiments the conditions for each requirement were complementary and it was necessary only to know that G/V did not exceed certain limits. In the experiments carried out here, following the attainment of thermal equilibrium with about 2/3 of a specimen molten, some 100 mm of alloy rod was solidified in about 1 week, and with furnace temperatures controlled to within ~$\pm 0.1°$ K it proved possible to avoid planar breakdown over the composition ranges concerned.

Thermal analyses
Sections 5 mm long were guillotined from solidified specimens and their freezing points determined in a multichamber DTA apparatus containing six crucible chambers surrounding a central reference block of pure aluminium or molten tin (Fig. 3). The apparatus was constructed from pure graphite with alumina crucibles and sheaths and 0·125 mm Pt/Pt–13% Rh thermocouples, and was supported in an evacuated silica tube, the thermocouples being taken to an external 'commutator' which allowed successive samples to be examined as the temperature fell along the liquidus trace, i.e. at increasing alloy concentrations. Thermal arrests were distinct and following considerable calibration appeared to be reproducible to within ±1 K.

Chemical analyses
Directionally frozen alloy rods were analysed in two ways. Semiquantitative results were obtained using a Cameca 46 electron microprobe analyser, simultaneously recording Cu and Mn Kα_1 radiations. For these measurements, alloy rods were sectioned into 20 mm lengths and a flat surface ground and polished along one side of each, these longitudinal sections were analysed on polished surfaces and subsequently etched for microscopical examination. Analyses were carried out by driving the samples parallel to the growth axis under the electron beam which was scanning a 300 μm line at right angles, i.e. across the rod axis.[11] Point

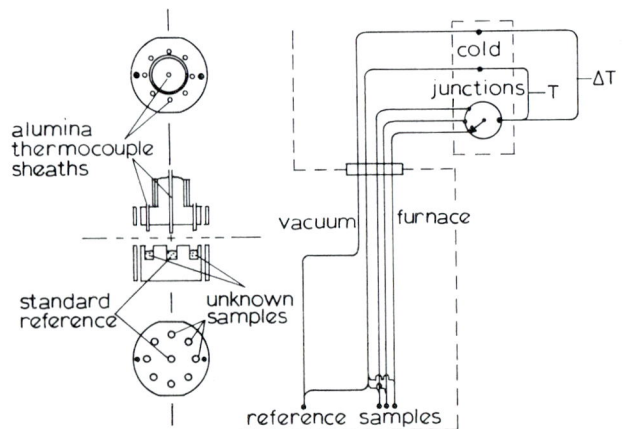

3   Elevation and plan of multichamber DTA apparatus, with thermocouple circuit and example; constructed in pure graphite with silica and stainless-steel supports

analyses were also made of relatively large ($>50\,\mu$m) particles of intermetallic compounds.

No attempt was made to correct the intensity of X-ray emissions in these multicomponent and polyphase samples, but a few smaller sections were dissolved and chemically analysed to provide checks of the actual compositions. In the alloys concerned, copper analyses were slightly higher and manganese slightly lower than the microprobe results.

For determination of solid–liquid distribution coefficients for regions of specimens within the aluminium solid solution, it was found more accurate to use atomic absorption spectrophotometry on dissolved samples. Whereas the microscopical examination and microprobe analyses were made on alloys of relatively high solute contents, (e.g. 10% Cu and 1% Mn) alloys for k_0 determinations were of lower concentrations, (*see* Fig. 1,) so that the greater part of a specimen froze as the primary solid solution. For this part of the work, calibration and trial runs were made first with binary Al–Cu, Al–Mn specimens, and then, from a number of ternary alloys so that the interdependence of k_0 for each element could be expressed in terms of the concentration of both solute elements. Concentration-distance curves were then fitted to that line best fitting the Scheil equation for given or variable values of k_0, and for small variations in the zero position. Estimates of the instrumental errors (e.g. solution concentrations, drift, calibration etc.) suggest that the analyses were well within $\sim \pm 5\%$ of the total in a range 0·05–5 wt-% Cu or Mn. Within these limits, the combined solute traces were self consistent, as shown below.

RESULTS
Microscopical and microprobe analyses
Figure 4 shows the type of data obtained for an alloy of initial concentration 10 wt-% Cu or 1 wt-% Mn, together with the total longitudinal section with microprobe traces for Cu and Mn, and selected optical micrographs, inset below.

At the beginning of the sample, on the left, is a region of high and variable composition which corresponds to that part of the rod which was partially molten during the initial temperature equilibration, – the copper concentrations being the more variable

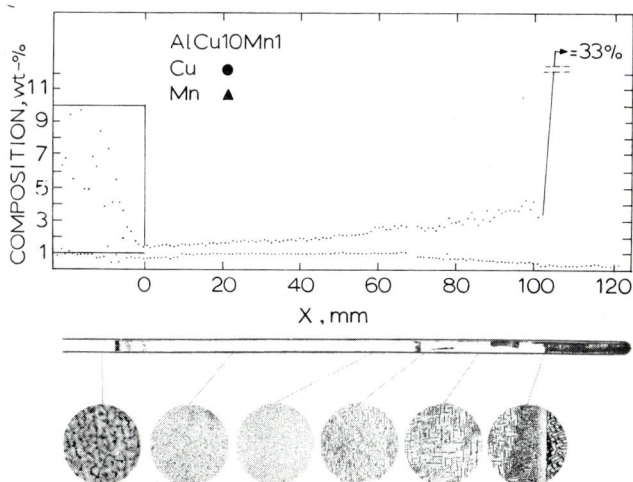

4   **Microprobe and microscopical analyses of directionally frozen alloy of nominal composition 10 wt-% Cu, 1 wt-% Mn**

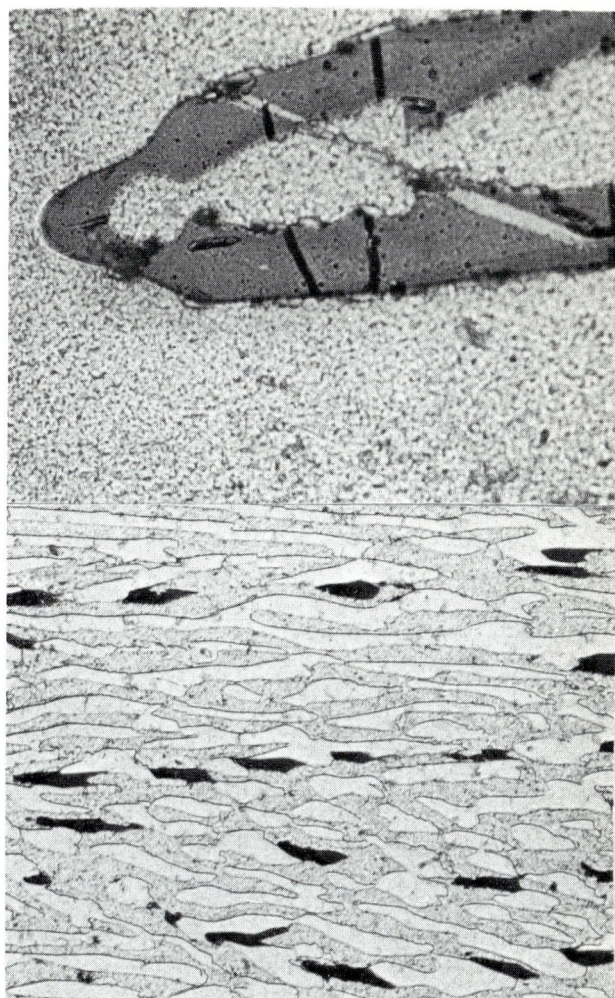

5   *a* **Two-phase region, Al solid solution + ternary compound;** *b* **three-phase terminal eutectic   [×216] (enlarged ×2)**

because of the wider freezing range and lower k_0 for this component.

Directional growth then began at position 0 mm and the composition moved across the solidus surface of the aluminium solid solution up to about 65 mm, the copper concentration rising slowly ($k_0 \sim 0.14$) and that for manganese imperceptibly ($k_0 \sim 0.85$). The limit of solid solubility terminates at ~ 2.5 wt-% Cu and 1.0 wt-% Mn, with the appearance of relatively massive flakes of the ternary compound T of approximate formula $Al_{12}CuMn_2$ and the liquid composition then moves along the monovariant eutectic trough (*see* Fig. 1), terminating at the ternary isothermal eutectic reaction $L \rightleftharpoons Al + CuAl_2 + T$ at ~ 33 wt-% Cu and 0·25 wt-% Mn and 547°C. The analyses recorded in the binary eutectic region correspond to the aluminium solid solution and do not include the massive compound particles; they therefore correspond to the solid solubility limit in that phase, and not to the monovariant trace; (*see* also Figs 5*a* and *b*).

Liquidus determination
Figure 6 shows liquidus data obtained for a specimen of the same composition as that shown in Fig. 4. The initial data falling from the freezing point of pure Al extend only up to the limit of solid solubility, while the

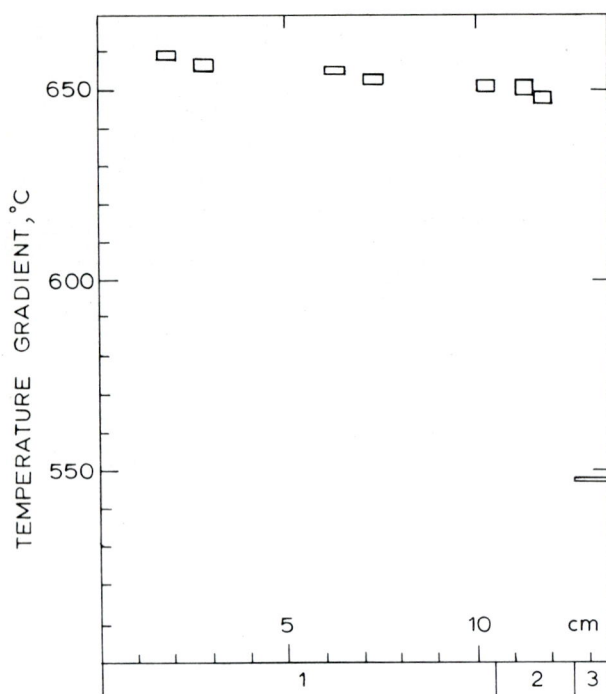

6   **Liquidus trace for 5Cu–1Mn alloy, (cf. Fig. 4)**

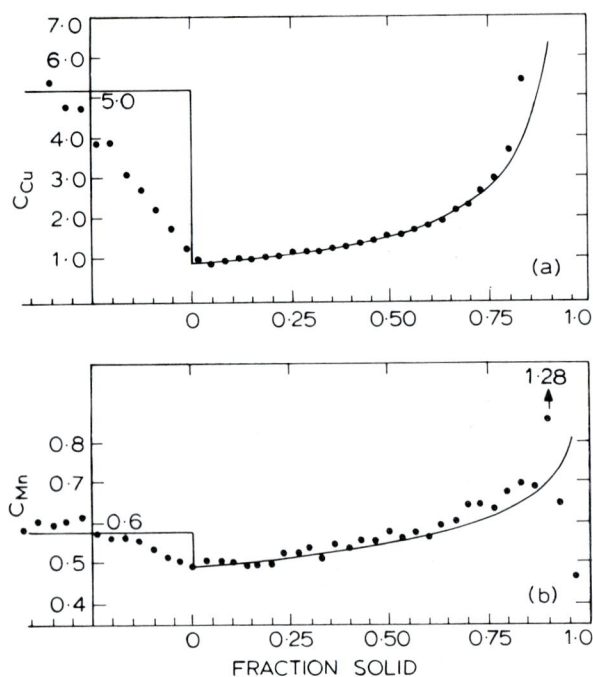

7   **Segregation traces for *a* Cu and *b* Mn in an alloy of nominal composition 5 wt-% Cu, 0·5 wt-% Mn, with the best fitting curves according to the Scheil equation**

subsequent data relate to the monovariant and isothermal parts of the specimen (system). These results are in good agreement with the published data, but necessarily, there is a wide discontinuity between the limit of solid solubility (~2·5 wt-% Cu and 1·0 wt-% Mn) and the point at which the alloy liquid concentration reached the monovariant eutectic at ~17 wt-% Cu and 1·2 wt-% Mn. This limitation is inherent in the method and data, because in any alloy solidified as described above (in the section on experimental requirements and procedures) the composition follows the solidus with an inevitable discontinuity at the solid solubility limit where the composition moves abruptly to that of the subsequent eutectic[8] or peritectic[12] reaction.

Determination of distribution coefficients

Figure 7(*a* and *b*) shows the analytical results for a ternary alloy of initial composition 5 Cu–0·5 Mn, with the Cu and Mn results relating to alternate segments, each 5 mm long. Correlation with the best fits with the Scheil equation are indicated by the solid lines, the initial solid + liquid regions being disregarded, i.e. the significant results relate to the position at which a specimen was completely molten and planar front growth began. Such results were obtained in duplicate for three initial compositions each containing 5 wt-% Cu and 0·3, 0·5, and 1·0 wt-% Mn. It was unnecessary to vary the copper contents since these varied significantly along any given specimen in any case, while the manganese segregation was much less pronounced.

Correlation of the results for each element, as functions of the concentration of each, then allowed the respective k_0 values to be interrelated, as shown below.

1   k Al–Cu $= (0·142 \pm 0·001) + (0·043 \pm 0·002)C_{Mn}$,
      wt-%

2   k Al–Cu $= (0·139 \pm 0·002) + (0·007 \pm 0·004)C_{Cu}$,
      wt-%

3   k Al–Mn $= (0·83 \pm 0·007) + (0·04 \pm 0·01)C_{Mn}$,
      wt-%

4   k Al–Mn, no dependence detected for C_{Cu}, wt-%

In this system the solid distribution coefficients are relatively insensitive to alloy composition, because the liquidus–solidus isotherms are relatively straight and parallel; thus the results are not spectacular. There are, however, cross terms for k_{Cu} and k_{Mn} and one must exist for k_{Mn} as a function of C_{Cu}, i.e. the solidus and liquidus surfaces are slightly curved as functions of both concentration terms. However, in a system such as Al–Cu–Ti,[13] or Cu–Ni–Sn,[3] where the respective liquidus slopes are of opposite sign, i.e. $k_{Al-Cu} < 1·0$ and $k_{Al-Ti} > 1·0$, the interdependence would be very marked and segregation traces curve across the liquidus/solidus surfaces with the concentrations of solutes changing rapidly in opposite senses.

DISCUSSION AND CONCLUSIONS

The present work has been undertaken to show that a general procedure based on directional solidification of alloy specimens can provide a feasible route to complete liquidus and solidus data in a multicomponent alloy system. The results described above relate to directionally solidified specimens of some six different compositions and conform closely in composition and temperature to the diagram of Day and Phillips (Fig. 1),[5] rather than the version given by Hannemann and Schräder.[6] The present results further refine the earlier data; thus, by taking the analytical results given in Fig. 7, and plotting these upon a ternary composition base (Fig. 8), it is possible to produce a segregation trace for an alloy of initial composition X. In fact, the Cu and Mn data are not exactly self consistent unless the limits of analytical accuracy of ~±5% are allowed, but the direction of the solidus trace is not in question. In a

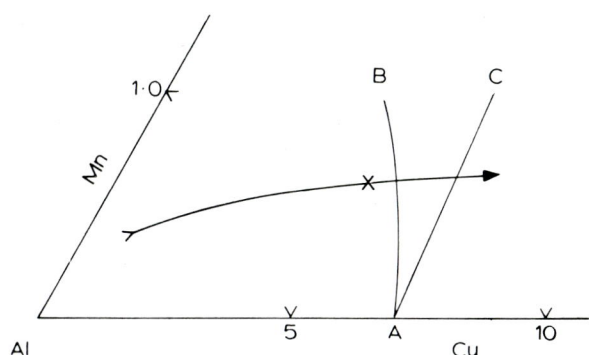

8 Ternary segregation trace corresponding to the data in Fig. 6 indicating the correct (*AB*) and previous[5] (*AC*) traces for 640 K liquidus isotherm

ternary diagram, the segregation trace must lie normal to the liquidus or solidus isotherms at any point, so that from Fig. 8, since the trace is almost linear, it can be seen that the inclination of the liquidus isotherms must be very similar to the solidus isotherms on that surface. The extra data obtained by the present authors then indicate that the liquidus isotherms are inclined (e.g. for 640°C) at a steeper gradient (i.e. towards a lower Mn content) than those given by Day and Phillips.[6] The traces *AB* and *AC* show the more probable and previous 640 K isotherm, respectively.

The results described here demonstrate the value and potential of the method. Clearly, there will be experimental limitations which will preclude the universal application of this approach, e.g. refractory problems, the inability to maintain planar front growth etc., but the method must nevertheless have wide potential application. In the present work, three research students were employed for separate periods of some six to nine months (including development work); D.

M. L. Bartholomew was responsible for k_0 determinations,[13] M. Jezuit for metallographic and microprobe analysis, and B. Watts for the thermal analyses.[14]

ACKNOWLEDGMENTS
The authors wish to acknowledge financial support from Alcan International (Research and Development) Ltd, the Science Research Council (D.M.L.B. and B.W.), and the British Council (M.J.), and are grateful to Professor Sir Peter Hirsch, F.R.S. for laboratory accommodation and facilities.

REFERENCES
1 W. HUME-ROTHERY *et al.*: 'The determination of metallurgical equilibrium diagrams', 1956, London, The Institute of Physics
2 M. G. DAY AND A. HELLAWELL: *J. Inst. Met.*, 1964, **93**, 276
3 B. D. BASTOW AND D. H. KIRKWOOD: *ibid.*, 1971, **99**, 277
4 W. A. TILLER: *Trans. AIME*, 1959, **215**, 555
5 M. K. B. DAY AND H. W. L. PHILLIPS: *J. Inst. Met.*, 1947, **74**, 833
6 H. HANEMANN AND A. SCHRÄDER: 'Ternäre legierungen des aluminiums', 1952, Düsseldorf, Verlag Stahleisen
7 W. B. PFANN: 'Zone melting', 1965, New York, Wiley
8 A. HELLAWELL: *Trans. AIME*, 1965, **233**, 1 516
9 R. M. SHARP AND A. HELLAWELL: *J. Cryst. Growth*, 1970, **8**, 29
10 J. A. BURTON *et al.*: *J. Chem. Phys.*, 1953, **21**, 1 987
11 W. A. TILLER, *et al.*: *Acta Metall.*, 1953, **1**, 423
12 N. J. W. BARKER AND A. HELLAWELL: *Met. Sci.*, 1974, **8**, 353
13 D. M. L. BARTHOLOMEW: D.Phil. thesis, 1974, University of Oxford
14 B. WATTS: Part II thesis, 1976, University of Oxford

In situ radiographic observations of solute redistribution during solidification

M. P. Stephenson and J. Beech

The conditions required for the radiographic examination of aluminium alloys during solidification are outlined. In the aluminium 2 wt-% gold alloys solute profiles which have, at times, been associated with a quenching artefact are shown to exist and to vary with growth rate. The profiles alter as the morphology of the solid/liquid interface is modified. As the interface becomes more cellular more solute is included and the profile diminished. The profile is very sensitive to growth direction and at a downwards growth rate of 0·012 mm/s none could be detected. This is not in agreement with work on aluminium–copper alloys but the discrepancy may be attributed to the difference in density between copper and gold. Growth orientation is shown to have a pronounced influence in the distribution of the Al₃Ni in an aluminium 7·5 wt-% nickel alloy.

M. P. Stephenson is now with the Tube Investments Ltd, Chesterfield, and J. Beech is at the Department of Metallurgy, Sheffield University

A great deal of the understanding of the solidification process has been gained using traditional methods of sectioning the solidified material. This presented some problems[1] and prompted workers to attempt direct observations of one kind or another or use analogs involving non-metallic materials. While a radiographic technique has been used previously[2] the present work is an attempt to improve resolution and to continuously record events using an X-ray TV camera.

DESCRIPTION OF THE TECHNIQUE
The apparatus
The specimen, which is about 300 μm thick, is sandwiched between two pieces of graphite which fit tightly into a slot in the 'graphite furnace'. The temperature of the graphite furnace is controlled by passing a current through it and the graphite sandwich can be withdrawn as indicated, so producing directional solidification. The whole furnace arrangement can be seen in Fig. 1. The X-ray camera tube is separated from the furnace assembly by a 5 μm thick aluminium foil which seals the hole in the water-cooled brass plate that forms the rear of the vacuum chamber. X-rays from a source operated at 35 kV and 10 mA are passed continuously through the sample as solidification proceeds. The sequences can be viewed directly on the TV monitor or tape recorded for subsequent viewing and quantitative assessment. The magnification on the viewing screen is about $\times 40$. The apparatus is shown schematically in Fig. 2 and this has been described in detail elsewhere.

Resolution
Since the film is not in contact with the specimen some geometric blurring (Ug) occurs

$$Ug = W \cdot \frac{L_2}{L_1}$$

where W is the focal width of the X-ray source, L_1 the source–specimen separation and L_2 the specimen–X-ray camera distance: with $L_1 = 350$ mm, $L_2 = 6 \cdot 0$ mm and $W = 0 \cdot 6$ mm a value of $Ug = 10^{-2}$ mm was obtained. This value compares favourably with that of the X-ray vidicon tube in the X-ray camera which is also quoted as 10^{-2} mm.

Contrast
This is governed by the absorption characteristics of the alloy and the effect of the resulting X-rays on the recording medium. Lamberts law $I = I_0 \exp -(\mu x)$, where I_0 is the intensity of the incident X-ray beam and I that of the transmitted beam after passing through material of thickness x and linear absorption coefficient μ.

$$\mu = KZ^3 \lambda^3$$

34

1 'Furnace' and specimen/graphite assembly

where Z is the atomic number of the element and λ the wavelength of the radiation.

Since the present technique demands a short exposure time all of the available energy is required and polychromatic radiation had to be used. Contrast is therefore developed by using alloys containing elements which have large atomic number differences. In addition a low value of K_0 is also useful in that this enhances the μ solid/μ liquid difference. For example an Al–30 wt-% Cu alloy with $K_0 \simeq 0.2$ initially separates solid of Al–6 wt-% Cu in equilibrium with a liquid of Al–30 wt-% Cu.

Experiments on a range of alloys showed that good contrast could be obtained if (μ liquid–μ solid) > 1 mm.

RECORDING MEDIUM
X-ray vidicon
The vidicon, which was X-ray sensitive, shifted the problems associated with movement blurring down to

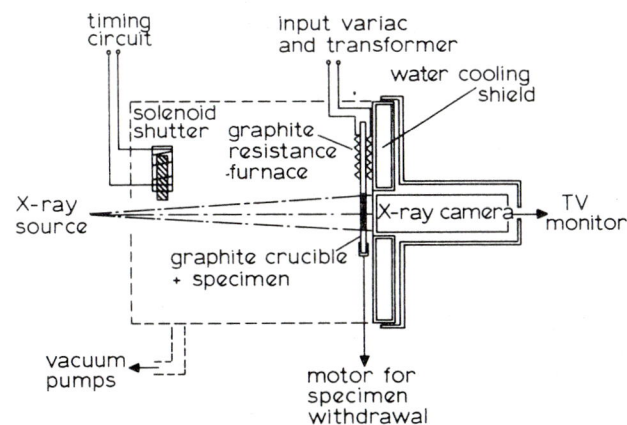

2 Schematic arrangement of the apparatus

the order of the scan time of $\simeq 0.03$ s, i.e. growth rates of up to about 0·3 mm/s could be used. The intensity used had to be sufficient to excite the photoconductive target material of the vidicon and this led to the choice of dimensions and X-ray characteristics already mentioned.

Density measurements
A Joyce–Loebl twin beam recording densitometer was used to examine the density of films which were obtained by photographing the image on the TV monitor.

Standardization
Calibration standards covering the range of compositions obtained in the alloys used were prepared. A number of alloys all of the same thickness were mounted in a single graphite slice, melted and radiographed. Thus a calibration curve of density v. composition was prepared and used to determine compositions in the experimental alloys used.

Growth rate and temperature gradient measurements
The growth rate was simply measured from the TV monitor and this could be done quite accurately. The temperature gradient could be estimated by placing thermocouples at intervals on the surface of the graphite sandwich. Two other built-in checks were available.

One check was provided by the distance between Al$_3$Ni needle tips, which grew as a primary phase, and the eutectic. Assuming that these interfaces grew at the equilibrium temperature the temperature gradient could be estimated. A second check was accidentally discovered when an Al–2 wt-% Au alloy was partially remelted and held at temperature and a band of low X-ray absorption developed (Fig. 3). The width of this was found to correspond to the freezing range of the alloy and again using equilibrium liquidus and solidus values the gradient could be measured. The 'band' formed under a temperature gradient and the intercellular liquid diffused into the bulk liquid until the

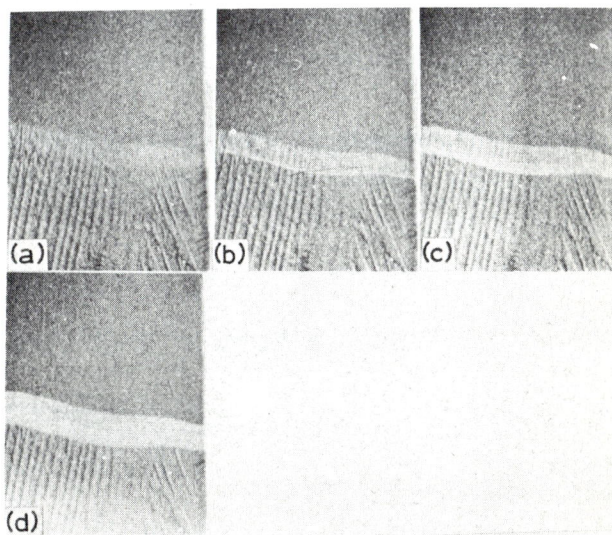

3 Process of the 'white band' development in an Al–2 wt-% Au alloy

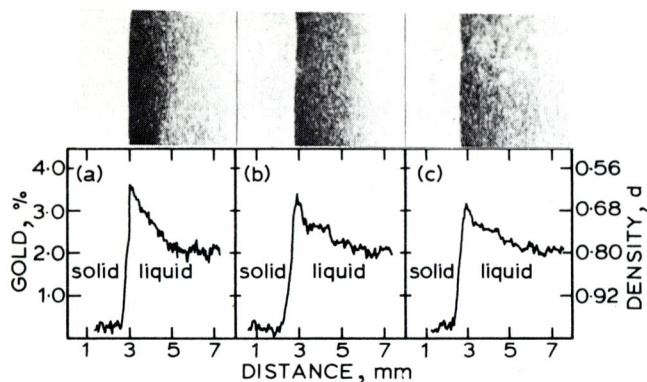

4   **The variation of the solute profile with growth rate**

eutectic temperature was reduced and solidification took place. Both of the checks supported the thermocouple measurements.

RESULTS AND DISCUSSION
Solute redistribution at the solid/liquid interface

Aluminium–gold alloys were used because they fulfilled the conditions outlined earlier and allowed the nature of the interface to be changed by altering the growth rate. The temperature gradient was constant at $\simeq$ 10°C/mm.

In Al–2 wt-% Au alloys, grown upwards at 5, 3, and 1·5 μm/s, the solute profile widths changed from 1 700 to 2 500 to 3 500 μm, respectively. These changes are recorded in Fig. 4, together with the density/concentration plots. It is clear that the quenching artefact to which attention has been drawn by other workers[1] cannot be present here since the observations were made at the solidification temperature.

At these growth rates the interface appeared to be planar on the radiographs and conventional metallography confirmed that this was the case. Certainly no cellular structure was found, but pores and ditches produced by transverse segregation are difficult to detect[3] and it cannot be said with certainty that they were not present in this case. A reasonable value of $7·1 \times 10^{-5}$ cm^2/s, the diffusivity of gold atoms in the liquid, is obtained if the measured widths are taken to be the 'characteristic' distances.

At higher growth rates the interface becomes cellular and the profile changes dramatically as solute is included in the solid. This is shown in Fig. 5 where interfaces obtained at 30 μm/s (Fig. 5a), and 10 μm/s (Fig. 5b) are compared with an apparently planar interface obtained at 3 μm/s, (Fig. 5c). The solute

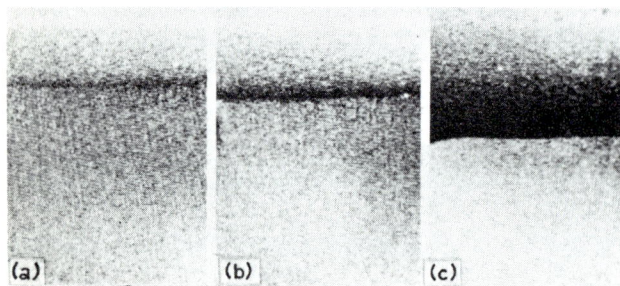

5   **The influence of the solid/liquid growth morphology on the solute profile**

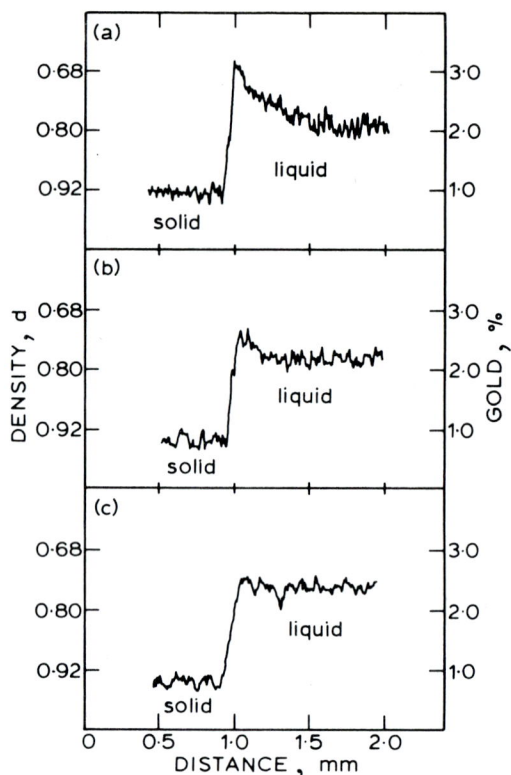

6   **Dependence of the solute profile on growth orientation with respect to gravity**

rejected ahead of the interface is now much less and the profile width is smaller than it would be with a planar interface.

The influence of growth direction with respect to gravity

The work described so far relates to the upward growth of solid into hotter liquid and hence to non-convective conditions. If growth is either downwards or sideways with ostensibly the same growth rate and interface type the solute profile changes, as shown in Fig. 6. It would be expected that in the downgrowth mode (Fig. 6c) the profile would be much reduced, and in fact using the present technique none could be detected. The sideways growth (Fig. 6b) shows a small profile. This is not in agreement with the findings of other workers who used Al–Cu alloys and obtained a profile width, in the downgrowth mode, of about 400 μm. In the present work a smaller profile would be expected in view of the greater density of gold, but it must be less than about 50 μm otherwise there would be no problem in detecting it.

Semiquantitative assessment of overall segregation

The change in composition of the bulk liquid during solidification could readily be determined and compared with a simple theoretical model. Since the interface was not always planar the modified Scheil equation[4] was used

$$\bar{C}_s = K_{eff} C_0 (1-g)^{K_{eff}-1} \qquad (1)$$

where $\bar{C}_s$ is the average concentration of the solid with incorporated solute, and $K_{eff} = \bar{C}_s/CL$. From this the concentration in the liquid is readily obtained

$$C_L = C_0 (1-g)^{K_{eff}-1} \qquad (2)$$

Table 1 K_{eff} **values for downward and upward growth**

Growth rate, μm/s	$K_{eff} = \bar{c}_s/c_L$ down	up
3	0·115	0·43
12	0·45	0·82
40	0·77	1·00
100	0·97	1·00

For a given alloy composition (C_0) and fraction solid (g) K_{eff} can be determined. The K_{eff} values given in Table 1 are for $C_0 = 2$ wt-% and $g = 0\cdot6$ and for down and up growth orientations. In both cases K_{eff} approaches k_0 at slow rates and unity at higher rates. The influence of convection is clearly illustrated, at the slower growth rates, by the lower K_{eff} values and consequently more rapid build up of solute in the liquid. The influence of growth rate is shown for the downward growth case in Fig. 7. The 3 μm/s rate shows a rapid rise as a function of fraction solid with, as shown earlier, less entrapped solute and at 100 μm/s little or no solute build up occurs.

It is clear that convection can occur within the sample despite its small dimensions and that the results obtained for the most part confirm observations made using other techniques. There is some doubt concerning the nature of the interface at 3 μm/s since K_{eff} is not equal to k_0. This either means that some solute is included and we cannot detect it, though this seems unlikely, or the inaccuracies in the measurements made have led to an over estimate of K_{eff}. This can easily occur, for if in equation (2) at g_s 0·6, $c_L = 4\cdot5$, $K_{eff} = 0\cdot115$ as shown in Table 1 but if c_L is 4·72, $K_{eff} = 0\cdot06$. The latter value is approximately equal to k_0. This difference in c_L values is within the estimated accuracy of the determination and so can in itself account for the discrepancy.

The influence of growth direction on crystal growth

Experiments designed to study the growth of a primary phase which depletes the liquid of solute revealed some interesting effects of orientation on primary crystal growth. An Al–7·5 wt-% Ni alloy was used since it satisfied the X-ray contrast requirements, grew in a faceted manner and depleted the liquid in nickel.

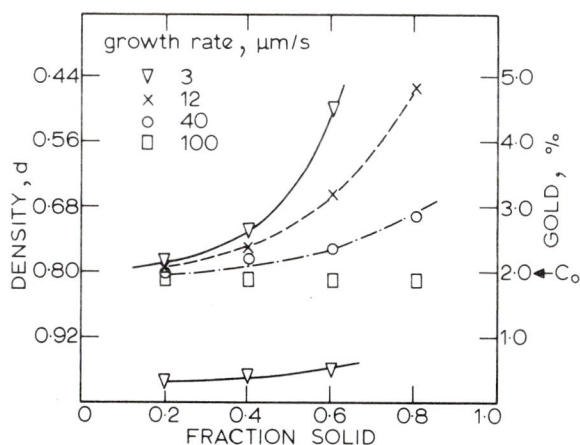

7 Concentration of solute as a function of fraction solidified

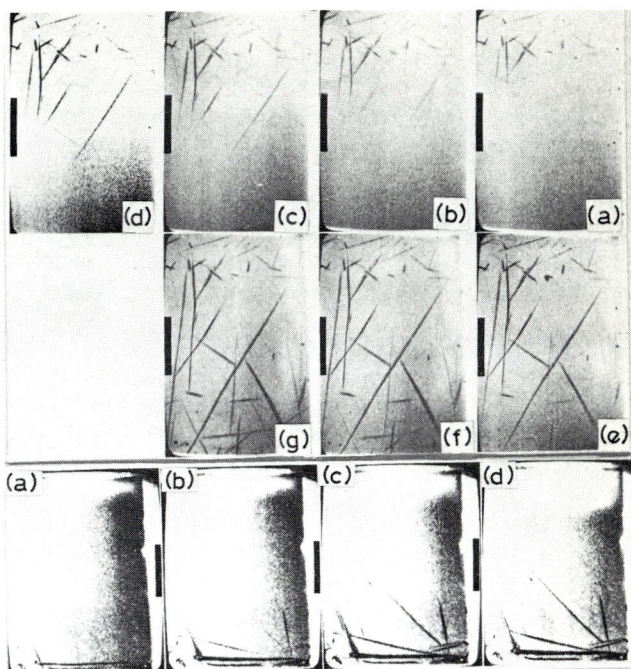

8 Formation of Al₃Ni in a downgrowth mode (top) and in an upgrowth mode (bottom)

Figure 8 shows the formation of Al$_3$Ni, grown in the downward (top picture) and upward (bottom picture) directions respectively, and the two show very marked differences. The primary growth now appears dark since the Al$_3$Ni is more dense. The downward growth situation gives a reasonably uniform distribution of the Al$_3$Ni but the latter phase is found only in the bottom section of the upward growth sequence.

This observation may be explained in terms of convection brought about by changes in chemical composition. The primary Al$_3$Ni crystal depletes the adjacent liquid of nickel and this liquid, being less dense will tend to rise. In the downward growth sequence no convection arises from this source and growth occurs uniformly. When the growth direction is upwards less dense solute-rich liquid rises and the more dense liquid descends. This produces a constant supply of solute-rich liquid for the growing primary phase at the bottom of the specimen. Therefore, as growth proceeds from the bottom the driving force for Al$_3$Ni growth decreases since its nickel content is decreasing. Hence coarse primary Al$_3$Ni needles grow at the bottom and the liquid rapidly falls to the eutectic composition and solidifies as such.

The eutectic is not always distinguishable from the liquid unless the conditions during the various stages of the process are ideal. These conditions are approached in Fig. 8a where the eutectic (not resolved) interface may be seen behind the primary Al$_3$Ni needle tips. This system again shows the importance of convection, and in particular the role of density differences associated with composition which outweigh those related to temperature differences in this case.

CONCLUSIONS

1 The radiographic technique allows solid–liquid interfaces to be studied at temperature and removes problems associated with quenching artefacts.

2 A solute profile, which changes in dimensions with growth rate, exists at 'planar' solid/liquid interfaces.

3 The profile is diminished as the interface morphology changes to a cellular one.

4 The orientation of growth with respect to gravity has a pronounced effect on solute segregation. In downgrowth sequences in Al–2 wt-% Au alloys no profile would be detected at a growth rate of 12 μm/s.

5 The overall solute distribution could be described by the modified Scheil equation.

6 Growth orientation can radically change the distribution of the Al$_3$Ni phase in Al–7·5 wt-% Ni alloys.

ACKNOWLEDGMENTS

The authors thank Professors G. W. Greenwood and B. B. Argent for the provision of research facilities, and M. P. Stephenson is grateful for the financial support received from the Science Research Council.

REFERENCES

1 A. HELLAWELL: 'The solidification of metals', 83, 1967, London, The Iron and Steel Institute
2 J. FORSTEN AND H. M. MIEKKOAJA: *J. Inst. Met.*, 1967, **95**, 143
3 J. FORSTEN AND H. M. MIEKKOAJA: *J. Inst. Met.*, 1971, **99**, 105
4 R. M. SHARP AND A. HELLAWELL: *J. Cryst. Growth*, 1971, **8**, 33

Temperature gradient zone melting and microsegregation in castings

D. J. Allen and J. D. Hunt

Experimental and theoretical work demonstrates that temperature gradient zone melting (TGZM) can cause extensive remelting during the later stages of dendritic solidification, and hence reduce microsegregation in castings. Observations on a transparent 'metal analog' material (succinonitrile) during dendritic solidification showed that the sidearms migrated several secondary arm-spacings up the temperature gradient. The mechanism of TGZM, in which a liquid pool in a temperature gradient simultaneously melts at its hot end and freezes at its cold end, is explained and shown to account for these observations satisfactorily. During rapid cooling, however, remelting may be suppressed. This is discussed in terms of an analytical model of the diffusion field in interdendritic pools. A comparison with published thermal data and dendrite spacing measurements suggests that a significant fraction of the original solid is remelted by TGZM during many, but not all, casting and welding processes. TGZM reduces microsegregation, since (for k < 1) the original, comparatively pure solid which remelts is replaced by less pure solid deposited at a lower temperature. Several observed features of microsegregation in castings may be explained on this basis.

J. D. Hunt is in the Department of Metallurgy and Science of Materials, Oxford University, and D. J. Allen is now with the CEGB Marchwood Engineering Laboratories, Southampton

Direct *in situ* observations of solidification have revealed several unexpected phenomena in recent years.[1] While most non-metallic crystals grow with faceted morphologies, a few organic materials behave like metals in that they undergo non-faceted dendritic solidification.[2] These materials have been extensively studied as transparent analogs for metals. A thin layer of organic alloy may be sandwiched between two glass slides to form a specimen cell, which rests across the gap between the hot and cold plates of a temperature gradient stage on an optical microscope.[3] In the present apparatus, solidification is accomplished by driving the hot and cold plates across the stage while the cell is held fixed, so that the same region of the cell is kept in the microscope viewfield throughout the course of solidification.[4] Figure 1 shows three successive photographs of a fixed region of a succinonitrile – 6 wt–% camphor specimen during steady-state dendritic solidification at a growth rate $R = 3\,\mu\text{m/s}$ and temperature gradient $G \sim 50°\text{C/cm}$. The labelled secondary dendrite arms were identified unambiguously from a sequence of intermediate photographs. Some of the smaller sidearms (e.g. number 7)

have remelted, due to coarsening processes which reduce the dendrite surface area.[5] Simultaneously, all the sidearms appear to have moved two or three secondary arm-spacings up the temperature gradient. Observations of stationary dust particles such as that marked by a vertical line in Fig. 1 confirmed that neither the specimen cell nor the bulk solid moved bodily during the experiment. The sidearm 'movement' is therefore attributed to a continuous process of melting and reprecipitation. If melting occurs at the hot sides of the liquid pools between secondary arms, while solidification occurs at the cold sides, all the sidearms will seem to 'march' together up the temperature gradient (Fig. 2).

TEMPERATURE GRADIENT ZONE MELTING

This apparently surprising behaviour may be understood as a form of temperature gradient zone melting (TGZM).[6] Consider a liquid droplet inside a solid specimen subjected to a temperature gradient (Fig. 3). The opposite ends of the droplet have different temperatures, and therefore have different compositions. Hence, there is a concentration gradient across

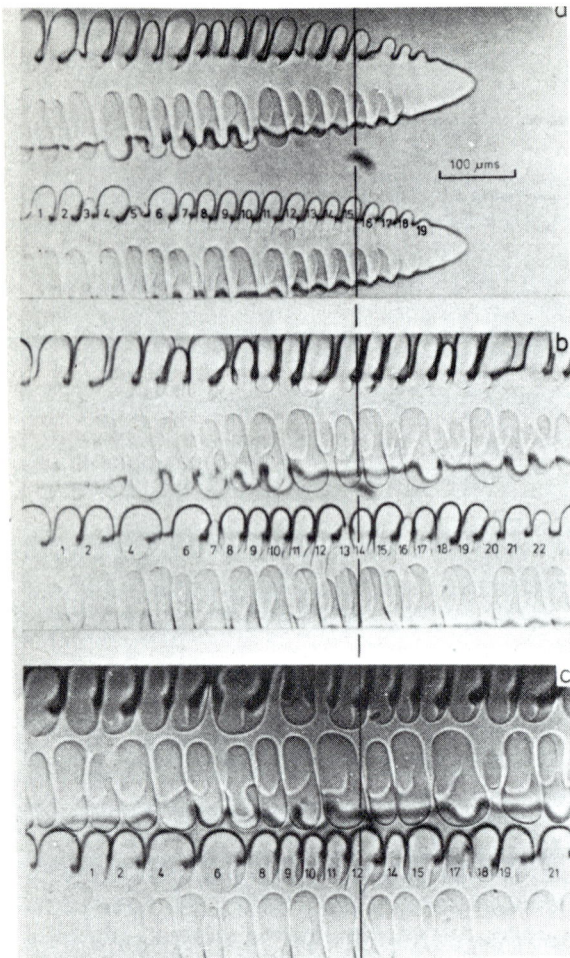

1   A fixed region of a solidifying succinonitrile–camphor specimen after, *a* 0 s; *b* 180 s; *c* 480 s [×98]

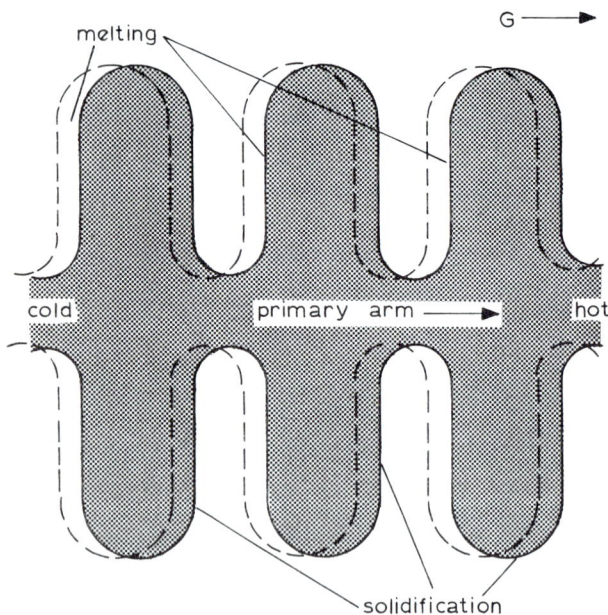

2   Sidearm 'movement' by TGZM

the droplet, and solute diffuses from cold to hot. This causes solidification at the cold end and melting at the hot end, so that the droplet migrates up the temperature gradient. TGZM is quite well known within the

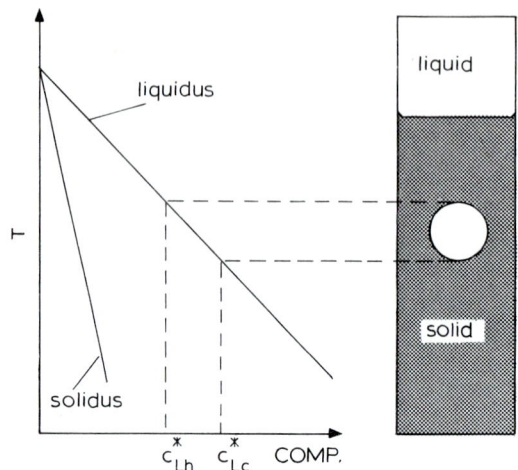

3   Mechanism of TGZM

fields of single crystal manufacture,[7] interface kinetics studies,[8,9] and nuclear fuel element degradation.[10] There have also been observations of 'solute trails' resulting from TGZM during cellular solidification.[11] However, until recently there has been little interest in the possible occurrence of TGZM during dendritic solidification. Indeed, it is by no means obvious that the model outlined above for droplet migration in a stationary thermal gradient ('static' TGZM) should have any relevance to dendritic solidification, where the morphology and thermal conditions are quite different.

Experimental evidence which shows that sidearm migration during solidification is similar to static TGZM has been fully described in a previous paper.[12] Using the same specimen as was used for the solidification experiment of Fig. 1, the migration velocities of liquid droplets undergoing static TGZM were measured over a range of temperatures. It was found that these velocities, when correctly averaged over the solidification range, equalled the average sidearm migration velocities within the experimental error of $\sim \pm 20\%$. As in static TGZM, the melting and freezing interfaces in Fig. 1 moved up the temperature gradient at very similar velocities.

Perhaps the most surprising aspect of these results is the extensive remelting which occurs despite continuous cooling of the specimen. The behaviour of the liquid pools between secondary arms may be discussed in terms of two simultaneous competing processes. TGZM acts to make both sides of the pool move up the temperature gradient. Cooling of the specimen, which causes an increase in local fraction solid, acts to make the two sides of the pool move towards each other. If TGZM is dominant, as in our experiments, the hot side of the pool melts and moves up the temperature gradient at a velocity approaching the velocity of the cold side. If solidification due to cooling is dominant, however, the hot side of the pool freezes and moves down the temperature gradient. As the thermal conditions in castings and weld pools may be somewhat different from our experimental conditions, further analysis is required to determine whether TGZM or solidification is dominant in these cases.

DIFFUSION IN INTERDENDRITIC LIQUID POOLS

Consider a binary alloy of liquidus slope $m = +dT/dc_L$ and distribution coefficient k which cools through its freezing range at a rate F in a linear temperature gradient G. Liquid composition is denoted by c_L, which becomes c_L^* at the solid/liquid interface. Liquid pools between neighbouring secondary dendrite arms may be simply modelled as planar-sided zones of width L perpendicular to the temperature gradient (Fig. 4). The origin of coordinates is positioned at the centre of the liquid zone, with the z axis parallel to G. If local equilibrium is assumed at the solid/liquid interface, c_L^* lies on the liquidus line, given by

$$\left(\frac{\delta c_L^*}{\delta z}\right)_t = \left(\frac{\delta T}{\delta z}\right)_t \Big/ \left(\frac{\delta T}{\delta c_L^*}\right) = \frac{G}{m} \tag{1}$$

as shown by the dashed line in Fig. 4. This only applies to the composition at the solid/liquid interface. As in all diffusion problems it is necessary to make further assumptions to determine the solute distribution elsewhere in the bulk material. When the secondary arms are first formed, there will be a transient solute distribution in the surrounding liquid corresponding to the rapid initial growth process. However, it has been shown[13] that the liquid pool width is generally very much smaller than the diffusion distance $\sim \sqrt{D_L t}$ through which a concentration disturbance is able to propagate during the local solidification time. It is, therefore, reasonable to assume that the initial transient rapidly relaxes to some form of quasi-stationary solute distribution. Now, as the specimen cools, c_L in the interdendritic liquid must continuously increase. We will assume that the rate of increase of c_L is the same at all parts of the liquid pool, thus obtaining a quasi-stationary solute distribution curve which moves

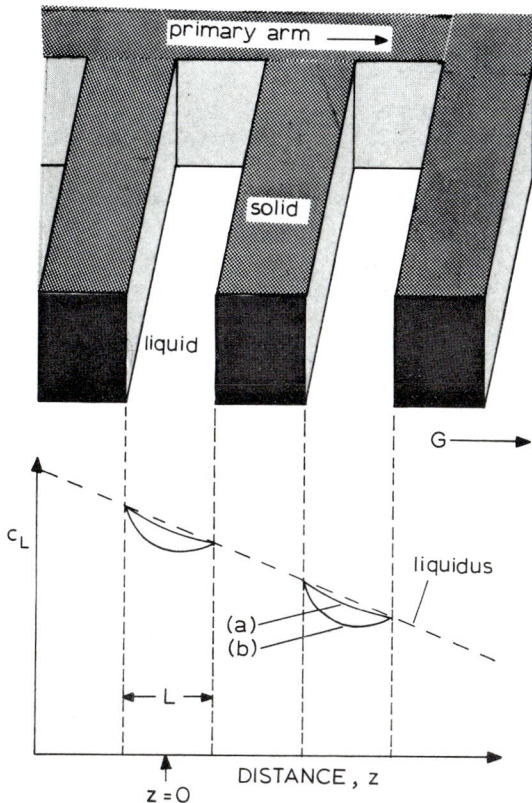

4 Model of interdendritic solidification, showing solute distributions in the liquid, *a* when TGZM is dominant; *b* when solidification is dominant

up the composition axis without changing shape. The rate of increase of c_L^* at a stationary segment of the solid/liquid interface is given (from the local equilibrium assumption) by

$$\left(\frac{\delta c_L^*}{\delta t}\right)_z = \frac{-F}{m} \tag{2}$$

Therefore, since c_L increases everywhere at the same rate,

$$\left(\frac{\delta c_L}{\delta t}\right)_z = \frac{-F}{m} \quad \text{throughout} \tag{3}$$

Hence, from Fick's second law,

$$\left(\frac{\delta^2 c_L}{\delta z^2}\right) = \frac{-F}{mD_L} \tag{4}$$

where D_L is the diffusion coefficient in the liquid. Equation (4) reflects the curvature of the diffusion profile required to cause an increase in c_L throughout the solidification process. Equation (1) reflects the 'tilt' of the diffusion profile due to the temperature gradient. A solution is obtained by integrating equation (4) twice, using equation (1) to put in boundary conditions at the interface on each side of the pool, and then redifferentiating to derive

$$\frac{\delta c_L}{\delta z} = \left(\frac{G}{m}\right)\left(1 - \frac{Fz}{D_L G}\right) \tag{5}$$

in the liquid pool. Neglecting solid-state diffusion, the interface velocity v (in the positive z direction) on either side of the liquid pool is given by

$$v c_L^* (1-k) = -D_L \left(\frac{\delta c_L}{\delta z}\right)_{\text{interface}} \tag{6}$$

Writing:

$$\frac{F}{G} = R' \tag{7}$$

where R' is the isotherm velocity along the temperature gradient direction (which equals R for the case of steady-state solidification), equations (5)–(7) may be combined to give

$$v_c = \left(\frac{-D_L G}{m c_L^* (1-k)}\right)\left(1 + \frac{R'L}{2D_L}\right) \tag{8}$$

$$v_h = \left(\frac{-D_L G}{m c_L^* (1-k)}\right)\left(1 - \frac{R'L}{2D_L}\right) \tag{9}$$

where the suffixes c and h refer to the cold and hot sides of the pool, respectively: note that m is negative when $k < 1$. Therefore remelting is suppressed if L exceeds a critical value

$$L_{\text{crit}} = \frac{2D_L}{R'} \tag{10}$$

This is represented by curve (*b*) in Fig. 4. However, v_h approximately equals v_c if

$$L \ll L_{\text{crit}} \quad \text{or} \quad \left(\frac{R'L}{2D_L}\right) \ll 1 \tag{11}$$

In this case, TGZM is dominant, as shown by curve (*a*) in Fig. 4. These equations provide a simple criterion to determine whether or not melting occurs. Melting is favoured by a high temperature gradient, low cooling rate, and low liquid pool width.

To estimate the fraction of the original solid in the secondary arms which is remelted by TGZM, f_m, we obtain the velocity of the pool centre

$$\bar{v} = \frac{-D_L G}{mc_L^*(1-k)} \qquad (12)$$

from equations (8) and (9). So the pool centre moves a total distance

$$d = \int \bar{v}\, dt \sim \int \bar{v}\left(\frac{-m}{GR'}\right) dc_L^*$$

$$\sim \left(\frac{D_L}{R'(1-k)}\right) \ln\left(\frac{c_2}{c_1}\right) \qquad (13)$$

where c_1, c_2 are the values of c_L^* in the liquid pool at the beginning and end of its solidification, respectively. Taking d as roughly equal to the distance moved by the hot side of each pool, and using secondary arm spacing on completion of solidification, λ_2, as a measure of the scale of structure

$$f_m \sim \frac{d}{\lambda_2} \qquad (14)$$

provided L is small by comparison with L_{crit}.

COMPARISON WITH EXPERIMENTAL DATA

In our experiments (Fig. 1), $R' = R = 3\ \mu m/s$, and $D_L \sim 3 \times 10^{-6}\ cm^2/s$ in succinonitrile alloys.[4] Therefore, from equation (10), $L_{crit} \sim 200\ \mu ms$, while from Fig. 1, $L \sim 5\ \mu ms$. Hence, from equations (8) and (9), v_c should not exceed v_h by more than about 5%. This prediction is in good agreement with the experimental observations.

Published data on typical casting and welding processes may be used to calculate L_{crit} and d from the above equations for comparison with measured λ_2 values, as shown in Table 1. Thermal data on small aluminium[14] and steel[15] sand castings are presented by Ruddle.[16] In the aluminium casting, an approximate average $G \sim 0.5°C/cm$ and $F \sim 3 \times 10^{-2}°C/s$, while D_L may be taken as $\sim 3 \times 10^{-5}\ cm^2/s$: λ_2 is obtained from the results of Bardes and Flemings[17] for the measured local solidification time of $\sim 3 \times 10^3\ s$. In the steel casting, $R' \sim 6 \times 10^{-3}\ cm/s$ and D_L may be taken as $\sim 10^{-4}\ cm^2/s$, while λ_2 is obtained from the results of Suzuki *et al.*[18] for the measured $F \sim 5 \times 10^{-2}°C/s$. In an aluminium weld, Lanzafame and Kattamis[19] measured

Table 1 TGZM during casting and welding

Material	Process	L_{crit}, μms	d, μms	λ_2, μms
Al–4% Cu	sand casting (cylinder, 13 cm dia. × 26 cm)	10	12	150
0.6% carbon steel	sand casting (18 cm square section)	300	250	450
2014 alloy (Al–4 Cu–1 Si–1 Mn)	TIG welding	6	7	14

L_{crit}, d, and λ_2 are defined in equations (10), (13), and (14) respectively, and discussed in the section concerning comparisons with experimental data

$\lambda_2 \sim 14\ \mu ms$ for a welding speed $\sim 4 \times 10^{-1}\ cm/s$. Since the shape of a weld pool and its associated isotherms is approximately independent of time, the isotherm velocity R' equals the component of welding travel speed resolved parallel to the local temperature gradient. So R' increases from the sides to the centre of the weld and may be taken as $\sim 10^{-1}\ cm/s$ in this case.

To find out whether or not melting occurs, L must be estimated from the measured final secondary spacing λ_2 and compared to the calculated L_{crit}. Since L is probably much smaller than the secondary spacing, which itself commonly increases several-fold by coarsening in the semi-solid region,[5] L may be an order of magnitude smaller than λ_2 (*see* Fig. 1). Therefore, using Table 1 and equation (10), it is suggested that melting probably occurs throughout the semi-solid region in the steel casting and aluminium weld, but may be restricted to the last stages of solidification (when L becomes very small) or suppressed altogether in the aluminium casting. Given these conclusions, the extent of remelting may be estimated from equation (14). This suggests that about half the original secondary arm material remelts in the steel casting and aluminium weld, but little or no remelting occurs in the aluminium casting. It may be noted that all the original secondary arm material remelted in the experiments of Fig. 1.

The difference in behaviour of the two castings may be attributed to the higher volumetric latent heat and lower thermal conductivity of steel, which promote a higher ratio of temperature gradient to cooling rate. The greater significance of TGZM in welding than in casting of aluminium alloys may be explained in terms of the external heat input and consequent high thermal gradient in welding processes.

MICROSEGREGATION

Remelting on this scale must be expected to affect microsegregation in the finished product. When the solidus slope dT/dc_S is negative, the effect of TGZM (and of dendrite coarsening) is to remelt comparatively pure solid deposited close to the growth front and replace it with less pure solid deposited at a lower temperature. Soluble impurities are thus incorporated in the primary phase, leaving less material to form interdendritic second phases in the final stages of solidification. Thus TGZM reduces microsegregation.

The 'complete mixing' or 'Scheil' equation[20] is commonly assumed to describe microsegregation in castings, although it generally overestimates the severity of segregation quite significantly. Solid-state diffusion,[13] coarsening,[21] and growth temperature depression[22] have all been invoked to explain the discrepancy. Since the secondary arm spacing commonly trebles during solidification,[4,5] coarsening processes alone may remelt about half the original solid. It is apparent that both TGZM and coarsening should often act to reduce microsegregation substantially.

The extent of TGZM, and the quantity of heat flow through a given region, both depend on the product of temperature gradient and time. In a casting, the quantity of heat extracted through a given region during solidification increases with distance from the thermal centre. These considerations suggest that more TGZM occurs near the sides of large ingots than in the small castings considered in this paper. Doherty and

Melford[22] observed that the severity of microsegregation in a large ingot increased gradually on moving away from the walls through the columnar zone, and increased sharply on entering the equiaxed zone. While other factors may also be involved, these results may plausibly be attributed to the greater homogenizing effects of TGZM in the columnar zone and away from the ingot centre.

ACKNOWLEDGMENTS

We thank Professor Sir Peter Hirsch, F.R.S., for provision of laboratory facilities. D. J. Allen acknowledges financial support from the Science Research Council.

REFERENCES

1 M. E. GLICKSMAN: 'Solidification', 155, 1971, Ohio, American Society for Metals
2 K. A. JACKSON AND J. D. HUNT: *Acta Metall.*, 1965, **13**, 1 212
3 J. D. HUNT *et al.*: *Rev. Sci. Instrum.*, 1966, **37**, 805
4 D. J. ALLEN: 'Dendritic solidification', D.Phil. Thesis, 1976, Oxford University
5 T. Z. KATTAMIS *et al.*: *Trans. TMS–AIME*, 1967, **239**, 1504
6 W. G. PFANN: *Trans. AIME*, 1955, **203**, 961
7 G. A. WOLFF AND A. I. MLAVSKY: 'Crystal growth: theory and techniques', vol. 1, (ed. C. H. L. Goodman), 193, 1974, New York, Plenum Press
8 W. A. TILLER: *J. Appl. Phys.*, 1963, **34**, 2 757
9 H. E. CLINE AND T. R. ANTHONY: *J. Appl. Phys.*, 1972, **43**, 10
10 F. A. NICHOLS: *J. Nuclear Mat.*, 1969, **30**, 143
11 W. BARDSLEY *et al.*: *Solid-State Electronics*, 1962, **5**, 395
12 D. J. ALLEN AND J. D. HUNT: *Met. Trans. A.*, **7A**, 767
13 T. F. BOWER *et al.*: *Trans. TMS–AIME*, 1966, **236**, 624
14 R. W. RUDDLE: *J. Inst. Met.*, 1950, **77**, 1
15 H. F. BISHOP *et al.*: *TAFS*, 1951, **59**, 435
16 R. W. RUDDLE: 'The solidification of castings', 2 ed., 1957, London, The Institute of Metals
17 B. P. BARDES AND M. C. FLEMINGS: *Mod. Cast.*, 1966, **74**, 100
18 A. SUZUKI *et al.*: *Nippon Kingaku Gakkai Shuho*, 1968, **32**, 1 301: also Brutcher translation 7 804, 1969
19 J. N. LANZAFAME AND T. Z. KATTAMIS: *Welding J. Res. Supp.*, 1973, **52**, 2 26s
20 E. SCHEIL: *Z. Metallkd*, 1942, **34**, 70
21 M. C. FLEMINGS *et al.*: *J. Iron and Steel Inst.*, 1970, **208**, 371
22 R. D. DOHERTY AND D. A. MELFORD: *J. Iron Steel Inst.*, 1966, **204**, 1 131

Microsegregation in Fe–Cr–C alloys solidified under steady-state conditions

B. A. Rickinson and D. H. Kirkwood

Fe–Cr–C alloys have been solidified under controlled steady-state conditions, and quenched during growth to investigate the development of dendrite morphology and microsegregation. No solute undercooling at the primary dendrite tips was measurable in any alloy examined, and the minimum chromium content at the centre of the primary arms did not change during solidification. The secondary arms formed initially at a composition close to the minimum in the primary cores, and developed by growth into discs with a cellular structure at their boundaries. These tertiary cells were transitory, annealing out rapidly and contributing to solute back diffusion into the discs. An asymmetric distribution of solute was measured through the thickness of the discs, and is believed to arise from the migration of the secondary arms under the temperature gradient causing significant homogenization. The secondary arm spacing increases directly with growth time, and the rate is sensitive to composition so that the more highly alloyed specimens possess markedly finer dendrite structures. The final microsegregation of chromium in the solidified alloys of $1\frac{1}{2}$% Cr–1% C and $1\frac{1}{2}$% C–2% C depends on the solidification conditions, and a minimum in segregation occurs at cooling rates around $100°C/min$. Increasing the carbon content initially worsens the segregation in $1\frac{1}{2}$% Cr alloys, producing a maximum at ~1·6% C as in previous work.

The authors are in Department of Metallurgy, University of Sheffield

Since the introduction of microprobe analysis, much work has been carried out on the problem of microsegregation in low-chromium steels,[1-4] especially on Fe–$1\frac{1}{2}$% Cr–1% C since they are particularly prone to the formation of banded microstructures which result in a deterioration of the mechanical properties. These investigations have involved the examination of fully solidified ingots, for which the cooling rates have been measured in most cases but not controlled.

The steady-state technique of solidification offers two important advantages. First, it allows the independent control of the velocity of the growth front, R, and of the temperature gradient in the liquid ahead of the front, G_L. Secondly, it allows the solidifying specimen to be quenched, and the development of the dendrite morphology and solute segregation to be examined, effectively during growth. This can give valuable information on the mechanism of microsegregation, and has recently been applied to aluminium–copper alloys,[5] to stainless steels[6] and other low-alloy steels.[7] The present investigation is the first attempt to apply this technique to the problem of microsegregation in low-chromium steels.

EXPERIMENTAL DETAILS

Alloys were made up from high-purity electrolytic iron and chromium, and from graphite, then hot rolled into rods and surface ground to fit alumina tubes of 6 mm i.d. and 250 mm length. The specimens contained in the alumina tubes were suspended within a vertical high-frequency induction furnace containing an argon atmosphere, and a thick graphite susceptor acted as the heat source. The specimens were driven at a predetermined rate down through the short hot zone where melting occurred. Below the hot zone solidification began at some point within the furnace, and at steady-state the velocity of the growth front equals the rate of traction of the specimen through the furnace. The temperature gradient was controlled by the power input and calibrated in runs employing a thermocouple within a specimen. During steady-state solidification the specimens were quenched by dropping into a brine bath beneath the furnace.

Both longitudinal and transverse sections were made on these specimens to study dendrite morphology and microsegregation. The degree of microsegregation in

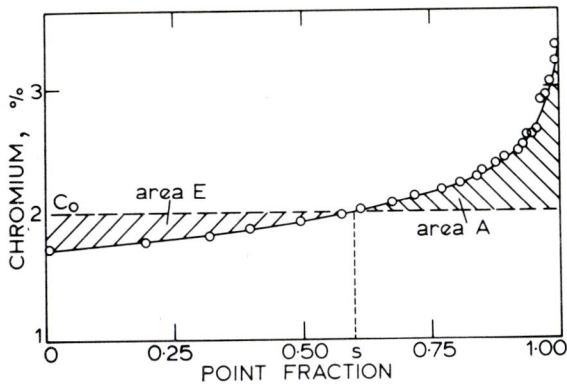

1 Cumulative solute profile for 1·9% Cr–0·47% C alloy; grown at $R = 6$ mm/min, $G_L = 11·7°C/mm$; sectioned at $\sim 1170°C$

any region was assessed in two ways; first, by taking the segregation ratio

$$S = \frac{\text{maximum conc. of Cr}}{\text{minimum conc. of Cr}}$$

and secondly using the segregation area index, E. The latter requires the construction of a cumulative solute profile, as in Fig. 1. It is obtained by taking random point analyses over a specimen and plotting the results as composition C against the cumulative volume fraction (the fraction of points whose composition is $\leqslant C$).[3] The total area under the curve in Fig. 1 corresponds of course to the average composition C_0. The area E in the figure has been used as a measure of microsegregation and tends to zero as segregation decreases. It will be noted that it does not depend on the form of the curve at high point fractions and is therefore useful in describing situations where some high solute liquid or carbides exist.

RESULTS
Growth morphology

Transverse sections of the rod specimens are normal to the primary dendrite arms of the growing solid, and in all specimens these primary arms were seen to be ordered in a close packed array (Fig. 2a). From the etching behaviour in the micrograph and from the distribution of solute in the section (*see* microsegregation), it appears that secondary arms grow from the primary stalks initially in the form of rods and subsequently thicken sideways breaking down into cellular tertiary arms before impinging on other growing arms. These cellular arrays are visible in Fig. 2a and their development is sketched diagrammatically in Fig. 3. This process is similar to that described by Subramanian *et al.*[8] for the solidification of iron alloys in small ingots, but in their case the tertiary arms were more developed and higher order arms were also observed. In the present work decreasing the temperature gradient in the liquid and increasing the alloy content were both found to favour the production of tertiary arms, and this may explain the difference in morphology of the ingot and steady-state specimens.

These tertiary cells were found to be transitory in the solidification of all the specimens. At temperatures greater than 20°C below the temperature of the primary dendrite tips, this structure had virtually disappeared leaving only secondary arms, often roughly hexagonal in outline (Fig. 2b) which approximate to discs with a diameter greater than their thickness. The disappearance of the tertiary cells can occur by two processes, (a) by diffusion of solute in the liquid from the cell walls as a result of high surface curvature leading to a coarsening of the solid structure;[9,10] or (b) by diffusion of the solute into the solid from the cell walls which results in an annealing out of the structure.[8] Presumably both occur, but annealing out by solid-state diffusion appears to be occurring at the roots of the cells in Fig. 2a, and this is confirmed by probe analysis

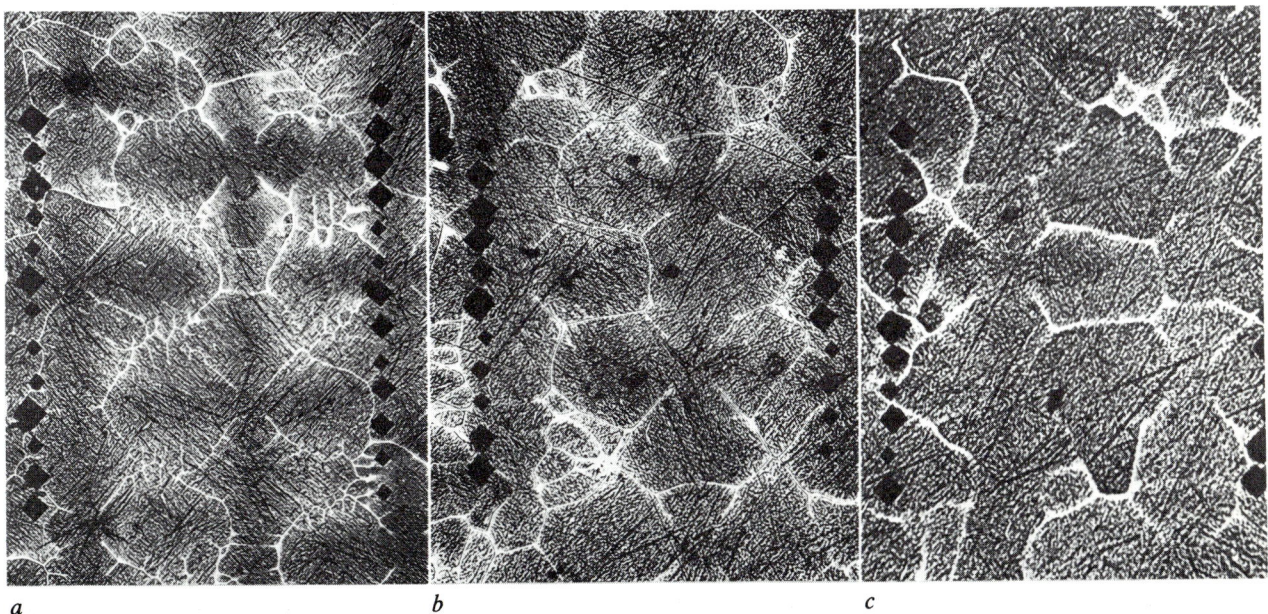

a 2 mm behind tips; *b* 12·8 mm behind tips; *c* 20·9 mm behind tips

2 Transverse sections of 1·4% Cr–1·0% C specimen, grown at $R = 15$ mm/min, $G_L = 6·7°C/mm$ [× 75]

3   Illustration of development of tertiary cells on secondary plates

of the region. It is a possibility that the cellular structure is an artefact introduced by quenching the solidifying specimen; however, this interpretation is not consistent with the observation of cell boundaries in the act of annealing. At around 200°C below the temperature of the tips the outline of the secondary discs is still clearly visible but more diffuse (Fig. 2c), and only a small amount of liquid as films at the boundaries appears to exist.

Primary arm spacing

The primary arm spacing was measured on transverse sections from the distance between neighbouring primary cores, taking an average of 10 such measurements. The spacing was found to be independent of the distance behind the primary tip, and therefore was measured only on one section for most specimens.

The effect of varying both the growth rate R and the temperature gradient in the liquid G_L on the primary arm spacing of the $1\frac{1}{2}$% Cr–1% C alloy was determined, and may be represented by the empirical relation

$$\lambda_1 = \text{const. } G_L^{-0.7} R^{-0.48}$$

However, the scatter in the results was rather large, and the exponents cannot be considered to be more accurate than ± 0.1. Figure 4 shows the logarithmic plot of λ_1 against cooling rate (measured as both $G_L . R$ and $G_{L+s} . R$ where G_{L+s} is the average temperature gradient within the liquid + solid region) and both produce a good linear relationship.

Secondary arm spacing

From longitudinal sections of specimens, one can reveal the long primary arms and obtain the secondary arm spacing at any point behind the tips of the primary dendrites. The secondary arm spacing was observed to increase with distance from the tip as in aluminium–copper alloys[5] as a result of dendrite coarsening. An example is given in Fig. 5 for the 1·4% Cr–1% C alloy solidified at a constant growth rate of 30 mm/min, and it is apparent that the temperature gradient has no effect on the spacing with distance from the tips in this series of experiments. The same conclusion was reached at the slowest growth rate, $R = 6$ mm/min. However, at the intermediate rate of $R = 15$ mm/min it does seem that the spacing is higher for some specimens grown under lower temperature gradients; this result is out of line with the rest and may possibly arise from an error in identifying the position of the growth front. The effect of increasing the growth rate is to decrease the spacing at a given point behind the tips, and it has been proposed that the spacing is a function only of the time of existence θ_g of the secondary arms. Plotting the spacing against θ_g in Fig. 6 shows that most points fall on a single line, if the results from $R = 15$ mm/min are excluded. Also included in Fig. 5 are the results from the steady-state solidification of a commercial steel AISI 52100 (Fe–1·7 Cr–0·96 C–0·43 Mn–0.34 Si) from which it is seen that small amounts of other elements can have a quite dramatic effect on reducing secondary arm spacings.

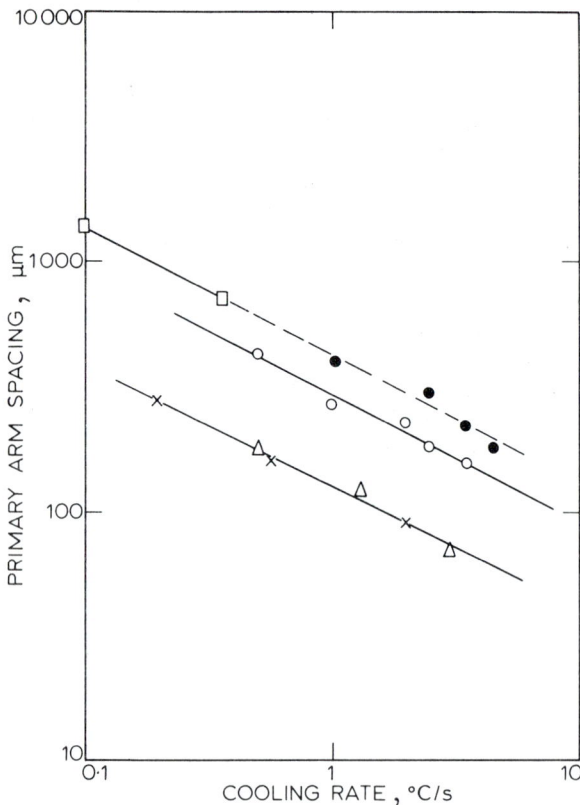

4   Primary arm spacing λ_1 against cooling rate; 1·4% Cr–1·0% C alloy; ● $G_L . R$, ○ $G_{L+s} . R$, △ ingot = present work; □ Refs. 1, 2; × Ref. 3

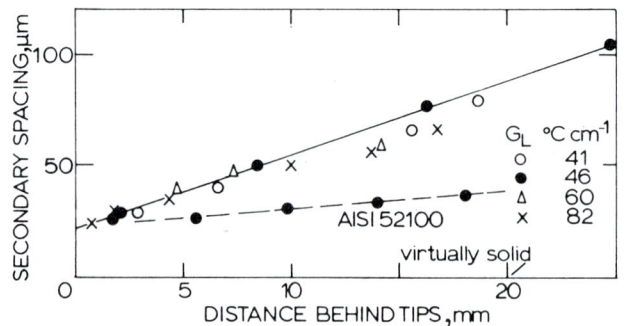

5   Secondary arm spacing λ_2 against distance behind growth front for specimens grown at 30 mm/min; 1·4% Cr–1·0% C alloy

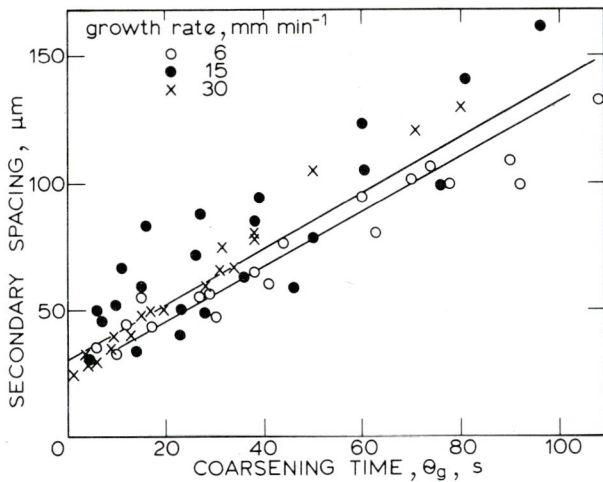

6 Secondary arm spacing λ_2 against growth time θ_g for all 1·4% C–1·0% C specimens

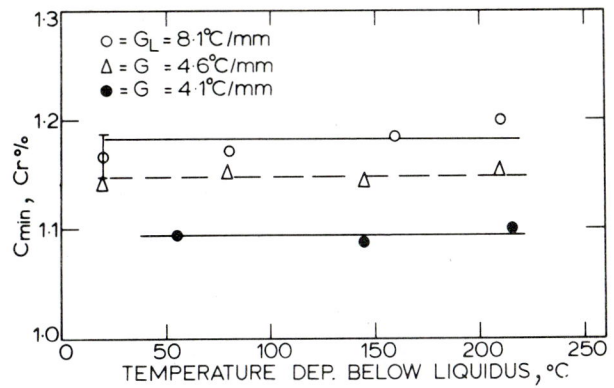

8 C_{min} plotted behind the growth front as temperature below that of tips, ΔT; 1·4% Cr–1·0% C alloy, specimens grown at $R = 30$ mm/min

Figure 7 is a plot of λ_2 against θ_g for a number of alloys of different composition. Again we obtain an approximately linear relationship in each case, and it is clear that the addition of carbon and of chromium both cause a considerable lowering of the coarsening rate.

Microsegregation during growth

In an attempt to follow the process of segregation during solidification, specimens of 1·4% Cr–1% C alloy grown under different conditions were sectioned transversely at different distances behind the growth front and the distribution of solute analysed.

From Fig. 8 it is seen that C_{min} at the primary cores (or within the secondary arms) does not change during solidification at $R = 30$ mm/min. This is true also for most other growth velocities within the accuracy of microprobe measurement. The possible exception is one specimen grown at $R = 15$ mm/min under a gradient of $G_L = 6°C$/mm where a small increase is apparently observed, and in view of the low value of final microsegregation recorded for this case the observation may be genuine. The value of C_{min} at the primary tips of Fig. 8 appears to show some effect of

G_L; however, this is not borne out by the results at other growth velocities, and the discrepancy between the values at $G_L = 4·6$ deg/mm and $G_L = 4·0$ deg/mm in Fig. 8 also indicates that the difference must arise from some other source, most probably a small variation in average composition. There is no clear evidence in this work that either G_L or R affect the value of C_{min} at the dendrite tips.

Although C_{min} does not change with distance from the tips, C_{max} as measured does, and gives rise to a change in the segregation ratio with fraction solidified. Figure 9 is a plot of segregation ratio as a function of temperature below that of the tips, ΔT, and is typical of the $1\frac{1}{2}$% Cr alloys. Under all growth conditions the ratio increases at first to a peak between $\Delta T = 60°C$ and $\Delta T = 80°C$, and thereafter falls to a constant value beyond $\Delta T = 200°C$, i.e. about 1 250°C. Sections taken at 1 100°C confirmed this constancy and revealed a complete absence of carbides in the 1·4% Cr–1% C alloy and in the commercial alloy under the growth conditions employed.

Many of the points above are also illustrated in the cumulative solute profiles presented in Fig. 10. It is seen that the level of chromium in the solute poor regions rises as solidification progresses although C_{min} remains unaffected. The high C_{max} for the 6 mm section corresponds to the segregation peak in Fig. 9, but it should be remembered that artefacts can be produced in quenching small volumes of liquid[8] and therefore C_{max} may be exaggerated. This would mean that part or all of the segregation peak arises also from a quenching artefact.

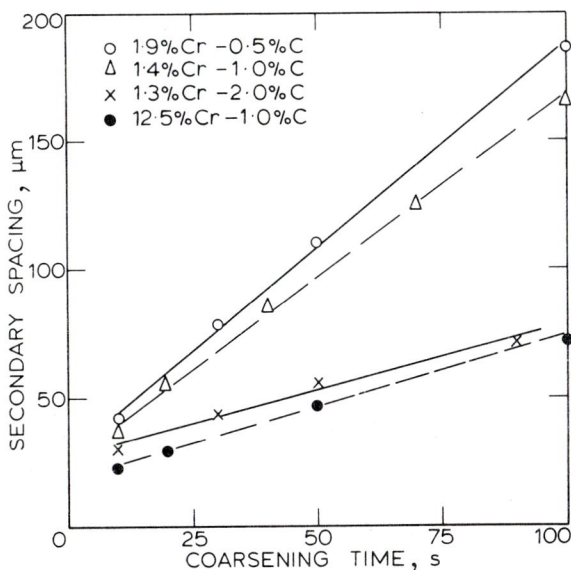

7 Change in secondary spacing with growth time for different alloys

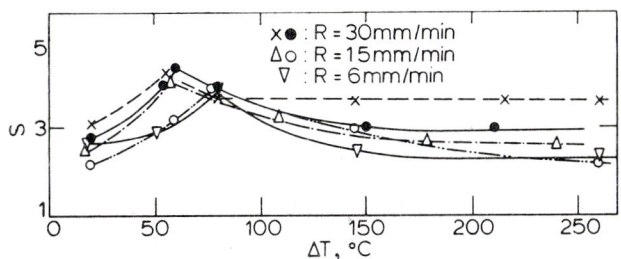

9 Segregation at different temperatures behind the growth front for 1·4% Cr–1·0% C alloy

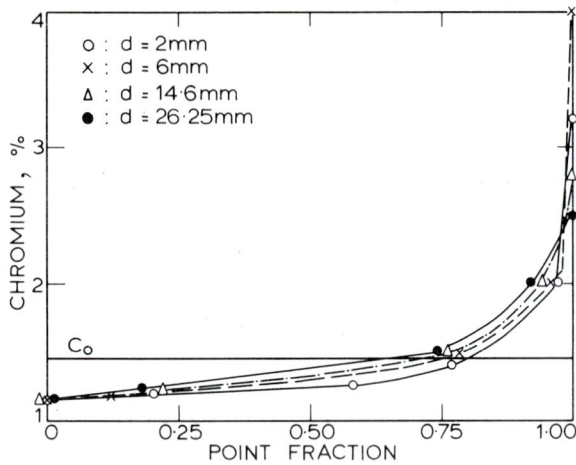

10   Cumulative solute profiles at different distances *d* behind tips for 1·4% Cr–1·0% C alloy; *R* = 6 mm/min; *G*$_L$ = 9·7°C/mm

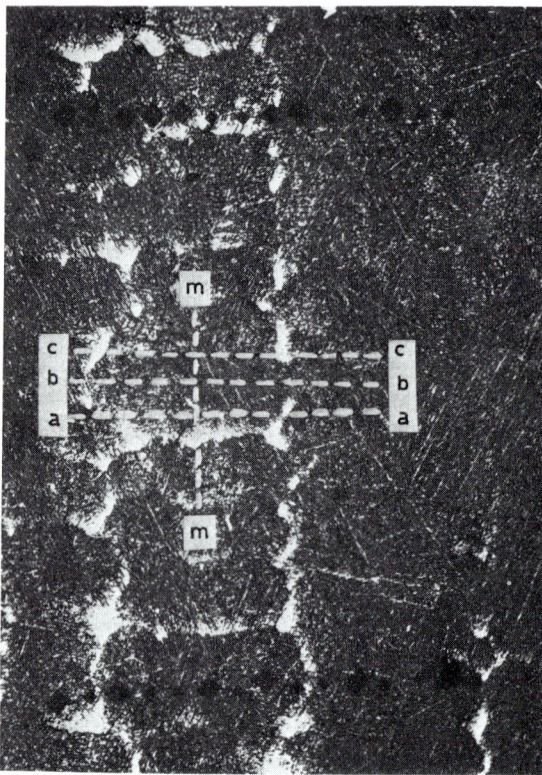

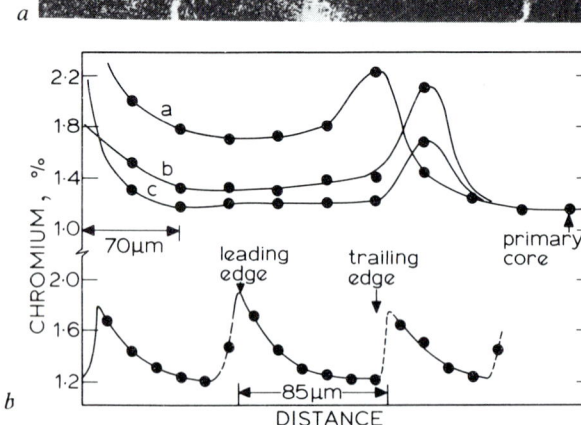

11   *a* Longitudinal section ~15 mm behind growth front for 1·4% Cr–1·0% C alloy; *R* = 15 mm/min; *G*$_L$ = 6·7°C/mm [×110]; *b* various solute profiles for chromium, scans indicated on Fig. 11*a*

A more detailed study of the solute distribution and its relation to the dendrite morphology was carried out in a few cases. The longitudinal section in Fig. 11*a* at about 15 mm from the tips shows a primary arm from which secondary arms, or discs as is clear from Fig. 2*b*, are attached. The profiles of chromium content are given for microprobe step scans across this section in different directions (Fig. 11*b*). The line scan *m*–*m* traverses through the centres of secondary dendrites parallel to the primary arms. The interesting fact is that the saw-tooth profile of solute is not symmetrical about the dendrite centre, but the surface of the disc facing towards the tips (the 'leading edge') has a high solute content which falls rapidly on moving into the dendrite and reaches a minimum near the other face (the 'trailing edge'). A series of microprobe traverses (*a*–*a*, *b*–*b*, and *c*–*c*) were also made along the length of a secondary disc from its edge to the primary core, and the composition profiles are given in Fig. 11*b*. These traverses correspond well with the thickness traverse *m*–*m*. It is clear that (*a*) the traverse *c*–*c* along the trailing edge indicates a composition in the disc close to that of the primary core; (*b*) in each section through a disc parallel to its face there is a large area fraction of the same composition, but this plateau composition increases rapidly on approaching the leading edge; and (*c*) the composition also rises within each section at the edges of a disc. It should be noted that the composition peak observed between the disc and the primary core occurs because the plane of section in the specimen is not exactly through the centre of the secondary disc.

This asymmetric distribution of solute was observed very clearly in the micrographs of high-chromium alloys solidified under steady-state conditions. Figure 12 is an example from a 12% Cr steel showing a non-martensitic rim high in solute at the leading edge of each plate. The same distribution has been detected in aluminium–copper[11] and other alloys,[12] and is believed to arise from 'temperature gradient zone melting' (TGZM) as described by Allen and Hunt.[13]

The effect of growth conditions on final segregation

To compare the resultant segregation produced by differences in growth conditions, sections were made where the alloy was considered to be fully solidified and beyond which no further changes in segregation were observed. Typically, this was in a region of the specimen which had been quenched from about 1200°C, where it has been established that the above conditions hold.

The segregation area index *E* was preferred for these comparative experiments since it does not rely on the extreme values of composition which may be exceptional in any given specimen, and also because it allows comparison between specimens containing carbides for which C_{max} may not vary. However, the segregation ratio has been used for comparison with other work.

No consistent effect of the growth conditions on final segregation could be detected in the 1·9% Cr–0·5% C alloy. The plot of *E* against the gradient *G*$_L$ for three different growth rates is given in Fig. 13 for the 1·4% Cr–1% C alloy. A minimum in the final segregation is evident for each growth rate, and the segregation at each minimum appears to be similar. Two results from the commercial steel were obtained which

12 Secondary arms in 12·5% Cr–1·0% C specimen, exhibiting martensite-free zone at leading edge [×250]

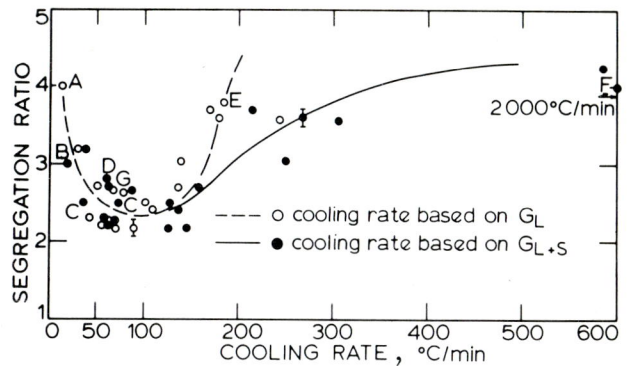

A, E unidirectional ingot[3]; B 2½ t ingot[1]; C 1% Cr steel[4]; D small ingot[3]; F brine-quenched sample[1](~2 000°C/min); G foundry casting, present work

14 Change in microsegregation with cooling rate for 1·4% Cr–1·0% C alloy;

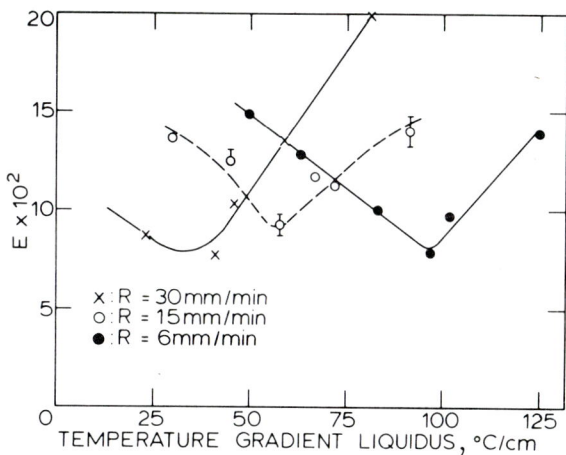

13 Change in microsegregation with growth conditions for 1·4% Cr–1·0% C alloy

cooling rate of around 100°C/min. The experimental results from previous work on similar alloys appear to fit on this curve quite well. An even better correlation is obtained on a plot of E against $G_L^2 . R$, producing a minimum at $G_L^2 . R = 90°C^2/(mm \; s)$, but no theoretical justification can be adduced for this.

Similar behaviour is obtained in the 1·3% Cr–2% C alloy in which carbides of the type M_3C appear, containing ~6% Cr. A plot of E against cooling rate for this alloy again indicates a minimum in segregation in the region of 100°C/min.

DISCUSSION

A systematic investigation into the equilibrium partition coefficients of chromium and carbon between solid and liquid in the Fe–Cr–C system has been carried out by the authors[14] in parallel with the solidification work. The results have been used in Table 1 to calculate $k_0 C_0$ for chromium, where C_0 is the average composition of the alloy obtained by microprobe analysis of the quenched liquid well away from the growth front, and this is compared with the average composition measured at the primary tips. The agreement is extremely good over a whole range of alloys, and it is clear that no solute undercooling is measurable at the growing dendrite tips under any of the growth conditions employed. The secondary arms formed near a primary tip probably possess a C_{min} about 5% higher than the tip, but this will result from growth into liquid

gave higher segregation for the same growth conditions than in the pure ternary alloy. Exactly the same pattern of results is produced using the segregation ratio in place of E and the minimum occurs at $S = 2·1$.

Since the minimum is achieved at a higher value of G_L for smaller R, it is instructive to plot segregation against cooling rate, $G_L R$. This has been done in Fig. 14, and in spite of some scatter a single curve may be drawn through all the experimental points which reveals that the minimum in final segregation associated with a

Table 1 Chromium concentration at dendrite tips

Alloy	C_0, % Cr	k_0	$k_0 C_0$	Average C_{min} at tips, %Cr
1·3% Cr–2·0% C	1·30	0·81	1·06	1·04
1·4% Cr–1·0% C	1·38	0·83	1·14	1·13
1·9% Cr–0·47% C	1·95	0·86	1·68	1·70
1·5% Cr–0·19% C	1·55	0·91	1·41	1·42
12·5% Cr–1·0% C	12·48	0·78	9·76	9·76
20% Cr–1·0% C	19·74	0·83	16·38	16·38
AISI 52100	1·70	0·77	1·31	1·28

of higher solute content rather than solute undercooling. It will be noted in Table 1 that the commercial steel (AISI 52100) has a significantly lower k_0 value compared with the pure alloy of similar composition, and this is no doubt responsible for the higher segregation observed in this steel. As we shall see, the partition coefficient is an important parameter in determining microsegregation in alloys in which back diffusion is limited, and its sensitivity to small compositional variations must contribute greatly to the variations in segregation between different industrial castings.

Consideration of the solute distribution in the longitudinal section shown in Fig. 11b can give some guidance in estimating relative contributions of solute back diffusion and solute incorporation during growth. At this section, which is ~15 mm back from the growth front, the first solid was formed about 1 min previously so that the maximum distance permitted for solid state diffusion is $2\sqrt{Dt} = 7\ \mu\mathrm{m}$ (using $D = 2 \times 10^{-13}\ \mathrm{m^2/s}$). Inspection of the solute profiles in Fig. 11b indicates that the solute enhanced rim around the discs penetrates ~70 μm and the profile extends from the surface at the leading edge into the thickness of the disc to ~60 μm. It is clear that solid-state diffusion can only play a part in lowering the solute maximum at the edge of the discs, but not in producing the extensive homogenization observed within the discs in these specimens.

Much of the solute in the rim of the disc will have been incorporated into the solid during growth in accordance with the Scheil equation. The predominant growth in the establishment of the secondary arms is along their length, followed by a widening into disc-like plates. It has been explained that this last stage involves the formation of cellular arrays at the perimeter of the plates which subsequently anneal out partly by back diffusion from the cell boundaries into the rim of the disc, and this must be a further reason for the extent of the solute profile.

Doubtless incorporation of solute during growth occurs to some extent in the thickening of the secondary discs, but to obtain the asymmetric saw-toothed profiles measured along their thickness direction, they would have to be produced first as thin plates and thicken by growth in one direction only. There is no evidence for this. A more probable explanation arises from some recent work by Allen and Hunt,[13] who have observed in transparent organic analogs that secondary arms migrate along the primary stalk under the influence of a temperature gradient. This occurs when the liquid separating two adjacent arms is at the same time in local equilibrium with the trailing edge of one interface and the leading edge of the other downstream where the local temperature is slightly lower. The maintenance of local equilibrium results in a composition gradient in the liquid and a flux of solute between the interfaces, and this in turn leads to a remelting of the trailing edge of the upper dendrite and solidification at the leading edge of the lower dendrite. Since this is occurring between all secondary dendrites the effect is to produce a migration of the arms up to the temperature gradient towards the primary tips. As solute-poor solid is being melted off and solidification is occurring from solute-enriched liquid a saw-tooth profile will develop in the arms and increasing amounts of solute will become incorporated into the solid.

This mechanism is believed to be the cause of the saw-tooth profile of solute in Fig. 11b. Allen and Hunt calculate that the distance migrated by an arm is given, approximately, by

$$d = \frac{D_L}{R(1 - k_0)} \ln\left(\frac{C_2}{C_1}\right)$$

where D_L is the diffusion coefficient in the liquid ($\sim 7 \times 10^{-7}\ \mathrm{m^2/s}$); R is the tip growth rate ($= 15$ mm/min); k_0 the partition coefficient ($= 0.83$); and C_1 ($\sim 1.2\%$) and C_2 ($\sim 1.8\%$) are the compositions of the solid at the tip and in the rim of the secondary arm at the point of interest, respectively. Using these values, which are for the section in Fig. 11a, we have $d = 67\ \mu$m in good agreement with the extent of the solute profile. If these estimates are correct, there must be little doubt that dendrite migration under a temperature gradient is significant during steady-state solidification. Since the section was taken at about half solidification time, the final migration throughout solidification could be greater than the arm spacing and must lead to considerable homogenization of the dendritic structure. This presumably does not apply to the solute rich regions at the circular boundaries of the discs which join to form walls of high solute liquid parallel to the temperature gradient. This is probably the explanation of the structure in aluminium–copper alloys[5] solidified under steep temperature gradients, where the secondary arms are observed to anneal out leaving long films of solute between the primaries.

The degree of microsegregation inherited by a solidified alloy therefore depends on the extent to which the segregation, developed by the rejection of solute from the growing solid, is mitigated by the two processes of back diffusion and homogenization through dendrite migration. Changes in these two processes with G_L and R must in some way account for the minimum in segregation found at certain solidification conditions.

It might be expected that increasing G_L would favour dendrite migration and therefore encourage homogenization. Allen and Hunt state that the total distance migrated by an arm is independent of G_L, since an increase of G_L reduces the time available for migration. However, their quantitative treatment of the problem is for a simple binary system, and their comment supposes that the last liquid to solidify is always of the same composition. It is not clear then that their comment applies to the more complex situation encountered in ternary alloys.

The variation of final secondary arm spacing λ_2^0 with solidification time θ_s may often be described by the expression: $\lambda_2^0 = \mathrm{const.}\ \theta_s^n$ where n is a constant determined by experiment. We have seen that the spacing during steady-state growth in Fe–Cr–C alloys is directly proportional to the time of growth, and if it is permissible to extrapolate these curves to complete solidification it would suggest that n = 1. Based on a mathematical analysis of back diffusion during solidification, Brody and Flemings[15] have argued that where $n > \frac{1}{2}$, microsegregation should be decreased by increasing the cooling rate. This might explain the initial fall in segregation with G_L observed in Fig. 13. The other factor which has been demonstrated to contribute to back diffusion in this study is the annealing out of the high solute regions in the tertiary cells which fringe the boundaries of the secondary discs. The development of these cells is restricted or prevented at high G_L (or cooling rate) presumably because the primary arms are too close to one another. It is suggested that when this happens more solute is rejected into the interdendritic regions between the discs, and if this cannot be accommodated by the thinner discs as G_L increases microsegregation must increase. The minimum in segregation is interpreted in this picture as a consequence of the complex morphology of the growing dendrite and the way it changes with growth conditions; the minimum will occur where the liquid is as highly dispersed throughout the structure as possible, allowing the maximum opportunity for the effect of back diffusion.

Doherty and Melford[1] were the first to attempt to measure the influence of cooling rate on the segregation ratio in these alloys by cooling small laboratory melts in different environments. Unfortunately most of the structures were equiaxed which are known to produce exceptional segregation. Their brine-quenched specimen ($\sim 2\,000°C/min$) of Fe–$1\frac{1}{2}$% Cr–1% C was columnar and gave a value of $S \sim 4.0$, but the cooling conditions are clearly far from the proposed optimum. Flemings *et al.*[3] measured the segregation ratio and cooling rates at different distances from the chill face in a unidirectional cast ingot of the same alloy and obtained values of $S \sim 4.0$ with no apparent variation. Some of the points are included in Fig. 14 and it is clear that they straddle the minimum. However, they also quenched a small ingot, which was cooled at a rate near to the conditions for minimum segregation, and obtained a value of $S \sim 2.8$ which lies nicely on the curve in Fig. 14. A directionally solidified ingot of this alloy was cast as part of the present investigation and sectioned at 13 mm from the chill face where the cooling rate was 78°C/min. The value of S at this point was found to be 2.65, and confirms that it is possible to obtain low values of microsegregation in foundry castings for certain well defined cooling conditions. The very low values of segregation ratio measured by Fredriksson and Hellner[4] in Fe–1% Cr–1% C steels are also explained by the fact that they fall within the minimum.

The influence of carbon on microsegregation in low-chromium alloys is indicated in Fig. 15, where the values of E have been normalized to 1.4% Cr. Because of the effect of cooling rate there can be no unique curve: one curve is a plot of minimum segregation and the other, above it, of a standard growth condition. In each case the segregation increases at first with carbon content as in previous work;[1,3,16] in particular the minimum segregation curve is in excellent agreement with Fredriksson and Hellner.[4] These authors attribute the increase to the change in partition coefficient of chromium with carbon. Fredriksson[6] has shown that Flemings' analysis of back diffusion during solidification can be used to estimate the segregation ratio where diffusion does not penetrate the dendrite core. The results of his calculation correlate reasonably well with experiment and it seems likely that k_0 is the dominant factor involved, particularly since it will also play an important role in dendrite migration. Both curves in Fig. 15 reach a maximum; the lower one occurs at ~ 1.6% C in close agreement with previous work.[1,3]

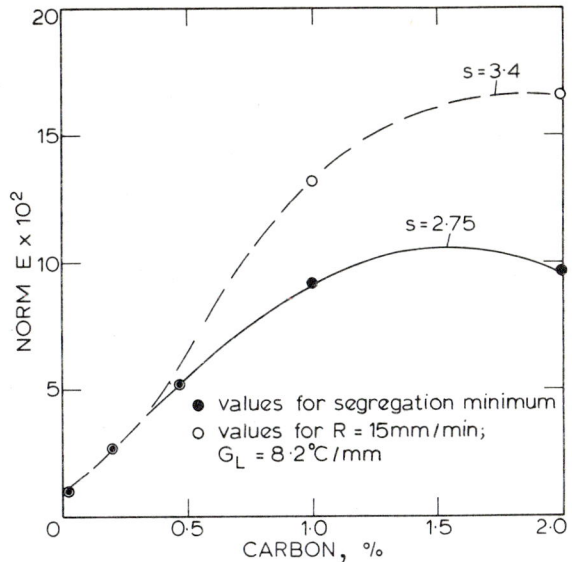

15   Change in microsegregation with carbon content for 1·4% Cr alloys

Despite the general agreement in the microsegregation results of the steady-state experiments and the previous work on ingots, some caution is required in making direct comparisons. An example is given in Fig. 4 where the primary dendrite spacing is plotted for different cooling rates. The results of Melford and co-workers[1,2] on a $2\frac{1}{2}$ t ingot appears to fit well with the steady-state values, but those of Flemings *et al.*[3] are distinctly lower. Our own results from the directionally-solidified ingot referred to above, are consistent with Flemings. It seems unlikely that such large discrepancies in primary arm spacing can be caused by minor compositional differences although small amounts of Si and Mn in the commercial alloy had significant effects on the chromium partition coefficient, on the rate of coarsening of the secondary arms and on the development of the tertiary cells. Other explanations for the discrepancy are, (*a*) the high values of G_L in the steady-state work as compared with the near zero gradients in the ingots; (*b*) the rapid change in cooling rate in the ingots which may not allow the primary arms to adjust to the steady-state value rapidly enough.

REFERENCES

1   R. D. DOHERTY AND D. A. MELFORD: *J. Iron Steel Inst.*, 1966, **204**, 1 131
2   D. A. MELFORD AND D. A. GRANGER: in 'Solidification of metals', 289, 1968, London, The Iron and Steel Institute
3   M. C. FLEMINGS *et al.*: *J. Iron Steel Inst.*, 1970, **208**, 371
4   H. FREDRIKSSON AND O. HELLNER: *Scand. J. Metall.*, 1974, **3**, 61
5   K. P. YOUNG AND D. H. KIRKWOOD: *Metall. Trans.*, 1975, **6A**, 197
6   H. FREDRIKSSON: *Metall. Trans.*, 1972, **3A**, 2 989
7   D. B. MOHARIL *et al.*: *Metall. Trans.*, 1974, **5**, 59
8   S. V. SUBRAMANIAN *et al.*: *J. Iron Steel Inst.*, 1968, **206**, 1 027
9   T. Z. KATTAMIS *et al.*: *Trans. AIME*, 1967, **239**, 1 504
10   M. KAHLWEIT: *Scripta Metall.*, 1968, **2**, 251

11 B. A. RICKINSON: Ph.D. Thesis 1975, University of Sheffield

12 F. WEINBERG AND E. TEGHTSOONIAN: *Metall. Trans.*, 1972, **3**, 93

13 D. J. ALLEN AND J. D. HUNT: *Metall. Trans.*, 1976, **7A**, 767

14 B. A. RICKINSON AND D. H. KIRKWOOD: *Met. Sci.*, 1978, **12**, 138

15 H. D. BRODY AND M. C. FLEMINGS: *Trans. AIME*, 1966, **236**, 615

16 P. M. ZHURENKOV AND I. N. GOLIKOV: *Met. Sci. Heat Treat. Metals*, 1964, (5–6), 293

Electric analog of eutectic growth

R. W. Series and J. D. Hunt

Using an electrical analog technique it has been possible to obtain numerical solutions to the diffusion equation relevant to the solute diffusion field ahead of a growing lamellar eutectic. By combining these solutions with the thermal conditions at the solid/liquid interface it is possible, by an iterative technique, to find self-consistent interface shapes. In the analog a plane wave term ahead of the interface is included. As a result, in contrast to a previous analog method,[1] it is possible to study any system. No ad hoc assumptions are made concerning the magnitude of this plane wave term and as a result we are able to find self-consistent interface shapes under conditions where Nash and Glicksman[2] predict no interface exists. The method can be used to study the behaviour of the solution to the diffusion equation under extreme conditions provided that $D/V > \lambda/4\pi$. Under normal conditions (for metallic systems) good agreement is found between results predicted by the analog and those calculated from the Jackson and Hunt theory.[3]

The authors are in the Department of Metallurgy and Science of Materials, Oxford University

LIST OF SYMBOLS

a_j curvature undercooling constant
B_0, B_n, A_0, A_n, V_0 Fourier coefficients
C^* reduced composition
C, C_e, C_i, C_s composition, eutectic, interfacial, solid
C_∞ $C_e + C_\infty$ is the bulk composition
D liquid diffusion coefficient
m_j signed liquidus slope
$\underline{n}$ normal to interface
r local radius of curvature of interface
S_α, S_β lamellar half width
$\underline{S}$ line of interface
ΔT interface undercooling
T, T_e, T_i temperature, eutectic, interface
v velocity
V_i mean potential along interface
$\Delta \bar{V}_\alpha = \bar{V}_i - (1/S'_\alpha)\int_0^{S'\alpha} V_i \, dx'$
x, z moving coordinate system
λ lamellar spacing
$\theta_\alpha, \theta_\beta$ *see* Fig. 1
$'$ primes denote the relevant quantity in the analog system

The analysis of the growth of a lamellar eutectic requires that solutions be found to the heat and solute flow equations. The shape of the solid/liquid interface is dictated by the interface equation which relates the temperature and composition of the material in contact with the interface to the equilibrium phase diagram. The interface equation links the temperature and composition fields at the interface. In this paper only two terms are included, a composition term and a capillarity or curvature term. Both terms make appreciable contributions to all systems during lamellar growth. Kinetic terms could be included if desired provided a suitable rate law can be given. Isotropy of surface energies is assumed although anisotropy could be incorporated if desired. The contact angles at the groove are specified by the balance of forces at the groove, and so are physically determined by the orientation of the two solid phases and the solid/liquid interfacial energies.

THEORY

Steady-state solutions to the heat and solute flow equations must be found. The equations are linked by the interface equation (equation 1) which determines the shape of the interface, which is not known *a priori*.

The interface equation, neglecting attachment kinetics, obtained by equating chemical potentials on both sides of the interface, is

$$\Delta T = T_e - T_i = m_j(C_e - C_i) + \frac{a_j}{r} \qquad j = \alpha, \beta$$

$$= m_j(C_e - C_i) + \frac{a_j[1 + (dz/dx)^2]^{\frac{3}{2}}}{d^2z/dx^2} \qquad (1)$$

The diffusion equation with moving coordinates x, y, z is

$$\nabla^2 C + \frac{v}{D}\frac{\partial C}{\partial z} = 0 \qquad (2)$$

53

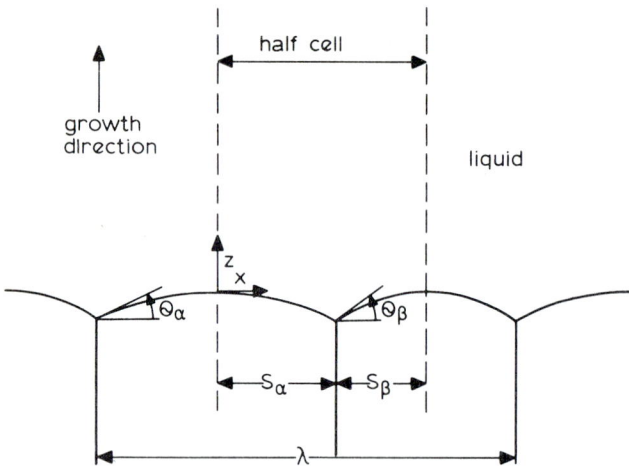

1 Lamellar eutectic interface

In metallic systems heat flow is very rapid compared with solute flow and to a good approximation we may assume the interface to be isothermal. The solute flow boundary conditions are[3]

(i) far field composition
(ii) composition is periodic with the lamellae
(iii) solute conservation at the interface.

The boundary conditions for the interface are

(i) interface is parallel to x axis at lamellae centres
(ii) contact angle at the groove is set by force balance at the groove.

The composition field may be written (for any interface shape)

$$C = C_e + C_\infty + \sum_{n=0}^{\infty} B_n \cos\left(\frac{n\pi x}{S_\alpha + S_\beta}\right)$$
$$\exp\left\{\left(\frac{-v}{2D} - \sqrt{\left(\frac{v}{2D}\right)^2 + \left(\frac{n\pi}{S_\alpha + S_\beta}\right)^2}\right)z\right\} \quad (3)$$

Under most experimental conditions

$$\frac{\pi}{S_\alpha + S_\beta} \gg \frac{v}{2D} \quad (4)$$

and equation (3) may be written (in the region $z \ll D/v$)

$$C = C_e + C_\infty + B_0 - B_0\frac{v}{D}z$$
$$+ \sum_1^{\infty} B_n \cos\left(\frac{n\pi x}{S_\alpha + S_\beta}\right)\exp\left(\frac{-n\pi z}{S_\alpha + S_\beta}\right) \quad (5)$$

This is a solution to the Laplace equation and so may be evaluated by studying the flow of electric current in a conducting medium.

Apparatus

The apparatus is shown in Fig. 2. Paper with a uniform resistivity was used as the analog of the liquid ahead of the growing interface. A current, representing the rejected solute, was injected through electrodes. An electrode, electrode γ, parallel to the x axis and placed a few lamellar spacings away from the interface provided a current to represent the term $B_0(v/D)$ in equation (5). The solid/liquid interface was divided into strips 5 mm wide (generally there were 20 strips) and

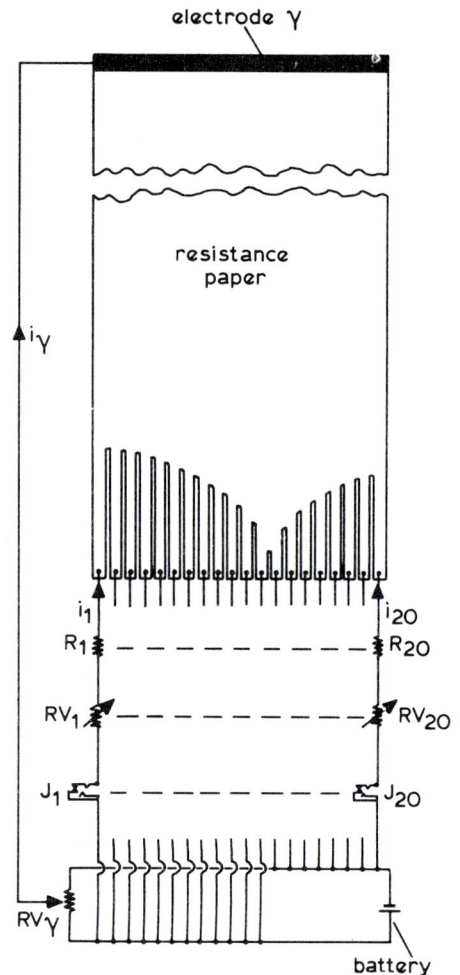

2 Schematic diagram of the apparatus

cuts were made between adjacent strips (on the 'solid' side) to prevent 'diffusion' in the 'solid'. The current through each strip could be adjusted by means of a variable resistor. Currents and voltages were measured with a potentiometric chart recorder which produced negligible loading to the circuit. Switched jack sockets were included to facilitate the measurement of currents.

Theory of the analog

The electric field in the analog obeys the Laplace equation and so, taking account of the geometry of the electrodes, we write

$$V = V_0 - A_0 z' + \sum_{n=1}^{\infty} A_n \cos\left(\frac{n\pi x'}{S'_\alpha + S'_\beta}\right)\exp\left(\frac{-n\pi z'}{S'_\alpha + S'_\beta}\right) \quad (6)$$

where the primes denote the analog equivalent to the various quantities. The coefficients A_n are determined by the boundary conditions. The coefficient A_0, as with the coefficient B_0, is determined by the requirement that the temperature of the two phases should agree with the heat flow part of the problem.

To determine the proportionality constant between potential and composition differences, the solute flux J (measured with respect to stationary coordinates) and the electrical current density J' are compared at the interface to give[4]

$$H = \frac{\Delta C}{\Delta V} = \frac{J\lambda}{J'\lambda'\rho D} \quad (7)$$

This shows that to set up the analog correctly we must adjust the currents injected into the resistance paper until they are in proportion to the solute flows which they represent. The current through the γ electrode is determined by the algebraic sum of the currents injected along the interface. To determine a known composition in the liquid we note that the algebraic sum of the currents flowing into a closed area of the resistance paper must be zero. Applying this across the 'interface' and the plane $z' = K$ (where K is a constant), writing the equations in terms of their solute flow equivalents we obtain

$$-\int_0^{\lambda/2} D\left(\frac{\partial C}{\partial \underline{n}}\right)_i \mathrm{d}S + \int_0^{\lambda/2} D\left(\frac{\partial C}{\partial z}\right)_{z=K} \mathrm{d}x = 0 \qquad (8)$$

Writing

$$-D\frac{\mathrm{d}C}{\mathrm{d}\underline{n}} = (C_i - C_s)v\frac{\mathrm{d}x}{\mathrm{d}\underline{S}}$$

and obtaining $\mathrm{d}C/\mathrm{d}z$ from equation (5) we have

$$\frac{2}{\lambda}\int_0^{\lambda/2}(C_i - C_s)\,\mathrm{d}x - B_0 = \bar{C}_i - \bar{C}_s - B_0 = 0 \qquad (9)$$

Note that this is only a property of the approximate solution and that it in effect defines $z = 0$. During steady stage growth $\bar{C}_s = C_e + C_\infty$ so we may relate the mean potential along the interface to the chosen values of C_e, C_∞, and B_0. Potential and composition at corresponding points are related by

$$C = C_e + C_\infty + B_0 + H(V - \bar{V}_i) \qquad (10)$$

Calculation scheme

Physical constants, bulk composition, v, λ, and phase diagram data, together with the contact angles at the groove, are all specified before an interface shape may be calculated. Starting with a planar interface and assuming the interface composition to be C_e, the analog was set up and a first approximation to the interfacial liquid composition profile determined. The value of B_0 is not specified by the solute flow but by the heat flow. To improve the estimate of B_0 equation (1) is integrated across each phase, the mean value of r being given by[3] $\sin\theta_j/S_j$. Equating the mean undercooling of each phase then gives the revised value of B_0

$$C_\infty + \text{revised } B_0 =$$

$$\frac{\dfrac{a_\beta \sin\theta_\beta}{S_\beta} - \dfrac{a_\alpha \sin\theta_\alpha}{S_\alpha} + \dot{m}_\beta H\Delta\bar{V}_\beta + m_\alpha H\Delta\bar{V}_\alpha}{m_\alpha + m_\beta} \qquad (11)$$

A revised interface shape is calculated by integrating equation (1) twice, keeping T_i constant and using the interfacial liquid composition measured from the analog. A very small adjustment to the volume fraction may need to be made. Using the new interface shape, B_0, interfacial liquid composition, and volume fraction, the analog is set up again. The process is repeated until a self-consistent interface is found (generally after the second iteration).

Fuller details of the calculation and iteration schemes used are given elsewhere.[4] All the numerical calculations required could be performed easily on a cheap programmable pocket calculator (Novus 4515).

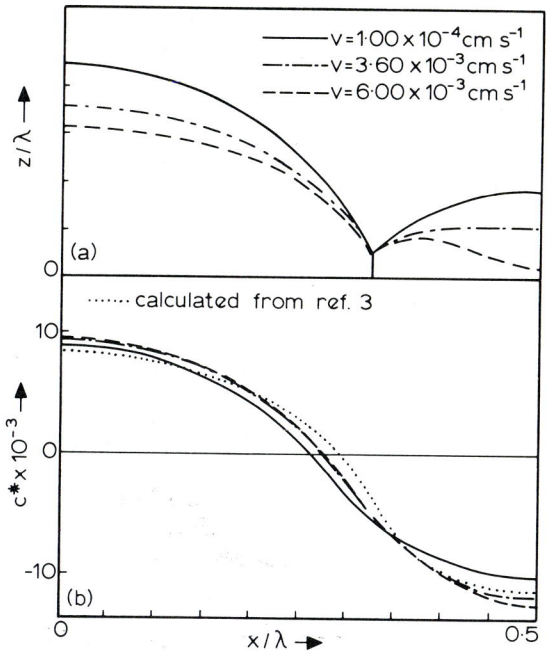

3 *a* **Self-consistent interface calculated for** $C_\infty = 0$, $\lambda = 10^{-4}$ **cm, and** v **as shown;** *b* **reduced interface composition for above together with** C^* **predicted by Jackson and Hunt theory**

RESULTS AND DISCUSSION

Figures 3, 4, 5 show self-consistent interface shapes calculated for a system with the following parameters: $a_\alpha = 1\cdot72\times10^{-5}$ cm K, $a_\beta = 6\cdot96\times10^{-6}$ cm K, solid α, β compositions: $0\cdot01$, $0\cdot71$ atom fraction (independent of T); $C_e = 0\cdot255$; m_α, $m_\beta = 100$, 400 K; $\theta_\alpha = 62\cdot5°$, $\theta_\beta = 42°$, and $D = 6\times10^{-6}$ cm^2/s. These values are similar to those chosen by Nash and Glicksman[2] for lead–tin. Compositions are expressed in reduced form C^* for convenience.

$$C^* = \frac{C - C_e - B_0 - C_\infty}{v\lambda} \qquad (12)$$

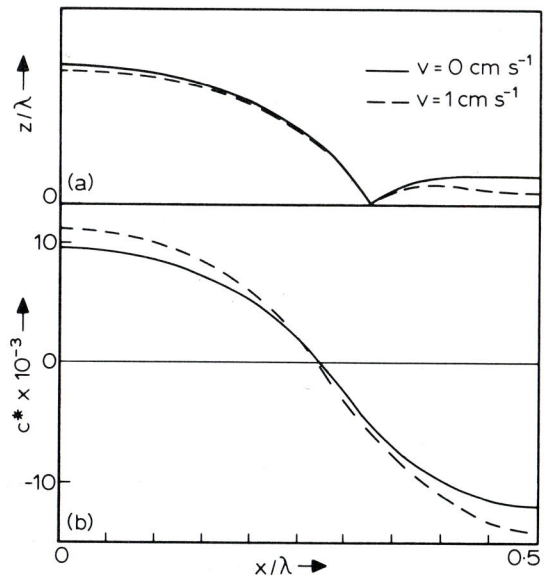

4 *a* **Self-consistent interface shape for** $C_\infty = 0$ $v\lambda^2 = v\lambda^2 = 4\times10^{-11}$ **cm^3/s, and** $v = 0$, **1 cm/s;** *b* **reduced interface composition for the above**

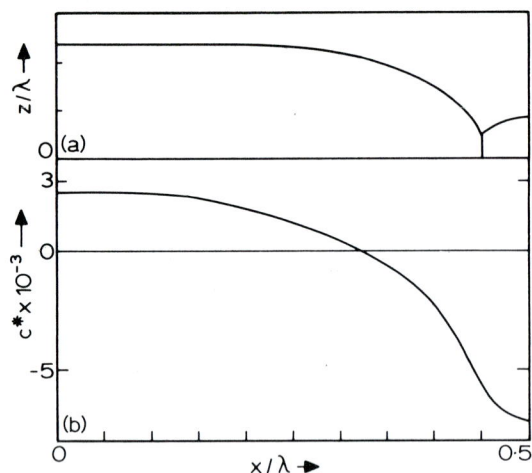

5   *a* Self-consistent interface shape for $C_\infty = 0.175$, $v = 2 \times 10^{-2}$ cm/s, $\lambda = 10^{-4}$; *b* reduced composition for the above

Table 1   Undercoolings calculated from the Jackson and Hunt theory compared with those predicted by the analog for the system described in the text under various conditions

C_∞	v, cm/s	λ, cm	ΔT Jackson and Hunt theory	ΔT, analog
0	1.00×10^{-4}	10^{-4}	0.110	0.110 ± 0.005
0	3.60×10^{-3}	10^{-4}	0.215	0.216 ± 0.005
0	6.00×10^{-3}	10^{-4}	0.286	0.289 ± 0.005
0	1	6.32×10^{-6}	3.52 (3.87)*	3.87 ± 0.1
−0.75	2×10^{-2}	10^{-4}	0.373	0.341 ± 0.01

* Undercooling calculated using modified theory (see text)

In some cases the compositions calculated by the Jackson and Hunt equations[3] are included. For a given system and composition the interface shape is to a large extent constant for a given value of $v\lambda^2$—changes in interface shape with v at constant $v\lambda^2$ being caused only by variation of the rejected solute flux along each interface, which only become appreciable at high velocities. As with the Jackson and Hunt theory an upper limit to $v\lambda^2$ is found, set by the formation of a pocket in the centre of one phase (not necessarily the wider). As $v\lambda^2$ is made smaller the interface tends to the arc of a circle. Values of the interface undercooling as predicted by the analog under various conditions are compared with those predicted by the Jackson and Hunt theory in Table 1. A more detailed discussion is published elsewhere.[4] Under normal experimental conditions good agreement is found. The interface shown in Fig. 5 is self-consistent, although under these conditions Nash and Glicksman[2] predict that no self-consistent interface exists. This discrepancy arises because in the latter theory the mean interface composition is arbitrarily restricted and θ_α, θ_β are treated as variables in the solution to the diffusion equation. In the theory outlined in this paper no restriction is placed on the interface composition (except that it must be between 0 and 1).

At high velocities composition variations in the liquid along the interface become considerable (a factor which is properly incorporated in the analog). This is the chief reason for the breakdown of the Jackson and Hunt theory at high velocity. As shown elsewhere[4] it is possible to modify the Jackson and Hunt theory by allowing the mean composition ahead of the α and β lamellae to have different (unrestricted) values rather than assuming the composition variation along the interface to be small. When properly carried out (*see* Ref. 4) this modification restores the agreement between the analog and the theory at high velocity (*see* Table 1).

REFERENCES

1   K. A. JACKSON *et al.*: *Trans. Met. Soc. AIME*, 1964, **230**, 1 547
2   G. E. NASH AND M. E. GLICKSMAN: Proceedings of the Conference on *in situ* composites II, 1975, Bolton Landing Lake George, N.Y.
3   K. A. JACKSON AND J. D. HUNT: *Trans. Met. Soc. AIME*, 1966, **236**, 1 129
4   R. W. SERIES *et al.*: *J. Cryst. Growth*, 1977, **40**, 221

Coupled zones in faceted/non-faceted eutectics

D. J. Fisher and W. Kurz

The coupled zone conveniently summarizes the solidification conditions which give rise to an entirely eutectic microstructure. Many widely used eutectic alloys, such as Al–Si and Fe–C are of f/nf type and are known, or expected, to exhibit a coupled zone which is skewed towards the faceted component. This can give rise to unexpected microstructures; e.g. Al dendrites in hypereutectic Al–Si alloys. It would be useful to be able to predict the shape of the coupled zone but this is not usually feasible for f/nf alloys because present analyses of dendrite and eutectic growth do not give correct results when applied to faceted morphologies. It is suggested here that this is because of the difficulties encountered by both faceted primary crystals and f/nf eutectics in optimizing their solid/liquid interface morphologies, due to the anisotropy of the growth kinetics of the faceted phase. In particular, it is found that the platelet dendrite model is better than the needle model for predicting Si 'dendrite' tip temperatures. The overall shape of the coupled zone for Al/Si was successfully predicted without invoking any undercooling due to interface attachment kinetics, the influence of which is deemed to be negligible in this case.

The authors are with the Swiss Federal Institute of Technology, Lausanne, Switzerland

It is often important to control the quantity of primary phase in an alloy close to the eutectic composition. Its presence may be beneficial, e.g. for the toughness of a brittle eutectic casting alloy, or detrimental, e.g. for the high-temperature strength of oriented *in situ* composites. The range of compositions and solidification conditions which give a fully eutectic microstructure are conveniently summarized by the coupled zone. Two types of coupled zone are observed in metallic alloys (Table 1[1-17]):

(i) Kofler's B-zone[18] (Fig. 1a, b) which always includes the eutectic composition, is typical of nf/nf eutectics, (where both phases have a low entropy of melting and, hence, non-faceted solid/liquid interfaces[19,20])

(ii) the A-zone, which is skewed towards the faceted component (Fig. 2), does not include the eutectic composition and leads to unexpected microstructures: e.g. Al dendrites in a hypereutectic Al–Si alloy; A-type zones are often observed in eutectic alloys which have a metalloid or intermetallic compound as a constituent phase (Table 1); these phases usually exhibit solid/liquid interfaces which are faceted;[19] this class of eutectic is referred to as being of faceted/non-faceted (f/nf) type[20] and includes important eutectic alloys such as Al–Si, Fe–graphite, and *in situ* composites reinforced by intermetallic phases.

The coupled zones described by Kofler were observed under conditions of high undercooling (high growth rate) (Fig. 1a). Mollard and Flemings[21] showed that the zone widened under conditions of high-temperature gradient and low growth rate. This gives a waisted appearance to the overall zone (Figs 1b, 2). This paper is concerned exclusively with the high speed, high undercooling region of the coupled zone.

There is considerable interest in theoretically predicting the shape and position of the coupled zone, but this is difficult because present analyses of crystal growth do not give even a rough approximation to the observed undercooling/growth rate relationships of many f/nf eutectics or faceted dendrites. Calculated values of undercooling for both are often too low by a factor of at least three, and usually much more. This paper explores some possible reasons for the discrepancy.

THE COUPLED ZONE

Figure 2 illustrates the origin of a skewed coupled zone under steady-state conditions. Each growth morphology, plane front eutectic or primary dendrites (of both phases), has its own growth rate/interface temperature relationship. In order to derive the coupled zone from such curves it is necessary to apply a competitive growth criterion, e.g. that the morphology having the highest interface temperature for a given growth rate, or the highest growth rate for a given

Table 1   Coupled zones in metallic alloys

Eutectic	Microstructure	Faceted phase	Faceted phase (structure)	Kofler zone*	Morphology of non-faceted, faceted dendrites		K_1 (equation 1)	Agreement with theory†
Al–Al$_9$Co$_2$	Regular	Al$_9$Co$_2$	Monoclinic	A[1]	Needle	platelet	—	Yes
Al–Al$_2$Cu	Regular	Al$_2$Cu	Tetragonal	B[2]	Needle	platelet	2·3–10·7[3,4]	*No*
Al–Al$_3$Fe	Irregular	Al$_3$Fe	Monoclinic	A[2,5]	Needle	platelet	—	Yes
Al–Al$_6$Fe	Regular	Al$_6$Fe	Orthorhombic	A[5]	Needle	platelet	—	Yes
Al–Al$_6$Mn	Irregular	Al$_6$Mn	Orthorhombic	A[6]	Needle	platelet	100[6]	Yes
Al–Al$_3$Ni	Regular	Al$_3$Ni	Orthorhombic	A[1], B[7]	Needle	platelet	—	Yes/*No*
Al–Si	Irregular	Si	Cubic(diamond)	A[8,9,10]	Needle	platelet	80–200[8,11]	Yes
Ni–Ni$_3$Nb	Regular	Ni$_3$Nb	Orthorhombic	A[12]	Needle	platelet	—	Yes
Sn–Bi	Irregular	Bi	Rhombohedral	A[13]	Needle	platelet	—	Yes
Zn–Zn$_{15}$Ti	Regular	Zn$_{15}$Ti	Complex	A[14,15]	Needle	platelet	—	Yes
Ag–Cu	Regular	—	—	B[16]	Needle	needle	—	Yes
Cd–Zn	Regular	—	—	B[17]	Needle	needle	—	Yes

* A; asymmetric (skewed), only includes eutectic composition at low growth rates B; symmetric, includes eutectic composition at all undercoolings

† Faceted (platelet-forming) phase leads to A-type zone

undercooling, will be the one observed. The range of temperature, as a function of composition, within which the eutectic fulfils this condition, i.e. the coupled zone, is then usually plotted on the phase diagram (hatched area, Fig. 2).

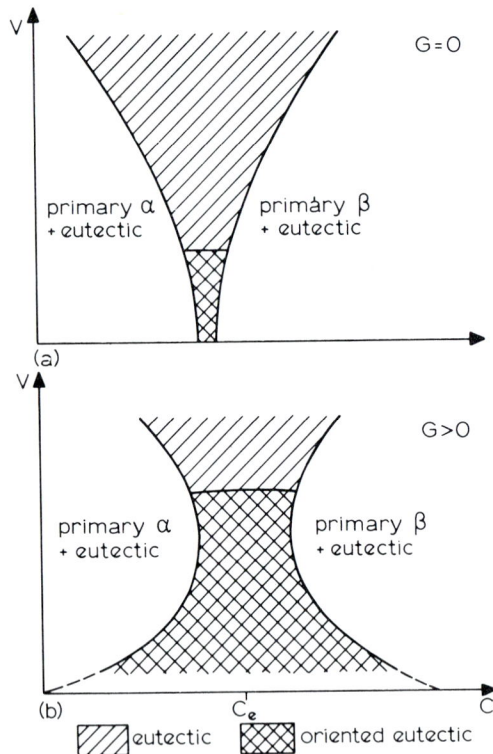

2   The construction of the coupled zone; the superposition of interface temperature/growth rate curves for the competing morphologies (plane front eutectic and dendrites) indicates the boundaries of the coupled zone at a given composition (circles) (α is non-faceted phase, β is faceted phase)

The interface temperature, T, of a non-faceted/non-faceted eutectic obeys a law of the form[22]

$$T = T_e - K_1 V^{\frac{1}{2}} \tag{1}$$

where T_e is the equilibrium eutectic temperature, K_1 is a function of the physical constants of the alloy, and V is the growth rate. Under conditions of controlled growth in a positive temperature gradient, the dendrite tip temperature is given by[23]

$$T = T_1 - K_2 V^{\frac{1}{2}} - \frac{GD}{V} \tag{2}$$

where T_1 is the (stable or metastable) liquidus temperature, K_2 is a function of the physical constants of the alloy, G is the temperature gradient at the interface, and D is the coefficient of solute diffusivity in the liquid. The curves defined by this expression exhibit a maximum so that the curve for the eutectic is cut twice: once at high speeds and once at low speeds (Fig. 2). This explains the waisted appearance of the overall zone. Since this paper is restricted to the discussion of high-speed growth, where the influence of the gradient

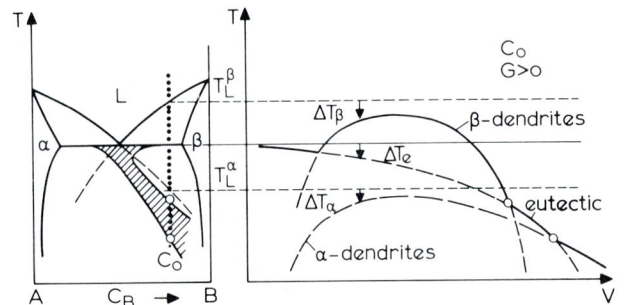

1   Schematic coupled zones for a non-faceted/non-faceted eutectic: the zone always includes the eutectic composition, a isothermal growth—the zone widens only with increasing growth rate; b growth in a temperature gradient—the zone also widens at low growth rates. Double hatched area represents oriented structures under directional solidification conditions

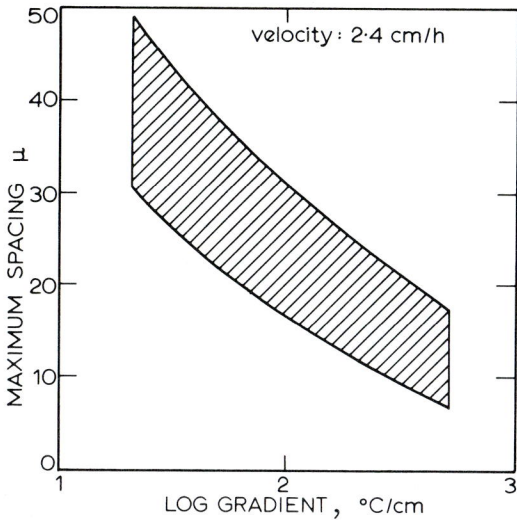

3 The maximum separation at branching of two adjacent borneol platelets as a function of temperature gradient (f/nf borneol/succinonitrile eutectic); the wide scatter is probably due to an effect of the detailed angular relationship of the adjacent platelets, and the perturbing effect of 'second-nearest' neighbours (*see* Fig. 4)

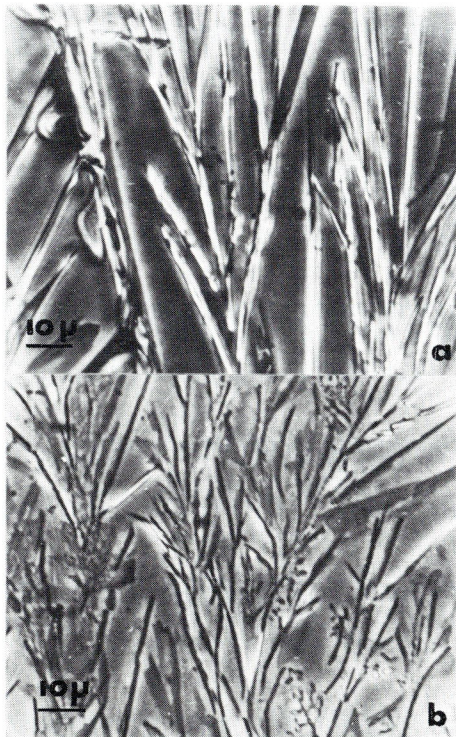

4 Longitudinal structures of borneol/succinonitrile eutectic grown at a growth rate of 2·4 cmh^{-1} under gradients of, a 14 K/cm, and b 395 K/cm; note the markedly increased branching tendency under high temperature gradients (*see* Fig. 3)

G (the third term in equation (2)) is negligible, one can rewrite equation (2) as simply

$$T = T_1 - K_2 V^{\frac{1}{2}} \tag{3}$$

Attempts have been made to predict the coupled zone of f/nf eutectics, using equations (1), (2), and (3).[24,25]

It is easy to see from Fig. 2, for example, that in order to give a skewed zone the rate of change of undercooling with velocity for the faceted primary phase (β) must be greater than that of the eutectic, whereas that of the non-faceted dendrite phase must be smaller. One is, in fact, faced with the double problem of explaining (*a*) why f/nf eutectic undercoolings (Table 1) can be anomalously high; and (*b*) why the undercooling of the faceted 'dendrites' can be yet higher.

THE UNDERCOOLING/GROWTH RATE RELATIONSHIP FOR FACETED/NON-FACETED EUTECTICS

In another paper[26] the authors have attempted to rationalize the observed behaviour of certain f/nf eutectics. Only the relevant points of the cited paper are pointed out here.

Faceted/non-faceted eutectics, such as Al–Si, are characterized by anomalously large interphase spacings and undercoolings (Table 1). It has also been observed that the microstructure[11,26] and the undercooling[11] can be influenced by an imposed temperature gradient (Figs 3, 4). The large spacings and undercoolings can be explained by non-optimum growth, which occurs because the f/nf eutectic spacing cannot change easily, whereas a nf/nf eutectic always grows with the spacing which gives the maximum growth rate or interface temperature (lower catastrophic limit, LCL, *see* Fig. 5).

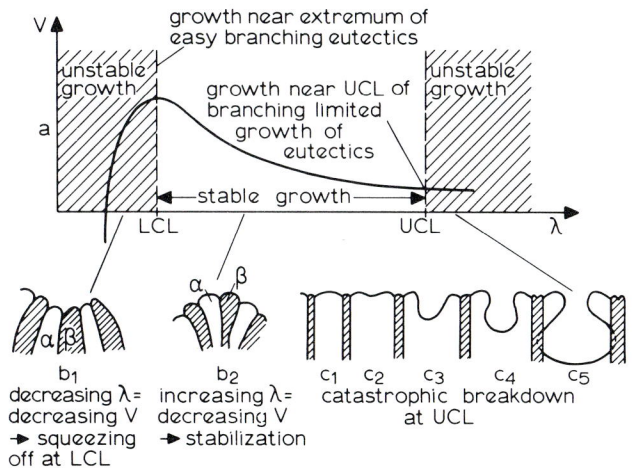

5 The range of theoretically possible eutectic spacings; in reality, two limits exist which reduce the range of physically observable spacings to the unhatched region; these are (i) the lower catastrophic limit (LCL), below which spacing fluctuations lead to 'squeezing off' of a platelet of one phase (b_1), thus increasing the spacing; and (ii) the upper catastrophic limit (UCL); on increasing the spacing, at a given growth rate, one phase falls back as shown in the sequence c_1 to c_5; the other phase will eventually become unstable (branch) at the UCL

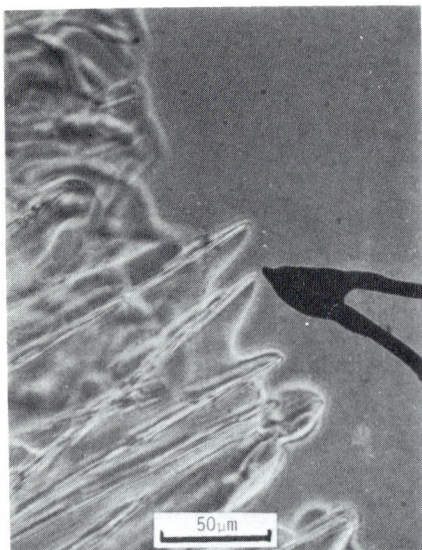

6 Solid/liquid interface of borneol/succinonitrile eutectic, showing large depressions in the succinonitrile/melt interface; similar interfaces are seen in quenched specimens of Al–Si and Fe/C eutectics[26]

The anisotropic growth kinetics of the faceted phase lead to what might be termed branching-limited growth. Therefore, the f/nf eutectic can only change its morphology via 'catastrophic' changes and, in order that these occur, the spacing must increase to the upper catastrophic limit (UCL, see c_4, Fig. 5). The observed spacings will be between the two limits, and the associated temperature or growth rate (Fig. 5) will be the ones measured. Since these will vary from place to place, the solid/liquid interface will be macroscopically irregular.

The effect of the temperature gradient is suggested to arise because its increase tends to prevent the formation of depressions (Fig. 6) in the non-faceted phase, thus increasing the solute concentration at the fibre/melt interface. This should cause the fibres to branch at a smaller separation (Fig. 4) and is equivalent to a lowering of the UCL. The observed spacings and undercoolings are, therefore, smaller under higher temperature gradients.

Transitions from faceted to non-faceted behaviour could conceivably occur in the faceted phase of f/nf eutectics (see next section: 'Test case'). This would increase the ease of branching, and could explain the discontinuity in eutectic spacings observed on increasing the growth rate in eutectics such as Al–Si[11] or Fe–C[41].

THE UNDERCOOLING/GROWTH RATE RELATIONSHIP FOR PRIMARY, DENDRITE-LIKE CRYSTALS

Instead of direct measurement, one can deduce dendrite undercoolings from known eutectic undercoolings and known coupled zone boundaries. Some idea of the undercooling of isolated Si dendrites was obtained from measurements[8] on the Al–Si coupled zone and eutectic undercooling. The constant, K_2, (equation (3)) found in this way,[24] was of the order of 350 $K s^{\frac{1}{3}}/cm^{\frac{1}{3}}$. Using the Zener type analysis of Burden and Hunt[23] (equation (3)), the value of the constant, K^2,

for Si was found to be equal to 72 $K s^{\frac{1}{3}}/cm^{\frac{1}{3}}$, and there was therefore a factor of five difference between theory and experiment. One must assume from this that there is some factor which is not taken fully into account by the usual theoretical models. Some tentative suggestions are interface attachment kinetics, time dependence, and non-optimum growth.

Interface attachment kinetics
Substances which grow with a faceted solid/liquid interface are expected to have greater difficulty in adding atoms to their (low index) faces than non-faceted materials, thus leading to a greater undercooling at a given growth rate. It has been popular, therefore, to ascribe anomalously high undercoolings, in faceted materials such as Si, to this cause. However, Steen and Hellawell[27] have estimated that for Si crystals growing from Al alloys near the eutectic composition, this undercooling is of the order of

$$\Delta T = 1 \cdot 5 V \qquad (4)$$

and such levels of undercooling are dwarfed by the other sources of undercooling (solute and curvature effects).

The fact that the kinetic contribution to undercooling must be separated from these other sources makes it very difficult to measure. Values for Si and Ge have been deduced from directional solidification and droplet migration experiments, but the results are very varied (Table 2), although this can partly be ascribed to different defect growth mechanisms. The directional growth experiments are obviously the more relevant and these indicate kinetic undercoolings even lower than those given by equation (4). This cause of undercooling, therefore, can certainly not be used to explain the anomalously high undercoolings of these materials.

Time dependence
This effect should play a rôle in non-faceted dendritic growth, since branches continually develop near the tip, and oscillation of the tip itself can occur.[28] Glicksman *et al.*[29] have ascribed a factor of five difference between theory and experiment, in non-faceted dendrite growth mainly to neglected time-dependent effects. Similar variations in geometry occur in faceted materials, but are here more erratic and have a greater effect on the dendrite growth since the changes in geometry are more radical, owing to the greater stability of faceted solid/liquid interfaces.[30]

By analogy with plane, single-phase growth,[31] repeated changes in growth rate could have a marked effect on the average interface solute concentration. At the present time it is not possible to make a meaningful estimation of the excess undercooling due to this cause.

Non-optimum growth
In all dendrite growth calculations there is a presumed optimization of the total undercooling or other similar criterion. This optimization is carried out with respect to chosen geometrical parameters, such as the radius of curvature at the dendrite tip. In effect, the dendrite is assumed to physically adjust its morphology in order to optimize some potential-like quantity and this optimization is modelled in the calculations. Two important

Table 2 Kinetic undercooling coefficients for Si and Ge alloys

Alloy	Method	Undercooling/growth rate law; $\Delta T = KV^n$		Reference
		K	n	
Ge–Au	TGZM	220	0·5	R. G. Seidensticker: *J. Electrochem. Soc.*, 1966, **113**, 152
Ge–Al	TGZM	45	0·5	W. A. Tiller: *J. Appl. Phys.*, 1963, **34**, 2 757
Si–Al	TGZM	52	0·5	V. N. Lozovskii and A. I. Udyanskaya: *Sov. Phys.-Cryst.*, 1968, **13**, 477
Si–Al	TGZM	~32	0·5	H. E. Cline and T. R. Anthony: *J. Appl. Phys.*, 1976, **47**, 2 325
Ge	Czochralski	35 (111)	0·5	
		23 (110)	0·5	J. C. Brice and P. A. C. Whiffin: *Solid-State Elect.*, 1964, **7**, 183
Ge	Czochralski	0·9	1	Y. M. Shashkov and V. P. Grishin: *Russ. J. Phys. Chem.*, 1964, **38**, 1 631
Ge	Czochralski	<0·3	1	D. R. Hamilton and R. G. Seidensticker: *J. Appl. Phys.*, 1963, **34**, 1 450
Ge	Czochralski	5	0·5	V. V. Voronkov: *Sov. Phys.-Cryst.*, 1973, **17**, 807
Si	Czochralski	6	0·5	V. V. Voronkov: *Sov. Phys.-Cryst.*, 1973, **17**, 807
Si	Czochralski	0·1	1	V. P. Grishin and Y. M. Shashkov: *Inorg. Mat.*, 1968, **4**, 1 252
Si	Czochralski	1·6	1	Y. M. Shashkov and V. P. Grishin: *Russ. J. Phys. Chem.*, 1964, **38**, 1 631

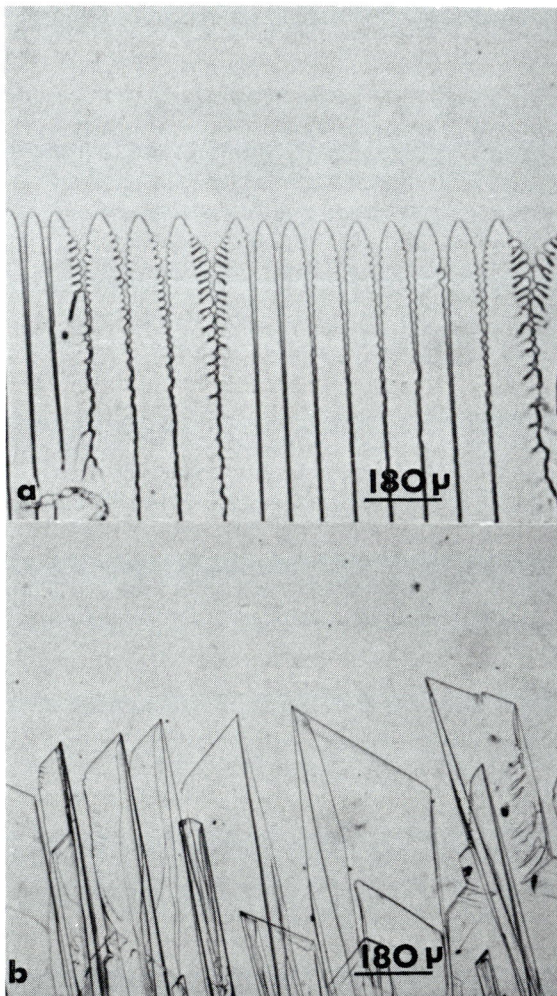

a non-faceted dendrites of pivalic acid; *b* faceted primary crystals of naphthalene (similar growth rates ~10 cm/h)

7 Typical non-faceted and faceted primary crystals

questions to be asked are, (*a*) whether the mathematical model includes enough 'free' morphological parameters to describe the real dendrite shape fully; and (*b*) what physical mechanism allows the dendrite to adopt a given morphology.

Existing analyses of dendrite growth generally treat one of two problems: the platelet model,[32] and the needle model.[33] However, Yost[34] has shown that if an elliptical parabolic dendrite model (which includes the other models as special cases) is taken, the optimum morphology will not have an exactly circular, but a slightly oval cross-section. This occurs even though isotropic interfacial energy is assumed. The effect should be even more marked in faceted materials where the surface energy and attachment kinetics can be expected to be anisotropic. In this respect, the anisotropy of the attachment kinetics is much more important than the absolute value. Faceted dendrites present blunt faces to the melt (Fig. 7*b*) and should be less efficient for solute rejection than the 'needle-like', non-faceted dendrites (Fig. 7*a*). For the moment, there exists no rigorous analysis for angular dendrites growing from the melt.[35] However, to test whether the assumption of a more anisotropic form for the Si primary crystals would result in a closer agreement with the experimental results, two simple Ivantsov-type calculations for needle- and platelet-like dendrites were compared. A similar analysis had already been carried out by Fredriksson[25] for the Fe–C system, but using two different mathematical methods. It was felt that this alone could affect the result. (Compare the results for two different 'needle' calculations; Fig. 8.)

In the present case the equations were: for platelets

$$T = T_1 - \frac{mC_0}{1 + (k-1)\pi^{\frac{1}{2}}P^{\frac{1}{2}}e^P \, \mathrm{Erfc}(P^{\frac{1}{2}})} + mC_0 - \frac{\Gamma}{r} \quad (5)$$

and for needles:

$$T = T_1 - \frac{mC_0}{1 + (k-1)Pe^P E_1(P)} + mC_0 - \frac{2\Gamma}{r} \quad (6)$$

where k is the distribution coefficient, C_0 is the initial alloy composition, P is the Péclet number ($= Vr/2D$), r

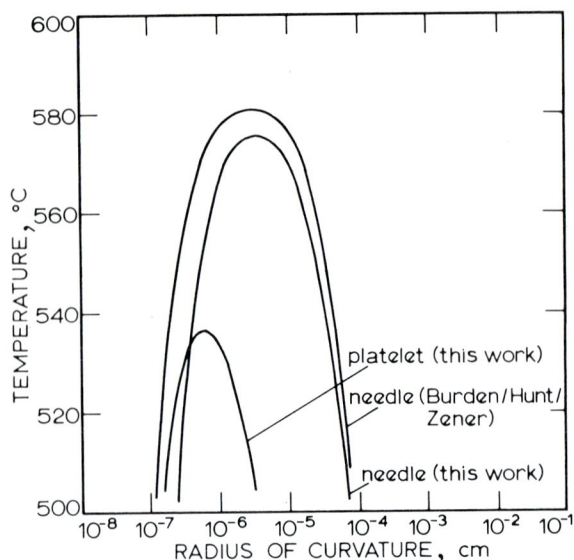

8   The calculated interface temperature as a function of radius of curvature, for various dendrite models (equations (5) and (6), concentration 13 wt-% Si, growth rate, 0·01 cm/s)

Table 3   Constants for Al–Si eutectic

D_{Al}	5×10^{-5} cm^2/s
D_{Si}	5×10^{-5} cm^2/s
k_{Al}	0·13
k_{Si}	0·05
m_{Al}	6·7°C (wt-%)$^{-1}$
m_{Si}	32°C (wt-%)$^{-1}$
Γ_{Al}	$1·06 \times 10^{-5}$°C cm
Γ_{Si}	$1·25 \times 10^{-5}$°C cm

is the tip radius, Γ is the Gibbs–Thompson coefficient, Erfc is the complementary error function, and E_1 is the exponential integral function. The values assumed for the physical properties are shown in Table 3. The maximum tip temperatures for platelets and needles (Fig. 8) are shown as a function of V in Fig. 9. For a composition, C_0, equal to 13 wt-% Si, the Si platelet undercooling (equation (5)) can be approximated by

$$\Delta T \simeq 240 V^{0.34} (10^{-5} < V < 10^{-1} \text{ cm/s}) \qquad (7)$$

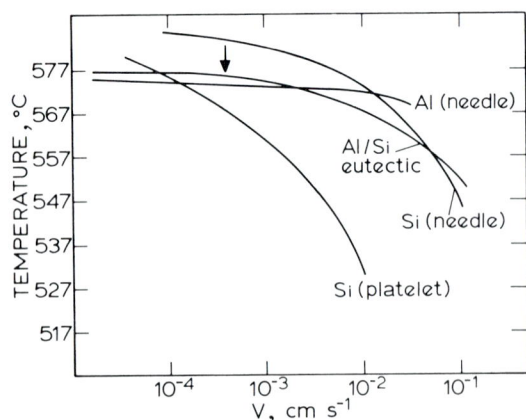

9   Maximum dendrite tip temperatures (13 wt-% Si) plotted as a function of growth rate, together with the measured[11] undercooling of Al–Si eutectic

and the Si needle undercooling by

$$\Delta T \simeq 114 V^{0.48} (10^{-5} < V < 10^{-1} \text{ cm/s}) \qquad (8)$$

As already mentioned, the Burden–Hunt analysis[23] for needles gives

$$\Delta T \simeq 72 V^{0.5} \qquad (9)$$

Optimization of the morphology

The mathematical solutions to growth problems are described by continuous curves but the system itself cannot always change in a continuous fashion in response to changing growth conditions or in order to reach the mathematically defined optimum. This phenomenon has been investigated using organic analogs which permit continuous observation of the morphology.[36]

The inability to adopt the optimum morphology is of great importance in the growth of faceted materials: at the same nominal growth rate one observes faceted shapes of widely different morphologies (Fig. 7b), whereas one can almost exactly superpose the images of different non-faceted dendrites tips (Fig. 7a). Figure 7b implies that the faceted dendrites are growing with different tip temperatures at the same growth rate. In fact, the local growth rates were different from dendrite to dendrite. The concept of steady-state growth is, basically, inapplicable to the growth of faceted crystals and f/nf eutectics, and the imposed growth rate in directional solidification is only maintained in an average sense by the solid/liquid interface. The large range of faceted dendrite shapes is analogous to the wide range of spacings in f/nf eutectics (*see* previous section: 'The undercooling/growth rate relationship') observed at one growth rate.

A TEST CASE; CALCULATION OF THE Al–Si COUPLED ZONE

The shape of the coupled zone was found in the way indicated schematically in Fig. 2. The temperature/growth rate curve for Al–Si eutectic was taken from the experimental results of Toloui and Hellawell[11] for a temperature gradient of 80°K/cm. The corresponding curves for primary Al dendrites (needles) and Si platelets were derived from equations (5) and (6). Figure 9 shows a typical set of curves (13 wt-% Si). The main effect of changes in composition is to shift the curves up and down according to the slopes of the liquidii. The form of the curves can also change due to the influence of the C_0 term in equations (5) and (6). By comparing such sets of curves, the coupled zone was obtained (hatched area, Fig. 10).

DISCUSSION

In Fig. 10 the calculated (hatched) and the experimentally determined (unhatched) coupled zones[8] are compared for Al–Si. One can see that without involving a kinetic undercooling the correct overall form of the zone is obtained. This is due mainly to the fact that the undercooling of the eutectic increases far more quickly with increasing growth rate than that of the Al dendrites. The high undercooling of the typical, irregular, f/nf eutectic has been suggested elsewhere[26] to originate from its non-optimum growth. Before the undercooling can be calculated it will be necessary to determine the conditions leading to instability and

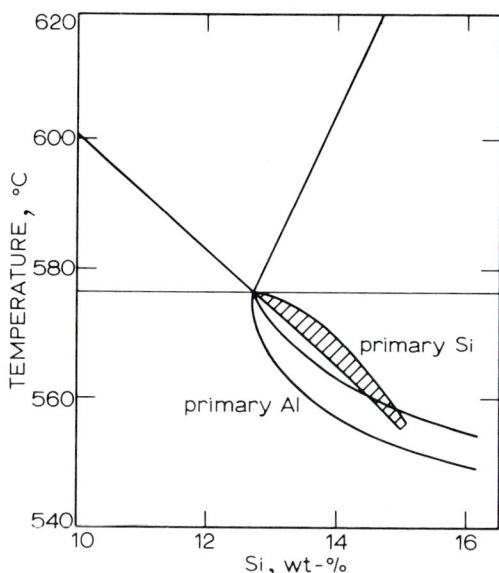

10 Coupled zones: carrying out the procedure indicated in Fig. 9 for various compositions gives the calculated Al–Si coupled zone (hatched area); also shown is the experimentally determined zone[8]

branching of the faceted phase in the eutectic. Therefore, the undercooling/growth rate relationship was taken from experiment.[11]

The second important factor required in order to obtain a skewed zone is that the undercooling of the faceted dendrites should increase faster than that of the eutectic with increasing growth rate; otherwise, the coupled zone will be very restricted or not exist at all. For instance, the undercoolings for needle growth given by equations (8) and (9) are far too small to permit primary Si to 'disappear' with increasing V, even at the eutectic composition. The infinite plate (equation (7)) seems to correspond better to the real growth behaviour of such primary crystals.

It is often stated that the coupled zone is skewed towards the faceted component, owing to the lower growth rate or higher undercooling of the latter when growing as primary crystals. This is not strictly true, since the position of the nonfaceted dendrite/eutectic boundary determines whether or not the zone is skewed (type-A), and the form of this boundary is, obviously, only influenced by the relative undercoolings of these two competing morphologies. The undercooling of primary crystals of the faceted phase is nevertheless very important since the other boundary must also be rotated towards the faceted component in order to avoid 'cutting off' the zone. Table 1 shows that platelet dendrites having higher undercoolings are almost always associated with skewed zones. Therefore, it seems that the anisotropic growth behaviour ('stiffness') of the faceted phase, leading to *non-optimum growth of eutectic as well as dendrites*, is the reason for the skewed form of the coupled zone.

As one can see from Fig. 10, quantitative agreement between theory and experiment is poor. This is due to two possible causes.

1 Real primary crystals of faceted phases normally have aspect ratios (width/thickness) of the order of 10. Therefore equation (7), which was derived for an infinite aspect ratio, overestimates the undercooling.

An intermediate form of the primary crystals should give the observed intersection with the eutectic curve deduced from the known coupled zone boundary[8] (arrow, Fig. 9). If the intermediate forms had morphologies which could be approximated by elliptical paraboloids, they could be treated using Yost's equation.[34]

2 The experimentally determined coupled zone is placed in doubt by later research: the undercoolings necessary to permit the coupled zone to be plotted were deduced from the value of K_1 (equation (1)) for the Al–Si eutectic.[8] This was taken to be constant but has lately been shown to depend on the temperature gradient.[11] As the temperature gradient may vary with growth rate, the actual eutectic undercooling may have changed in an unknown fashion in reference 8. This would modify the real form of the coupled zone (unhatched, Fig. 10).

Finally, there is a possibility that the 'faceted' primary and eutectic phases could switch to non-faceted behaviour because the 'roughness' of the atomic interface may depend on growth rate and on the concentration of the melt with which it is in contact.[37,38] This may explain why Al_2Cu dendrites are faceted while the Al_2Cu phase in the eutectic is apparently not, although both have essentially the same solid composition. It could also explain why Al–Al_2Cu eutectic constitutes an exception to Table 1 in that the dendrites are usually observed to be faceted[39] but the zone is of B-type. There is, in fact, a reported transition from faceted (platelet) to non-faceted (needle) growth of Al_2Cu primary crystals with increasing growth rate.[40]

CONCLUSION

Prediction of the coupled zone (within which an entirely eutectic microstructure is obtained) is desirable. This can be done by comparing the undercooling/growth rate relationship of competing growth forms. Theoretical models developed so far for the growth of nf/nf eutectics and non-faceted primary crystals are incapable of explaining the observed undercooling/growth rate relationship of f/nf eutectics and faceted primary crystals, and give predicted undercoolings which are too small. It is proposed that the main difference between faceted and non-faceted growth, for both eutectics and primary crystals, is linked to difficulties encountered by the system in optimizing its morphology, owing to the higher stability of faceted interfaces, and anisotropy of the growth rates of different faces. In particular, owing to a limited ability to branch, this leads to anomalously high undercoolings in eutectics such as Al–Si and Fe–C. Non-optimum growth also occurs in faceted dendrites but, also, owing to anisotropy of the interface attachment kinetics of faceted materials, the form should be better represented by a platelet than by a needle shape in many cases. Simple calculations show that the use of the platelet shape for the faceted phase and the needle shape for the non-faceted phase, together with experimentally found growth laws for the eutectic, gives a predicted coupled zone for Al–Si which is reasonably comparable to the experimentally derived one.

A wide range of growth behaviours can thus be explained simply on the basis of details of growth morphology without invoking an undercooling due to interface attachment kinetics.

ACKNOWLEDGMENTS
The authors wish to thank Dr H. Jones, Sheffield University, for reading the paper critically. The work was supported by the 'Commission pour l'Encouragement des Recherches Scientifiques', Bern, the Brown Boveri Research Centre, Baden, and Sulzer Brothers, Winterthur, Switzerland.

REFERENCES
1 R. S. BARCLAY et al.: *J. Mat. Sci.* 1971, **6**, 1 168
2 E. SCHEIL AND J. MASUDA: *Aluminium*, 1955, **31**, 51
3 A. MOORE AND R. ELLIOTT: 'The solidification of metals', 167; 1968, London, The Iron and Steel Institute
4 M. TASSA AND J. D. HUNT: *J. Cryst. Growth*, 1976, **34**, 38
5 I. R. HUGHES AND H. JONES: *J. Mat. Sci.*, 1976, **II**, 1 781
6 J. A. EADY et al.: *J. Aust. Inst. Met.*, 1975, **20**, 23
7 E. SCHEIL AND R. ZIMMERMANN: *Z. Metallkd.*, 1957, **48**, 509
8 H. A. H. STEEN AND A. HELLAWELL: *Acta Metall.*, 1972, **20**, 363
9 M. F. X. GIGLIOTTI AND G. A. COLLIGAN: *Metall. Trans.*, 1972, **3**, 933
10 D. C. JENKINSON AND L. M. HOGAN: *J. Cryst. Growth*, 1975, **28**, 171
11 B. TOLOUI AND A. HELLAWELL: *Acta Metall.*, 1976, **24**, 565
12 F. D. LEMKEY: United Technologies Report, NASA Cr-2278
13 M. F. X. GIGLIOTTI et al.: *Metall. Trans.*, 1970, **1**, 1 038
14 G. L. LEONE AND H. W. KERR: *J. Cryst. Growth*, 1976, **32**, 111
15 S. GOTO et al.: *J. Jap. Inst. Met.*, 1973, **37**, 466
16 B. L. JONES et al.: *J. Cryst. Growth*, 1971, **10**, 313
17 M. SAHOO et al.: *J. Mat. Sci.*, 1976, **11**, 1 680
18 A. KOFLER: *Monatsh. Chem.*, 1949, **80**, 441
19 K. A. JACKSON: 'Liquid metals and solidification', 174, 1958, Cleveland, Ohio, American Society for Metals
20 J. D. HUNT AND K. A. JACKSON: *Trans. Met. Soc. AIME*, 1966, **236**, 843
21 F. R. MOLLARD AND M. C. FLEMINGS: *Trans. Met. Soc. AIME*, 1967, **239**, 1 526
22 K. A. JACKSON AND J. D. HUNT: *Trans. Met. Soc. AIME*, 1966, **236**, 1 129
23 M. H. BURDEN AND J. D. HUNT: *J. Cryst. Growth*, 1974, **22**, 109
24 D. J. FISHER AND W. KURZ: Proc. 6th Int. Conf. Light Metals, 135, 1975, Düsseldorf, Aluminium-Verlag
25 H. FREDRIKSSON: *Metall. Trans.*, 1975, **6A**, 1 658
26 D. J. FISHER AND W. KURZ: Proc. Symp. 'The quality control of engineering alloys and the role of metals science', Mar. 1977, Delft
27 H. A. H. STEEN AND A. HELLAWELL: *Acta Metall.*, 1975, **23**, 529
28 L. R. MORRIS AND W. C. WINEGARD: *J. Cryst. Growth*, 1967, **1**, 245
29 M. E. GLICKSMAN et al.: *Metall. Trans.*, 1976, **7A**, 1 747
30 A. A. CHERNOV: *J. Cryst. Growth*, 1974, **24/25**, 11
31 W. A. TILLER et al.: *Acta Metall.*, 1953, **1**, 428
32 W. P. BOSZE AND R. TRIVEDI: *Metall. Trans.*, 1974, **5**, 511
33 R. TRIVEDI: *Acta Metall.*, 1970, **18**, 287
34 F. G. YOST: Ph.D. Thesis, 1972, Iowa State University, USA
35 M. E. GLICKSMAN: Japan–US joint seminar on solidification of metals and alloys, 98, 1977, National Science Foundation, USA
36 D. J. FISHER AND W. KURZ: unpublished work
37 J. W. CAHN: *Acta Metall.*, 1960, **8**, 554
38 B. MUTAFTSCHIEV: 'Adsorption et Croissance Cristalline', 231, 1965, Paris, CNRS
39 J. P. RIQUET et al.: *Comptes Rend.*, 1974, **279**, 497
40 H. FREDRIKSSON: *Jernkontorets Ann.*, 1971, **155**, 571
41 M. HILLERT AND V. V. SUBBA RAO: 'The solidification of metals', 204, 1968, London, The Iron and Steel Institute

Relative contributions of capillarity and temperature gradient effects on the degradation of eutectic composites

M. McLean

Evidence of post-solidification coarsening following the directional freezing of three eutectic alloys (73C, $\gamma-\gamma'-Cr_3C_2$ and Al/Al_3Ni) is presented. The data for $\gamma-\gamma'-Cr_3C_2$ are not compatible with a simple correction to account for solid state capillarity motivated coarsening. A model is developed showing that the presence of a temperature gradient parallel to the solidifying fibres can modify the rate of solid-state coarsening significantly.

The author is with Division of Materials Applications, National Physical Laboratory, Teddington, Middlesex

Control of the microstructural scale of eutectic composites can be achieved by directional solidification under defined conditions, in particular by selection of an appropriate solidification rate R. The extensive experimental data (e.g., Chadwick[1]) have been largely consistent with the theories of eutectic solidification by Tiller[2] and by Jackson and Hunt[3] (TJH) which predict that $\lambda^2 R = K$ where λ is a measure of the microstructural scale (e.g. mean fibre radius or spacing) and K is a constant independent of the magnitude of the applied temperature gradient G. However, although the variation of λ with R at a constant G has been extensively studied, there is relatively little experimental confirmation that K is indeed independent of G.

It is also well established that the microstructural dimensions λ are increased by aging at temperatures where high diffusivities can lead to relatively rapid mass transport. The most common driving force for such phase coarsening derives from the reduction in the total interfacial energy of the system (Ostwald ripening) and, for the case of fibres growing by volume diffusion through a matrix, the variation of λ with time t is given by a variant of the Lifschitz–Slyozov–Wagner theory[4,5]

$$\lambda^3 = \lambda_0^3 + \frac{4}{9}\frac{c\Omega^2\gamma D}{kT}t \qquad (1)$$

where λ_0 is the value of λ at $t=0$; Ω is the atomic volume; γ is the fibre/matrix interfacial energy; c, D are the concentration and diffusivity of the rate controlling species, and kT is the thermal energy. More recently it has been noted that the rate of phase coarsening can be greater in the presence of a temperature gradient than in isothermal conditions[6,7] and McLean[6] and Jones and co-workers[7] have proposed models for the coarsening of two-phase materials in a transverse temperature gradient. Although interpretation of the experimental data in terms of the rival models is still a subject of debate there is now strong experimental evidence that increased rates of coarsening can occur in the presence of a temperature gradient.

The TJH theories account for the structures produced at the point of solidification. In practice the microstructures are observed after the material has been subjected to a rather complex post-solidification heat treatment in cooling to room temperature. Consequently some microstructural degradation analogous to the aging processes described above can modify the predictions of the theories of eutectic solidification. Indeed, some evidence for post-solidification spheroidization[8] and coarsening[9] has been reported. In this paper evidence of post-solidification coarsening during the directional solidification of three eutectic alloys will be presented and indications of the relative importance of capillarity and temperature gradients on these processes will be made.

EXPERIMENTAL PROCEDURES

Precast ingots of eutectic composition were direction-ally solidified at various rates in two modifications of a Bridgman apparatus having quite different temperature gradients. The alloys were melted by radiation from a graphite susceptor heated by a 0·5 MHz radio genera-tor in a constant power mode. The two variations of this apparatus are shown schematically in Fig. 1. In the first, designated CH, the ingot was supported on a water-cooled copper hearth which was withdrawn at a uni-form rate from the heated susceptor. Since the temperature gradient during solidification can vary as the distance between the hot zone and the chill increases, the apparatus was modified to produce a constant, higher temperature gradient by withdrawing the ingot into a liquid metal coolant as shown in Fig. 1b (LMC).

Solidification of the Al–Al$_3$Ni eutectic was observed directly by high voltage electron microscopy. Thin foils of eutectic composition were loaded in a resistively heated specimen stage in an AEI EM7 electron micro-scope. When melted the specimen was maintained as a thin foil between the supporting copper grids and, possibly, by the coherent alumina scale. On reducing the temperature of the hot stage the progression of the solid/liquid interface in the radial temperature gradient, produced by the additional heating of the electron beam, could be observed directly.

EXPERIMENTAL RESULTS
(Co, Cr)–(Co, Cr)$_7$C$_3$

The (Co, Cr)–(Co, Cr)$_7$C$_3$ eutectic, described by Thompson and Lemkey[10] and referred to as 73C, was

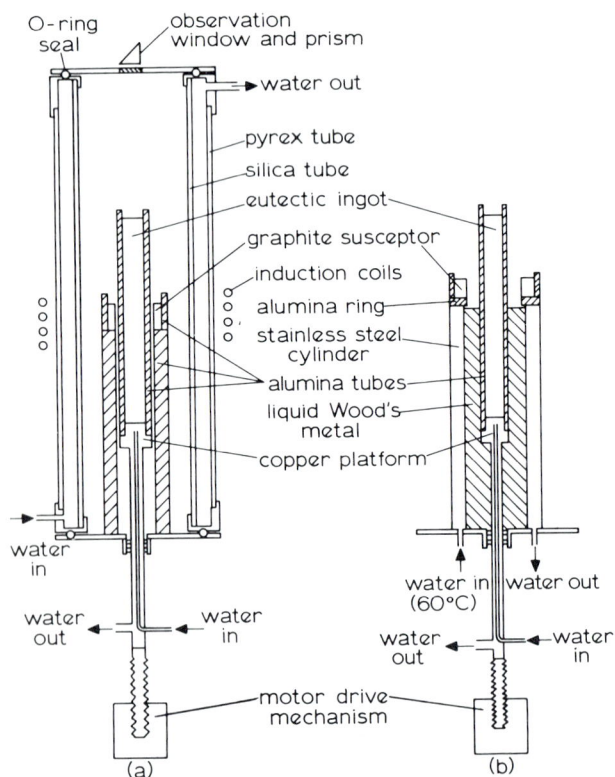

a water chilled hearth (CH); b liquid metal cooling (LMC)

1   Schematic illustrations of the two modifications of the directional solidification apparatus

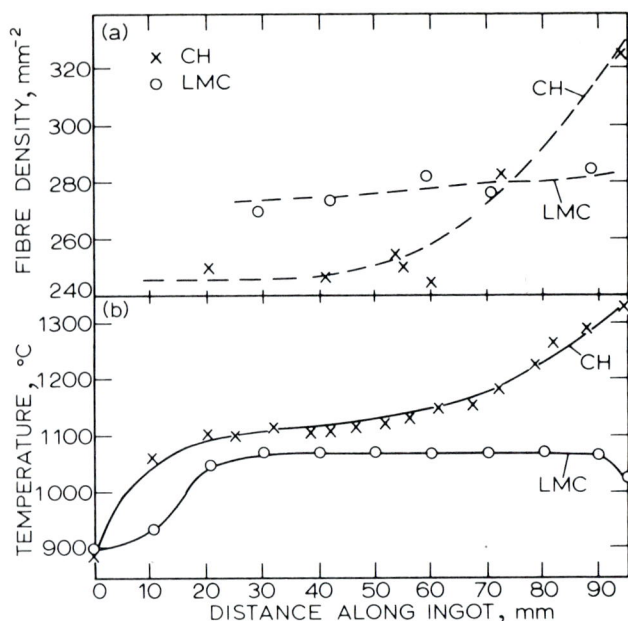

2   a Variation in the density of carbide fibres along ingots of 73C directionally solidified at 130 mm/h in CH and LMC; b variation in the temperature of ingots of 73C at the exit of the hot zone during directional solidification in CH and LMC, as a function of the length of ingots solidified

directionally solidified at 130 mm/h in both the CH and LMC rigs under constant power conditions. The result-ing ingots were sectioned at various points and the density of fibres on transverse sections measured. The results, shown in Fig. 2a, indicate a relatively uniform fibre density along the LMC ingot but considerable variation along that grown in CH. The LMC micro-structure is noticeably finer than that of the CH material.

The temperatures of the ingots leaving the hot zone produced by a copper concentrator were monitored using optical pyrometry in a series of parallel experi-ments using the same power input. The temperature distributions during the solidification of ingots 100 mm in length are shown in Fig. 2b. Although the tempera-tures were not measured at the solid/liquid interface but at a cooler part of the ingot they clearly indicate the important differences between the two rigs. In LMC the temperatures measured are consistently lower than in CH because of the more efficient heat extraction which, in turn, leads to higher temperature gradients. Moreover, the LMC temperature profile is flatter, being essentially constant over some 70 mm. In CH, however, there is much more scatter and a steady significant rise in temperature towards the end of the ingot. This probably indicates that the major contribu-tion to heat loss in CH is by radiation from the seg-ments of the ingot projecting above and below the susceptor rather than by conduction to the chilled hearth. There is therefore a significant rise in the melt temperature, and hence G, towards the end of the ingot when the amount of radiation is essentially halved.

There is a clear correlation between the spatial vari-ations in fibre density and temperature shown in Figs 2a and b, respectively, the finer microstructures being associated with the higher temperature gradients.

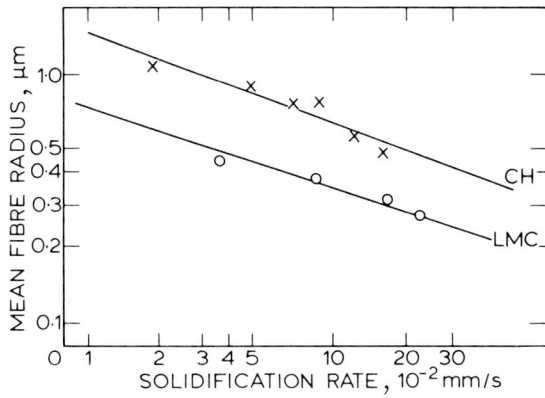

3 Log (mean fibre radius) versus log (solidification rate) for $\gamma-\gamma'-Cr_3C_2$ directionally solidified in CH and LMC

4 a *In situ* HVEM observations of the solidification of the Al–Al₃Ni eutectic; b comparison of the *in situ* observations of solidification microstructures with the data of Livingston *et al.*[14]

$\gamma-\gamma'-Cr_3C_2$

The $\gamma-\gamma'-Cr_3C_2$ eutectic, identified by Tidy[11] and Chadwick,[12] has been directionally solidified in both CH and LMC at a wide range of solidification rates (67 mm to 1 500 mm/h) and the fibre densities measured at equivalent positions in each ingot. The results are shown in Fig. 3 as a log–log plot of the mean fibre radius *v.* solidification rate R. The data obtained in each apparatus are by themselves consistent with the predictions of the TJH theories giving straight lines with slopes of approximately $\frac{1}{2}$. However, there is a significant displacement of the curves obtained at the two different gradients estimated to be

$$G_{CH} = 5 \text{ K/mm and } G_{LMC} = 13 \text{ K/mm}$$

These data provide clear evidence that the scale of eutectic microstructure can be sensitive to the magnitude of the temperature gradient. Indeed in the $\gamma-\gamma'-Cr_3C_2$ system the change in mean fibre radius obtained by altering G by less than a factor of three is greater than by changing R by a factor of ten. It is noteworthy that the 73C system was less sensitive to changes in G.

Al–Al₃Ni

The advance of the solidification interface of the Al–Al₃Ni eutectic can be followed in the sequence of HVEM micrographs shown in Fig. 4a. The direction of the fibres clearly follows the temperature gradient in the plane of the specimen. In the sequence shown the solidification rate is 200 mm/h and the mean fibre diameter is 0.09μm. The solid/liquid interface is apparently planar, indicating that G must be relatively high but it is not possible to measure G in this experiment. McLean and Loveday[13] have previously estimated $G \sim 123$ K/mm in this system. The data obtained in the *in situ* observations of eutectic solidification, shown in Fig. 4b, fall noticeably below the extensive results of Livingston *et al.*[14] This is probably because the present observations refer to the point of solidification whereas previous microstructures, observed after cooling to room temperature, have been subject to post-solidification coarsening.

DISCUSSION

Bullock *et al.*[15] have previously shown that the mean fibre radius, $\bar{\lambda}$, resulting from eutectic solidification followed by capillarity motivated coarsening (Ostwald

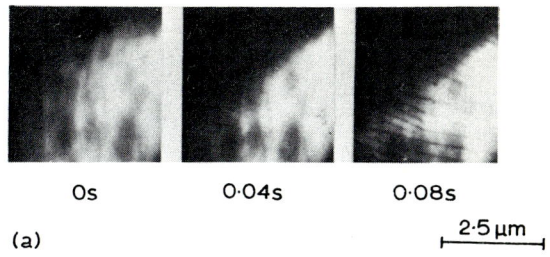

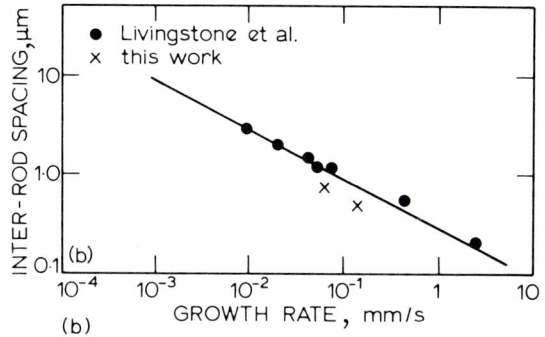

ripening) in cooling by traversing a constant temperature gradient G at a rate R, is given by

$$\bar{\lambda}^3 = AR^{-\frac{3}{2}} + \frac{B}{GR} \qquad (2)$$

where A and B are constants. Of the present data, those for $\gamma-\gamma'-Cr_3C_2$ are most suitable for quantitative analysis. Taking the values of $\bar{\lambda}$ for $R = 315$ mm/h in CH and LMC, equation (2) places a *lower* limit on the ratio of the temperature gradients. Rearranging equation (2),

$$\frac{G_1}{G_2} = \frac{\bar{\lambda}_2^3 - AR^{-\frac{3}{2}}}{\bar{\lambda}_1^3 - AR^{-\frac{3}{2}}} > \left(\frac{\bar{\lambda}_2}{\bar{\lambda}_1}\right)^3$$

The present data would predict $G_{LMC}/G_{CH} > 8$.

It is possible to obtain estimates of the ratio of temperature gradients by more direct arguments:

(i) by imbedding thermocouples in certain ingots direct, albeit approximate, measurements of the temperature gradients were made; the results, $G_{LMC} = 13$ K/mm and $G_{CH} = 5$ K/mm, gave a value of the ratio $G_{LMC}/G_{CH} = 2.6 \pm 1$

(ii) the breakdown in well aligned composite microstructure occurs when the ratio G/R is less than a critical value which varies from system to system; for $\gamma-\gamma'-Cr_3C_2$ the breakdown occurs for $315 < R < 400$ mm/h in CH and $600 < R < 750$ mm/h in LMC; this leads to the estimate $1.5 < G_{LMC}/G_{CH} < 2.4$, which is consistent with the direct measurements.

Although it is clear that some post-solidification coarsening has taken place it cannot be explained quantitatively on the basis of capillarity-motivated coarsening leading to equation (2).

It is of interest to consider the effect of the temperature gradient on the post-solidification coarsening

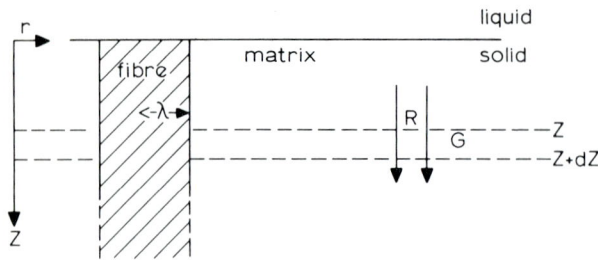

5 Schematic illustration of a solidifying eutectic defining the parameters in the model

kinetics. Previous theories have considered the case where G is applied transverse to the axes of the fibres. Here G is parallel to the fibres. Consider the idealized form of the solidifying eutectic shown in Fig. 5 where cylindrical co-ordinates are defined. Capillarity-motivated coarsening arises because of a radial concentration gradient within the matrix, due to the solute concentration adjacent to the fibre matrix interface c_r being different from the mean solute concentration $c_{\bar{r}}$ (r is the fibre radius and $\bar{r}$ is the mean fibre radius); c_0 is the concentration adjacent to a planar interface.

$$c_r = c_0 \exp\left(\frac{2\gamma\Omega}{kTr}\right) \tag{3}$$

The presence of the temperature gradient G leads to a solute distribution within the matrix[16]

$$\frac{dc}{dz} = \frac{cQ^*G}{kT^2} \tag{4}$$

where Q^* is the heat of transport. Recognizing that $c(z, r)$ this may be expressed

$$\frac{d}{dz}\ln c(z, r) = \frac{Q^*G}{kT^2} = \frac{Q^*G}{kT_E^2(1 - Gz/T_E)^2} \tag{5}$$

where T_E is the eutectic temperature. Integration of equation (5) with the boundary condition that equation (3), i.e. local equilibrium, is satisfied at the solid/liquid interface gives

$$c(z, r) = c_0 \exp\left[\left(\frac{2\gamma\Omega}{kTr}\right) + \frac{Q^*Gz}{kT_E^2(1 - Gz/T_E)}\right] \tag{6}$$

The capillarity driving force is therefore modified by the second term of the exponential factor. The analysis described by Bullock et al.[15] is extended in the Appendix to give a corrected form of equation (2).

$$\bar{r}^3 = AR^{-\frac{3}{2}} + \frac{B}{GR}\exp\left(-\frac{Q^*}{kT_E}\right) \tag{7}$$

Thus the longitudinal concentration gradient due to G biases the solute concentrations within the matrix to either amplify or attenuate the driving force for Ostwald ripening depending on the sign of Q^*. When Q^* is negative, as it is reported to be for carbon diffusion in nickel,[17] within the limits $4 < Q^* < 40$ kJ/mol the presence of the temperature gradient, irrespective of its magnitude, accelerates capillarity coarsening by a factor of between 1·4 and 20. On the other hand a positive Q^* would reduce the effects of Ostwald ripening by about the same factors. It is noteworthy that Q^* is reported to be positive for carbon diffusion in cobalt[17] and that the dependence of r on G

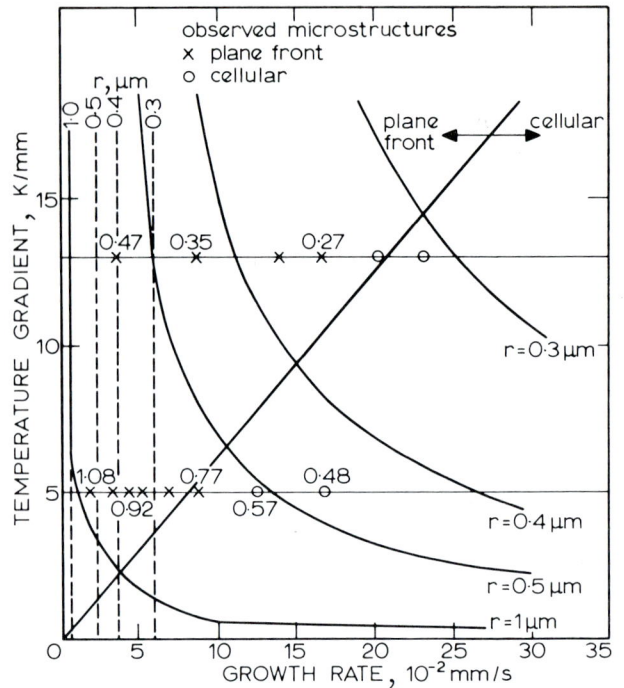

6 Map showing the regimes of composite and cellular microstructures for γ–γ'–Cr_3C_2 solidified in various G and R; the numbers adjacent to the data-points are the mean radii of the carbide fibres; approximate contours of constant fibre radii are indicated and the dashed vertical lines are the predictions of the TJH theory

in the alloy 73C is much less pronounced than for γ–γ'–Cr_3C_2.

The microstructural features of the γ–γ'–Cr_3C_2 solidified in various G and R are mapped in Fig. 6. The diagonal line through the origin indicates the boundary between plane front solidification, giving a well aligned microstructure, and cellular freezing. The present data are consistent with hyperbolic contours of constant fibre radius predicted by the present theory. The TJH theory would give contours normal to the R axis as indicated by the broken lines.

CONCLUSIONS

1 The microstructural scale of three directionally solidified eutectics depends on both the temperature gradient and the solidification rate.

2 The results are not consistent with an estimate of the contribution from solid state Ostwald ripening.

3 A model is developed showing that the rate of Ostwald ripening can be modified by the presence of a temperature gradient.

ACKNOWLEDGMENTS
The author thanks Dr P. Quested and Mr E. Bullock for comments and assistance. Parts of the work were funded by the Ministry of Defence (Procurement Executive) and others were within the COST 50 collaborative programme.

REFERENCES
1 G. A. CHADWICK: *Prog. Mater. Sci.*, 1963, **12**, (2)
2 W. A. TILLER: 'Liquid metals and solidification', 1958, Cleveland, Ohio, ASM

3 K. A. JACKSON AND J. D. HUNT: *Trans. AIME*, 1966, **236**, 1 129

4 I. M. LIFSCHITZ AND V. V. SLYOZOV: *J. Phys. Chem. Solids*, 1961, **19**, 35

5 C. WAGNER: *Z. Elektroken*, 1961, **65**, 581

6 M. McLEAN: *Scripta Metall.*, 1975, **9**, 439

7 D. R. H. JONES: *Metal Science*, 1974, **8**, 37

8 S. MARICH: *Metall. Trans.*, 1970, **1**, 2 953

9 G. GARMONG AND T. H. COURTNEY: *Metall. Trans.*, 1975, **6A**, 1 945

10 E. R. THOMPSON AND F. D. LEMKEY: *Metall. Trans.*, 1970, **1**, 2 799

11 D. C. TIDY: Ph.D. thesis, Department of Metallurgy, University of Cambridge, 1970

12 G. A. CHADWICK: British Patent Specification No. 38115/73

13 M. McLEAN AND M. S. LOVEDAY: *J. Mater Sci.*, 1974, **9**, 1 104; *and* 1974, **9**, 2 069

14 J. D. LIVINGSTON *et al.*: *Acta Metall.*, 1970, **18**, 399

15 E. BULLOCK, *et al.*: *Acta Metall.*, 1977, **25**, 333

16 P. G. SHEWMON: 'Diffusion in Solids', 1963, N.Y., McGraw-Hill

17 J. G. SHAW AND W. A. OATES: *Metall. Trans.*, 1971, **2**, 2 127

APPENDIX

Consider a fibrous eutectic solidifying at a rate R in a temperature gradient G. The mean fibre radius on solidification r_0 is given by the TJH theory

$$r_0 = AR^{-\frac{1}{2}}$$

where A is a constant. The rods will coarsen in the solid state by Ostwald ripening. Following Bullock *et al.*[15] and modifying the equations to take account of the effect of the longitudinal temperature gradient (equation (6))

$$r_{z+dz}^3 = r_z^3 + \frac{K}{T_E - Gz} \exp\left[-\frac{Q}{k(T_E - Gz)} + \frac{Q^* Gz}{kT_E(T_E - Gz)} \right] \frac{1}{R}\,dz$$

where

$$K = \frac{4}{9} \frac{c_0 \Omega^2 \gamma D_0}{k}$$

and the diffusivity has been expressed as $D = D_0 \exp(-Q/kT)$.

The equation

$$\frac{dr^3}{dz} = \frac{K}{R} \cdot \frac{1}{T_E - Gz} \exp\left[-\frac{Q}{k(T_E - Gz)} + \frac{Q^* Gz}{kT_E(T_E - Gz)} \right]$$

can be reduced to a standard form by the substitution $u = Q/k(T_E - Gz)$. Integration with the boundary condition that $r = r_0$ when $z = 0$ gives

$$r^3 = r_0^3 + \frac{B}{GR} \exp\left(-\frac{Q^*}{kT_E} \right)$$

where B is a constant.

Instability of graphite crystal growth in metallic systems

A. Munitz and I. Minkoff

The instability behaviour of graphite growth from Ni–C solution in the presence of Pb is described. The experiments were performed using a Bridgman-type apparatus and the relative instability of steps and surfaces was examined, dependent on variation of the solute concentration and rate of crucible extraction. The instability of the (0001) graphite surface is characterized by the development of pyramidal hillocks, while instability of growth steps leads to the formation of elongated laths. The experimental results are discussed in terms of the distribution of solute at the interface and the related undercooling. For small interface undercooling the (0001) graphite surface advances in a macroscopically planar manner. Pyramidal hillocks develop with increased Pb concentration and steps become unstable at a value of solute concentration intermediate between planar and pyramidal behaviour. The intermediate range of unstable behaviour of steps can be correlated with some observed crystal morphologies for freely growing graphite from solution, taking into account the variation of solute supersaturation over a growing surface.

The authors are with the Department of Materials Engineering, Technion, Haifa, Israel

The theory of instability in growth has been well developed for surfaces having isotropic surface energy and interface kinetics.[1,2] For faceted crystals, different instability phenomena are observed for which a theory has been suggested.[3,4] The stepped nature of the faceted crystal surface is important and instability analysis takes this into account.

The instability of a faceted crystal surface can be analysed in its departure from planarity by examining the time dependence of the growth of a perturbation. The instability can show itself in several different ways. Two cases which have been examined are

(i) formation of pyramids
(ii) change of surface slope resulting in a change of crystal form, e.g. an hour-glass form.

In this paper, some experimental observations are given of instability related to a solute boundary layer for graphite growth. The relative instability of steps and surfaces are discussed as related to solute distribution and supersaturation. In particular, step unstable behaviour is discussed as an intermediate process between planar surface growth and pyramidal instability. In this intermediate range, new growth morphologies may arise.

INSTABILITY THEORY APPLIED TO NON-FACETED AND FACETED SYSTEMS

For non-faceted crystals the conditions for instability are described by the constitutional supercooling theory,[1] or by the approach using perturbation analysis adopted by Mullins and Sekerka.[2] In the constitutional supercooling theory, the instability of a surface is described by the expression

$$G \leqslant \frac{mC_0(1/k - 1)}{D/V}$$

This expression relates the velocity of advance of the interface V and the solute distribution ahead of it to the slope of the liquidus m, the distribution coefficient k, and the temperature gradient in the liquid G.

In the theory of Mullins and Sekerka[2] the instability relationship acquires the form

$$\tfrac{1}{2}(\mathcal{G}_S + \mathcal{G}_L) < mG_{CS}G(\omega)$$

G_{CS} is the constitutional supercooling gradient. In the constitutional supercooling theory G is replaced by the weighted function $\tfrac{1}{2}(\mathcal{G}_S + \mathcal{G}_L)$. $G(\omega)$ contains terms involving the interfacial energy.

For faceted crystals, isotropic surface energy and interface kinetics do not apply. However, identical factors related to the solute effects in the above theories still describe the solute unstabilizing influence. The actual behaviour of the surface must be examined from the point of view of its structure, step source behaviour, and kinetics of growth.

The stepped nature of the faceted surface plays a role in surface attachment kinetics and the growth rate is related to the slope and surface structure. Pyramidal instability can be related to a change in surface slope, p, which gives a change in attachment coefficient, b. In a theoretical analysis of hour-glass forms Chernov[4] related the change of surface slope, p, over a crystal face to the supersaturation variation over it. The initially planar face adopts changes of slope so that p is smallest at the edge where the supersaturation is the largest and highest at the face-centre, where the supersaturation is the smallest. This allows the crystal face to conserve its slope during the crystallization process.

The inter-relationship between step and surface instability has been less well researched and is discussed in this paper.

EXPERIMENTAL METHOD

The experiments reported here were part of a more extensive programme utilizing directional solidification techniques in order to study the instability of faceted crystals. Ni–C solution was used, and some results are reported of Pb addition to study graphite instability.

The apparatus used is shown in Fig. 1. This consists of a Bridgman-type furnace heated by a Crusilite element. The graphite crucible could be lowered or raised at different speeds. Two speeds, 10^{-2} cm/s and 10^{-3} cm/s were employed for the present experiments. The temperature distribution along the axis of the tube, without the inserted crucible, is shown in Fig. 1.

The furnace was held at $1\,550°C$ to equilibrate the charge, which consisted of Ni pellets plus Pb. These materials were of the following purity: Ni (Baker analysed reagent) -0.003Pb, 0.002Co, 0.005Fe, remainder Ni; Pb (Matheson reagent) 99.999% purity. The crucible was machined from boron-free semiconductor grade graphite, the dimensions of which are given in Fig. 1. According to the Ni–C equilibrium diagram, the nickel dissolves approximately 2.5% C at $1\,550°C$.

The experiments described in this paper were performed while raising the crucible through the furnace. At the end of each experiment, the nickel alloy was

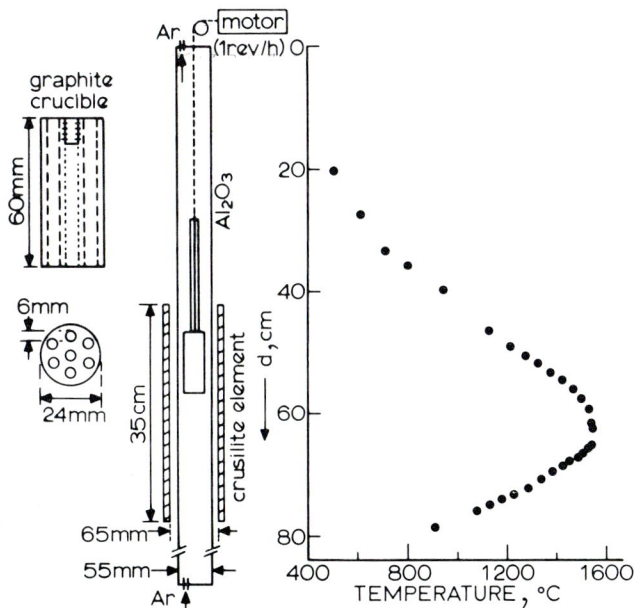

2 Elongated steps on surface of graphite crystal; crucible withdrawal rate 10^{-3} cm/s: Pb concentration 0·35%

sectioned vertically and examined both metallographically and by microprobe analysis on the half. The second half was utilized for graphite dissolution experiments. The graphite was extracted by dissolving the matrix in $1:1$ HNO_3 at $50°C$. After careful washing and drying, the crystals were examined in the scanning electron microscope.

OBSERVATIONS

The observations of graphite surfaces were to enable examination of two types of instability

  (i) step instability (Fig. 2)
  (ii) surface instability (Fig. 3).

At a traverse speed of 10^{-3} cm/s, the steps become unstable at an estimated Pb concentration of 0.35%,

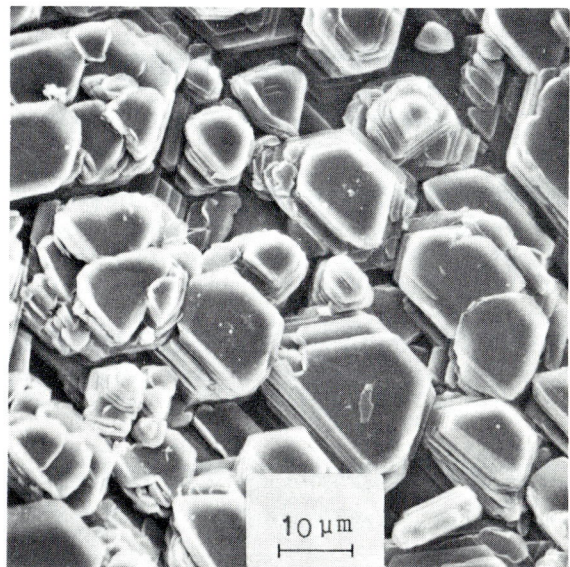

1 Diagram of apparatus employed showing dimensions of graphite crucible and temperature distribution along axis of Al_2O_3 tube

3 Pyramids on surface of graphite crystal; crucible withdrawal rate 10^{-3} cm/s: Pb concentration 3·5%

4   Etch pit distribution on graphite crucible with pyramidal development; etched O_2–Ar mixture, 700°C; 30 min etch pit distribution 5×10^7 cm^{-2} [$\times$3 000]

and the surface becomes unstable at an estimated Pb concentration of 0·5%. As the crucible traverse speed is increased to 10^{-2} cm/s, the step becomes unstable at a higher Pb concentration of 1·4%.

Concentration corrections were made by estimating the Pb loss by evaporation from the alloy during the performance of the experiment. At speeds of less than 10^{-4} cm/s, the evaporation losses became too great for satisfactory experimentation.

The instability of a step is characterized by long laths appearing on the surface of approximate height $2\,\mu$m and width $5\,\mu$m. The instability of a surface is characterized by pyramids. The slope of these varies with the concentration of Pb in solution, the average slope measured being 20° at 0·7% Pb and 60° at 3·5% Pb.

Etching experiments, performed by heating the crystals for 30 min at 700°C in an O_2–Ar mixture, (2·6 mbar O_2–728 mbar Ar) gave an etch pit distribution of 5×10^7 cm^{-2} (Fig. 4). These etch pits have been related to emerging screw dislocations.[5] Each pyramid is from a multiple dislocation source.

ANALYSIS OF OBSERVATIONS
Solute supercooling at interface due to Pb
The influence of the Pb is to increase the liquidus temperature gradient at the boundary and to influence constitutional supercooling. The liquidus temperature gradient is equal to m multiplied by the solute concentration gradient, where m, taken from the Ni–Pb binary published in Ref. 6, has a value $\simeq 50$°C/wt-% Pb.

Growth at small supercooling
The theory of growth of a faceted crystal for small supersaturation or supercooling relates the advance of an interface to a screw dislocation mechanism. In the absence of solute elements other than carbon in the Ni–C alloy, the interface undercooling is entirely kinetic given by $\Delta T = Te - T_K$, where T_K is the growth temperature required for the interface process. From

published literature,[7] this value of ΔT is assumed to be about 4°C. Growth hillocks under these conditions should have surface slopes given by $p = \tan^{-1} a/\lambda \simeq 1°$–2°, where a is the height of a surface step and λ is the spacing between steps.

Pyramidal growth on a surface
The spacing between steps λ, created at the source, varies inversely with supersaturation or supercooling. The hillock slope given by $p = \tan^{-1} a/\lambda$ therefore increases with supersaturation, and hillock instabilities should exceed this value. The slope p is, therefore, approximately zero at $\Delta T \simeq 0$°C and becomes roughly 60° at 3·5% Pb. This is related to undercooling at the surface, which is roughly estimated from the Pb concentration in solution to be 10°C. The slope p of the pyramids should vary approximately linearly with undercooling.[8]

Step instability on a surface
At an intermediate supersaturation between that at which the surface advances uniformly and the higher value leading to pyramid formation, the steps may become unstable and develop arms along the surface in a dendritic manner. Seeger[9] calculated the supersaturation over a step for very dilute solutions relating this to the angle made by the step with the surface.

Steps may become unstable in the following ways:

(i)   macrosteps may break down into microsteps[10]
(ii) the step faces may buckle[9]
(iii) the steps may show the typical dendritic breakdown noted here.

It is not clear which factors determine solute distribution at the surface and which influence the supersaturation at a step. Seeger[9] considered supersaturation for the geometry of a step and ignored surface migration. Where the instability effects are related to supercooling due to solute, concentration profiles normal to and in the plane of the surface might be considered. At a small supercooling, small p and large λ, concentration gradients of solute should exist between steps at the surface as well as across the solute boundary layer, normal to the surface. As the step spacing decreases, the solute distribution between steps becomes smoothed out, so that only solute gradients normal to the surface are of importance. Smoothing out of the solute distribution between steps as the spacing decreases is due both to the overlapping of diffusion fields and to the dependence of the maximum concentration between steps on λ. As λ decreases, the value of the maximum solute concentration decreases. These variations of solute distribution must be considered in analysing relative step and surface instability phenomena (*see* Ref. 12).

Change of velocity
An increase in the crucible withdrawal velocity requires a corresponding increase in the concentration of added Pb to produce the instability effects noted for a surface or a step. This is related to the effect of fluid flow at the surface of the growing crystal and to the change in solute profile which results.[11] The analysis for the present experiments is given in Ref. 12.

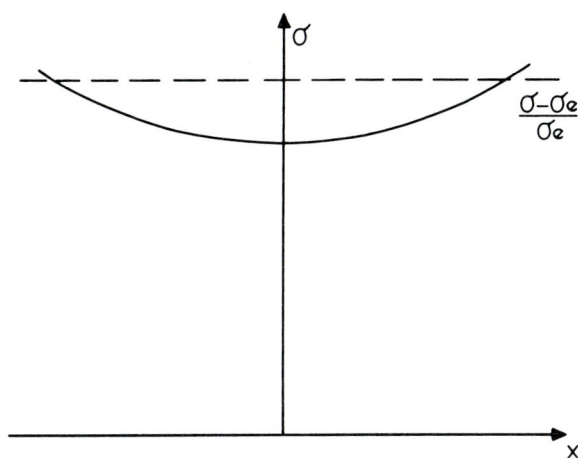

5 Theoretical supersaturation over a growing crystal surface,[3] expressed as $\bar{\sigma} = (\sigma - \sigma_e)/\sigma_e$, where σ is the actual concentration at surface and σ_e is the equilibrium concentration

SUPERSATURATION VARIATION OVER A SURFACE AND GROWTH FORMS

The supersaturation, calculated over a growing surface,[9] has the form shown in Fig. 5. In the experiments reported in this paper, there was no related variation in the observed instability between the crystal edge and centre, although supersaturation effects have been noted in graphite crystals growing from solution.[13]

Figure 6 shows two effects related to step instability and the variation of supersaturation over a growing crystal surface. In Fig. 6a, the ledges formed by the step instability depart from the surface when approaching the crystal edge, apparently under the influence of the high supersaturation existing there. In Fig. 6b, a spherical form of graphite is shown with marked ledges traversing the crystal. This indicates a solute supersaturation driving force for growth. Both crystals grew in Ni–C–B alloys.[13] The spherical crystal probably owes

its form to ledges being driven round the crystal edge. Two forms, a cylinder or a sphere, or sometimes a combination, have been observed.

SUMMARY

Experiments have been described which demonstrate different regimes of growth observed for graphite crystals in Ni–C solution containing an impurity, Pb. At zero impurity content, a macroscopically planar surface of the graphite grows and at high Pb concentrations this surface develops pyramids. In the intermediate region of impurity concentration the steps become unstable, taking the form of long laths traversing the crystal surface.

Increasing the crucible withdrawal rate requires an increase in the solute concentration to produce the effects noted. Some aspects of the theory of growth of faceted crystals have been correlated with these observations. While no effects of the variation of supersaturation over a growing face were noted, such effects have been noted in other experiments on freely growing crystals. A short discussion has been given correlating this with some observed growth forms.

In the present paper results are described for one added element, Pb. A more extensive analysis of the results for different elements and combinations of elements is given in Ref. 12.

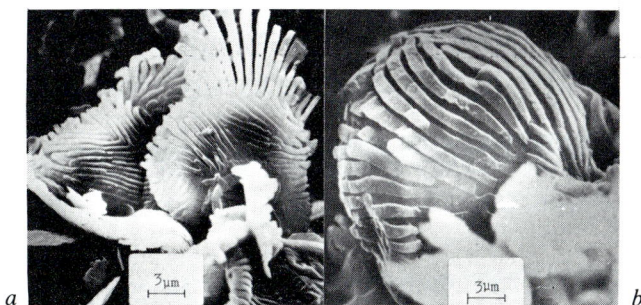

6 Step instabilities of Ni–C–B alloy a on graphite crystal departing from edge under driving force of high supersaturation[13]; and b traversing a spherical graphite crystal[13]

REFERENCES

1 W. A. TILLER *et al.*: *Acta Metall.*, 1953, **7**, 428
2 W. W. MULLINS AND R. F. SEKERKA: *J. Appl. Phys.*, 1964, **35**, 444
3 A. A. CHERNOV: *Sov. Phys. Crystallogr.*, 1972, **16**, (4), 734
4 A. A. CHERNOV: *J. Cryst. Growth*, 1974, **24/25**, 11
5 C. ROSCOE AND J. THOMAS: *Proc. Roy. Soc.*, 1966, **A297**, 397
6 M. HANSEN: 'Constitution of binary alloys', (2 Ed); 1958, New York, McGraw-Hill
7 W. A. TILLER: 'Recent research on cast iron', (ed. H. D. Merchant), 129; 1968, London, Gordon and Breach
8 A. A. CHERNOV: 'Crystal growth and characterisation', (eds. R. Ueda and J. B. Mullin), 33, 1975, North-Holland
9 A. SEEGER: *Phil. Mag.*, 1953, **44**, 1
10 A. A. CHERNOV: *Sov. Phys. Uspekhi (English Transl.)*, 1961, **4** (1), 116
11 R. L. PARKER: 'Solid state physics', 151; 1970, New York, Academic Press
12 A. MUNITZ: D.Sc. Thesis (Hebrew, English Abstract) Dept. of Materials Eng., Technion, Haifa (to be submitted 1977)
13 I. MINKOFF AND B. LUX: *Nature*, 1970, **225**, 540

DISCUSSION

Session 1: Growth morphology, solute distribution, and ripening

Chairman: H. Morrogh (Director, British Cast Iron Research Association)

The following papers were discussed; *Keynote Address: Cellular and primary dendrite spacings* by J. D. Hunt; *Morphological instability at a grain-boundary groove* by J. D. Ayres and R. J. Schaefer; *Dendrite arm spacings in hypoeutectic Pb–Sb alloys directionally solidified under steady and non-steady conditions* by J. A. Spittle and D. M. Lloyd; *Aberrations observed in the relationship of dendrite size-alloying elements for low-alloy steel* by D. J. Hurtuk and A. A. Tzavaras; *Segregation and the determination of phase equilibria in multicomponent systems: Al–Cu–Mn* by D. M. L. Bartholomew, M. Jezuit, B. Watts, and A. Hellawell; *In-situ observations of solute redistribution during solidification* by M. P. Stephenson and J. Beech; *Temperature gradient zone melting and microsegregation in castings* by D. J. Allen and J. D. Hunt; *Microsegregation in Fe–Cr–C alloys solidified under steady-state conditions* by B. A. Rickinson and D. H. Kirkwood; *Electric analog of eutectic growth* by R. W. Series and J. D. Hunt; *Coupled zones in faceted/non-faceted eutectics* by D. J. Fisher and W. Kurz; *Relative contributions of capillarity and temperature gradient effects on the degradation of eutectic composites* by M. McLean; *Instability of graphite crystal growth in metallic systems* by A. Munitz and I. Minkoff

DR H. MORROGH (Director, British Cast Iron Research Association): It is my privilege and honour to open this Conference on Solidification and Casting. I am sure that the organizers are greatly encouraged by the response of delegates from many countries. I understand that over 50% are from overseas.

Ten years ago I had the privilege of chairing a committee under the auspices of the then Iron and Steel Institute (now The Metals Society), the Institute of Physics and the Physical Society, and the Institution of Metallurgists, to organize a conference on the solidification of metals which was held in December 1967 at Brighton. Eventually our proceedings were published in a very modest little volume. I compared it with today's proceedings; 10 years ago we had a very fine conference and I am sure those attending enjoyed it and found much of interest. Our declared aim was to bring together solidification scientists and technologists who were concerned with the problems of the solidification of metals in industrial practice. When I look back, I think we failed in our aim, although we had a very interesting time, and I hope this present conference will succeed where we failed. From the efforts of the committee, they deserve to succeed.

It is perhaps trite to remind you that the process of solidification influences and determines the properties of metals in many applications. It certainly determines the properties of castings, and it is pleasing to note that the word casting appears in the title of this conference. It influences the properties of segregation in ingots, it influences continuously cast materials, the properties of fusion welds, and new modes of solidification lead to the possibilities of the development of new engineering materials.

Indeed, it is an important topic and I am pleased to say that it has almost become a fashionable one. It is my general experience that with a few exceptions technology is usually way ahead of the ability of scientists to explain what the technologist achieves. The technologist has to work with impure materials and he complicates the situation by additions arrived at for their effect empirically. Great progress has been made in the fundamental study of 'pure' materials. I use the word pure in inverted commas because I am not really sure that you always achieve the high purity that is claimed. We certainly need a greater understanding of the effect of impurities on growth mechanisms and on heterogeneous nucleation. I find some encouragement in the attention which is evidently being paid, judging by the programme, to impurities in solidification processes in this conference. I am encouraged by the practical note in Session 4 and with increased emphasis in Sessions 5 and 6. I have noted that Professor Nutting has the responsibility of closing the conference, and I am sure he will do this in his usual most efficient way. However, I am certain also that he will have insufficient time to do justice to the conference, and I do hope that when the proceedings are published, the Organising Committee find it possible to develop some sort of concensus of the conclusions which it is possible to draw from this conference.

PROFESSOR M. C. FLEMINGS (Massachusetts Institute of Technology): I have a comment on Dr Hunt's paper. I think he is to be complimented on the excellent analysis; he called it crude but I think the point is that it brings out for the first time the basic factor that leads to the establishment of primary dendrite arm spacing, namely that the primaries do what they can to at least approach, if not attain, a minimum undercooling position. Dr Ardell in my group at MIT has been doing somewhat similar work and has come up with similar results. The only thing to be added here is that Ardell has also shown that the composition dependence in his analysis, as in John Hunt's, is around a $\frac{1}{4}$ power, and that also agrees with experiment. The data of Young and Kirkwood is replotted in the figure. The points move as the theory says they ought, that is the point.

DR D. H. KIRKWOOD (Sheffield University): I would like to ask Dr Hunt why the Trivedi analysis was not used? He used this in a previous paper with Burden. Would this not have given a better agreement?

DR HUNT: I do not think very much would be gained by using the Trivedi analysis. I wanted to keep the analysis as simple as possible. There is also some question whether the difference between the Trivedi analysis and straightforward (Ivantsov) analysis is real.

MR C. MASCRE (Centre Technique des Industrie, Fonderie): I have a comment for Mr Ayres. The method of measuring surface energy with the groove is of paramount importance. My question concerns the error occurring from impurities. From theoretical evaluation the solid–liquid surface energy can range from 100% to 0·2% of the solid vapour surface energy. Considering measurement, the dihedral angle method collapses because the contact angle is often zero. Considering the homogeneous nucleation method, in our laboratory not many materials nucleate homogeneously so heterogeneous nucleation is used and a correction of 10^{-10} to 10^{-20} can be included in the constant. Fortunately there is the groove method, but it is very important to consider the errors. In the preceding paper on succinonitrile there is a very careful discussion on this point. Nevertheless, it seems to me that you do not consider the possibility of impurities with direct segregations concentrating in a small volume close to the boundary of solid crystals. Such an enrichment can explain a large part of the groove. Do you take into account this possibility of enrichment, by diffusion or convection? Have you made experiments on the effect of impurities added at known levels, or on the effect of the distance between the grooves on the measurement of solid–liquid surface energy?

MR AYRES: I am afraid that I cannot give you much in the way of theoretical analysis, but for those of you who are not really familiar with our work, our previous paper on grooves dealt with stationary grooves in a stable thermal gradient situation. In this work we would measure the exact shape of the groove and determine the surface energy from this. These measurements were also carried out on high-purity materials. We are concerned about the effect of impurities. It is not the sort of experiment where you can get in and probe the impurity level at the interface. In the previous work we made every effort to eliminate impurity effects. One way in which we think the effects would have shown up if they were important would be that of shifting the position of the interface by altering the gradient. The interface moves, assumes a new stable position with a new groove depth. When we did this, it was a very short time before we achieved a stable shape which was retained for a matter of hours and if there had been any sort of partitioning other than those demanded by equilibrium effects, this groove should have altered its shape with time. So we think in our previous experiments that we have not had any serious effects because of impurities. In the results which I described in the paper we are working with a migrating boundary and there is certainly every possibility that, even though we had pretty pure material, we could have trapped impurities at that boundary, and those impurities could affect the result.

DR L. M. HOGAN (University of Queensland, Australia): I have a small point to make about the Spittle and Lloyd paper. The authors refer to our suggestion that primary dendrite arm spacing is determined by changes before the establishment of steady-state growth. In subsequent experiments with a transparent analog we have observed the initial transient stage. Starting with an initially stationary plane interface, fine perturbations developed as growth commenced; these coarsened with partial remelting so that some crystals were isolated just ahead of the interface and could change orientation by rotation. As growth continued, in a positive temperature gradient, the isolated crystals and remaining projections developed into branched dendrites with varying orientations which competed so that some were eliminated. By the time steady-state growth was established an average primary arm spacing was established which remained constant throughout steady-state growth. The length of the transient period determined the final spacing and was related to the freezing conditions, i.e. growth rate, temperature gradient, and liquidus–solidus interval. It can be said that this is also a coarsening process, though different from that responsible for secondary spacing.

PROFESSOR F. WEINBERG (University of British Columbia, Canada): Considering Dr Hurtuk's paper with regard to columnar growth in continuously cast billets, on the basis of his report I looked at 11-strand continuously cast mild-steel billets and measured the columnar length. The carbon content ranged from 0·18 to 1%. The length of the columnar zone was plotted as a function of carbon content. It was noted that the length of columnar zone is a pretty sloppy measurement. The length varies quite a lot along the billet for no apparent reason, but in any event, I did not see any correlation with carbon as he did. We suggest that the effect is probably quite small.

DR HURTUK: It is very difficult to get good measurements of columnar growth on an etched cross-section. We took measurements from four sides of an etched bloom and averaged these, and correlated that with the average carbon content for the cast, which is not a precise measurement either. As one of the figures showed, despite the many other variables present, the effect of carbon still showed through. We then verified it in the laboratory and we think it is real.

DR E. SELCUK (Middle East Technical University, Turkey): I would like to comment on Dr Hurtuk's paper. I do not think his model simulates continuous casting correctly because the tundish in continuous casting, which functions as a reservoir of liquid metal, regulates the metal flow and the heat as well. His ingots did not seem to be perfectly directionally solidified because I saw some dendrites appearing in the centre of the ingots, growing from the side walls.

DR HURTUK: On the question of whether our model reproduced the conditions of continuous casting, we did not attempt to do that. We really wanted to give trends so that we could use that information and then apply it to the caster and completely redefine the whole process on the caster itself. We realise that it is hard to reproduce those conditions not only with the tundish but with the copper chill we are using and the eventual air gap that forms. It is very difficult to reproduce the thermal gradients that we get on the caster. We wanted to identify trends and apply that to the caster.

You mentioned the question of side growth in the ingots. It is true that on some we did have growth from the side. The slides I showed were early ingots we produced. We refined the technique as time went on, with better insulation for one thing, and for most of the ingots produced we were able to obtain a flat interface and we know there was no interference from side growth.

MR R. WIDDOWSON (BSC, Teesside Labs): I have two questions for Dr Hurtuk. First, did he find that the extent of the columnar structure was sensitive to secondary cooling intensity? Does he consider regulation of secondary cooling intensity a practical means of regulating this type of structure? Secondly, to which point in the process is superheat referred, i.e. ladle, tundish, or mould?

DR HURTUK: On secondary cooling and its effect on columnar growth, that would certainly have a tremendous effect on the extent of columnar growth. One of the things we are trying to institute in our caster is control of secondary cooling to control the extent of columnar growth. We feel that coupling controlled secondary cooling with the superheat would give us the best control of columnar growth without going to different types of dynamic method. The measurements of superheat in our caster were taken both in the tundish and verified later with measurements in the mould itself.

DR J. CAMPBELL (Fulmer Research Institute Ltd): I would like to draw attention to the interpretation of the results from Dr Hurtuk's work, because it seems to me that in a lot of the work of this nature, one can see discontinuities in the results. It is worth looking for such discontinuities and drawing attention to them, particularly in the case of the length of the columnar zone and per cent carbon. With the 0·5% carbon, which is the peritectic composition, one would expect two quite separate curves depending whether it is solidifying to gamma or delta iron.

DR HURTUK: Concerning the plot of per cent carbon *v.* columnar growth, I think you are quite correct in that we may have two separate sets of data. At this time, we were just starting this business and our data was limited. The data only extended up to 0·60% carbon, where in fact the other set of data may have just begun, so we drew a smooth curve through all the points—being very naïve about the whole thing. However, you may have a good point there.

DR D. H. KIRKWOOD (University of Sheffield): On the paper by Hellawell *et al.*, in which they were determining partition coefficients, it is clearly important that we get accurate values for partition coefficients. If we look at the values in Table 1, it appears that they are claiming a fairly high accuracy for the partition coefficients. They are claiming a value of about 0·14 for the pure binary aluminium–copper. It is the accuracy of this technique that I wish to question. Clearly, it is very nice to have shortcuts but shortcuts with wrong answers are no good to anyone. It so happens that we have been working on the partition coefficients with Al–Cu–Ni by rather more conventional techniques of quenching and analysing with the microprobe and we get rather different results. I should say that these rapid quenches were followed by equilibration for very different times. The variation in time was taken into account, and also for particle size, and we believe that we have got very reliable values, therefore, for partition coefficients for Al–Cu. The liquidus we get is exactly the same as the previous work of Burden and Hunt, which we believe to be very accurate. The solidus is very different and gives us an equilibrium partition coefficient of 0·21 as against the value of 0·14, and clearly we believe our result is right.

If you determine the partition coefficient from the theoretical equation from the heat fusion, again you get a value of 0·21. I am rather puzzled about this very large discrepancy of 50% between our values. Did the authors ever quench during their solidification and examine the assumptions which are the basis of their analysis? Was there any sign of microsegregation towards the outside, the gap between the crucible tube and the solidified material itself? Was there any sign and what was the extent of the build-up of solute at the interface?

I also plotted the values of Hellawell *et al.* as a log concentration against log fraction liquid; according to the Scheil equation this should be a straight line giving a slope of $(1 - k_0)$. If you do this, you get a very good straight line and it turns out that k_0 is 0·24 so that there is an internal inconsistency within their results. I think the only explanation is that, somehow, they have got the wrong average composition for the liquid initially. I would like to ask them how they determine this composition and what accuracy they claim.

PROFESSOR HELLAWELL: In reply to Dr Kirkwood, I do agree that the discrepancy between his results and our interpretation of our results should be resolved. The experimental problem complicating interpretation of segregation curves arises because when a specimen of given C_0 is placed in a furnace above its melting point, so that one-half or two-thirds of it is molten, there is part of the specimen between the solidus and liquidus temperature. During and after the time for attaining thermal equilibrium we cannot precisely define C_0 or the actual starting point for growth. Bartholomew therefore interpreted the results for best fit allowing C_0, $g = 0$, and k_0 to vary. This would account for the inconsistency in different ways of interpreting our results, but not the discrepancy between our results and those of Burden and Hunt with Dr Kirkwood's higher value of k_0.

In reply to the query about the planarity of the interface—yes, we did confirm this on quenched specimens.

PROFESSOR F. WEINBERG (University of British Columbia, Canada): Concerning the movement of secondary dendrite branches, I was surprised to see that the shape of the secondary branches was maintained in the process of moving. I think the secondary branches take up this narrow neck, broadening out, because the branch breaks through the solute layer on the primary dendrite showing that it is very sensitive to local solute thermal conditions and yet later on, when you get to the melting–freezing process, the morphology does not change, which I find very surprising. I would like to ask the authors to comment about that.

I have no feeling at all for the coefficient of thermal expansion of the organic material with which we are dealing. Is it possible that, as a solid cools, there are small movements of the dendrite with respect to the glass walls of the cell which might add to the kind of transverse movement we see?

DR ALLEN: First of all, Professor Weinberg talked about the structure not changing while the arms were migrating. If you look at our pictures you will see that the structure does change, it does all the things you would expect it to do in an isothermal situation. In Fig. 1, arm number 7, which is a small one and thinner than number 6 or 8 on either side of it, is fairly well developed in the top picture; it is melting off in the second picture and it is not there in the third, so one does get these changes. Over the surface of a dendrite arm, the composition varies basically on two counts. It varies because you have variations in temperature that lead to temperature gradient zone melting, and it varies because there are variations in interfacial curvature. That explains the various coarsening effects explained by Flemings *et al*. It is an interesting question as to whether these two effects interact in any way. All we can say just from looking at it is it looks as if they do not actually interact, they both go on at once, but it would be a good question on which to do further research.

As regards the second question on the thermal expansion, we did worry about this. In fact, if it happened it would detract from our measurements and

tend to make the arms go the other way. There is semi-solid throughout the temperature range of the pictures; further back there is fully solidified material which anchors itself to the glass cell walls and keeps everything fixed. If you imagine all of the cell cooling, the dendrite is anchored at the back and as you cool it, the primary length which is cooling will be shrinking, and that could drag all the secondary arms back down the temperature gradient towards lower temperatures so that this would counteract the migration effect we were observing. We therefore looked at this to ensure that movement was not underestimated, by growing some reasonably coarse dendrites, solidifying them very quickly, and then looking in the microscope to see whether this rather rapid cooling would drag everything backwards. We could not see it moving at all so it gave us the indication that thermal expansion effects were negligible as far as our measurements were concerned.

MR J. D. AYERS (Naval Research Laboratories, USA): I also have a comment directed to Drs Allen and Hunt. I have done a lot of work on succinonitrile. One of the studies we did was on dendrite growth and we were using very high-purity material specifically to get the sort of compositional effects they are studying. In these studies where we had isolated dendrites of succinonitrile going into a supercooled melt, we also found that the side branches pointed forward and would appear to migrate towards the tip region. We believe in this instance it was strictly a thermal effect, that there was a greater gradient in a diagonal direction away from the core, and that the solidification was occurring preferentially on the forward faces of the side branch and there was no reason for remelting to have occurred because in the high purity situation we had, there was no driving force for remelting.

DR ALLEN: First of all, however high the purity, as long as you are still in a region where the liquidus line has some sort of slope and is not absolutely flat, then there will be a composition difference between the two ends of the liquid region, which are at different temperatures. Therefore, there will still be a remelting effect for the reasons we explained.

You talked about the side arms growing forward. This certainly does happen and we had to be rather careful in the measurements we took to avoid getting this into our results. For instance, if we compared the positions of a sidearm tip in Fig. 1*a* and *c*, our measured displacement in the temperature gradient direction would only be partly due to temperature gradient zone melting. The growth of the sidearm towards the temperature gradient direction would also contribute to the displacement measured. It is rather difficult to avoid structural features which are not affected by that. The best we thought we could do was the junction between the secondary arm and the primary stalk. We thought that would be a reasonably reproducible place to take measurements and we took all our measurements from here when doing the quantitative results in our previous paper. If we had taken them all from different places on the sides of the dendrites, it would have looked as if they migrated further and the extra effect would have been due to the fact that they point forwards. It is interesting that they should point forward. This again is related to differences of composition, due to the temperature gradient which biassed the growth in the forward direction, or there may also be a thermal effect. It would be interesting to look at that further. We think that in our system with the solute content of about 6 wt-% the reason why the sidearms point forwards is a solute effect. Considering three sidearms, such as numbers 9, 10, and 11 in Fig. 1, the liquid at the tip of 9 will be least pure and the liquid at the tip of 11 will be most pure, so arm 10 will reject solute most efficiently by growing somewhat towards 11 and away from 9, that is pointing forwards. Thus the effect of temperature gradient in biassing the direction of sidearm growth somewhat resembles temperature gradient zone melting except that melting is not involved. In super-pure systems, however, it is probably a thermal effect rather than a solute effect which makes sidearms point forwards.

PROFESSOR HELLAWELL: I would like to remark on some aspects of the paper by Fisher and Kurz on faceted/non-faceted eutectics. We have also concluded that the kinetic undercooling, e.g. for Si, is very small and does not significantly contribute to the observed large undercoolings in alloys such as Al–Si, Fe–C, etc. The problem in such a system is that the faceted phase cannot easily adjust its dimensions laterally to the heat flow direction. In the eutectics this relative inflexibility of one phase leads to variable and large spacings and, consequently, to large solute undercoolings for the non-faceted (metallic) phase. This in turn produces a non-planar growth front (faceted phase projecting forwards into the liquid, albeit wetted by metal close to its growing edge/point) and a sensitivity to the imposed temperature gradient. It must follow that since temperature gradient is an important variable, there is no unique coupled region in these systems but rather a whole range of such regions.

Session 2

Eutectic and peritectic solidification

Chairman: W. Kurz (Ecole Polytechnique Federale de Lausanne, Switzerland)

Editors: I. G. Davies (BSC Swinden Laboratories, Rotherham)
R. Jordan (Alcan International, Banbury, Oxon)

Keynote Address: Eutectic and peritectic solidification

M. Hillert

Some fundamental aspects relating to the eutectic and peritectic reactions in binary alloys are reviewed. The usual eutectic-like and peritectic-like modes of growth are discussed for both types of reaction. New growth-rate equations are given for some of the cases. The effect of a ternary addition in promoting a dendritic growth of a eutectic aggregate is analysed and it is shown that the tip radius of such a dendrite is controlled by a completely different mechanism from that on an ordinary dendrite. The transition between eutectic and peritectic reactions in ternary systems is discussed in relation to the diffusivities of the various elements in the various phases.

The author is with the Royal Institute of Technology, Stockholm, Sweden

Some fundamental problems related to eutectic and peritectic reactions will be dealt with in this paper. The author's main field of interest is solid-state transformations and an attempt will be made to look at solidification problems with eyes that have been trained to examine solid-state problems. In particular, in the solid-state field it is not difficult to establish isothermal conditions experimentally and consequently the theoretical models developed in that field are usually strictly isothermal. Diffusion controlled reactions are thus regarded as the result of some supersaturation. In the solidification field the approach is often quite different and one tends to regard supercooling as the driving force. Complex situations have been analysed, for instance, by splitting up the total supercooling in various ΔT contributions, even when the various levels of temperature thus defined have no physical significance. Another example of this attitude is the widespread use of the concept of constitutional supercooling, a concept that seems very strange to one so accustomed to thinking that diffusion controlled reactions always have a constitutional driving force.

The author's personal view is that in the solidification field one tends to make things more complicated than they are by considering the temperature gradient together with concentration gradients. Admittedly, there are phenomena that can be understood only if both types of gradient are considered, the stability of a planar solidification front in a temperature gradient being a good example. However, it would be a useful strategy always to attempt to explain a phenomenon through an isothermal model before including specific effects of the temperature gradient. This method is used in the present paper.

COOPERATIVE EUTECTIC GROWTH

Modern research in the field of metals started with the introduction of the X-ray technique. When applied to phase transformations this technique revealed an orientation relationship between growing plates of a precipitate phase and the parent phase. This discovery led to a tendency to emphasize the role of crystallography; as an example from the solid-state field, an orientation relationship between the parent phase and one of the growing phases in a eutectoid structure was postulated and received considerable attention but was never confirmed experimentally. In the solidification field there could of course be no orientation relationship to the parent phase which is liquid. Instead, the possible orientation relationship between the two eutectic phases has attracted much attention and has often been reported in experimental studies. However, the possible effect of such an orientation relation on the growth was neglected when the theory of diffusion controlled eutectic growth was worked out in some detail for lamellar structures by the author[1] and extended to rod-like eutectics by Jackson and Hunt.[2] This theory, which is isothermal in nature, seems capable of explaining the main features of the eutectic reaction, thus indicating that the orientation relationship does not have a dominating influence. Among other things, the theory predicts the existence of a liquid layer of changed composition ahead of the solidification front. This layer has an approximate thickness of D/v and its existence can be confirmed by changing the growth rate v in directional solidification experiments.

It is well known that the existence of an enriched layer ahead of a single phase growing in a temperature gradient tends to break down the planar front and make it cellular or dendritic. It has been suggested[3] that the layer formed ahead of a eutectic structure should have a similar effect and thus play an important role in breaking down the smooth solidification front of a eutectic and in the development of a primary precipitation of single-phase dendrites ahead of the eutectic front. One has even attempted to use the criterion, developed for the transition from planar to cellular single-phase solidification in a temperature gradient. The author sees this as an example of the temperature gradient being used to explain a phenomenon that is essentially isothermal in nature. The fact that a completely eutectic structure can often be obtained over a considerable range of composition, but not outside, is mainly an isothermal problem although the range may be increased by the application of a temperature gradient. This phenomenon cannot be a simple function of G/R. Furthermore, the layer should not have the postulated effect because the theory of diffusion controlled eutectic growth indicates that the eutectic growth rate is rather insensitive to the alloy composition. As pointed out long ago,[1] the main question is whether the eutectic reaction gives a higher growth rate than the separate growth of either one of the phases. This principle has recently been applied successfully in a calculation of the coupled zone in the Fe–C system[4] and the effect of a temperature gradient should be introduced as a modification of this model.

Before leaving the theory of eutectic growth, it is worthwhile taking a close look at the predicted interface (Fig. 1). The numbers refer to the case of pearlite but the general shapes apply to any lamellar eutectic. The calculation of the shape started from the three-phase junctions where the three surface energies were assumed to balance each other, thereby fixing the angles. The calculation then yields the shape. It is interesting that the curvature is different for the two phases and varies along the edges. There will thus be differences in pressure between the phases and within each one. It is not known how and to what degree the phases can support such differences, nor how the growth conditions would change if the pressure differences are not supported. It has been suggested that this might result in an extra loss of free energy.[5] It should be interesting to examine how the theory is modified if one assumes that the pressures as well as the surface tensions are balanced at the three-phase junctions.

DEGENERATE EUTECTIC GROWTH

The well organized growth of two eutectic phases side by side has been described as a result of their cooperation,[1] the reason for the cooperation being that in order to grow each phase needs the element of which the other phase must rid itself. However, if one examines the growth problem in more detail, one may conclude from the same evidence that each phase should prefer to grow along the interface between the other one and the parent phase. This could result in a constant battle between the phases, each one trying to isolate the other from contact with the parent phase. Indeed, this tendency is often very pronounced, especially if one of the phases is present as a pro-eutectic precipitation. At the start of the eutectic reaction this phase will be covered with a halo of the other phase. In some cases like spheroidal cast iron, the halo formation will dominate the whole eutectic reaction. The resulting microstructures of a eutectic reaction, where the two phases have not been able to establish a cooperation, are coarser and are often called degenerate (or divorce) eutectic structures. Usually it takes some time after the nucleation of the two phases for them to establish cooperation and grow in the well known eutectic fashion. Secondary phases, which often form out of some impurity towards the end of the solidification in commercial alloys, usually form in such a small volume fraction that there is insufficient time for cooperation to be established. Consequently, the final microstructure may fail to reveal that they formed by a eutectic reaction.

The process by which the second eutectic phase to form grows along the surface of the primary precipitation will now be examined in detail. The assumption of a balance of surface tensions at the three-phase junctions and the neglect of any diffusion in the solid phases results in a picture like that in Fig. 2. The diagram resembles what one would achieve by cutting out half a lamella of each type in Fig. 1. The main difference appears to be that the diffusion distances have shortened to half. Thus it may be suggested that the growth rate v can be estimated from the theory for the normal eutectic reaction. The following form should be preferred[6]

$$v = \frac{2D^L(^ex^{L/\alpha} - {}^ex^{L/\beta})}{f^\alpha f^\beta (x^\beta - x^\alpha)S} \tag{1}$$

where f^α and f^β are the mole fractions of α and β, respectively. According to the maximum growth rate hypothesis, the quantity S should be twice the critical value S_c.

By analogy to the peritectic case[7] the above growth process may be classified as the degenerate eutectic reaction. The subsequent thickening owing to solid-

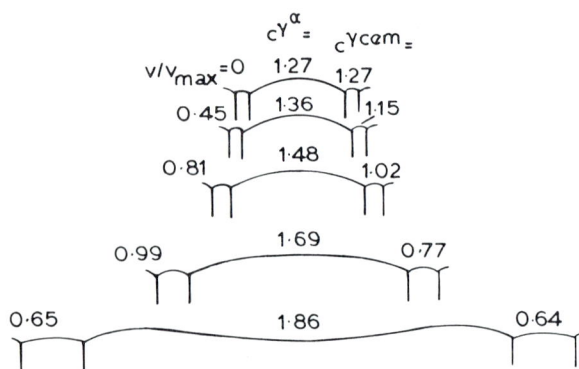

1   Predicted shape of solidification front for lamellar eutectic (from Ref. 1)

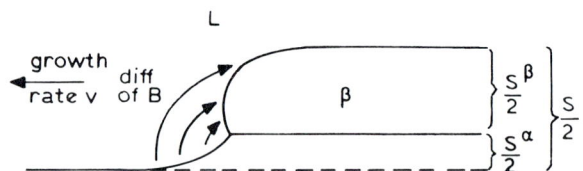

2   Degenerate eutectic reaction by which second eutectic phase β grows along surface of primary phase α

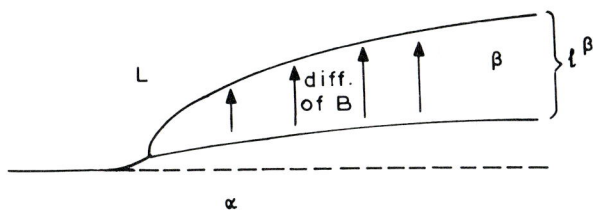

3 Degenerate eutectic transformation occurring after primary α-phase has been covered by layer of second eutectic phase β

state diffusion should then be classified as the degenerate eutectic transformation (Fig. 3). Its rate has recently been derived[8] and is given by the expression

$$(l^\beta)^2 = \frac{2(x^L - x^\alpha)(x^{\beta/\alpha} - x^{\beta/L})D^\beta t}{(x^\beta - x^L)(x^\beta - x^\alpha)} \qquad (2)$$

This growth process will continue until interrupted by the cooperative eutectic growth which is much faster.

PERITECTIC REACTION AND TRANSFORMATION

The peritectic case has attracted relatively little attention, but some interest has been shown quite recently[7] in papers concerned mainly with the behaviour of peritectic alloy systems in directional solidification.[9-11] We shall first discuss the behaviour of a peritectic system under natural cooling conditions.

In a peritectic system the solidification starts by the primary precipitation of a phase α. Below the peritectic temperature, a new solid phase β will form, with a strong tendency to grow along the α/L interface and thus to isolate the primary phase from contact with the liquid. The reason for this tendency is that the phase diagram predicts the following type of reaction

$$\alpha + L \to \beta$$

Thus it may seem natural that β should grow where both the parent phases are present. However, the β-phase may also grow as a primary precipitation into the liquid, in particular if the temperature is gradually decreasing. So there may be a competition between the two modes of growth and it should be interesting to estimate the growth rate of β along the α/L interface.

As in the eutectic case we shall first discuss the geometric shape and assume that there is a balance of surface tensions at the three-phase junction. If this condition is combined with an assumption that there is no diffusion in the solid phases and the α/β interface is thus fixed behind the three-phase junction, one obtains the shape shown in Fig. 4. The arrows indicate the diffusion of the alloying element *B*. The diagram

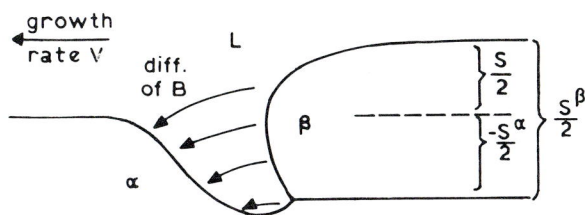

4 Peritectic reaction by which second solid phase β grows along surface of primary phase α

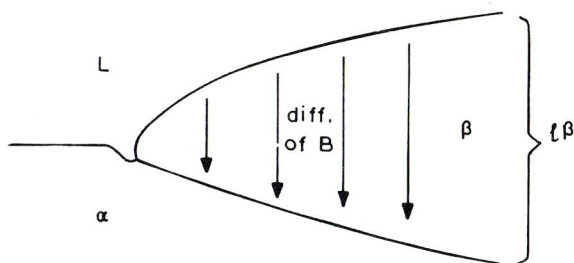

5 Peritectic transformation occurring after primary α-phase has been covered by layer of second solid phase β

demonstrates that the α-phase goes into solution simultaneous to the growth of β. A puzzling feature is that some new α forms just before the three-phase junction. This occurs because of the strong negative curvature of the α/L interface. Furthermore, it should be pointed out that we are again faced with the problem related to the different pressures in the α- and β-phase, which was mentioned in connection with the eutectic case.

The diagram resembles the one constructed for the degenerate eutectic reaction (Fig. 2) and it seems reasonable to apply the same equation. Since the α-phase is now even closer to the β/L interface one should perhaps further increase the numerical factor. The equation takes the following form if the definitions of f^α and f^β are maintained from the eutectic case, which means that f^α is now negative, because α dissolves, and f^β is larger than unity

$$v = \frac{2D^L({}^e x^{L/\beta} - {}^e x^{L/\alpha})}{-f^\alpha f^\beta (x^\beta - x^\alpha)S} \qquad (3)$$

According to the maximum growth rate hypothesis, *S* should have twice the critical value.

The reaction can also take place by diffusion in the solid phases. We shall consider the effect of diffusion only in the growing β-phase, by which the thickness of the β-layer may grow after its initial formation (Fig. 5). This has been called the peritectic transformation as distinct from the peritectic reaction.[7] It resembles the degenerate eutectic transformation and the same rate equation can be applied. It now takes the form[8]

$$(l^\beta)^2 = \frac{2(x^L - x^\alpha)(x^{\beta/L} - x^{\beta/\alpha})D^\beta t}{(x^L - x^\beta)(x^\beta - x^\alpha)} \qquad (4)$$

ALIGNED PERITECTIC GROWTH

By the application of a temperature gradient it is possible to make the solidification proceed in a certain direction. If the temperature gradient is strong enough, the development of primary dendrites can be suppressed and a planar solidification front may even develop. When applied to peritectic alloys of a composition that falls within the α + β two-phase field, this technique can suppress the formation of dendrite side-arms and the result will be an aligned α + β two-phase structure.[9]

It was pointed out by Chalmers[12] that a temperature gradient of sufficient strength should produce a planar solidification front for peritectic as well as other types of alloys, and Chalmers suggested that this should lead to a coupled growth of the two solid phases. Two recent

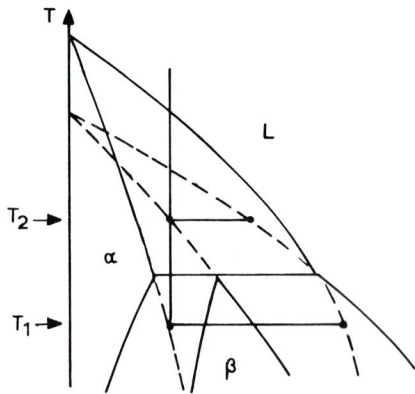

6   **Peritectic phase diagram showing temperatures for stationary planar growth of α-phase (T_1) and β-phase (T_2) at indicated alloy composition**

attempts to confirm this suggestion have failed.[9,10] In both cases alternating bands of α and β were obtained.

It is possible that this result can be explained with reference to the phase diagram. Figure 6 shows a typical peritectic diagram and it is evident that an alloy of the indicated composition can solidify to pure α at a temperature T_1. If the temperature gradient is successively increased, the temperature of the solidification front will thus decrease continuously and the dendritic morphology will change to cellular and finally it will become planar at T_1. However, the extrapolated phase boundaries for the β + L equilibrium indicates that, in the absence of α, it should be possible to make the same alloy solidify to pure β with a planar solidification front at T_2 which is higher than T_1. If the β-phase is nucleated it may thus form a complete layer ahead of the planar α-front. Furthermore, the β-phase in contact with the liquid is not stable with respect to α at the temperature T_2 because it is above the peritectic temperature. As a consequence, if α is nucleated it may grow along the β/L interface and again form a planar α-front.

Using the above explanation it is easy to understand that Boettinger[9] observed temperature fluctuations in his experiments. If this explanation is correct, it should be possible to obtain a planar one-phase solidification if the experimental combination of gradient and rate is such that there is insufficient time for nucleation or

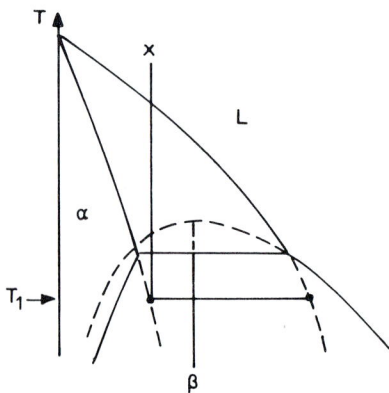

7   **Peritectic phase diagram allowing planar growth of α but not β at indicated alloy composition**

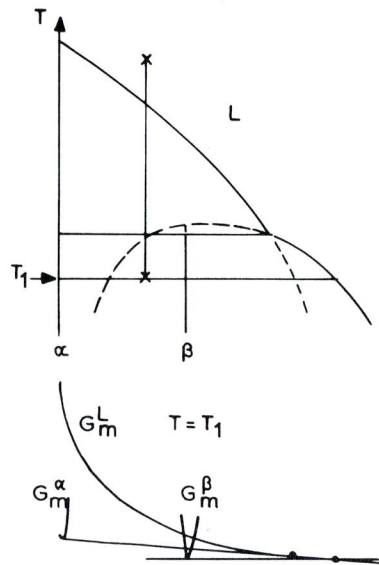

8   **Peritectic phase diagram which may allow eutectic growth of α + β for indicated alloy composition, Gibbs energy diagram constructed for temperature T_1**

growth of a new phase ahead of the planar solidification front. In this particular case of directional solidification, the result may not depend solely upon the well known type of combination G/R. An easier way to prevent the formation of β ahead of the planar α-front would be to select an alloy system where the β-phase has a more constant composition (Fig. 7). This diagram demonstrates that a planar growth of β is now impossible at the alloy composition under consideration, because the β-phase cannot have that composition. For the same reason, rapid growth of β is not possible and it should be easy to prevent β from forming ahead of the planar α-front. On the other hand, this type of alloy system offers a new possibility. If one attempts to solidify an alloy of a slightly higher B-content with a planar α-front, the initial liquid will fall below the liquidus lines of both α and β and an ordinary eutectic reaction may start. This is better demonstrated by a system where both the α- and β-phases have constant compositions. Figure 8 shows such a case and the related Gibbs energy diagram. The diagram demonstrates that the β-phase can be in equilibrium with an A-rich liquid as well as a B-rich liquid. In the former case, the situation resembles the eutectic case and it should be possible for B to diffuse away from the α/L interface and reach β/L interface. A necessary requirement for an aligned eutectic structure to form is that local equilibrium can be attained between the three phases at the three-phase junctions. As in the eutectic reaction, this may be accomplished by means of the increased pressure owing to the curved α/L and β/L interfaces (*see* Fig. 9). The essential feature is probably the point where the liquid composition at the three-phase junctions falls. It may be that the eutectic reaction requires this composition to fall between those of the α- and β-phases and not to the right of the β-phase. In order to achieve this, one must apply a larger pressure on the α-phase than on the β-phase and this can be accomplished by simply choosing an alloy composition such that the volume fraction of α is small, i.e. a composition close to that of

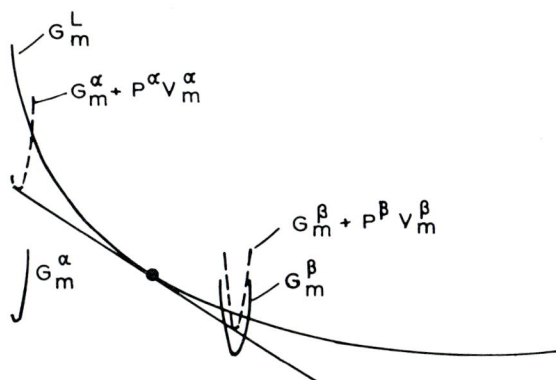

9 **Gibbs energy diagram showing condition at three-phase junctions in eutectic-like growth. The Gibbs energy curves for solid phases have been lifted by amount PV_m where P is given by curvature of solid/liquid interface, point of tangency on L-curve should fall between α and β**

the β-phase. It should be emphasized that this type of reaction should not require a strong temperature gradient but simply a low enough temperature.

If a B-rich layer of liquid has been established ahead of the solidification front, one should expect an ordinary peritectic reaction. With an alloy system where the α- and β-phases have constant compositions, it should not be difficult to produce aligned structures, and with a sufficiently strong temperature gradient one should be able even to obtain a rather planar solidification front. However, this is not sufficient justification for using the term 'coupled growth' and it does not seem probable that the coarseness of the aligned structure will be determined by the maximum growth rate hypothesis or the equivalent hypothesis for directional solidification which uses minimum undercooling. The solidification front cannot be even approximately isothermal and it makes no sense to talk about a particular undercooling.

EUTECTIC-LIKE REACTIONS
In systems with a liquid miscibility gap one may have a monotectic reaction:

$$L_1 \rightarrow \alpha + L_2$$

This resembles the eutectic reaction and it is interesting to examine the effects of exchanging one of the solid phases for a liquid. Two factors should be pointed out. First, there can be no orientation relationship between the two growing phases, and the fact that the monotectic reaction frequently results in aligned eutectic-like structures indicates that the orientation relationship often reported for eutectics does not play a major role in eutectic reaction. However, the fact that the monotectic structures all seem to be of the rod-like type indicates that the orientation relationship in eutectics may be of importance for the development of lamellar eutectics. Secondly, the growing liquid phase cannot maintain any pressure differences. As a result, the theory of the eutectic reaction cannot be applied directly to the monotectic case. The L_1/L_2 interface must have a constant curvature and, in addition, the pressure that this curved interface creates in the growing liquid phase is transmitted into the interior of

the monotectic two-phase structure. The consequence of this effect on the theory should be examined.

Similar to the eutectic reaction, the monotectic reaction may occur by cooperative growth or it may degenerate and result in droplets of the new liquid phase. The realization of this simple fact led to the suggestion that one should differentiate between four kinds of MnS inclusions in steel: normal eutectic, degenerate eutectic, normal monotectic, and degenerate monotectic.[13] The old classification system did not realize the possibility of having a normal monotectic structure.

Quite a number of systems can give rise to a metatectic reaction:

$$\alpha \rightarrow \beta + L$$

This reaction has attracted some interest recently.[14] Typically, it occurs soon after an alloy has solidified completely to an α-phase and is a result of the transformation of α to a new solid β-phase which has a lower solubility of some element that stabilizes the liquid phase. It is thus evident that the volume fraction of the L-phase is usually rather small. In spite of the fact that the reaction formally resembles the eutectic type, it seems natural not to expect normal eutectic growth but rather a degenerated eutectic reaction, yielding a structure where the new liquid phase is situated along the previous solid/solid interfaces. This case has been observed.[15]

An attempt has been made to produce an aligned eutectic-like structure by this reaction.[16] A single-phase β specimen from a peritectic alloy system was heated into the $\alpha + L$ two-phase region and the metatectic reaction $\beta \rightarrow \alpha + L$ was indeed observed and gave a rod-like eutectic structure that could be preserved by quenching (Fig. 10).

EFFECT OF TERNARY ADDITION ON EUTECTIC GROWTH
It is well known that an alloying addition has about the same effect on eutectic growth as it has on primary solidification. A ternary addition to a binary eutectic alloy may thus cause the eutectic colonies to break down into cells and the cells may even branch in a

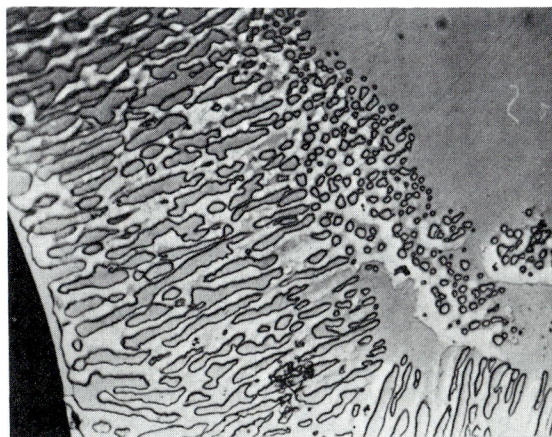

10 **Rod-like two-phase structure obtained by metatectic reaction of Γ-phase in FeZn system when heated into $\alpha + L$ region, [×800]**

dendritic fashion.[17] By directional solidification one can stabilize a planar eutectic solidification front and the criterion for *G/R* developed for single-phase solidification has been applied successfully.[18] There seems to have been relatively little interest in the study of these two-phase cells or dendrites, although one may find interesting problems in this area. One may wonder, for instance, how it is possible for the ternary addition to escape the growing dendrite when the two main elements have time to redistribute themselves only between the two growing phases. This problem is accentuated by the fact that the two-phase dendrite is very thick and contains a large number of lamellae. Furthermore, one may ask why the two-phase dendrite is so thick. The same amount of alloying addition to a one-component specimen should produce very thin dendrites. In an attempt to stimulate work in this area, a very rough theoretical treatment of these problems is presented in the Appendix. The result shows that the width of the two-phase dendrite is not in any way controlled by the same factor that operates in ordinary dendritic growth. In contrast with ordinary dendrites, the thickness of the two-phase dendrite is predicted to decrease with increasing alloy content.

CRITERIA FOR EUTECTIC AND PERITECTIC REACTIONS IN TERNARY SYSTEMS

In ternary systems it often occurs that a three-phase reaction which is eutectic on one binary side is peritectic on another side. Somewhere inside the ternary system there is a transition. The criterion used in many textbooks to predict exactly where the transition takes place is based on the assumption that there is no diffusion in the solid phases. Long ago, a completely different criterion was derived for the case where there is such slow cooling that all differences in chemical potentials can be eliminated by diffusion.[18,20] Of course, the truth falls somewhere between these two extremes. This problem has been largely neglected although the equivalent problem has been discussed for transformations in the solid state.[16,21] However, in a recent paper Fredriksson[22] has opened up this area. It is not sufficient to look at the overall reaction. One must analyse in detail what reaction takes place at different locations in the system and one must consider the different diffusivities of various elements in all the phases present. As a result of his analysis, Fredriksson could predict that a reaction which had an overall peritectic character could locally produce a eutectic or a metatectic reaction on one of the sides of the growing solid phase.

REFERENCES

1 M. HILLERT: *Jernkontorets Ann.*, 1957, **141**, 757
2 K. A. JACKSON AND J. D. HUNT: *Trans. AIME*, 1966, **236**, 1 129
3 F. R. MOLLARD AND M. C. FLEMINGS: *ibid.*, 1967, **239**, 1 526, 1534
4 H. FREDRIKSSON: *Metall. Trans.*, 1975, **6A**, 1 658
5 M. HILLERT: 'The mechanism of phase transformations in crystalline solids', 231, 1969, London, Institute of Metals
6 M. HILLERT: *Metall. Trans.*, 1972, **3**, 2 729
7 H. W. KERR, *et al.*: *Acta Metall.*, 1974, **22**, 677
8 M. HILLERT: *Scri. Metall.*, 1972, **11**
9 W. J. BOETTINGER: *Metall. Trans.*, 1974, **5**, 2 023
10 A. P. TITCHENER AND J. A. SPITTLE: *Acta Metall.*, 1975, **23**, 497
11 D. H. ST. JOHN AND L. M. HOGAN: *ibid.*, 1977, **25**, 77
12 B. CHALMERS; 'Physical metallurgy', 271, 1959, New York, John Wiley
13 H. FREDRIKSSON AND M. HILLERT: *J. Iron Steel Inst.*, 1971, **209**, 109
14 S. WAGNER AND D. A. RIGNEY: *Metall. Trans.*, 1974, **5**, 2 155
15 H. FREDRIKSSON AND J. STJERNDAHL: *Metall. Trans.*, 1975, **6B**, 661
16 M. HILLERT AND B. UHRENIUS: *Scand. J. Metall.*, 1972, **1**, 223
17 J. E. GRUZLESKI AND W. C. WEINGARD: *Trans. AIME*, 1968, **242**, 1 785
18 M. D. RINALDI *et al.*: *Metall. Trans*, 1972, **3**, 3 133, 3 139
19 O. S. IVANOV: *Dokl. Akad. Nauk.*, 1949, **49**, 349
20 M. HILLERT, 1958, **189**, 224; and 1960, **195**, 201
21 M. HILLERT: 'Physical chemistry in metallurgy', (ed. Fisher, Oriani, Turkdogan), 445, 1976, Pittsburgh, Pa, US Steel Corporation
22 H. FREDRIKSSON: *Met. Sci.*, 1976, **10**, 77
23 M. HILLERT: *Metall. Trans.*, 1975, **6A**, 5

APPENDIX
Dendritic growth of eutectic colony in presence of impurity

We shall consider the dendritic growth of a eutectic resulting from the presence of an impurity *C* to a binary eutectic alloy *A–B*. The impurity is less soluble in the solid phase α and β than in the liquid phase.

Owing to the curvature of the α and β edges, there will be a displacement of the phase boundaries because of the Gibbs–Thomson effect (Fig. 11). The various quantities are defined in the diagram. For small undercoolings the triangles will have the same shape and we find

$$\frac{\Delta x_C}{\Delta x_{AB}^1 - \Delta x_{AB}} = \frac{\Delta x_C^0}{\Delta x_{AB}^0} = F \qquad (5)$$

where $F = \sqrt{3}/2$ if α and β are pure *A* and *B* and the liquid phase is ideal. Δx_{AB} is the concentration difference that drives the diffusion of *A* and *B* in front of the eutectic colony. According to the binary theory[6] we have

$$v = E \Delta x_{AB}/S \quad \text{where} \quad E = 2D_{AB}/f^\alpha f^\beta (x^\beta - x^\alpha) \qquad (6)$$

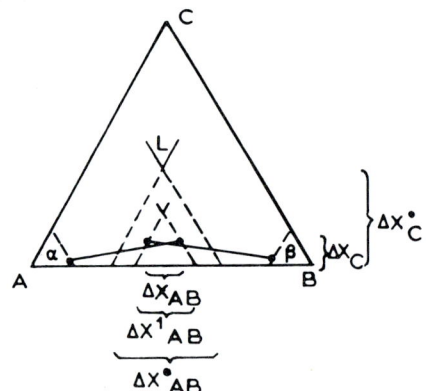

11 Ternary phase diagram showing effect of impurity *C* on eutectic reaction in *A–B* alloy

where S is the interlamellar spacing and Δx_C is the concentration within the liquid phase that drives the diffusion of C away from the growing $\alpha + \beta$-aggregate. For the diffusion of C we shall apply the theory of dendritic growth. If the impurity content is low, the supersaturation of $\alpha + \beta$ is high and we can use a simple, approximate rate equation[23]

$$v = \frac{D_C \Delta x_C}{2r \Delta x_C^{i/e}} \tag{7}$$

where $\Delta x_C^{i/e}$ denotes the difference in C-content of the initial liquid and the eutectic structure. This difference is equal to the initial impurity content if C is insoluble in the two solid phases. We shall treat $\Delta x_C^{i/e}$ as a constant. Let us introduce the critical spacing for the binary case. It is related to the distance between the two solubilities

$$S_c = k/\Delta x_{AB}^0 \tag{8}$$

Assuming that the eutectic front is planar, the spacing is given by the distance between the two solubility lines at the impurity level that is maintained at the front

$$S_1 = S_c/(1 - \Delta x_{AB}^1/\Delta x_{AB}^0) \tag{9}$$

However, if the eutectic colony is shaped like a dendrite, the actual spacing must be larger and is related to the curvature of the front. The relationship is complicated but for small curvatures it seems reasonable to apply the following type of equation

$$r(S - S_1) = S_1^2 \tag{10}$$

where r is the macroscopic radius of curvature at the tip of the dendritic eutectic colony.

We can immediately eliminate Δx_C and r by inserting equations (5) and (10) into equation (7). By combining equations (6) and (7) we obtain

$$v = \frac{D_C(\Delta x_{AB}^1 - \Delta x_{AB})F(S - S_1)}{2S_1^2 \Delta x_C^{i/e}} = \frac{E \Delta x_{AB}}{S} \tag{11}$$

This yields

$$\frac{\Delta x_{AB}^1}{\Delta x_{AB}} = 1 + \frac{L^2 S_1^2}{(S - S_1)S} \tag{12}$$

where $L^2 = 2E\Delta x_C^{i/e}/FD_C$ (12a)

By inserting equation (12) into equation (6) we find

$$v = \frac{E \Delta x_{AB}^1}{S + L^2 S_1^2/(S - S_1)} \tag{13}$$

We have here v as a function of S and Δx_{AB}^1 only, because S_1 is a function of Δx_{AB}^1. Notice that

$$\frac{dS_1}{d\Delta x_{AB}^1} = \frac{S_c/\Delta x_{AB}^0}{(1 - \Delta x_{AB}^1/\Delta x_{AB}^0)} = \frac{S_1^2}{S_x \Delta x_{AB}^0} \tag{14}$$

In the binary theory of dendritic growth and eutectic growth one finds v as a function of the coarseness and in order to predict the actual growth rate and spacing one must use some additional criterion, such as Zener's maximum growth rate hypothesis. In the present case there are two degrees of freedom because we have both types of growth. We shall apply Zener's hypothesis and make the derivatives with respect of S and Δx_{AB}^1 equal to zero. We thus obtain

$$S/S_1 = 1 + L \tag{15}$$

$$\Delta x_{AB}^1 = \tfrac{1}{2}\Delta x_{AB}^0 \tag{16}$$

By inserting these values into the previous equations we find

$$S = 2S_c(1 + L)$$

$$\Delta x_{AB} = \frac{1 + 2L}{1 + L}\Delta x_{AB}^1 = \frac{1}{2} \cdot \frac{1 + 2L}{1 + L}\Delta x_{AB}^0 \tag{17}$$

$$v = v_{binary}^{max} \cdot \frac{1 + 2L}{(1 + L)^2} \tag{18a}$$

where

$$v_{binary}^{max} = \frac{D_{AB}\Delta x_{AB}^0}{2S_c f^\alpha f^\beta (x^\beta - x^\alpha)} \tag{18b}$$

L is a correction which is small if the impurity content is low,

$$L = \sqrt{\frac{4D_{AB}\Delta x_C^{i/e}}{FD_C f^\alpha f^\beta (x^\beta - x^\alpha)}} \tag{19}$$

Finally, the radius of the tip of the dendritic eutectic colony will be

$$r = 2S_c/L \tag{20}$$

Complex-regular growth in the bismuth–lead eutectic

D. Baragar, M. Sahoo, and R. W. Smith

Eutectic alloys of the bismuth–lead systems have been solidified unidirectionally in order to determine the extent to which the temperature gradient at the solid/liquid interface, and the thermal conductivity anisotropy of the faceted phase influence the growth morphology of complex-regular eutectics. The results show that the structural changes observed in this complex-regular eutectic can be described adequately in terms of constitutional undercooling, and are, therefore, analogous to the changes in the cell structure known to occur in single-phase binary alloys. Measurements of the size of the complex-regular cells λ of the lead–bismuth eutectic as a function of growth rate were also made. It was found that the cell size λ varies as a function of growth rate and obeys the relationships predicted by the theory for the growth of single-phase binary alloys, i.e. $\lambda^2 \propto R^{-2/3}$ (small constitutional undercooling) and $\lambda^2 \propto R^{-1}$ (large constitutional undercooling).

The authors are in the Metallurgical Engineering Department, Queens University, Kingston, Ontario, Canada

The bismuth–lead eutectic forms a complex-regular cellular structure during solidification. In previous work[1] it has been shown that the shape of the complex-regular cell is defined by the faceted bismuth phase and that the orientation of the bismuth and, as a result, the cell shape change as the growth rate is varied. It was suggested[1] that the predominant orientation of the bismuth phase under any particular growth conditions is determined, both by the degree of constitutional undercooling at the solid/liquid interface and by heat flow due to the thermal anisotropy of bismuth. The purpose of the present investigation was to establish the importance of thermal conductivity anisotropy by studying the way in which the cell shape was influenced by the magnitude of the temperature gradient at the solid/liquid interface.

COMPLEX-REGULAR CELLULAR GROWTH

Hunt and Jackson[2] have suggested that for regular lamellar eutectics to grow it is necessary that (a) a lamellar space-change mechanism must be available to accommodate local fluctuations in growth rate; and (b) rapid variation in the lamellar half-width must be possible in order to accommodate fluctuations in composition. These two processes are not readily available for faceted–non-faceted eutectics and, consequently, a compositional boundary layer is formed. It was proposed[3] that an irregular structure forms if the boundary layer is small and a complex-regular structure forms when the boundary layer is large.

Croker *et al.*[1] have built a model of the skeletal bismuth phase of the complex-regular cell formed in the lead–bismuth system, (Fig. 1). The figure is essentially a near right-angled pyramid defined by three main {100} bismuth plates which join in the ⟨111⟩ axis. The secondary arms from these plates have {100} faces and all plates are terminated by {100} macrofacets forming the three cube faces of the pyramid. It was found that all the microstructural forms in this eutectic could be generated by successive sections through this model. These structures are shown in Fig. 2 along with the respective quenched solid/liquid interfaces. The change in orientation of the cell was thought to occur as a result of constitutional-undercooling effects and heat-removal considerations.

At low growth rates, constitutional undercooling is small and the depth the cell can grow into the melt is small. Under these conditions, the cell is oriented such that the 'fish-spine' cell is produced, (Fig. 2a). As the growth rate is increased, constitutional undercooling is increased, and the trigonal cell is now the favourable orientation, (Fig. 1). It was suggested that the trigonal cell (Fig. 2b) is more stable than the fish-spine cell because it projects further into the undercooled layer for a given cell width. The appearance of the cubic cell at higher growth rates was thought to occur in order to increase the thermal conductivity of the bismuth phase in the growth direction. The conductivity of bismuth

1 Model cell of skeletal bismuth phase in lead–bismuth eutectic; trigonal axis of bismuth is perpendicular to plane of paper

perpendicular to the trigonal axis ($\langle 111 \rangle$) is 1·73 times greater than that parallel to $\langle 111 \rangle$, (Fig. 2b). As the growth rate is increased it is presumed that heat removal becomes more important and the cell orientation which has a greater thermal conductivity in the growth direction is stable, resulting in the appearance of the cubic structure (Fig. 2c).

EXPERIMENTAL

In order to determine whether or not the argument given above was correct, the eutectic alloy was grown over a range of growth rates and temperature gradients. The alloys were prepared by placing the weighed components into a Pyrex tube and sealing the tube under a partial pressure of argon. The charge was then melted and shaken vigorously in order to ensure homogeneity. Growth of weighed amounts of alloy was carried out in a vertical 'Bridgman' crystal-growing apparatus. The melt was moved downward through the furnace into a water-cooled quenching bath.

Temperature gradients were measured using a 0·01 cm wire chromel–alumel thermocouple, which was arc welded and subsequently insulated with a thin coat of alundum cement before being positioned at the centre of the growth tube through two diametrically opposed holes, which were later sealed with alundum cement. The output of the thermocouple was attached to a recording potentiometer in order to obtain a continuous plot of temperature v. distance. The temperature gradient was varied by changing the furnace setting.

RESULTS AND DISCUSSION

The results of this work are summarized in Fig. 3 which is a plot of the growth rate R and temperature gradient G at which a particular structure occurs. As can be seen, at a given growth rate the fish-spine cell is found at higher temperature gradients, the trigonal cell at intermediate values, and the cubic cell at lower gradients. However, the ratio of temperature gradient to growth rate G/R at which a particular structure occurs is not constant. At a growth rate of 8×10^{-5} cm/s, the cubic structure is stable to G/R values approximately seven times greater than that at a growth rate of $4 \cdot 5 \times 10^{-3}$ cm/s. At a growth rate of $1 \cdot 3 \times 10^{-3}$ cm/s, the cubic structure occurs at about the same G/R values as at $4 \cdot 5 \times 10^{-4}$ cm/s. On the basis of the picture of complex-regular growth given by Hunt and Harle[3], at higher growth rates the compositional boundary layer should be larger than at lower growth rates because the system will have less time to adjust to growth rate or compositional fluctuations. Consequently, as the growth rate is increased two effects may be expected to increase the constitutional undercooling, namely, the increase in the bismuth content at the interface and the increase in the sharpness of the solute profile. If this were the case, the cubic structure would be expected to be stable at higher G/R values as the growth rate is increased. This interpretation is inconsistent with the experimental values reported above. Also, close observation (*see* Fig. 4) of the bismuth lamellar shows that the width varies markedly along its length, which suggests that the bismuth plates do not have difficulty in adjusting their width. Thus, it is necessary to seek another explanation of the phenomenon.

Such an explanation of complex-regular growth can be made in terms of growth rate v. undercooling relationships of the two phases involved. Faceted materials require a much larger driving force for the interface attachment of atoms to occur at the same rates as for a non-faceted material. Therefore, the composition at which the two phases of a faceted–non-faceted eutectic can grow at the same temperature in a coupled manner[4,5] is richer in the faceted component than the equilibrium value. This results in an increased driving force for growth of the faceted material and a decrease for the non-faceted material. Thus, a boundary layer will develop, the size of which will depend on the difference in the undercooling required for the growth of the two phases. This boundary layer permits the establishment of a zone of constitutional undercooling with respect to the faceted phase, which responds by forming macrofaceted cellular projections.

In the lead–bismuth eutectic, if it is assumed that each type of cell appears at a set value of constitutional undercooling (as in the case of single-phase materials), then the G/R value at which that cell appears will not necessarily be constant or increasing as the growth rate is increased. For example, if the growth rate–undercooling relationship for the faceted and non-faceted phases are shown in Fig. 5, then the boundary layer resulting from interface attachment kinetics will be relatively large at a velocity V_1, but small and constant for growth velocities exceeding V_2. In this situation, the concentration of bismuth in the liquid at the solid/liquid interface would decrease and then become constant. Because of this, lower G/R values would be required for the cubic structure to be stable. The observations reported here support this argument.

Once a boundary layer has developed, the faceted phase can form cells. The shape of these cells will be

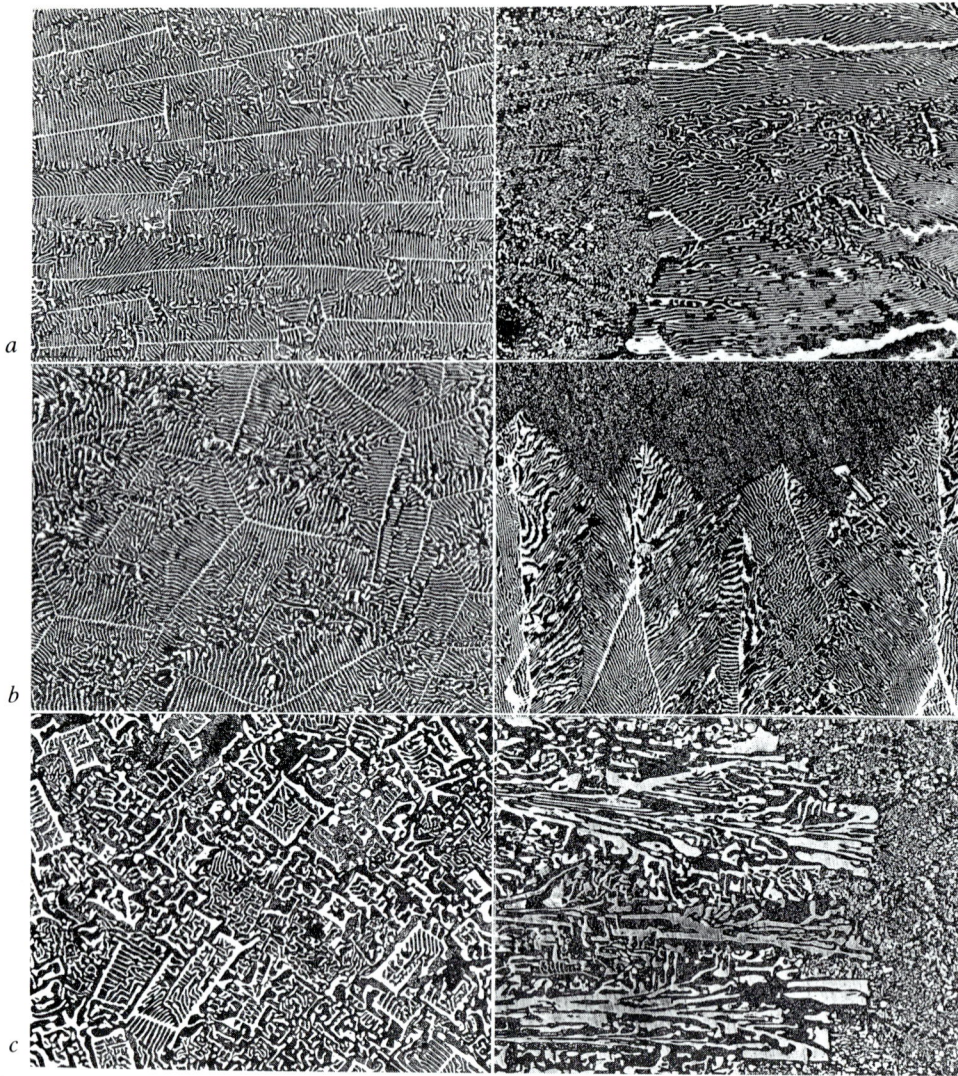

2   **Transverse section (left) and quenched interface (right) of the lead–bismuth eutectic showing;** *a* **fish-spine cell, growth rate = 4·5 × 10^{-4} cm/s, temperature gradient = 140°C/cm;** *b* **trigonal cell, growth rate = 4·5 = 10^{-4} cm/s, temperature gradient = 90°C/cm;** *c* **cubic cell, growth rate = 4·5 × 10^{-4} cm/s, temperature gradient = 50°C/cm;   [× 95]**

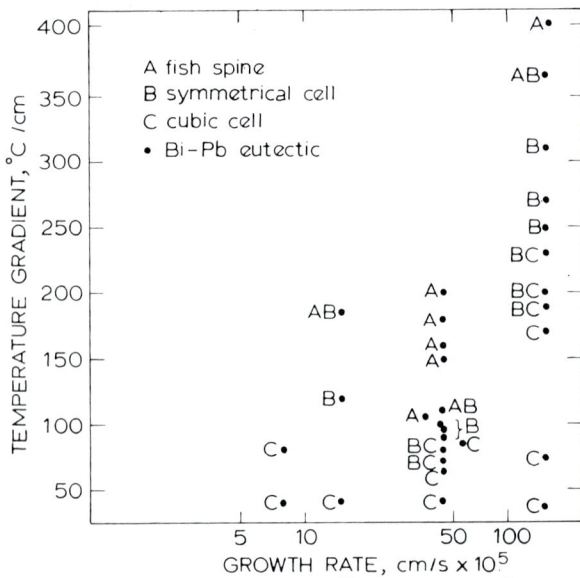

3   **Plot of temperature gradient *v.* growth rate showing structures found for bismuth–lead eutectic**

4   **Transverse section showing large variation in lamellar width of bismuth phase; growth rate = 8 × 10^{-5}, temperature gradient = 200°C/cm;   [× 1 100]**

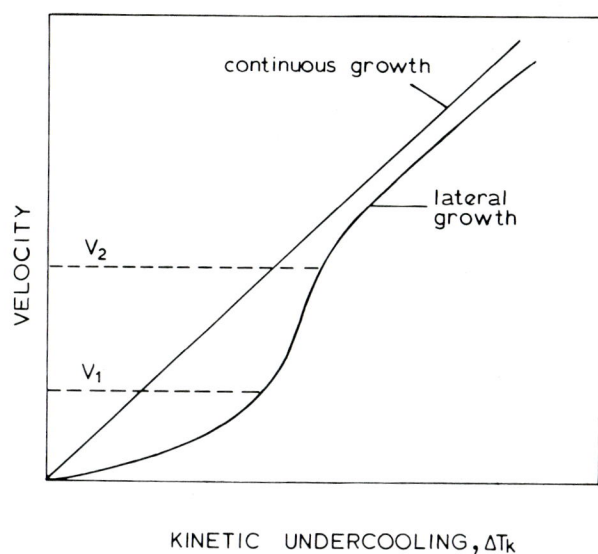

5 Schematic illustration of probable growth rate *v*. undercooling curves for continuous growth (non-faceted materials) and lateral growth (faceted materials)

governed, as is the case for single phase materials, (*a*) by the degree of constitutional undercooling; (b) by crystallographic effects; and (c) by heat-removal and surface-energy considerations.[6,7,8] When growth occurs with a planar interface, that is at high G/R values, a single-phase material will grow in the direction of maximum heat flow.[9] As G/R is lowered, constitutional undercooling develops and diffusion of solute becomes important. Elongated cells are found to occur in this region.[10,11] As G/R is lowered further, a greater degree of constitutional undercooling is present ahead of the solid/liquid interface. Thus, the distance from the cell base to the cell tip is increased, and this necessitates a lower tip concentration, as has recently been demonstrated by Jin and Purdy.[12] Therefore more solute must diffuse from the tip to the base, and as Kramer *et al.*[13] have shown, using a simple model of solute conservation, the area in which solute is incorporated must increase. Thus the regular shaped cell is developed which has more area available for solute incorporation per area of cell on the solid/liquid interface than the elongated cell.[13]

As the degree of constitutional undercooling is increased beyond that at which the regular-shaped cells are stable, the cells change shape and acquire the characteristics of dendrites.[14,15] The cell tips for a square pyramidal cell, are markedly different from those of the regular or elongated cell and grow in the characteristic dendrite direction. Chalmers[15] refers to them as cellular dendrites because they are intermediate between regular cells and dendrites. It is suggested[16] that the cellular dendrite is formed because, as the constitutional undercooling is increased, more solute flows to the boundary. Eventually, the cell boundary can no longer absorb this additional solute and the congestion is relieved by branching of the cells to form the cellular dendrite.

Although the cellular growth of faceted materials is similar to that of non-faceted phases,[17] the shapes of

the cells will be determined by the orientation of the faceting planes with respect to the growth direction. This will also be true for the complex-regular growth of the bismuth–lead eutectic. At the onset of constitutional undercooling, the bismuth orientation will be determined by a combination of thermal conductivity and diffusion considerations. The stable cell is thus the fish-spine type which has an orientation between the direction of maximum ($\perp \langle 111 \rangle$) and minimum ($\| \langle 111 \rangle$) thermal conductivity. The fish-spine cell also has a ratio of cell boundary area to cell area in the solid/liquid interface which is 1/3 of that for cells growing in the $\langle 111 \rangle$. As G/R is lowered further, thermal conductivity becomes less important and diffusion becomes more important. The stable cell is now the trigonal cell which is presumed to be more efficient than the elongated cell in terms of accommodating the redistribution of solute.

In the case of a faceted material such as bismuth, the width of the cell is determined by its height. Consequently, there comes a point when further segregation cannot be accommodated with a trigonal cell structure and, therefore, the system forms a cellular structure for which the lateral dimension is not determined by the projection distance into the liquid. In the case of the bismuth–lead eutectic, the cubic cell develops which has this characteristic. It also grows in the characteristic dendrite direction of bismuth and has a cell shape different (Fig. 2*c*) from that of the fish-spine cell on the trigonal cell. It is concluded that the cubic cell is a form of cellular dendrite found in single-phase materials. The fact that the cubic cell is found at the lowest values of temperature gradient at a given growth rate does not support the theory that the cubic cell appears as a result of the system's attempt to bring the direction of highest thermal conductivity into the growth direction. The complex–regular cellular growth in the lead–bismuth eutectic is thus found to be closely analogous to cellular growth in single-phase non-faceted materials.

In order to learn more about the growth behaviour of the complex–regular eutectic, the cell size was measured as a function of growth rate. The results are shown in Fig. 6, where cell area is plotted as a function of growth rate for the fish-spine, trigonal, and cubic cells. The least-squares method was used to determine the slope of these plots and it was found that cell area varied as $(1/R)^{2/3}$ for the fish-spine cell and $1/R$ for the trigonal cell.

The only quantitative analysis of cell growth available is for non-faceted cellular growth of a single-phase dilute binary alloy.[18,19] From this, the cell area λ^2 may be calculated in two ways. The first is to find the value of the wavelength[20] of a sinusoidal perturbation which causes the interface to become unstable and put that equal to the corresponding cell size, namely

$$\lambda^2 \propto R^{-2/3} \qquad (1)$$

This relationship should be relevant at the onset of instability.[20] Alternatively, the frequency can be found by maximizing the growth rate of a pertubation with respect to frequency. It is thought that this method will be more relevant at large degrees of instability. The results show that

$$\lambda^2 \propto R^{-1} \qquad (2)$$

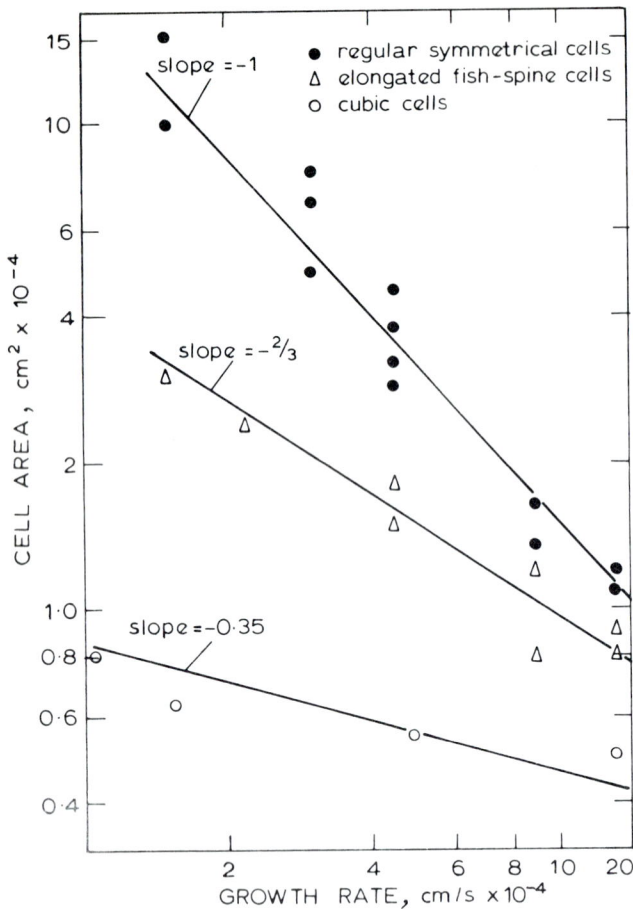

6 Cell area of fish-spine symmetrical (trigonal) and cubic cells of the Pb₂Bi–Bi eutectic as function of growth rate

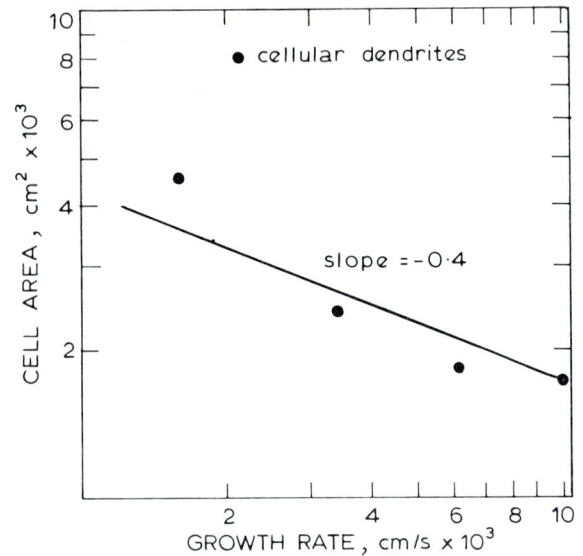

7 Area of cellular dendrites of a Fe–8Ni alloy as function of growth rate; after Jin and Purdy[12]

effects of growth velocity on the solidification structure of cellular dendrites. Measurements of the area of cellular dendrites (taken from their Fig. 1 of Ref. 12) have been plotted in Fig. 7 as functions of growth rate. A comparison of this curve with the curve for the cubic cell (Fig. 6) shows that these two cell forms obey similar relationships between cell area and growth rate. This supports the proposition made earlier that the cubic cell is a form of cellular dendrite as found in single-phase material. The results of the cell-size measurements reported here support the observation that complex-regular cellular growth in the lead–bismuth system is analogous to the cellular growth of single-phase non-faceted materials.

More recent experimental evidence has provided new information on the dependence of cell size on growth rate. Morris and Winegard[21] studied cell development in non-faceted material and found that the cell spacing of elongated cells, which occur at small degrees of instability, increased with decreasing growth rate. Coriell et al.[20] replotted this data and found that λ^2 varies with $R^{-2/3}$ in agreement with equation (1). The growth-rate dependence of the hexagonal form of the cellular interface which is found when there is a large degree of undercooling present, has been studied by Plaskett and Winegard.[22] They found that cell area varied inversely with the growth rate which is the relationship predicted by Coriell et al.[20] for growth in this region.

The results reported here show that the fish-spine and the elongated cell of a single-phase, non-faceted material have the same growth rate dependence on cell area. In addition, the present authors have found from measurements of the temperature gradient that the fish-spine cell appears near the onset of instability, as does the non-faceted elongated cell.[21,23,24] Similarly, the symmetrically shaped hexagonal cell and the tri-gonal cell have the same dependence of cell area on growth rate and both occur at larger degrees of instability than the elongated cell.

The growth of cellular dendrites has not received the attention given other growth forms. However, Jin and Purdy[12] have recently carried out controlled solidification experiments in order to investigate the

REFERENCES

1 M. N. CROKER et al.: *J. Cryst. Growth*, 1971, **2**, 121
2 J. D. HUNT AND K. A. JACKSON: *Trans. AIME*, 1966, **236**, 843
3 J. D. HUNT AND D. T. J. HURLE: *ibid.*, 1968, **242**, 1 043
4 L. M. HOGAN: *J. Aust. Inst. Met.*, 1964, **9**, 228
5 A. KOFLER: *Z. Metallkd.*, 1950, **41**, 221
6 G. F. BOLLING AND W. A. TILLER: *J. Appl. Phys.*, 1960, **31**, 2 040
7 M. C. FLEMINGS: 'Solidification processing', 67, 1974, Toronto, McGraw-Hill Inc
8 L. MORRIS AND W. WINEGARD: *J. Cryst. Growth*, 1969/70, **69**, 61
9 D. T. J. HURLE: *Prog. Mater. Sci.*, 1962, **10**, 88
10 D. WALTON et al.: *Trans. AIME*, 1955, **203**, 1 203
11 H. BILONI et al.: *ibid.*, 1966, **236**, 930
12 I. JIN AND G. PURDY: *J. Cryst. Growth*, 1973, **23**, 39
13 J. J. KRAMER et al.: *Trans. AIME*, 1963, **227**, 374
14 H. BILONI AND B. CHALMERS: 'Principles of solidification', Ref. 39, 1964; New York, Wiley
15 B. CHALMERS: *ibid.*, 104, 1964; New York, Wiley
16 J. O. COULTHARD AND R. E. ELLIOT: 'The solidification of metals', 61; 1968, London, The Iron and Steel Institute
17 W. BARDSLEY et al.: *Solid-State Electron.*, 1962, **5**, 395

18 W. MILLINS AND R. SEKERKA: *J. Appl. Phys.*, 1964, **34**, 323; *ibid.*, 1964, **35**, 444
19 R. SEKERKA: *J. Cryst. Growth*, 1968, **3**, 71
20 S. R. CORIELL *et al.*: *ibid.*, 1976, **32**, 1
21 L. MORRIS AND W. WINEGARD: *ibid.*, 1969, **5**, 361

22 T. S. PLASKETT AND W. WINEGARD: *Can. J. Phys.*, 1959, **37**, 361
23 W. A. TILLER: 'The art and science of growing crystals', 295, 1963, New York, Wiley
24 F. MOLLARD AND M. FLEMINGS: *J. Cryst. Growth*, 1969, **5**, 361

Orientations and interfaces in directionally solidified eutectic systems

I. G. Davies and D. D. Double

The various types of orientation relationships that can develop during directional solidification of binary eutectic crystals are discussed with reference to a number of representative alloy systems. It is shown how lattice-coincidence considerations often determine the phase–phase and habit plane orientations; however, the adoption of these particular epitaxies depends on the conditions under which solidification is carried out. Their effect in determining the final morphology of the duplex crystal is described.

I. G. Davies is with the Sheffield Laboratories, BSC, and D. D. Double is in the Department of Metallurgy and Science of Materials, University of Oxford

In this paper, the authors consider the crystallographic orientation relationships that are developed in binary eutectic alloys under conditions of directional solidification.[1] Since, in these duplex crystals, three degrees of freedom are required to specify the relative phase–phase orientation and two to specify that of the habit plane, it is clear that a number of configurations are possible in which the orientation relationship is fixed to a greater or lesser degree. The situation is summarized in Table 1 and Fig. 1; this figure shows, on stereographic projections, how the phase–phase orientations can be (Case A) completely random, (Case B) specified along a common axis but free to rotate about this axis, and (Case C) uniquely defined by mutual parallelism along two non-collinear axes. In these three situations, the various possibilities that exist for the development of preferred habit planes and growth textures are indicated. Further reference will be made to Table 1 and Fig. 1 in the examples which follow, in order to examine the extent to which such orientations are adopted by different alloy systems during growth.

LAMELLAR MORPHOLOGIES
The aligned lamellar and rod-like microstructures that are characteristic of metallic eutectic alloys reflect the closely coupled and cooperative mode of growth of

Table 1 Phase-phase orientation relationships

Case	Phase–phase orientation* relation-ships	Possible growth textures	Habit plane(s)
A	None, no directions or planes in either crystal show preferred alignment	Only a single fibre texture could develop	One phase could assume preferred solid/liquid facets
B	A common pole $X_a \parallel X_b$ is shared about which each phase is free to rotate	A single fibre texture is possible (cf case A) except when the growth axis $\parallel X_a \parallel X_b$ when a double fibre texture would occur	Solid/liquid facets on one phase (cf case A) or stabilization of the plane $X_a \parallel X_b$ from partially preferred epitaxy; the former possible with single or double fibre texture, the latter only with single fibre texture
C	Fully preferred epitaxial orientations $X_a \parallel X_b$, $Y_a \parallel Y_b$ and therefore $Z_a \parallel Z_b$	If preferred, must always be a double-fibre texture without freedom to rotate mutually	Whether preferred from solid/liquid faceting or solid/solid coincidence, will always relate to both phases

* See Fig. 1 for stereographic projections of phase–phase orientation relationships

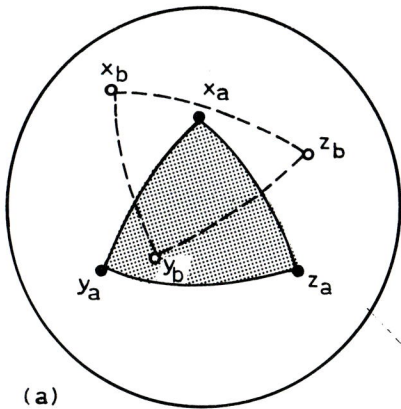

(a)

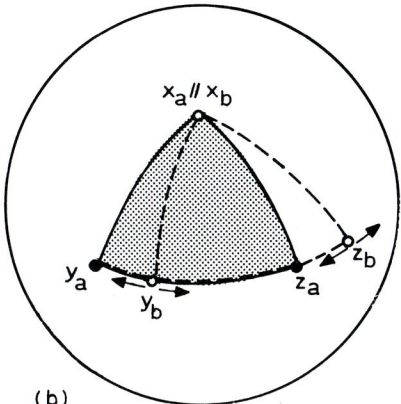

(b)

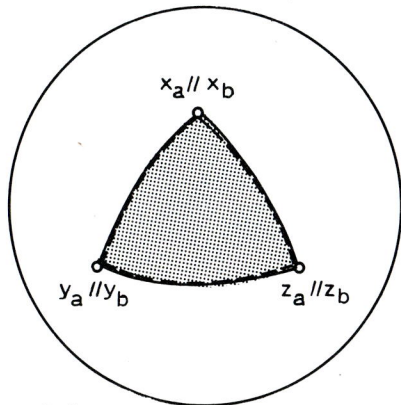

(c)

a Case *A*; *b* Case *B*; *c* Case *C*

1 Stereographic projections of phase–phase orientation relationships; three mutually perpendicular axes (X_a, Y_a, Z_a) of phase (*a*) are fixed in the stereographic projection; alternative orientations of phase (*b*) are represented by poles X_b, Y_b, and Z_b, which are also 90° apart

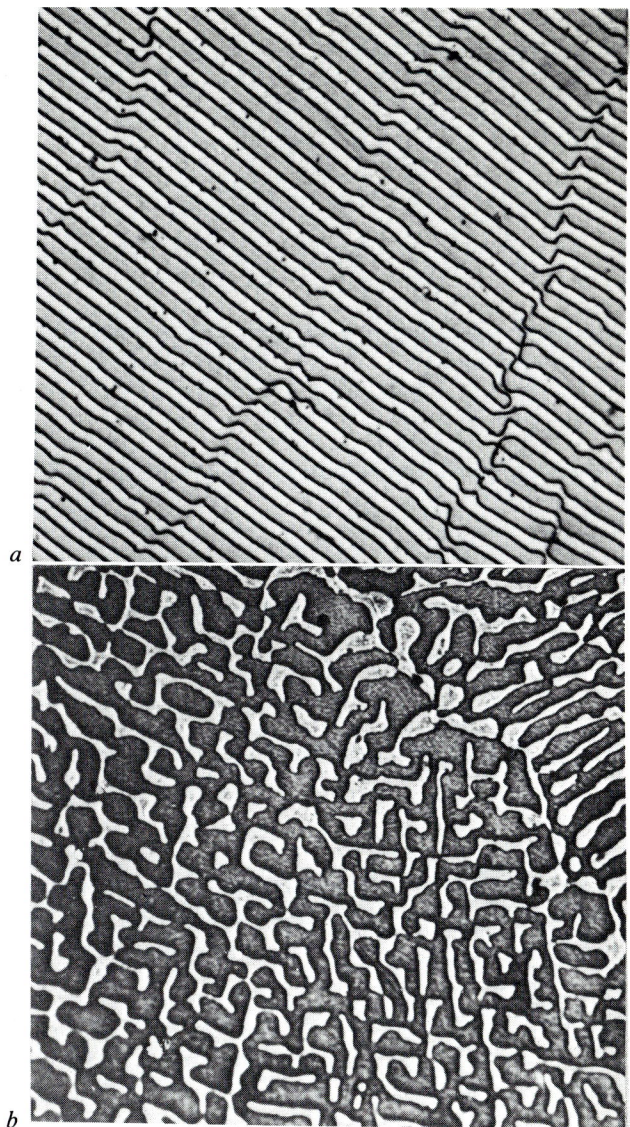

a

b

a regular lamellar microstructure, section transverse to growth axis [×350]; *b* degenerate morphology in eutectic Al–CuAl$_2$ [×400]

2 Optical micrographs of microstructure in eutectic Al–CuAl$_2$

which these two-phase crystals are capable under controlled conditions.[2] The establishment of these regular structures is associated with the development of particular phase–phase orientations which allow the formation of low-energy interface planes between the phases, usually closely parallel to the growth axis.

The eutectic Al–CuAl$_2$ system has been extensively studied and, although it does not represent the simplest example in this category, it shows a number of features that are probably of general relevance. A number of orientation relationships have been quoted but the one

most commonly found in lamellar grains (Fig. 2*a*) may be written[3,4,5]:

$$(111)\text{Al}\|(2\bar{1}1)\text{CuAl}_2 \quad \text{and} \quad [1\bar{1}0]\text{Al}\|[120]\text{CuAl}_2$$

(1)

This orientation represents a uniquely fixed phase–phase configuration such as that represented in Case *C* in Table 1. The preferred growth direction is usually around $[\bar{1}12]$Al, i.e. parallel to $(111)\text{Al}\|(2\bar{1}1)\text{CuAl}_2$. This texture is probably dictated by the growth anisotropy of the CuAl$_2$ phase since it represents an extension in the plane of the ordered $\{2\bar{1}1\}$ layer lattice of CuAl$_2$ along an axis which is a compromise between two close-packed directions.

Kraft[3] has explained this orientation relationship in terms of the crystallographic similarities that exist in the atomic densities and interplanar spacings of the (111)Al and $(2\bar{1}1)$CuAl$_2$ planes. A simple two-dimensional model which illustrates the partial coincidence

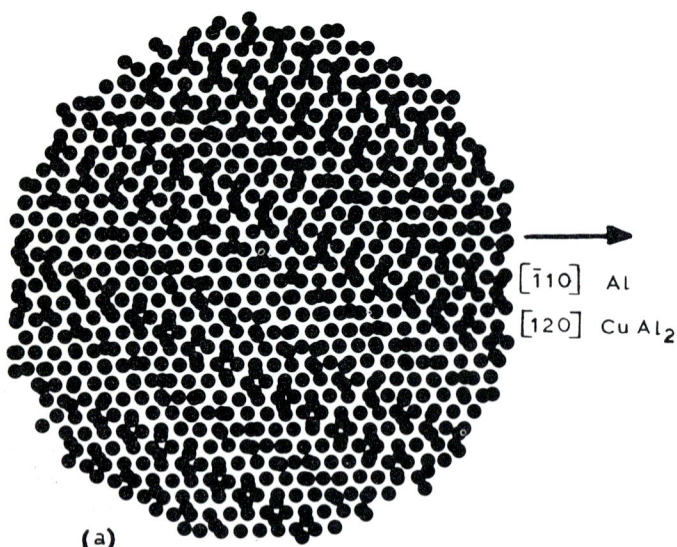

(a)

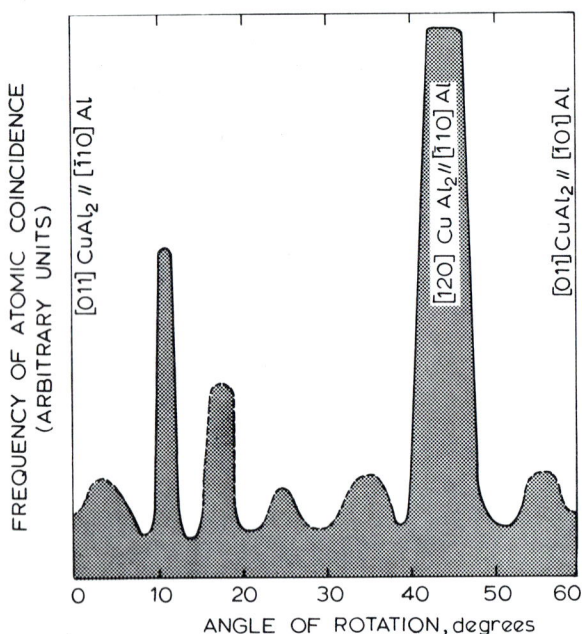

3 *a* Coincident site pattern obtained by superimposition of two nets representing atomic positions in the (111) Al plane and the (2̄1̄1) CuAl₂ plane, respectively; the axes [1̄10]Al and [120]CuAl₂ are parallel; *b* frequency of atomic coincidence in patterns such as that shown in Fig. 2*a* plotted as function of relative angle of rotation about the normal to the (111)Al‖(2̄1̄1)CuAl₂ planes

4   Transmission electron micrograph of lattice misfit dislocations imaged at lamellar interface in eutectic Al–CuAl₂   [×50 000]

when [1̄10]Al‖[120]CuAl₂—the orientation (1) actually observed.

Figure 3*a* is useful since it shows the semicoherent nature of the interface. In practice, the misfit between the two lattices at the interface is accommodated by regular arrays of dislocations (Fig. 4), the density of which relates to the difference in interplanar spacings between the phases and the orientation of the interface.[4,7]

The curve in Fig. 3*b* predicts that a number of subsidiary coincidence peaks should occur at other relative rotations of the two planes. This is of some interest because it has been observed in degenerate morphologies of Al–CuAl₂, (i.e. those preceding the evolution of the regular lamellar structure during growth (Fig. 2*b*), that limited epitaxies persist in which (111)Al‖(2̄1̄1)CuAl₂ but with no apparent parallelism between low-index directions within these planes.[4,5] That is to say that the phases have a degree of freedom to rotate about the normal to their common planes, which is the partially preferred situation shown in Case B in Table 1 and Fig. 1. Thus, it seems that configurations like that shown in Case B may be quite common as intermediate stages during the development of more rigidly defined epitaxies such as Case C. In some cases, they may represent the best sort of compromise that is achievable when growth is non-directional. In this context, one may note the alternative orientation relationship quoted by Lawson *et al.*,[8] who studied the cellular morphologies of Al–CuAl₂ (i.e. those resulting from growth at a curved solid/liquid interface). They found that the approximate epitaxy could be given by:

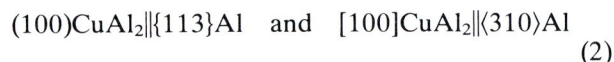

$$(100)CuAl_2\|\{113\}Al \quad and \quad [100]CuAl_2\|\langle310\rangle Al$$
(2)

Although it is not immediately obvious, this relationship can be reproduced fairly accurately by rotating the lamellar orientation (1) through some 33° about the normal (111)Al‖(2̄1̄1)CuAl₂. Significantly perhaps, this rotational operation coincides with the second largest peak in the coincident site pattern curve shown in Fig. 3*b*.

possible between these planes is shown in Fig. 3*a*. Two nets, one representing the atomic arrangement in the (111)Al plane and the other that in the (2̄1̄1)CuAl₂ plane (aluminium atoms only), are superimposed to produced a moiré pattern effect, or coincident site pattern.[6] As the two nets are rotated relative to one another about their plane normals, the periodicity of the pattern varies. The frequency of this periodicity, (which gives a measure of the density of atomic coincidences), is plotted as a function of rotational angle in Fig. 3*b*. Although the method is, at best, only semiquantitative, a number of fairly well defined peaks are obtained, the greatest of which, clearly, occurs

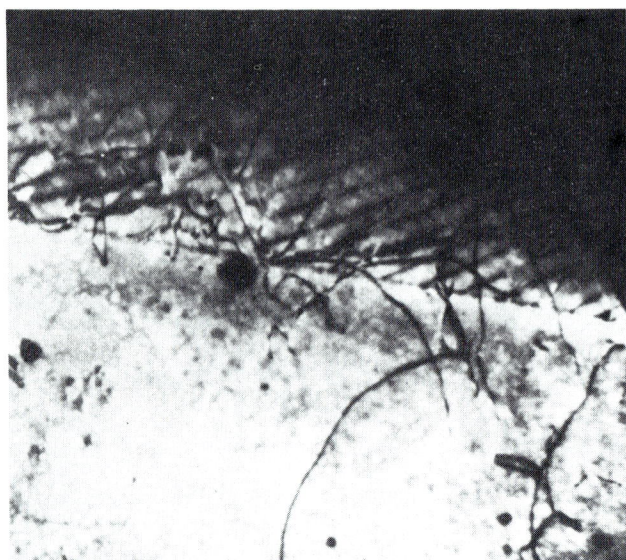

A puzzling feature of lamellar Al–CuAl$_2$ is that the macroscopic habit plane, which might be expected to lie parallel to (111)Al∥(2$\bar{1}$1)CuAl$_2$, has been found to deviate some 12° from this orientation and appears irrational.[4,5] This is not a usual occurrence in lamellar crystals although it has been observed in another system, Al–Zn.[9] The reason for it is not clear, but the detailed electron microscopy work of Garmong and Rhodes[10] may be relevant to this problem since these authors have shown that the interfaces in Al–CuAl$_2$ are not planar but consist of a series of small steps or facets. Thus, it is possible that although on a microscopic scale the habit plane contains ledges parallel to (111)Al∥(2$\bar{1}$1)CuAl$_2$, the macroscopic interface appears irrational because of stepped facets of some alternative habit plane(s).

ROD-LIKE MORPHOLOGIES

From considerations of interfacial energy, it can be predicted qualitatively that a rod-like morphology should be preferred over a lamellar morphology in systems that contain a low volume fraction of one of the phases.[11] In general, a correlation of this type is valid, but because the formation of a rod structure requires the adoption of more than one set of habit planes (unlike the lamellar structure), the transition often depends on the particular crystallography and low-energy interface possibilities of the duplex crystal.

The effect of decreasing the volume fraction of one of the phases in promoting a lamellar to rod transition has been demonstrated in off-eutectic composition alloys of Pb–Sn and Al–CuAl$_2$ produced under 'coupled growth' conditions where primary dendrite growth was suppressed.[12,13] Figure 5 shows a micrograph of an example in hypoeutectic Al–CuAl$_2$, consisting of a lamellar grain adjacent to a grain with a rod-like structure of CuAl$_2$. It has been determined[14] that the phase–phase orientations in the rod grain were the same as those in the lamellar grain, as in orientation (1), except

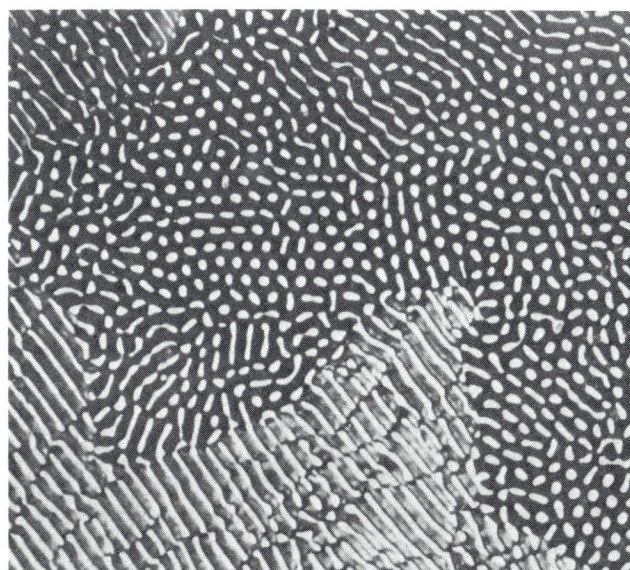

that the growth direction no longer corresponded to [$\bar{1}$12]Al but approximated to [100]Al. It is, therefore, clear that in this case the structural transition has been effected not only by the reduction in volume fraction of CuAl$_2$ (to about 25%) but also because the crystal orientations are no longer compatible with the formation of a lamellar habit, i.e. because the growth direction no longer lies parallel to the (2$\bar{1}$1) layer planes of CuAl$_2$.

The rods of CuAl$_2$ in Fig. 5 have rounded cross-sections with no obvious macroscopic interface facets; in other rod eutectic structures, this is not usually the case. In the system Al–Al$_3$Ni (Fig. 6a), the low symmetry of the orientation relationship is reflected in the irregular, multiple-faceted cross section of the Al$_3$Ni rods. The orientation most often observed is[15,16,17]:

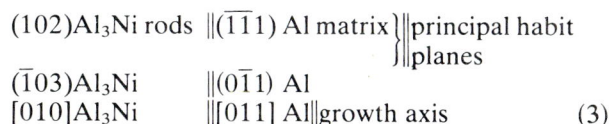

(102)Al$_3$Ni rods ∥($\bar{1}$11) Al matrix ⎫∥principal habit
⎬∥planes
($\bar{1}$03)Al$_3$Ni ∥(0$\bar{1}$1) Al ⎭
[010]Al$_3$Ni ∥[011] Al∥growth axis (3)

Again, the formation of the principal facets can be related to the fairly close periodic coincidence that is possible between the two crystal lattices along certain planes; this is illustrated in Fig. 6b for the case of (102)Al$_3$Ni plane. Secondary, less well defined habit planes also occur. Clearly, these various planes are not equivalent and there is a tendency to expand those that are of the lowest interfacial energy. In the extreme, this anisotropy can result in the development of a lath-like or broken lamellar-type microstructure (Fig. 6c).

In radial growth at cellular solid/liquid interfaces of Al–Al$_3$Ni, Lawson et al.[8] have shown that although the primary (102)Al$_3$Ni∥($\bar{1}$11)Al interface is retained, the aluminium phase is, to a certain extent, free to rotate about the [$\bar{1}$11] axis. In a more extreme situation where the imposed growth direction is continually changing, Jaffrey and Chadwick[18] have found that no specific orientation relationship is maintained between the phases, although the fibres of Al$_3$Ni retain their preferred [010] growth texture. Thus it seems that, as in lamellar crystals, a specific epitaxy such as Case C (Table 1 and Fig. 1) is obtained only under conditions of directional solidification. Non-directional growth leads to a progressive loss in the orientation relationship through configurations like Case B, and finally to Case A.

IRREGULAR MORPHOLOGIES

Irregular microstructures are most commonly associated with metal–non-metal eutectic systems (namely, Al–Si, Fe–C, and Al–Ge (Fig. 7)) in which the difference between the growth kinetics of the two phases (non-faceting and faceting) does not allow the cooperative mode of growth that is possible in metal/metal combinations. The microstructure of these alloys is determined largely by the particular growth anisotropies of the faceting phase and the occurrence of crystal defects (i.e. twins etc.) which facilitate its growth and allow some measure of accommodation with the metallic phase at the duplex solid/liquid interface.[19-22] Even in directionally frozen samples, preferred orientation relationships between the phases are generally absent (though the faceting phase may show a preferred growth texture) and the habit planes that occur arise

5 Optical micrograph of rod-like morphology of CuAl$_2$ in Al matrix produced in off-eutectic composition alloy under 'coupled growth' conditions; section transverse to growth axis [×350]

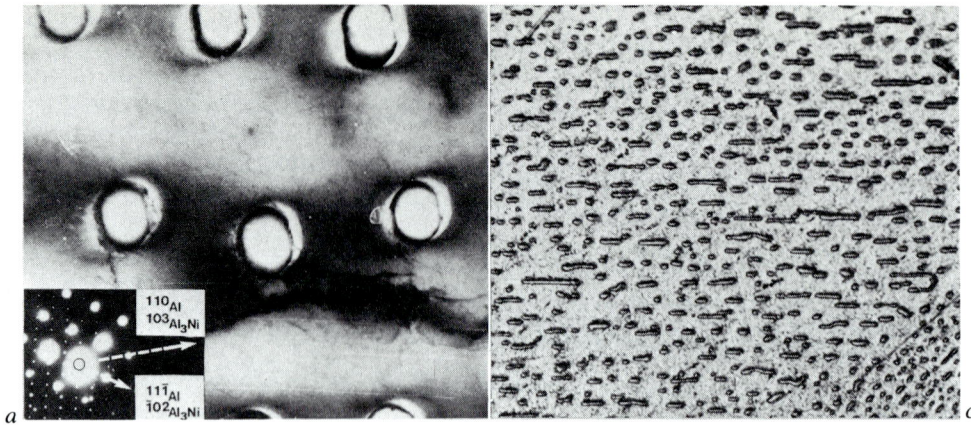

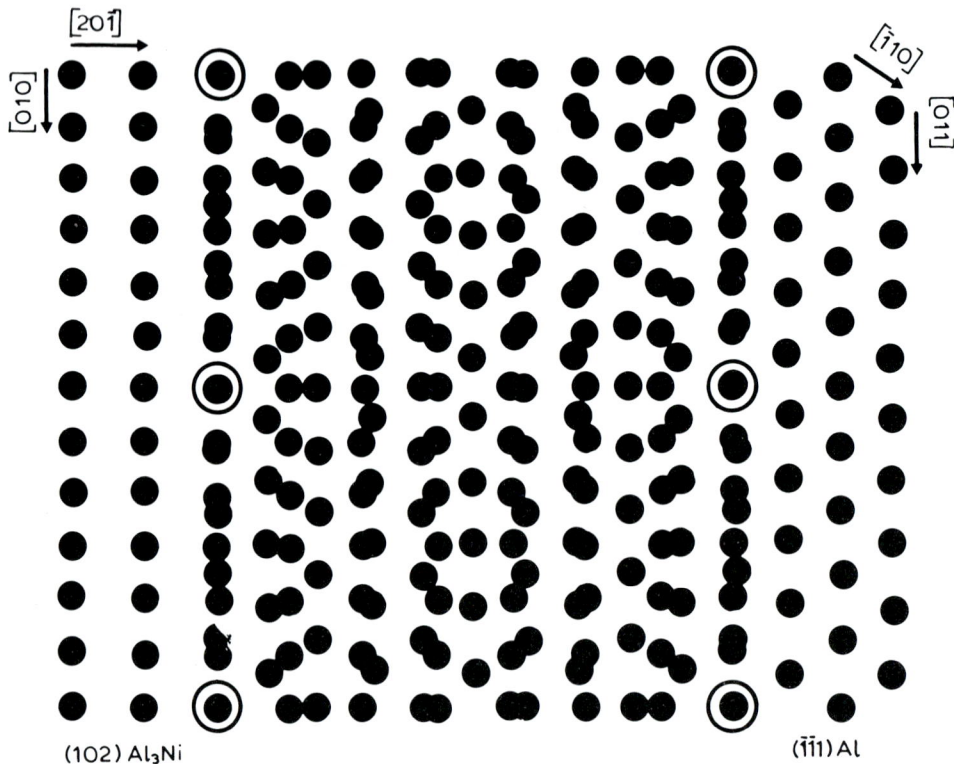

6   *a* Transmission electron micrograph and associated diffraction
patterns from transverse section of eutectic Al–Al₃Ni, [×2 500]; *b*
superimposed patterns of the (102)Al₃Ni and (111)Al lattice planes
illustrating the periodic coincidence that occurs when
[010]Al₃Ni∥[011]Al; *c* optical micrograph of broken lamellar structure
obtained in eutectic Al–Al₃Ni, section transverse to growth axis, [×225]

from solid/liquid faceting behaviour of the non-metallic phase. In general then, the lack of preferred epitaxies in these systems is best represented by Case A in
Table 1.

However, the distinction between such systems and
metallic (non-faceting) eutectic alloys is not rigid and
inevitably examples occur with intermediate characteristics; the pseudo-binary alloy $Co(15\%\,Ni)$–$Cr_{23}C_6$ is
a case in point. The microstructure of a section transverse to the growth axis (Fig. 8a) is almost wholly
irregular and typical of a faceting/non-faceting alloy.
Nevertheless, from studies using the electron microscope (Fig. 8b), it transpires that a specific and
reproducible orientation relationship exists:[23]

$(100)Co(Ni)\|(100)Cr_{23}C_6$

and   (4)

$[010]Co(Ni)\|[010]Cr_{23}C_6$

In this simple cube–cube epitaxy, the common prefer-
red growth direction lies parallel to $\langle 111\rangle$, and although
the habit planes seem haphazard and wholly irrational
on a macroscopic scale, in the electron microscope they
are seen to consist of a number of straight seg-
ments/facets which lie parallel to the common $\{\bar{1}11\}$
planes of the two phases.

The symmetrical epitaxy between the phases can be
attributed to the fact that the lattice parameter of the
Co(Ni) solid-solution phase (fcc, $a = 3\cdot52$–$3\cdot56$ Å) is
almost exactly one third of that of the unit cell of the
carbide phase (complex cubic, $a = 10\cdot6$ Å). This size

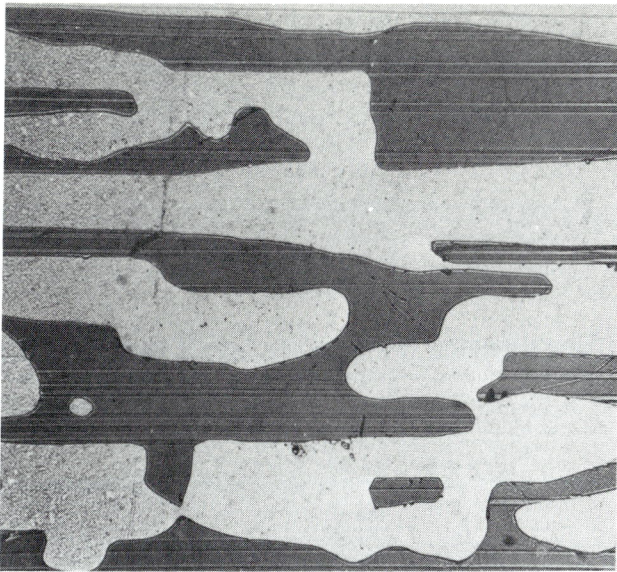

7 Optical micrograph of microstructure in eutectic Al–Ge, section parallel to growth axis; fine striations in the Ge phase are traces of {111} twins [×500]

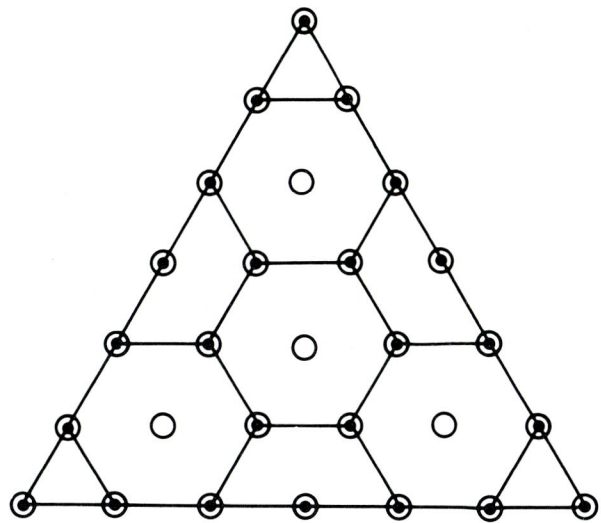

9 Diagram showing close atomic coincidence across (111) interface plane between Cr atoms in carbide phase (small filled circles) and Co/Ni atoms in matrix (large open circles)

relationship allows very good matching between the atoms of the two lattices, in the cube–cube orientation, especially across a {111} interface plane (Fig. 9).

The irregularity of the microstructure can, to a certain extent, be related to certain obvious incompatibilities in the crystallographic growth preferences. With a [111] growth texture and a tendency towards the formation of {$\bar{1}11$} interface planes, (*a*) the interface planes cannot lie parallel to the growth axis but are inclined at about 20° to it; and (*b*) three possible sets ($\bar{1}11$), ($1\bar{1}1$), and ($11\bar{1}$)) of such planes exist in a symmetrical configuration about the growth axis. In the first instance, there will be a continual tendency for the phases to grow in an edgewise manner at the solid/liquid interface, and in the second they have the freedom to alternate in morphology between the multiplicity of interface planes that are available. Thus in these circumstances, even ignoring the faceting character of the carbide phase, it seems unlikely during directional freezing of this alloy that a stable steady-state profile could be maintained at the growth front, as would be necessary for the development of a more regular microstructure.

Occasionally in this alloy, quasi steady-state 'complex–regular' cells develop (a rather badly formed example may be seen at the edge of Fig. 8*a*). The geometrical arrangement of the spines and side plates in these cells is such that they consist of lamellar arrays parallel to intersecting {$\bar{1}11$} planes (Fig. 10). Similar complex–regular cells have been observed in hypereutectic Al–Si and Al–Ge alloys[19,20] and in the same

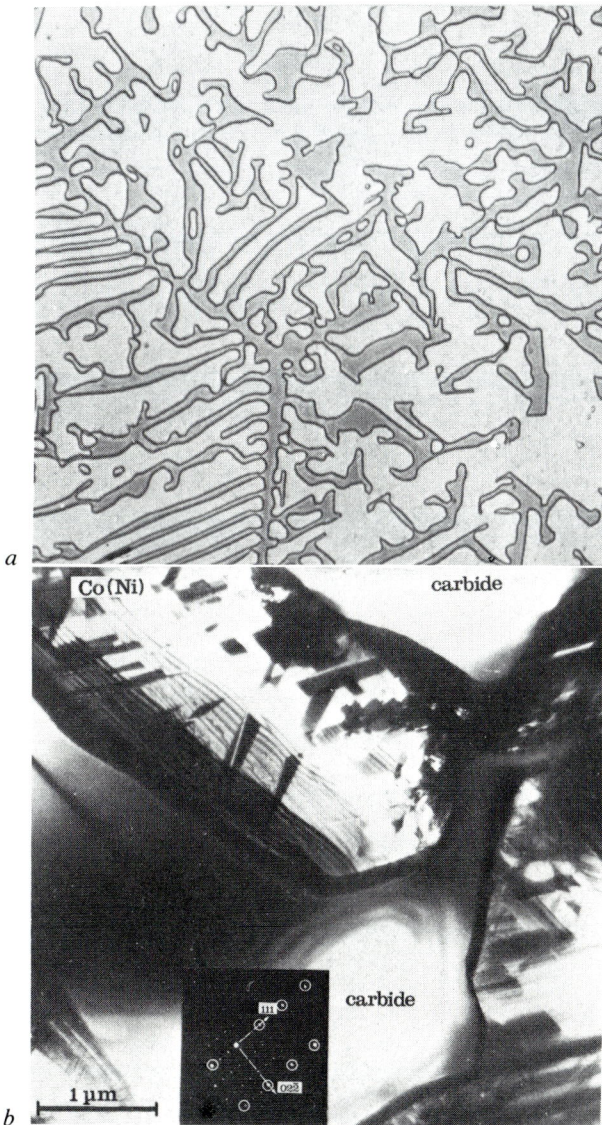

8 *a* Optical micrograph of microstructure in eutectic Co(Ni)–Cr$_{23}$C$_6$, section transverse to growth axis, [×500]; *b* transmission electron micrograph of section of eutectic Co(Ni)–Cr$_{23}$C$_6$ parallel to growth axis; superimposed diffraction patterns (Co(Ni) spots circled) illustrate the simple cube–cube orientation of the phases

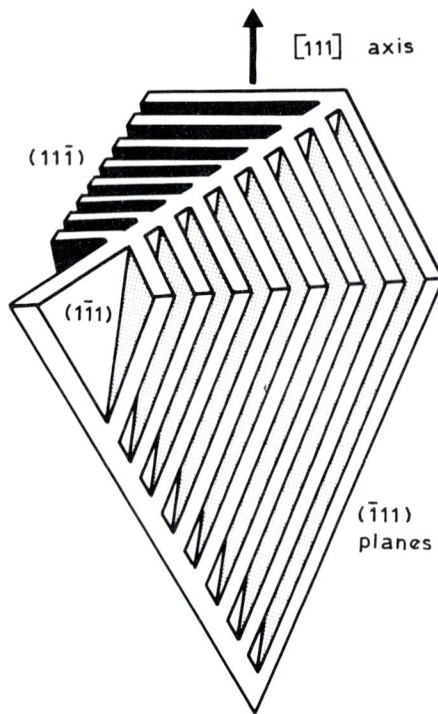

10   Diagram showing intersecting {111} geometry of a 'complex-regular' cell in eutectic Co(Ni)–$Cr_{23}C_6$

Labels on diagram: [111] axis, (11$\bar{1}$), (1$\bar{1}$1), ($\bar{1}$11) planes

way it has been shown that their morphologies can be correlated with the crystallographic arrangement of the phases.

ACKNOWLEDGMENTS

This work has been sponsored in part by the Science Research Council. D. D. Double would like to acknowledge the cooperation from members of the Département des Materiaux of the Ecole Polytechnique Federale de Lausanne where some of the electron microscopy work was carried out. I. G. Davies acknowledges support from the French Ministry of Industry and Scientific Development, and from members of the Département de Métallurgie, Ecole Nationale Superieure des Mines de Saint Etienne, where much of this work was conducted.

REFERENCES

1   L. M. HOGAN *et al.*: 'Advances in materials research', Vol. 5, 83, 1970, Wiley Interscience
2   K. A. JACKSON AND J. D. HUNT: *Trans. AIME*, 1966, **236**, 843 and 1 129
3   R. W. KRAFT: *ibid.*, 1962, **224**, 65
4   I. G. DAVIES AND A. HELLAWELL: *Philos. Mag.*, 1969, **19**, 1 285
5   I. G. DAVIES AND A. HELLAWELL: *Philos. Mag.*, 1970, **22**, 1 255
6   W. BOLLMANN: *Philos. Mag.*, 1967, **16**, 363 and 383
7   G. C. WEATHERLEY: *Met. Sci.*, 1968, **2**, 25
8   W. H. S. LAWSON *et al.*: *J. Cryst. Growth*, 1972, **12**, 209
9   D. D. DOUBLE *et al.*: *ibid.*, 1968, **2**, 191
10 G. GARMONG AND C. G. RHODES: *Acta Metall.*, 1974, **22**, 1 373
11 D. J. S. COOKSEY *et al.*: *Philos. Mag.*, 1964, **10**, 745
12 F. R. MOLLARD AND M. C. FLEMINGS: *Trans. AIME*, 1967, **239**, 1 526 and 1 534
13 R. M. JORDAN AND J. D. HUNT: *Metall. Trans.*, 1971, **2**, 3 401
14 I. G. DAVIES: D. Phil. Thesis, 1969, University of Oxford, Department of Metallurgy and Science of Materials, p. 79
15 F. D. LEMKEY *et al.*: *Trans AIME*, 1965, **233**, 334
16 W. K. TICE *et al.*: 'Fifty years of progress in metallographic techniques', 239, 1966, ASTM STP **430**
17 G. R. ARMSTRONG *et al.*: Unpublished work 1970, University of Oxford Department of Metallurgy and Science of Materials
18 D. JAFFREY AND G. A. CHADWICK: *Trans. AIME*, 1969, **245**, 2 435
19 A. HELLAWELL: *Prog. Mater. Sci.*, 1970, **15**, 1
20 M. G. DAY AND A. HELLAWELL: *Proc. R. Soc.*, 1968, **A305**, 473
21 D. D. DOUBLE AND A. HELLAWELL: *Acta Metall.*, 1969, **17**, 1 071
22 M. ORON AND I. MINKOFF: *Philos. Mag.*, 1969, **9**, 1 059
23 G. ZAMBELLI AND D. D. DOUBLE: to be published

Crystallographic study on eutectic microstructure of Al–Si system

K. Kobayashi, P. H. Shingu, and R. Ozaki

Direct structure observations and the electron diffraction analysis of thin films prepared from Al–Si melts have been performed, using the electron microscope, to study the crystallographic characteristics of individual crystals which compose the typical irregular eutectic microstructure. The eutectic silicon crystals are to a certain extent continuous and the directional preference of growth to $\langle 112 \rangle$ is noted, interpreted as being due to the repeated nucleation on a re-entrant corner formed by the multiple twins. Branching and growth direction change in eutectic silicon crystals, whenever the $\langle 112 \rangle$ direction becomes difficult due to the restriction of heat flow direction and/or solute silicon distribution, are found to occur by the creation of a new twin on the growth habit faces to keep the direction preference, $\langle 112 \rangle$, unchanged. The aluminium grains have diameters comparable with the spacing between eutectic silicon plates, indicating that the aluminium phase nucleates repeatedly on eutectic silicon plates. The apparently random crystallographic orientation relationships between the eutectic silicon and the adjoining aluminium crystals, measured by the selected area electron diffraction analysis, have been classified into simple relationships by taking the growth habit and the twinning in the silicon crystals into consideration. The most frequently measured orientation relationship is $[001]Al\|[110]Si$, $[100]Al\|[\bar{1}1\bar{1}]Si$.

The authors are in the Department of Metal Science and Technology, Kyoto University, Japan

The Al–Si system, which is similar to the Fe-graphite system, has a complex eutectic structure and is usually classified as a typical irregular type. Because of the apparent irregularity, a detailed investigation of this eutectic microstructure requires the crystallographic studies of each individual crystal which composes the eutectic structure. The use of X-ray diffraction technique is not suitable for this purpose because the area from which the crystallographic information is obtained cannot be smaller than $50\mu m$ in dia., even by the use of the micro-focus method, the size which barely compares the size of typical eutectic silicon in the case of the slow solidification.

In this work, direct structure observations and the electron diffraction analysis, by electron microscope, of thin films prepared from Al–Si melts have been performed to obtain such structural information as, (*a*) growth habit, branching and growth direction change of eutectic silicon crystals; (*b*) size and shape of aluminium grains; and (*c*) crystallographic orientation relationships between the eutectic silicon and the adjoining aluminium crystals.

EXPERIMENTAL PROCEDURE

Al–Si master alloys with the composition range of 8–20 wt-% Si (the eutectic composition is Al–12·7 wt-% Si[1]) were prepared using pure aluminium (99·99%) and high purity silicon (better than 99·999%). After polishing the surface, these master alloys were immersed in aqua regia to remove oxides and impurities on the surface. Thin films of this eutectic were prepared in an argon atmosphere by dipping small elliptical loops of iron wire 0·3 mm in dia. into the molten alloy in an alumina crucible and withdrawing it at a rate of 0·1–1 cm/s. Cooling rate was measured by immersing a 0·1 mm dia. looped alumel–chromel thermocouple into the melt and withdrawing it at the same rate as in the case of sample preparation. A proper area of the thin film was chosen under the optical microscope, cut out, and mounted on a specimen holder of the electron microscope. The apparatus used to prepare the thin films is shown in Fig. 1.

RESULTS

Growth habit of eutectic silicon crystal

A typical transmission electron micrograph of a directionally solidified thin film specimen is shown in Fig. 2*a*. In the same figure, a schematic drawing of the micrograph, the electron diffraction pattern obtained from the area marked by a circle and the indexed reciprocal lattice points for the Si[E] and Si[F] crystals in the micrograph are also shown in Fig. 2*a*, *b*, and *c*, respectively. The diffraction pattern (*C*) tells that the incident

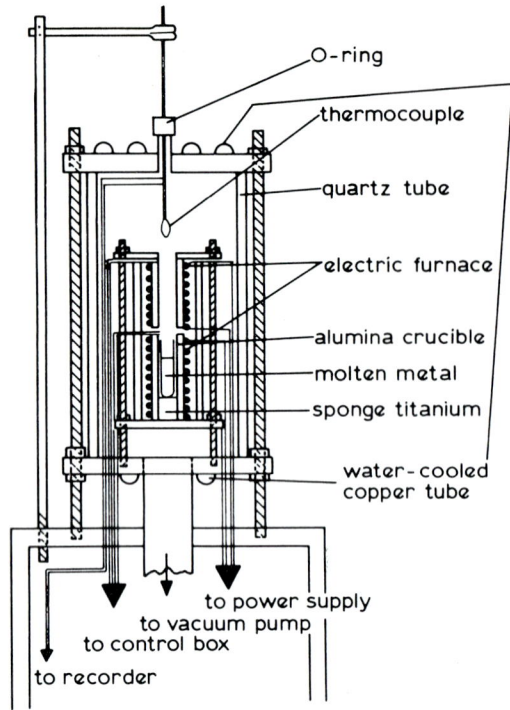

1 Diagram of experimental apparatus for thin film

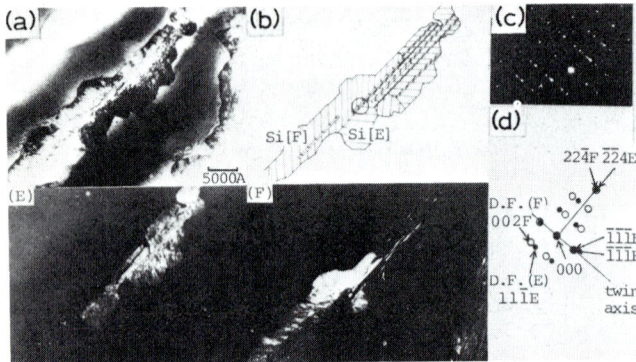

a transmission electron micrograph of thin film of Al–20 wt-% Si alloy and its schematic drawing *B*, *C*, and *D* diffraction patterns taken from circled areas in *a*.

3 Branched eutectic silicon crystal

The streaks which are seen on the diffraction pattern in the direction parallel to the direction of twin axis indicates the existence of multiples of thin plate crystals[2] perpendicular to [111]E, F direction. The dark field images using (11$\bar{1}$)E and (002)F indicates that these thin plates are twinned crystals which have the identical orientation with either Si[E] or Si[F] as shown in Fig. 2*e* and *f*.

The surface planes of the flat portion of silicon crystals are {111} as it is seen in the micrograph; however, there exist parts of silicon crystal surfaces with complex shapes indicating the existence of surface planes other than {111}.

Branching and growth direction change of eutectic silicon crystal

A transmission electron micrograph of a branched eutectic silicon crystal is shown in Fig. 3, together with the schematic explanations of branching and the electron diffraction patterns obtained from three different parts of the branched silicon crystal.

When the incident beam direction and the trace of the boundary of Si[C] and Si[D] are defined as [$\bar{1}$01]C and [$\bar{1}$2$\bar{1}$]C of Si[C] crystal respectively, the twin plane between Si[C] and Si[D] crystals is ($\bar{1}$11)C. When the twin plane is taken as (111)D for Si[D] crystal, the incident beam direction is [$\bar{1}$10]D and the preferred growth direction becomes [11$\bar{2}$]D. Thus it is seen that the twin plane runs parallel to the growth direction of Si[D] and perpendicular to the plane of the micrograph. This fact can also be confirmed from the diffraction pattern Fig. 3*d*. The fact that the directions of streaks of diffraction spots of Fig. 3*b* and *d* differ by an angle of 70·5° indicates that the Si[C] crystal shares a twin plane ($\bar{1}$11)C with Si[D] crystal and also shares another twin plane (1$\bar{1}$1)C with Si[B] crystal. When the (1$\bar{1}$1)C twin plane is defined as ($\bar{1}$11)B plane for the Si[B] crystal, the incident beam direction is [$\bar{1}$10]B and the growth direction becomes [$\bar{1}$12]B. The twin plane between Si[C] and Si[B] crystals is thus seen to run parallel to

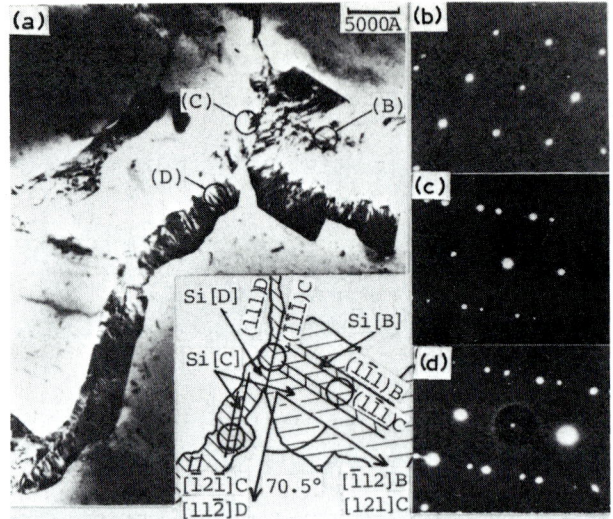

a transmission electron micrograph of thin film of Al–12 wt-% Si alloy; *b* schematic drawing of (*a*); *c* diffraction pattern taken from circled area in (*b*); *d* explanation of (*c*); *e* dark field image by the (111)E reflection; *f* dark field image by the (002)F reflection

2 Eutectic silicon crystal containing multiple twins with twin planes parallel to preferred growth direction [11$\bar{2}$]F(= [11$\bar{2}$]E)

beam is parallel to [$\bar{1}$10]F direction, where the subscript F refers to Si[F] crystal, that the preferred growth direction is [11$\bar{2}$]F, and that the plane which is shared with Si[E] crystal is ($\bar{1}$1$\bar{1}$)F. Si[F] and Si[E] are related by a twin with ($\bar{1}$1$\bar{1}$)F and [111]F as the plane and axis of twinning.

When the twin plane for Si[E], which is shared with Si[F], is defined (111)E, the incident electron beam is [$\bar{1}$01]E and the growth direction for Si[E] is [112]E. Thus it is seen that the twin plane between Si[F] and Si[E] is parallel to the growth direction and perpendicular to the plane of the micrograph.

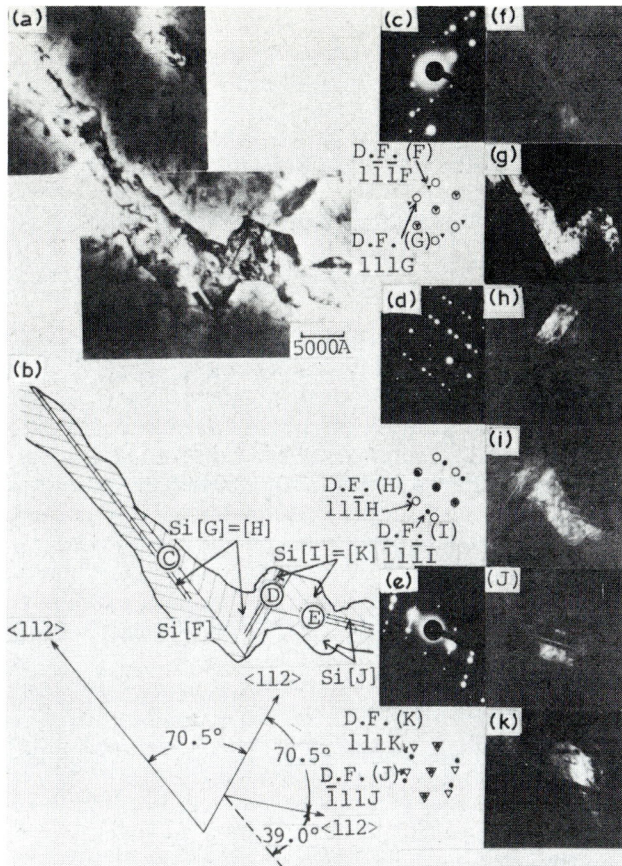

a transmission electron micrograph of thin film of Al–20 wt-% Si alloy; *b* schematic drawing of *a*; *c, d,* and *e* diffraction patterns taken from circled areas in *b*; *f–k* dark field images

4 Growth direction change of eutectic silicon crystal

the direction of growth and perpendicular to the plane of the micrograph.

The twinning may take place on any {111} type plane, hence the growth direction changes via twinning are observed in many different ways. The diffraction patterns shown in Fig. 4*f–k* reveal that the crystal [F], [G](=[H]), [I](=[K]) and [J], shown in the micrograph Fig. 4*b*, are four differently oriented crystals all oriented with one of their ⟨110⟩ axes parallel to the incident beam. The crystals [F] and [G], [G] and [I], and [I] and [J] are related to each other by twinning and all the twin planes for these twinnings are parallel to the incident beam; also the twin traces are parallel to ⟨112⟩ direction, hence the angles between these traces are 70·5°. Figure 4 thus shows an example of a 39° growth direction change of a growing eutectic silicon crystal by repeating two twinnings keeping the crystallographic growth direction preference of ⟨112⟩ unchanged.

These results clearly demonstrate that the branching and the growth direction change of a growing eutectic silicon crystal takes place by the twinning mechanism. The growth direction after twinning resumes ⟨112⟩ and the existence of multiple twinning in both original and branched silicon crystal is frequently observed.

Size and shape of eutectic aluminium grains

The bright and dark field images of eutectic aluminium grains shown in Fig. 5 revealed that the grain size of

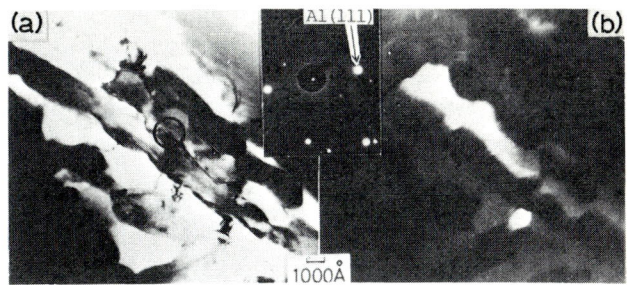

a transmission electron micrograph of thin film of Al–12 wt-% Si alloy; *b* dark field image by the Al(111) reflection

5 Size and shape of eutectic aluminium grains

eutectic aluminium crystals is of the order of 1 000–10 000 A which is comparable to the spacings but much less than lengths of continuous eutectic silicon crystals. Such repeated nucleation of aluminium is due to the relatively small temperature gradient ahead of the solidification front during the unidirectional solidification of a thin film specimen. The measured cooling rate in this case was 0·1–10°C/s and the growth rate[3] estimated from the eutectic silicon spacing was 0·1–1·0 cm/s.

Crystallographic orientation relationships between the eutectic silicon and the adjoining aluminium crystals

When the twinning in the eutectic silicon crystal is taken into consideration,[4,5] most specimens, out of 130 measured cases, were classified into three types of simple relationship as follows;

$$[001]Al\|[110]Si, \quad [100]Al\|[\bar{1}1\bar{1}]Si \quad (1)$$

$$[001]Al\|[110]Si, \quad [100]Al\|[00\bar{1}]Si \quad (2)$$

$$[\bar{1}10]Al\|[\bar{1}10]Si, \quad [111]Al\|[110]Si \quad (3)$$

The measured frequency of occurrence for each type of relationship is given in Table 1. In the table, the classification into one of these three types is made whenever the measured relationship falls within 15° single axis rotation from the exact relationship.

DISCUSSION

For the detailed crystallographic study of eutectic silicon crystals in Al–Si alloys the use of X-ray micro Laue or ECP (electron channelling pattern) diffraction techniques, which were useful in the demonstration of the evidences of TPRE mechanism[6,7] of growth of plate-like primary silicon crystals reported previously,[4,5] is

Table 1 Classification of the orientation relationships between eutectic silicon and aluminium crystals

	Type (1)	Type (2)	Type (3)	Others
As measured	45	13	27	14
After twinning operation	23	4	4	—
Total	68	17	31	14
Percentage	52·3	13·1	23·8	10·8

not possible owing to the small size of crystals. The electron microscope is, therefore, extensively used in the present work though the accuracy of the orientation determination is not as high as in the case of the previous methods.

The evidence which supports the TPRE mechanism, the existence of multiple twinnings which run parallel to the direction of growth, is frequently observed for the eutectic silicon crystals oriented with their ⟨110⟩ axis, hence also the twin planes, parallel to the incident beam as shown in Figs 2, 3, and 4. Branching or growth direction change of eutectic silicon crystals may take place by the dendritic growth of a single crystal, but the growth direction is largely restricted in such a case if the TPRE mechanism of growth is to be followed. The incorporation of twinnings allows the growth direction change to all directions which are the integer multiplication of 70·5°; the three dimensional geometry of this mechanism may best be understood as being based on the repeated stacking of tetrahedrons bounded by {111} planes at the position of direction change. The evidence of such cases is given in Fig. 4 and is discussed in the previous section.

Eutectic silicon crystals are known to project into the liquid to grow ahead of the eutectic aluminium phase.[8,9] Since the free surfaces of eutectic silicon provide the nucleation sites for the eutectic aluminium, the surfaces of eutectic silicon are most frequently {111}. However, the occurrence of exposure of the surfaces with the habit planes other than {111} are sometimes observed,[10] which may be a consequence of the two-dimensional nucleation mechanism of growth. The fraction of the surface area of particular crystallographic indices and the effectiveness as the nucleation site of that surface determines the frequency of the occurrence of a particular crystallographic orientation relationship between the eutectic silicon and aluminium crystals. The formation of corners bounded by {111} planes as a result of the growth direction change by twinning, as shown in Fig. 6, may be providing the easy nucleation sites for eutectic aluminium crystals.

The most frequently (52·3%) measured crystallographic orientation relationship between the eutectic silicon and aluminium crystals, which is identical with the measured relationship in the case of plate-like primary silicon and aluminium,[5] is consistent with the nucleation of aluminium on the {111} silicon plane.

The relatively high proportion of other observed types of relationships to the most frequent type compared to the case for the primary silicon and aluminium crystals, may be due to the occurrence of free eutectic silicon surfaces other than {111}, as shown in Fig. 2.

CONCLUSIONS

Direct structure observations (and the electron diffraction analysis by electron microscope) of their films prepared from Al–Si melts have been performed. The following results concerning the microstructure and the growth characteristics are obtained.

1 The eutectic silicon crystals are continuous to a certain extent and the preferred growth direction of ⟨112⟩ with the habit plane of {111} is noted, and is interpreted as being due to the repeated nucleation event on a re-entrant corner formed by the multiple twins parallel to the growth direction.

2 Branching and growth direction change of eutectic silicon crystals, which may be caused by the restriction of heat flow direction and/or solute silicon distribution, are found to take place by the creation of a new twin on the growth habit faces. The growth direction preference, ⟨112⟩, does not change after the branching or growth direction change.

3 The aluminium grains have diameters comparable to the spacing between eutectic silicon plates, indicating that the aluminium phase nucleates repeatedly on eutectic silicon plates.

4 The apparently random crystallographic orientation relationships between the eutectic silicon and adjoining aluminium crystals, measured by the selected area electron diffraction analysis have been classified into simple relationships by taking the growth habit and the twinning in the silicon crystals into consideration. The most frequently measured orientation relationship is

[001]Al∥[110]Si,   [100]Al∥[Ī1Ī]Si

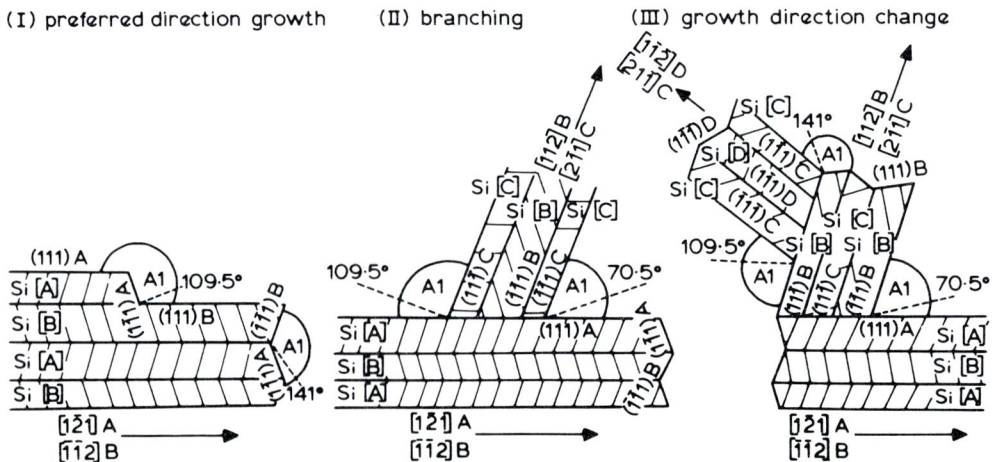

6 **Diagrams of possible cases of nucleation of a eutectic aluminium crystal on a twinned eutectic silicon crystal; note that crystals (A) and (C), and (B) and (D) are not related by twinning**

ACKNOWLEDGMENTS

The authors wish to thank H. Esaka, M.Sc., for his assistance in the experiments.

REFERENCES

1 R. A. MEUSSNER: US Naval Research Labs. Report, 1959, No. 5331, 6
2 P. B. HIRSCH *et al.*: 'Electron microscopy of thin crystals', 82, 1965, London, Butterworths
3 M. H. BURDEN AND H. JONES: *J. Inst. Met.*, 1970, **98**, 249
4 K. KOBAYASHI *et al.*: *J. Mater. Sci.*, 1975, **10**, 290
5 *ibid.*: 1976, **11**, 399
6 D. R. HAMILTON AND R. G. SEIDENSTICKER: *J. Appl. Phys.*, 1960, **31**, 1 165
7 R. S. WAGNER: *Acta Metall.*, 1960, **8**, 57
8 A. HELLAWELL: *Prog. Mater. Sci.*, 1970, **15**, 20
9 A. J. McLEOD *et al.*: *J. Cryst. Growth*, 1973, **19**, 301
10 H. FREDRIKSSON *et al.*: *J. Inst. Met.*, 1973, **101**, 285

Study of the structure of Al–Al$_3$Ni composites grown in non-uniform cylindrical moulds

N. A. El-Mahallawy and M. M. Farag

Structural changes caused by axisymmetric variation in the cross-sectional area of growing Al–Al$_3$Ni eutectic composites have been studied for different angles of divergence and growth rates. The experimental results showed that slower growth rates and coarser structures exist in the divergence area and at nucleation sites of new grains that start on divergence shoulders. Fibre misalignment, with respect to the specimen axis, was also observed in the divergence area. These structural irregularities are more pronounced for larger angles of divergence and faster growth rates.

N. A. El-Mahallawy is in the Department of Production Engineering, Ain Shams University, Cairo, and Professor M. M. Farag is in the Department of Materials Engineering, American University in Cairo, Egypt

Previous work has shown that directional solidification of alloys can lead to large improvements in mechanical and physical properties. Unfortunately, most of the published information is obtained from straight cylindrical specimens while practical applications involve more complicated shapes. Recent studies on directionally solidified heat-resistant alloys[1–4] and aluminium-base alloys[5,6] have shown that variations in shape can cause large variations in the structure. In the present work the authors describe a study of structural changes caused by axisymmetric variation in the cross-sectional area of cylindrical specimens of Al–Al$_3$Ni eutectic alloy.

EXPERIMENTAL PROCEDURE

Use of the directional-solidification apparatus selected for the present work involves withdrawing a crucible containing the molten alloy from a resistance-heated furnace through a water-cooled copper chill at the desired growth rate. In order to obtain the desired changes in the composite shape, the crucible was machined as shown in Fig. 1. The crucible was made of high-purity graphite and was held in an alumina tube which was connected to the drive system. The temperature gradient at the solid/liquid interface of the straight part of the crucible was kept constant at about 130 K/cm throughout the experimental work and was measured using a Pt–Pt/10Rh thermocouple contained in an alumina tube, 1 mm in diameter. The nominal growth rates used were 2, 2·5, 4, 6, and 12 cm/h.

The material used for composites was Al–Al$_3$Ni eutectic alloy prepared from pure aluminium (99·99%) and nickel (99·9%) with a composition of 6·06 wt-% Ni. The grown composites were sectioned longitudinally and transversely for macroscopic and microscopic examination. For examination in the scanning electron microscope, the specimens were etched electrolytically using a solution of 3 cm^3 HCl, 50 cm^3 H$_2$O, 50 cm^3 alcohol.

EXPERIMENTAL RESULTS AND DISCUSSION
Growth rate variations in divergence

The present composites exhibited fibrous Al$_3$Ni phase in the straight parts of the specimens at all growth rates except for 2 cm/h, where some tendency to form ribbons was observed. The structure was finer at the faster growth rates and the inter-fibre spacing varied according to the relationship:

$$\lambda^2 R = 20·5 \qquad (1)$$

where λ is inter-fibre spacing in μm and R is growth rate in cm/h.

Although the speed at which the crucible was withdrawn from the furnace was constant during a given experiment, it was noticed that the inter-fibre spacing λ varied within the divergence region of the specimens. The λ values were converted into growth rates using equation (1) after correcting for fibre inclination to the cross-section. The fibre inclination was determined from longitudinal sections. The local values of growth

106

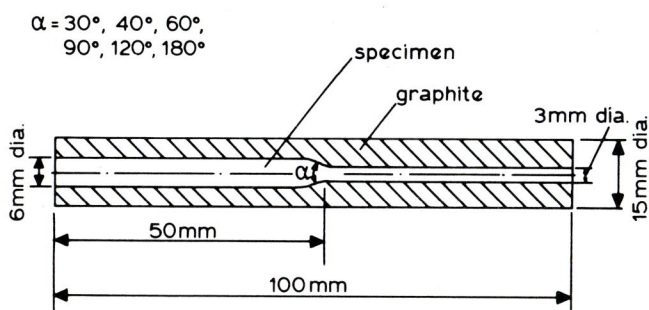

1 **Specimen and crucible arrangement**

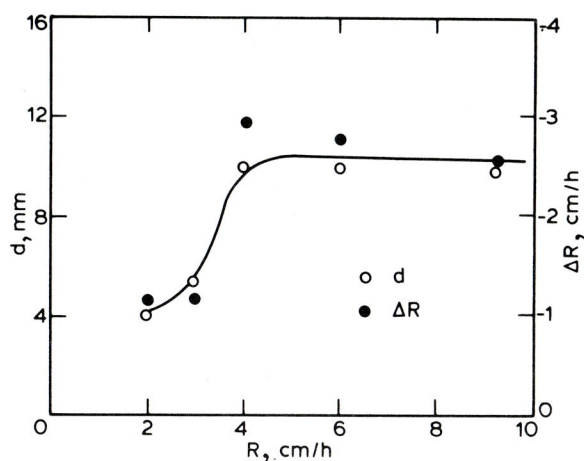

2 **Effect of nominal growth rate on actual growth rate for α = 120°**

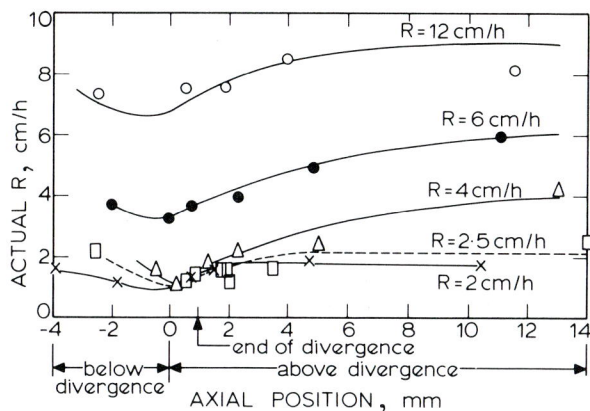

4 **Effect of nominal growth rate on *d* and Δ*R* for α = 120°**

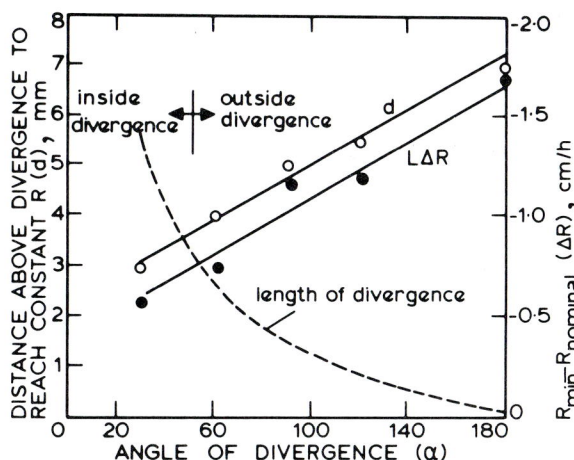

5 **Effect of α on *d* and Δ*R* for nominal growth rate = 2·5 cm/h**

rate within the divergence region are shown in Figs 2 and 3 for different nominal growth rates and divergences angles, respectively. The results show that the local growth rates decrease below the nominal value (reaching a minimum around the start of divergence), then increase again to the nominal value after a distance *d* above the start of divergence. Figure 4 shows the variation of the distance *d* with the nominal growth rate *R*. The figure also shows that Δ*R*, which is the magnitude of the decrease in growth rate below the nominal value, varied in the same way as *d*. The variation of both *d* and Δ*R* with the divergence angle is shown in Fig. 5. The values of Δ*R* and *d* can be taken as a measure of the extent to which divergence disturbs the structure. The results given in Figs 4 and 5 indicate that wider angles of divergence at higher growth rates cause more extensive damage to the structure.

Nucleation of new grains in divergence
Longitudinal sections of specimens grown under different divergence angles and growth rates were

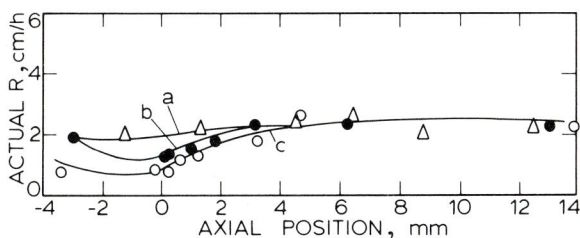

a α = 30°; *b* α = 90°; *c* α = 1

3 **Effect of divergence angle on actual growth rate for nominal growth rate = 2·5 cm/h**

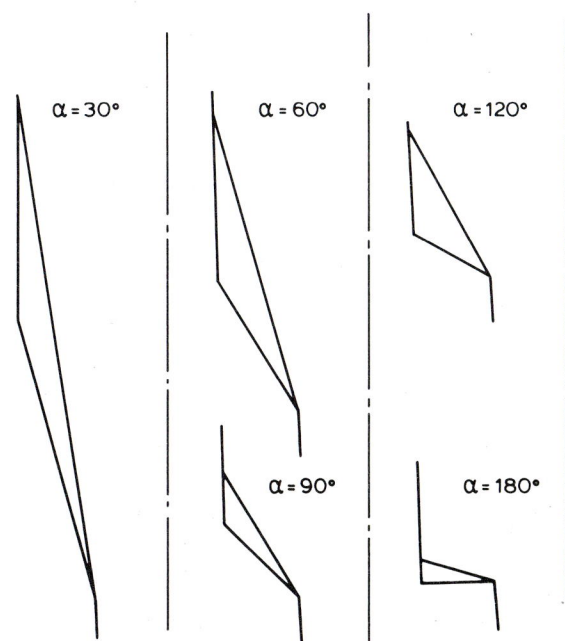

6 **Effect of divergence angle on shape of new grains; *R* = 2 cm/h**

a $\alpha = 60°$ [×100]; b $\alpha = 120°$ [×50]; c $\alpha = 180°$ [×50]

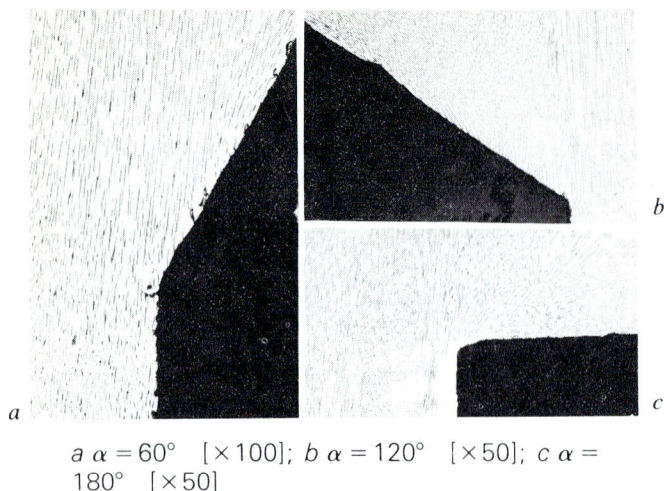

7 Different types of observed microstructure $R = 4$ cm/h

anodized and examined under polarized light. New grains were observed at the mould wall in the divergence area as shown diagramatically in Fig. 6. Although the divergence angle had a strong influence on the shape and size of the new grains, the effect of nominal growth rate was negligible.

Figure 7 shows a variety of structures that were observed at different divergence angles. Scanning electron microscopic examination showed that Al_3Ni phase in the new grains started in the lamellar form, as shown in Fig. 8. The Al_3Ni lamellae changed to the fibrous form by multiple branching and the change was sufficiently well defined to permit a contour to be drawn separating the lamellae from the fibres. The size of the lamellar region of the structure was measured by its maximum thickness normal to the mould wall t_{max} and by its length along the mould wall l. Figure 9 shows that the size of the lamellar region of the structure increased as the angle of divergence was increased, but was not affected by the growth rate.

Preliminary measurements of interlamellar and fibre spacings in an alloy grown at 2·5 cm/h show that the growth rate was slow in the lamellar region (about 1·5 cm/h), then increased rapidly in the lamellar fibre transition region to exceed its nominal value. The growth rate then rapidly decreased to the nominal value.

Fibre orientation in divergence

Fibre orientation in the longitudinal section was measured with respect to the axis of the cylindrical part of the specimen. Areas of the longitudinal section with equal fibre orientations were joined with contours, as shown in Figs 10 and 11. The equiorientation contours generally show that the presence of divergence has no significant effect on the fibre orientation in the central part of the specimen. The contours also show that the change in fibre orientation away from the axial direction takes place gradually, so that the maximum misorientation is reached shortly before the corner corresponding to the end of divergence. After this corner, the fibres gradually rotate back to the axial orientation. The results given in Figs 10 and 11 show that higher values of misorientation are caused by larger angles of divergence and faster growth rates.

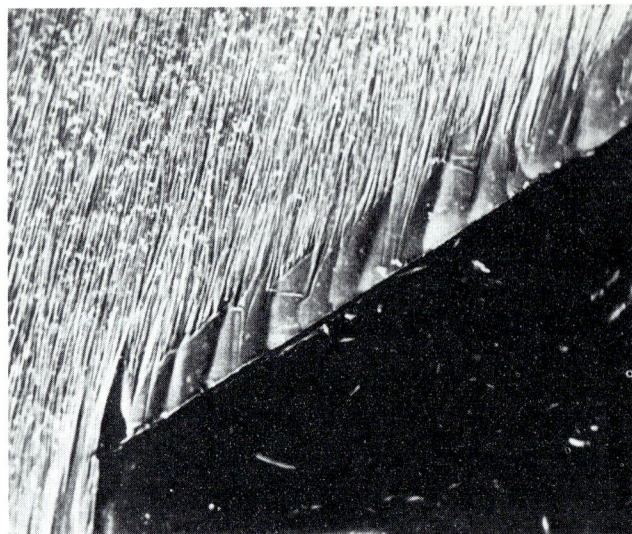

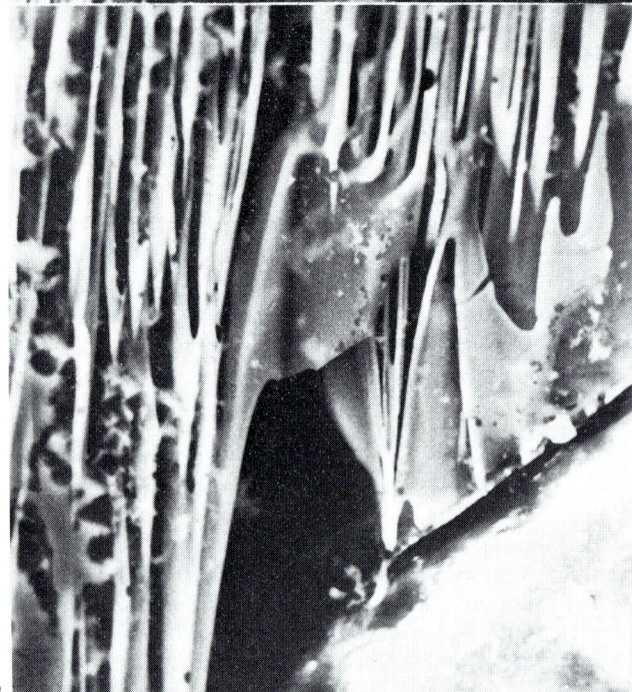

a [×100]; b [×500]

8 Formation and breakdown of Al_3Ni lamellae in the new grains; $R = 2·5$ cm/h

Chadwick et al.[7,8] have shown that Al_3Ni fibres always grow in the direction of heat extraction, i.e. at right angles to the solid/liquid interface. By applying this principle to Figs 10 and 11, it can be deduced that the solid/liquid interface gradually increases its curvature as it progresses within the divergence area and then straightens back beyond the corner corresponding to the end of divergence. Larger angles of divergence and faster growth rates result in more curved solid/liquid interface which can lead to breakdown of composite structures in some cases.[9] The main reason for the increased curvature of the solid/liquid interface is expected to be the difference in conductivity between the alloy and graphite crucible. This view is supported by analytical calculations using a heat-transfer model.[10]

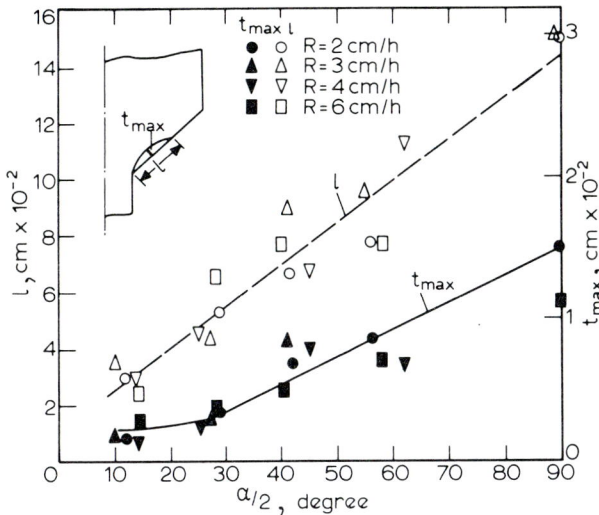

9 Effect of angle of divergence on *l* and *t*ₘₐₓ for different growth rates

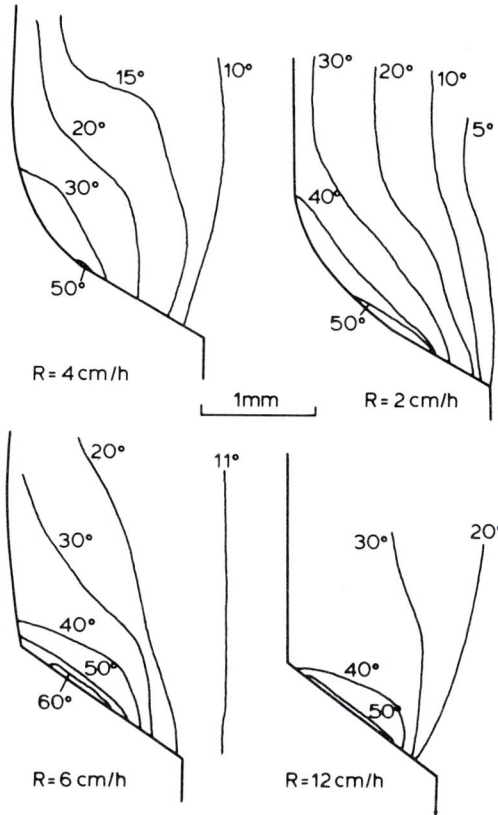

10 Effect of α on equi-orientation contours *R* = 2·5 cm/h

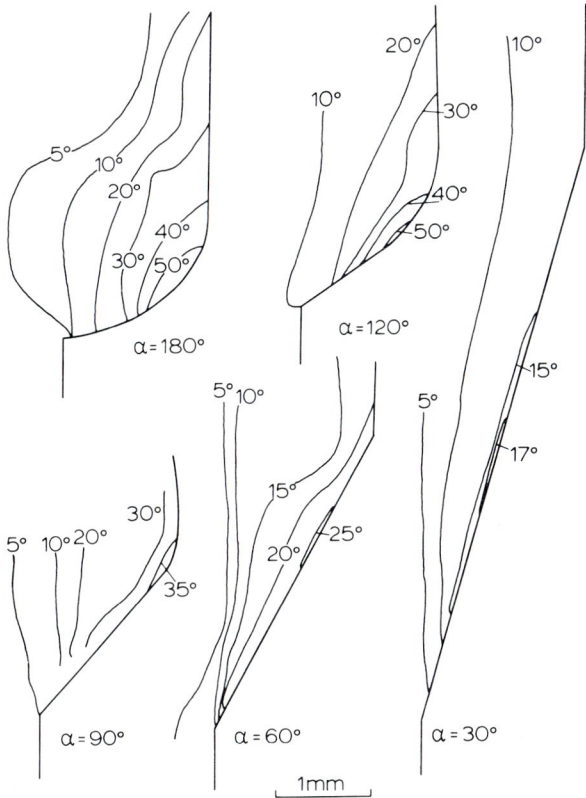

11 Effect of growth rate on equi-orientation contours for divergence angle 120°

behaviour of the component. Wider angles of divergence, corresponding to more sudden changes in cross-sectional area, and higher growth rates yield more severe conditions. The expected deterioration of the mechanical behaviour would be aggravated by the lack of alignment of the reinforcing phase in the divergence area.

ACKNOWLEDGMENT

N. A. El-Mahallawy gratefully acknowledges the help given by Professor Dr K. Schwerdtfeger and Dr H. Jacobi of the Metallurgy Department, Max Planck Institute, Dusseldorf.

REFERENCES

1 L. D. GRAHAM: Proc. of the Conf. on *In Situ* Composites, Vol. 1, 107; 1973

2 F. L. VER SNYDER AND M. E. SHANK: *Mater. Sci. Eng.*, 1970, **6**, 213

3 F. L. VER SNYDER *et al.*: *Mod. Cast.*, 1967, **50**, 360

4 F. STAUB *et al.*: *Foundry Trade J.*, 1973, 227

5 M. M. FARAG AND M. C. FLEMINGS: *Metall. Trans.*, 1976, **7A**, 215

6 L. BUCHAKJIAN: M.Sc. thesis, Massachusetts Institute of Technology, 1974

7 G. A. CHADWICK: 'Solidification of metals', 138, 1968, London, The Iron and Steel Institute

8 D. JAFFREY AND G. A. CHADWICK: *Metall. Trans.*, 1970, **1**, 3 389

9 J. D. HUNT AND J. P. CHILTON: *J. Inst. Met.*, 1962–63, **91**, 338

10 N. A. El-MAHALLAWY AND M. M. FARAG: *J. Cryst. Growth*, (to be published)

DISCUSSION

From the observed changes in the structure of a growing composite when the cross-sectional area increases, the following general conclusions can be drawn.

The observed slower growth rates in the bulk of the divergence area and at the nucleation sites of new grains yield coarser structures and can even cause a change in the morphology of the reinforcing phase. These changes can adversely affect the mechanical

Influence of various elements on Al–Al₆Fe eutectic system

P. G. Keong, J. A. Sames, C. McL. Adam, and R. M. Sharp

The metastable Al–Al$_6$Fe fibrous eutectic has tensile strengths in the range 300–500 MPa, but low ductility. By precipitation hardening the matrix of these alloys it has been possible to increase both the strength and the ductility of the fibrous eutectic structure. In this investigation, ternary Al–Fe–Cu, Al–Fe–Zn, Al–Fe–Mg, and Al–Fe–Mn alloys have been studied, and a quinary fibrous Al–Fe–Mg–Zn–Cu alloy developed. The authors describe the interaction of the alloying elements copper, magnesium, manganese, and zinc with the stable Al–Al$_3$Fe and metastable Al–Al$_6$Fe systems, and outline the effect of solid/liquid interface growth velocity on the development of cellular eutectic microstructures. The solidification behaviour of these alloys is complicated by the interaction of the stable–metastable eutectic transitions with the dendrite–eutectic–cellular transitions. The final precipitation-hardenable eutectic alloy can be grown over a wide range of interface growth velocities and temperature gradients typical of commercial direct chill casting practice.

The authors are in the Department of Chemical and Materials Engineering, University of Auckland, New Zealand

The aluminium–iron binary alloy system has an equilibrium eutectic reaction at 1·8 wt-% iron and 655°C forming a eutectic of aluminium and 4·5 vol.-% Al$_3$Fe.[1] However, it has been shown[2] that at high solid/liquid interface growth velocities this eutectic system is suppressed in favour of a metastable Al–Al$_6$Fe eutectic, occurring at 2·8 wt-% iron and having up to 20 vol.-% Al$_6$Fe rods.

Al$_6$Fe rods in a matrix of aluminium have been grown unidirectionally by Adam *et al.*[3] and tensile tests reported in their work show strengths in the range 300–500 MPa. Fracture strains were, however, low and were in the range 0·004–0·015. Highest strengths were obtained with alloys of the highest iron content. The minimum growth rate necessary for suppression of the Al$_3$Fe phase increases with iron content, so that to grow the strongest alloys requires extremely high (2 mm/s) growth rates.

In this work the influence of various third elements on the Al–Fe binary eutectic systems has been examined for two reasons. First, it might be expected that the addition of a third element would alter the stability of one eutectic system in relation to the other, making it easier or more difficult to grow the desirable Al$_6$Fe rods. Secondly, the presence of a third element in the aluminium matrix gives rise to the possibility of precipitation hardening as well as fibre reinforcement.

Successful precipitation hardening of eutectic aluminium alloys has been carried out in the Al–Cu system.[4] However, in that case, as in any binary system, the precipitate particles were the same phase as the eutectic phase. In ternary systems, such as those described in this paper, it is possible to have precipitate particles which are different from the eutectic rod phase.

The choice of third elements was influenced by these two considerations. Copper and manganese, for example, form compounds with iron which are isostructural with Al$_6$Fe: they might, therefore, be expected to stabilize that compound. Thirdly, elements which would be suitable for precipitate formation would be those which form no ternary intermetallic compound with aluminium and iron (e.g. not silicon) but which form suitable binary or ternary compounds with aluminium.

EFFECT OF VARIOUS ELEMENTS ON STABILITY OF Al–Al₆Fe EUTECTIC
Copper
Figure 1 shows the aluminium-rich end of the Al–Fe–Cu phase diagram,[5] Fig. 1*a* being a horizontal projection of the liquidus surfaces, and Fig. 1*b* a vertical section through the liquidus valley. The compound α is Al$_6$(Cu, Fe) which is a stable compound, isostructural

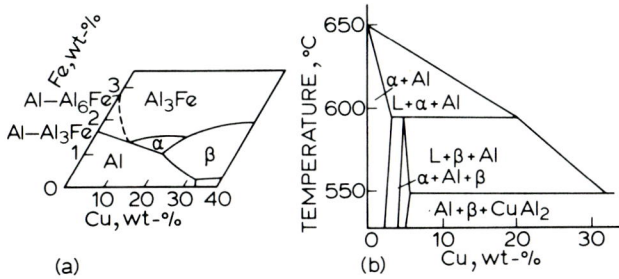

a horizontal projection of liquidus surfaces; *b* vertical section through liquidus valley

1   Al–Fe–Cu phase diagram[5]

with the metastable Al₆Fe. Because these structures are identical, the authors have extrapolated the stable monovariant trough up to the binary composition Al–2·8 wt-% Fe to represent the metastable reaction. The compound β is Cu₂Fe–Al₇. If the formation of β is to be avoided, Fig. 1*b* shows that the copper content must be less than 3 wt-%.

Alloys containing 3 wt-% copper and iron in the range 2·5–3·2 wt-% were made and unidirectionally solidified in alumina tubes. After solidification the specimens were examined under an optical microscope in order to determine the microstructure. The theoretical solidification behaviour of a ternary stable eutectic system competing with one which is metastable and coupled with the possibility of breakdown to either dendrites or two-phase cells is somewhat complicated. It is possible, however, to recognize an Al₆Fe–Al₆(Cu, Fe) structure, which in these alloys was always rod-like and cellular. Figure 2 shows the growth rates over which Al₆Fe–Al₆(Cu, Fe) was stable. The line on the diagram represents data from the binary Al–Fe system from the work of Adam *et al.*[3] It is clear that the addition of copper only marginally increases the stability of the metastable system, presumably because the

monovariant trough is at a lower iron content than the binary eutectic. Thus, were it not for the stabilizing influence of the α-phase, the line representing the limit of stability of Al₃Fe would be shifted to lower iron contents.

Manganese

Manganese forms a compound Al₆Mn with aluminium, structurally identical to Al₆Fe.[6] It might be expected, then, that additions of manganese to Al–Fe alloys might stabilize the Al–Al₆Fe eutectic reaction. Comparison with aluminium–iron alloys is difficult, however, since the extended solid solution of Al₆Fe and Al₆Mn means that the volume fraction of the fibres is increased. The most satisfactory method of comparison is to compare the stability of Al₃Fe at various growth rates in an Al–Mn–Fe alloy with Al₃Fe in an aluminium–iron alloy whose iron content is the sum of the iron and manganese contents.

Table 1 (section (i)) gives the results for two alloys in the Al–Fe–Mn system. Consider the Al–0·5 Mn–2·8 Fe alloy. Primary Al₃Fe is present when the growth rate is less than 400 μm/s. This compares with a growth rate of 440 μm/s necessary for suppression of primary Al₃Fe in the equivalent aluminium–iron alloy.[3] In the Al–1 Mn–2·8 Fe alloy no primary Al₃Fe was observed at growth rates above 200 μm/s. Additions of manganese do, therefore, stabilize the Al–Al₆Fe reaction as expected. However, as Table 1 (section (i)) indicates, primary Al₆Mn is produced at growth rates higher than those necessary for suppression of primary Al₃Fe.

There appears, therefore, to be no advantage in adding manganese to Al–2·8 Fe alloys in order to increase the volume fraction of the fibres, although there would be some point in substituting some of the 2·8% Fe with manganese. Large substitutions of manganese are not, however, desirable as the mechanical properties of aluminium–manganese alloys have been shown to be inferior to those of aluminium–iron alloys.[7]

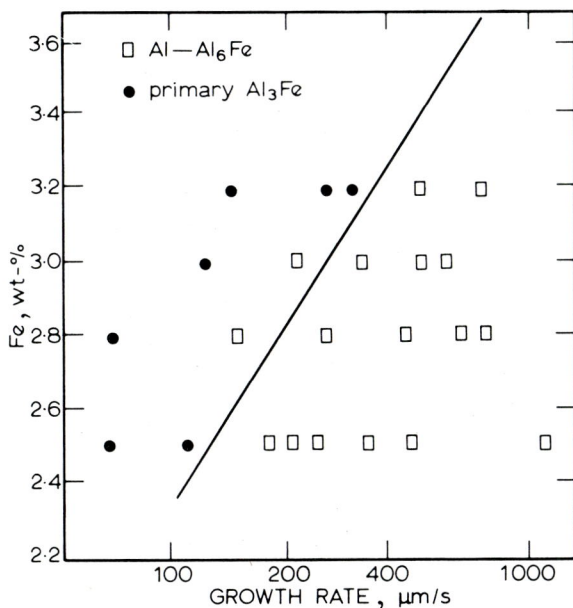

2   Effect of iron concentration on interface growth velocity necessary for Al–Al₆Fe eutectic structures in Al–3 wt-% Cu–Fe alloys: line represents data for binary Al–Fe alloys[2]

Table 1   Effect of alloy composition on relative stability of Al₆Fe and Al₃Fe

Alloy composition	Minimum growth rate (μm/s) necessary for suppression of	
	Al₃Fe	Al₆Mn
(i) Al–Mn alloys		
Al–2·8 Fe–1 Mn	<200	775
Al–3·8 Fe	750	—
Al–2·8 Fe–0·5 Mn	400	550
Al–3·3 Fe	440	—
(ii) Al–Mg alloys		
Al–2 Mg–2·8 Fe	450	—
Al–3·3 Fe	440	—
Al–4 Mg–2·8 Fe	1800	—
Al–3·5 Fe	600	—
(iii) Al–Zn alloys		
Al–5 Zn–2·8 Fe	415	—
Al–3·2 Fe	370	—
Al–7·4 Zn–2·8 Fe	600	—
Al–3·4 Fe	510	—

Magnesium

Magnesium forms no compound similar to $Al_6(Cu, Fe)$ or Al_6Mn and its effect on the stability of Al_6Fe might, therefore, be expected to result only from its lowering of the iron content of the monovariant trough. The $Al-Al_3Fe$ monovariant trough passes through the compositions 0 Mg–1·8 Fe, 2 Mg–1·3 Fe, and 4 Mg–1·1 Fe.[8] The authors, therefore, compare the structure of these alloys with the structure of aluminium–iron alloys whose compositions are 0·5 and 0·7% higher than those of the ternary alloys. Table 1 (section (ii)) gives the results. It appears from these that high magnesium contents have a considerable stabilizing effect on the $Al-Al_3Fe$ eutectic reaction.

The detrimental effect of magnesium in promoting primary Al_3Fe stability can be countered by selective manganese substitutions, which would tend to 'restabilize' eutectic $Al-Al_6Fe$. Under conditions of increasing growth rate (Fig. 3) or increasing magnesium content, primary Al_3Fe can be effectively converted into eutectic $Al-Al_6Fe$ by small manganese additions. The shaded regions of Fig. 3 represent compositions which produce primary Al_3Fe. By using 0·5% Mn, for instance, at a growth velocity of 1 000 μm/s, up to 6% Mg can be incorporated into the $Al-Al_6(Fe, Mn)$ eutectic structure without nucleation and growth of primary Al_3Fe.[9]

Zinc

Zinc, like magnesium, forms no ternary compound with iron and aluminium and again its effect on the stability of Al_6Fe might only be expected to result from the lowering of the iron content of the monovariant trough. This trough passes through the compositions of 0 Zn–1·8 Fe, 5 Zn–1·4 Fe, and 7·4 Zn–1·2 Fe.[10] Table 1 (section (iii)) summarises the stability of Al_6Fe in the

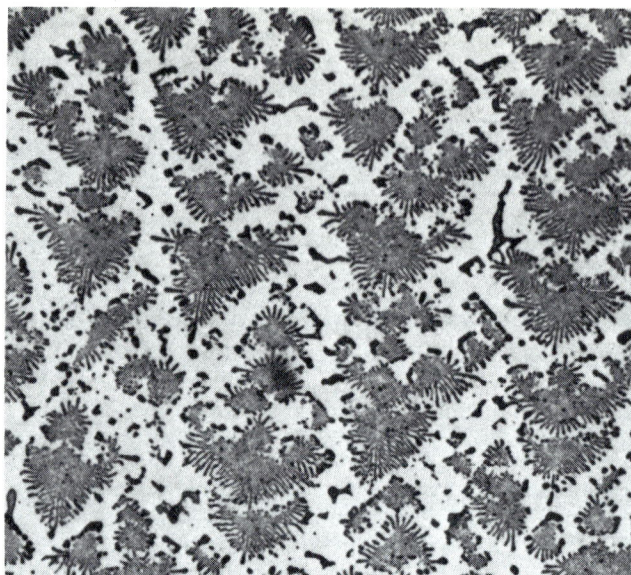

4   Al–2·8 wt-% Fe–7·4 wt-%: growth velocity 1 310 μm/s; transverse section   [× 600]

Al–Fe–Zn alloys investigated compared with equivalent Al–Fe alloys. It is apparent that zinc has very little effect on the relative stability of Al_6Fe and Al_3Fe.

INFLUENCE OF ALLOYING ELEMENTS ON EUTECTIC SOLIDIFICATION BEHAVIOUR

Binary $Al-Al_6Fe$ eutectic alloys solidify with an essentially planar solid/liquid interface to produce a regular array of parallel fibres. Introduction of a third element produces a cellular interface, when the following well known relationship is satisfied

$$\frac{G}{R} \leqslant \frac{mC_O(1-k_O)}{D_L-(k_O)}$$

where the symbols have their usual meanings.

Ternary $Al-Al_6Fe$ alloys containing copper, magnesium, manganese, or zinc solidify as cellular eutectic colonies. The colony structure produced by each element, however, is quite different. Zinc appears to have least effect on interface curvature (Fig. 4) producing equiaxed cells with little evidence of coarse cell-boundary Al_6Fe. Copper appears to have a considerable effect on cellular morphology (Fig. 5) producing $Al_6(Fe, Cu)$ plates at the cell extremities rather than curved fibres (Fig. 6). These plates appear to be similar to those observed by Hughes and Jones[11] in binary $Al-Al_6Fe$ alloys solidified at high growth velocities. Copper also produces sufficient lateral microsegregation for growth of β ($Cu_2, FeAl_7$) in intercellular liquid grooves. The morphology of $Al-Al_6(Fe, Mn)$ cells is similar to that observed by Eady *et al.* for binary $Al-MnAl_6$ alloys.[7] These cells appear to develop crystallographic plates not unlike those produced at cell boundaries by $Al-Al_6(Fe, Cu)$ alloys. The effect of magnesium is predominantly to produce very fine cells, with the cell extremities producing Al_6Fe plates rather than fibres. This plate-like morphology may simply reflect a high degree of interface curvature, rather than modification of the interfacial energy anisotropy which could produce a low energy interface.

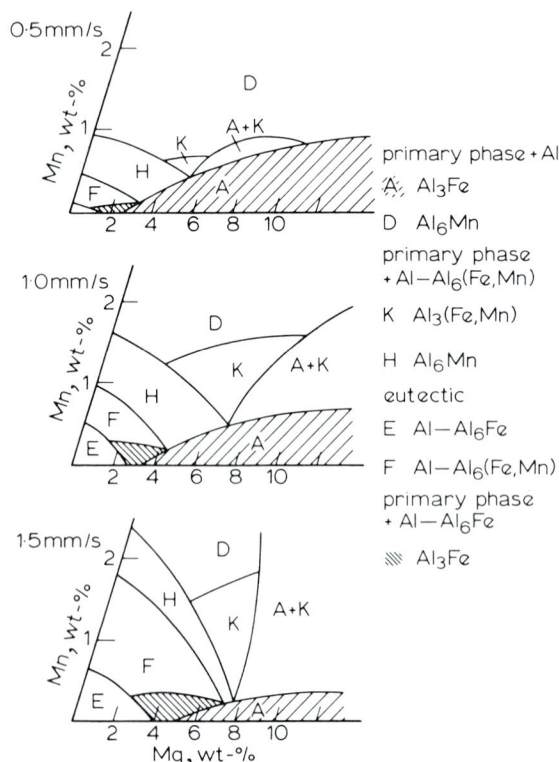

3   Al–Fe–Mg–Mn quaternary phase fields at 2·8 wt-% Fe

Legend for Fig. 3:

primary phase + Al

A Al_3Fe

D Al_6Mn

primary phase + $Al-Al_6(Fe, Mn)$

K $Al_3(Fe, Mn)$

H Al_6Mn

eutectic

E $Al-Al_6Fe$

F $Al-Al_6(Fe, Mn)$

primary phase + $Al-Al_6Fe$

Al_3Fe

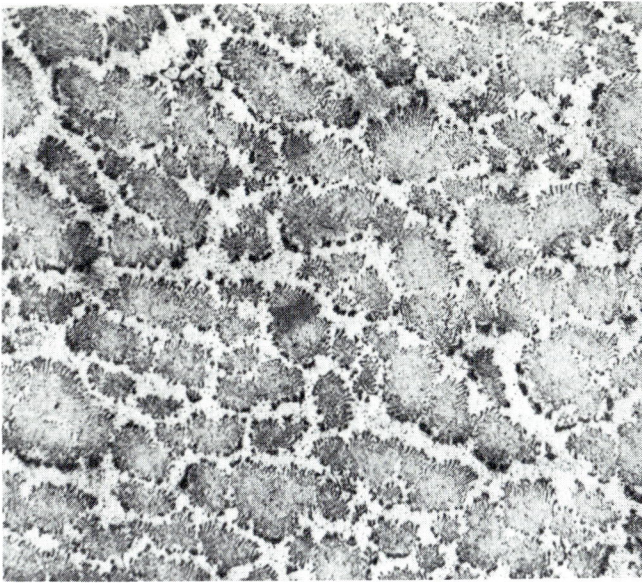

5   Al–2·8 wt-%Fe–3 wt-% Cu: growth velocity 750 μm/s; transverse section   [×600]

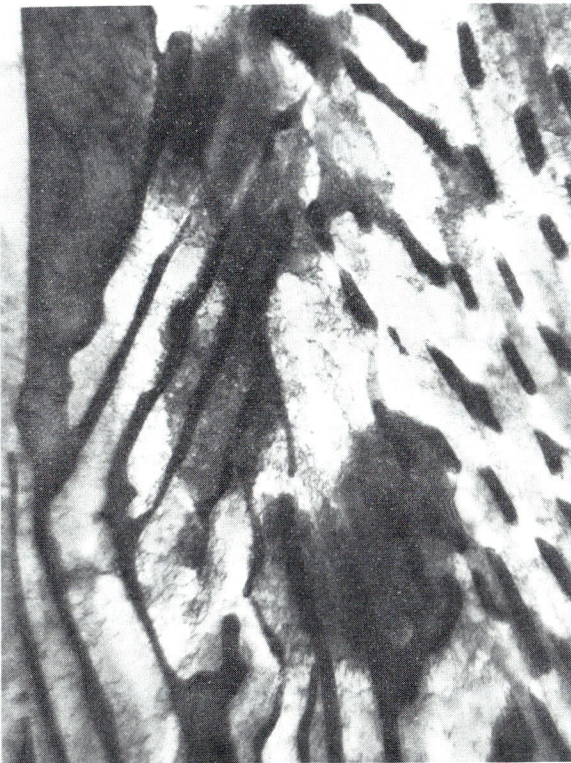

6   Transmission electron micrograph of cellular Al₆(Fe, Cu) rods and plates growing at cell extremities in Al–2·8 wt-% Fe–3 wt-% Cu alloys   [×25 000]

In these alloys there is a general trend for the eutectic cell size to decrease with increasing growth rate, and the relationship between these parameters is shown in Fig. 7. The alloys were solidified at a constant temperature gradient of 8·5°C/mm, and the relationship between cell size and growth velocity is different to that obtained by Rumball for eutectic Al–Zn alloys.[12]

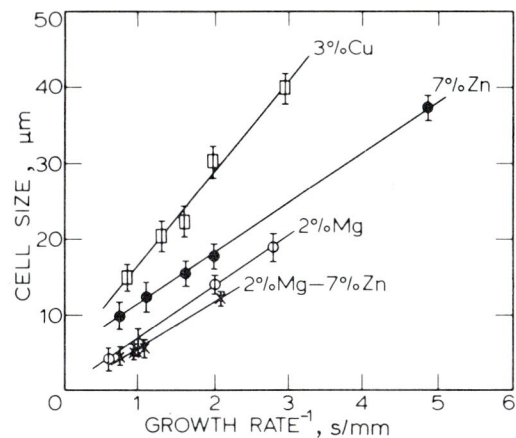

7   Effect of growth rate on cell size for various ternary alloys containing 2·8 wt-% Fe

The relationship between eutectic cell spacing and solidification parameters can be derived in a similar way to the single phase cell spacing relationship.[13]

The functional relationship between eutectic cell spacing y and growth rate is

$$y = \frac{mD_L}{GR}\left(\frac{\partial C}{\partial y}\right)_{y=0}$$

where D_L is the liquid diffusion coefficient and the y-axis is transverse to the growth direction.

The gradient of the cell spacing $v.$ (growth rate)$^{-1}$ plot then yields data about the transverse concentration gradient, in terms of normal solidification parameters:

$$d\left(\frac{dy}{1/R}\right) = \frac{mD_L}{G}\left(\frac{\partial C}{\partial y}\right)_{y=0}$$

Substitution of data for ternary Al–Al₆Fe alloys yields the values for transverse concentration gradients given in Table 2. This table also lists values of the distribution coefficient parameter $((1-k_O)/k_O)$, and the parameter $((C_S/k_O)-C_O)(m/G)$, which is a measure of the depth of the cell-boundary groove. The authors have compared the transverse concentration gradients with phase-diagram variables for the ternary eutectic alloy systems, but no successful correlation relating eutectic cell size to these variables appears possible.

INFLUENCE ON MATRIX ORIENTATION

Eutectic Al₆Fe fibres always have the [001] direction normal to the solid/liquid interface. The orientation of the aluminium matrix, however, appears to be variable. In binary Al–Al₆Fe alloys solidifying with a planar solid/liquid interface the following relationship has been reported[2]

Al₆Fe[001]∥Al[110]∥growth direction

In ternary alloys solidifying with a cellular microstructure the observed orientation of the aluminium matrix is [100]∥growth direction.

At the centres of eutectic cells, the orientation relationship observed for ternary alloys is independent of the solute element, and is

Al₆Fe[001]∥Al[100]∥growth direction

This orientation relationship changes toward the cell boundaries, but for cellular alloys the aluminium matrix

Table 2   Cell size related to system parameters for various solute elements in the Al–Al$_6$Fe cellular eutectic

Element	$\dfrac{mD}{G}$	$\dfrac{dy}{d(1/R)}$	$\left(\dfrac{\partial C_y}{\partial y}\right)_{y=0}$	$\dfrac{1-k_O}{k_O}$	$\left(\dfrac{C_s}{k_O}-C_O\right)\dfrac{m}{G}$, cm
Zn	$\dfrac{1{\cdot}76\times6{\cdot}25\times10^{-5}}{85}$	$6{\cdot}65\times10^{-5}$	51·4 wt-%/cm	1·26	3·74
Mg	$\dfrac{5\times2{\cdot}7\times10^{-5}}{85}$	$7{\cdot}1\times10^{-5}$	44·7	1·50	2·14
Cu	$\dfrac{2{\cdot}5\times3\times10^{-5}}{85}$	$12{\cdot}0\times10^{-5}$	136	6·2	1·13

phase always has the dendritic growth orientation, $\langle100\rangle$. The variable orientation relationships[14] observed in other fibrous eutectic alloy systems with cellular eutectic microstructures may simply reflect different degrees of solid/liquid interface curvature and different stages in the progressive rotation of the aluminium matrix orientation from $\langle110\rangle$ towards $\langle100\rangle$.

A FIBROUS EUTECTIC PRECIPITATION-HARDENED ALLOY

The incorporation of magnesium, zinc, and copper within the matrix of a fibre-reinforced alloy makes precipitation hardening of the matrix possible. The stability of the Al–Al$_6$Fe eutectic depends on low magnesium contents; the undesirability of β Cu$_2$FeAl$_7$ intercellular compound limits the copper content to 3%; low manganese contents stabilize the Al–Al$_6$Fe eutectic but high manganese contents produce primary Al$_6$Mn. These considerations restrict possible alloy compositions to low copper and manganese contents and thus favour precipitation hardening by MgZn$_2$ and MgZn$_{11}$ species rather than Mg$_5$Al$_8$, Mg$_3$Zn$_3$Al$_2$, or CuAl$_2$. Two fibrous eutectic alloy compositions were selected, and Fig. 8 shows the precipitation-hardening response of these alloys compared to data obtained by Polmear for a standard Al–8 Zn–2 Mg alloy.[15]

A detailed discussion of the mechanical properties of these alloys is outside the scope of this paper.[9] The hardness, however, of the Al–Al$_6$Fe eutectic depends

on growth velocity,[11] and is approximately HV 50 for 2·8 wt-% Fe unidirectionally solidified Al–Al$_6$Fe alloys. The stronger of the two selected alloys in the solution-treated condition has a hardness of HV 160 and the same alloy in the precipitation-hardened condition about HV 230. This peak hardened condition corresponds to a yield strength of 540 MPa, an ultimate tensile strength of 635 MPa, and 7% elongation at failure. These tensile properties are extremely sensitive to the presence of primary Al$_3$Fe which causes premature tensile instabilities above the yield stress. Optimum mechanical properties for these alloys require the absence of primary phases and well developed cellular eutectic microstructures.

ACKNOWLEDGMENTS

The authors are indebted to Professor A. L. Titchener, Dean of the School of Engineering, University of Auckland, for the provision of laboratory facilities and to Comalco Ltd for the provision of financial assistance and a post-graduate scholarship for J. A. Sames. The work reported in this paper is currently subject to an Australian and US Patent application by Comalco Ltd, Melbourne, Australia.

REFERENCES

1 L. E. MONDOLFO: 'Aluminium alloys, structure and properties', 283, 1976, London, Butterworths
2 C. MCL. ADAM AND L. M. HOGAN: *J. Aust. Inst. Met.*, 1972, **17**, 81
3 C. MCL. ADAM *et al.*: 'Proc. int. conf. on in-situ composites', (Vol. II), 309, 1973, National Materials Advisory Board
4 C. G. RHODES AND G. GARMONG: *Metall. Trans.*, 1972, **3**, 1 861
5 H. W. L. PHILLIPS: *J. Inst. Met.*, 1953–4, **82**, 197
6 H. W. L. PHILLIPS: 'Annotated equilibrium diagrams', 48, 1959, London, Institute of Metals Monograph No. 25
7 J. A. EADY *et al.*: *J. Aust. Inst. Met.*, 1975, **20**, 23
8 H. W. L. PHILLIPS: *J. Inst. Met.*, 1941, **67**, 275
9 J. A. SAMES: M.E. Thesis, 1977, University of Auckland
10 G. V. RAYNOR *et al.*: *Acta Metall.* 1953, **1**, 629
11 I. R. HUGHES AND H. JONES: *J. Mater. Sci.*, 1976, **11** 1 781
12 W. M. RUMBALL: *Metall.*, 1968, **77**, 141
13 M. C. FLEMINGS: 'Solidification processing', 83, 1974, New York, McGraw-Hill
14 G. GARMONG *et al.*: *Metall. Trans.*, 1973, **4**, 707
15 I. J. POLMEAR: *J. Inst. Met.*, 1957–58, **86**, 113

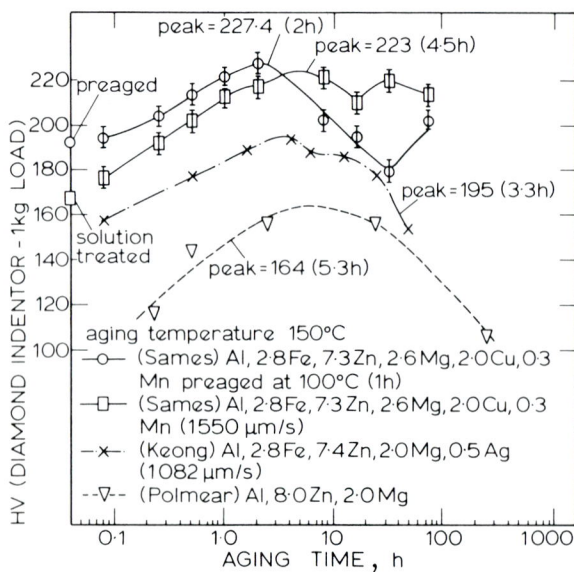

8   Precipitation hardening response

Influence of silicon and aluminium on the solidification of cast iron

T. Carlberg and H. Fredriksson

The influence of silicon and aluminium on the solidification of eutectic cast iron has been studied by unidirectional solidification experiments. In particular, the influence of these elements on the transition from grey to white structure has been studied. The graphite morphology in the different alloys has also been studied. Both aluminium and silicon favour the grey structure. Aluminium gives a very coarse flake graphite structure but silicon gives a very fine, undercooled, graphite structure. The transition from grey to white is discussed from the point of view of a different growth mechanism of graphite and from the point of view of the Fe–Al–C and Fe–Si–C phase diagrams.

The authors are in the Department of Casting of Metals, The Royal Institute of Technology, Stockholm, Sweden

Commercial cast iron alloys solidify as either white or grey iron, depending on the content of alloying elements and the rate of cooling. One usually tries to prevent or counteract the formation of white iron. It is well known and well documented[1,2] that this can be achieved by the addition of silicon. In recent years, cast iron has also been produced in which the silicon is replaced by aluminium.[3] This also reduces the risk of white iron formation. Silicon and aluminium both result in a completely different morphology of the precipitated graphite. It is therefore of interest to compare the effects of these alloying elements on the solidification process, in order to cast some light on the origin of the different graphite morphologies. Since very little is known, in any case, about the solidification process in cast iron alloys containing aluminium, a comparison and study of these alloys has been carried out in the present investigation. Many workers have investigated the transition from white to grey solidification,[4,5] which displays a hysteresis effect with respect to the cooling rate. This hysteresis effect and a number of differences in the origin of the structure are explained, taking calculated phase diagrams for Fe–Al–C and Fe–Si–C as a starting point.

EXPERIMENTAL METHOD

The solidification experiments were carried out in a gradient solidification apparatus consisting of a silite rod furnace into which a water-cooled copper probe was partially inserted from above. A crucible containing the alloy was introduced from beneath and placed at a well defined height where the temperature was 1 270°C. An aluminium oxide tube of 1·7 mm dia. was then passed through the copper probe. Molten alloy was sucked into the tube to a height of about 10 cm. The water-cooled probe produced a temperature gradient along the sample, and at a certain level the melt solidified. The aluminium oxide tube was connected to a motor which drew the tube upwards through the probe at a constant speed, so that a constant solidification speed was obtained. By using a gear mechanism, it was possible to study speeds from 0·0017 to 1·67 mm/s. After a certain time at constant speed, the sample was drawn as quickly as possible out of the furnace and quenched in salt water. The test rods were then ground longitudinally and studied under an optical microscope. A thermocouple was used to check that the temperature gradient in the furnace remained constant throughout the experiments. The furnace atmosphere consisted of argon, which was introduced from below. The alloys and growth rates used in the gradient solidification tests are listed in Tables 1 and 2, respectively.

In order to facilitate evaluation of the results and to make the investigation as complete as possible, differential thermal analysis (DTA) measurements were carried out on three of the alloys. In DTA, the sample is allowed to pass back and forth through a temperature range from below A_1 to above the melting point several times at varying rates. At the same time a reference sample is passed through the same temperature cycle. The difference in temperature between the test and reference samples is measured and recorded on a plotter. When a transformation occurs in the test sample, energy is released or absorbed, and the temperature difference between it and the reference sample changes. In this way, transformations occurring in the test sample can be studied by reading off the

Table 1   Composition of the alloys

Element, %	Alloy no.								
	1	2	3	4	5	6	7	8	9
C	4·26	4·20	4·14	4·11	4·03	4·3	4·3	4·2	3·9
Al	—	0·50	1·00	1·96	2·90	—	—	—	—
Si	—	—	—	—	—	0·21	0·54	1·29	2·27

Table 2   Growth rates in experiments, mm/s

Experiment									
A	B	C	D	E	F	G	H	K	L
1·664	0·834	0·417	0·167	0·083	0·042	0·017	0·0083	0·0042	0·0017

1   **Longitudinal section of solidification front of eutectic Fe–C alloy: $V = 0.0017$ mm/s   [×20]**

2   **Longitudinal section of solidification front of eutectic Fe–C alloy with 1·0 wt-% Al: $V = 0.0017$ mm/s   [×20]**

temperature at which they occur and the amount of energy absorbed or released. The cooling rates used in the DTA measurements were 20°, 12°, 10°, and 6°C/min, and the alloys studied were Nos. 1, 4, and 5.

OBSERVATIONS ON SOLIDIFICATION PROCESS

The samples that were allowed to solidify at the lowest speed, 0·0017 mm/s, show major alterations with increasing aluminium content. Figure 1 shows the unalloyed sample and Figs 2 and 3 show samples containing 1 and 2·9% aluminium, respectively. As the aluminium content rises, the solidification front progresses from a quite smooth front with flaky graphite solidification to an irregular front in which the two phases cannot cooperate at all. Table 3 gives the lengths of the solidification interval for alloys of different aluminium content. It may also be noted that in places the graphite grows freely out into the melt almost as far as the austenite. This is best seen in Fig. 3.

Table 3   Solidification range, mm

	Alloy no.			
	1	2	4	5
Length, mm	0	0·5	3·6	6·6

3   **Longitudinal section of solidification front of eutectic Fe–C alloy with 2·9 wt-% Al: $V = 0.0017$ mm/s   [×25]**

4 Longitudinal section of solidification front of eutectic Fe–C alloy with 0·54 wt-%Si: $V = 0·001\,7$ mm/s [×20]

5 Longitudinal section of solidification front of eutectic Fe–C alloy with 2·27 wt–%Si: $V = 0·001\,7$ mm/s [×20]

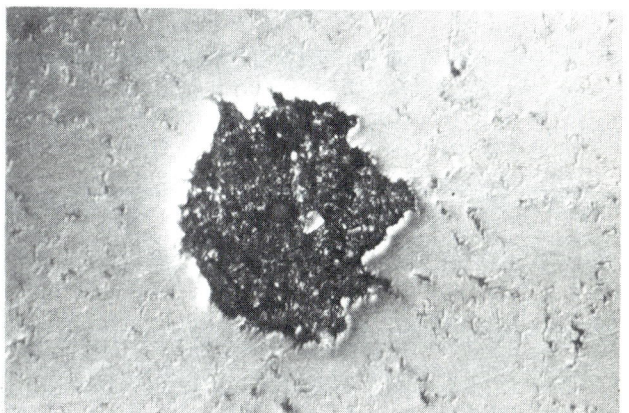

a

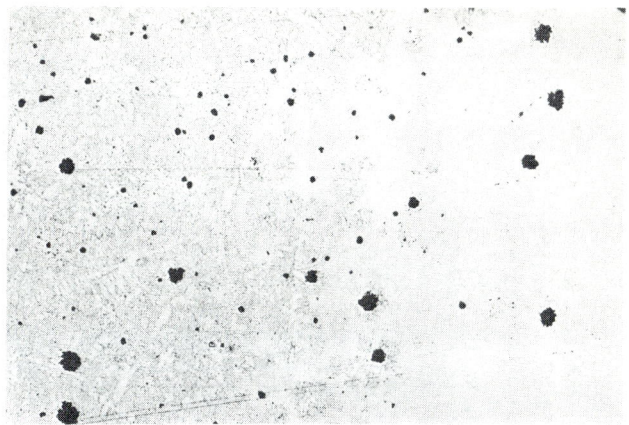

b

a [×28]; *b* [×115]

6 Longitudinal section of solidification front of eutectic Fe–C alloy: $V = 0·083$ mm/s

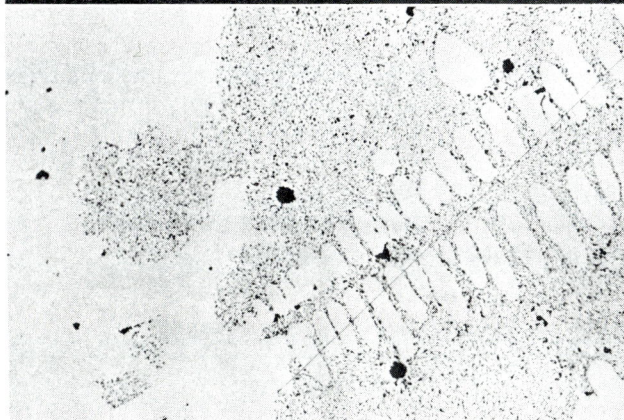

7 Longitudinal section of solidification front of eutectic Fe–C alloy with 0·54 wt-%Si: $V = 0·083$ mm/s [×56]

With a rising silicon content, the solidification front does not change in the same way as with a rising aluminium content. Figures 4 and 5 show samples containing ∼0·5 and ∼2·3% silicon, respectively. The front remains smooth and the length of the solidification interval does not increase. On the other hand, the structure seems to be less clearly oriented along the length of the sample, and graphite begins to be precipitated ahead of the solidification front because the sample was somewhat hypereutectic. At the next higher speed, a transition to supercooled graphite was obtained in the unalloyed samples and in the samples containing silicon. In the latter samples, supercooled graphite was also obtained at all higher speeds.

Figures 6 and 7 show the structure in the samples containing 0 and 0·54% silicon, respectively, that solidified at the speed of 0·083 mm/s. In these samples, and especially in those with high silicon content, isolated graphite nodules were observed to be precipitated some way ahead of the eutectic solidification front. Figure 8 shows the appearance of these nodules under high magnification. They consist of closely packed aggregates of very fine flakes with numerous branchings. This morphology is presumably the same as that found by Morrogh[6] in rapidly solidified hypereutectic cast iron alloys.

8 Graphite nodule at high magnification [×715] from specimen with 0·54 wt-%Si: $V = 0·083$ mm/s

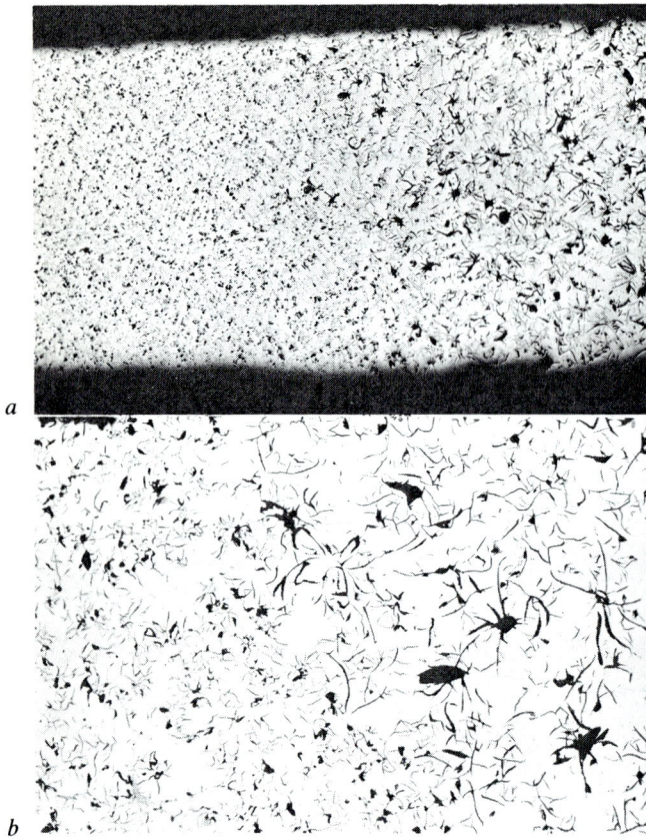

a [×28]; *b* [×715]

9 Longitudinal section of solidification front of eutectic Fe–C alloy with 2·9 wt-%Al: V = 0·083 mm/s

With rising Al content, the eutectic reaction showed an increasing tendency to produce a flaky graphite structure. In the alloy with the highest aluminium content, this structure was obtained at all speeds. Figure 9 shows the structure of the sample of 2·9% aluminium alloy that solidified at 0·083 mm/s.

The mean spacing of the graphite flakes was determined by placing a line of known length at a number of points of eutectic structure in the samples and counting the flakes that intersected the line. Figure 10 shows the graphite flake spacing plotted against aluminium and silicon content at various speeds. It is clear that an increased aluminium content results in a coarser structure, especially at the lowest speed. The silicon content has no effect at all on the graphite flake spacing at the higher speeds, while at the lowest speed there is a very slight tendency towards a coarser structure with rising silicon content.

If the dependency between the coarseness and the solidification speed in the individual alloys is investigated, it is found that the speed has less effect on the coarseness of the supercooled structure than on that of the flaky graphite structure. Similar results have been reported previously.[7] In summary, the graphite flake spacing changes considerably less with the solidification speed and the content of alloying elements in supercooled graphite than in flaky graphite.

The results of the DTA measurements are given in Table 4 and shown in Fig. 11. They show that the eutectic reaction begins at increasingly higher temperatures with increasing aluminium content. This is in good

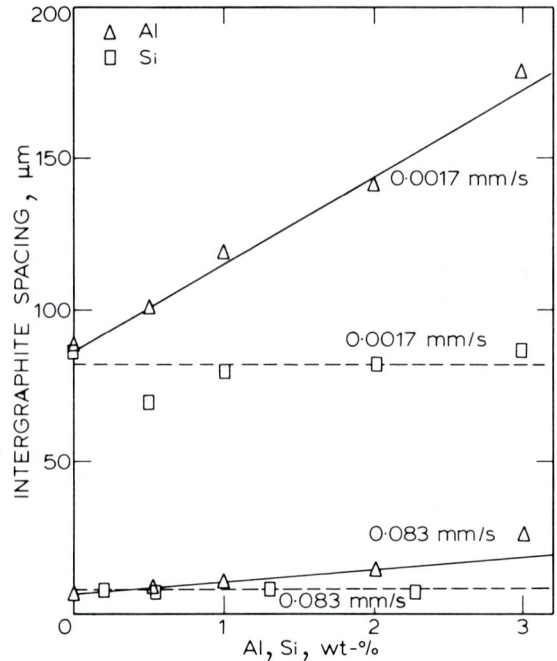

10 Intergraphite spacing as a function of Al and Si contents at two different growth rates

agreement with earlier observations.[8,9] On the other hand, the temperature of the eutectic reaction seems to rise very little with rising silicon content. The DTA measurements also show that the solidification interval increases as the aluminium content rises. The solidification interval is less affected by the silicon content of the samples.[10]

TRANSITION FROM GREY TO WHITE IRON FORMATION

After the sample was sucked up into the tube, it was allowed to remain still for 5 min to give the solidification front time to stabilize. At the same time, the solidified material nearest the melt became graphitized, so that the experiments began with grey iron

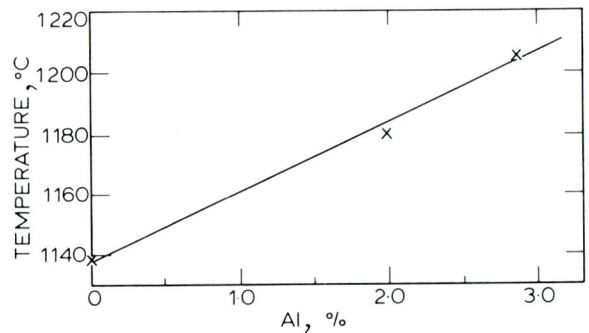

11 Eutectic temperature as a function of Al contents measured by DTA

Table 4 Result of DTA measurements

Alloy no.	Eutectic arrest temperature, °C
1	1 138
4	1 181
5	1 204

formation. In some samples there was an immediate transition to white, or mottled iron formation, while others solidified as grey iron throughout. In some samples, solidification began with a grey zone which then changed over to a white, or mottled structure. At the highest speed it could be observed that the length of the grey zone increased with the aluminium·content, and at 2·9% the entire sample solidified as grey iron. At lower speeds, the entire sample solidified as grey iron for all aluminium contents. Similar results were obtained in the silicon samples, but the tendency was weaker. Thus, at the highest speed, the 2·3% sample did not solidify entirely as grey iron. The results are shown in Figs 12 and 13. In these figures, completely filled dots represent samples that solidified entirely as grey iron; unfilled dots are samples that solidified entirely as white iron or switched to white iron after a time; and partly filled dots are samples that switched to a mottled structure after a time. It is evident that aluminium and silicon favour the formation of grey iron and that the effect of aluminium is somewhat more marked.

PHASE DIAGRAMS FOR Fe–Si–C AND Fe–Al–C

The Fe–Al–C phase diagram was calculated for the following temperatures: 1 160°, 1 180°, and 1 200°C. The two-phase γ/L area was calculated using the following equation[11]

$$X_C^L = {}^\circ X_C^L + \frac{k_{Fe}^{\gamma/L} - k_{Al}^{\gamma/L}}{1 - k_{Fe}^{\gamma/L}} X_{Al}^L \cdot X_C^L$$

The values used for the distribution constants are assumed to be independent of the composition. The term ${}^\circ C_C^L$ was evaluated at the temperatures concerned on the basis of the binary Fe–C phase diagram in Ref. 9. The constants $k_{Fe}^{\gamma/L}$ and $k_C^{\gamma/L}$ were evaluated from the Fe–C diagram at 1·10 and 0·51, respectively. Since it is reported[12] that aluminium segregates negatively,

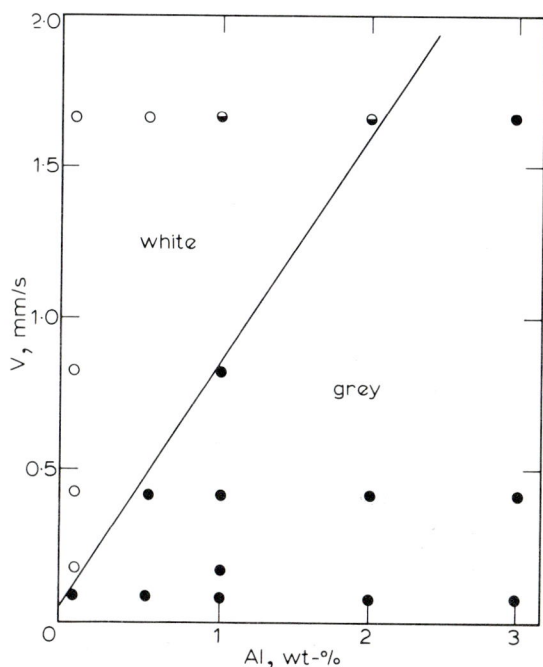

13 Transition from grey to white solidification as a function of growth rate and Si content

i.e. the aluminium content is higher in the austenite than in the melt at equilibrium, the constant $k_{Al}^{\gamma/L}$ should be greater than unity, and it was assigned the value of 1·1. The following equation, from Ref. 13, was used to calculate the line between the liquid phase and liquid phase + graphite

$$X^{L/gr} = {}^\circ X^{L/gr} + mX_{Al}$$

where $m = -0·52$, according to Schenk.[13] For the equilibrium between liquid phase and cementite the equation

$$X_C^{L/cem} = {}^\circ X_C^{L/cem} - 3X_{Al} {}^\circ X_C^{L/cem} + mX_{Al}$$

is used, which is obtained from the condition of equilibrium $-a_{Fe}^3 \cdot a_C = $ constant, assuming that the carbon activity in the melt in equilibrium with cementite is affected in the same way as in equilibrium with graphite.

The calculated liquids are shown in Fig. 14. The triple points $\gamma|L|gr$, $\gamma|L|cem$, and $L|gr|cem$ are joined.

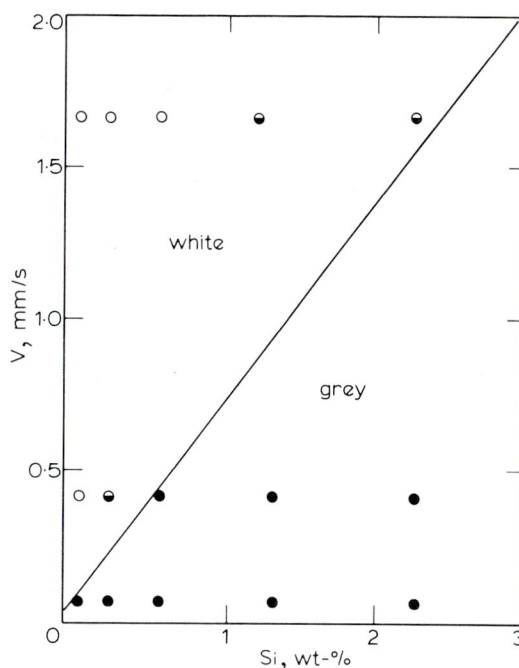

12 Transition from grey to white solidification as a function of growth rate and Al content

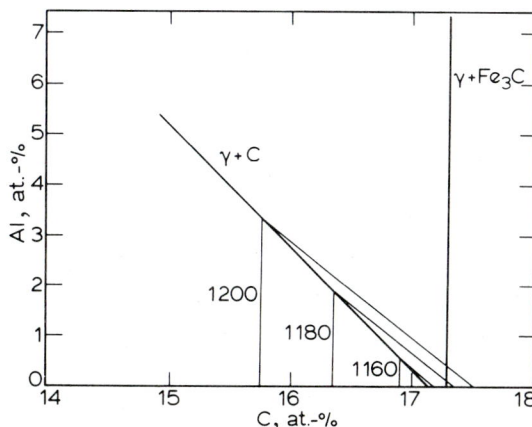

14 Calculated eutectic lines for stable and metastable Fe–C–Al system

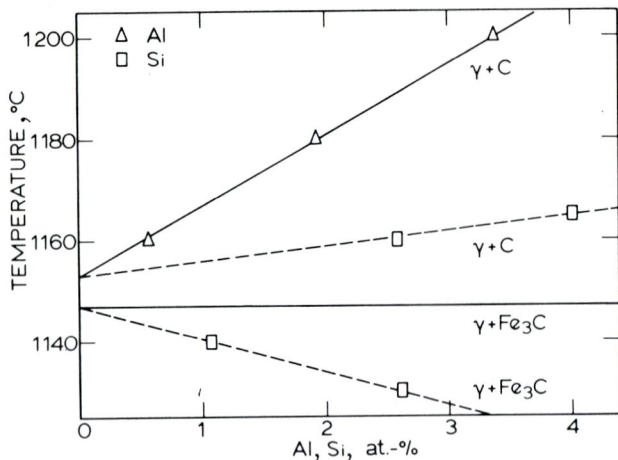

15 Calculated eutectic temperatures as a function of Al and Si contents for stable and metastable Fe–C–Al and Fe–C–Si systems

The eutectic temperature obtained from this graph is plotted against aluminium content for the cases of grey and white iron formation in Fig. 15. In the case of grey iron, the graph shows the eutectic temperature rising over 20°C for every unit percentage weight of aluminium added. The DTA measurements (Fig. 11) also showed a rise of 20°C per unit percentage weight, and the same is shown in a phase diagram by Löhberg and Ueberschaer.[8]

The ternary Fe–Si–C phase diagram (Fig. 16) was calculated in a similar way using the above equations and the constants given above. The distribution constants $k_{Si}^{\gamma/L}$ was assigned the value 1·04 to obtain agreement between the calculations and the phase diagram according to Schürmann and von Hirsch.[10] The value chosen appears to agree with the distribution constants reported in the literature. The *m* value used was −0·11, according to Schenk.[13] Figure 15 shows the influence of Si on the temperature of the metastable and the stable eutectic reaction.

DISCUSSION

It is known from the work of Hultgren *et al.*[14] Hillert and Subba Rao,[4] and Fredriksson[15] that flaky graphite can be regarded as a primary precipitation of graphite, and that supercooled graphite can be regarded as the result of a normal eutectic reaction. The transition from flaky to undercooled graphite might be affected by a

number of factors, such as surface tension between the graphite and the melt, or the phase-boundary kinetics at the graphite/melt interface.[5,15] In the present case the authors have found that cast iron alloys containing aluminium precipitate flaky graphite, and alloys containing silicon give supercooled graphite. It is difficult to decide whether this is due to the cooperation between the growing phases being affected by the diffusion field ahead of them, or to the effect of the two elements on the phase-boundary kinetics. The authors will attempt below to assess the effect of Si and Al on the rate of growth of the graphite and the austenite, in order to investigate whether the difference in the morphology of the graphite can be explained in terms of differing diffusion conditions.

The literature contains no simple mathematical models of growth in ternary alloy systems. In the present analysis a simplified model of growth conditions in ternary alloy systems will be applied based on the analysis by Zener[16] and Hillert.[17] This modified model of the growth of graphite austenite and a normal eutectic is described in the Appendixes.

The growth rate was calculated for the three structures in eutectic alloys containing no additive, 2% aluminium and 2% silicon, respectively. The calculations were made using the values given in Ref. 4. The results of those calculations are shown in Figs 17–20. The undercooling for the different reactions varies with the composition. In Fig. 21 are shown the regions where different reactions occur in the fastest way in alloys without and with, respectively, 2% Al and 2% Si. The figure shows that the region for coupled growth (undercooled graphite) decreases in size with the increasing Al and Si content and decreases somewhat faster in alloys with Al.

It is interesting to note that the experiments also show this fact, although it is more pronounced. The

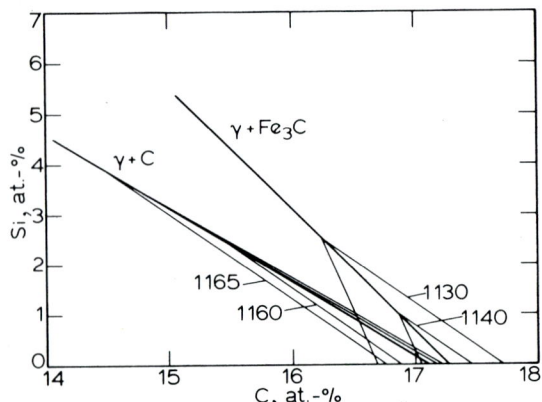

16 Calculated eutectic lines for stable and metastable Fe–C–Si system

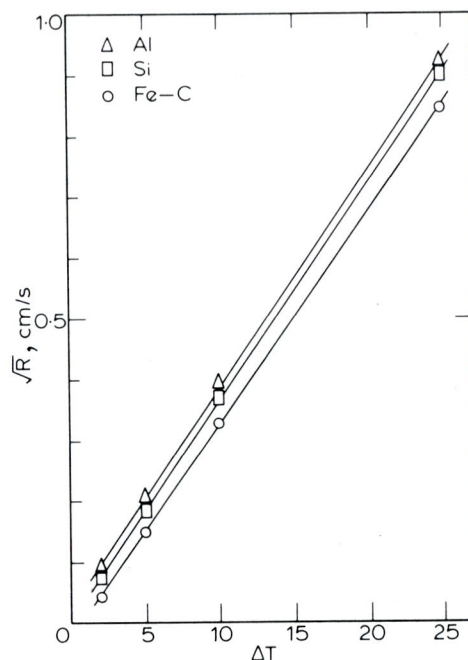

17 Calculated growth rate for austenite as a function of undercooling: calculations for pure Fe–C, Fe–C with 2 wt-% Al, and Fe–C with 2 wt-% Si

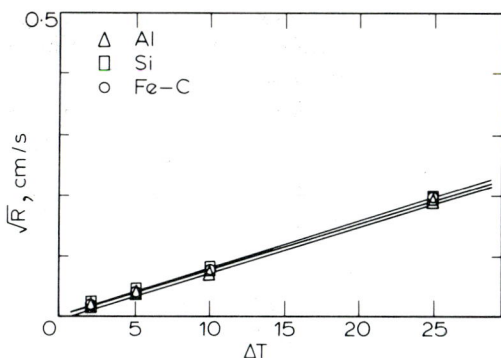

18 Calculated growth rate for graphite as a function of undercooling: calculations for pure Fe–C, Fe–C with 2 wt-% Al, and Fe–C with 2 wt-% Si

author of one investigation[15] discusses the effect of phase-boundary kinetics on the precipitation of super-cooled graphite. It was found that an increase in the phase-boundary-controlled growth of graphite favours the formation of flaky graphite. In the present case this would imply that aluminium has a marked effect on the kinetics and thereby favours the precipitation of flaky graphite.

THE AUSTENITE BORDER ROUND THE GRAPHITE

Figures 1–3 show the effect of aluminium on the solidification process. They show that with a rising Al content there is a decrease in the tendency for austenitic growth round the graphite. It is well known that the zone at the graphite/melt phase boundary is depleted in carbon, and this would favour precipitation of austenite around the graphite. However, this border of austenite forms with great difficulty in the alloy

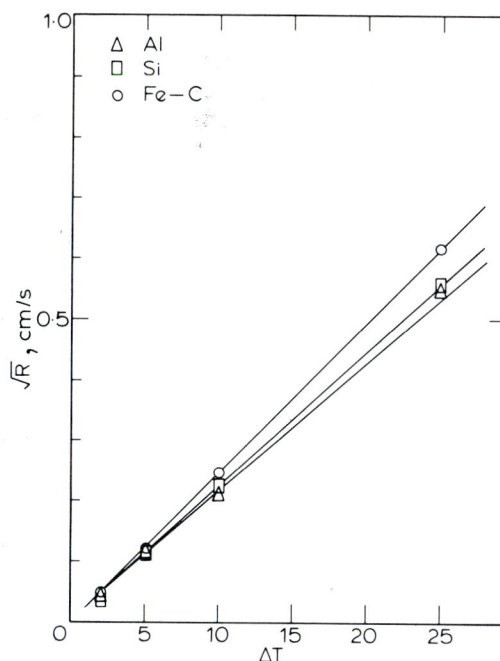

20 Calculated growth rate for cooperative growth of austenite + graphite eutectic as a function of undercooling: calculations for pure Fe–C, Fe–C with 2 wt-% Al, and Fe–C with 2 wt-% Si

containing 2·9% aluminium (Fig. 3). A possible explanation for the failure of austenite to precipitate round the graphite might be that the distribution constant $k_{Al}^{\gamma/L}$ is greater than unity. However, a theoretical analysis based on the phase diagram shows that this cannot be the explanation. A more likely explanation is that the phase-boundary kinetics at the graphite/melt interface are so slow for high aluminium contents that the carbon content of the phase-boundary graphite/liquid is equal to the carbon content at the phase-boundary austenite/liquid, and therefore the driving force behind the formation of new surfaces is not enough to enable growth of austenite round the graphite.

LENGTH OF SOLIDIFICATION INTERVAL

The lengthening of the solidification interval with rising aluminium content is easily explained if it is noted that the distribution constant $k_{Al}^{\gamma/L}$ is greater than unity. In areas of molten metal that have been enclosed in the

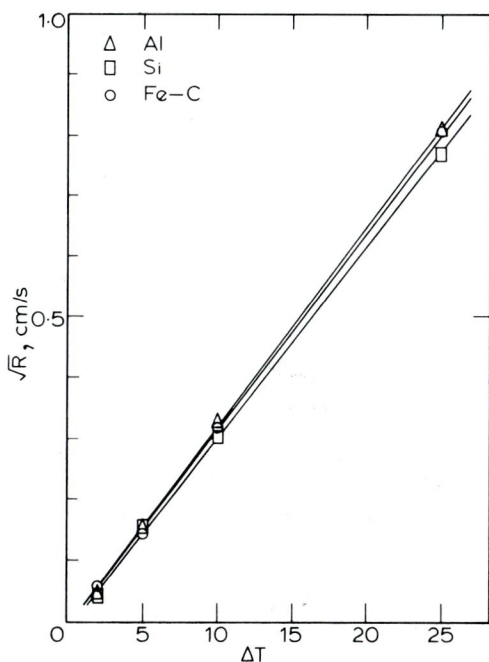

19 Calculated growth rate for cementite as a function of undercooling: calculations for pure Fe–C, Fe–C with 2 wt-% Al, and Fe–C with 2 wt-% Si

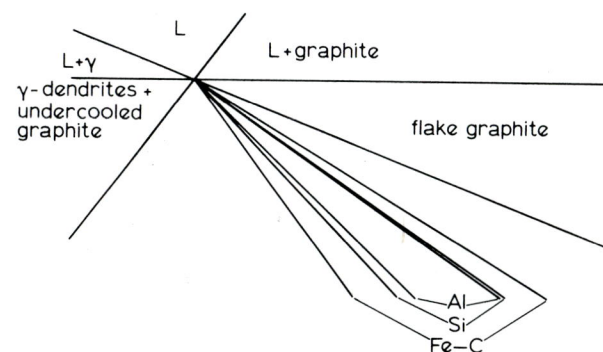

21 Calculated extension of coupled zone for pure Fe–C, Fe–C with 2 wt-% Al, and Fe–C with 2 wt-% Si

course of solidification, the aluminium content will therefore decrease owing to the negative segregation of aluminium. If the aluminium content falls, the freezing point will also fall, implying that molten metal will still be present in these areas quite low down in the temperature range. If the solidification interval for alloy no. 5 (2·9% Al) is taken, which is 6·6 mm, and that distance is measured off on the gradient curve from the liquidus temperature, 1 205°C, it can be seen that the temperature drops by 60°C over the solidification interval. If it is assumed that all the aluminium has been removed from the areas that solidify last, the solidification temperature should have fallen by around 60°C, because the liquidus temperature rises 20°C per percentage unit of aluminium, as noted above. It should, therefore, be possible to explain the lengthening of the solidification interval in this way.

Silicon does not cause the same alterations of the solidification front as does aluminium (*see* Figs 4 and 5 for 0·54% and 2·3% silicon, respectively). Silicon does not segregate as markedly as aluminium,[8] nor does it give an elongated solidification interval. Further, the addition of silicon results in a hypereutectic melt. This means that the graphite sometimes tends to grow into the melt ahead of the solidification front (Fig. 4). At the highest silicon content (Fig. 5) the formation of graphite nuclei ahead of the solidification front can also clearly be seen. This also has the effect of reducing the degree of longitudinal orientation of the structure.

TRANSITION FROM GREY TO WHITE IRON FORMATION

The experiments show that aluminium and silicon both have a strong tendency to favour stable solidification of cast iron. It was also observed that aluminium had a more marked graphite stabilizing effect than silicon.

If the Fe–C–Al system is examined and the slopes of the eutectic lines in the stable and the metastable diagrams are compared, (Fig. 14) it is clear that the line curves off more markedly towards lower carbon contents in the case of stable solidification. If Fig. 15 is also examined, it is clear that the eutectic temperature increases more in the case of grey iron than white iron formation as the aluminium content rises. Both these changes in the phase diagram naturally contribute to an increased tendency towards grey iron formation.

In the presence of a silicon additive, the eutectic line curves off even more markedly towards lower carbon contents than it does in aluminium alloys, and this resulted in the silicon alloys becoming hypereutectic. In spite of this, the silicon is less effective than aluminium in stabilizing the graphite. The reason for this is to be found in Fig. 15, where the eutectic lines $L|gr|\gamma$ and $L|cem|\gamma$ do not separate as markedly as in the Fe–C–Al system.

To illustrate this, the authors carried out a series of calculations for eutectic alloys containing no additive, 2% silicon, and 2% aluminium, respectively. It was assumed that solidification began with precipitation of austenite, and the growth rates of graphite and cementite were then calculated as a function of supercooling. The constants used in the calculations are taken from Ref. 4. The results are shown in Figs 22–24. It can be seen that in the pure Fe–C alloy, precipitation of cementite begins at 13°C below the starting temperature for graphite precipitation. In the case of the silicon

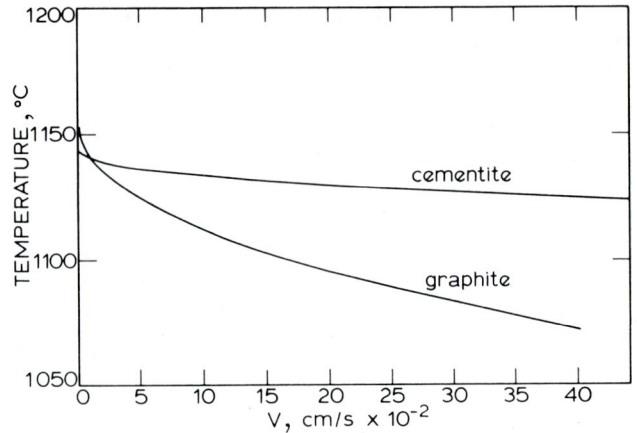

22   Calculated growth rate for cementite and graphite as a function of temperature for eutectic Fe–C alloy

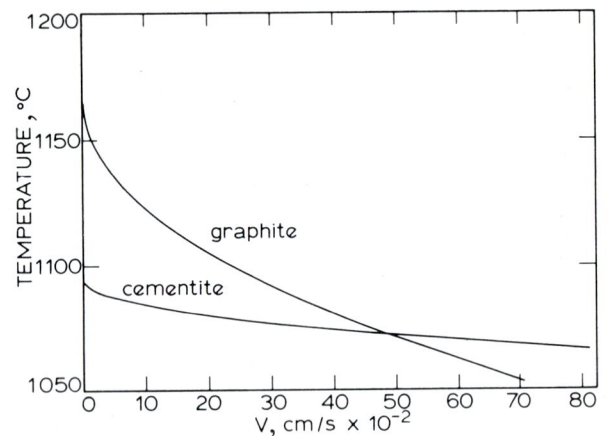

23   Calculated growth rate for cementite and graphite as a function of temperature for eutectic Fe–C alloy with 2 wt-% Si

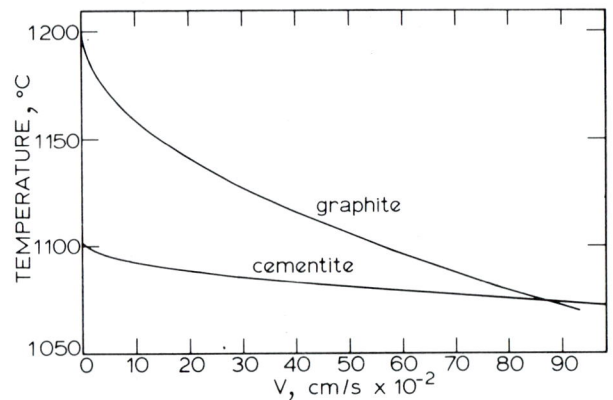

24   Calculated growth rate for cementite and graphite as a function of temperature for eutectic Fe–C alloy with 2 wt-% Al

and the aluminium alloys, the temperature differences are 99° and 131°C, respectively. It can further be seen that the transition to white iron formation occurs for alloys containing no additive, 2% silicon, and 2% aluminium at graphite growth rates of $0·9 \times 10^{-2}$, 49×10^{-2}, and 87×10^{-2} cm/s, respectively. Hence, it is concluded that the probability of white iron formation is considerably reduced in the silicon and aluminium alloys as compared with the unalloyed samples. It can also be seen that the effect of aluminium is definitely stronger. At the same time, however, it must be noted that no allowance was made in these calculations for the possibility of a phase-boundary controlled reaction in connection with the precipitation of graphite.

TRANSFORMATION DIAGRAM

The transitions from grey to white iron formation and from white to grey iron formation have been discussed previously by Hillert and Subba Rao[4] and Fredriksson and Remaeus.[5] In these studies, it was noted that there is a large difference in rate between the two transformations. The same hysteresis effect was observed in the present study in the case of unalloyed cast iron. This hysteresis has previously been attributed to the increasingly unfavourable conditions for graphite and cementite nucleus formation. Earlier discussions have not taken into account the fact that the composition of the cast iron melt will be affected during a eutectic reaction by the growth rate ratio of the two precipitated solid phases. Owing to its rapid growth rate, precipitation of cementite will cause a displacement of the carbon content of the melt towards the L/cementite liquidus, thus reducing the ability of the graphite-precipitation process to compete with it. Similarly, the growth of cementite can be affected by the precipitation of graphite, and it should be noted that precipitation of graphite affects the carbon content of the melt.

CONCLUSIONS

It has been shown that aluminium stabilizes grey solidification of cast iron more than silicon. Aluminium favours precipitation of coarse flake graphite, due to an interphase control reaction. The phase diagrams Fe–C–Al and Fe–C–Si are calculated and the experimentally made observations are explained from those diagrams.

REFERENCES

1 I. C. H. HUGHES: 'The solidification of metals', 184, 1968, London, The Iron and Steel Institute
2 W. OLDFIELD: Recent research on cast iron', Paper presented at the ASM Seminar, 1964, Detroit
3 J. N. DEFRANCQ et al.: *Giesserei–Praxis*, 1970, (3), 34
4 M. HILLERT AND V. V. SUBBA RAO: 'The solidification of metals', 204, 1968, London, The Iron and Steel Institute
5 H. FREDRIKSSON AND B. REMAEUS: 'Second conference on cast iron'; 1975, Geneva, Delta
6 H. MORROGH: *H. BCIRA Res. Dev.*, 1955, **5**, 655
7 P. SAYMA et al.: *J. Crystal Growth*, 1974, **22**, 272
8 K. LÖHBERG AND A. UEBERSCHAER: *Giessereiforschung*, 1969, **21**, (4), 171
9 M. HANSEN: 'Constitution of binary alloys', (2 ed.), 353, 1958, New York, McGraw-Hill
10 E. SCHÜRMANN AND J. VON HIRSCH: *ibid.*, 1966, **18**, 398.
11 M. HILLERT: 'Calculation of phase equilibria': Paper presented at the ASM symposium on phase transformations, October 1968, Detroit
12 J. CHARBONNIER AND J. C. MARGERIE: 'Recent research on cast iron', 389; 1968, New York, Gordon and Breach
13 H. SCHENK et al.: *Giessereiforschung*, 1960 **47**, (2), 25
14 A. HULTGREN et al.: *J. Iron Steel Inst.*, **176**, 1954, 365
15 H. FREDRIKSSON: *Metall. Trans.*, 1975, **6A**, 1 658
16 C. ZENER: *Trans AIME*, 1946, **167**, 550
17 M. HILLERT: *Jernkontorets Ann.*, 1957, **141**, 757

APPENDIX 1

The following equations can be derived for the growth rate of a dendrite tip under the assumption that the growth rate is controlled by diffusion of the elements C and M, respectively.

$$V = \frac{D(X_C^{L/\alpha} - X_C^L)}{R(X_C^{L/\alpha} - X_C^{\alpha/L})} \tag{1}$$

$$V = \frac{D(X_M^{L/\alpha} - X_M^L)}{R(X_M^{L/\alpha} - X_M^{\alpha/L})} \tag{2}$$

In these equations the driving force for the diffusion is $(X_C^{L/\alpha} - X_C^L)$ and $(X_M^{L/\alpha} - X_M^L)$, respectively. The total driving force $({}^\circ X_C^{L/\alpha} - X_C^L)$ and $({}^\circ X_M^{L/\alpha} - X_M^L)$, respectively, where ${}^\circ X_C^{L/\alpha}$ and ${}^\circ X_M^{L/\alpha}$ are equilibrium values from the phase diagram, must be shared between surface energy and energy for diffusion. In a system with three components the total driving force can be made by using the equation

$$X_C^{L/\alpha} - X_C^L + X_M^{L/\alpha} - X_M^L$$
$$= \left(1 - \frac{R^*}{R}\right)({}^\circ X_C^{L/\alpha} - X_C^L + {}^\circ X_M^{L/\alpha} - X_M^L) \tag{3}$$

where R^* is the critical radius of the dendrite tip, when all the driving force is used for surface energy.

Using equations (1)–(3) above, the equation

$$V = \frac{D}{R} \frac{[({}^\circ X_C^{L/\alpha} - X_C^L) + ({}^\circ X_M^{L/\alpha} - X_M^L)]}{[(X_C^{L/\alpha} - X_C^{\alpha/L}) + (X_M^{L/\alpha} - X_M^{\alpha/L})]} \left(1 - \frac{R^*}{R}\right) \tag{4}$$

is obtained. The maximum value of V is obtained at $R = 2R^*$, where

$$R^* = \frac{\sigma V_m}{(-\Delta G_m)} = \frac{\sigma V_m T_O}{\Delta H_m \Delta T} \tag{5}$$

and this gives

$$V_{max} = \frac{D \Delta H_m \Delta T}{4 \sigma V_m T_O} \frac{({}^\circ X_C^{L/\alpha} - X_C^L + {}^\circ X_M^{L/\alpha} - X_M^L)}{(X_C^{L/\alpha} - X_C^{\alpha/L} + X_M^{L/\alpha} - X_M^{\alpha/L})} \tag{6}$$

APPENDIX 2

For eutectic growth in a system with the three components, Fe, C, and M, the driving force for diffusion can be obtained in the following way

$$X_C^{L/\alpha} - X_C^{L/\beta} + X_M^{L/\alpha} - X_M^{L/\beta}$$
$$= \left(1 - \frac{s^*}{s}\right)({}^\circ X_C^{L/\alpha} - X_C^{L/\beta} + {}^\circ X_M^{L/\alpha} - {}^\circ X_M^{L/\beta}) \tag{7}$$

where s^* is the critical interlamellar spacing, when all the driving force is used for surface energy.

The following equations can be derived for the growth rate controlled by diffusion

$$V_C^\alpha = D^L \frac{V_m^s}{V_m^L} \frac{X_C^{L/\alpha} - X_C^{L/\beta}}{X_C^{L/\alpha} - X_C^{\alpha/L}} \frac{1}{s^\alpha} \tag{8}$$

and

$$V_M^\alpha = D^L \frac{V_M^s}{V_M^L} \frac{X_M^{L/\alpha} - X_M^{L/\beta}}{X_M^{L/\alpha} - X_M^{\alpha/L}} \frac{1}{s^\alpha} \qquad (9)$$

Also, the following equations must be valid during the solidification

$$f^\alpha (X_C^{L/\alpha} - X_C^{\alpha/L}) = f^\beta (X_C^{\beta/L} - X_C^{L/\beta})$$
$$= f^\alpha f^\beta (X_C^{\beta/L} - X_C^{\alpha/L}) \qquad (10)$$

$$f^\alpha (X_M^{L/\alpha} - X_M^{\alpha/L}) = f^\beta (X_M^{\beta/L} - X_M^{L/\beta})$$
$$= f^\alpha f^\beta (X_M^{\beta/L} - X_M^{\alpha/L}) \qquad (11)$$

$$f^\alpha = \frac{s^\alpha}{s} \qquad (12)$$

Inserting equations (10) and (12) into equation (8) gives

$$V_C^\alpha = D \frac{V_M^s}{V_M^L} \frac{(X_C^{L/\alpha} - X_C^{L/\beta})}{(X_C^{\beta/L} - X_C^{\alpha/L})} \frac{1}{f^\alpha f^\beta s} \qquad (13)$$

and in the same way inserting equations (11) and (12) into equation (9) gives

$$V_M^\alpha = D \frac{V_M^s}{V_M^L} \frac{(X_M^{L/\alpha} - X_M^{L/\beta})}{(X_M^{\beta/L} - X_M^{\alpha/L})} \frac{1}{f^\alpha f^\beta s} \qquad (14)$$

The growth rate V must equal $V_C^\alpha = V_M^\alpha$, which means that V can be derived from equations (7), (13), and (14)

$$V = D \frac{V_M^s}{V_M^L} \frac{(^\circ X_C^{L/\alpha} - ^\circ X_C^{L/\beta} + ^\circ X_M^{L/\alpha} - X_M^{L/\beta})}{(X_C^{\beta/L} - X_C^{\alpha/L} + X_M^{\beta/L} - X_M^{\alpha/L})} \frac{1 - (s^*/s)}{f^\alpha f^\beta s} \qquad (15)$$

The maximum growth rate is derived at $s = 2s^*$, where

$$s^* = \frac{2\sigma V_M^L}{(-\Delta G_M)} = \frac{2\sigma V_M^L T_O}{\Delta H_M \Delta T} \qquad (16)$$

and this gives

$$V_{max} =$$

$$\frac{D \Delta H_M \Delta T}{8\sigma V_M^L T_O} \frac{V_M^s}{V_M^L} \frac{(^\circ X_C^{L/\alpha} - ^\circ X_C^{L/\beta} + X_M^{L/\alpha} - ^\circ X_M^{L/\beta})}{(X_C^{\beta/L} - X_C^{\alpha/L} + X_M^{\beta/L} - X_M^{\alpha/L})} \frac{1}{f^\alpha f^\beta} \qquad (17)$$

Role of interphase boundary adsorption in the formation of spheroidal graphite in cast iron

W. C. Johnson and H. B. Smartt

The interphase boundaries in grey and ductile cast iron were studied with a scanning Auger microprobe. Sulphur and oxygen were found to be adsorbed at the flake/metal interfaces in the grey iron, while the nodule/metal and intercrystalline graphite interfaces in the ductile iron were free of foreign elements. The only magnesium detected in the magnesium-modified ductile iron was combined with phosphorus and sulphur as a compound. A model is presented which proposes that Fe–C–Si eutectic alloys in the absence of surface-active impurities (e.g. in vacuum casting of high-purity materials) produce nodular graphite owing to the inherent instability of the graphite/melt interface. The sulphur and oxygen always present in commercial alloys adsorb at the graphite/melt interface, effectively 'stabilizing' the active sites on the graphite basal planes, and preventing spherulitic growth. The purpose of modifiers is to getter these impurities.

W. C. Johnson is with Physical Electronics Ind., Inc., Minnesota, and H. B. Smartt is with the Ford Motor Company, Michigan, USA

Although the modification of cast iron from grey iron to ductile iron has been studied extensively,[1-4] the mechanism by which the graphite morphology is altered by magnesium and/or rare earth additions is still not well understood. Most of the extant theories assume subtle compositional changes in either the graphite/metal or intercrystalline graphite boundaries,[1-3,5-12] but the analysis techniques used previously, such as the electron microprobe,[5] autoradiography,[11] or sessile drop measurements[9] are not capable of successfully detecting the spatially limited chemical variations presumed to be responsible for the changes. Recently, however, a scanning Auger microprobe has been used to determine directly the interfacial compositions of grey and ductile irons.[13-15] In this paper, the authors intend to demonstrate the utility

of the Auger electron spectroscopy (AES) technique in studying the interphase boundary composition of cast iron; and to propose a modification model which is consistent with the observed impurity adsorption.

EXPERIMENTAL

Both the grey and ductile irons examined (as-cast microstructures in Fig. 1) were produced from induction-melted base iron, stream inoculated with FeSi,[16] and cast into greensand moulds. The ductile iron was modified with a commercial Mg–Fe–Si alloy. Bulk chemical analysis of both alloys (Table 1) indicated that they were typical commercial cast irons.

AES has been applied before to the study of grain-boundary segregation by examining intergranular fracture surfaces.[17-19] If the whole fracture surface is of

Table 1 Composition of cast iron

Type of cast iron	Composition, wt-%											
	C	Si	Mn	S	P	Mg	O	Cr	Ni	Cu	Mo	V
Ductile iron	3·28	2·2	0·58	0·004	0·020	0·09	0·0016	0·02	0·05	0·01	0·01	0·03
Grey iron	3·64	2·5	0·55	0·04	0·032	—	0·013	0·02	0·09	0·1	0·02	0·02

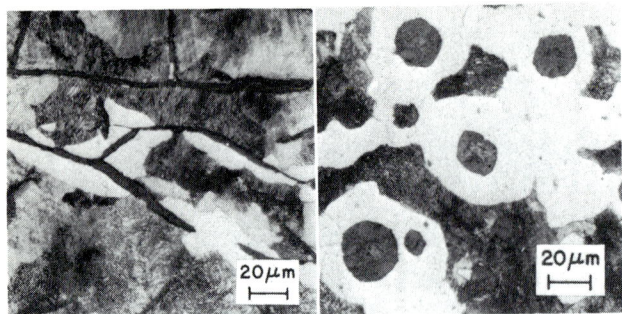

a microstructure of grey iron; *b* microstructure of ductile iron

1   Structure of cast iron

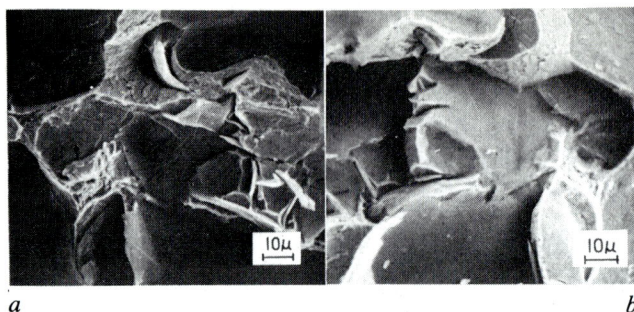

a SEM image of fracture surface; *b* SEM image of opposite fracture surface

3   High magnification SEM micrographs of grey iron fracture

uniform intergranular character, the average surface composition obtained over the area covered by the exciting electron beam ($>200 \mu$m in dia.), would represent the grain-boundary composition. However, in order to study the composition of the graphite/metal interfaces in cast iron, it is apparent from Fig. 2 that the fracture surface has to be probed with a surface analysis technique with adequate spatial resolution to resolve the areas of interphase boundary fracture. By using a highly focused electron beam for analysis, the scanning Auger microprobe or SAM, was developed.[20] Analogous to the standard electron microprobe, the SAM can be used to obtain images, chemical maps, and spot analyses of a surface with a spatial resolution of 5 μm, but utilizes Auger electrons instead of X-rays for chemical analysis, and therefore analyzes only the top 2–3 atomic layers.[20–22]

After *in situ* fracture in an ultra-high vacuum system (pressure less than 5×10^{-10} torr to prevent contamination of the freshly exposed surfaces), the exposed graphite/metal interfaces were thoroughly analyzed with the SAM, including chemical mapping and spot analysis. Inert-ion sputter etching was used to determine the depth extent of the chemical varia-

tions[23,24] and a scanning electron microscope (SEM) was used to provide the high-resolution micrographs necessary for interpretation of the SAM data.

RESULTS
Grey iron
Figure 3*a* shows a typical graphite/metal interfacial fracture (region marked with an arrow) from a grey iron specimen. The metal above and below the interfacial fracture area exhibits a torn appearance, indicating ductile fracture. A micrograph of the opposite fracture surface, (Fig. 3*b*), shows the cleavage step in the graphite flake corresponding to the cleavage portion of the flake in Fig. 3*a*, (Figs 3*a* and *b* fit together as mirror images).

Analyses of this fracture surface using SAM demonstrates the necessity of good spatial resolution in a study of this type. The SAM image, (Fig. 4*b*), although not as sharp as the SEM image of the same area, (Fig. 4*a*) owing to the larger beam, still shows the fracture details. The Auger maps for iron (Fig. 4*c*), and carbon, (Fig. 4*d*), fit together as positive and negative, while the sulphur, (Fig. 4*e*), and the oxygen maps (Fig. 4*f*), indicate that these elements are located in discrete patches on the surface.

It is evident from the Auger spot analyses, (Fig. 5), that the interfacial fracture region, (Point 1), has both sulphur and oxygen on the surface. A rough calculation of the impurity coverages at this interphase boundary yields 20at.-% oxygen and 5at.-% sulphur, and inert-ion sputter etching equivalent to removing about 5 atomic layers of iron removed all traces of the impurities, indicating that they were highly localized. The spot analysis from the adjacent ductile fracture region, (Point 2), establishes that the impurities detected at Point 1 do not result from contamination occurring after fracture, but rather are characteristic of the interphase boundary before fracture. The cleaved graphite flake, (Point 3), showed just carbon. When a similar sample was annealed for 5 h at 1 100°C (48°C below the Fe–C eutectic temperature) and quenched, the interphase boundaries showed similar amounts of sulphur and oxygen, indicating that the impurity adsorption took place during solidification, rather than during cooling.

Ductile iron
Figures 6*a*–6*g* show a SEM–SAM composite analysis for a ductile iron fracture surface. The Auger maps for

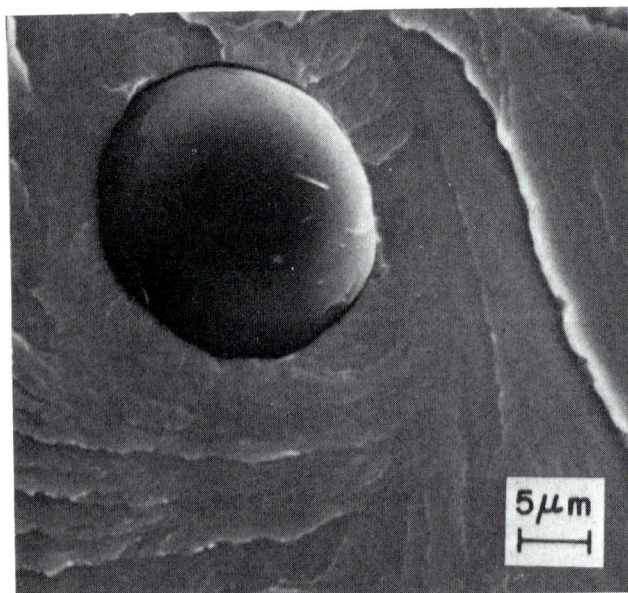

2   SEM micrograph of ductile iron fracture surface

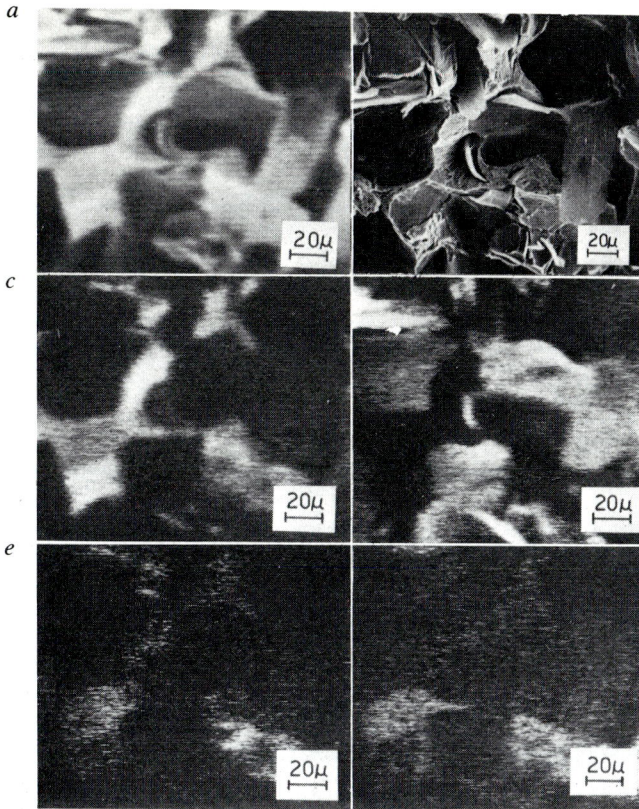

a SAM absorbed current image; *b* SEM image; *c* iron Auger map; *d* carbon Auger map; *e* sulphur Auger map; *f* oxygen Auger map

4 SEM/SAM analysis of grey iron fracture surface

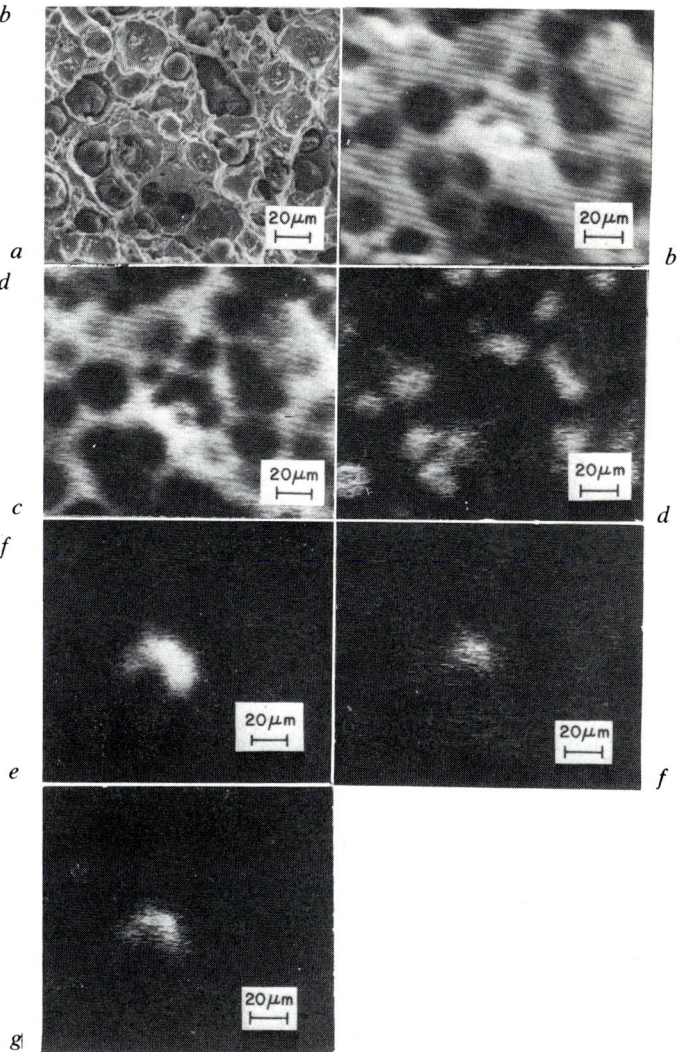

a SEM image; *b* SAM absorbed current image; *c* iron Auger map; *d* carbon Auger map; *e* phosphorus Auger map; *f* sulphur Auger map; *g* magnesium Auger map

6 SEM/SAM analysis of ductile iron fracture surface

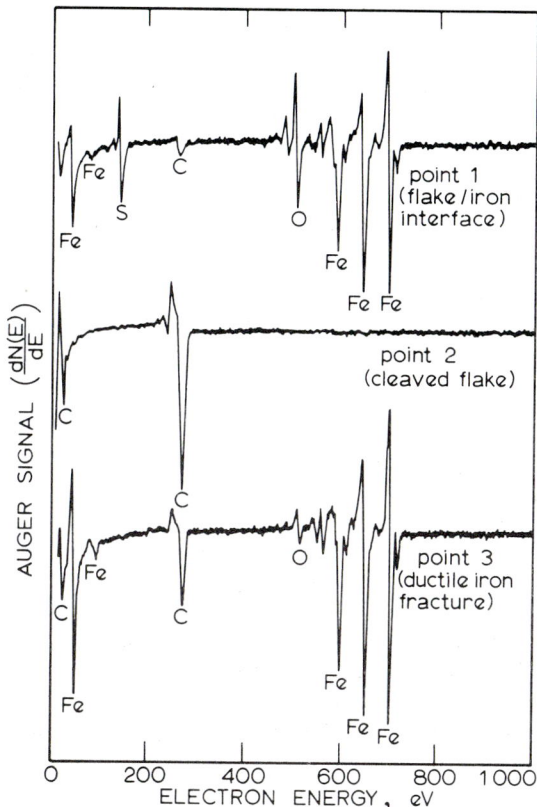

5 Auger spot analysis of grey iron fracture surface

iron, (Fig. 6*c*), and carbon, (Fig. 6*d*), again appear as positive and negative. The phosphorus, (Fig. 6*e*), sulphur, (Fig. 6*f*), and magnesium, (Fig. 6*g*), Auger maps show that these elements occur together in a spot at the centre of the maps.

Point 1 from the spot analyses, (Fig. 7), corresponding to the spots rich in phosphorus, sulphur and magnesium showed a high concentration of these three elements. Inert-ion sputter etching equivalent to 200 layers of iron removed did not alter the surface composition of this area, and a high-magnification SEM micrograph, (Fig. 8*a*), reveals that a Mg–P–S precipitate is responsible. The magnesium Auger peak was found to be shifted from its customary energy of 1 186 eV for elemental magnesium, to 1 182 eV, which reflects the formation of a magnesium compound.[25]

Point analysis 2, from an area with fractured nodules and nodules which pulled loose cleanly (Fig. 8*b*), shows carbon and a little iron, but no impurities or modifying elements. The other half of the graphite/metal interphase boundary (the crater left

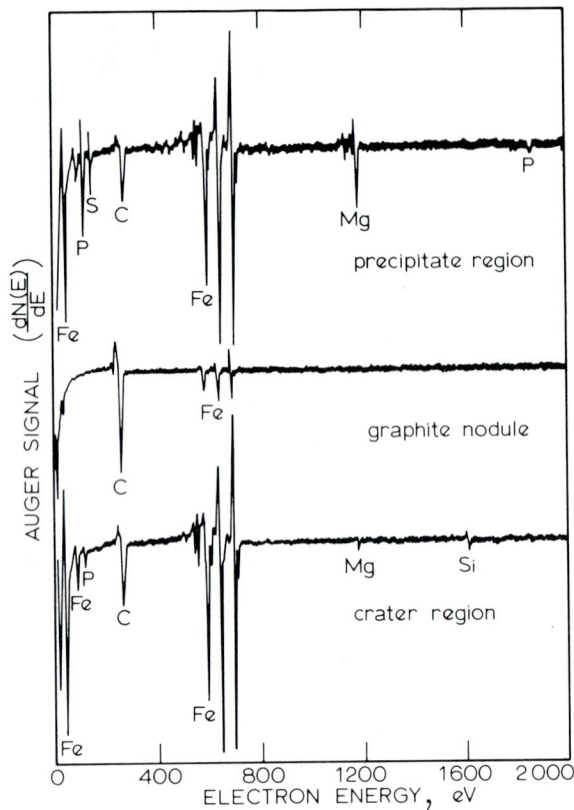

7 Auger spot analysis of ductile iron fracture surface

when the nodule pulled loose), (Fig. 8c), showed a little carbon residue from fracture, as well as some fine Mg–P precipitate particles. Once again, the magnesium Auger peak had the 4 eV chemical shift characteristic of compounds. The only magnesium detected on ductile iron fracture surfaces was always combined with phosphorus and/or sulphur, as a compound.

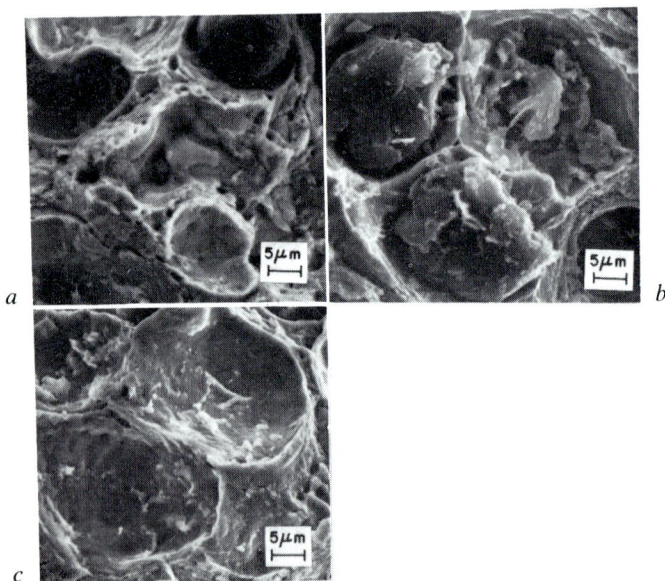

a high magnesium, phosphorus, and sulphur region; *b* broken graphite nodule region; *c* metal half of graphite/metal interphase boundary

8 SEM micrographs of ductile iron spot analysis areas

The origin of the iron signal in the Auger spot analyses was at first confusing, since all nodules showed it, but flakes from grey iron did not; however, back-scattered electron SEM images clearly showed that it was caused by metal inclusions trapped in the nodules during solidification.[26,27]

DISCUSSION
Spherulitic growth model

Spherulitic growth is fairly common in many systems, particularly polymers,[28–30] as well as the Fe–C, Ni–C, and Co–C metal systems. In all instances, the nucleation of the nodules takes place in the liquid, the growing particles remain thermally isolated, and the non-metallic phase is the one that forms nodules.[2] Owing to this thermal isolation, spherically symmetric solute and thermal fields are set up around the nucleus, and the only sink for the latent heat of solidification is the melt. The growing particle can then become hotter than the ambient liquid and a negative temperature gradient will exist at the growth interface.[30]

Under these conditions, perturbations on the surface will encounter cooler liquid and will be favoured relative to the rest of the interface for continued growth.[31] For small particles, capillary forces will damp out these perturbations. Mullins and Sekerka[32] have derived an expression for the critical radius R_C beyond which a particle such as this becomes unstable

$$R_C = \frac{2(\gamma_{SL}/L_V)(7 + 4k_S k_L)}{(T_M - T_\infty)/T_M} \quad (1)$$

where

γ_{SL} = graphite interfacial free energy
L_V = latent heat of solidification per unit volume
k_S = thermal conductivity of the solid
k_L = thermal conductivity of the liquid
T_M = solidification temperature
T_∞ = ambient temperature

The fractional undercooling, ΔT_U is defined as

$$\Delta T_U = (T_M - T_\infty)/T_M \quad (2)$$

Typical values for the parameters in equation (1) at 1 000°C are

$\gamma_{SL} = 1\,500$ erg/cm^2 = $3\cdot6 \times 10^{-5}$ cal/cm^2 (Ref. 9)
$L_V = 8\,831$ cal/mole = 1 660 cal/cm^3 (Ref. 33)
$k_S = -0\cdot08$ to $-0\cdot12$ cal/s cm K (Refs 34 and 35)
$k_L = -0\cdot03$ to $-0\cdot07$ cal/s cm K (estimated from Refs 36–38)

These values yield

$$R_C = \frac{7 \pm 2 \times 10^{-7}\,\text{cm}}{\Delta T_U} = \frac{70 \pm 20\,\text{Å}}{\Delta T_U} \quad (3)$$

For a typical fractional undercooling of 15% ($T_\infty = 1\,273$°K and $T_M = 1\,473$°K), particles with radii greater than about 300–500 Å will no longer be stabilized by capillary forces and perturbations on these particles will grow rather than decay. Since graphite nucleates heterogeneously from 100 Å carbon aggregates in the melt,[5] capillary forces will not act as a stabilizing factor for very long.

Since the graphite particles are small and thermally isolated, the perturbations that grow out into the melt will be at the same temperature as the rest of the particle and the Fe-rich liquid between protrusions will solidify fairly slowly. This will lead to long protrusions from the particle which, due to spherical symmetry, would produce a star-shaped spheroid; however, eventually the cavities fill in owing to branching and the star-shaped particle becomes a macroscopically smooth spherulite. This rapid growth and branching traps pockets of liquid within the spheroid, which solidify as metal inclusions, as observed in this study and others.[26,27]*

Continued growth defects during graphite growth make renucleation of basal plane surfaces unnecessary. Owing to the large difference in size between the metal atoms and the carbon atoms, non-carbon atoms incorporated in the lattice generate substantial stresses, which produce a high density of screw dislocations ($>10^3/cm^2$)[39] along with splitting and tearing of the growing structure. The screw dislocations produce growth spirals on the basal planes, which can rapidly extend the crystal in the [0001] direction by growth in the [10$\bar{1}$0] direction,[40,41] while splitting and branching of the graphite,[42–44] combined with its high flexibility,[44] allows the relative orientation of the basal planes to change during growth. Thus, it is possible for a pseudopolycrystalline nodule to grow from a single graphite nucleation event, with the basal planes tangential to the surface of the spheroid.[45]

Effect of impurity adsorption

The area covered by the active sites on a basal plane surface is only a small fraction of the total area; hence, the growth ledges are more sensitive to a small amount of adsorbed impurity than the edge of the crystal, which consists entirely of active sites.[41] Impurity adsorption, such as the oxygen and sulphur observed in the present study, can easily 'poison' the growth ledges on the basal planes. Blazinakov[46] and Chernov[31,41,47,48] have shown how impurities adsorbed on active sites can have a strong effect on the growth rate. By slowing the propagation of the basal plane growth defects, the interface is essentially stabilized by the adsorption (in contrast with the stabilization induced by capillary force observed for very small particles). The growth perturbations cannot utilize the thermal gradient at the interface to develop into long protrusions, since the impurity drag effect reduces the difference between the growth rates of the active sites and the rest of the interface.[46]

The edges of the crystal perpendicular to the basal plane continue to grow, even with impurity adsorption, since renucleation, growth spirals, and instability are not important for the growth of these planes. Carbon atoms can be added directly on to these planes, since the whole plane is active, and therefore a decrease in growth rate due to impurity adsorption does not change the basic growth mechanism. The graphite particle then extends in the [10$\bar{1}$0] direction, and thickens in the [0001] direction, producing a flake.

---

* Such long protrusions do form in certain polymer systems.[28–30] In metal–graphite systems, the graphite basal planes are tangential to the spheroidal surface, and thus the fast-growing [10$\bar{1}$0] direction develops parallel to the surface of the nodule and the slow-growing [0001] direction develops perpendicular to this surface. Thus, the surface fills in rapidly between the protrusions, and the resulting spheroid has a surface smoother than those typically found in polymers

Modification and vacuum casting

This study suggests that the normal form of graphite in cast iron is as spherulites. This is consistent with the results of investigations which have shown that spherulitic graphite can be obtained from ultra-high purity vacuum melting without the use of modifiers.[49–51] Although it may be argued that the starting materials contained just enough magnesium and/or rare earth elements to act as modifiers, the fact that no traces of modifier adsorption were detected in this study, indicates that the direct intervention of these elements at the growth interface is not supported by the experimental evidence.

This model is also consistent with the observed effects of modifiers. These additives chemically combine with the oxygen and sulphur in the melt, preventing them from adsorbing on the graphite. The affinity of a particular scavenger element for oxygen and/or sulphur as well as the stability of the compound formed determines the effectiveness of a modifier, which explains why magnesium–rare earth alloys modify better than magnesium or rare earths alone.[52,53]

Another factor which determines the effectiveness of a modifier is its resistance to 'fade' (the loss of modifying potency as a molten alloy is held before casting).[5] The combination of atmospheric contamination of the molten alloy and vaporization of the modifier, proceeds until the magnesium present is insufficient to combine chemically with all of the impurities, and the alloy then solidifies as grey iron. Some of the particles formed by the modifiers serve as graphite nuclei, while the remainder either float out of the melt or are retained as inclusions. The fact that only Mg–P–S precipitates were detected in this study suggests that the other compounds formed are of such a size or specific gravity that they float out of the melt.

Vacuum-cast alloys rely on the purity of the starting materials and the degassing effect of the vacuum environment to produce low impurity concentrations in the melt, and hence, spheroidal graphite. Because this process cannot be controlled as easily as gettering the impurities, mixed morphologies can develop.

ACKNOWLEDGMENTS

The authors would like to thank N. G. Chavka, G. L. Parrott, and F. E. Alberts for technical assistance, and B. V. Kovacs for many helpful suggestions. In addition, they are indebted to N. A. Gjostein, C. L. Magee, K. R. Kinsman, A. Hellawell, P. P. Wynblatt, R. W. Clark, and W. L. Winterbottom for critical evaluations of the manuscript.

REFERENCES

1 B. LUX: *Giessereiforsch.* (*in English*), 1970, **22**, 158
2 B. LUX: *ibid.*, 1970, **22**, 65
3 S. BANNERJEE: *Br. Foundryman*, 1965, **58**, 344
4 H. MORROGH: *Trans. Am. Foundryman's Soc.*, 1948, **56**, 72
5 H. D. MERCHANT: 'Recent research on cast iron', (ed. H. D. Merchant), 33, 1968, New York, Gordon and Breach
6 F. H. BUTTNER *et al.*: *Am. Foundryman*, 1951, **20**, 49
7 I. MINKOFF AND B. LUX: *AFS Cast Metals Res. J.*, 1970, **6**, 181
8 J. C. SAWYER AND J. F. WALLACE: *ibid.*, 1969, **5**, 83

9 R. H. McSWAIN *et al.*: *ibid.*, 1974, **10**, 181

10 H. M. WELD *et al.*: *J. Met.*, 1952, **4**, 738

11 W. A. SPINDLER AND R. A. FLINN: *Trans. Am. Foundryman's Soc.*, 1962, **70**, 1 017

12 G. T. VANROOYEN AND G. PAUL: *Met. Sci.*, 1974, **8**, 370

13 W. C. JOHNSON *et al.*: *Scripta Metall.*, 1974, **8**, 1 309

14 W. C. JOHNSON AND H. B. SMARTT: *ibid.*, 1975, **9**, 1 205

15 W. C. JOHNSON AND H. B. SMARTT: *Metall. Trans. A*, 1977, **8A**, 553

16 H. B. SMARTT *et al.*: Unpublished research

17 W. C. JOHNSON AND D. F. STEIN: *Metall. Trans.*, 1974, **5**, 549

18 D. F. STEIN *et al.*: *Can. Met. Q.*, 1974, **13**, 79

19 A. JOSHI AND D. F. STEIN: *Metall. Trans.*, 1970, **1**, 2 543

20 N. C. McDONALD *et al.*: *Res. Develop.*, 1976, **27**, (8), 42, 48, 50

21 C. C. CHANG: *Surf. Sci.*, 1971, **25**, 53

22 W. C. JOHNSON *et al.*: *Can. J. Spectroscop.*, 1972, **17**, 88

23 G. K. WEHNER AND D. J. HAJICEK: *J. Appl. Phys.*, 1971, **42**, 1 145

24 G. K. WEHNER: 'Methods and phenomena I–methods of surface analysis', (ed. A. W. Czandrna), 1975, New York, Elsevier Scientific Publishing Co

25 C. D. WAGNER: *Anal. Chem.*, 1975, **47**, 1 201

26 I. MINKOFF AND W. C. NIXON: *J. Appl. Phys.*, 1966, **37**, 4 848

27 VON PETER WARBICHLER AND W. GEYMAYER: *Giessereiforsch.*, 1973, **25**, 29

28 H. D. KEITH AND F. J. PADDEN JR.: *J. Appl. Phys.*, 1964, **35**, 1 270

29 H. D. KEITH AND F. J. PADDEN JR.: *ibid.*, 1964, **35**, 1 286

30 H. D. KEITH AND F. J. PADDEN JR.: *ibid.*, 1963, **34**, 2 409

31 A. A. CHERNOV: *Sov. Phys. Cryst.*, 1972, **16**, 734

32 W. W. MULLINS AND R. F. SEKERKA: *J. Appl. Phys.*, 1963, **34**, 323

33 R. HULTGREN *et al.*: 'Selected values of the thermodynamic properties of binary alloys', 485, 1973, Metals Park, Ohio, American Society for Metals

34 A. R. UBBELOHDE AND F. A. LEWIS: 'Graphite and its crystal compounds', 58, 1960, Oxford, Clarendon Press

35 A. GOLDSMITH *et al.*: 'Handbook of thermophysical properties of solid materials–revised edition–volume I: Elements', 105, 115, 1961, New York, The Macmillan Co.

36 *ibid.*, 357

37 A. GOLDSMITH *et al.*: 'Handbook of thermophysical properties of solid materials, revised edition–Vol. II: Alloys', 64, 1961, New York, The Macmillan Co.

38 Y. S. TOULOUKIAN *et al.*: 'Thermal conductivity-metallic elements and alloys', 1 128, 1970, New York, Plenum Press

39 C. ROSCOE *et al.*: *J. Mater. Sci.*, 1971, **6**, 998

40 F. C. FRANK: *Disc. Faraday Soc.*, 1949, **5**, 48

41 A. A. CHERNOV: *Sov. Phys. Uspekhi*, 1961, **4**, 116

42 B. LUX: *AFS Cast Met. Res. J.*, 1972, **8**, 49

43 B. LUX *et al.*: 'The metallurgy of cast iron', (ed. B. Lux, I. Minkoff, and F. Mollard), 475, 1975, St. Saphorin, Switzerland, Georgi Publishing Co.

44 A. HELLAWELL: personal communication

45 M. J. HUNTER AND G. A. CHADWICK: *J. Iron Steel Inst.*, 1972, **210**, 117

46 G. BLIZNAKOV: *Z. Phys. Chem. (Leipzig)*, 1958, **209**, 372

47 A. A. CHERNOV: *Sov. Phys. Dok.*, 1960, **5**, 654

48 A. A. CHERNOV: *Sov. Phys. Dok.*, 1960 **5**, 470

49 P. M. THOMAS AND J. E. GRUZLESKI: *J. Iron Steel Inst.*, 1973, **211**, 426

50 M. HONMA AND A. MINATO: Imono, *J. Jpn Foundryman's Soc.*, 1959, **31**, 1 064

51 J. P. SADOCHA AND J. E. GRUZLESKI: 'The metallurgy of cast iron', (ed. B. Lux, I. Minkoff, and F. Mollard), 443, 1975, St. Saphorin, Switzerland, Georgi Publishing Co.

52 M. J. LALICH: 'The metallurgy of cast iron', (ed. B. Lux, I. Minkoff, and F. Mollard), 561, 1975, St. Saphorin, Switzerland, Georgi Publishing Co.

53 W. G. WILSON *et al.*: *J. Met.*, 1974, **26**, 14

Transition from peritectic to eutectic reaction in iron-base alloys

H. Fredriksson

Steels solidify either to ferrite or to austenite depending on the contents of different alloying elements. The addition of elements which stabilize austenite gives rise to a peritectic reaction during the solidification process. A primary precipitation of austenite occurs when the content of austenite stabilizing elements is high. The addition of ferrite stabilizer to such an alloy produces a eutectic reaction. This very complex reaction in alloyed steels is discussed, and experimental examples will be taken from alloys in the Fe–Ni–Cr–Mo system.

The author is with the Department of Casting of Metals, The Royal Institute of Technology, Stockholm, Sweden

LIST OF SYMBOLS

f = volume fraction
$^{\circ}G^{\alpha} - {}^{\circ}G^{\beta}$ = free energy difference between α- and β-phase; for the different phases α, γ, L these have been taken from Refs. 1, 2
X = mole fraction
$X^{\alpha/\beta}$ = mole fraction at phase boundary α/β in the α-phase

The solidification process of steel normally starts with a primary precipitation of ferrite or austenite. In the former case many steels go through a peritectic reaction during the solidification process. Alloying elements in steel either stabilize ferrite or austenite. The choice and amount of alloying element may then influence the solidification process. The binary phase diagram of iron-base alloys, with an austenite stabilizing element such as carbon, nitrogen, nickel or manganese, always shows a peritectic reaction. Steels normally contain one or two of these elements. Most iron-base alloys also contain ferrite stabilizing elements such as chromium, molybdenum, silicon, and/or tungsten. Those ferrite stabilizing elements have a large influence on the peritectic reaction as recently shown by the author.[3,4] Stainless steels mainly contain Ni, Cr, N, and sometimes Mo, i.e. two austenite stabilizing elements and one or two ferrite stabilizing elements. Stainless steels offer a unique possibility to study the influence of ferrite stabilizing elements on the peritectic reaction. An investigation of the peritectic reaction in stainless steels has been made in this work using the controlled solidification technique. The result will be reported and theoretically analysed.

EXPERIMENTAL

The controlled solidification experiments were performed in the following way. The alloy was melted in an alumina crucible and then sucked into a thin-walled alumina tube of 1·7 mm i.d. The tube was placed in a hot furnace with a steep temperature gradient and because of this the alloy in the tube was partly remelted. The tube was removed at a constant rate through the gradient in which solidification took place in a reasonably well defined zone. The experiments were interrupted suddenly by quenching the tube in brine and the remaining melt then solidified to a very fine structure which could be distinguished metallographically from the solidification structure formed at the constant rate.

To distinguish between austenite and ferrite in the microstructure some of the samples were colour etched. The samples were first well prepared metallographically and then heat treated in air at 480°C for 30 min. By this treatment the austenite achieves a red–brown colour and ferrite light yellow.

In order to investigate the segregation profile of the alloying elements, microprobe analyses were made in some samples. The specimens were then polished before being analysed. The measured intensity was corrected with well developed computer programs.

In this report the experimental results from six different alloys will be discussed (Table 1). The alloys

Table 1   Experimental results from six alloys

Alloy	Cr	Ni	Mo	N
1	8·8	8·8	0·55	0·06
2	17·7	11·7	2·65	—
3	18·0	13·8	2·61	—
4	20·4	15·7	2·88	—
5	20·4	16·0	6·26	0·09
6	20·2	20·4	9·8	—

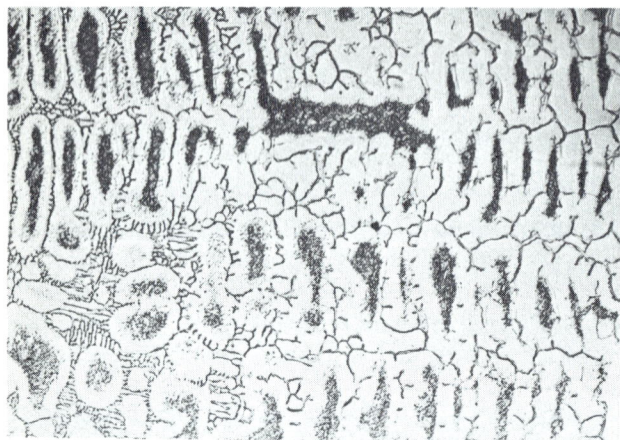

1   Start of austenite precipitation in 18/8 steel unidirectionally solidified at 1·2 cm/min[1]

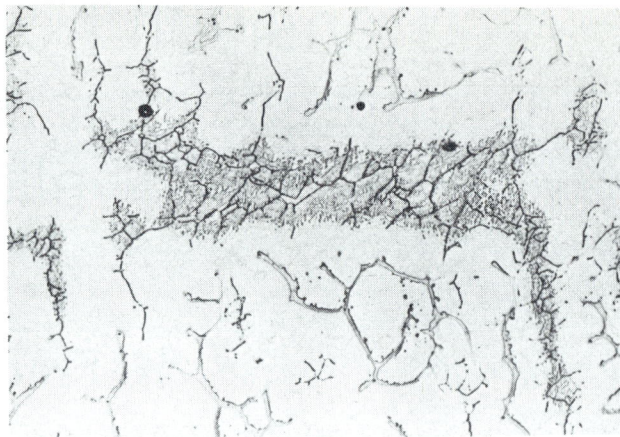

2   Microstructure at very last solidified part of sample shown in Fig. 1[1]

were chosen from three different investigations made at the department of Casting of Metals at the Royal Institute of Technology, Stockholm.[3,5,6]

RESULTS
0% Mo
In normal 18/8 steel the solidification process starts with a primary precipitation of ferrite followed by a secondary precipitation of austenite. The start of the austenite precipitation will be determined either by the content of Ni and N in the sample, or by the solidification rate.[3]

Figure 1 shows the start of the austenite precipitation in an 18/8 steel with 0·06% N (Table 1). The sample has solidified at a rate of 1·2 cm/min. The dark etching two-phase structure represents ferrite dendrite arms transformed to Widmanstätten austenite during the quench; around them an austenite border has been formed. The austenite is light etching. To the left, a fine dendrite structure is shown between the large dendrite arms. This fine structure has been formed during the quench and indicates that one-third of the volume in this part of the sample was liquid at the moment of quenching. To the right in Fig. 1 this fine dendrite structure cannot be observed, which means that the solidification process has been finished within a small temperature range.

It can be seen in Fig. 2, which shows an area in the right-hand side of Fig. 1 at a larger magnification, that in the last solidified parts of the sample a network of ferrite has been formed. This structure can be followed much further away from the solidification front and thus it may safely be concluded that it formed before the quench. Evidently, there have been two transitions

during the solidification, the first from ferrite to austenite, the second back to ferrite. It may be suggested that the first one is mainly a result of the increased stability of austenite relative to ferrite as the temperature drops, and that the second is owing to segregation of a ferrite stabilizing element during the growth of austenite. At the same moment as austenite appears it also starts to grow into the primary ferrite and the volume fraction of the ferrite is drastically lowered.

The concentration profiles of chromium and nickel were measured across a dendrite stem after the start of the austenite formation (Fig. 3). The diagram shows that the ferrite remaining before quenching has fairly

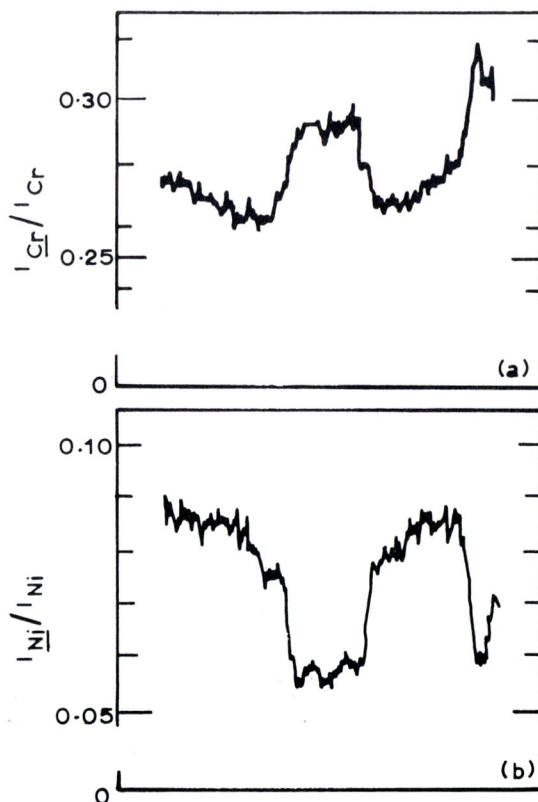

3   Distribution of *a* chromium, *b* nickel over dendrite cross just behind end of solidification process; arrows indicate centre of dendrite cross[1]

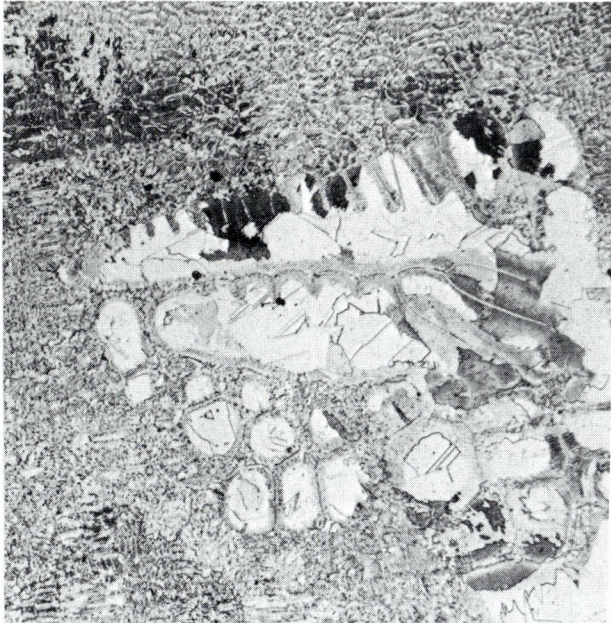

4 **Solidification front in unidirectionally
solidified stainless steel, 0·05 cm/min (alloy 2,
Table 1) [×100]**

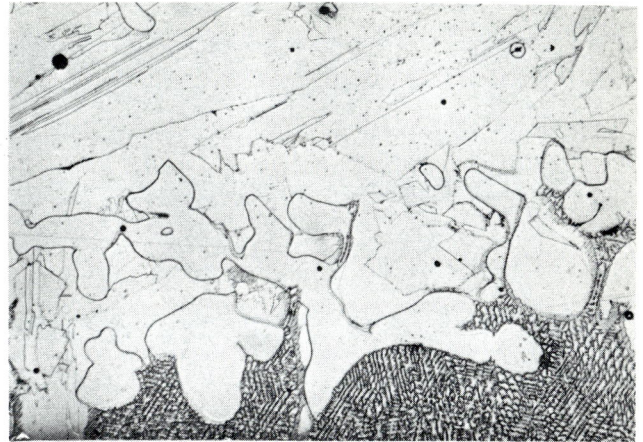

5 **Start of austenite precipitation in same
sample as in Fig. 4 [×65]**

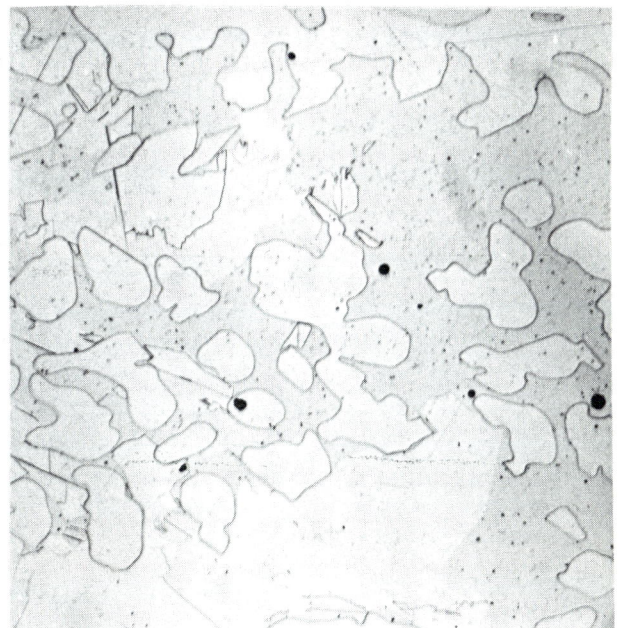

6 **Structure at very last solidified parts in same
sample as in Fig. 4 [×100]**

constant nickel and chromium contents. The further
growth of austenite into the ferrite, which has occurred
during the quench, has not affected the distribution
markedly. The austenite present before the quench
shows a strong variation of nickel as well as chromium.
It is difficult to make a detailed interpretation of these
profiles because the austenite has grown in two
opposite directions, into the primary ferrite and into
the remaining melt. The latter process should lead to an
enrichment of nickel and chromium in the melt and,
consequently, to higher alloy contents in those regions
which solidified last. This tendency is evident for both
alloying elements in Fig. 3, and the variation of the
alloy contents within the austenite in Fig. 3 should thus
reflect the segregation of both elements into the
remaining melt. The first process will lead to an
enrichment of chromium and to a depletion of nickel in
the regions which solidified first.

3% Mo

Three alloys with 3% Mo and with different Ni, Cr, and
N contents have been investigated at different
solidification rates (Table 1). Figures 4 and 5 show the
solidification sequence at a solidification rate of
0·05 cm/min. The sample solidified with a primary
precipitation of ferrite to the right in Fig. 4. Roughly
5 mm (30°C) behind the solidification front, austenite
starts to precipitate (Fig. 5). The micrograph shows that
austenite and ferrite precipitate simultaneously from
the melt. At the same time the austenite is growing into
the ferrite. The ferrite in Fig. 5 contains transformation
twins formed during the quenching. Figure 6 shows the
amount of ferrite and austenite 12 mm (70°C) behind
the solidification front.

At the higher solidification rate, 0·5 cm/min, the
austenite precipitation did not start before 10 mm
(60°C) behind the solidification front. The reaction
occurred in a way similar to that at the lower
solidification rate.

Figures 7–9 show the solidification sequence in alloy
3. The solidification rate was 0·05 cm/min. The alloy
has, with the exception of nickel, the same composition
as alloy 2. The solidification sequence starts with a
primary precipitation of austenite (Fig. 7) and is almost
immediately followed by a secondary precipitation of
dendritic ferrite between the austenite dendrite arms
(Figure 8). From now on, ferrite and austenite precipitate together from the melt (Fig. 9). The
solidification at the higher solidification rate,
0·5 cm/min, also starts with a primary precipitation of
austenite, and the precipitation of ferrite does not start
before the very end of the solidification sequence.

The solidification sequence of alloy 4 was quite
similar to that of alloy 3. In this sample microprobe
measurements have been made. The concentration of
Cr, Ni, and Mo in the centre of a dendrite arm was
determined, and the effective distribution coefficient
calculated. The results are shown in Table 2.

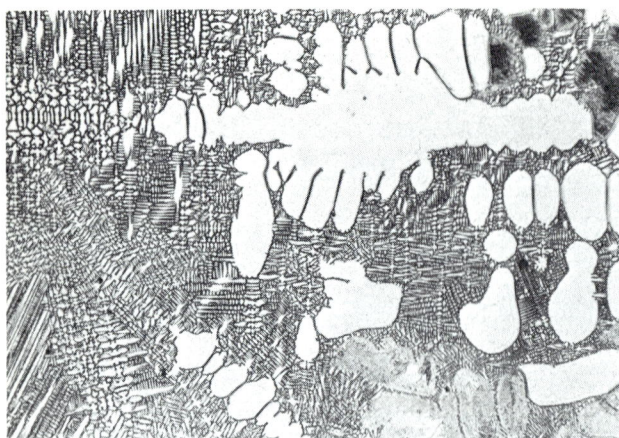

7 Solidification front with simultaneous precipitation of austenite and ferrite (alloy 3, Table 1): solidification rate 0·05 cm/min

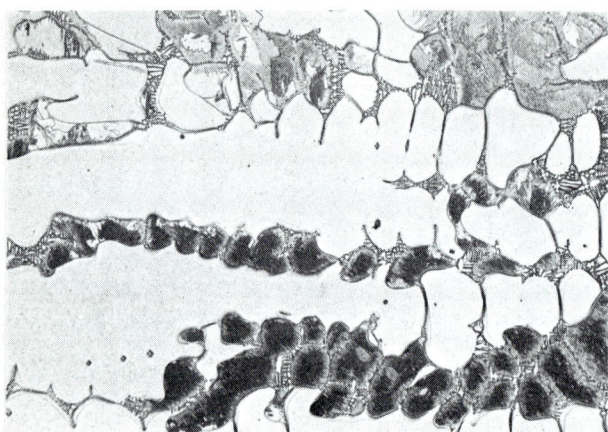

8 Further precipitation of ferrite and austenite in same sample as in Fig. 7, structure 15°C behind solidification front

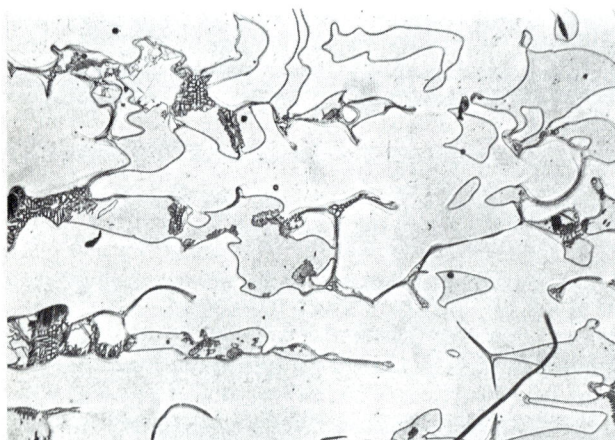

9 Structure in last solidified parts of sample shown in Fig. 7

Table 2 Effective distribution coefficient

Alloy	k_e^{Cr}		k_e^{Ni}		k_e^{Mo}	
4	0·91	(γ)	0·95	(γ)	0·65	(γ)
5	0·97	(δ)	0·84	(δ)	0·80	(δ)

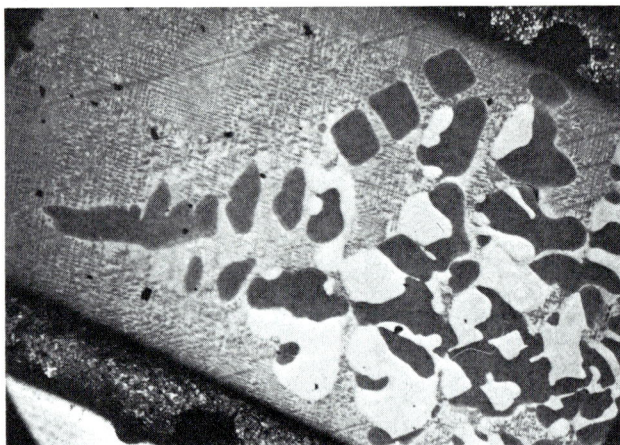

10 Solidification front in alloy 6 (Table 1); solidification rate 0·01 cm/min

6 and 9% Mo

The solidification sequence of alloy 5 was very similar to that of alloy 2. In this sample microprobe measurements were also made and the effective distribution coefficient was calculated (Table 2).

Alloy 6 was studied at three different solidification rates, 0·5, 0·01, and 0·002 cm/min. Figure 10 shows the solidification front at a solidification rate of 0·01 cm/min. The solidification starts with a primary precipitation of austenite (dark etching, Fig. 10), which is followed immediately by a secondary precipitation of austenite (light etching, Fig. 10). Ferrite and austenite both precipitate from the liquid during the subsequent cooling. With increasing solidification rate the secondary precipitation of ferrite starts at a lower temperature (Fig. 11). About 50% liquid is left to solidify. The micrograph was taken from the sample solidified with 0·5 cm/min. After a decrease in the solidification rate to 0·002 cm/min the reaction was very similar to a normal eutectic reaction (Fig. 12).

PERITECTIC REACTION

In iron-base alloys the peritectic reaction is normally $\delta + L \rightarrow \gamma$. A quantitative treatment of this reaction can be made in the following way.[4] The amount of ferrite precipitated before the start of the peritectic reaction can be estimated by the lever rule according to the high diffusivity in ferrite.[7] At the start of the peritectic

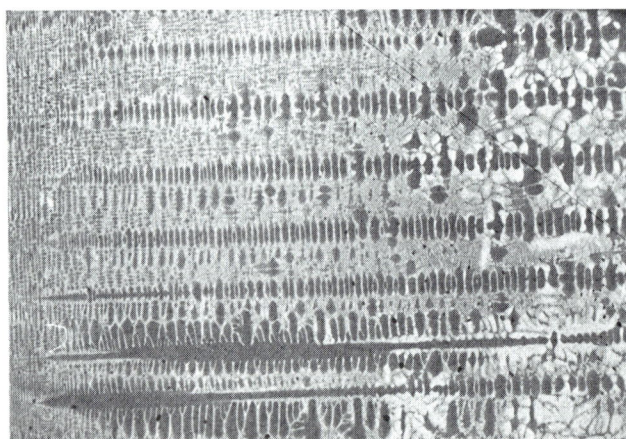

11 Solidification front in alloy 6 (Table 1); solidification rate 0·5 cm/min

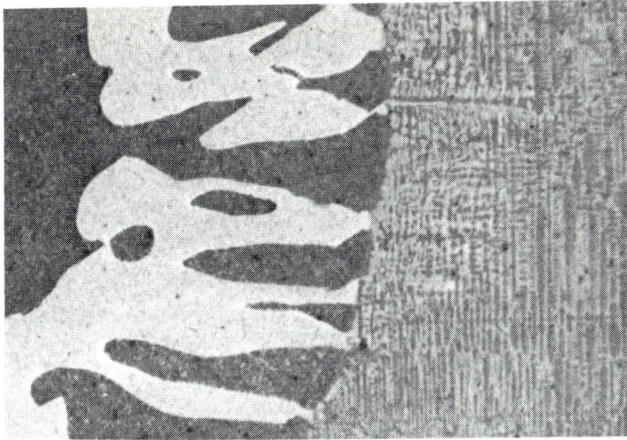

12 Solidification front in alloy 6 (Table 1); solidification rate 0·002 cm/min

reaction austenite is formed at the interface between ferrite and liquid and completely surrounds the ferrite (*see* Fig. 1). During the subsequent reaction austenite will grow into the liquid as well as into the ferrite. The concentration distribution in the austenite can be calculated with Scheil's segregation equation[8,9] for these two reactions.[4] Such a calculation has been made for an alloy with 9·8% Ni and 13·3% Cr with the constants given in Table 2 and according to the directions given in Appendix 1. The results of those calculations are shown in Fig. 13. A comparison with Fig. 3 shows that the experiment gives a similar result.

TRANSITION FROM PERITECTIC TO EUTECTIC REACTION
Fe–Cr–Ni alloys
During the austenite precipitation ferrite stabilizing elements will segregate more heavily to the liquid than will austenite stabilizing elements. Because of this segregation the liquid can be supersaturated with ferrite and a transition from a pure peritectic to a eutectic/peritectic reaction occurs (Figs 1 and 2). In the latter case ferrite and austenite are simultaneously precipitated from the melt and at the same time austenite is growing into the ferrite as a solid state transformation. This transition can be explained from the phase diagram Fe–Cr–Ni.

Very little is known about the phase diagram Fe–Cr–Ni. In this paper the diagram given by Bain and Paxton[10] will be used. Figure 14 shows a modification of this diagram. The three-phase line extends into the diagram from the binary Fe–Ni diagram parallel with the Cr-axis. At a Cr-content larger than roughly 15%, the three-phase line bends more and more towards higher Ni-contents. The three-phase triangle is shown at some temperatures. The diagram shows that the three-phase reaction changes from a pure peritectic reaction to a pure eutectic reaction. According to the treatment of the peritectic reaction in the previous paragraph, the segregation direction of liquid can be calculated during the peritectic reaction. This will be obtained by differentiating the Scheil segregation equation, when the following result is obtained

$$\frac{dx_{Ni}^{L}}{dx_{Cr}^{L}} = \frac{x_{Ni}^{L}(1 - k_{Ni}^{\gamma/L})}{x_{Cr}^{L}(1 - k_{Cr}^{\gamma/L})} \qquad (1)$$

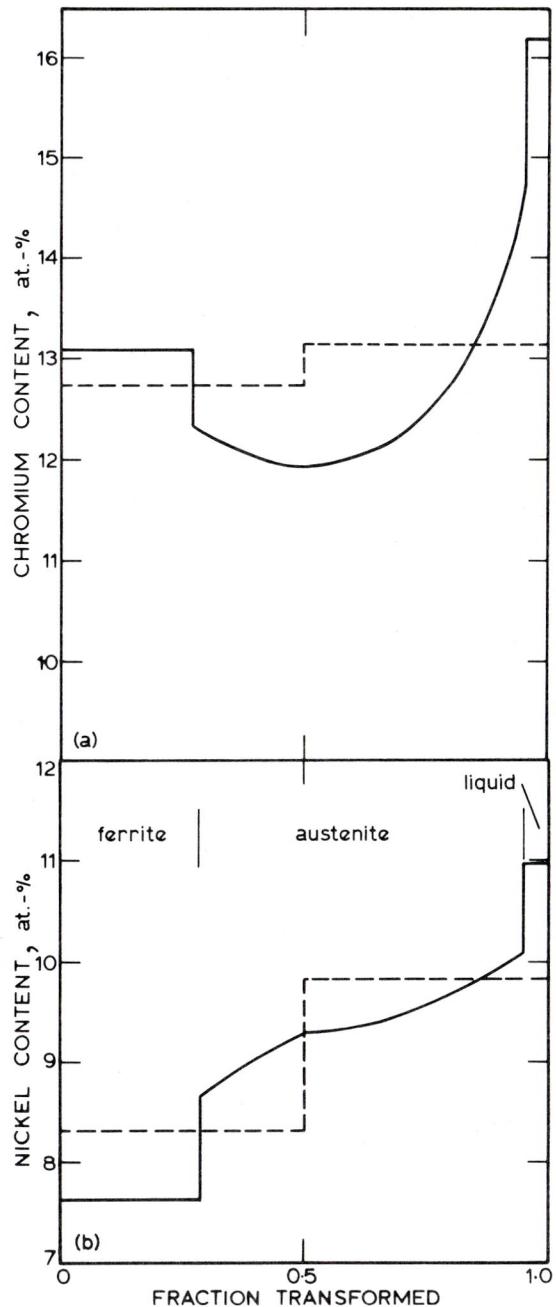

13 Distribution of *a* chromium and *b* nickel during peritectic reaction, assumed to start when 50% ferrite has been formed

This is the same as the direction of the tie line between austenite and liquid. If the slope of the tie line is smaller than the slope of the three-phase line, ferrite must precipitate at the same time as austenite. This occurs in Fig. 14 when the Ni and Cr contents are roughly 12 and 19%, respectively.

INFLUENCE OF Mo ON TRANSITION FROM PERITECTIC TO EUTECTIC REACTION
There is no information in the literature about the influence of molybdenum on the three-phase equilibrium $\alpha/\gamma/L$ in Fe–Cr–Ni–Mo alloys. In order to estimate the influence of molybdenum on the three-phase reaction the phase diagrams Fe–Cr–Ni and Fe–Cr–Ni–Mo were calculated in a rather simple way (Appendix 2). The calculations are based on the

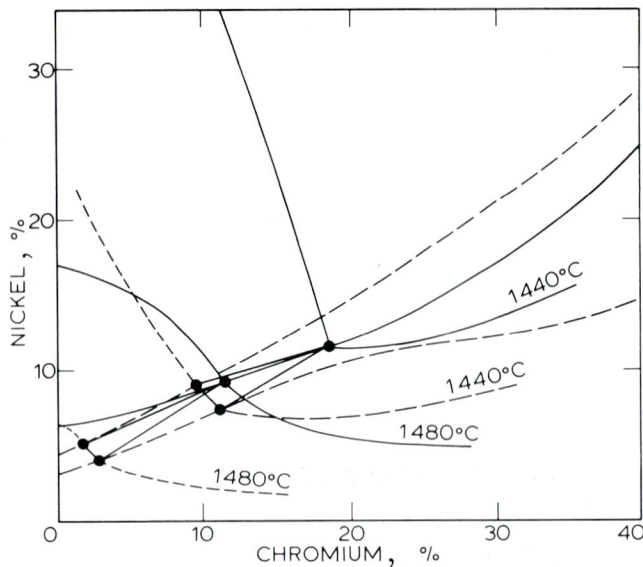

14   Phase diagram Fe–Cr–Ni according to Bain and Paxton[8]

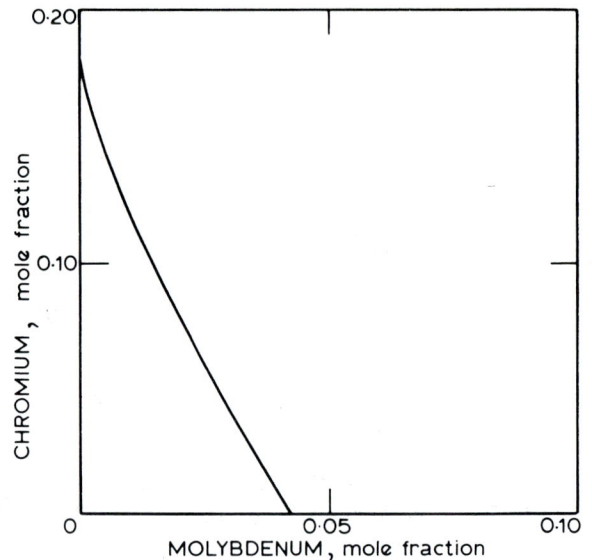

16   Transition from peritectic to a eutectic reaction as function of Cr and Mo content in Fe–Cr–Ni–Mo alloy with 11·9% Ni

assumption that the thermodynamics of iron can be described by Rault's law. This is an approximation which had been used earlier by Hume–Rothery[11] and seems to describe the phase diagrams quite adequately. The partition coefficients between γ/L and α/L for Cr, Ni, and Mo have been assumed to be constant and independent of temperature and alloying content. The partition coefficients have been chosen from the microprobe measurements shown in Table 2. The calculated phase diagrams are shown in Fig. 15, where the lines represent the projection of the three-phase equilibrium at different Mo-contents in Fe–Cr–Ni–Mo alloys. Following the same principles described in the previous paragraph, the transition from a peritectic to a eutectic/peritectic reaction occurs at 11·9% Ni and 17·2% Cr in alloys without molybdenum. However, the quaternary system Fe–Cr–Ni–Mo has to be analysed if the tie line cuts the liquidus surface of ferrite in order to describe the transition. This can be done by vector analyses. This has been done, and the result of the calculation is shown in Fig. 16 for an alloy with 11·9% Ni. It is worth noting that the transition is very strongly influenced by an increase of the Mo content in the alloy.

QUANTITATIVE TREATMENT OF EUTECTIC REACTION

Figs 2, 5, 7, 11, and 12 illustrate the eutectic reaction. Ferrite and austenite is growing into the ferrite. Figures 17a and b illustrate the concentration distribution of an element which stabilizes ferrite and austenite, respectively, during such a reaction. The diagram has been drawn by assuming that the diffusion rate in the liquid as well as in the ferrite is so fast that no concentration gradients exist in those phases. The diffusion in the austenite has been assumed to be so low that no back diffusion occurs in this phase. A mass balance at a small change at the volume fraction of liquid gives the following expression

$$df^L x_M^{L/\gamma}(1 - k_m^{\gamma/L}) + df^\alpha x_M^{\alpha/\gamma}(1 - k_M^{\gamma/\alpha})$$
$$= dx_M^L(f^\alpha/k_M^{\alpha/L} + f^L) \qquad (2)$$

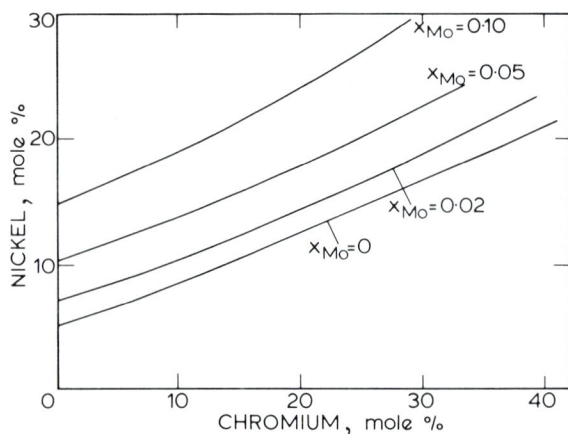

15   Projections of phase diagram Fe–Cr–Ni–Mo, at constant Mo content

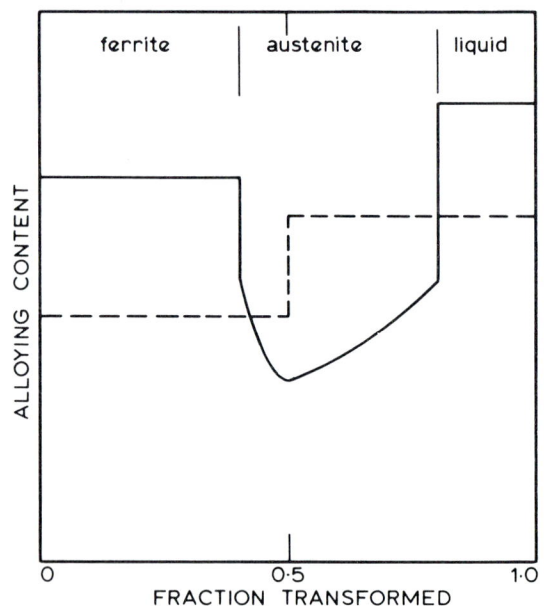

17   Eutectic three-phase reaction in Fe–Cr–Ni alloys

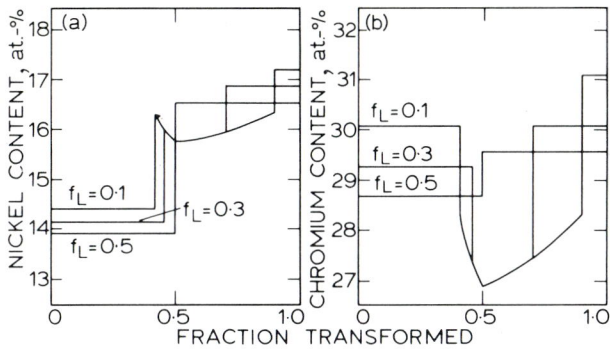

18 Distribution of, *a* nickel and *b* chromium in Fe–Cr–Ni alloy during eutectic reaction, assumed to start when 50% ferrite has been formed

Such an equation is given for each element. Those equations should be solved simultaneously, together with the equilibrium equations (6)–(11). This can only be done numerically by stepwise integration. Such calculations have been performed for an alloy with 15% Cr, and the results are shown in Fig. 18. The calculations have been made only for a ternary alloy Fe–Cr–Ni, but it may be assumed that molybdenum strongly favours the ferrite precipitation from the liquid and that austenite grows faster into the ferrite in the alloy without molybdenum than in the alloy with molybdenum. Strong evidence for this was given when calculations were performed for ternary Fe–Mo–Ni alloys.

DISCUSSION
The above calculations have been made with a large number of simplifications. One of the most important factors is the diffusion rate in the different phases. The experiments show that an increased solidification rate displaces the start of the secondary reaction to a lower temperature. This is true for a primary precipitation of ferrite as well as for a primary precipitation of

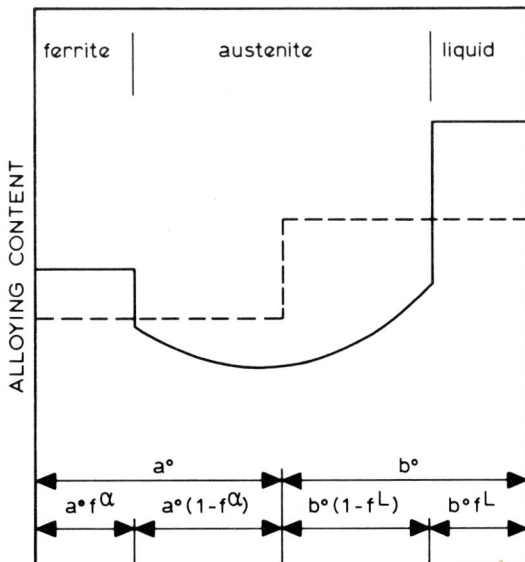

19 Schematic diagram of distribution of alloying elements during peritectic transformations

austenite, and can be explained only by an accentuated segregation at an increased solidification rate. This implies that Scheil's segregation equation does not fully explain the concentration distribution at a primary precipitation of austenite. It also shows that the lever rule is not fully valid for a primary precipitation of ferrite. Nevertheless, the calculations are capable of describing the transition from a peritectic to a eutectic reaction quite adequately.

SUMMARY
The three-phase reaction in Fe–Cr–Ni and Fe–Cr–Ni–Mo alloys has been studied by unidirectional solidification experiments. The transition from a peritectic to a eutectic reaction has been verified experimentally in these alloys. The three-phase reaction has also been theoretically analysed, and the experimental results have been verified.

REFERENCES
1 R. ORR AND J. CHIPMAN: *J. Iron Steel Inst.*, 1967, **299**, 630
2 H. HARVIG: *Jernkontorets Ann.*, 1971, **155**, 157
3 H. FREDRIKSSON: *Metall. Trans.*, 1972, **3**, 2 989
4 H. FREDRIKSSON: *Met. Sci.*, 1976, **10**, 77
5 H. FREDRIKSSON AND I. SVENSSON: Report to the Swedish Board for Technical Development, Dec. 1974
6 C-Å. DÄCKER AND H. FREDRIKSSON: Report to the Swedish Board for Technical Development, Mar. 1976
7 H. FREDRIKSSON: *Scand. J. Metall.*, 1976, **5**, 27
8 E. SCHEIL: *Z. Metallk.*, 1942, **34**, 70
9 H. O. BRODY AND M. C. FLEMINGS: *Trans. AIME*, 1966, **236**, 615
10 'Metals handbook', 1948, Ohio, ASM
11 W. HUME-ROTHERY AND R. A. BUCKLEY: *J. Iron Steel Inst.*, 1964, **202**, 531

APPENDIX 1
Equation (1) shows the concentration distribution during the peritectic reaction. A material balance gives the concentration distribution before the start of the peritectic reaction

$$x_{Ni}^{O} = a \cdot x_{Ni}^{\alpha/L} + bx_{Ni}^{L/\alpha} \tag{1}$$

$$x_{Cr}^{O} = a \cdot x_{Cr}^{\alpha/L} + bx_{Cr}^{L/\alpha} \tag{2}$$

$$1 = a + b \tag{3}$$

where a and b are fractions of α and L. The distribution in the three phases during the reaction are described by Scheil's segregation equation

$$x_{M}^{\gamma/L} = k_{M}^{\gamma/L} x_{M}^{L0} (f^{L})^{-(1-k_{M}^{\gamma/L})} \tag{4}$$

$$x_{M}^{\gamma/\alpha} = k_{M}^{\gamma/\alpha} x_{M}^{\alpha0} (f^{\alpha})^{-(1-k_{M}^{\sigma/\gamma})}. \tag{5}$$

where f^{L} and f^{α} are the fractions of L and α remaining, and M stands for Cr, Ni, and C. At each interface there is local equilibrium. Rault's law gives the composition at the interphase as a function of temperature. For the γ/L equilibria one gets

$$\frac{1 - x_{Ni}^{L/\gamma} - x_{Cr}^{L/\gamma}}{1 - x_{Ni}^{\gamma/L} - x_{Cr}^{\gamma/L}} = \exp - \frac{{}^\circ G_{Fe}^{L} - {}^\circ G_{Fe}^{\gamma}}{RT} \tag{6}$$

For the γ/α equilibrium one gets

$$\frac{1 - x_{Ni}^{\gamma/\alpha} - x_{Cr}^{\gamma/\alpha}}{1 - x_{Ni}^{\alpha/\gamma} - x_{Cr}^{\alpha/\gamma}} = \exp - \frac{{}^\circ G_{Fe}^{\gamma} - {}^\circ G_{Fe}^{\alpha}}{RT} \tag{7}$$

and

$$x_{Ni}^{\gamma/L} / x_{Ni}^{L/\gamma} = k_{Ni}^{\gamma/L} \tag{8}$$

$$x_{Ni}^{\alpha/\gamma} / x_{Ni}^{\gamma/\alpha} = k_{Ni}^{\gamma/\alpha} \tag{9}$$

$$x_{Cr}^{\gamma/L} / x_{Cr}^{L/\gamma} = k_{Cr}^{\gamma/L} \tag{10}$$

$$x_{Cr}^{\alpha/\gamma} / x_{Cr}^{\gamma/\alpha} = k_{Cr}^{\gamma/\alpha} \tag{11}$$

Putting equations (4)–(5) and (8)–(11) into equations (6) and (7) gives f^α and f^L as functions of temperature. Equations (4) and (5) then give the concentration distributions.

APPENDIX 2

Assuming that Rault's law is valid for iron, the equilibria γ/L and α/L give

$$\frac{1 - k_{Cr}^{\alpha/L} \cdot x_{Cr}^{L/\alpha} - k_{Ni}^{\alpha/L} \cdot x_{Ni}^{L/\alpha} - k_{Mo}^{\alpha/L} \cdot x_{Mo}^{L/\alpha}}{1 - x_{Cr}^{L/\alpha} - x_{Ni}^{L/\alpha} - x_{Mo}^{L/\alpha}}$$
$$= \exp \frac{{}^\circ G_{Fe}^{L} - {}^\circ G_{Fe}^{\alpha}}{RT} \tag{12}$$

$$\frac{1 - k_{Cr}^{\gamma/L} \cdot x_{Cr}^{L/\gamma} - k_{Ni}^{\gamma/L} \cdot x_{Ni}^{L/\gamma} - k_{Mo}^{L/\gamma} x_{Mo}^{L/\gamma}}{1 - x_{Cr}^{L/\gamma} - x_{Ni}^{L/\gamma} - x_{Mo}^{L/\gamma}}$$
$$= \exp \frac{{}^\circ G_{Fe}^{L} - {}^\circ G_{Fe}^{\gamma}}{RT} \tag{13}$$

Solving $x_{Cr}^{L/\gamma}$ and $x_{Cr}^{L/\alpha}$ from these two equations and putting them equal at the three-phase equilibrium produces one equation giving

$$x_{Ni}^{L/\gamma} = x_{Ni}^{L/\alpha}$$

for a constant value on x_{Mo}^{L}.

Unidirectional solidification of peritectic alloy

A. Ostrowski and E. W. Langer

This paper presents an analysis based on the coupled growth of eutectic alloys and shows that a great difference exists between the undercooling v. growth rate and lamellar spacing relationships for the eutectic and peritectic cases, respectively. This difference causes instabilities in the solidification front and thus precludes coupled growth of the two-phase peritectic alloys. Solidification experiments, where the temperature gradient/growth rate ratio is sufficiently large to suppress non-planar freezing, do not result in coupled growth of the two phases but give a periodic separation in the form of bands. In the present investigation of Ag–Zn alloys a unidirectional solidification technique under stationary conditions has been employed.

A. Ostrowski is Research Assistant and E. W. Langer is Professor in the Department of Metallurgy, The Technical University of Denmark, Lyngby, Denmark

LIST OF SYMBOLS

a_α, a_β	constants
c_i, c_i	concentration of component i at the interface of α- and β-phases
C_0	equilibrium concentration of element i
C_∞	difference between the peritectic composition and the actual composition far from the interface
D	diffusion coefficient
L_α, L_β	heats of fusion per unit volume
m_α, m_β	slope of liquidus curve
S_α, S_β	half widths of α and β lamellae respectively
$\sigma_\alpha^L, \sigma_\beta^L$	liquid/solid surface energies per unit area angles which characterize the curvature of the interface
T_p	peritectic temperature
T_i	temperature of the solidification front

The phase relationships in peritectic systems have been investigated extensively during the last few years. The possibility of coupled growth of primary and secondary (peritectic) phases has been of special interest, but this problem has not yet been solved. Uhlmann *et al.*[1] have analysed the process of unidirectional solidification of Ag–Zn alloys, and they found that the steady-state solidification of the peritectic melt with a planar solid/liquid interface cannot occur. Later Boettinger[2] was able to show that Zn–Cd alloys of peritectic composition, which are solidified with a planar liquid/solid interface, have a characteristic structure, containing bands of primary and peritectic phases lying perpendicular to the growth direction. Titchener *et al.*[3]

have investigated the Sn–Sb and Cu–Zn systems and reached the same conclusions as Boettinger.

EXPERIMENTAL PROCEDURE

The object of the experiments was to determine the structure obtained in a two-phase peritectic alloy, solidified with a plane solidification front, and so it was necessary to determine the G/R value (temperature gradient/solidification rate) resulting in a plane solid/liquid interface in agreement with the theory for constitutional undercooling. Unidirectional solidification experiments were performed on a Ag–Zn alloy with 10 wt-% Ag. A value of $G/R = 8 \cdot 10^6 [°C \cdot s/cm^2]$ results in a plane solidification front. Details of the experimental technique employed are described elsewhere.[6]

Alloys were prepared from 99·98% purity Ag and 99·99% purity Zn. Quartz tubes with a 3 mm i.d. and 100 mm long were filled with the molten alloy, which was then allowed to freeze unidirectionally by pulling the tube out of a furnace. A steep temperature gradient was maintained by means of a watercooled chill. Electrically insulated thermocouples inserted in the alloy were used for measuring the temperature gradient in the liquid and solid alloy.

RESULTS

No coupled eutectic-like growth was observed of ϵ and η after plane solidification. Instead, a banded structure of alternating layers of $\eta + \epsilon$ and η was obtained, the bands lying perpendicular to the growth direction, Figs 1–4.

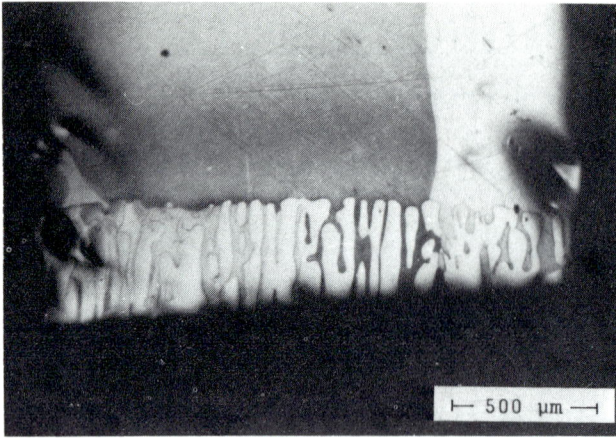

1   The first layer

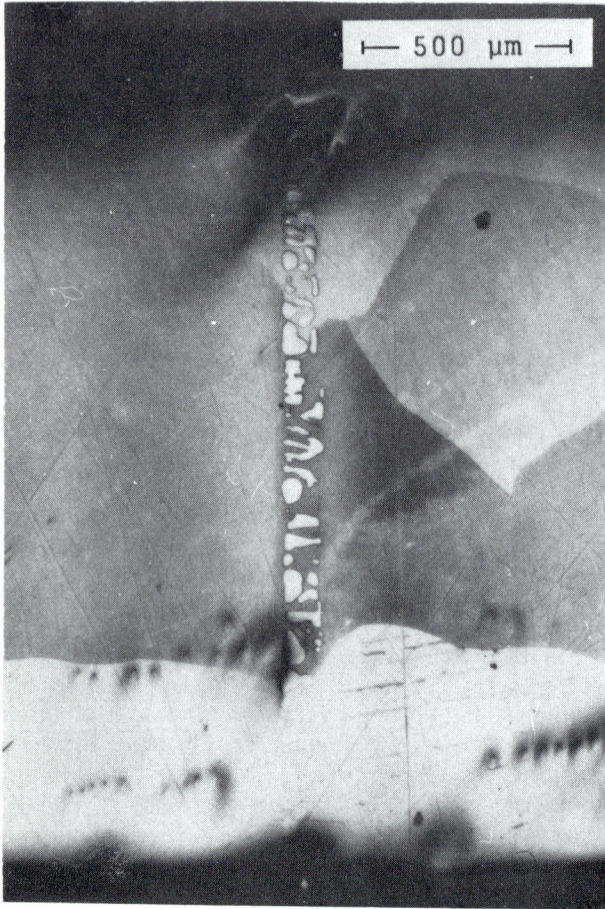

2   The second layer

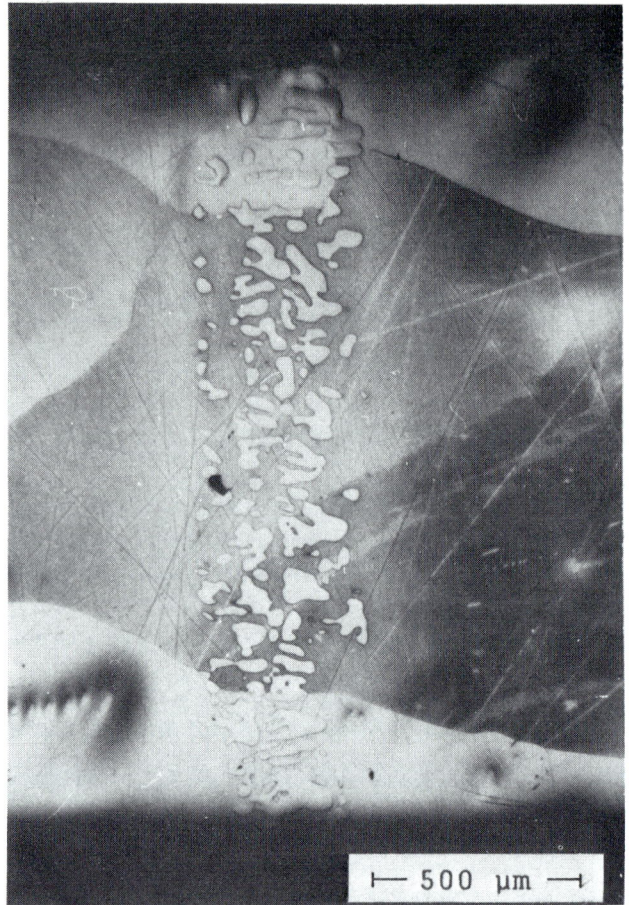

3   The third layer

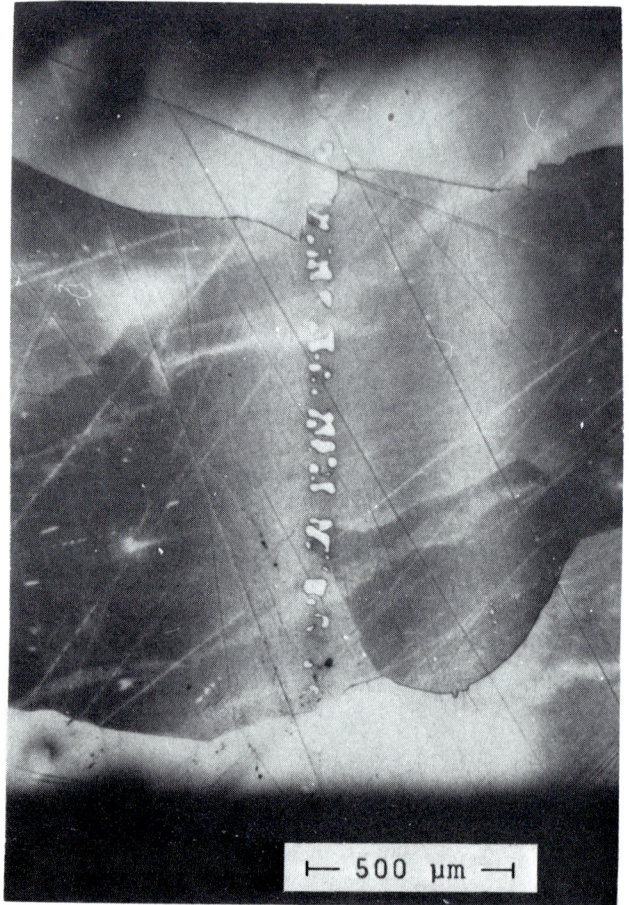

4   The fourth layer

In the first specimen three layers of this kind were observed. Microprobe analysis of the ϵ-phase showed a Zn content of 83 wt-%, which is in accordance with the equilibrium diagram (Fig. 5); similarly, the η-phase was found to contain 92 wt-% Zn along the whole length of the specimen. Duplicate experiments with the same G/R value resulted in the same structure—well defined layers of cellular ϵ-phase in a η-matrix (Fig. 6).

It was first thought that this banded structure might be caused by irregularities in the drive mechanism or in the furnace temperature control unit. As the banding was not observed in subsequent experiments with lower G/R values which resulted in dendritic structures, and

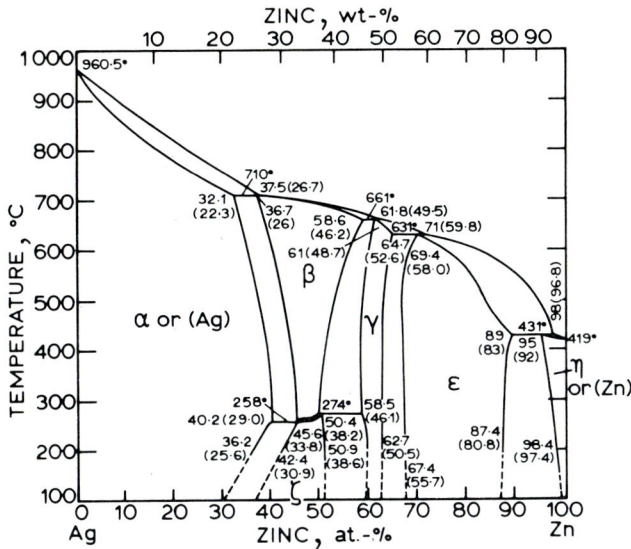

5 Ag–Zn system

where the two mentioned external sources for disturbances would also have been effective, it was concluded that the banded structure must be caused by some other factor.

DISCUSSION

Having rejected changes in temperature and solidification velocity as being the causes of this one must look for other experimental factors, such as the large difference in the coefficients of thermal expansion

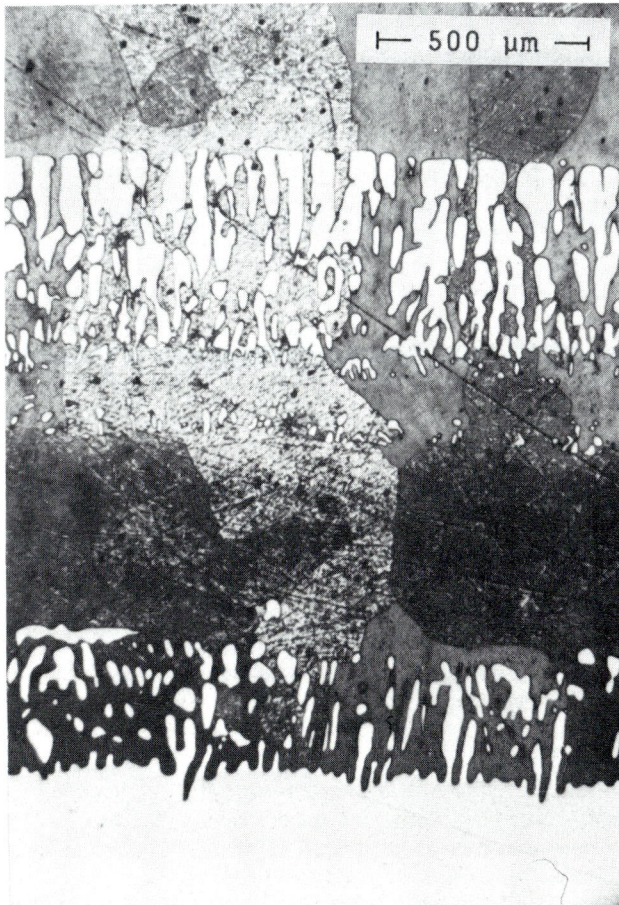

6 The solidification front and the layer of ε-phase

between the quartz tube and the alloy. This gives rise to the presence of a small space between the quartz and metal surface. If the same conditions are maintained during the entire solidification process, it will not be possible to observe any effect of this fact, but a steep decline in temperature will occur if contact is suddenly established between the tube and the alloy.

This results in an increasing degree of constitutional undercooling which will be sufficient for nucleation of the primary phase. This phase is now able to grow and a decrease in undercooling in front of the primary phase conversely stops its growth because the contact between the alloy surface and the quartz tube will cease shortly afterwards as a result of thermal contraction on cooling. Therefore, the temperature gradient in the melt begins to increase again, resulting in a one-phase structure.

This repeating process gives rise to bands of the primary phase, and thus the observed structure can be explained on the basis of the differences in thermal coefficients of expansion between the alloy and the quartz tube, and the question as to whether or not coupled growth can occur in a peritectic system is unresolved.

Analysis of the possibility of coupled growth in a peritectic system
Stationary solidification with undercooling of primary phase

Figure 7a depicts a phase diagram containing an ideal peritectic reaction, and it is indicated which phases will be present under equilibrium conditions at any temperature. Assuming that the melt with a composition C_0 is undercooled to the temperature T below the peritectic temperature, the resulting situation can be analyzed by means of free energy/composition curves in Fig. 7b.

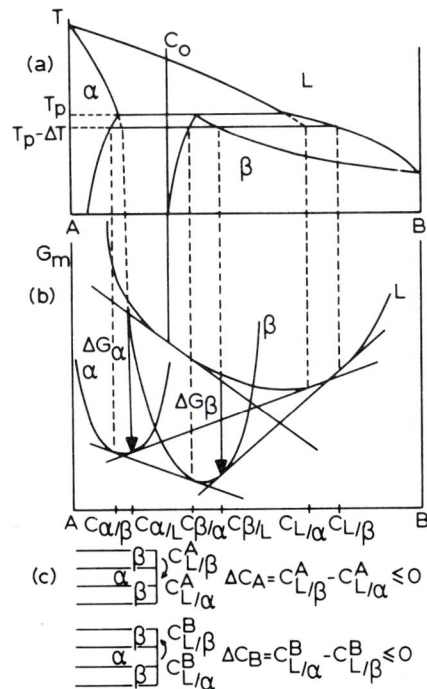

7 Analysis of the possibilities of coupled growth by means of free energy curves

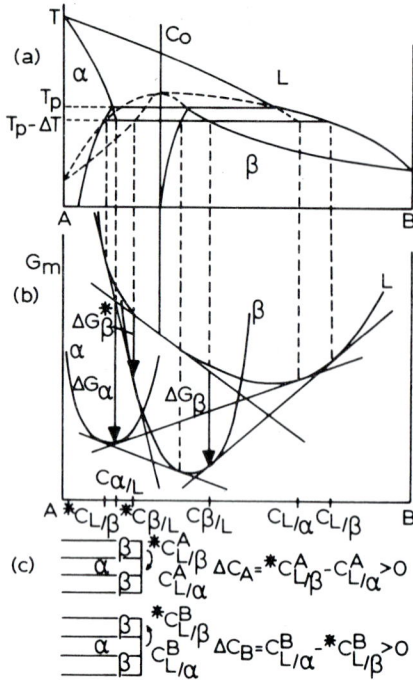

8   **The metastable case of the peritectic equilibrium diagram *C corresponds to the metastable concentration**

Drawing a tangent to the molar free energy curve at the composition C_0 shows negative free energy changes ΔG, both for the α- and β-phases, and thus indicates a positive driving force for phase separation. In spite of the negative ΔG_β the β-phase is not able to grow because the condition must be fulfilled that equilibrium prevails in the phase boundary.

The next step is to investigate the possibility of cooperative growth of the two phases. From Fig. 7c it can be seen that the concentration changes ΔC_A and ΔC_B are negative, which implies that cooperative growth of α and β is not possible.

Solidification with undercooling of both phases

After having analysed the solidification case above, the metastable solidification of alloy C_0 will be examined by means of free energy curves (Fig. 8). The dashed lines show extensions of equilibrium lines for β and L above the peritectic temperature T_p. The β-phase rejected along these metastable lines will be named β^*. The metastable liquidus and solidus merge in a maximum point and have a common tangent here. As in the case above ΔG_α and ΔG_β will be negative; this also applies to ΔG_β^*.

The analysis of the possibility for cooperative growth of α and metastable β^* phases yields positive values for ΔC_A and ΔC_B which permits a possible flux for both A- and B-atoms ahead of the solidification front by lamellar coupled growth as shown in Fig. 8c. This implies that cooperative growth of the two phases is possible.

It can be concluded that the existence of a metastable β^*-phase makes it possible to obtain coupled growth in an alloy system with a peritectic reaction in the same way as in a eutectic system. From free energy curves it can be predicted whether coupled growth is thermodynamically possible at a slight undercooling below the peritectic temperature.

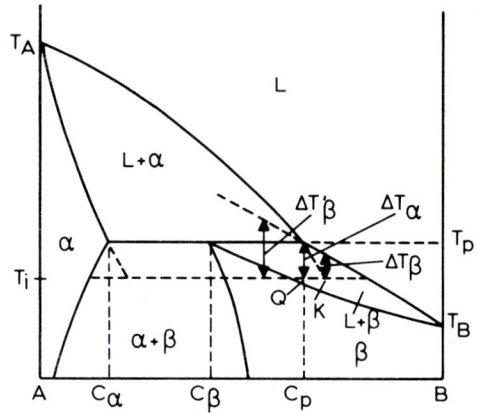

9   **Metastable extrapolation of liquidus and solidus lines and undercooling for α- and β-phases**

It is, however, also of interest to examine the thermal conditions prevailing during isothermal solidification of alloys of near-peritectic composition. During solidification under metastable conditions, isothermal solidification after quenching an alloy of composition C_p is considered (Fig. 9).

The temperature in the solid/liquid interface is T_i, and the undercooling is defined as

$$\Delta T = T_L - T_i$$

where T_L is the liquidus temperature for α- or β-phase according to the equilibrium diagram.

If it is assumed that the liquid alloy of composition C_p is undercooled a finite amount to the temperature T_i (below T_p) to point Q, it will be supersaturated with respect to both α and β, since Q is situated below the liquidus curves for both the α- and β-phases. If a particle of α-phase is formed in the liquid, B-atoms are rejected into the surrounding liquid phase, changing the composition of the melt from Q towards point K, and thus the undercooling with respect to α, ΔT_α, is decreasing towards zero, whereas the undercooling with respect to β is changing to ΔT_β.

At the point K, ΔT_α equals zero, which means that only β crystals are formed. By their growth A-atoms are rejected and the composition of the surrounding liquid will change towards point Q. This change of composition will result in a higher degree of undercooling with respect to the β-phase, and the melt will follow the metastable liquidus line for β, until the liquid has totally solidified.

The metastable case thus gives undercooling in front of the growing α-phase and superheating ahead of the β-phase, both with respect to the peritectic temperature. The β-phase stops its growth after a certain value for ΔT_β is reached, and special circumstances are required to retain the whole process.

An analysis based on the theory for growth of lamellar and rod-shaped eutectics[4] (see Appendix) results in three curves (Fig. 10) connecting undercooling ΔT with lamellar spacing λ. It is seen that for a eutectic growing with a given velocity, the undercooling ($\Delta T = T_E - T_i$) is positive for large values of λ. The two lower curves represent the peritectic case where the undercooling ($\Delta T = T_p - T_i$) is negative (superheating) for large λ values. A change of sign for ΔT_β, (see Appendix) in agreement with Fig. 10 gives a curve corresponding to

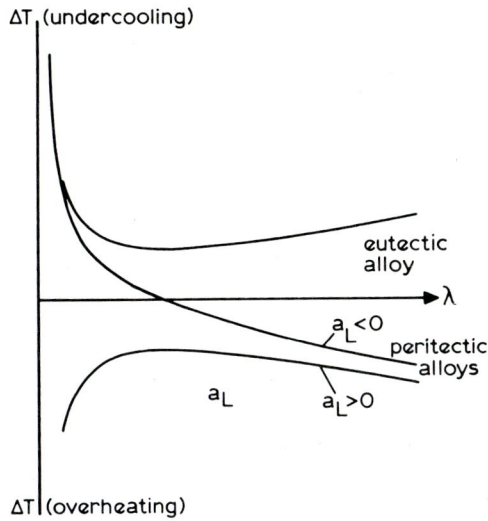

10 Variations of interface undercooling ΔT, with lamellar spacing λ

the eutectic case, and consequently makes a stable, coupled growth of two phases in the peritectic system theoretically possible, although this situation has not yet been achieved in practice.

REFERENCES

1 D. R. UHLMANN AND G. A. CHADWICK: *Acta Metall.*, 1961, **9**, 835

2 W. J. BOETTINGER: *Metall. Trans.*, 1974, **5**, 2 023

3 A. P. TITCHENER AND J. A. SPITTLE: *Acta Metall.*, 1975, **23**, 497

4 K. A. JACKSON AND J. D. HUNT: *Trans. Metall. Soc. AIME*, 1966, **236**, 1 129

5 L. V. SYDOW: 'Isoterm Stelnande av Peritektiska Legeringssytem', 1968, Stockholm, Kungliga Tekniska Högskolan

6 F. R. MOLLARD AND M. C. FLEMINGS: *Trans. Metall. Society AIME*, 1967, **239**, 1 534

APPENDIX

A calculation of the undercooling in front of the solidification front is presented, using a method similar to that used by Boettinger,[2] in the case of the coupled growth of two phases in the peritectic system

$$\Delta T = T_p - T_i$$

$$\Delta T_\alpha = m_\alpha(C_i^\alpha - C_p); \qquad m < 0;$$

$$\Delta T_\beta = m_\beta(C_i^\beta - C_p); \qquad m < 0;$$

$$C_i^\alpha = C_p + C_\infty + B_0 + \frac{2(S_\alpha + S_\beta)^2}{S_\alpha}\frac{R}{D}C_0 P$$

$$C_i^\beta = C_p + C_\infty + B_0 - \frac{2(S_\alpha + S_\beta)^2}{S_\beta}\frac{R}{D}C_0 P$$

$$P = \sum_{n=1}^{\infty} \left(\frac{1}{n\pi}\right)^3 \sin^2\left(\frac{n\pi S_\alpha}{S_\alpha + S_\beta}\right);$$

$$B_0 = \frac{S_\alpha(C_p - C_\alpha) - (C_\beta - C_p)S_\beta}{S_\alpha + S_\beta}$$

$$= C_p - \frac{C_\alpha S_\alpha + C_\beta S_\beta}{S_\alpha + S_\beta}$$

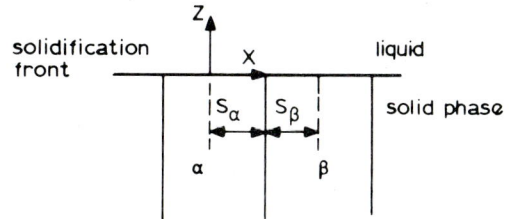

$$a_\alpha = \frac{T_p}{L_\alpha}\sigma_\alpha^L \sin\theta_\alpha^L;$$

$$a_\beta = \frac{T_p}{L_\beta}\sigma_\beta^L \sin\theta_\beta^L;$$

$$\Delta T_\alpha = m_\alpha\left(C_p + C_\infty + B_0 \right.$$
$$\left. + \frac{2(S_\alpha + S_\beta)^2}{S_\alpha}\frac{R}{D}C_0 - C_p\right) + \frac{a_\alpha}{S_\alpha}$$

$$\Delta T_\beta = m_\beta\left(C_p + C_\infty + B_0 \right.$$
$$\left. - \frac{2(S_\alpha + S_\beta)^2}{S_\beta}\frac{R}{D}C_0 P - C_p\right) + \frac{a_\beta}{S_\beta}$$

$$\frac{\Delta T_\alpha}{m_\alpha} - \frac{\Delta T_\beta}{m_\beta} = \frac{2(S_\alpha + S_\beta)^2}{S_\alpha}\frac{R}{D}C_0 P$$

$$+ \frac{2(S_\alpha + S_\beta)^2}{S_\beta}\frac{R}{D}C_0 P$$

$$+ \frac{a_\alpha}{m_\alpha S_\alpha} - \frac{a_\beta}{m_\beta S_\beta}$$

for $\Delta T_\alpha = \Delta T_\beta$

$$\frac{\Delta T}{m} = Q\lambda R + \frac{a_L}{\lambda};$$

$$\lambda = 2(S_\alpha + S_\beta)$$

$$\frac{1}{m} = \frac{1}{m_\alpha} - \frac{1}{m_\beta}; \qquad Q = \frac{P(1+\zeta)^2}{\zeta D}(C_\beta - C_\alpha)$$

$$a_L = 2(1+\zeta)\left(\frac{a_\alpha}{m_\alpha} - \frac{a_\beta}{\zeta m_\beta}\right); \qquad \zeta = \frac{S_\beta}{S_\alpha}$$

Controlled solidification of peritectic alloys

H. D. Brody and S. A. David

Pb–Bi and Sn–Cd as models of alloys from peritectic systems were directionally solidified under controlled conditions for moderate and high values of the temperature gradient/growth rate ratio G/R. A variety of microstructures were obtained. Alloys which would solidify as two phases under equilibrium conditions, when solidified under conditions of plane-front growth at high values of G/R, formed single-phase metastable solid solutions of the peritectic product phase. At moderate values of G/R the alloys were grown with well aligned two-phase morphologies. However, because growth of the two solid phases is not coupled, the spacing between phases is much coarser than in eutectic systems. An analysis of the growth conditions that lead to the observed microstructures is presented.

H. D. Brody is in the Department of Metallurgical and Materials Engineering, University of Pittsburgh, and S. A. David is at the Oak Ridge National Laboratory Tennessee, USA

In recent years, controlled solidification of peritectic alloys has been receiving a great deal of attention and has been the subject of detailed analytical and experimental studies.[1-5] Much of the recent work was carried out in order to resolve the several conflicting hypotheses that have been proposed to describe the structures expected to form on directional solidification of peritectic alloys.[6-8] Several authors have discussed the directional freezing of peritectic alloys under the conditions of small values of the temperature gradient/growth rate ratio G/R.[6,9] However, initial results on directional freezing of peritectic alloys under conditions of moderate and high values of G/R have produced surprising results that are not in complete agreement with the existing hypotheses.[5]

For an alloy of composition C_0 between the compositions above pure A and up to C_L in the peritectic phase diagrams of Fig. 1, the freezing process for small G/R may be described, schematically, as in Fig. 2. The temperature gradient is shown in Fig. 2a, where X_t, X_p, and X_r represent the instantaneous positions of the liquidus, peritectic, and effective solidus isotherms, respectively. The effective solidus, as shown in Fig. 1, is the temperature where the last liquid C_L^f freezes as the average composition of the cored, often two-phase solid reaches C_0. As in previous studies,[10-11] it is assumed (a) that the liquid is uniform in composition along an isotherm, (b) that equilibrium applies to the solid/liquid interface*, and (c) that there is negligible

undercooling before nucleation of solid phases. Then solid-plus-liquid coexist between X_t and X_r and the liquidus curve may be used to generate the curve of solute distribution in the liquid within the solid-plus-liquid zone (Fig. 2b). For low values of G/R, it has been assumed that diffusion across the mushy zone is negligible and the tips of the dendrites are not undercooled owing either to constitutional or kinetic effects.[10,12-14] If, for simplicity in sketching, the dendrites are considered to be plates, Fig. 2c represents the distribution of phases; S_2 first nucleates at the peritectic temperature and quickly envelopes S_1. From the point of complete envelopment, two-phase equilibrium is maintained at two interfaces: S_2/L and S_1/S_2. In some systems S_2 has been found to nucleate and grow epitaxially on S_1.[15] In other systems, S_2 has been reported to nucleate heterogeneously in the liquid and then to grow together until it surrounds S_1.[2] In directional freezing, nucleation of S_2 near T_p is not considered to be a difficulty owing to the presence of previously grown S_2 in regions of lower temperature. Figure 2d shows a volume element taken between two dendrite arms at a time when the element is at a temperature below T_p and above T_s'. At the L/S_2 interface, S_2 forms by solidification with concomitant rejection of solute at the interface and an increase in the liquid composition, e.g. at T'

$$L(C_L') \rightarrow S_2(C_{2L}') + \text{solute } [B] \qquad (1)$$

Diffusion occurs across the S_2 rim owing to the composition gradient in the S_2, as shown in Fig. 2e. The difference $(C_{2L}' - C_2')$ increases as the temperature decreases below the peritectic temperature. As a result

* The equilibrium interface compositions are given by the equilibrium partition ratio $K = C_S^*/C_L^*$, where C_S^* and C_L^* are the solid and liquid compositions given by the tie line at the temperature of a particular location. Herein K_1 refers to the equilibrium between S_1 and L and K_2 to equilibrium between S_2 and L

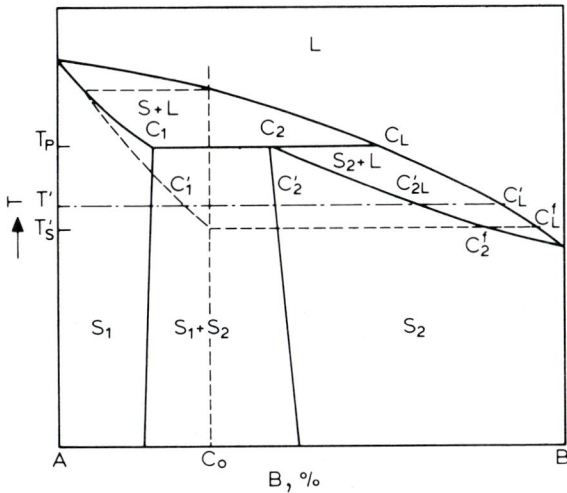

1 Binary peritectic phase diagram

of this diffusion flux, solute is brought to the S_1/S_2. interface. At the latter interface, solute is incorporated by the transformation of S_1 to S_2, e.g. at T'

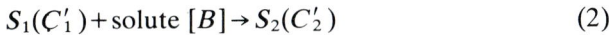

$$S_1(C'_1) + \text{solute } [B] \rightarrow S_2(C'_2) \qquad (2)$$

It is by the combination of these two reactions that the peritectic transformation proceeds

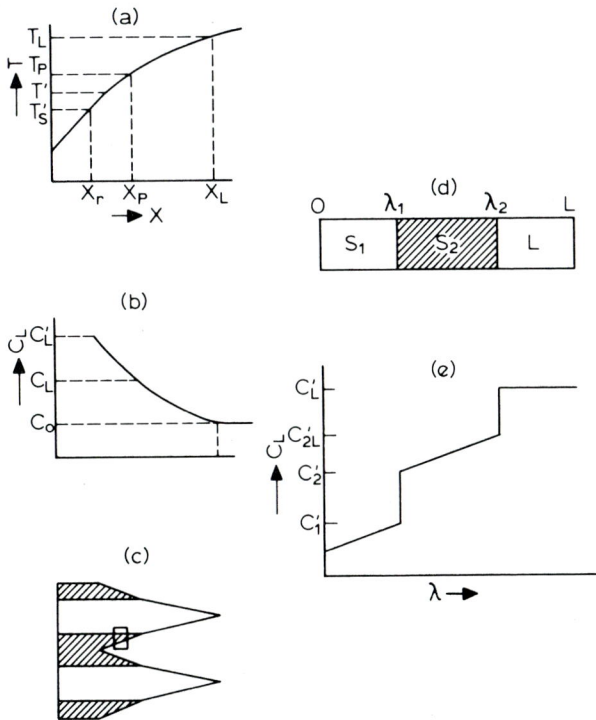

$$L + S_1 \rightarrow S_2 \qquad (3)$$

a temperature distribution during unidirectional solidification; *b* liquid composition in solid and liquid and in liquid regions; *c* illustration of sequence of formation of solid phases indicating position of characteristic volume element; *d* sketch of volume element considered in solid and liquid region; *e* concentration profile along length of volume element

2 Model for directional solidification of peritectic alloys at low values of *G/R*

However, the ratio of L and S_1 that are consumed is not fixed by the phase diagram, but is determined by the extent of diffusion through the S_2 rim. In the limit of the case of negligible diffusion in S_2 during freezing, no S_1 would be consumed, freezing would not be complete until the volume element reached the melting point of B, and the last solid to form would be pure B.

As the G/R ratio is increased, the diffusion flux along the concentration gradient in the liquid (Fig. 2*b*), increases and the temperature of the dendrite tips T_t is lowered below the equilibrium liquidus T_L by an amount of ΔT, which is given approximately by[10-13]

$$\Delta T = T_L - T_t = \frac{D_L G}{R} \qquad (4)$$

where D_L is the diffusion coefficient of the solute in the liquid. Thus, S_1 first forms at a lower temperature and the composition of the first S_1 to form is increased above $K_1 C_0$ according to the solidus curve. Also, it is more likely that as G/R increases, the dendrites will become simpler in morphology and eventually side branches will not form. (An unbranched rod interface would be termed cellular by Flemings.[12]) At a sufficiently high value of G/R, the constitutional undercooling will be sufficient to suppress the dendrite tip temperature to the equilibrium solidus. For compositions between pure A and C_1, this condition is the same as that for plane front growth of S_1

$$G/R \geqslant \frac{m_1 C_0 (1 - K_1)}{D_L K_1} \qquad (5)$$

where m_1 (defined positive) is the slope of the liquidus curve for S_1 at $C_L = C_0/K_1$. This constitutional supercooling criterion for plane-front freezing of S_1 is plotted as $\overline{oa}$ in Fig. 3.

The condition for suppressing the S_1 dendrite tips below the peritectic temperature, which is the equilibrium solidus for compositions between C_1 and C_2 is given by

$$G/R \geqslant \frac{m_1 (C_L - C_0)}{D_L} \qquad (6)$$

This relationship is plotted as $\overline{abd}$ in Fig. 3.

As the diffusion flux across the mushy zone causes an increase in the solute concentration of the dendrite tips, it also causes a decrease in solute concentration of the liquid at the end of freezing. For compositions above C_2, a planar S_2/L interface will result when the composition at the dendrite roots is C_0/K_2 and T'_s equals the equilibrium solidus temperature as in the constitutional supercooling criterion written for S_2

$$G/R \geqslant \frac{m_2 C_0 (1 - k_2)}{D_L} \frac{}{k_2} \qquad (7)$$

where m_2 (defined positive) is the slope of the liquidus curve for S_2 at $C_L = C^f_L$. For compositions between pure A and C_2 a planar interface between S_2 and L can be achieved when values for the metastable extension of the S_2 liquidus and solidus are substituted in equation (7). In this region, the S_2/L interface would be at a temperature higher than the peritectic temperature. This criterion is plotted as $\overline{obc}$ in Fig. 3.

Region A represents the normal freezing conditions at small values of G/R. Here, dendrites of the primary phase S_1 would grow ahead of the peritectic isotherm

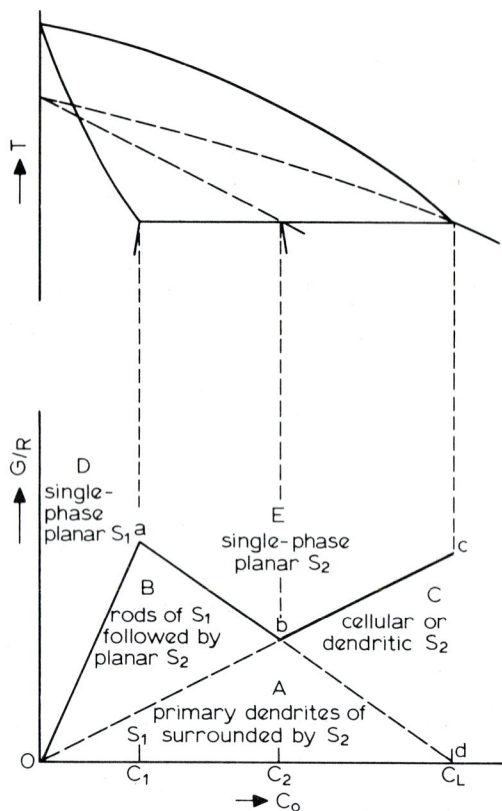

3 Critical G/R v. C_0 plot to obtain planar solid/liquid interface as defined by equations (5)–(7); expected microstructures and interface morphology are also shown

and be surrounded by gradually thickening S_2 below the peritectic isotherm, as shown in Fig. 2. For higher values of G/R in Region B primary dendrites or cells of S_1 will grow with higher undercoolings at the tips and S_2 will freeze at a plane front at the peritectic temperature. For higher initial solute contents in Region C, S_2 will grow with either a dendritic or cellular interface. For still higher values of G/R in Regions D and E, singular-phase material will grow at a plane front. For low solute contents in Region D, S_1 will grow. For higher solute contents in Region E, S_2 will form. For compositions between C_1 and C_2 at values of G/R in Region E, single-phase S_2 will grow at the metastable S_2 solidus, even though two phases would be stable. Since the metastable solidus is higher than the peritectic temperature, *the requirement for suppressing the formation of S_1 would not be given by equation (6) but would fall on a line slightly below ab in Fig. 3.* For simplicity of presentation, the revised criterion is not shown.

Figure 3 represents, clearly, the initial growth forms under the assumptions listed above, namely:

  (i) no undercooling due to kinetic or curvature effects

  (ii) equilibrium at all solid/liquid interfaces

  (iii) no undercooling before nucleation of solid phases

  (iv) perfect mixing in the liquid along any isothermal surface.

However, the structure that forms first would not be a final equilibrium structure and so changes in structure

might occur before the end of freezing. For example, in Regions A and B up to initial solute contents of C_1 single-phase S_1 would be the equilibrium product of solidification. Diffusion in S_1 during freezing would result in a reduction in the extent of coring in primary S_1 and a reduction in the amount of S_2 that formed.[10] For substantial diffusion in S_1 during freezing, S_2 would not form. Similarly in Region A, S_1 will tend to convert to S_2 by the reaction described in equation (2) as a result of diffusion through the rim of S_2. The interface between S_1 and S_2 might move first in one direction then another due to the difference in diffusion rate within S_1 and S_2 and the slopes of the solvus lines.

In Region E for compositions between C_1 and C_2, single-phase S_2 would form at the metastable solidus temperature, which would be above the peritectic temperature. After forming, the S_2 would cool through two regions of two-phase equilibrium, $S_1 + L$ and then $S_1 + S_2$, which are the equilibrium phase distribution. If S_1 is slow to nucleate and/or diffusion in the solid is sluggish, metastable S_2 expected after freezing would persist to room temperature. If S_1 were to nucleate and grow sufficiently below the peritectic temperature, no interference with the liquid/solid interface would occur. However, if S_1 nucleated close to the interface and diffusion were rapid, the metastable S_2/L interface could shift to an equilibrium S_1/L interface which would be cellular or dendritic at these conditions.

EXPERIMENTAL

Alloys from the Pb–Bi and Sn–Cd system were directionally solidified at a variety of G/R values in order to establish the growth conditions for peritectic systems in which diffusion in the solid phases is relatively slow. All samples were made from pure metals with initial purities greater than 99·99%. Rods of the desired composition were formed by first melting an alloy of composition C_0 in a crucible, and drawing the molten alloy into a Pyrex tube by means of a partial vacuum. Before drawing the sample, a 100 μm chromel–alumel thermocouple in a 0·1 cm double-bore mullite protection tube was inserted in the Pyrex tube. The thermocouple tip was coated with boron nitride slurry.

Samples were placed in a vertical resistance-zone melting and freezing apparatus, shown in Fig. 4. The distance between the heating element and the following chills was set at 0·5 cm. Gradients of up to 360° and 340°C/cm were achieved in Pb–Bi and Sn–Cd, respectively. At least one zone-melting pass was made before the final directional-freezing pass. In the last pass, the solid/liquid interface was quenched by switching off the power to the heater and increasing the water flow through the chills. Owing to the highly oxidizing nature of the Pb–Bi alloys a special polishing technique involving a combination of mechanical and electro-polishing techniques was developed and used. The electrolyte consists of 40 ml of glycerol, 10 ml of acetic acid, and 10 ml of nitric acid. Samples of Sn–Cd were polished to 0·01 μm alumina and etched with a solution containing 5 ml of nitric acid and 95 ml of lactic acid.

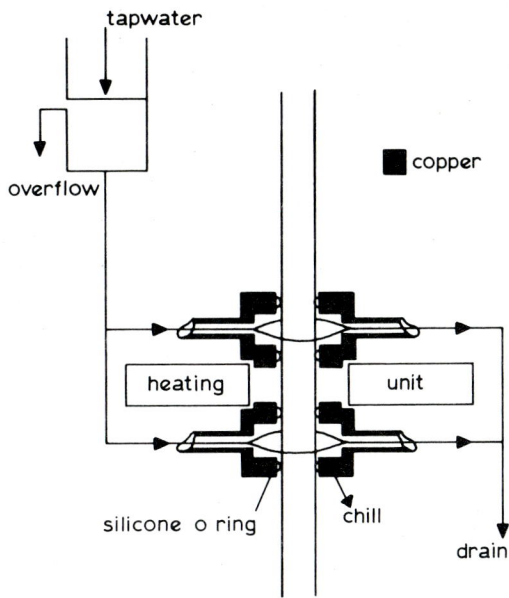

4 Low-temperature vertical zone-melting unit showing heating and cooling arrangements

5 Constitution diagram of Pb–Bi system[16] with extended metastable solidus; data points represent actual interface temperatures measured at solid (plane front β)/liquid interface

Pb–Bi system

As indicated in the phase diagram for Pb–Bi (Fig. 5), the peritectic reaction at 184°C is

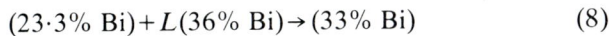

$$(23{\cdot}3\% \text{ Bi}) + L(36\% \text{ Bi}) \rightarrow (33\% \text{ Bi}) \qquad (8)$$

Samples were directionally solidified in the zone-refining apparatus with bismuth compositions ranging from 24 to 35 wt-% and for values of G/R ranging from $0{\cdot}48 \times 10^6$ to $4{\cdot}58 \times 10^{6}$°C s/cm². Table 1 lists the samples made and the microstructural observations. Figure 6 shows the range of microstructures observed. Photomicrographs at the bottom are of transverse

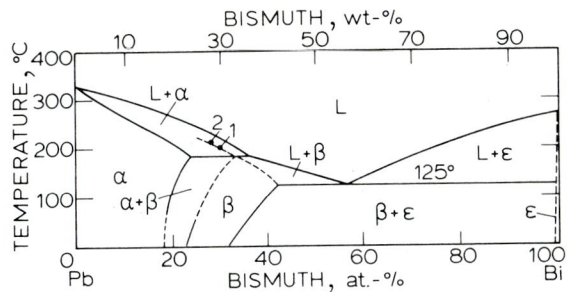

sections, perpendicular to the heat flow. Photomicrographs at the top are of longitudinal sections, parallel to the heat flow. The longitudinal sections were taken at the position of the quenched solid/liquid interface.

At low values of G/R, Fig. 6a, the α-phase leads the β-phase. The dendrites are branched and are seen to be highly plate-like at low values of G/R, (see Fig. 7). At higher values of G/R (Fig. 6b) the α-phase still leads the β, but the rods of α are not branched and the β appears to form at a plane front. However, the latter observation is tenuous. If there were a liquid groove extending below the region where β first forms, it would not be readily seen as the liquid and β in this system are nearly the same composition, i.e. $k_2 \sim 1$. In the transverse section, the structure is similar to a directionally solidified eutectic, except that the spacing is coarser. The spacing between the rods of α is 60 μm. At high values of G/R (Fig. 6c), the α rods are completely suppressed and β grows at a plane front. If the conditions are close but not sufficient for a planar

Table 1 Composition, growth conditions, and observed microstructures in Pb–Bi alloys

C_0, wt-% Bi	G, °C/cm	R, cm/s × 10⁴	G/R, °Cs/cm² × 10⁻⁶	Structure	Quenched S/L interface shape
24	220	4·58	0·48	$\alpha + \beta$	Branched α dendrite leading
24	244	2·83	0·86	$\alpha + \beta$	Rods of α leading
24	298	1·39	2·1	$\alpha + \beta$	Rods of α leading
26	250	3·47	0·72	$\alpha + \beta$	Branched α dendrite leading
26	360	2·83	1·27	$\alpha + \beta$	Rods of α leading
26	350	2·22	1·57	$\alpha + \beta$	Rods of α leading
26	255	1·39	1·84	$\alpha + \beta$	Rods of α leading
26	275	1·39	1·98	$\alpha + \beta$	Branched α dendrite leading
28	200	1·94	1·03	$\alpha + \beta$	Rods of α leading
28	260	1·66	1·55	$\alpha + \beta$	Rods of α leading
28	338	0·83	4·87	β	Plane front β
30	230	4·58	0·50	$\alpha + \beta$	Rods of α leading
30	283	4·39	0·64	$\alpha + \beta$	Rods of α leading
30	283	1·67	1·70	$\alpha + \beta$	Rough interface
30	220	1·02	2·17	β	Plane front β
30	226	0·83	2·72	β	Plane front β
32	277	4·58	0·60	$\alpha + \beta$	Rods of α leading
32	225	1·39	1·60	β	Cells of β
32	275	1·39	1·98	β	Plane front β
33	152	5·55	0·27	$\alpha + \beta$	Rods of α leading
33	195	4·17	0·47	$\alpha + \beta$	Rods of α leading
33	258	1·39	1·86	β	Plane front β
33	266	2·22	1·20	β	Cells of β
35	354	5·55	0·64	β	Cells of β
35	340	4·58	0·71	β	Cells of β
35	255	3·06	1·10	β	Cells of β

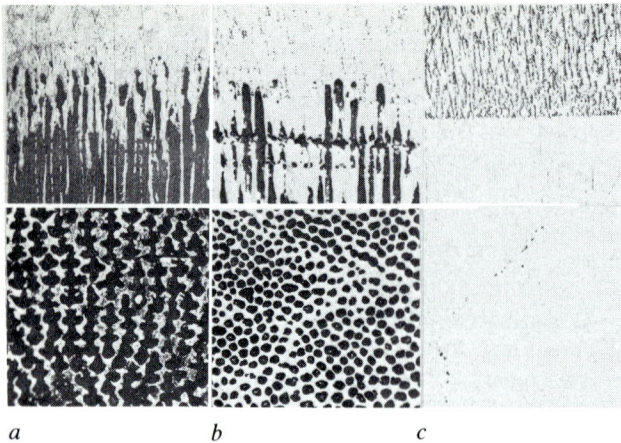

a $C_0 = 28$-wt-% Bi; $G/R = 1.03 \times 10^{6}$°Cs/cm^2; *b* $C_0 = 28$ wt-% Bi, $G/R = 1.55 \times 10^{6}$°Cs/cm^2; *c* $C_0 = 28$ wt-% Bi, $G/R = 4.87 \times 10^{6}$°Cs/cm^2

6   Typical aligned structures showing effect of *G/R* ratios; in each case the top micrograph is a longitudinal section including the quenched interface and the bottom is a transverse section   [×20]

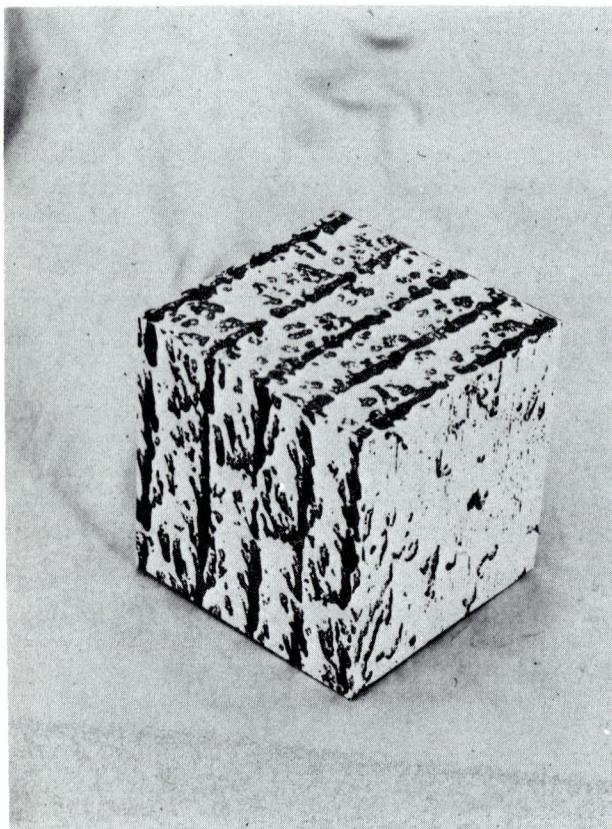

7   Three-dimensional composite photomicrograph showing branched plate-like dendrites of *α*-phase surrounded by *β*-phase

β/L interface to be stable, a cellular interface develops, as shown in Fig. 8.

Figure 9 summarizes the results for Pb–Bi. Lines representing the criteria in equations (5)–(7) are drawn in Fig. 9. These lines are surprisingly good fits for the

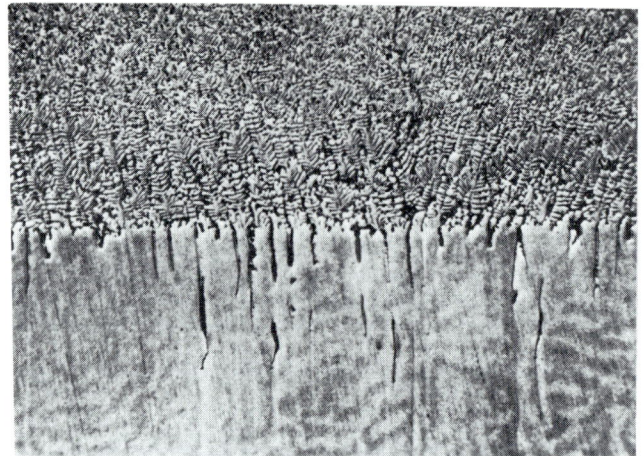

8   Photomicrograph of quenched interface which is cellular *β*   [×50]

boundaries separating the morphological behaviour of the alloys.

If the solidification behaviour of Pb–Bi supports the theory presented above, then the temperature of the β/L interface for conditions of single-phase growth at compositions between 24 and 33% Bi should be above the peritectic temperature. Two experiments were run to determine the interface temperature. In each rod, two thermocouples were inserted at a distance of 0·5 cm apart. The sample was directionally solidified until the bottom thermocouple indicated a temperature below the peritectic; then the sample was quenched. The distance of the interface from both thermocouples was measured and the temperature of the interface estimated by assuming a linear temperature gradient in

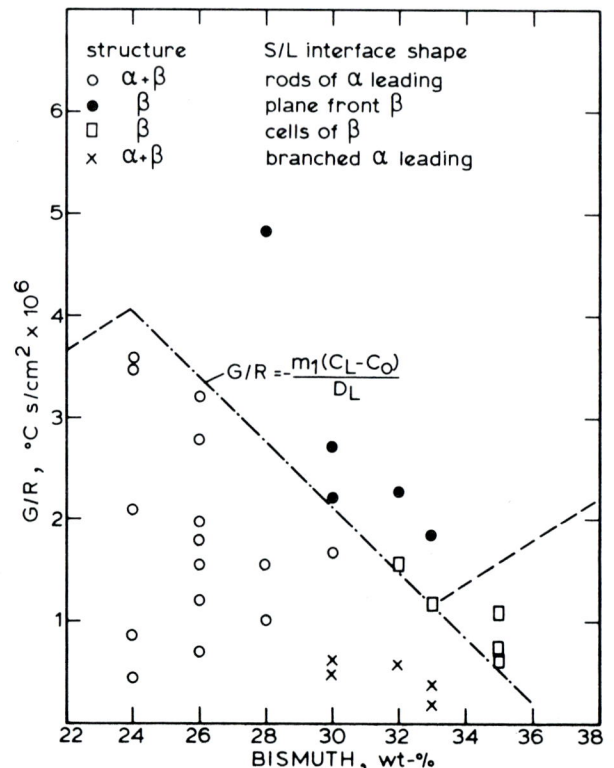

9   Structures of directionally solidified Pb–Bi alloys and interface shapes as functions of *G/R* and C_0

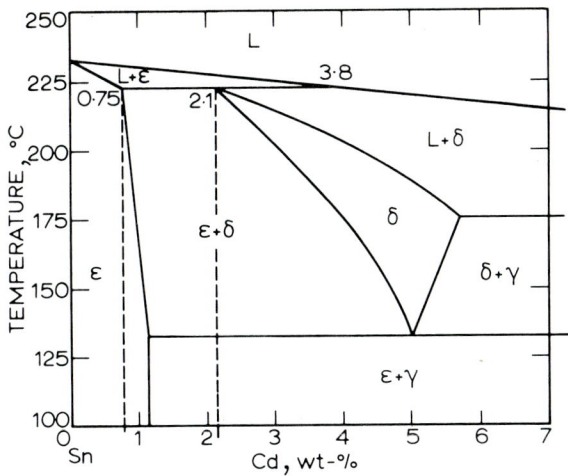

10 Constitution diagram of Sn–Cd system

a $C_0 = 1.5$ wt-% Cd, $G/R = 0.9 \times 10^5$°Cs/cm^2; *b* $C_0 = 1.5$ wt-% Cd, $G/R = 1.03 \times 10^6$°Cs/cm^2

11 Typical longitudinal and transverse micrographs showing effect of G/R ratios [×65]

the region of the two thermocouples. The temperatures measured for Pb–28 Bi and Pb–30 Bi alloys were on a line that is consistent with the metastable extension of the β solidus. These points are plotted as (1) and (2) in Fig. 5.

Sn–Cd system

As indicated in the phase diagram for Sn–Cd, (Fig. 10) the peritectic reaction at 221°C is reported as[17]

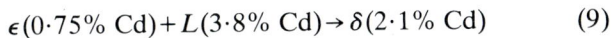

$$\epsilon(0.75\% \text{ Cd}) + L(3.8\% \text{ Cd}) \rightarrow \delta(2.1\% \text{ Cd}) \qquad (9)$$

Samples were directionally solidified in the zone melting and freezing apparatus with cadmium compositions ranging from 1.0 to 1.75 wt-% and for values of G/R ranging from 0.5×10^5 to 1.5×10^6°C s/cm^2. Table 2 lists the experiments carried out and the microstructural observations. The range of microstructures observed in Sn–Cd was similar to that in Pb–Bi with one major additional microstructural feature. Figure 11 indicates the range of microstructures that parallel Pb–Bi. Figure 11*a* shows the composite-like microstructure obtained at moderate values of G/R as rods of ϵ lead what appears to be a planar interface of δ. At high values of G/R, single phase δ grows at a planar interface. Additionally, some samples at high values of G/R

exhibited a banded structure (Figs 12 and 13). The bands generally alternate between ϵ and δ and run transverse to the growth direction, parallel to the isotherms. Occasionally a band will consist of ϵ and δ growing side by side.

In Fig. 14 the results for Sn–Cd are summarized. The results given by Boettinger[1] are also plotted in Fig. 14. With the exception of the region of banding, the results are consistent with the theory presented above. The lines based on the criteria of equations (5)–(7) are not in as good agreement with the boundaries of morphological behaviour as was the case for Pb–Bi.

There may be several reasons for the banding phenomenon. In one explanation of the banding the δ/L interface which is expected to form in region B would be

Table 2 Composition, growth conditions, and observed microstructures in Sn–Cd alloys.

C_0, wt-%	G, °C/cm	R, cm/s × 10^4	G/R, °Cs/ cm^2 × 10^{-5}	Structure	Quenched S/L interface shape
1·0	241	16·66	1·45	$\epsilon + \delta$	Rods of ϵ leading
1·0	230	8·33	2·76	Banded ϵ and δ	—
1·0	200	4·58	4·30	Banded ϵ and δ	—
1·0	152	2·22	6·84	Banded ϵ and δ	—
1·5	152	16·66	0·90	$\epsilon + \delta$	Rods of ϵ leading
1·5	234	11·11	2·10	Banded ϵ and δ	—
1·5	150	4·58	2·76	Banded ϵ and δ	—
1·5	143	1·39	10·30	δ	Plane front δ
1·5	340	2·70	12·2	δ	Plane front δ
1·75	82	16·66	0·49	δ	Cells of δ
1·75	251	26·28	0·95	δ	Cells of δ
1·75	127	4·58	1·52	δ	Cells of δ
1·75	95	4·58	2·10	δ	Cells of δ
1·75	179	3·47	5·20	δ	Plane front δ

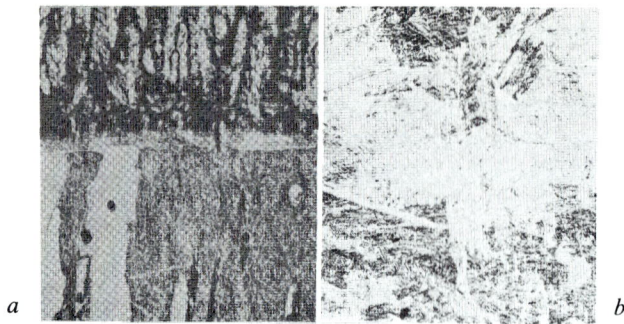

12 *a* **Photomicrograph of longitudinal section including quenched interface showing ε-phase nucleating and growing along with δ-phase without any interference with planar solid/liquid interface;** *b* **transverse micrograph of same sample**

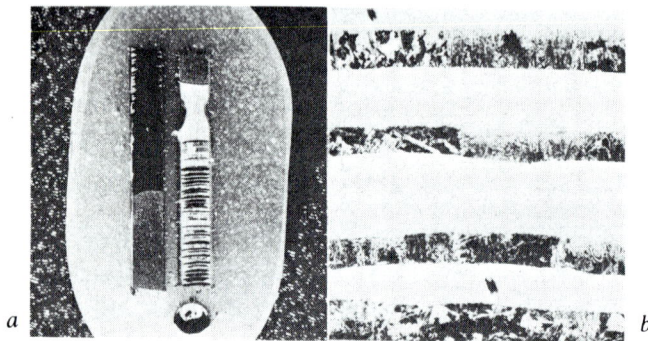

a photomacrograph of directionally solidified sample, [×2]; *b* photomicrograph of same sample showing alternate layers of ε and δ [×25]

13   Banding in directionally solidified Sn–Cd alloys

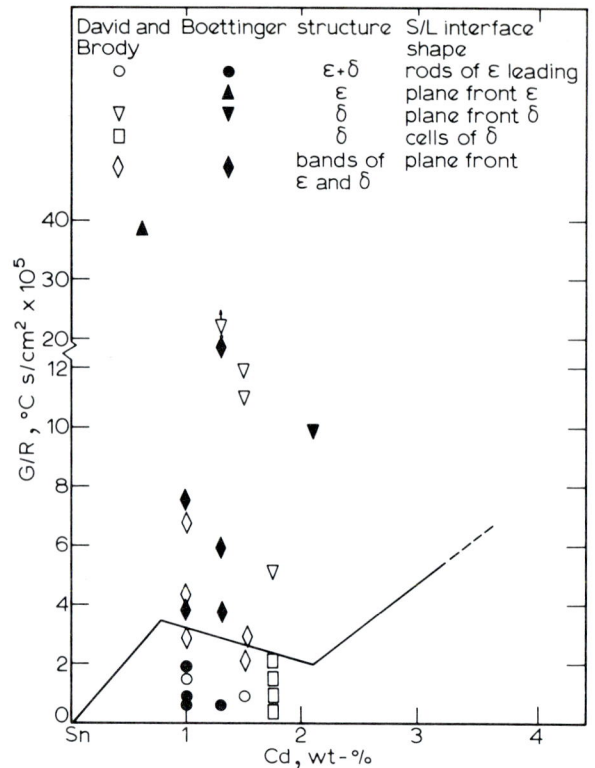

14   Structures of directionally solidified Sn–Cd alloys and interface shapes as functions of *G/R* and *C*₀

disrupted by nucleation of ε behind the interface. This ε would grow up to the interface and grow for a short time until the solute built up to a point to nucleate δ again. This process is repeated every time ε is nucleated at high temperature. For higher values of G/R, banding is extinguished. This can be explained because the solidified δ cools more quickly to a temperature where ε will not nucleate owing to the higher temperature gradients.

SUMMARY

A theory has been proposed and experiments with Pb–Bi and Sn–Cd have given reasonable confirmation that peritectic alloys can be directionally solidified to give a variety of microstructures.

1 At low values of G/R, branched dendrites of the pro-peritectic phase will be surrounded below the peritectic isotherm by the peritectic product phase.

2 At moderate values of G/R, simple growth forms (such as rods) of the pro-peritectic phase grow with high undercooling ahead of a planar interface of the peritectic product phase. The structure resembles a eutectic composite except it is coarser.

3 At high values of G/R, formation of the pro-peritectic phase is suppressed and the peritectic product phase grows at a plane front. For those compositions

that would be two-phase according to the equilibrium phase diagram, the temperature of the planar interface is above the peritectic temperature and is the temperature on the metastable extension of the solidus of the peritectic product phase. The phase that forms will be metastable (supersaturated with solvent, for $k < 1$). The planar interface can be disrupted by nucleation of the pro-peritectic phase behind the interface.

ACKNOWLEDGMENTS

The authors wish to thank Dr G. I. Scherbakov for his contribution during the initial stages of this investigation. The authors are grateful to the Office of Naval Research (ONR Contract N00014-67-A-0402-0003) for support of this work.

REFERENCES

1 W. J. BOETTINGER: *Metall. Trans.*, 1974, **5**, 2 023

2 N. J. W. BAKER AND A. HELLAWELL: *Met. Sci.*, 1974, **8**, 353

3 A. P. TITCHENER AND J. A. SPITTLE: *Acta Metall.*, 1975, **23**, 297

4 D. H. ST. JOHN AND L. M. HOGAN: *ibid.*, 1977, 77

5 G. I. SCHERBAKOV *et al.*: *Scripta Metall.*, 1974, **8**, 123

6 D. R. UHLMANN AND B. A. CHADWICK: *Acta Metall.*, 1961, **9**, 835

7 B. CHALMERS: 'Physical metallurgy', 271, 1959, New York, John Wiley and Sons

8 M. C. FLEMINGS: 'Solidification processing', 117, 1974, New York, McGraw-Hill Book Company

9 J. A. SARTELL AND D. J. MACK: *J. Inst. Met.*, 1964, **93**, 19

10 H. D. BRODY AND M. C. FLEMINGS: *Trans. AIME*, 1966, **236**, 615

11 T. F. BOWER *et al.*,: *ibid.*, 1966, **236**, 624

12 M. C. FLEMINGS: 'Solidification processing,' Chapter 3, 1974, New York, McGraw-Hill Book Company

13 R. M. SHARP AND M. C. FLEMINGS: *Metall. Trans.*, 1974, **5**, 823

14 R. M. SHARP AND M. C. FLEMINGS: *ibid.*, 1973, **4**, 997

15 A. B. GRENINGER: *Trans. Amer. Inst. Min. Met. Eng.*, 1937, **124**, 379

16 M. HANSEN AND K. ANDERKO: 'Constitution of binary alloys', 1958, New York, McGraw-Hill Book Company

17 W. T. PELL-WALPOLE: 'Metals handbook', 1 189, 1948, ASM, Cleveland

DISCUSSION

Session 2: Eutectic and peritectic solidification

Chairman: W. Kurz (Ecole Polytechnique Federale Lausanne, Switzerland)

The following papers were discussed *Keynote Address: Eutectic and peritectic solidification* by M. Hillert; *Complex–regular growth in the bismuth–lead eutectic* by D. Baragar, M. Sahoo, and R. W. Smith; *Orientations and interfaces in directionally solidified eutectic systems* by I. G. Davies and D. D. Double; *Crystallographic study of the eutectic microstructure in Al–Si system* by K. Kobayashi, P. H. Shingu, and R. Ozaki; *Study of the structures of Al–Al₃Ni composites grown in non-uniform cylindrical moulds* by N. A. El-Mahallawy and M. M. Farag; *Influence of various elements on Al–Al₆Fe eutectic system* by P. G. Keong, J. A. Sames, C. M. Adam, and R. M. Sharp; *Influence of silicon and aluminium on the solidification of cast iron* by T. Carlberg and H. Fredriksson; *Role of phase boundary adsorption in the formation of spheroidal graphite in cast iron* by W. C. Johnson and H. B. Smartt; *Transition from peritectic to eutectic reactions in iron-base alloys* by H. Fredriksson; *Unidirectional solidification of peritectic alloys* by A. Ostrowski and E. W. Langer; *Controlled solidification of peritectic alloys* by H. D. Brody and S. A. David

DR B. CANTOR (University of Sussex): I would like to ask Professor Hillert about the slide of the eutectic dendrite that he showed at the end of his paper. Which alloy system was it in and under what sort of conditions was it prepared? I have seen similar dendritic-type eutectic structures in aluminium alloys, but in all the cases that I have seen, it has been on the surface of a chill-cast specimen so that the solidification rate has probably been extremely high. I have never been able to prepare such a structure in any sort of controlled way.

PROFESSOR HILLERT: I apologise if I gave the wrong impression. This was not a picture taken by myself. I took it from the literature, a publication by Weingard, and these structures were seen in the aluminium–copper system containing a third element.

DR R. M. JORDAN (Alcan International Ltd, Banbury): You can grow eutectic dendrites in the Al–Fe–Mn and Al–Fe–Ni systems at growth rates greater than about 10^{-2} cm/s.

In their paper on cellular growth in eutectic alloys Barager *et al.* show a transverse section through a directionally frozen sample of Pb-Bi eutectic in Fig. 4. They cite the variation in lamellar spacings shown in this micrograph as evidence that the Bi plates do not have difficulty in adjusting their width during growth.

I do not understand how plate width variations seen on a single transverse section can be evidence that the widths change easily in the longitudinal direction during growth. Perhaps the authors would like to comment further on this.

PROFESSOR R. W. SMITH: Increasingly, we find evidence to suggest that faceting materials are able to grow normal to what we presume to be the faceting planes. For instance, eutectic specimens in which the non-faceting phase has been removed by deep etching or sublimation frequently show protrusions. In broken lamellar eutectics these protrusions may even space the gap between adjacent perforated lamellae. We are looking closely at such events and are trying to determine the associated kinetic undercooling, which we believe to be small.

DR B. CANTOR: In the Al–CuAl$_2$ eutectic Dr Double has shown a graphic and elegant model of atomic matching between the (111) Al and (211) CuAl$_2$ planes. This he uses to explain the preferred orientation relation in this eutectic. However he notes in the text of his paper that in many cases the lamellar habit plane is not composed of the two planes. This is a very curious situation. The preferred orientation relationship is explained by good atomic matching between the two phases although the material frequently prefers to exhibit poorer atomic matching on quite different planes and the high atom-matching planes are not in fact observed.

A similar situation exists for the Al–Al$_3$Ni eutectic where Garmong and Rhodes have argued that good atom matching on two planes, I think the (102) Al$_3$Ni and (111) Al, determines the crystallography of the structure. Once again, there are reported cases where these planes are parallel in the eutectic orientation relation, but this particular facet is not observed on the Al$_3$Ni rods. Dr Double points out in his paper that for Al–CuAl$_2$, the lamellar interface can contain steps so that there may be microfacets of the (111) Al∥(211)/CuAl$_2$ type. If indeed the atom-matching across these planes is an important feature leading, say, to a low-energy interface, then I cannot see why the plane of the interface should be made up of such microfacets. I would expect one of two results. First, a strong tendency for a low energy (111) Al∥(211)CuAl$_2$ interface would produce a macroscopic facet, i.e. the lamellar interface would be parallel to these planes regardless of other effects. Alternatively, a second, strong crystallographic effect such as anisotropic interface kinetics might also be present. In this case, why does the interface still not exhibit a macroscopic facet, possibly no longer parallel to the heat flow direction? This type of effect seems to occur in the Al–Ag eutectic system. Without a good explanation of this, I think we may be driven to conclude that atomic matching to produce a low-energy interface is at best a second order effect.

DR DOUBLE: I admit there is a problem. The Al–CuAl$_2$ system is rather an unfortunate example; it is certainly not the simplest system. Kraft was able to explain the orientation relationship found between the faces in terms of the similarity and the atom-matching between (111) Al and (211) CuAl$_2$ planes. There is an awkwardness however, in that the interface plane does not in fact coincide with these planes but consistently deviates about 11° away from them. I do not know the reason why this occurs. The work of Garmong showed that the interface plane in the Al–CuAl$_2$ system contained a microscopically stepped interface profile and it seems likely that these steps would contain (111)Al∥(211) CuAl$_2$ facets, but this has not so far been proved.

This type of irrationality of habit plane is also observed in the Al–Zn system. It is perhaps significant that both these eutectics also display macroscopic helical twisting of the lamellar planes during growth. There seems to be some connection between these effects, but to date the reason for the spiral rotation phenomenon is not clearly understood.

On talking about the Al–Ni system, I assume we are talking about the Al–Al$_3$Ni rod eutectic system. In a rod system there is a different situation, where unlike the lamellar system one must have more than one set of interface planes to form the rod-like structure. Some argument exists about the identity of the interface planes in the Al–Al$_3$Ni system but basically one can get fairly reasonable atom-matching between the Al matrix and the (103)Al$_3$Ni and (102)Al$_3$Ni planes. Such matching appears to be reasonably consistent with the overall morphology of the rods.

DR J. HUNT (University of Oxford): I wonder whether the question on pressure differences across the α–β boundary which Professor Hillert raised is of importance to Dr Cantor's question. A possible way of resolving the pressure difference problem is to have a curved α–β boundary just behind the cusp followed by some plastic deformation which straightens the α–β boundary as it grows (see diagram).

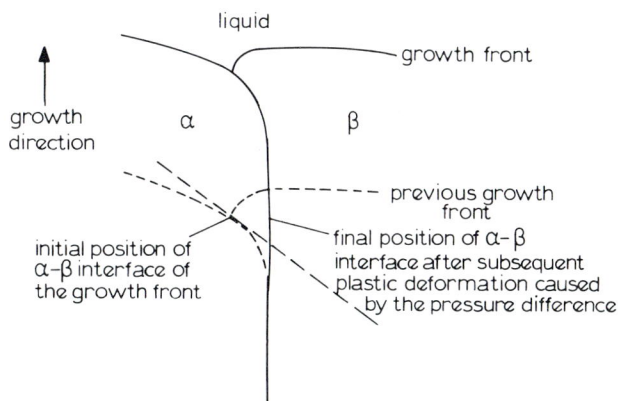

The low energy boundary selected by growth would then not be the final lamellar boundary but would be a few degrees off this orientation.

PROFESSOR HILLERT: The mechanism that Dr Hunt suggests may certainly play some role but it is difficult to say whether it can provide the complete answer to the balancing of forces between the three phases. Unfortunately, it might be very difficult to work out this model quantitatively and thus to make an experimental test possible.

DR L. M. HOGAN (University of Queensland): Dr Kobayashi has shown that the aluminium grains in the Al–Si eutectic have a favoured orientation relationship with the silicon. Since the silicon changes orientation frequently during growth, this suggests that the aluminium in the eutectic must be polycrystalline with a fairly substantial range of orientations. In other words, there should be large grain boundaries visible in the aluminium phase between the silicon plates. I wonder if Dr Kobayashi could say if these have been observed. They are very difficult to observe.

DR KOBAYASHI: Within a single 'eutectic grain' the aluminium displays a number of high angle boundaries. These are frequently observed experimentally by electron diffraction, and one example is shown in Fig. 5.

DR G. HAOUR (Battelle Research Centre, Geneva): It is indicated in Dr Kobayashi's paper that in the Al–Si eutectic, silicon grows by the TPRE mechanism. It would be interesting to learn whether as a consequence of this, Dr Kobayashi favours any given explanation for the modification mechanism of Al–Si by sodium?

DR H. JONES (University of Sheffield): Drs Sharp and Adam describe the effects of precipitation hardening alloying additions in producing strength and ductility increases in Al–Fe eutectic alloys. I would like to suggest an alternative explanation for the increases in ductility that they set out to achieve. Clearly the effect of introducing precipitation hardening additions will be to increase strength but I find it difficult to understand how this could also increase ductility and my explanation for that would be that these additions might change the proportion of the eutectic and the matrix phases. Do the authors observe such an effect on making their alloying additions and might this be a more plausible explanation of the improvement in ductility?

DR ADAM: The increased ductility of Al–Al$_6$Fe eutectic on addition of solute elements such as Cu, Mg, Zn, and Mn has been attributed to precipitation hardening but the microstructure of the alloys shows that fibres are curved into the transverse direction, so that some resistance to the propagation of a deformation band ahead of a crack is effectively provided. There is also some contribution from the volume fraction effect mentioned by Dr Jones. We believe that the predominant effect, however, is due to the precipitate species effectively locking dislocations in the intercellular areas, so that on fibre failure within the cell, the released stored energy from the fibres is absorbed in work-hardening the inter-cellular areas. In precipitation hardened alloys without fibres, deformation would simply cause work hardening of the matrix at increasing stresses. In these alloys Al–Fe–Mg–Zn, we think that the work-hardening rate is very low, at effectively a constant stress, when fibres commence to fail.

DR V. KONDIC (University of Birmingham): I would like to address a question to Professor Fredriksson on the subject of grey cast irons. Whenever non-isothermal cooling conditions exist three main groups of variables have been found to effect the structures of grey cast iron: impurities, basic composition, and cooling rate conditions.

In all cases a characteristic freezing temperature is associated with a particular metastable or stable graphite eutectic structure. Has Professor Fredriksson measured the freezing temperature of various alloys in his experiments? I believe that there could be a similar relationship between freezing temperature and eutectic structures of grey iron in the unidirectional freezing experiments as in the multi-directional cooling work.

PROFESSOR FREDRIKSSON: No, we have not measured the temperatures at the solidification front. It seems possible, however, to relate different morphologies in cast iron to the growth rate and/or undercooling with the help of known growth laws if one considers the flake graphite as a primary precipitate.

DR D. A. MELFORD (Tube Investments Research Laboratories): I should like to ask Dr Johnson a question about the role of sulphur in the formation of graphite nodules. First, I would like to say that what

seems to happen when a nodule forms is that calcium–magnesium sulphide forms from the melt and a magnesium–aluminium silicate spinel grows around it epitaxially. The graphite grows epitaxially on the spinel.

If the sulphide/spinel nucleus forms from the melt at a temperature above the graphite liquidus, it can become poisoned by adsorbed sulphur as the melt cools. When examined in a transmission microscope with X-ray facilities, the combined sulphur signal, from the calcium–magnesium sulphide remains stable, whereas a more widely distributed but fainter X-ray emission from the adsorbed sulphur disappears over the course of about an hour. Is it possible that, in the higher vacuum of the Auger equipment used by Dr Johnson, electron emission from adsorbed sulphur was not detected because of rapid evaporation of the adsorbed material?

DR JOHNSON: My experience indicates that electron stimulated desorption of volatile species is more of a problem than vacuum desorption. I assume that the TEM experiments described use beam energies of at least 100 kV, while the Auger experiments use 5–10 kV electron beams; hence the beam desorption would be less of a problem in the Auger situation! The fact remains that elemental sulphur was never detected in ductile cast irons in our experiments. Perhaps the elemental sulphur detected in the TEM studies was contamination due to sample preparation rather than adsorption of elemental sulphur during solidification.

Another point to consider is that it is necessary to overdope with magnesium to obtain ductile iron. Hence it is difficult to believe that excess elemental sulphur could be available after modification. I also question whether spinel is necessary for nucleation. It may be sufficient but I doubt if it is necessary.

PROFESSOR HELLAWELL (University of Wisconsin, USA): I also question whether spinel is necessary for nucleation. I was somewhat surprised to hear Dr Melford state, rather as a matter of fact, that the spherulitic structures occurred when graphite was able to nucleate on minute particles of a spinel. Do we know that for a fact? Is there always a nice piece of spinel in the middle of every one?

DR MELFORD: I cannot answer that question. It certainly can happen as a fact in chill cast magnesium treated iron.

PROFESSOR HELLAWELL: We know, for example, from the work of Sadocha and Gruzleski that spherulites can also be produced in a grey iron by vacuum purification of the melt. How would this be compatible with nodules nucleating—always—on spinel particles?

DR C. R. LOPER (University of Wisconsin, USA): Following the remarks of Dr Hellawell, it should be noted that the literature does contain several reports indicating the presence of magnesium containing spinels in the centre of spheroidal graphite particles. Work in process at the University of Wisconsin however, indicates that those compounds are just as likely to be found distributed throughout the matrix with no association whatsoever with the graphite spheroids. This may indicate a predominance of these compounds throughout the structure, and that these compounds could act as heterogeneous nucleation sites. But these compounds may not be the controlling factor in causing the spheroidal form of graphite.

DR E. SELCUK (Middle East Technical University, Turkey): I remember a paper by Hunter and Chadwick concerning a careful examination of thin slices of grey cast iron and spherulitic cast iron and I think they reached the conclusion that in spherulitic graphite cast iron and grey cast iron, the centre midline was a small flake. I think this is in line with Professor Hellawell's suggestion that, on keeping grey cast iron under vacuum for a long time, one would get spherulitic graphite. I do not think spinels are always there to nucleate spherulitic cast iron.

DR J. E. GRUZLESKI (McGill University, Montreal): I have two questions for Dr Johnson. First, is it possible to quantify your results so that you can actually tell us what the concentration of impurity elements at the boundary is?

The second question relates to the oxygen which you found at the interface between the flake and the iron matrix. I presume that you feel that this oxygen arose during the solidification process but I wonder if in fact some, or maybe even all of this oxygen may have got to the boundary by interface boundary diffusion in the time between when the sample was solid and when you actually looked at it? The reason I mention this is because in our work on vacuum treated cast irons, we have tried very often to measure the oxygen content and we find that in fact the only way in which you can get any reproducible results is to chill the sample, and have no graphite present. The reason why we think this arises, although it is really only a presumption, is because of oxygen diffusion in the boundary between the graphite and the iron matrix. Perhaps some of the oxygen may have arisen from there. Obviously the other elements you found were placed there during the solidification process.

MR M. A. HORSFALL (BCIRA): Referring to the paper by Drs Johnson and Smartt in view of the large differences obtained in oxygen content by AES at flake and nodule/matrix interfaces, it would be interesting to know if the authors have examined the total oxygen contents of chilled samples of the modified and unmodified irons by fusion techniques and if so whether the results were consistent with those obtained by AES.

DR JOHNSON: The grey iron examined had been held for 10 hours at a temperature within 50°C of the eutectic temperature and then quenched to ensure that the impurity segregation observed had not occurred during cooling. Since the samples were typically cut from the centre of 5 cm dia. castings, it is difficult to believe that interphase boundary diffusion from the outside of the casting could account for the observed oxygen—i.e. the diffusion distances necessary are enormous compared to the time at temperature.

The amount of sulphur and oxygen appeared to be roughly 20 at.-% oxygen and 5 at.-% sulphur at the interface. It disappeared after about 5 atomic layers of iron were removed with inert ion sputter etching indicating that it was highly localized at the interface.

DR GRUZLESKI: If the oxygen was coming in from outside, I would expect to find it in the ductile iron as

well. We would expect to find some oxygen in those samples.

DR JOHNSON: I do not think you would expect to find it in the ductile because of graphite.

PROFESSOR H. W. KERR (University of Waterloo, Canada): In his paper on the eutectic to peritectic transition, I was concerned by Professor Fredriksson's definition of a eutectic. In view of the small specimen widths compared to the dendrite sizes in Professor Fredriksson's specimens, it seems possible that the appearance of one or both of austenite and ferrite may be sensitive to the nucleation frequency of each. Could he comment on the reproducibility of the appearance of the two phases, and are the two phases appearing as dendrites truly growing as a 'eutectic' in the normal sense of diffusive coupling.

DR P. R. SAHM (Brown, Boveri and Cie): In trying to explain the simultaneous growth of the two dendritic phases in Professor Fredriksson's work, I also wondered about the applicability of the term 'eutectic reaction' which, by conventional definition, should be non-variant in nature. However, there is an interesting example where one phase apparently grows in a dendritic manner during a eutectic reaction. This is in the cobalt–chromium carbide eutectic, where the cobalt base material, (Co–30% Cr), grows apparently ahead as the leading phase, and if it is quenched, it has a dendritic appearance, while the carbide grows a little behind. This may not be quite the same as Professor Fredriksson's 'eutectic', where he just seems to have two dendrites growing at different places but not bound by a diffusion reaction.

PROFESSOR FREDRIKSSON: In answer to the second part of Professor Kerr's comment, there was no nucleation problem for the two phases. The two dendrites always grew side by side for quite long distances, and our experiments were quite reproducible. With regard to the definition of a eutectic reaction—I define this as simply the simultaneous growth of two or more solid phases from the liquid. There are different types of eutectic reaction giving different microstructures, and the most common of these are the normal coupled eutectic structures where two phases grow side by side in a coupled manner. In my case, the two phases clearly grow simultaneously from the liquid, but there is no coupled growth in the normal sense, and the dendrites are not bound by a diffusion reaction as pointed out by Dr Sahm. I should also like to point out that as well as in the cobalt–chromium–carbide system described by Dr Sahm, there are still further examples of this type of growth. For example, in the iron–carbon–aluminium system, two primary phases can grow simultaneously—flake graphite and austenite. Another example is in the cast iron system, where there is a eutectic reaction without co-operative growth.

THE CHAIRMAN: Where do the compositional fluctuations come from to enable γ and δ dendrites to grow together in Professor Fredriksson's experiments?

PROFESSOR FREDRIKSSON: The austenite dendrites shown in the paper formed in the spaces between the ferrite dendrites. The composition of austenite and ferrite can be very similar, and composition fluctuations are not necessary for these structures to occur.

DR D. J. ALLEN (CEGB, Marchwood): I would like to come back to Professor Fredriksson's paper and suggest a possible reason why it is possible to get this 'abnormal' sort of eutectic growth where there appears to be no coupling. The coupling in a normal eutectic system occurs because the solute that the beta phase wants is being rejected by alpha, and the solute that the alpha phase wants is being rejected by beta, and so the two phases grow alongside each other in a coupled fashion. Since the compositions of ferrite and austenite are not very different, the tendency to coupling is weak in this case. Now, it is well known that the addition of a third component with low solubility in both alpha and beta phases can make a planar eutectic growth front break down into colonies. The impurity thus seems to be trying, albeit unsuccessfully, to prevent coupled growth in such cases, so I would suggest that an impurity such as sulphur might be sufficient to prevent coupled growth in the ferrite–austenite eutectic system.

PROFESSOR FREDRIKSSON: Yes, in the Fe–Cr–Ni system the shape of the three-phase triangle shows that the two solid phases, ferrite and austenite, have very similar compositions and they differ from the liquid. As a consequence the tendency for coupling is low and this may explain why the two dendrites do not show much tendency to cooperate during growth.

DR S. A. DAVID (University of Pittsburgh, USA): Dr Ostrowski and Professor Langer attribute the banded structures obtained during the directional solidification of their peritectic alloys to the difference in coefficient of thermal expansion between the alloy and the container material. Could they explain briefly what the difference in coefficient of thermal expansion has got to do with the type of alternate microstructures obtained?

DR B. A. RICKINSON (Osborn Steels): I would like to follow on the question made by Dr David to Dr Ostrowski and Professor Langer. For plane front growth of their alloy G/R is very high, and from personal experience with Al- and Ni-based alloys it is extremely difficult to remove mechanical banding at very low growth rates. I personally would be far more agreeable to accept the authors' explanation of banding if they had supplied evidence to confirm that it occurred at much higher growth velocities (increasing G at the same time to keep G/R sufficiently high for plane front growth).

DR OSTROWSKI: We believe that the difference in thermal expansion between the alloys and container material normally produces an air gap between the container inner wall and the solid metal. Sometimes the air gap disappears and the temperature at the solidification front drops sharply, changing G/R. This gives rise to the nucleation of ϵ-phase which grows in the η-matrix. If the air gap is re-established, the temperature rises again, and ϵ stops growing. We believe this results in the ϵ-banding that we observe.

MR D. J. FISHER (Ecole Polytechnique Federale de Lausanne): I would like to comment on the controversy over whether coupled peritectic growth is possible. Professor Hillert said in the opening paper that he thought that it was possible, but Dr David has disagreed with this and Ostrowski and Langer put forward theoretical arguments to this effect as well. In some unpublished work carried out by Fisher and Kurz on the Sn–Sb system, microstructures indistinguishable from coupled fibrous eutectic microstructures were observed. The results were not reproducible and the eutectic-like structures neither occupied the whole specimen cross-section nor extended in a continuous fashion along the whole length of the specimen.

DR J. A. SPITTLE: Dr David suggests that the presence of banding in directionally frozen peritectic alloys, or planar growth of the secondary phase with high G/R conditions, depends on whether or not the primary phase nucleates behind the interface. Professor Hillert in his explanation of banding says that it occurs due to nucleation of the primary phase ahead of the interface because the secondary phase is less stable than the primary above the peritectic temperature.

These are very different explanations—nucleation behind the interface and nucleation ahead of it. Would either Professor Hillert or Dr David like to comment on this. In particular, I would like to ask Dr David where he gets planar growth of the secondary phase, in the case of very high G/R. Is there definitely no primary phase present at the very beginning of growth?

DR DAVID: As to the first question, I think we have clear evidence for our mechanism in the case of the tin–cadmium system. In this case we have observed at a quenched solid/liquid interface that the primary beta phase does in fact nucleate behind the interface and then starts to grow. It grows for a while until the solute builds up allowing the delta phase to nucleate again; and so it goes on beta and delta, beta and delta.

On the second question, whenever we grow the material at very high G/R ratios, to begin with there is a two-phase structure and later on, it transforms to a single phase structure with peritectic growth.

DR SPITTLE: I assume when you say it is a single phase, it has been a single phase from the very commencement of solidification?

DR DAVID: It could be a little bit confusing in a sense. If you look at the start of solidification there is a transition region that occurs between the as-cast material and the single phase material. There is some sort of transition structure that exists before the growth of the single phase.

DR SPITTLE: I was wondering whether in fact, by using very low growth rates to achieve high G/R ratios you are in fact creating a composition which will always freeze out beta anyway. In fact, you may be below the peritectic temperature.

DR DAVID: The second phase was not growing below the peritectic temperature. In the case of the tin–cadmium system, we never made temperature measurements, but in the case of lead–bismuth we did, and interface temperatures definitely proved to be much higher than the peritectic temperature.

PROFESSOR HELLAWELL: I am becoming a little confused by these discussions of alternative peritectic reactions. Could this be clarified? Much depends on whether convective mixing occurs in the liquid close up to the growth front. Are the various results and observations for steady-state growth with no convection, or for growth with efficient liquid mixing?

DR DAVID: If you are talking about the effect of macrosegregation during growth there are two things which will contribute to such an effect. One is fluid flow, the other is thermal transport. Fluid flow would not have been important in the Sn–Cd system because samples were grown vertically in narrow tubes (3–5 mm) in a positive temperature gradient, and the solute rejected (cadmium) was heavier than the bulk liquid. In the Pb–Bi system, Bi is lighter than lead and so there might have been some convection in this case. Thermal transport should not have been a problem under the growth conditions that we used.

The following discussion was subsequently received from Professor Hillert as a written contribution to clarify some of the points raised on peritectic growth.

PROFESSOR HILLERT: In the discussion the question was raised as to how the mechanisms of banding in peritectic alloys, proposed in the paper by Brody and David and in my paper, compare with each other. It seems to me that we are in essential agreement. I based my arguments on isothermal considerations only, whereas Brody and David gave a more detailed analysis and considered the effect of the temperature gradient G and the velocity R. Their result as presented in their Fig. 3 is somewhat puzzling, however. The construction of their diagram depends strongly on the three equilibrium compositions at the peritectic temperature. This seems unnatural since the peritectic equilibrium can only be established at an infinitely slow growth rate. The conditions which are now under discussion are radically different.

A more realistic diagram can be derived from the same basic equations if one compares the temperatures at the solidification front for the various modes of growth. For planar growth the well known theory of constitutional supercooling yields; for S_1:

$$D_L G/R = m_1 C_0 (1-K_1)/K_1 \qquad (1)$$

For S_2:

$$D_L G/R = m_2 C_0 (1-K_2)/K_2 \qquad (2)$$

These lines are plotted as Oa and Obc in Fig. A using values of m_1, m_2, K_1 and K_2 which are defined by the schematic phase diagram. It is important to note that the G/R value which is required in order to establish planar growth is lower for S_2 than for S_1. Both modes of growth are possible above the line for S_1. It seems natural to expect that the phase which grows at the highest temperature will win in competitive growth. One should thus compare the temperature at the planar growth fronts. They can be derived for any alloy composition C_0 from the phase diagram; for S_1:

$$T - T_p = m_1(C_P - C_0/K_1) \qquad (3)$$

For S_2:

$$T - T_p = m_2(C_P - C_0/K_2) \qquad (4)$$

C_P is the peritectic composition. The boundary between the regions for planar growth of S_1 and S_2,

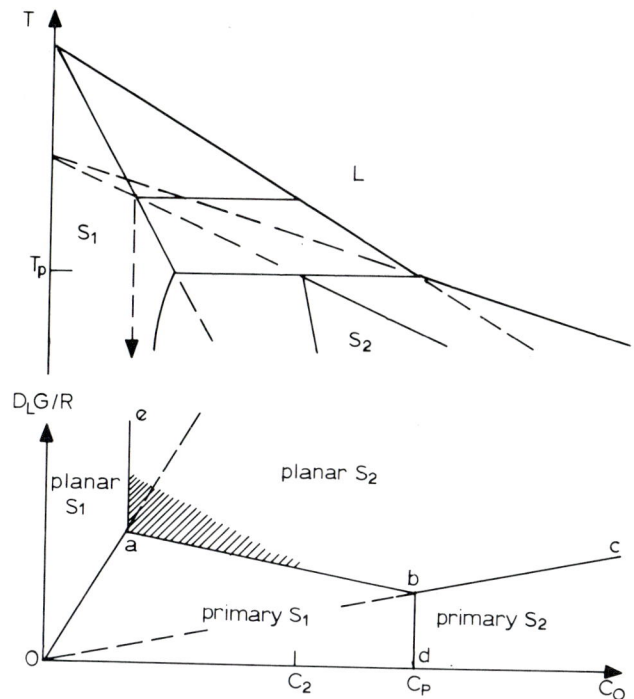

A Planar solid/liquid interface based on comparison of temperatures at solidification front for various modes of growth

respectively, is obtained by combining these two equations and putting their T values equal. The result is

$$C_0 = C_P K_1 K_2 (m_1 - m_2)/(K_2 m_1 - K_1 m_2) \qquad (5)$$

This vertical line is plotted as ae in Fig. A. It can of course be obtained directly from the phase diagram.

For the temperature at the tip of cells or dendrites of S_1, Brody and David used the following approximate equation,

$$D_L G/R = m_1(C_L - C_0) \qquad (6)$$

C_0 is the alloy composition and the temperature enters through the liquidus composition C_L. The following expression can be obtained directly from the phase diagram.

$$C_L = C_P + (T_P - T)/m_1 \qquad (7)$$

A combination of equations (6) and (7) yields, for S_1

$$T - T_P = m_1(C_P - C_0) - D_L G/R \qquad (8)$$

In the same way we obtain for S_2

$$T - T_P = m_2(C_P - C_0) - D_L G/R \qquad (9)$$

Again it seems natural to expect that the phase which grows at the highest temperature will win. The boundary between the regions for primary precipitation of S_1 and S_2 respectively, is then found by combining equations (8) and (9) and putting their values equal. The result is

$$C_0 = C_P \qquad (10)$$

This yields the vertical line db in Fig. A, quite different from the sloping line given by Brody and David.

The boundary between the region of planar growth of S_2 and cellular or dendritic growth of S_1 is obtained by combining equations (4) and (8) and putting their values equal.

$$D_L G/R = (m_1 - m_2)(C_P - C_0) + m_2 C_0 (1 - K_2)/K_2 \tag{11}$$

It is interesting and satisfactory that this relation yields a straight line between the points *a* and *b* in Fig. A. Four well-defined regions have thus been obtained. In the region of planar growth of S_1 the solidification will yield 100% S_1 but some S_2 may subsequently precipitate as the temperature is lowered far below the peritectic one. In the region of planar growth of S_2 the solidification will yield 100% S_2. This phase will be metastable if the alloy composition is to the left of C_2, the peritectic composition of the S_2 phase. It can immediately start to precipitate S_1 if that phase is nucleated and it can even transform completely by a eutectic type of reaction.

$$S_2 \rightarrow S_1 + L$$

This may lead to an interruption of the solidification to S_2 as suggested in the papers presented at the conference. However, this can only occur before the temperature has fallen below the peritectic one. As pointed out by Brody and David, the phenomenon of banding should thus grow less probable at increasing values of G/R. It should be most likely in the hatched region in Fig. A.

The solidification is expected to yield 100% S_2 in the region where it starts with a primary precipitation of S_2. However, in the region of primary precipitation of S_1, one should expect a subsequent formation of S_2 from the liquid phase, in particular if diffusion is slow in the S_1 phase. This formation of S_2 could be planar if G/R falls well above the line *Ob*. The exact condition is difficult to derive because of the change in composition of the liquid phase owing to the primary precipitation of S_1.

Session 3

Nucleation and grain refinement

Chairman: R. D. Doherty (University of Sussex)

Editors: A. Hellawell (University of Wisconsin, USA)

J. O. Ward (Henry Wiggin and Company, Hereford)

Keynote Address: Heterogeneous nucleation and grain refinement in aluminium castings

A. Hellawell

The action of heterogeneous nucleating agents is considered for a cooling melt containing a dispersion of particles, and it is shown how the combined nucleation and growth processes can describe a cooling curve and the eventual grain size which might be expected in a casting. The separate influences of the nucleating potential of a substrate, of cooling rate, alloy constitution, and particle size are considered, and the results are related to the Al–Ti, Al–Zr and Al–Cr systems. The model for nucleation and grain refinement is then extended inasmuch as it might apply to grain refinement in ternary Al–Ti–B alloys and a number of contradictory effects are discussed.

The author is in the Department of Materials, University of Wisconsin, USA

At the previous conference, held in 1967 in Brighton, Glasson and Emley[1] reviewed the additions and treatments which can be made to magnesium and aluminium-base alloys to promote grain refinement during solidification, drawing attention to the wide range of reports in the literature and the often contradictory explanations which have been suggested for the mechanisms involved. Inevitably, the present paper will relate to some of the same problems, but a somewhat different approach will be made. In the first place, the action which nucleating particles have is considered with respect to the shape of a cooling curve immediately below the freezing point of a metal or alloy melt, and secondly, the factors which influence this action are related, where possible, to grain-refining behaviour in aluminium alloys with titanium, zirconium and chromium and to ternary alloys containing boron.

The author begins with the following premises

(i) grain refinement is desirable during casting, (*a*) to facilitate and ensure reproducible properties for subsequent working, in particular to produce a uniform surface finish; (*b*) to reduce internal stresses in alloy castings which contain brittle constituents, by dispersing the latter; (*c*) to allow rapid solidification during continuous or semi-continuous casting; and (*d*) in small and intricate castings to promote a uniform solidification pattern and sound product

(ii) in foundry practice we are never concerned with homogeneous nucleation, and, indeed, there is considerable doubt[2] whether this has been observed under strict laboratory conditions. Rather, we are concerned with relatively rapid nucleation at relatively low undercoolings of a few degrees, catalysed by the presence of surfaces, either solid–liquid or liquid vapour.[3]

We know that in the absence of any intentional additions to a melt a very wide degree of control can be exercised upon grain size and shape through the casting conditions, notably the 'chill' and rate of heat removal by the mould, and by the superheat or casting temperature, combinations of these determining the growth rate and prevailing temperature gradients. Thus, grains are nucleated by fragments of solid metal which are either swept from the chilled surface, as in 'big bang' nucleation at low superheats, or find their way into the bulk liquid from the columnar region or upper surface of a melt. However, these are only a special case of heterogeneous nucleation in which the catalytic substrate particles happen to be the same material as the bulk casting, and it is therefore appropriate to consider the general situation where liquid contains a dispersion of solid particles, from whatever source(s), as it cools below its freezing point.

MODEL FOR GRAIN REFINEMENT DURING SOLIDIFICATION

Consider a volume of metal liquid of uniform temperature, cooling at a given rate, and containing a dispersion of particles which the solid metal tends to wet. It will be necessary to make some arbitrary but reasonable assumptions if the situation is to be put on any quantitative basis, and the following were made recently by Maxwell and the present author.[4] We assumed that particles were spherical and of given radius R_0, and that one particle could promote only a single nucleation event. We considered particles in the range, $R_0 = 0.3$ to 3.0 µm; TiB_2 particles found in Al–Ti–B alloys are at the lower and Al_3Ti particles at the upper end of this range.[5] In practice, particles are rarely spherical—TiB_2 crystallises as flat platelets—but this is not a serious objection because growth around them at early stages would occur spherically.[1,4] Again, evidence has been presented[6] to suggest that a particle, e.g. of Al_3Ti, may nucleate more than one grain of aluminium, possibly because it was twinned or polycrystalline, but this does not seriously modify the argument, involving only a small numerical factor (like $\times 2$) which it transpires is small by comparison with other factors.

A typical dispersion of particles in practice might be of density, N_V^P, 10^3 mm^{-3}, i.e. having a mean separation of ~ 100 µm. As the melt cools below the freezing point, two processes take place concurrently; solid nucleates upon the available substrate surface at a rate which rises exponentially with the undercooling, and when the temperature has fallen below that required to exceed the curvature term (given by the Gibbs–Thompson expression, $\Delta T_c 2\gamma_{CL}/R_0\Delta S$, where ΔT_c is the curvature undercooling, γ_{CL} the metal solid–liquid surface energy, R_0 as before, and ΔS the entropy of fusion) nucleated particles will begin to grow and evolve latent heat. At first, the temperature will continue to fall almost uniformly, but as nucleation and growth accelerate the cooling rate decreases until the temperature reaches a minimum and recalescence occurs. Although substrate particles are being 'used up' during this stage, the nucleation rate per unit area of available substrate rises to a maximum at the minimum temperature and thereafter falls off very rapidly. This sequence takes place in a short time interval, < 5 s, and such nucleation as can take place is almost complete just beyond the minimum temperature—after that the process is entirely one of growth, although there may still be many substrate particles present which never had time (undercooling) to promote a nucleation event. One of the things we need to know is what proportion of the particles were active as catalytic sites for grain refinement. If every particle nucleated one (or two etc.)

grain(s) the situation would be rather simple, but intuition dictates that this is unlikely and that for any substrate there must be a particle density beyond which further additions are superfluous. Estimates of the optimum or maximum useful particle density are important because they identify several factors which influence the efficacy of a 'grain refiner' and throw considerable light on what must be happening in actual cases.

The model is then as depicted in Fig. 1 and it is necessary to outline briefly how the nucleation and growth processes can be treated quantitatively.

Nucleation

To estimate a free energy, ΔG^*, of a critical or stable embryo we use the classical theory of Volmer *et al.*[17] which gives an expression of the form

$$\Delta G^* \propto \frac{\gamma_{CL}^3}{(\Delta S . \Delta T)^2} \tag{1}$$

where γ_{CL}, ΔS and ΔT are as before, and the nucleation rate per unit time and volume, $\dot{N}$,

$$\dot{N} \propto e^{-\Delta G^*/kT} \tag{2}$$

For heterogeneous nucleation there will be three surface energy terms to consider, γ_{CL}, γ_{CS} and γ_{SL}, where C = crystal (metal), S = substrate, and L = liquid (metal). We do not, in general, know these quantities, so that as a convenience it is usual to take the model of Volmer[7] and describe the 'wetting' of the substrate by an included contact angle, θ, for a hemispherical cap upon a flat surface. This is almost certainly not physically realistic because embryo nuclei are more probably monolayers or multilayers of atoms, having no geometrical shape which simply relates to a sphere. However, it is the best available compromise, satisfactory as long as it is remembered that it is simply a convenient number which can be used to describe the potential of a substrate to act as a surface for nucleation, where $\Delta G^* \propto f.(\theta)$, where $f(\theta)$ is a cosine function such that $\Delta G^* \to 0$ as $\theta \to 0$.

Using the above expressions it is then possible to calculate $\dot{N}$ at each successive time interval and undercooling, ΔT, and hence from $\Sigma_0^t \dot{N}$, the total number of such events and the number of grains nucleated at successive times/temperatures below the freezing point, when $t = 0$. If the cooling rate were linear this would be a simple calculation, leading to exhaustion of sites, but it is not linear and the growth and resulting evolution of latent heat must be considered.

Growth

Maxwell and Hellawell assumed that in the early stages (< 5 s) growth would be spherical, and the microstructure of refined alloys (Fig. 2) suggests that this is a valid assumption. Therefore, for each successive time interval, δt, the volume freezing and heat evolved is $\propto 4\pi R^2 . \delta R . \Delta H$, where R is the radius of a particle at that instant, δR the incremental change, and ΔH the latent heat of fusion; this quantity of heat must be summed for each successive interval (typically 10^{-2} s) and added to the temperature in order to correct the initially constant cooling rate.

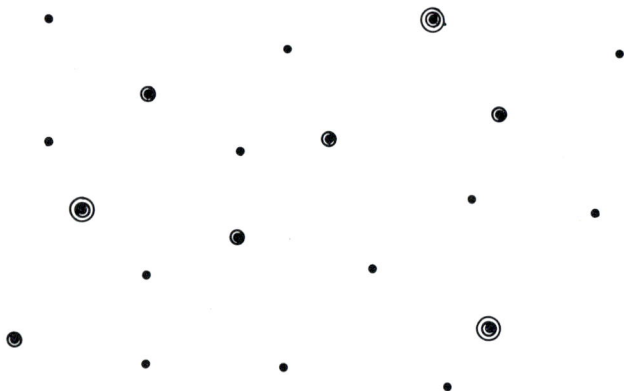

1 Depicting a dispersion of potential heterogeneous particles, some of which have nucleated and grown spherical shells of metal incrementally

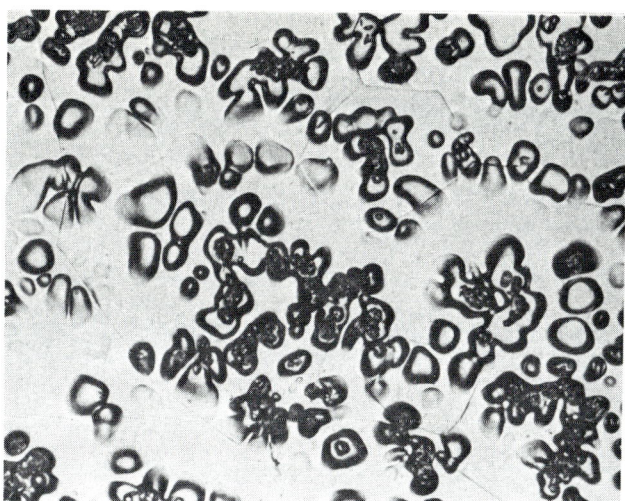

2 Alloy of Al–0·6 wt-% Ti chill cast, showing Ti-rich regions of spherical outline surrounding Al₃Ti particles [× 85 (enlarged × 3)]

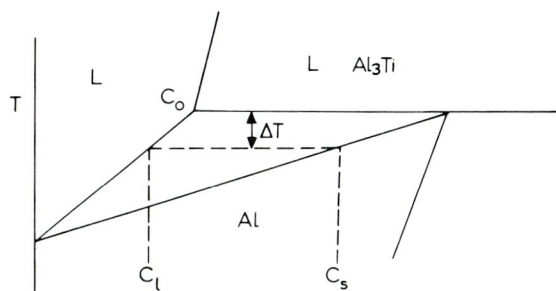

3 The Al–Ti phase diagram at low Ti concentrations, peritectic temperature 665°C, peritectic liquid concentration 0·15 wt-% Ti, aluminium-rich solid and liquid interfacial concentrations C_s and C_L at undercooling ΔT

At these early stages of growth, with particles of substrate typically ~ 100 μm apart, we are not concerned with solid impingement, nor with overlapping solute fields (soft impingement[17])—the particles and their occasional metal envelopes are still far apart because the ratio of latent heat/specific heat is of the order $10^3:1$ and very little solid is needed to reverse the cooling curve. The subsequent events are controlled only by latent heat evolution and its conduction into the bulk liquid. For a thermal conductivity, D_{th}, the thermal fields around a growing crystal are of the order $\sqrt{D_{th}t}$ in width, and in a metal $\sqrt{D_{th}}$ for one second is 1–10 mm, i.e. more than an order of magnitude greater than the particle separations with which we are concerned. It is therefore a reasonable approximation to assume that the evolution of latent heat is uniform throughout the melt, particularly if there is simultaneous stirring or convective mixing.

Growth from such particles as have nucleated solid metal will not begin until the undercooling exceeds that for curvature (cf. p. 162) and for aluminium on particles, $R_0 = 1$ μm, this would correspond to $\Delta T \gtrsim 0.25$ K. Thereafter growth will proceed increasingly rapidly with ΔT, and, were there no other obstacle recalescence would be rapid at very small undercoolings and further nucleation inhibited, i.e. grain refinement would be poor, even were the substrate potentially active, $\theta \to 0$, $= n.^3$

As is generally recognized, however,[8] the situation is more complicated because the growth temperature, and hence the ambient temperature of the liquid, is also very sensitive to solute rejection at the growing solid–liquid interface. If the solute term is large the growth temperature is depressed and a longer time at a lower temperature encourages further nucleation events.

Solute rejection and undercooling
To assess the magnitude of this contribution we consult first the relevant phase diagram between metal and the nucleating substrate. It so happens that most systems in which grain refinement occurs are of peritectic form, such as Al–Ti (Fig. 3). There is nothing particularly significant about this feature, it is simply that if there is to be a dispersion of intermetallic compound which is

stable above the freezing point of the metal in question, then a peritectic system is most probable. The existence of the peritectic reaction itself, e.g. Al₃Ti + liquid ⇌ Al (Fig. 3) is probably also incidental, because it does not follow that the primary phase, Al₃Ti in this case, is necessarily a good nucleating agent for the second solid, aluminium solid solution in this case, and indeed there are probably as many cases where the primary solid is not a preferred substrate as there are when it is preferred.[9] However in Al–Ti, as in some of the other systems between aluminium and transition metals (Nb, Zr, and Cr), the presence of preferred orientations between substrates and metal[5,10,11] would seem to indicate that compounds such as Al₃Ti are preferred substrates, although there may be some disagreement as to which relative orientations are actually preferred.

Referring then to Fig. 3, for growth of aluminium upon a dispersion of Al₃Ti particles which have precipitated from above the peritectic temperature, we are concerned with a limiting concentration C_0 (assuming precipitation of Al₃Ti was complete), and growth at some ΔT at which the solid and liquid interfacial concentrations would be C_s and C_L, i.e. there is a solute accumulation at the growth front (Fig. 4) and the growth temperature is depressed by an amount

$$\Delta T_{solute} \propto C_0 m(k-1) \qquad (4)$$

where m is the liquidus slope and k the solid–liquid distribution coefficient.

For spherical growth, this solute accumulation increases with $R^{\frac{1}{2}}$, where R is the radius at any instant ($R > R_0$), but whatever the growth model the relative solute undercoolings from system to system depend upon C_0, m and k, above ($= n$).[4]

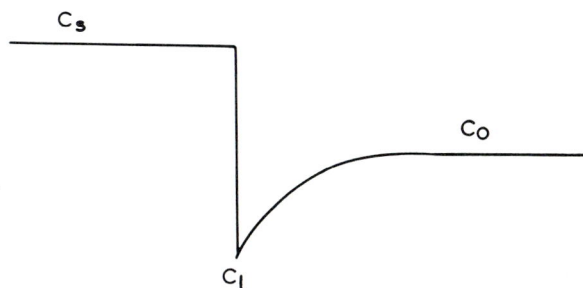

4 Solid–liquid concentration profile during growth of aluminium solution upon Al₃Ti (ref. Fig. 3)

The cooling curve and associated events

Taking the model outlined above, Maxwell[4] computed the form of a cooling curve for an alloy such as Co in Fig. 3, combining the nucleation and growth expressions for a variety of cooling rates, phase equilibria, substrate particle radii and over a small range of the cryptic contact angle θ—summing the events in time intervals of 10^{-2} s. The only quantity which was entirely unknown was the number, θ, although simple estimates lead one to expect $\theta < 10°$ for most cases of efficient heterogeneous nucleation.

The form in which the results appear is in Fig. 5a and b. Figure 5a shows a cooling curve for an Al–Ti alloy, computed for an initial rate of 0·5 K/s, Co = 0·15 wt-% Ti, $R_0 = 1$ μm and containing a particle density, N_V^P of 10^3 mm^{-3}. Good agreement with experiment was obtained if θ was assumed to be $\sim 7°$, i.e. this number was arrived at by comparison with observed results.

Negligible nucleation and growth occurs for some 1·2 K of undercooling, or 2·4 s after the melt passes the freezing point, but in the following second the reaction is rapid, and as far as nucleation is concerned it is virtually complete. What happens in this small time interval is summarized in Fig. 5b which is taken from the inset frame in Fig. 5a. Corresponding to the detailed cooling curve, it is instructive to look at (a) the nucleation rate, $\dot{N}$, per unit volume of melt; (b) the sum of events, $\sum_0^t \dot{N} . \delta t$, to give the total number of grains nucleated at any time; and (c) the volume fraction, V_F, of solidified metal, and R, the mean radius of growing metal spheres at any instant. Taking these in turn we see that:

(i) $\dot{N}$ is negligible up to 2·4 s when it is still only 10^{-3} mm^{-3} s^{-1}, but is rising steeply to a maximum rate of nearly 10^4 mm^{-3} s^{-1} at a temperature just before the minimum, thereafter decreasing very rapidly as particles are exhausted and particularly as recalescence begins. Nucleation has virtually finished by ~ 3.5 s

(ii) N_V^G, the grain density, is similarly small at 2·4 s but rises with $\dot{N}$ to a final level of $\sim 10^3$ mm^{-3} by the time recalescence begins, i.e. under the conditions chosen, nearly every substrate particle would nucleate a metal crystal

(iii) meanwhile, as noted earlier, the growing particles are still relatively far apart during this critical phase, and the volume fraction of solid metal, V_F, is still $< 10^{-3}$ by the time nucleation events have finished. This emphasises the point that there is no question of solid or solute impingement during the relevant stages of the process and heat evolution is the dominant influence

(iv) finally, confirming the latter point, the mean particle radii, R, are still very small during this period and do not exceed 5 μm until well after nucleation is complete. Up to this size, growth is essentially spherical (c.f. Fig. 2) but thereafter undoubtedly becomes dendritic as the entire melt begins to solidify.

This then is the background against which the action of a potential nucleating agent must be considered, and the question next arises as to how this situation can best be represented for different variables.

Factors influencing the process

One useful way to summarize nucleation behaviour is to compare the final grain density, N_V^G and the initial substrate density, N_V^P. This has been done for aluminium alloys for four alternative factors (Figs 6–9). In all cases we are concerned with how substrate activity compares with an ideal ratio of one grain per substrate particle,

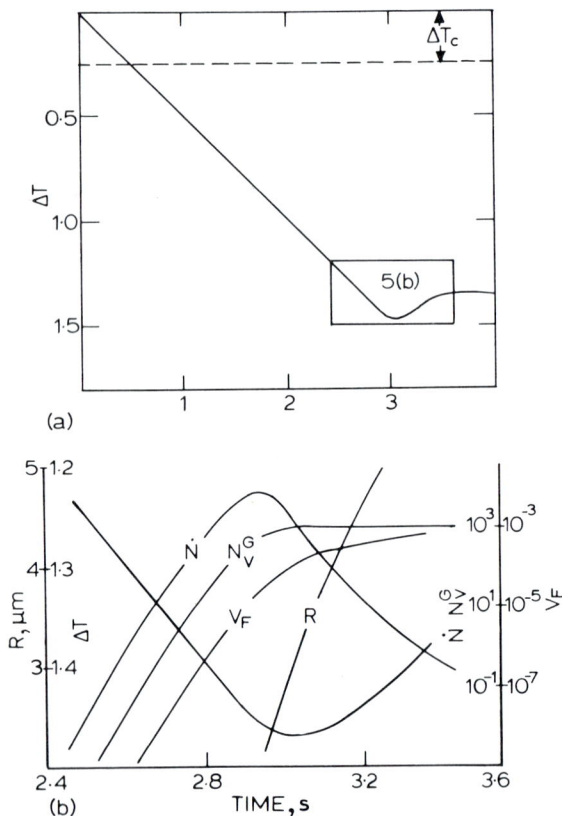

5   *a* Cooling curve computed for Al–Ti alloy containing 10^3 mm^{-3} particles of Al$_3$Ti, radius $R_0 = 1$ μm, initial cooling rate 0·5K/s and $\theta = 7°$; *b* detail taken from inset Fig. 5a showing corresponding variations of $\dot{N}$ = number of nucleation events mm^{-3}/s, N_V^G = total number of nucleation events mm^{-3}, V_F = volume fraction of solid metal, and R = mean radii of solid metal spheres

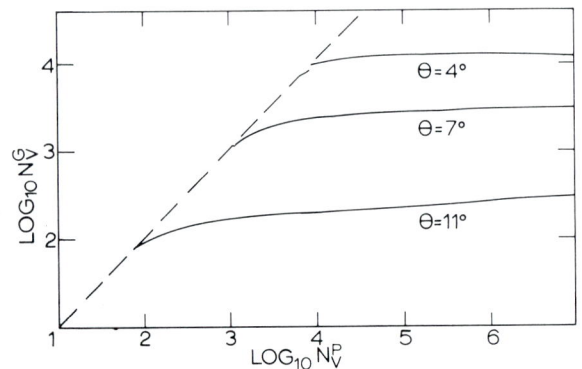

6   Plotting number of grains mm^{-3} v. number of particles mm^{-3} for Al–Ti system, Al$_3$Ti radius 1 μm and cooling rate 0·5K/s; for three values of the hypothetical contact angle $\theta = 4°$, $\theta = 7°$, and $\theta = 11°$

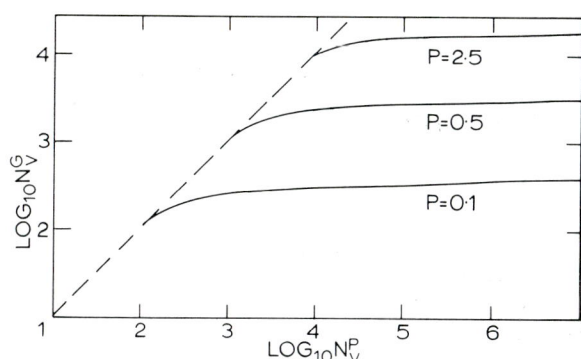

7 As Fig. 6 with $\theta = 7°$ and for cooling rates $P = 0.1$K/s, $P = 0.5$K/s and $P = 2.5$K/s

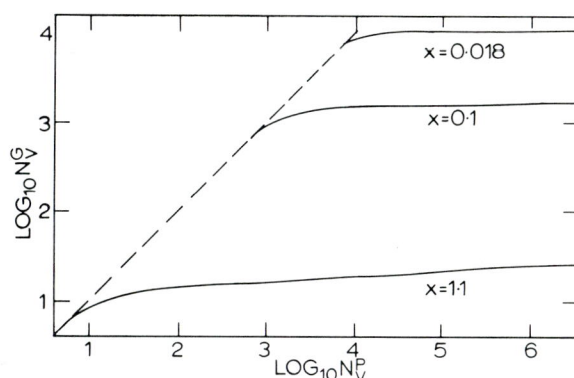

8 As Figs 6–7, with $\theta = 4°$, $P = 0.5$K/s, and for phase diagram data, X, from AlTi ($X = 0.018$), Al–Zr ($X = 0.1$) and Al–Cr ($X = 1.1$), where $X = \frac{1}{2}(C_0 m(k-1))$

which appears on Figs 6–9 as straight lines of unit slope. We may take these figures separately.

Figure 6 uses the data for Al–Ti and shows the effect of different assumed values of θ, at a given cooling rate of 0.5 K/s. The middle example, $\theta = 7°$, corresponds to the composite diagram of Fig. 5a and b, and gives the closest approximation to both an observed cooling curve and the corresponding grain size. In the range shown, a change of θ by $\sim \pm 3°$ changes the limiting grain density by approximately an order of magnitude. In the case of Al–Ti, with $R_0 = 1$ μm, the deviation from 1:1 takes place fairly abruptly at $N_v^P \approx 10^3$ mm^{-3}, i.e. greater particle density then produces negligible further grain refinement and there would be only an increasing density of excess substrate particles which never became active, and it would be pointless to exceed that 'optimum' addition of grain refiner. This is discussed further in the next section.

In Figure 7 we consider the same example, for $\theta = 7°$, but for different cooling rates, P, from 0.1K/s to 2.5K/s. This small range of cooling rates covers much of those encountered in foundry practice. Thus, in semicontinuous casting, with a traction rate of ~ 20 cm/min (33 μm/s) and a prevailing temperature gradient of 1–2K/cm, the local cooling rate of liquid metal approaching the freezing point is between 0.25–0.5K/s. In smaller individual castings the rate will, of course, be very variable but rarely by more than an order of magnitude of that for continuous casting. Again, the range chosen demonstrates order of magnitude changes in the optimum grain refinement and shows how the grain refining action is curtailed by the casting conditions.

Figure 8 introduces perhaps the most striking and new information in this present context, that is, the influence of solute undercooling as it varies from system to system. In the discussion for the corresponding session, ten years ago,[8] this aspect was mentioned but its importance was not expanded.

Figure 8 concerns variations of a quantity, X, where $X \propto 1/C_0 m(k-1)$, the measure of solute rejection in any system. To facilitate comparison, the plots in Fig. 8 relate to $\theta = 4°$, with $P = 0.5$K/s as in Fig. 6, and the values of X are those for the systems Al–Ti ($X = 0.018$), Al–Zr ($X = 0.1$), and Al–Cr ($X = 1.1$), in which the relevant substrate compounds are Al$_3$Ti, Al$_3$Zr and Al$_7$Cr—all of which were assumed to be equally good (or better than) potential catalysts, i.e. lower values of θ than appropriate to Al$_3$Ti, which might be equated with

good lattice matching between compound-metal, high substrate–liquid surface energy.[10,11]

As has been shown[1,11,12] there is a dramatic variation in the grain refining action in these three systems, such that Al$_3$Ti is much more efficient than Al$_3$Zr, which in turn is better than Al$_7$Cr, the action of which is slight or negligible.[1] Figure 8 shows simply that this variation can be accounted for by differences in the phase diagrams without further recourse to any other factors. Discussions of relative lattice disregistry are quite superfluous—Al$_3$Zr is isomorphous with Al$_3$Ti and has similar lattice parameters. All these compounds may be equally good *potential* catalysts, but only in one system, Al–Ti, is the ambient temperature depressed sufficiently to develop a rapid nucleation rate before recalescence occurs. It follows, that however good a nuclent might be *potentially*, viz. $\theta \to 0$ etc., if it is insoluble in a pure metal—perhaps even Al$_2$O$_3$ in Al—its grain refining *activity* will be precluded if the growth temperature is not significantly depressed below the freezing point.

Figure 9 presents a further practical variable, that of substrate particle size. Particle sizes were briefly discussed earlier (see **Nucleation**) but we note that grain refining additives, such as those based on master alloys of the type Al–5 wt-% Ti, 1 wt-% B, contain compound dispersions of sizes depending on the mode of precipitation during chemical reaction (involving alumino- and boro-fluorides,[1]) casting conditions and subsequent extrusion or rolling treatments. In Fig. 9, fixing all the other variables, the effect of an order of magnitude

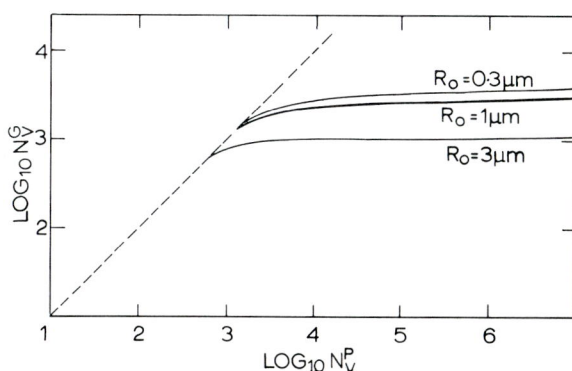

9 As Figs 6–8, with $\theta = 7°$, $P = 0.5$K/s for $X = 0.018$ (AlTi) and particle radii, $R_0 = 3$ μm, 1 μm and 0.3 μm.

change in R_0 from 0.3 μm to 3.0 μm can be observed. The results relate to the Al–Ti system, where typically the size of Al$_3$Ti particles is more likely to approach 3 μm than the lower sizes. The comparison is dominated by heat production, so that, while one might have expected that as R_0 increased the available surface area and nucleation rate per particle would increase, this effect is swamped by the latent heat term—$4\pi R^2\delta R.\Delta H$—which prevents the melt temperature for the larger particles falling as low as it can for the smaller. The effect is not as great as shown in the previous comparisons of Figs 6–8, but this result is for given particle densities, N_V^P, and not for given concentration levels which is quite different (*see* later section **Size distribution**).

FURTHER APPLICATION OF THE MODEL

In the previous sections a model was presented which describes the general reaction of a heterogeneous nucleant. The treatment has, so far, been somewhat abstract although values of parameters have been related as closely as possible to real examples. The major piece of information which has come out of these basic considerations is the importance of the phase equilibria and the way in which *potentially* as opposed to *actually* active substrates may be distinguished, as between the Al–Ti, Al–Zr, and Al–Cr systems.

It is desirable now to consider the application of these principles to more complicated practical examples. There are three areas of interest; (*a*) is it possible to use the model to estimate the limiting effective concentration of a solute element, in practice?; (*b*) how does the information relate, if at all, to the unexplained activity of ternary Al–Ti–B additives, compared with binary Al–Ti grain refiner?; and (*c*) is it possible to make any reasonable comments upon the time for which a refining agent remains efficacious following its additions to a melt—the problem of 'holding' or 'retention' times?

Optimum concentrations

Let us take the Al–Ti system (Fig. 3) and consider the addition of a binary alloy containing 5wt-% Ti, the mean particle radius of Al$_3$Ti particles being 3 μm. Assuming that the particles of compound do not have any time to dissolve between the time of addition to the melt and the moment of casting (but *see* this volume[13]) it should be possible to calculate N_V^P quite simply and hence arrive at the necessary dilution of additive to reach the 'optimum' useful composition in Fig. 9 (lower line).

Referring to the binary Al$_3$Ti + liquid immediately above the peritectic temperature (Fig. 3) we find that 1 wt-% Ti in excess of Co (0.15 wt-% Ti) is equivalent to ~ 2 vol.-% of Al$_3$Ti or, in a 5 wt-% Ti alloy to some 10% by volume of potential substrate (1 wt-% Ti $\equiv$ 3 wt-% Al$_3$Ti, ρAl$_3$Ti $= 3.13$. 10^3 kg m^{-3} and ρ liquid Al $= 2.35$ kg m^{-3}). On Fig. 9 we see that the critical N_V^P is $\sim 10^3$ mm^{-3}, which with $R_0 = 3$ μm corresponds only to $\sim 2.10^{-3}$ excess wt-% Ti, i.e. $0.15 + 2.10^{-3}$ wt-% Ti. Hence using a master alloy of 5 wt-% Ti, with those conditions of cooling rate and particle size, the dilution of the addition need not exceed 1 : 2500; more concentrated additions should not significantly improve grain refinement, the limiting grain size then being ~ 100 μm. In practice, with $P \sim 0.5$ K/s, this is in fact an observed lower limit in grain size, but the calculation is immediately suspect because we do not know the effective

solubility limit for Al$_3$Ti in molten aluminium, and, if it were necessary to reach equilibrium at $C_0 + 2.10^3$ wt-% Ti the necessary addition would be nearly 2% of master alloy. This is the unknown factor which presently prevents our making useful quantitative predictions. Evidence from thermal analysis[6] suggests that the precipitation of Al$_3$Ti is incomplete at rapid cooling rates, so that a chill-cast alloy of Al-5 wt-% Ti for example, will contain a slightly supersaturated Al solid solution, by perhaps a fractional percentage.[18]

Size distribution and Al–Ti–B alloys

The observation is[1,11,12] that in certain concentration ranges and with certain cooling rates, and addition of 1 wt-% B to an Al-5 wt-% Ti additive will increase the grain refining effect by an order of magnitude. The ternary master alloy then contains relatively coarse particles of Al$_3$Ti and very many fine TiB$_2$ crystals.[5] The immediate inference might be to suppose that TiB$_2$ rather than Al$_3$Ti is now the responsible agent, but there is contradictory evidence in the microstructure of the master alloy. In the refined metal, in which the addition was diluted by 1 : 500 for example, occasional TiB$_2$ crystals are to be found, although whether they nucleated the grains within which they occur is an open question. But in the master alloy, Al$_3$Ti particles occur within grains, and are often related in a preferred crystallographic sense to the matrix, while the vast majority of TiB$_2$ particles occur intergranularly or interdendritically, i.e. they appear not to be wetted by aluminium and their orientations are random. One must ask, to what is the enhanced activity due, if the extra particles are not easily wetted by the metal? Such an uneven distribution of TiB$_2$ flakes is not that which would be expected from an excess of nucleating particles (i.e. far beyond optimum N_V^P), because a substrate which was wetted by the metal, a potentially good nucleant, would be enveloped uniformly during growth whether or not individual particles became active nucleating sites. We know further[5,16] that in ternary alloys richer in boron than the Al–TiB$_2$ tie line are relatively inactive, i.e. the enhanced activity occurs only with Al$_3$Ti present in the master alloy addition.

Figure 10 shows an outline of the ternary liquidus in the Al–Ti–B system, and although many features of the precise binary and ternary reactions are uncertain, it is clear that the TiB$_2$ liquidus rises very steeply from the Al-rich corner of the system.[1,11,14] In the 5 wt-% Ti, 1 wt-% B alloy, the general conclusion must be that TiB$_2$ (having a melting point of $\sim 2\,800$°C) will either precipitate first from the liquid, on cooling, or will already be present at temperatures below the liquidus if formed by chemical reaction with salts, and further that Al$_3$Ti will probably not precipitate until the temperature falls close to the binary Al–Ti liquidus. We have no information as to whether TiB$_2$ nucleates Al$_3$Ti. However, suppose that Al$_3$Ti did nucleate upon at least a proportion of TiB$_2$ crystals—perhaps in the course of some monovariant or ternary reaction—how would this affect the situation, i.e. with Al$_3$Ti effectively dispersed more finely?

To consider this, we can refer again to Fig. 9 and the influence of particle size. In Fig. 9, size effects are compared for given particle densities, N_V^P, but in an alloy of given concentration, e.g. 1 wt-% Ti, this is not quite the relevant quantity because in addition to the isolated

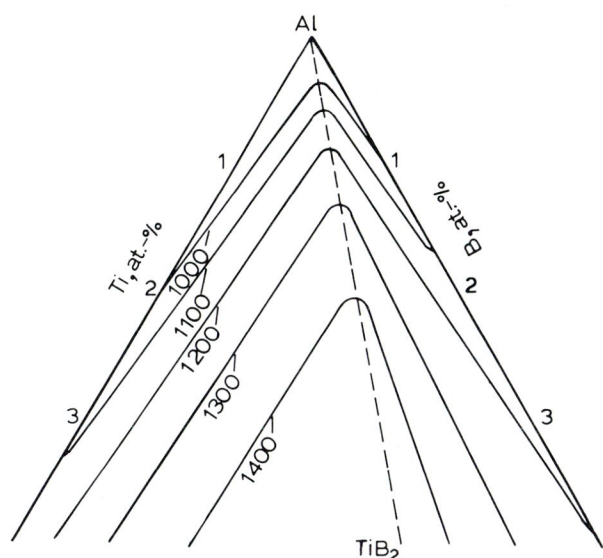

10 Outline form of Al-rich liquidus surface in Al–Ti–B alloys

effect of size per particle we must also consider how the particle density changes as the size decreases in a given alloy, and this makes a big difference. For a given volume % of substrate, the effect of splitting up the particles into many finer ones also implies an increase in N_V^P and this is shown in Fig. 11 where the information of Fig. 9 is replotted for wt-% Ti in excess of the effective liquidus limit, C_e, such that $C_e = C_0$ for equilibrium but may be higher in practice, as noted earlier. The effect of size can now be seen more clearly and produces some dramatic changes in low concentration ranges. Thus, taking a very low excess Ti content of 10^{-5} wt-%, whereas 3 μm particles would produce negligible grain refinement, the same amount of Al_3Ti as 0·3 μm particles will increase N_V^G by some three orders of magnitude and reduce the grain size from ~1 mm to ~100 μm.

I do not know what definite evidence exists for Al_3Ti wetting or being absorbed upon the finer TiB_2 crystals, although Bäckerud[15] has presented some micrographs which might support this hypothesis. However, the model described here does illustrate that a change in the size distribution would itself account for the enhanced activity of a substrate, and moreover that the concentration levels involved need be only very low.

Again, as in the previous estimate, we are not able to make a precise estimate of the effective solubility limit, C_e, and on cooling we do not expect that all the Al_3Ti will necessarily precipitate. Some further knowledge of dissolution rates[13] will be very relevant in this context.

Finally, it is instructive to refer to the detail of the Al-rich corner of the ternary system. The concentrations at which the liquidus surfaces lie are so exceedingly low that we can only guess at the reactions involved.[14,16] There are perhaps two points which can be made.

First, the TiB_2 liquidus must be steeper and lie at lower solute contents than that for Al_3Ti. If this is so, it may be that the Al_3Ti solubility limit is depressed to lower Ti levels in the ternary system, i.e. C_e may be lower than it is in the binary system. Such an effect would be very relevant here, quite apart from the size effect, because the excess Ti content may well increase dramatically; if the estimates of Fig. 11 are realistic, increasing the Al_3Ti particle density from as little as 10^{-5} to 10^{-3} mm^{-3} would itself account for much of the enhanced activity associated with boron. It should be emphasized that we are dealing with astonishingly small solute concentrations.

Secondly, TiB_2 particles are to be found in both dilute and slowly cooled samples. This implies that whatever ternary reactions may take place, as between liquid–TiB_2–Al_3Ti, the reaction must be incomplete or produce TiB_2 as a product, such fine particles would be expected to dissolve easily during a peritectic reaction. In this context three possible quasi-peritectic or peritectic reactions would seem to be indicated, all involving liquid of very low boron concentrations. These are

(i) $L + Al_3Ti = TiB_2 + aluminium$
(ii) $L + TiB_2 = Al_3Ti + aluminium$
(iii) $L + Al_3Ti + TiB_2 = aluminium$.

We have no obvious method of distinguishing between these alternatives and it may never be possible to do so experimentally, but their form might have some relevance to reaction mechanisms and rates, and have some bearing on the following complication.

Retention time

When a relatively concentrated dispersion of particles, *viz.* as an alloy Al–5 wt-% Ti, 1 wt-% B, is added to molten aluminium at a temperature well above the binary or possible ternary peritectic reactions (the temperatures of which we do not know), perhaps at 700°C or higher, the first expectation might be that a significant proportion of the lower melting Al_3Ti will begin to dissolve. In the absence of any cross reactions involving TiB_2, one might therefore expect a steady deterioration in the efficacy of the addition to promote grain refinement, assuming, that is, that Al_3Ti and not TiB_2 is the active substrate. In practice, however, the above is not the case and for some time after the addition is made the grain refining action improves, reaching a maximum potency, e.g. after a 10 min holding period. Thereafter the efficiency decreases and the refining effect fades gradually.

We have at present no explanation for this effect, beyond the simple thought that a short interval is needed to 'suitably' disperse the active ingredient, the precise identity of which we still do not know! Metallographic and crystallographic examinations must indicate Al_3Ti, but there is something about the combination with

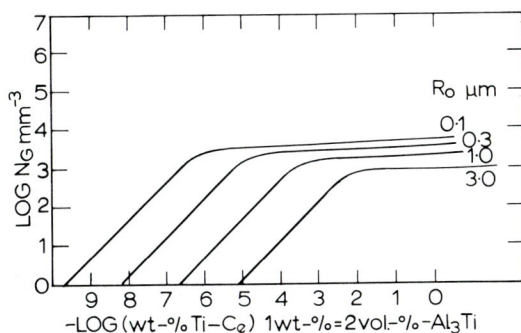

11 Data from Fig. 9 replotted for different particle sizes as functions of excess Ti concentration beyond the effective liquidus concentration C_e

boron which enhances the activity. One question which must be answered, is whether the cross reaction which takes place, perhaps during the 'holding' time prior to casting, can be described in terms of a possible ternary reaction of the alternatives suggested previously. It is worth while noting that all those reactions involve both the intermetallic compounds, and if these compounds do not associate in the liquid, i.e. remain widely dispersed, the reaction rates will be relatively slow. This reaction rate will be particularly slow for the true ternary peritectic reaction, (point (iii) above), which involves liquid and both compounds simultaneously.

It is worth noting again that in the chill-cast master alloy we can expect some supersaturation with respect to Ti, because precipitation of Al_3Ti is unlikely to be complete according to the equilibrium diagram. Remelting such an alloy, just above the peritectic temperature would actually involve further precipitation of Al_3Ti.

In practice, by careful quality control and trial and error in use, it seems that the aluminium foundry has stumbled upon a remarkable product in Al–Ti–B, by accident rather than by design. Like many other discoveries of this kind, involving minor additions which produce dramatic effects (sodium in Al–Si alloys, magnesium and cerium in cast irons) we still seek the full explanation of the reaction mechanism.

SUMMARY OF CONCLUSIONS

1 Application of a relatively simple model for nucleation and growth upon a dispersion of substrate particles shows how cooling curves can be analysed and grain sizes calculated with relatively few assumptions. It is particularly noteworthy that all grain refining action is virtually complete immediately following recalescence.

2 The model can be used to separately assess the influences of nucleating potential, cooling rate, relevant phase equilibria and particle size. The influence of solute undercooling is particularly striking in a comparison between the grain refining action in Al–Ti, Al–Zr and Al–Cr systems, and stresses the difference between *potential* and *actual* substrate activity.

3 In any system, the difficulty of making quantitative predictions hinges primarily upon lack of effective solubility data, but simple calculations do emphasize that remarkably low excess solute concentrations will be involved.

4 The enhanced catalytic activity occurring in ternary aluminium alloys with boron, Al–Ti–B system, remains something of an enigma. It is noted, however, that any cross reaction, as between Al_3Ti and TiB_2, which dispersed the active Al_3Ti more finely could account for the observed enhancement.

5 The grain refining action of an additive is time dependent and passes through an optimum some time after addition to the melt. We do not presently know whether this is simply a problem of mixing, or whether it relates to time dependent ternary phase reaction.

ACKNOWLEDGMENTS

The author wishes to acknowledge support for much of this work from the Science Research Council, and particularly from Alcan International Research and Development Ltd, Banbury, and the London and Scandinavian Metallurgical Co. Ltd, Rotherham, both of which have been very helpful in drawing attention to aspects of foundry practice in the use of grain refining agents.

REFERENCES

1 E. L. GLASSON AND E. F. EMLEY: 'The solidification of metals', 110; 1967, London, The Iron and Steel Institute

2 C. R. LOPER: University of Wisconsin, personal communication; *see also* this conference, p. 169

3 J. CAMPBELL: 'The solidification of metals', 18; 1967, London, The Iron and Steel Institute

4 I. MAXWELL AND A. HELLAWELL: *Acta Metall.*, 1975, **23**, 229

5 I. G. DAVIES *et al.*: *Met. Trans. AIME*, 1970, **1**, 275

6 J. CISSÉ *et al.*: *Met. Trans. AIME*, 1971, **2**, 1 733

7 M. VOLMER: *Z. Elektrochem.*, 1929, **35**, 555

8 'The solidification of metals', Discussion I, 119; 1967, London, The Iron and Steel Institute

9 N. J. W. BARKER AND A. HELLAWELL: *Met. Sci.*, 1974, **8**, 353

10 F. A. CROSSLEY AND M. F. MONDOLFO: *Trans. AIME*, 1951, **191**, 1 143

11 A. CIBULA: *J. Inst. Met.*, 1949–50, **76**, 321

12 G. W. DELAMORE AND R. W. SMITH: *Met. Trans. AIME*, 1971, **7**, 1 744

13 J. H. PEREPEZKO *et al.*; this conference, p. 169

14 I. MAXWELL AND A. HELLAWELL: *Met. Trans. AIME*, 1972, **3**, 1 487

15 L. BÄCKERUD: *Jernkontorets Ann.*, 1971, **155**, (8), 422

16 I. MAXWELL AND A. HELLAWELL: *Acta Metall.*, 1975, **23**, 901

17 J. W. CHRISTIAN: 'Theory of phase transformations in metals and alloys', 2nd ed., 1975, Chap. 10, Pergamon Press

18 I. MAXWELL AND A. HELLAWELL: *Acta Metall.*, 1975, **23**, 895

Undercooling of low-melting metals and alloys

J. H. Perepezko, D. H. Rasmussen, I. E. Anderson, and C. R. Loper, Jr.

An emulsification procedure has been developed to generate stable droplets of metals and alloys with melting temperatures below about 450°C. The droplets are supported in aromatic oils and stabilized by reaction with organic peroxides and organic acid modifiers. The emulsification procedures were developed to segregate nucleants present in the bulk metal into a small fraction of the droplets and to permit measurement of the nucleation undercooling in the presence of different types of surface layers. A maximum undercooling would be limited not by a surface catalysis of nucleation, but by a homogeneous nucleation of crystallization in the undercooled liquid metal. It has been believed that the undercooling to homogeneous nucleation of crystallization is generally equivalent to about $0.2T_m$. Application of the improved droplet technique has resulted in a substantial increase in the maximum undercooling values for Bi and Sn from about $0.2T_m$ to $0.32T_m$ and $0.34T_m$, respectively. These results indicate that the previously reported undercoolings represent the onset of a heterogeneous nucleation process. Moreover, this implies that the values of σ_{LS} calculated from previous undercoolings underestimate the actual value. Similar large undercoolings to crystallization have been achieved in Sn–Bi alloys. The depression of the undercooling temperature for Bi by the addition of Sn solute is proportional to the depression of the melting temperature of bulk alloys by Sn, suggesting a colligative solute–solvent behaviour. For Sn-rich alloys, X-ray diffraction examination of the initial solidification structure has demonstrated the formation of a single-phase Sn solid solution. This composition-invariant solidification reaction results in a considerable metastable extension of primary solid solubility. These observations indicate that the droplet technique is of value in the study of metastable phases and structural modifications produced during solidification at high undercooling.

The authors are in the Department of Metallurgical and Mineral Engineering, University of Wisconsin, USA

The undercooling of liquid metals and alloys before solidification is a fairly common observation. Indeed, with reasonable care, substantial undercoolings approaching 200°–250°C ($0.1–0.2T_m$) may be obtained in bulk-size liquid samples of Fe, Ni, and Co.[1-3] The controlling factor which limits the possible undercooling in bulk samples is the effect of heterogeneous nucleants. For this reason, most studies of undercooling and solidification refer to the catalytic effects of nucleants, usually of uncertain identity and potency.

By examining smaller sample volumes that are obtained by dispersing a liquid metal into fine droplets, the maximum undercooling values may be extended appreciably.[4,5] At high degrees of dispersion, nucleants which may be present in the bulk liquid are isolated into a small fraction of the droplet population so that the majority of the droplets may solidify without the influence of nucleants. Most of the applications of the droplet concept have involved the observation of drop- let solidification on an inert substrate,[6-10] but in a few cases a droplet emulsion contained within a suitable carrier fluid has been used.[4,5,11]

In pure metals the maximum undercooling temperature is considered to be represented by the temperature at which homogeneous nucleation proceeds at an appreciable rate. As a result of numerous studies applying the droplet-substrate method to a variety of pure metals, maximum undercoolings were obtained equivalent to about $0.2T_m$.[6,7] The observation of a general relationship between maximum undercooling and the absolute melting temperature tended to support the assertion that in most metals the onset of homogeneous nucleation required an undercooling of $0.2T_m$. However, the complete demonstration of homogeneous nucleation has been carried out only for two pure metals, Hg[11] and Ga.[12] In both cases the onset of homogeneous nucleation was equivalent to $0.33T_m$.

Recently, the droplet emulsion technique has been modified and improved so that studies of solidification at high undercooling on metals and alloys that melt below about 450°C are now possible with this approach.[4,5] Following the new procedures, large undercoolings have been obtained for several pure metals and alloys which exceed by large amounts the previously reported limits for maximum undercooling. This development permits a quantitative assessment of the undercooling and nucleation of liquid alloys and a detailed study of the effect of large undercoolings on the structure and stability of liquid and solid phases.

EXPERIMENTAL DETAILS

Droplet dispersions of low-melting metals and alloys were produced by emulsifying a mixture of a liquid-metal sample and carrier fluid with a high-speed shearing device. During the emulsification process the liquid metal droplets were stabilized by a chemical reaction that deposits a surface layer on each droplet. The procedure followed in this work was derived from the earlier work of Turnbull,[11] (who used it to prepare emulsions of mercury droplets) but the present authors modified and developed the technique so that it could be applied to metals and alloys that melt below about 450°C.

The formation of a stable metal-droplet emulsion requires a suitable carrier fluid, an appropriate oxidant, and an acid catalyst. Droplet emulsions, which undercooled to the greatest extent, were produced with surfactant reagents which are combinations of an organic peroxide and an aromatic dicarboxylic acid in a thermally stable oil. Further details of the procedures have been described elsewhere.[4]

After preparation, the emulsions were transferred to a sample tube of a differential thermal analysis (DTA) apparatus. A standard heating and cooling rate of approximately 10°C/min was employed throughout the study. The melting temperature of each metal during heating served as a calibration point and the undercooling to nucleation was determined by the difference in temperature between the melting point and the peak of the exothermic reaction during cooling. After emulsification and undercooling, the resulting droplets with diameters between 5 and 20 μm are in a suitable form for X-ray diffraction studies. Powder diffraction patterns of samples were used to determine lattice parameters and crystal structures of as-crystallized and annealed droplet dispersions.

RESULTS

An important requirement in the initial evaluation of the undercooling behaviour of a droplet emulsion of a pure metal is the observation of a uniform undercooling to a single well-defined exothermic crystallization event. This type of undercooling behaviour is illustrated in Figs 1 and 2, which present DTA thermograms for pure Sn and pure Bi droplet emulsions. The nucleation temperatures recorded in the thermograms of Figs 1 and 2 which are 57°C for Sn and 98°C for Bi, represent undercoolings of 175° and 173°C, respectively. It should be noted that the maximum undercoolings are equivalent to $0.34 T_m$ and $0.32 T_m$ for Sn and Bi, respectively. It can be established that these new undercooling

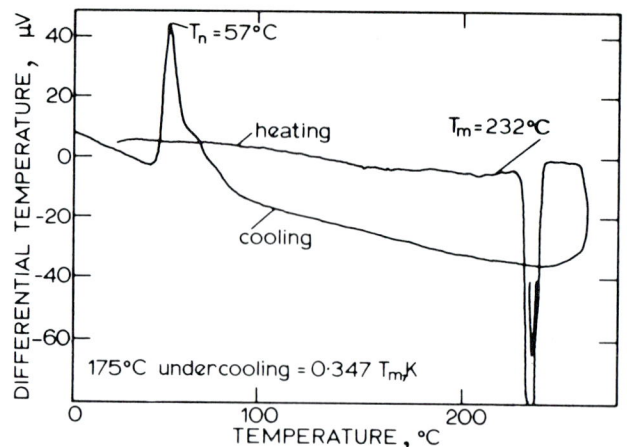

1   DTA thermogram for pure Sn droplets: cooling curve indicates uniform undercooling of 175°C to nucleation temperature of 57°C

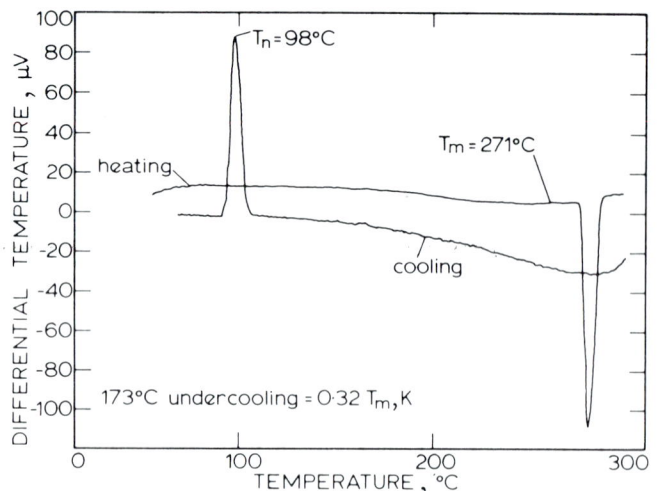

2   DTA thermogram for pure Bi droplets: cooling curve indicates uniform undercooling of 173°C to nucleation temperature of 98°C

values represent those limited by homogeneous nucleation of crystallization by measurements of nucleation kinetics.

The large undercoolings achieved for pure metals may be maintained for alloy droplet emulsions as demonstrated by the thermogram shown in Fig. 3 for a Sn-85Bi alloy. In fact, Sn–Bi alloys represent a model system to examine undercooling and nucleation in a eutectic alloy system. The nucleation temperatures determined for the entire composition range in the Sn–Bi system are indicated in the equilibrium phase diagram shown in Fig. 4. There are several features in the pattern of behaviour of the compositional dependence of the nucleation temperature that are worthy of comment. In Bi-rich alloys the nucleation temperature is depressed by Sn solute, in the same manner as Sn depresses the liquidus temperature. This suggests a colligative solute–solvent interaction in which solute affects the properties of microscopic clusters in the same way that it affects bulk properties. For Sn-rich alloys, a different trend is observed for the dependence of nucleation temperature upon composition. The nucleation temperature is depressed from that for pure Sn by solute additions, but the amount of depression is smaller than the depression of the

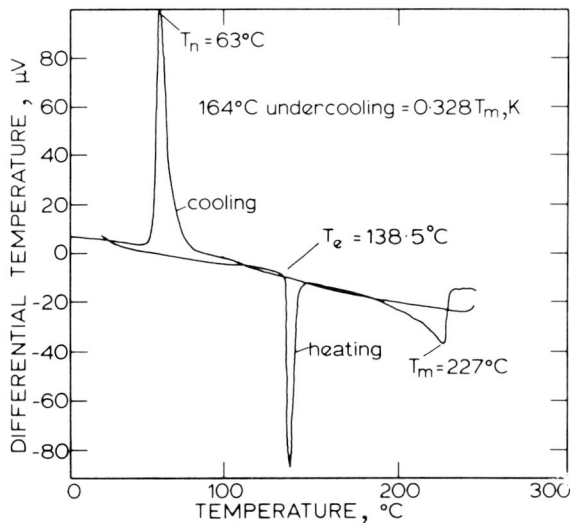

3 DTA thermogram for an emulsion of a Sn–85 wt-% Bi alloy; heating curve indicates temperatures of eutectic melting and final melting at 138·5° and 227°C, respectively; cooling curve indicates uniform undercooling of 164°C to crystallization temperature of 63°C

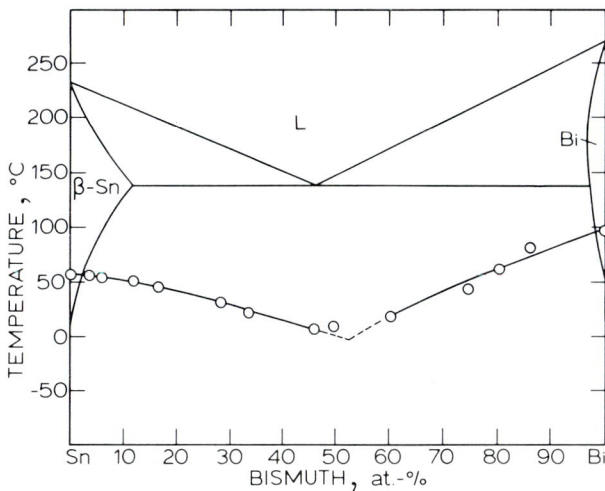

5 Lattice spacings determined at 25°C for Sn–Bi alloys crystallized at high undercooling; maximum equilibrium solid solubility is indicated by vertical dashed line; circled points refer to previous measurements[13] and square points represent present results

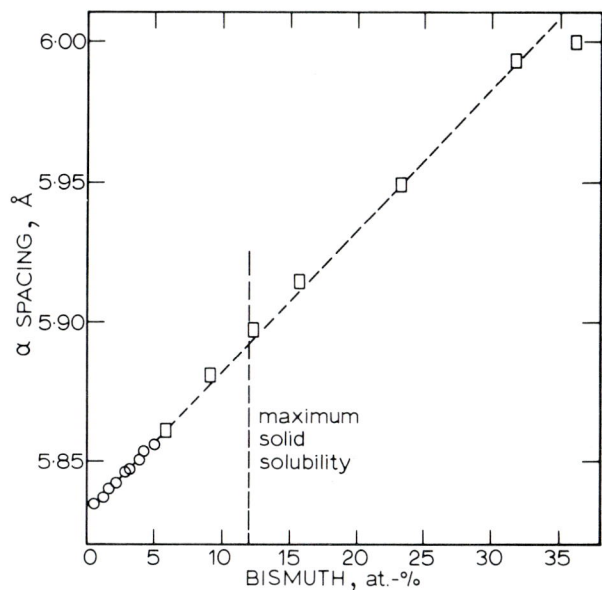

4 Summary of composition dependence of nucleation undercooling in Sn–Bi system

liquidus temperature, so that the magnitude of the undercooling measured from the liquidus decreases with increasing Bi concentration.

X-ray diffraction studies on droplets immediately after solidification have provided useful structural information on the initial solidification products. Lattice parameter measurements for the Sn-rich alloys are plotted in Fig. 5. At low solute contents, the composition dependence of the a parameter of the tetragonal Sn solid solution is in good agreement with previously reported values.[13] The continued increase of the a parameter with increase in Bi content, beyond the maximum equilibrium solid solubility for the Sn-rich primary solid solution, demonstrates that the solid solubility has been extended out to at least 35 at.-% Bi. The solubility extension probably extends past 35 at.-% Bi up to about the eutectic composition; however, in this composition range, thermal decomposition occurs more rapidly at room temperature than in the alloys with a

lower Bi content. The existence of an initial Sn solid solution has also been confirmed by metallographic techniques. That the initial solidification product for compositions in excess of the equilibrium solubility is metastable has been established by the observation that the equilibrium eutectic microstructure develops in droplets after aging at room temperature, as shown in Fig. 6. For the Bi-rich alloys, the X-ray diffraction results on droplets immediately after solidification indicate the presence of the two-phase eutectic structure.

DISCUSSION

The present results for the nucleation temperature in pure Sn and Bi are substantially lower than the values that have been reported previously and have been believed to represent the limits to maximum undercooling. With the droplet-substrate procedure, the maximum undercoolings for Bi and Sn have been reported to be 90° and 105°C,[6] respectively, while the droplet–emulsion approach has yielded undercoolings of 173° and 175°C, respectively. On the basis of this comparison, it would appear that nucleation catalysts are present in the droplet-substrate samples which are removed effectively by the droplet–emulsion technique. In fact, a similar conclusion has been stated by Paull et al.[14] in their recent discussion of past work on nucleation undercooling of small liquid drops in contact with a substrate. The current studies have demonstrated clearly that the previous maximum undercooling values which have been taken to represent homogeneous nucleation actually represent a heterogeneous nucleation process. However, this observation should not be taken as proof that the new maximum undercooling values represent solidification by a homogeneous nucleation process. Instead, the present findings emphasize that before homogeneous nucleation conditions can be claimed with assurance, it is absolutely

6 **Optical micrograph of sectioned Sn–25 at.-% Bi alloy droplet; partial decomposition of metastable Sn solid-solution into equilibrium eutectic phases has occurred after annealing at 25°C for 50 h   [× 1900]**

where ΔG_V is the liquid/solid free energy difference, k is Boltzmann's constant, T is the absolute temperature, n is the number of atoms per unit volume, J is the nucleation rate, h is Planck's constant, and ΔG_A is the activation free energy for atom addition to a cluster. Following Turnbull,[6] values of $\exp(-\Delta G_A/kT) = 10^{-2}$, $J = 2 \times 10^{14}\,\text{s}^{-1}\,\text{m}^{-3}$, and $\Delta G_V = \Delta H_m \Delta T/T_m$ where ΔH_m is the heat of fusion and $\Delta T = T_m - T$ were used to evaluate equation (1). For bismuth, with $\Delta H_m = 5\cdot29 \times 10^8\,\text{J/m}^{3,15}$ and $n = 2\cdot82 \times 10^{28}\,\text{atoms/m}^3$, equation (1) yields a value for σ_{LS} of $79\cdot3\,\text{mJ/m}^2$. Similarly, for tin, with $\Delta H_m = 4\cdot32 \times 10^8\,\text{J/m}^{3,15}$ and $n = 3\cdot7 \times 10^{28}\,\text{atoms/m}^3$, equation (1) yields a value for σ_{LS} of $70\cdot6\,\text{mJ/m}^2$. A comparison of the present calculated values for σ_{LS} with the values of $54\cdot5\,\text{mJ/m}^2$ for Sn and $54\cdot4\,\text{mJ/m}^2$ for Bi calculated from previous undercooling determinations[6] indicates that the current undercooling measurements yield about a 50% increase in σ_{LS}. These results are in agreement with the assessments of Miller and Chadwick[16] who suggested that Turnbull's values for σ_{LS} should be increased by a factor of about 3/2, and represent about the same magnitude of correction for Sn and Bi as reported by Stowell[17] for submicron lead particles.

In judging the effect of the amount of undercooling to crystallization upon the magnitude of σ_{LS}, it should be noted that σ_{LS} as given by equation (1) is related to the value for ΔG_V at the crystallization temperature. The maximum undercooling, in turn, can be related to the free energy change by the first-order approximation, $\Delta G_v \simeq \Delta H_m \Delta T/T_m$. While this approximation for ΔG_v is valid for low undercoolings, it is necessary to include a correction for the temperature dependence of ΔH_m and ΔS_m, involving ΔC_p, the specific heat difference between undercooled liquid and solid, at high undercoolings. Heretofore, ΔC_p has been taken as zero[11] or equal to the value at the melting point,[12] but recent measurements[18,19] have noted that a difference between the liquid and solid specific heat exists at the melting point, and that this difference continues to increase with undercooling. If the ΔG_v value at the nucleation temperature is modified by a ΔC_p correction, which has been determined recently for droplet emulsions of Sn and Bi,[19] the revised ΔG_v value is decreased by 2–3% below that given by the first-order approximation. This correction of ΔG_v results in a slight reduction in the value of σ_{LS} to $77\cdot5\,\text{mJ/m}^2$ for Bi and $69\cdot6\,\text{mJ/m}^2$ for Sn.

Independent measurements of σ_{LS} are not generally available for most metals, but for pure Bi, Glicksman and Vold[20] have determined the value of $82 \pm 4\,\text{mJ/m}^2$ for σ_{LS} at the melting point. This measured value compares favourably with the value calculated from the present nucleation-undercooling results. However, it should be noted that while the different determinations of σ_{LS} for Bi agree, they refer to different temperatures, spanning the undercooling range. Based upon this comparison, for bismuth, it appears that σ_{LS} changes only slightly in value between the nucleation temperature, T_n (98°C) and the melting point. Although it is possible that the limit to maximum undercooling in bismuth may be greater than 173°C, this would increase the calculated σ_{LS} value based upon nucleation undercooling. Recent theoretical attempts[21,22] to model the liquid/solid interface attribute σ_{LS} mainly to configurational entropy factors. Based upon this model, σ_{LS} is

necessary to demonstrate that the nucleation rate depends on droplet volume, rather than on droplet surface area.[11] Furthermore, the present results cast doubt on the correlation that the onset of homogeneous nucleation occurs at an undercooling equivalent to $0\cdot2T_m$. Based upon the two studies[11,12] in which homogeneous nucleation conditions were established, it appears that the limit to maximum undercooling may be $0\cdot33T_m$, but again this apparent correlation must be substantiated by further work.

Until recently, measurements of the homogeneous crystallization temperature provided a major source of information concerning the magnitude of the liquid/solid interfacial energy, σ_{LS}. Values of σ_{LS} may be calculated by employing the maximum undercooling temperature and classical homogeneous nucleation theory. However, since the maximum undercooling values used in the past calculations have been shown to represent heterogeneous nucleation, at least in the case of Bi and Sn, the previously calculated values for σ_{LS} are expected to underestimate the true value. Although the true magnitude for σ_{LS} cannot be set until homogeneous nucleation conditions are regarded as established, it is of interest to ascertain the magnitude of the correction to the previous calculations implied by the present undercooling results.

The maximum undercooling values may be related to an effective minimum value of σ_{LS} by the nucleation-rate equation employed by Turnbull[6]:

$$\sigma_{LS} = [(3 \times 2\cdot303\Delta G_v^2 kT/16\pi)$$
$$\times \log(nkT \exp\{-\Delta G_A/kT\}/Jh)]^{1/3} \quad (1)$$

expected to scale with the reduced temperature (T/T_m). If the measured value of σ_LS for Bi at T_m is scaled to T_n, a predicted σ_LS of about 56 mJ/m^2 would be expected at T_n. This result is contrary to the present calculated value for σ_LS at T_n. A more detailed assessment of the magnitude of σ_LS and its temperature variation will be possible as homogeneous nucleation kinetics measurements become available for more metals.

The results on the undercooling of alloys in the Sn–Bi system indicate that a significant modification of the solidification products may be achieved through nucleation under high undercooling conditions. In the Sn–Bi system, although the equilibrium structure at room temperature for alloys containing between 12 at.-% Bi and 35 at.-% Bi is given by a Sn-rich and Bi-rich two-phase mixture, the initial solidification structure is a supersaturated single-phase Sn-rich solid solution. In the composition range from the maximum equilibrium solubility in the β-Sn phase to almost the eutectic composition, a composition-invariant solidification reaction occurs which results in a substantial metastable extension of primary solid solubility. Previous work[23] involving a 'splat cooling' technique in which liquid Sn–Bi alloys were rapidly cooled to $-190°$C has indicated also that a solubility extension may be obtained out to about the eutectic composition. The similarity of solubility extension values obtained with the slow cooling droplet technique and with the rapid cooling splat cooling technique is interesting and suggests that the magnitude of the cooling rate is of secondary importance to the ability to achieve a state of high undercooling before nucleation. Evidently, at high undercoolings, kinetic factors permit a Sn-rich solid solution to form, even though thermodynamic factors such as the maximum reduction in free energy of the system favour the formation of a (Sn-rich + Bi-rich) two-phase mixture.

In connection wtih the presence of a metastable solid-solution phase as the initial solidification structure in the Sn–Bi system between 12 and 35 at.-% Bi, it is important to consider some of the thermal effects associated with the nucleation of crystallization in a highly undercooled liquid-alloy droplet emulsion. After nucleation of a liquid-metal droplet, subsequent crystallization is very rapid at high undercooling. The rapid solidification of a droplet creates an essentially adiabatic system so that the liberated latent heat of solidification will increase the temperature of the droplet. An estimate of the magnitude of this effect can be made by considering the fraction of solid phase X_s that must solidify in order to liberate sufficient heat to raise the temperature of the liquid surrounding the solid to the melting point. If the liquid specific heat C_p^l and heat of fusion ΔH_m are assumed to be independent of temperature, X_s may be approximated[24] by $X_\mathrm{s} \simeq C_\mathrm{p}^\mathrm{l} \Delta T/\Delta H_\mathrm{m}$. On the basis of this relation, it is expected that for pure bismuth, where $C_\mathrm{p}^\mathrm{l} = 30.5$ J/mole-K,[15] $\Delta T = 173$ K and $\Delta H_\mathrm{m} = 11.3$ kJ/mole, rapid crystallization of an undercooled droplet would cause about 45% of it to solidify initially before recalescence could heat the droplet up to the melting point. Subsequent completion of the crystallization process requires an exchange of heat between the system and the surroundings. Clearly, this thermal behaviour should be of great concern in any study of the structural characteristics of products of solidification at high undercooling. For example, it is possible that if the

rise in temperature is sufficient, the initial solid that is nucleated from the melt may undergo a structural modification or perhaps, even a partial remelting, especially if the initial structure is not an equilibrium structure, but rather a metastable phase. The severity of the recalescence annealing effect would be expected to depend upon the degree of metastability of the initial solidification products. Although the extended solid-solution phase in the Sn–Bi system is metastable to prolonged annealing at room temperature as demonstrated by the decomposition into the equilibrium two-phase mixture given in Fig. 6, this structure appears to be resistant to decomposition during the recalescence heating following nucleation. The relatively sluggish decomposition kinetics of the metastable structure formed during crystallization at high undercooling by the droplet technique suggests that the inert non-catalytic surface coating that was introduced during emulsification to prevent surface nucleation of crystallization may not be a very favourable site for the nucleation of any subsequent solid-state transformation.

It is noteworthy that when the splat cooling technique is utilized to extend the β-Sn phase solid solubility in the Sn–Bi system, thermal decomposition of the metastable structures takes place below room temperature.[23] When the extension of solid solubility in Sn–Bi alloys is accomplished by crystallization at high undercooling during a slow cooling by the droplet technique, thermal decomposition is relatively sluggish at room temperature. This difference between the decomposition kinetics of the metastable phases produced by the two techniques is likely to be related to the condition of the specimen surface. The surface of a foil prepared by splat cooling is expected to be fairly rough and probably coated with a simple metal oxide which can serve as a viable heterogeneous nucleation site for solid-state transformation. The surface of a fine droplet, on the other hand, is fairly smooth and coated with a complex surfactant film. In addition, owing to the rapid cooling rates involved, splat-cooled foils contain a relatively high density of imperfections,[25] while droplets cooled slowly are single crystals and can be expected to show a relatively low density of imperfections. This suggests that the droplet technique can offer important advantages in the study of metastable structures formed during crystallization at high undercooling.

CONCLUSIONS

A modified droplet-emulsion technique has been employed in order to obtain large undercoolings before crystallization in pure Sn and Bi and Sn–Bi alloys. The undercoolings observed for pure Sn and Bi are 175° and 173°C, respectively, and are substantially larger than the maximum undercoolings reported previously. The current studies have demonstrated clearly that the previous maximum undercooling values which have been taken to represent the onset of homogeneous nucleation actually represent the onset of a heterogeneous nucleation process. Determinations of maximum nucleation undercooling have been used extensively to calculate values of σ_LS, based upon classical nucleation theory. By using the increased nucleation-undercooling values determined in this study for Sn and Bi, it has been possible to recalculate values for σ_LS; these revised values are about 50% greater than the values based upon previous undercoolings. For the case of bismuth, the value of σ_LS

calculated from the nucleation undercooling measured in this study is in fair agreement with the experimental determination of σ_{LS} at the melting point.

For Bi-rich alloys, of the Sn–Bi system, the nucleation temperature is depressed by solute, in nearly the same manner that the liquidus temperature is also depressed by solute. This behaviour suggests a colligative solute–solvent interaction in which solute affects the properties of microscopic clusters in the same way that it affects bulk properties. For Sn-rich alloys, a composition-invariant solidification reaction occurs and results in a metastable extension of solid solubility out to at least 35 at.-% Bi. Although the metastable Sn-rich solid solution decomposes thermally at room temperature, the decomposition kinetics are sufficiently sluggish for the metastable structure not to be altered by recalescence annealing effects. Observations of metastable alloy structure crystallization in Sn–Bi alloys suggest that the thermal decomposition of this structure proceeds less rapidly when the metastable phase is formed by the droplet technique than when it is produced by rapid quenching from the melt. These findings indicate that a significant modification of alloy solidification products may be achieved through nucleation under high undercooling conditions by the use of the droplet-emulsion technique.

ACKNOWLEDGMENT

The authors very much appreciate the partial support of this work by the US Army Research Office. Some of the high-purity metals used were kindly provided by the American Smelting and Refining Company Central Research Laboratory.

REFERENCES

1 J. L. WALKER: 'Physical chemistry of process metallurgy, Part 2,' (ed. G. R. St. Pierre), 845, 1961, New York, Interscience Publishers
2 G. A. COLLIGAN AND B. J. BAYLES: *Acta Metall.*, 1962, **10**, 895
3 T. Z. KATTAMIS AND M. C. FLEMINGS: *Trans. AIME*, 1966, **236**, 1 523
4 D. H. RASMUSSEN AND C. R. LOPER: *Acta Metall.*, 1975, **23**, 1 215
5 D. H. RASMUSSEN et al.: 'Rapidly quenched metals, Part 1', (ed. N. J. Grant and B. C. Giessen), 51, 1976, Cambridge, Massachusetts, MIT Press
6 D. TURNBULL: *J. Appl. Phys.*, 1950, **21**, 1 022
7 D. TURNBULL AND R. E. CECH: *ibid.*, 1950, **21**, 804
8 B. E. SUNDQUIST AND L. F. MONDOLFO: *Trans. AIME*, 1961, **221**, 157
9 P. B. CROSBY et al.: 'The solidification of metals', 10, 1968, London, The Iron and Steel Institute
10 T. TAKASHASHI AND W. A. TILLER: *Acta Metall.*, 1969, **17**, 651
11 D. TURNBULL: *J. Chem. Phys.*, 1952, **20**, 411
12 Y. MIYAZAWA AND G. M. POUND: *J. Cryst. Growth.*, 1974, **23**, 45
13 J. A. LEE AND G. V. RAYNOR: *Proc. Phys. Soc. (London)*, 1954, **B67**, 737
14 J. PAULL et al.: *Scripta Metall.*, 1976, **10**, 845
15 R. HULTGREN et al.: 'Selected values of the thermodynamic properties of the elements,' 1973, Metals Park, Ohio, ASM
16 W. A. MILLER AND G. A. CHADWICK: *Acta Metall.*, 1967, **15**, 607
17 M. J. STOWELL: *Philos. Mag.*, 1970, **22**, 1
18 H. S. CHEN AND D. TURNBULL: *Acta Metall.*, 1968, **16**, 369
19 D. H. RASMUSSEN AND J. H. PEREPEZKO: To be published
20 M. E. GLICKSMAN AND C. L. VOLD: *Scripta Metall.*, 1971, **5**, 493
21 F. SPAEPEN: *Acta Metall.*, 1975, **23**, 729
22 F. SPAEPEN AND R. B. MEYER: *Scripta Metall.*, 1976, **10**, 257
23 R. H. KANE et al.: *Acta Metall.*, 1966, **14**, 605
24 M. E. GLICKSMAN AND R. J. SCHAEFER: *J. Chem. Phys.*, 1966, **45**, 2 367
25 G. THOMAS AND R. H. WILLENS: *Acta Metall.*, 1964, **12**, 191

Nucleation in peritectic systems

P. G. Boswell, G. A. Chadwick, R. Elliott, and F. R. Sale

The droplet technique devised previously for the study of heterogeneous nucleation in eutectic systems has been extended to the case of peritectic alloys. Specimens containing droplets of the peritectic liquid entrained in a matrix of the primary solid phase were prepared by heat treating chill-cast alloys. The nucleation temperatures of the droplets have been obtained using differential thermal analysis. Measured undercoolings below the liquidus of the low-temperature peritectic phase ranging from 0–16 K have been obtained in a variety of systems including $Al–Al_3Ti$; $Cd–Ag$; $Pb–Pb_3Bi$; $Cu–Sn$; $Ag–Sn$; and $Cu–Fe$. The undercoolings are discussed in relation to the classical description of the peritectic reaction.

P. G. Boswell and G. A. Chadwick are in the Department of Mining and Metallurgical Engineering, University of Queensland, Australia. R. Elliott and F. R. Sale are in the Department of Metallurgy, University of Manchester

The classical description of the peritectic 'reaction'[1] invokes heterogeneous nucleation at the peritectic temperature T_p, of the secondary (II) solid phase on the primary (I) solid phase/liquid L interface (*see* Fig. 1 for a schematic phase diagram). Lateral growth of II then occurs to give I surrounded by a rim of II and thereby isolating I from the liquid. Subsequent *transformations* at the I/II and L/II interfaces then proceed via mass transfer within II and these transformations have been examined experimentally[2] and analysed theoretically in some detail. On the other hand, the 'reaction' stage involving the formation of the II rim via a three-phase 'reaction' $(I + L \rightarrow II)$ has received little attention. However, it is customarily assumed that zero undercooling below T_p is required for the heterogeneous nucleation of II at the I/L interface.[3,4]

Southin[5] has recently used the internal droplet technique, originally developed by Wang and Smith,[6] to measure undercoolings for secondary phase nucleation in several binary eutectic systems. The technique involves annealing samples containing finely distributed, solute-rich interdendritic material at a temperature T_A just above the eutectic temperature in the two-phase (I/L) field. The solute-rich material melts and equilibrates with the matrix to give a reasonably uniform distribution of small droplets entrained in large grains of the primary (I) phase. Following the annealing treatment at T_A, the alloy is allowed to cool at a slow rate (~ 0.2 K/s) and the nucleation of the secondary (II) phase is monitored by thermal analysis. By using this method both Wang and Smith[6] and Southin[5] demonstrated that a significant undercooling below the II-phase liquidus was invariably required to nucleate the II-phase in the uncontaminated liquid droplets, although the liquid was in perfect contact with solid

material (namely the I-phase). In view of these results, it might be expected that in the absence of impurity particles, an undercooling below T_p would be required to nucleate II in a peritectic 'reaction'. A study, using the internal droplet technique, of the peritectic 'reaction' was therefore undertaken in order to establish the undercoolings for II-phase nucleation in several commonly investigated peritectic systems.

EXPERIMENTAL

A total of 14 peritectic systems were investigated by the internal droplet technique and details of each transformation are given in Table 1. The alloys were prepared from high-purity ($> 99.999\%$) component elements by melting and casting appropriate quantities of the components (having rod or shot form) in evacuated (10^{-5} torr) quartz tubes (11 mm dia.). Other alloy fabrication procedures involved vacuum induction melting high-purity components in an alumina crucible followed by casting into cast iron moulds (for Fe–Cu alloys); powder methods (for Bi–Ni alloys); and argon-arc melting (for Al–Ti alloys).

Alloy compositions were chosen so that, on the basis of published phase diagrams, samples equilibrated at $T_A = T_p + 5$ K contained 7 ± 3 vol.-% of liquid. In several cases it was found that the available phase diagrams were in error because either too little or too much liquid was found to be present at T_A. Corresponding adjustments were therefore made to the alloy compositions in order to give suitable volume fractions of the liquid phases.

Cylindrical billets (3 mm dia. × 6 mm in height) were cut from the chill-cast alloys and examined by differential thermal analysis using a Rigaku–Denki DTA unit operated at maximum sensitivity (5 μV full-scale

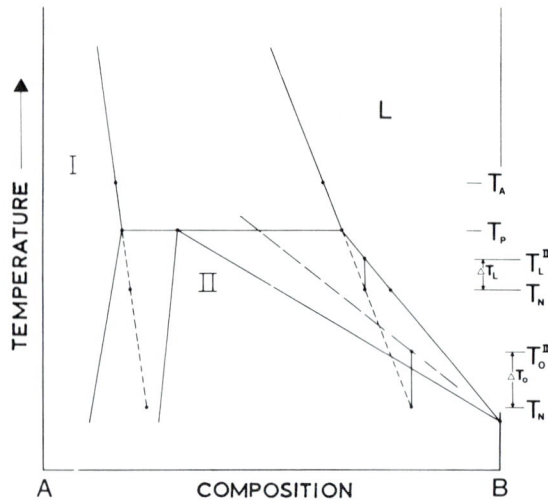

1 Schematic phase diagram for a peritectic system

Table 1   Undercoolings for II-phase nucleation in peritectic systems

System	Reaction L+I→II		T_p, K	ΔT_p, +0·5 K	ΔT_L^{II}, K	$\Delta T_L^{II}/T_L^{II}$
Ag–Sn						
(a)	α(Ag)	ζ	997	non-detected (0?)	—	—
(b)	ζ	ϵ	753	4	2±2	0·003
Al–Ti	Al$_3$Ti	α(Al)	938	7	5±2	0·005
Cd–Ag	ϵ	η	616	14	6±2	0·010
Cd–Cu	γ	Cd$_3$Cu	670	non-detected (0?)	—	—
Cd–In	(Cd)	Cd$_3$In	469	6	3±1	0·006
Cu–Fe	γ(Fe)	α(Cu)	1367	15	13±1	0·010
Cu–Ge	α(Cu)	ξ	1096	28	4±2	0·004
Cu–Sn	ϵ	η	688	30	3±1	0·005
Ni–Bi	α(Ni)	NiBi	928	non-detected (0?)	—	—
Pb–Bi	α(Pb)	ϵ	457	19	16±1	0·035
Pb–Ca	α(Pb)	Pb$_3$Ca	60	17	16±1	0·027
Sb–Sn	(Sb)	β	698	9	4±2	0·006
Sn–Sb						
(a)	β	β'	598	14	2±1	0·003
(b)	β'	(Sn)	519	14	12±2	0·025
Sn–Cd	(Sn)	β	496	non-detected (0?)	—	—

deflection; chromel/alumel thermocouples). Both pan- and block-type sample holders were used and nickel billets, equal in size to the specimen billets, were employed as standards. The samples were heated to T_A at 1–2 K/s and held to within 0·5 K of T_A for times varying from 10 min to several days. Following the annealing treatments, the alloys were cooled at 0·02 K/s and the exothermic solidification transformations were monitored. At the end of continuous cooling, the samples were heated to T_A at 0·167 K/s and the melting temperatures were investigated. The anneal–cool–reheat cycle was repeated until no appreciable changes could be detected, on cooling, between successive

thermograms. The specimens were maintained under an inert (argon) atmosphere during the thermal analysis experiments in order to minimize oxidation of the alloys.

Characteristic thermograms, comprising strong and weak peaks, corresponding to solidification of the grain boundary and droplet liquids, were obtained from most of the alloys investigated. In order to facilitate interpretation of the as-solidified microstructure, each specimen was quenched during the final cooling sequence when the specimen attained a temperature that was ~10 K below the temperature of the droplet peak.

RESULTS
Differential thermal analysis

Figure 2 shows typical DTA thermograms obtained using Cd–Ag, Cu–Sn, Pb–Bi, Sn–Sb, Cu–Ge, and Cu–Fe peritectic alloys. The first large exothermic peak corresponds to II-phase formation in the grain-boundary liquid and the second smaller peak corresponds to II-phase formation in the liquid droplets. The grain-boundary peak was, in some cases, sharp and well defined after equilibration at T_A (*see* the traces for Cd–Ag and Cu–Sn in Fig. 2). In other systems (for instance, Cu–Ge and Cu–Fe) there appeared to be a continuous but steadily decreasing exothermic reaction on cooling following a reasonably sharp grain-boundary peak. It is likely that the continuous evolution of heat corresponds to the diffusion-controlled growth of the secondary phase into the grain-boundary liquid whereas the initial peak corresponds to the nucleation and growth of the II-phase rim around the grain-boundary liquid. A droplet peak arose during the continuous reaction.

Undercooling for nucleation of the secondary phase

Following Southin,[5] the observed peritectic temperature T_p^{obs} was taken to be the temperature at which the differential temperature trace departed from the baseline owing to the heat evolution accompanying the formation of the II-phase in the grain-boundary liquid. The mean nucleation temperature T_N of the droplets was taken to be the temperature corresponding to the maximum of the droplet peak (*see* Fig. 2). The experimental T_p^{obs} values agreed to within ±2 K with the

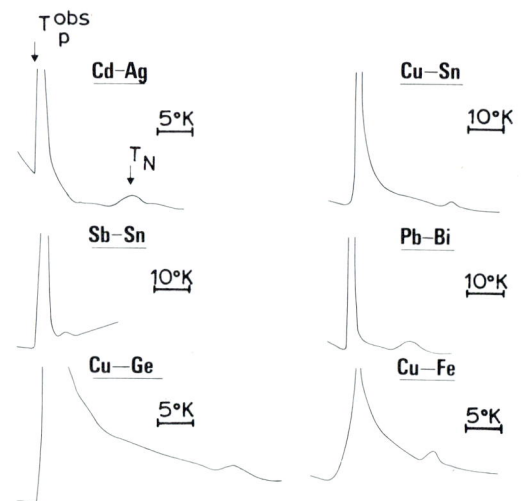

2 Typical DTA thermograms observed on slow cooling peritectic alloys from T_A

accepted values of T_p for the systems investigated (*see* Table 1). However, efforts were not made to specify accurately any small differences between T_p and T_p^{obs}. Instead, the primary concern was the temperature difference $\Delta T_p = T_p^{obs} - T_N$ and this difference was measured to within ± 0.5 K.

The experimentally determined T_N and ΔT_p values are tabulated in Table 1 together with the undercooling ΔT_L^{II}, below the II-phase liquidus (*see* Fig. 1). This latter undercooling is a measure of the driving force for the nucleation of the II-phase because it can be assumed that the liquid in the droplets remains in equilibrium with the primary (I) solid phase on slow cooling from T_A to T_N. Hence the composition of this liquid follows the extrapolated I-liquidus while the composition of the precipitated I-solid follows the extrapolated I-solidus (*see* Fig. 1). Owing to inaccuracies in the reported positions of the I and II-phase liquidi, even for the relatively common peritectic systems investigated, the errors in the ΔT_L^{II} values are generally large (20%) and in some instances are thought to approach 50%.

Optical Microscopy

The equilibrium droplet shapes in alloys quenched for T_A demonstrated most of the morphological features described by Miller and Chadwick.[7] For instance, non-spherical shapes, attributable to anisotropy of the (I) solid/liquid surface energy[7] were observed in many of the annealed alloys and several alloys showed a tendency to faceting of elongated droplets. The shapes of droplets quenched from just below T_N were not significantly different to the shapes obtained on quenching from T_A.

Resolution of the droplet microstructures by optical microscopy was rarely accomplished. However, several types of microstructures could be identified, namely those displaying a droplet rim of II-phase with a fine-scale two-phase structure in the remainder of the droplet (e.g. Cu–Sn); coarse, dendritic II-phase grains with an unresolved interdendritic structure (e.g. Ag–Sn(a)); several non-dendritic II-phase grains located around the droplet periphery (e.g. Cu–Ge); and finally droplets with no fine-scale structures (e.g. Cu–Fe). The first three types represent microstructures resulting from the diffusion-controlled growth of the II-phase in the droplet liquid such that solute partitioning led to rimmed or dendritic II-phase morphologies. The featureless Cu–Fe droplet structure probably arose as a result of 'massive', non-partitioning solidification of the liquid. This result was confirmed by electron microscopy by which it was found that the solidified Cu–Fe droplets contained a single II-phase (α(Cu)) crystal.

DISCUSSION
Nucleation of the secondary phase

According to classical nucleation theory, a critical nucleus of the II-phase will be characterized by its composition, location (homogeneous or heterogeneous), and orientation relationship with the surrounding I-phase. The nucleation site and the orientation relationship will be interrelated to some extent because the repeated observation of a specific orientation relationship is indicative of heterogeneous nucleation (orientated growth without epitaxial nucleation being improbable). However, the lack of an observed orientation relationship does not necessarily imply homogeneous nucleation because heterogeneous nucleation without a specific nucleus–substrate orientation relationship can arise.

The composition of the critical nucleus cannot be measured experimentally so the composition must be estimated from thermodynamic considerations. For nucleation at variable composition, the maximum driving force for nucleation is given by the 'parallel-tangent' construction on a free-energy diagram. The corresponding II-phase nucleus will be slightly poorer in B than the equilibrium liquidus composition at T_N and the driving force for nucleation will be approximately equal to $L^{II} \Delta T_L^{II}/T_L^{II}$, where L^{II} is the latent heat of solidification of II-phase from the liquid. At temperatures below T_O^{II}, nucleation at constant composition is thermodynamically possible. T_O^{II} is the locus of points defining the temperature and composition at which II-phase solid and liquid have the same free energy. The driving force for nucleation at constant composition is approximately equal to $L^{II} \Delta T_O^{II}/T_O^{II}$, where ΔT_O is the undercooling below T_O^{II} (i.e. $\Delta T_O = T_O^{II} - T_N$, shown in Fig. 1). On assuming that T_O^{II} lies mid-way between the II-phase liquidus and solidus we note that of the peritectic systems investigated here, only Cu–Fe and Sn–Sb(b) were observed to undercool below T_O^{II} with $\Delta T_O^{II}/T_O^{II}$ values of 0.008 and 0.005, respectively. The errors in these values are relatively large (~ 50%), owing to the rather arbitrary selection of T_O^{II}.

Reference to Table 1 shows that none of the observed $\Delta T_L^{II}/T_L^{II}$ nor the two $\Delta T_O/T_O^{II}$ undercooling coefficients approached the previously reported coefficient of ~ 0.2 required for homogeneous nucleation.[8] Hence, the II-phases in the systems investigated probably nucleated heterogeneously on the I/L droplet interfaces. It is perhaps surprising that the large undercooling coefficients observed by Southin[5] in his eutectic droplet experiments were not obtained in the present study of peritectic droplets. For example, Southin[5] reported $\Delta T_L/T_L$ values of 0.20, 0.33, and 0.19 for Cu(Ag), Al(Ge), and Cd(Bi) alloys, respectively, whereas the largest $\Delta T_L^{II}/T_L$ coefficient measured here was 0.04 for Pb(Bi). There appears to be no fundamental reason for the observation of generally low $\Delta T_L^{II}/T_L$ values in the peritectic system and the relatively frequent observation of large values in eutectic systems.

Nevertheless, it has been shown in this investigation that in the absence of impurities particles, an undercooling below T_L^{II} is required for the nucleation of II. The classical description[3] of the peritectic 'reaction' is therefore inaccurate since it postulates a zero undercooling below T_p for the heterogeneous nucleation of II on the I/L interface.

Droplet solidification

An exhaustive analysis of all the possible droplet solidification mechanisms and micromorphologies will not be attempted. Instead, only those aspects that can be used to clarify the understanding of the nucleation stage will be described. Broadly speaking, two types of II-phase growth from the melt can be identified, namely, 'massive' growth without solute partitioning and long-range, diffusion-controlled growth with solute partitioning. The former only arises at temperatures below T_O^{II} and gives largely featureless droplets upon complete solidification. Few data describing the II-phase nucleation site or the I–II orientation relationship would be

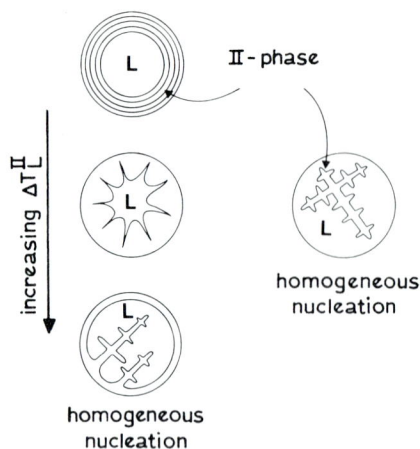

3 Schematic illustrations of partially solidified structures obtained during cooling from T_N

gathered from a metallographic examination of the massively solidified droplet, so crystallographic techniques (such as microfocus X-ray, selected-area electron diffraction, and polarized light methods) must be used. The identification of an orientation relationship between the massive II-phase and the I-phase would constitute evidence for epitaxial nucleation and the absence of any relationship would imply (but not necessarily demand) nucleation in the liquid away from the I-phase substrate.

Schematic representations of the types of continuously cooled droplet structures resulting from II-phase growth, with solute partitioning, following epitaxial and non-epitaxial nucleation are given in Fig. 3. It is assumed that negligible dissolution of the I-solid occurs during cooling from T_N and that complete solidification is accomplished by the direct transformation of the liquid to the II-phase. The second assumption implies that additional phases that may arise at temperatures below T_p are not nucleated and able to grow at undercoolings below T_N. Under these circumstances, and with solute partitioning, the liquid composition follows, at low cooling rates, the II-liquidus at temperatures below T_N and the volume fraction of II-phase solidified within the droplet is given by the lever-rule construction. Solidification to II is complete at the temperature at which the composition of the II-solidus is equal to the liquid composition at T_N.

Comparisons between the observed and predicted droplet microstructures support the interpretation of the undercooling results based upon heterogeneous nucleation of the II-phase on the I/L interface. For example, the rimmed structures described briefly above are indicative of heterogeneous, substrate nucleation. Attempts to determine whether the observed dendritic morphologies developed as a result of II-phase growth from a centrally located or peripherally located site were unsuccessful. The structures of droplets with dendritic morphologies therefore require further investigation. Work is currently being directed to establishing whether or not there is an orientation relationship between the massively solidified α(Cu) phase in the Cu–Fe droplets. Preliminary results indicate there is a relationship so it is likely that the α(Cu) nucleated heterogeneously on the γ(Fe)/L interface and grew at temperatures below T_O^{II}.

CONCLUSIONS

1   The internal droplet technique has been used to demonstrate that small, but significant, undercoolings below the peritectic temperature are required for the heterogeneous nucleation of the secondary (II) phase in several peritectic systems. Thus, the classical description of the peritectic 'reaction' is inaccurate.

2   The undercooling coefficients ($= \Delta T_L^{II}/T_L^{II}$ where T_L^{II} is the II-phase liquidus temperature and ΔT_L^{II} is the recorded undercooling below T_L^{II}) for heterogeneous nucleation of the II-phase on the I/liquid interface are generally small (< 0.04).

REFERENCES

1 J. A. SARTELL AND D. J. MACK: *J. Inst. Met.*, 1964, **93**, 19

2 A. P. TITCHENER AND J. A. SPITTLE: *Met. Sci.*, 1974, **8**, 112

3 F. N. RHINES: 'Phase diagrams in metallurgy', 84, 1956, New York, McGraw-Hill

4 B. CHALMERS: 'Physical metallurgy', 271, 1962, New York, John Wiley

5 R. T. SOUTHIN: Ph.D. Dissertation, University of Cambridge, 1970

6 C. C. WANG AND C. S. SMITH: *Trans. AIME*, 1950, **188**, 156

7 W. A. MILLER AND G. A. CHADWICK: *Proc. R. Soc.*, 1969, **313A**, 257

8 D. A. TURNBULL: *J. Appl. Phys.*, 1950, **21**, 1022

Nucleation in steel castings by interstitial compounds

J. V. Wood

Various theoretical approaches to the problem of grain refinement are reviewed in terms of their applicability to the production of effective nucleants for use in steel castings. A variation of the lattice disregistry model is outlined and preliminary data from electron optical studies are presented in order to test this theory.

The author is in the Department of Metallurgy and Materials Science, University of Cambridge

Since the publication of the lattice disregistry theory proposed by Turnbull and Vonnegut[1] in order to explain the heterogeneous nucleation of a solid from its melt by innoculants, several papers have been published which have expanded both the theoretical and experimental knowledge of this subject. While it has become apparent that a number of important variables (e.g. lattice matching, reduced interfacial energies, surface electron distribution) influence the effectiveness of a foreign nucleant, it is unfortunately true that most of these variables cannot be measured in any fundamental way. Hence, the time-honoured technique of the empiricist is still used to a large extent in the practical application of innoculants to castings. In the present paper the author intends to outline how modern electron investigative techniques (both optical and spectrographic) may be employed to examine the behaviour of pure compounds in nucleating steel castings.

BACKGROUND

In two important papers[2,3] the current knowledge on the addition of numerous compounds to liquid iron is described. Bramfitt[2] added transition metal carbides and nitrides to superheated pure iron (~ 110 K superheat), and measured the amount of supercooling in the liquid before the onset of solidification. Six compounds were found to be effective under these conditions (TiC, TiN, SiC, ZrN, ZrC, and WC). A lattice disregistry model was proposed using a modified form of the Turnbull–Vonnegut model which gave a number of simple orientation relationships based on the (001) and (0001) planes for cubic and hexagonal compounds, respectively. Campbell and Bannister[3] added a number of carbides, oxides, and borides to an Fe-3% steel. Again titanium compounds proved to be the most effective, with TiB_2 as the best. However, it was proposed that this compound decomposed in the liquid steel according to the reaction

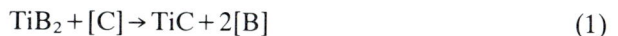

$$TiB_2 + [C] \rightarrow TiC + 2[B] \qquad (1)$$

The boron in solution acts as a grain growth inhibitor. Such results allow us to define the three main areas of grain refinement as follows

(i) creation of sufficient nuclei (either by direct addition or by stirring) to allow heterogeneous nucleation to occur throughout the melt

(ii) addition of surface-active elements or compounds (e.g. boron or metalloids) which encase grains that have started to grow (by solute rejection at the surface) and thus allow further stable nuclei to develop and grow

(iii) achieve sufficient supercooling within the melt to allow as many nuclei as possible to become stable.

Previously, (i) and (ii) have generally been amalgamated with a preferred division between chemical and mechanical (including electromechanical) techniques for inducing grain refinement.

IMPORTANT PARAMETERS

In view of the above classification, it is necessary to define accurately the area that is being studied in order to isolate the main variables involved. However, in practice it is generally true that all three occur in many situations concurrently. For the present case, classes (i) and (ii) were examined, while class (iii) is peculiar to a number of rapid casting techniques.

In order to predict the effectiveness of adding any compound to liquid iron, the following parameters must be known

(i) the solubility of the compound at the addition temperature

(ii) nucleation rate of the compound from a super-saturated liquid (i.e. assuming that the compound will dissolve by a significant amount and will recombine near the equilibrium liquidus temperature to act as an effective nucleant)

(iii) lattice parameters of the compound and iron at the nucleation temperature

(iv) the interfacial energies associated with liquid-nucleant system.

It is probable that most systems of interest will have large solubilities in liquid steels and that the nucleation of particles from the supercooled liquid is the most significant single variable in their effectiveness. Evidence to support this assumption comes from the work of Campbell and Bannister[3] who demonstrated that the total weight of TiB_2 (rather than the total number of particles) added during solidification influenced the final grain size. As outlined in equation (1), these authors presumed that TiB_2 was not found again on cooling, but that TiC was formed instead. Aronsson[4] has summarized solubility data for a number of carbides and nitrides at the melting point of pure iron, and most compounds lie between 0·1–1·0 at.-%.

Bramfitt correctly assumes that to compare the lattice parameters of both the compound and iron at the melting temperature is only correct if a cube–cube orientation relationship between the two is preserved. While this relationship is found in the solid state in austenitic steels (e.g. $M_{23}C_6$, NbC etc.), it is hardly ever found for simple compounds which have precipitated in solid ferrite. If both the lattice parameters of ferrite and the compound behaved ideally on heating, a direct comparison of lattice planes at ambient temperature would be justified. However, most compounds under investigation are defect structures and have large variations at any one temperature.[5] Thus, it is necessary to predict both a reasonable parameter at the liquidus temperature and the magnitude of divergence allowable to make planar matching a possibility. Bramfitt assumed orientation relationships for three simple planes of ferrite and compared the disregistries. In the technique described below, the author attempts to predict which orientation relationships are most likely from a planar matching argument and then to determine whether or not the disregistry falls within the error band of the compound at the liquidus temperature. In practice, this is about 10% for most compounds.

In the present case, an iron-base alloy containing 3·3 wt-% silicon was used, and assumed to have a lattice parameter at its melting point of 2·936 Å. Taking a consensus of values for TiC from the work of Pearson[5] and Goldschmidt,[6] the planar matching (given below) lies within the 10% limit (in fact between 3 and 5% in this case)

Fe		TiC
(110)	—	(200)
(200)	—	(220)
(211)	—	(222)
(220)	—	(133)

From these data, an expected orientation relationship can be calculated for any compound in any matrix. Data for all the compounds listed in Table 1 have been obtained in this way for both the silicon steel and austenitic iron with a value of $a_0 = 3\cdot604$ Å at its melting

Table 1 Compounds added to liquid steels

Cubic	Hexagonal	Cubic	Hexagonal
V_4C_3	M_7C_3	ZrC	Nb_2C
TiC	Si_3N_4	ZrN	Zr_2B
TaC	Cr_2N	TiN	TiB_2
NbC	Mo_2C	NbN	—

point. From these data, detailed orientation relationships can be predicted and compared with those found by experiment.

In the case of TiC, the Baker–Nutting orientation relationship is normally found between cubic precipitates and bcc iron

$$\{100\}_{ppte}\|\{100\}_\alpha$$
$$\langle100\rangle_{ppte}\|\langle110\rangle_\alpha$$

whereas, the following relationship is predicted

$$\{100\}_{TiC}\|\{110\}_\alpha$$
$$\langle1\bar{1}0\rangle_{TiC}\|\langle100\rangle_\alpha$$

and for austenitic steel above the orientation relationship that would be expected is

$$\{100\}_{TiC}\|\{111\}_\gamma$$
$$\langle100\rangle_{TiC}\|\langle\bar{1}10\rangle_\gamma$$

as opposed to the cube–cube relationship normally found. If the austenite subsequently transforms to ferrite then the Kurdjumov–Sachs relationship must be superimposed on this relationship. The experiments described below were designed to determine the validity of this approach.

EXPERIMENTAL DETAILS AND RESULTS

Two steels were chosen, namely: Fe–3·3Si and Fe–0·55C. Both were made in a vacuum furnace using electrolytic iron, 99·99% Si, and spectrographically pure graphite as additions, respectively. The apparatus for melting was similar to that used by Bramfitt with a rf induction heater coupling into a graphite susceptor. The charge of 100 g was melted in an alumina crucible after the chamber had been evacuated to 10^{-2} torr and backfilled with pure argon to a pressure of 400 torr. After the steel had been melted, the compound was added by means of plunging an envelope containing the material, made of pure iron, into the melt and then withdrawing the iron rod as soon as dissolution had taken place. Exact temperatures were not monitored, but a superheat of at least 50 K was maintained before innoculation. In general 500 mg samples of compound were added.

Table 1 contains a list of the compounds that were added in this investigation. Powders were bought from commercial firms and ranged in purity from 99–99·9%. The size of the particles varied but most were within the range 2–5 μm. After solidification had taken place, the ingots were sectioned and etched to examine the grain size. Only TiC and ZrB_2 significantly altered the grain size of the Fe–3·3Si steel, and Si_3N_4 (not in the form of powder but as a solid substrate) was the only compound to have any effect on the Fe–0·55C steel. Exact grain size values are generally relevant only to the specific experimental situation, but the above compounds reduced the grain size by half to a third of that without additions.

The longitudinal sections were examined in an electron microprobe analyser to determine the amount that the additives segregated in the final casting. A number of compounds affected the distribution of silicon, but these were not detectable in interdendritic regions. However, in the case of TiC and ZrB_2, large particles ($\sim 2-5$ μm in dia.) were observed which were rich in Ti and Zr, respectively. The exact siting of these particles appeared to be random in that they were found in the centre and at the edge of dendrites.

Transverse slices from the ingots were taken and chemically polished to 5 μm thickness; 3 mm discs were then punched from these sheets and polished in 5% pechloric acid dissolved in 2-butoxyethanol at 0°C using a 75 V-potential. Resulting foils were examined in a 200 kV electron microscope.

Figure 1a shows a general view of Fe-3·3Si to which 0·5 g of TiC has been added. Small particles can be observed by their strain fields and these are TiC particles which have come out of solution during the solid-state cooling. They have the Baker-Nutting orientation relationship with the matrix. Figure 1b is a detail taken from

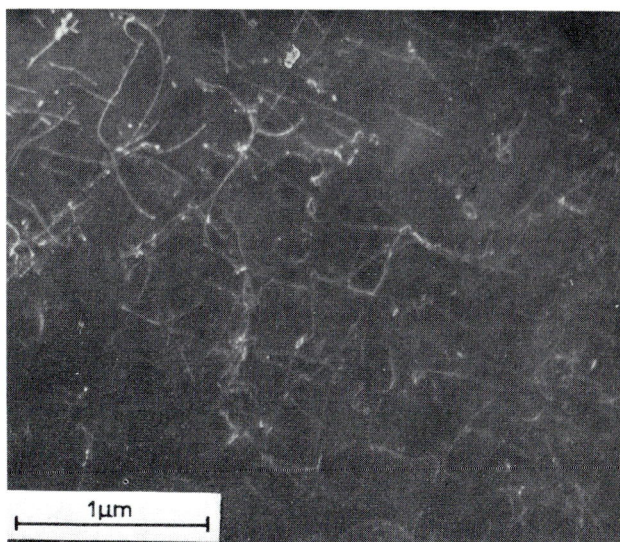

2 **Same area as in Fig. 1a using weak-beam conditions to show dislocation configurations**

Fig. 1a showing the much larger TiC particles, (they appear light in the micrograph). These appear to be rounded off at the edges and, as yet, all attempts to assign an orientation relationship have been unsuccessful. That they do not light up with the small particles in dark field is further evidence to demonstrate that these are original undissolved particles. Normally, only one grain is found in each foil and more work is needed to characterize this relationship unambiguously. From Fig. 1a it can be seen that there are a number of these undissolved particles in the field of view, and this demonstrates clearly that not every particle acts as a nucleus, even if it remains out of solution.

Weak-beam microscopy can be utilized to demonstrate the nature of dislocation structures at particle/matrix interfaces. Figure 2 is a weak-beam micrograph of the area shown in Fig. 1a. From this it can be seen that a general dislocation tangle is predominant throughout the whole area. This has been caused by the internal stresses produced during cooling. A number of particles of the reprecipitated TiC appear to have dislocation loops around them, whereas there appears to be no indication of the dislocation structure at the large particle interfaces. The large particles are very thin (as a result of preferential etching) and this could account for the lack of dislocation structure; however dislocations would have had time to climb away at temperatures just below the solidification temperature.

Out of ten discs examined which contained ZrB_2, there was no evidence of large particles which had been preserved from the liquid state. Figure 3 shows the general dispersion of reprecipitated ZrB_2 particles. A number of large holes are found throughout the thin areas, which indicates that large particles have dropped out during thinning. Evidence from the microprobe analyser has clearly shown the existence of large undissolved particles and it is concluded that ZrB_2 acts very much like TiC, although boron which has dissolved can act as a grain growth inhibitor. As a ferrite grain grows, a boron layer will be built up at the liquid/solid interface. Figure 4 shows the expected boron concentration profile across this interface (the maximum solubility in pure α iron is 60 ppm[7] and in Ni-Cr austenite is about 100 ppm[8]). The iron-boron system has a deep eutectic

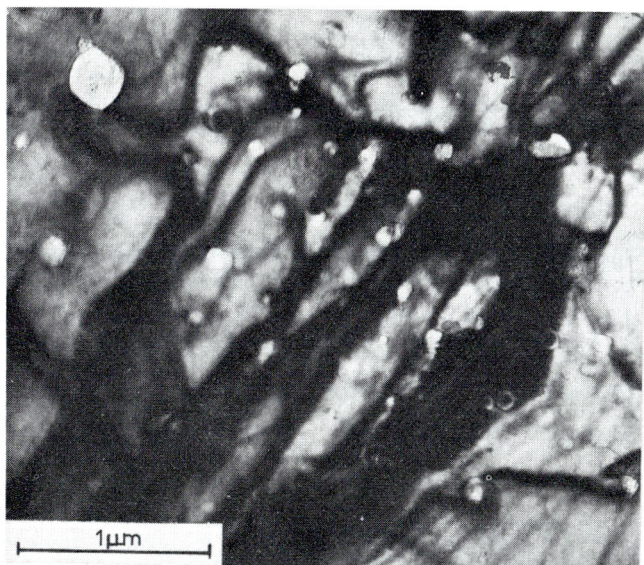

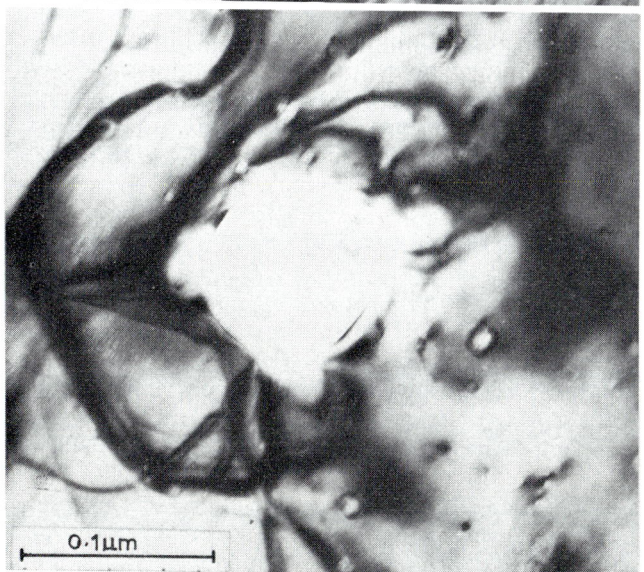

1 **Transmission electron micrograph of Fe-3·3Si casting with 0·5 g of TiC added; *b* is detail of *a* showing undissolved TiC particle**

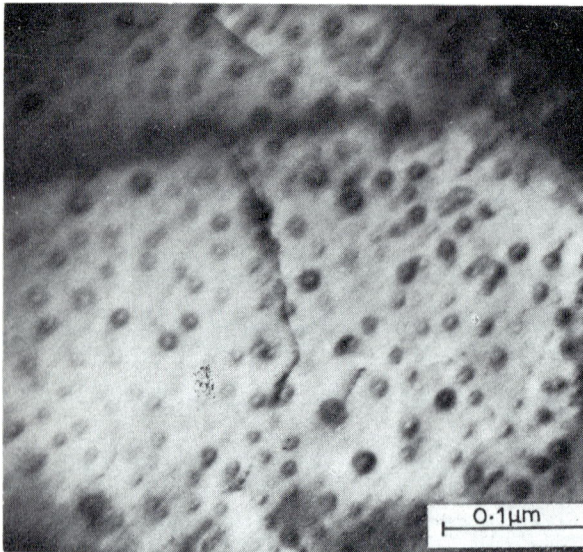

3   Transmission electron micrograph of silicon steel with 0·5 g ZrB₂ added; all particles have been precipitated in solid state

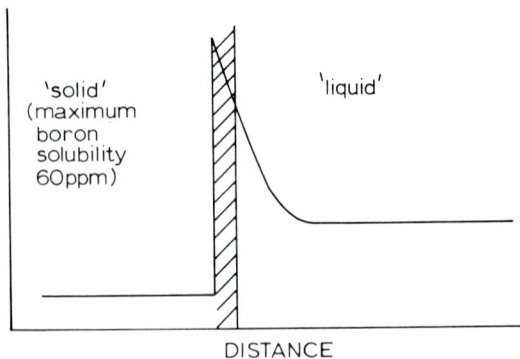

4   Concentration profile of boron across solid/liquid interface of ferrite; hatched region indicates low melting point liquid which could be quenched out to give amorphous structure

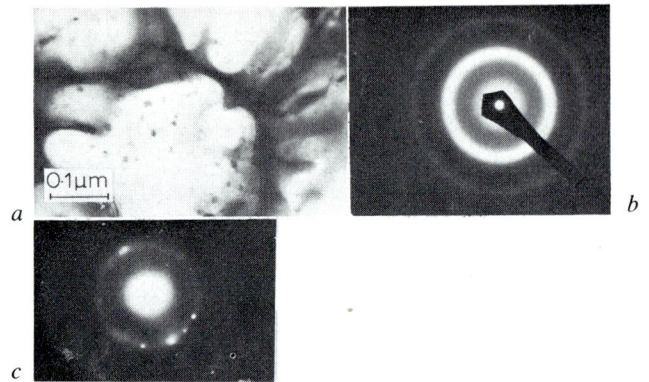

5   *a* Transmission electron micrograph of splat-quenched ferrite grain with diffuse boundary; *b* diffraction pattern from *a*; *c* diffraction pattern from amorphous iron–nickel boron alloy

well which is used to advantage in making amorphous metals. Hence, this layer will have a very low melting point compared with that of the growing ferrite and thus grain growth will be actively inhibited. If the system could be frozen at this point with negligible diffusion then it should be possible to observe such an amorphous layer. An iron–nickel chromium steel which solidifies as ferrite with large titanium and boron additions (5 and 1 wt-%, respectively) has been splat quenched under controlled conditions.[9] Figure 5*a* shows such a solidified structure with a crystalline (ferrite) centre and a diffuse boundary. Selected-area electron diffraction from this grain (Fig. 5*b*) shows both distinct crystalline spots and a diffuse ring characteristic of an amorphous or microcrystalline phase. Figure 5*c* is a diffraction pattern from an amorphous alloy of similar composition. It can be seen that the first ring is at exactly the same spacing. In the duplex-phase steel, this ring is not caused by the presence of oxide or precipitate, as the ring disappears on aging above 400°C with precipitation of distinct particles of M₂B in ferrite.

Thus, although TiB₂ does not appear to nucleate ferrite, dissolved boron in liquid iron can actively inhibit grain growth in iron for a short period while other nuclei

become effective. If no other nuclei are present then TiB₂ will appear to be ineffective as the present results confirm.

Si₃N₄ only reacts with carbon steels if melting is performed under a partial pressure of argon (i.e. it does not occur under normal atmospheric conditions). By examining sections through Si₃N₄-steel interfaces, it can be seen that the liquid has reacted along Si₃N₄ grain boundaries extremely quickly. If Cr and Ni are added to preserve the austenitic structure down to room temperature, the morphology of the M₂₃C₆ particles near the interface change from being in a fine eutectic dispersion to forming large hexagonal particles. The following reaction is proposed to account for this reaction in the absence of air: $Si_3N_4 + 3[C] \rightarrow 3SiC + 4[N]$. It is not clear whether the dissolved N would combine to form a compound or N₂ gas.

DISCUSSION AND CONCLUSIONS

The present research confirms that TiC is effective as a grain refiner in silicon steels, though more evidence is needed from electron microscopy to confirm any lattice matching argument; TiB₂ was not effective, and if equation (1) is assumed to be correct, then this can be directly attributed to the lack of carbon impurities in the present system. ZrB₂ (hexagonal) has lattice parameters 3·170 and 3·533 Å and must be considered to be an active nucleant (as opposed to TiB₂) as large particles are found (either undissolved or recombined) in the solid ingot.

Dissolved boron in liquid steel can actively prevent grain growth by building up a layer of low melting point liquid ahead of the solid interface. In general, this slows down growth allowing further nuclei to become active if present.

Bramfitt[2] has shown that SiC is relatively effective in nucleating liquid iron and this would account for the behaviour in nucleating the apparently austenitic steel. Fe–0·55C is just past the peritectic point and as carbon is extracted to form SiC, the immediate area will be deficient in carbon and will start to solidify as ferrite. Thus, SiC has not shown itself to be effective at nucleating austenite, but ferrite which subsequently transforms as diffusion occurs.

ACKNOWLEDGMENTS

The author wishes to thank Professor R. W. K. Honeycombe for advice and the provision of laboratory facilities. This work was undertaken on a post-doctoral fellowship supported by Foseco International.

REFERENCES

1 D. TURNBULL AND R. VONNEGUT: *Ind. Eng. Chem.*, 1952, **44**, 1 292
2 B. L. BRAMFITT: *Metall. Trans.*, 1970, **1**, 1 987
3 J. CAMPBELL AND J. W. BANNISTER: *Met. Technol.*, 1975, **2**, 409
4 B. ARONSSON: 'Steel strengthening mechanisms', **77**, 1969, Climax Molybdenum Corporation
5 W. B. PEARSON: 'Handbook of lattice spacings and structures of metals and alloys, Vol. 2, 1967, Oxford, Pergamon Press
6 H. J. GOLDSCHMIDT: 'Interstitial alloys', 1967, London, Butterworth
7 A. BROWN: Ph.D. Dissertation, 1973, University of Cambridge
8 H. J. GOLDSCHMIDT: *J. Iron Steel Inst.*, 1971, **209**, 900
9 J. V. WOOD AND R. W. K. HONEYCOMBE: to be published

Influence of mould wall micro-geometry on casting structure

A. Morales, M. E. Glicksman, and H. Biloni

The formation of the chill zone during casting involves complex interactions of liquid metal flow, metal/liquid mould heat transfer, nucleation catalysis, and dendritic growth. Chill-zone grain size depends on both independent nucleation events and crystal multiplication processes induced by melt turbulence, so that grain refinement may arise from quite diverse phenomena. This investigation was carried out to establish how the chill-zone microstructure and columnar-zone growth rates depend on mould wall characteristics. A flow channel fluidity test, similar to that used by Prates and Biloni, was used to relate the overall heat-transfer rate to the mould wall roughness and to interpret the mechanisms of grain formation in binary Al–Cu alloys. The nucleation sites of the dendritic grains were observed microscopically to correspond uniquely with the distribution of mould wall asperities, where high local heat-transfer rates are expected. An interesting observation made during the fluidity test was that the overall heat-transfer rate varied substantially as the surface roughness was varied, whereas the local heat-transfer rate – as estimated from the distribution of nucleation sites – remained virtually constant. Thus, factors influencing the overall solidification rate, such as the thermal diffusivity of the mould wall and the presence of surface coatings, can be controlled relatively independently of the factors affecting the local heat-transfer rate. To confirm the effects of local and overall heat-transfer rates, a series of vertical mould castings were poured. Alloys of $99.99\%\,Al$, $0.5\%\,Cu$, and $Al–1\%\,Cu$ were bottom poured and cast against a variety of mould walls composed of graphite, copper, steel, and aluminium, each either polished, machined to a prescribed microgeometry, or coated with alumina or lamp black. The macroetched structures show clearly that the chill zone and its columnar extension always responded in the same manner: i.e. the grain size was influenced by the surface roughness and local heat-transfer rates, whereas the columnar zone extension was controlled only by the overall heat flow through the mould wall. Thus, the mould wall microgeometry, as influenced by machining, polishing, and coating has a profound influence on the grain structure of these castings by effects manifested in the local and average heat-transfer rates. Moreover, it appears to be possible to transfer results obtained from fluidity tests to actual mould surfaces with similar microgeometries, since it is the highly localized supercooling of the melt at mould wall asperities that appears to be responsible for the predendritic nucleation and grain refinement.

A. Morales is at the University of Bogota, Columbia, M. E. Glicksman is in the Rensselaer Polytechnic Institute, New York, USA, and H. Biloni is at the CNEA Buenos Aires, Argentina

In previous works by Bower and Flemings[1] and Biloni and Morando,[2] the thermal conditions existing at the chill zone were simulated. The formation of the chill zone involves complex interactions of liquid metal flow, metal/mould heat transfer, nucleation catalyst, and dendritic growth.[1-5] Chill-zone grain size depends both on independent nucleation events and crystal multiplication processes induced by melt turbulence, so that grain refinement may arise from quite diverse phenomena.

Recent work by Prates and Biloni[5] presented a method based on a linear fluidity test by which a measure of the engineering heat-transfer coefficient at the metal/mould interface h_i, is obtained when the fluidity length L_f is plotted as a function of the liquid superheat ΔT in °C. Essentially, the test simulates the heat-extraction conditions at the chill zone as do Refs. 1 and 2. Prates and Biloni[5] considered the effect of the solute in Al–Cu alloys as well as the effect of different coatings and mould materials on the h_i parameter. They also considered the effect of the h_i value (for a given mould microgeometry) and the mould microgeometry (for a bare copper substrate) upon the surface structure, by means of the detection of the predendritic nuclei.[6] However, the method used was essentially similar to that employed by Biloni and Chalmers,[6] that is, the solidification of a liquid drop entering into contact with a cold substrate. Although the thermal conditions could be quite similar to those existing at the chill zone, the dynamic conditions of fluid flow do not correspond to those existing when the mould is filled by the melt, a condition better simulated by the fluidity test. In any case, however, the observation of the as-cast structure, without any metallographic preparation, can give considerable information about the formation of pre-dendritic, dendritic, and multiplied structures.[7] Even in the case of 99·993% pure aluminium, the existence of predendritic nuclei can be detected and related to the origin of the grains.[6] Consequently, a relationship between micro- and macrostructure may be established in a similar way for pure metals (essentially very dilute alloys[8]) and alloys.

The present work was carried out to establish how the chill-zone macrostructure and columnar-zone rates depend on wall characteristics. Fluidity tests will be performed as well as castings on laboratory scale. It is expected to obtain (*a*) a more exact deposition and control of the coating thickness than prior works[5]; (*b*) a more accurate detection of the relationships existing between the mould surface microgeometry and the resulting structure in bare and coated moulds; and (*c*) a better understanding of the surface structure formation and its influence on the columnar zone when real castings are considered.

EXPERIMENTAL TECHNIQUES
Materials
99·99% aluminium, Al–0·5% Cu, and Al–1% Cu alloys were used. The copper purity was 99·99%.

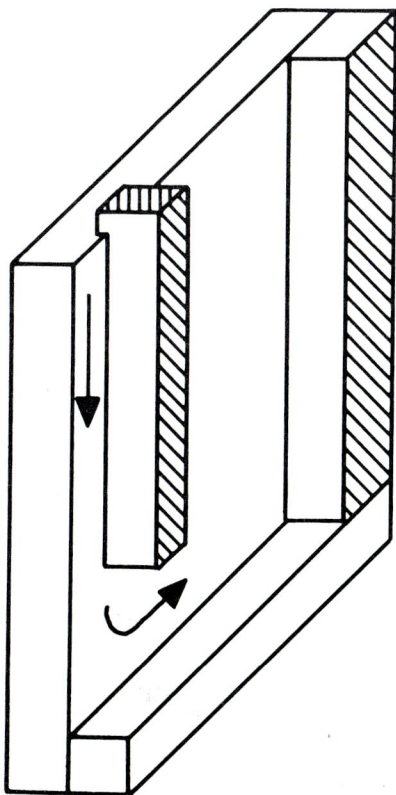

1   Schematic diagram of mould Type *A*

Fluidity test
The fluidity test was performed in a device essentially similar to that employed by Prates and Biloni.[5] The metallostatic head used was 8·5 cm, and the channel material was copper with semi-infinite walls. When the fluidity test was performed on the bare copper surface, two different microprofiles were obtained: one by polishing the surface with 600 emery paper, and the other by making perpendicular mechanical grooves on a surface with the same mechanical preparation. The network obtained in the latter case was 250 μm wide.

Laboratory-scale experiments
Two types of mould were used in the experiments involving laboratory-scale ingots; these were designated *A* and *B*.

Type A
Figure 1 shows the mould design where the melt is poured from the bottom. The heat-extraction zones of the mould were two 2·5×18×24 cm, assuring the bidirectional extraction of heat. The other parts of the mould shown schematically in Fig. 1 were thermally isolated with a FOSECO coating. The mould materials were steel (SAE 1010), aluminium, and graphite. In each case, one of the plates was polished with 600 emery paper and the other was covered with 200 μm thickness of alumina. 99·99% Al was poured in this type of mould and ΔT was 100°C in all cases.

Type B
Figure 2 shows schematically the device used. The Pyrex window was useful in order to observe the solidification front advance as well as to provide a very flat and isolated surface. The lateral copper plates had different surfaces conditions for different experiments as follows:

(i) polished with 600 emery paper
(ii) polished with 600 emery paper and machined in a similar way to the fluidity channel
(iii) coated with a 200 μm thickness of alumina
(iv) covered with a 100 μm thickness of lamp black.

With this type of mould 99·99% Al and Al–0·5Cu and Al–1%Cu alloys were used with ΔT equal to 100°C.

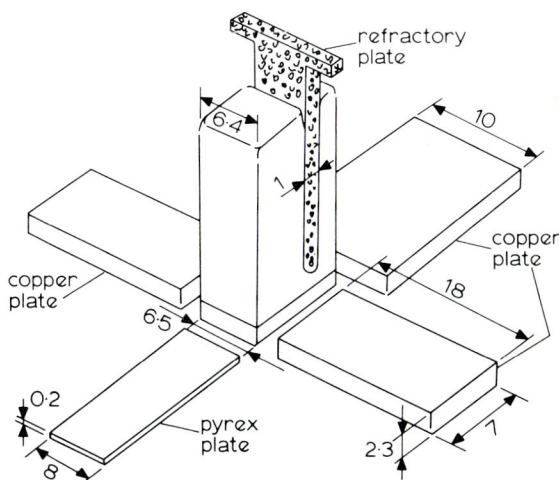

2   Schematic diagram of mould Type *B*: all dimensions in cm

Coating procedure
Alumina

The alumina was deposited on the fluidity channel surface and the different mould walls, previously specified, through a pressure pistol. The alumina had a granulometry of 450 mesh. The procedure represented a compromise between (a) a complete adherence to the mould surface, homogeneous thickness and good physical and mechanical properties of the coating in order to resist the temperature changes during the test as well as the erosion caused by the liquid flow; and (b) a minimum percentage of the agglomerative, $SiNO_3Na$ in the present case, used in order to obtain a good intergranular binding of the alumina. After several experiments the conditions adopted were:

   (i) alumina: 1 part, mesh 450
   (ii) water: 5 parts
   (iii) $SiNO_3Na$: 4 wt-% for the alumina
   (iv) borax (used as activator): 0·05%.

The optimum temperature of the plates to be covered was between 120° and 140°C during the coating operation. The deposition was made through automatic passages of the pistol along the surface of the plates. The thickness was measured with the micrometer of a microscope focusing the bare surface (protected with adhesive tape) and the coated surface.

Lamp black

The lamp black deposition was made as described elsewhere[9] and the thickeness was controlled as in the case of alumina.

Microprofile of the surfaces

Figure 3 shows the microprofile measured by a rugosimeter for a copper surface polished with 600 emery paper. It is representative of all the surfaces of different materials having this type of surface preparation. Figure 4 corresponds to the microprofile obtained with the

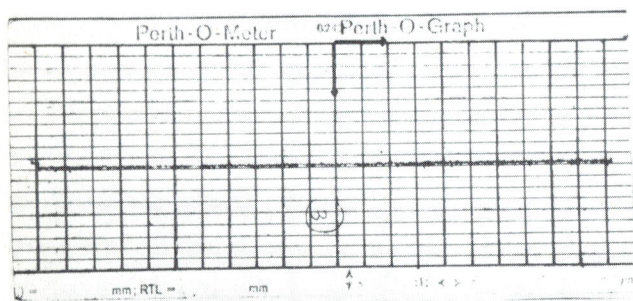

3   Microprofile corresponding to a 600 emery paper polished surface: Ra = 0·2 μ, ESC = 10 mm − 10 μ

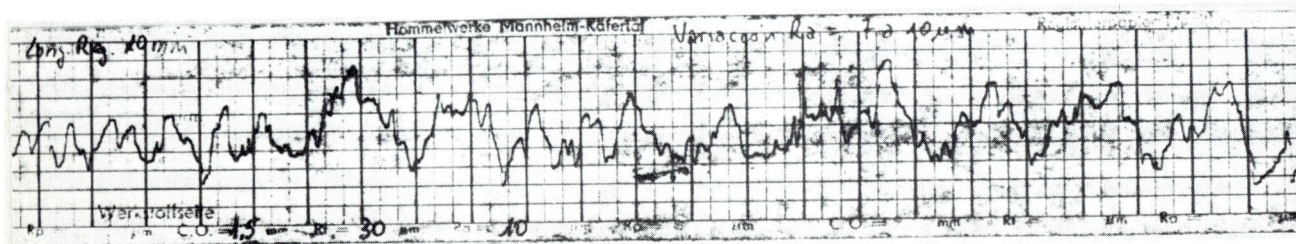

rugosimeter when alumina coating was used. In order to prevent accidents on the apparatus, a film of aluminium was deposited on the surface using a vacuum camera. The film, a few microns thick, permits a good measure of the microprofile, protecting the device and assuring reliable results.

When lamp black was used, the microprofile could not be obtained with the rugosimeter but a microscopic inspection showed a very flat surface, better than a 600 emery paper preparation.

Metallographic observations

The observations were made on macroscopic scale without etch and after standard aluminium macroetching. In the latter case the macrostructure is easily and better revealed in the case of 99·99%Al surfaces. Microscopic observations were made on the as-cast surfaces without metallographic preparation, a method which gives extensive structural information.[7]

EXPERIMENTAL RESULTS
Fluidity experiments

The relationships existing between the fluidity experiments, under different mould surface conditions, and the surface structures can be summarized as follows:

   (i) for a 600 emery paper finishing (see profile in Fig. 3) the predendritic density is higher at the entrance of the fluidity channel and scarce along the channel: elongated grains are observed along the channel, presenting a cellular dendritic substructure and eventually multiplication at the 'meniscus marks'
   (ii) for machined channels the predendritic distribution corresponds to the edges and nodes of the machined network: Fig. 5 is a typical structure observed
   (iii) when an alumina coating was used with a surface microprofile corresponding to Fig. 4, the surface structure along the sample was as shown in Fig. 6: a dendritic structure starts from predendritic nuclei without signs of multiplication mechanisms.

On the other hand, when the fluidity channel presents two different bare surface conditions, polished with 600 emery paper and machined surface, the fluidity length is different. In this case L_f is 36% longer than the case where the bare copper channel surface is 600 emery paper polished (Fig. 7).

Casting Experiments
Type A

Independently of the mould material, the results were essentially similar. The ingot surface in contact with the 600 emery paper polished mould wall presents small,

4   Microprofile corresponding to alumina-coated surface: Ra = 10 μ Rt = 30 μ

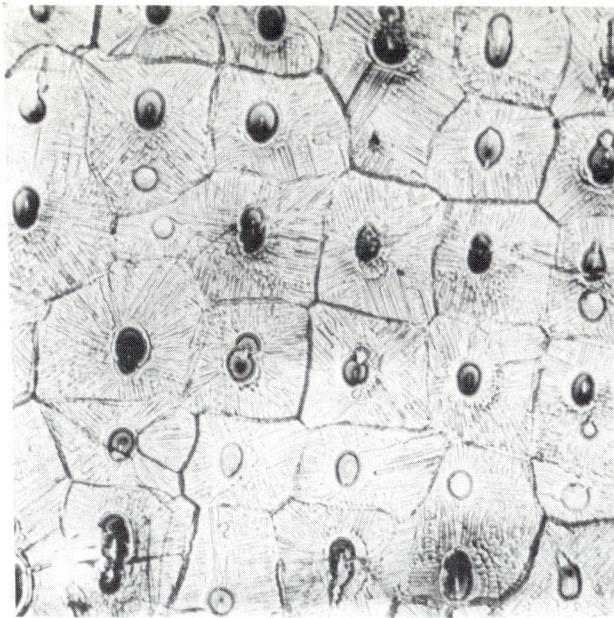

5 Typical structure when mould wall shows perpendicular grooves machined on 600 emery paper polished surface: predendrite nuclei at edges and corners of the grooves; as-cast structure, without etching [×80]

6 Typical structure corresponding to an Al–Cu alloy solidified in contact with an alumina-coated surface presenting a microprofile of the Fig. 4 type: as-cast structure without etching [×100]

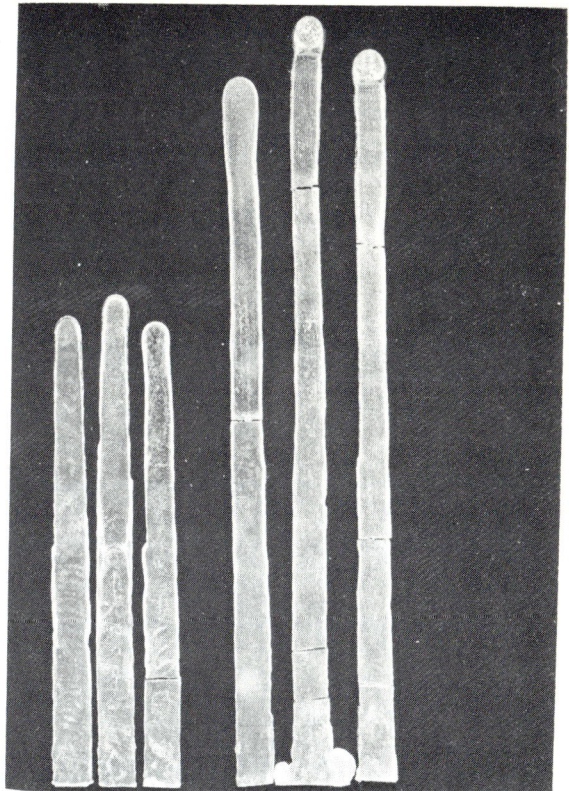

7 Samples of Al–1%Cu obtained with a fluidity test and $\Delta T = 100°C$: samples on the left correspond to those poured into a channel with bare copper walls polished with 600 emery paper; samples on the right correspond to those poured into a channel with bare copper walls polished with 600 emery paper and machined

a   b

8 Type *A* casting: macrostructure of 99·99%Al ingot surface in contact with *a* 600 emery paper polished wall surface; and *b* alumina-coated wall surface [×0·50]

equiaxed grains at the bottom from where the larger and elongated grains grow. These grains are eventually interrupted by a band of small grains situated at the meniscus marks. From these grains new large and elongated grains grow: Fig. 8a is a representative structure.

On the other hand, when the mould wall is coated with alumina, the ingot surface presents a fine equiaxed grain: Fig. 8b is a representative structure.

A longitudinal section of the three series of experiments, involving different mould materials, is shown in Fig. 9. In all cases, the columnar zone starting from the alumina-coated wall is much finer but the grain size is only about ⅓ of the ingot thickness.

Type B

Several experiments were performed with the device shown in Fig. 2. Table 1 summarizes the experiments performed and gives the representative figures which

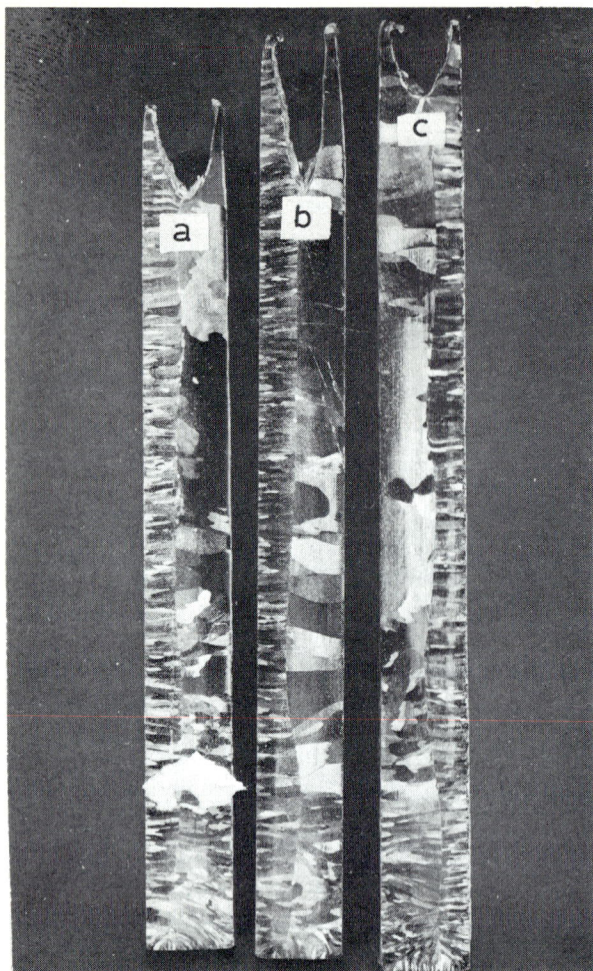

a left side: steel mould covered with alumina, right side: steel mould 600 emery paper polished; *b* left side: graphite mould covered with alumina, right side: graphite mould 600 emery paper polished; *c* left side: alumina mould 600 emery paper polished; right side: alumina mould covered with alumina

9   Type *A* experiments (99·99%Al); longitudinal reactions of ingots with different wall mould conditions   [×0·5]

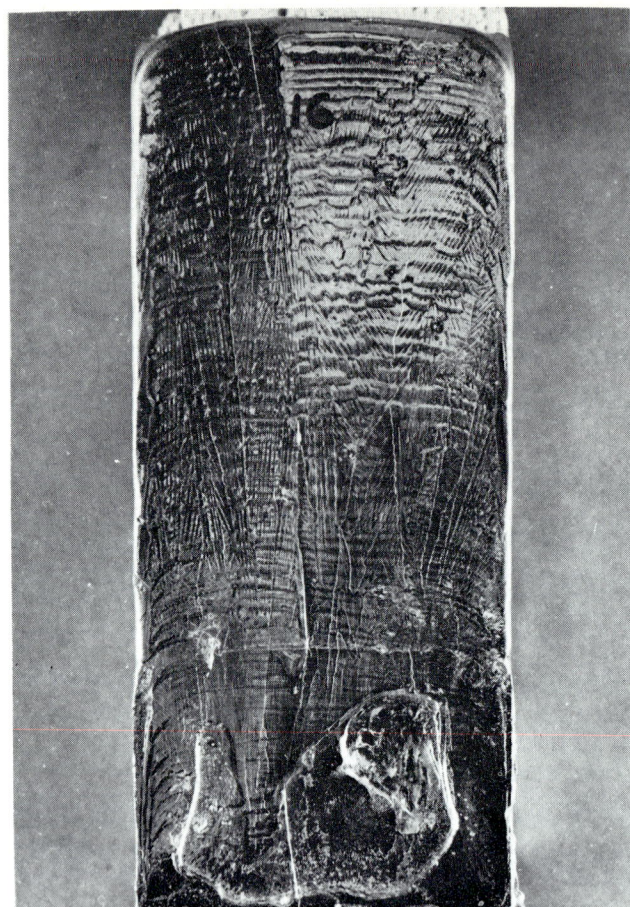

10   Type *B* experiments (99·99%Al); as-cast surface solidified in contact with the Pyrex window. Note meniscus marks   [×0·75]

show the structures and substructures observed in each case. The following results will be analysed.

1   Comparing the macrostructures obtained through different experiments (*see* Table 1) the same mould surface condition gives similar results independently of the material used, 99·99%Al, Al–0·5%Cu, or Al–1%Cu.

2   When the melt solidified in contact with a flat surface, which was polished with 600 emery, coated with

Table 1   Experiments with casting type B

Experiment No.	Alloy	Surface condition on left mould wall	Surface condition on right mould wall	Representative figures
I	99·99%Al	Polished with 600 emery paper	Coated with alumina	3, 4, 10, 11, 8*b*
II	99·99%Al	Coated with alumina	Coated with alumina	4, 8*b*
III	99·99%Al	Coated with alumina	Coated with lamp black	4, 10, 12, 14
IV	99·99%Al	Polished with 600 emery paper and machined	Coated with alumina	4, 8*b*, 13
V	Al–0·5%Cu	Polished with 600 emery paper	Polished with 600 emery paper	3, 11, 10,15
VI	Al–1%Cu	Polished with 600 emery paper	Polished with 600 emery paper	3, 11, 10, 15
VII	Al–1%Cu	Polished with 600 emery paper and machined	Polished with 600 emery paper and machined	14, 5
VIII	Al–1%Cu	Coated with alumina	Coated with alumina	4, 8*b*, 6

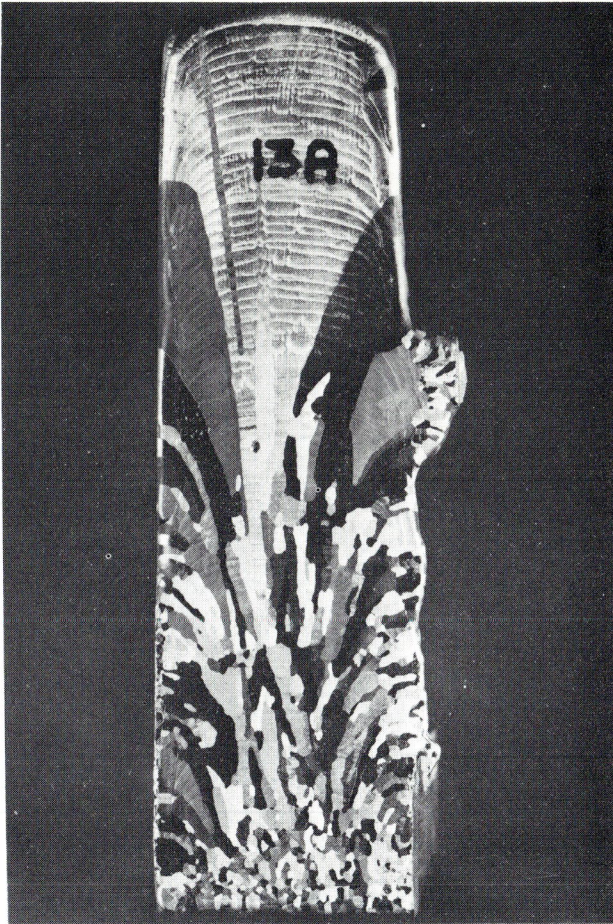

11 Type *B* experiments (99·99%Al); metal solidified in contact with left wall mould (*see* experiment I, Table 1); macrostructure after standard macroetching [×0·75]

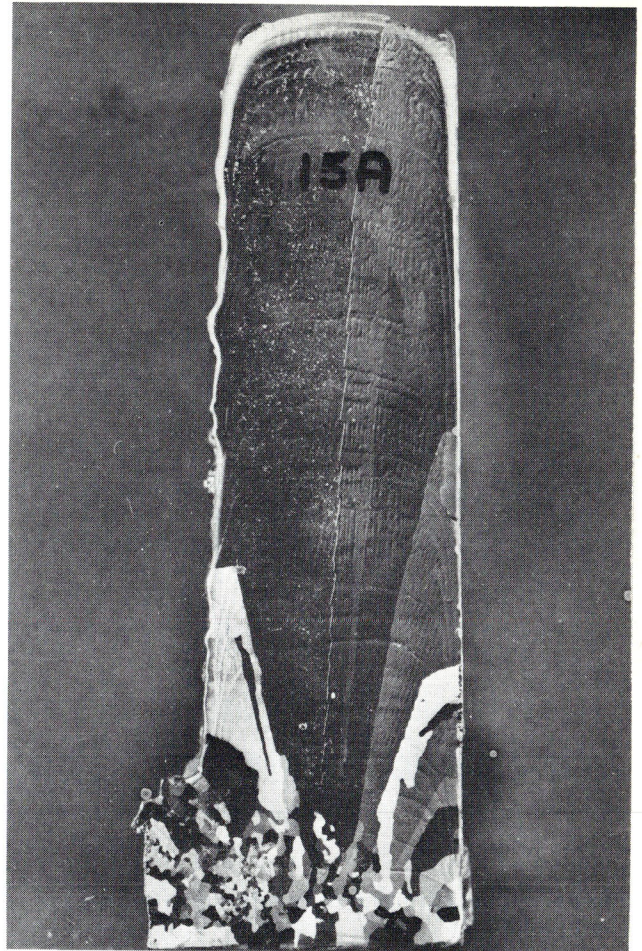

12 Type *B* experiment (99·99%Al); metal solidified in contact with right wall mould, (*see* experiment III Table 1)—macrostructure after standard macroetching, showing that practically all the surface is covered by a single grain [×0·75]

lamp black, or the Pyrex window, the successive positions of the solid/liquid interface are shown by the meniscus marks. Figure 10 corresponds to 99·99% Al solidifying in contact with the Pyrex window. The macrostructure is quite similar to Fig. 5 from Ref. 7, where an Al–5%Cu alloy solidified in contact with a polished copper mould.

3 The macrostructure corresponding to a liquid solidified in contact with a flat surface presents similar characteristics to those analyzed in casting Type *A*, when the same surface characteristic arises. Figures 11 and 12 correspond to those mould surfaces polished with 600 emery paper and covered with lamp black, respectively, and could be compared with Fig. 8*a*.

4 The macrostructure corresponding to a liquid solidified in contact with a surface covered with alumina presents an aspect quite similar to Fig. 8*b* and a microstructure quite similar to Fig. 6. When the surface mould has been polished and machined the macrostructure obtained is as shown in Fig. 13. The microstructure is quite similar to Fig. 5.

5 As in the case of casting Type *A*, the surface structure determines the bulk structure of the ingots; for example, Fig. 14 corresponds to a longitudinal section of an ingot corresponding to experiment III, Table 1.

6 The substructure corresponding to the meniscus marks, especially in the Al–Cu alloys, presents a substructure that strongly suggests that a multiplication

mechanism operates locally. Figure 15 corresponds to the as-cast aspect of the meniscus mark. A careful inspection of the substructure revealed similar results to those shown in Fig. 6 from Ref. 7: the elongated grains have predendritic regions as their origin, meanwhile in the meniscus marks a multiplication mechanism is the leading process.

7 When a mould surface presents asperities, given by the alumina coating, or a machined surface the meniscus marks practically disappear.

DISCUSSION
Fluidity Test
The mould microgeometry plays an important role in the type of structure obtained. As is shown schematically in Fig. 16, the liquid enters into contact with the cusps of the surface. At these asperities a high rate of heat extraction occurs on the laminar film of the liquid and as a consequence the mechanism proposed by Biloni and Chalmers[6] can operate, and a predendritic region will form. The network produced by the surface machining provides a given distribution of cusps and as a consequence has an important effect on the resultant structure. As is shown schematically in Fig. 16, when the meniscus of the liquid enters into contact with the cusps

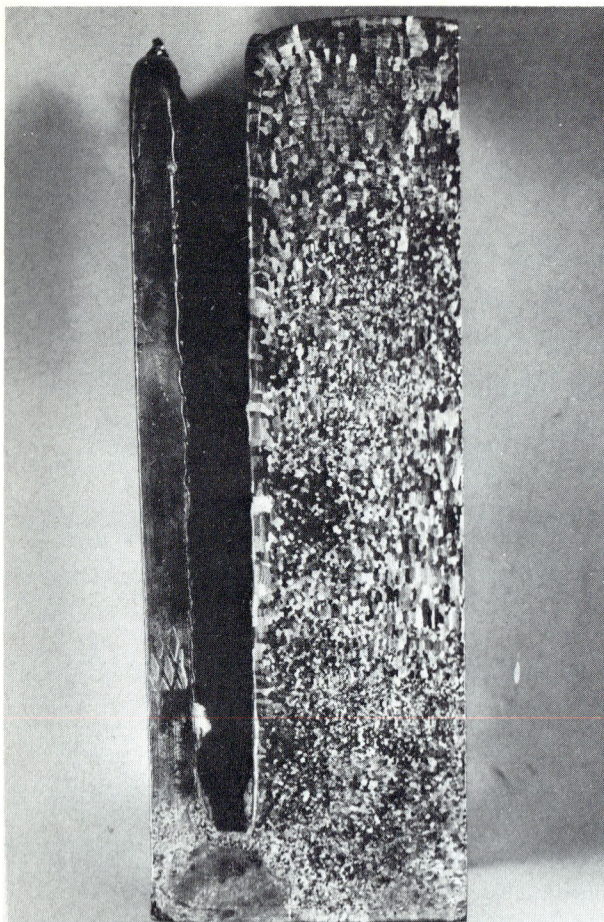

13   Type *B* experiment (99·99%Al); metal solidified in contact with left wall mould (*see* experiment IV, Table 1); macrostructure after standard macroetching, showing fine grains   [×0·75]

14   Longitudinal section corresponding to experiment III, Table I, after macroetch. Note the different grain size having as origin the mould walls with different microgeometries   [×0·75]

provided by the grooves a predendritic nucleation occurs, followed by a rapid cellular dendritic growth[6] on the portion of the flat surface among grooves. New grooves are reached by the tips of the dendrite cells, nucleating new predendrites, etc. The result is a structure of the type shown in Fig. 5. When the surface is covered with alumina the facts are essentially the same. In this case the alumina grains deposited on the surface produce the microprofile shown in Fig. 4, providing nucleation cusps. Although the heat diffusivity of the alumina is poor the small volume of the laminar layer supercools producing predendritic nuclei. The structure of Fig. 6 is the result.

For a very flat surface with a microprofile of the type shown in Fig. 3 the finer grains at the beginning of the fluidity channel may be explained as a result of the liquid supercooling and activation of independent nucleation centres from which points the grains grow along the mould walls of the channel. Because no significant asperities exist on the surface, the repetitive nucleation does not occur frequently, and as a result the cellular substructure persists along the channel. The difference in flow length L_f shown in Fig. 7 can be explained by taking into consideration the amount of metal surface in contact with the mould walls. The surface tension of the liquid metal impedes total wetting with the surface microprofile of the wall. When surface rugosity becomes

smaller, the effective contact between liquid and the mould surface and the *overall* heat-transfer rate grow and as a consequence the h_i parameter is higher, and thus L_f is smaller. Then, for two different microgeometries, the *local* heat-transfer rate at the points where the metal is in contact with the mould walls is the same, and the *overall* heat-transfer rate will depend on the rugosity of the surface. The results discussed above seem to contradict the results presented by Prates and Biloni,[5] who consider the relationship between the mould rugosity and the density of predendritic nuclei as observed on the Al–5% Cu solidified surface. From Fig. 6,[5] it appears that a very flat surface will have the higher density of predendritic nuclei. However, it is necessary to take into account the different types of experiments. Prates and Biloni used the method employed by Biloni and Chalmers, i.e. a liquid drop entering into contact with a cold surface. In this case all the metal supercools and the situation is different from that presented schematically in Fig. 16. It seems logical that when this condition arises and no relevant asperities on the mould surface exist, continuous growth is more economical from the point of view of energy than repetitive nucleation. It is probable that critical rugosity would be necessary in order to produce repetitive nucleation, and, consequently, grain refinement, but it has not been investigated in the present work.

15 **Substructure corresponding to a meniscus mark in an Al-1%Cu alloy solidified in contact with a flat mould surface: no etching [×80]**

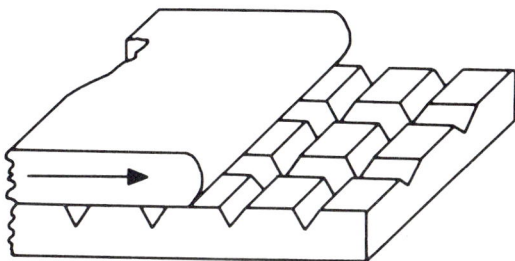

16 **Schematic diagram of repetitive nucleation mechanism on a polished and machined surface**

Ingots

The preceding discussion can be extrapolated to the case of the ingots obtained and may help the understanding of the structures and substructures obtained.

In the Type *A* experiments, for a polished mould surface, the bottom part of the ingot surface shows small, equiaxed grains, probably as a result of a mechanism of copious nucleation when the liquid makes contact with the mould walls. The large and elongated grains following this fine region (Fig. 8*a*) may be explained as the growth of the grains parallel to the mould wall. Taking into account that asperities of critical size do not exist on the mould wall surface no repetitive nucleation occurs, as discussed above. The only source of grain production is through a multiplication mechanism at the meniscus as a result of the inevitable variations of pouring speed. On the other hand, when the mould wall is coated with

alumina the metal enters into contact with the asperities corresponding to the microprofile of Fig. 4 and, consequently, a repetitive nucleation arises in a similar way to the mechanism discussed in the fluidity test.

The extension of the columnar region growing from the surface is controlled by the overall heat flow through the mould walls. Taking into account the thickness of the casting (2 cm), the h_i value practically controls the solidification rate.[10] In Fig. 9, the grains growing from the bare polished surfaces are twice as long as those starting from the coated surface. This result agrees with the measurements obtained by Morales.[11] Using an accurate measure of the liquid speed in the fluidity channel through a resistometry method,[12] rather than the water model used by Prates and Biloni,[5] Morales determined for Al-Cu alloys that the h_i values corresponding to bare copper and steel surfaces are twice those of the same surface covered with alumina deposited by a pistol, as in the present work.

In Type *B* experiments the results can be discussed in a similar way. The macrostructure clearly shows that the chill zone and its columnar extension always responded in the same manner; namely, the grain size is influenced by the surface roughness and local heat-transfer rates, whereas the columnar zone is controlled only by the overall heat flow through the mould wall. In this sense, Fig. 15 is a dramatic example. The control of the local heat transfer through a given geometry proved to be capable of eliminating the meniscus marks, a casting defect that has localized segregation as a result of the multiplication mechanism arising in that region.[7] The repetitive nucleation mechanism provoked by a suitable microgeometry prevents the formation of long dendritic branches able to be multiplied,[1-2] a fact stressed by Prates and Biloni[5] even in conditions of turbulent liquid flow. As a consequence, the ingot presents a better and brighter surface aspect.

CONCLUSIONS

The mould wall microgeometry, as influenced by machining, polishing, and coating has a profound influence on the grain structure of the casting by effects manifested in the local and average local heat transfer. Moreover, it appears to be possible to transfer results obtained from fluidity tests to actual mould surfaces with similar microgeometries, since it is the highly localized supercooling of the melt at mould wall asperities that appears to be responsible for the predendritic nucleation and grain refinement.

ACKNOWLEDGMENTS

The present work was partially supported by the Multinational Program of Metallurgy OAS-CNEA. The authors wish to acknowledge the general comments in the works of Professors M. Hillert, R. Doherty and M. C. Flemings during the post-doctoral seminars held in Buenos Aires under the auspices of the OAS-CNEA Program.

REFERENCES

1 T. F. BOWER AND M. C. FLEMINGS: *Trans. Met. Soc. AIME*, 1967, **239**, 216
2 H. BILONI AND R. MORANDO: *ibid.*, 1968, **242**, 1 121
3 B. CHALMERS: 'Principles of solidification', 1964, New York, John Wiley

4 M. C. FLEMINGS: 'Solidification processing', 1974, New York, McGraw-Hill

5 M. PRATES AND H. BILONI: *Metall. Trans.*, 1972, **3**, 1 501

6 H. BILONI AND B. CHALMERS: *Trans. Met. Soc. AIME*, 1965, **233**, 373

7 H. BILONI: 'The solidification of metals', 74, 1968, London, The Iron and Steel Institute

8 H. BILONI *et al.*: *Trans. Met. Soc. AIME*, 1966, **236**, 930

9 M. PRATES: Universidad Nacional del Sur, Bahía Blanca, Argentina, Thesis, 1971

10 O. S. PIRES *et al.*: *Z. Metallkund.*, 1974, **65**, 143

11 A. MORALES; Universidad Nacional de Rosario, Argentina, Thesis, 1975

12 G. O. HIRA: *Bull. Jpn. Inst. Met.*, S1 B14 1950, **6**, 27

Aluminium grain refinement by crystal multiplication mechanism stimulated by hexachloroethane additions to the mould coating

N. L. Cupini and M. Prates de Campos Filho

An experimental analysis was made to demonstrate the possibility of grain refining aluminium cast in chill and sand moulds using hexachloroethane additions to the insulating mould coating material. The results showed an improvement in the grain-refining power in comparison with conventional boron–titanium inoculation techniques. It was possible to establish correlations between the percentage of hexachloroethane in the coating and solidification time, final grain size, and the extent of the equiaxed zone. It is suggested that the hexachloroethane (a thermally effervescent compound) stimulates a crystal multiplication mechanism that continuously disrupts the growing structure during the solidification process.

The authors are in the Mechanical Engineering Department, Fabrication Process and Materials Division, State University of Campinas (UNICAMP), Campinas–SP, Brazil

It is well known in the technology of casting aluminium that grain refinement of the solidified macrostructure improves the mechanical behaviour and other characteristics of the cast ingot or piece.[1,2] Three basic methods can be used to grain refine the macrostructure: thermal, chemical, and mechanical.[3] The thermal method is based essentially on control of heat transfer conditions at the metal/mould interface. However, the efficiency is rather low.[4] The chemical method is based on the action of an inoculating agent that produces efficient heterogeneous nucleation sites finely dispersed in the melt.[5–7] This is the most useful method for industrial applications because of its high efficiency in reducing the as-cast grain size.

Both thermal and chemical methods utilize nucleation mechanisms. The mechanically-induced grain refinement, however, results from a crystal multiplication mechanism, which consists of extensive fragmentation of previously nucleated crystals. This is normally obtained through the introduction of turbulent stirring in the melt by gas-bubble agitation, or by vibration, or by other means of movement of the metal/mould system.[8–10] This method of industrial application, despite its efficiency in reducing grain size, is restricted to small ingots because of the high cost of introducing mechanical energy in metal/mould systems with high inertial mass and size. In the case of interrupted spin-

ning, for instance, only a minor portion of the mechanical energy is used to agitate the melt during the solidification.[11]

The present work was carried out to demonstrate the possibility of grain refining aluminium by a modification of the mechanical method, without wasting mechanical energy and without the need for complicated mechanical apparatus. This was achieved by the mixing of hexachloroethane with a mould coating. The hexachloroethane is a commercial substance that sublimes at low temperatures (about 180°C), resulting in effervescent evolution of gas-bubbles in the melt during solidification and consequent turbulent flow in the liquid metal.

EXPERIMENTAL PROCEDURE
Experimental work was carried out with commercial purity aluminium of a grade normally used in Brazilian foundries. Chemical analysis showed it to be of 99·44% purity, with 0·14% Fe, 0·32% Mg, and less than 0·10% Si.

The aluminium was heated in a Salamander crucible to 60°C above the liquidus temperature and cast into cylindrical sand and mild-steel moulds of 60 mm i.d., 15 mm wall thickness, and 120 mm height. A few experiments were made in sand moulds with 38 mm and 22 mm i.d., the external dimensions being the same in all cases.

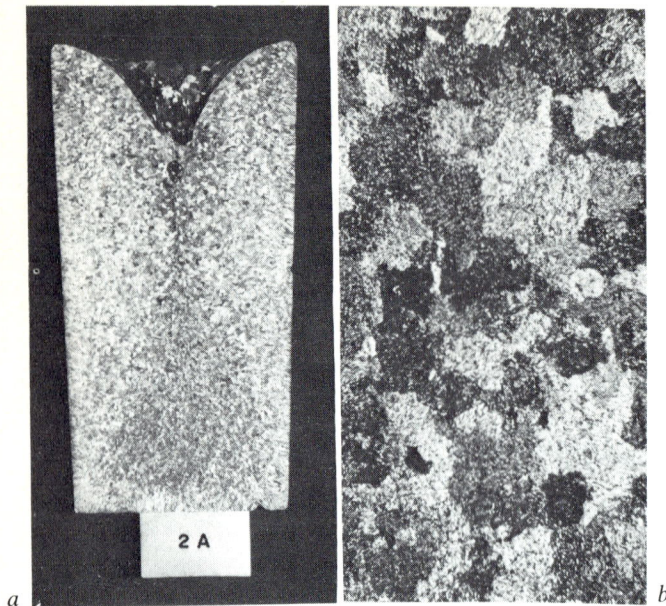

1 *a* Macrostructure [×0·5] and *b* microstructure [×10] for boron–titanium conventionally inoculated ingot, solidified in steel, mould coated with Dycote wash (solidification time 52 s)

In all experiments the solidification time was measured by examination of the time–temperature plot obtained from a Yokogawa precision recorder connected to a chromel/alumel thermocouple.

The mould coating was prepared by mixing Dycote 39 powder (FOSECO), water and a proportion of granular hexachloroethane varying from 0 to 70 wt-%. This mould wash was applied by spraying the internal surface of the mould which was preheated to 80°C in an electrical oven.

In order to assess the efficiency of this method in reducing the grain size reference experiments were carried out. These consisted of inoculating with a traditional boron–titanium grain refiner (Nucleant 2/FOSECO) in both types of mould washed internally with Dycote 39 without hexachloroethane.

RESULTS
Reference samples
In order to compare the additional grain refining power produced by using hexachloroethane, reference samples were examined after inoculation with boron–titanium in the conventional way using both types of mould. A typical result is shown in Fig. 1. To confirm that the hexachloroethane was not itself acting as an inoculant, it was added in the conventional way. This gave only a columnar structure, as shown in Fig. 2.

EFFERVESCENT COATING ON STEEL MOULD
The effect of the hexachloroethane in the coating on the final grain size is shown in Fig. 3, indicating the greater grain refining that results as compared to conventional boron–titanium inoculation. It can be clearly seen that (at least in quantities greater than 10%), the amount of hexachloroethane has no sensible effect on the final grain size, or on the extent of the equiaxed zone. This is

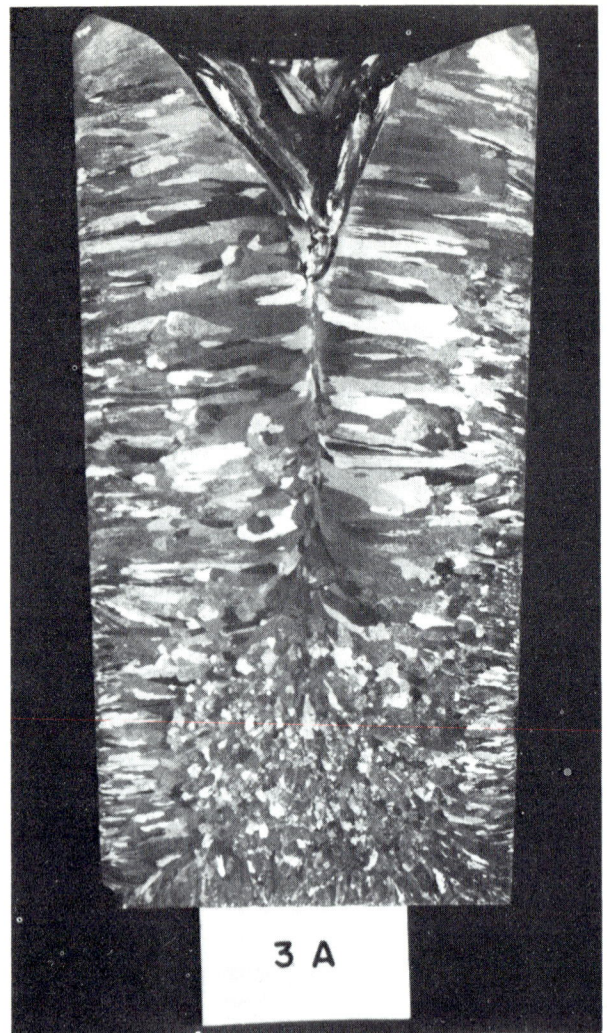

2 Macrostructure [×1] of the ingot conventionally inoculated with pure hexachloroethane and solidified in steel mould (solidification time 53 s)

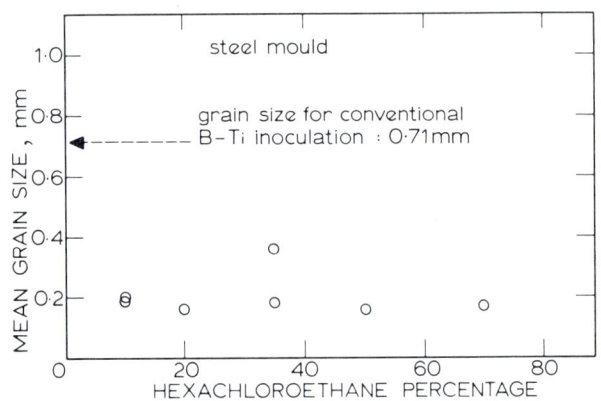

3 Influence of hexachloroethane percentage of the Dycote-base wash on the final grain size for steel mould

apparent from Figs. 4, 5 and 6. This means that (for the steel mould) a small quantity of hexachloroethane is sufficient to exert a powerful grain refining effect. It was noted that increasing the percentage of hexachloroethane in the coating resulted in an increased total solidification time, and this is illustrated in Fig. 7.

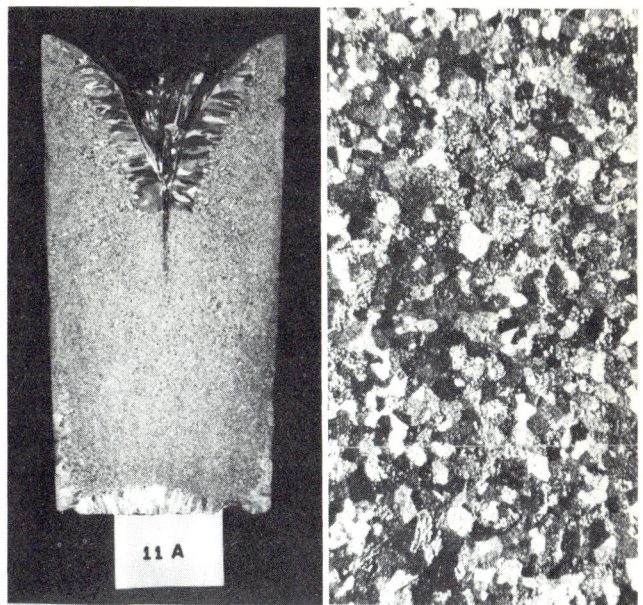

6   **Same as Fig. 4 for 70% hexachloroethane (solidification time 92 s)**

4   **Macrostructure [×1] for steel mould without hexachloroethane in the Dycote wash (solidification time 52 s)**

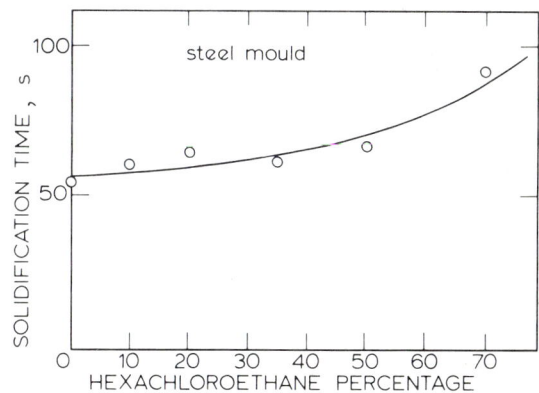

7   **Influence of hexachloroethane percentage of the Dycote wash on the solidification time for steel mould**

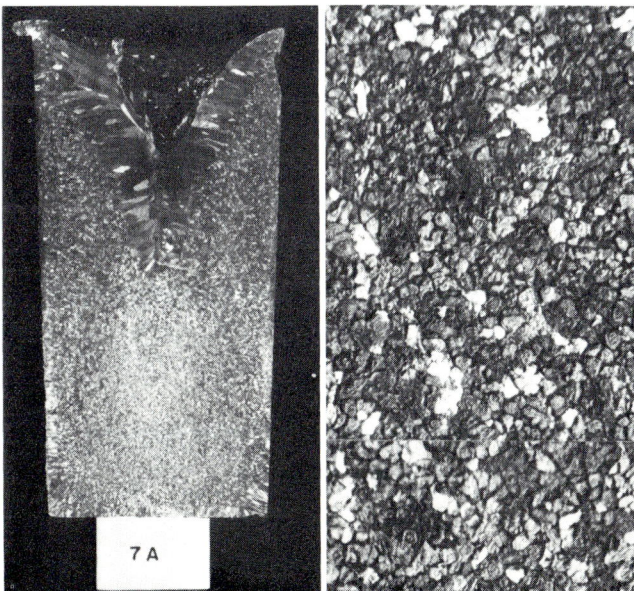

a   *b*

5   ***a* Macrostructure [×0·5] and *b* microstructure [×10] for the same as in Fig. 4, but with 10% hexachloroethane in Dycote-base wash (solidification time 61 s)**

Effervescent coating on sand mould

For the large sand moulds (60 mm dia.), no change was observed in the grain size for any percentage addition of hexachloroethane, as shown characteristically in Fig. 8. On the other hand grain refinement did result when smaller diameter (38 mm and 22 mm) moulds were used as shown in Fig. 9. These results reveal the importance of the total solidification time on the grain refining efficiency of the hexachloroethane. The efficiency is dependent on the relationship between the total solidification time and the time during which the hexachloroethane acts. This observation can be deduced by comparing Fig. 8 with Fig. 9 which shows that diminishing solidification time leads to increases in the extent of grain refinement.

DISCUSSION

Bearing in mind the result of the reference sample test of Fig. 1, as well as the results shown in Fig. 3, it can easily be concluded that in the steel mould (Figs. 5 and 6), the effervescent coating method has a greater grain refining

8 Macrostructure [×1] for sand mould with 10% hexachloroethane in the Dycote-base wash (solidification time 330 s)

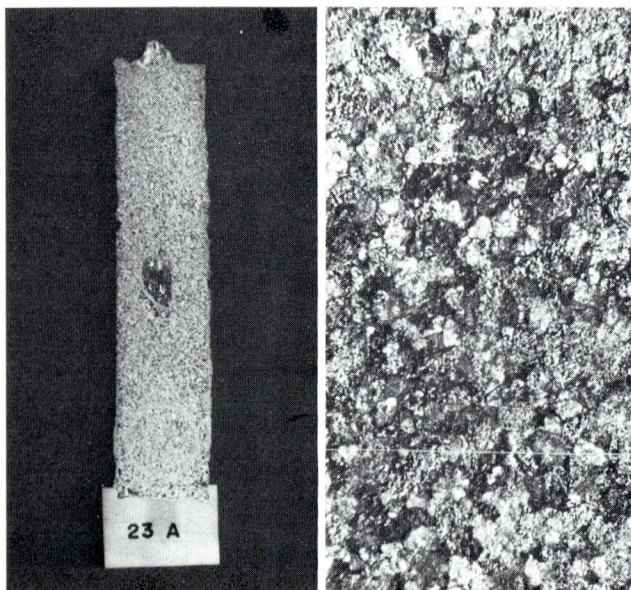

a *b*

9 *a* Macrostructure [×0·5], and *b* microstructure [×10] for 20% hexachloroethane in 22 mm i.d. sand mould (solidification time 90 s)

power than the conventional boron–titanium inoculation procedure. Even for small proportions of hexachloroethane in the Dycote base wash, the improved efficiency is clear.

For the sand moulds the method is only effective for the smaller internal diameter moulds as can be seen by comparing Fig. 8 with the results shown in Fig. 9. This suggests the existence of a correlation between the efficiency of the method and the total solidification time. This is made more clear by the results shown in Table 1. This can be accounted for by the fact that the effervescence initiated by the hexachloroethane occurs only for a limited period of time. This would imply that to have the maximum degree of grain refinement, the total solidification time has to be equal to or less than the activation time of the hexachloroethane.

It was observed that, if all other variables are fixed, increasing the hexachloroethane percentages in the wash causes an increase in the total solidification time (Fig. 7). Although this effect is of minor importance, it may be noted that it is probably due to the promotion by gas evolution of an enhanced gap at the metal/mould interface during the solidification process which lowers the heat transfer coefficient appreciably.

The observed grain-refining action produced by the effervescence resulting from the use of hexachloroethane in the Dycote wash, together with the fact that the low purity aluminium used grows in a dendritic or cellular/dendritic fashion, supports the proposed grain multiplication mechanisms of Jackson *et al.*[12], and Tiller and O'Hara.[13] Thus the volatilization of the hexachloroethane provokes a sufficiently turbulent fluid motion to produce thermal fluctuations and remelting to disrupt the solid structure that protrudes into the liquid during solidification, particularly during the formation of the chill zone.[14] The resulting detached solid fragments are dynamically misaligned and carried by the fluid body forces. Subsequently, if these particles are removed into the bulk of liquid, they may survive in the local almost isothermal conditions and give rise to the fine-grained structure.

The results obtained are not sufficiently conclusive to suggest that this mechanism is the only one which would result in the grain-refining action observed with hexachloroethane. Other grain refining mechanisms,

Table 1 The influence of the casting conditions* on the efficiency of grain refinement of the effervescent (hexachloroethane) coating

Mould material	Mould dia., mm	Hexachloro-ethane in mould wash, %	Total solidification time, s	Observed grain refining power
steel	60	10	61·0	high
steel	60	35	62·0	high
steel	60	50	66·0	high
sand	22	20	90·0	high
steel	60	70	92·0	high
sand	38	20	240	partial
sand	60	10	330	none
sand	60	35	351	none

* All castings poured with 60°C superheat

probably on a minor scale, could also contribute; for example the 'showering of crystals' from the surface, as pointed out by Southin.[10] Further work needs to be carried out to quantify the relative importance of the different grain refining mechanisms present in the system.

CONCLUSION

The results of the present work permit the proposition of a simple and economic method to grain refine commercial aluminium castings by the addition of hexachloroethane in insulating mould coatings. The efficiency of the method depends on a compromise between the mould heat transfer and the cross-sectional area of the casting, in order to ensure an adequate solidification time, i.e. the greater the heat extraction rate of the mould, the greater can be the cross-sectional area of the casting (for the solidification time to be equal to or less than the residence time of the hexachloroethane in the system). If these conditions exist, the proposed method shows a greater grain refining power than conventional boron–titanium inoculation. The results suggest that the hexachloroethane promotes, by its effervescence, an intensive turbulence flow in the liquid metal that stimulates strong crystal multiplication by fragmentation of the growing structure during the solidification process.

ACKNOWLEDGMENTS

The authors are pleased to express their appreciation to Drs G. J. Davies and T. W. Clyne for helpful and stimulating discussion, to Mr N. Scarpato and to Mr J. W. C. Carvalho for helping in several of the experiments, and to Financiadora de Estudos e Projetos—FINEP—for their indispensable financial support.

REFERENCES

1 A. CIBULA AND R. W. RUDDLE: *J. Inst. Met.*, 1949–50, **76**, 361
2 A. L. MINCHER: *Metal Ind.*, 1950, **76**, 435
3 E. A. FEEST: *Metal. ABM*, 1974, **30**, 19
4 O. S. PIRES et al.: *ibid.*, 1972, **28**, 779
5 G. W. DELAMORE AND R. W. SMITH: *Metall. Trans.*, 1971, **2**, 1 733
6 J. CISSE et al.: *Metall. Trans.*, 1974, **5**, 633
7 A. J. CORNISH: *Met. Sci.*, 1975, **9**, 477
8 R. T. SOUTHIN: *J. Aust. Inst. Met.*, 1965, **10**, 115
9 G. S. COLE AND G. F. BOLLING: *Trans. Met. Soc. AIME*, 1969, **245**, 725
10 R. T. SOUTHIN: 'The solidification of metals', 305, 1968, London, The Iron and Steel Institute
11 F. A. CROSSLEY et al.: *Trans. Met. Soc. AIME*, 1961, **221**, 419
12 K. A. JACKSON et al.: *ibid.*, 1966, **236**, 149
13 W. A. TILLER AND S. O'HARA: 'The solidification of metals', 27, 1968, London, The Iron and Steel Institute
14 M. PRATES AND H. BILONI: *Metall. Trans.*, 1972, **3**, 1 501.

Influence of composition on intrinsic grain refining in binary copper alloys

L.O. Gullman and L. Johansson

Copper alloys of varying concentrations were solidified from the upper surface which was covered by a transparent salt flux. Depending on alloy concentration, the melt surface was observed to solidify in one of two ways: as a continuous flake (low concentration), or under formation of numerous small crystals (high concentration). When the concentration was increased transition from flake to small crystal growth occurred within a narrow interval. The lowest concentration of solutes which gave small crystals was found to be related to partition coefficients such that $c \cdot (1-k)/k \sim constant$. Melt flow rate and undercooling were low when solidification started. The grain refining is assumed to be an effect of dendrite disintegration by the coarsening mechanism.

The authors are with Gränges Aluminium Research Laboratory, Finspång, Sweden

Not every grain in a solidification structure has its origin in an individual nucleation event. It has been suggested that some grains may be formed by fragmentation of the growing solid. The extent to which such a crystal multiplication process contributes to the number of grains in a particular cast structure is still, however, undetermined.

The concept of crystal multiplication has received support from the work of Jackson et al.[1] who showed that secondary dendritic branches may remelt partially and thereby form new grains.

Possible mechanisms for dendrite fragmentation have been investigated theoretically by Tiller and O'Hara.[2] They found that the probability of formation of new grains by fragmentation would depend upon alloy composition, undercooling, and rate of melt flow during solidification.

The effect of composition on grain refining was studied experimentally by Tarshis et al.[3], who plotted the grain size data for alloys as a function of a parameter P comprising concentration and constitution characteristics. The influence of undercooling on crystal multiplication was studied by Southin et al.[4] who found that refining effects were dependent upon alloy composition and undercooling.

The grain size data in the experiments referred to have been obtained from experiments where solidification took place under intense turbulent motion in the liquid alloy[3] or after high undercooling.[4] It would be of great practical significance, however, if the grain refining effects for a particular alloy could be further clarified under less extreme conditions. The solidification experiments in the present work have been carried out in such a way that undercooling preceding nucleation of solid and melt flow during solidification were kept to a minimum.

EXPERIMENTAL PROCEDURE

Binary copper alloys were melted in an apparatus shown in Fig. 1. The melts were kept in a graphite crucible (ECV, top inner diameter 80 mm). The crucible was heavily insulated around its outer walls and under the bottom to direct the heat losses upwards. The melt surface was covered by a layer of molten sodium tetraborate (borax) about 5 mm thick, which was dehydrated by fusion before use.

The copper used was of OFHC quality. The alloying elements (excluding oxygen) were added through the salt in elementary form (Ag, Sn, Zn) or as master alloys of technical purity (Cu–10%Si, Cu–15%P). Oxygen was introduced by exposing the copper to air during melting.

Copper or prealloyed ingots were melted under salt and heated to 25–35°C above liquidus temperature; furnace heating was then shut off and the melts were left to solidify. During cooling and solidification, the melt surface was illuminated at an angle by a high-intensity lamp (see Fig. 2). By looking at the melt surface at a small angle (5°–10°) from the reflected light rays, small perturbations caused by the formation of solid could be observed easily. Some solidification sequences were recorded with a cine camera.

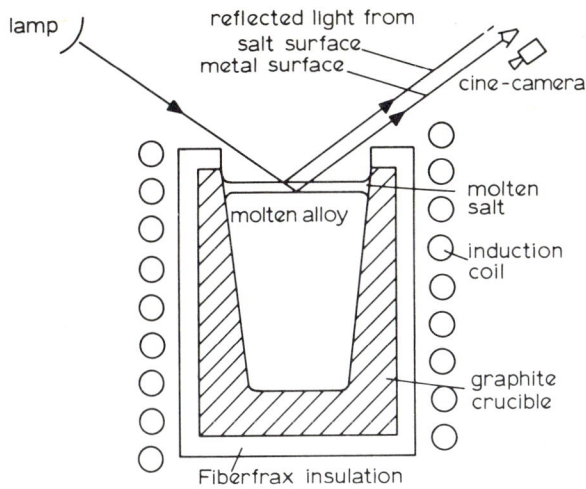

1 Experimental apparatus

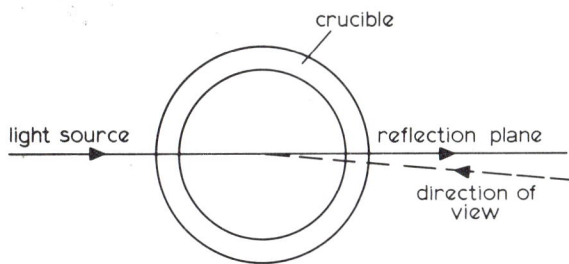

2 Illumination arrangement, viewed from above

The alloy series was studied as described above after successive additions of alloying elements, followed by melting and solidifying after each addition. The appearance of the first solid, rapidly growing in the interface between salt and alloy melts, was observed and characterized.

Temperature measurements
A thermocouple protected by an Al_2O_3 tube was immersed continuously in the melts to permit a reproducible temperature cycle. The Al_2O_3 tube was corroded by the borax melt and had to be exchanged at intervals.

Attempts were made to measure the undercooling of the alloy surface before nucleation of the solid. To achieve this the thermocouple bead was protected from the salt by a thin-walled graphite tube (*see* Fig. 3). The flat bottom of the graphite tube was immersed through the salt until it just touched the alloy surface.

RESULTS
The first solid visible at the interface could be formed in one of two ways; either as a coherent shell or subdivided into numerous small crystals.

Shell growth
In unalloyed copper (with low oxygen content, *see* below) or in dilute alloys, the interface solidified by the formation of a thin shell of solid phase. Solidification started from one or two nucleation sites and continued until the whole interface had solidified. At the beginning the shell was very thin, which was demonstrated by touching it with a ceramic rod. A few seconds after complete solidification of the interface, the shell had

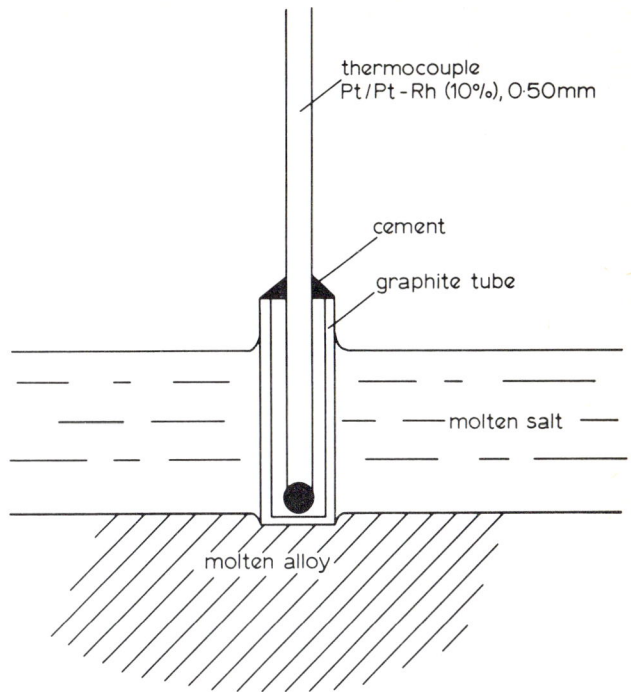

3 Apparatus for measuring surface undercooling

acquired such a rigidity by downward growth that it was no longer possible to break it with the rod.

Primary nucleation often took place at the crucible wall or at the alumina tube protecting the thermocouple. The exact nucleation site was in many cases impossible to localize because part of the interface was in the shade of the crucible wall.

The solidification front advanced at a rate which, from a frame-by-frame analysis of the cine film, was estimated to be 9 ± 1 mm/s.

Formation of small crystals
In concentrated alloys the interface solidified by the formation of a large number of small crystals originating from a few nucleation centres. There was a more or less gradual change from shell formation to small crystal formation when the concentration of the solute was increased. While growing radially, the crystals rapidly moved away from the nucleation site. New small crystals formed from those already existing. This continued until the interface was completely covered by solid. These secondary crystals moved away from the solidification fronts where they had originated and they also appeared to be repelled by each other.

Some small crystals disappeared from the interface before a coherent crust had formed. It was found that these crystals settled to the bottom under the influence of gravity (*see* below).

Influence of composition
The gradual change from shell to small crystals took place within a concentration interval (different for different solute elements). The results from all the experimental series are shown in Fig. 4, where the alloying elements have been characterized by their equilibrium partition coefficients, k. The figure shows that the solidification process changes in a systematic manner such that transition intervals of different alloys lie on a straight line.

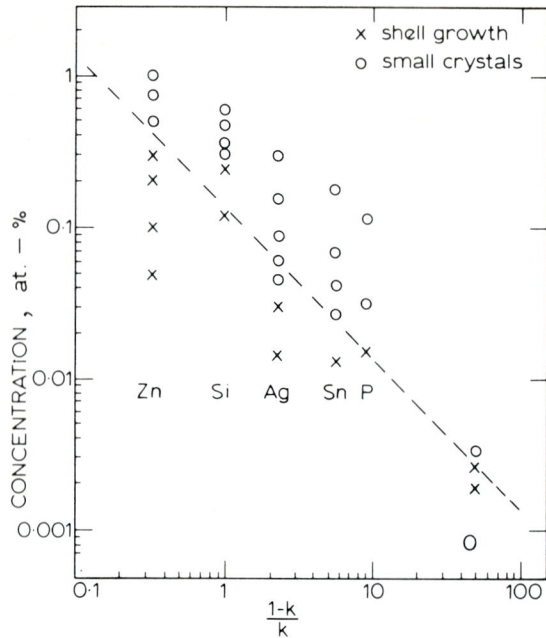

4 **Character of solidification at different alloy constitution and concentration**

The first unalloyed melt in each series appeared to solidify in an unsystematic manner. Analysis of some unalloyed melts showed that they had different oxygen contents. These results are shown in Fig. 4, and it appears that a change in the structure characteristic occurs with a change in the concentration of absorbed oxygen in the melt.

Thermocouple readings
Before solidification, the cooling rate of the melts was 0·3°–0·4°C/s. Undercooling of the interface just before nucleation was found to be 1·7 ± 0·5°C.

By comparing the readings of two simultaneously immersed thermocouples at the interface, it was observed that the temperature at the centre of the interface was slightly lower than that at the periphery, indicating that, as anticipated, heat was lost from the melt surface.

Convective currents could be seen easily in the salt by observing minute dirt particles which were carried along by the melt flow. The convective pattern is shown schematically in Fig. 5. The flow directions show that cooling is most intense at the centre.

Before solidification, the thermocouple readings from the interface oscillated, indicating that convective movements had developed in the alloy melts.

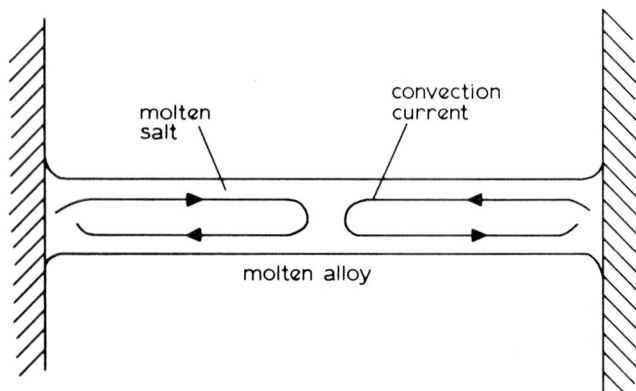

5 **Convection currents in molten salt**

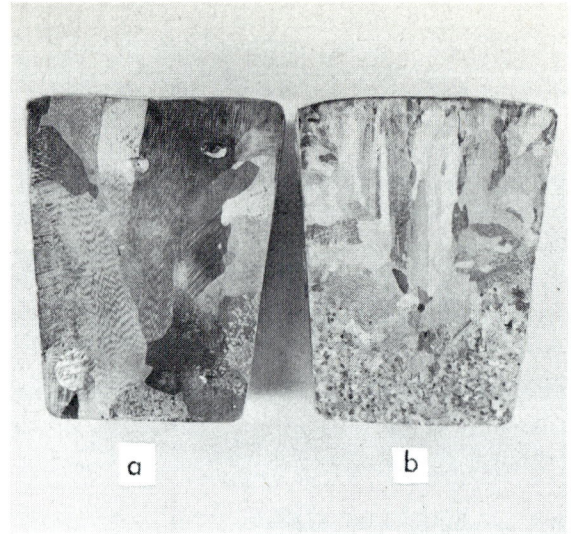

6 **Typical macrostructure *a* after shell growth, and *b* after crystal formation   [×0·5]**

Structure of solidified alloys
Alloys which had solidified by shell formation had a coarse columnar structure throughout the entire ingot (*see* Fig. 6*a*). From the structure it is obvious that a considerable part of the latent heat was lost upwards.

Microsegregation patterns indicated that the first solid had formed from the melt with very weak, if any, tendencies for branching.

Ingots which had solidified by formation of small crystals had fine equiaxed grains at the bottom, the rest being columnar (Fig. 6*b*). The crystals visible on the top were small and equiaxed (Fig. 7), but the same crystals had changed to columnar during their further growth downwards.

The volume of the equiaxed part of the ingot was proportional to the solute concentration above the minimum value necessary for small crystal formation. The data given in Fig. 8 were obtained from separate experiments. Extrapolation along the lines to volume

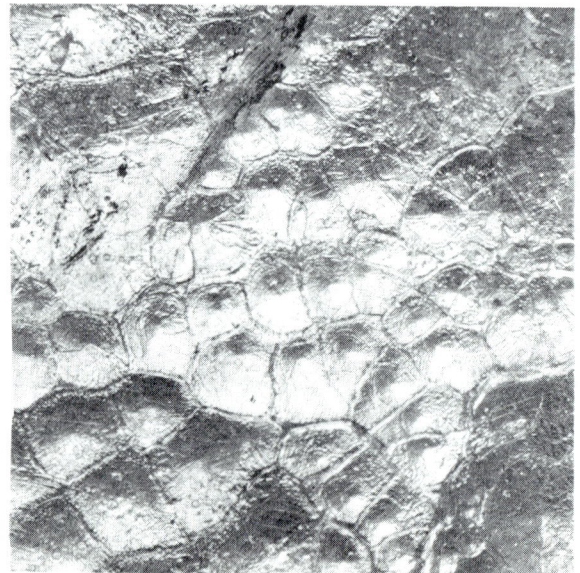

7 **Example of ingot surface solidified by small crystal formation, illumination from above [×10]**

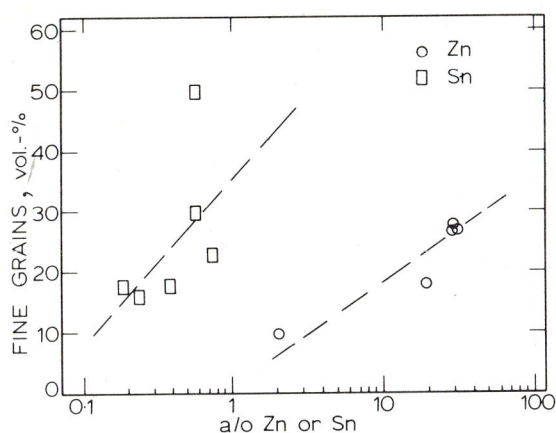

8 Relation between volume of fine grains and alloy content

part = 0 yield concentration values for the onset of grain refining in reasonable agreement with those found by observing the change of interface solidification.

The solid alloys having equiaxed crystals at the bottom showed a completely dendritic microstructure.

DISCUSSION

Grain refining effects in the present work depend on constitution and concentration. Observations made during experiments suggest that the grain refining is the result of a crystal multiplication process rather than of increased nucleation. The multiplication proceeded at small undercooling and is not restricted to the recalescence period.

Composition effect

Experimental observations were correlated to constitution and concentration by plotting them (Fig. 4). Alloys solidified by the formation of small crystals are generally found above the line

$$\log c + \log [(1-k)/k] = -0.85 \quad (1)$$

while those with a coarse grain are on the other side.

There is no sharp boundary between the two areas. The basic condition for dendrite disintegration is the development of secondary branches. This will occur gradually with increasing concentration provided the thermal conditions are unchanged. Consequently, the onset of grain refining also will be gradual, a fact which makes the classification uncertain for alloys of composition near the borderline. Also, small errors in the values chosen for respective partition coefficients will affect the positions significantly in the diagram of the different alloy series.

Rewriting equation (1), the conditions for grain refining can be formulated

$$c \cdot (1-k)/k > 0.14 \quad (2)$$

The constant refers to actual base metal and solidification conditions.

The effect of composition was similar to that found by Tarshis *et al.*[3] However, correlation of the present experimental observations with *P* or partition coefficient alone, as carried out by Southin and Weston,[4] have yielded an inferior fit.

Undercooling

By performing the experiments on a comparatively large laboratory scale (5 kg) and by using metals of normal technical quality, solid started to grow after a very slight undercooling, in contrast to the experiments by Southin and Weston.[4] Crystal multiplication was observed to proceed for a few seconds after the recalescence period, a fact which renders dendrite remelting a less probable explanation for the grain refining. A more probable mechanism, which is capable of operating isothermally at low undercooling, is fragmentation by dendrite coarsening, studied by Kattamis *et al.*[5] According to calculations by these authors, the time necessary for loosening secondary arms of $\sim 10~\mu m$ spacing by the coarsening mechanism is shorter or at least of the same order of magnitude as the time available in the present experiments. Of course the same mechanism could operate during the later stages of solidification of the experimental melts, and it is not necessary to assume that all grains at the bottom of the crucible have originated at the salt/metal interface. It is probable, however, that the first dendrites that grow into an undercooled melt will form an extended network that can be disintegrated easily by coarsening. Hence fragmentation can be expected to be most intensive during the early stages of solidification.

Convective currents

It has been argued that fragmentation can be effected by drag forces from the melt flowing past growing dendrites.[2] In the present experiments, melt flow was restricted to natural convection only. If convective currents in the salt had exerted any significant frictional drag on the growing solid, then detached particles would have been carried along from the centre towards the periphery. As this was generally not found to be the case, it can be concluded that the flow of molten salt had no direct effect on crystal multiplication.

The flow rate of convection currents in the molten alloy could not be observed in the same simple way, but can be assumed to be low as only small temperature differences were found to exist within the melt. Also the rate of feeding liquid flowing into the dendrite network is slow at the low growth rates that prevailed during the present experiments. Thus it is improbable that these slow currents of liquid alloy should contribute significantly to the grain multiplication by the mechanisms put forward by Tiller and O'Hara.[2]

The repulsion of small crystals by each other can be explained as an effect of interfacial tension forces (the Marangoni effect[6]). Considering the probable order of magnitude of these interfacial forces (estimated to 10^{-1} N/m) in comparison to the tensile strength of copper at its melting point (about 5 N/mm²), it is realised that fragmentation of copper dendrites by the same forces is highly improbable.

CONCLUSIONS

1 In binary copper alloys intrinsic grain refining effects can be expected provided $c \cdot [(1-k)/k] >$ constant; the value of the constant depends on actual solidification conditions.

2 The observed grain refining is probably due to isothermal fragmentation of dendrites, growing at low undercooling.

3   Under the experimental conditions dendrites may be fragmented by the coarsening process.

ACKNOWLEDGMENT
The authors wish to thank Gränges Metallverken for permission to publish this paper.

REFERENCES
1 K. A. JACKSON *et al.*: *Trans AIME*, 1966, **236**, (2), 149

2 W. A. TILLER AND S. O'HARA: 'The solidification of metals', 27, 1968, London, The Iron and Steel Institute

3 L. A. TARSHIS *et al.*: *Metall. Trans.*, 1971, **2**, (9), 2 589

4 R. T. SOUTHIN AND G. M. WESTON: *J. Aust. Inst. Met.*, 1973, **18**, (2), 74

5 T. Z. KATTAMIS *et al.*: *Trans. AIME*, 1967, **239**, (10), 1 504

6 L. E. SCRIVEN AND C. V. STERNLING: *Nature*, 1960, **187**, 186

Formation mechanism of eutectic grains

A. Ohno, T. Motegi, and K. Ishibashi

In order to determine the origin of eutectic grains in eutectic alloys, the macrostructures and microstructures of ingots (of Al–Cu, Al–Ni, Al–Zn, Pb–Sn, and Sn–Zn), solidified under static and vibration conditions, were examined. In alloys not of eutectic composition and which were either hypoeutectic or hypereutectic, primary crystals separated from the mould wall in the initial stage of solidification and tended to appear in the central region of the ingots. When the primary crystals were not the same as the leading phase of the eutectic, the eutectic which formed on the mould wall grew in a columnar manner independent of the existing primary crystals in the liquid. When the primary crystals were of the leading phase, the formation of equiaxed eutectic grains occurred from the primary crystals in the liquid. It was also found that eutectic grains on the mould wall could be separated by turbulence in the liquid.

The authors are in the Department of Metallurgical Engineering, Chiba Institute of Technology, Narashino, Chiba-Ken Japan

It is known that the boundaries of the eutectic grains influence the strength of eutectic system alloys. However, the mechanism of formation of equiaxed eutectic grains in an alloy which consists entirely or partly of eutectic is not well understood. Generally, it has been accepted that equiaxed eutectic grains nucleate in the liquid at the growing interface of primary crystals. For instance, in order to explain the formation of eutectic grains of Fe–C alloys, Lakeland[1] has proposed that austenite dendrites and eutectic grains are regenerated in a cyclic manner, as illustrated in Fig. 1. Eutectic grains, A, B, and C nucleate on or near projecting dendrites and, as the growth of these grains nears completion, dendrites are regenerated in the intergranular positions (a) and (b) in Fig. 1. The regenerated dendrites grow into liquid which is constitutionally supercooled because of the rejection of sulphur and other impurities. As the austenite grows, carbon is rejected into the melt, and when the carbon content of the surrounding liquid is high enough, eutectic grains nucleate on a dendrite surface (at D and E). The cycle is repeated as grains D and E near completed growth and a dendrite is regenerated at (d).

The present authors[2] have observed the solidification phenomena of Sn–Bi, Sn–Pb, and Sn–Sb alloys and found that the equiaxed crystals in cast metals were formed by the separation of necked-shape crystals from the mould wall in the initial stage of solidification before the formation of a stable solid skin, as illustrated schematically in Fig. 2.[3]

This suggests that the primary crystals of eutectic system alloys can also be separated from the mould wall before the eutectic reaction takes place. If the separated crystals are of the leading phase of a eutectic alloy system eutectic grains can originate at the surface of the separated crystals. The eutectic grains themselves may also be separated from the mould wall where eutectic reaction first occurs. In order to investigate these suggestions and to determine the factors which control the eutectic grain size, the macrostructures and microstructures of alloy ingots (of Al–Cu, Al–Ni, Al–Zn, Pb–Sn, and Sn–Zn) solidified under static and vibration conditions were studied.

EXPERIMENTAL METHOD

Hypoeutectic and hypereutectic alloys having compositions close to the eutectic composition of the systems, were melted in graphite crucibles using an electric furnace. The molten metal was poured into a metallic or sand mould 30 mm in diameter and 65 mm deep under static and vibration conditions. In the latter case, a frequency of 50 Hz and 0·3–1·0 mm in amplitude was used. The structures of the vertical and horizontal sections of ingots were observed macroscopically and microscopically using a polarization microscope.

DISTRIBUTION OF PRIMARY CRYSTALS AND EQUIAXED EUTECTIC GRAINS

In both hypo- and hypereutectic alloys primary crystals always tended to appear in the central region of ingots,

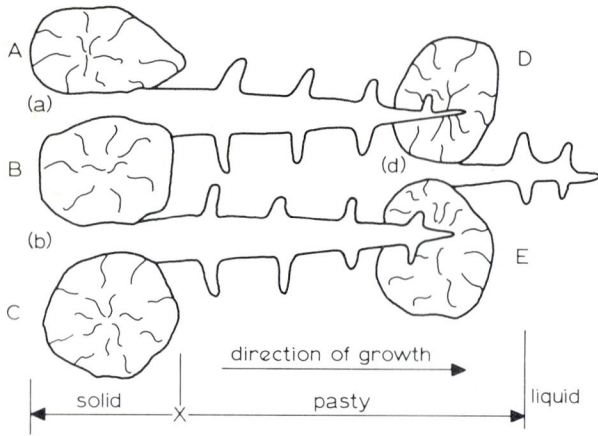

1   Schematic representation of growth of irregular interface by cyclic mechanism of alternate precipitation of eutectic grains, austenite dendrites, and eutectic grains

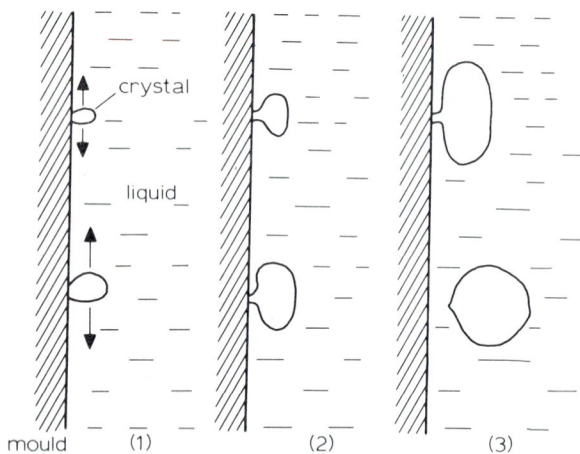

2   Schematic illustration of separation of necked-shape crystals from mould wall

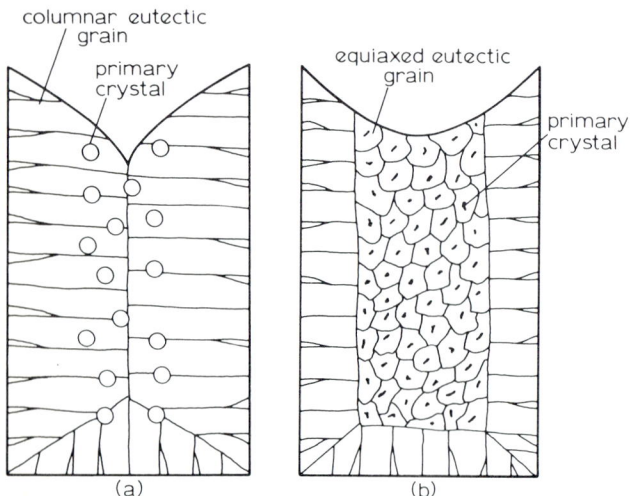

a primary crystals of secondary phase; *b* primary crystals of leading phase

3   Schematic illustration of macrostructures of eutectic system alloys

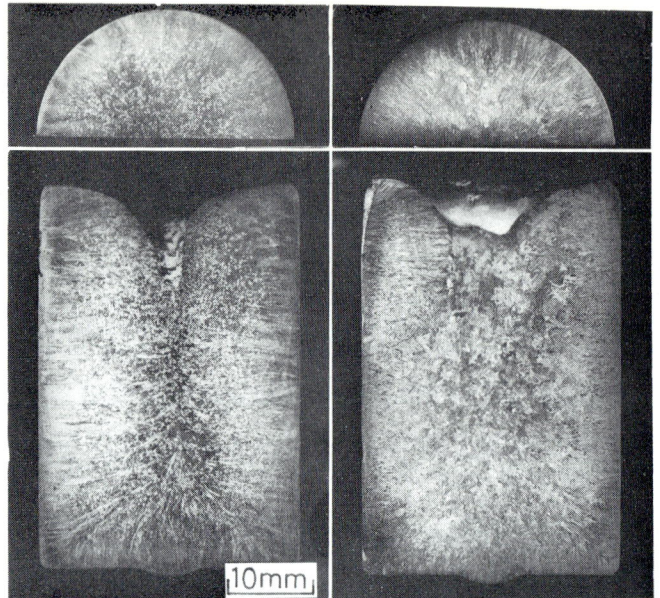

a hypoeutectic alloy (32%Cu) ingot; *b* hypereutectic alloy (34%Cu) ingot of Al–Cu system, cast at 700°C

4   Typical macrostructures of eutectic system alloys

as shown schematically in Fig. 3, indicating that the separation of primary crystals occurred from the mould wall before the start of the eutectic reaction. The region of primary crystals in hypo- and hypereutectic alloy systems increased as the alloy composition was moved further away from the eutectic composition. It was also observed that eutectic grains tended to grow from the mould wall, forming columnar eutectic grains (Fig. 3a). If the primary crystals were the same as the leading phase of the eutectic, an equiaxed zone of eutectic grains always appeared in the central region of the castings, as shown schematically in Fig. 3b. Figure 4 shows the typical macrostructures of hypoeutectic and hypereutectic alloy ingots of the Al–Cu system which were cast at 700°C in metallic moulds. In Fig. 4a, the macrostructure of Al–32%Cu alloy (hypoeutectic) shows columnar eutectic grains in the entire area of the ingot and primary Al particles in the central region, while the macrostructure of Al–34%Cu (hypereutectic) alloy showed columnar eutectic grains in the outer zone, and equiaxed eutectic grains associated with primary $CuAl_2$ particles in the central region (Fig. 4b). Similar macrostructures were observed in all of the alloys used in this work.

PRIMARY CRYSTALS AND EUTECTIC GRAINS

The macrostructures of eutectic alloys showed that the formation of eutectic grains started on the mould wall where the temperature of the liquid first reached the eutectic temperature. It was also found that if the primary crystals were the same as the leading phase of the eutectic in the alloys, the eutectic structures originated on the surface of the primary crystals. However, if the primary crystals were of the other phase of the eutectic, the eutectic structures did not originate on the primary crystals.

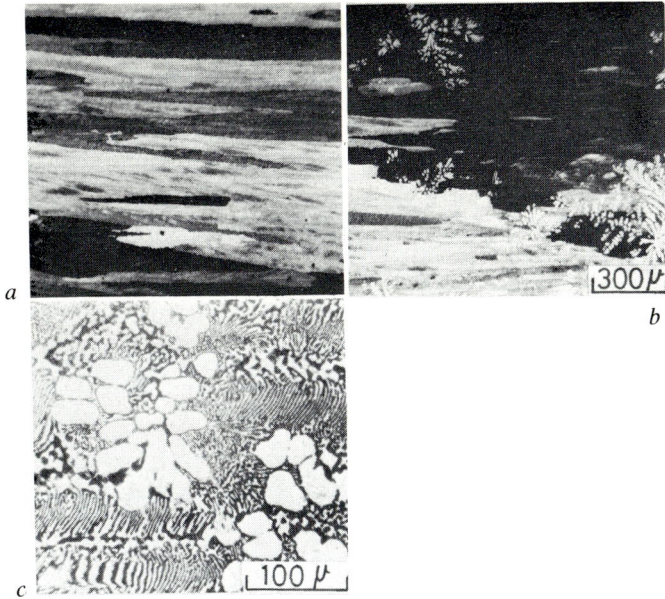

a outer region; b and c central region

5   Microstructures of Al–32%Cu (hypoeutectic) alloy ingot

Figure 5 shows the microstructures of the outer region and of the central region of Al–32% Cu alloy (hypoeutectic). Columnar eutectic grains consisting of lamellae of CuAl$_2$ and Al were observed over the whole of the ingot and primary Al dendrites were observed mostly in the central region. Figure 6 shows the microstructures of the outer region and of the central region of Al–34%Cu (hypereutectic) ingot. Columnar eutectic grains were observed in the outer region, while equiaxed eutectic grains were associated with primary CuAl$_2$ particles in the central region. Similar relationships between the primary crystals and the eutectic matrix were observed in all the alloy systems studied in this work.

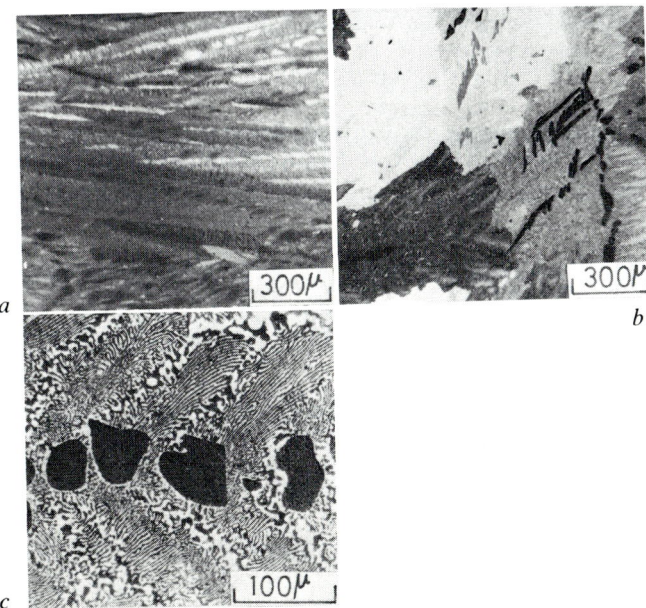

a outer region; b and c central region

6   Microstructures of Al–34%Cu (hypereutectic) alloy ingot

It is known that leading phases in the eutectic separation of Al–Cu, Al–Zn, Pb–Sn, and Sn–Zn are CuAl$_2$, Al, Sn, and Zn, respectively.[4] When these leading phases existed as the primary crystals, equiaxed eutectic grains were always obtained. However, when the primary crystals were of the other phase in the eutectic, the grains were columnar. In the case of solidification of Al–Ni alloys, NiAl$_3$ was found to fulfil the requirements of the leading phase.

FORMATION OF EQUIAXED EUTECTIC GRAINS ON PRIMARY CRYSTALS

From the results of observations of the structures of eutectic alloys, the growth process of eutectic grains is illustrated in Fig. 7. Both the leading and the secondary phase grow from the mould wall into the remaining liquid, forming columnar grains. Since the degree of undercooling for the growth of a crystal in a liquid is smaller than that for the nucleation of the crystal, if the primary crystals are not of the leading phase, columnar eutectic grains which grow from the mould wall can grow independently from the primary crystals, as shown in Fig. 7a. However, if the primary crystals of the leading phase which separate from the mould wall in advance exist in the region ahead of the growing columnar eutectic grains, projections grow from the surface of the primary crystals when the temperature of the liquid at the surface of the primary crystals reaches the eutectic temperature. They finally form independent eutectic grains (see Fig. 7b).

The formation of the eutectic structure from the primary crystal in the case of the solidification of Al–Si hypereutectic alloys was directly observed by Löhberg and his co-workers.[5] Their work clearly showed that eutectic formed from the primary Si crystal, which was the leading phase of the Al–Si system.

SEPARATION OF EUTECTIC GRAINS FROM MOULD WALL

Equiaxed eutectic grains were often observed in ingots of the alloys which had no leading phase as primary

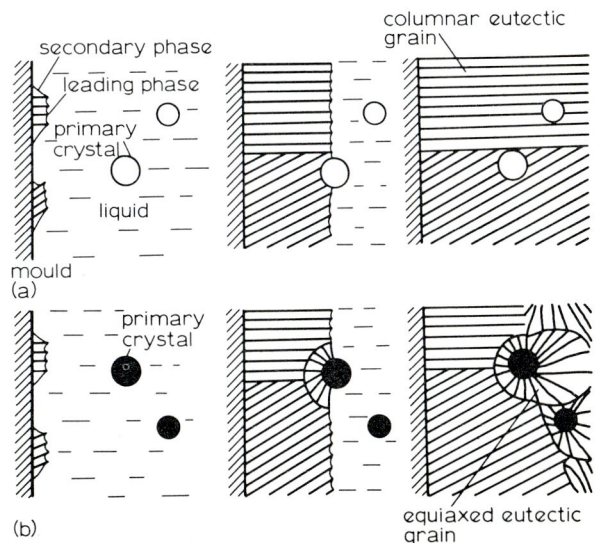

a without leading phase as primary crystal; b with leading phase as primary crystal

7   Schematic illustration of formation of eutectic grains

206   *Ohno et al.*

crystals when the molten alloy was poured with high superheat into a mould, or when a mould which had small cooling capacity was used. In order to demonstrate this, hypoeutectic alloys and hypereutectic alloys of the Al–Cu, Al–Ni, Al–Zn, Pb–Sn, and Sn–Zn system were solidified under vibration conditions of 0·3–1·0 mm in amplitude and 50 Hz.

Vibrations greatly promoted the grain refining of primary crystals in alloys of both hypoeutectic and hypereutectic composition, and expanded the region of primary crystals. When the primary crystals were of the leading phase of the eutectic, the eutectic grains were refined and the equiaxed eutectic zone also expanded. Even when the primary crystals were of the other eutectic phase, vibration caused the separation of equiaxed eutectic grains from the mould wall. Figures 8 and 9 show typical structures which illustrate the effect of vibration on the solidified structure of a hypoeutectic and a hypereutectic alloy of the Al–Cu system.

Separation of eutectic grains occurred more readily in sand than in metallic moulds. Because the sand mould has less cooling capacity than the metallic mould, it is considered that the formation of a stable solid skin of eutectic grains on the mould wall takes place much more slowly than in the metallic mould. In other words, the separation of the eutectic grains occurs much more easily from a sand mould wall than from the wall of the metallic mould.

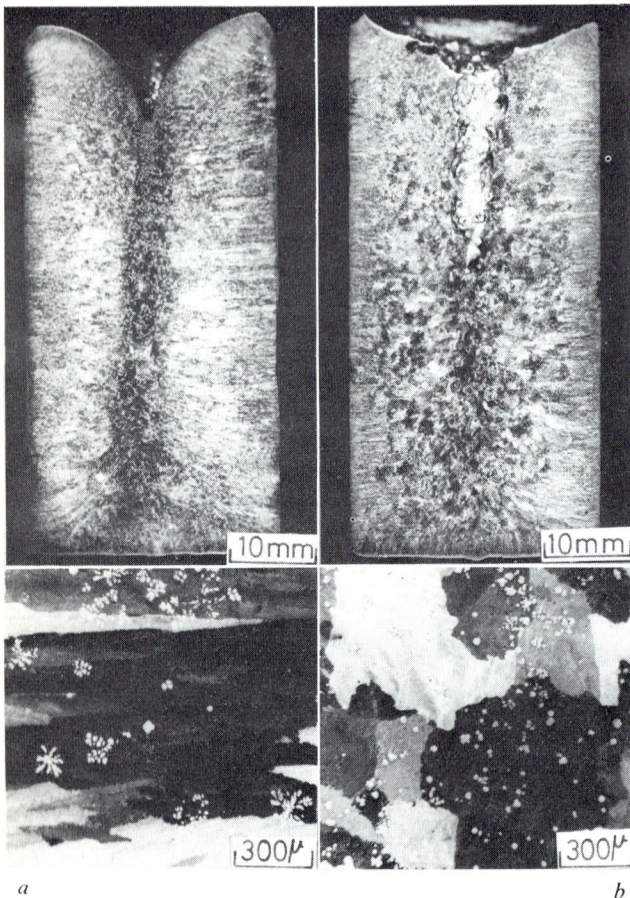

a static; *b* vibration of 0·5 mm in amplitude, cast at 600°C

8   Influence of vibration on macrostructure and microstructure of Al–32%Cu alloy (hypoeutectic) ingot

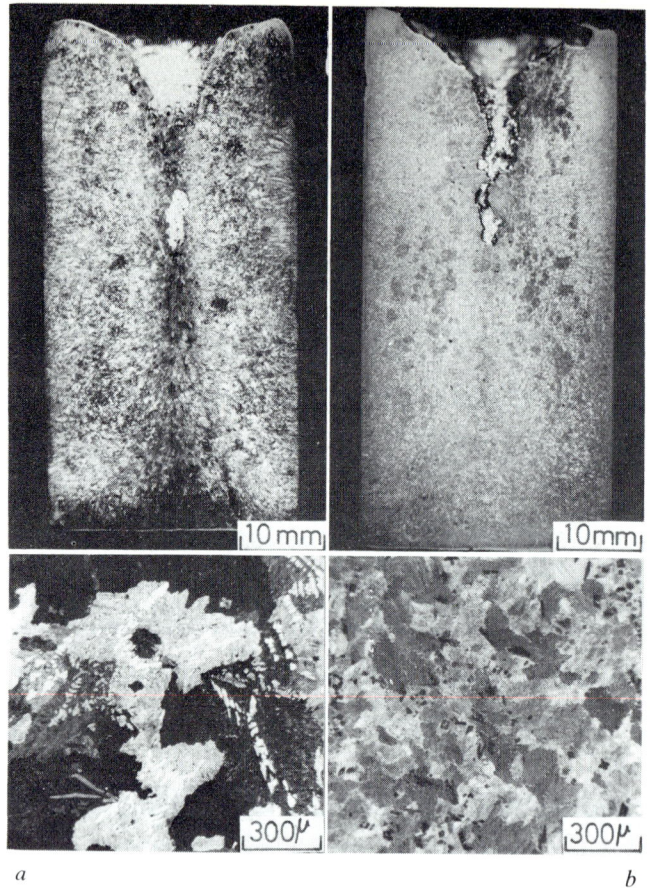

a static; *b* vibration of 0·5 mm in amplitude, cast at 600°C

9   Influence of vibration on macrostructure and microstructure of Al–34%Cu alloy (hypereutectic) ingot

CONTROL OF EUTECTIC GRAINS

Since the surface of the primary crystals of the leading phase of a eutectic alloy system is the place of origin of eutectic grains, the increase in the number of the primary crystals of the leading phase results in an increase in the number of eutectic grains. If the structure with fine eutectic grains of an alloy system is desired, the alloy having primary crystals of the leading phase must be chosen. Agitation and vibration of molten metal during the solidification will be effective for the distribution of as many as possible of the primary crystals in the molten metal before the eutectic reaction takes place. However, if the presence of equiaxed eutectic grains in an ingot is undesirable, an alloy having primary crystals of the leading phase should be avoided, and the molten metal should be cast in the mould under static condition and without turbulence.

CONCLUSION

It is concluded that in Al–Cu, Al–Ni, Al–Zn, Pb–Sn, and Sn–Zn system alloys, the primary crystals can be separated from the mould wall before the start of the eutectic reaction. Eutectic grains first begin to form on the mould wall and have the tendency to form columnar grains starting on the mould wall. Eutectic grains also can be separated from the mould wall before the formation of the stable skin of eutectic grains. The primary crystals of the leading phase of a eutectic are the place of origin of eutectic grains.

REFERENCES

1 K. D. LAKELAND: *J. Aust. Inst. Met.*, 1965, **10**, 55

2 A. OHNO *et al.*: *Trans. Iron Steel Inst. Jpn*, 1971, **11**, 18
3 A. OHNO AND T. MOTEGI: *AFS Cast Met. Research J.*, Jun. 1975, 45
4 V. DE L. DAVIES: *J. Inst. Met.*, 1964–65, **93**, 10
5 V. S. JUSTI *et al.*: *Giesserei*, 1972, **24**, 37

Grain refinement of electroslag remelted ingots

J. Campbell and P. D. Caton

The results of a theoretical and experimental appraisal are set out for a simple grain refining process, using the addition to the melt of particles (not powder) of the same composition as the melt. Experiments on 75 mm dia. ingots of electroslag remelted (ESR) specimens of an austenitic stainless steel, INCONEL ALLOY* 600, and NIMONIC ALLOY* 90 are described. Particles are added to the liquid metal pool via the slag layer by simply feeding them into the mould during the course of the melt. Particles act as (a) nuclei for grain refinement, reducing grain size by up to a factor of 100, and (b) internal chills, increasing the rate of solidification in the central regions of the ingot most inaccessible to conventional heat extraction. In this way the rate of solidification has been increased by a factor of about 10, dendrite arm spacing reduced by a factor of about 3, and production rates increased by up to about 2 times without any additional power requirement. Concurrent electromagnetic stirring at moderate rates aids the process by (a) decreasing power requirements by up to 20%; (b) providing an improved distribution of added particles; (c) producing an improved surface finish; and (d) seemingly aiding a grain multiplication effect whereby the number of grains in the ingot exceeds the number of particles added by a factor of 3×10^5. The mechanical properties of NIMONIC ALLOY 90 were substantially improved, and response to heat treatment was increased: homogenized material was softer, and aged material was harder than control material. Tensile strength was raised without loss of ductility.

The authors are with the Fulmer Research Institute Ltd, Slough

The electroslag remelting (ESR) process, shown schematically in Fig. 1, has proved valuable for the production of high-quality steels and nickel-base alloys by its action in reducing elements, such as sulphur, which are deleterious in many high-temperature and high-strength situations, and in its action in reducing the number of large inclusions: it often compares favourably with vacuum arc remelting since the ESR materials are usually more easily forged.

The advantages of ESR over conventional ingot casting are, however, gradually eroded as the ingot size increases. This is because the smaller surface/volume ratio of larger ingots results in increased solidification times for the centre of the ingot. Thus, grain size increases, dendrite arm spacing coarsens, and various segregation mechanisms have time to develop to levels more typical of conventional ingots. For instance, macrosegregation from end to end and from the centre to the outside of the ingot becomes noticeable, and channel segregates (the equivalent of 'A' and 'V' segregates in conventional ingots) appear.

The problem of grain size has been tackled by two methods: (a) grain refinement by the addition of foreign nucleants;[1] and (b) grain refinement by electromagnetic stirring.[2] When applied to steels, the first method usually results in contamination of the metal and a consequent reduction in mechanical properties.[3,4]

The second method is certainly capable of reducing the grain size of steels when properly applied. The metal, of course, is not contaminated, but a serious disadvantage is the enlargement of the dendrite arm spacing: the coarsening kinetics of the arm spacing is accelerated by the bulk motion of the melt, since the rate is no longer solely controlled by bulk diffusion.

Both these attacks on the grain-size problem have left the central problem untouched: that of the slow cooling of the centre of the ingot. Thus, micro- and macrosegregation are at best unaltered, and possibly increased.

The objective of this work was to test an idea for the simultaneous solution to all these problems by the use of a simple process. Particles of identical composition to

* Trademark.

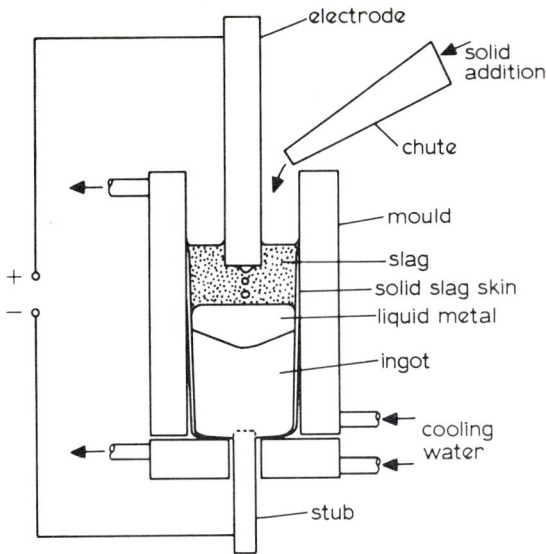

1 Schematic view of ESR process, showing stub instead of normal baseplate connection and vibratory feeder and chute

that of the electrode were added through the slag layer during the course of a melt. The particles were designed to be added at such a rate that they did not entirely melt. In this condition they served two important roles:

(i) they acted as nuclei for the initiation of new grains: as nuclei they had the advantage over the addition of foreign nuclei since (*a*) they did not contaminate the metal; and (*b*) they were guaranteed to be effective for any system, as opposed to the foreign nucleants which are usually highly specific to one alloy or class of alloys

(ii) they acted as minute internal chills: in this way the heat in the centre of the ingot could be extracted quickly, only having to diffuse to the nearest solid particle as it was warmed up, instead of a distance many orders of magnitude further to the mould wall; thus, the time-dependent phenomena, such as dendrite arm spacing and the segregation processes, would all be reduced to values characteristic of a small ingot.

HEAT- AND MASS-BALANCE RELATIONS
To obtain an approximate idea of the relative quantities of solid which may be added to the liquid to produce solid nuclei there are two regimes to consider which represent limiting conditions.

Slow heat transfer
When the rate of transfer of heat to the added solid occurs at a rate which is slow compared to the rate at which solid is added, then the maximum rate of addition is limited by the density of packing of the solid particles. For equal spheres the density of random close packing is close to 64%. Other geometries of solids and different degrees of closeness of packing can give figures which vary widely from this value. This regime is probably not applicable for metals, but might be true for some ceramics.

Rapid heat transfer
When the thermal conductivities are sufficiently high for the redistribution of heat to require negligible time, then the solution to the problem of the maximum levels of addition reduces to a simple heat balance. There are two cases to be considered.

1 When the solid is added in such a small quantity that the heat available from the liquid can cause a fraction f_1 of the added particles to melt, then the relative quantities of solid and liquid now are

$$f = \frac{m_1}{m_2} = \frac{\displaystyle\int_{T_2}^{T_m} Cp_2\, \mathrm{d}T}{\displaystyle\int_{T_1}^{T_m} Cp_1\, \mathrm{d}T + f_1 L_1}$$

where the subscripts 1 and 2 refer to the solid and liquid, respectively, m is the mass, Cp is the specific heat, L is the latent heat of fusion, and T_m and T are the melting point and initial temperature of the appropriate phase, respectively. When $f_2 = 1$, this relationship defines the extreme lower limit of the solid addition, below which no nuclei will survive, thus nullifying one of the principal aims of this work.

When $f_1 = 0$, then the added particles do not melt at all (but no liquid freezes). For this pseudo-equilibrium situation we have the approximation

$$f = \frac{m_1}{m_2} = \frac{T_m - T_2}{T_m - T_1}$$

This approximation holds for the case of $Cp_1 \simeq Cp_2$, which is tolerably true when the solid and liquid phases are metals or alloys of the same compositions, and for the case when the specific heats do not change significantly with temperature (which is not a well founded assumption).

2 When the solid is added in amounts which are sufficient to freeze a fraction f_2 of the liquid, the weight fraction of solid to liquid is easily shown to be

$$f = \frac{m_1}{m_2} = \frac{\displaystyle\int_{T_2}^{T_m} Cp_2\, \mathrm{d}T + f_2 L_2}{\displaystyle\int_{T_2}^{T_m} Cp_1\, \mathrm{d}T}$$

This relationship also reduces considerably for conditions of interest in this work: taking average values of Cp_2 over the limited temperature range in the liquid, and noting that

$$Cp_2(T_m - T_2) \ll f_2 L_2,$$

we obtain

$$f = \frac{m_1}{m_2} = \frac{f_2 L_2}{Cp_1(T_m - T_1)}$$

When all the liquid freezes as a result of the solid addition, then putting $f_2 = 1$ we obtain values of m_2/m_1 of 28% for pure nickel and 21% for pure iron, (the corresponding values for Cu and Al are 45 and 50%, respectively) when adding solid at room temperature ($T_1 \simeq 20°C$). These relations are shown graphically for Ni and Fe in Figs 2 and 3, respectively.

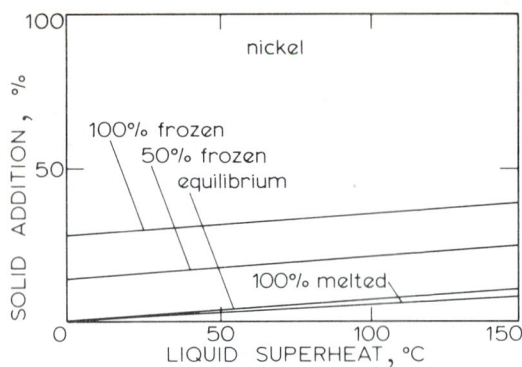

2 Graphical view of heat balance for cold additions of solid nickel to liquid nickel at various temperatures above its melting point

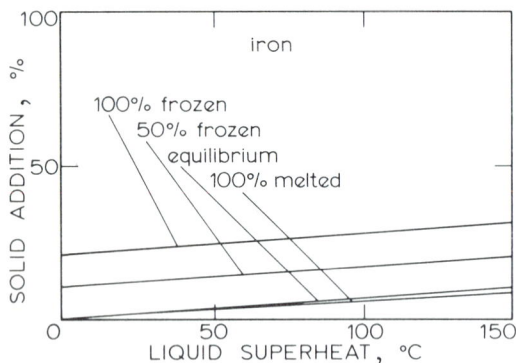

3 Graphical view of heat balance for cold additions of solid iron to liquid iron at various temperatures above its melting point

Effect of slag layer

When solid metallic particles are introduced into the top of the mould, with the aim of reaching the liquid metal pool, they are heated during their passage through the slag, which, under normal operating conditions, is at temperatures in excess of 1 700°C. It is possible that a solid slag envelope may freeze around the particles as they sink through the layer. If so, then they are certain to float on the slag/metal interface since the solid slag is poorly wetted by the metal (this is clearly revealed by the rounded form of the crowns of ESR ingots). The presence of much floating debris at the interface has been observed in transparent model studies by one of the authors.[5] Thus, the earliest moment at which the particle could enter the metal pool will be that instant when the particle has shed its solid shell of slag. The particle would then be at the melting point of the slag.

It is possible, however, that the particle may continue to lie on the metal/slag interface even after the slag shell has been melted. This is because boundary layers, probably of oxides, will prevent assimilation of the solid. Again, this can be observed in transparent analog studies[5] where even liquid metal droplets are sometimes observed to rebound from the metal/slag interface, roll around for several seconds, and then suddenly be assimilated after the boundary layers have had time to diffuse away. If assimilation is delayed in this way then the particle may have already started to melt before it enters the metal pool.

In all probability, therefore, the particles are heated to a temperature somewhere between the melting points of

the slag and metal phases before entering the metal pool. The effect of this large degree of preheating has been calculated, and is shown in Fig. 4. Average temperatures of the metal pool are in the region of 25°C above the liquidus, but the effects of higher and lower superheats are indicated by the length of the bar lines in the figure. Clearly, the maximum allowable additions are very much increased by the preheating effect of the slag.

Mass balance

Consider the mass transfer in unit time. A mass w of solid particles has been added, during which time a mass $\pi d^2(v + V)\rho/4$ of the electrode has been melted and been added to the ingot, whose additional mass is $\pi D^2 V\rho/4$, where v and V are the velocities of the tip of the electrode and top of the ingot, (or, more exactly, the metal pool), respectively, ρ is the density of the solid, and d and D are the diameters of the electrode and ingot, respectively (taking the slightly reduced diameter characteristic of steady-state melting, since the slag temperatures are lower, and the slag skin on the ingot is, consequently, slightly increased). The relationships are easily adjusted for electrodes and ingots of non-circular cross-section. Therefore, a balance is obtained under steady-state conditions:

$$\pi d^2(v + V)\rho/4 + w = \pi D^2 v\rho/4$$

but, in addition, there is the fractional addition to the metal pool

$$f = \frac{4w}{\pi D^2 V\rho}$$

These two equations lead to a relationship for f in terms of operationally measurable parameters.

$$f = 4w[1 - (d/D)^2]/(\pi\rho d^2 v + 4w)$$

This has been the equation used to work out the values of f which have been found to be experimentally desirable in this work. Typical values of w were 0, 50, 100, 150, and 200 g/min. Values of electrode feed v were in the range 15–25 mm/min.

EXPERIMENTAL METHOD

The electroslag melting furnace on which the work was carried out had available a maximum power of 40 kW, dc, with the electrode usually positive, and a maximum of 2 000 A; it could take electrodes up to 1·5 m long. The mould used in this work was a double-wall, water-cooled type, of 75 mm i.d. and height 300 mm, constructed of copper.

A molten slag start was used. The base electrical connection was made not via the baseplate but via a stub which projected through a central hole in the baseplate, and which became welded to the ingot (Fig. 1). The stub was made from the same material as the ingot, and could be subsequently cut off and re-used.

A standard weight of 500 g of slag was used of composition (wt-%) 80CaF$_2$, 10CaO, and 10Al$_2$O$_3$. The slag cap on the ingot after solidification was typically 30–40 mm thick.

An electromagnetic stirring coil was wound on the outside of the mould. It consisted of 80 turns of average diameter 130 mm over a length of 250 mm. The magnetic field intensity B at its centre may be obtained approximately from the relationship for a long solenoid:

$$B = \mu_O NI/l$$

where B is in units of Tesla (or N/A/m or Wb/m^2), $\mu_O = 4\pi \times 10^{-7}$ N/A^2 (or J/c^2 or H/m), N/l = number of turns per metre, and I = current through the coil in amps. For the conditions in this investigation, B is calculated to be $4 \cdot 0 \times 10^{-4}$ T for each amp through the coil. Thus, for $I = 10$ A, $B = 4 \cdot 0$ mT, equivalent to 3 200 A turns per metre. The preliminary experiments revealed that even this modest field was found to be in excess of normal requirements by a factor of 2·5 to 10.

When considering the field intensity B in units of N/A/m, it is clear that the product of B and current density in the liquid metal/liquid slag system will give the stirring force per unit volume in the melt:

$$\text{force per unit volume} \propto BI_2/D^2$$

$$\propto NI_1 I_2/lD^2$$

where I_1 is the current in the coil and I_2 is the current in the melt. The formula is simplified for dc melting because of the absence of any skin effect concentrating the field and the current in the outer layers of the melt.

A vibratory feeder was arranged near the furnace (Fig. 1) so that metal particles could be delivered at accurately metered rates down a chute and into the top of the mould. The angle of delivery was about 30° to the horizontal, so that particles tended to ricochet around the annular gap between the electrode and the mould before falling into the slag. Thus, some degree of dispersion of the particles around the circumference of the melt was achieved in this way.

Additions could only be made after the upper surface of the slag had remelted. This was usually 8–10 min after the start of the melt, compared to a total melt time of 22–30 min. The corresponding fraction of the length of the ingot from the base was only 0·15–0·20 of the length of the ingot, as a result of the slow beginning of the melt during the warm-up period, and also of the extra material introduced from the additions after this period.

The nickel-base alloys chosen for this work were INCONEL ALLOY 600 of nominal composition (wt-%): 15·5Cr–8Fe–1·0Mn–0·50Si–0·50Cu–0·15C, 0·015S, and NIMONIC ALLOY 90 of nominal composition (wt-%): 20Cr–18Co–5Fe–2·5Ti–1·5Al–0·13C.

RESULTS
Preliminary studies
Preliminary melts were carried out with En 58B stainless steels of nominal composition (wt-%): 18Cr–8Ni, stabilized with titanium. The purpose of these melts was simply to obtain experience with the production of approximate values for a steady melting rate, particle addition rate, and electromagnetic stirring rate in a relatively inexpensive and easily obtainable material, while efforts were proceeding to obtain nickel-base alloys in bar and wire form (the wire to be cut to diameter lengths so as to form compact particles of about 3 mm dia.).

These preliminary exercises produced valuable experience before the main programme of nickel–base work was begun. The main findings are listed below.

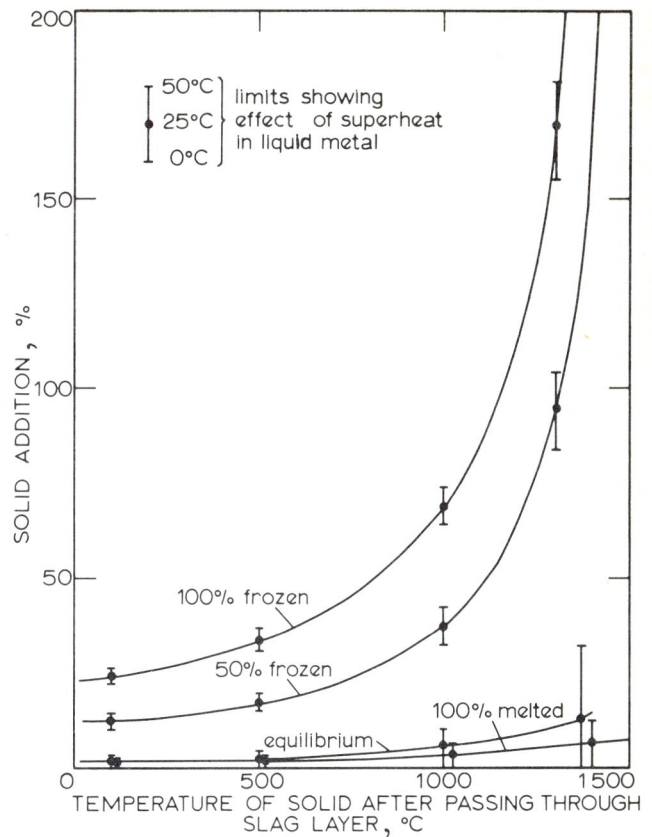

4 **Graphical view of heat balance for addition of solid iron, preheated to various temperatures into liquid iron at 25°C above its melting point: bar lines indicate the effect of small changes never far above its melting point**

1   Several methods of starting the melt were tried. The only consistently successful method was obtained using a liquid slag start.

2   The melt rate when using a mould of 75 mm dia. and an electrode of 32 mm dia. could be varied from a minimum of approximately 120 g/min (electrode speed 15 mm/min, power 17 kVA) to approximately 270 g/min (electrode speed 34 mm/min, power 32·4 kVA).

3   The rate of addition of particles required accurate metering by a vibratory feed device. Values of up to 53 vol.-% addition were found to be satisfactory from the point of view of unimpaired surface finish so long as the power dissipation was adequate (25 kVA). At these rates of addition but lower power dissipation (15–19 kVA) the slag layer was partially frozen, causing the additions to pile up in the annular gap between the electrode and mould wall. This led to arcing and damage to the mould. At intermediate conditions, where the slag layer was severely cooled but not frozen, the slag skin thickened from its usual value of 0·5 mm to about 4 mm. The ingot was therefore necked in during the continuation of these conditions, and the surface finish was poor, containing partially welded-in particles. Also, bright unmelted particles of chopped wire were found embedded in the thickened slag layer. This melting condition also led, of course, to an unacceptable loss of slag, which, in turn, affected the subsequent stability of melting conditions.

Table 1 Summary of melt conditions

Ingot	Rate of addition, g/min	%	Stirring A/turn/m	s^{-1}	Volts	Amps	Electrode speed, mm/min
INCONEL ALLOY 600							
ES 9	0	0	0	0	26	1 000	11
	0	0	0	0	26	1 250	19
	0	0	1 600	1–2	26	1 300	24
	0	0	3 200	< ~3	unstable	—	—
ES 10*	0	0	1 280	1–2	26	1 250	22·0
	50	15	1 280	1–2	26	1 250	18·5
	100	27	1 280	1–2	26	1 250	16·5
	150	36	1 280	1–2	26	1 250	14·7
ES 11*	0	0	1 440	2	26	1 200	25
ES 12	100	24·3	1 280	2	27·5	1 200	18·3
ES 13	200	43	1 280	2	25·0	1 250	12·0
ES 14	0–200	0–43	640–1 600	0–2	25	1 250	18·3
ES 15	200	42	500	1	25	1 400	14
ES 16	0	0	1 280	2	26	1 250	22
NIMONIC ALLOY 90							
ES 18	100	29	1 600	2	20	880	28
ES 19	0	0	1 600	2	21	790	28
ES 20	200	48	1 600	2	20	1 000	22

* 2g FeS added at 2 min intervals

4 Microexamination of early ingots revealed that it seemed impossible to find any trace of the remnants of added particles which had formed the nuclei of new grains: the exact match of chemical composition and the apparent perfections of the interface between the nucleus and the matrix caused the nuclei to be invisible.

The problem was overcome on later ingots by adding sulphur to the matrix in the form of 2 g additions of FeS at 2 min intervals throughout the melt. A typical microstructure is shown in Fig. 5. The added nuclei are revealed as sulphur-free regions up to 0·5 mm dia. This compares with their original size of approximately 3 mm dia., showing that most of the particle had melted away even under conditions of high (52%) addition, although at moderately high power (25 kW). The sulphur additions at regular intervals also helped to define the posi-

tion of the solidification front at each stage of the melt. In this way, deliberately introduced changes in melt conditions at various stages throughout the melt could be linked accurately to the corresponding location in the ingot.

5 The effectiveness of the electromagnetic stirring coil was investigated. It was found that a current I_1 of 5 A was adequate for stirring rates up to 2 rev/s, when the melting current I_2 was about 1 200 A. When I_1 was reduced to 1·5 A, the stirring rate fell to 1 rev/s. Higher rates of stirring caused the slag to be centrifuged up the inside of the mould and caused loss of contact between the electrode and slag, leading to arcing and instability of melting. The moderate rates of stirring in the range 1–2 rev/s were found to be beneficial from several points of view:

(i) in the absence of additions, the melt rate automatically increased by a significant amount, about 20%, without any increase in power requirement: thus the melting efficiency of the process was improved

(ii) the additions were only made to one side of the mould, causing some asymmetry of the ingot structure in the absence of stirring: stirring at moderate rates improved the distribution of additions and the consequent symmetry of the ingot structure

(iii) moderate rates of stirring were found to improve surface finish by preventing the incorporation of particles in the slag skin: in the absence of stirring, such particles were often partly welded to the ingot, leading to a most unsatisfactory rough surface, particularly on the side at which the additions were made.

6 The apparent disadvantage arising from the impossibility of adding grain refining additions during the early stages before the slag was fully molten was found to be less important than was feared. This non-grain-refined length was only 15–20% of the length of

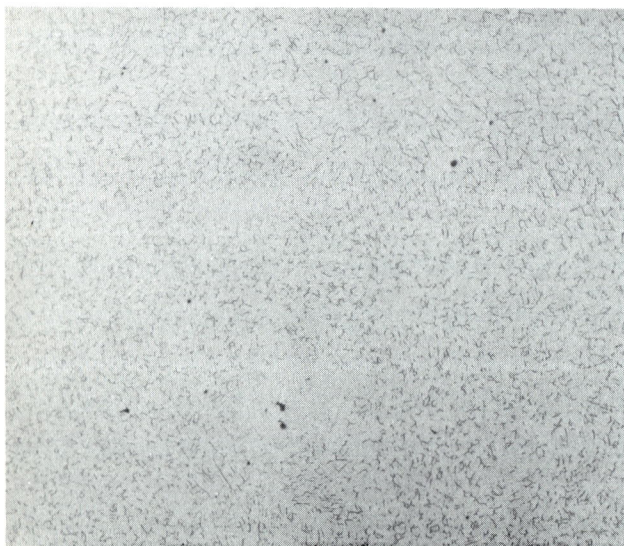

5 **Typical microstructure of stainless steel to which approximately 0·1 wt-% sulphur had been added, revealing the sites of unmelted nuclei as sulphur-free regions   [×20]**

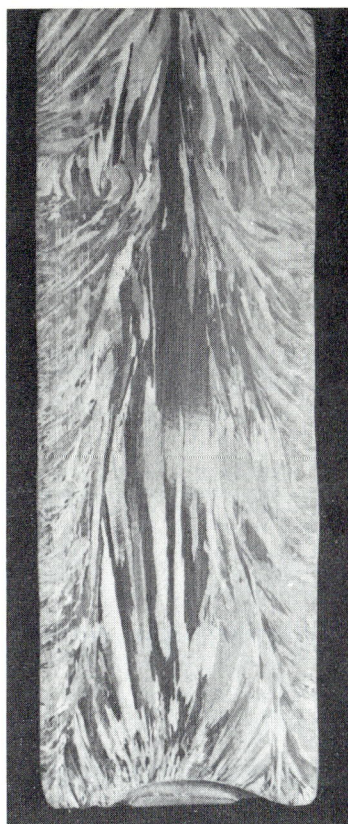

6 Macrostructure of INCONEL ALLOY 600 ingot subjected to alternate conditions of zero stirring, and very rapid stirring, showing little effect on crystal structure: crystals growing inwards near the top are a result of loss of electrical contact during a burst of extra rapid stirring [×0·5 etched in Marbles reagent]

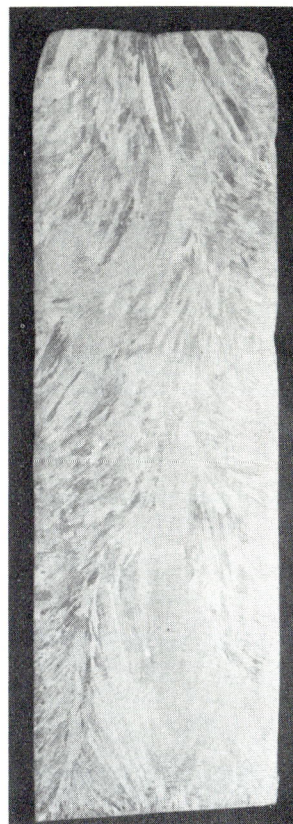

7 Macrostructure of INCONEL ALLOY 600 ingot subjected to slow stirring and addition rates of 200 g/min at intervals throughout melt [×0·5], etched in Marbles reagent]

the ingot (corresponding to a height of about half the ingot dia.), and in any case consisted of reasonably fine grains and fine dendrite arm spacing as a result of its proximity to the water-cooled baseplate.

7 The rate of ingot production was capable of increase by a factor commonly around 1·3, but over short periods as high as 3·3 with no extra power required. It seems, therefore, that a doubling of the production rate might be attainable under carefully optimized conditions, with no increase in melting power costs. All these major findings were confirmed for nickel-base alloys in the subsequent programme.

INCONEL ALLOY 600 results
Figure 6 shows an initial attempt to explore the range of melt rates and stirring rates which were possible, all along the length of a single ingot of INCONEL ALLOY 600, allowing close and detailed comparisons to be made. Rapid stirring of the melt at widely spaced intervals has clearly not produced corresponding changes in grain size in the central portions of the ingot (the inward curving crystals $\frac{1}{3}$ of the way down from the top of the ingot are a consequence of loss of contact between the electrode and the slag due to extremely rapid stirring).

Figure 7 represents a direct comparison with Fig. 6 showing the effect of intermittent additions at 200 g/min. In those zones where additions had been made the structure consisted of fine, equiaxed grains.

Where the additions were discontinued, the structure reverted to fine, columnar grains. Comparative grain sizes deduced from line counts (perpendicular to the grain direction for columnar grains) are given in Table 2.

Table 2 Grain size determinations from line counts in central areas of ingots, away from edge effects at mould wall

Ingot	Number of intercepts	Distance, mm	Grain size, mm	mm
ES 9	13	35	2·7	—
	5	30	6·0	—
	5	25	5·0	—
	15	35	2·3	—
	—	—	average 4·0	—
ES 14 (Coarsest areas, where grains revert to columnar)				
	29	25	0·86	—
	24	20	0·83	—
	24	25	1·04	—
	22	20	0·91	—
	—	—	average 0·91	—
(Finest areas of equiaxed grains)				
	25	10		40 µm
	27	10		37
	19	10		53
	22	10		45
	—	—		average 44

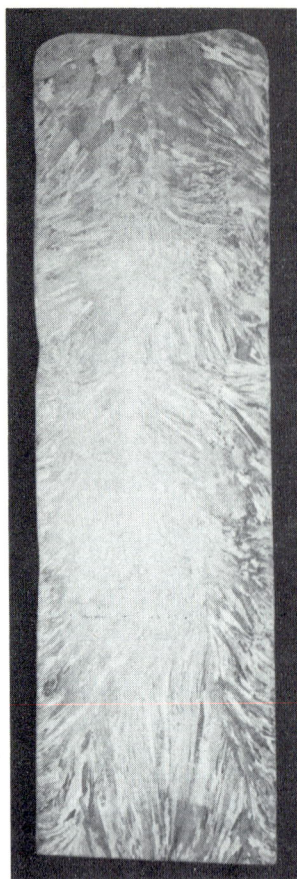

8   Macrostructure of INCONEL ALLOY 600 ingot subjected to rapid stirring, plus 2 g FeS additions at 2 min intervals   [×0·5, etched in Marbles reagent]

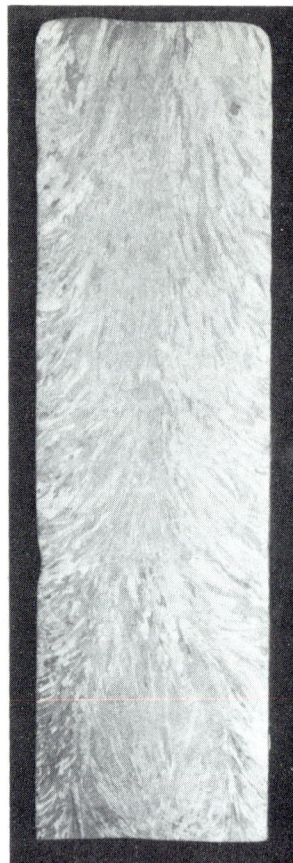

9   Macrostructure of INCONEL ALLOY 600 ingot, conditions as for Fig. 8, but with additions varying from 0, 50, 100, to 150 g/min during progress of melt   [×0·5, etched in Marbles reagent]

The two ingots were particularly featureless when polished and etched sections were subjected to micro-examination. Although grain refinement had occurred, no nuclei could be observed. Subsequent ingots were therefore made with additions of 2 g FeS at 2 min intervals throughout the melt. These are discussed below.

Figure 8 shows that the addition of FeS has increased the tendency of the alloy to form finer grains when compared with Fig. 6. This is to be expected, to some extent, since the distribution coefficient k for sulphur is so low (i.e. is much less soluble in the solid than in the liquid) that the dendrite morphology will be altered to a more fragile form,[6] thus facilitating dendrite fragmentation.

Figure 9, nevertheless, illustrates the considerable grain refinement effect which was achieved by the addition of nuclei. It is also clear that the grain-size distribution is also narrower (i.e. the grain size is more uniform).

The microstructures shown in Figs 10 and 11 were taken from equivalent locations in these two ingots, from a point half way from the base (actually close to 0·4 fraction of the height in the macrographs, since part of the base was machined off to facilitate holding the ingot during sectioning and surfacing). Figure 10 shows a region consisting of three subgrains, whereas the evidently polycrystalline region of Fig. 11 does not reveal anything that can be positively identified as a nucleus. Some of the lighter etching grains might

indicate the remnants of an original nucleus. The difficulty of detecting nuclei emphasises the perfection with which the particles are incorporated into the ingot.

Dendrite arm spacings taken from these micrographs are listed in Table 3.

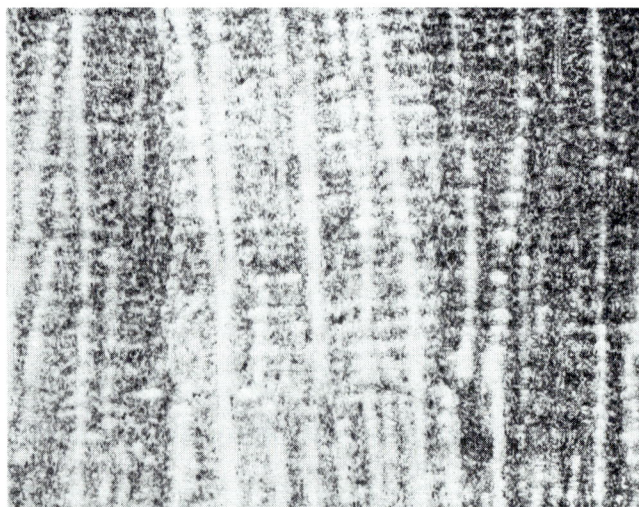

10   Micrograph of longitudinal section taken from centre of the INCONEL ALLOY 600 ingot shown in Fig. 8: the whole area consists of 3 subgrains   [×15, etched in ferric chloride in ethanol]

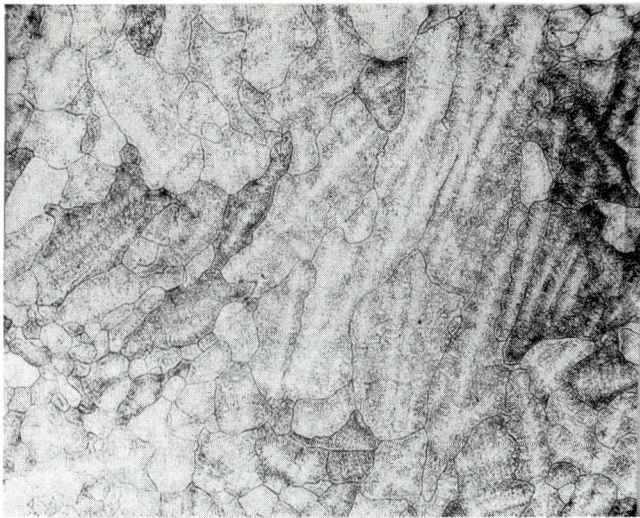

11 Micrographs of longitudinal section taken from the centre of the INCONEL ALLOY 600 ingot shown in Fig. 9: some of the lighter etching regions (relatively sulphur-free) may indicate remnants of added nuclei [×15, etched in ferric chloride in ethanol]

Table 3 Average dendrite arm spacings, with standard mean deviations

Ingot	Primary spacing, μm	Secondary spacing, μm
ES 11	225 ± 15	100 ± 10
ES 10	195 ± 15	40 ± 4
Ratio 11/10	1·15	2·5

12 Macrostructure of NIMONIC ALLOY 90 ingots with solid additions of 0, 100, and 200 g/min (from left to right) [×⅓, etched in HCl plus hydrogen peroxide]

NIMONIC ALLOY 90 results

Macrostructures of ingots with various levels of solid additions, zero, 100, and 200 g/min are shown in Fig. 12. The fine chevron central grains surrounded by long columnar grains are typical of NIMONIC ALLOY 90 ESR ingots.

The unetched microstructure revealed no evidence of the added particles. The etched microstructures in Fig. 13 do show significant differences, although no nuclei could be identified with any certainty.

The apparently improved homogeneity indicated by the microstructure of the grain-refined material is reinforced by the hardness response to heat treatment. Table 4 shows how the material homogenized at the standard 1 150°C for 12 h is softer than the control. In contrast, the material subjected to the standard aging treatment of 700°C for 16 h was harder than the control.

The stress-rupture tests carried out at 870°C and 140 MPa were satisfactory.

The room-temperature tensile results were high on ductility and rather low on strength. This was almost certainly the result of the standard aging treatment now being unsuitable for this rather homogeneous material – the material had been effectively overaged. Prior knowledge of the shape of the aging curves for this material suggests that an optimized homogenization and aging treatment will yield a superior strength without sacrificing existing levels of ductility.

DISCUSSION

All the initial objectives of this programme with regard to improved ingot structure appear to have been met. These include grain and dendrite arm refinement, with

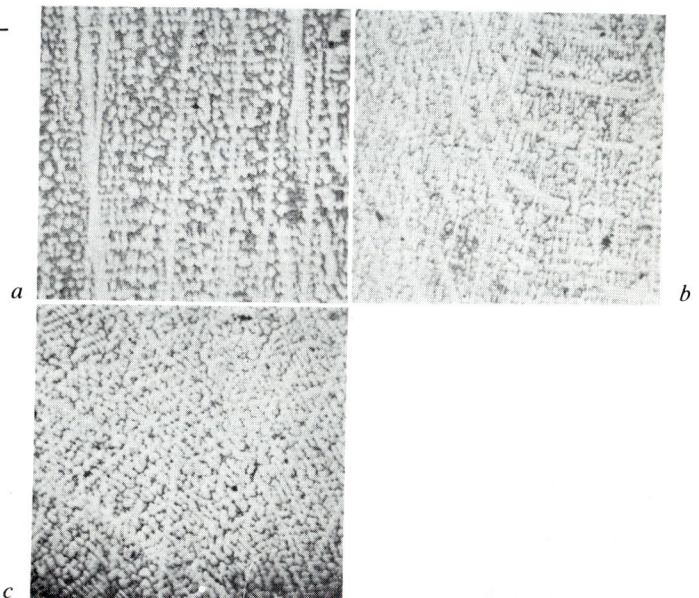

a zero addition; *b* medium rate of solid addition; *c* high rate of solid addition

13 Microstructures of longitudinal sections taken from the mid-radius position of the ingots shown in Fig. 12, showing increased randomness and seemingly improved homogeneity of grain-refined material [×12, etched in Marbles reagent]

Table 4 Mechanical properties of NIMONIC ALLOY 90 ESR ingots

Property	Condition*	Typical conventional results on wrought ESR material	Experimental results Solid addition rate, g/min		
			0	100	200
Brinell hardness	AC		309	306	297
	H	230–300	226	219	205
	A	340–400	240	265	271
0·2% proof stress, MPa	AC		542	545	—
	H		520	538	—
	A	770	599	620	—
Tensile strength, MPa	AC		867	866	—
	H		877	904	—
	A	1 200	966	940	—
Elongation, %	AC		31	29	—
	H		32	35	—
	A	25	33	35	—
Red area, %	AC		23	28	—
	H		30	48	—
	A	35	45	40	—
Stress rupture, hours to failure	AC			68	—
	H			89	—
	A	150		179	—

* AC = as cast; H = homogenized 1 150°C for 12 h; A = aged 700°C for 16 h. Hardness results are averages of 10 readings, standard mean deviation ± 2HB.

no detrimental features such as slag or porosity entrapment or poor surface finish. An additional, unexpected benefit was the increase in effective melt rate for no increase in power requirement.

The absence of any increase in power requirement is probably the consequence of (*a*) the electromagnetic stirring, which seems to increase heat transfer to the electrode, and so promotes more rapid melting; and (*b*) the reduction in temperature of the slag layer, by internally heating particles in a useful manner, so that less heat is wasted by loss to the water-cooled mould.

The ingots have been too small to determine whether macrosegregation effects can be reduced. However, the indications are promising. The increase in freezing rate can be deduced from the reduction in secondary dendrite arm spacing. The secondary arm spacing x increases with solidification time t according to the approximate coarsening law

$$x \propto t^{0·35}$$

Thus, since x has been observed to be reduced by a factor of 2·5, then t will be reduced correspondingly by more than an order of magnitude. This should significantly reduce segregation processes which take time to become established and time to build up to appreciable levels.

Grain counts on the stainless steel ingots have revealed that the number of grains in the grain refined ingot exceeds the number of added nuclei by a factor of approximately 3×10^5. It is quite evident from the macrostructures that stirring of the liquid alone does not cause dendrite damage in such ingots, where the solidification range is relatively small and the temperature gradient high. It can be speculated that grain multiplication therefore arises as a result of the mechanical damage inflicted on the dendritic front by the arrival of a circulating and tumbling mass of relatively heavy nuclei. The disparity between the numbers of added nuclei and the final grains also helps to explain why the nuclei are so hard to find: they are the 'needles in the haystack' which they have created.

CONCLUSIONS

1   Grain refinement of ESR ingots has been achieved for stainless steels and the nickel-base alloys INCONEL ALLOY 600 and NIMONIC ALLOY 90. The grain size in the centre of ingots was reduced by a factor of up to nearly 100.

2   The secondary dendrite arm spacing in the centre of ingots was reduced by a factor of up to 2·5, indicating an increase in freezing rate of approximately an order of magnitude.

3   The rate of ingot production (measured for instance, in g/s) was increased by 20–40% with no increase in power requirements. There are indications that much higher increases might be attainable.

4   Rotation of the melt at rates of approximately 1 rev/s was found to be necessary to obtain symmetrical distribution and good surface finish.

5   Incorporation of the added particles into the ingot appeared to be so perfect that traces of nucleating particles could only be found by deliberate contamination of the ingot with sulphur, and not always then. Furthermore, detrimental features such as porosity, entrapped slag, and poor surface finish were found to be easily avoided when using the correct conditions of addition rate and stirring rate.

6   The response to heat treatment and the mechanical properties of NIMONIC ALLOY 90 were significantly improved.

ACKNOWLEDGMENTS

Part of this work was supported by Arbed, Luxembourg, under the 'COST 50' scheme.

REFERENCES

1 J. CAMPBELL AND J. W. BANNISTER: *Met. Technol.*, 1975, **2**, (9), 409

2 J. CAMPBELL AND D. I. DAWSON: 'Magneto-hydrodynamics in casting processes', BISRA Open Report MG/58/71, British Iron and Steel Research Association

3 N. CHURCH *et al.*: *Trans. Amer. Foundryman's Soc.*, 1966, **74**, 113; *J. Met.*, 1967, **19**, (Jun.), 44

4 P. STULER: *Schweissen Schneiden*, 1967, **19**, (6), 270

5 J. CAMPBELL: *J. Met.*, 1970, **22**, (7), 23

6 K. A. JACKSON *et al.*: *Trans. AIME*, 1966, **236**, 149

Structure of continuously cast eutectics used in the manufacture of fine-grained aluminium sheet

L. R. Morris

This paper describes the relationship between the cast structure and final properties of a recently developed type of aluminium alloy that is strengthened by a grain size of about 1 μm. The grain boundaries are stabilized by a uniform dispersion of second phase particles, about 0·2–0·5 μm dia., formed by breaking up a rod-like eutectic. In the Al–Fe–Mn, Al–Fe–Si, and Al–Si eutectic systems the phases, in direct chill (DC) cast ingot, are coarse and faceted. The necessary fine rod-like structure is produced by casting 0·7 cm thick strip between water-cooled steel rolls. Growth is almost unidirectional and the solidification velocity of 1–1·5 cm/s produces branched rods of about 0·2 μm dia. The final properties of the sheet depend upon the diameter of the rods and the uniformity of the cast microstructure because these defects affect the particles which control the grain size and thus the strength. Coils of sheet 140 cm wide and 5000 kg in weight have been produced.

The author is with Alcan Research Centre, Kingston, Ontario, Canada

It is well known that the grain size in commercial aluminium alloys can be controlled by second-phase particles; however, the minimum grain sizes achieved are in the order of 25 μm and there is no appreciable grain-boundary strengthening. In aluminium, and other fcc metals, there are a large number of slip systems and little dislocation locking at normal purity; therefore, grain-boundary strengthening does not become important until the grain size is reduced to about 1 μm.[1]

This paper describes the cast structure, processing, and tensile properties of a recently developed series of aluminium alloys in which a grain size of about 1 μm is stabilized by a uniform dispersion of second-phase particles generated by breaking up a rod-like eutectic structure. The eutectics that are of use commercially have cheap ingredients, a relatively small volume fraction (less than 10%) of brittle rod-like phases, and reasonable corrosion resistance. A large volume of second-phase particles, large particles, or a non-uniform dispersion reduces the ductility.

The properties of sheet rolled from large direct chill (DC) cast eutectic ingots have been discussed previously.[2,3] The processing of this DC ingot will be reviewed first and then more recent work on eutectic, or near eutectic, alloys cast at high freezing rates on a commercial twin-roll strip caster will be presented.

PROCESSING OF DC INGOT

In aluminium ingot 10–50 cm thick the DC casting process imposes a solidification velocity of 0·05–0·2 cm/s, normal to the macroscopic freezing front and, usually, a low temperature gradient (essentially zero in the central portions of large ingot). In general, regular rod-like eutectics can be cast by the DC process to give a cellular eutectic structure with average rod diameters of 0·1–0·2 μm, which can be broken up, by hot rolling, to produce particles of less than 0·5 μm average dia.

The main processing steps, illustrated for the Al-6 wt-%Ni eutectic in Fig. 1 are

  (i) rapid solidification to form a fine rod-like structure
  (ii) hot deformation to segment the rods and disperse the particles
  (iii) cold work to generate a dense dislocation cell structure
  (iv) final annealing to form fine grains or subgrains—the high misorientation boundaries thus formed are pinned by the particles; the grain diameter is approximately equal to the local particle spacing.

The final grain size depends primarily on (a) the rod diameter; (b) the amount of plastic deformation; and (c) the final annealing temperature. The influence of the final annealing temperature on the tensile properties of 1 mm thick sheet is illustrated in Fig. 2. The yield

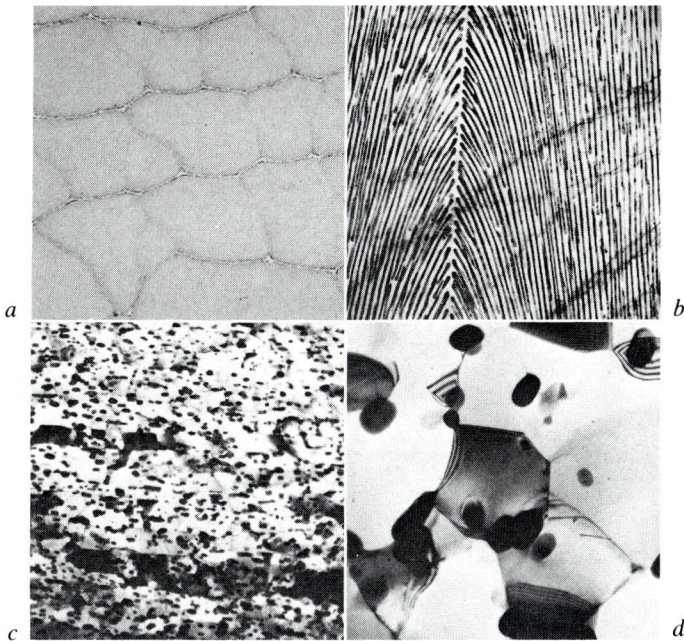

a DC cast eutectic cells [×250]; *b* rod structure in cells, TEM [×2700]; *c* hot worked at 500°C, TEM [×2200]; *d* cold worked plus anneal at 300°C, TEM [×7 800]

1 Structure of Al–6Ni alloy showing steps in processing leading to fine grain stabilized by NiAl₃ particles

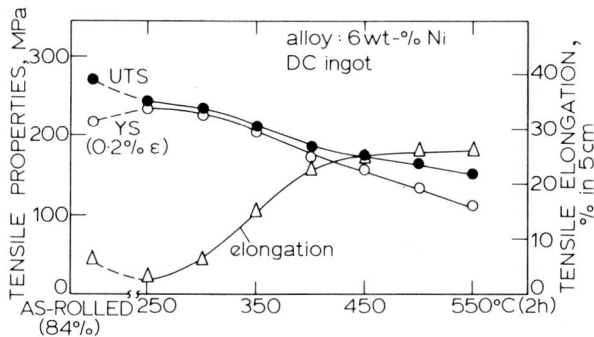

2 Room temperature tensile properties of Al–Ni eutectic alloy after various isochronel anneals; material was DC cast, hot worked, cold rolled (84% reduction) to 1 mm sheet

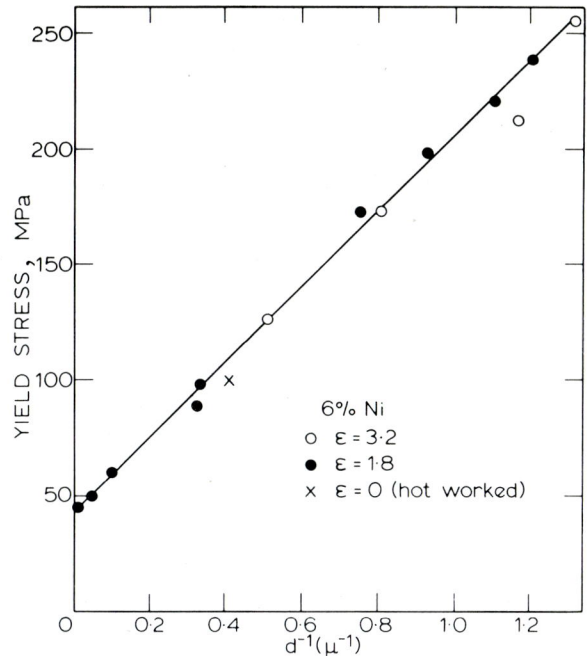

3 Yield strength *v.* grain size data for Al–6Ni; grain size defined as $d = \sqrt{A}$, where *A* is the average grain area; different grain sizes produced by varying amount of cold work and final anneal temperature

strength decreases (the grain size increases) gradually and continuously as the final annealing temperature is increased. There is no definable recrystallization step; the elongation increases rapidly at about 400°C, not because of a sudden increase in grain size, but because the sheet with a grain size of less than about 1 μm is plastically unstable in a manner similar to that observed in ultrafine-grained steels.[4]

The influence of grain size on the yield strength (0·2% offset strain) is shown in Fig. 3. The yield strength (YS) at a grain size of 20 μm is 50 MPa while the YS at a grain size of 0·8 μm is 250 MPa. The optimum combination of strength and ductility for the stretch forming of sheet, for instance, would occur after a 400°C anneal which produces a grain size of about 1·2 μm. The YS of this material is less than that of a fibre strengthened Al–NiAl₃ composite, as measured in the direction of the fibres.[5,6] The segmented particles contribute only about 20 MPa to the YS of the sheet. The advantage in using the particles to stabilize a fine grain size is that the ductility is much higher and the strength is isotropic in the plane of the sheet.

HUNTER CAST STRIP

The production of ultrafine-grain aluminium wrought products via large DC ingot is limited to regular eutectics or monovariant alloys which grow in a coupled manner with a rod-like structure. A number of other eutectics, such as those containing Fe, Mn, and Si are attractive, because they are cheap and compatible with present alloys, but they cannot be used in large DC ingot form because the faceted second phase is then too coarse or unevenly distributed.

The Hunter type twin-roll caster[7] produces coils of strip, about 0·7 cm thick × 150 cm wide × 5 000 kg, at a very high solidification rate. The Hunter caster is illustrated in Fig. 4. The basic operational principle of this type of casting machine is that both the solidus and liquidus fronts must be located between the feeder tip and the point where force is first exerted on the slab by the rolls; the process becomes unstable when pressure is exerted on the strip when it is semisolid. This distance is about 1 cm. The process is therefore limited to alloys with low freezing ranges and is generally used to cast alloys with low solute contents. It is evident, however, that alloys close to a eutectic composition would also, as a class, be suitable for this type of casting process.

The freezing front, for steady-state casting, is very flat; growth is almost unidirectional along the length of the slab, and the solidification rate is normally in the range 1–1·5 cm/s. A description of the phases found in Hunter slab and 12 cm thick DC ingot is given in Table 1.

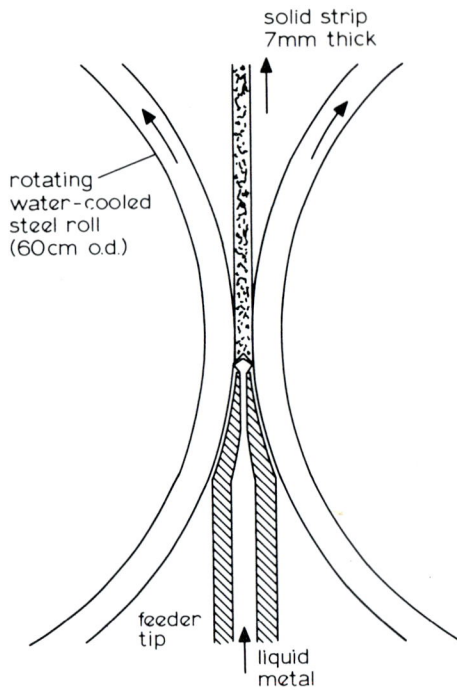

4 **Twin-roll (Hunter type) strip caster (schematic)**

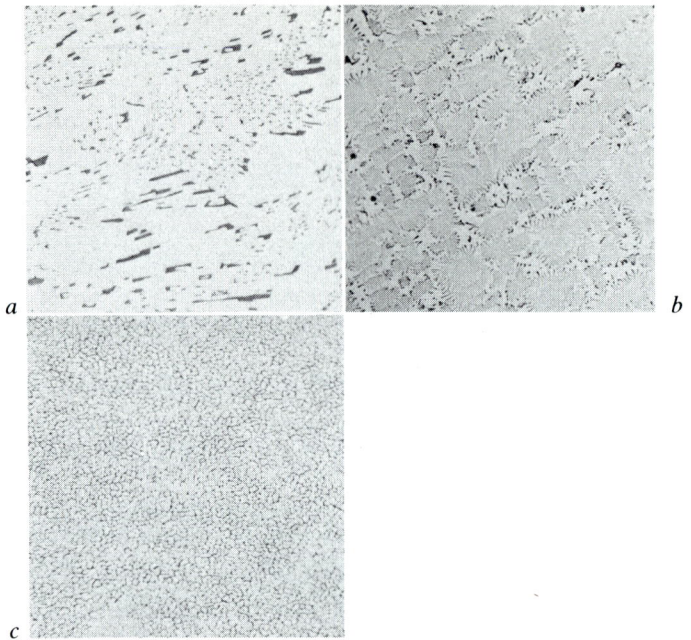

a DC ingot (45 cm thick), growth rate approximately 0·05 cm/s; *b* DC ingot (12 cm thick), growth rate approximately 0·12 cm/s plus higher temperature gradient in liquid; *c* Hunter cast strip (0·7 cm thick), growth rate approximately 1·2 cm/s (note aluminium dendrites with 2·5 μm cell dia.)

5 **Commercial purity Al–Fe–Mn eutectic cast by three techniques [×250]**

Structure of Hunter strip

A comparison of the as-cast structures of the Al–Fe–Mn ternary eutectic, cast by three different methods, is shown in Fig. 5. The $(Fe,Mn)Al_6$ phase (the other equilibrium phase, $FeAl_3$, does not form) solidifies as faceted plates in large DC ingot; as a cellular rod-like eutectic in small DC ingot; but as a fine interdendritic network in the Hunter cast strip. At higher magnification (Fig. 6) the intermetallic phase in the Hunter strip is seen to consist of a branched rod-like growth form that envelops the aluminium dendrite cells.

A high density of misorientation boundaries was found in the as-cast intermetallic phase. These geometric growth defects can be seen from the changes in diffraction contrast along the rods in Figs 6 and 7. The defect boundaries possibly originate from the autonucleation of one crystal orientation on another as these complex phases are forced to grow at high velocity

along the irregular channels left by the freezing dendrite cells.

Thermal stability

The irregular intermetallic phase network is thermally unstable and segments after annealing at 350°C for 2 h. The rod-like eutectics that have been grown unidirectionally and are fairly clean of substructure are much more stable.[8,9] The instability is initiated by the growth of a groove at the intersection of the growth defect and the interface with the aluminium matrix. The mechanism is similar to that observed in the lamellar Al–CuAl$_2$ eutectic after plastic deformation.[10] In Fig. 6

Table 1 Phases in as-cast Al alloy ingot

Alloying element, wt-%	Casting technique	Growth rate, cm/s	Phase	Volume, %	Approximate dia. or thickness, μm
6Ni	DC	0·12	NiAl$_3$	10	0·1
1·7Ni–1·7Fe	DC	0·09	(Fe, Ni)$_2$Al$_9$	8	0·18
1·7Ni–1·7Fe	Hunter	1·1	(Fe,Ni)$_2$Al$_9$	8	0·06
9·5Si	DC	0·12	Si	9	1–5
9·5Si	Hunter	1·0	Si	9	0·3
1·7Fe	DC	0·1	FeAl$_6$+FeAl$_3$	~4	1–2
1·7Fe	Hunter	1·3	Tet Al–Fe*	—	0·2
1·8Fe–1·1Si	DC	0·1	α–AlFeSi	~4	1–2
1·8Fe–1·1Si	Hunter	1·1	β–AlFeSi	—	0·15
1·8Fe–0·9Mn	DC	0·12	(Fe,Mn)Al$_6$	~7	0·3
1·8Fe–0·9Mn	Hunter	1·2	Tet Al–Fe*		0·2

* Tet Al–Fe: metastable Al–Fe phase which can incorporate a small amount of Mn, stoichiometry not known, bc Tet crystal structure; α: nominally Fe$_3$SiAl$_{12}$ (cubic); β: nominally FeSiAl$_5$ (Tet); (Fe, Mn)Al$_6$: isomorphous with MnAl$_6$(Ortho).

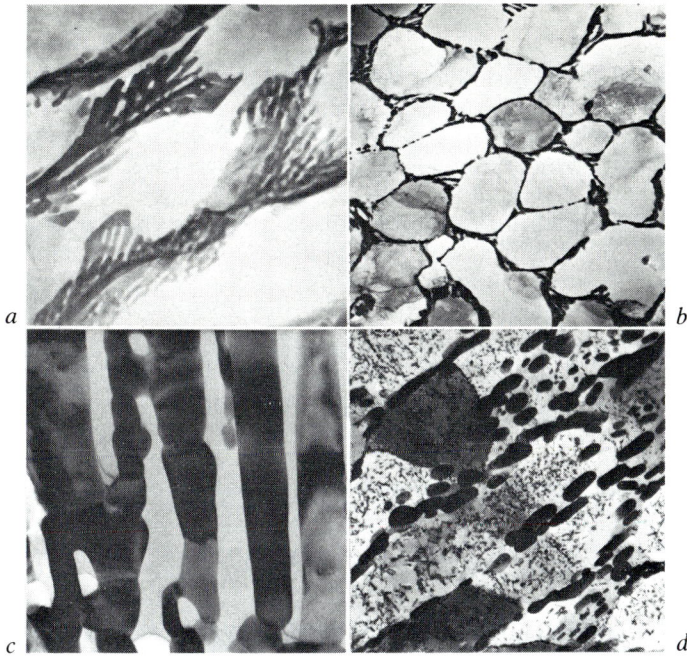

a as-cast (Fe, Mn)Al$_6$ structure, longitudinal, [× 4 600]; *b* as-cast structure, transverse [× 2 700]; *c* annealed at 300°C (2 h), [× 23 000]; *d* annealed at 500°C (2 h), [×2 700] the precipitate in Al dendrite cells is MnAl$_6$

6 Structure of Hunter cast Al–1·8Fe–0·9Mn eutectic, TEM micrographs

7 Hunter cast Al–1·7Fe alloy showing growth defects in intermetallic phase, phase is not FeAl$_3$ or FeAl$_6$ but a metastable phase with bc Tet crystal structure, TEM [× 46 000]

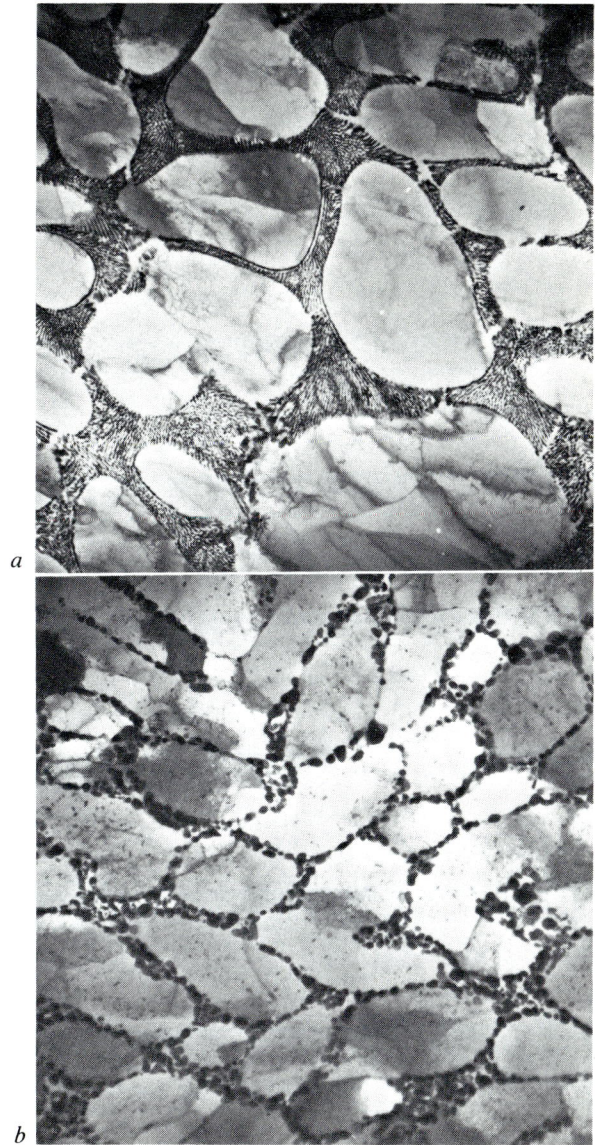

a as-cast structure [× 3 200]; *b* annealed at 500°C (2 h) [× 3 200]

8 Structure of Hunter cast Al–1·6Ni–1·6Fe– 0·5Mn monovariant alloy; intermetallic phase (Fe, Ni, Mn)$_2$Al$_9$; TEM micrographs

the triple point grooves are quite evident after a 300°C anneal and the structure has spheroidized after a 500°C anneal. The same transition can be seen in Fig. 8 for the (Fe,Ni)$_2$Al$_9$ phase in an Al–Fe–Ni–Mn alloy.

The production of equiaxed particles by hot rolling a rod-like eutectic in DC ingot also depends upon the generation of substructure in the intermetallic phase. When a fine NiAl$_3$ rod is deformed at 500°C, an array of almost equiaxed particles is formed (Fig. 1). When the same material is deformed at room temperature the NiAl$_3$ rods deform plastically and fracture to some extent but the average aspect ratio remains greater than 100 after 80% reduction. The influence on the substructure produced by plastic deformation can be seen in Fig. 9 where the material has been hot rolled at 350°C and the segmenting of the rods at crystal defects has not gone to completion.

Phase transformations

Because of the high growth rate, the intermetallic phase found in Hunter strip is often not the equilibrium phase. As is shown in Table 2, the metastable as-cast phase transforms during annealing to the stable phase, often with a second metastable transitional phase at intermediate temperatures. In the four alloys tabulated here the morphological breakdown to particles occurs at a

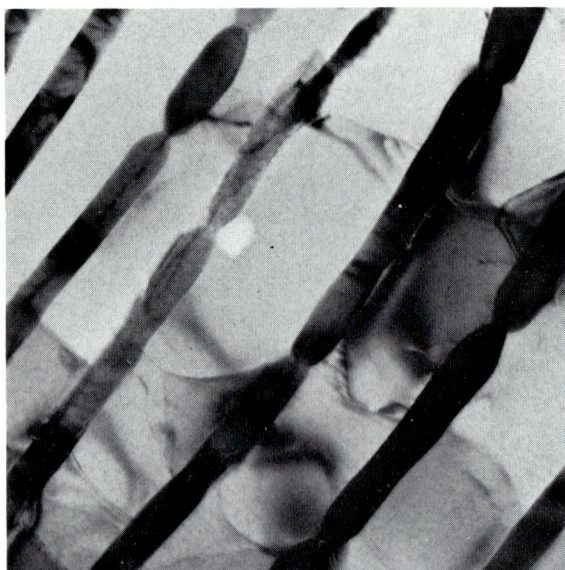

9 DC cast Al–6Ni eutectic hot rolled, 80% reduction, at 350°C; note formation of grooves at (strain induced) substructure in NiAl₃ rods

lower temperature than the first detectable phase change.

PROPERTIES

A comparison of the tensile properties of 1 mm sheet rolled from the three (Fe,Mn)Al₆ structures shown in Fig. 5 is given in Table 3. The coarse faceted structure gave a YS of 60–70 MPa, which is only slightly above the

base YS of 45 MPa measured in sheet with a 30 μm grain size. Both the coupled growth microstructure in the small DC ingot and the fine interdendritic network in the Hunter strip produced a YS in the range 110–175 MPa. Similarly, in Table 4, it can be seen that the YS of sheet rolled from Hunter strip is more than double that obtained from the same alloy cast as DC ingot.

The increased YS of sheet rolled from the finer eutectic structure is a result, principally, of the smaller grain size that is stabilized by the finer particle spacing. The particles themselves contribute only about 10–20 MPa to the YS.[3] The tensile properties of a Hunter cast Al–Fe–Mn eutectic is plotted in Fig. 11, and can be compared with the properties given in Fig. 2 for sheet rolled from the Al–NiAl₃ eutectic. The yield strength of the Al–Fe–Mn alloy tends to be lower for a given anneal but the combinations of strength and tensile elongation are comparable.

In the Hunter cast strip the eutectic structure is broken up into particles by annealing. Initially, this leaves a network of particles at the boundaries of the aluminium dendrite cells. The particle distribution becomes more uniform as the strip is rolled (Fig. 10). After 15% reduction the particles are segregated in bands and the grain size is approximately equal to the band spacing (dendrite cell size) rather than the average particle spacing. After 87% reduction the bands of particles are no longer distinguishable and the grains are smaller and more uniform in size.

SUMMARY

Ultrafine-grained aluminium sheet, with a good combination of yield strength and formability, can be

Table 2 Phase transformations* during annealing of Al alloys rapidly solidified by Hunter process

Annealing temperature °C (2 h)	System			
	1·7Fe–0·8Mn	1·7%Fe	1·6Fe–0·3Cr	1·8Fe–1·1Si
As-cast	Tet Al–Fe	Tet Al–Fe	Tet Al–Fe	β
350	Tet Al–Fe	Tet Al–Fe	Tet Al–Fe	β
400	Tet Al–Fe + α	Tet + tr.α	Tet + α	β
450	(Fe, Mn)Al₆	Tet + α	Tet + α	β
500	(Fe, Mn)Al₆	Tet + α	Tet + α	β
550	(Fe, Mn)Al₆	α	α + FeAl₃	β + α
600	(Fe, Mn)Al₆	FeAl₃	FeAl₃	α

* Tet Al–Fe: metastable Al–Fe phase which can incorporate a small amount of Mn, stoichiometry not known, bc Tet crystal structure; α: nominally Fe₃SiAl₁₂ (cubic); β: nominally FeSiAl₅ (Tet); (Fe, Mn)Al₆: isomorphous with MnAl₆(Ortho).

Table 3 Tensile properties of 1 mm Al–Fe–Mn eutectic sheet produced with structures shown in Fig. 5

Ingot	300°C (2 h)			400°C (2 h)		
	UTS, MPa	YS, MPa	Elongation, %	UTS, MPa	YS, MPa	Elongation, %
DC ingot (45 cm thick)	145	70	33	140	60	33
DC ingot (12 cm thick)	185	170	22	155	110	30
Hunter strip (0·7 cm thick)	180	175	22	165	130	31

Table 4 Tensile properties of 1 mm sheet produced from Hunter strip and 12 cm thick DC ingot

Alloy	Ingot	Annealing temperature					
		300°C (2 h)			350°C (2 h)		
		UTS, MPa	YS, MPa	Elongation, %	UTS, MPa	YS, MPa	Elongation, %
9·5%Si	DC Hunter	140	55	33	145	50	28
		200	145	16	185	125	20
1·7%Fe	DC Hunter	145	60	30	140	55	32
		160	130	22	150	110	21
1·8%Fe +1·1%Si	DC Hunter	130	55	30	—	—	—
		230	125	22	—	—	—

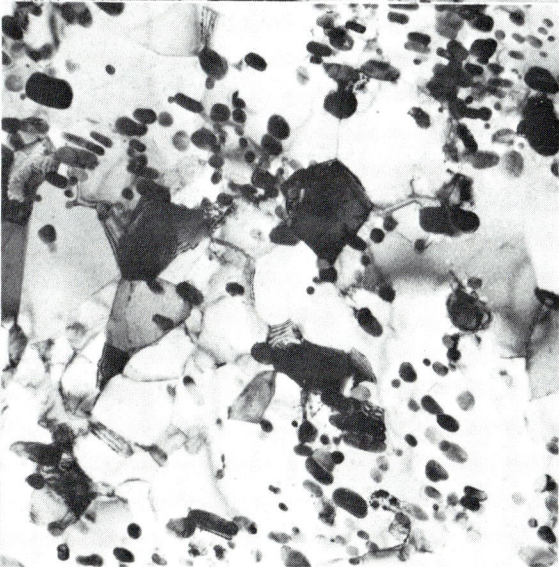

a 15% reduction; true strain 0·16; *b* 87% reduction; true strain 2·04

10 Effect of amount of reduction by cold rolling on particle distribution and grain size in Hunter cast Al–1·7Fe–1·7Ni alloy; TEM micrographs; phase is (Fe,Ni)$_2$Al$_9$; both samples annealed at 350°C (2 h)

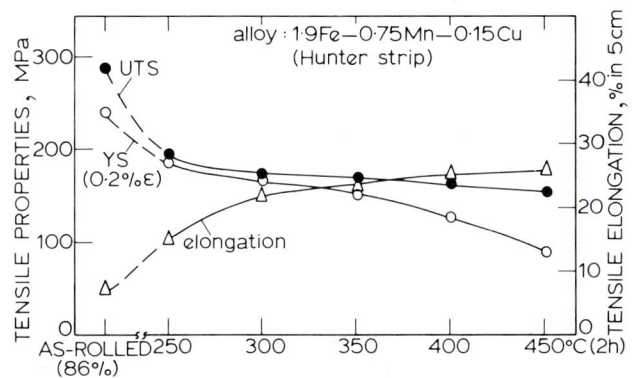

11 Room temperature tensile properties of Al–Fe–Mn ternary eutectic alloy after various isochronal anneals; alloy cast as 0·7 cm strip, annealed at 500°C, then cold rolled (86%) to 1 mm sheet

manufactured commercially by the two methods outlined below.

1 Regular rod-like eutectics that grow in a coupled fashion can be cast as large DC ingot, which at the imposed growth rates of about 0·1 cm/s produces rods of about 0·1 μm dia.. A uniform dispersion of particles, 0·2–0·5 μm dia., can be generated by hot rolling or hot extrusion; the break-up of the structure is promoted by a combination of plastic deformation and the formation of grooves at the strain-induced substructure in the rod-like phase.

2 Eutectics in which the second phase tends to be faceted must be cast at much higher growth rates, using a Hunter-type caster, to produce the required particle size. Since the coupled zone tends to be skewed away from the aluminium phase, the structure of these eutectic alloys, when grown at about 1 cm/s, usually consists of primary aluminium dendrites surrounded by a fine eutectic network. The particles are most conveniently produced, in the as-cast strip, by thermal treatments since the as-cast structure is very unstable because of the irregular morphology and high density of growth defects.

REFERENCES

1 R. J. MCELROY AND Z. C. SZKOPIAK: *Int. Metall. Rev.*, 1972, **17**, 175

2 L. R. MORRIS: 4th Proc. international conference on strength of metals and alloys', Vol. 2, 649, 1976, Nancy, France

3 L. R. MORRIS *et al.*: *ibid.*, 131

4 W. B. MORRISON AND R. L. MILLER: 'Ultrafine-grain metals', 183, 1970, New York, Syracuse University Press

5 F. D. LEMKEY *et al.*: *Trans. AIME*, 1965, **233**, 334

6 W. H. S. LAWSON AND H. W. KERR: *Metall. Trans.*, 1971, **2**, 2 853

7 E. F. EMLEY: *Int. Met. Rev.*, 1976, **21**, 75

8 G. C. WEATHERLY: 'Treatise on Materials Sciences and Technology, Vol. 8, 121, 1975, New York, Academic Press

9 I. R. HUGHES AND H. JONES: *J. Mater. Sci.*, 1977, **12**, 323

10 E. HO AND G. C. WEATHERLY: *Met. Sci.*, 1977, **11**, 109

DISCUSSION

Session 3: Nucleation and grain refinement

Chairman: R. D. Doherty (University of Sussex)

The following papers were discussed: *Keynote Address: Heterogeneous nucleation and grain refinement* by A. Hellawell; *Undercooling of low-melting metals and alloys* by J. H. Perepezko, D. H. Rasmussen, I. E. Anderson, and C. R. Loper; *Nucleation in peritectic systems* by P. G. Boswell, G. A. Chadwick, R. Elliot, and F. R. Sale; *Nucleation in steel castings by interstitial compounds* by J. V. Wood; *Influence of mould wall microgeometry on casting structure* by A. Morales, M. E. Glicksman, and H. Biloni; *Aluminium grain refinement by crystal multiplication mechanism stimulated by hexachloroethane additions to the mould coating* by N. L. Cupini and M. Prates de Campos Filho; *Influence of composition on intrinsic grain refining in binary copper alloys* by L. O. Gullman and L. Johansson; *Formation mechanism of eutectic grains* A. Ohno, T. Motegi, and K. Ishibashi; *Grain refinement of electroslag remelted ingots* by J. Campbell and P. D. Caton; *Structure of continuously cast eutectics used in the manufacture of fine-grained aluminium sheet* by L. R. Morris.

MR C. MASCRE (Centre Technique des Ind. Fonderie): I have a question for Professor Hellawell. What happens when $\theta = 0$ in Volmer's theory? We remember that $\theta = 0$ when there is a tendency for bonding between A and B, as would be normal between nucleant and nucleated phases.

PROFESSOR HELLAWELL: Dr Mascre asks what happens when θ goes to zero, implying no barrier to nucleation. This would be equivalent, for example, to adding pure solid A to liquid A as it passed through the freezing point.

I suppose that the solid particles would not begin to grow until the undercooling exceeded that for curvature. Then they would all start to grow more or less spontaneously and there would be recalescence and subsequent solidification at a very low undercooling. Any further small particle additions after recalescence would dissolve if they were not entrapped in the growing solid.

If, however, the liquid were impure, the solid would grow at a lower temperature such that further solid particle additions might well survive and nucleate further grains.

As in my paper, p. 162, I entirely agree that the Volmer model is only a mathematical convenience. I am

sure that nobody now believes that such nuclei are hemispheres and in any case, the angular construction for θ is not a true equilibrium.

DR V. KONDIC (University of Birmingham): Referring to the very interesting paper from Cambridge studying experimental nucleation, we have used a method in Birmingham which I could not claim to be new, but some might find interesting to use.

We immerse swarf from the metal we are studying in a suitable organic lacquer, then coat it with the nucleant that we are studying. This overcomes the great difficulty of mixing light nuclei of small size with the melt. The metal which is being studied is poured at a given temperature so that the swarf melts, leaving the nuclei suitably dispersed in the metal. In this way we found, using a similar method to that used at Wisconsin (by emulsifying nuclei), that we could also study the effects of surface energy and wetting. I think those who are studying nucleation phenomena experimentally might find this useful, if you have not seen it before.

DR K. G. DAVIS (Department of Energy, Canada): I would like to congratulate Professor Cupini on a very interesting development. Something that would worry me about this bubbling from the hexachloroethane wash

225

would be the possible production of either porosity in the casting or else degradation in surface quality. Would he comment on that?

PROFESSOR CUPINI: We did not observe macro-porosity. The extent of microporosity has not yet been investigated. The surface quality of the ingots was good, indicating that the action of the gas had been beneficial.

DR E. P. EMLEY (British Aluminium Co. Ltd): Hexachloroethane, and certain chlorides, have been tried experimentally for mould washes for magnesium-base alloys and in that case there was no doubt that the surface quality of the castings was improved.

I should think, however, that there wasn't enough hexachloroethane in the washes to outlast the solidification process. If one is going to arrange for enough hexachloroethane to be still active by the time the casting is going solid, I would have thought there could be risk of another kind of defect, particularly in alloys with a longish melting range; perhaps aluminium alloys containing magnesium and a fair melting range would run the risk of inclusions of magnesium chloride being trapped in the casting. I wonder if inclusions of that kind have been looked for at all?

PROFESSOR CUPINI: No microstructural examination was made to try to detect magnesium chloride. It is anticipated that hexachloroethane treatment will not be effective in large sections that have long solidification times.

DR J. BEECH (University of Sheffield): I found Dr Wood's paper very interesting, particularly the latter part. I did not fully understand about the titanium diboride, whether that was supposed to be going into solution and titanium carbide precipitated. I ask this because thermodynamic data show that titanium carbide is not stable and so it is unlikely to precipitate from a liquid. I can understand that if it is put into solution as titanium carbide, it may hang around long enough to be a nucleant, but if nitrogen is used in the calculation, titanium nitride can be stable in this sort of condition and it is known that this is a nucleant in certain types of steel.

DR WOOD: This was a prediction put forward by John Campbell that titanium diboride dissolves and forms titanium carbide. We did not rely on this. We added pure titanium carbide or pure titanium diboride. Sometimes we put both in. There was no obvious benefit from adding titanium carbide and titanium diboride simultaneously. We did not rely on that reaction and I agree it is probably very unlikely indeed.

THE CHAIRMAN: Would Mr Campbell like to comment on the question whether titanium carbide can form in respect of his previous publication on electroslag refinement of steels?

DR CAMPBELL: There is one small point in its favour. If particles are actually being added, these do take a finite time to go into solution, and during that time there is a diffusion zone rich in titanium surrounding the particles in which the solubility product of the carbide as well as the nitride might easily be exceeded.

Both titanium and carbon will be concentrated ahead of the solidification front, and may precipitate titanium carbide there. Such particles might be displaced into the bulk liquid by turbulence associated with the arrival of

droplets in the metal pool. The arrival of dissolving titanium diboride particles at the solidification front will combine these effects and thus further enhance the heterogeneity of the local environment.

DR H. A. DAVIES (University of Sheffield): I was interested in Dr Wood's proposition that possibly iron–boride formed as a barrier to crystal growth in the case he cited. Is this what was proposed?

DR WOOD: Something along that line, yes. I do not know the exact mechanism.

DR DAVIES: Is this liquid iron–boride that forms? Yet you reported that on coolings the phase was amorphous. Would the cooling rate have been sufficient to vitrify this alloy? It would require a cooling rate of about 1 million deg/s to vitrify.

DR WOOD: I did say that this was a splat-quenched material and therefore that there was amorphous material round it. It is just taking that result from the splat-quenching study and it is not necessarily right to extrapolate it back, but it is a possible mechanism. That is what I was trying to say.

DR DAVIES: Did you, on the other hand, find crystalline iron–boron in the slow-cooled samples?

DR WOOD: One does not find iron–boron just like that. Obviously most of the boron comes out of solution and comes out as a boride of some kind.

DR DAVIES: Did you find an iron boride?

DR WOOD: Not an iron boride, no. We would give it the designation M_2B because the diffraction pattern we found does not fit Fe_2B or one of the other boride phases exactly.

PROFESSOR R. W. SMITH (Queens University, Canada): I have a question for Dr Campbell. It would appear that cold metallic particles are tipped through the flux layer into the melt. If, as is suggested, they are to extract the heat, then they could well chill off the flux. If they do this, then they take some flux into the melt and the end results will be flux inclusions. Is this a problem?

DR CAMPBELL: I am glad you asked that question. There is no doubt that the flux runs at a lower temperature because of the addition of cold particles. This is what contributes in fact to the greater overall thermal efficiency of the process, in that the heat is now no longer disappearing into the water cooling. It is actually being used up in heating the particles on their way through this flux layer, but as the particle arrives with its shell of solid flux at the liquid metal, it will float because of surface tension. It can fairly easily be shown that a particle up to about 10 mm in diameter will continue to float there until the solid flux has started to melt, and when it starts to melt and a small 'window' appears through which the liquid metal can 'see' the metal particle, then there is direct metal-to-metal contact. At the point at which that occurs, there is a reasonable degree of wetting, and then, of course, the surface tension changes and the particle is wetted and it moves straight through. I believe this is why slag inclusions are not seen in the steel. It is a self-correcting process that the particles are not allowed to go through until they are wetted by the interface. They are arrested by the interface until they are wetted, which means that they must be arrested until the slag shell disappears.

PROFESSOR SMITH: If you add a substantial amount of solid particles you may find some of the slag-covered particles will be pushed under the metal pool surface.

DR CAMPBELL: Yes, it is possible to overload the process thermally, but there is no point in doing that. It will not work. The idea is to add just sufficient particles to get a useful degree of cooling and grain refinement in the bulk of the ingot without chilling off the slag layer so entirely that the whole process comes to a stop. In fact there is rather a wide range of process conditions which will continue to work quite satisfactorily. It took us some time to find them originally, but having found them, we now know what to do and we can make it work every time without chilling off the slag layer entirely.

PROFESSOR SMITH: You claim that this technique is useful to extract heat from within the melt. How much have you got to take out in order to freeze it? It must be a very small proportion since if large proportions of added solid are used, then presumably the slag layer will be over-chilled and the process will stop.

If the shell of flux does not melt before the particle enters the liquid metal pool, then little surface cleaning of the particle will take place and so oxides will be added to the metal pool.

DR CAMPBELL: By carefully measuring dendrite arm spacings in the ingot, we get a reduction of about a factor of three in dendrite arm spacings which, if the normal coarsening law is taken as being approximately a one-third power law, would predict roughly a factor of 10 increase in local solidification speed within the centre of the ingot. In spite of the fact that we lose some of the chilling power of the particles on their way through the slag, we are still getting a factor of 10 improvement in solidification speed.

PROFESSOR A. A. TZAVARAS (Aristotelian University of Salonika, Greece): I have a question for Dr Campbell. I would agree with him that the same base material would be the ideal nucleating agent for a melt. Would he care to comment on the problem of uniform distribution of the nucleating agent inside the melt?

DR CAMPBELL: This is an important point and it gives me my only cause for fears about scaling the process up. This is the problem of distributing these particles over the entire surface of, say, a large ingot of which most of the surface, in any case, is obstructed by the presence of the electrode, but that is just engineering, basically, and I think that this is something that we can overcome. On our very small ingots, that has been an easy problem to solve. We have achieved it by a certain amount of electromagnetic stirring, very comfortably and very easily, but I do not expect success to come so easily as we scale up in size. We are in the process of scaling up so I hope we will soon have the answers.

DR N. B. BRYSON (Aluminium Company of Canada Ltd): Could I ask Dr Campbell what percentage of the total mass flow rate is added in the form of particles?

DR CAMPBELL: We find that we can add up to about 50% particles by the electroslag route. I think if the normal thermal balance of just adding particles to metals

is done, it will be found that the optimum ought to be about 20%. It is because of the preheating of the particles on their way through the flux layer that they lose some of their chilling power, to put it rather crudely, so the thermal balance turns out that in fact up to about 50% can be added and possibly even more before the whole process becomes thermally overloaded and will not take it.

DR G. HOYLE (BSC, Sheffield Laboratories): I was interested to read that Dr Campbell found an increasing melt rate by stirring, without the addition of any particles. I am wondering where the extra energy comes from to provide this melting, or alternatively if there is no extra energy, where is the energy actually being saved that enables the higher melting rate to be achieved?

I have done one or two rough calculations on Dr Campbell's figures, and it seems to me that the values for specific energy consumption that he has been getting are rather high, between $2\frac{1}{2}$ and 5 kWh/kg, according to the values in his Table 1. If this is correct, and I make quite a few assumptions in getting the figures, it means that stirring is not in fact causing a great improvement on what would be expected to be normal conditions in ESR, but it may be removing some of the inefficiency that might be associated with the small scale of operation. In fact, in Table 1 the melt No ES 19, which seems to be the best one, appears to be about the right value for electroslag.

I would like to ask Dr Campbell if he has determined any figures for specific energy consumption. If it is high, in his normal experiments, where is the waste heat going and if it is not high, where is the heat being saved? This is without additions. I would also be interested to see some figures for specific energy consumption for melts in which particles are added, because it seems that a very big saving can be made in this case. For example, the melt ES 18 appears to be about the same electrode speed and power as ES 19, and in the latter case there is about 29% addition, so I would be interested to see the actual figures.

DR CAMPBELL: You have asked a great many questions, Mr Hoyle, and frankly I do not have the answers to all of them. In particular, we have not looked at specific energy consumption because I was very well aware that these are very small scale melts, in which this was really just a feasibility study, and had I worked out any figures from them at all, I would have been a little sceptical. I am very hopeful, though, that we shall soon be in possession of some better facts and figures than we have at present.

When we are doing a normal melt, we select the amps and the volts and set it all going, stand back for about 20 minutes for these very small ingots, and the ingots are complete. With the particle additions and without much noticeable difference in power settings, we discovered that the melts were over in double quick time. Of course, we were adding 50% additions, or approximately that weight of addition, and it does seem that we have got a much better chance of using the energy more efficiently in the process by putting very cold particles directly through the slag. These particles are absorbing heat most efficiently from the slag because they are so cold, so that the slag is running at a lower temperature, and therefore heat losses to the mould wall are reduced, and most of the energy that is put into the electroslag unit does not go into useful melting energy at all. Most of it disappears straight out through the water cooling and so this is a method of reducing the very great heat loss from the water cooling.

DR H. A. DAVIES (University of Sheffield): I do not want to overburden Dr Campbell with questions, but what part does having to provide up to 50% of particulate material, presumably of the previously electroslag refined material, play in the overall economics of the process?

DR CAMPBELL: I am glad you asked that question. Let me clarify one or two points before I answer. We added chopped wire particles because this is the only material we could get hold of sufficiently quickly in order to carry out this work. No back-of-envelope calculations are required to tell you that that is extremely uneconomical. Our work was a feasibility study just to see whether it could be done or not, and we were so delighted with the results which showed that it could be done, that we are now working very hard on a method of producing particles, of the size required for the process, economically.

DR DAVIES: And of the composition required, free of impurities!

DR CAMPBELL: I am sorry, I missed your point. I thought you were making a proposition to me and had something tucked up your sleeve.

DR DAVIES: If you are using any sort of chopped wire, this itself contains impurities, but you are implying that on the way down this is itself purified?

DR CAMPBELL: No, I am not suggesting that. I think this is a complete red herring. If an electrode is taken from an electroslag unit suddenly and quenched, sliced through, and examined, as was done at Sheffield several years ago, it is found that for several millimetres back from the melted front, practically all the inclusions have disappeared. They have disappeared simply because they have gone into solution in the solid state, which means that most of the inclusion removal that occurs in electroslag actually happens before it is ever melted and it is just that the temperature is so high, very close to the solidification point, that those very small inclusions have plenty of time in which to go into solution. They have to be quite macroscopic inclusions to avoid going into solution. We have been using chopped wire in which the inclusions are very fine, so even if those particles never melt, whatever sulphides and oxides might have been there would in any case have gone into solution and have then got to reprecipitate in some form if they have the chance. I think a lot of them do not reprecipitate and those that do, reprecipitate rather finely and this is one of the major benefits of the process.

A lot has been done on the chemistry of the process to show that sulphur is removed by as much as 50%, or something like that, but the amount of inclusions which are removed or redistributed is out of all proportion to that rather small 50%. In my view, the chemistry is rather secondary and it is the solidification which is important.

DR E. SELCUK (Middle East Technical University, Turkey): I have always been suspicious of solid material additions for grain refining. Our experiments have shown that sometimes it works and sometimes it does not. It is true for titanium diboride and titanium dioxide.

In the case of electroslag refining, I think that we have a fairly good chance of making use of the liquid motion within the slag and within the liquid pool. In ESR, we have the water-cooled mould, the electrode, the slag pool, and the metal pool. There is an electrodynamic pressure created by the current flowing through the electrode into the slag pool, which is a function of current drawn and the ratio of the cross-sectional area of the electrode and the copper mould.

Since we have reasonably directional solidification and columnar grains are growing at an angle to the mould wall, we have some evidence that this liquid motion within the liquid pool might cause a fragmentation of the tips of the columnar grains and might push them into the centre of the solidifying ingot.

The point is to catch an optimum point for the current drawn and the ratio of the cross-sectional area of the electrode and the mould. I think if one can catch the optimum point it will be much easier to grain-refine the electroslag ingots without the addition of solid particles of which, as I have said, I have my suspicions.

DR CAMPBELL (*written reply*): In reply to Dr Selcuk's remarks concerning the use of naturally occurring electromagnetic stirring in electroslag refining, he was kind enough to refer at length to my work on the naturally occurring fluid flow processes in ESR (*J. Metals*, 1970, July, 1) where I derived the approximate relation for the velocity, v, of stirring, with a current of I amps and electrode and ingot wear of A_1 and A_2, respectively:

$$V = K\left(\frac{1}{A_1} - \frac{1}{A_2}\right)a_1 z a_n (1-za)/3 - (1+a)/3 \quad (1-za)$$

where k is constant, L is the diameter of the mould and the index 'a' is probably somewhere between 1 and 2.

The difficulty with any attempt to utilize the relation in any useful way is that A_1 and A_2 are normally fixed to be as close as possible to avoid radiation losses from the surface of the slag bath, and to reduce the number of electrode changes if a long ingot is required. The only other important adjustable parameter is the current, I, which is normally limited by considerations of ingot quality.

DR E. F. EMLEY (British Aluminium Co. Ltd): I would like to go back to Professor Hellawell's paper and ask three questions, and offer a comment.

As he says, it is a very different approach from that of Glasson and myself, but I think it is none the worse for that, and personally, I would always favour a simple explanation which looks as if it might have general validity.

Now in the three cases I know of very strong grain refinement, there is a proven epitaxial relationship between the matrix and the supposed nucleating species. First of all, I would like to ask if there are any cases of strong grain refinement where it is believed that there is not such an epitaxial relationship? The three cases are Al by Ti, Mg by Zr, and Mg–Al by C.

The second point is that, presumably, the existence of an epitaxial relationship in the solid state must virtually guarantee a value of theta less than 10°, and in fact I would have imagined that any peritectically formed phase in a binary system would be pretty well wetted by the matrix almost by definition. Perhaps he would comment on that?

My third point is to ask if Professor Hellawell has looked at the grain refinement of magnesium by manganese? I ask that because I think it might be a good test case. The system is a peritectic one so I think there will be very good wetting of the solid phase. The diagram is very much like magnesium–zirconium or aluminium–titanium, and I would have thought that it would suit Professor Hellawell's model. The only snag is that manganese does not grain refine magnesium, so if Professor Hellawell's model would correctly predict that manganese should not grain refine magnesium, then I think this would be a very strong point in favour of his views.

Finally, he indicates in his paper that it is difficult to explain the induction period between when you add the grain refining agent to aluminium, this being aluminium–titanium–boron hardener alloy, and when you actually develop maximum grain refining effect, and it seems to me that one can perhaps explain this. Professor Hellawell mentioned the probability, which I would think is almost a certainty, that boron has the effect of steepening the liquidus and pushing the peritectic point of the aluminium–titanium binary system over towards the lower titanium contents, and this is the basis of the mechanism of aluminium–titanium–boron grain refining suggested by Cornish two years ago (A. J. Cornish: *Metal Sci.*, 1975, 9, 477).

If this picture is correct it will be necessary, in order to be able to produce Ti–Al$_3$ nucleated phase at low Ti content, to establish in the melt a finite solubility of boron of several ppm. TiB$_2$ is a very insoluble kind of substance and it is sure to take some time to establish this solubility. It is an experimental fact that induction periods for grain refinement titanium–boron hardener of the usual 5:1 ratio are quite short, but if you move to the 30:1 ratio they are very much longer—about 30 min—and it would seem to me that it would be possible to predict this qualitatively from Cornish's model because it ought to be quicker to obtain saturation with the addition of a 5:1 hardener than a 30:1 hardener, having the same total titanium content each time.

PROFESSOR HELLAWELL: On the first point, I do not know of any examples of nucleation without preferred orientation between substrate and the material which is being nucleated. It goes against the grain to believe that that would be the case. That is to say, if the substrate is an active one and is a crystalline material, one would expect that there would be a minimum in the free energy for certain orientations, but I do not know of any cases.

With regard to the second point, about the peritectic systems, assuming that the primary phase is always going to be a good nuclear for the second phase, I do not really know whether I think that that should be so. I do not see why it should follow that, in a peritectic system, the primary solid phase should necessarily be a good substrate for the secondary solid. How important this may be to the final microstructure will, of course, depend upon the relative volume proportion of the primary phase because, even if the second solid to nucleate was elsewhere than on the primary, the reaction proceeding by liquid diffusion will soon bring the two solids into

contact—albeit without any preferred epitaxial relationship. In a case like that, where the primary phase is widely dispersed as a minor constituent, the reaction rate might be quite different, i.e. slower.

Dr Emley asked if our nucleation model would say anything about the lack of grain refinement in the Mg–Mn system. I think it does. In that system the peritectic αMn + L = Mg is only 2°C above the freezing point of Mg and the solute undercooling would be negligible. Thus, in the context of my paper, αMn may well be a 'potential' nucleant for Mg, but the absence of significant undercooling would effectively preclude its ever becoming 'active'.

DR P. R. BEELEY (University of Leeds): I would like to refer to the paper by Cupini and Prates on the effect of hexachloroethane in producing grain refinement.

It might be of interest to note that a few years ago, BCIRA developed a very effective method for refining the structures obtained in white iron castings intended for malleabilizing. They thereby managed to suppress cracking in these castings locally. The method of doing that was to surface-coat with a compound, of which one of the many constituents was zinc. It could well be that there is a similar mechanism operating since zinc is volatile.

As regards that technique, it is to be noted about this paper that they did not succeed in achieving the mechanism on their larger ingot which was, by its own standards, a fairly small ingot— about 60 mm diameter. The work at BCIRA was certainly on castings of very thin section.

The other type of commercially viable grain refining process based on surface treatment is that used in the gas turbine blade field where coatings such as cobalt aluminate are very effective, but they are only very effective on castings of extremely thin section.

I would suggest that any method that is to be based on a surface treatment will only be viable on castings of relatively small section thickness, and that for heavy castings we shall have to look for a method in which foreign nuclei are introduced or *in situ* solid is used.

In relation to Dr Campbell's additions of solid to the liquid bath, I would suggest that a far better approach is perhaps to be found in some of his other work in which the solid is already there, i.e. the solid which is actually developing in the casting, and what we really want, if we are to follow that approach, is much more effective and controllable electromagnetic stirring devices. I would submit that there lies the route—it is either that or heterogeneous nuclei which can be distributed by some means round the castings.

PROFESSOR N. L. CUPINI: For the mould techniques, we have a short total solidification time so it is easy to establish a compromise between the hexachloroethane addition and the total solidification time. It is anticipated that hexachloroethane treatment will not be effective in large sections that have long solidification times.

DR J. CAMPBELL (*written reply*): In reply to Dr Beeley's statement that he would prefer to see the solid 'already present' in the pool being utilized by means of some kind of electromagnetic stirring, this is an avenue that I have already explored, with little success. The reasons seem to be as follows.

1 The secondary arm spacing is coarsened by stirring, even though grain size might be a little reduced in some steels.

2 In most steels, grain size is not reduced at reasonable rates of stirring. This seems to be the result of the dendrite form of most steels, which does not have secondary arms with narrow roots, as a result of the relatively large effective distribution coefficients. Additions of solutes such as boron change this, of course. Dendrites merely change direction, growing into the direction of flow, and thus giving a spiral grain form to the ingot. High rates of stirring centrifuge the slag up the sides of the ingot, leading to sporadic loss of contact with the electrode, and consequent arcing and instability of melting. The ingot is characterized by somewhat ragged grain but its poor surface makes it useless.

3 Electromagnetic stirring is, in practice, difficult to carry out on an industrial scale as a result of both fundamental and practical problems.

4 Grain refinement in itself is of less importance than other more serious problems encountered in ESR ingots, all of which are associated with the problem of extracting heat from the centre of the ingot. Electromagnetic stirring has, of course, little influence on this fundamental problem.

DR F. D. LEMKEY (United Technologies Research Centre, USA): I have a question for Dr Morris. Did he investigate the properties of fully coupled eutectic structures of Al–Fe–Mn, where no primary aluminium dendrites occurred, in sheet prepared by the Hunter process? If so, did the ductility remain acceptable to sheet application?

DR MORRIS: It has been found that with compositions that are much above the eutectic composition, there are primary crystals like faceted crystals. This does not do much good for the ductility of the sheet or product. It has to be remembered that the temperature gradients in most continuous-casting processes are quite low. Also, at a growth rate of 1 cm/s the S/L interface temperature is well below the equilibrium liquidus temperature.

Another consideration that I did not go into is that there are optimum grain sizes; hardening coefficients change with the grain size because the path for dissipation of motion is being changed. At very small grain sizes the work hardening is low, and for stretch-forming operations this does not perform very well, so the grain size is in fact purposely increased up to about $1 \cdot 5$ μm to give the best combination of properties, strength and formability.

DR D. ST. JOHN (Queensland University, Australia): I have three questions on the paper by Boswell *et al.* If T_N is heterogeneous nucleation of phase II then what is T_pobs if it is not heterogeneous? Is it not reasonable to expect that the primary phase will compete with the grain-boundary junction as both have excellent nucleating abilities?

Secondly, is there evidence that the secondary peak, T_N, corresponds to phase II nucleation on phase I?

Has it been noted that in the Al–Al$_3$Ti system, an alloy of 5%Ti cooled at about 1°/min* show two arrests, one at T_pobs and one at approximately 660°? The difference is 5°C.

* D. St. John, Ph.D. Thesis; H. W. Kerr *et al.*: *Acta Metall.*, 1974, **22**, 677

DR ELLIOTT: On the first point you mentioned—why is there a difference between the temperature recorded for the nucleation—the interpretation that has previously been given to this is that grain boundaries are much more effective nucleation sites than the interior of the droplet and that is why there is this difference.

In answer to the second question, I do not think the configuration will necessarily be the same on the grain boundary as on the interior of the droplet.

DR ST. JOHN: The first was about nucleation on the grain boundary, the second one was what evidence do you have that the second drop is nucleating on the primary phase.

DR ELLIOTT: This is one of the difficulties with the stage that the work is in at the moment. I really think that both your second and third questions will come out of the structural work that is done on the solidifying droplets. I am sorry I cannot be more specific on this but I am not conversant with the details at the moment.

DR SALE (*written reply*): The plateaux on the cooling curve cannot be the same events that gave the two peaks on the DTA curves. In fact the temperature difference between the two plateaux, about 5°, is reduced in the case of the big DTA peak. Also the cooling curves are produced with the small energy event occurring first (small plateau) followed by the large thermal event (large plateau). DTA shows one very large event. (Nucleation and subsequent growth of grain-boundary liquid) followed by a small thermal effect (nucleation and growth of liquid in droplets), the peak area being approximately equal to two different forms of liquid phase.

DR M. H. BURDEN (Tube Investments Research Laboratories): I would like to ask Dr Wood a question. In his paper he gave a long list of compounds with which he tried to nucleate. Most of these compounds are soluble in liquid steel. Did he take this into account in his studies and did he do any work on this because in dissolving in liquid steel it would in effect drop the superheat and therefore could alter the grain size. Because of this his results could be interpreted in different ways depending on whether or not the compound dissolves.

DR WOOD: This is an interesting point. We were not really very concerned with getting a very fine grain size. We were much more concerned to look at particles which did not dissolve in the liquid steel and which we could later observe to see how they orientated themselves with respect to as-solidifying ferrite. As such, powder was added until particles of undissolved carbide and nitride were found (for bodies, distinct observation by electron probe microscopy of the transition metal element). Naturally these additions could affect superheat conditions and the conditions put forward in the paper should not be used as rigorous parameters under which such powders worked. Indeed the results on exact grain size using these conditions were very disappointing.

DR V. KONDIC (University of Birmingham): Two years ago, Professor Ohno paid us a visit at Birmingham University and gave us a most interesting talk on his ideas of solidification. In that talk he introduced the principle of crystal detachment from the mould wall that he showed on the slide this morning. According to this, crystals have their 'necks' cut off and then they rapidly move inwards, still moving around eventually to settle in the middle of the mould.

I would like to comment on the results presented this morning. I believe there is an alternative explanation to account for the presence of primary dendrites in the middle of the ingots that are used in his experiments on hypereutectic phases in hypereutectic alloys.

Professor Ohno will recall that numerous experiments have been done on examining inverse segregation in aluminium–copper alloys. If you repeat the experiments of the kind you carried out with an alloy containing, say, 6% copper in aluminium, you will observe that a eutectic near to the mould wall would be about 8% and in the centre about 4%. If you extrapolate these experiments to the near-eutectic systems, you could plausibly suggest that the mechanism to account for the presence of primary crystals in the middle would be based on invoking the principles of inverse segregation. It does not necessitate movement from the mould wall. It would assume nucleation in the centre of the ingot. The trouble with this sort of hypothesis would be to explain what has happened to the hypereutectic. It might be simply that a different kind of segregation occurs because of the differences in solidification, the inverse segregation of hypoeutectic alloys and normal segregation of hypereutectic alloys. According to the well known theories of such segregation, the results he obtained could be accounted for by independent nucleation in ingot centres as opposed to the idea of crystal transfer from mould walls.

PROFESSOR OHNO (*written reply*): I think that the inverse segregation of the hypoeutectic alloy of an Al–Cu system is caused by the precipitation of primary crystals of α-phase in the centre region of the ingot, which were separated from the mould wall in the initial stages of solidification.

In the case of the solidification of a hypereutectic alloy of this system, the primary crystals are of Cu–Al$_2$ phase; therefore, inverse segregation does not occur in the central region of the ingot. I do not think that the primary crystals of hypo- or hypereutectic systems are produced by independent nucleation in the ingot.

DR J. SPITTLE (University College, Cardiff): Following Dr Kondic's remarks, we have observed a large number of eutectic systems, looking at this sort of thing with regard to the distribution of primaries. In most cases it is certainly not a fact that the primaries are always located towards the centre. Often there is a uniform distribution of the primaries throughout the ingot cross-section and not located towards the centre, so perhaps this is a bit spurious in that an attempt is being made to describe an explanation for a particular distribution. I have never seen a marked distribution of the primaries towards the centre, and personally I feel that perhaps they are in fact uniformly distributed and directly nucleated throughout the bulk liquid in an undercooled liquid.

DR L. M. HOGAN (University of Queensland, Australia): With regard to Professor Ohno's paper, I looked at CuAl$_2$ some years ago and was interested in

the same phenomenon that Professor Ohno was observing in relation to the microstructures rather than the position of the dendrites. We are inclined to agree with the last comment, that certainly the dendrites do move and are found towards the centre of an ingot. I think it is worth mentioning also that we found that observing the very surface of the ingot (small ingot) there was always a primary particle at the surface to nucleate every eutectic grain, that is the columnar eutectic grain growing in the hypereutectic alloy. This was always CuAl$_2$. In the alloy eutectic composition I have found both primary dendrites right on the surface, the chilled crystals, and CuAl$_2$ was also the active nucleant and the aluminium dendrites were apparently unaffected in these circumstances. On the hypereutectic side, the effect of the CuAl$_2$ was obvious and the eutectic preferred to grow from the CuAl$_2$ dendrites giving the equiaxed effect, whereas on the hypoeutectic side, as was observed, the primary dendrites were completely independent of the eutectic growth.

THE CHAIRMAN: In the paper by Boswell *et al.*, the authors remarked on the smaller undercoolings that they found compared to the larger values found in some cases by Southin[1] in his studies of nucleation in eutectic systems. In this context, I would like to draw attention to some ideas that I am about to publish[2] on Southin's data as re-analysed by Cantor.[3] Cantor obtained, from Southin's undercooling results, values for the heterogeneous nucleation parameter cos θ:

$$\cos \theta = \frac{AL - \sigma Ab}{\sigma BL} \quad (1)$$

where σAL is the solid liquid interfacial energy for the primary substrate, A. σBL that of the nucleated phase B, and σAB the solid/solid interfacial energy.

Cantor[3] obtained 22 values of cos θ for different $A - B$ pairs of which four were considered by Cantor as suspect due to the approximation used. Of the remaining 18, the nine from systems where both elements had fully metallic, fcc, cph, or bcc structures, were selected for analysis (Table 1).

Since solid–liquid interfacial energies of pure metals have been correlated with the melting points of the metals, as $\sigma AL \, Tm(A)$, it was decided to see if the values of cos θ could be correlated[4] with the melting point ratios $Tm(A)/Tm(B)$. It was found[2] that the cos θ

Table 1

System A–B	Cos θ	$Tm(A)/Tm(B)$
Ag–Pb	0·97	2·06
Cu–Pb	0·96	2·26
NiAl$_3$–Al	0·95	1·90*
Al–Pb	0·59	1·56
Zn–Al	0·49	1·15
CuAl$_3$–Al	0·38	1·25**
Ag–Cu	0·21	0·91
Cu–Ag	−0·22	1·1
Pb–Ag	−1·0	0·49

* NiAl$_3$ is a peritectic phase so $Tm(A)$ was taken as 1 773 k, about 140k less than the highest melting compound NiAl
** $Tm(A)$ taken as 1 173k for similar reason

values could be fitted, by linear regression analysis, to equation (2), with a correlation coefficient of 0·9

$$\cos \theta = \frac{Tm(A)}{Tm(B)} - 1$$

A full discussion of this conclusion will be given in a forthcoming note[2] but in the context of the observation by Boswell *et al.* it may be of interest to point out that in peritectic systems, the primary phase inevitably has a higher melting point than the secondary peritectic phase. Therefore, equation (2) suggests that, as reported by Boswell *et al.*, low undercoolings would be expected for peritectic droplet nucleation.

REFERENCES

1 R. T. SOUTHIN: Ph.D. dissertation, University of Cambridge, 1970
2 R. D. DOHERTY: to be published
3 B. CANTOR: to be published
4 D. TURNBULL, *J. Appl. Phys.*, 1950, **21**, 1 022

DR G. HAOUR (Battelle, Geneva (*written contribution*)): It has been observed in the past that a preliminary sequence of cooling and heating the liquid enhanced the undercooling subsequently attainable. Would you expect that such thermal cycling would also have an effect on the undercoolings obtained in your small particles experiments? This refers to the paper by Rasmussen *et al.*

MR D. A. GRANGER (Aluminium Company of America (*written contribution*)): In his paper, Professor Hellawell talks of the 'unexplained' behaviour of the Al–5Ti–1B grain refining master alloy. Dr Emley has given an explanation for the manner in which the grain refiner acts. I should like to offer another interpretation based on the following experimental observations.

When an aluminium alloy melt with a small amount of titanium (0·004%) in solution is treated with a boron-containing gas, such as BF_3, to form fresh TiB_2 particles, there is no grain refinement. But the addition of another small increment of titanium will immediately produce a very effective grain-size reduction. This indicates that the TiB_2 particles act as nuclei for titanium in solution which in turn nucleate α-Al grains. This sequence of events can be used to understand the behaviour of the Al–Ti–B grain refiners. When the refiner is added to a melt, the matrix dissolves away releasing TiB_2 and Al_3Ti particles. The Al_3Ti phase is soluble in molten aluminium at the small addition levels in general use. It will take several minutes for these particles to dissolve, but when dissolution is complete the grain refiner is at its peak effectiveness. From then on 'fade' sets in because the dense TiB_2 particles settle out under gravity.

DR R. P. H. FLEMING (City of London Polytechnic): In the presentation of his paper, Professor Hellawell stated that, according to the 'Turnbull theory of nucleation', good nucleating agents should have a similar crystal structure to that of the metal to be nucleated; and further in discussion he stated that he was not aware of any literature or results in which the nucleating agent did not have epitaxy with the metal to be nucleated.

In fact, there is a considerable literature summarized by CIBULA in '*Grain Control*' (Institution of Metallurgists' publication, 1969) in which it is shown that a variety of nucleating agents have been investigated in aluminium alloys; nucleating agents included TiC, TaC, MbC, ZrC, TiB_2, Zr_2B etc. Of these compounds, those that have the greatest mismatch factor (δ) were found by experiment to be the most effective nucleating agents. These observations of nucleating agents in aluminium alloys is in complete contrast to Turnbull's theory of nucleation and so it is perhaps true to say that the selection of nucleating agents is still largely a 'try-it-and see' process for most metals and alloys.

Session 4

Solidification and quality control; continuous casting

Chairman: D. Melford (Tube Investments Ltd)

Editor: A. Nicholson (BSC, Swinden Laboratories)

Keynote Address: Continuous casting

F. Weinberg

The continuous casting of both ferrous and non-ferrous metals has been investigated extensively over the last twenty years, reflecting the major position this casting technique has now assumed in the industry. Taylor[1] in the Howe Memorial Lecture of 1975 reviewed the results of recent investigations related to the continuous casting of steel. Emley[2] has recently reviewed the continuous casting of aluminium, including both the semicontinuous DC casting process generally used in the industry and the continuous casting machines currently in use or being developed. No attempt will be made in this paper to cover the subject in a comprehensive way, for it is much too large. Instead, particular areas will be discussed, primarily those relating to solidification, mathematical modelling, and cracking, which are areas dealt with in the other papers presented in this session.

The author is in the Department of Metallurgy, University of British Columbia, Vancouver

CONTINUOUS CASTING OF ALUMINIUM ALLOYS

Some of the continuous casting machines and processes for aluminium alloys are listed in Table 1 and are described in Ref. 3. In general the processes listed divide into two groups, one producing sheet or wire directly from the melt, with a small reduction in section by rolling, and the second producing slabs and billets by horizontal casting. In both cases it is claimed that equivalent or better material can be produced using these processes, as compared to the semicontinuous DC vertical casting process, with significant savings in capital and operating costs.

It is reported that the nearly pure aluminium alloys, used for electrical and thermal conductors, are cast with the least difficulty. There is little tendency for cracking in these alloys, low porosity, good surface quality, and a small cell size. For the Alusuisse Caster I,[3] which is representative of the twin roll strip casters, for 7 mm strip the cell or interdendritic spacing is reported to be approximately 5 μm, considerably smaller than that normally found in DC castings. With higher alloy content casting can become more difficult. Casting these alloys with the Propezi process,[3] for example, the low and medium mechanical strength alloys are readily castable into rod. The high-strength mechanical alloys cannot be cast at the present time. High-strength alloys cast by the Cegedur–Pechiney–Secim process,[3] can exhibit internal cracks, internal porosity, external cracks, surface roughness, and solute segregation, if the casting conditions are not properly selected and applied.

Table 1 Aluminium continuous casting systems

	Alloys cast
Sheet, plate and foil	
Hunter continuous casting process	Al
Pechiney 3C process	Al
Mann rotary strip casting and rolling line	Al
Alusuisse caster I and caster II	Al
Hazelett twin belt caster	Al, Zn, Pb, Cu, steel
Rod, bar and wire	
Properzi process for continuously cast and rolled rod	Al, Pb, Zn, Cu
Cegedur–Pechiney–Secim continuous casting and rolling process	Al
Southwire Aluminium SCR systems	Al, Cu
Ingot, slab and billet	
Clark single strand horizontal casting system	Al
Wagstaff horizontal casting machine	Al
Reynolds horizontal casting process	Al
Kaiser aluminium process for horizontal continuous casting	Al
Alcoa horizontal continuous casting process	Al

These effects are also present in semicontinuous DC casting of the high-strength alloys.

These results indicate that modelling studies defining optimum conditions for the casting of the high-strength alloys with minimum internal defects would be valuable. In DC casting attention has been directed primarily to the water spray region in considering internal cracking. Using both modelling and experimental measurements, Bryson[4] has developed water spray practices which can reduce or eliminate cracking at higher casting speeds.

With a closed head horizontal casting system, the mould region as part of the casting head is more significant in a heat transfer analysis. An analysis of such a system is presented in another paper in these proceedings. By considering the heat transfer in the mould in more detail, it becomes apparent that volume shrinkage of the solid shell during solidification can affect the heat transfer to the mould. As the solid cools, shrinkage occurs, the amount depending on the alloy composition and the temperature distribution. The shell separates from the mould, reducing the heat transfer between the shell and the mould. However, hydrostatic pressure from the liquid pool can force part of the shell back against the mould in a complex way. This also occurs in the continuous casting of steel to be discussed later, and is of major significance in relation to surface cracking. In aluminium alloys the effect can result in the generation of surface defects.

Investigations have been reported of the development of a fine grain structure in horizontally cast billets by controlling the fluid flow, where normally large grains would be obtained.[5] A fine grained structure has also been reported when casting in an electromagnetic field.[6] For level pouring of vertically cast billets Bergmann[7] describes a surface macrostructure which is attributed to macrosegregation associated with fluctuations in the metal flow. All of these observations pertain to specific systems; it is not clear how relevant they would be in other systems.

Non-ferrous alloys other than aluminium are cast continuously on a large scale. Some of the machines listed in Table 1 are used to cast a range of materials as shown. Other machines not listed here have been developed particularly for casting copper.[8] In general, the lower melting point alloys can be grouped with aluminium; the higher melting point materials, e.g. brass, are cast in a manner similar to that used for steel. Accordingly there is no need to consider the casting of these materials as a separate group.

CONTINUOUS CASTING OF STEEL

Steel is currently being cast continuously in the form of billets and slabs by vertical (stick) casting or low head curved mould systems, using water cooled, oscillating, copper moulds. The billets and slabs are normally square or rectangular in section. Beam blanks are being cast in curved moulds[9] and the development of billets of cylindrical section for seamless pipe is currently being carried out.[10] A casting machine for producing high quality billets of cylindrical section for pipe has been developed in which the billet and mould are continuously rotated during vertical casting.[11] It is not known how the quality of round billets produced in this manner compares to the round billets cast without rotation.

Horizontal casting systems of unique designs are currently being developed for the Watts process and by Davy–Loewy.[12] A mathematical model of the heat transfer for the Watts process has been published recently.[13]

From a solidification point of view, both vertical casting of slab and billet and curved mould casting can be considered as similar. Differences between various installations will be associated with (*a*) the size of slab or billet cast; (*b*) lubricating practice, either rape-seed oil or powdered flux being used; (*c*) deoxidation practice; and (*d*) pouring practice using open pour or immersed nozzles of different designs. Each installation will have individual operating parameters including water flows and pressures in the mould and spray regions, withdrawal rates, and other parameters, all of which will be adjusted for the particular steel being cast. Accordingly, any general analysis of the continuous casting process will have to be sufficiently flexible to allow for these variations.

Solidification

The solidification process during continuous casting reaches a steady state which, when combined with the large scale of the operation, offers an attractive system for consideration as far as solidification is concerned.

Solidification occurs initially at the mould wall and meniscus, forming a thin shell which progressively thickens to approximately 1·3 cm when it leaves the mould. The details of how the shell forms and is distorted are discussed later. Below the mould, water sprays cool the exposed slab or billet surface. The characteristics of the sprays are important as they are related directly to internal cracking of the casting. Following the water sprays, the billet, still containing a liquid core, is cooled by radiant heating. Pool depths can vary from 500 cm to 3 000 cm depending on the casting installation and casting conditions. Normally casting will continue for 5×10^3 s, but with multiple heats and interchangeable nozzles, steel slab has been continuously cast without interruption for $3·9 \times 10^5$ s.[14]

Fluid flow in the liquid pool, which can influence surface quality, cast structure, and segregation, can be considered in two parts: within the mould and in the submould region. In the mould region the liquid pool is extensively mixed by the momentum of the input stream, to a depth close to the bottom of the mould. With open pour, the flow is turbulent; with closed pour using immersed bifurcated nozzles flow is lamellar. This has been demonstrated by adding radioactive tracers to the liquid pool and examining the resultant distribution of the tracer in autoradiographs of the sectioned casting.[15,16] Turbulent mixing in the mould with open pour enables aluminium to be added direct to the pool for deoxidation of the steel. The controlled lamellar flow with immersed nozzles markedly improves the surface quality of slabs used for the production of sheet. A mathematical model of mixing in the mould region has been developed by Szekely and Yadova[17] for different input stream configurations.

Below the mould, fluid flow is markedly reduced, tending to be confined to the central part of the pool and decreasing with distance down the pool.[16] Details of the flow pattern are not clear. The lower part of the pool contains a significant amount of solid dendritic 'debris', possibly filling the lower quarter of the pool, as determined from the position of tantalum pellets dropped in the liquid pool during casting (Fig. 1). In the final stages of freezing there is appreciable downward movement of liquid resulting from volume shrinkage. This was demonstrated clearly in some experiments carried out

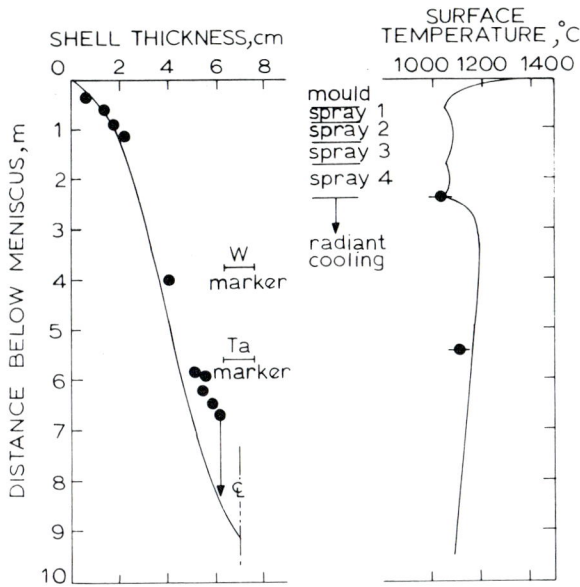

1 **Shell thickness and surface temperature of billet as function of distance below meniscus: solid lines from model calculations, points are observed data[26]**

by the present author in which tantalum pellets containing radioactive gold encased in mild steel were dropped into a 17×17 cm vertically cast mild steel billet. After the pellet stopped falling, the gold was released from the pellet and drawn downwards to fill the void resulting from volume shrinkage. The results are shown in Fig. 2. The downward movement of liquid extended over a distance of 75 cm along the billet axis. This movement could play a major role in the development of centre line porosity and segregation in the castings.

Cast structure

The cast structure of continuously cast steels, as with static castings, consists of both columnar and equiaxed grains. The extent of each structure present depends on the pouring temperature, low pouring temperatures tending to produce equiaxed structures, and higher temperatures columnar grains.[18] At intermediate pouring temperatures both structures are present, the transition from columnar to central equiaxed grains occurring sharply at a uniform thickness around the billet. The presence of columnar grains is important when considering internal cracking as it has been shown that cracking occurs predominantly between columnar grains.[18]

With curved mould casting, the structure may not be symmetrical about the central axis of the billet. Equiaxed grains falling down the liquid pool may tend to pile up in the lower half of the curved billet, resulting in a shift of the centreline towards the inner curved surface of the billet. The structure can consist of columnar grains between the central line and the inner surface and equiaxed grains in the outer half of the billet section.[16] The curvature may also affect the inclusion distribution in the billet. If the radius of curvature is small, inclusions which form during solidification when the strand is nearly horizontal can clump and float upwards until they meet the advancing upper interface. A high-density band of inclusions can develop as shown in Fig. 3, and

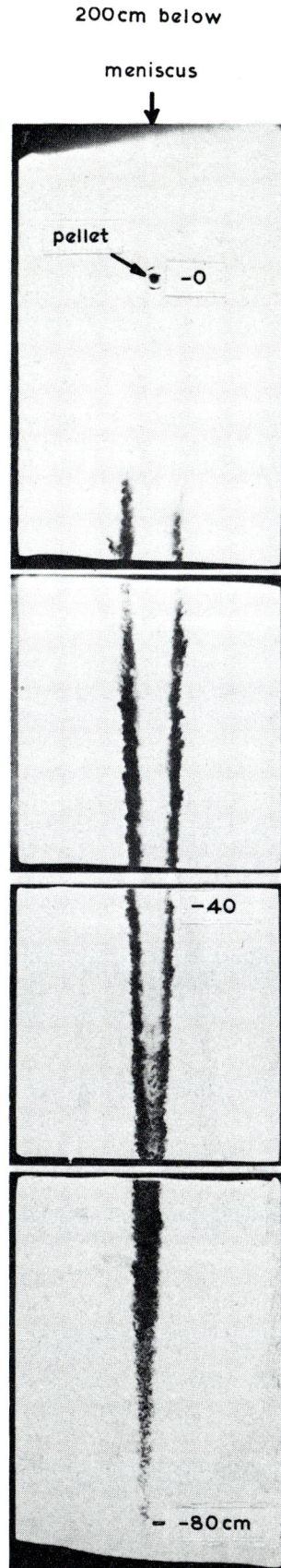

2 **Downward flow of liquid near bottom of pool owing to volume shrinkage**

can markedly affect the physical properties of the cast billet.[19]

The mechanism for nucleation of the central equiaxed grains is not clear. If dendrite remelting in the solid–

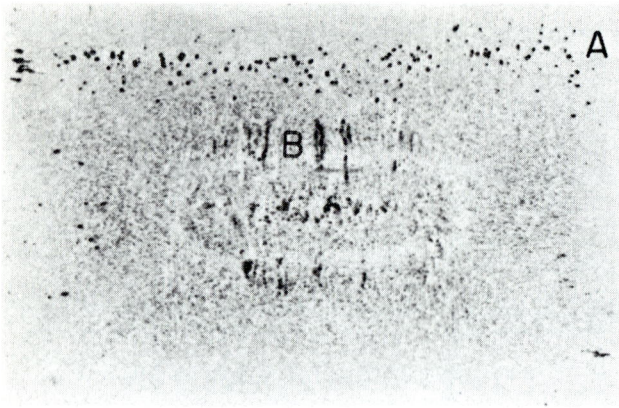

3   Sulphur print of cross-section of billet cast
    with small radius of curvature: band of MnS
    inclusions at *A*, halfway cracks at *B*[19]

liquid region is assumed to be the dominant mechanism,
then it cannot be effective below the mould since there is
no evidence of flow of the interdendritic liquid into the
liquid pool. At the bottom of the mould, where turbulence owing to the input stream stops, the dendrites are
very small and the shell is thin. If all the nucleation
occurs at this point, it is difficult to envisage a sharp
columnar-equiaxed transition occurring after more than
half the billet section has solidified, as is observed.

The cast structure is relatively insensitive to casting
rate, other factors being equivalent. The present author
carried out experiments on two vertical billet casting
machines for mild steel billets 18×18 cm in section, one
casting at a rate of $2 \cdot 2$ cm/s and the second at $5 \cdot 8$ cm/s.
The structure of the resultant billets was essentially the
same in both cases.

In the central part of the billet, centreline porosity is
often observed as well as V-segregate patterns. In addition, a periodic bridging of the centreline by large
dendritic grains is observed. These effects are probably
associated with the extensive downward movement of
the loosely packed solid dendrites in the lower part of
the pool resulting from volume shrinkage. Interlocking
of the larger dendrites could account for the porosity and
V-segregation. It is not clear what mechanism produces
the periodic bridging; but it is not directly associated
with the mould oscillation.

The solid shell in the mould region

Initially it was important to determine the shell thickness
of the strand as it left the mould under given operating
conditions, to prevent break-outs from occurring. As
more experience is gained for each installation, this
information becomes less important.

In practice it is found that the thickness of the solid
shell can fluctuate appreciably in the mould region. This
is illustrated in the autoradiographs in Fig. 4 which show
that fluctuations in the shell thickness can develop
immediately below the meniscus. Thin regions of the
shell are associated with depressions on the outside
surface of the billet. The extent of the fluctuations will
differ for different casting machines and different lubricating practices. In addition, the fluctuations are
strongly dependent on the carbon content of the steel. In
the region of $0 \cdot 1\%$ C, cast billets exhibit a rough outside
surface and extensive fluctuations in shell thickness.[20]
This is accompanied by a sharp drop in heat transfer

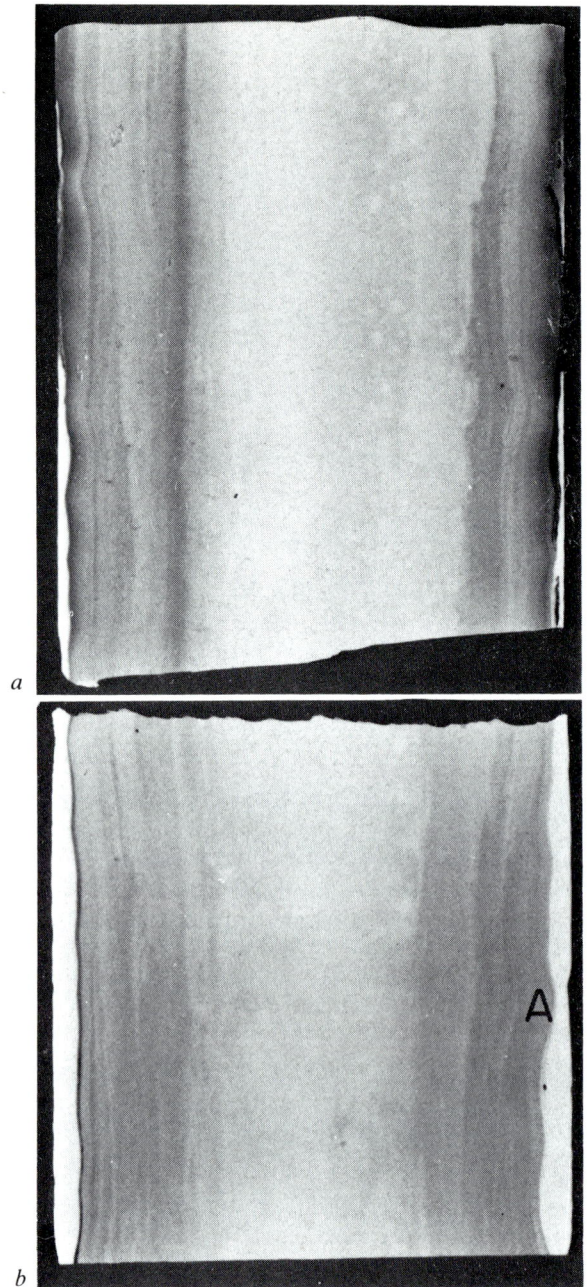

4   Autoradiography showing solid shell contour
    near meniscus (*a*) and in mould region (*b*);
    note thin shell and depressed outer surface at
    A[16]

between the steel and the mould, as shown in Fig. 5.
Above $0 \cdot 25\%$ C the shell is smooth and the heat transfer
constant. An examination of heat transfer in the mould
as a function of carbon content in an operating billet
caster showed similar results,[21] except that a drop in heat
transfer was also observed at high carbon levels
($0 \cdot 7\%$ C).

The variations in heat transfer and resultant variations
in shell thickness are associated with distortions of the
solid shell owing to tearing or buckling. Assuming the
shell sticks periodically at the meniscus and tears on the
downward stroke of the mould, a buckling mechanism
can be evolved which accounts for the separation of the
shell from the mould.[22,16] However, this does not
explain the effect of the carbon concentration. A
mechanism to account for this has been proposed by

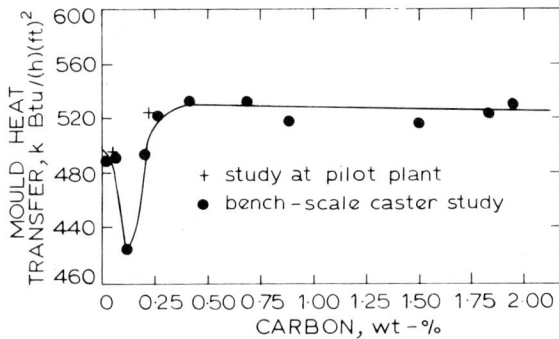

5 Mould heat transfer as a function of carbon content of steel[20]

Grill and Brimacombe[21] based on buckling resulting from the δ–γ-phase transformation of the solid shell during cooling.

In addition to variations in shell thickness parallel to the billet axis, variations occur in the horizontal plane, perpendicular to the billet axis. These variations are of major concern when considering corner and longitudinal surface cracks. In billets, on a horizontal plane, the shell tends to become thin at the corners. This results from shrinkage, which separates the shell from the mould, and the hydrostatic pressure of the liquid pool, which pushes the large faces back to the mould surfaces, leaving the corners separated from the mould. In slabs, the distortion of the shell is more complex particularly in the corner region, as shrinkage of the wide face will tend to move the shell corner along the face as well as towards the slab axis. Separation can cause shell reheating leading to cracking. In general, the shell distortion in both billets and slabs depends on the radius of curvature of the corners of the mould and the mould taper. Some investigators are examining these effects both experimentally and by using modelling techniques.[23]

Below the mould for smaller billets, the shell is strong enough to retain the liquid pool without retaining rolls. For larger billets and slabs retaining rolls are required to prevent bulging of the strand. An example of the solid shell which bulged and was subsequently pushed back to its original position by retaining rolls is shown in the autoradiograph of Fig. 6. The solid shell (A) along the side face of the beam blank has a flat, smooth outer surface produced by the retaining rolls. The solid–liquid interface (B), delineated by the addition of radioactive gold to the liquid pool, has a wavy contour which shows the shell bulged and then buckled when it was pushed back by the retaining rolls. The sharp corner in the shell at (C) results from a misaligned roller. Note the internal cracks (D) associated with the deformed shell.

Mathematical modelling of the casting process

As outlined by Taylor,[1] a series of models has been developed for the continuous casting process based on an analysis of the heat balance in the system. The models differ in the manner of solution of the appropriate equations but primarily differ in the assumptions, boundary conditions, and physical parameters used in the solution. The applicability of the models is therefore primarily dependent on how accurately they describe a full-scale operative casting system.

A comparison of measured and calculated shell thicknesses has been made using a variety of experi-

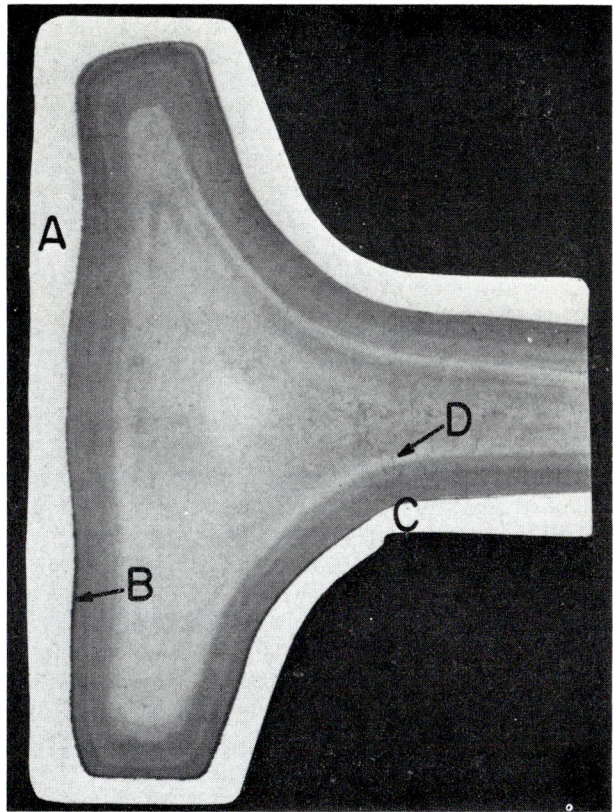

6 Autoradiograph showing solid shell contour of beam blank: solid shell *A*, interface contour *B*, misaligned roller *C*, cracks at *D*

mental techniques. These include measuring the residual shell when breakouts occur, radioactive tracer additions to the liquid pool, sulphur additions, and others. Surface temperatures in the submould region have also been measured and compared to predicted values calculated with a mathematical model.

The most recent models, including the one to be discussed in this session, use finite difference methods to calculate the temperature distribution in the solid shell. One such model, developed by Brimacombe *et al.*,[24] compares pool profiles calculated using the best data available for several casting systems and operating conditions, with shell thickness measurements made using radioactive tracer additions to the pool. In general, good agreement is obtained between calculated and measured values of the shell thickness in the mould region. Taylor[1] considers that tracer additions to the liquid pool do not delineate sharply the solid–liquid interface because of inadequate mixing. The observations of the present author indicate that, contrary to the above statement, extensive mixing occurs in the mould region, with the tracer addition sharply defining the interface.

Differences between the calculated and observed shell contours were observed in a stainless-steel slab and high-carbon mild steel heats.[25] In both cases the calculated values gave thinner shells. This difference is believed to result from the relatively wide separation of the solidus and liquidus lines in these alloys, compared to normal mild steel. The calculated values of the shell thickness are based on the solidus position; the tracer additions show the liquidus positions, which could account for the observed difference.

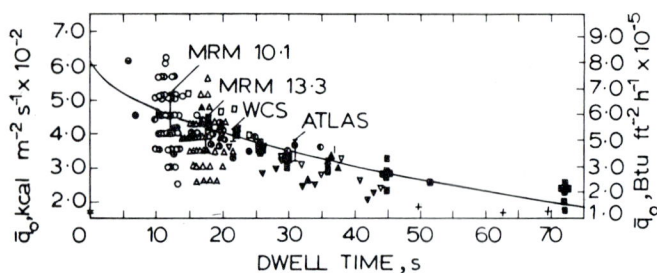

7 Average flux of heat extracted by mould as a function of dwell time[25]

In applying the model to different casting machines, the value used for the heat transfer coefficient between the steel shell and the mould as a function of casting conditions and mould parameters is critical. On the basis of an extensive examination of heat transfer in the mould at a number of installations using different lubricants, Lait *et al.*[25] have shown that the values determined by Savage and Pritchard[22] are generally applicable to many systems. This is illustrated in Fig. 7,[25] which plots the average mould heat flux as a function of dwell time in the mould for a large number of heats at a number of installations. The scatter appears large as the figure shows the widely diverging data most prominently. Eighty per cent of the data fell within the bars indicated, showing good agreement with the calculated curve.

Accordingly, the model can be applied to the mould region in most installations with confidence. The model cannot be applied directly when significant buckling of the shell occurs, without specific assumptions concerning the buckling and the resultant local variation in heat transfer between the shell and the mould being made.

In the water spray and radiant cooling regions below the mould, Taylor[1] considers that the finite difference models adequately define the temperature of the billet both at the surface and throughout the shell. The present author would agree with this, with the reservations that better data for the water spray heat transfer coefficients are desirable and that some consideration should be given to the effect of the retaining rolls in slabs and large billet casting on the heat transfer process.

The evidence that the models are satisfactory is based on both surface temperature measurements, described by Taylor,[1] and tracer additions to the liquid pool. In experiments with which the present author was associated lead additions containing radioactive gold were made near the mould wall to the liquid pool during casting. The lead addition left a residual trail on solid dendrite projections at the interface which was detected in the autoradiographs. The results[26] are shown in Fig. 1 where the solid line is the calculated shell thickness and the points indicate the measured shell thickness. Agreement is good in the upper half of the pool. Near the bottom the measured thickness is greater than the calculated value, presumably because of the presence of solid dendritic debris. The observed surface temperatures on the right also agree well with the calculated values. Note the position of the tantalum pellet, which gives some indication of the extent of the debris filled region at the bottom of the pool.

Segregation

Generally all continuously cast steels exhibit interdendritic microsegregation similar to that observed in normal castings. The amount and distribution of the segregation depends on the steel composition and the dendrite size.

The extent of macrosegregation in billets and slabs is not clearly defined. In general, macrosegregation outside the central zone of the slab or billet is not significant. In the central region, particularly along the central axis, appreciable segregation of various constituents has been reported.[27] The present author found no segregation along the central axis of a stainless-steel slab other than normal microsegregation, when using electron probe microanalysis. Enhanced centreline segregation of carbon and phosphorus has been reported in slabs which have bulged during casting.[28] The enhanced centreline segregation is probably associated with the downward flow of enriched liquid resulting from volume shrinkage, but the details of how this occurs are not clear, nor is it known why it occurs in some cases and not in others.

Cracking

The formation of cracks during the continuous casting of steel billets is currently of major concern to the industry. This is reflected in this section of these proceedings in which half of the papers presented deal with this subject.

Different types of cracks are observed in continuously cast slabs and billets. On the surface, corner, longitudinal, and transverse cracks can be present. In the interior, cracks occur near the surface, halfway between the surface and centreline, and along the centreline. The crack lengths vary from fine cracks to very long longitudinal surface cracks. In general, the cracks result from strains generated in the solid shell during solidification. The crack paths tend to be related to the cast structure and microsegregation.

A detailed analysis of the formation of cracks is complex. It consists of several major components: (*a*) an evaluation of the local change in temperature distribution causing thermal strains; (*b*) calculation of the resultant strain and stress owing to the temperature changes; and (*c*) a comparison of these results with some criterion which will indicate whether local cracking will occur. The analysis would require data describing the mechanical properties of steel having the same structure and segregation as that occurring in the continuously cast material, at temperatures near the melting point. In addition, the effects of recovery as part of the mechanical properties should be considered as well as the phase transformations which occur when the steel is cooled. The latter may be significant when using tests specimens heated up to the test temperature as an indication of material behaviour which has cooled from the melt.

As solidification occurs during continuous casting, the solid shell shrinks and distorts as it cools, causing local strains. If the cooling is relatively slow and continuous, it is reasonable to assume that the resultant strains are relieved by recovery at the high temperatures being considered. Accordingly, for cracking in the interior of the billet, the important thermal stresses are those generated through rapid changes in temperature. These occur when the strand leaves the mould, when it enters and leaves the spray zones in the cooling system, and when solidification stops.

Grill *et al.*[19] have analysed halfway cracking in billets resulting from surface reheating when the billet leaves the spray zone. The experimentally verified mathematical model was used to determine the temperature distribution in the shell at the end of the spray zone and the beginning of the radiant cooling zone. Reheating of the surface, when spraying is terminated, causes compressive stresses on the surface and tensile stresses in the interior, with the latter causing cracking.

Assuming that no significant strains were present in the strand at the end of the spray zone and that no recovery occurred during the subsequent surface reheating period of 20–30 s, calculations were made of the strain distribution throughout the solid shell. Values of the elastic modulus, elastic limit, and slope of the plastic stress–strain curve were estimated from available data. The calculated values of stress throughout a billet section were then compared to a critical value for cracking estimated from the data of Adams,[29] and the presence of cracks were predicted in the halfway region for a given reheating. These results are compared to the observed position of halfway cracks in Fig. 8 in a billet cast under the conditions used in the calculations. Good agreement is obtained between the calculations and the observed cracks.

The above analysis depends heavily on the mechanical properties of the steel used to calculate the strain and stress. More recent data by Wray[30] indicate that the yield stress, the slope of the plastic component of the stress–strain curve, and the stress required for cracking to occur used in the above analysis are too high. A recalculation has been done by Sorimachi and Brimacombe using more recent data.[31] The agreement between calculations and experiment remains good in their recalculation.

Mechanical properties

The tendency for cracking to occur in slabs and billets is influenced by the structure and composition of the steel. For example, halfway cracks tend to occur in columnar structures along the interdendritic grain boundaries.[18]

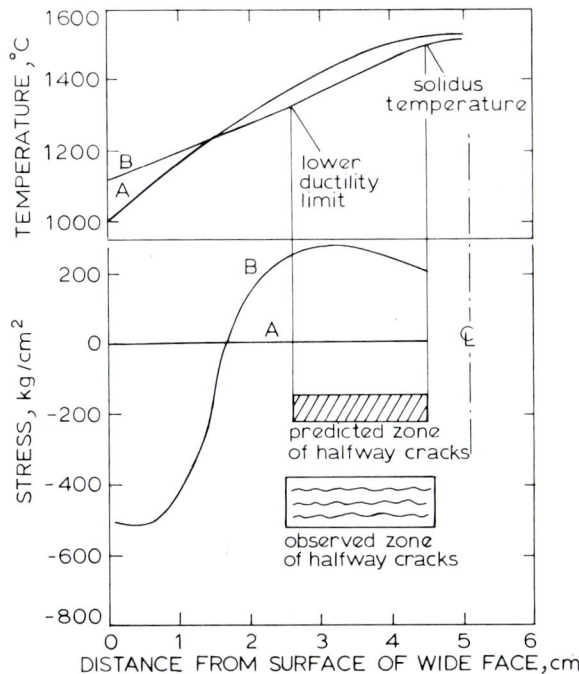

The effect of composition was investigated in a general way by the present author by examining the extent of halfway cracking in 15×15 cm billets of different composition, cast under normal operating conditions in the same machine. The extent of halfway cracking was rated from 0 for no cracking to 4 for extensive cracking. Eleven strands were examined and the cracking rated from sulphur prints of transverse sections of the billets. The results are shown in Fig. 9, in which the concentrations of the elements in the steel are plotted vertically and the extent of cracking horizontally. Considering the bottom set of points for carbon, each point corresponds to one strand. Going from no cracking on the left to extensive cracking on the right, it is evident that there is no systematic dependence of cracking on the carbon content.

Considering the other elements plotted in Fig. 9, the phosphorus level is observed to increase with increasing cracking systematically for eight of the billets. There is some increase in copper associated with more extensive cracking and a decrease in Mn/S at the low cracking levels. These results suggest that phosphorus at concentrations above 0·02% could be a primary factor in cracking, but other factors, such as excessive water sprays and abnormally high concentrations of other elements, could result in cracking in conjunction with lower phosphorus levels.

A number of detailed studies of the mechanical properties of steel at high temperatures have been carried

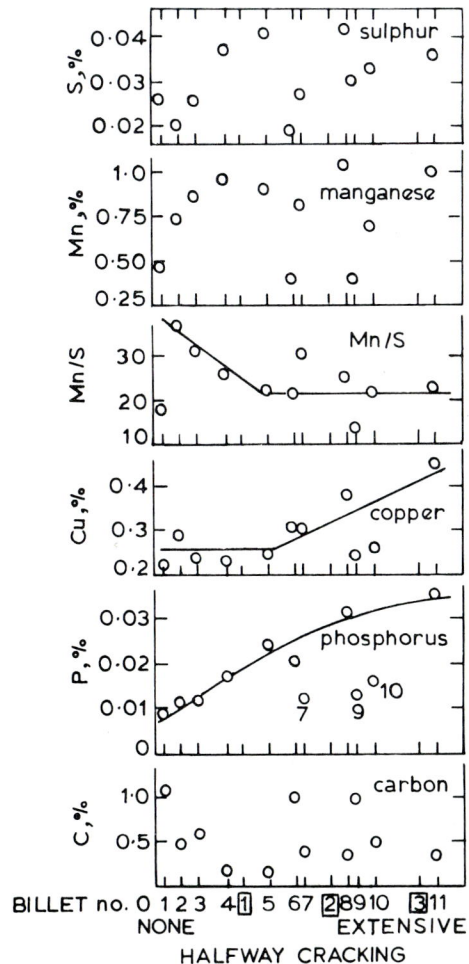

8 Profiles of temperature and stress perpendicular to wide face at centre plane of billet[19]

9 Composition of 11 billets plotted against tendency for halfway cracking

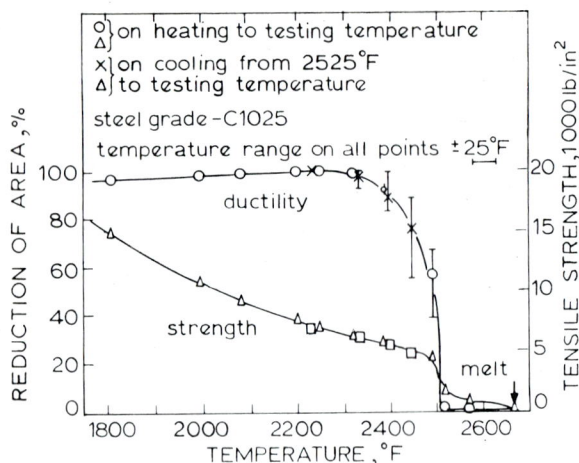

10 Ductility and strength *v.* temperature for tensile test samples cut from mild steel continuously cast billet[29]

out. Adams[29] and Fuchs[32] machined test samples from continuously cast billets and measured their strength and ductility as a function of temperature and composition with a Gleeble testing machine. This uses resistance heating to heat the sample, and is characterised by steep temperature gradients from the centre to the outside of the sample and by generally rapid pulling rates. In both investigations it was found that there was a sharp drop in strength and ductility with increasing temperature at temperatures well below the liquidus, as shown in Fig. 10. The temperature at which this drop occurred decreased with increasing carbon content of the steel. The brittle behaviour of the steel near the melting point would account for the cracking observed in continuously cast billets and slabs.

Adams[29] proposed that the brittle behaviour was caused by the presence of a high concentration of phosphorus at the recrystallized austenitic grain boundaries. This would result in a liquid layer forming at the grain boundaries at temperatures well below the liquidus, causing the observed drop in strength and ductility. That high levels of phosphorus can be segregated at grain boundaries under equilibrium conditions has been proposed by Hondros *et al.*[33]

Recent experiments by the present author, which measured the temperature at which continuously cast steel samples separated by melting under a small stress, showed that separation occurred at the solidus temperature, essentially independent of the phosphorus concentration in the steel. Also, separation often occurred at interdendritic grain boundaries which were as-cast and not recrystallized. In these experiments a ductile–brittle transition was observed to occur at approximately 45°C below the solidus temperature for continuously cast steel having a wide range of composition. This subject is dealt with in several of the papers in this session.

A second ductile–brittle transition in steel has been examined and reported by Lankford[34] and others. In this case, if steel is sensitized by heating to 1 430°C and then cooled, it becomes brittle when tested at temperatures below about 1 100°C. The same behaviour is observed in samples which have been cooled direct from the melt. The brittle behaviour is related by Lankford to low Mn : S ratios in the steel, the low ratios allowing some

FeS to form at grain boundaries, making them brittle. The relevance to cracking of this ductile–brittle temperature in continuously cast billets is unclear. The transition temperature is too low for the effect to be relevant to cracking in the mould and water spray regions. Cracking owing to withdrawal roll pressures does not appear to be a problem under normal casting operations.

The plastic component of the stress–strain curve at high temperatures is difficult to define as the plastic deformation has been shown to be cyclic[35] as a result of recrystallization. The present author has observed that cyclic stress–strain curves are obtained with test specimens made from continuously cast billets as well.

The procedure for selecting a cracking parameter based on stress or strain data is not clear. Perhaps a fracture-mechanics approach might prove useful. The ultimate tensile stress for continuously cast mild steel decreases progressively with increasing temperature, the rate of decrease being small at high temperatures. The present author has found that the ultimate stress for samples which fail in a ductile or brittle manner is the same. This indicates that a simple stress level criterion for cracking is not satisfactory. This problem is discussed in the papers dealing with cracking in this section of these proceedings.

Clearly, considerable effort is still required to clarify the complex factors contributing to the cracking of continuously cast steels.

REFERENCES

1 C. R. TAYLOR: *Metall. Trans. B*, 1975, **6**, 359
2 E. F. EMLEY: *Int. Met. Rev.*, 1976, **21**, 75
3 Continuous Casting Seminar Papers, Kansas City Seminar 1975, Aluminum Association, New York, USA
4 N. B. BRYSON: 'Light metals', (ed. W. G. Rotsell), 429, 1972, New York, AIME
5 R. E. SPEAR *et al.*: *J. Met.*, 1971, Oct., 42
6 Z. N. GETSELEV: *ibid.*, 1971, Oct., 38
7 W. J. BERGMANN: *Metall. Trans.*, 1970, **1**, 3 361
8 A. I. NUSSBAUM: 'Continuous casting', 73, and D. Barnes *et al.*, 93; 1973, New York, Met. Soc. AIME
9 G. S. LUCENTI: *Iron Steel Eng.*, 1969, July, 83
10 H. SCHREWE AND J. GLASER: 'Continuous casting of steel', 239, 1977, London, The Metals Society; *also* Concast Inc. with Dalmine Siderca SAIC Campana, Argentina, private communication
11 M. BABEL: 'Continuous casting', 17; 1973, New York, AIME; J. Basilis and G. Jego: 'Continuous casting of steel', 201, 1977, London, The Metals Society; S. Wakabayashi: *ibid.*, 209
12 H. E. ALLEN *et al.*: 'Continuous casting of steel', 215; 1977, London, The Metals Society; J. Marsh and D. Toothill: *ibid.*, 221
13 V. STANEK AND J. SZEKELY: *Metall. Trans. B.*, 1976, **7**, 619
14 C. H. BODE *et al.*: 'Continuous casting of steel', 75, 1977, London, The Metals Society
15 S. K. MORTON AND F. WEINBERG: *J. Iron Steel Inst.* 1973, **211**, Jan., 13
16 J. E. LAIT *et al.*: *Ironmaking Steelmaking*, 1974, **1**, 35
17 J. SZEKELY AND R. T. YADOVA: *Metall. Trans.*, 1972, **3**, 2 673; *ibid.*, 1973, **4**, 1 379
18 G. VAN DRUNEN *et al.*: *Ironmaking Steelmaking*, 1975, **2**, 125

19 A. GRILL *et al.*: *Ironmaking Steelmaking*, 1976, **3**, 38
20 S. N. SINGH AND K. E. BLAZEK: *J. Met.*, 1974, Oct., 17
21 A. GRILL AND J. K. BRIMACOMBE: *Ironmaking Steelmaking*, 1976, **3**, 76
22 J. SAVAGE AND W. H. PRITCHARD: *J. Iron Steel Inst.*, 1954, **178**, 269
23 A. GRILL *et al.*: *Metall. Trans. B*, 1976, **7**, 177
24 J. K. BRIMACOMBE *et al.*: 'Mathematical process models in iron and steelmaking', 174, 1973, London, The Metals Society
25 J. E. LAIT *et al.*: *Ironmaking Steelmaking*, 1974, **1**, 35
26 F. WEINBERG: *Metall. Trans. A*, 1975, **6**, 1971
27 W. R. IRVING AND A. PERKINS: 'Continuous casting of steel', 107, 1977, London, The Metals Society; S. Myoshi: *ibid.*, 286
28 K. SASAKI *et al.*: Sumitomo Metal Industries, Osaka, Japan, private communication
29 C. J. ADAMS: 'AIME Open Hearth Proc.', 290, 1971, Pittsburgh, Pa, USA
30 P. J. WRAY: *Metall. Trans. A*, 1975, **6**, 1 197
31 K. SORIMACHI AND J. K. BRIMACOMBE: *Ironmaking Steelmaking*, 1977, **4**, 240
32 A. FUCHS: Estel-Ber. Forschung Entwicklung Unserer Werke No. 3, 1975, (3) 127
33 E. D. HONDROS *et al.*: *Met. Mater.*, 1976, Jan., 26
34 W. T. LANKFORD: *Metall. Trans.*, 1972, **3**, 1 331
35 P. J. WRAY AND M. F. HOLMES: *Metall. Trans. A*, 1975, **6**, 1 189

Simulation of heat flow and thermal stresses in axisymmetric continuous casting

J. Mathew and H. D. Brody

Thermal stresses in continuous casting play a role in the development of casting defects such as hot tearing, cold cracking, and surface cracking. Hence, an ability to predict the level of thermal stresses in an ingot in relation to the casting conditions is important for the optimum design of continuous casting processes. The model described in this paper is a finite-element analysis of heat flow and the development of thermal stresses in axisymmetric continuous casting applicable to the simulation of both ferrous and non-ferrous alloys. The model is capable of simulating a wide range of boundary conditions, and the program is simple to use. The model takes into account the heat flow and displacements in the axial and radial directions in the casting and accounts for the temperature dependence of material properties. The stress simulation uses a thermoelastic-plastic creep analysis and can compute both the elevated-temperature stresses and the residual stresses in the ingot at room temperature. The analysis is found to be useful in studying the formation of various thermal stress induced defects such as hot tearing, surface cracking, and cold cracking in continuously cast ingots. A parameter called maximum normalized stress – obtained by dividing the maximum (algebraic) stress by the yield stress of the material at the temperature of the region of interest – is found to correlate well with the observed relationship between the hot tearing tendency of continuously cast aluminium and the casting rate. The analysis has been applied to predict the dependence of temperature profile and freezing (sump) profile in the casting on the casting rate. The predicted effect of the casting rate on the mould heat flux is found to be consistent with experimental measurements.

J. Mathew is with Metallurgical Research Laboratories of Combustion Engineering Inc., Tennessee, and H. D. Brody is Professor of the Department of Metallurgical and Materials Engineering, University of Pittsburgh, Pennsylvania, USA

ANALYSIS

Details of the analysis can be found in the references.[1–3] Only a brief description is given here. The heat flow simulation uses a finite element method to take into account the steady-state heat transfer in the axial and radial directions in a cylindrical ingot. This makes the analysis applicable to the study of ferrous and non-ferrous continuous casting. Unlike steel, non-ferrous metals such as aluminium and copper have fairly good conductivity and are cast at relatively low speeds. This makes the axial heat transfer relatively important in continuous casting of these alloys. The axial heat transfer is not appreciable in continuous casting of steel except at the bottom of the mould, where the radial resistance to heat transfer due to the air gap makes the axial heat transfer relatively important.

The model takes into account the variation of material properties with temperature and can handle both equilibrium and non-equilibrium solidification modes. (The results presented in this paper assume non-equilibrium solidification.) Hot tearing is believed[4–8] to originate in the 'mushy' zone of the alloy near the effective solidus temperature. The fraction of liquid which is in equilibrium with the solid at any temperature within the mushy zone of the alloy and the effective solidus temperature of the alloy depend very much on whether equilibrium or non-equilibrium solidification is assumed.[9] The latent heat liberated during solidification can be incorporated into the specific heat or it may be incorporated into the enthalpy of the material.

The model has the flexibility to account for virtually all boundary conditions encountered in continuous casting. This includes, and is not limited to, a refractory header at the top of the mould (as customarily used in aluminium DC casting) and a delayed quench. In the mould region where an air gap generally forms, the position and extent of air gap formation can be computed by the analysis. Provision to compute the air gap formed is very important in the analysis of thermal stresses. The location and extent of air-gap formation have a strong effect on the temperature field in the casting, and hence on the thermal stresses, but have only a weak effect on the sump profile.[10]

The thermal stress analysis is also an axisymmetric finite-element analysis. It takes into account the deformation in the axial as well as radial directions in the casting. (Previous analyses[11,12] of thermal stresses in continuous casting assumed a plane strain deformation. A plane strain assumption is not valid when there are thermal gradients in the withdrawal directions as in the continuous casting of aluminium.) The present analysis takes into account the elastic, plastic, creep, and thermal deformations of the material. The analysis has the flexibility to account for the variation of the material properties (stress–strain relationships and creep law) with temperature. Virtually all boundary conditions encountered in continuous casting of cylindrical ingots can be accounted for in the analysis, including the effect of the metallostatic head and mould friction forces. The analysis computes the stresses in the mushy zone of the alloy below the 'coherency temperature', which is defined as the temperature above which the alloy behaves like a liquid and below which it behaves like a solid. It has been proposed[14] that the coherency temperature of an alloy plays a major role in deciding its hot tearing sensitivity.

As the details of the validation of the model are given in Refs. 1–3, only a brief description is given here for completeness. The heat-transfer analysis was checked by comparing the model-predicted temperature profiles and freezing profiles with measured quantities. (Both laboratory and industrial casting conditions for ferrous and non-ferrous continuous casting were checked.) The agreement between the measured and calculated sump profiles and temperature fields was found to be very good. The stress analysis was checked by two methods. First, the displacement field was checked by comparing stress fields predicted by the model with analytical results for simple, ideal cases (not continuous casting) where analytical solutions are available. The agreement was found to be very good. Secondly, the application of the model to continuous casting was checked by applying the analysis to the DC casting of Al–Si alloy. The residual stresses in a DC cast ingot were measured by Roth *et al.*[13] The residual stresses and positions of stress reversal predicted by the model were found to be in good agreement with those reported by Roth *et al.* The results obtained from the analysis are as follows:

(i) temperatures and cooling rates at different positions in the casting
(ii) position of the liquidus and solidus isotherms in the casting
(iii) heat flux from the casting to the mould at different heights in the mould
(iv) axial, radial, hoop, shear, and equivalent stresses at different locations in the casting
(v) principal stresses and principal planes corresponding to stresses in (iv)
(vi) maximum normalized stress and hydrostatic stress at different locations in the casting: maximum normalized stress is obtained by dividing the maximum (algebraic) stress by the

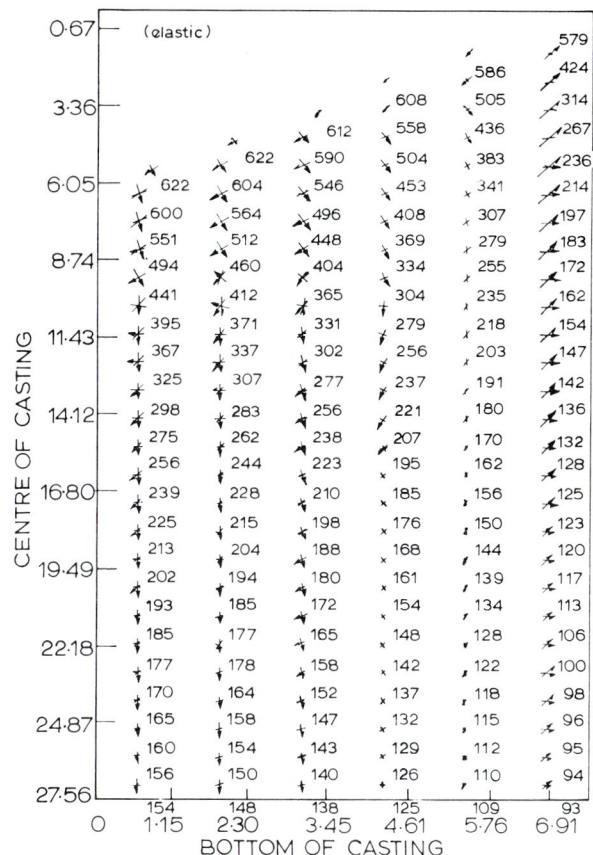

1 Elastic stress distribution in a 38 cm dia. ingot cast at 5 cm/min (stress scale: 1 cm = 65 kg/mm²)

yield stress of the material at the temperature of the location where the stress is computed
(vii) radial and axial deformations of the ingot
(viii) plastic strains at different locations in the casting.

The analysis computes the elastic stresses in the absence of relaxation as well as the stresses after plastic flow and creep.

APPLICATION OF THE MODEL TO AN Al–Mg ALLOY

Figure 1 shows the magnitudes and directions of the elastic thermal stresses in a 38 cm dia. 5082 aluminium–magnesium alloy ingot cast at 5 cm/min. (As the casting conditions assumed in the simulation may not be encountered in actual practice, this example is given only for illustration.) The principal stresses are plotted to scale as vectors normal to the principal planes. The temperatures at positions where the stress is computed are indicated by numbers next to the stress vectors in Fig. 1. To account for the coherency temperature of the alloy, the elements above the coherency temperature have been removed from the assembly and proper boundary conditions due to metallostatic head have been applied. The stresses after plastic flow and creep have been taken into account are shown in Fig. 2; note that the magnitude and direction of the principal stresses have changed considerably after plastic flow and creep. The principal stresses are normal to the solidification front in the casting near the top of the ingot. This is to be expected as the metallostatic head acts normal to the solidification front. However, the principal planes rotate as the ingot is withdrawn and become aligned with the

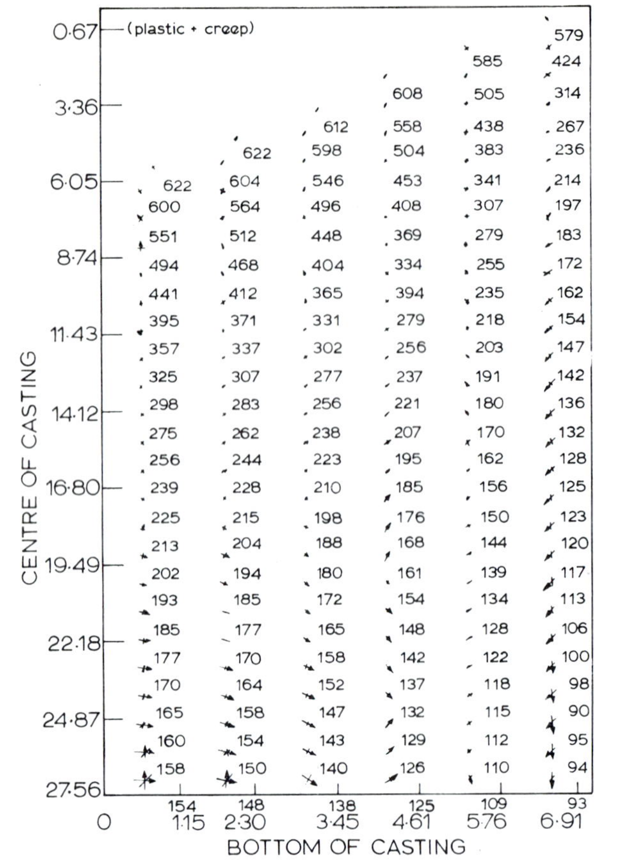

Figure (left column):

Axis label (vertical, left): CENTRE OF CASTING
Axis label (bottom): BOTTOM OF CASTING

Vertical scale values: 0.67, 3.36, 6.05, 8.74, 11.43, 14.12, 16.80, 19.49, 22.18, 24.87, 27.56

Bottom scale values: 0 · 1.15 · 2.30 · 3.45 · 4.61 · 5.76 · 6.91

(plastic + creep)

```
0.67                                              579
                                        585       424
3.36                          608       505       314
                    612       558       438       267
          622       598       504       383       236
6.05  622 604       546       453       341       214
      600 564       496       408       307       197
8.74  551 512       448       369       279       183
      494 468       404       334       255       172
      441 412       365       394       235       162
11.43 395 371       331       279       218       154
      357 337       302       256       203       147
      325 307       277       237       191       142
14.12 298 283       256       221       180       136
      275 262       238       207       170       132
      256 244       223       195       162       128
16.80 239 228       210       185       156       125
      225 215       198       176       150       123
      213 204       188       168       144       120
19.49 202 194       180       161       139       117
      193 185       172       154       134       113
22.18 185 177       165       148       128       106
      177 170       158       142       122       100
      170 164       152       137       118        98
24.87 165 158       147       132       115        90
      160 154       143       129       112        95
27.56 158 150       140       126       110        94
      154 148       138       125       109        93
```

2 Stress distribution after plastic flow and creep in a 38 cm dia. ingot cast at 5 cm/min (stress scale: 1 cm = 100 kg/mm²)

Table 1 (right column):

Table 1 Maximum normalized stress in a 38 cm dia. 5082 Al–Mg alloy ingot cast at 8·9 cm/min*

(Axis of ingot — right-hand column; Bottom of ingot)

1	3·075					
2	5·333	−0·698				
3	2·913	−0·207				
4	1·185	0·489	−0·285			
5	0·666	0·608	−0·086			
6	0·567	0·751	−0·305	−0·411		
7	0·519	0·794	−0·925	−0·696		
8	0·461	0·798	−0·595	−1·578	−1·376	
9	0·393	0·816	0·421	−2·795	−2·255	
10	0·312	0·817	0·966	−2·154	−4·214	−8·348
11	0·214	0·816	1·057	−0·708	−6·178	−7·375
12	0·149	0·750	0·933	0·401	−3·709	−9·200
13	0·120	0·688	0·811	1·119	−1·258	−6·839
14	0·113	0·691	0·829	1·476	0·517	−2·984
15	0·112	0·698	0·838	1·354	1·681	−0·263
16	0·113	0·680	0·871	1·237	1·716	1·517
17	0·113	0·651	0·908	1·164	1·451	1·495
18	0·110	0·611	0·946	1·133	1·330	1·273
19	0·104	0·566	0·961	1·179	1·282	1·160
20	0·092	0·521	0·966	1·265	1·323	1·174
21	0·078	0·483	0·972	1·344	1·472	1·319
22	0·081	0·456	0·991	1·439	1·640	1·552
23	0·081	0·450	1·035	1·542	1·828	1·813
24	0·070	0·458	1·108	1·673	2·035	2·095
25	0·074	0·476	1·201	1·829	2·259	2·385
26	0·090	0·503	1·302	2·001	2·494	2·674
27	0·111	0·524	1·387	2·164	2·725	2·957
28	−0·042	0·433	1·438	2·288	2·887	3·216

Bottom of ingot

* Axial and radial divisions are equally spaced: axial and radial divisions are 2·5 cm and 3·8 cm, respectively; negative signs indicate compresive stresses

coordinate axes at lower temperatures. The existence of a neutral plane between the surface and the centre of the casting also can be seen from the stress plots.

It is obvious from Figs 1 and 2 that the magnitude of the stresses change considerably at different locations in the casting. However, since the temperature also changes, the severity of the stresses cannot be obtained readily through the absolute stresses. In order to make the stress values more meaningful, the stress is normalized by dividing the computed stress values by the yield stress of the material at the temperature considered. The maximum (algebraic) value of this quantity is selected and is called the maximum normalized stress.

Table 1 gives the maximum normalized stress at different positions in a 38 cm dia., 5082 Al–Mg alloy ingot poured at 8·9 cm/min. (Negative signs before the stress values indicate compressive stresses. The axis of the ingot is at the right-hand column of the stress values.) The maximum normalized tensile stress increases within the mushy zone and reaches a maximum near the solidus. The peak normalized stress near the solidification front can be related to the hot tearing tendency of DC cast aluminium alloys leading to 'centreline' or 'star' cracks.

Away from the solidification front, the maximum normalized stress decreases. However, near the bottom of the casting, it increases again and reaches another maximum. The peak in the maximum normalized stress at the bottom of the ingot may be related to the formation of cold cracks in non-ferrous alloys. Some as-cast alloys (for example, 7075 aluminium alloys) have very poor ductility at room temperature. Ingots of these alloys, if the residual stresses developed during the casting are high enough, will crack during cooling. These cracks, also, will be symmetrical with respect to the axis of the ingot (since the peak of the maximum normalized stress occurs near the axis of the ingot). Since the cracks form after the ingot has been completely solidified, the cracks will be open (i.e., will not be healed by flow of liquid metal).

A third zone where a peak in the maximum normalized stress occurs is at the surface of the ingot. Referring to Table 1, a peak in the maximum normalized stress is found to occur near the mould exit. If these stresses exceed the strength of the material, surface cracks may form in the ingot.

Table 2 gives the maximum normalized stress in an ingot identical to that in Table 1 except that the ingot is cast at 10·2 cm/min. The effect of increasing the casting rate on the maximum normalized stress can be studied by comparing Tables 1 and 2; note that the maximum normalized stress near the solidification front of the ingot increases from 1·7 to 3·5 as the casting rate is increased from 8·9 to 10·2 cm/min. Thus, according to the model, the hot tearing (star cracking) tendency of the ingot should increase with increase in casting rate. This is confirmed by industrial experience.[1,15–18] From results of experiments available to the authors, the ingot simulated in Tables 1 and 2 can be cast without cracking at 8·9 cm/min, whereas it would exhibit fine centreline cracks when cast at 10·2 cm/min.

From Tables 1 and 2 it is also seen that the maximum normalized stress near the surface of the ingot increases with decrease in the casting rate. This is to be expected, as the air gap formed will be smaller and the resultant

Table 2 Maximum normalized stress in a 38 cm dia. 5082 Al–Mg alloy ingot cast at 10·2 cm/min*

1	1·801					
2	3·512	−0·233				
3	2·240	−0·103				
4	0·974	0·446	−0·434			
5	0·641	0·146	0·515			
6	0·534	0·468	−0·456			
7	0·464	0·814	−0·979	−0·647		
8	0·403	1·004	−1·294	−0·871		
9	0·340	1·001	0·062	−1·909	−1·610	
10	0·274	0·969	0·954	−3·137	−2·606	
11	0·208	0·906	1·304	−1·991	−4·635	−9·370
12	0·131	0·859	1·101	−0·414	−6·553	−7·820
13	0·089	0·767	1·050	0·731	−4·404	−9·265
14	0·078	0·659	1·030	1·480	−1·637	−8·439
15	0·080	0·676	0·990	2·389	0·729	−4·510
16	0·089	0·690	0·970	2·341	2·229	−0·517
17	0·097	0·688	0·985	1·870	3·465	2·122
18	0·101	1·668	1·012	1·655	2·661	3·304
19	0·098	0·631	1·056	1·541	2·138	1·186
20	0·087	0·581	1·084	1·513	1·947	1·876
21	0·072	0·529	1·092	1·562	1·896	1·802
22	0·073	0·486	1·097	1·625	1·975	1·893
23	0·070	0·466	1·114	1·684	2·118	2·081
24	0·056	0·462	1·152	1·757	2·258	2·320
25	0·055	0·469	1·209	1·848	2·402	2·558
26	0·070	0·485	1·275	1·952	2·546	2·779
27	0·101	0·514	1·335	2·049	2·684	2·976
28	−0·041	0·546	1·417	2·121	2·767	3·146

Bottom of ingot

Axis of ingot

* Axial and radial divisions are equally spaced: axial and radial divisions are 2·5 cm and 3·8 cm, respectively

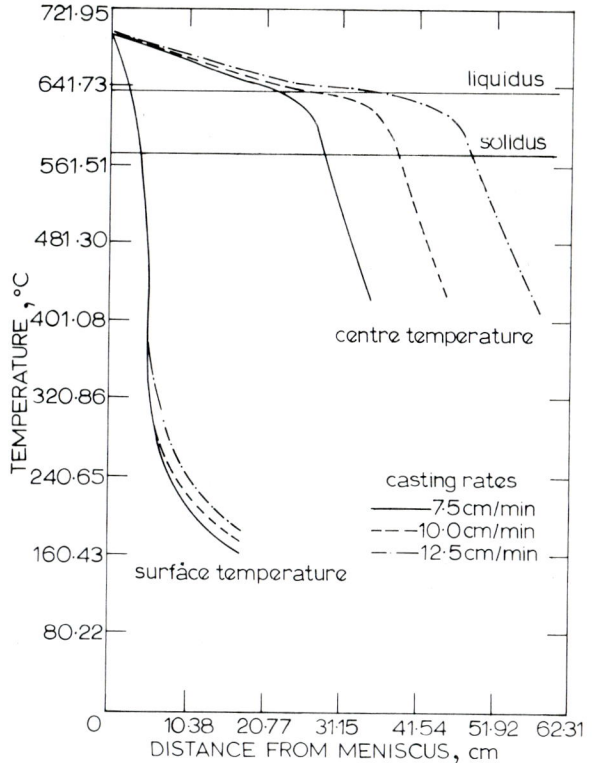

3 Effect of casting rate on temperature profile in 5084 Al–Mg alloy ingot cast at 7·5, 10, and 12·5 cm/min

reheating of the ingot will be less for lower casting rates. It is also of interest to note that the residual stresses (at the end of cooling of the ingot) decrease with increase in casting rate. The residual stresses are useful in studying the tendency of the ingot to cold cracking. However, it should be pointed out that cold cracking is influenced by both the stresses and the plastic strains. The present simulation gives both the stresses and the plastic strains for use in studying the cold cracking of continuously cast ingots.

EFFECT OF CASTING RATE ON TEMPERATURE PROFILE

Figure 3 shows the effect of casting rate on the surface and centre temperature profiles in a 5082 Al–Mg alloy ingot. The casting rate is found to have a considerable effect on the centre temperature profile. However, the surface temperature profile is practically insensitive to casting rate. (These observations are in agreement with measurements of thermal profile in continuous casting using thermocouples, etc.[15,18]) Since increasing the casting rate is found to increase the hot tearing tendency of a casting, one way to explain the hot tearing tendency of the casting is to explain it with reference to the differences in surface and centre cooling rates. Bryson[15] has done this for an A6063 alloy. He defined a 'hot cracking parameter', the difference in cooling rates between the centre and the surfaces of the casting, which he used successfully in explaining the hot tearing sensi-

tivity. Table 3 gives the centre and surface cooling rates in 16 and 38 cm dia. ingots cast at different speeds. The hot cracking parameter is indeed found to increase with an increase in casting rate. However, as can be seen from Table 3, this parameter fails to explain the size effect on hot tearing sensitivity.

EFFECT OF CASTING RATE ON FREEZING PROFILE

The freezing profile in a casting is very important as this has a predominant effect on the formation of defects. For example, a long sump in a casting may give problems

Table 3 Effect of casting rate on cooling rates at the time the centre of the casting solidifies

Casting rate, cm/min	Surface cooling rate, °C/s	Centre cooling rate, °C/s	Difference between centre and surface cooling rates
16 cm dia. casting			
10·0	1·84	8·23	6·39
17·5	2·14	13·40	11·26
25·0	2·20	14·40	12·20
30·5	1·91	18·22	16·31
37·5	2·03	19·79	17·76
38 cm dia. casting			
5·0	0·17	1·96	1·80
7·5	0·10	2·90	2·80
8·9	0·18	3·14	2·96
10·0	0·14	3·46	3·32
12·5	0·17	3·73	3·56

RADIAL DISTANCE, cm

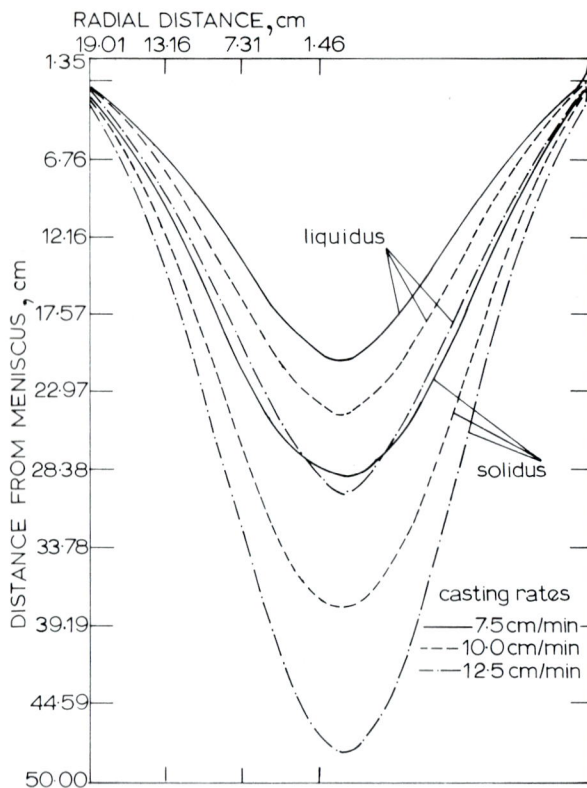

4   **Effect of casting rate on freezing profile in a 38 cm dia. 5084 Al–Mg alloy ingot cast at 7·5, 10, and 12·5 cm/min**

Table 4   Effect of casting rate on the heat flux (cgs) in a 38 cm dia. casting

Distance from meniscus, cm	Casting rate, cm/min			
	5·0	7·6	8·9	12·7
1·25	22·33	29·38	31·54	85·06
3·75	2·39	2·48	2·64	2·90
6·25	56·92	63·85	67·59	75·48
8·75	28·39	34·07	36·29	43·35
11·25	21·19	26·32	22·85	33·93
13·75	16·76	21·47	19·39	28·25
16·25	13·71	18·14	16·90	24·38

such as hot tears, surface cracks, and cold cracks in continuously cast ingots. The analysis is found to predict correctly the effect of casting rate on the hot tearing tendency of the casting. Thus, the analysis can be calibrated through experiments to determine the maximum casting rate at which an ingot may be cast without cracking. Once this information is available for typical conditions, the analysis may be used to optimize casting conditions taking into account stress-induced defects. Thus, the present simulation provides a powerful tool to the operators of continuous casting installations.

ACKNOWLEDGMENTS
The authors are grateful to the Office of Naval Research (Contract N0014-67-A-0402-0003) and the ALCOA Foundation for contributing to the support of this project. The computations were performed at the Computer Center, University of Pittsburgh, on the DEC-10 system.

with fluid flow induced defects such as segregation, etc.[19,20] The sump depth also has a great effect on the feeding of the casting. For example, a bowed sump[21] may result in the formation of porosity and associated defects.

Figure 4 shows the effect of casting rate on the position of the solidus and liquidus isotherms in a 38 cm dia. casting. Increasing the casting rate is found to increase the sump depth considerably. The strong influence of casting rate on the sump depth has been observed by many investigators[10,19,22–34] who, by means of techniques such as the breakout method, adding radioactive tracers and pellets, inserting metal rods into the sump, etc., have measured the sump depth in continuous casting. However, the striking feature of Fig. 4 is that with increase in casting speed the separation between the solidus and liquidus isotherms increases. This increases the size of the mushy zone in the casting. One of the main reasons for the increased tendency of the casting to hot tear at high casting speeds is probably its effect on the cooling rates existing in the casting and on the freezing profile in the casting.

HEAT FLUX FROM SURFACE OF INGOT
Table 4 gives the influence of casting rate on the heat flux from the surface of a 38 cm dia. casting. The heat flux is found to increase with the increase in casting speed. This is consistent with the observations of Singh and Blazek[35] who measured the heat flux in the mould of a continuous caster at different casting speeds.

SUMMARY
The analysis presented in this paper is useful in studying the formation of various thermal stress induced defects

REFERENCES
1 J. MATHEW: Ph.D. Dissertation, Department of Metallurgical and Materials Engineering, University of Pittsburgh, Pittsburgh, Pa. 15261, 1977
2 J. MATHEW AND H. D. BRODY: 'Nuclear metallurgy', (ed. R. J. Arsenault *et al.*), Vol. 20, Pt 2, 1138; 1976
3 J. MATHEW AND H. D. BRODY: *ibid.*, Vol. 20, Pt 2, 978
4 T. MORINAGA *et al.*: 'Fundamental investigation of hot tearing in light metals', paper presented at the 31st International Foundry Congress, 1964, Amsterdam
5 J. VAN EEGHEM AND DESY: *Trans. AFS*, 1965, **73**, 282
6 J. A. WILLIAMS AND A. R. E. SINGER: *J. Inst. Met.*, 1968, **96**, 5
7 S. A. METZ AND M. C. FLEMINGS: *Trans. AFS*, 1969, **77**, 329
8 S. A. METZ AND M. C. FLEMINGS: *ibid.*, 1973, **78**, 453
9 T. F. BOWER *et al.*: *Trans. Met. Soc. AIME*, 1966, **236**, 624
10 J. K. BRIMACOMBE AND F. WEINBERG: *J. Iron Steel Inst.*, 1973, **211**, 24
11 J. MORICEAU: 'Light metals', 119; 1975, New York, AIME Metallurgical Society
12 A. GRILL *et al.*: *Ironmaking Steelmaking*, 1976, **3**, 38
13 A. ROTH *et al.*: *Aluminium*, 1942, **24**, 206
14 R. J. CLAXTON: 'Continuous casting', (ed. K. R. Olen), 341, 1973, New York, AIME Metallurgical Society

15 N. B. BRYSON: 'Light metals', 429, 1972, New York, AIME Metallurgical Society

16 Definitions and Causes of Continuous Casting Defects: Nomenclature of Continuous Casting Defects Goup of the Steelmaking Division, British Iron and Steel Research Association, 1967, London, The Iron and Steel Institute

17 A. G. SHAKESPEARE: 'Continuous casting of ferrous and non-ferrous metals" 1, 1966, National Trade Press

18 D. G. HARRINGTON: Personal communication, Kaiser Aluminum, Mt. Pleasant, California

19 D. M. LEWIS AND J. SAVAGE: *Metall. Rev.*, 1956, **1**, 65

20 W. M. WOJCIK: 'Continuous casting', (ed. K. R. Olen), 217, 1973, New York, AIME Metallurgical Society

21 D. D. BEATTIE *et al.*: Proc. Conf. on modelling and simulation for applied control systems, 57; 1973, University of Bath

22 D. A. PEEL AND A. E. PENGELLY: 'Mathematical models in metallurgical process development', 186, 1970, London, The Iron and Steel Institute

23 D. A. PEEL *et al.*: 'Pilot plant studies of heat transfer, solidification and resultant structure of continuously cast aluminum', Paper presented at the AIME annual meeting, Denver, 1970

24 E. A. MIZIKAR: *Trans. Met. Soc. AIME*, 1967, **39**, 1747

25 J. W. DONALDSON AND M. HESS: 'Continuous processing and process control', (ed. T. R. Ingraham), 299, 1968, New York, Gordon and Breach

26 D. J. P. ADENIS *et al.*: *J. Inst. Met.*, 1962–63, **91**, 395

27 A. D. AKIMENKO *et al.*: 'Continuous casting of steel', 59, 1962, London, The Iron and Steel Institute

28 J. J. GAUTIER *et al.*: *J. Iron Steel Inst.*, 1970, **208**, 1053

29 R. D. PEHLKE: *ASM Metals Engr. Q.*, 1964, **4**, (2), 42

30 R. D. PEHLKE: 'Computer applications in metallurgical engineering', (ed. R. D. Pehlke and M. J. Sinnott), 75, 1964, Metals Park, Ohio, American Society for Metals

31 L. SAROFF: 'Continuous casting', (ed. K. R. Olen), 323, 1973, New York, AIME Metallurgical Society

32 J. LAIT *et al.*: *ibid.*, 171

33 J. SZEKELY AND R. T. YADOYA: *Metall. Trans.*, Oct., 1972, **3**, 2673

34 S. K. MORTON AND F. WEINBERG: *J. Iron Steel Inst.*, 1973, **211**, 13

35 S. N. SINGH AND K. E. BLAZEK: 'Heat transfer profiles in continuous casting mould as a function of various casting parameters', Paper presented at the AIME NOH BOSC Meeting, St. Louis, 28–31 March 1976

Heat-transfer characteristics in closed head horizontal continuous casting

R. Hadden and B. Indyk

Horizontal continuous casting has obvious practical advantages over vertical or curve mould casting as it makes unnecessary the deep casting pits and curve mould equipment, and, owing to the floor mounting, gives greater safety and accessibility. However, the asymmetry, an unavoidable aspect of horizontal continuous casting, introduces factors which must be considered in design and operation of the equipment. A series of thermocouples embedded in strategical positions allowed accurate measurements of point-by-point flow from the metal to the mould and cooling system, one of the most important aspects of continuous casting. It has been found that the bottom part of the mould is more effective in heat extraction from the metal than the top in all cases. However, at higher speeds a temperature inversion occurs as a result of which the heat flow through a large section of the top part of the mould is larger than that through the corresponding section of the bottom part. This has an important effect on the grain structure of the metal, and therefore influences the quality of the finished product.

The authors are at the University of Strathclyde

LIST OF SYMBOLS

A = heat-transfer area
h = heat-transfer coefficient
h_1 = heat-transfer coefficient at hot end
h_2 = heat-transfer coefficient at cold end
$LMTD$ = logarithmic mean temperature difference
$\bar{T}$ = mean temperature
$\bar{T}_{M_1}$ = mean temperature of metal at hot end
$\bar{T}_{M_2}$ = mean temperature of metal at cold end
$\bar{T}_{W_1}$ = mean temperature of water at cold end
$\bar{T}_{W_2}$ = mean temperature of water at hot end
V = casting speed
$\dfrac{dQ}{dt}$ = rate of transfer of heat
$\dfrac{dQ_W}{dt}$ = rate of heat absorption by water

θ = temperature of the mould
θ_1 = the surface temperature of the mould adjacent to metal
θ_2 = the surface temperature of the mould adjacent to water jacket
$\Delta\theta = \theta_1 - \theta_2$ = the temperature difference across the mould wall

In recent years there has been an increasing interest in horizontal continuous casting. There are obvious advantages: there is no requirement for a tall building since the headroom requirements are minimal, and deep casting pits are dispensed with, the component units being floor mounted. These features contribute to a significant saving in capital investment. In addition a horizontal caster is safer than the vertical one since, in the event of a breakout or mould overflow, the molten metal can be collected in suitably placed sand boxes with minimal damage to the machine components. It can further be argued that the floor-level location of the controls and casting mechanism are more accessible for routine maintenance or in the event of a major breakdown.

THE HORIZONTAL CASTER

The general layout of the caster is shown schematically in Fig. 1. A more detailed view of the casting head in relation to the withdrawal mechanism is shown in Fig. 2.

The tundish is heated by an air/gas burner, the flame of which is directed against the lower part of the crucible. A water-cooled composite graphite mould is used. The first section of the mould, which is cement sealed to the tundish, has a designed thick section to allow for wastage by burning and is easily replaced. The second section, which is made from a fine-grained graphite and consequently has a superior machine finish, is enclosed on its upper and lower faces by a book-end-type steel

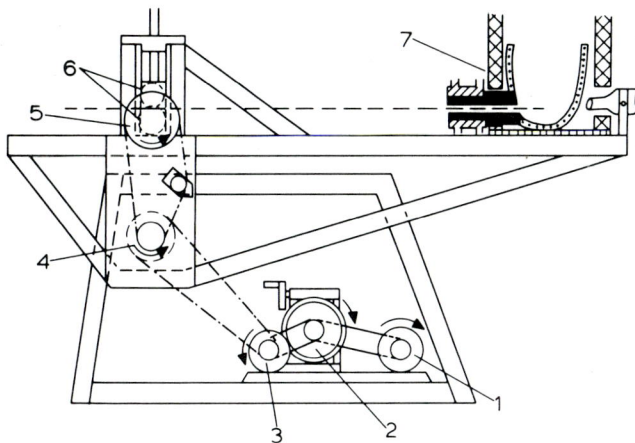

1 electric motor; 2 variator; 3 spur reduction gear; 4 chain sprocket; 5 magnetic plate clutch; 6 main drive rolls; 7 casting head

1 Schematic layout of horizontal caster

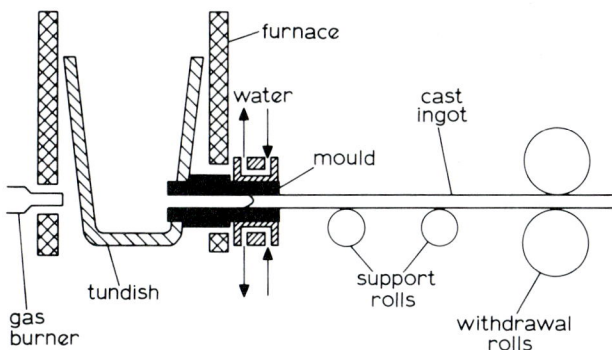

2 Casting head and withdrawal mechanism

water jacket. The upper and lower sections of the water jacket have independent inlet and outlet coolant channels.

The charge is prepared in a hf induction furnace and lip poured into the preheated tundish. A keyed starter bar, which has the same dimensions as the mould orifice, plugs the mould exit and allows the first metal to be poured. After a suitable delay of about 10 s to allow freezing into the key, the starter bar is extracted laterally by the withdrawal rolls. The casting is withdrawn intermittently and supported by carrier rolls before passing through the withdrawal rolls. Metal temperature in the tundish is continually monitored throughout the casting operation by an immersed chromel/alumel thermocouple. Any necessary temperature adjustment is by the gas burner control. The tundish is replenished with molten charge to keep the level of the molten metal constant.

The heat characteristics of the horizontal mould were studied by examining the increase in mould cooling water temperature and temperature distribution within the graphite mould during the casting. Aluminium, brass, and gunmetal were used to cast billets of various breadths and thicknesses. The temperature of the molten metal was controlled to within $\pm 5°C$. The water flow was maintained constant to each mould and the temperature of incoming and outgoing water was constantly monitored.

A range of casting speeds was investigated by gradually increasing the withdrawal speed at regular intervals. Preliminary trials established that the inlet and outlet water temperatures stabilized after 60 s from a change in casting speed.

The temperature distribution within the mould was measured by means of chromel/alumel thermocouples apportioned throughout the length and thickness of the upper and lower broad face sections of the graphite mould, the junctions being embedded in the mould and jammed tightly with graphite powder to ensure good thermal contact. The temperature recording was by a high-speed scanning system using digital voltmeter, the output of which was recorded on the punch tape.

HEAT TRANSFER

Consideration of the complete energy balance of the continuous casting process can be subdivided into three distinct sections: the tundish, the solid billet when it leaves the mould, and the mould.

The main purpose of the tundish is to direct the liquid metal into the mould at a steady rate and temperature, and maintain a steady hydrostatic pressure of the metal at the entrance to the mould. In commercial casters where the volume of liquid metal in the tundish is large, the steady state is achieved fairly quickly owing to the fact that heat losses in the tundish constitute only a small fraction of the heat content of the molten metal. Where the heat losses in the tundish are a significant fraction of the heat content of the molten metal it is necessary to supply an additional source of heat in order to achieve a steady state in this section of the caster. Interest in heat transfer here is usually limited to economic consideration and the achievement of a uniform temperature of the liquid metal. Further, owing to the different ways in which this can be attained and the different geometry of the tundish, little purpose would be served in concentrating on this section.

Heat transfer in the second section is important where the rate of cooling may influence the ultimate grain structure of the finished product. In particular, it warrants detailed consideration where, owing to the casting conditions and physical properties of the metal, the liquid sump in the billet may be extending beyond the mould. However, this section is easier to tackle from both the theoretical and practical points of view.

Heat transfer in the third section, i.e. the mould, is of primary importance as it is the overriding factor for the ultimate success of the whole process. For this reason, the authors concentrate on the mould.

The overall heat balance in the mould could be simply represented by considering the mould as a countercurrent recuperator with the metal flowing in one direction and the coolant in the other. Thus, the following equation can be written:

$$\frac{dQ}{dt} = A \cdot \frac{h_2(\bar{T}_{M_1} - \bar{T}_{W_2}) - h_1(\bar{T}_{M_2} - \bar{T}_{W_1})}{\ln\frac{h_2(\bar{T}_{M_1} - \bar{T}_{W_2})}{h_1(\bar{T}_{M_2} - \bar{T}_{W_1})}} \qquad (1)$$

Equation (1) allows for the fact that the heat-transfer coefficient at the entrance to the mould is significantly different from that at the outlet, but does not describe the manner in which it changes along the mould. A further simplification could be introduced if an average heat-transfer coefficient is used instead of the actual one; thus,

$$\frac{dQ}{dt} = A \cdot h \cdot (LMTD) \tag{2}$$

where

$$LMTD = \frac{(\bar{T}_{M_1} - \bar{T}_{W_2}) - (\bar{T}_{M_2} - \bar{T}_{W_1})}{\ln \dfrac{\bar{T}_{M_1} - \bar{T}_{W_2}}{\bar{T}_{M_2} - \bar{T}_{W_1}}} \tag{3}$$

dQ/dt is evaluated from the heat gained by the coolant, and A is known from the geometry of the mould. The only difficulty is presented by the mean temperature of the metal at either end, although in some cases where the conductivity of the metal is high and casting speed is low it may be assumed to be approximately constant across the billet.

This approach is useful in assessing general energy requirements, but gives little detailed information about local solidification processes having a major bearing on the structure of the finished product. For this a detailed investigation of local heat-transfer phenomena is necessary.

A typical relationship between the rise in the coolant temperature and casting speed, where other variables are kept constant, is shown in Fig. 3. It is useful to note that the relationship is not linear, as might be expected.

The shape of the solidification front is to a certain extent indicated by the surface appearance of the billets (Fig. 4). In this respect the rectangular cross-section of the billet is an advantage, since a similar visual observation is not possible on a cylindrical billet.

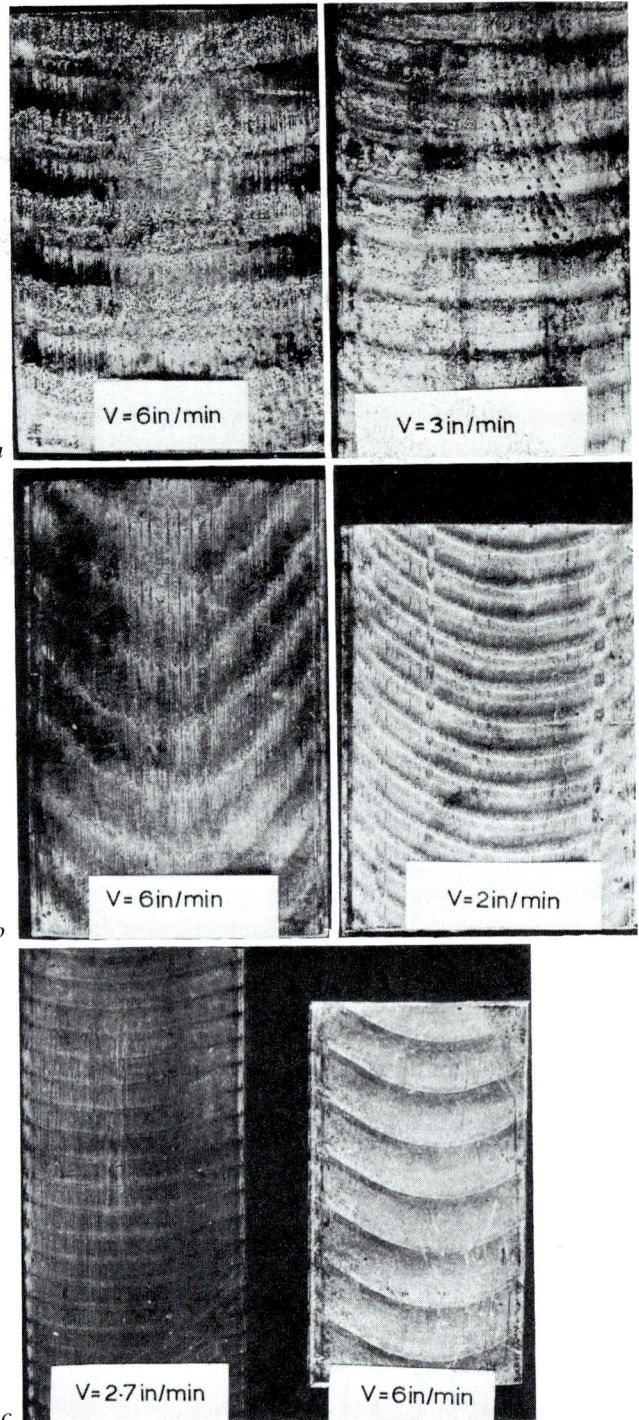

a aluminium, 7.5×2.0 cm; b gun metal, 7.5×2.0 cm; c brass, 7.5×0.5 cm

4   Typical surface appearance of billet

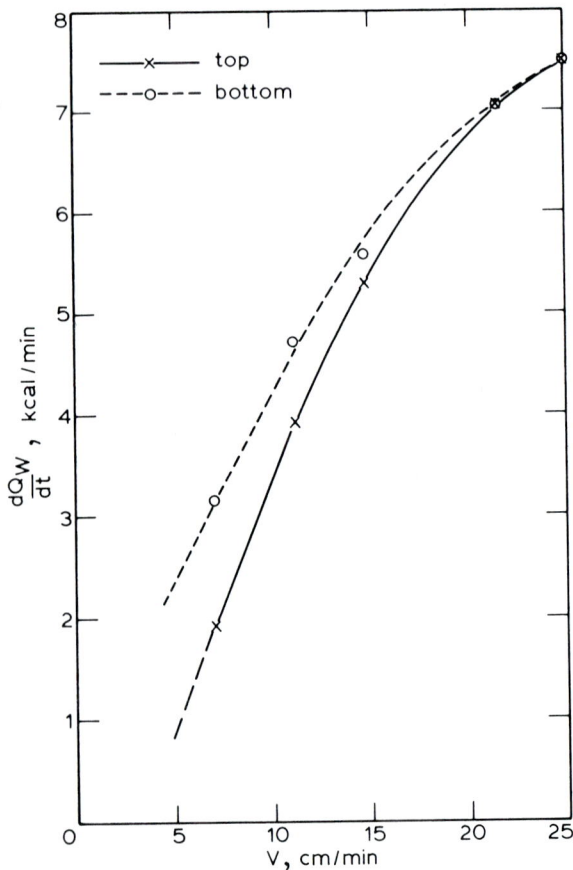

3   Relation of heat extraction rate by water to casting speed

A typical macrostructure of slabs at various casting speeds is shown in Fig. 5. This clearly indicates that solidification, and therefore heat extraction by the mould, is highly asymmetric in vertical direction at low casting speed, but as the speed increases symmetry is approached.

The typical temperatures of the inside and outside surface of the mould were obtained by extrapolation of the values recorded by pairs of adjacent thermocouples situated close to either surface (Figs 6 and 7). By separating the records of the top and bottom parts of the

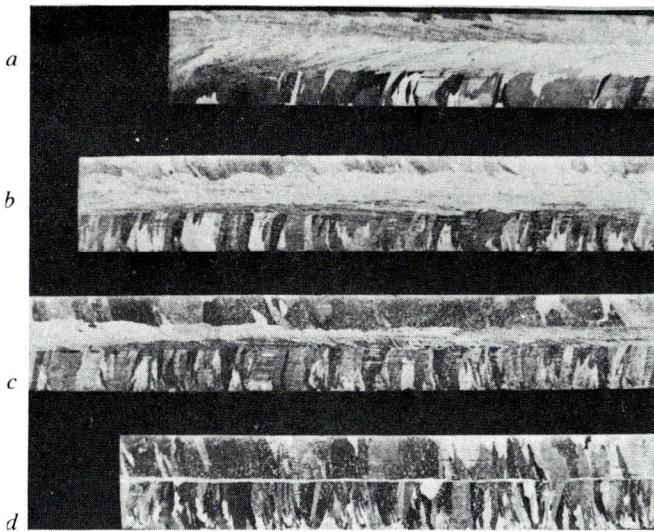

a 5 cm/min; *b* 7·5 cm/min; *c* 10 cm/min; *d* 15 cm/min

5 Typical macrostructure of slabs at various casting speeds

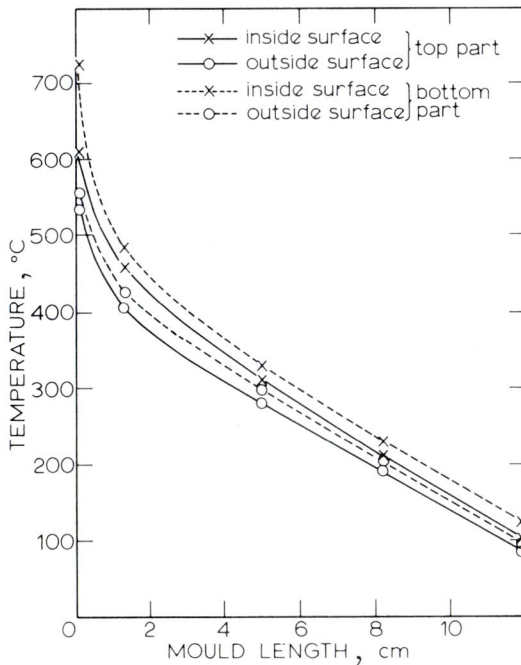

6 Surface temperature of mould: *V* = 7·0 cm/min

mould it was possible to determine local heat transfer through each part.

DISCUSSION

Physical properties such as thermal conductivity, surface tension, angle of contact, viscosity, and UTS of metals and alloys in the neighbourhood of the melting point are virtually unknown. The authors place no trust in values of thermal conductivity deduced from the Lolenz relation, which is approximately valid for ambient temperatures but has not been thoroughly checked at higher temperatures. The subsequent discussion therefore avoids problems which incorporate these properties and is thus limited in its scope. A method of measuring the temperature profile, in the metal billet in both liquid and

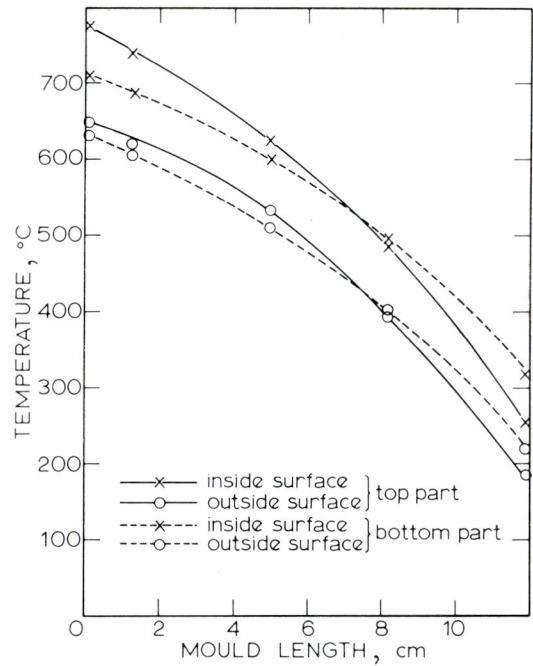

7 Surface temperature of mould: *V* = 25 cm/min

solid state, has been developed by the authors but so far this has been applied only to vertical continuous casting and is the subject of a further paper.

For the purpose of detailed analysis of heat transfer the mould has been divided into ten equal segments and heat-transfer rate has been evaluated for both the top and bottom parts of each segment for various casting speeds. Typical examples of such evaluations are given in Tables 1 and 2. It is seen that the total heat transfer through the bottom part of the mould is higher than that through the top in both cases, but the proportion changes. While at low speed 55% of the total heat extracted from the billet goes through the bottom part, at a higher speed it amounts to 53%. Although this asymmetry seems small, it is evident from Fig. 5 that it has a profound effect on grain structure.

Table 1 Heat flow through horizontal broad face mould walls for the cast: *V* = 7·5 cm/min

Segment No.	Top				Bottom			
	θ_1	θ_2	$\Delta\theta$	$\frac{dQ}{dt}$, cal/s	θ_3	θ_4	$\Delta\theta$	$\frac{dQ}{dt}$, cal/s
1	551	460	55	82	570	480	90	134
2	420	375	45	67	440	395	45	67
3	380	310	40	60	365	330	35	53
4	290	260	30	45	305	275	30	45
5	235	210	25	38	253	225	30	45
6	185	170	15	22	205	185	20	30
7	145	130	15	22	163	145	20	30
8	100	85	15	22	125	105	20	30
9	80	65	15	22	90	70	20	30
10	50	35	15	22	55	35	20	30

Total $\frac{dQ}{dt}$ = 402   Total $\frac{dQ}{dt}$ = 494

$\frac{dQ_w}{dt}$ = 410 cal/s $\frac{dQ_w}{dt}$ = 500 cal/s

Table 2   Heat flow through horizontal broad face mould walls for the cast: *V* = 24 cm/min

Segment No.	Top θ_1	θ_2	$\Delta\theta$	$\dfrac{dQ}{dt}$, cal/s	Bottom θ_3	θ_4	$\Delta\theta$	$\dfrac{dQ}{dt}$, cal/s
1	760	640	120	180	700	635	65	96
2	715	605	110	165	670	600	70	105
3	660	565	95	143	630	555	75	112
4	600	500	100	150	585	500	85	127
5	530	440	90	135	530	440	90	134
6	450	360	90	135	475	370	105	157
7	360	280	80	120	410	295	115	172
8	260	210	50	75	330	220	110	165
9	170	135	35	53	280	100	100	150
10	85	60	25	38	160	75	85	127

Total $\dfrac{dQ}{dt}$ = 1 194   Total $\dfrac{dQ}{dt}$ = 1 345

$\dfrac{dQ_w}{dt}$ = 1 200 cal/s $\dfrac{dQ_w}{dt}$ = 1 350 cal/s

A closer inspection of Tables 1 and 2 reveals another significant difference. At low casting speeds the temperature of the mould is always higher at the bottom than at the corresponding position of the top part of the mould. At higher speeds, however, over a considerable part of the mould the situation is reversed. It is also noticeable that both the temperature gradient and the heat-transfer rate undergo a similar inversion. The heat flux through the bottom part of the mould shows a maximum, while no such maximum exists in the case of the top part.

The probable explanation of this temperature inversion may be found by considering the flow pattern of the liquid metal (Fig. 8), and the nature of heat transfer at the metal/mould interface.

At low speeds, the periodical flow of liquid metal is more orderly, resembling laminar flow in spite of a sharp bend. There is a relatively small amount of mixing of metal between the hotter central layers and the cooler layers at the boundaries, and a steady temperature gradient is established to a larger degree than at high speeds. The flow lines are more congested at the bottom, resulting in higher temperature and a steeper temperature gradient close to the bottom interface. At high

speeds the higher degree of turbulence, due to the vortex at the bend, results in the transfer of hot metal from the centre to the neighbourhood of the top interface.

Further along the mould the metal begins to solidify, forming a skin of solid metal directly in contact with the inner face of the mould. At low speeds, solidification begins sooner and the liquid sump is much shallower. Steady convective currents within the metal have a chance to develop so that hotter metal gradually migrates upwards, delaying the solidification in the upper region. The weight of the metal is supported by the bottom part of the mould ensuring much better thermal contact and thus higher heat transfer than at the top. When the solid skin is thick enough not to yield to hydrostatic pressure exerted by the column of liquid metal, a small gap develops between the metal and mould surfaces, markedly reducing the heat-transfer rate. Again, owing to the weight of the metal, better contact is maintained at the bottom than at the top.

At higher speeds, owing to increased depth of the liquid sump, the solidified skin is thinner and weaker and consequently yields to the hydrostatic pressure, ensuring good thermal contact over much larger range of the

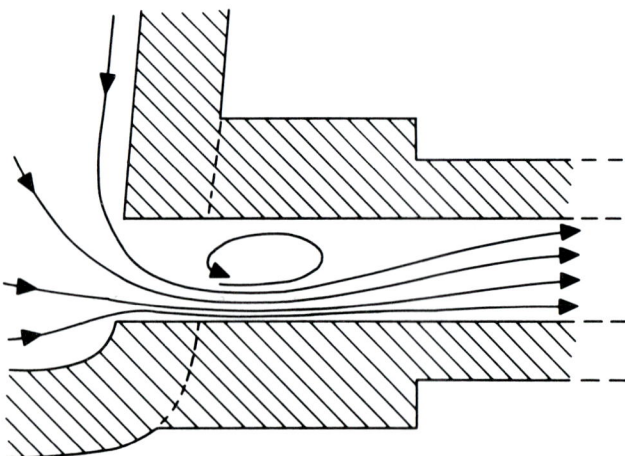

8   Flow patterns at entrance of mould

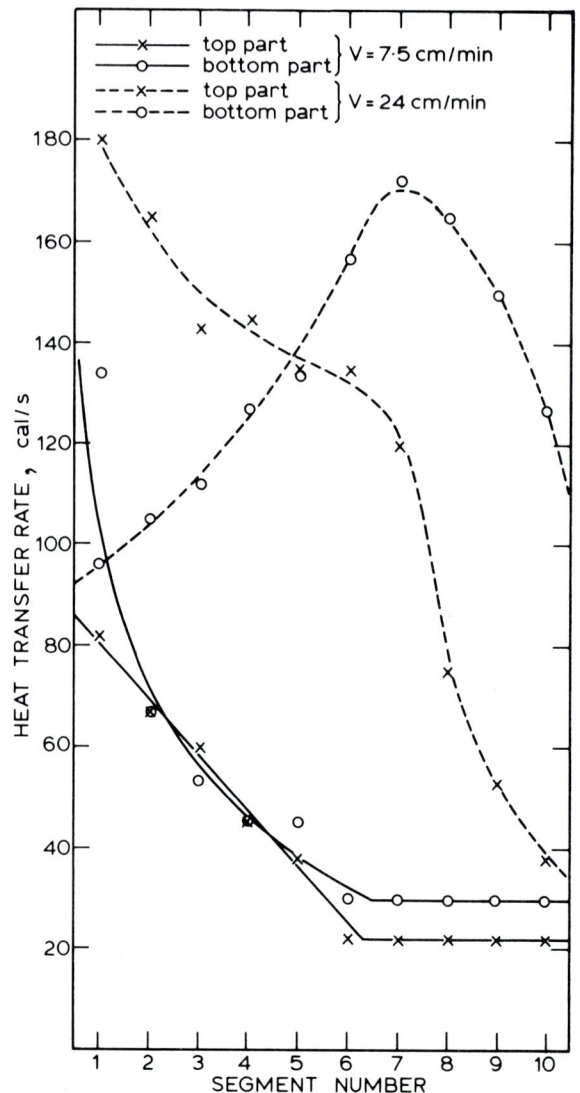

9   Heat transfer through mould

upper part of the mould. At the same time, the convective currents within the liquid metal result in a larger fraction of heat being transferred by this section of the top part of the mould. At the point where a significant gap develops, the situation reverts to the one at low speeds, thus producing the temperature inversion.

It is perhaps worth noting that the curvature of the surface temperature profiles in Figs 6 and 7 is reversed. However, it is necessary to point out that although this tendency is present in most cases, the curvature can be altered to a marked degree by the rate of flow of water.

CONCLUSIONS

1 A marked asymmetry in the heat extraction rate which is a function of the casting speed is present in horizontal continuous casting. This strongly influences the crystalline structure of the finished product.

2 At low casting speeds the rate of heat transfer through the top part of the mould is lower than that through the bottom at all points along the mould. However, as the casting speed increases, the temperature inversion occurs, where over a large part of the mould the temperature and consequently the heat-extraction rate is higher at the top than at the bottom.

Thermal analysis of solidification of aluminium alloys during continuous casting

I. Jin and J. G. Sutherland

An improved mathematical model for the distribution of temperature and solid fraction in a dendritically solidifying aluminium alloy has been developed, based on non-equilibrium solidification theory. Using the model, the time-dependent variation of temperature distribution, or for thin ingot casting the geometric variation, in various aluminium alloys during solidification is calculated for both uniform boundary heat-transfer conditions and a perturbed thermal boundary condition. It is shown that a perturbation of the heat transfer at the boundary causes a distortion of the solidifying shell which increases the non-uniformity of the heat transfer. This mechanism is dependent on the alloy composition and on the initial casting conditions. A shorter freezing range alloy and a higher initial heat-transfer coefficient produce a stronger shell distortion tendency, and hence a less stable heat-transfer condition.

The authors are with Aluminium Company of Canada Ltd, Kingston Research Centre, Ontario, Canada

Considerable theoretical analyses on the heat-transfer problem during alloy solidification have been developed in recent years[1-4] in order to predict the effect of casting conditions and materials properties on the temperature distribution in a solidifying alloy. A correct analysis of the mathematical boundary value problem involving a phase change must be composed of three essential parts: the heat equation, the imposed initial and boundary conditions, and the correct mechanism of solidification. In previous work it has been a common practice to assume that an alloy solidifies over the temperature range between the liquidus and solidus temperatures. This assumption may be reasonable for casting conditions involving such slow cooling that the solidification process is not dendritic, but for dendritic solidification, which is the case for most continuous casting conditions, these previous models fail to give realistic temperature and solid fraction distributions in the liquid–solid region.

The purpose of the present work is to develop a mathematical model for the temperature distribution in a solidifying alloy during high-speed thin-slab continuous casting, and to examine the effect of alloy properties and casting conditions on the thermal behaviour of solidifying alloys. This is done with the ultimate aim of clarifying, in particular, why the tendency for shell distortion is greater in a short freezing range alloy than in a long freezing range alloy for an identical casting condition, and why a faster cooling condition enhances the shell distortion tendency for a given alloy. Until now,

both of these questions have remained essentially unanswered, although the effects have been experienced for some considerable time.

In the present work, interest is restricted to the solidification of ideal binary aluminium alloys.

MATHEMATICAL MODEL

To represent the physical situation of continuous casting, the cartesian coordinate system (x, z), whose origin is at the wall of the meniscus, is chosen; positive x is taken toward the slab centre and positive z toward the casting direction. At $z = 0$, the melt is continuously supplied at a constant temperature T_p and the cooling alloy is moving at a constant speed v in the z-direction. At the slab surfaces, $x = 0$ and $x = L$, heat is removed at a rate characterized by a heat-transfer coefficient h which is a function of z. For high-speed casting, the heat flow is practically one dimensional. The heat balance on a differential element which moves at the casting speed in the z-direction gives the partial differential equation governing the alloy temperature T as a function of distance x and time t:

$$\rho(C + \alpha \cdot \Delta H)\frac{\partial T}{\partial t} = \lambda \frac{\partial^2 T}{\partial x^2} + \beta\left(\frac{\partial T}{\partial x}\right)^2 \quad (1)$$

where ρ is the density, C is the specific heat, α is the amount of solid formed per unit temperature drop of solid–liquid mixture, ΔH is the latent heat of fusion, λ is the thermal conductivity, and β is a parameter $(\partial\lambda/\partial T)$ defining the variation of thermal conductivity with temperature.

The initial and boundary conditions are:

$$t = 0, \qquad T = T_p, \qquad 0 < x < L \tag{2}$$

$$t > 0, \qquad \lambda\frac{\partial T}{\partial x}\bigg|_{x=0} = h(T_S - T_W), \quad \text{at } x = 0 \tag{3}$$

$$\frac{\partial T}{\partial x} = 0, \qquad\qquad \text{at } x = \frac{L}{2} \tag{4}$$

where T_S is the slab surface temperature and T_W is the cooling water temperature.

The numerical solution of equation (1) for the initial and boundary conditions can be obtained if the temperature-dependent parameters are determined. The crucial thing to note is that α, λ, and β are dependent on the temperature as well as on the solidification process at hand. The cooling rate is assumed to be sufficiently fast to result in dendritic solidification, and, as is reasonable for most substitutional alloys, the solute diffusion in the solid is assumed to be negligible compared with that in the liquid. Then, for alloy systems with a nearly constant partition coefficient K_O, the interface concentration of solid C_S at temperature T is[5,6]:

$$C_S = K_O C_O (1 - f_s)^{K_O - 1} \tag{5}$$

where f_s is the fraction of solid and C_O is the bulk alloy composition. The interface composition also has to satisfy the local equilibrium condition, i.e.

$$C_S = \frac{1}{m_S}(T - T_m) \tag{6}$$

where T_m is the melting temperature of pure aluminium and m_S is the solidus slope. From equation (5),

$$f_s = 1 - \left(\frac{C_S}{K_O C_O}\right)^{1/(K_O - 1)} \tag{7}$$

By differentiating equation (7) with respect to T, we obtain

$$\alpha = \frac{1}{K_O C_O}\frac{1}{m_S}\frac{1}{K_O - 1}\left(\frac{C_S}{K_O C_O}\right)^{(2 - K_O)/(K_O - 1)} \tag{8}$$

The thermal conductivity in the liquid–solid mixture may be taken as:

$$\lambda = \lambda_s f_s + (1 - f_s)\lambda_1 \tag{9}$$

where λ_S and λ_1 are thermal conductivities of solid and liquid, respectively. By differentiating equation (9) with respect to T,

$$\beta = (\lambda_s - \lambda_1)\frac{\partial f_s}{\partial T} + f_s\frac{\partial \lambda_s}{\partial T} + (1 - f_s)\frac{\partial \lambda_1}{\partial T} \tag{10}$$

When the temperature reaches the eutectic temperature, it is assumed that the eutectic reaction occurs linearly in the temperature range of 0.1°C, i.e.,

$$\alpha = \frac{1 - f_s^*}{0.1} \tag{11}$$

where f_s^* is the fraction of solid at the beginning of the eutectic temperature.

For the numerical solution of the heat equation, an implicit finite-difference approximation method was

Table 1 Values of casting and physical parameters used in computation

Parameter		Al–1·3 wt-% Mn	Al–4·5 wt-% Mg
L	Slab thickness, cm	1·27	1·27
T_p	Pouring temperature, °C	680	670
T_W	Water temperature, °C	20	20
h	Heat-transfer coefficient, W/cm² °C	0·209 0·418	0·209 0·418
v	Casting speed, cm/s	11·43 22·86	11·43 22·86
λ	Thermal conductivity, W/cm °C		
	Solid	2·09	2·09
	Liquid	0·92	0·92
ρ	Density, g/cm³		
	Solid	2·7–0·00023T	2·7–0·00023T
	Liquid	2·36	2·36
C	Specific heat, J/g °C		
	Solid	0·89 + 0·000468T	0·89 + 0·000468T
	Liquid	1·08	1·08
ΔH	Latent heat of fusion, J/g	393	393
K_O	Partition coefficient	0·93	0·28
T_m	Melting temperature of Al, °C	660	660
$T_{liq.}$	Liquidus temperature, °C	659	642
$T_{eut.}$	Eutectic temperature, °C	658·5	451

used. The computed results by the mathematical model described above are those which simultaneously satisfy the principle of heat flow and the dendritic solidification mechanism idealized by the classical non-equilibrium solidification theory.

UNIFORM HEAT TRANSFER
The effect of heat-transfer coefficient and freezing range of an alloy on the macroscopic steady-state temperature distribution may best be examined by comparing isotherms under uniform heat-transfer boundary conditions. For this purpose the alloy systems, Al–1·3 wt-% Mn and Al–4·5 wt-% Mg, have been chosen. In these systems the liquidus and solidus slopes are nearly constant. Also, these alloys should provide a good comparison between short and long freezing range systems, respectively.

Values of parameters used for the present computation are shown in Table 1. (Most of the physical and thermodynamic constants are obtained from Ref. 7.) Isotherms from the computed results are shown in Figs 1 and 2. From the isotherms, it is noted that, for a constant heat-transfer condition, the region between the liquidus and eutectic isotherms is very narrow in an Al–1·3 wt-% Mn alloy, and is wide in an Al–4·5 wt-% Mg alloy. This indicates that the solid surface formation distance of a short freezing range alloy is small, whereas a long freezing range alloy forms its surface over a long distance. From a structural point of view, a wide scattering of microporosity from centre to surface would thus be expected in a long freezing range alloy slab, while a concentrated porosity would be expected in the centre in a short freezing range alloy slab. For a given alloy, a

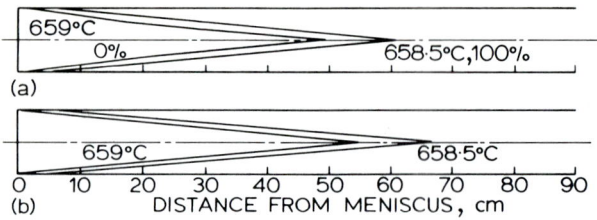

a $v = 11.43$ cm/s, $h = 0.209$ W/cm^2 °C;
b $v = 22.86$ cm/s, $h = 0.418$ W/cm^2 °C

1   Isotherms for two different casting conditions: Al–1.3 wt-% Mn; slab thickness = 1.27 cm

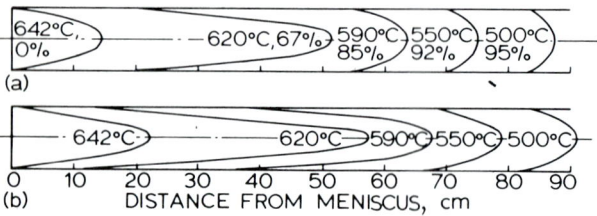

a $v = 11.43$ cm/s, $h = 0.209$ W/cm^2 °C;
b $v = 22.86$ cm/s, $h = 0.418$ W/cm^2 °C

2   Isotherms and fractions solidified for two different casting conditions: Al–4.5 wt-% Mg; slab thickness = 1.27 cm

higher heat-transfer coefficient should produce less surface porosity; however, this effect would be minimal in an extremely short freezing range alloy such as Al–1.3 wt-% Mn.

NON-UNIFORM HEAT TRANSFER

Experimentally, the behaviour of a solidification system does not generally follow the course predicted by the simple idealized theoretical model. For instance, solidifying an alloy on a rigid flat uniformly cooled mould does not generally result in uniform heat transfer, and, with virtually the same macroscopic isotherm configuration, a variety of different casting results are obtained. In this paper the instability of the thermal boundary condition in a solid mould cooling system will be examined.

The initial heat-transfer coefficient uniquely determines the temperature gradient near the surface in a given solidification system, and the time-dependent temperature distribution determines the stress and strain distributions in the solidifying alloy. The strain development (or the shell distortion) in turn usually changes the heat-transfer coefficient; thus, the intrinsic non-uniformity which exists in all real casting systems propagates in time as well as geometrically.

The driving force for the shell distortion ψ is the absolute value of the gradient of the cooling rate in a solidifying alloy, i.e. for one-dimensional cooling

$$\psi = \left| \frac{\partial \dot{T}}{\partial x} \right| \qquad (12)$$

To examine this function numerically, a small perturbation of heat-transfer coefficient (20% reduction) was introduced to an idealized uniform heat-transfer boundary condition, and the corresponding change of the temperature field in time was calculated. By choosing an

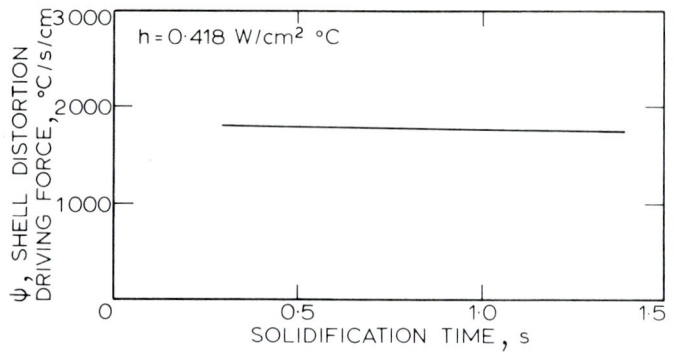

3   Variation of the shell distortion driving force ψ for Al–1.3 wt-% Mn alloy

arbitrary small distance near the surface (0.0635 cm) and an arbitrary small time increment (0.01 s), the variation of the shell distortion driving force (rate of change of thermal gradient near the surface) was calculated as a function of solidification time. Computed results of the function during the shell formation period are shown in Fig. 3 for Al–1.3 wt-% Mn and in Fig. 4 for Al–4.5 wt-% Mg.

It is noted that, for a constant heat-transfer coefficient, the shell distortion driving force in Al–1.3 wt-% Mn alloy is nearly the same throughout the solidification period, whereas that in Al–4.5 wt-% Mg alloy starts from a small value and increases in time, and, after a significant time, reaches nearly the same level as in Al–1.3 wt-% Mn alloy. Thus, when a short freezing range alloy solidifies, shell distortion starts as soon as the shell is formed and the non-uniformity of the thermal boundary condition becomes stronger in time. However, with a long freezing range alloy, the shell distortion driving force at the early stage is minimal. At later stages, when the shell distortion driving force is significant, the shell is thick and therefore the driving force presumably remains in the form of thermal stress rather than distortion in the shell. The effect of heat-transfer coefficient is evident in Fig. 4. The shell distortion driving force and its rate of increase decrease with decreasing heat-transfer coefficient, indicating that a higher heat-transfer coefficient enhances shell distortion and hence non-uniformity.

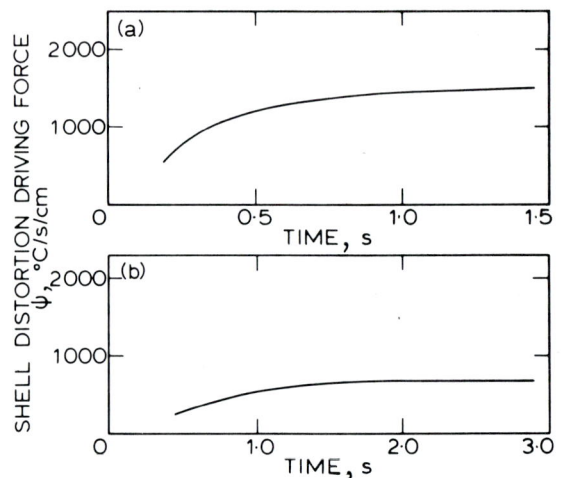

a $h = 0.418$ W/cm^2 °C;   *b* $h = 0.209$ W/cm^2 °C

4   Variation of the shell distortion driving force ψ for Al–4.5 wt-% Mg alloy

The physical background for the perturbation results is not explicitly shown because the overall analysis has been carried out by the numerical method. However, a close examination of the computational equation indicates that, in a short freezing range alloy, because the solidified shell is already 100% solid, a perturbed thermal boundary condition contributes initially only to a surface temperature change, whereas, in a long freezing range alloy, the solidifying shell is a liquid–solid mixture and a perturbed heat flux causes a change of surface solid fraction (which requires latent heat) and hence it causes a smaller initial change of surface temperature. Therefore, the amount of surface temperature change in a long freezing range alloy is smaller than that in a short freezing range alloy.

The present analysis was concerned mainly with the thermal behaviour of solidifying alloys to clarify the basic difference in their intrinsic thermal properties. It is well understood that a complete analysis of shell distortion is a combination of thermal analysis and mechanical analysis and in some circumstances the mechanical aspects may be the most important. However, the mechanical properties of a solidifying alloy are very complex functions of temperature, alloy composition and of the solidification process (or cooling rate). Because of the limited knowledge of such properties at this time, the results presented here are those from a partial analysis. Even alone, however, this thermal analysis shows that the shell distortion tendency in solidifying alloys is decreased by increasing freezing range and by decreasing cooling rate.

SUMMARY AND CONCLUSIONS

1 A mathematical model, based on the classical non-equilibrium solidification theory, has been developed to calculate the temperature and solid fraction distributions in a dendritically solidifying binary alloy.

2 Computed isotherms of a thin slab under a uniform heat-transfer condition indicate that, in a short freezing range alloy, solid surface formation distance is short and porosity is concentrated near the centre line, whereas, in a long freezing range alloy, solid surface formation distance is long and porosity is diffuse.

3 The shell distortion driving force at the early stage of solidification is greater with higher heat-transfer coefficients and with alloys which have a short freezing range.

REFERENCES

1 D. J.-P. ADENIS *et al.*: *J. Inst. Met.*, 1962–63, **91**, 395

2 E. A. MIZIKAR: *Trans. Met. Soc. AIME*, 1967, **239**, 1 747

3 E. Y. KUNG AND J. C. POLLOCK: *Simulation*, 1968, **1**, 29

4 J. SZEKELY AND V. STANEK: *Metall. Trans.*, 1970, **1**, 119

5 W. G. PFANN: *Trans. AIME*, 1952, 747

6 H. D. BRODY AND M. C. FLEMINGS: *Trans. Met. Soc. AIME*, 1966, **236**, 615

7 'Aluminium: Vol. 1, Properties, Physical metallurgy and phase diagrams', (ed. K. R. Van Horn), 1967, Metals Park, Ohio, American Society for Metals

Continuous observation of hot crack-formation during deformation and heating in SEM

H. Fredriksson and B. Lehtinen

A new technique for studying hot crack formation has been developed. Small cast tensile test samples were placed, heated and strained, in a scanning electron microscope. The deformation process was recorded on a TV monitor. The temperature and the stress–strain curve were registered at the same time. Different alloys were studied (Al–Cd, Al–Sn, and Pb–Sn). The transition from ductile to brittle fracture at the moment of melting was registered for the different alloys. The fracture surfaces were studied. The movement of the melt between the dendrite arms was also registered. The observation is discussed from the point of view of different brittle theories.

H. Fredriksson is with the Royal Institute of Technology, Stockholm, and B. Lehtinen is with the Institute of Metal Research, Stockholm, Sweden

One of the most difficult problems to master on casting is the formation of hot cracks. A lot of work has been done to explain the formation process of these cracks. During the first part of this century the theory was proposed that hot cracks were formed in a brittle region roughly 100°C below the very last solidification temperature. Between 1940 and 1950 the hypothesis was put forward[1-2] that hot cracks were formed at the end of the solidification process, when some melt still remained between the dendrite arms. This hypothesis was verified by Pellini,[3] who showed by X-ray analysis and simultaneous temperature registration that hot cracks were formed at the very end of the solidification sequence. Since then reasons have been discussed for the embrittlement at a temperature near the very end of the solidification process. Borland and co-workers[4,5] have shown that a grain-boundary film of liquid metals during the very end of the solidification process is a possible reason for the embrittlement. However, there are several proposals[6] which suggest that the embrittlement is caused by penetration of the crack tip by the liquid. In order to study the crack growth at a temperature near the very end of the solidification process a number of tension tests have been performed in a scanning electron microscope at this temperature.

EXPERIMENTAL TECHNIQUE

A scanning electron microscope was equipped with a tension tester for small flat samples, Figs. 1 and 2. The tension tester could create a tension force of 2 kN max.

The drawing speed could be varied from 0·05 to 1·0 mm/min. The force was measured by a strain gauge and registered by a recorder. Around three faces of the sample a heat-resisting furnace was placed, Fig. 2. The furnace permitted heating of the sample to 600°C max. Between the furnace and the sample a thermocouple was placed, Fig. 2, and the temperature was registered by a temperature recorder.

Simultaneous recordings of the image, the stress–strain curve, and the temperature were made by a video technique: the arrangement is shown in Fig. 3. The simultaneous heating–straining test was carried out in two different ways. In one case the sample was heated to a temperature just above the melting point and then the tension test was started, and in the other, the tension was started during the heating process and the melting occurred during the deformation process.

Two aluminium-base alloys with 4% Sn and 4% Cd respectively, have been examined. Some lead alloys with varying Sn contents up to 15% have also been investigated. The specimens were cut from the columnar zone in small ingots, mainly perpendicular to crystal boundaries. About 50 different experiments were performed. Three of these experiments will be discussed in detail.

THE ALLOYS

Two factors determined the selection of the alloys, the solidification temperature range and the surface tension conditions of the liquid against solid crystals. Figures 4–6 show the phase diagrams for the three alloys

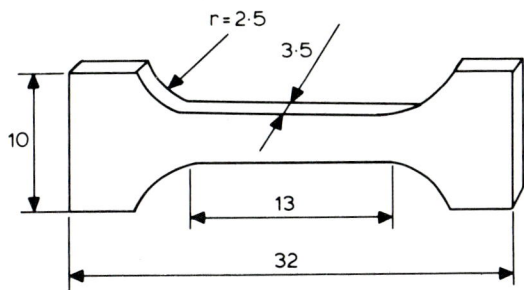

1 Tensile test specimen

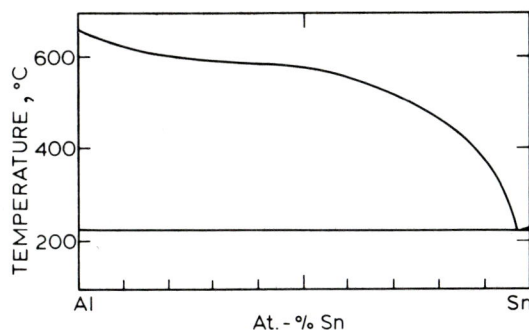

4 Al–Sn phase diagram[7]

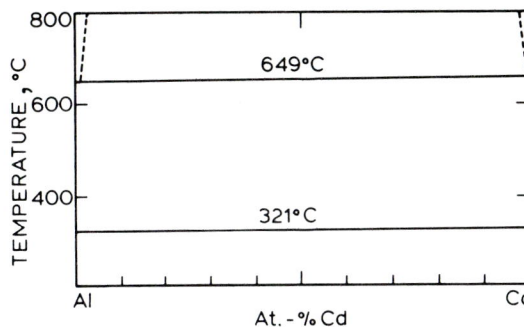

5 Al–Cd phase diagram[7]

A sample; *B* furnace; *C* thermocouple;
D tension blocks

2 Specimen and furnace arrangement in tensile test equipment

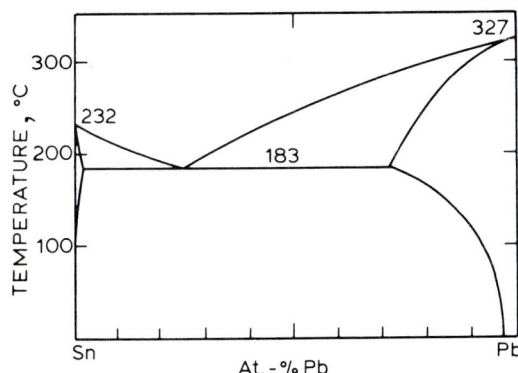

6 Sn–Pb phase diagram[7]

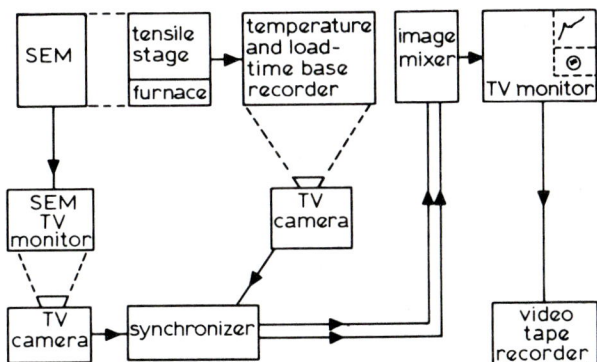

3 Key figure of registration arrangement

studied. Al–Sn and Al–Cd alloys both have a very large temperature solidification range. The system Al–Cd has a two-phase region, L_1/L_2, with a monotectic reaction at 649°C. At this reaction the precipitated liquid contains very small amounts of aluminium. This liquid solidifies at a temperature of 321°C. The Al–Sn system has a very wide two-phase region α(Al)/L and a low solubility of Sn in Al. The system contains a eutectic reaction at 228°C. In the Pb–Sn system, the lead phase can solve a large quantity of Sn. The solidification temperature range is smaller than in the two systems described above. In this system a eutectic reaction occurs at a temperature of 182°C.

The solidification process normally starts with a primary precipitation of dendrites with segregation of alloying elements to the last solidified parts. The distribution of the very last solidified melt will be determined not only by the solidification conditions but also by the surface tension between the liquid and the solid phase. A low value of the ratio of the liquid/solid surface

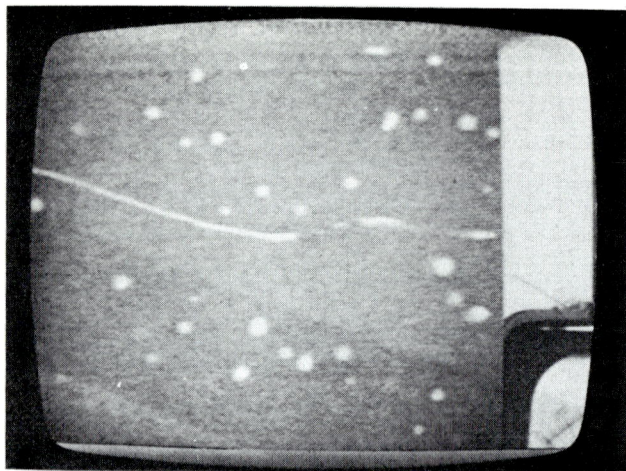

7   Casting structure in Al–Sn sample

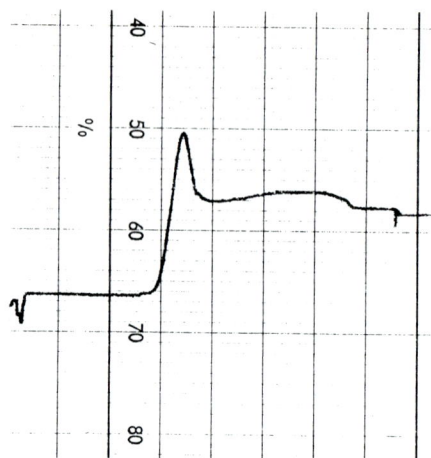

9   Tension–tensile curve in the experiment with the Al–Sn alloy

tensions $\gamma^{L/S}$ and solid/solid $\gamma^{S/S}$ gives rise to a spreading of liquid along the crystal boundaries.[8] Measurements of the ratio $\gamma^{L/S}/\gamma^{S/S}$ have been performed by Smith[8] in the Al–Sn system and by Rogerson and Borland[5] in the Al/Sn and Al–Cd systems. These investigations show that the Sn-rich liquid in the system Al–Sn has very good wetting properties of the aluminium crystal boundaries. The Cd-rich liquid in the Al–Cd system are not wetting the boundaries. A crystal-boundaries film of Sn-rich liquid is to be expected in the Al–Sn samples, but not in the Al–Cd samples.

The Pb–Sn system shows a different kind of character to the other two systems, and therefore alloys from this system were studied. Very little is known about the wetting properties of the crystal boundaries of the eutectic liquid in this system. However, the casting structure shows that the liquid acts in a very similar way to the Sn-rich liquid in the Al–Sn system.

RESULTS
Al–4% Sn
The experiment with the Al–Sn alloys was made in the following way. The furnace and the sample were heated to 240°C and the tension experiment started immediately afterwards.

Figure 7 shows the casting structure. Along the horizontal centre line a crystal boundary can be obser-

ved. It is well delineated by a 1 mm wide path of a Sn-rich phase (white areas in the figure). Inside the Al crystals small, round inclusions of a Sn-rich phase can be observed. At a temperature of about 230°C these inclusions and the Sn-rich phase in the crystal boundary were melted. This was shown by the fact that small, round Sn drops formed along the crystal boundary at the sample surface, Fig. 8.

At a temperature of 240°C the tension experiment was started. At the beginning the tension rose linearly, Fig. 9. There was a pure elastic deformation sequence. Just before the maximum stress force has been reached a crack starts to form along the crystal boundary. Figure 10a–e shows the process of nucleation and growth of the crack. At the start of the crack formation the tension will be drastically reduced. The growth of the crack was

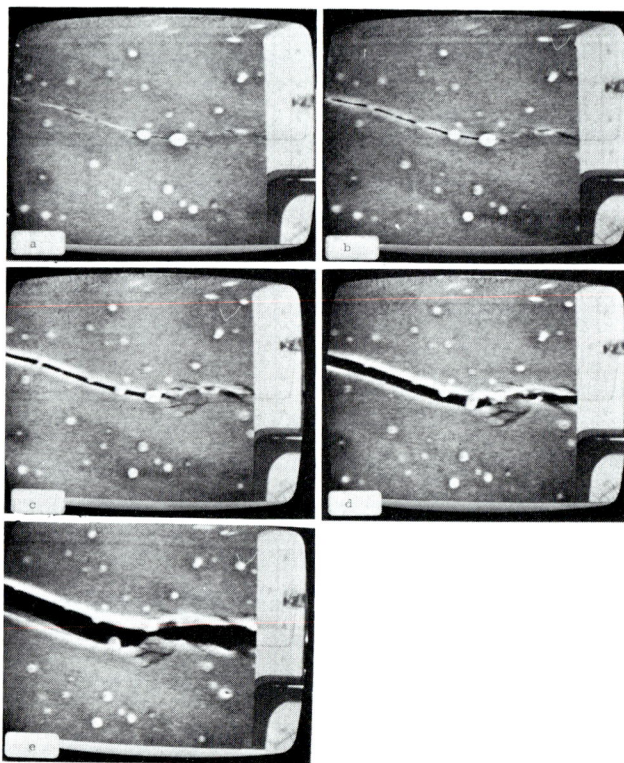

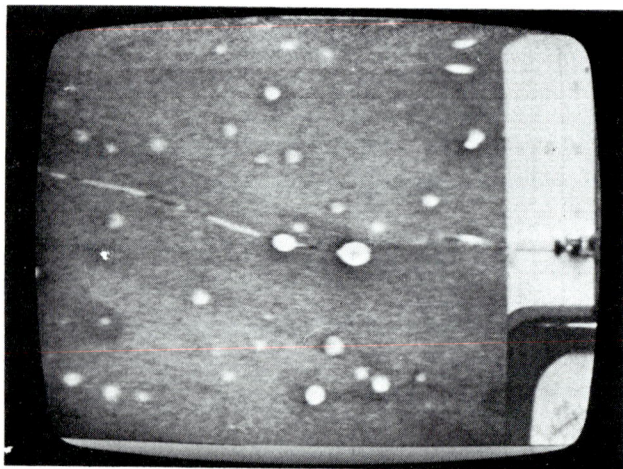

8   Melting of Sn–Al eutectic in the sample shown in Fig. 7

10   Different stages of the crack growth in the sample shown in Fig. 7

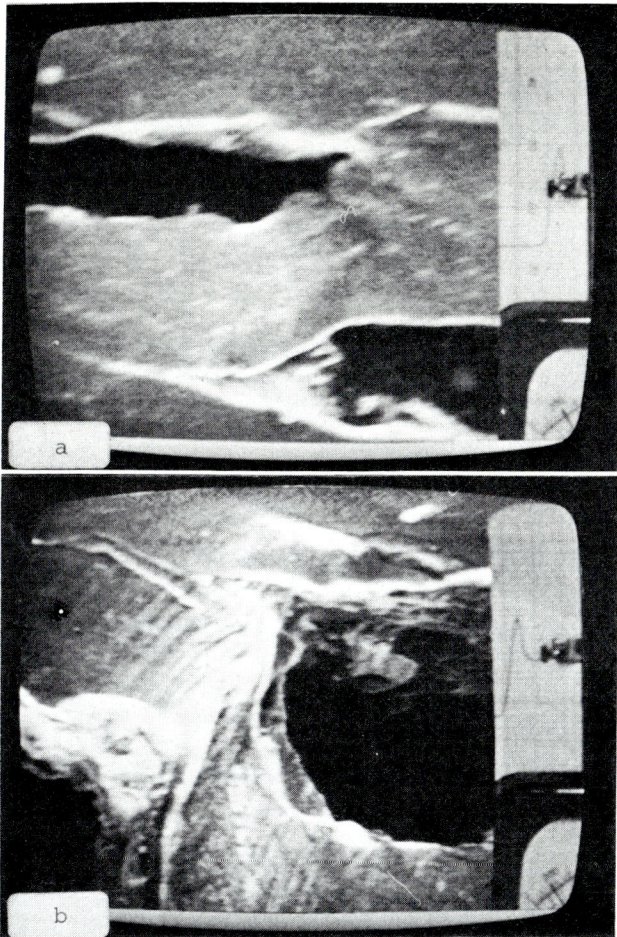

11 Plastic deformation of bridges of solid material

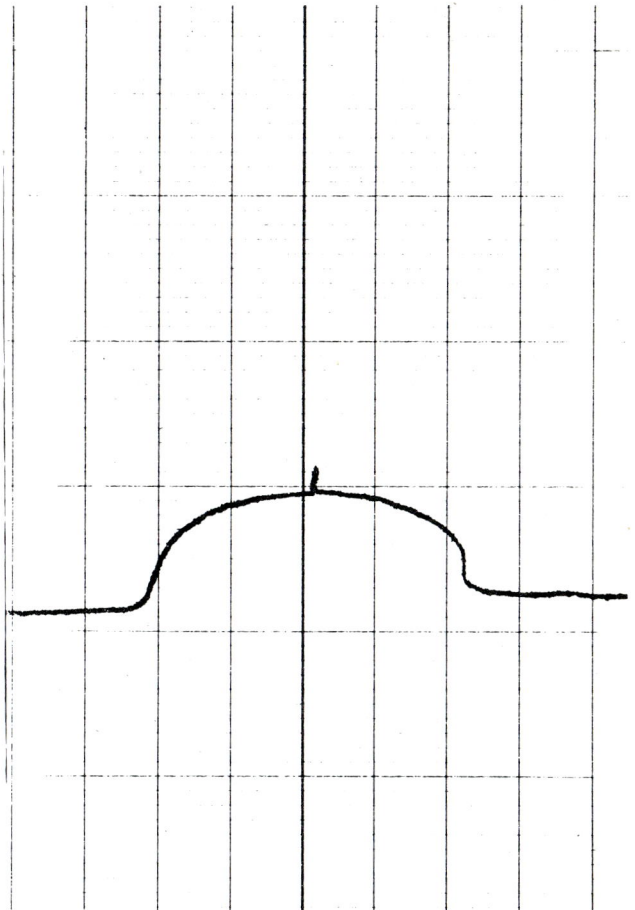

12 Stress–elongation curve in the experiment with the Pb–Sn alloy

stopped at the end of the grain boundary. Figure 11*a* shows that the crack growth has been stopped at the end of a grain boundary and a bridge of solid material has been formed between two grains. This bridge will be deformed plastically and within this region a normal ductile fracture will occur, Fig. 11*b*.

Pb–15% Sn

About 15 experiments were performed with different lead–tin alloys. The experiment which will be discussed here was made in the following way. The sample was heated to 150°C and the tension was started at this temperature. The tension and heating then occurred simultaneously. The stress–elongation curve, Fig. 12, shows an elastic–plastic deformation sequence. When the maximum stress was reached, the temperature in the sample had increased to 180°C, just below the melting point. Figure 13 shows the structure at this point. Through a horizontal line in the figure a crystal boundary is situated. The boundary is marked with a 3 mm wide path of Pb/Sn eutectic. Perpendicular to this boundary in the upper part of the figure two interdendritic regions can be seen. Between the matrix and the eutectic structure along the grain boundary a series of cracks have been formed. The sample fractures some seconds before the eutectic start to melt. A crack is nucleated and grows quickly through the whole sample along the crystal grain boundary, Fig. 14*a*. In the figure the eutectic has not been melted. This can be seen by the sharp edges at the crack against the top surface of the

sample. About 5 s after the fraction formation, melting occurs. At this moment the fraction surface was more even. At the same time the edges against the sample surface were rounded. Figure 14*b* shows the structure just after melting. Figure 14*c* shows the crack across the sample.

Figures 15*a* and *b* shows the fracture surface. The figures show that the crack has gone along a crystal boundary with a continuous film of Pb/Sn eutectic. The fracture surface shows a dendritic pattern.

In the area near the fracture the eutectic melt has been sucked away from the interdendritic regions, Figs 14*c*

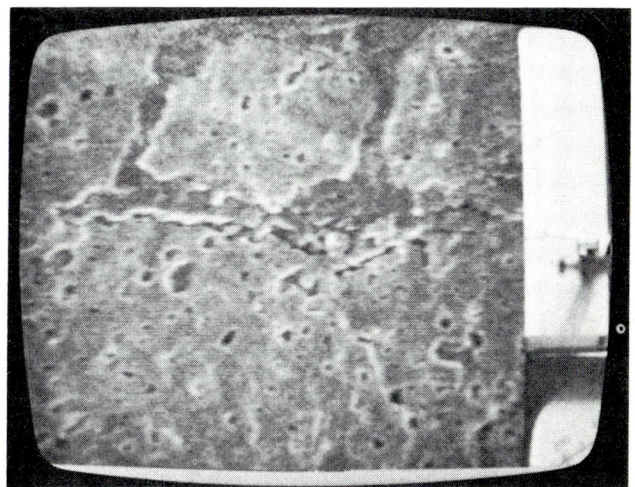

13 Casting structure in Pb–Sn alloy

15   **Fracture surface in the Pb–Sn alloy at two different magnifications**

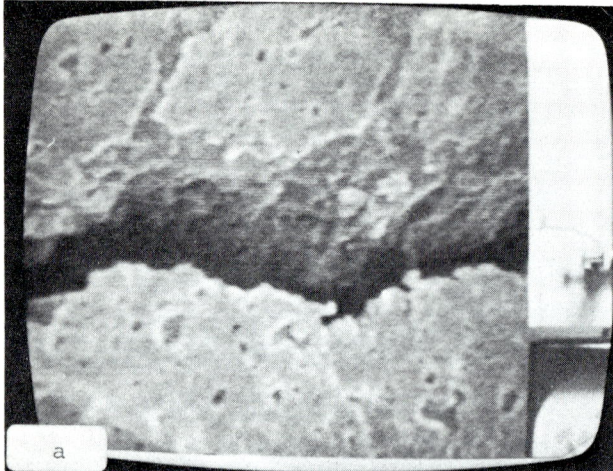

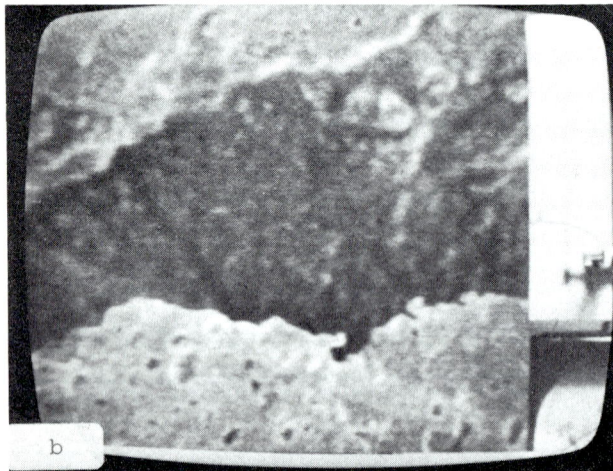

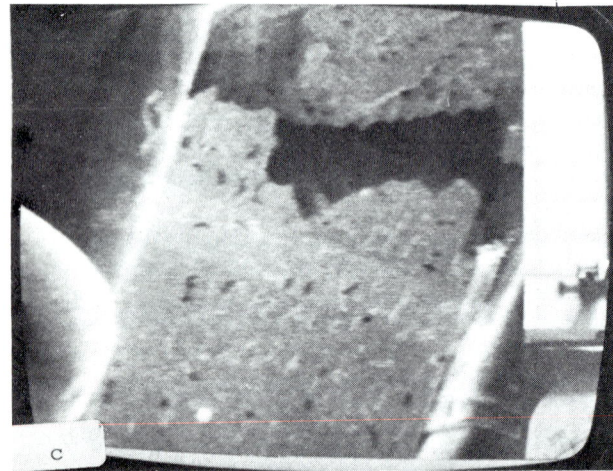

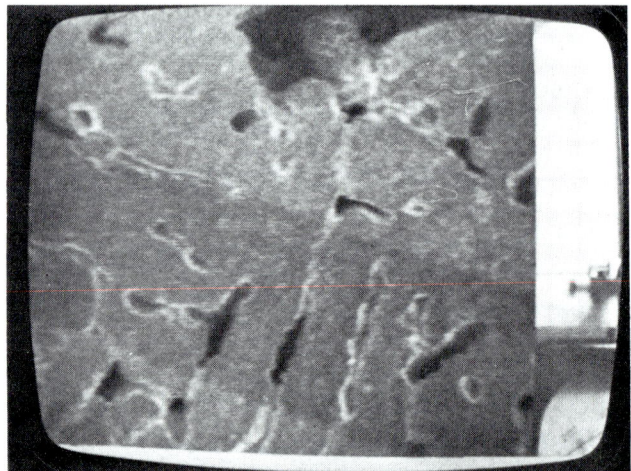

a crack growth has started; *b* melting starts; *c* fractured sample

14   **Different stages during the crack formation in the Pb–Sn alloy**

16   **The area around the fracture in the sample shown in Fig. 14c**

and 16. This process could be carefully followed in a couple of similar experiments. In this case a selected area was observed close to the fracture region. Figures 17a–c illustrate the sucking sequence. Figure 17a shows a region with two interdendritic regions just before fracture and melting. At the moment of fracturing and melting, the melt is sucked away from the interdendritic regions. Figure 17b shows a hole formed when the eutectic liquid was removed from the lower left interdendritic region in Fig. 17a. Figure 17c shows the

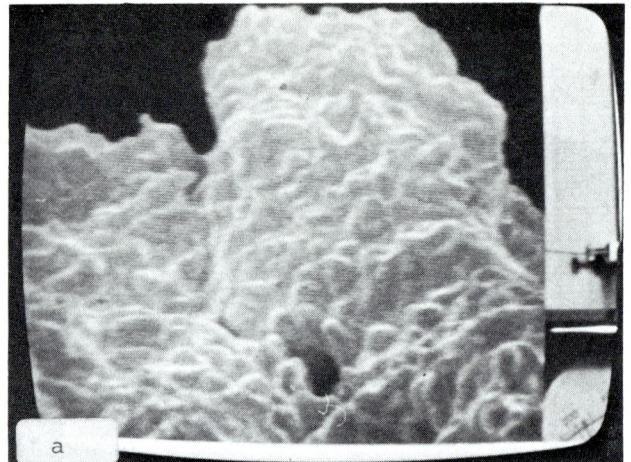

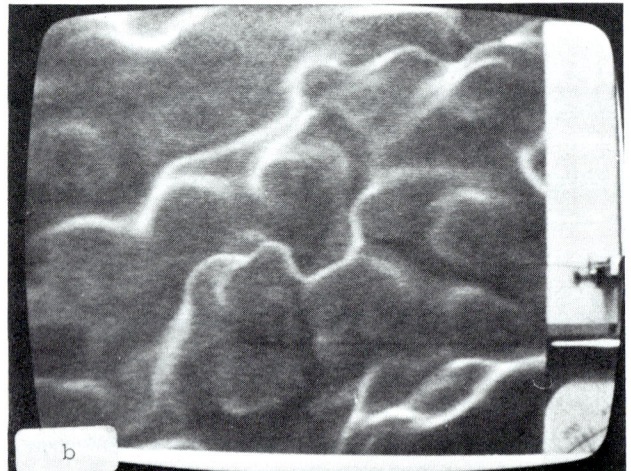

structure roughly 5 s later. The eutectic melt has now been removed from the upper left interdendritic region in Fig. 17a.

Al–4% Cd

About 10 experiments were performed with different heating and deformation sequences. One of the experiments gave the following result. The sample was

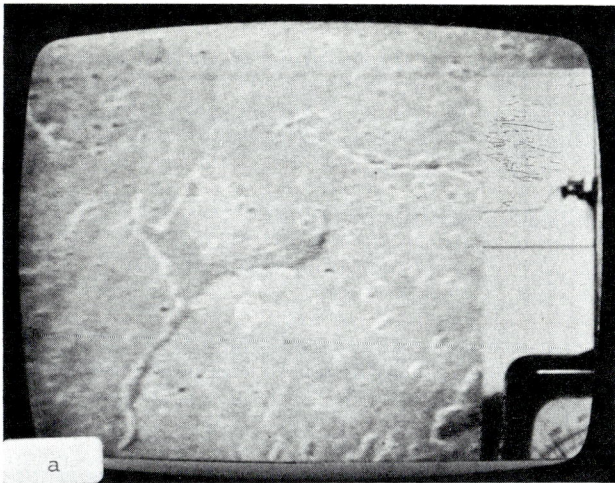

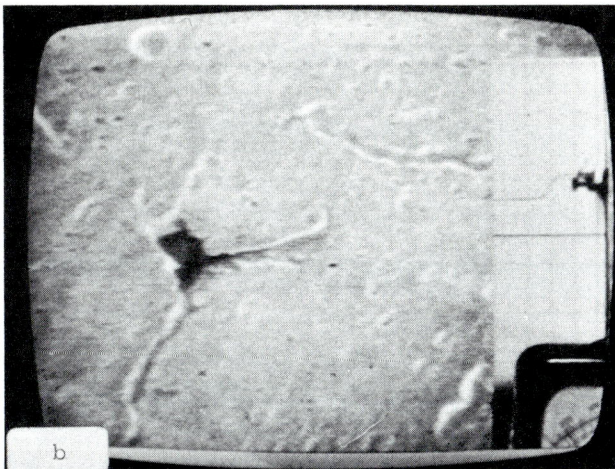

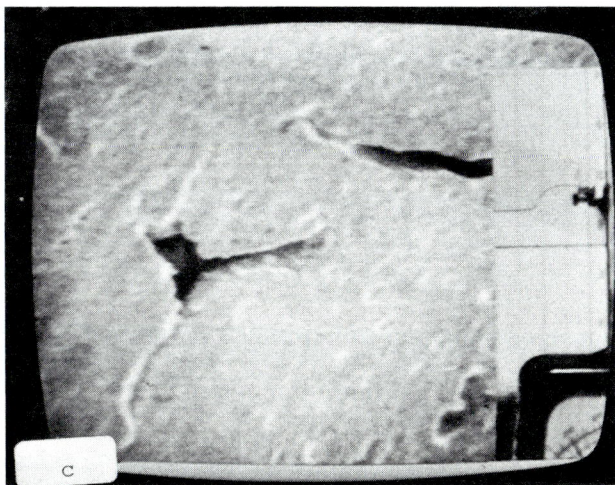

17 **Different stages of removing eutectic liquid from the interdendritic regions during the crack growth**

18 **Tension–tensile diagram in the experiment with the Al–Cd alloy**

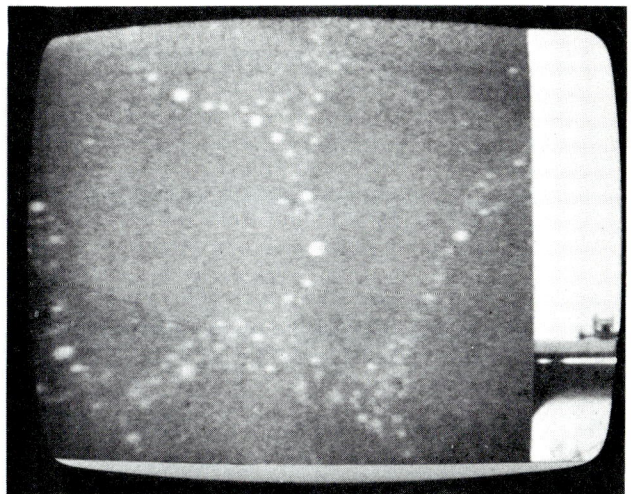

19 **Casting structure in Al–Cd sample**

heated to a temperature just above 400°C and the deformation process was then started. The temperature was kept between 350° and 400°C during the whole deformation sequence. This means that the temperature is always above the melting point. Figure 18 shows the tension–tensile diagram of the whole deformation sequence.

Figure 19 shows the structure in the sample just before the start of the deformation sequence. The figure

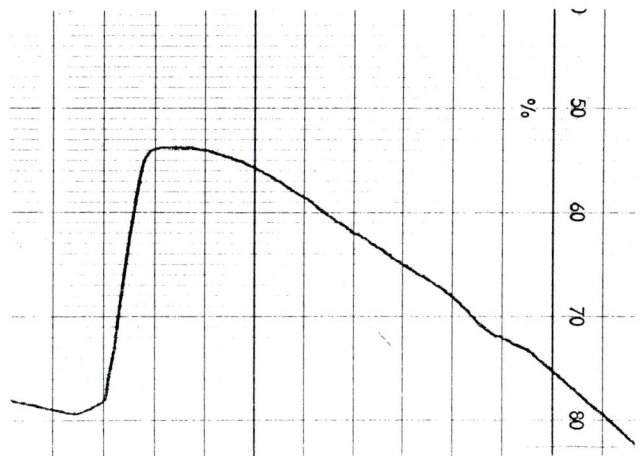

shows a dendritic pattern with drops of Al–Cd mono-tectic in interdendritic regions. The monotectic structure consists of small Cd drops in an Al matrix.

At the start of the tension experiment an elastic deformation occurred. At the end of the elastic deformation process, cracks could be observed along the crystal boundaries (Fig. 20*a*). These cracks propagate during the continuous deformation process without giving rise to fracture, Fig. 20*b* and *c*. Figure 21*a–c* shows a series of views at increasing magnifications at the end of the tension experiment. Also, at the very end of the deformation process no embrittlement could be observed.

Figure 22 shows the surface of the sample just before the specimen fractures. Large cracks have been formed along the crystal boundaries across the sample. Figure 23*a* and *b* shows the fracture surface at the large crack in the centre of Fig. 22. Figure 23*a* and *b* shows that the fracture in this sample is ductile.

DISCUSSION

The two aluminium alloys studied show quite a different fracture behaviour. The Al–Sn alloys show a brittle fracture along crystal boundaries. However, the Al–Cd alloy shows a ductile fracture. Also, in the latter case the fracture follows the grain boundaries. The drops of Cd

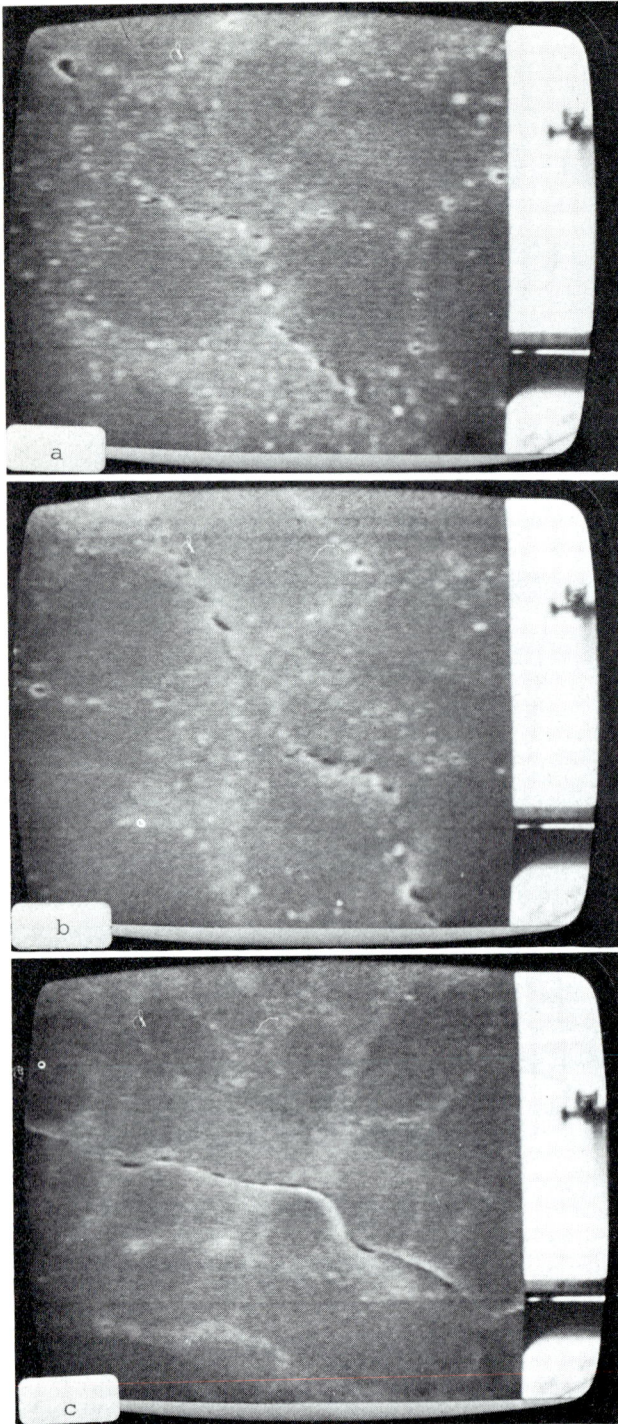

20   Initial state of crack growth in Al–Cd sample

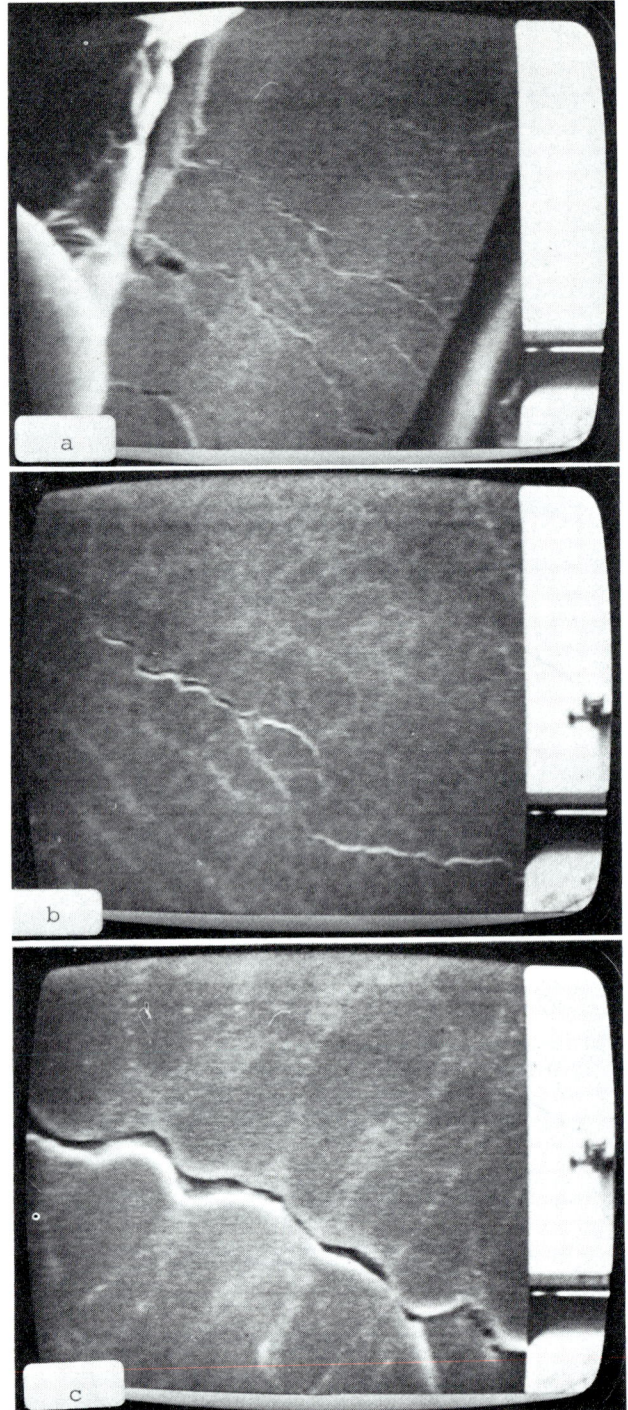

21   Cracks formed along grain boundaries in the Al–Cd sample: increasing magnification from *a* to *c*

liquid in the grain boundaries have no influence on the crack growth. During the deformation process they act in a very similar way to spherical solid inert slag inclusions. The difference in fracture behaviour between the two alloys seems to be the wettability of the crystal boundaries by the liquid.

It is interesting to note that exactly the same embrittlement behaviour as in Al–Sn alloys occurs in Pb–Sn alloys. This embrittlement does not occur if the crystal boundaries go along the sample parallel with the drawing direction. Such an experiment was performed. In this case, brittle fracture could not be observed. The liquid did not have any influence on the fracture. No penetra-

tion of the crack tip by the liquid could be observed. This phenomenon was demonstrated in Figs. 11*a* and *b*. In this case the cracks first grew along the crystal boundary, and the growth was stopped at the end of the boundary. When this occurs, the deformation process changes. The crack growth is no longer affected by the present liquid metal and the deformation continues in a plastic mode.

CONCLUSIONS

A technique for deformation studies at high temperatures has been developed. The technique has been used

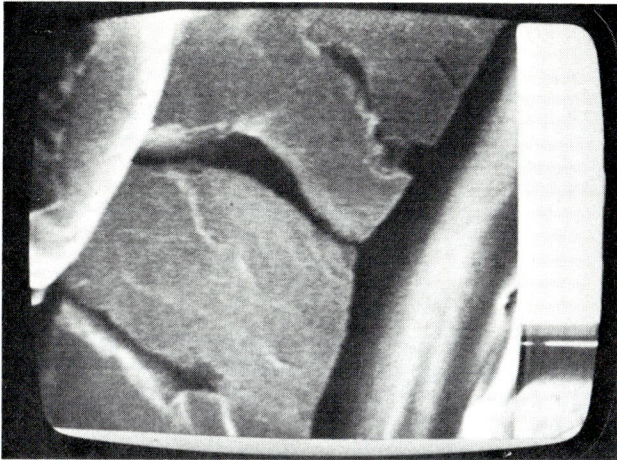

22 Al–Cd specimen just before fracturing

for studies of hot crack formations in cast alloys. The deformation sequence has been registered on video tapes. The experiment shows that hot cracks are formed just above or just below the melting point of the alloys if the alloys contain a eutectic liquid with very good wetting conditions of the crystal boundaries.

ACKNOWLEDGMENT
Financial support by the Swedish Board for Technical Development is gratefully acknowledged.

REFERENCES

1 HULTGREN AND G. HELMER: *Jernkontorets Ann.*, 1944, **128**, 457
2 A. R. E. SINGER AND S. A. COTRELL: *J. Inst. Met.*, 1947, **73**, 33
3 W. S. PELLINI: *Foundry*, 1952, **80**, 124
4 J. C. BORLAND: *Brit. Weld. J.*, 1960, **7**, 508
5 J. H. ROGERSON AND J. C. BORLAND: *Trans. Met. Soc. AIME*, 1963, **227**, 2

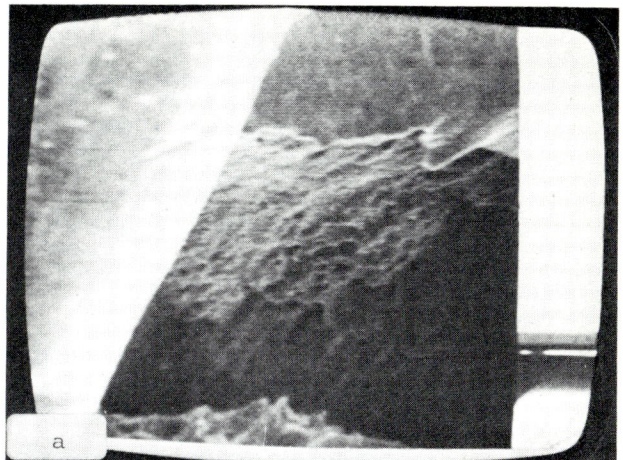

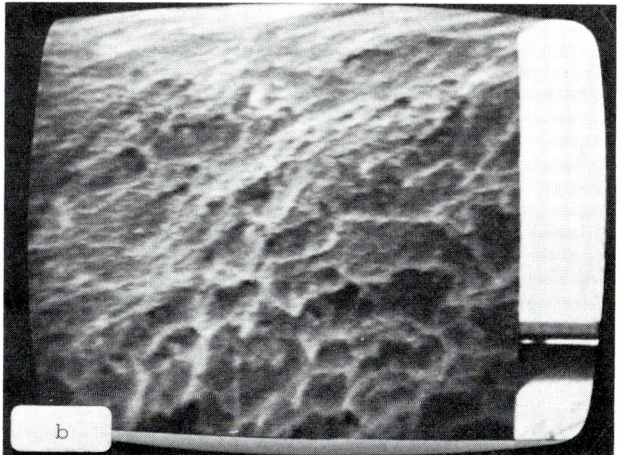

23 Fracture surface at two different magnifications in the Al–Cd sample

6 N. S. STOLOFF AND T. L. JOHNSON: *Acta. Metall.*, 1963, **11**, 251
7 M. HANSEN: 'Constitution of binary alloys', 1958, New York, McGraw-Hill
8 C. S. SMITH: *Trans. AIME*, 1948, **175**, 15

Crack formation and tensile properties of strand-cast steels up to their melting points

K. Kinoshita, G. Kasai, and T. Emi

The facial and internal crack susceptibility of steel solidifying in a strand casting machine has been investigated for Si and Al–Si killed steel slab specimens reheated or in situ *solidified on an Instron-type testing machine. The work-hardening parameter as determined on hot tensile stress–strain curves is shown to be a good measure of the facial crack susceptibility, whereas the zero ductility point, determined in the vicinity of the solidus temperature of the corresponding steel, is found to give a satisfactory interpretation of the characteristics of the internal crack susceptibility. Metallurgical factors influencing the two susceptibilities are also discussed.*

The authors are with the Kawasaki Steel Corporation, Japan

Longitudinal or transverse facial cracks occurring in the skin of a solidifying steel shell in the mould or the secondary cooling zone of the strand casting machine are caused mainly by the imbalance between the thermal stress imposed upon, and hot tensile properties of, the skin. Internal cracks arising in the shell in the upper and pool end portion in the secondary cooling zone are also due to ferrostatic bulging stress and supplementary thermal stress overcoming the strength of the shell. To prevent the formation of these cracks, therefore, heat extraction from the shell must be controlled to minimize the thermal stress and reinforce the strength of the shell. The strength, primarily a function of the shell temperature, is strongly influenced by dendritic structure, redistribution, and precipitation of solutes.

For the facial cracks, earlier studies[1,2] examined mostly reduction of area which is of not so much relevance to predicting crack susceptibility. The influence of thermal history, and hence solute precipitation, of a specimen on the crack formation has not yet been fully discussed. For the internal crack which propagates along interdendritic region in the shell, previous investigations[3,4] dealt only with reheated specimens, the structure of which is totally different from the solidified specimen. One exception was the work by Wilber et al.[5] on in situ solidified specimens, but detailed consideration was not given on the susceptibility to internal cracking in the high-temperature region.

It is the objective of this paper to determine the susceptibilities of strand-cast steels to facial and internal cracks by making detailed investigation into hot tensile properties up to their melting points with special reference to the dendritic structure and solute precipitation in the solidifying shell.

EXPERIMENTAL
Material
Rod-shaped specimens were machined, from the outer 30 mm (i.e. from within the outer columnar dendritic shell) of strand cast, spray-quenched slabs, with their axes parallel to the width of the slabs. The slabs (see Table 1) are of Si, Al–Si killed or pipeline grade steels.

HOT TENSILE PROPERTIES
A specimen of 8 mm dia. × 25 mm gauge length was placed as shown in Fig. 1 on an Instron-type testing machine, and heated in situ under Ar by an induction energized Mo susceptor to examine hot tensile properties. The susceptor enabled even distribution of

Table 1 Chemical composition of materials (wt-%)

Steel No.	C	Si	Mn	P	S	Al	N (ppm)	O (ppm)
1	0·16	0·24	0·81	0·021	0·021	0·004	38	—
2	0·14	0·24	0·69	0·021	0·005	0·036	58	20
3	0·16	0·40	1·36	0·021	0·016	0·020	46	24
4*	0·10	0·28	1·56	0·026	0·007	0·028	57	33
5	0·15	0·20	0·82	0·021	0·014	0·006	44	33
6	0·15	0·20	0·89	0·017	0·015	0·041	57	21

* 0·032Nb, 0·026V.

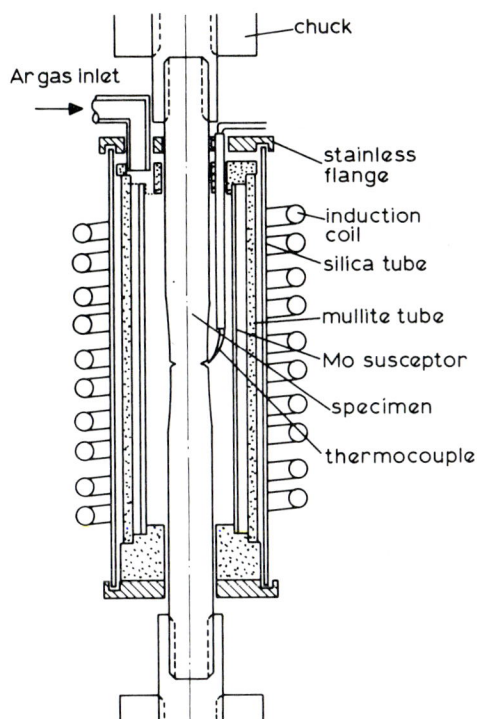

1 Sectional view of apparatus used for as-cast and pre-annealed hot tensile tests

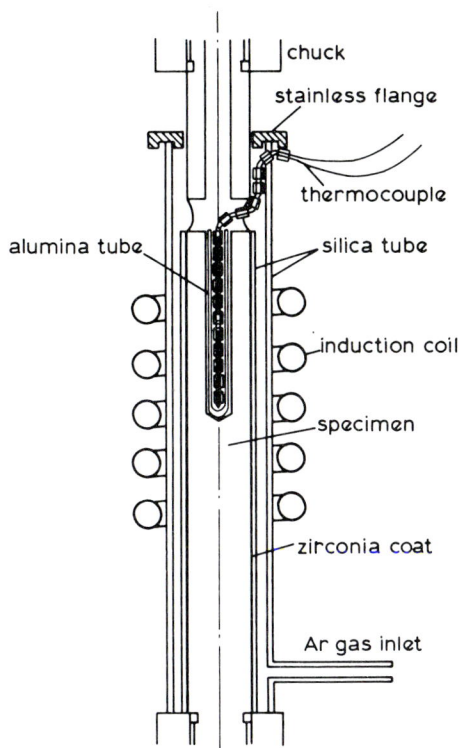

2 Sectional view of apparatus used for *in situ* solidified hot failure test

temperature of less than ± 5 K over the gauge length even after necking of the specimen. Care was taken to minimize automatically temperature variations during a run to be less than ± 1 K with a PR–13 thermocouple attached to the heated surface of the specimen.

Hot tensile properties were examined in two series of tests: first, to keep the specimen at test temperature for 900 s, then to deform it at 0·017 mm/s to fracture. This is designated the hot tensile test for the as-cast specimen.

The second test is to anneal the specimen at 1 543 K (or 1 623 K for Nb–V type pipeline steel, No. 4) for 3·6 ks, then cool at 0·4 K/s to test temperature before deforming. This is denoted the hot tensile test for a pre-annealed specimen.

Susceptibility to facial crack formation
A V-notched pre-annealed specimen of 12 mm o.d. and 8 mm i.d., 0·25 mm notch-tip radius, and $\pi/4$ rad. flank angle, was employed. The specimen was deformed up to 1 mm elongation, unloaded, and quenched to room temperature, under experimental conditions which were otherwise as described above. Then the specimen underwent microscopic observation of (a) structure; and (b) total length of cracks (if any).

Susceptibility to internal crack formation
A 20 mm dia. specimen was closely fitted into a zirconia-lined fused silica tube, as shown in Fig. 2, installed on the testing machine, and melted under Ar by direct induction heating. The lining was specifically placed with fine zirconia cement to avoid deformation of the silica tube upon melting of the specimen. It also helped to prevent dissolution of silica into the melt, and hence contamination of the melt leading to sticking of the silica tube to metal upon solidification. Temperature was controlled by a PR–13 thermocouple inserted into the centre of the hot zone of the specimen.

Hot ductility was determined by melting the specimen at 1 833 K for a few minutes, and Al added through the thermocouple hole if required, cooling it at 0·3 K/s to test temperature, straining the solidified specimen at 0·083 mm/s to fracture, and quenching it to room temperature. This is designated the hot failure test for the *in situ* solidified specimen.

Brittle fractured and sectional surfaces of these specimens were examined with a scanning electron microscope (SEM).

RESULTS AND DISCUSSION
Hot tensile properties of Si and Al–Si killed steels
Stress (σ) – strain (ϵ) curves have been obtained for Si killed (No. 1) and Al–Si killed steels (Nos. 2, 3, and 4) at 1 050–1 700 K.

An example for Al–Si killed plate steel (No. 3, tensile strength 50 kg/mm^2) is shown in Fig. 3. Intermittent decreases in stress observed for a larger strain at higher temperatures are obviously attributable to the concurrent recrystallization for the plastic deformation of austenite.[6] The stress for pre-annealed specimens (solid line) is much lower than that for as-cast specimens (broken line) in the lower strain range ($\epsilon < 0.1$) in both the lower austenitic range (1 173 $< T <$ 1 473 K) and the ferritic range ($T <$ 1 173 K). The above softening, due to the preliminary annealing, is also observed for Al–Si-killed conventional plate steel (No. 2, tensile strength 40 kg/mm^2) and Nb–V type Al–Si killed pipeline steel (No. 4).

On the other hand, the softening for Si killed steel (No. 1, tensile strength 40 kg/mm^2) is observed only in the ferritic range, but not in the austenitic range, as shown in Fig. 4 as roughly the same stress for as-cast and pre-annealed specimens at the lower strain range.

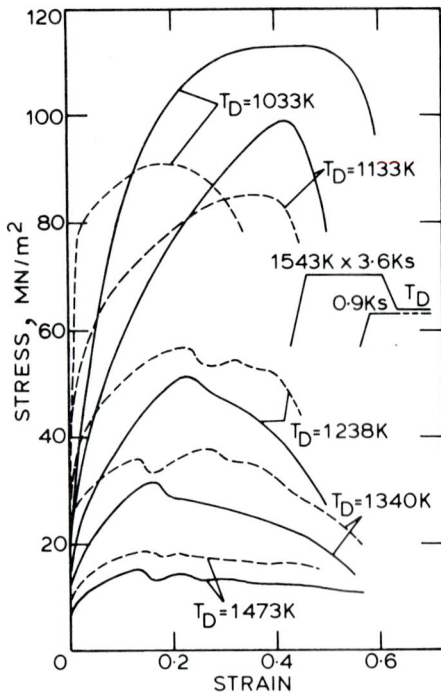

3 Stress–strain curves for No. 3 plate steel

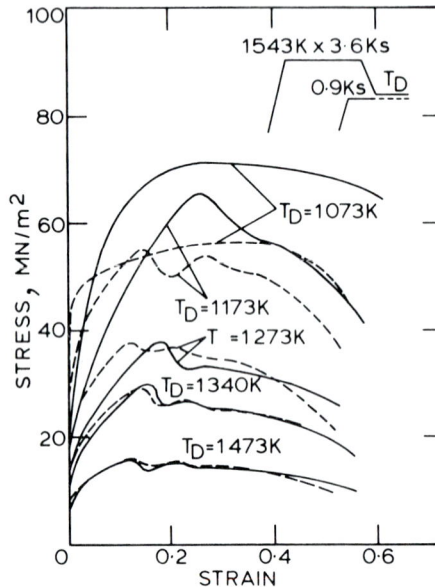

4 Stress–strain curves for No. 1 conventional plate steel

The influence of deformation temperature and heat treatment of specimens on the flow stress at $\epsilon = 0.004$ is summarized in Fig. 5.

Work hardening of the steels examined is determined in terms of a parameter n as defined by $\mathrm{d}(\log\sigma)/\mathrm{d}(\log\epsilon)$ $\{ = (\epsilon/\sigma)\,(\mathrm{d}\sigma/\mathrm{d}\epsilon)\}$ at $\epsilon = 0.02$ to 0.10 as the slope of $\log\sigma - \log\epsilon$ curves shown in Fig. 6. The work-hardening parameter is equal to the uniform strain ϵ_u, which is the strain at the onset of necking.[7] The effect of deformation temperature and heat treatment of specimens on n is shown in Fig. 7. Comparison of Fig. 5 with Fig. 7 indicates that the larger the flow stress $\sigma_{0.004}$, the smaller the value of n.

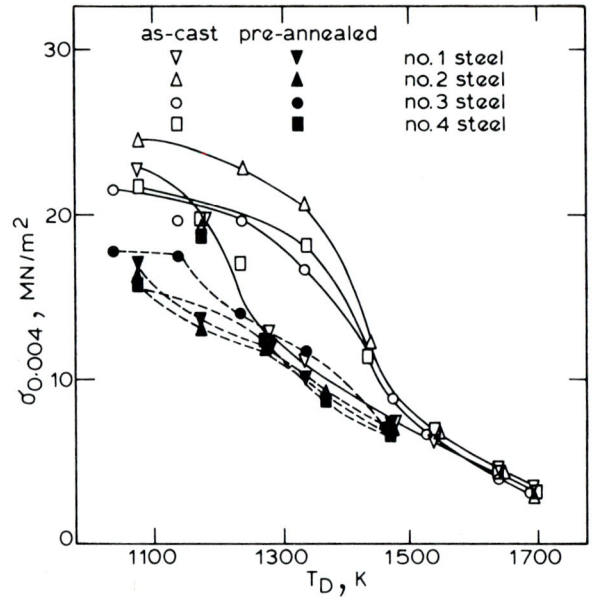

5 Variation of flow stress $\sigma_{0.004}$ with deformation temperature

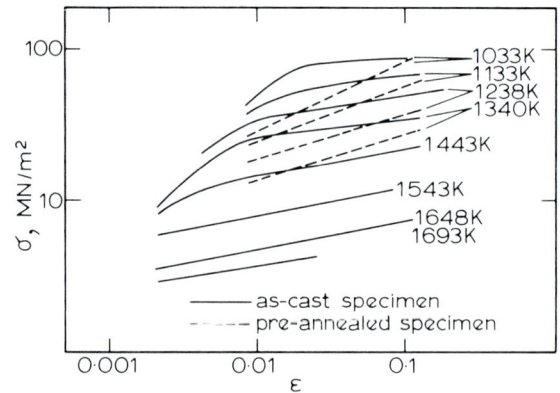

6 Stress–strain curves for No. 3 plate steel

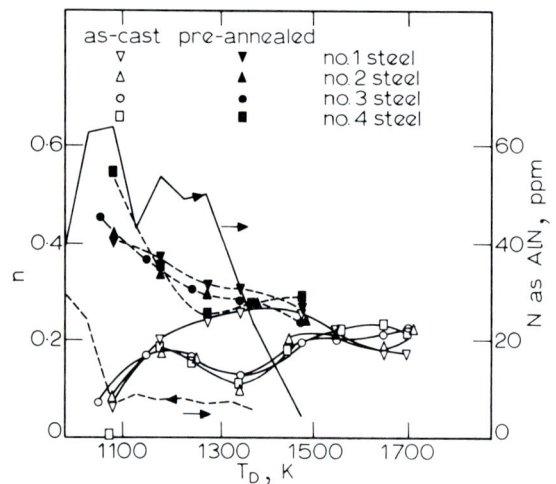

7 Variation of work-hardening parameter and precipitation AlN with temperature

In the lower austenitic temperature range, n for Al–Si killed steels is, as shown by Fig. 7, smaller for as-cast than for pre-annealed specimens. As shown elsewhere,[8] the decrease of ductility, as represented by the smaller n, cannot be interpreted by the corresponding change in the austenite grain size. The observation that *n* for Si killed steel in this temperature range is the same for as-cast and pre-annealed specimens and for pre-annealed Al–Si killed specimens gives further confirmation of the foregoing argument. The presence of Al in the steel seems to be responsible for the decrease in ductility. Regarding this, isochronal precipitation of AlN in an Al–Si killed steel has been determined at different temperatures, the result being shown in Fig. 7. It is interesting to note that the formation of AlN in pre-annealed specimens (lower line) is almost insignificant in contrast to that in as-cast specimens (upper line). Increasing formation of AlN in the latter specimens corresponds reasonably to decreasing *n*, indicating that precipitation hardening[9] by AlN may possibly account for the decreasing ductility of as-cast Al–Si killed specimens in the lower austenitic temperature range.

In the ferritic temperature range, however, *n* for as-cast specimens is much smaller than that for pre-annealed ones for both Si and Al–Si killed steels. This may be due to the difference in the texture of the two classes of specimens.

Facial crack formation on Nb–V type pipeline steel castings

Cracks formed at 1 173 K on the notch root of V-notched specimen and facial cracks observed in corresponding strand-cast slab are shown in Figs 8 and 9,

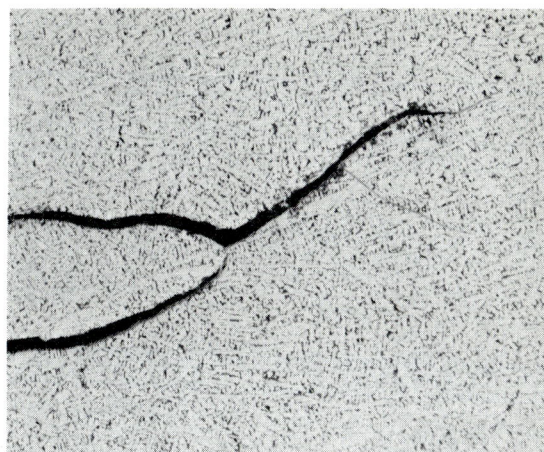

9 Facial crack formed on Al–Si killed strand-cast slab [×10]

respectively. The cracks on the specimen and slab propagate along austenitic grain boundaries. Total crack length found on the notch root is plotted in Fig. 10 as an index for crack susceptibility against deformation temperature. For solution treated specimens (1 623 K × 3·6 ks), the crack susceptibility reaches its maximum at 1 223 K, whereas it decreases with decreasing temperature from 1 273 to 1 073 K for incompletely solution treated specimens (1 623 K × 0·6 ks).

Also shown in Fig. 10 is the amount of acid insoluble niobium formed in the solution-treated (lower line) and incompletely solution treated (upper line) specimens at varying temperatures. The crack susceptibility compared with the amount of acid insoluble Nb seems to indicate that niobium precipitates, mostly carbides, are responsible, like AlN, for the increasing crack susceptibility of the incompletely solution treated specimens at 1 250–1 300 K.

To prevent the formation of facial cracks, therefore, bending and unbending of Al–Si killed and Nb–V type steel castings must be done at temperatures below 1 050 K. Also, recalescence of the steel surface to the lower austenitic temperature range, which enhances precipitation of AlN and/or niobium carbides, should be avoided during strand casting.[10]

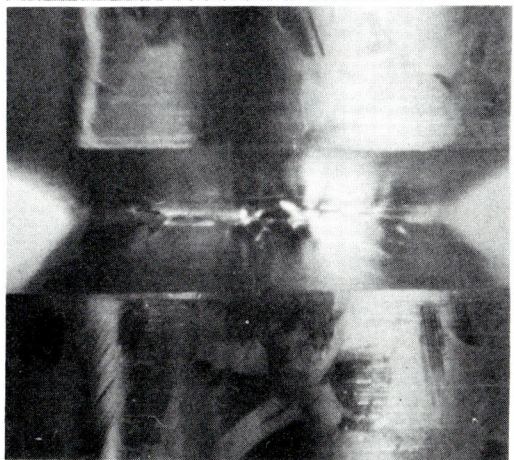

a sectional view. [×100]; *b* external view, [×6·5]

8 Crack formed at 1 173 K on notch root of V-notched specimen

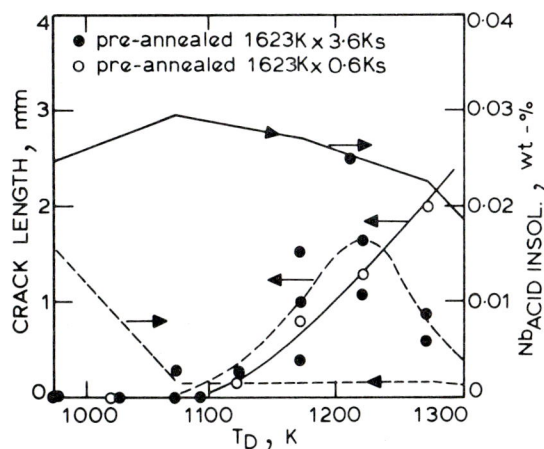

10 Variation of crack susceptibility and acid insol. Nb with temperature

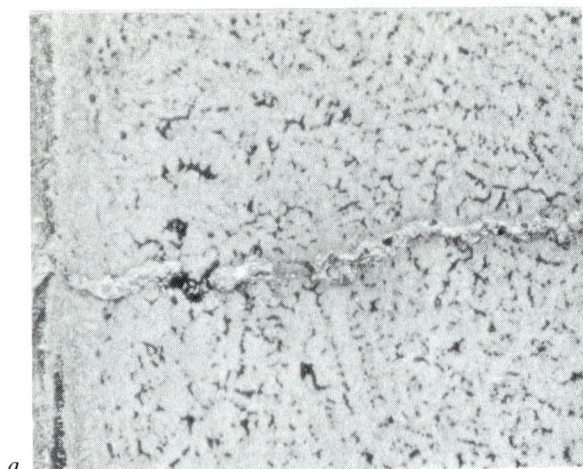

a

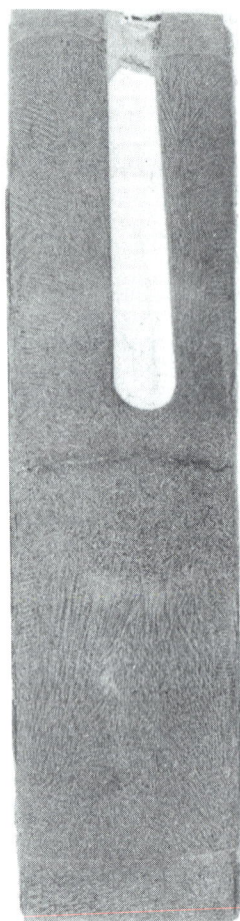

b

a [×100]; b [×1·5]

11   Longitudinal sectional view of *in situ* solidified Al–Si killed specimen

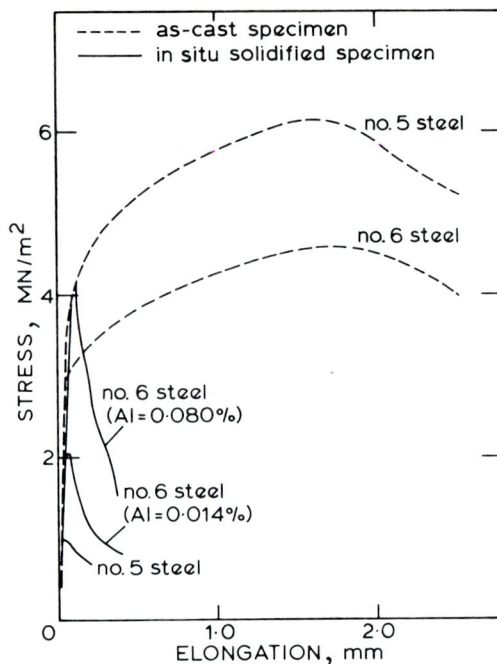

12   Stress–elongation curves for steels No. 5 and 6

appreciable difference in dendrite arm spacings is observed between Si and Al–Si killed specimens.

Stress–elongation curves for *in situ* solidified Si (No. 5) and Al–Si killed (No. 6) specimens are shown in Fig. 12 together with those for as-cast specimens of corresponding chemistry. The Si killed specimens exhibit much lower stress than the Al–Si killed ones at the same temperature. The dependence of the tensile stress on temperature and Al content is replotted from Fig. 12 in Fig. 13.

Internal cracking has been shown to occur along the interdendritic region in the 'mushy' zone and adjacent solidified shell, with solute enriched liquid penetrating into the cracks.[11] This is conceivable on the basis of the present observation, i.e., the tensile strength given in

Internal crack formation in Si killed steel castings

Only those specimens solidified by completely filling the silica tube, without shrinkage cavities in and around the hot zone, have been subjected to further investigation.

A longitudinal sectional view of an *in situ* solidified Al–Si killed specimen (*see* section on 'Susceptibility to internal crack formation', above), is shown in Fig. 11. Cracks are found to extend mostly along the inter-dendritic region, intersecting only a few principal arms of dendrites. This also applies to Si killed specimens. No

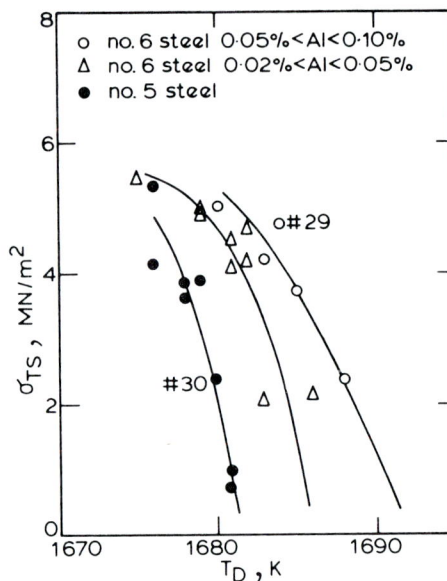

13   Variation of tensile strength of *in situ* solidified specimen with deformation temperature

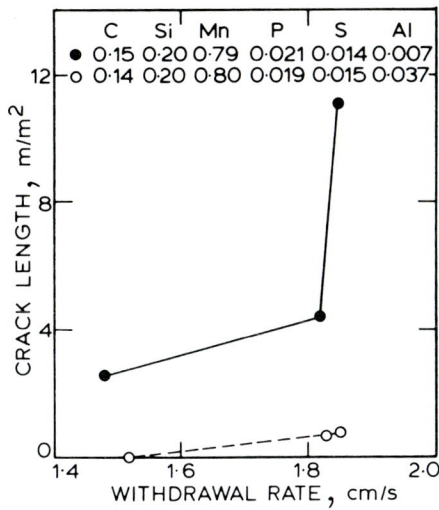

14 Variation of crack susceptibility with withdrawal rate

Fig. 13 for sub-solidus temperature is in the order of 2 MN/m², easily susceptible to the initiation and propagation of cracks. However, it is increasing steeply with decreasing temperature, preventing further extention of the cracks into lower temperature region in the outer shell.

The tensile strength at a given temperature decreases, and hence the zero ductility point, as defined by the

intersection of the curves at the abscissa, decreases with decreasing Al content. This gives a reasonable explanation of the phenomenon observed in commercial strand casting that Si killed steel slabs are, as shown in Fig. 14, more susceptible to internal crack formation than Al–Si killed steels (in particular, at higher withdrawal rates).[12]

SEM observation of fractured surface of Si killed specimen (No. 30 in Fig. 13) is compared with that of Al–Si killed specimen (No. 29 in Fig. 13) in Fig. 15. The former exhibits protruded steel tips on the fractured surface which is covered with film identified by EPMA to be MnS. The latter (No. 29) comprised smooth, round surfaces, indicative of dendrite arms to which type II MnS is adhered. The observed lower tensile strength for Si killed steel may have been caused by the existence of the MnS film. Since only type I MnS and no film-like MnS is found in the bulk of the Si killed specimen (Fig. 16), however, the above argument cannot yet be fully confirmed.

The zero ductility point of *in situ* solidified and as-cast specimens is shown in Fig. 17 as a function of

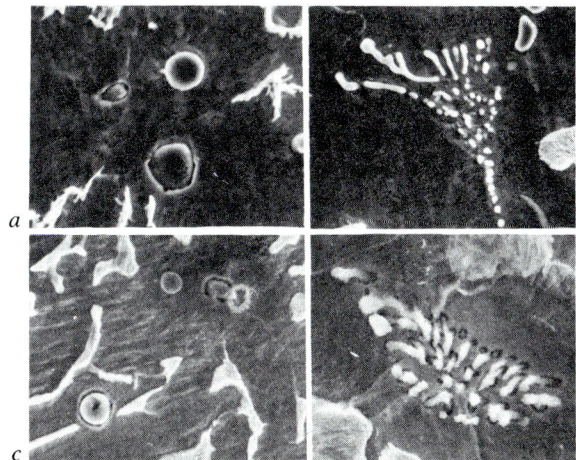

a Si killed slab, [×500]; *b* Al–Si killed slab, [×500]; *c* Si killed specimen, [×500]; *d* Al–Si killed specimen [×1 000]

16 Manganese sulphide in bulk of specimen and strand-cast slab

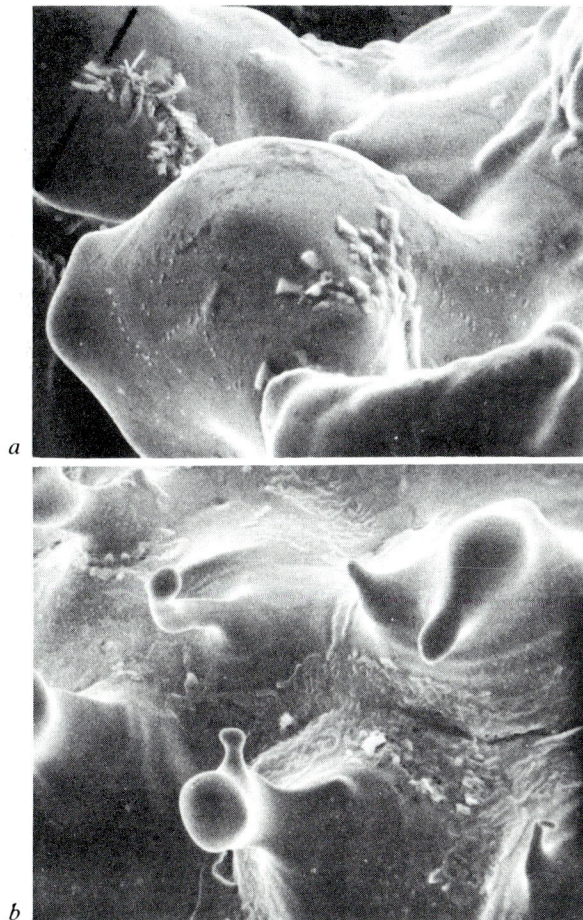

a Al–Si killed (No. 29), [×1 000]; *b* Si killed (No. 30), [×500]

15 Fractured surfaces of *in situ* solidified specimen

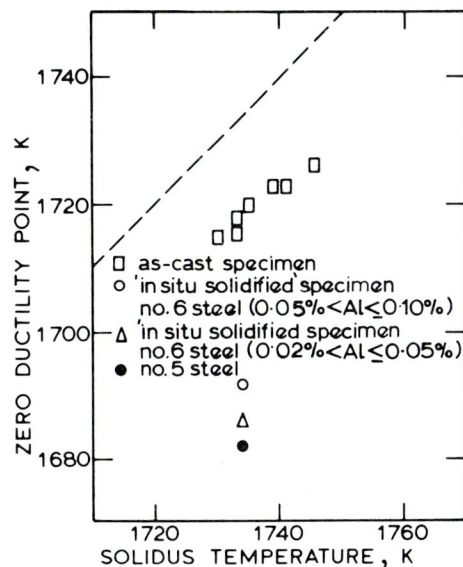

17 Zero ductility point *v.* solidus temperature

corresponding solidus temperature. The zero ductility point is 43–53 K lower than the solidus temperature for *in situ* solidified specimens in contrast to only about 15 K for as-cast specimens. This clearly indicates, as stated elsewhere,[3] that *in situ* solidified experiments are essential in order to obtain relevant data for the study of high-temperature mechanical properties of the solidification structure of steels.

CONCLUSIONS

Hot tensile properties have been determined with an Instron-type testing machine for specimens cut from the outer shell of strand-cast slabs and reheated or *in situ* solidified on the machine. The results obtained are:

(i) hot ductility, as a good measure of facial crack susceptibility of strand-cast steel castings, is evaluated in terms of the work-hardening parameter determined on hot tensile stress–strain curves for the reheated specimens

(ii) the hot ductility in the ferritic temperature range is influenced by steel texture, whereas in the austenitic temperature range it is decreased by the precipitation of aluminium nitride (in Al–Si killed steel) and niobium carbides (in Nb–V type Al–Si killed steel)

(iii) the zero ductility point of *in situ* solidified specimen is about 50 K lower than the solidus temperature of the corresponding steel, decreasing markedly for Si killed steel. This is considered to be due to the formation of film-like MnS in Si killed steel in place of type II MnS in Al–Si killed steel, and gives a reasonable explanation of the higher internal crack susceptibility of Si killed steel in commercial strand casting.

REFERENCES

1 S. HASEBE: *Tetsu-to-Hagané*, 1962, **48**, (6), 761
2 A. FUCHS: ESTEL Berichte aus Forschung u. Entwicklung uns. Werke, 1975, Heft 3, 127–135
3 C. J. ADAMS: Proceedings NOH–BOS Conference (AIME), 1971, **54**, 290
4 H. FUJII *et al.*: *Tetsu-to-Hagané*, 1976, **62**, (4), S93
5 G. A. WILBER *et al.*: *Metall. Trans.*, 1975, **6A**, 1 727
6 P. J. WRAY AND M. F. HOLMES: *ibid.*, 1975, **6A**, 1 189
7 W. B. MORRISON AND R. L. MILLER: Ultra Fine Grain Materials, 16th Sagamore Army Materials Res. Conf., 183, 1970, NY, Syracuse University
8 T. TANAKA AND T. ENAMI: *Tetsu-to-Hagané*, 1972, **58**, (13), 1 775
9 M. F. ASHBY: 'Strengthening methods in crystals', (ed. A. Kelly and R. B. Nicholson), 137, 1971, London, Elsevier
10 T. NOZAKI *et al.*: *Tetsu-to Hagané*, 1976, **62**, (12), 1 503
11 H. FUJII *et al.*: *ibid.*, 1976, **62**, (14), 1 813
12 S. MORIWAKI: personal communication

Comparison between experimental data and theoretical predictions relating to dependence of solidification cracking on composition

T. W. Clyne and G. J. Davies

A major requirement for the complete understanding and control of the phenomenon of solidification cracking during casting is comprehension of the mechanisms by which changes in solute level influence the observed incidence of cracking. However, previous attempts to introduce a rationale, which would enable the cracking susceptibility to be related to the composition, have produced predictions which, even in simple binary eutectic systems, give poor correlation with experimental data. The present work proposes that the most significant effects produced by a variation in solute level result from changes in mass transport during the vulnerable two-phase (liquid–solid) regime. It is postulated that this is a consequence primarily of the rate of change of local liquid fraction as determined by changes in solute redistribution. Analysis founded on this postulate is presented which enables predictions of cracking susceptibility to be made from the characteristics of the phase diagram. The predictions are shown to be in agreement with experimental data relating cracking susceptibility to composition for the Al–Mg, Al–Si and Al–Zn systems.

T. W. Clyne is at the Departamento de Engenharia Mecanica, UNICAMP, Campinas Brazil. G. J. Davies was in the Department of Metallurgy and Materials Science, University of Cambridge and is now in the Department of Metallurgy, University of Sheffield

Some of the salient features of solidification cracking have been apparent for a considerable time, although clarification of the importance of the different parameters has been hampered by experimental difficulties. Simple characteristics, such as the origin of the contraction stresses, the approximate solidification sequence leading to interdendritic separation, and the importance of liquid feeding to vulnerable regions, are now well understood and have been outlined by several authors.[1-7] It remains necessary, however, to define precisely the mechanisms involved in cracking. Early work[8-12] established that there is a connection between crack susceptibility and freezing range.

Assessment of the validity of theories concerning susceptibility–composition variations have been hampered by the absence of a reliable cracking test. The majority of previous tests[1,2,4,5,8-16] have either failed to ascribe a 'severity' to the cracking produced under particular conditions (classifying castings merely as 'cracked' or 'uncracked'), or else have ascribed a value in a simplistic and unjustifiable way (such as surface crack length). A recently developed test,[17] however, makes a direct assessment of cracking using an electrical method.

In binary eutectic systems, attempts[1,2,5,9,15,18] to predict variations in cracking susceptibility with composition from the appropriate phase diagram have concentrated mainly on the concept of a large solidification interval (freezing range) being deleterious, because it would render the last solidifying region (the 'hot-spot') susceptible to contraction stresses over a greater temperature interval. Usually the freezing range was represented by the vertical distance between the liquidus and solidus, but even when this idea is modified by the introduction of a 'temperature of coherence', the agreement with experimental results is not good. For example, with binary Al alloys, the predicted susceptibility–composition curves are at variance with experimental data in the majority of common systems.

No previous theoretical attempts have been made which take into account the *time* interval of solidification. In the present work, the basic theory involves consideration of the time during which processes related to crack production may take place. In particular, the changes in local liquid fraction with time are considered to define regimes in which particular processes are constrained to occur. By taking a ratio of the time available for interdendritic separation (cracking) to that for other processes which are acting against the production of cracks, a figure can be produced which may be taken as indicative of the cracking susceptibility. The analysis is shown to lead to predicted susceptibility–composition curves which can be correlated most satisfactorily with experimental data for the Al–Mg, Al–Si and Al–Zn systems.

THEORETICAL

The contraction stresses to which a solidifying alloying is subjected arise primarily from the volume contraction β associated with the phase change (usually about 6%) although thermal contraction may reinforce the stress build-up. In general it can be assumed that β is effectively independent of composition within the normal range of composition of binary alloys.[19,20] In addition it is assumed that the local liquid fraction f_L in any volume element decreases continuously up to the time of complete solidification. This need be valid in the 'hot-spot', which will be the region most vulnerable to cracking, only in the range where $f_L < 0.5$.

In the vicinity of the hot-spot the following processes can occur:

(i) *strain accommodation by solid movement*: termed *mass feeding*, this involves cooperative rearrangement of the growing solid dendrites under the influence of the contraction stress. The grain structure and the dendrite structure will have an influence and it is expected that there will be a lower limit below which mass feeding is geometrically impossible

(ii) *strain accommodation by liquid movement*: interdendritic liquid flow might be expected to assist stress relief both by liquid healing of cracks and, probably more importantly, by continuous interdendritic feeding, allowing slight relaxation between adjacent dendrites. A lower limit would be expected corresponding to a marked reduction in the permeability of the porous dendrite array before which the volume flow will be proportional to f_L^2[21-23]

(iii) *interdendritic separation*: this can occur only below a certain value of f_L since it requires the liquid to be distributed as an interdendritic film

(iv) *interdendritic bridging*: at a late stage in the solidification sequence, the integrity of the interdendritic film will become impaired and regions of solid cohesion will be established between adjacent dendrites. If this can occur to a sufficient extent, the strength of the casting will be greater than that required to prevent interdendritic separation; the value of f_L below which this becomes dominant would be expected to be dependent on dendrite morphology and also on the surface energy of the liquid–solid interface.

The cracking susceptibility coefficient (CSC) can be defined as

$$\mathrm{CSC} = \frac{t_V}{t_R} \qquad (1)$$

where t_V is the vulnerable time period and t_R is the time available for stress relaxation processes. In order to predict the variation of CSC with composition it is necessary to predict the variation of f_L with time for a range of compositions. It is difficult to fix the limits of f_L to correspond with the various processes described above. Using the results of previous work concerned with liquid–solid rheology and interdendritic fluid mobility[12,22,24,25] approximate limits can be defined as,

mass and liquid feeding $0.1 < f_L < 0.6$

interdendritic separation $0.01 < f_L < 0.1$

interdendritic bridging $f_L < 0.01$

For the solidification conditions encountered in normal castings the Schiel equation[26] holds locally.[27,28] Thus

$$C_L = C_0 f_L^{k-1} \qquad (2)$$

where C_L is the liquid composition after a fraction $(1 - f_L)$ of an alloy with initial composition C_0 and distribution coefficient k has solidified. This can be rewritten as

$$f_L = \left(\frac{T_M - T}{T_M - T_L} \right)^{+1/(k-1)} \qquad (3)$$

where T_M is the melting temperature of the pure base metal and T_L is the liquidus temperature of the alloy of composition C_0. In practice two conditions of heat flow need to be considered:

(i) $\dfrac{dQ}{dt} = \text{constant}$

(ii) $\dfrac{dQ}{dt} \alpha t^{-0.5}$

where $dQ = (\rho H \cdot df_L + \rho C\, dT)$. The former should apply where heat transfer is interface controlled and the latter when heat transfer is conductivity controlled. ρ is the density (assumed the same for solid and liquid), H the latent heat, and C the specific heat. Integrating the above and incorporating equation (3) and the boundary conditions $t = 0$, $T = T_L$ and $t = 1$, $T = T_E$ the eutectic temperature, leads to the two relationships[19]

$$\text{(i)} \quad t = \frac{\frac{H}{C}(1 - f_L) + (T_M - T_L)(f_L^{k-1} - 1)}{\frac{H}{C}\left(1 - \left[\frac{T_M - T_E}{T_M - T_L}\right]\right)^{-1/(1-k)} + (T_L - T_E)} \qquad (4)$$

and

$$\text{(ii)} \quad t = \left(\frac{\frac{H}{C}(1 - f_L) + (T_M - T_L)(f_L^{k-1} - 1)}{\frac{H}{C}\left(1 - \left[\frac{T_M - T_E}{T_M - T_L}\right]\right)^{-1/(1-k)} + (T_L - T_E)} \right)^2 \qquad (5)$$

Figure 1 shows the predicted variations in the liquid fraction with fractional time for the Al–Mg system for the two cases treated above. The method of determination of t_R and t_V from plots such as those in Fig. 1 is shown in Fig. 2. A computer program was used to

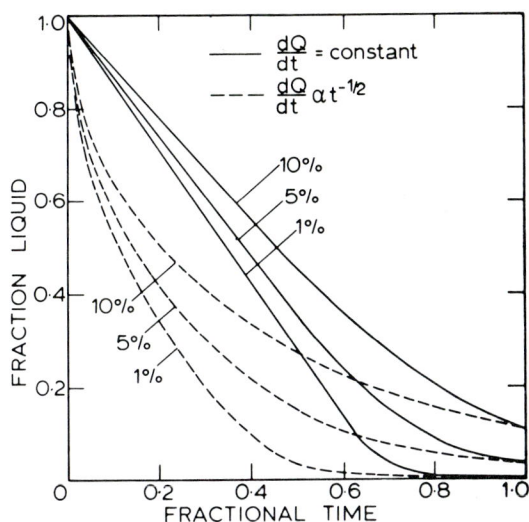

1   Predicted variation of liquid fraction with fractional time for aluminium alloys containing 1, 5, and 10% Mg, respectively (*see* key)

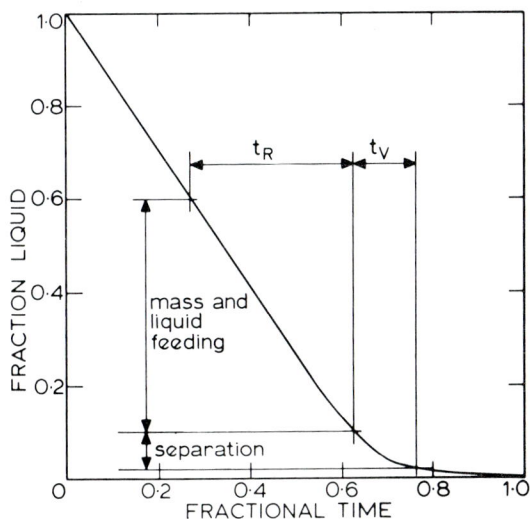

2   Method of determination of time of stress relaxation t_R and vulnerable time t_V from a plot of the variation of liquid fraction with fractional time

calculate the variation of the cracking susceptibility coefficient (CSC, equation (1)) with composition for various systems using equations (4) and (5) and the appropriate limits on f_L as defined above. Predictions for the Al–Mg system are shown in Fig. 3.

The boundary conditions used in making the above predictions imply that the solidification process needs to be considered only up to the point at which eutectic liquid appears. This is validated by the use of the *ratio* of times. Certainly the existence of eutectic liquid with its plane–front mode of solidification will stabilise the dendrite structure if it is still in the relaxation range, and promote feeding if it is in the vulnerable range.

COMPARISON OF THEORY AND EXPERIMENT

Theoretical predictions were compared both with experimental values of the extent of cracking obtained

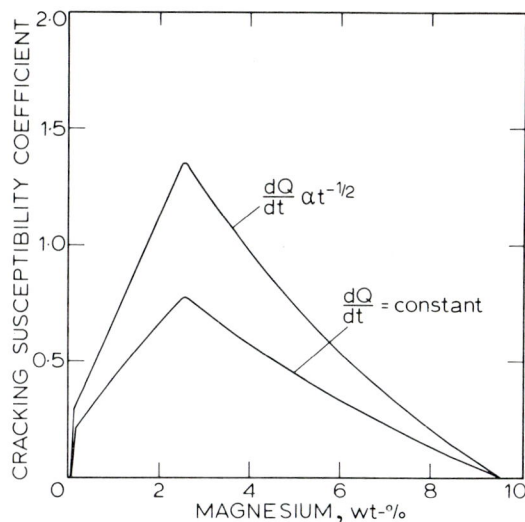

3   Predicted variation of cracking susceptibility coefficient with composition for Al–Mg system

using the test facility previously described by the authors[17] and with results from ring-cracking tests.[2,16] The results are shown for Al–Mg, Al–Si, and Al–Zn in Figs. 4, 5, and 6, respectively. In these figures only the prediction from equation (5) is given. Included in these figures is a plot of the cracking susceptibility as predicted using as a criterion the solidification temperature interval, as determined from the phase diagram.[2,5,18] In making these comparisons it is important to note that the influence of process variables, such as superheat, is mainly on the overall extent of cracking and not on the composition of maximum susceptibility.[16,19] Examination of the results shows a good correlation between prediction using equation (5) and experiment, and little or no correlation with the solidification interval criterion.

The present analysis has application to the study of changes in susceptibility as a result of compositional variations associated with both alloying and impurity levels. Quantitatively, the initial severity of the gradients of the susceptibility–composition curves for various solutes gives good correlation with observed crack-promotion potency for impurities in pure aluminium.[19]

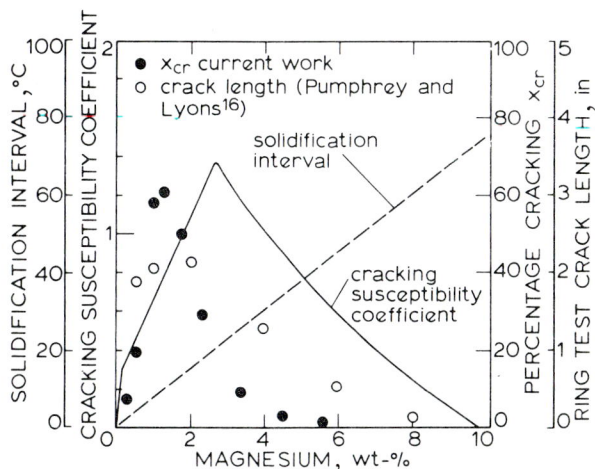

4   Comparison of predicted cracking susceptibility with experimental data for Al–Mg alloys

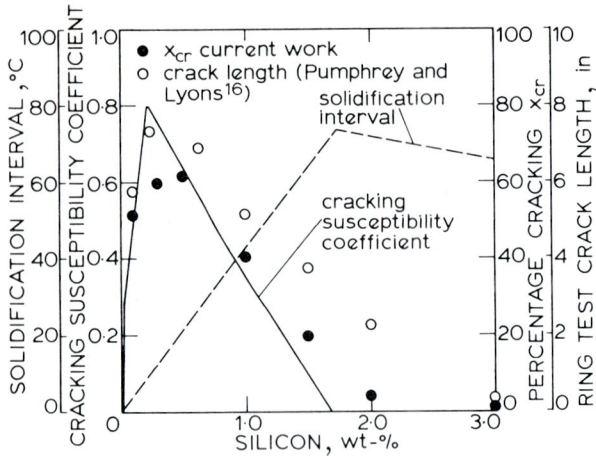

5 **Comparison of predicted cracking susceptibility with experimental data for Al–Si alloys**

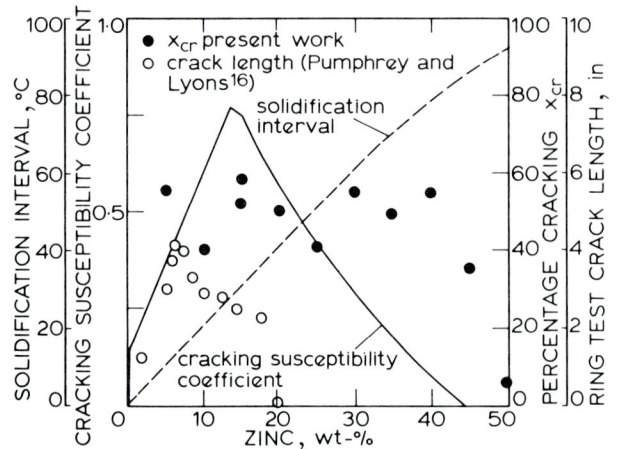

6 **Comparison of predicted cracking susceptibility with experimental data for Al–Zn alloys**

The following order is observed (in *decreasing* tendency to promote cracking): Sn, Si, Cu, Mg, Zn.

Taking into account the present assumptions together with the possibility of unusual dendrite morphologies or substantial variations in liquid properties, the existing correlation is quite acceptable. It would be of interest to examine the application of this approach to more complex multicomponent systems.

ACKNOWLEDGMENTS
The authors would like to thank Professor R. W. K. Honeycombe for the provision of laboratory facilities and the Science Research Council (TWC) for financial support.

REFERENCES
1 D. C. G. LEES: *J. Inst. Met.*, 1946, **72**, 343
2 A. R. E. SINGER AND P. H. JENNINGS: *J. Inst. Met.*, 1947, **73**, 197
3 W. G. PELLINI: *The Foundry*, 1952, **80**, 124
4 H. F. BISHOP *et al.*: *Trans. Amer. Found. Soc.*, 1952, **60**, 818
5 R. A. ROSENBERG *et al.*: *Trans. Amer. Found. Soc.*, 1960, **68**, 518
6 S. A. METZ AND M. C. FLEMINGS: *Trans. Amer. Found. Soc.*, 1970, **78**, 453
7 P. R. BEELEY: *Met. Technol.*, 1976, **3**, 128
8 S. L. ARCHBUTT *et al.*: *J. Inst. Met.*, 1928, **40**, 219
9 J. VERO: *The Metal Industry*, 1936, **48**, 431
10 A. A. BOCHVAR AND Z. A. SVIDERSKAYA: *Bull. USSR Acad. Sci.* (Scientific and Technical division), 1947, 349
11 E. J. GAMBER: *Trans. Amer. Found. Soc.*, 1959, **67**, 237
12 H. F. BISHOP *et al.*: *Trans. Amer. Found. Soc.*, 1958, **65**, 247
13 H. C. HALL: *The Metal Industry*, 1937, **51**, 9
14 H. F. HALL: Iron and Steel Inst. Special Report No. 23, 1938, 73
15 A. R. E. SINGER AND P. H. JENNINGS: *J. Inst. Met.*, 1947, **73**, 273
16 W. I. PUMPHREY AND J. V. LYONS: *ibid.*, 1948, **74**, 439
17 T. W. CLYNE AND G. J. DAVIES: *Br. Foundryman*, 1975, **68**, 238
18 J. C. BORLAND: *Brit. Weld. J.*, 1960, **8**, 508
19 T. W. CLYNE: Ph.D. Thesis, University of Cambridge, 1976
20 V. KONDIC: 'Metallurgical principles of Founding', 1968, 49, London, Arnold
21 T. S. PIWONKA AND M. C. FLEMINGS: *Trans. Met. Soc. AIME*, 1966, **236**, 1157
22 R. MEHRABIAN *et al.*: *Met. Trans.*, 1970, **1**, 1209
23 M. J. STEWART AND F. WEINBERG: *J. Cryst. Growth*, 1972, **12**, 217
24 M. DECROP: *Fonderie*, 1966, **249**, 467
25 D. B. SPENCER *et al.*: *Met. Trans.*, 1972, **3**, 1925
26 E. SCHEIL: *Z. Metallke.*, 1942, **34**, 70
27 H. D. BRODY AND M. C. FLEMINGS: *Trans. Met. Soc. AIME*, 1966, **236**, 615
28 T. F. BOWER *et al.*: *ibid.*, 1966, **236**, 624

Origins of defects in continuously cast blooms produced on a curved mould machine

M. H. Burden, G. D. Funnell, A. G. Whitaker, and J. M. Young

This paper is concerned with the cast structure and defects which occur in steel blooms continuously cast at Round Oak Steel Works. The cast structure is described and related to flow patterns within the mould. Results showing the distribution of inclusions in bloom and bar are presented. Studies of surface cracking and corner cracking are discussed and related to casting practice.

Mr Burden, Mr Funnell, and Mr Whitaker are with TI Research Laboratories, Saffron Walden, Essex.
Mr Young is with Round Oak Steel Works, Brierley Hill, West Midlands

This paper is concerned with the cast structure and defects occurring in steel blooms continuously cast at Round Oak Steel Works on a 4-strand curved mould machine.

The casting machine at Round Oak was commissioned in 1976. It has a 12 m radius and is designed to cast large blooms with a cross section of 355×317 mm mainly for subsequent rolling to rounds. A diagram of the machine is shown in Fig. 1. The liquid steel is supplied by two 120 t electric-arc furnaces and is argon stirred in the ladle to reduce temperature stratification and provide accurate temperature control. From the ladle to the tundish the liquid metal stream is unprotected, but from the tundish to the mould a submerged pouring practice is used. The copper moulds are 700 mm long, have a 0·6% taper, and are oscillated sinusoidally. Below the mould, in the secondary cooling zone, water is sprayed on to the surface at a rate of $0·3$ dm^3/kg over a length of 8 m. At the casting speed of 0·6 m/min, which is used at present, the steel is completely solidified before the straightening rolls. The steel grades cast on the machine to date have been plain carbon and carbon manganese steels mainly for seamless tubemaking.

THE CAST STRUCTURE

At high superheats ($>20°C$) the cast grain structure of the blooms has been found to be asymmetric, as shown in the transverse section (Fig. 2*a*) and this is thought to be characteristic of curved mould machines. The columnar region extends further to the centre on the inner radius and the equiaxed zone is formed towards the outer radius side. This is attributed to equiaxed

grains falling under gravity towards the outer radius because the strand has curved significantly away from the vertical before solidification is complete. This mechanism of formation of the asymmetric structure can be observed easily using ammonium chloride analog castings. The asymmetry of the grain structure can still be seen in transverse sections of round bar rolled from the continuously cast blooms (Fig. 3) and this can be used to relate the orientation of the bar to the original as-cast bloom. The grain structure can change significantly throughout a cast because of variations in superheat. Fig. 2*a* and *b* are transverse sections taken from the front and back, respectively, of the same cast. Figure 2*b* shows a larger, finer grained equiaxed zone than 2*a* and illustrates the effect of a drop in tundish temperature of 10°C. The corresponding longitudinal sections are shown in Fig. 2*c* and *d*. These figures also illustrate the effect of superheat on the extent of central porosity and segregation.

The orientations of columnar dendrites are thought to reflect closely the flow patterns in the liquid steel ahead of the growing interface, the dendrites being angled towards the flow of liquid.[1] Flow patterns in the mould region of the strand have been studied using a water modelling technique, and have been found to be affected by the outlet design of the submerged pouring tube. A number of different pouring tube designs were examined by modelling with the aim of reducing the penetration of inclusions into the strand. One of the most promising designs was a four-port-outlet pouring tube with the ports directed upwards at 15° to the horizontal. This design was chosen for use on the Round Oak machine

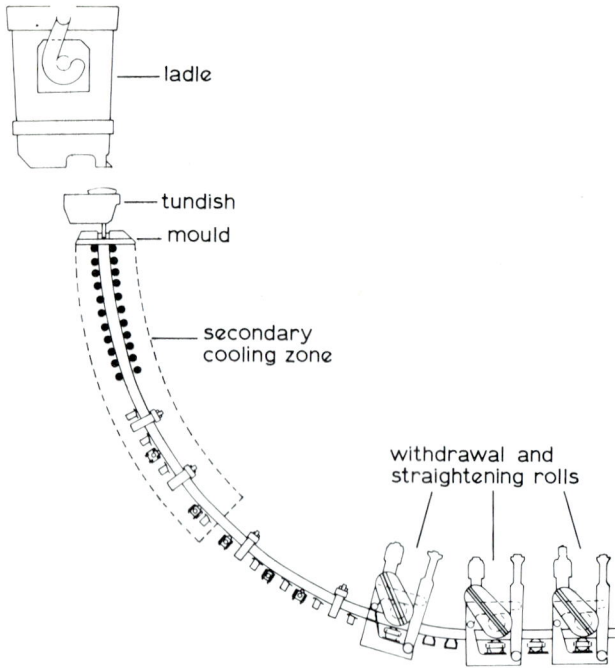

1   Diagram of Round Oak Steel Works continuous casting machine

inner radius

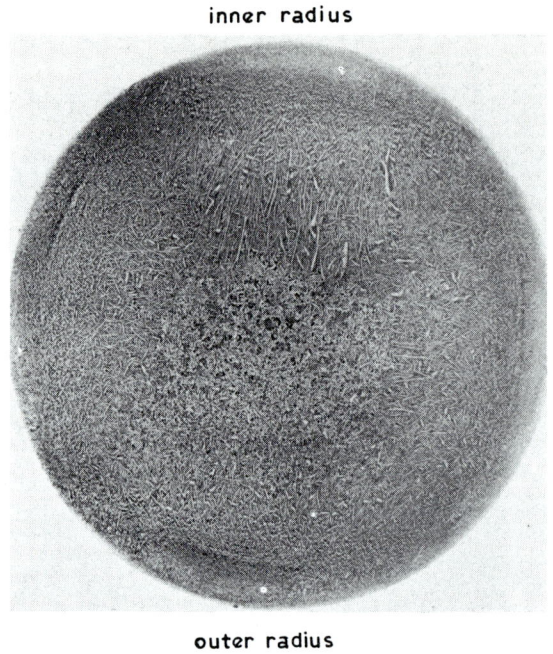

outer radius

3   Macroetched section of round bar rolled from continuously cast bloom showing retention of asymmetric structure   [×0·45]

and the flow patterns which it produced are shown in Fig. 4. There is good agreement between the orientations of the columnar dendrites in the cast structure and the flow pattern. In the centres of the bloom faces immediately opposite the ports, the columnar dendrites are angled downwards at approximately 20° for the first 5–10 mm. At this shell thickness the dendrites abruptly change direction to an angle about 20° upwards to the horizontal. The columnar dendrites usually remain angled upwards for a significant distance towards the equiaxed zone, suggesting that flow remains predominantly down

the interface. In a number of the longitudinal sections, especially on the upper inner radius side of the bloom, the angle of columnar dendrite growth has been observed to alter direction once or twice during growth towards the centre, suggesting that some form of recirculatory flow could be occurring deep in the liquid pool.[2]

DISTRIBUTION OF INCLUSIONS

The steelmaking practice at Round Oak is based on aluminium deoxidation, the soluble aluminium content being within the range 0·01–0·04%. This results in the

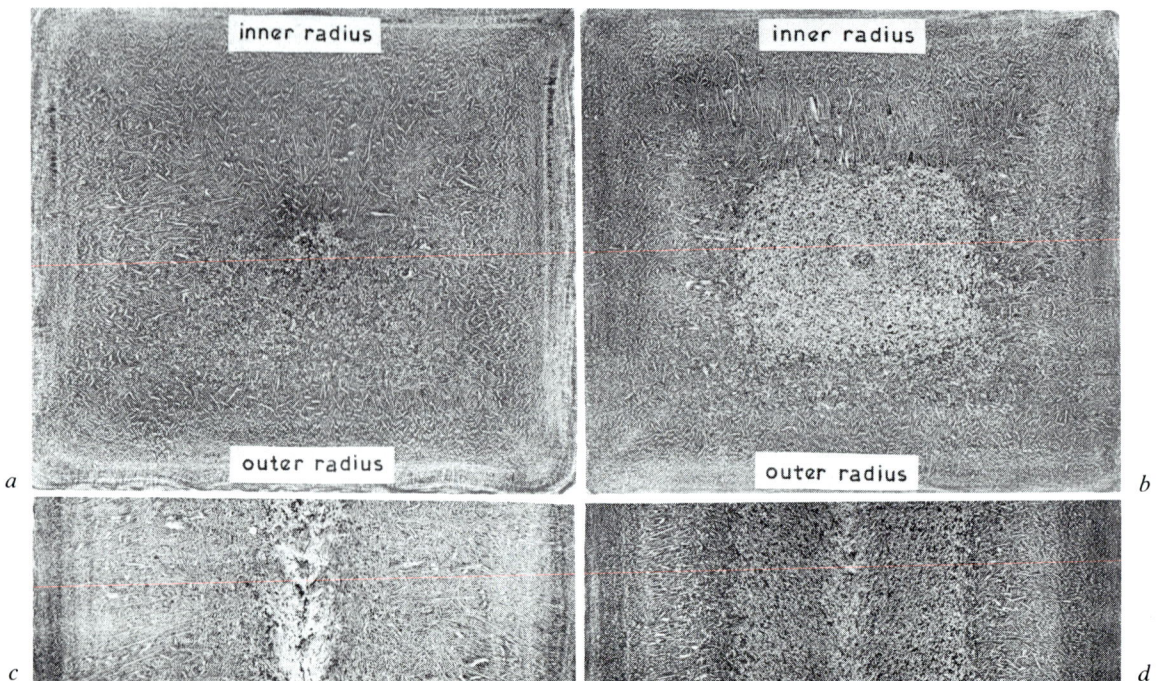

2   Macroetched sections of as-cast bloom: *a* and *b* transverse sections from front and back of cast, respectively [×0·2]; *c* and *d* corresponding central longitudinal sections [×0·2]

measurement and automatic inclusion counting on transverse sections of round bar rolled from the continuously cast blooms, as shown in Fig. 6. The orientation of the bar in relation to the cast bloom was determined by macroetching adjacent sections.

4   Water model showing liquid flow patterns within mould   [×0·17]

formation of alumina inclusions. The distribution of clusters (greater than $50\,\mu$m in diameter) of these inclusions across the bloom has been found to be asymmetric (Fig. 5). The inclusions are concentrated towards the inner radius section of the bloom. This is thought to occur as a result of flotation of the inclusions within the liquid pool, the rising inclusions being caught on the inner radius interface. The asymmetric distribution of inclusions can also be detected by oxygen

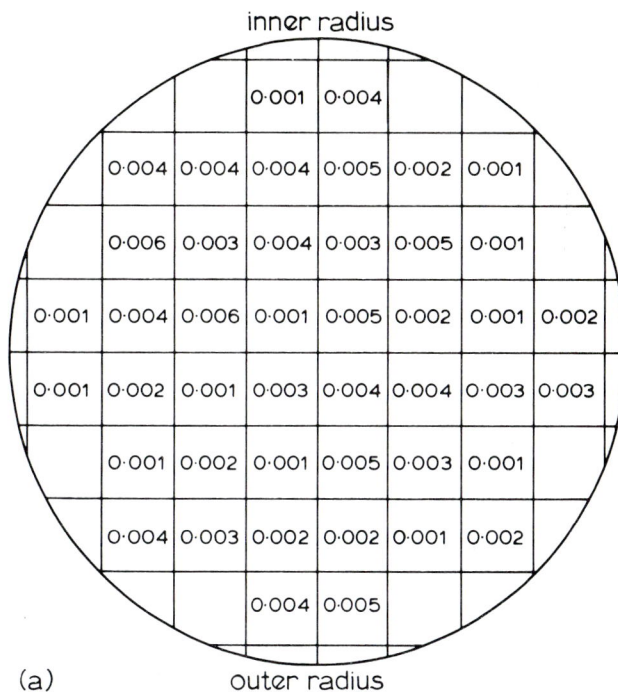

(a)

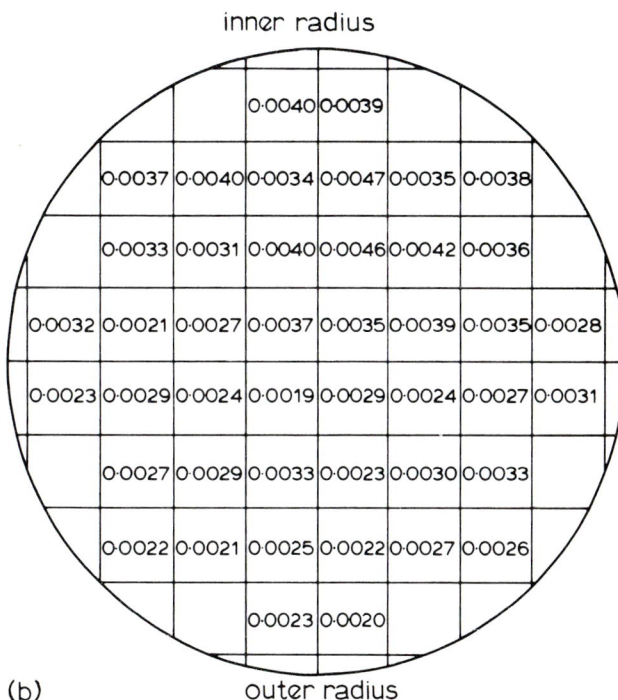

(b)

a volume fraction of oxide inclusions determined at various positions across transverse section of bar using Quantimet 360; *b* oxygen concentration measurements at corresponding positions across same bar

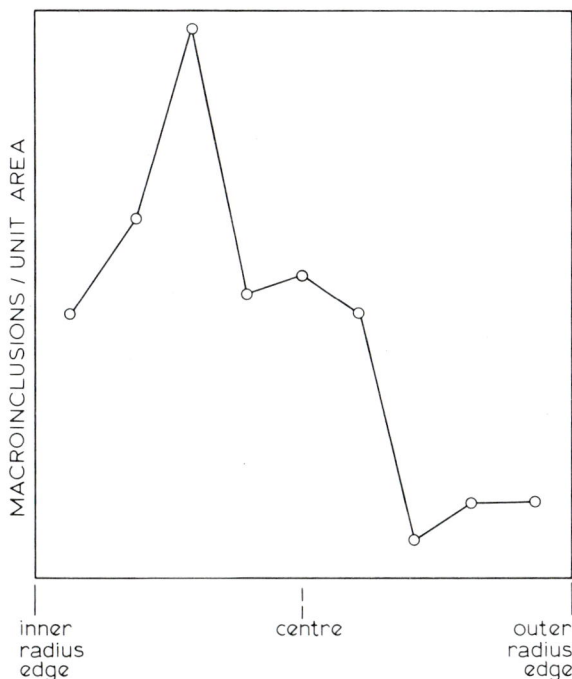

5   Positions of macroinclusions across centre of transverse section of bloom

6   Transverse sections of round bar rolled from continuously cast blooms

7   Cracks revealed on as-cast bloom surface by macroetching   [×0·22]

CRACKS

One of the major problems in the continuous casting of steel is the formation of cracks. Quality problems can arise from both surface and corner cracking. The following two sections describe work on these defects.

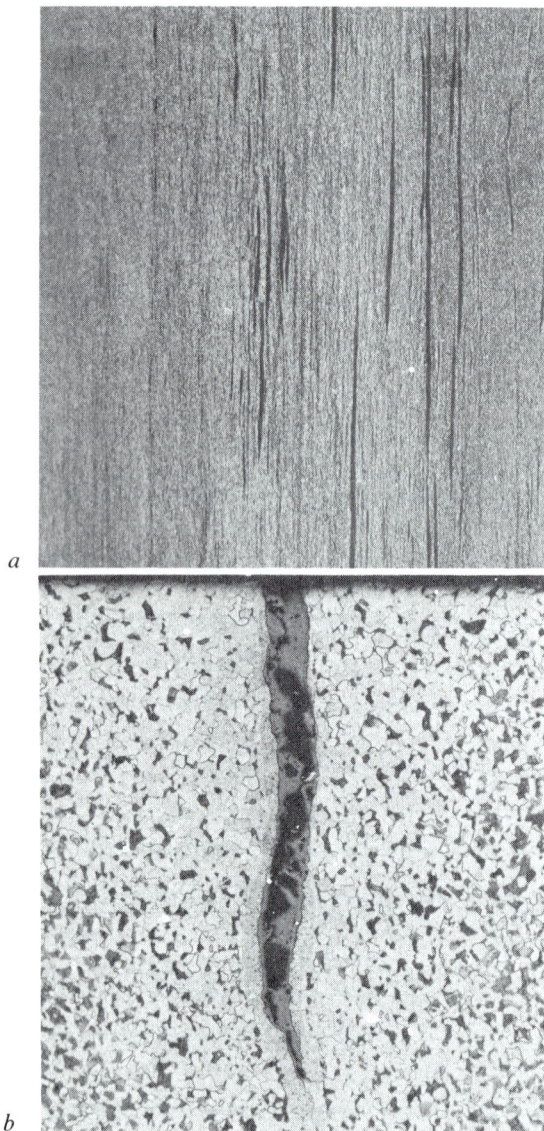

a macroetched surface of rolled bar showing cracks [×1·2]; *b* micrograph of surface crack in rolled bar [×56]

8   Defects on round bar rolled from blooms

Surface cracking or checking

During the first two months of operation of the caster at Round Oak, a form of surface cracking occurred which was revealed by deep etching (Fig. 7) or scarfing runs. The cracks were not visible on the surface of the as-cast bloom and could be up to 5 mm in depth. They occurred mainly at the centres of bloom faces and were similar to hot checking reported previously in other studies to be the result of Cu pick-up from the mould.[3,4] The cracks were not removed by scaling in the reheat furnace and could produce serious defects on round bar (Fig. 8) rolled from the blooms.

The cracking was found to be intergranular between austenite grains. The austenite grain boundaries were coincident in many cases with the original delta ferrite solidification structure (Fig. 9). Microanalysis of the surface regions showed that there was considerable enrichment of Cu, Ni, Sn, and Sb in and around the cracks (Fig. 10). In many places areas of Cu-rich phase were observed. This Cu-rich phase was analysed and found to contain about 85% Cu, 5% Sn, 4% Ni, 5% Fe, and 1% Sb. An alloy of this composition would have been molten at a temperature of about 1 050°C. The presence of such large amounts of Cu-rich phase suggested that the cracking was probably caused by liquid films of Cu penetrating down the grain boundaries. If the presence of Cu at the surface of the strand had been caused by pick-up from the mould, then enrichment of Ni and Sn would not have been expected. It was thought likely, therefore, that the increased residual concentrations were caused by preferential oxidation of the iron. This mechanism is well known[5-8] for steels which are heated in an oxidizing atmosphere to a temperature greater than 1 000°C for periods longer than one hour. It occurs because residuals such as Cu,

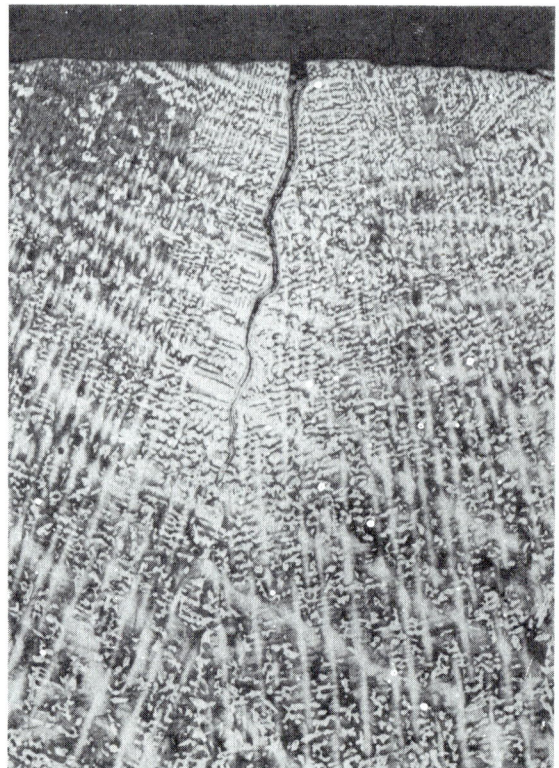

9   Micrograph showing intergranular crack in as-cast bloom   [×35]

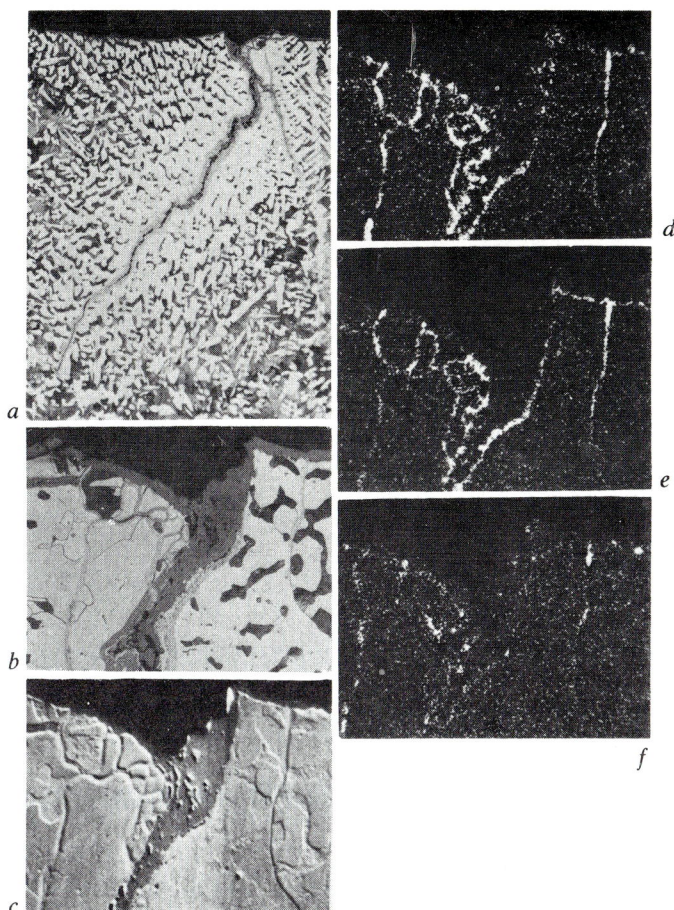

a optical [×33]; b optical [×145];
c electron image; d Ni; e Cu; f Sn

10 Optical, electron, and X-ray images showing residual enrichment of Cu, Ni, and Sn around crack at surface of as-cast bloom

11 Simulation apparatus

to the secondary cooling zone. A specimen which was cooled without any water spray showed no Cu enrichment. However, a specimen which was sprayed intermittently with water showed considerable enrichment of Cu, Ni, and Sn similar to that which had been observed at the surface of the continuously cast strand.

After it had been determined that water was enhancing residual enrichment, a more detailed study was undertaken to investigate the effect of water spray volume, temperature, and some compositional variables on surface cracking. A hot bending test method was used, in conjunction with the simulation technique shown in Fig. 11, for assessing the surface hot shortness under different conditions. The bend specimens were induction heated rapidly to 1 350°C and then cooled to 1 200°C or 1 000°C and held at this temperature for 20 min. During the 20 min holding period water was sprayed intermittently on to the surface of the specimen every 20 s. The quantity of water sprayed on to the surface was varied from 1·5 ml to 15 ml per injection. After the 20 min spraying period, the specimens were withdrawn from the induction coil, cooled rapidly to 1 000°C in the case of the sample sprayed at 1 200°C, and then bent through 180° at 1 000°C. The central sprayed region of the specimen was sectioned and the maximum and total crack depth determined. The specimens were produced from a number of commercial and laboratory casts to give variations in Cu, Ni, Si, and Al concentrations for carbon steels within the range 0·1–0·3%C. The results were found to fall into two categories.

Figure 12 shows results which were typical of steel containing less than 0·01% Al. The maximum crack

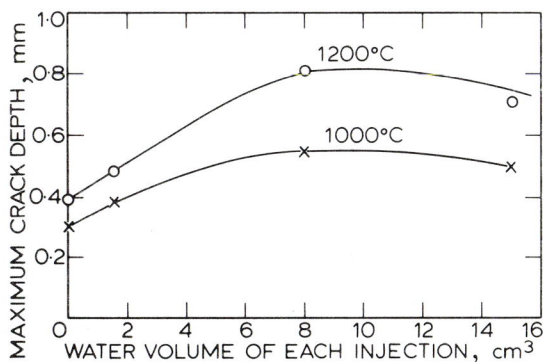

Ni, and Sn have only a small tendency to oxidize in the presence of iron. Iron is removed preferentially from the surface layers by oxidation, while the residual elements are left behind in the subscale layer.

At TI Research Laboratories a considerable amount of work has been carried out[6–8] into this type of enrichment and associated cracking produced by reheating and deformation during the bar- and tube-making processes. The amount of Cu found on the surfaces of the continuously cast strands, however, was greater than that generated during reheat operations, even though the time at high temperature during continuous casting was considerably shorter than for the reheating processes. There was some factor, therefore, which was enhancing the formation of Cu enrichment. Because of the positions of the cracks at the centres of the bloom faces, it was considered possible that water sprays might be contributing to the enhanced enrichment of the residual elements.

This possibility was examined in the laboratory using the technique shown in Fig. 11. A steel specimen, $10 \times 10 \times 100$ mm, within a silica tube was induction heated rapidly to 1 350°C and then cooled at a rate similar to that which occurred at the surface of the strand. Water could be sprayed intermittently on to the surface of the specimen during the temperature range corresponding

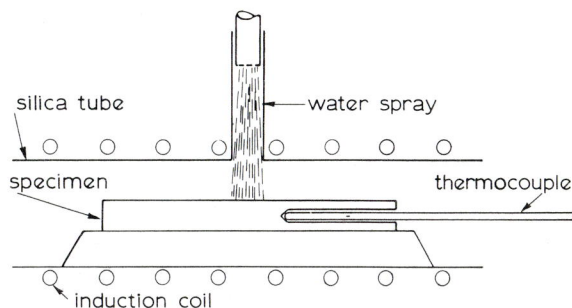

12 Variation of maximum crack depth with water volume for specimens of 0·2% C, 0·30% Cu, 0·007% Al steel sprayed with water at temperatures of 1 200° and 1 000°C, deformed at 1 000°C

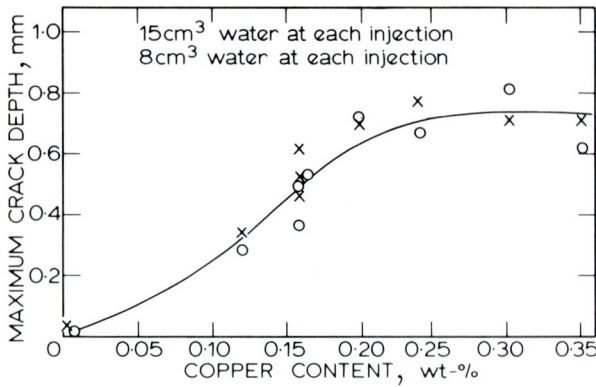

13   Variation of maximum crack depth with copper content for specimens sprayed at 1 200°C with 8 or 15 cm³ of water for each injection

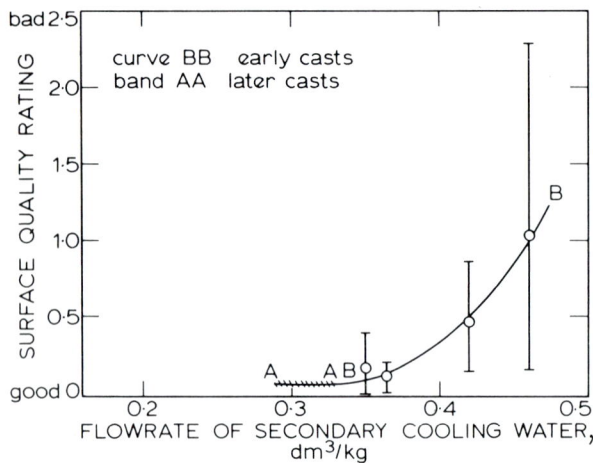

15.   Variation of surface quality rating with aluminium concentration of steel: surface quality rating is a combination of the number of rejected and dressed bars

depth increased with increasing spray water volume but levelled off at the high water flow rates. The specimens sprayed at 1 200°C gave more severe cracking than those sprayed at 1 000°C. Microscopic examination of the cracks indicated that they were caused by liquid films of Cu, and the severity of cracking at the higher water flow rates was found to be related to the Cu content (*see* Fig. 13).

For steels containing higher Al concentrations, however, the crack depths became considerably larger for specimens sprayed at the lower temperature of 1 000°C. Chemical analyses of the bend specimens in the vicinity of the spray showed that the severe cracking at 1 000°C and high water flow rates was associated with high levels of AlN ($\sim$40 ppm). A high concentration of AlN ($>$25 ppm) is known[9,10] to reduce significantly the ductility of steel in the temperature range 800°–1 100°C.

It appears from the laboratory experiments, therefore, that there are at least two mechanisms of crack formation: (*a*) for low-aluminium steels sprayed at high and low temperatures, and high-aluminium steels sprayed at high temperature, the cracking is caused by liquid films of copper produced by preferential oxidation; and (*b*) a second concurrent mechanism which occurs in high-aluminium steels sprayed at low temperatures produces very severe cracking and is caused by

AlN precipitated by the thermal cycling action of the sprays.

Support for the role of cooling water and aluminium content revealed by the laboratory studies was obtained by the analysis of plant data. These indicated that the surface quality improved considerably when the secondary cooling water spray levels were reduced (Fig. 14). At the high water flow rates the surface quality appeared to be influenced by the aluminium concentration of the steel (Fig. 15). Furthermore, analysis for AlN in the surface layers of an as-cast bloom which had exhibited bad surface cracking showed the presence of a high concentration ($\sim$80 ppm). This concentration decreased to $\sim$20 ppm at a depth of 4 mm into the bloom.

As for the role of copper, observations on the plant indicated that it contributed mainly to the formation of small cracks, which did not present quality problems. Copper, however, may also play a part in the cracking associated with AlN by acting as a defect initiator through the residual enrichment process.

Reducing the secondary cooling water flow levels on the casting machine to 0·28–0·3 dm³/kg and adjustments to the spray nozzles have been successful in virtually eliminating this defect.

Corner cracking

Figure 16 shows a typical corner crack revealed by etching a transverse section of a bloom. Usually in the as-cast bloom the cracks are subsurface. In the reheat

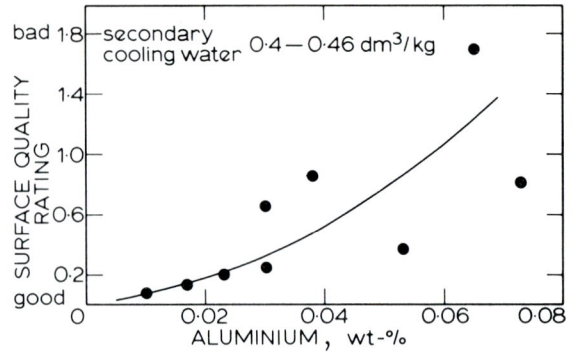

14   Variation of surface quality rating with volume flow rate of secondary cooling water: surface quality rating is a combination of number of rejected and dressed bars

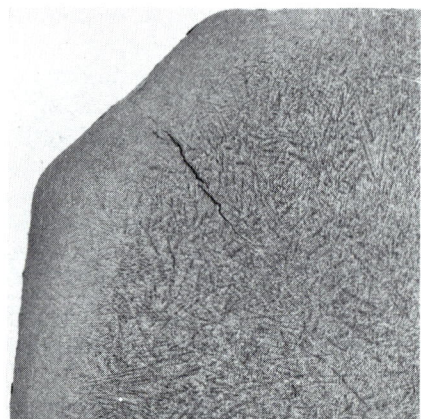

16   Corner crack in macroetched as-cast bloom   [×1·8]

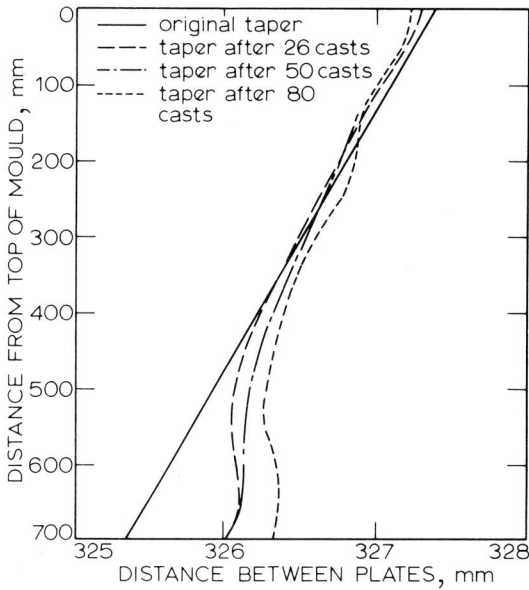

17 Variation of mould taper with number of successive casts

18 Experimental technique for measuring tensile strength and ductility of steel in as-cast condition

furnace, however, as a result of scaling the crack can be exposed and become oxidized, giving rise to a defect in the subsequent rolled bar.

Analysis of plant data has shown that carbon levels in the range 0·15–0·24% and high superheats encourage cracking, a conclusion which is in agreement with the results of other studies.[11] Mould taper and machine alignment have also been found to be very important factors. The gradual wear of the mould taper during successive casts (Fig. 17) progressively causes the corner cracking frequency to increase, limiting the mould life in some cases to less than 100 casts.

Studies of a shell remaining after a breakout showed that a corner crack first started to form 250 mm below the liquid steel meniscus. Optical microscopy of the

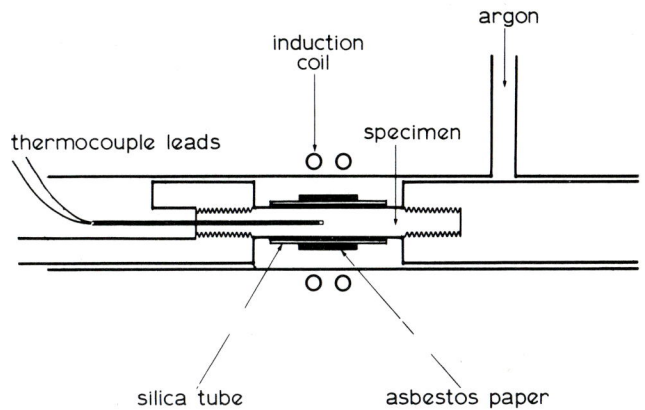

cracks in the breakout shell indicated that they started to form at or near the solid/liquid interface and then propagated towards the outside of the bloom. Near the solid/liquid interface, the cracks were infilled with sulphur-rich liquid.

The morphology of the fracture faces of the cracks has been studied by scanning electron microscopy to obtain a better understanding of the mechanism of their formation. One of the problems in trying to look at the characteristics of the fracture faces which existed at the high temperature of formation is that during the relatively slow cooling of the bloom considerable surface diffusion can take place. In addition, precipitation of sulphides and changes in morphology of the sulphides can also occur. Studies are continuing on these aspects. In order to understand more clearly the fracture processes a laboratory study was undertaken to determine the tensile strength, ductility, and fracture-face morphology of as-cast steel fractured at different temperatures. The experimental technique used is shown in Fig. 18. About 2 cm of the gauge length of a

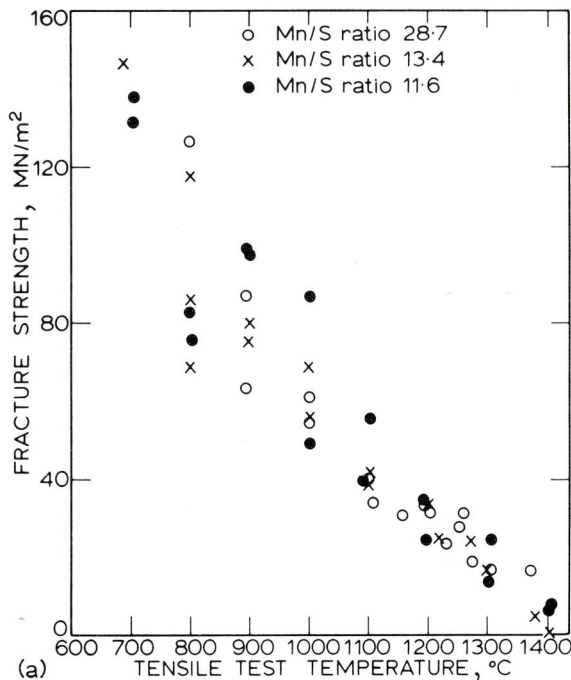

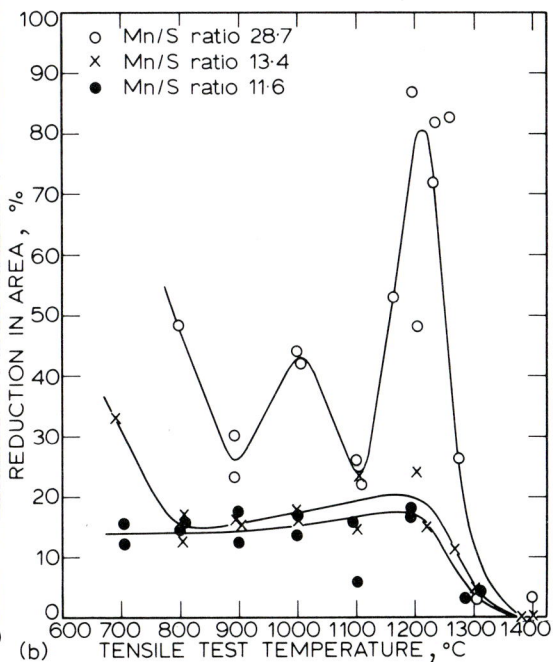

a tensile strength; *b* reduction in area to fracture

19 Variation of mechanical properties with temperature in as-cast steels having different Mn/S ratios

20 Fracture face morphologies of; *a* corner crack in breakout shell [×160]; and *b* tensile specimen tested at 1 450°C   [× 300]

tensile specimen was melted using induction heating. The molten region was contained by a silica tube. The specimen was then resolidified, cooled to the test temperature, and tensile tested at a strain rate of approximately 0·2 s. Tests were carried out on a number of 0·2% C steel casts having varying Mn/S ratios.

The results of the tests shown in Fig. 19*a* and *b* indicate that while the Mn/S ratio does not affect the fracture stress significantly, it does alter the reduction in area at temperatures from 900° to 1 400°C. The fracture face observed in a corner crack from a breakout shell has been compared with those of laboratory tensile specimens fractured over a range of temperatures from 800° to 1 450°C. Figure 20 illustrates the close similarity between the bloom fracture and a laboratory specimen tested at 1 450°C, indicating that corner cracks are formed at around this temperature.

SUMMARY
This paper has described laboratory and plant studies which are leading to a better understanding of the mechanisms by which defects are initiated during continuous casting. In many cases appropriate practical remedies for these can be suggested and, overall, the quality of the continuously cast material is proving markedly superior to ingot steel.

REFERENCES
1 B. CHALMERS: 'Principles of solidification', 270, 1964, New York, John Wiley
2 A. KOHN: 'The solidification of metals', 416, 1967, London, The Iron and Steel Institute
3 H. MORI: *Tetsu-to-Hagané (J. Iron Steel Inst. Jpn)*, 1974, **60**, (no. 7), 784
4 T. NOZAKI *et al.*: *ibid.*, 1971, **57**, (no. 11), S673
5 A. D. NICHOLSON AND J. D. MURRAY: *J. Iron Steel Inst.*, 1965, **203**, 1 007
6 D. A. MELFORD: *J. Iron Steel Inst.*, 1962, **200**, 290
7 D. A. MELFORD: 'Proc. Third International Symposium on X-ray Optics and X-ray Microanalysis', 557; 1963
8 D. A. MELFORD: *J. Iron Steel Inst.*, 1966, **204**, 495
9 P. A. PORTEVIN: *Rev. Métall.*, 1962, **59**, (no. 11), 915
10 Unpublished research, Tube Investments Research Laboratories
11 H. VON ENDE AND G. VOGT: *J. Iron Steel Inst.*, 1972, **210**, 889

Solidification of steel in continuous-casting moulds

M. Wolf and W. Kurz

The initial seconds of solidification in a continuous-casting mould govern the entire subsequent strand formation. Since surface quality and shell growth are decisively influenced at such an early stage of operation, disturbances in skin solidification could result in breakouts. At present, there exists only limited detailed knowledge of the initial stage solidification in the mould. Therefore, an attempt has been made to correlate macroscopic shell thickness (which follows a square root of time equation), with the casting variables influencing the structure. Using the deduced functions in combination with the parabolic rate constant for the increase in the thickness of the solidifying zone, the secondary dendrite arm spacing was calculated, resulting in a reasonably good agreement with experimental results. From this analysis, the authors draw some conclusions with respect to increasing the growth rate of the solidified skin in the mould.

M. Wolf is with Concast AG, Toedistrasse, Zurich, and W. Kurz is in the Swiss Federal Institute of Technology, Lausanne, Switzerland

LIST OF SYMBOLS

A = material constant ($= -\rho_s \Delta H / \kappa_s$), Ks/cm^2
a = heat diffusivity ($= \kappa / \rho c$), cm^2/s
B = constant, equation (18), μm K$^b s^{-b}$
b = exponent, equation (18)
c = specific heat, J/K. g
G = temperature gradient, K/cm
ΔH = latent heat of fusion, J/g
K = solidification constant, equation (1), cm/$\sqrt{s}$
K' = solidification constant, equation (2), cm/s^n
l = subscript for liquid
$M(K)$ = function of K equation (13)
n = exponent, equation (2)
R = casting rate, m/min
s = subscript for solid
t = time, s
T = temperature, K and °C
$\dot{T}$ = cooling rate, K/s
T_{li} = liquidus
T_l° = initial temperature of liquid
T_m = melting point $\simeq (T_{li} + T_{so})/2$
T_{so} = solidus
T_s^0 = surface temperature of solid
V = growth rate of solid/liquid interface in y direction, cm/s
y = coordinate perpendicular to mould surface
Y = solid/liquid interface position on y axis (shell thickness), cm
θ = local solidification time, equation (17), s
κ = thermal conductivity, W/cmK
Λ = coefficient in equation (6)

λ = dendrite arm spacing, μm
ρ = density, g/cm^3

In the continuous casting of steel, a water-cooled copper mould is used in the first stage of solidification. The mould has to fulfil two important functions:

(i) at the beginning of the solidification process, the surface of the strand is formed in contact with the mould wall just below the free liquid surface, (often called the meniscus)
(ii) when the strand leaves the mould, the solidified shell must be sufficiently thick to contain the liquid metal safely; in addition, since the mechanical properties of any casting are strongly influenced by the solidification conditions,[1] close control of the shell growth in the first seconds of strand formation is of considerable importance.

The aim of this work was to analyse experimental measurements of the skin thickness in the mould region and to use this information to determine the solidification parameters and finally, the microstructure of the cast product.

EXPERIMENTAL OBSERVATIONS ON SHELL GROWTH

As has been shown,[2,3] the heat flow in the continuous-casting mould is a function of the distance along the length of the mould (Fig. 1). There is a small heat flow at the start followed by a rapid increase, up to a maximum at a solidification time of about 2 s, which corresponds to

287

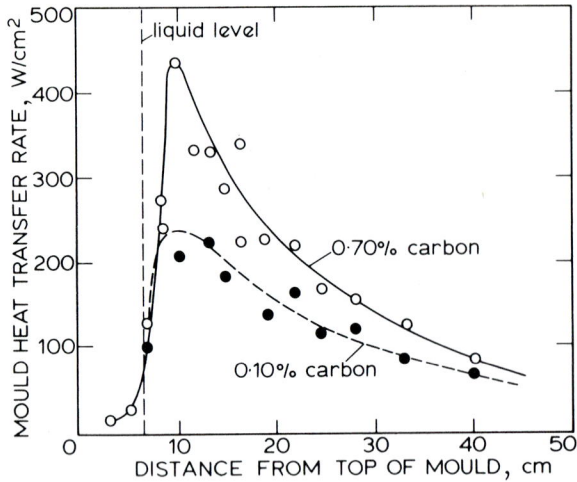

1 Heat flux as function of position in mould, for two steels[3]; casting speed = 2·1 cm/s

a point 3–10 cm below the meniscus. The heat flow then decreases steadily to a low value at the end of the mould. Singh and Blazek[3] argue that these three zones of heat flow might be rationalized in the following way:

   (i) at the beginning, the steel/mould contact is very localized but increases with time because of the formation of a continuous solid skin (Fig. 2) and increasing ferrostatic pressure
  (ii) the observed maximum in heat flow may be due to the good thermal contact of the hot solid shell resulting from its being pressed against the copper mould by the internal pressure
 (iii) once the solid skin becomes sufficiently self-supporting, it contracts, forming an air gap; this, together with the increasing shell thickness, decreases the heat flow.

The heat flow is also strongly dependent on composition (Fig. 1). On studying the mean heat flow of the mould as a function of the carbon concentration in the steel (Fig. 3), a sharp minimum is observed at about 0·1 wt-% C.[3,4] The constant of proportionality K of the square root of time law

$$Y = K\sqrt{t} \tag{1}$$

shows a similar tendency (Fig. 3); K is referred to as

3 Mean heat flux in mould[4] and parabolic rate constant calculated from Figs 12 and 13 of Ref. 4 as functions of carbon content of the steel; (mould section 83 × 83 mm, casting speed 1·3 m/min)

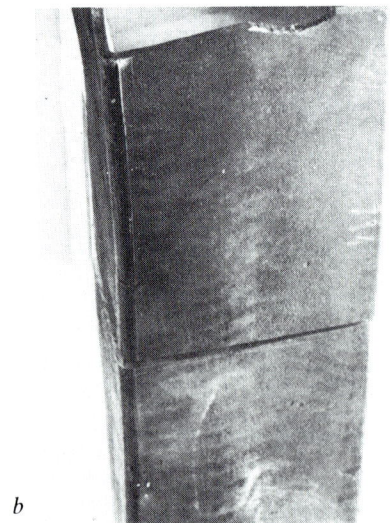

2 Breakout shell of 0·07% C steel, showing degree of partial solidification underneath the meniscus; (section 120 × 120 mm, R = 2·7 m/min)

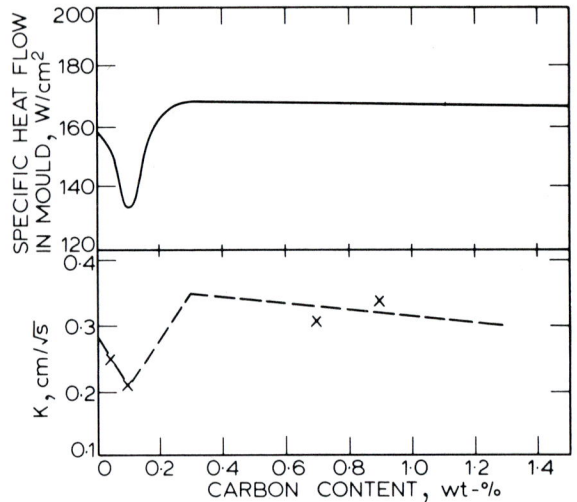

a

b

4 Surface quality of breakout shells (a) 0·11% C-steel (section 108 × 108 mm, R = 3·3 m/min); (b) 0·62% C-steel (section 115 × 115 mm, R = 2 m/min)

Table 1 Characteristic data for shell growth in the mould derived from various investigations

Reference	Method*	Carbon content of steel, wt-%	Cross-sectional dimensions, mm	Pouring type	R, m/min	n‡	K'‡
4	BO	various, see Fig. 3	various, see Fig. 3	open	—	—	—
7	BO	0·25	various, see Fig. 6	open	—	—	—
8	RI	0·17	150×300	open	0·70	0·65	0·17
9	BO	0·11	115×115	open	1·9	0·65	0·18
10		0·10	200×200	open	—	0·86	0·15
11	RI	0·06	slab†	closed†	—	0·5	0·25
12	S/Pb	0·60	110×110	open	—	0·73	0·27
13	calc.	0·20	150×930	closed†	—	0·86	0·08
14	calc.	(0.15)	200×240	open	0·6	0·73	0·11
				closed	0·6	0·73	0·11
15	RI	0·06	220×1 250	closed	1·0	0·5	0·21
					1·4	0·5	0·32
					2·0	0·5	0·21
16	RI	0·20	bloom†	open	0·9	0·75	0·12
17	RI	0·18	140×140	open	1·4	0·65	0·14
	RI	0·85	140×140	open	1·5	0·73	0·16
18	RI	0·13	133×133	open	2·10	0·70	0·12
	RI	0·13	133×133	open	2·65	1·0	0·075
	RI	0·15	101×101	open	2·8	1·0	0·11
19		0·10	200×200	closed†	1·0	0·56	0·36
20	BO	(<0·20)	152×152	open	1·52	0·52	0·22
21	RI	0·14	200×2 050	closed	0·65	0·85	0·06
					0·85	0·70	0·08
					1·05	0·77	0·11
22	calc.	0·06	slab†	closed	1·0	0·74	0·08
23	feeler	0·06	225×1 320	closed	0·4	0·49	0·29
					0·6	0·52	0·23
					0·8	0·55	0·18
24	RI	0·14	200×1 500	closed	0·65	0·74	0·07
25	S/Pb	0·12	200×1 600	closed†	0·67	0·63	0·16

* Method of shell growth determination: BO ≐ breakout; RI = radioactive isotope; S/Pb = tracing with S and Pb addition; feeler = direct sensing during casting; calc. = calculated from heat-transfer measurements
† Condition not explicitly stated
‡ See equation (2)

the parabolic rate constant and, typically, takes values of 0·2–0·35 cm/√s (16–27 mm/√min) in the mould. The very different surface qualities of low- and high-carbon steels are shown in Fig. 4. The rough and wavy surface of the 0·11% C composition is typical of this grade[4] of steel and is also visible at the solid/liquid interface (Fig. 4a). In comparison with this, a 0·62% C steel shell has very smooth outer and inner surfaces, which is consistent with the very different heat-flow and K values. The reason for the difference in surface quality is not clear but might be connected with the peritectic reaction.[5] A better understanding of the first stages of solidification in the mould is needed.

A summary of most of the relevant measurements of shell thickness in the mould Y is shown in Table 1[4-25] and Fig. 5. All the profiles for breakout shells should be typical of those for a solid fraction of about 0·5,[6] a value which is also found with tracer techniques. It has been observed that equation (1) is rarely obeyed and the measured growth law is usually of the form

$$Y = K't^n \qquad (2)$$

where n lies in the range 0·5–1; K' decreases as n increases. It can also be seen from Fig. 5 that for the same value of n, K' is higher in high-carbon steels. There is no visible difference between the results for open and

closed pouring.* The straight line between K' and n (Fig. 5) cuts the ordinate for $n = 0·5$ at $K \simeq 0·25$ for C < 0·20% and $K \simeq 0·35$ for C > 0·20%. This result agrees well with the values in Fig. 3.

Another very important factor which influences the values n and K' in equation (2) is the superheat of the liquid steel (Fig. 6). The higher the melt temperature, the lower the value of K' and the higher that of n.[7] This tendency can be attributed to the retardation of solidification, which is caused by the necessity of decreasing the temperature of the liquid steel.

ESTIMATION OF SHELL GROWTH RATE AND SOLIDIFICATION PARAMETERS
Theoretical consideration
The heat equation with the boundary conditions: planar solid/liquid interface at the melting point, T_m, constant melt temperature far from the solid/liquid interface T_i° and constant surface temperature T_s° can be solved analytically.[26] (A similar solution has already been applied by Jones[27] to continuous casting of steel). One obtains:

$$Y = Kt^{0.5} \qquad (3)$$

* Open pouring is generally used for billet sections with oil lubrication, while closed pouring is used mainly for slab and bloom casting with submerged nozzles and a slag on the free liquid surface

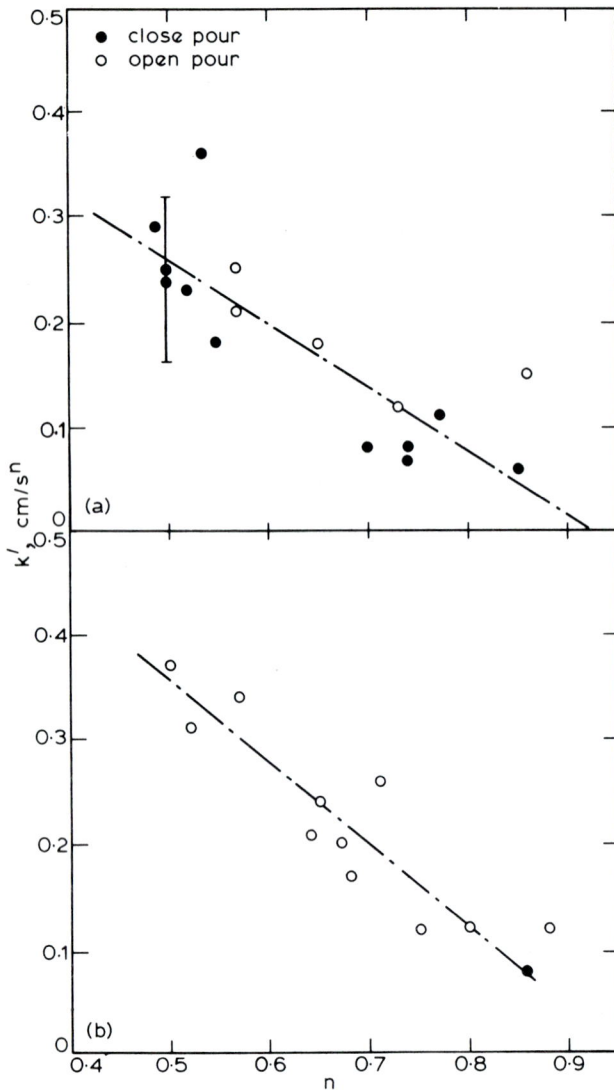

a low-carbon steels, C < 0·2 wt-%; *b* high-carbon steels, C > 0·2 wt-%

5 Relationship between K' and n values of equation (2) for shell growth in mould, for low- and high-carbon steels (cf. Table 1)

$$V = \mathrm{d}Y/\mathrm{d}t = 0.5Kt^{-0.5} \qquad (4)$$

$$G_s = \frac{\partial T_s}{\partial y} = \frac{(T_m - T_s^\circ)}{\mathrm{erf}\,\Lambda\sqrt{\pi a_s t}}\exp(-K^2/4a_s) \qquad (5)$$

$$K = 2\Lambda a_s^{0.5} \qquad (6)$$

The value of Λ can be obtained from:

$$\frac{(\kappa_1/\kappa_s)(a_s/a_1)^{0.5}(T_1^\circ - T_m)}{\Lambda\,e^{(a_s/a_1)\Lambda^2}\,\mathrm{erf}\,c[(a_s/a_1)^{0.5}\Lambda]} - \frac{(T_m - T_s^\circ)}{\Lambda\,e^{\Lambda^2}\,\mathrm{erf}\,\Lambda} = \frac{-\Delta H\pi^{0.5}}{c_s} \qquad (7)*$$

For the case of zero superheat in the liquid, equation (7) simplifies to

$$\Lambda\,e^{\Lambda^2}\,\mathrm{erf}\,\Lambda = -c_s(T_m - T_s^\circ)/\Delta H\pi^{0.5} \qquad (8)$$

ΔH is negative for solidification. In this case, the temperature gradient in the solid can be directly

* The authors are grateful to Dr. T. Emi, Chiba, Japan, for pointing out an error in equation 10.1.40 of Ref. 26.

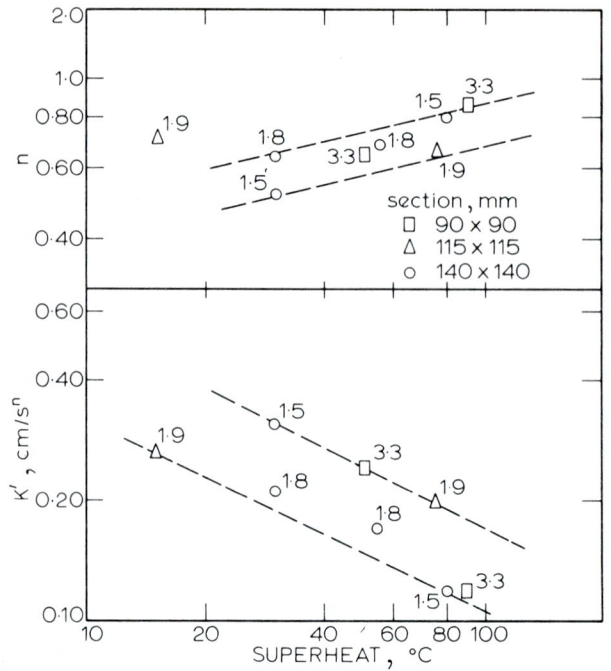

6 Effect of steel superheat (difference in tundish and liquidus temperature) on shell growth in the mould for 0·25% C steel (derived from the results given in Ref. 7; figures along curves refer to casting rate in m/min)

obtained by use of the boundary condition ($\kappa_1 G_1 - \kappa_s G_s = \rho_s\,\Delta HV$):

$$G_s = \partial T_s/\partial y = AV \qquad (9)$$

$$A = -\rho_s\,\Delta H/\kappa_s \qquad (10)$$

Using equation (6) and the correct value of a_s for steel (Table 2), one obtains $K = 0.5\Lambda$. With this and equation (8), one then obtains a relationship between K and $(T_m - T_s^\circ)$, the total temperature difference across the solidified steel shell (Fig. 7). If K is in the range 0·2–0·4, the complicated equation (8) may be approximated by:

$$(T_m - T_s^\circ) \simeq 7.1 \times 10^3 K^{2.5} \qquad (11)$$

Using this relationship, the temperature gradient in the solid at the solid/liquid interface G_s may be easily obtained *via* equation (5) and reduces, for steel, ($a_s = 0.063\ \mathrm{cm^2/s}$,[28,29] $0.2 < K < 0.4$) to

$$G_s = M(K)t^{-0.5} \qquad (12)$$

$$M(K) = \frac{1.6 \times 10^4 K^{2.5}\,e^{-4K^2}}{\mathrm{erf}(2K)} \simeq 8 \times 10^3 K^{1.5}\,e^{-4K^2} \qquad (13)$$

In a similar way, one could derive an analogous equation for G_1. However, unless convection could be taken into account, this would not be satisfactory.

Assuming zero superheat in the liquid pool, which is a reasonable assumption in the case of low casting

Table 2 Properties of steel close to the melting point[28,29]

ΔH, J/g	c_s, J/gK	ρ_s, g/cm³	κ_s, W/cm K	κ_1, W/cm K	a_s, cm²/s	a_1, cm²/s
260	0·7	7·4	0·335	0·35	0·063	0·075

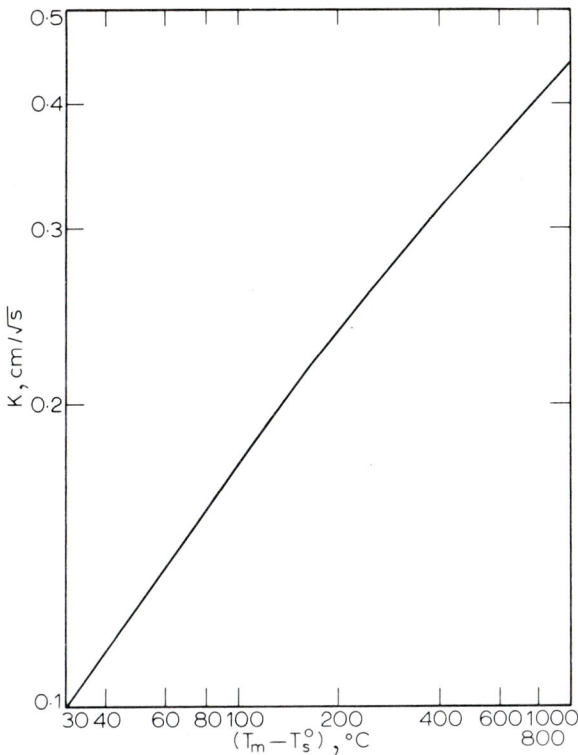

7 Parabolic rate constant *K* for carbon steel as function of temperature difference across solid shell; (*see* equations (6) and (8); no superheat)

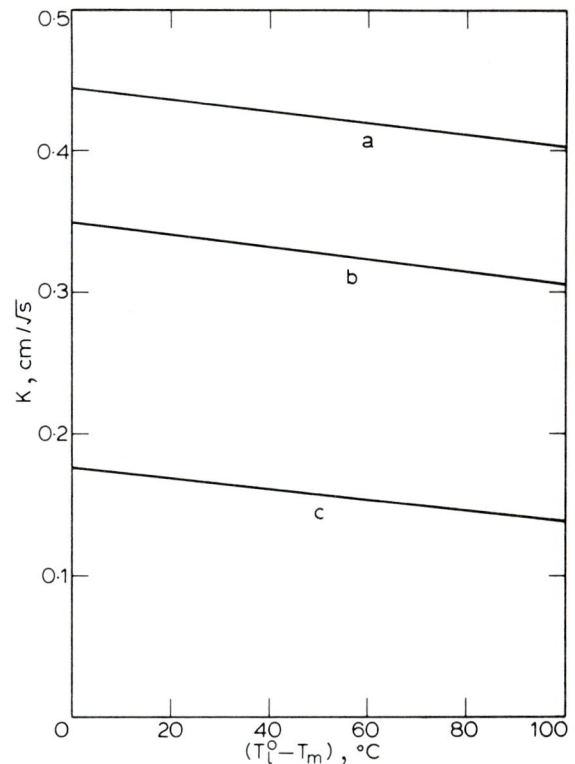

a $(T_m - T_s^\circ) = 1\,000°C$; *b* $(T_m - T_s^\circ) = 500°C$; *c* $(T_m - T_s^\circ) = 100°C$

8 Parabolic rate constant *K* for carbon steel as function of true superheat in liquid; (temperature difference in melt between *y* = ∞ and *y* = *Y*)

temperatures, if can be seen from Fig. 7 that the maximum temperature difference across the skin $(T_m - T_s^\circ)$ is 130°–500°C for the observed range of $0·2 < K < 0·35$. These are very plausible values.

The effect on *K* of an initial superheat in the melt has been calculated using equations (6) and (7), (Fig. 8). Again, direct use of this relationship is not possible because of convection effects, but it can be seen that a superheat in the liquid decreases *K* only slightly. In Fig. 6, *K'* decreases quite rapidly with increased superheat. However, this change is accompanied by an increase in the value of *n*. This behaviour is due to the finite volume of the melt, which leads to a decrease in $T_l^\circ - T_m$ with increasing time (*see* below).

Growth conditions and structure of 0·62% C steel

With these facts in mind, an estimation of the parameters influencing solidification in the continuous-casting mould can be attempted. Because of its more regular growth (Fig. 4*b*) and its well defined primary structure, the 0·62% C steel was chosen as an example. The thermal properties given in Table 2[28,29] were used for the calculations. The simplifying assumptions made, which are especially crude for the first moments of solidification, were (Fig. 9):

(i) that during the first 1·5 s, corresponding to the region of increasing heat flux (Fig. 1), only a mean value $\bar{K}$ (= 0·2) is assumed for a very rapidly changing temperature difference in the solid, without any superheat in the liquid; this is equivalent to a mean temperature difference across the solid shell of 130°C (Fig. 7)

(ii) that between 1·5 and 16 s (supposed to correspond to the end of the mould) the temperature of the solid skin surface exposed to the mould is assumed to remain constant and *K* is assumed to be 0·30; this leads to a surface temperature of ~ 1 100°C.

These assumptions permit the calculation of the skin thickness *Y* as a function of time (Fig. 9*c*). Regression analysis of the points in Fig. 9*c* (experimental values) with respect to equation (2) leads to the relationship:

$$Y = 0·24t^{0·58} \tag{14}$$

A similar regression analysis of the points on the full line from 0·5–16 s in Fig. 9*c*, which represent the model described above, gives:

$$Y = 0·21t^{0·64} \tag{15}$$

This shows that even with such a simple model, a reasonable fit with experimental results is possible. The result enables one to calculate the parameters needed for predicting the solidification structure (Fig. 10). The cooling rate in the solid close to the solid/liquid interface was obtained by using the relationship:

$$\dot{T}_s = -G_s V = -AV^2 \tag{16}$$

For a given alloy, the local solidification time θ is found using:

$$\theta = (Y_{li} - T_{so})/\dot{T}_s \tag{17}$$

but only when G_s represents the mean gradient in the mushy zone with reasonable accuracy. All values, except

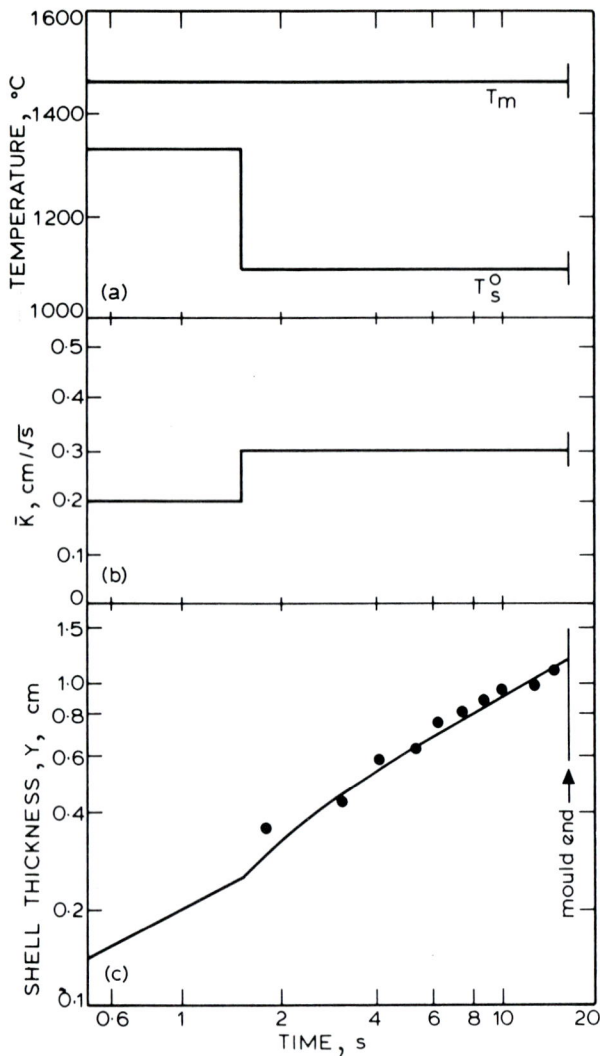

9   **Calculated shell thickness Y as function of time (full line), compared with experimental points (c); temperatures and assumed K values as indicated in (a) and (b)**

V, have been calculated for $t > 1\cdot5$ s because of the uncertain and varying conditions at the start.

Growth rate fluctuations (coupled with temperature gradient fluctuations) of the type shown (Fig. 10) will be caused by sudden surface temperature changes due to varying thermal contact with the mould especially at the beginning of solidification (cf. Figs 2 and 4a). These lead to the familiar banded structure of low-carbon steel (Fig. 11). Furthermore, Fig. 11 indicates that the shell thickness for low-carbon steels varies markedly with position in the mould at a given time.

Knowing $\dot{T}$, one can use the established empirical relationship for dendrite arm spacing:

$$\lambda = B\dot{T}^{-b} \tag{18}$$

Using the present model, one can only determine the secondary dendrite arm spacing λ_2, because the growth conditions at the dendrite tip cannot be obtained and convection, which is always present in the mould, seems to have a large effect on the primary dendrite spacing.[30,31]

Suzuki *et al.*[32] determined the spacing–cooling rate relationship for $0\cdot6$% C steels and found:

$$\lambda_2 = 145\dot{T}^{-0\cdot39} \tag{19}$$

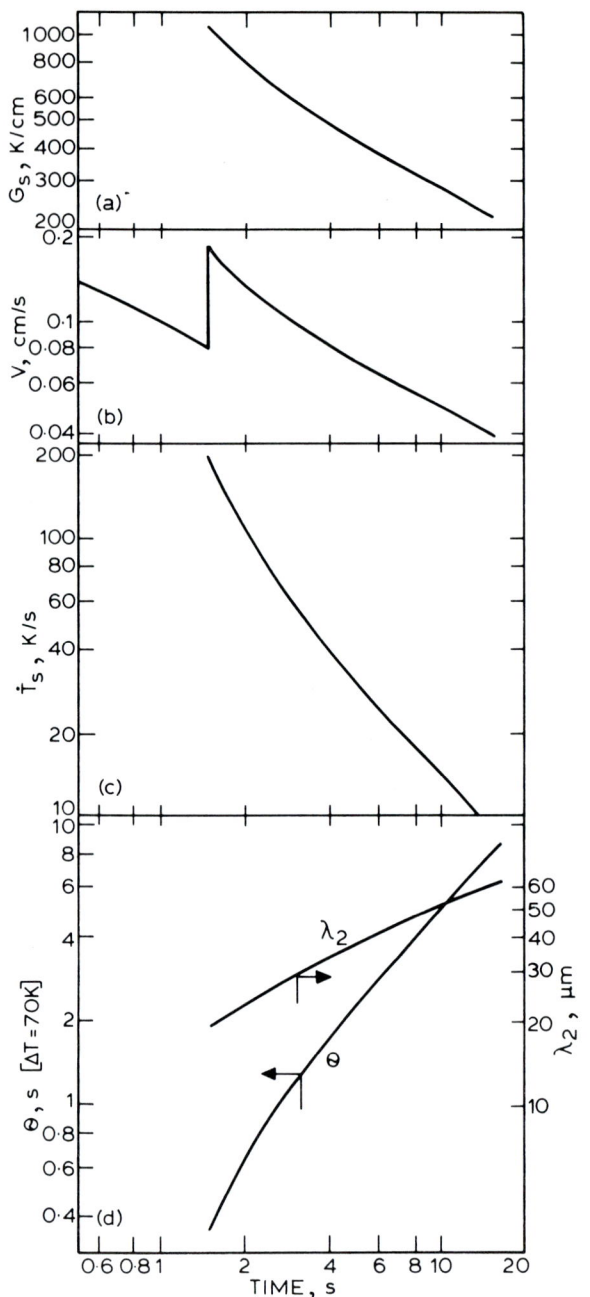

10   **Important solidification parameters calculated on the assumptions of Fig. 9b; (G_s has been calculated using equation 9)**

surface of casting   1 mm

11   **Microstructure of 0·07% C steel at 55–60 cm from meniscus (Oberhoffer's reagent); same shell as that in Fig. 2**

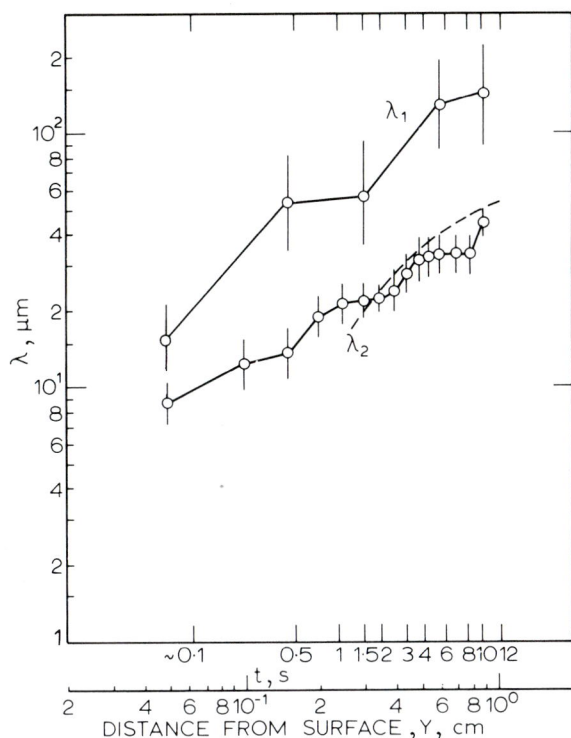

12 **Measured primary λ_1 and secondary λ_2 dendrite arm spacing for 0·62% C steel as function of distance from outside of casting (full line); broken line represents λ_2 obtained with cooling rate from Fig. 10 and equation (19)**

(where λ is in μm and $\dot{T}$ is in K/s). This equation gives the λ_2 values indicated (Fig. 10) when one assumes that $\dot{T}_s \simeq \dot{T}$.

Figure 12 shows the experimentally determined values of λ_1 and λ_2 as functions of the distance from the outer surface compared with the calculated ones; as can be seen, the agreement is good.

DISCUSSION

The solidification constants K or K' (equations (1) and (2)) are generally accessible parameters in continuous casting. Measurements of shell growth in the mould usually indicate growth according to equation (2) where $n > 0·5$. However, theory shows that under constant boundary conditions, n should reduce to $0·5$. The high and rapidly varying surface temperature of the solid skin and the superheat of the liquid at the beginning of solidification lead to initially small and rapidly varying K values. Regression analysis (equation (15)) of the calculated $Y(t)$ points, assuming two different K values and $n = 0·5$ (Fig. 9b), leads to mean values of n greater than $0·5$. This is one way to explain the difference between equations (1) and (2).

The present analysis of a simplified model is applicable because of the special conditions in the continuous-casting mould which may be assumed to be constant in the region of the decreasing heat flow, which covers nearly the whole length of the mould (Fig. 1). According to the equations given above, K of the square-root law can be easily related to important solidification parameters, e.g. $V = 0·5K/\sqrt{t}$; $G_s = M(K)/\sqrt{t} \simeq 0·5AK/\sqrt{t}$; $\dot{T} \simeq AK^2/4t$; $\theta \simeq 4(T_{li} - T_{so})t/AK^2$. This permits the determination of an important structural

parameter, the secondary dendrite arm spacing, since $\lambda_2 \simeq B(4t/AK^2)^b$. Therefore, K (with $n = 0·5$) is a very important and useful value for the macroscopic, as well as for the microscopic characterization of the solidification process.

Many factors have an effect on K including the cooling efficiency of the mould, mould geometry, lubrication, casting speed, etc. In view of shell growth, K should be uniform and high in order to obtain high casting rates and to reduce the probability of breakouts. A high value of K also leads to a fine dendritic structure having good mechanical properties. High K values are connected with low casting temperatures and steels containing less than about 0·02% C or more than 0·3% C.

At the start of solidification in the mould, the temperature gradient, and the growth rate of the solid/liquid interface perpendicular to the mould surface are high ($G_s > 10^3$ K/cm; $V > 1$ mm/s). This leads to a fine dendritic structure at the surface of the casting ($\lambda_2 \simeq 10\ \mu$m).

Owing to the simplifying assumptions employed (no convection, no superheat of the liquid, and a plane solid/liquid interface), the calculated values of G_s should be taken as an order of magnitude estimation only. In spite of this, the model is able to furnish essential and new information concerning this solidification process which, owing to its importance in the steel industry, strongly influences the quality of many steels.

CONCLUSION

On the basis of a well known limiting solution of the one-dimensional heat equation, it is possible to propose a model which permits the rapid determination of the solidification parameters, V, G, and $\dot{T}$, once the macroscopic growth law of the solid shell in the continuous-casting mould is known. Furthermore, analysis of the heat-flow problem and of experimental data on shell growth indicates that, for the major part of the length of the continuous-casting mould, the square root of time law is adequate for describing the thickening of the shell. The constant K of this law should be as high as possible in order (a) to increase operating safety; (b) to permit casting at higher rates; and (c) to promote finer microstructure and better surface quality. The value of K increases with decreasing surface temperature and with decreasing casting temperature. Steels with C contents less than about 0·02 wt-% C and greater than 0·3 wt-% C have K values 40–50% higher than those of 0·1% C steels. The solidification parameters of a 0·62% C steel have been calculated. A comparison of estimated and measured secondary dendrite arm spacings allows one to conclude that the proposed method might be generally useful for the prediction of primary structures in continuously cast steel.

ACKNOWLEDGMENT

The authors are grateful to Dr H. Jones of Sheffield University for critically reading the manuscript.

REFERENCES

1 W. KURZ: Proc. 5th Int. Conf. Vacuum Metallurgy and ESR, 5, 1976, Munich

2 W. R. IRVING *et al.*: 'Thermal control requirements for continuous casting', paper presented at the Int. Iron and Steel Congress, 1974, Düsseldorf

3   S. N. SINGH AND K. E. BLAZEK: *OH-Proc.*, 1976, St. Louis

4   S. N. SINGH AND K. E. BLAZEK: *J. Met.*, 1974, **26**, 17

5   H. JACOBI: *Arch. Eisenhüttenwes*, 1976, **47**, 345

6   J. ASAI AND J. SZEKELY: *Ironmaking Steelmaking*, 1975, **2**, 205

7   B. TARMANN AND G. FORSTNER: *Radex-Rundsch.*, 1966, 51

8   B. N. KATOMIN AND V. S. RUTES: *Izv. Akad. Nauk SSSR Met.*, 1957, Jan., 123 (Brutcher translation No. 3944)

9   K. USHIJIMA: *Tetsu-to-Hagané*, 1962, **48**, 747

10 N. G. GLADYSHEV *et al.*: *Stal in English*, 1969, Sept., 788

11 H. NEMOTO AND T. KAWAWA: *OH-Proc.*, 13, 1969

12 A. SUZUKI *et al.*: *Tetsu-to-Hagané*, 1970, **56**, 95; (Brutcher translation No. 8398)

13 M. KUBA *et al.*: *ibid.*, 1971, **57**, 648; (Brutcher translation No. 8729)

14 G. VOGT AND K. WÜNNENBERG: *Klepzig Fachber.*, 1972, **80**, 491

15 K. SAITO AND M. TATE: *OH-Proc.*, 238; 1973

16 J. LAIT *et al.*: *OH-Proc.*, 269; 1973

17 S. K. MORTON AND F. WEINBERG: *J. Iron Steel Inst.*, 1973, **211**, 13

18 J. K. BRIMACOMBE *et al.*: 'Mathematical process models in iron and steel making', 174, 1975, London, The Metals Society

19 F. ESSER AND K. KRUSE: *Neue Hütte*, 1973, **18**, 705

20 L. SAROFF: Proc. 102nd AIME Annual Meeting, 323; 1973, Chicago, AIME

21 T. OHASHI *et al.*: *Trans. Iron Steel Inst. Jpn.*, 1975, **15**, 571

22 R. ALBERNY *et al.*: *Rev. Métall.*, 1976, July–Aug, (7/8), 545

23 E. BACHNER AND M. USSAR: *Stahl Eisen*, 1976, **96**, 185

24 T. KAWAWA: Paper presented at the 2nd Japan-Germany Seminar, Tokyo, 1976

25 T. KAWAWA *et al.*: *Can. Met. Quart.*, 1976, **15**, 129

26 J. SZEKELY AND N. J. THEMELIS: 'Rate phenomena in process metallurgy', 1971, New York, Wiley

27 H. JONES: *J. Inst. Met.*, 1969, **97**, 38

28 F. RICHTER: 'Die wichtigsten physikalischen Eigenschaften von 52 Eisenwerkstoffen', Sonderbericht No. 8, Düsseldorf, 1973, Verlag Stahleisen

29 C. J. SMITHELLS: 'Metals reference book', (5th ed.) 1976, London, Butterworths

30 T. OKAMOTO *et al.*: *J. Cryst. Growth*, 1975, **29**, 131

31 M. H. BURDEN AND J. D. HUNT: *Met. Sci.*, 1976, **10**, 156

32 A. SUZUKI *et al.*: *Nippon Kinzoku Gakkai-shi*, 1968, **32**, 1 301; (Brutcher translation 7804)

Secondary cooling and product quality in continuous casting

A. Etienne and A. Palmaers

Crack formation in continuously cast blooms is discussed in relation to the cooling conditions. The discussion is based on a synthesis of plant experience, laboratory measurements, and model calculation results. A criterion for crack formation is proposed.

A. Etienne is with C. R. M. Liege, and A. Palmaers is in the Department of Metallurgy, University of Liege, Belgium

The importance of the amount and distribution of water cooling on the quality of continuously cast slabs, blooms, and billets has been shown in many works.

In slab casting, inappropriate water cooling is most often associated with surface defects such as longitudinal, transverse, or star cracks.[1-3] These defects appear to be caused by mechanical stresses applied to the strand but are magnified by poor secondary cooling. It has been reported that the amount of cooling also affects central segregation.[4,5]

Some steel grades cast as blooms or billets are prone to internal cracking and appear to be highly sensitive to the distribution and amount of spray water. Formation of halfway cracks has already been discussed and is thought to be, in many instances, produced by thermal stresses, associated with important surface reheating.[6,7,8]

This paper attempts to describe cracking mechanisms induced in continuously cast blooms by thermal stresses. The present work makes a synthesis of plant experiments, laboratory measurements, and calculation results to formulate a reasonable cracking criterion which can be used to control or design spray cooling.

INDUSTRIAL EXPERIENCE IN BLOOM CASTING

Experience gained with the curved mould type bloom caster at Cockerill–Seraing[9] showed that a fair proportion of blooms, made of St 52 and St 45 grades, were marked by heavy axial cracking, known as star cracks, when the spray water was distributed to a length of 8·5 m below the mould. These defects were more frequent with the smaller sections. Reducing the total amount of sprayed water did not suppress the cracking. The cracks disappeared, however, when the sprayed length was shortened to 1·5 m. Similar procedures have been carried out in a number of bloom casters.

Following these first observations, experiments were carried out to study the effect of the cooling intensity on the internal quality in the case of the short spray cooling. Table 1 gives the experimental conditions achieved during the trials and Table 2 lists the composition of the cast steels.

The higher water flow rates caused crack opening in the 187-square (Fig. 1) and 237-octagonal blooms whereas the 288-octagonal bloom remained free of cracks for all spray conditions. The length and severity of cracks increased markedly with the water flow rates.

Table 1   Experimental conditions

Size, mm	Average casting speed, m/min	Water flow rates used in secondary cooling, l/min on 1·5 m	on 8·5 m
288 octagonal	0·7	200	400
		360	
237 octagonal	1·00	230	380
		310	
		415	
187 square	1·35	220	380
		310	
		400	

Table 2   Composition of cast steels, wt-%

Size, mm	C	Mn	Si	S	P
288 octagonal	0·200	1·217	0·257	0·015	0·014
237 octagonal	0·160	1·210	0·270	0·013	0·015
	0·170	1·300	0·280	0·013	0·019
187 square	0·175	0·641	0·277	0·016	0·014
	0·190	0·524	0·245	0·010	0·012

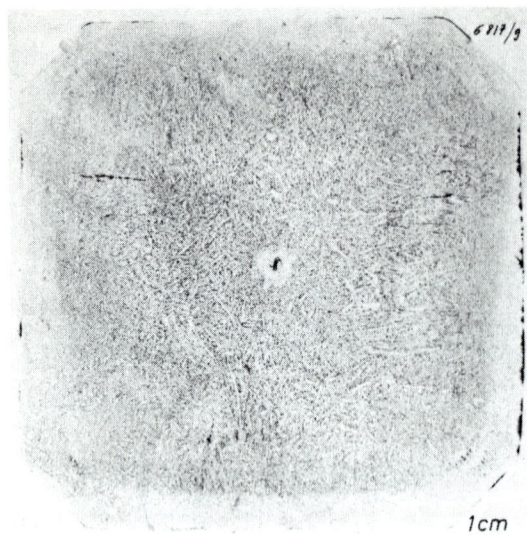

1 Sulphur print of cracked bloom

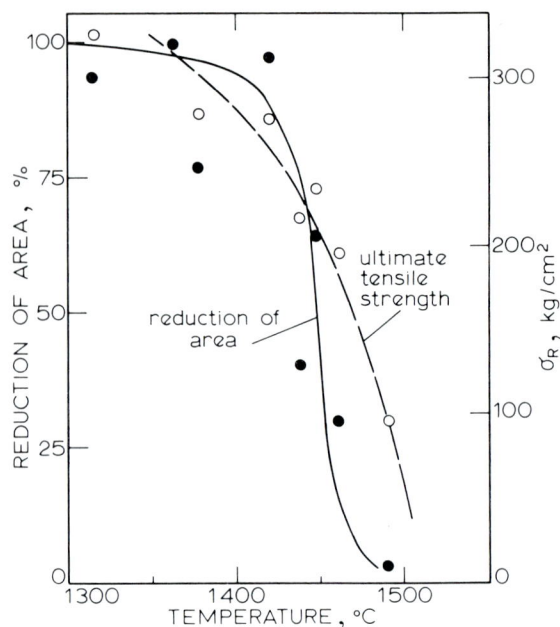

2 Gleeble test results obtained with 0·185% C steel

MECHANICAL PROPERTIES OF CARBON STEELS AT ELEVATED TEMPERATURE

A few authors[8,10] attempted to explain cracking during continuous casting on the basis of a pure elastic–plastic behaviour of steel. These data, which are obtained by extrapolation, may lead to inaccurate assessments.

An extensive measurement programme was undertaken in the CRM laboratories in order to obtain reliable information on the mechanical properties of carbon steel at high temperatures in strain conditions similar to what is achieved in continuous casting.

Two techniques were worked out. The first made use of a Gleeble machine which allowed tensile tests to be carried out up to the melting point but which applied a minimum strain rate of $0·3 \, s^{-1}$. The second one made use of an Instron machine which was allowed to cover a wide range of strain rates extending from $2·8 \times 10^{-3}$ to $5·6 \times 10^{-5} \, s^{-1}$ but where the tensile test temperature was kept below 1 350°C.

The specimens used for these tests were machined in the skin of continuously cast blooms: the specimen axis was taken parallel to the bloom axis and so was perpendicular to the primary dendrite direction. So far, tests have been carried out on four types of steel, the compositions of which are given in Table 3.

The Gleeble tests were conducted to investigate the ductility loss which takes place in the vicinity of the melting point.[11–13] As is common practice in this kind of test, the steel ductility was measured by the reduction of area and the ultimate tensile stress. The results are given in Fig. 2.

Table 3 Composition of steels tested in the laboratory, 10^{-3}%

C	Mn	Si	S	P	Al$_{total}$
84	600	230	14	11	10
98	450	248	12	7	8
88	450	224	12	7	11
81	450	226	11	7	10
88	525	202	25	19	6
185	1 175	222	15	22	7
412	618	247	14	14	12

Heat treatments of the specimen before the tensile test (melting or sensibilizing) did not appear to affect the steel behaviour. The results can be summarized as follows:

(i) in the case of low sulphur steel, the ductility loss appears some 40°C below solidus temperature in 0·09 and 0·185% C steels, and some 65°C below solidus temperature in 0·4% C steel; increasing the sulphur content to 0·025% promotes the ductility loss to some 80°C below solidus temperature

(ii) the level of the ultimate tensile stress does not vary significantly for these four grades of steel.

The results of these Gleeble tensile tests provide no clue to explain the differences observed in casting behaviour between 0·09 and 0·185% C steels, though they provide indications which assist in interpreting cracking problems encountered with a steel having a high sulphur or a high carbon content. Tests were carried out with an Instron machine in a range of temperatures from 900° to 1 350°C. Figure 3 gives an example of the measured stress–strain curves. The relative lack of precision in determining the elastic parameters and the strong dependence of the work hardening coefficient upon strain rate, elongation, and temperature make it impossible to formulate a simple rule to describe all the results. These are best presented in tables. The data obtained with 0·185% C steel are summarized in Table 4.

Differences between the three carbon steels listed in Table 1 are noted in their tensile behaviour but it will be shown below that they cannot account for the cracking sensitivity of the 0·185% C steel.

THERMAL STRESS MODEL

A simplified thermal stress model was developed with a view to assessing a satisfactory cracking criterion applicable to bloom and billet casting.

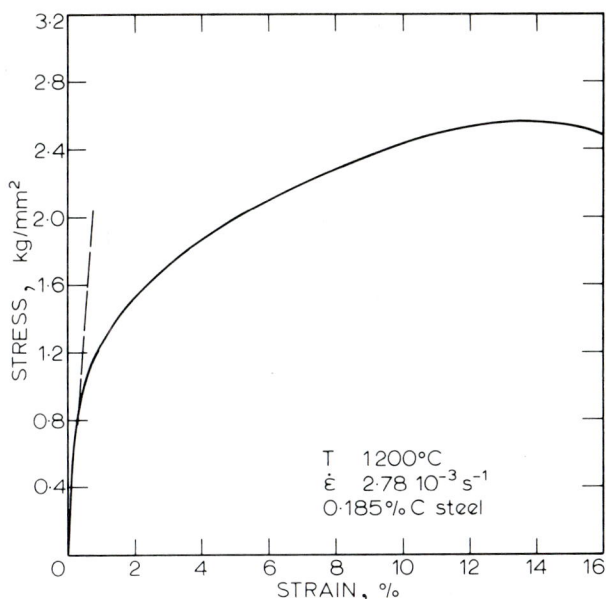

3   Instron tests: stress–strain curve

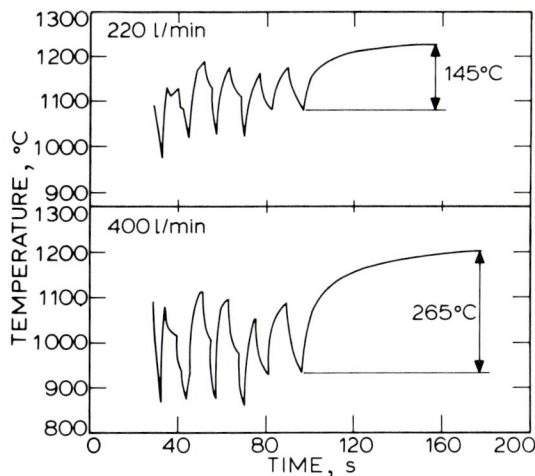

4   Surface temperature profiles calculated for 187-square bloom cast at 1·35 m/min

This model handles a uniaxial state of stress in the bloom cross section, assimilating one side to a beam. The following assumptions are made:

(i) the corners prevent deflection of the beam, the length of which varies according to internal stresses

(ii) the effect of thermal gradients along the beam is not considered

(iii) the beam is assumed to be free of stress at the beginning of calculations.

The model is based on a classical type of reasoning by which the overall beam deformation is calculated in order to obtain a balanced internal state of stress, these stresses being induced by the impeded thermal deformation.[14]

The model uses finite difference calculus. The solidified thickness is divided in n-slices; each slice n_i is characterized at time t_i by its average temperature T_i. The temperature profiles in the solidified shell given at two instants t_i and t_{i+1} are compared and the internal stresses and induced strains are calculated. The temperature profiles are calculated by a classical solidification–heat transfer model.[15]

It was found necessary to deal with short time intervals (5 s) and to repeat this calculating procedure to cover any given period of time, accumulating stresses, and strains.

The results reported in this paper do not take into account stress relaxation though the calculation procedure gives full opportunity to do so. Tables, such as Table 4 for 0·185% C steel, are used to relate stress and strain as a function of temperature and strain rate. Above 1 350°C, stresses are interpolated between 0-value at the solidus temperature and table value at 1 350°C. An average expansion coefficient of 22×10^{-6} was used.[16]

The calculations were first carried out at the end of the secondary cooling in the various conditions listed in Table 1. To illustrate these results, Fig. 4 gives the calculated surface temperature profile in a 187-square bloom sprayed with 220 and 400 l/min of water, respectively. Figure 5 illustrates strains induced across the shell during surface reheating. Table 5 summarizes the maximum strains calculated to take place across the beam in relation to the temperature existing at that particular point. It should be noted that the stress level remains fairly low, the maximum value being smaller than 30 kg/cm².

Though the authors are aware of the extreme crudeness of the beam assumption, similar calculations were made at the end of the liquid pool in cases of a short (1·5 m) and a long (8·5 m) secondary cooling. The liquid pool depth for each case reported in Table 1 was greater than 14 m. This meant that the small surface reheating (50°C) observed at the end of the long secondary cooling was well damped down when the metal solidified

Table 4   Instron tensile tests: stress*–strain relationships measured in 0·185% C steel

T, °C	ϵ%($\dot{\epsilon} = 2.78 \times 10^{-3}$s^{-1})						ϵ%($\dot{\epsilon} = 2.78 \times 10^{-4}$s^{-1})						ϵ%($\dot{\epsilon} = 5.55 \times 10^{-5}$s^{-1})					
	0·2	0·4	0·6	0·8	1·0	1·5	0·2	0·4	0·6	0·8	1·0	1·5	0·2	0·4	0·6	0·8	1·0	1·5
950	1·11	2·04	2·62	3·04	3·20	3·78	0·81	1·48	2·09	2·45	2·56	2·91	0·60	1·09	1·72	2·04	2·12	2·30
1 000	0·98	1·80	2·13	2·43	2·61	2·97	0·71	1·29	1·64	1·87	2·02	2·26	0·52	0·94	1·30	1·48	1·60	1·76
1 050	0·89	1·38	1·59	1·78	1·91	2·22	0·65	1·02	1·22	1·37	1·46	1·64	0·48	0·79	0·96	1·08	1·14	1·25
1 100	0·81	1·06	1·27	1·45	1·60	1·81	0·59	0·81	0·99	1·11	1·20	1·34	0·44	1·64	0·79	0·88	0·92	1·02
1 150	0·71	0·90	1·10	1·22	1·33	1·49	0·52	0·69	0·84	0·92	0·98	1·08	0·40	0·54	0·66	0·71	0·74	0·80
1 200	0·62	0·80	1·0	1·08	1·25	1·37	0·47	0·61	0·76	0·81	0·88	0·97	0·36	0·48	0·60	0·64	0·65	0·71
1 300	0·56	0·70	0·78	0·86	0·94	1·03	0·42	0·51	0·59	0·64	0·66	0·70	0·32	0·38	0·41	0·42	0·44	0·47
1 350	0·46	0·60	0·65	0·69	0·74	0·82	0·32	0·40	0·45	0·48	0·50	0·55	0·24	0·27	0·31	0·34	0·35	0·36

* Stress in kg/mm²

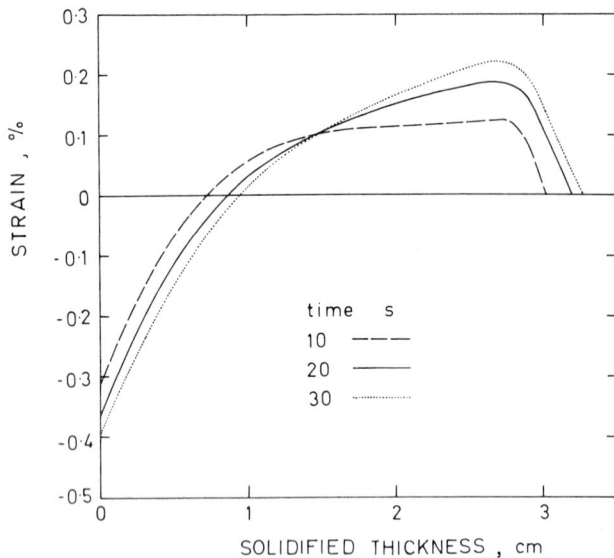

5 Strains induced across solidified shell at end of spray cooling: 187-square bloom, 400 l/min

completely. Figure 6 shows calculated strains induced after complete solidification of a 187-square bloom.

DISCUSSION

The results given in Table 5 lead one to conclude that the level of strain calculated at the end of the secondary cooling depends on bloom size and is a function of cooling intensity.

Similar calculations carried out with the mechanical properties pertinent to 0·09% C steel give similar results. Thus it appears that the measured mechanical properties cannot be used to explain the differences in cracking behaviour between 0·09 and 0·185% C steel. The reason must be looked for in the shrinkage linked to the peritectic transformation which superimposes extra strains on the above calculated values. The maximum linear shrinkage resulting from the $\delta - \gamma$ transformation

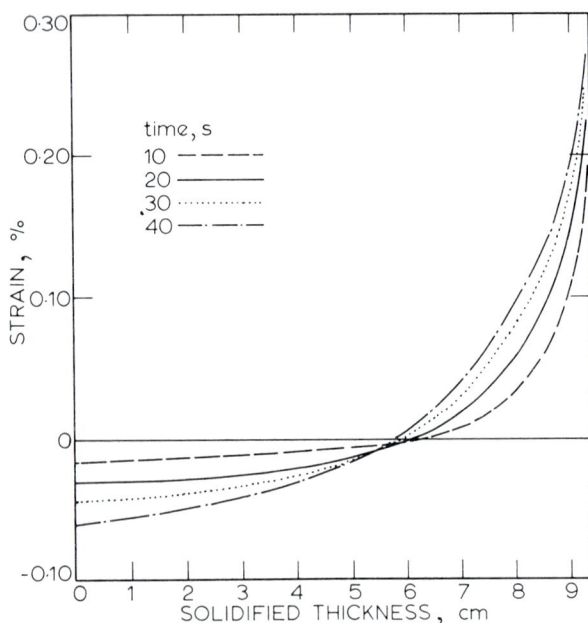

6 Strains induced across solidified shell at end point of solidification: 187-square bloom, long spray cooling

Table 5 Maximum tensile strains, %, calculated in the vicinity of the solidification front in relation to the temperature at that point

| Bloom size, mm | Water flow rates, 1/min | Time elapsed after the end of spraying, s | |
		20	30
288 octagonal	360	0·13	0·17
		1 444°C	1 432°C
	200	0·9	0·12
		1 464°C	1 450°C
237 octagonal	415	0·17	0·20
		1 445°C	1 435°C
	230	0·11	0·14
		1 457°C	1 448°C
187 square	400	0·19	0·22
		1 466°C	1 449°C
	220	0·13	0·15
		1 463°C	1 457°C

around the peritectic temperature is roughly estimated to about 0·25%.[17] It would be interesting to know the effect of the Si and Mn contents on peritectic temperature and compositions. On the basis of the Fe–C diagram, the range of C-content from 0·09 to 0·17% C is critical since the $\delta - \gamma$ transformation occurs entirely in the low ductility region, between 1 440°C and solidus temperature.

A similar analysis was done at the end point of solidification. It showed that the probability of cracking increased when the cast size decreased. A longer spray cooling favoured cracking because it achieved a lower surface temperature which promoted internal cooling and stronger shell.

CONCLUSIONS

An analysis of all these results leads the authors to think that cracking occurs in the vicinity of the solidification front in the low ductility range when the total tensile strains go beyond some limiting value. The results obtained in our work indicate that the ultimate strain might be around 0·3%. Cracking sensitivity depends on steel composition, which affects the low ductility range and acuteness of peritectic transformation.

ACKNOWLEDGMENTS

The authors wish to express their gratitude to the Cockerill Company which allowed the completion of the industrial trials. In particular, they express their deep appreciation to R. Pesch for his cooperation in this work.

REFERENCES

1 W. R. IRVING AND A. PERKINS: 'Continuous casting of steel', 107, 1977, London, The Metals Society
2 T. NAZAKI et al.: Tetsu-to-Hagané (J. Iron Steel Inst. Jpn), 1975, **2**, A17
3 W. RESCH et al.: Stahl Eisen, 1976, **9**, 432
4 S. MYOSHI: 'Continuous casting of steel', 286, 1977, London, The Metals Society
5 N. GROSSKURTH et al.: Stahl Eisen, 1974, **15**, 673
6 Y. AKETA AND K. USHIJIMA: Tetsu-to-Hagané Overseas (J. Iron Steel Inst. Jpn), 1962, **4**, 334
7 G. VAN DRUNEN et al.: Ironmaking Steelmaking, 1975, **2**, 125

8 A. GRILL *et al.*: *Ironmaking Steelmaking*, 1976, **1**, 38

9 R. MAAS *et al.*: C.R.M. Metall. Rep., 1974, **38**, 11

10 K. A. FEKETE: *Radex Rundsch.*, 1974, **3**, 135

11 G. A. WILBER *et al.*: Metall. Trans. *A*, 75, **6**, 1 727

12 C. J. ADAMS: *Proc. Nat. Open Hearth and Basic Oxygen Steelmaking*, 1971, **54**, 290

13 A. FUCHS: *Estel Ber. Forschung Entwicklung unsere Werke*, **75**, 3, 127

14 B. E. GATEWOOD: 'Thermal stresses', 9, 1957, New York, McGraw Hill

15 A. ETIENNE AND P. MIGNON: Rapport interne du C.R.M., 74, S7

16 Y. DARDEL: *Rev. Métall.*, 1964, 220

17 P. COHEUR: Notes du cours de Connaissance des Matériaux Métalliques, ch. III, 19, Université de Liège

Quality of continuously cast slabs

R. J. Gray, A. Perkins, and B. Walker

The effects of steel composition and casting parameters on slab quality are discussed in the light of the BSC's experience in the casting of carbon–manganese steels. The casting parameters considered include casting temperature, casting speed, section size, mould lubricant, secondary cooling, strand support, roll alignment, and wear. These parameters are related to surface cracking, internal cracking, and segregation in the slab. In order to reduce defects, production experience and plant trials have been combined with a mathematical model of the casting process to give some operating guidelines and to develop the understanding of the process.

R. J. Gray and A. Perkins are with Teesside Laboratories, BSC, and B. Walker is with Teesside Division, BSC

In the past six years, the proportion of world crude steel production put through the continuous casting route has increased from 10 to 20%. A recent forecast predicts that this trend will continue and many new steelmaking units will rely entirely on continuous casting. Approximately half of the continuously cast tonnage in the world today is from wide slab machines (typically producing 1 500–2 000 × 180–300 mm sections), one-quarter from multistrand bloomcasters and one-quarter from billet casters. The dominance of slab casters and multistrand bloom casters is due largely to the widespread use of BOS with production rates of the order of 300 t/h. Twin-strand slabcasting is most compatible at this tonnage rate but it introduces special problems in the design and operation of the caster in order to achieve a high product quality. Specific quality defects which can arise in cast slabs are

(i) segregation of alloying elements and impurities in the centre of the strand
(ii) internal cracking, particularly along the centre-line of the strand
(iii) surface cracking, which may be either longitudinal or transverse.

The occurrence of these problems varies with the design of the casting machine, the casting conditions used, and the grade of steel being cast. The authors refer principally in this paper to plate grades; in casting strip grades, steel cleanness is also imperative. The results presented below are based on samples from a total BSC production of over 3 Mt since 1973.

INTERNAL QUALITY

The macrostructure of continuously cast slabs has some similarity with that of ingots, in that a small chill zone (about 3 mm thick), formed in the mould, is followed by a columnar zone which continues to grow during spray cooling until solidification is completed with a central equiaxed zone. The principal deviations from ingot structure arise because

(i) the slabs are thin (180–305 mm) compared with ingots: thus, the 2 columnar zones may comprise most of the slab thickness
(ii) limitations in the strand support system together with large ferrostatic pressures in the liquid core towards the bottom of the casting machine give rise to mechanical factors which can greatly influence internal defects and segregation.

Macrostructure

Figure 1 shows the effect of superheat temperature (excess over liquidus) in the tundish on the thicknesses of columnar and equiaxed zones in the as-cast slabs for 10 000 t of carbon–manganese plate steel (0·08–0·12 C, 1·15–1·30 Mn). This figure shows that a high temperature promotes columnar growth and in order to obtain a wide equiaxed zone it is necessary to impose an upper limit on the tundish temperature. Other authors have similarly reported that temperature is the major factor affecting equiaxed zone width.[1]

Casting speeds for slabs in Fig. 1 ranged from 0·6 to 1·1 m/min with total spray cooling water between 0·7 and 0·95 l/kg, being almost independent of the casting speed. Multiple regression analysis showed that over these ranges the speed produced only a 6% change in equiaxed zone width while the spray water had no significant effect at all. The following empirical relationship has been derived for 180–230 mm slabs:

$$\text{Equiaxed zone (\%)} = 42 \cdot 9 - 0 \cdot 58 \, (°C \text{ superheat})$$
$$- 12 \cdot 9 \, (\text{m/min casting speed}).$$

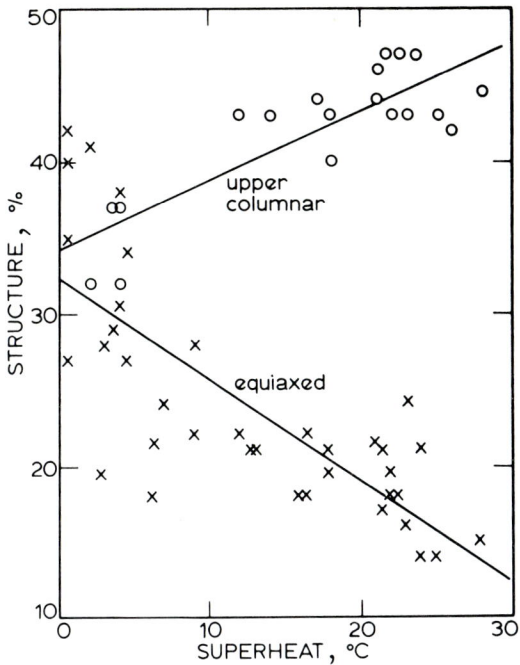

1 Effect of superheat on thickness of equiaxed and columnar zones

These results are in accordance with a mathematical model of the caster,[2] which shows the solidification rate to be independent of casting speed at constant spray water (Fig. 2). The model also shows that even doubling the spray water produces only a small increase in solidification rate, because the rate of heat extraction is limited mainly by the thermal conductivity of steel and is, therefore, not greatly influenced by variations in surface conditions. Thus, the widths of the columnar and equiaxed zones are only marginally influenced by casting speed and spray cooling water; liquid steel temperature has the dominant effect.

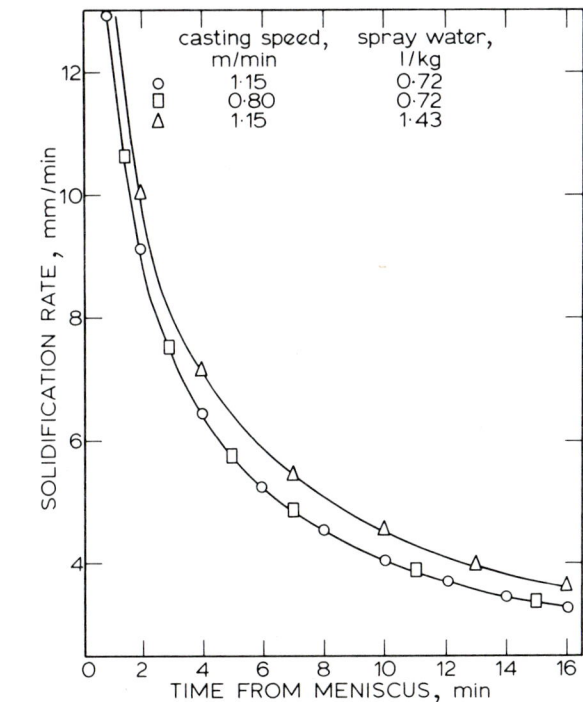

2 Calculated solidification rates for various casting conditions

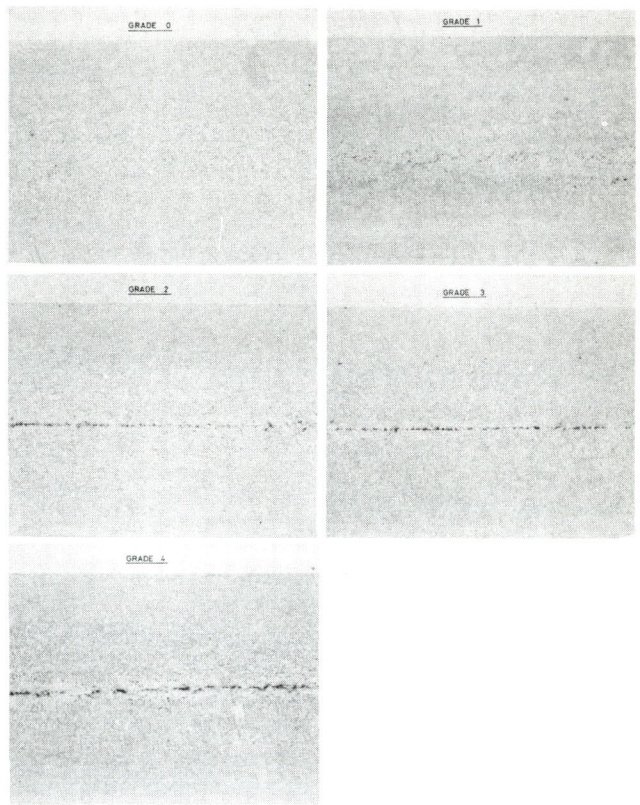

3 Internal quality standard: sulphur segregation [×0·19]

Segregation
Metallurgical factors

The build up of solute elements ahead of the solidification front is encouraged by the columnar growth mechanism, while an equiaxed growth present near the centre of the slab distributes the segregated liquid interdendritically throughout the equiaxed zone. A grading system based on standard sulphur prints from transverse sections of the slab (Fig. 3) provides a semiquantitative means of assessing segregation. Grades 2–4 illustrate the tendency for the segregate to lie along the centreline of the slab when casting conditions are adverse.

Figure 4 shows for the casts shown in Fig. 1 the manner in which sulphur segregation increases as the width of the equiaxed zone is reduced. The equation,

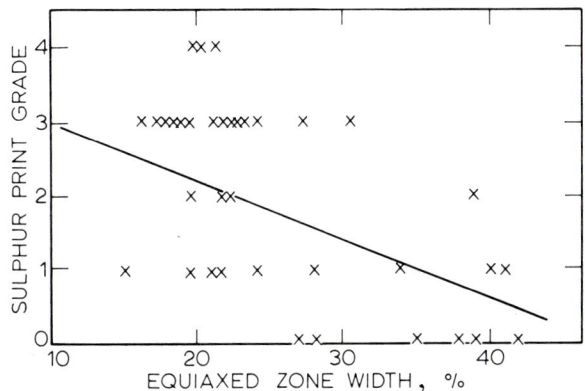

4 Relationship between equiaxed zone width and sulphur print grade

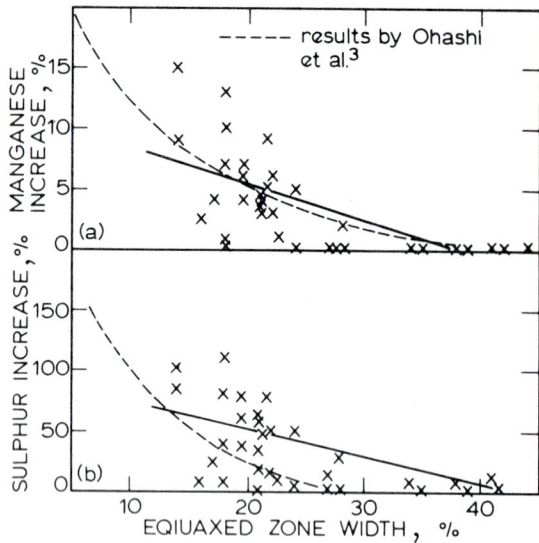

5 Effect of equiaxed zone width on segregation of *a* manganese and *b* suiphur

sulphur print grade $= 3 \cdot 8 - 0 \cdot 08$ (% equiaxed zone), fits the data shown.

Centreline segregation has also been assessed by comparing the analysis of drillings taken from the segregated region with those from the bulk of the slab well away from the segregate. A drill size of 3 mm was used and the degree of segregation was calculated as the percentage change in each alloying element from the bulk to the central region. Figures 5*a* and *b* show that the segregation of manganese and sulphur increases with decreasing equiaxed zone width according to the equations:

$$\text{\% increase in Mn} = 11 \cdot 3 - 0 \cdot 31 \text{ (\% equiaxed zone)}$$
$$\text{\% increase in S} = 96 - 2 \cdot 09 \text{ (\% equiaxed zone)}$$

Also shown in Figs 5*a* and *b* are the corresponding results of Ohashi *et al.*[3] for a similar steel quality. The two sets of results agree well, indicating that differences between casting machines have not, as expected, greatly affected the relationship between equiaxed zone width and segregation. Thus, the casting temperature, through its effect on macrostructure, has an important influence on the quality of slabs, which is reflected in the final product. Figure 6 illustrates directly the effect of superheat on the percentage of acceptable plates (assessed from sulphur prints) for sulphur levels in the range $0 \cdot 012 - 0 \cdot 023\%$. Compared to the superheat, the sulphur level has a small effect on acceptability for this grade of plate. A maximum permissible superheat of 15°C is now imposed on casts for more demanding applications, thus ensuring a wide equiaxed zone and the minimization of segregation.

Current developments include electromagnetic stirring of the solidifying strand to produce fine, detached dendrites which sink into the central region of the liquid core, giving a broader equiaxed zone.

Non-metallurgical factors

The slab casters referred to in this paper have curved moulds and strand support systems of 9·8 and 12·2 m radius. The ferrostatic pressure in the liquid core of the strand causes some bulging of the solidifying shell between, and at, the strand support rolls. The ferrostatic pressure increases with distance below the meniscus, but so also does the strength of the shell as it becomes thicker and cooler. The amount of bulging between rolls can be roughly estimated using the theory of thin plate bending, and assuming a mean value of Young's modulus for that part of the shell below 1 000°C and assuming zero shell strength above 1 000°C. On this basis, the

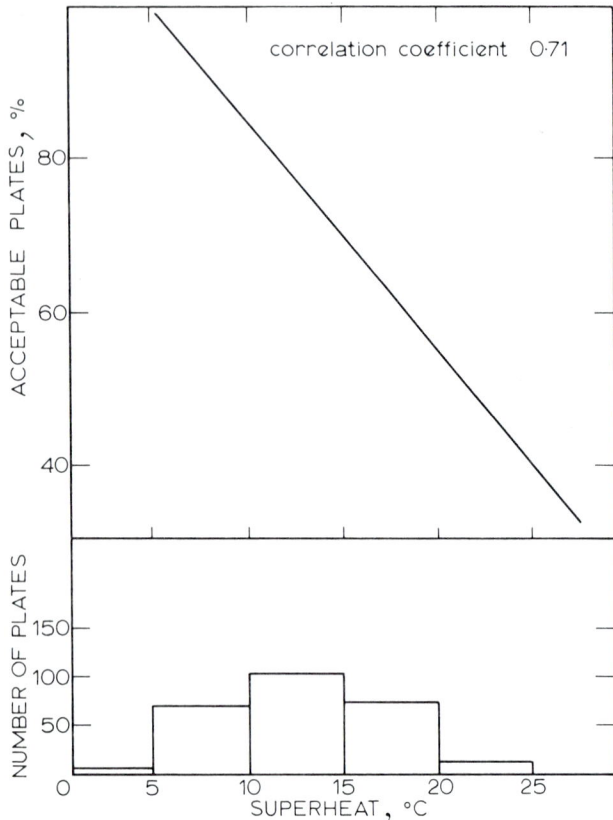

6 Effect of superheat on plate quality

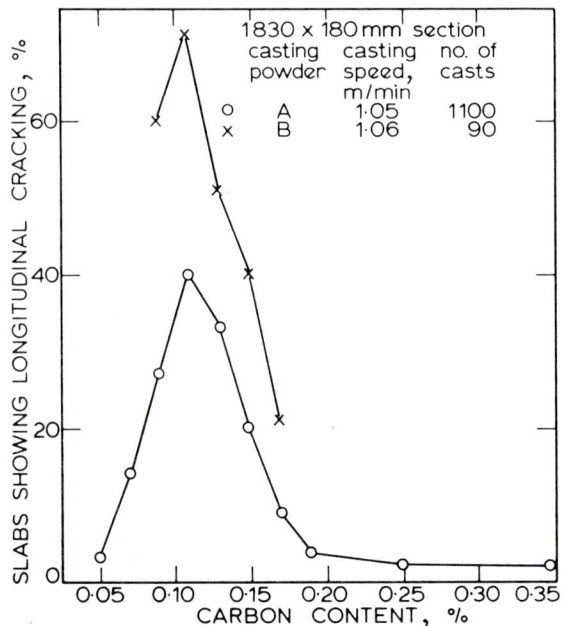

7 Effect of carbon content on longitudinal surface cracking

maximum bulge in a $1\,525 \times 305$ mm section, near the point of complete solidification, is estimated to be 0·7 mm at a casting speed of 0·72 m/min and standard spray cooling rates. If the cooling is increased by 60%, the mean shell temperature is reduced from 850° to 835°C; the shell strength increases and the bulging is reduced to 0·4 mm. These calculations are currently being compared with values measured on the casting machine, showing them to be of the correct magnitude. However, the lack of high-temperature data is a considerable handicap in the development of a comprehensive bulging theory.

Bulging can also occur at the rolls due to flexing of the rolls and this has been measured to be up to 3 mm during casting.

Bulging can also occur due to roll wear. The likelihood of bulging is increased by some machine design features which employ 'floating-roll' extraction systems where traction is provided by ferrostatic loading of the rolls by the shell. With such a design, variations in operating conditions may require compensating changes in the applied roll-clamping pressures to maintain a constant pass line.

The result of bulging near the point of complete solidification is to allow the segregated liquid to flow into pockets which may remain cut off from liquid feed until solidification is complete, thus giving a high segregation index and poor sulphur print. If the casting speed is increased from 0·7 to 1·0 m/min for 180 and 230 mm thick slab keeping the amount of spray water in l/kg constant, the sulphur print (Fig. 3) deteriorates by about 1 grade. Calculations show that this increase in speed moves the point of complete solidification about 4 m further along the strand, thereby increasing the ferro-static pressure at this point, and increasing bulging. Miyoshi[4] showed that by increasing the spray water from 1·01 to 1·30 l/kg, the segregation index was reduced from 20–35% to 10–20%, owing to the reduction in bulging. The increase in spray water may have been accompanied by a small reduction in equiaxed zone width, but the effect of this on segregation was evidently insignificant in comparison with the effect of reduced bulging.

Centreline cracking

Centreline cracking occurs midway between (and parallel with) the broad faces of the slab and is associated with the solidification shrinkage of trapped pockets of liquid and deformation of the strand near the point of complete solidification. It is thus intimately related to casting speed, spray cooling rate, and bulging, in a similar manner to segregation. It has also been found to be correlated with carbon content, cracking increasing with increase in carbon. Table 1 gives results for three

Table 1 Effect of carbon content on centreline cracking

Section size, mm	Cracked slabs in carbon range, %				
	0·04–0·07	0·08–0·11	0·12–0·15	0·16–0·19	0·20–0·24
1830×180	4·0	3·0	4·4	6·7	—
1830×230	0·0	0·9	1·4	3·4	7·0
1525×305	—	0·2	2·1	0·8	—

Sample: 1 Mt.

slab thicknesses, covering a total sample size of 1 Mt, equivalent to 32 000 slabs. It is possible that the explanation of this effect is the higher tendency for columnar grain growth and larger solidification shrink-age at increasing carbon contents. Other metallurgical factors, such as superheat and sulphur content, exert relatively little influence. A detailed discussion of the measures adopted in order to minimize centreline cracking has been given previously.[5]

SURFACE CRACKING

The main cracking problems in continuously cast slabs have been due to two types of surface cracking: longi-tudinal, occurring mainly within the central 50% of the broad face, and transverse, often occurring in the reciprocation marks near the edges of the broad face at the slab corner.

Longitudinal cracking

Figure 7 shows the effect of carbon content on longi-tudinal cracking in $1\,830 \times 180$ mm mild steel slabs for two different mould lubricants (casting powders). There is a pronounced maximum in cracking at about 0·11% carbon with powder A; the same pattern is followed by powder B, but the level of cracking is higher. Figure 8 illustrates the effect of carbon content on the mean linear coefficient of shrinkage over the first 25°, 50°, and 100°C of cooling below the solidus temperature, cal-culated from the phase diagram. The mean shrinkage coefficients were calculated using published[6] data and by taking a value of 0·38% for the linear shrinkage due to the $\delta-\gamma$ transformation.[7] The similarity in position and shape of the peaks in Figs 7 and 8 indicates that cracking probably occurs within 50°C of the solidus owing to the additional shrinkage accompanying the $\delta-\gamma$ trans-formation in the steel shell. Grill and Brimacombe[7] have proposed that this additional shrinkage is responsible for

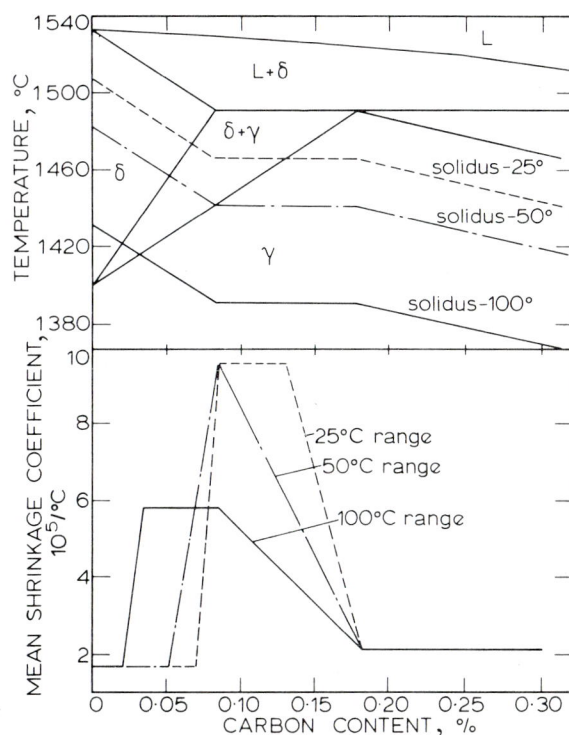

8 Effect of carbon content on shrinkage after solidification

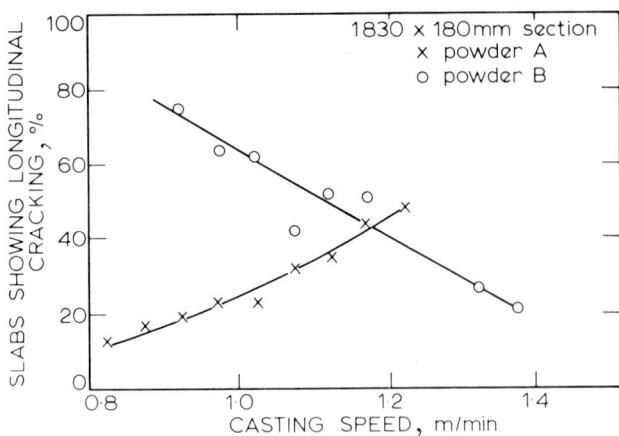

9   Effect of casting speed on longitudinal cracking for two casting powders

the minimum in mould heat transfer which is seen at 0·1% carbon and which is accompanied by pronounced horizontal rippling on the inner shell surface.

Figure 9 shows the effect of casting speed on the incidence of cracking for the two types of casting powder. Laboratory tests have shown that powder *A* melts to form a slag of high viscosity (0·8 Ns/m² at 1 300°C) which produces less cracking at lower casting speeds. Powder *B*, however, forms a low viscosity slag (0·15 Ns/m² at 1 300°C) which performs better the higher the casting speed. These results indicate that the slag viscosity required for a uniform thickness of insulating, lubricating slag film between the strand and mould increases with decreasing casting speed. Figure 8 shows that the effect of carbon level is independent of the effect of slag viscosity. Figure 10 shows the effect of section size and casting speed on cracking using powder *A*. The cracking peak is present at each section size and cracking increases with decreasing section thickness. The level of cracking at 180 mm is double that at 230 mm; Figure 9 shows that this difference could be entirely due to the difference in casting speeds since at a constant thickness of 180 mm and using powder *A*, cracking increases from 15 to 29% when the casting speed increases from 0·85 to 1·05 m/min. The results shown in Fig. 10 cover an additional 600 000 t of slab over the results previously published.[5] Longitudinal cracking is also affected by sulphur level; levels above 0·02% lead to increased cracking which is attributed to

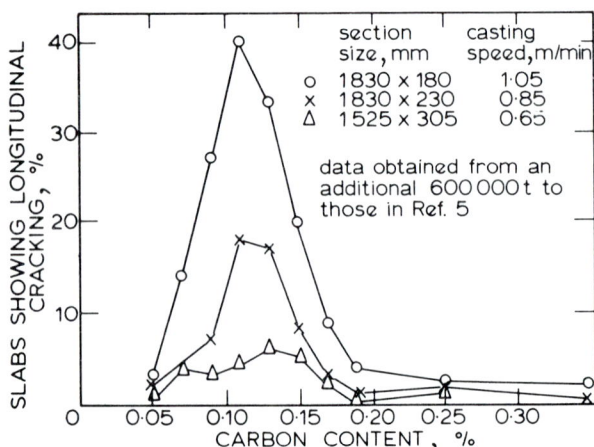

11   Stress distribution across broad face

the tendency of sulphur to form low melting point segregates which reduce the cohesive strength of the shrinking steel shell at temperatures below the theoretical solidus.

While longitudinal cracking normally arises in the mould, the cracks can be propagated or new cracks formed by excessive or non-uniform cooling immediately below the mould. Thus, cracking has been reduced by control of heat transfer in the top zone.[5]

Transverse cracking

Transverse cracks are usually shallow and can give severe problems because they are not normally visible until the slab surface is dressed or until the slab has been rolled to plate. The cracks, when they occur, are found mainly in high-strength steels having manganese levels above 1% which contain 0·03% niobium or vanadium. Stresses computed using a finite difference thermal stress model (Fig. 11) show that peaks in longitudinal tensile stress σ_z occur near the edges of the broad faces, owing to the large temperature gradients produced by the intense cooling of narrow water sprays. The peaks in σ_z correspond approximately with the location of transverse cracking, indicating that thermal stress is a likely cause of this defect. There is evidence that the cracking propagates at the straightener, since the defect is more extensive on the top surface of the slabs, where straightening induces tensile stresses, than on the bottom surface, where the straightening stresses are compressive. Furthermore, hot tensile tests carried out during cooling

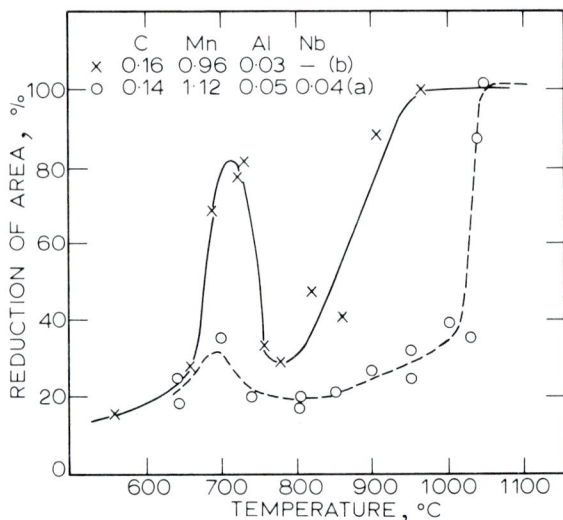

10   Effect of section size on longitudinal cracking

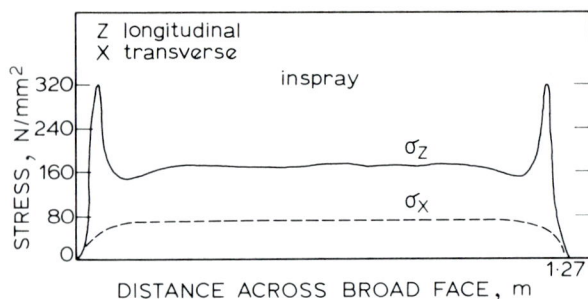

12   Effect of steel composition on hot ductility

from 1 300°C show that the ductility (% reduction in area at fracture) of the crack-susceptible qualities (*see* e.g. Fig. 12(*a*)) deteriorates rapidly at 100°C above the normal straightening temperature of about 900°C. The non-susceptible qualities (Fig. 12(*b*)) show good ductility at temperatures down to 900°C or below. The deterioration is attributed to the precipitation of nitrides and carbides of aluminium and niobium/vanadium.

CONCLUSION
During the production of carbon–manganese continuously cast slabs within BSC, various quality problems have been encountered. From the experience gained on production plant and with the assistance of mathematical modelling and hot ductility testing, a number of these problems have been significantly reduced. Segregation is governed by equiaxed zone width and bulging of the cast slab; a wide equiaxed zone is obtained by minimizing the superheat of the liquid steel, while bulging is reduced by reducing casting speed and increasing spray cooling water. Centreline cracking also increases with bulging and increasing carbon content, but is little affected by superheat. Machine design features and support roll condition play a critical role in the achievement of good slab quality.

Longitudinal surface cracking depends strongly on carbon content, and comparison with thermal shrinkage data indicates that it occurs within 50°C of the solidus. The viscosity of the slag lubricant used in the mould is also important, the optimum value decreasing with increasing casting speed. Transverse cracking appears to be related to thermal stresses arising during secondary cooling, and to the hot ductility of the particular steel quality at the straightening temperature.

ACKNOWLEDGMENTS
The authors are indebted to Works colleagues and to senior management for permission to publish this paper.

REFERENCES
1 H. NEMOTO: *Trans. Iron Steel Inst. Jpn*, 1976, **16**, 51
2 A. PERKINS AND W. R. IRVING: 'Mathematical process models in iron- and steelmaking', 187, 1975, London, The Metals Society
3 T. OHASHI *et al.*: *Trans. Iron Steel Inst. Jpn*, 1975, **15**, 571
4 S. MYOSHI: 'Continuous casting of steel', 286, 1977, London, The Metals Society
5 W. R. IRVING AND A. PERKINS: *ibid.*, 107
6 G. TAMMANN AND G. BANDEL: *Arch. Eisenhüttenwes*, 1934, **7**, 571
7 A. GRILL AND J. K. BRIMACOMBE: *Ironmaking Steelmaking*, 1976, **3**, (2), 76

Low pressure die casting of aluminium alloys

E. Selçuk and L. Meriç

A pilot-plant scale, low pressure die casting unit has been designed and constructed which is capable of producing Al-alloy castings weighing $4 \cdot 5 \times 10^{-1}$ to $4 \cdot 5$ kg. Process parameters have been investigated and it has been found that to obtain a laminar flow of liquid metal into the mould cavity, 3×10^{4} Nm^{-2} gauge pressure is required for a stalk diameter of 3×10^{-2} m. Solidification characteristics of the cast parts have been investigated from macrostructure, porosity, and feeding viewpoints. The process has been found to be readily applicable to pure Al as well as Al–Si alloys of hypoeutectic composition and has many metallurgical advantages when compared with conventional high pressure die casting. An approach has been made to analyse the thermal behaviour of low pressure die casting in terms of the average temperature of the mould, duration of the cycle, and degree of superheat. It has been shown that solidification time can be calculated, thus information can be fed into the automation system for pressurizing, mould opening, and ejection times.

The authors are at the Middle East Technical University, Ankara, Turkey

The subject of pouring liquid metals into permanent moulds with pressure has received attention since the Second World War. Developments in aviation and the automotive industry have demanded cast parts of aluminium alloys in large quantities with accurate and reproducible dimensions. This has led to the development of die casting. The casting of metals with low melting points such as aluminium, its alloys, and others into metal dies has usually been performed under high pressures varying between 3×10^{7} and 7×10^{7} Nm^{-2}. This inevitably means a high overall investment. The pressure generators, dies, and casting unit itself are all expensive parts.

The aim with low pressure die casting is to reduce the highly sophisticated instrumentation and costly installation involved in high pressure die casting and to replace it with a process that is simple yet gives high yield and good quality products.

RELATIVE THEORIES OF GRAVITY, HIGH AND LOW PRESSURE DIE CASTING

Die casting was developed to enable permanent metal moulds consisting of two or more parts to be used repeatedly, in place of the usual one-off sand moulds. In gravity die casting liquid metal entering the mould by gravity provides the hydrostatic pressure within the molten alloy. Gravity die casting is suitable for high volume production of small, simple castings that have a fairly uniform wall thickness and no undercuts or intricate coring. It is difficult to achieve very low castable section thickness because of the low hydrostatic pressure

of the liquid. Metals like aluminium undergo a high shrinkage of 6–8% during solidification; because of this, and also in order to provide additional pressure during feeding, risers are necessary for gravity die casting.

High pressure die casting is closely related to gravity die casting, reusable metal moulds being employed in both processes. There is a difference, however, in the mould filling methods. The success of gravity die casting is controlled by the fluidity of molten alloy which enables the die to be filled. Pressure die casting utilizes pressure instead of gravity to fill the die much more rapidly and force the alloy against the mould surface; thus by pressure die casting more complex shapes can be produced than by other processes.

Flow of liquid metal

The flow of liquid metal in the die casting can be investigated by the application of Bernoulli's equation. Bernoulli's equation is applicable only when the pressure, velocity, viscosity, and density of the fluid are constant for all finite periods under consideration; the flow obeys the law of continuity; and all energy losses from the system are irreversible. The energy losses in the system are generally irreversible because the heat generated by the friction at the wall is primarily lost to the die rather than the flowing stream. The law of continuity is obeyed when the system is pressurized. The possible density fluctuations in the metal caused by the entrapment of air and other gases in the flowing stream can be averaged to approach the condition of a constant density. The small increase in the viscosity near the melting point will not affect the calculation because of the relative insignificance of its contribution to the total energy losses at high Reynolds numbers.

The pressure and velocity are not constant during a die casting cycle as the piston accelerates from zero at the beginning and decelerates again at the end. An average can be taken over a suitable time interval; hence for the theoretical evaluation, the conditions of Bernoulli's theorem are assumed to be satisfied.

The flow of liquid metal in the pressure die casting as it enters the die, during the filling stage, and then finally in the pressurizing cycle, has been investigated by Frommer and Lieby[1] in terms of the behaviour of a single jet stream starting from the gate, extending through the die cavity, and hitting the opposite wall. According to Frommer and Lieby, as the jet stream hits the wall it causes a violent turbulence and forms a pool. As the pool becomes bigger a relatively quiet zone is formed at the back of the pool and metal begins to flow out along the two side walls. This gradually fills the cavity by the extension of the pool and the side stream, and the turbulence eventually dies. Frommer and Lieby have calculated the injection velocity in terms of the injection pressure and specific gravity.

A different approach has been made by Brandt.[2] He proposed that the incoming stream spreads out immediately towards the sides and the cavity is filled when both of these side wall streams hit the upper wall. Later Stuhrke and Wallace[3] showed that when the cavity is filled through an ingate producing a jet, Frommer's theory is valid. When the ingate joins the cavity with a wide angle, metal will rise in the cavity with a solid front. This is in line with the theory put forward by Brandt.

AN ALTERNATIVE APPROACH: LOW PRESSURE DIE CASTING

Permanent mould gravity casting requires bulky feeders and risers, as does high pressure die casting. In high pressure die casting, depending on the casting weight and projected area, tremendous locking forces are necessary to keep the die halves together because of the enormous fluid pressure exerted by the liquid metal. Careful engineering and robust construction are required to withstand the repeated hammer blows delivered by the injection action.

In gravity die casting oxidation and dross formation will take place during pouring, owing to the possible turbulence and unavoidable exposure of the liquid metal. This is valid for cold chamber pressure die casting.

In pressure die casting, high velocity liquid metal causes various problems. Flow lines and surface porosity will develop and seams will result when two divided streams of liquid metal freeze without adequate welding. As a result of premature solidification of thin forerunners, foliation will destroy the surface quality. The cavitation phenomenon occurring in the rapid flowing liquid will create a porous texture, and the leakage resistance of pressure die cast metals will be very low. Another consequence is that high pressure die cast parts are not suitable for heat treatment.

A new technique is necessary, therefore, for casting light metals in metallic moulds to produce heat treatable and weldable parts with better properties and high production yield, yet involving no major investments in equipment.

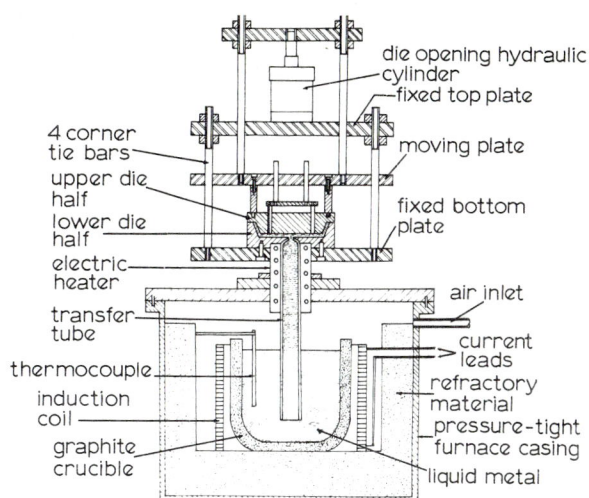

1 Low pressure die casting unit

The simple physical principle behind low pressure die casting is the rise of a fluid in a tube immersed in a pool when excess pressure is applied on the surface, and the return of the fluid to the pool on release of the pressure.

EQUIPMENT

Figure 1 shows the low pressure die casting unit developed at the Middle East Technical University. The arrangement consists of two parts: a hydraulic casting unit, and a furnace which is a pressure tight vessel containing an induction-heated crucible.

The hydraulic casting unit is a simple four bar hydraulic press. The lower and middle plates contain the lower and upper die halves respectively. The lower plate has an opening through which the transfer tube and heating coil passes and makes a direct liquid metal contact with the lower die half. The upper plate contains the hydraulic actuator and its reverse acting mechanism. A safety catch against the free fall of the moveable parts is also located here. Moving plate is guided by four sliding type bearings.

Multifunctional design enables the members of the casting unit to be used for different purposes. It is possible to open the furnace lid by the moving plate. Four vertical tie bars can serve as locations for the attachment of auxiliary hydraulic actuators which will operate in case the dies are multipieces or if there is a moveable core.

The furnace is a pressure tight, flat end cylindrical vessel designed according to the ASTM unfired vessel code. The top plate is bolted to the vessel. A 0·4 m (40 cm) diameter hole on the top plate is used for cleaning, inspection, and adjustment of the crucible, liquid metal level indicator, and thermocouples. The transfer tube is heated with an electrical heating element embedded in a ceramic sleeve which covers the part of the tube from the liquid metal level to the nozzle. An induction-heated graphite–SiC type crucible of 60 kg capacity is used for melting and as a holding furnace. Coil leads go through a pressure tight flange located on the side of the vessel. A separate hydraulic piston is used to lift the furnace and lock it in position during operation.

2   **Mould and core assembly for hollow cylinder**

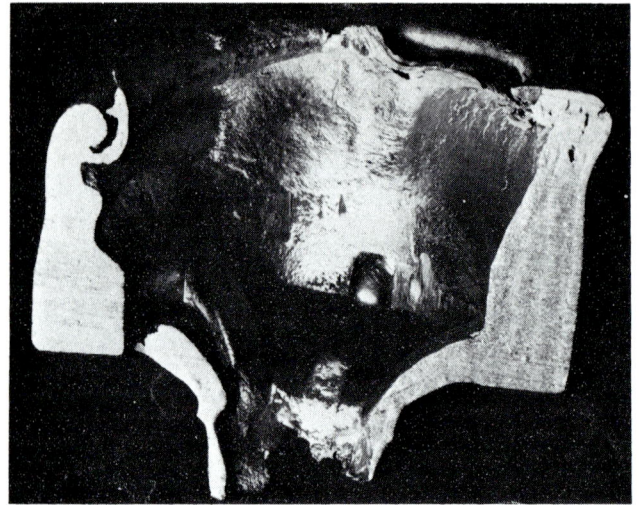

4   **Incomplete cast**

EXPERIMENTS AND PRODUCTION

Experiments were performed before full scale production to determine the process parameters. A hollow cylinder was cast. Figure 2 shows the mould and core assembly together with thermocouples used for the experiments. The change of air pressure in the furnace chamber and the degree of superheat and resulting macrostructures have been examined. The temperatures of the mould surface, mould metal interface, core tip, and core surface were recorded by chromel/alumel thermocouples. The temperature of the mould was stabilized by making at least 20 shots and the data from the thermocouples were recorded and evaluated for the next 20 shots.

Commercially pure aluminium and Al–6·5% Si alloy have been used throughout the experiments. Metal is melted usually in a fuel-fired furnace and then transferred into the vessel, induction heating being used only to adjust the liquid metal temperature. However, an induction furnace is suitable also for starting with a cold charge.

Cast parts have been sliced vertically into two parts to investigate the voids and porosities, and have been

subjected to macrometallographic examination. A pilot-plant low pressure die casting unit was then used for full scale production of 5 kW electrical motor cover plate. Figure 3 shows a successful casting.

RESULTS

Figure 4 shows an incomplete cast caused by improper pressurizing. Figure 5 shows the voids and air pockets resulting from turbulent filling and improper venting. Figure 6 shows the hot tears observed, improper ejection timing, and details of the runner system. Figure 7 shows the macrostructure, the photograph having been

3   **Successful casting: electric motor cover plate**

5   **Air pockets and voids resulting from turbulent filling and improper venting**

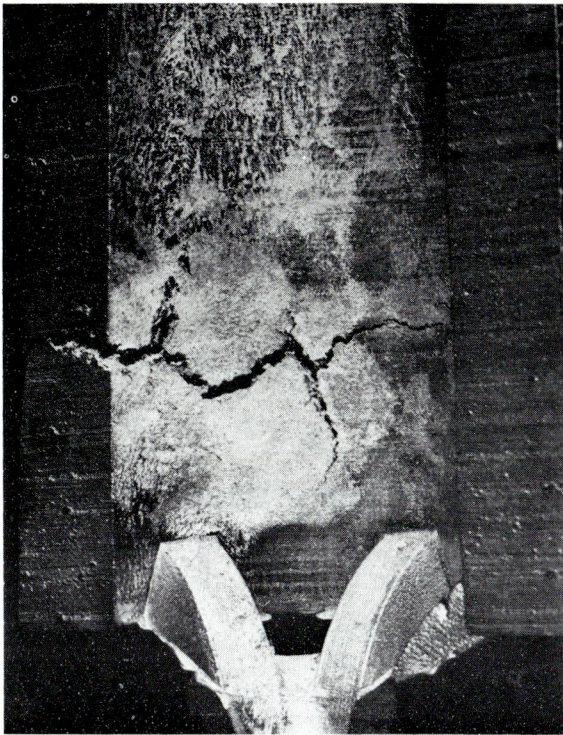

6 Hot tears formed owing to early ejection

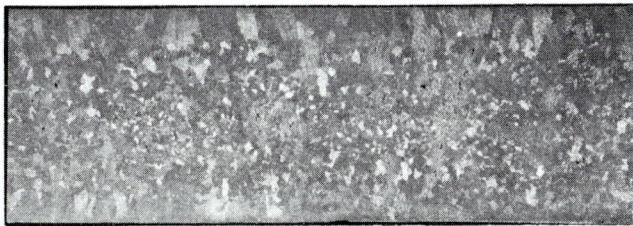

7 Macrostructure: etchant Tuckers reagent

8 Section of pressure die cast cover plate showing heavy porosity

taken from a full scale piece, and the extent of soundness can also be observed. Figure 8 is the section of a pressure die cast cover plate showing heavy porosity.

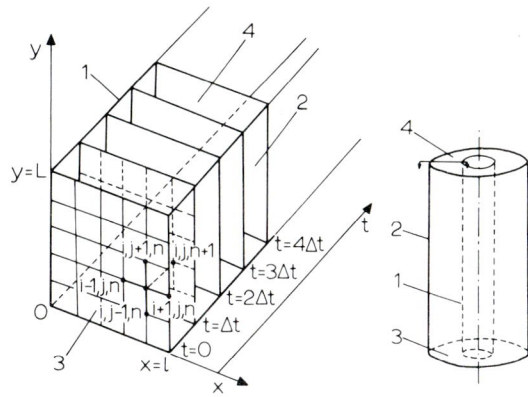

9 Schematic representation of hollow cylinder as rectangular prism

DISCUSSION
Fluid flow
The gauge pressure required to raise the liquid metal in the transfer tube can be calculated from the simple relation

$$P = h \cdot \rho \tag{1}$$

where P = the gauge pressure
h = height of the liquid column
ρ = density of the liquid.

However, Bernoulli's equation when adapted for this case will appear as

$$P = h \cdot \rho + \frac{\rho}{2g} V^2 + \Delta P_f \tag{2}$$

where V = velocity of liquid metal flowing through the transfer tube
ΔP_f = equivalent pressure drop owing to frictional resistance exerted by the transfer tube. All other symbols are the same as in equation (1).

Equation (2) shows that slightly higher pressure is required to fill the mould cavity than that given by equation (1). It is desired that the flow of liquid metal should be in the laminar regime. The Reynolds number is given as 2 100 for the end point of the laminar regime and 4 000 for the starting point of the turbulent regime. Therefore, one can calculate two boundary velocity values. Equation (3) gives the Reynolds number

$$\frac{VD\rho}{\nu} = N_r \tag{3}$$

where ν = viscosity of the liquid, taken at 1·8 cps at 700°C for aluminium
ρ = density of the liquid; for aluminium at 700°C it is equal to 2·3 g/cm^3
D = gate diameter which is equal to 3 cm (3 × 10^{-2} m).

Velocities of the liquid metal entering the cavity are calculated as $5·47 \times 10^{-2}$ m/s and 1×10^{-1} m/s for Reynolds numbers 2 100 and 4 000, respectively. The operating pressure should be controlled to give metal filling velocities within these limits. Otherwise, the mould cavity will be filled with liquid metal in turbulent flow, which will cause defective casting.

The transfer tube can be considered as a sand roughened pipe where the mechanical energy loss in a flow of fluid is computed by the Darcy–Weisbach equation

$$h_f = f\frac{L}{D}\frac{V^2}{2g}\qquad(4)$$

where h_f = the head (energy) loss
L = length of the pipe
D = inside diameter
V = velocity of flow
f = friction factor taken as 0·03.

The gauge pressure necessary has then been calculated for the upper boundary velocity as $2·4\times10^4\,\mathrm{N/m^2}$. Experiments have shown that when the pressure is gradually raised to $3\times10^4\,\mathrm{N/m^2}$ liquid metal will penetrate into the intricate points of the mould and will push the air inside the die through the vent holes.

The liquid metal should rise in the transfer tube quickly until it reaches the ingate, thus the heat losses in the transfer tube will be minimum. Filling of the mould cavity should take place with a non-turbulent rise of the liquid metal. This requires careful adjustment of the rate of pressurizing, otherwise large air pockets and heavy porosity will develop in the cast parts. However, this does not mean the complete stopping of air admittance. When this happens, unsolidified metal in the mould will flow back to the crucible. Therefore, metal should travel as fast as possible in the transfer tube until it reaches the gate, when it should slow down. The velocity values calculated refer to the velocity of liquid after this stage when the system reaches an acceptable steady state.

Heat transfer

Hollow cylindrical shapes present rather complicated heat transfer behaviour. The heat flow from the centre core is restricted and the casting as a whole has a lower cooling rate than a plate of the same cross-section. The simplest approach is to consider the shape as a plate, but to correct for an 'effective plate thickness'. If T is the true wall thickness and Te the effective plate thickness, then the following equation[5] may be used:

$$Te = kT\qquad(5)$$

Figure 9 shows a schematic representation of the hollow cylinder as a rectangular prism.

As the liquid metal enters the mould cavity with a considerable superheat, the heat transfer analysis can be performed in two steps. The first step involves the heat transfer with superheat. Performing a heat balance on a differential element gives the differential equation governing the alloy temperature T as a function of distance and time:

$$\frac{\partial T}{\partial t} = \frac{k}{\rho C_p}\left(\frac{\partial^2 T}{\partial x^2} + \frac{\partial^2 T}{\partial y^2}\right)\qquad(6)$$

where T represents the superheat temperature, and ρ, k, C_p are the same for both liquid and solid. The boundary and initial conditions required for the solution of equation (6) are:

$$t = O:\quad T = T_s\quad\text{for } O < x < 1$$
$$O < y < L\qquad(6a)$$

$$t > O:\quad T = T_1\quad\text{at } x = O\qquad(6b)$$

$$T = T_2\quad\text{at } x = 1\qquad(6c)$$

$$\frac{dT}{dy} = O\quad\text{at } y = O\qquad(6d)$$

$$T = T_4\quad\text{at } y = L\qquad(6e)$$

This differential equation is then written in implicit finite-difference form as

$$\frac{T_{i,j}^{n+1} - T_{i,j}^{n}}{\Delta t} = \frac{k}{\rho_p^C}\left(\frac{T_{i-1,j}^{n+1} - 2T_{i,j}^{n+1} + T_{i+1,j}^{n+1}}{\Delta x^2}\right.$$
$$\left. + \frac{T_{i,j-i}^{n+1} - 2T_{i,j}^{n+1}\,T_{i,j+1}^{n+1}}{\Delta y^2}\right)\qquad(7)$$

The above form of the difference equation is next modified if Δx is made equal to Δy and the constant terms are grouped into a single constant B defined as

$$B = \frac{k\,\Delta t}{\rho C_p\,\Delta x^2}(1 + 4B)T_{i,j}^{n+1}$$
$$= T_{i,j}^{n} + B(T_{i-1,j}^{n+1} + T_{i+1,j}^{n+1} + T_{i,j-1}^{n+1} + T_{i,j+1}^{n+1})\qquad(8)$$

Solving for $T_{i,j}^{n+1}$ the equation takes the form

$$T_{i,j}^{n+1} = C_{i,j} + D(T_{i-1,j}^{n+1} + T_{i+1,j}^{n+1} + T_{i,j-1}^{n+1} + T_{i,j+1}^{n+1})\qquad(9)$$

where

$$C_{i,j} = \frac{T_{i,j}^{n}}{1 + 4B}$$

$$D = \frac{B}{1 + 4B}$$

By applying equation (9) to all grid points $i = 2, 3, \ldots, m - 1$ and $j = 2, 3, \ldots, p - 1$ in turn, and by employing a Taylor's expansion to obtain the approximation for use at the point on the insulated boundary, $j = 1$, i.e.

$$T_{i,1}^{n+1} = C_{i,1} + D(T_{i-1,1}^{n+1} + T_{i+1,1}^{n+1} + 2T_{i,2}^{n+1})\qquad(10)$$

Temperatures at all grid points are computed at time intervals of Δt. The time elapsed until the temperatures at all grid points are equal to the liquidus temperature is calculated.

Solidification will start at a temperature just below the liquidus, i.e. $T < T_{\text{liq}}$. The second step of the heat transfer analysis involves latent heat effects. Performing a heat balance on a differential element and a generation term arising from latent heat gives the following differential equation:

$$\frac{\rho}{k}(C_p + \alpha\,\Delta H)\frac{\partial T}{\partial t} = \frac{\partial^2 T}{\partial X^2} + \frac{\partial^2 T}{\partial y^2}\qquad(11)$$

where ΔH is the latent heat of fusion and α is the kg of solid formed per °C fall per kg of solid–liquid mixture which was determined from the phase equilibrium diagram: $\alpha = \alpha(T)$ which is a known function of temperature. The boundary and initial conditions required for the solution of equation (11) are:

$$t = O:\quad T = T_{\text{liq.}}\quad\text{for } O < x < 1$$
$$O < y < L\qquad(11a)$$

together with equations (6b)–(6e).

The implicit finite-difference approximation of equation (11) is

$$\frac{\rho}{k}(C_p + \alpha \, \Delta H)\frac{T_{i,j}^{n+1} - T_{i,j}^{n}}{\Delta t} = \frac{T_{i-1,j}^{n+1} - 2T_{i,j}^{n+1} + T_{i+1,j}^{n+1}}{\Delta x^2}$$
$$+ \frac{T_{i,j-1}^{n+1} - 2T_{i,j}^{n+1} + T_{i,j+1}^{n+1}}{\Delta y^2} \quad (12)$$

Using a similar subdivision of the cast and taking Δx equal to Δy and grouping the constant terms into a single constant E defined as

$$E = \frac{\rho}{k}(C_p + \alpha \, \Delta H)\frac{(\Delta x)^2}{\Delta t}$$

equation (12) can be put into the following form:

$$T_{i-1,j}^{n+1} - 2T_{i,j}^{n+1} + T_{i+1,j}^{n+1} + T_{i,j-1}^{n+1} - 2T_{i,j}^{n+1} + T_{i,j+1}^{n+1}$$
$$= E(T_{i,j}^{n+1} - T_{i,j}^{n}) \quad (13)$$

which is to be solved together with the equation modified for the insulated end condition, *i.e.*

$$T_{i-1,1}^{n+1} - 2T_{i,1}^{n+1} + T_{i+1,1}^{n+1} + 2T_{i,2}^{n+1} - 2T_{i,1}^{n+1}$$
$$= E(T_{i,1}^{n+1} - T_{i,1}^{n}) \quad (14)$$

at a succession of time periods. Computation is ended when the temperature next to the insulated end has fallen to 400°C.

Summing up the time elapsed to obtain this temperature with the corresponding time to achieve the liquidus temperature by the first computer program gives information about the solidification time that can be fed into the electronic control system for pressurizing, mould opening, and ejection times.

Macroexamination

Macroexamination of cast parts showed mainly equiaxed grain structure and minimum porosity when compared with high pressure die castings. Porosity formation in high pressure die casting can be explained referring to the theory put forward by Frommer and Lieby.[1] The jet stream entering the cavity with high turbulence will stir the liquid in the mould violently and the solidification taking place in such a heavily stirred liquid will result in many entrapped voids. The high pressure applied during the process might be enough to squeeze up the voids, although this is not a welding as in the case of hot forging. The squeezed up voids will create high stress fields in their neighbourhood which will be active on heating. This is the main reason why high pressure cast pieces are usually listed as non-heat treatable objects.

Solidification in low pressure die casting takes place in a non-turbulent liquid, the growth of columnar grains is almost perpendicular to the mould wall, and the equiaxed zone presents a fairly homogeneous grain cohesiveness.

CONCLUSIONS

The following conclusions can be drawn.

1 Low pressure die casting proves itself to be a versatile technique in which casting size and plan area are not limiting factors.

2 Negligible locking forces are required to keep die halves together because of the low hydrostatic pressure of the liquid metal in the die.

3 Low pressure die casting requires no feeders and additional risers.

4 Laminar flow of the liquid metal into the mould cavity produces sound, heat treatable castings with homogeneous grain cohesiveness.

5 Heat transfer analysis of sample castings provides valuable information for automatic control of the operation.

ACKNOWLEDGMENTS

The authors wish to thank Dr N. Selçuk of the Chemical Engineering Department, Middle East Technical University, for help in preparing computer programs. The support of ELSAN A. S. of Ankara is also acknowledged.

REFERENCES

1 L. FROMMER AND G. LIEBY: *Giesserei*, 1952, **H13**, 311
2 W. BRANDT: *Tech. Zentr. Prakt. Metallearbeit*, 1937, **47**
3 W. W. STUHRKE AND J. F. WALLACE: *AFS Trans.*, 1965, **73**, 569
4 T. P. YAO AND V. KONDIC: *J. Inst. Met.*, 1952–53, **81**, 17
5 R. A. FLINN: 'Fundamentals of metal casting', 49, 1963, Addison–Wesley Publishing Co.

DISCUSSION

Session 4: Solidification and quality control; continuous casting

Chairman: D. Melford (Tube Investments Ltd)

The following papers were discussed: *Keynote Address: Continuous casting* by F. Weinberg; *Simulation of heat flow and thermal stresses in axisymmetric continuous casting* by J. Mathew and H. D. Brody; *Heat-transfer characteristics in closed head horizontal continuous casting* by R. Hadden and B. Indyk; *Thermal analysis of solidification of aluminium alloys during continuous casting* by I. Jin and J. G. Sutherland; *Continuous observation of hot crack-formation during deformation and heating in SEM* by H. Fredriksson and B. Lehtinen; *Crack formation and tensile properties of strand-cast steels up to their melting points* by K. Kinoshita, G. Kasai and T. Emi; *Comparison between experimental data and theoretical predictions relating to dependence of solidification cracking on composition* by T. W. Clyne and G. J. Davies; *Origin of defects in continuously cast blooms produced on a curved mould machine* by M. H. Burden, G. D. Funnell, A. G. Whittaker, and J. M. Young; *Solidification of steel in continuous-casting moulds* by M. Wolf and W. Kurz; *Secondary cooling and product quality in continuous casting* by A. Etienne and A. Palmaers; *Quality of continuously cast slab* by R. J. Gray, A. Perkins, and B. Walker; *Low pressure die casting of aluminium alloys* by E. Selcuk and L. Meric

DR A. GRILL (University of Ben Gurion, Israel): I would like to compliment Dr Etienne for trying to give some more mechanical data on the properties of the steel but I cannot agree with her completely. The two-dimensional model for stress evaluation in continuously cast steel, developed by myself with Professor Brimacombe and Professor Weinberg, takes into consideration the plastic behaviour of the material and is no more 'pure' than the one-dimensional model of Dr Etienne and her co-workers. We also tried to develop a two-dimensional elasto-plastic stress calculation model. We included in the model the available high temperature mechanical property data that I agree are not too reliable, but we still succeeded in explaining the formation of internal cracks that occurred in several experimental installations that we investigated.

The model has since been modified by Mr Sorimachi and Professor Brimacombe to include recent mechanical data published by Dr Wray from US Steel, and I feel that the two-dimensional stress model is still the best approach to calculate thermally induced stresses.

Referring now to the paper by Jin and Sutherland; even though the paper was on aluminium it would be possible to use two-dimensional stress models to evaluate the coupled behaviour of the heat flow and the formation of the solidified shell since these two

phenomena, as Dr Southin pointed out, are not related. The heat flow influences the gap formation between the solidified shell and the mould and this gap will influence the heat flow in the next stages of the mould. We investigated the behaviour of the gap in the mould for the continuous casting of slabs and it is possible to get an understanding of the coupled behaviour using the elasto-plastic model for stress calculations. With the corresponding mechanical properties of aluminium it is possible to use the same approach in order to calculate the behaviour in the casting of thin slabs of aluminium.

As regards the paper by Dr Burden, I should like to know what mechanism the author suggests for explaining the enrichment of copper at the grain boundaries during continuous casting. I would suggest that the water sprays enhance the cracking by increasing the tensile stresses at the surface, and these stresses produce cracks at the copper rich regions.

DR ETIENNE: I think we would agree on the necessity of having high temperature data on the mechanical properties of steel; the model is just as good as the data fed in.

The second point is that I wanted to develop this model as a tool to control cracking and in cooling

312

systems design. I think that the one-dimensional model is still adequate to do this job, though a two-dimensional model would be more complete and precise.

MR SUTHERLAND: I think the point that Dr Grill made is not entirely applicable to the paper presented which covered quantitatively only thermal analysis. I discussed mechanical analysis only in a qualitative sense. This was not included since such an analysis involves some parameters and some assumptions which are not well established, and hence the quantitative data so generated would be somewhat questionable. Without this full quantitative analysis, but having the thermal analysis including the effects of heat transfer perturbations, it is possible to gain a fairly good understanding of the mechanisms which operate to produce shell distortion, and to see why alloys with different freezing ranges should behave differently. Unreliable quantitative data would be of little value.

DR BURDEN: I did not have time to go into full details about preferential oxidation. At the surface, the iron is oxidized away and leaves the copper, tin, and antimony underneath the scale layer. The action of the water sprays speeds up this process considerably by intermittently removing the oxide layer and increasing the rate of oxidation. Although some of the residual enrichment is removed in the scale, this is not sufficient to prevent a large increase in the concentration of residuals, particularly copper, at the surface. I agree with Dr Grill that the thermal cycling action of the sprays will produce the stresses which give rise to most of the cracking. However in the absence of copper enrichment and AlN precipitation the thermal cycling alone would not give rise to cracking.

MR JIN: I have a comment on Dr Grill's question, in relation to aluminium continuous casting. The important objective of our paper is to establish a mathematical method which predicts the degree of non-uniformity generated during continuous casting as a function of the initial heat transfer coefficient, and alloy composition, by defining and calculating the shell distortion driving force at an early stage of shell formation.

PROFESSOR F. WEINBERG (University of British Columbia): I wish to refer to the paper by Dr Burden, and in particular the appearance of the surface of the tensile test and the temperature at the cracking face. At high temperatures there were nice smooth round faces associated with dendrites and at low temperatures it was quite different. In my own work I have found something similar. When I get smooth round fracture surfaces at any temperature I can generally associate these with internal cracks present in the sample before it was tested. [The same thing can be seen locally, for example, if the tensile test pattern is deformed and taken from a continuously cast ingot with liquid nitrogen.] Is it possible that the smooth surfaces seen at high temperature are associated with internal cracks present in the sample before they were pulled, or are they really associated with test temperature?

Moving on to the paper by Dr Wolf and his mathematical model, I suggest that a simple power model using the square root function of thickness is not a very suitable kind of formulation for stress calculations in terms of the complexity being dealt with.

Taylor, in his review, claims that no one has defined the thickness of the shell inside the mould and therefore no mathematical model can be proved. I think this is wrong. The radiotracer experiments we have done show the position of the shell if the solidus/liquidus gap is small. However, the model does not fit very well with stainless and high carbon steels because it predicts the solidus, and any delineation of the shell shows the liquidus.

With regard to Dr Etienne's work on stress/strain curves at high temperature, these are particularly complicated by the fact that recovery occurs. Experiments by Wray showed that the stress rises and starts dropping immediately in a series of cycles; we have also observed this phenomenon in our own experiments at high temperatures. We have never observed significant work hardening so it is difficult to develop any kind of stress/strain curve with confidence.

DR BURDEN: The figure I showed of surface fracture at 1 450°C (Fig. 20) is characteristic of fracture at this temperature; it does not show a contraction cavity already present in the specimen before testing. The internal surfaces of contraction cavities have a dendritic morphology, but they can be distinguished from those surfaces fractured at 1 450°C, which have a better developed dendrite structure, more like growing dendrites. They also show deeper grooves and re-entrant angles.

In the tensile tests we performed the specimens were rapidly cooled after fracture, and little diffusion could have occurred which would change the surface morphologies.

DR B. INDYK (University of Strathclyde): Professor Weinberg has mentioned that no one has measured the thickness of the shell. I know of a number of cases where the actual solidification profile has been measured.

I would like to ask Mr Sutherland whether he has checked his calculations against the practical data available?

MR SUTHERLAND: We have not compared experimental and analytically predicted shell distortion because we have not quantitatively determined such characteristics of shell distortion as amplitude or wave length. Indeed it may not be practicable to do this, since there could be many complex interface effects, any one of which might cause a variation in the heat transfer sufficient to initiate a local shell distortion, which will become stronger with time. We have made many observations of simple cast surfaces which have shown significant shell distortion and we want to determine the basic reason for this. Why is there more turbulence in short freezing range alloys? The magnitude of the shell distortion and its wave length depends on the source of the perturbation that initiates the disturbance, and in some cases even the way in which the metal flows off into the mould. From observation of cast materials we have correlated this break in the oxide skin with the shell distortion which occurs.

PROFESSOR D. W. KURZ (Ecole Polytechnique Federale de Lausanne): I would like to reply to Professor Weinberg's remarks. We did not try to fit the $\sqrt{t}$ law to the measured shell thickness values. During the initial period of solidification we used a different constant in the square root law. Equations (14) and (15) in our paper show that it is possible to get something experimentally which differs from the square root law. We wished to use the large amount of information available on shell thickness measurements together with an analytical solution of the heat equation to obtain a first approximation to the conditions prevailing at the solid/liquid interface. Figure 12 of our paper shows clearly that this can be achieved reasonably well.

PROFESSOR WEINBERG: I want to correct a misconception. There is a good fit between theory and experimental measurement of shell thickness, contrary to the statement by Taylor (C. R. Taylor: *Met. Trans. B.*, 1975, **6B**, (9), 359).

PROFESSOR A. A. TZAVARAS (Aristotelian University of Salonika, Greece): I have a question for Drs Kurz and Wolf. It was mentioned that high carbon steels have a smooth interface and low carbon steels have a rough interface. Is this observation general, or is this something that has been seen only in one or two cases?

My second question is to Mr Jin and Mr Sutherland. Mr Sutherland indicated one way for the slab to 'self-destruct'. Has he also considered a bridging effect inside the shell leading to segregation, shrinkage etc.?

DR WOLF: The difference in surface smoothness between low and high carbon steels is firmly established by a large number of observations on both breakout shells and mould heat flux. The strong tendency to shell distortion of low carbon steels might be explained by related shrinkage behaviour when undergoing the delta/gamma transformation as proposed by Grill and Brimacombe (*Ironmaking Steelmaking*, 1976, **3**, (2), 76). However, this model does not account for high carbon steels. Therefore we think that shell strength is also an important factor which appears to decrease with increasing carbon content. Thus the strand shell cannot shrink away from the mould wall, and becomes smooth. More investigations are necessary in order to understand the mechanism fully.

MR SUTHERLAND: 'Bridging' through the thickness of a slab during casting can obstruct feeding and, due to the volume change caused by solidification, it is possible that this could bring about another form of shell distortion, i.e. local shell collapse. This distortion could, of course, also affect the heat transfer. We believe that shell distortion starts at an early stage of shell formation in narrow freezing range alloys, and that it does not generally occur at all in wide freezing range alloys, because the initial driving force is lower. If the heat transfer is reduced, the fraction solid in these alloys is also reduced. This rapidly lowers the shell bending strength until even a small metallostatic pressure will restore good ingot-to-mould contact.

PROFESSOR WEINBERG: With regard to Professor Tzavaras's point on the 0·1% carbon steels and the surface roughness, it is my understanding that this is a major problem in the industry and if someone could come up with a neat solution as to why the shell buckles and how to control it, this would be great. The mechanism proposed by Dr Grill is that it is a delta/gamma phase transformation which causes an additional strain. US Steel do not believe this and point

out the bump in the curves in other alloys besides steel. They have a solution to the problem. They know the mechanism, but they would not tell me what it was. It is very intriguing.

PROFESSOR H. FREDRIKSSON (RIT, Stockholm): I would also like to comment about this rough shell. We have seen the same effect in low carbon and 18/8 steels. We made careful metallographic analyses, and found that the thicker areas had solidified directly to ferrite and the thinner areas had solidified directly to austenite. We think this would also be the case for low carbon steels.

DR N. B. BRYSON (Aluminium Company of Canada Ltd): First, I would like to congratulate Professor Brody on an excellent paper. The value of this type of work would be greatly enhanced if people in industry would give freely of any experimental data which they may possess.

Secondly, I have a few comments on the use of computer-generated empirical cracking parameters. The one quoted in Professor Brody's paper was developed by me six years ago. Recently, my colleague Mr Jin has derived a parameter which is non-dimensional. This improved parameter is the ratio of the cooling rate at the centre to the cooling rate at the surface. Using this non-dimensional parameter, it is possible to predict the cracking rate for an ingot of any diameter, knowing the cracking rate of an ingot of one specific diameter.

DR G. B. BARKRAY (Israel Institute of Metals): How did Dr Mathew establish his cracking criteria? Also, was the stress calculation done once for the whole length of the DC cast billet or was it done incrementally?

PROFESSOR BRODY: We have not proposed a cracking criterion. We calculate the stresses (strains and strain rates) which would be present in the absence of cracking. Whilst awaiting the development of a cracking criterion we suggest that that normalized stress which would cause cracking in a particular alloy under a specific set of conditions be used to extrapolate/interpolate for other conditions using the same material.

DR BARKRAY: My next question asked whether the length was constant during your calculation of stress?

PROFESSOR BRODY: As can be seen in Figs 1 and 2 of our paper, we calculated temperature and stress distributions from the regions of the superheated metal in the mould to a region about 28 in. down the spray zone where the metal had cooled to 100°–150°C.

DR BARKRAY: So you used incremental lengths in your calculations.

PROFESSOR BRODY: No, we performed the calculation over the whole ingot using a fixed coordinate system.

DR BARKRAY: How did you calculate your thermal load? What temperature did you use?

PROFESSOR BRODY: We have to put in a thermal expansion coefficient for stresses that result from thermal expansion. This is a steady-state calculation, not a transient calculation, and it is a steady-state stress that is being calculated.

DR BARKRAY: The thermal stresses have to be calculated as a result of changes in temperatures, not as temperatures themselves. What temperature changes do you consider?

PROFESSOR BRODY: The temperatures change from point to point so the temperature changes are a function of the γ direction, not a function of time. Two other points: we have compared our heat flow calculations with reported measurements of skin thickness and on cooling rate. We have made our comparisons for steel, magnesium, and stainless steel and found good agreement. These comparisons are reported in reference 2. Incidentally, validation of the stress model is discussed in reference 3.

Dr Grill mentioned that it would be possible to calculate the air gap using a mechanical model. Our model uses an iterative process to compute the position of the air gap and its heat transfer coefficient. The heat transfer coefficients are first assumed, and a temperature distribution computed. Then an elastoplastic computation is used to calculate the position and extent of the air gap. The heat transfer coefficients are then recalculated and the process repeated until acceptable convergence is found. Finally, the calculated heat transfer coefficients are used for the steady-state calculation of the temperature distribution.

DR B. MINTZ (BSC, Scottish Labs): I would like to make a number of comments on the paper by Kinoshita *et al.* I was interested in the view that n-values are a better measure of the tendency to surface cracking than reduction of area in hot ductility tests and find it difficult to follow the reasoning behind this. The evidence I have found is that transverse cracks on the surface of aluminium killed steel slabs are propagated on straightening at the tangent point by a deformation-induced precipitation of AlN at the austenite grain boundaries which prevents grain-boundary sliding, and I am not clear how this relates to work hardening rates. Whatever the detailed mechanism is, one is interested in localized failure at grain boundaries and not in the uniform elongation of the sample. It would be very interesting if you would present your reduction of area curves against test temperature for comparison with the n value curves, and also to see evidence of how these relate to commercial transverse surface cracking in concast slabs.

The second point is the value to be placed on notched sample results. It certainly seems a good idea to notch the hot tensile test specimen but I was rather surprised that a V-notch had been used, since this may concentrate the stress too locally. Thus with the coarse austenite grain sizes present and the fact that failure is intergranular, the tendency for cracks to propagate would not be related to the inherent ductility but more to geometrical effects, i.e. whether the grain boundary was oriented favourably with respect to the stress system. If the root of the notch went along a grain boundary, then crack propagation would be easy. Figure 8 in the paper shows a very good example of this. Conversely, if the root of the notch was nowhere near an austenite grain boundary, I wonder whether crack propagation would be easy.

Only if a large number of samples had been tested would the results be representative of the true cracking trend and I wonder whether sufficient data has been accumulated to establish this with any certainty.

The third comment concerns the apparent lack of correlation between cracking index and n-values for the niobium–vanadium steel given a full solution treatment (Fig. 8). The cracking behaviour was very similar at 1 100 K and 1 300 K but the n-value at 1 100 K was twice that at 1 300 K. Can any explanation be given for this?

Finally, the cracking index data for a C/Mn/Nb/V/Al steel suggest that raising the straightener tangent point temperature above 850° increases the tendency to form cracks, this reaching a maximum at about 1 000°C. This seems to be very different to commercial practice currently being used which, granted, is based on hot ductility reduction of area measurements. These criteria suggest that for C/Mn/Nb/Al steels the ductility minimum occurs in the temperature range 750°–850°C and the tendency to transverse cracking is reduced by going to temperatures outside this range. Commercially, transverse cracking has been overcome either by soft cooling to ensure that the temperature at the straightener tangent point is in the range 900–950°C, or by lowering the temperature to well below 800°C using hard cooling conditions. The former practice does not accord with your results but is in agreement with the behaviour to be expected from changes in reduction-of-area with temperature, given by hot ductility tests. How do you reconcile your results with this?

MR KINOSHITA (written reply): In reply to your first query, at austenitic temperatures, the reduction of area levels off close to 100%; thus, this criterion is incapable of detecting differences in deformability between different steels. Our aim was to find an index which can be used at austenitic temperatures. In this respect, the n-value criterion seemed promising since it gave decreasing values with increasing amount of AlN as shown in Fig. 7. The higher crack sensitivity of Al–Si killed steels compared with Si killed steels is clearly demonstrated by the smaller n-values for the former steels. Nevertheless this dependency remains qualitative until further investigations are made into the size distribution of AlN, and grain size effects. The correlation between n-value and crack formation is at present only indirect, because the n-value as well as the reduction of area only represent the deformability of the specimen until it fractures at the grain boundaries, i.e. the less sensitive is the specimen to the crack, the larger the value of n or reduction of area. Since the n-value criterion can better represent hardening by AlN precipitation at austenitic temperatures than does reduction of area, we preferred the former.

The reproducibility of the crack length obtained by hot tensile tests has been satisfactory at austenitic temperatures, although the number of tests at ferritic temperatures is at present too small to deduce any plausible conclusion.

The good reproducibility can be attributed to the following fact: since the austenite grain size was about 200–400 μm in diameter in reheated specimens (either 'as-cast' or 'pre-annealed', *see* Fig. 8) a total length of more than 24 mm of grain boundaries, was situated within the area in which crack formation was observed (notch root width 0·5 mm × circumferential length 24 mm). The total length of fractured boundaries, however, as shown in Fig. 10 was less than 2·5 mm, only one tenth of the grain-boundary length available.

As regards your third question, this may have been caused by the difference in the size distribution of precipitated NbC and AlN between 1 100 K and 1 300 K. Further investigation will be made to confirm this.

Finally, our observations indicate that transverse cracks on the surface of slabs of Al–Si killed steels, particularly those containing Nb occur on straightening at the tangent point by AlN (and NbC) precipitated during the course of repeated cooling and recalescence in the preceding secondary cooling zone. If the surface temperature in this region is reduced to avoid recalescence by an improved spray water distribution, the surface temperature at the straightening point has little influence, at least between 750° and 950°C, on the formation of cracks under otherwise optimized casting conditions. However, if the roll gap and/or path line have been in misalignment, crack formation has been experienced at the straightening point both at austenitic and ferritic temperatures.

DR N. MCPHERSON (BSC, Scottish Division): Dr Mintz raised the question of soft and hard cooling of continuously cast slab material. Work carried out at BSC Ravenscraig Works has shown that increasing the specific secondary cooling water over the range 1·01/kg to 1·61/kg increased the severity of some surface defects. Has Dr Gray found that hard cooling resulted in a deterioration in surface quality? Has Dr Gray any evidence to back this up?

DR R. J. GRAY (BSC, Teesside Laboratories): We have confirmed that reducing the secondary cooling decreases transverse cracking, as I indicated in the paper. We have not yet tried increasing the cooling rate beyond the standard water flows that are recommended for the machines so that we cannot confirm or deny that, at the moment.

DR D. J. ALLEN (CEGB): I have a question for Dr Clyne. You have a reasonable correlation between solidification cracking and (t_v/t_r). But have you tried plotting your results against the simpler ratio (t_v/t), where t is the total solidification time?

DR CLYNE: As Dr Allen suggests it is the time for which the solidifying structure is vulnerable to interdendritic separation, (t_v), which is the crucial parameter. Dividing this by total time rather than strain accommodation time, the theoretical predictions remain virtually unchanged. It might be argued, however, that the ratio used is marginally more justifiable because the accommodation time affects the stress operating during the vulnerable period, whereas the bridging period (which would, of course, be included in a total time denominator) has no significance beyond representing the end of the vulnerable period.

As may be inferred from this, the upper f_L limit chosen for liquid and mass feeding is not critical and may even be taken as 1·0 without sensibly changing the predictions. It is the upper and lower cracking limits which are the important parameters.

Session 5

Solidification and quality control; macrosegregation, porosity, and simulation of solidification

Chairman: J. A. Reynolds (Steel Castings Research and Trade Association)

Editor: P. Beeley (Department of Metallurgy, University of Leeds)

Keynote Address Solidification and aspects of cast metal quality

P. R. Beeley

Certain aspects of quality, such as structure, homogeneity, and soundness are affected to a large extent by the solidification stage, and hence the relationship between solidification and quality has been receiving increasing emphasis. This paper briefly reviews past work as a background to recent research into segregation, porosity, and allied topics relevant to both ingots and castings. The use of computer and simulation techniques as an aid to quality is also reviewed.

The author is in the Department of Metallurgy, University of Leeds

The quality of a cast product is defined by a number of characteristics but central to the final performance of the material are the attributes of structure, homogeneity, and soundness. These aspects of quality are all dependent to a large extent on the solidification stage—hence its importance as a field of research. A notable feature of the work of recent years has been the increasing emphasis on material quality. This applies particularly to shaped castings, in which preoccupation with the prevention of local defects of macroscopic dimensions had previously relegated quality in the metallurgical sense to a subsidiary role. This past situation is reviewed briefly as a background to the more recent research into segregation, porosity, and allied topics relevant to the quality of both ingots and castings. Structure itself has been largely dealt with in previous sessions, although it has a significant bearing on the problems with which the author is concerned here.

SOUNDNESS AND THE FEEDING OF CASTINGS

The most common major casting defect was the massive shrinkage cavity or pipe, resulting from failure to compensate for liquid and solidification contraction. To prevent this and ensure adequate feeding required the manipulation of gating practice and of feeder head size and position to produce suitable temperature gradients in the system. Many earlier solidification studies were concerned with this macroscopic aspect of freezing: the now familiar idea of controlled directional solidification was originally conceived in the same context. The publications of Briggs[1] and of Ruddle[2] made early contributions to the development of interest in this area; other early studies[3,4] were concerned with the causes and elimination of hot tears in castings.

Meanwhile, the cast iron field had provided an outstanding metallurgical development based on struc-

ture control in cast material: the production of the ductile spheroidal graphite irons,[5] which were to give a much needed impetus to the cast metals industry. With this and a few other exceptions, however, the structural aspects were largely neglected. This may have been partly attributable to the fact that the steels, as the main family of high-strength materials, undergo transformations which reduce the significance of initial crystallization relative to final properties. Most of the other exceptions were cases in which there exists a strong interrelation of structure and soundness; hence the early interest in grain refining treatments for aluminium alloys, which enhance feeding and reduce the incidence of hot tears, apart from benefits to mechanical properties.

In the period after 1950 the use of accumulated data on solidification behaviour and the availability of modern aids such as mouldable exothermics and insulators enabled macroscopically sound shaped castings to be produced with progressively higher standards of reliability. In the 1950s and early 1960s there was the development and widespread adoption of practical quantitative approaches to the problem of feeding, although mainly in relation to steels and other skin forming alloys: these relied on empirical observations rather than knowledge of thermal properties. Much of the work was carried out in the USA,[6–9] although rooted in the earlier theoretical and experimental work of Chworinov[10] correlating the geometry and total freezing time of cast blocks. Chworinov's rule relates the total freezing time of a cast shape to the volume/surface area ratio according to the equation

$$t = k\left(\frac{V}{A}\right)^2$$

Its validity, derived theoretically, was confirmed by

experimental determinations on a variety of cast shapes ranging from plates to spheres.

On the basis of this and subsequent findings, a fully developed quantitative system for the design and location of feeder heads and for the use of chills, exothermics, and other aids to soundness was published by Wlodawer[11]: the work of Jeancolas[12] followed similar principles. By virtue of this and other work, especially the major contributions of Pellini and associates,[13] the feeding of sand castings in skin forming alloys was now well understood. It remained, however, to solve feeding problems associated with alloys freezing in the 'pasty' manner and susceptible to discrete porosity. Solutions are also needed in the different thermal situation of gravity die casting. Data on feeding behaviour and techniques for individual types of alloy are reported in the literature, for example, in Refs. 14–19: the alloys include SG and white cast irons and various aluminium and copper alloys.

Progress in the elimination of major casting defects, as outlined above, was assisted by parallel developments in non-destructive testing, in respect of range, sensitivity, and interpretation.

SOLIDIFICATION AND STRUCTURE

The main focus of research attention had meanwhile shifted to the development of structure in cast alloys. Chalmers and associates were among those who undertook the more systematic investigation of solidification phenomena long recognized qualitatively.[20] Attention was principally centred on the influence of thermal conditions in freezing, and a period which saw AIME and ASM symposia,[21,22] a substantial review by Winegard,[23] and a monograph by Chalmers[24] was notable for numerous investigations of growth processes, solid/liquid interface morphology and solute redistribution under conditions of controlled unidirectional freezing of small specimens.

The Brighton Conference of 1967[25] reflected the widespread interest in this field, which has since developed further. The approach previously confined to the study of solid-state phenomena had come to be applied to solidification. The work of Flemings and associates at the Massachusetts Institute of Technology was already represented at that conference and has occupied a central position since. There has been intense international activity in the field, while specialized practical applications of structure control have also emerged: VerSnyder, in his review of the development of gas turbine materials,[26] summarizes the work in one such area. The title of Flemings's recent work, 'Solidification processing',[27] also used for his Howe Memorial lecture,[28] aptly reflects the modern situation, in which freezing is no longer seen as a fortuitous event but as a process to be controlled to attain closely defined metallurgical objectives.

INGOTS AND SEGREGATION: EARLY WORK

Much of the above discussion on feeding was concerned with shaped castings. Turning more specifically to ingots, problems of soundness have received much less attention, partly because the relative simplicity of shape reduces the emphasis on feeding technique and partly because subsequent mechanical working increases the tolerance for porosity: in many cases this extends to the deliberate use of controlled gas evolution to improve wrought metal yield.

In the manufacture of ingots, more attention has been devoted to the problem of segregation, with its major influence on the quality and properties of the wrought product. Early concern with this aspect was expressed in the institution, within the ambit of The Iron and Steel Institute, of the Committee on the Heterogeneity of Steel Ingots, which for some years coordinated activity relating to the problem in steel.[29] It was appreciated that macrosegregation, defined as zonal segregation on a scale larger than that of the normal microstructure, could reduce hot workability and produce brittle zones and severe anisotropy in the wrought material.

Much of the early research into ingot segregation involved laborious sectioning and drilling for chemical analysis, while conflicting hypotheses were advanced for the formation of the familiar *A*- and *V*-segregate patterns. Similarly, in the non-ferrous field, attention was given to 'inverse segregation' and 'tin-sweat', as encountered in the casting of bronze ingots.

MODERN RESEARCH ON SEGREGATION

Detailed attention to basic mechanisms of segregation arose in relation to its positive application for zone refining, developed primarily for the production of high-purity materials for the electronics industry. The conditions in such a process are again those of unidirectional solidification and thus amenable to analysis and controlled experiment. Pfann's monograph[30] dealt in depth with the quantitative aspects of solute redistribution as influenced by temperature gradient, solidification rate, and mixing conditions in the liquid fraction. This and contemporary work were, however, mostly concerned with conditions of sharp separation of liquid and solid fractions, rather than with the deeply incised solid structure obtained in dendritic growth, the mode most commonly encountered under practical casting conditions.

Dendritic structure and segregation

Dendritic freezing had long been qualitatively familiar to metallurgists, as had dendritic segregation or coring. In the earlier quantitative work on solidification, however, the focus of attention had been the relatively modern observations of cellular interfaces and the conditions leading to successive transitions from one mode of growth to another. A revival of interest in dendritic structure soon followed with the greater appreciation of the importance of dendrite arm spacing as a metallurgical parameter with profound influence on mechanical properties. Such influence frequently transcends that of grain size and arises mainly through the scale of microsegregation and associated concentrations of solutes and second phases. This effect had been quantified by various groups of workers in the early 1960s[31,32] and witnessed in the results of investigations on section sensitivity in cast high-strength alloys.[33]

Clarification of numerous aspects of dendritic structure and segregation emerged from work by Flemings and associates with various alloys including steel,[34] aluminium–copper[35], and bismuth–tin.[36] Many of the ideas developed have been presented by Flemings[25] and not least in importance was the elucidation of terminology. Pictures were developed based on iso-concentration surfaces determined by probe analysis, while the plate-like morphology of some dendrites was also emphasized.

Conditions during dendritic segregation were subject to some dispute, mainly in relation to the composition of the interdendritic liquid.[25] In subsequent work[37,38] it has become accepted that mixing in this liquid is essentially complete and that dendritic segregation corresponds well with the Scheil equation

$$C_s = k_0 C_0 (1 - g)^{k_0 - 1}$$

where

C_s = the local solute concentration in the solid
g = the fraction solidified
k_0 = the equilibrium distribution coefficient.
(This assumes diffusion in the solid to be negligible.)

Segregation in ingots was also examined,[25] under conditions of both directional and normal freezing and including radioactive tracer work.

Later contributions on dendritic structure and segregation[39,40] emphasized the further mechanism of Ostwald ripening as a factor influencing final spacings and accounting for some discrepancies between earlier theoretical and experimental work: coarsening occurs by dissolution under the influence of interface curvature on local melting point. An alternative mechanism has recently been proposed for the coarsening effect involving coalescence of adjacent dendrite arms: in detailed studies under steady state growth conditions, arm spacings were related to thermal conditions and solute content.[41] Other work has extended considerations of microsegregation to ternary alloy systems,[42] while the influence of constitutional factors on dendrite spacing has also been further examined.[43]

Fluid motion in segregation

In much of the research in the period so far discussed, the main emphasis had been placed on thermal conditions and diffusion, although work on segregation had considered mixing in the liquid fraction. If one were to look for a fresh characteristic since 1967 it might well be the much greater emphasis on dynamic effects on both structure and segregation: some of these effects had received limited attention at Brighton.

The role of convection and thermal perturbation in the development of cast structures was reviewed on that occasion by Tiller and O'Hara,[25] with particular reference to the origin of the equiaxed zone in ingots and castings. The importance of fluid motion was illustrated by the work of Jackson *et al.* with transparent analogs,[44] showing dendrite fragmentation, while Cole and Bolling presented further experimental evidence for the effects of buoyancy forces and convective flow on ingot structure[25,45]; such flow could be modified artificially as a technique of structure control. Cole subsequently reviewed in more detail the nature of transport processes and fluid flow, again with the main emphasis on the structural aspects.[46] Southin's work on the shower mechanism and on the operative factors in dynamic nucleation techniques, for example, vibration, stirring, and gas agitation,[25,47] contributed further to this developing interest in dynamic effects.

The interrelation of dendritic structure and segregation has already been referred to, but it is in the field of macrosegregation that the effects of fluid motion are to be most readily observed. Several influences can generate such motion in a solidifying system; the most evident is simple thermal convection in the bulk liquid. Less obvious is the density variation produced by local differences in solute content, giving rise to constitutional or 'thermosolutal' convection. Superimposed on these effects are the feeding mechanisms produced by liquid and solidification shrinkage and by solid contraction in the freezing zone. In some cases, gas precipitation provides added pressure for the displacement of residual liquid.

Various forms of macrosegregation have been re-examined in the light of these influences and new explanations offered in some cases. The recent studies have been based on theoretical analysis, transparent analogs, and experimental metallic systems. The greater complexity imposed by variations of geometry and by the numerous long-range thermal and constitutional influences on motion make analysis difficult: as recently observed by Weinberg,[48] there is little quantitative information on the overall effects of fluid flow. Much attention is nevertheless being given to the subject.

Characteristic forms of macrosegregation

Among the effects in which fluid flow plays a critical part are inverse segregation and the still more common segregation patterns familiar in ingot structures. Long-range outward flow of low-melting-point liquid through interdendritic spaces was emphasised by Youdelis and co-worker[25,49] in their expositions of inverse segregation. This process, mainly under the influence of contraction in the outer solidifying zone, has been further and more recently analysed by Flemings *et al.*[50-52] with particular attention to the reheating influence of the air gaps formed at the metal/mould interface. Exudation of solute-rich liquid at the surface has also been treated by Kaempffer and Weinberg,[53] again broadly in accord with the classical theory of back flow consequent on local reheating.[54,55]

In the case of *A*- and *V*-segregates, much early work had been carried out without agreement on basic explanations. Recent researches have had a major impact on this problem, with general acceptance of the central role of the movement of liquid through the solidification zone. A frequent feature of these patterns is the concentration of segregates in local channels within which preferential flow of low-melting-point liquid has occurred: *A*, channel, and freckle segregation are manifestations of the same effect.

Blank and Pickering[25] associated *A*-segregation with channels or 'ropes' of liquid within the columnar dendritic region but tentatively attributed these to solute entrapment and further nucleation and growth beyond. Several researches have since provided evidence for flow along such channels. McDonald and Hunt[56] used transparent analogs with local dyeing to show the effect of lower density liquid in producing thermosolutal convection and channelling in the solidification zone, and the same authors demonstrated the reversal of flow direction induced by density reversal.[57] This work was later reinforced by findings from metal systems, notably those of Streat and Weinberg[58] and Hebditch and Hunt.[59]

Using autoradiographic techniques Streat and Weinberg observed the development of 'pipes' in lead-rich Pb–Sn alloys by upward flow from density inversion in the liquid; they later demonstrated the influence of solute-induced density variations on flow direction by the contrasting behaviour of tin-rich alloys in the same system.[60] Conditions for reducing segregation were postulated as including dendritic structure refinement, increasing growth rate and decreasing temperature gradient. Hebditch and Hunt[59] showed the development of macrosegregation in a Sn–6% Zn alloy, in which gravity flow occurred despite little change in density with composition. These authors[61] later stressed the increased importance of the convective interdendritic flow mechanism in slow freezing, as against the volume change mechanism with rapid freezing, or the diffusion layer mixing mechanism facilitated by near-planar interfaces. Further work in this general area is described in the paper by Fisher and Hunt in this volume.

'Freckling' was the term initially used to describe channel segregation encountered in directionally solidified superalloys. The nature of freckles was investigated in depth by Giamei and Kear[62]: they were found to consist of elongated aggregates of fine, randomly oriented, equiaxed grains, associated with alloy segregation and porosity and consistent with the late freezing of low-melting-point liquid channels. The convective mechanism and consequent upward jets in preferentially eroded channels were observed by Copley et al. in a transparent analog[63]: from analysis of this and other evidence they concluded that the freckles originated in the same phenomenon of density inversion in the pasty zone.

Continuing earlier work on the role of liquid movement in segregation,[50–52] a full quantitative analysis of gravity induced interdendritic flow and of conditions producing channel, freckle, and other forms of macrosegregation was presented by Mehrabian et al.[64] and further reviewed by Mehrabian and Flemings.[37] The analysis was based on the horizontal steady state solidification of aluminium–copper alloys and experimental results were later found to agree well with the theoretical model developed.[65] An explanation was advanced for the preferential remelting which results in channelling, based on the relation between the velocity and direction of convective flow and the movement of isotherms through the solidification zone: if the flow velocity exceeds the isotherm velocity the necessary conditions for instability are created. Channel segregates were subsequently reproduced in laboratory ingots to confirm the theoretical predictions.

Experiments have since been performed[66] to explore further the geometry and permeability of the porous mesh existing during solidification, using removal of the liquid fraction under applied pressure.

Other current work on segregation includes that shortly to be described by Takahashi et al. in which variations in fluid flow velocity have been experimentally induced, and by Weinberg et al. who examine the influence of ingot shape on A-segregate formation.

There is thus continued interest in this technologically important field. The economic significance of the problem has been stressed by Singh et al.,[67] who demonstrated the favourable heat treatment kinetics produced by fine dendrite arm spacings. As emphasized elsewhere, however,[68,69] heat treatment can never be adequate to eliminate the influence of macrosegregation, which in the long term must be suppressed at the outset by solidification control.

Discussion of the topic will be concluded on a positive note by reference to the adaptation of the recently developed stir-casting technique to the separation of the liquid fraction in differential crystallization.[70] As in zone refining, we thus have a further example of a modern technique brought to the aid of a very old metal-refining principle.

POROSITY IN CAST METALS

In contrast to the large volume of work on cast structure and segregation, research on porosity has been relatively scarce. The previously cited work on feeding techniques for shaped castings was concerned primarily with skin-forming alloys, which develop massive cavities rather than discrete porosity. The latter is associated with alloys of longer freezing range and its form and distribution are determined by alloy constitution, thermal conditions, and gas content. The problem of interdendritic fluid flow is again of obvious importance, in this case to the feeding of shrinkage in the later stages of solidification. It is usually beneficial to grain refine such alloys to prolong the mass feeding stage,[71] while if sufficient surface chill can be obtained to produce an intact skin, internal pore formation becomes difficult and feeding is assisted by atmospheric pressure.

Early contributions to the study of porosity were made by Piwonka and Flemings,[72] and by Campbell[25]: these dealt with the mechanism of void formation in cast material and with the interaction of gas and shrinkage. Subsequent papers by Campbell[73,74] reviewed the full range of feeding processes in castings and paid special attention to 'burst feeding', involving repeated collapse of the dendritic mesh under pressure created by shrinkage, atmosphere, and metallostatic head. The importance of effective interdendritic feeding in the production of premium castings was stressed, especially in association with the fine dendrite arm spacings essential to their high mechanical properties. Conditions giving rise to surface and internal porosity have since further reviewed by Flemings[27] and the nature of the cavities examined experimentally by Walker.[75]

The partial dependence of soundness on structure arises from the influence of growth morphology on the geometry of residual passages available for transmission of feed liquid to shrinkage sites. The same morphology governs the dispersion of residual porosity, segregates, and inclusions. The influence of these discrete minor defects can be regarded as a general one on material properties, as distinct from the local role of a single major defect.[16]

The interrelation of structure and porosity also accounts for the association of columnar structures with high levels of mechanical properties. The strongly directional freezing conditions which give rise to such structures are ideal for feeding, so minimizing the level of microporosity. This is the dominant influence on properties and is, except in special cases, more important than that of preferred orientation: this applies particularly to alloys undergoing subsequent transformations.

Under such ideal conditions, structure sensitivity of properties in cast material is also greatly reduced, indicating that this characteristic depends mainly on the dispersion of microporosity: this has been demonstrated in both high-strength steels[76] and copper alloys.[77]

The incidence of porosity is similarly evident in the phenomenon of section sensitivity. With the exception of grey cast iron, in which the effects of cooling rate on pearlite content and graphite form are paramount, the most important factors are the amount and distribution of microporosity and segregates, usually both coarser and more prevalent in thick sections and central regions. Evidence is seen in the marked reduction of the mass effect in unidirectionally as compared with conventionally solidified blocks.[78] Properties will be considered in a subsequent session and will not be further discussed here.

The respective roles of gas content and shrinkage in determining the nature and distribution of porosity have been previously highlighted by Cibula.[79] This aspect is to be further treated in a paper in this volume on porosity in aluminium alloys by Entwistle *et al.* while another paper, by Feurer and Wunderlin, reports on the direct monitoring of pore formation by acoustic emission measurement.

While the primary aim must be to prevent pore formation, Coble and Flemings[80] have demonstrated that shrinkage or hydrogen pores of size up to $\sim 1\,\mu$m can be annealed out in reasonable times, while the feasible size can be increased by a factor of ten under external pressures of $\sim 2\,\text{MN/m}^2$.

COMPUTATION AND SIMULATION TECHNIQUES AS AN AID TO QUALITY

Throughout the years during which solidification has been under investigation, various groups have applied mathematical techniques to derive data useful to the solution of casting problems. Much of the early work in this field was reviewed by Ruddle,[2] while Flemings[27] more recently surveyed the principles employed in the prediction of heat flow in refractory moulds, in which the thermal properties of the mould itself are critical, and in ingot and die casting, in which resistance at the metal/mould interface becomes a major controlling factor. The importance of the derived data in relation to feeding (width of the pasty zone) and structure (local solidification time) was stressed.

Most studies of this type have involved assumptions which reduce complexities due to shape, superheat, changing thermal properties, and interface conditions, but computer simulation has enabled some of these factors to be taken into account in later numerical work.

Eisen and Campagna[81] used computer simulation to develop the molten pool configuration in the casting of VAR slabs, while Durham and Berry[82] successfully employed the technique for prediction of the unidirectional freezing of a pure metal, taking into account the air gap, variation of thermal properties with temperature, and varying initial conditions. The latter paper contains a substantial review and tabulation of earlier numerical work in the field.

The factor of shape was embodied in numerical heat-transfer work by Pehlke and associates,[83] who were able to compute solidification profiles for *L*- and *T*-shaped sand cast sections in low-carbon steel which agreed well with previous experimental data determined by the Pellini group: work in France has been devoted to similar shapes.[84] In later work by Pehlke[85] the inward progress of solidification in square sand cast bars in aluminium and in a long-freezing-range silicon brass was numerically simulated, again in reasonable conformity

with the author's own experiments. This group reviews its own and other simulation studies in shaped castings in some detail in another paper in this volume.

In recent work by Davies,[86] 2-dimensional solidification profiles of plate castings were computed as a further development of earlier work[87] and were used to trace the progress of solidification along the centreline. Feeding distances were estimated and calculations for various alloys were found to agree well with practical feeding experiments. The same approach is extended to gravity die castings in today's contribution by Davies and Roe, while further simulation work, involving variations of several process parameters, will be reported by Hansen.

These ambitious developments still leave open the question of whether they will eventually find a direct place in industrial process control or whether their main use will be to provide back-up to more empirical techniques.

A further application of computer prediction of heat transfer to the development of cast structure is described by Ballantyne and Mitchell: their paper deals with solidification in the specialized remelting processes for the production of refined ingots in special steels and alloys.

CONCLUSIONS

One thus has an overall picture of a great and growing pool of knowledge of solidification, much of it relevant to the continuing task of perfecting quality in cast products, whether shaped castings or primary forms for the wrought metal industries. In this search for quality one can also look forward to help from advanced developments such as those to be discussed in the final session of this conference. Some of these bear on all three aspects of cast metal quality referred to at the beginning of the present paper: in relation to structure they offer refinement and control, to soundness reduced porosity, and to homogeneity the prospect of diminished segregation. Given such progress, the future of casting as a means of building shape and structure seems assured.

REFERENCES

1 C. W. BRIGGS: 'The metallurgy of steel castings', 1946, New York, McGraw Hill
2 R. W. RUDDLE: 'The solidification of castings', 1957, London, The Institute of Metals
3 H. F. BISHOP *et al.*: *Trans. Amer. Foundrymen's Soc.*, 1957, **65**, 247
4 R. A. ROSENBERG *et al.*: *ibid.*, 1960, **68**, 518
5 H. MORROGH AND W. J. WILLIAMS: *J. Iron Steel Inst.*, 1947, **155**, 321
6 J. B. CAINE: *Trans. Amer. Foundrymen's Soc.*, 1948, **56**, 492
7 J. B. CAINE: *ibid.*, 1952, **60**, 16
8 C. M. ADAMS AND H. F. TAYLOR: *ibid.*, 1953, **61**, 686
9 H. F. BISHOP *et al.*: *ibid.*, 1955, **63**, 271
10 N. CHWORINOV: *Giesserei*, 1940, **27**, 177
11 R. WLODAWER: 'Directional solidification of steel castings', 1966, Oxford, Pergamon Press
12 M. JEANCOLAS: *Foundry Trade J.*, 1966, **120**, 255
13 W. S. PELLINI *et al.*: *Trans. Amer. Foundrymen's Soc.*
14 N. P. SINHA AND V. KONDIC: *Brit. Foundryman*, 1974, **67**, 155
15 N. P. SINHA AND V. KONDIC: *ibid.*, 1975, **68**, 9
16 N. S. MAHADEVAN *et al.*: *ibid.*, 1971, **64**, 60

17 L. VON RICHARDS AND R. W. HEINE: *Trans. Amer. Foundrymen's Soc.*, 1973, **81**, 571
18 G. SCIAMA AND M. JEANCOLAS: *Cast Metals Res. J.*, 1973, **9**, (1), 1
19 H. DEVAUX AND M. JEANCOLAS: *Cast Metals Res. J.*, 1972, **8**, (3), 103
20 W. A. TILLER *et al.*: *Acta Metall.*, 1953, **1**, 428
21 'The solidification of metals and alloys', 1951, New York, AIME
22 'Liquid metals and solidification': 1958, Cleveland, Ohio, American Society for Metals
23 W. C. WINEGARD: *Metall. Rev.*, 1961, **6**, 53
24 B. CHALMERS: 'Principles of solidification', 1964, New York, Wiley
25 'The solidification of metals', 370, 1968, London, The Iron and Steel Institute
26 F. L. VERSNYDER AND M. E. SHANK: *Mater. Sci. Eng.*, 1970, **6**, 213
27 M. C. FLEMINGS: 'Solidification processing', 1974, New York, McGraw-Hill
28 M. C. FLEMINGS: *Metall. Trans.*, 1974, **5**, 2 121
29 Reports 4–9, Committee on Heterogeneity of Steel Ingots, 1932–1939, London, The Iron and Steel Institute
30 W. G. PFANN: 'Zone melting', (1 ed.); 1958, New York, Wiley
31 M. C. FLEMINGS: *Foundry*, 1963, March **91**, 72; April, 69
32 R. E. SPEAR AND G. R. GARDNER: *Trans. Amer. Foundrymen's Soc.*, 1963, **71**, 209
33 P. J. AHEARN *et al.*: *ibid.*, 1962, **70**, 1 154
34 T. Z. KATTAMIS AND M. C. FLEMINGS: *Trans. Met. Soc. AIME*, 1965, **233**, 992
35 T. F. BOWER *et al.*: *ibid.*, 1966, **236**, 624
36 P. J. AHEARN AND M. C. FLEMINGS: *ibid.*, 1967, **239**, 1 590
37 R. MEHRABIAN AND M. C. FLEMINGS: 'Solidification', 311, 1971, Cleveland, Ohio, American Society for Metals
38 F. WEINBERG AND E. TEGHTSOONIAN: *Metall. Trans.*, 1972, **3**, 93
39 T. Z. KATTAMIS *et al.*: *Trans. Met. Soc. AIME*, 1967, **239**, 1 504
40 T. Z. KATTAMIS *et al.*: *J. Inst. Met.*, 1967, **95**, 343
41 K. P. YOUNG AND D. H. KIRKWOOD: *Metall. Trans.*, 1975, **6**, 197
42 R. MEHRABIAN AND M. C. FLEMINGS: *ibid.*, 1970, 455
43 E. A. FEEST: *J. Inst. Met.*, 1973, **101**, 279
44 K. A. JACKSON *et al.*: *Trans. Met. Soc. AIME*, 1966, **236**, 149
45 G. S. COLE AND G. F. BOLLING: *ibid.*, 1966, **236**, 1 366
46 G. S. COLE: 'Solidification', 201, 1971, Cleveland, Ohio, American Society for Metals
47 R. T. SOUTHIN: *Trans. Met. Soc. AIME*, 1967, **236**, 220
48 F. WEINBERG: *Metall. Trans.*, 1975, **6**, 1971
49 J. S. KIRKALDY AND W. V. YOUDELIS: *Trans. Met. Soc. AIME*, 1958, **212**, 833
50 M. C. FLEMINGS AND G. E. NEREO: *Trans. Met. Soc. AIME*, 1967, **239**, 1 449
51 M. C. FLEMINGS *et al.*: *ibid.*, 1968, **242**, 41
52 M. C. FLEMINGS AND G. E. NEREO: *ibid.*, 1968, **242**, 50
53 F. KAEMPFFER AND F. WEINBERG: *Metall. Trans.*, 1971, **2**, 2 477
54 L. NORTHCOTT: *J. Inst. Met.*, 1946, **72**, 31
55 D. E. ADAMS: *ibid.*, 1948–49, **75**, 809
56 R. J. MCDONALD AND J. D. HUNT: *Trans. Met. Soc. AIME*, 1969, **245**, 1 993
57 R. J. MCDONALD AND J. D. HUNT: *Metall. Trans.*, 1970, **1**, 1 787
58 N. STREAT AND F. WEINBERG: *ibid.*, 1972, **3**, 3 181
59 D. J. HEBDITCH AND J. D. HUNT: *ibid.*, 1973, **4**, 2 008
60 N. STREAT AND F. WEINBERG: *ibid.*, 1974, **5**, 2 539
61 D. J. HEBDITCH AND J. D. HUNT: *ibid.*, 1974, **5**, 1 557
62 A. F. GIAMEI AND B. H. KEAR: *ibid.*, 1970, **1**, 2 185
63 S. M. COPLEY *et al.*: *ibid.*, 1970, **1**, 2 193
64 R. MEHRABIAN *et al.*: *ibid.*, 1970, **1**, 1 209
65 R. MEHRABIAN *et al.*: *ibid.*, 1970, **1**, 3 238
66 D. APELIAN *et al.*: *ibid.*, 1974, **5**, 2 533
67 S. N. SINGH *et al.*: *ibid.*, 1970, **1**, 1 383
68 G. F. BOLLING: 'Solidification', 341, 1971, Cleveland, Ohio, American Society for Metals
69 G. S. COLE: *Metall. Trans.*, 1971, **2**, 357
70 R. MEHRABIAN *et al.*: *ibid.*, 1974, **5**, 785
71 A. F. CIBULA: *Foundry Trade J.*, 1967, **122**, 337
72 T. S. PIWONKA AND M. C. FLEMINGS: *Trans. Met. Soc. AIME*, 1966, **236**, 1 157
73 J. CAMPBELL: *Cast Metals Res. J.*, 1969, **5**, (1), 1–8
74 J. CAMPBELL: *Brit. Foundryman*, 1969, **62**, 147
75 S. J. WALKER: *Cast Metals Res. J.*, 1971, **7**, (1), 13
76 R. F. POLICH AND M. C. FLEMINGS: *Trans. Amer. Foundrymen's Soc.*, 1965, **13**, 28
77 Report of T.S. 63, *Brit. Foundryman*, 1971
78 P. J. AHEARN AND F. C. QUIGLEY: *Trans. Amer. Foundrymen's Soc.*, 1964, **72**, 435
79 A. CIBULA: 'Gases and metals', 23, 1970, London, Butterworths
80 R. C. COBLE AND M. C. FLEMINGS: *Metall. Trans.*, 1971, **2**, 409
81 W. B. EISEN AND A. CAMPAGNA: *ibid.*, 1970, **1**, 849
82 R. DURHAM AND J. T. BERRY: *Trans. Amer. Foundrymen's Soc.*, 1974, **82**, 101
83 R. E. MARRONE *et al.*: *Cast Metals Res. J.*, 1970, **6**, (4), 184
84 G. SCIAMA: *ibid.*, 1972, **8**, (4), 145
85 R. D. PEHLKE *et al.*: *ibid.*, 1973, **9**, (2), 49
86 V. DE L. DAVIES: *ibid.*, 1975, **11**, (2), 33
87 V. DE L. DAVIES *et al.*: *Brit. Foundryman*, 1973, **66**, 305

Observations on the nature and extent of gravitational interdendritic fluid flow

K. M. Fisher and J. D. Hunt

The progress of macrosegregation in model castings has been followed using a quenching technique. Ingots of Sn-5Pb were grown under identical conditions and quenched after different lengths of time. Samples were taken from identical positions in each casting for composition analysis. The reproducibility of the method was confirmed by comparing the solute distributions determined in three castings quenched after similar times of growth. The observed macrosegregation was consistent with the operation of the gravitational interdendritic fluid-flow mechanism. A picture of the way in which composition changed with time at different positions in the castings was built up. This showed that the composition at a given position fell rapidly in the first 2 or 3 min after the dendrite tip front had passed the position. It also showed that there was a more gradual, yet significant, fall in composition up to 9 or 10 min after the front had passed a position. These time values corresponded to distances of 1 and 4 cm into the semi-solid region; the liquid volume fraction values at these positions were calculated to have been ∼0·70 and 0·30, respectively.

The authors are in the Department of Metallurgy and Science of Materials, University of Oxford

Fluid flow in the semi-solid zone due to gravitational forces has been recognized as a mechanism for macrosegregation for a number of years.[1] Although the operation of this mechanism has been clearly established,[1-5] little has been done to discover where in the semi-solid the majority of flow occurs, or what magnitudes of segregation can be caused by fluid flow at different points in the semi-solid. Hebditch and Hunt[5] used a model casting system in order to grow a number of thin section Sn–Pb, Pb–Sn, and Sn–Zn ingots, and study the macrosegregation present in them. The system allowed the growth processes in the ingots to be terminated by quenching at any stage. They inscribed a fine-scale grid across the dendrite-tip interface in several of their quenched ingots, and took samples from each grid point for composition analysis. Their results showed that the solute composition gradually decreased with distance behind the interface, up to a distance of about 2 cm into the semi-solid.

This fine-scale work was later improved by the present authors.[6] The same basic model casting system was used, and Sn–Pb ingots of various alloy compositions (0·5, 2, and 5 wt-% Pb) were quenched during growth. Pieces were cut from these ingots across the dendrite tip interface, as shown in Fig. 1a. Skims were then taken off the end of each specimen parallel to the plane of the interface at 0·5 mm intervals (Fig. 1b). The swarf from each

skim was collected and analysed. Figure 1c shows the composition v. position plots for two specimens cut from a Sn–5 Pb casting. Similar results were obtained in specimens cut from Sn–Pb and Sn–0·5 Pb castings. The graphs clearly indicate that fluid flow in the region immediately behind the interface (up to about 6 mm behind in these Sn–5 Pb ingot specimens) has a considerable effect on the macrosegregation of solute. The technique allowed observations on the structure of the semi-solid to be made between skims. These showed that the composition began to fall only after the dendrite tips could be seen on the skimmed section, and also that the region of rapid composition change corresponded to that in which large interdendritic liquid channels existed.

By means of these two experiments[5,6] it was possible to determine the composition v. position pattern in a single ingot quenched after a fixed time of growth. This is not the best way of observing where fluid flow is important in causing macrosegregation. Since the growth process in the ingots examined was not at steady state, no information could be obtained from the results concerning any subsequent variations in composition at a given position. Therefore, although both experiments indicate that an important region for fluid flow lies just behind the interface, they give no indication of whether or not flow in other parts of the semi-solid is important.

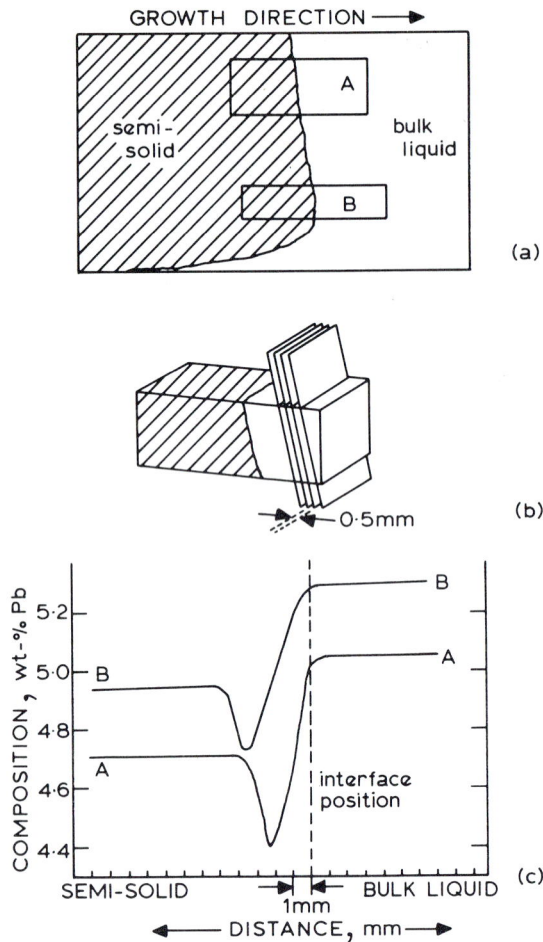

GROWTH DIRECTION ⟶

(a)

(b)

(c)

a positions from which specimens were cut; *b* sampling technique; *c* composition *v.* position profile determined in each specimen

1   Experiment performed on model casting[6]

The best method of determining the relationship between flow and segregation in different parts of the semi-solid is to study the variation in composition of a given small volume in an ingot with time. This variation can then be tied in with the growth process in the ingot to give the required information. Unfortunately, there are substantial experimental difficulties involved in measuring the changing composition of a given volume in an ingot during solidification. It was therefore decided that a quenching technique should be used. A number of ingots were grown under identical conditions and quenched after different times of growth. Samples were then taken from identical positions in each ingot and their compositions determined. In this way composition *v.* time plots at fixed positions in the ingots could be produced.

EXPERIMENTAL PROCEDURE

The model ingot system mentioned previously[5,6] was used. In order to grow castings reproducibly many improvements and modifications were made to the basic system; the furnace is shown in Fig. 2. There was a central slot around which the main heating element was wound. A second heater was wound inside an insert which fitted up into the bottom of the slot and which could be removed during quenching. The thin-section

2   Slot furnace

mould shown in Fig. 3 fitted down inside the slot furnace. The mould was kept well insulated from its surroundings by using a ceramic fibre blanket.

Alloy specimens were grown unidirectionally away from the vertical copper chill. The ingots produced were 108 mm long, 65 mm tall, and 13 mm thick. The stainless steel box prevented metal leakage, and the top asbestos insert helped to stop oxidation of the casting surface as well as ensuring a constant mould volume. A third heating element was wound around the copper cooling pipes leading from the chill block in order to stop these acting as a heat sink; this was switched off during the freezing process.

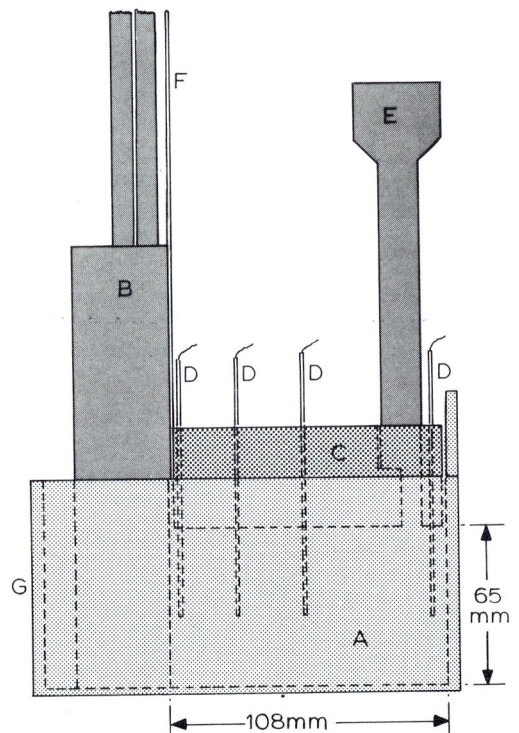

A mould cavity for alloy specimen; B copper chill; C top asbestos insert; D sample thermocouples; E copper filling funnel; F stainless steel tube for nucleant insertion; G stainless steel box mould

3   Schematic representation of thin section box mould

The run procedure is given below. The components of the mould were painted with colloidal graphite, and the mould was then assembled and placed into the furnace slot; this was then well packed with insulating material. The three heaters were switched on and the mould heated above the alloy liquidus temperature. A known quantity of carefully homogenized, superheated alloy was then poured into the mould cavity down the funnel, and the funnel was removed. After 2–3 h the majority of the superheat had been lost, and the three heaters were then individually adjusted until the temperatures of the mould and furnace enclosure were about 2°C above the alloy liquidus temperature. Monitoring thermocouples had been positioned around the mould so that an isothermal temperature enclosure could be obtained. Four sample thermocouples within the mould registered the temperature of the molten metal at fixed distances from the chill (5, 28, 55, 100 mm). When the temperature had stabilized the top heater was switched off and air was passed through the cooling block. The air flowrate was monitored on a flowmeter and kept identical for each casting. The melt was nucleated by passing a length of cold tin wire down a stainless steel tube fixed to the cooling block. The progress of the dendrite tip front could be followed using the sample thermocouples. When the interface passed a thermocouple the temperature it registered began to fall. Estimates of growth rates and temperature gradients were therefore obtained for each casting. Growth was terminated by removing the bottom heater insert, and pushing the mould down through the furnace slot into a quench bucket of water below.

After quenching, each ingot was weighed and its dimensions measured, in order to ensure that there was uniformity between ingots. A skim of exactly 2·5 mm was removed from one of the large faces of the ingot using a flycutter. This face was then ground on 600 grade SiC paper and etched for about 10 min in aqua regia in order to show the macrostructure.

Samples were taken from identical positions in each ingot using a 1 mm thick circular saw. Cuts were made through the full thickness of the ingot at fixed distances from its chill end and at fixed heights from its bottom face. Figure 4 shows the appearance of an ingot after sampling, and the positions of the cuts. Great care was taken to collect every piece of swarf removed from each position. The composition value of each sample was then determined by atomic absorption spectrophotometry. The analysis technique was developed and refined until accuracies of around ±1% of the composition values could be achieved.

RESULTS

Figure 5 shows the macrostructure of a Sn–5 Pb ingot quenched 15 min 24 s after cooling began. The interface shows a characteristic cut-back towards the bottom of the ingot, which is the result of macrosegregation caused by convective interdendritic fluid flow. Solute atoms of Pb are rejected by the growing dendrites, causing the interdendritic liquid to become enriched in solute. This liquid is denser than the bulk liquid since it is colder and contains more Pb.[7] Therefore, it has a tendency to move downwards. Consequently, in order to complete the convection cycle, liquid must move towards the chill at the top of the casting and away from the chill at the bottom; this is illustrated schematically in Fig. 6.

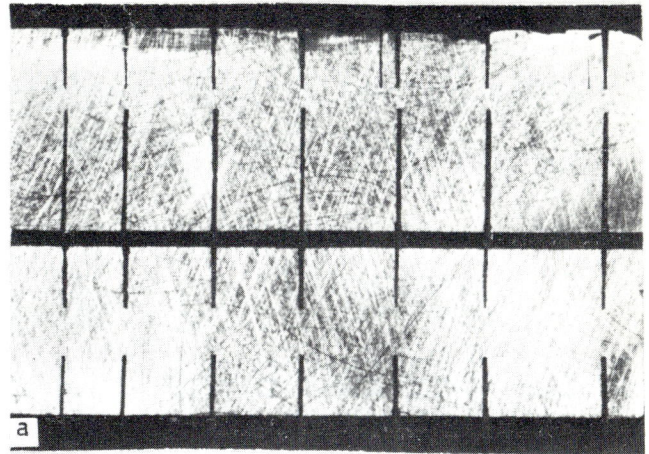

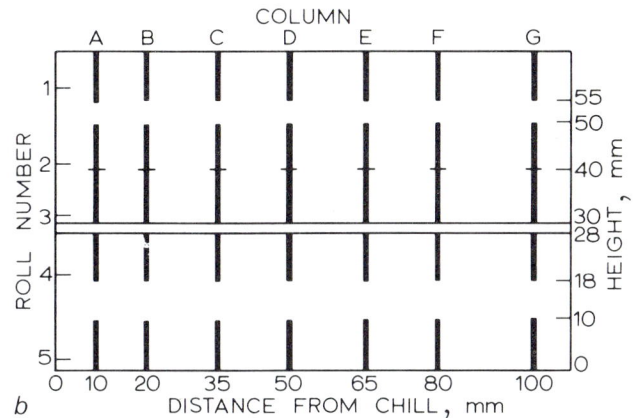

4 *a* Appearance of a model ingot after sampling, and *b* position of cuts

5 Typical macrostructure of a Sn–5 Pb ingot, quenched after 15 min 24 s; growth from left to right. The dendrite tip interface is clearly visible, and is cut-back at the bottom of the ingot; channels can be seen at the bottom of the semi-solid region

Liquid flow towards the chill leads to an increase in the amount of solid being deposited on the dendrites, and so causes negative segregation.[1,4] Liquid flow away from the chill, up the temperature gradient, should ideally lead to a uniform partial melting of the surrounding dendrites, and hence to positive segregation. In practice, however, such flow is unstable and preferential channels for liquid flow tend to form.[1–6] Such channels can be

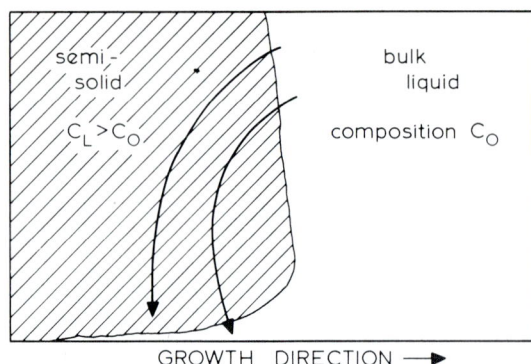

6   Schematic illustration of the expected fluid flow cycle in Sn–5 Pb ingots

seen towards the bottom of the semi-solid region of the ingot shown in Fig. 5. Non-equilibrium liquid rich in Pb can emerge from the semi-solid region through these channels, and because it is denser than the bulk liquid it tends to remain at the bottom of the ingot. The liquidus temperature of material in this region is therefore lower, and so the dendrite tip interface becomes cut back.

In order to test the reproducibility of the model casting system, three ingots of Sn–5 Pb were grown under identical conditions for similar lengths of time: 15 min 19 s, 15 min 24 s, and 15 min 28 s. Figure 7 shows the composition v. position plot determined for

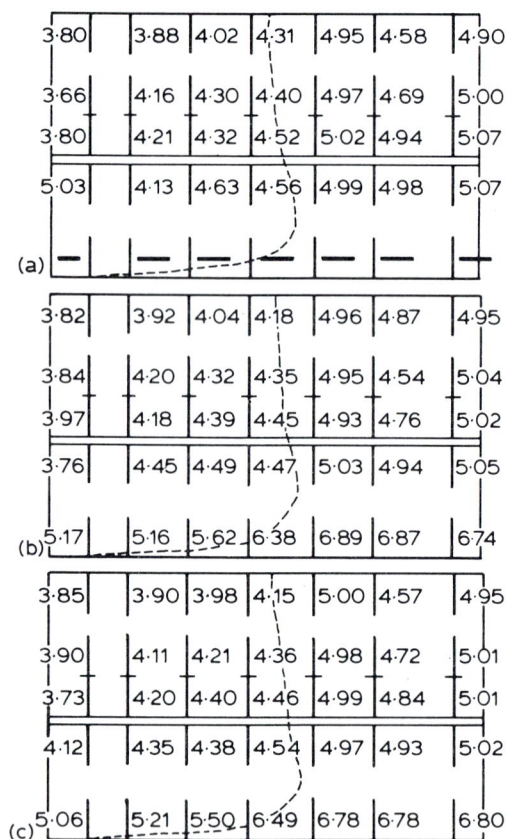

a 15 min 19 s; *b* 15 min 24 s; *c* 15 min 28 s

7   Composition v. position diagrams for three ingots of Sn–5 Pb quenched after similar times of growth. The dendrite interface position in each ingot is shown by a dashed line; composition values are given in wt-% Pb and are correct to within ±0·05 wt-% Pb

each ingot. The composition values quoted are considered to be correct to within ±0·05 Pb. A number of important observations can be made from the results and these are summarized below:

(i) at the majority of positions in the top portions of the three ingots the composition values are remarkably consistent
(ii) the agreement between values in the bottom portions is less convincing; this is thought to be due to the instability of the fluid flow in these regions, which leads to channel formation (*see* Figs 5 and 6); the positions of these channels within the ingot depend on the original microstructure, which obviously will not be identical in different ingots; consequently, these solute-enriched channels may intersect a sampling position in one ingot while not intersecting it in another
(iii) the build-up of Pb-rich liquid at the bottom of the ingot, which gives rise to the cut-back in the interface, can be clearly seen; there is very little solute enrichment in the remainder of the bulk liquid
(iv) the dendrite tip interfaces found in the three ingots are shown in Fig. 7; these are very similar both in position and shape for all three ingots
(v) there is a large fall in composition across the interface, followed by a gradual decrease further back into the semi-solid
(vi) the apparent low composition values of bulk liquid at positions 1F, 2F, and 3F are thought to have been caused by feeding of liquid from this region during quenching.

Having established that reproducible castings could be obtained from the model system, two further ingots were grown for the longer time of 22 min 54 s before quenching. Composition analysis performed on these ingots confirmed that the technique gave reproducible results and showed that there was a significant change in the composition values at many of the sampling positions. A series of ingots quenched after different times of growth was therefore assembled. The relevant information concerning these ingots is given in Table 1.

The progress of the macroscopic dendrite tip interface in the ingots is shown in Fig. 8. It can be seen that the cut-back became increasingly pronounced as growth

Table 1   Characteristics of ingots quenched after different growth periods

Run number	Quench time, min, s	Casting weight, g	Melt composition, wt-% Pb	Initial holding superheat, °C
17	8, 32	643	4·95	2·5
21	10, 40	639	4·99	2·25
24	15, 19	639	5·02	2·0
11	15, 24	642	—	4·0
12	15, 28	645	—	3·5
23	18, 39	643	5·07	1·5
15	22, 54	645	4·99	2·0
16	22, 55	638	4·96	2·0
20	30, 35	643	4·98	2·0
22	75, 6	643	4·97	2·5

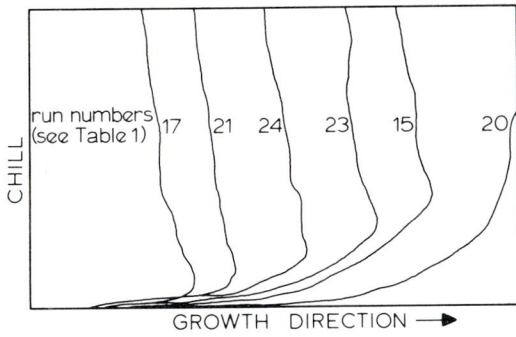

8 Progress of the macroscopic dendrite tip interface in the ingots with time; ingot number is given for each interface position (*see* Table 1 for times of growth)

proceeded. Figure 9 indicates how the composition pattern in the ingots changed with time of growth. Composition *v.* position diagrams for three ingots quenched after 10 min 40 s, 22 min 54 s and 75 min 6 s are shown. The continued accumulation of solute at the bottom of the castings can be clearly seen. The sharp fall in composition after the interface has passed a sampling position is also clearly shown, even at positions near the bottom of the ingots. Figures 10 and 11 show the composition *v.* time plots obtained at certain of the standard positions in the ingots. The dotted lines on the

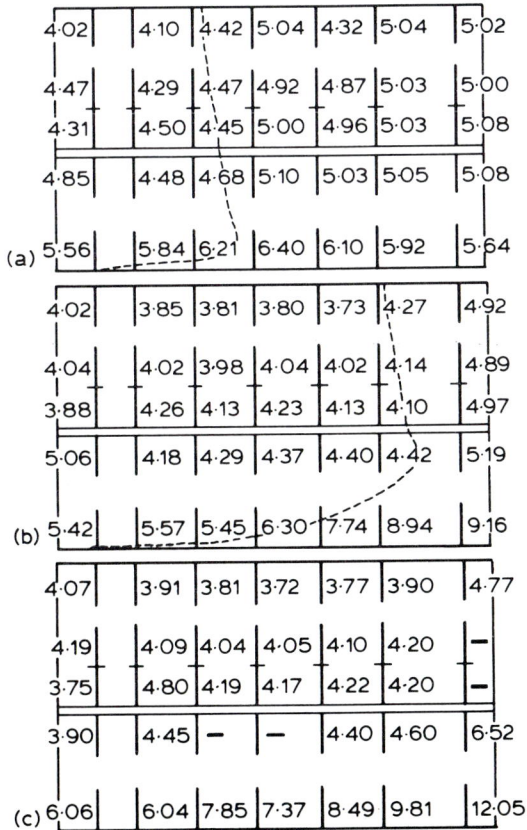

10 Composition *v.* time plots for the standard positions in row 2, columns A–G. Calculated times at which dendrite interface passed the positions are shown by vertical dashed lines; times are taken from the start of cooling

graphs indicate the times at which the interface passed each position. These times were calculated from the growth velocity data obtained from the thermocouples in the ingots. The important features of the curves are:

(i) at every position the composition fell rapidly in the first 2 min after the front of the dendrite tip had passed; the magnitude of this fall was, at most positions, about 10% of the nominal composition value

(ii) following this initial sharp fall there was a more gradual decrease in composition over a further time period of 7 to 8 min (i.e. about 9–10 min after the front had passed); after this time the composition remained roughly constant for the remainder of the freezing process

(iii) the composition data determined at position 3A, and to a lesser extent 2A, (*see* Fig. 4) exhibit

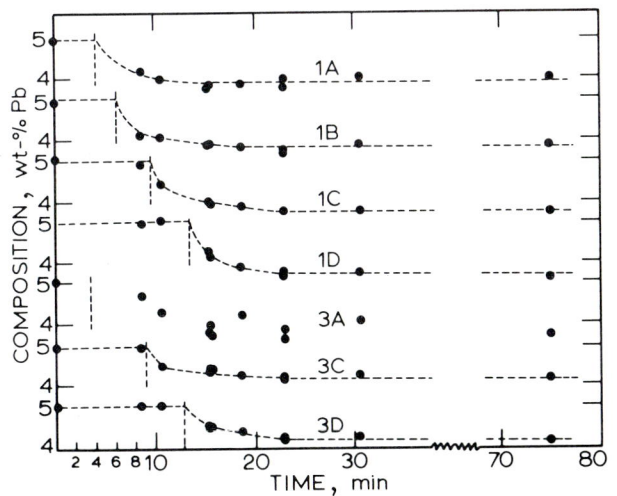

a 10 min 40 s; *b* 22 min 54 s; *c* 75 min 6 s

9 Composition *v.* position diagrams for three ingots of Sn–5 Pb quenched after various times; dendrite interface shown by dashed line; composition values in wt-% Pb, correction to ±0·05 wt-% Pb

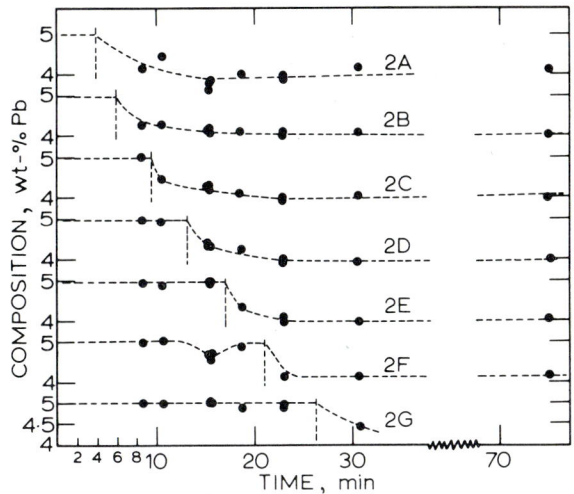

11 Composition *v.* time plots for the standard positions 1A, 1B, 1C, 1D, 3A, 3C, and 3D. Calculated times at which dendrite interface passed the positions are shown by vertical dashed lines; times are taken from the start of cooling

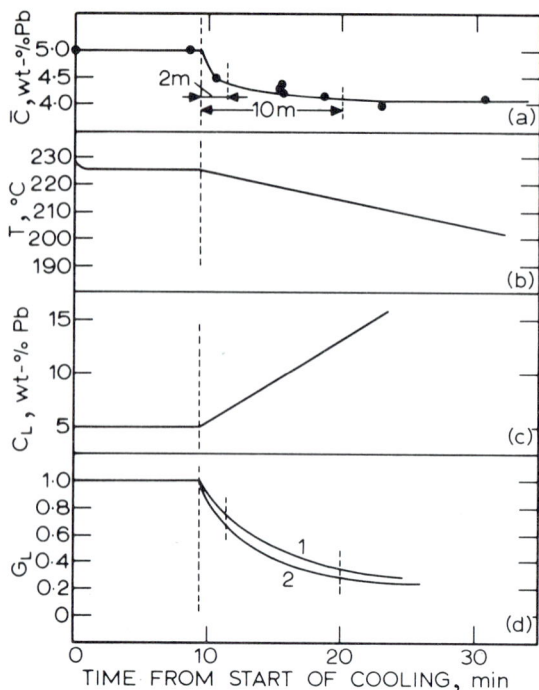

a average solute composition $\bar{C}$ v. t; *b* temperature T v. t; *c* liquid composition c_L v. t; *d* liquid volume fraction G_L v. t, calculated by two methods

12 Change of certain variables with time at position 2C

inconsistencies; the composition seems to have risen and fallen with time; this type of behaviour was also found to occur at the majority of positions in rows 4 and 5; as explained above, it is thought to be due to the non-reproducibility of channel formation within the ingots. Channels were definitely visible in the macrostructures of certain ingots in the region of positions 2A and 3A (*see* Fig. 4).

The rate of advance of the dendrite tip front in the ingots was measured by the sample thermocouples to be 7×10^{-3} cm/s. The time periods given in points 1 and 2 above therefore correspond to distances behind the interface of 1 and 4 cm, respectively.

Figure 12 shows the variation of several parameters with time at position 2C in the ingots; Figure 12*a* gives the experimentally determined average composition ($\bar{C}$) variation. Figure 12*b* shows the temperature variation calculated from the known temperature gradients in the ingots, and Fig. 12*c* shows the variation in liquid composition (C_L) calculated using the T v. t curve, together with the liquidus slope of the equilibrium phase diagram. Finally, Fig. 12*d* gives the variation in volume fraction liquid (G_L), for which two curves are shown. Curve 1 was drawn by substituting the C_L values from Fig. 12*c* into the Scheil equation:

$$G_L = \left(\frac{C_0}{C_L}\right)\frac{1}{1-k} \qquad (1)$$

where C_0 is the nominal solute composition, 5 Pb, and k is the equilibrium distribution coefficient. This gives an overestimate of G_L at all times, since the Scheil equation does not allow for any fluid flow. Curve 2 was drawn by substituting the values of $\bar{C}$ and C_L from Figs 12*a* and 12*c* into the equation:

$$\bar{C} = \bar{C}_S(1-f_L) + C_L f_L \qquad (2)$$

where f_L is the weight fraction of the liquid, and the average solid composition $\bar{C}_S$ is assumed to be 0.5 wt-% Pb. From these curves it can be seen that fluid flow continued to lead to macrosegregation until the volume fraction liquid fell below a value of about 0.30. The experiment has, therefore, shown that significant amounts of macrosegregation can be caused by gravitational fluid flow occurring deep into the semi-solid region, and not solely by flow close to the dendrite tip front.

REFERENCES

1 R. J. MCDONALD AND J. D. HUNT: *Trans. AIME*, 1969, **245**, 1 993
2 R. J. MCDONALD AND J. D. HUNT: *Metall. Trans.*, 1970, **1**, 1 787
3 S. M. COPLEY *et al.*: *ibid.*, 1970, **1**, 2 193
4 R. MEHRABIAN *et al.*: *ibid.*, 1970, **1**, 1 209
5 D. J. HEBDITCH AND J. D. HUNT: *ibid.*, 1974, **5**, 1 557
6 K. M. FISHER: Part II Thesis, Oxford University, 1974
7 H. R. THRESH *et al.*: *Trans. AIME*, 1968, **242**, 819

Effect of fluid flow on macrosegregation in steel ingots

T. Takahashi, K. Ichikawa, and M. Kudou

It is possible to obtain a quantitative understanding of macrosegregation in a steel ingot if the flow velocity of bulk liquid during solidification can be determined. The flow velocity of the bulk liquid can be determined by applying Taylor's vortex flow to the solidification of carbon steel. As a result, the following relations are introduced: $Ke = 1 - (1 - k_0) \, S_h$ and $U/V = 7500 \, S_h/(1 - S_h)$ where Ke is the effective distribution coefficient, defined as $\overline{Cp}/C_{L_0}$, $\overline{Cp}$ is the average solute concentration of the dendrite and the interdendritic region, C_{L_0} is the solute concentration in the bulk liquid, k_0 is the equilibrium distribution coefficient, S_h is the fraction solid, depending on the washing depth in the solidification zone, which is composed of solid and liquid, U is the flow velocity of the bulk liquid, and V is the solidification rate. The flow velocity of the bulk liquid is also evaluated by both the deflection angle of dendrites and the solidification rate. Furthermore, the macrosegregation Ke is theoretically explained on the basis of mass transfer with turbulent mixing in the solidification zone.

The authors are in the Department of Metallurgical Engineering, Hokkaido University, Sapporo, Japan

Fluid flow during solidification strongly influences micro- and macrosegregation, the stability of growing crystals, and the formation and floating separation of non-metallic inclusions. In the present investigation Taylor's vortex flow is applied to the solidification of carbon steel so that the effect of the flow of bulk liquid on macrosegregation behaviour and solidification morphology is examined in relation to the solidification zone, which is composed of solid and liquid.

EXPERIMENTAL PROCEDURE

When only the inner one of two concentric cylinders is rotated, Taylor's vortex flow[1] is initiated in the bulk liquid between them. The temperature and concentration of bulk liquid become uniform through the vortex flow. From Taylor's experimental results, it is found that the flow velocity of the bulk liquid U is determined by the peripheral velocity of the inner cylinder U_s: U is equal to about half U_s.[2,3] Using this method, the flow velocity can be changed by controlling the rate of rotation. Figure 1 shows the experimental apparatus. Solidification proceeds from the inner to the outer side by water cooling of the inner cylinder, so that the flow velocity of the bulk liquid can be quantitatively obtained from the peripheral velocity at the solidification front.

EXPERIMENTAL RESULTS
Deflection angle of dendrites

The primary dendrites in the columnar zone are deflected against the flow direction of the bulk liquid. Figure 2 shows the relationship between the deflection angle of the dendrites θ and the solidification rate V at different flow velocities of the bulk liquid U. The deflection angle increases with increasing flow velocity and decreasing solidification rate. The relationship is shown by the following equation,

$$\theta = 22 \cdot 49 U^{-1 \cdot 77 \times 10^{-1}} \log 3 \cdot 72 \times 10^{-3} U^{2 \cdot 08}/V \qquad (1)$$

If the deflection angle of the dendrites and the solidification rate are known, the flow velocity of the bulk liquid during solidification can be evaluated.

CONCENTRATION DISTRIBUTION OF SOLUTE

In solidification without rotation the concentrations of C, Si, Mn, P, and S in the solidified shell remain unchanged regardless of position. In solidification with rotation, the solute concentration in the solid decreases with increasing flow velocity of the bulk liquid and with decreasing solidification rate. This tendency implies that the solute-rich liquid between the dendrites within the solidification zone is washed out by the flow of bulk liquid.

332 Takahashi et al.

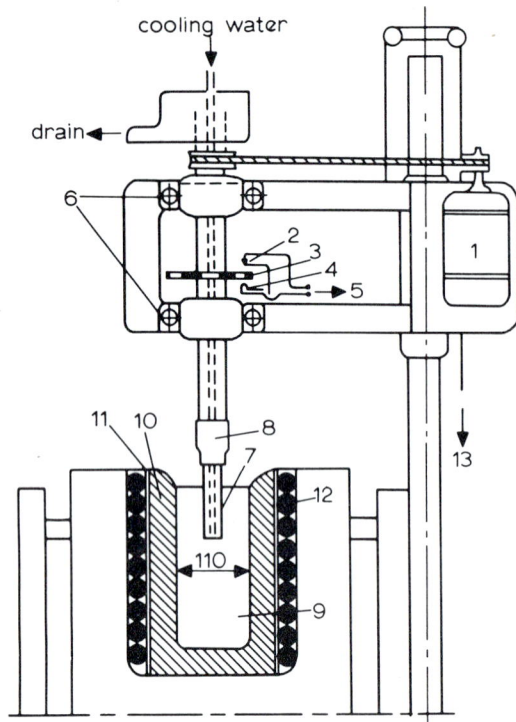

1 variable motor; 2 photodiode; 3 rotating plate; 4 light source; 5 digital counter; 6 pilobrock; 7 cooling pipe; 8 socket; 9 molten steel; 10 magnesia clinker; 11 coil cement; 12 high-frequency coil; 13 balance weight

1 Schematic illustration of experimental apparatus

EXPERIMENTAL CONSIDERATIONS RELATING TO THE EFFECTIVE DISTRIBUTION COEFFICIENT
Washing effect of the flow of bulk liquid in the solidification zone

The solidification zone is divided into p and q zones according to the morphology of coexistence of the solid and liquid. In the p zone, that having the higher fraction of solid, the liquid is entrapped by the solid. The average solute concentration of the p zone thus remains unchanged after solidification, and the p zone can be

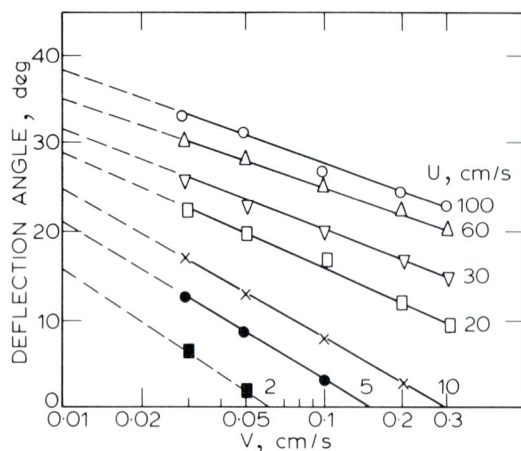

2 Relationship between deflection angle of dendrites θ and solidification rate V at different flow velocities of bulk liquid U

dealt with as a virtual solid. The q zone, which has the lower fraction of solid, can be subdivided into q_1 and q_2 zones. The q_1 zone contains a dendritic network, and the liquid in it can flow through the dendrite arm spacings. The q_2 zone is the so-called 'pasty' zone containing free dendrites. Only the q zone will be considered for macrosegregation as affected by the flow of bulk liquid.

Assuming that the fraction solid is a linear function of the distance in the solidification zone, the washing depth can be converted into the fraction solid. If the solidification zone is washed until the fraction solid reaches some value S_h, the effective distribution coefficient Ke is expressed approximately as follows,

$$Ke = \overline{Cp}/C_{L_0} = 1 - (1-k_0)S_h \qquad (2)$$

where $\overline{Cp}$ is the average solute concentration of the dendrite and the interdendritic region, C_{L_0} is the solute concentration in the bulk liquid, and k_0 is the equilibrium distribution coefficient. In equation (2), Ke decreases with increasing S_h. The maximum washing depth is reached at the boundary between the p and q zones. Inserting S_h^*, as the fraction solid at the boundary, into equation (2) gives the minimum effective distribution coefficient. The values of the minimum effective distribution coefficient were calculated by substituting $0.67^{2,3}$ for S_h^*, and the value of the equilibrium distribution coefficient for each alloying element for k_0, into equation (2): these calculated values were compared with observed values in the rim zone of a rimming ingot with violent flow by rimming action. It was seen that the calculated values agreed well with the minima observed.[3]

Solidification rate, flow velocity, and effective distribution coefficient

The fraction solid corresponding to the washing depth S_h is controlled by the solidification rate V and the flow velocity of the bulk liquid U. Using equation (2), the relationships between S_h, U, and V for sulphur are obtained empirically as follows:

$$U/V = 7\,500S_h/(1-S_h) \qquad (3)$$

where the equilibrium distribution coefficient of sulphur is 0.02.[4] Considering the role of the washing effect, the relationship given in equation (3) must be realized even for the other elements coexisting with sulphur.

From equations (2) and (3) the effective distribution coefficient Ke for $0 \leqslant S_h < 0.67$ is expressed as follows:

$$Ke = 1 - 1.33 \times 10^{-4}(1-k_0)(1-S_h)(U/V) \qquad (4)$$

Figure 3 shows the relationships between flow velocity, solidification rate, and the effective distribution coefficients of carbon and phosphorus, in solidification with rotation. The solid and dashed lines calculated from equation (2) at $S_h = S_h^*$ and equations (3) and (4) show good agreement with the observed values of carbon and phosphorus respectively. Reasonable results are obtained for other alloying elements as well as for C and P. The effective distribution coefficient for each alloying element decreases with increasing flow velocity and decreasing solidification rate. The equilibrium distribution coefficients of carbon and phosphorus are 0.34^5 and 0.20,[6] respectively. When U and V are constant, the smaller the equilibrium distribution coefficient, the smaller becomes the effective distribution coefficient.

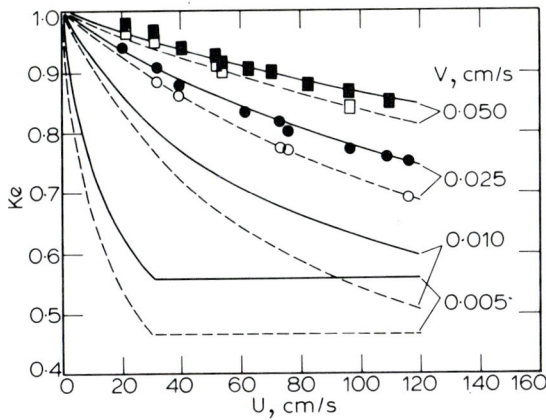

3 Comparison of the calculated and observed effective distribution coefficients of carbon and phosphorus Ke as dependent on flow velocity of bulk liquid U at different solidification rates V; the solid and dashed lines show calculated values of carbon and phosphorus respectively (●, ■ and ○, □ show the observed values of carbon and phosphorus respectively)

THEORETICAL EXAMINATION OF THE EFFECTIVE DISTRIBUTION COEFFICIENT

When turbulence is caused in the q zone by stirring, the quantity of solute which is transported per unit time by the turbulent mixing at $x = h$ is equivalent to the balance quantity of $V(C_{L_0} - \overline{Cp})$:

$$V(C_{L_0} - \overline{Cp}) = -(1 - S_h)A_D\left(\frac{dc}{dx}\right)_{x=h} \qquad (5)$$

where VC_{L_0} is the quantity transferred into the solidifying zone, $V\overline{Cp}$ is the quantity after solidification, A_D is the mixing diffusivity, and x is the distance along the direction of solidification.

Considering the solidification of the remaining liquid in the region without the flow of bulk liquid, the concentration gradient at $x = h$ is shown as follows:

$$\left(\frac{dc}{dx}\right)_{x=h} = -C_{L_0}(1 - k_0)/L(1 - S_h) \qquad (6)$$

where L is the thickness of the solidifying zone.

On the other hand, the mixing diffusivity A_D which is in proportion to both the turbulent flow velocity U,[7] and the mixing length Δh[8], for example the magnitude of the

vortex,[9] is represented by the following relationship:

$$A_D = B\Delta hU \qquad (7)$$

where B is an experimental constant. It is assumed that Δh is the spacing of liquid between the primary dendrites at $x = h$: $\Delta h = l(1 - S_h)$, where l is the primary dendrite arm spacing.

Substituting equations (6) and (7) into equation (5), the following relationship is obtained

$$Ke = 1 - B(l/L)(1 - S_h)(U/V) \qquad (8)$$

Equation (8) is based on the concept of mass transfer with turbulent mixing in the solidifying zone. The effective distribution coefficient in equation (8) must be equivalent to that in equation (4). As a result, $B(l/L)$ is equal to 1.33×10^{-2}. The ratio l/L remains unchanged at 1.65×10^{-2} except in the initial and the later stages of solidification in this investigation, since l increases with increasing L. Thus, the experimental constant B which is necessary to determine the mixing diffusivity is calculated to be 0.81×10^{-2}. This value is approximately equal to Kármán's value[10] of 1.19×10^{-2} in the field of hydrodynamics.

CONCLUSION

Using a modified Taylor's vortex method to measure the flow velocity of the bulk liquid, the mutual relationships between flow velocity, solidification rate, solute concentration distribution, and dendrite deflection angle are established, and the effect of fluid flow on macrosegregation is examined quantitatively in relation to the solidification zone.

REFERENCES

1 G. I. TAYLOR: *Proc. R. Soc.*, 1935, **151**, 494
2 T. TAKAHASHI AND I. HAGIWARA: *J. Jpn. Inst. Met.*, 1965, **29**, (12), 1 152
3 T. TAKAHASHI et al.: *Trans. Iron Steel. Inst. Jpn.*, 1972, **12**, 412
4 C. E. SIMS: 'Electric furnace steelmaking, Pt. 2, 1 ed., 99, 1963, New York, John Wiley & Sons
5 T. WADA AND H. WADA: Preprint Abstract at 61st Meeting of Japan Institute of Metals, 1967, **53**, 174
6 R. E. SMITH AND J. L. RUTHERFORD: *J. Met.*, 1957, **9**, 478
7 H. SCHLICHTING: 'Boundary layer theory', 4 ed., 476, 1960, New York, McGraw-Hill
8 H. SCHLICHTING: *ibid.*, 481
9 V. G. LEVICH: 'Physicochemical hydrodynamics', 1 ed., 172, 1962, Englewood Cliffs, Prentice-Hall
10 E. R. G. ECKERT AND R. M. DRAKE, JR: 'Heat and mass transfer', 2 ed., 226, 1959, New York, McGraw-Hill

Solidification of high-carbon steel ingots

F. Weinberg, J. Lait, and R. Pugh

Fluid flow, carbon segregation, and cast structure were examined in 0·7% C steel ingots of 550 kg and 2500 kg sizes, cast in big-end-up and big-end-down moulds. The fluid flow and cast structures were determined from autoradiographs of the sectioned ingots, to which Au^{198} was added during solidification. It was found that the fluid flow differed for the two mould configurations considered, and in some respects differed from that in previous reports. No significant interaction between liquid in the pool and in the solid–liquid region was observed. A regular array of A-segregate pipes was observed, rich in Au^{198} and C, which is attributed to upward flow of low density liquid. The A-segregate patterns were different in the two mould configurations considered. The major part of the macrosegregation of carbon was associated with the A-segregates.

F. Weinberg is with the Department of Metallurgy, University of British Columbia, Vancouver, and J. Lait and R. Pugh are with the Steel Company of Canada, Ontario, Canada

High-carbon steel ingots tend to have a high concentration of carbon in their upper parts. This is attributed to the upward flow of carbon-rich interdendritic liquid, owing to buoyancy forces resulting from temperature and composition gradients. The purpose of this investigation was to examine the fluid flow in high-carbon steel ingots during solidification for two mould geometries and to relate this to the observed macrosegregation of carbon in the ingots.

PROCEDURE

Three high-carbon steel ingots were cast at the Stelco Research Centre in Hamilton, Ontario, Canada. The size, pouring temperature, and composition of the killed steel are given in Table 1. All the ingots were hot-topped, using insulated sideboards, and covered with an exothermic topping compound. During solidification small pellets of radioactive gold attached to a steel rod were inserted rapidly into the centre of the melt, from the top, at various times during solidification. The pellets were 0·30 cm dia. and 0·65 cm long, and had an activity of approximately 100 mCi. The pellets melted within an estimated 20 s and the radioactive gold spread rapidly in the liquid pool. In addition, small pellets of lead containing radioactive gold were dropped down the side of the liquid pool at the same time as the gold pellet was added.

After the ingots had solidified and cooled, they were removed from the cast iron moulds and flame cut longitudinally and transversely. The sectioned surfaces were then milled flat and autoradiographed to show the distribution of radioactive gold in the plane of section. The same surfaces were also sulphur printed following the autoradiography.

The temperature distribution in the ingots during solidification was calculated using a finite-difference heat-transfer analysis. This analysis was carried out by

Table 1   Ingot size and composition

Test no.	Ingot weight, kg	Ingot size, cm	Temperature, °C		Composition, wt-%				
			Tap	Ladle	C	Mn	S	P	Si
1	550 BEU*	23×41×127	1610	1577	0·67	0·77	—	—	—
2	2720 BEU	41×64×168	1610	1571	0·68	0·73	0·023	0·010	0·66
3	2268 BED†	43×45×170	1632	1588	0·70	0·77	0·024	0·012	0·63

* Big end up.
† Big end down.

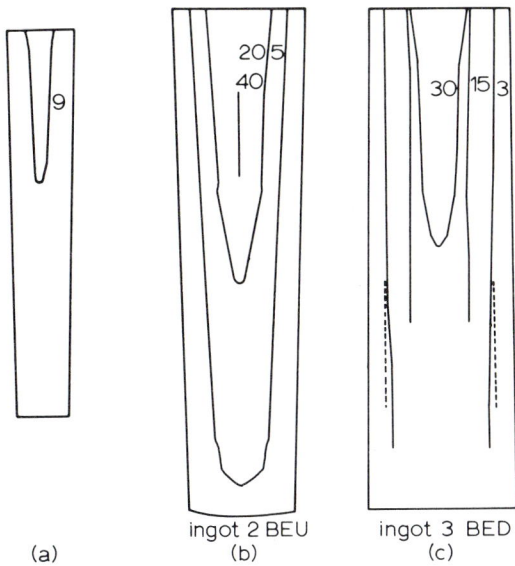

a, ingot 1; *b* ingot 2; *c* ingot 3

1 Contours of liquid pool at times (min) indicated after teeming

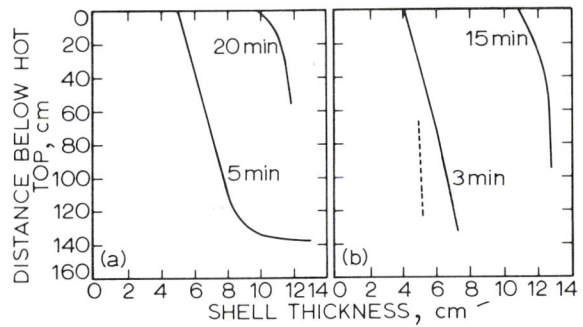

a ingot 2; *b* ingot 3

2 Shell thickness as function of distance below hot top at times (min) indicated after teeming: solid lines obtained from pool contours, dashed lines from lead addition

J. K. Brimacombe* at the Stelco Research Laboratory. Temperatures predicted by the analysis were verified at several points in an ingot by thermocouple measurements.

The carbon distributions in the ingots were measured from drillings taken from the billet, using standard analytical procedures.

RESULTS AND DISCUSSION
Fluid flow in the liquid pool

The distribution of the radioactive gold in the liquid pool indicates the extent of fluid flow in the pool at the time of addition.

For the small ingot 1, the gold addition was made 9 min after teeming. The autoradiographs showed that fluid flow was confined to a relatively narrow central zone in the upper half of the ingot at the time of addition. The contour of the zone is shown in Fig. 1*a*.

In the 3 t BEU ingot 2, the zone contours (Fig. 1*b*) indicate that extensive fluid flow was present in the liquid pool 5 min after teeming. In 20 min flow was confined to the central region of the upper half of the ingot, and after 40 min the flow had effectively stopped. The contours for the BED ingot 3 (Fig. 1*c*) show that more extensive flow occurs in the lower half of the ingot, 15 and 30 min after teeming, and that there is no clearly defined lower pool boundary.

The shell thickness delineated by the gold added to the pool increases with increasing distance down the ingot (*see* Fig. 2*a* and *b*). The lead additions containing radioactive gold, in ingot 3, delineated a constant narrower shell thickness, as shown in the dashed line of Fig. 2*b*. The lead additions produced small dark areas in the autoradiographs, caused, it is believed, by lead being caught by projecting dendrites along the interface. These results indicate that fluid flow in the liquid pool in the lower part of the ingot does not extend to the solid–liquid region.

* J. K. Brimacombe is in the Department of Metallurgy, University of British Columbia, Vancouver, Canada

The heat transfer analysis also indicates that there is little penetration of the solid–liquid region by liquid from the pool. The calculated temperature distributions during solidification of ingots 2 and 3, on a plane midway up the ingots, are shown in Fig. 3*a* and *b*. The thickness of the shell delineated by the radioactive gold added to the liquid pool is shown by the short vertical bars, for the times indicated. Comparing the position of the vertical bars with the corresponding temperature distribution at the time of addition, the bars are observed to be positioned well away from the solidus temperature and close to the liquidus temperature in all cases. In the vicinity of

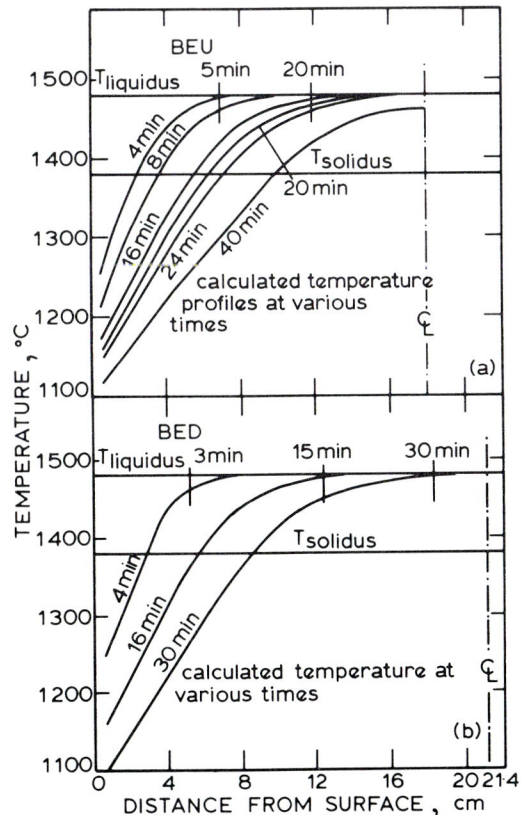

a, ingot 2; *b* ingot 3

3 Temperature distribution on horizontal plane midway up ingot at indicated times after teeming: vertical lines are interface positions determined from pool profiles

the liquidus, the calculated temperatures are very approximate. Accordingly the positioning of the vertical lines in Fig. 3*a* and *b*, inside the liquidus, is not considered significant. The calculated temperatures at the solidus are well defined.

Andrews and Gomer,[1] in their investigation of fluid flow in large castings, observed that the lower contour of the liquid pool exhibited upward streaks, particularly during the latter part of solidification. This suggested that small induced currents were present in the lower part of the ingot and rose slowly during solidification. This flow pattern was consistent with the model proposed by Blank and Pickering[2] to account for their observations of the distribution of inclusions in large castings. The present results did not exhibit any upward streaks or give any clear indication of the presence of induced currents in the lower half of the ingot. It is still possible, however, that induced currents at the bottom of the ingot were present, but did not interact with the general flow in the upper part of the billet and therefore did not pick up any of the radioactive gold. That two adjacent flows in a metal can exist with little interaction between the two was demonstrated by Stewart and Weinberg.[3]

Cast structure

Further evidence of extensive fluid flow in the liquid pool during the early part of solidification comes from observations of the growth direction of the columnar dendritic grains as observed in the sulphur prints of the ingot sections. In all the ingots, the columnar dendritic grains were observed to grow at an average angle of 20° to the horizontal, pointing upwards. Experiments on systems rotated during solidification have shown a change in growth direction with rotation, the dendrites pointing upstream.[4] The relationship between growth direction and flow velocity has not been established for steel. The present results indicate that a strong downward flow of liquid was present at the solid/liquid interface, at least in the early part of solidification.

The extent of the columnar region exhibiting the upward growth direction differed markedly for the two large billets (*see* hatched regions in Fig. 4*a* and *b*). The columnar grains extend to near the bottom of the ingot, and grow an appreciable distance into the ingot; in the case of ingot 3 (Fig. 4*b*) they grew nearly to the centre. If there is a direct correlation between flow velocity and columnar dendrite grain-growth direction, then this would indicate that extensive flow occurred during most of the solidification period, contrary to previous observations. The present authors do not believe that there is a direct correlation between flow velocity and dendrite growth direction. The imposition of fluid flow across an advancing interface causes nucleation to occur and new dendrites to grow in a direction pointing into the flowing stream, as observed. As long as nucleation does not recur, owing to a sudden perturbation of the system, the dendrites will continue to grow in the same direction regardless of changes in the flow velocity or temperature gradient. Accordingly, for the present ingots, the columnar grain direction was established early in the solidification process, when extensive flow was present, and continued in the same direction as the flow velocity decreased to zero. It is not clear why the columnar zone increased in width from top to bottom of the ingot (Fig.

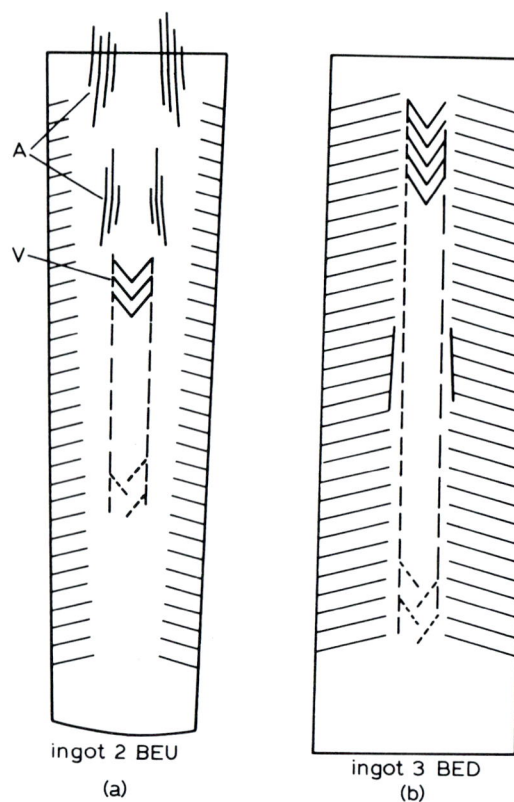

a ingot 2; *b* ingot 3

4   Cast structure: columnar region with upward pointing columnar dendrites (hatched area), A-segregate area and V-segregate area

4*a*), nor why the zone extended far into the billet (Fig. 4*b*).

The size and distribution of the equiaxed grains in the central part of the ingot were most clearly resolved in the autoradiographs of the small ingot (Table 1). Examples of the structure shown in Fig. 5 demonstrate a number of points:

(i)   very large and small dendrites are found adjacent to one another (Fig. 5*a*). Since dendrite size is controlled by local thermal conditions, appreciable movement of dendrite grains occurred in the equiaxed zone during solidification

(ii) in the equiaxed regions large dendrites are observed, with fine dendrite branches. The dendrites can interact physically to produce barriers to solid movement around them, producing V-segregates or porosity. An example of interlocked dendrites above a V-segregate 'cavity' is shown in Fig. 5*b*

(iii) large dendritic arrays can project from the general interface. The projections can be distorted or broken (*see* Fig. 5*c*)

(iv) there is no indication of grains which nucleate and grow in the upper part of the ingot continuously falling to the ingot bottom to form a mass of debris adjacent to the interface.

A-segregation

A-segregates were prominent in both of the BEU ingots (1 and 2), in the autoradiographs and sulphur prints, but were much less prominent in the BED ingot 3. Their

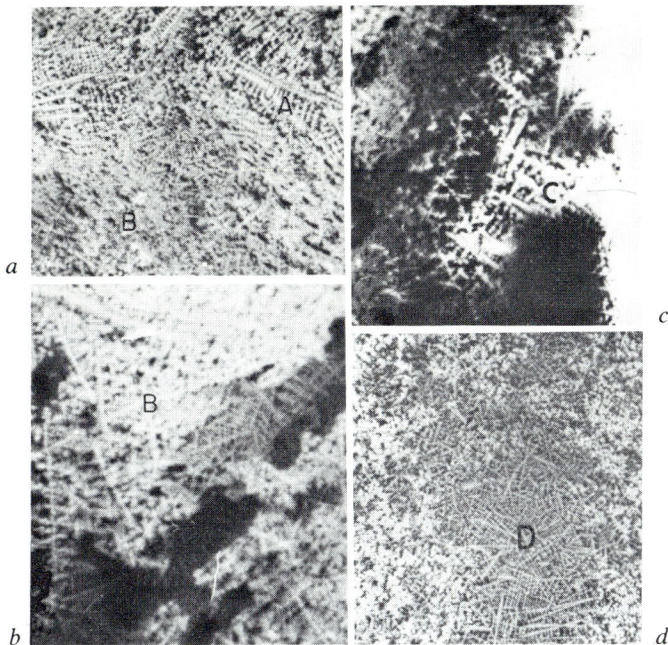

5 **Autoradiograph of ingot 1: *a* adjacent large *A* and small *B* dendrites indicate dendrite movement during solidification; *b* meshed dendrites at *B* form bridge; *c* projecting dendritic array *C* at interface pushed downward**

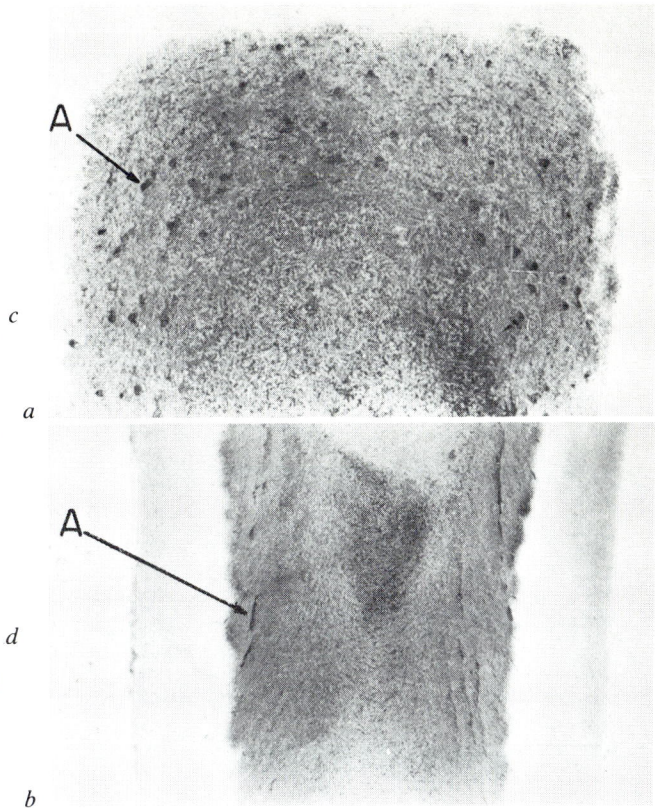

a central region of horizontal section; *b* vertical section

6 **Autoradiographs of ingot 2 showing A-segregates**

position in the ingot is shown schematically in Fig. 4*a* and *b*. In Fig. 4*a* two groups of A-segregates are shown at the upper and lower limits of the area containing A-segregates. The appearance of the A-segregates in the autoradiographs is shown in Fig. 6*a* and *b*. In Fig. 6*a* the A-segregates appear as small dark circular areas, roughly equidistant from one another, in a region midway between the outside surface and the centre of the ingot. Sulphur prints of the same surface showed the same circular areas, which were rich in sulphur. The dark circular areas were observed often to have a sharp outline, consistent with the hypothesis that they were associated with a remelting process. In some cases the dark areas were present in gold-free regions. The A-segregates did not appear to be significantly changed by changes in ingot size or grain structure in the BEU ingots, but their number was markedly reduced in the BED ingot.

The present results are consistent with a mechanism in which segregated solute-rich liquid of low density flows upward through the dendritic array in the solid–liquid region during solidification.[5-7] In the present case, carbon segregates to produce low density liquid. The A-segregate pipes form along the best open channel available in the solid–liquid region, remelting dendrite branches as they form. The observed relatively uniform distribution of pipes on the transverse section indicates that each pipe provides a path for flow from the surrounding region. The low incidence of A-segregates in the BED ingot 3 is attributed to the inward taper of the walls, which thus block the upward flow of the pipes.

A more detailed examination of the pipes was made, using the scanning electron microscope, on sections perpendicular to the pipe axes. The pipes were found to be porous and to contain a high density of inclusions. In some cases the irregular pores had smooth walls,

indicating that the porosity resulted from inadequate feeding in the final stages of solidification. The inclusions in the pipes were often large and were identified as primarily MnS and silicates. The MnS inclusions were often of Type III with large facets.

V-segregation

The regions containing V-segregation in the large ingots are shown in Fig. 4*a* and *b* by the V-pattern. They occur in the centre of the ingot and tend to be most pronounced in the upper half. Volume shrinkage results in a downward movement of the dendritic grains. Some of the larger dendrites mesh and lock (Fig. 5*b*), leaving a void below them, which generally fills with liquid forming a V-segregate pattern.

Macrosegregation

The distribution of carbon in the two large ingots is shown in Figs 7 and 8. In Fig. 7*a* the carbon levels are highest near the top of the ingot and lowest at about 40 cm below the top. The BED ingot (Fig. 8) shows less variation in carbon levels. These results are similar to those reported by Smith.[8]

Estimates of the carbon levels in the A-segregate pipes were made by taking drillings from a transverse section of the BEU ingot 2, each drilling containing a pipe. The carbon level determined would be an average of the pipe concentration and that of the surrounding material included in the drillings. The results from three pipes are given in Table 2. Taking the pipe diameter as 0·32 cm (from the maximum size of the pipes in the autoradiographs), and a carbon level of 0·75% for the

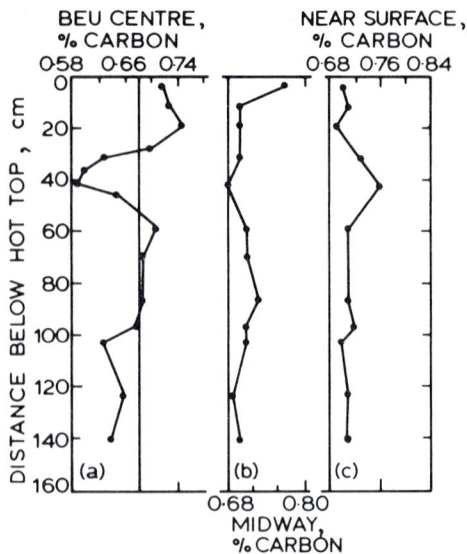

a along central axis starting below hot top; b midway between central axis and surface; c near surface

7 Carbon concentration in ingot 2; ladle concentration 0·68% C

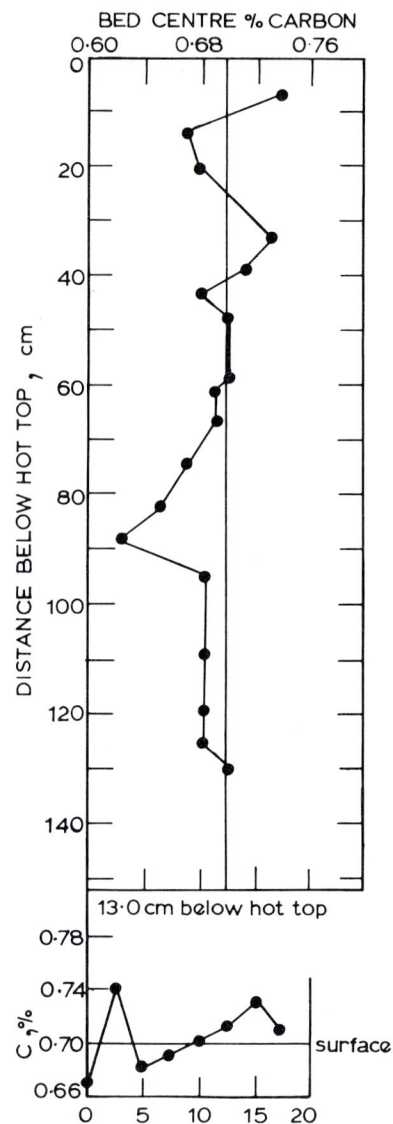

8 Carbon concentration in ingot 3 along central axis and perpendicular to central axis 13·0 cm below hot top: ladle concentration 0·70% C

remainder, the results in Table 2 indicate a carbon level of about 2% in pipes 2 and 3 and of a much higher value in pipe 1, which might have contained several pipes not discernible in the sulphur print.

The high carbon levels in the upper central region of the ingot could be owing to the A-segregate pipes and upward flow of liquid through the coarse central dendritic array.[9] The present results indicate that the high carbon levels at the top are largely a result of the A-segregates. In the BED ingot 3, fewer A-segregate pipes resulted in lower carbon levels at the top.

CONCLUSIONS

1 Strong convection currents are present in the early part of solidification. This is followed by limited flow in the upper half of the ingot, and finally by no flow in the liquid pool.

2 There is little or no interaction of the liquid pool with the solid–liquid region at the solidifying interface.

3 During solidification, more extensive flow occurs in the lower part of a BED ingot than in a BEU ingot.

4 No induced currents were observed in the lower part of the ingot as postulated in previous investigations. Also no grains were observed to fall from the upper to the lower part of the ingot to form a layer of debris on the bottom.

5 The upward growth direction of the columnar grains is established early during solidification by the downward flow of liquid at the interface. The columnar growth direction does not change as the rate of fluid flow subsequently decreases.

6 There is appreciable movement of grains in the equiaxed zone during solidification. Dendrites can project well ahead of the interface and can be bent. Dendrites can mesh and form bridges leading to V-segregates or porosity.

7 A-segregates result from the upward flow of carbon-rich liquid in the solid–liquid zone. Fewer A-

segregates are formed in BED ingots than in BEU ingots. The concentration of carbon in the segregates is high, as is the inclusion density. A large component of the high carbon level in the upper part of the ingot results from the presence of A-segregates. The remainder is owing to upward flow of carbon-rich liquid through the coarse dendritic array in the equiaxed zone.

Table 2 Carbon concentration in volume containing A-segregate pipe

	% C measured	Diameter of drill used, cm
Pipe 1	1·17	1·9
Pipe 2	1·12	0·63
Pipe 3	1·09	0·63
Ladle analysis	0·68	

REFERENCES

1 K. W. ANDREWS AND C. R. GOMER: 'The solidification of metals', 369, 1967, London, The Iron and Steel Institute

2 J. R. BLANK AND F. B. PICKERING: *ibid.*, 370

3 M. J. STEWART AND F. WEINBERG: *J. Cryst. Growth*, 1972, **12**, 228

4 B. CHALMERS, 'Principles of solidification', 270, 1964, New York, J. Wiley.

5 R. L. MCDONALD AND J. D. HUNT: *Trans. AIME*, 1969, **245**, 1 993

6 S. M. COPLEY *et al.*: *Metall. Trans.*, 1970, **1**, 2 193

7 N. STREAT AND F. WEINBERG: *ibid.*, 1972, **3**, 3 181

8 S. J. SMITH, *J. Iron Steel Inst.*, 1970, **208**, 247

9 N. STREAT AND F. WEINBERG: *Metall. Trans.*, 1976, **7B**, 417

Observation of porosity formation during solidification of aluminium alloys by acoustic emission measurements

U. Feurer and R. Wunderlin

Total counts of acoustic emission during solidification of aluminium alloy samples were found to increase with increases in the hydrogen content of the melt. It was therefore assumed that the acoustic events were somehow related to the formation of porosity in the samples. This assumption was verified both by basic experiments on bubble formation in aqueous solutions and a comparison of the pore fraction in the solidified samples with total acoustic emission. A simplified theory on hydrogen precipitation is qualitatively in fair agreement with acoustic emission during solidification. It is concluded that at least two different types of pores are formed, one between dendrite arms and the other between the single dendrite grains.

The authors are with Swiss Aluminium Ltd, Neuhausen, Switzerland

Microshrinkage or porosity formation during casting and solidification of aluminium alloys often causes severe problems in the fabrication of high-quality products. Although it is well known that porosity is initiated both by hydrogen in the melt and by insufficient feeding of solidification shrinkage, few experimental data on the mechanism of pore formation are available. Tentative experiments on acoustic emission during controlled solidification of aluminium alloy samples have shown that the total acoustic emission counts are proportional to the hydrogen content of the melt. Density measurements were in good agreement with these findings. It was therefore assumed that acoustic emission must be somehow related to the formation of hydrogen bubbles in the solidifying material and that acoustic measurements could thus provide useful data on both total porosity and the mechanism of its formation. Some of the results of the authors' recent investigations are presented below, mainly to indicate the possible future application of acoustic emission measurement to problems of porosity formation during solidification.

EXPERIMENTAL PROCEDURE

The experimental arrangement is shown schematically in Fig. 1. The solidification experiments were carried out in a small steel crucible of about 30 cm^3 volume. The crucible was adapted by an extension at the bottom to the acoustic emission transducer (*see* Fig. 2) which had a nominal resonance frequency of 140 kHz, a sensitivity of -75 dB (Ref. 1 V/μ bar) and which was connected to the electronic equipment (*see* caption for Fig. 1). As shown in Fig. 2, a thermocouple could be immersed in the melt for an optional determination of the solidification conditions in the centre of the sample. The experimental procedure is described below. The hydrogen content of the molten alloy was first determined by the 'Alu-Melt-Tester'. After carefully skimming the melt, a portion of it was poured into the crucible and the measuring equipment started. Usually, acoustic emission counts (ring-down counting) and temperature were recorded simultaneously *v.* time. After the sample had completely solidified, it was taken either for density determination or metallographic investigation. The basic experiments on bubble formation in aqueous solutions were carried out in a similar manner; however, the crucible was in this case made of transparent plastic material.

BASIC EXPERIMENTS

If it is really pore formation that is detected in these experiments, there must be some basic phenomenon which is creating a sound wave in the solidifying material. In crack formation within solid materials the creation of the acoustic events can be explained by a sudden release of stored elastic energy. A detailed description of this theory was given by Pollock.[1] A similar model of the supersaturation of gas in liquid aluminium could describe the origin of acoustic emission

340

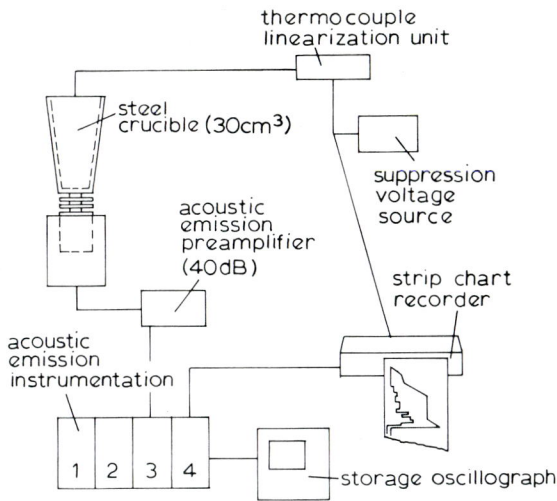

1 Schematic diagram of experimental equipment for acoustic-emission measurements during solidification

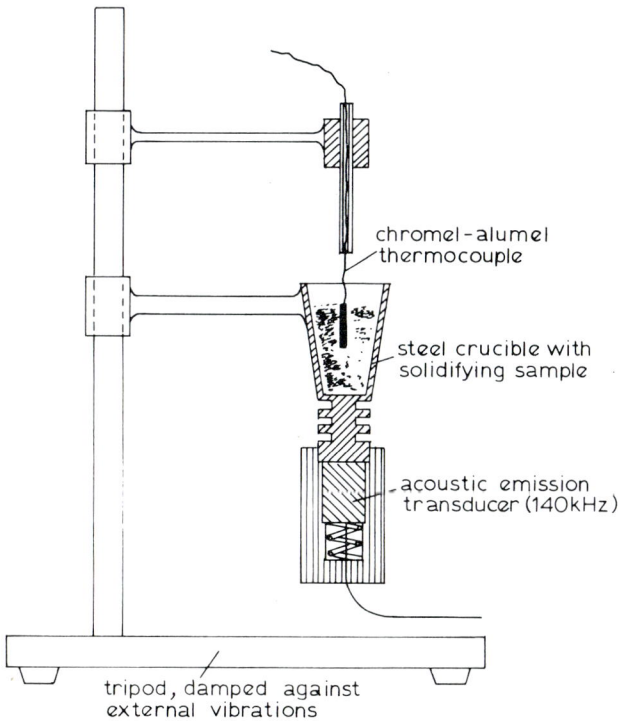

2 Sketch of arrangement for solidification experiments

by bubble formation. In the initial equilibrium condition, a given amount of hydrogen solute is present in the liquid aluminium. Owing to several causes, described later, the local hydrogen content may increase until an unstable condition is reached in which, by a release of 'supersaturation energy', a bubble of hydrogen can be formed. The acoustic emission event would be caused by a sudden displacement and compression of the liquid aluminium around the bubble surface, creating a shock pressure wave which travels through the solidifying material to the mould wall, where it initiates a secondary sound wave which is finally detected by the acoustic emission transducer.

In order to form a bubble the partial pressure of hydrogen in the melt has to exceed the local pressure by

$$\Delta P = P_H - P_x = \frac{2\sigma_{LV}}{r} \qquad (1)$$

where

P_H = partial pressure of hydrogen
P_x = local pressure, based on atmospheric, metal-lostatic, and other pressure terms
σ_{LV} = specific liquid/gas surface energy
r = bubble radius.

Therefore, it can be reasonably assumed, that acoustic emission amplitudes would be proportional to the surface energy and inversely proportional to the bubble radius. This assumption was checked by several experiments on bubble formation in aqueous solutions. One of them, the so-called 'beer experiment', was modified slightly by using $NaHCO_3$ solutions. A measured volume of solution (15 ml) was poured into a plexiglass 'crucible' which was connected to the acoustic emission transducer in a manner similar to that shown in Fig. 2. To this solution 1·5 ml 10% HCl solution was added and CO_2 bubbles of about 0·5–2 mm diameter were immediately formed. The acoustic emission was recorded by continuous ring-down counting; the results of these experiments are shown in Fig. 3. When the chemical reaction was finished, 15 mg of fine-alumina powder was added to the mixture, causing a much more intensive formation of numerous CO_2 bubbles of smaller size, about 0·2–0·5 mm in diameter. Typical acoustic emission oscillograms, corresponding to points 1 and 2

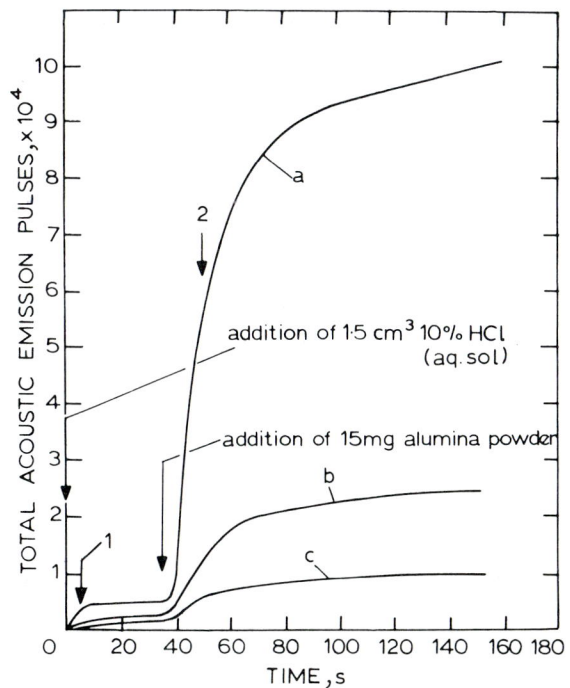

a $\sigma_{LV} \cong 72$ erg/cm^2; *b* $\sigma_{LV} \cong 65·5$ (1% C_2H_5OH); *c* $\sigma_{LV} \cong 55·5$ (5% C_2H_5OH)

3 Basic experiments on acoustic emission from bubble formation in aqueous $NaHCO_3$ solutions (20 g/l); arrows 1 and 2 mark positions where oscillograms of Fig. 4 were taken

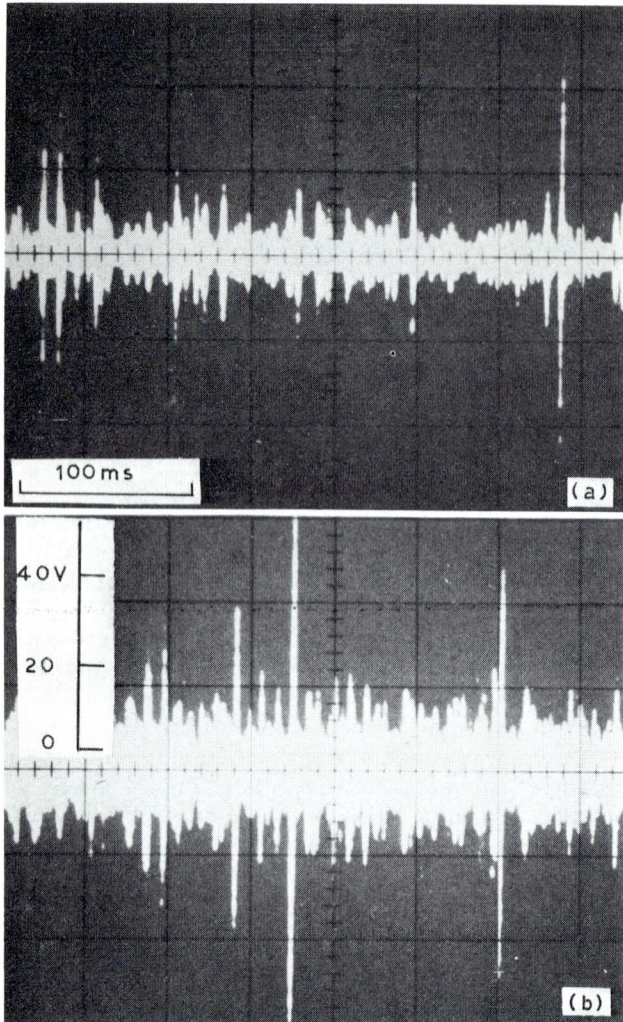

a oscillogram taken at point 1 in Fig. 3;
b oscillogram taken at point 2 in Fig. 3

4   Typical oscillograms of acoustic emissions during experiment a of Fig. 3

in Fig. 3, are represented in Fig. 4, demonstrating the influence of bubble size on acoustic emission amplitude. The influence of surface energy, which was varied by additions of ethanol to the hydrocarbonate solution, can be seen by comparison of curves *a*, *b*, and *c* in Fig. 3. The non-linear relationship can be attributed to several possible reasons, e.g. poisoning of the nucleating additive, variation of the amplitude-threshold ratio, and increased multiplication of higher amplitude acoustic emission events. Qualitatively, however, these experiments support the preceding assumptions and indicate that acoustic emissions result from pore formation in the solidification experiments.

RESULTS OF SOLIDIFICATION EXPERIMENTS

The results reported here were obtained from experiments with an Al–4·5 Cu–0·2 Ti alloy. Similar results were, however, obtained with pure aluminium and other alloys. In Fig. 5, three typical acoustic emission curves recorded during solidification of the grain-refined Al–4·5 Cu alloy are shown. Curve *a* is from the degassed material, whereas for curves *b* and *c*, the melt was artificially enriched with hydrogen. The difference

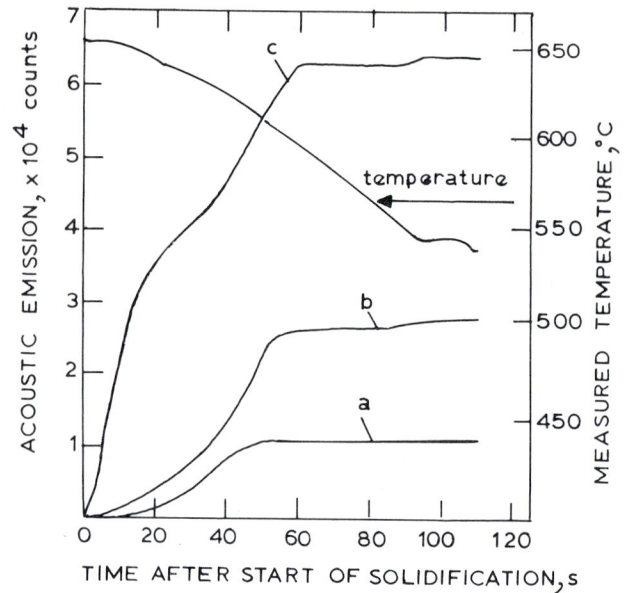

a 0·05 cm³H₂/100 g; *b* 0·17 cm³H₂/100 g;
c 0·23 cm³ H₂/100 g, 0·1% Al₂O₃

5   Examples of acoustic emission v. time curves during solidification of the alloy Al–4·5 Cu–0·2 Ti

between the measured hydrogen contents of *b* and *c* is attributable to the fact that the addition of alumina powder to the melt usually gives higher values in the first bubble test, owing to improved bubble nucleation conditions. It can therefore be assumed that the material for the two experiments had approximately the same hydrogen content. The addition of the fine alumina powder to the melt in experiment *c* obviously created optimal heterogeneous nucleation conditions for bubble formation, in agreement with the results of the basic experiments described above. In Table 1, the pore fractions of the solidified samples are compared with the measured total acoustic emission counts and the measured hydrogen contents in the melt; mean values of

Table 1   Summary of results of acoustic emission experiments for the alloy Al–4·5 Cu–0·2 Ti

Designation remarks	H₂ content, cm³ STP/100g	Total acoustic emission, 10⁴ counts	Pore fraction*, %
A	0·05	1·05	0·19
B	0·17	2·75	0·46
C (*see* Fig. 5)	0·23	6·35	0·63
No alumina	0·05	0·95 ± 0·21	0·21 ± 0·03
addition	0·10	2·35 ± 0·25	0·36 ± 0·03
	0·17	2·98 ± 0·65	0·37 ± 0·03
With alumina	0·08	4·23 ± 0·45	0·40 ± 0·01
addition	0·13	4·50 ± 0·57	0·43 ± 0·02
	0·23	5·90 ± 0·46†	0·65 ± 0·05

* Pore fractions based on a maximum density of 2·790 g/cm³
† Standard deviation of mean value
Linear regression: $y = a_0 + a_1 x$, where y = pore fraction, %, x = total acoustic emission, 10^4 counts, $a_0 = 0.15\%$, $a_1 = 0.0767\%/10^4$ counts, r^2 = coefficient of determination = 0·88.

30 additional experiments are also given. The linear regression of pore fraction v. total acoustic emission shows a fair agreement between both groups of experimental data (*see* notes for Table 1). A similar correlation was found in experiments with 99·85% purity grade aluminium (grain-refined with 0·3% TiB-hardener). The regression coefficients obtained from 27 data pairs were: $a_0 = 0·04\%$; $a_1 = 0·0204\%/10^4$ counts; $r^2 = 0.95$.

DISCUSSION

From the results presented above it is possible to make the following inferences:

(i) total acoustic emission depends not only on the hydrogen content of the melt but also on the bubble-nucleation conditions and alloy composition

(ii) several processes may operate in the porosity formation mechanism, e.g. enrichment of hydrogen in the liquid at the solid/liquid interface, heterogeneous or homogeneous bubble nucleation, supersaturation of hydrogen in the solidified structure, bubble agglomeration, or porosity coarsening owing to insufficient feeding conditions

(iii) several types of porosity must be formed during solidification, especially in cases involving heterogeneous bubble nucleation, causing different stages in the acoustic emission v. time curves (Fig. 5, curve c).

The case of optimal heterogeneous nucleation, neglecting surface-energy effects and diffusion processes, may be described by the following theoretical model, presented schematically in Fig. 6. Consider a volume element of liquid metal. In a grain-refined material, solidification starts simultaneously at different nucleation sites and, in general, small dendrites are formed which grow continuously until the end of solidification. Between the dendrite arms, the liquid metal at the solid/liquid interface is enriched not only in solute elements but also in hydrogen. Therefore at a certain time after the start of solidification, bubbles of hydrogen are formed between the dendrite arms, probably at nucleating oxide particles. Later, after the fraction liquid in the volume element has decreased, the hydrogen content in the spaces between the single dendrites is increased sufficiently to allow pore formation. Towards the end of solidification, feeding of solidification

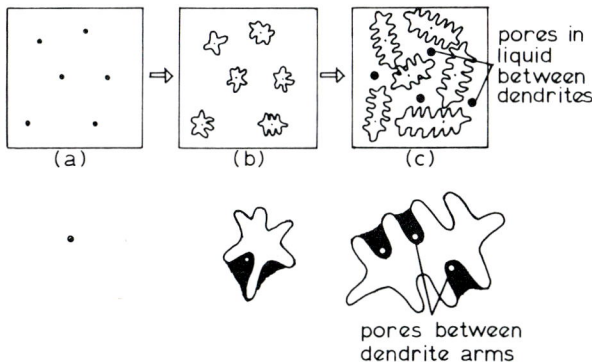

a $g_L = 1·0$; *b* $g_L = 0·9$; *c* $g_L = 0·5$

6 Theoretical model of formation of gas porosity during solidification; lower diagrams show enlarged views of developing dendrites and pores

shrinkage becomes increasingly difficult, causing a drop in local pressure in the liquid and so providing an additional driving force for a third type of porosity. Since this type seems to create rather small acoustic emission events, only the first two types of porosity are considered in the following analytical treatment.

The hydrogen content of a solidifying volume element can be described by

$$S_0 = S_L g_L + S_S (1 - g_L) \tag{2}$$

where

S_0 = initial hydrogen content of the liquid metal, cm^3 STP/100 g
S_L = hydrogen content of the liquid phase
S_S = hydrogen content of the solid phase
g_L = volume fraction liquid.

For a microscopic element in the space between two dendrite arms, S_L can be replaced by S_E and g_L by g_E, assuming that fraction liquid equals fraction eutectic. Furthermore, assuming that nucleation is heterogeneous and that pore formation is possible if the hydrogen content in the liquid exceeds the equilibrium solubility (thus neglecting surface-energy effects), the amount of hydrogen precipitated between the dendrite arms can be approximated by

$$S_{ai} \cong (1 - g_L)\{(S_0 - S_S) - g_E(S_L^0 - S_S)\} \tag{3}$$

whereby

$S_{ai} \geqslant 0$
S_{ai} = amount of hydrogen precipitated between dendrite arms
S_L^0 = equilibrium hydrogen solubility in the liquid phase
g_E = volume fraction eutectic.

In writing the term $(1 - g_L)$, it is assumed that bubbles formed between the dendrite arms are approximately proportional to the fraction solid of the macroscopic volume element (Fig. 6). A similar equation can be derived for the hydrogen precipitation between the single dendrites:

$$S_{aL} \cong g_L\{(S_0 - S_S) - g_L(S_L^0 - S_S)\} \tag{4}$$

whereby

$S_{aL} \geqslant 0$
S_{aL} = amount of hydrogen precipitated between the dendrites

S_{ai} and S_{aL} must both be positive to give a contribution to the total amount of porosity formation:

$$S_a(\text{total}) = S_{ai} + S_{aL}$$
$$\text{if } S_{ai} \geqslant 0; S_{aL} \geqslant 0 \tag{5}$$

Fraction liquid (or fraction solid) can be roughly calculated from a cooling curve of the sample, assuming that fraction solid is proportional to the temperature difference between the actual cooling curve and a theoretical curve with zero heat of fusion. The result of such a calculation is presented in Fig. 7*a*, based on the temperature measurement made in experiment *c* in Fig. 5. Fraction eutectic is readily obtained by the non-equilibrium lever rule. Finally, S_L^0 can be described by

$$S_L^0 = A_S \exp\left(-\frac{B_S}{T}\right) \tag{6}$$

344 *Feurer and Wunderlin*

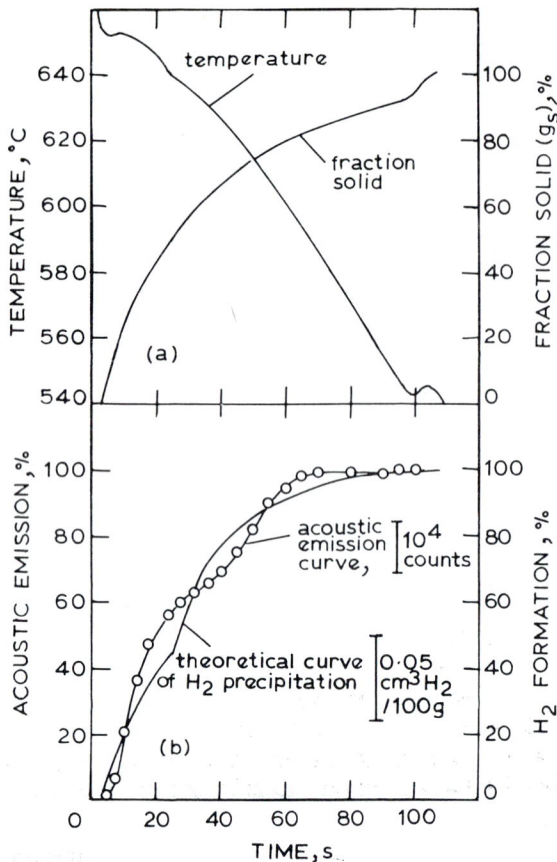

7 Calculation of *a* fraction solid and *b* hydrogen precipitation during experiment *c* of Fig. 5; comparison with acoustic-emission data

where A_S and B_S are constants depending on alloy composition and T is the absolute temperature. Taking data from the work of Opie and Grant[2] and assuming a constant solid solubility of hydrogen of $S_S = 0.03$ cm³/100 g, the amount of hydrogen precipitated *v.* time in experiment *c* was calculated by using equations (3)–(6). A comparison with the acoustic emission data is given in Fig. 7*b*. Although the theoretical curve does not fit completely to the acoustic emission curve, it at least shows the two-stage behaviour. Also, as can be seen from equation (1), acoustic emission depends on bubble size and it can be assumed that pores formed in the first stage are smaller than those formed in the second, thus causing higher acoustic emission intensities. It can

therefore be stated that pore formation is detected in these experiments.

This statement does not correspond with a theory proposed by Tensi,[3] who interpreted acoustic emission as friction noises between single crystallites in the solidifying material. However, these findings were made in rather different experimental conditions, e.g. much longer solidification times, so that comparison with the present results is difficult.

Finally, in order to give some indication as to future work, the simplified theory presented above could be improved considerably by taking into account effects of local pressure (i.e. interdendritic fluid flow), and influences of diffusion, local solidification time, and surface energy. For example, the difference between the first stages of curves *b* and *c* in Fig. 5 could be attributed to a certain amount of hydrogen supersaturation in the solidified material of *b*, owing to the fact that bubble formation between the dendrite arms is impossible under homogeneous nucleation conditions. This is one of several mechanisms which require investigation in detail. Also, a complete mathematical analysis of the origin of the acoustic-emission events will be a most interesting topic. Considerable contributions to this future work will be available from basic investigations carried out by several authors, including Campbell[4-7] and Piwonka and Flemings.[8] The ultimate objective of further investigations is obviously the solution of the practical problems caused by porosity formation, i.e. the control of solidification conditions or alloy composition in order to obtain cast material with a minimum amount of porosity. One might even be tempted to think of a quality-control method using acoustic emission to determine the susceptibility of an alloy to porosity formation.

REFERENCES

1 A. A. POLLOCK: *Non-Dest. Test.*, 1973, 265
2 W. R. OPIE AND N. J. GRANT: *Trans. AIME*, 1950, **188**, 1 237
3 H. M. TENSI: Proc. Second Acoustic Emission Symposium, 46, 1974, Tokyo
4 J. CAMPBELL: *Trans. AIME*, 1967, **239**, 138
5 J. CAMPBELL: *ibid.*, 1968, **242**, 1 464
6 J. CAMPBELL: *Br. Foundryman*, 1969, 147
7 J. CAMPBELL: *Trans. AIME*, 1969, **245**, 2 325
8 T. S. PIWONKA AND M. C. FLEMINGS: *ibid.*, 1966, **236**, 1 157

Development of porosity in aluminium-base alloys

R. A. Entwistle, J. E. Gruzleski, and P. M. Thomas

Microporosity distributions have been investigated in cast ingots of an Al–4·5% Cu alloy and an Al–8% Si alloy using a density technique. A definite pattern of microporosity is found within the ingot over a wide range of casting conditions and hydrogen concentrations. This pattern can be interpreted in terms of the solidification of the ingot. At very high gas levels when free gas pores exist in the liquid, the pattern disappears and porosity becomes uniform throughout the ingot. Under some conditions the amount of microporosity is related to the local solidification time.

The authors are in the Department of Mining and Metallurgical Engineering, McGill University, Montreal, Canada

The results of several investigations into the formation of microporosity in cast metals are available in the published literature, although, as Kondic[1] has recently pointed out, there has been a great imbalance between the amount of research effort expended on the nucleation and growth aspects of solidification and that devoted to porosity and other casting defects. A significant body of theoretical work describing the physics of pore formation can be found[2-6] as well as several experimental investigations into porosity formation.[7-17]

The conditions which favour the development of a pore in a solidifying metal are best described by the following equation:

$$P_g + P_s > P_{atm} + P_{s-t} + P_H \qquad (1)$$

where

P_g = equilibrium pressure of dissolved gases in the melt

P_s = negative pressure owing to shrinkage caused by feeding difficulties

P_{atm} = pressure owing to the atmosphere over the solidifying system

P_{s-t} = pressure required to overcome surface tension at the pore/liquid interface

P_H = pressure owing to the metallostatic head over the pore.

The two factors which promote pore formation, therefore, are dissolved gas and solidification shrinkage. These two factors interact with each other and in most casting situations it is difficult to state which is predominant in causing porosity.

In the present work the authors have examined how the porosity is distributed within laboratory scale ingots of an Al–4·5% Cu alloy and an Al–8% Si alloy. Many investigators have treated ingot porosity only in terms of the mean porosity; however, as will be shown, porosity can deviate considerably from the mean in different locations within the ingot. Thus mean values of porosity not only give a poor picture of ingot porosity but they may also be in considerable error if the samples from which the mean values are taken are not chosen carefully with relation to the ingot geometry and the freezing pattern. This work therefore provides information on where porosity is located within an ingot and also on how the amount and the distribution of porosity vary with the solidification conditions and gas content in two aluminium base alloys.

The Al–4·5% Cu alloys were cast into steel moulds about 10 cm dia.×10 cm high. The casting variables examined were melt superheat and mould temperature. The Al–8% Si alloys were cast into a water-cooled steel mould 10 cm dia.×30 cm high. Water cooling of the bottom third of this mould allowed higher rates of solidification than were possible in the Al–4·5% Cu experiments. Dissolved gas (H_2) content was measured by casting samples in graphite Ransley moulds[18] and analysing them for hydrogen by a subfusion technique.

Porosity distributions within ingots were determined by cutting a 10 mm thick vertical centreline slice from the ingot. This slice was sectioned into 40 cubes and the vol.-% porosity was determined on each by a density method.[19] From this grid of porosity values it was possible to draw lines of equal porosity by joining the appropriate points on the grid in much the same way as a

Table 1 Experimental conditions and mean porosity values

Alloy	Pouring temperature, K	Superheat, degrees	Mould temperature, K	Hydrogen content, ml H_2 (STP)/100 g Al	Mean porosity, vol.-%
Al–4·5% Cu	973	50	298	0·19	0·9
	973	50	298	0·30	0·8
	973	50	343	0·19	0·8
	973	50	513	0·19	0·8
Al–4·5% Cu	1 098	175	298	0·46	1·1
	1 098	175	343	0·46	0·9
	1 098	175	523	0·46	1·1
	1 098	175	673	0·46	2·9
Al–4·5% Cu	1 223	300	298	0·45	3·1
	1 223	300	343	0·45	2·7
	1 223	300	523	0·45	3·5
Al–8% Si	1 023	125	H_2O cooled	0·10	0·1
	1 023	125	H_2O cooled	0·30	0·8
	1 023	125	H_2O cooled	0·62	2·6
	1 023	125	H_2O cooled	0·66	2·3
	1 023	125	H_2O cooled	0·82	3·0

topographic map is constructed. Samples bordering the top surface of the ingot often contained internal macropores and considerable open porosity. Density measurements on these samples were therefore difficult to interpret and are not included in the figures given in this paper.

POROSITY DISTRIBUTIONS WITHIN INGOTS
The experimental conditions employed for the ingots discussed here are given in Table 1, along with the hydrogen content and the mean porosity of the centre-line slice. A major conclusion of the work is that the porosity within an ingot usually follows a definite pattern which in most cases can be interpreted in terms of the freezing pattern of the ingot. The porosity pattern, however, disappears at abnormally high gas concentrations when free gas pores exist during solidification, and also at very low gas concentrations.

Figures 1–3 illustrate typical porosity distributions obtained from the solidification of the Al–4·5% Cu ingots. Casting under conditions of low superheat produces the simplest distribution, relatively independent of mould temperature. In this type of distribution (Fig. 1) lines of equal porosity (isopores) lie more or less parallel to the mould walls and increase in magnitude to the ingot centre. As conditions are changed so as to promote slower rates of solidification, the porosity distributions also change. Distributions found by pouring at 1 098 K (superheat of 175°) are illustrated in Fig. 2. At

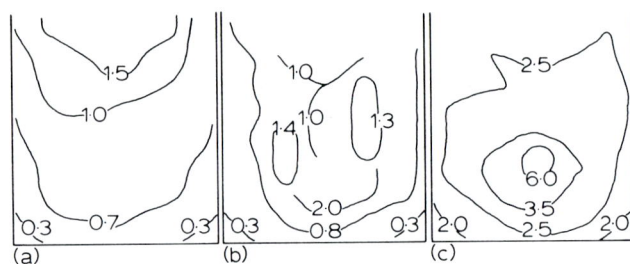

a mould temperature 298 K; *b* mould temperature 523 K; *c* mould temperature 673 K

2 Porosity profiles in ingots cast from 1 098 K (superheat 175°)

low mould temperatures (298 K) the distribution resembles that found by pouring with less superheat; however, as the mould temperature is increased the distribution changes to one with roughly concentric isopores and pockets of very high central porosity. This trend continues as the superheat is increased still further to 300° (pouring temperature 1 223 K) (Fig. 3) and under these conditions many isolated pockets of porosity are found.

The various distributions can be readily associated with the structure of the ingot. Distributions of the type shown in Fig. 1 were always found if the ingot contained a fine equiaxed grain structure. The isopores outline the interface between the mass of fine settling grains and the

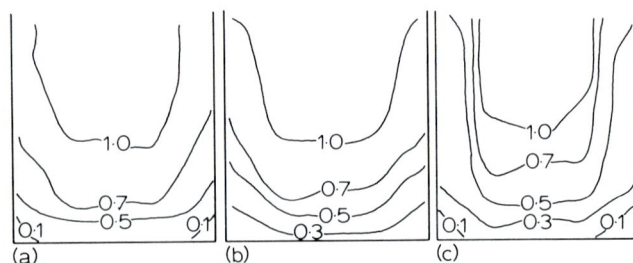

a mould temperature 298 K; *b* mould temperature 343 K; *c* mould temperature 513 K

1 Porosity profiles in ingots cast from 973 K (superheat 50°)

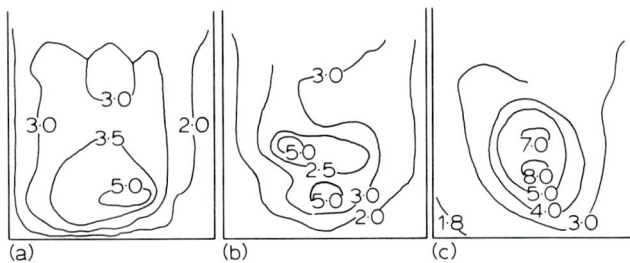

a mould temperature 298 K; *b* mould temperature 343 K; *c* mould temperature 523 K

3 Porosity profiles in ingots cast from 1 223 K (superheat 300°)

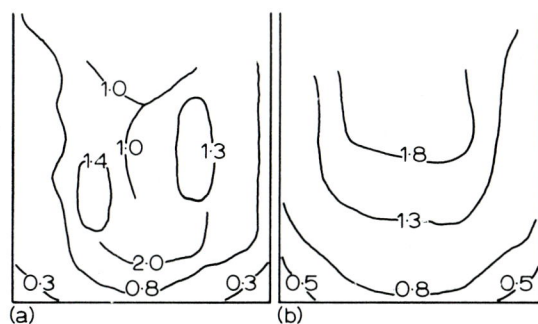

a non-grain refined; *b* Ti–B grain refined

4 Effect of grain refinement on porosity distribution: pouring temperature 1 098 K, mould temperature 523 K

bulk liquid at various stages of solidification. More concentric distributions found at higher superheats were associated with coarse columnar–equiaxed structures or coarse columnar structures. They reflect the difficulty of feeding long interdendritic channels. In some cases (e.g. pouring temperature of 1 223 K, mould temperature 523 K) it is likely that the top of the ingot froze over before the centre owing to the enhanced radiant heat transfer at high temperatures. Hence the isopores which follow the solid/liquid interface are approximately concentric, and very high central porosity exists because of the impossibility of feeding the centre.

A good example of the influence of structure and solidification mode is shown in Fig. 4. Two ingots were cast under the same conditions except that one was grain refined with a Ti–B grain refiner. When the structure was changed to one of fine equiaxed grains, the porosity distributions also changed, although the mean porosities remained about the same for each ingot: 1·4% for the grain-refined ingot and 1·1% for the non-grain-refined ingot.

INFLUENCE OF HYDROGEN CONTENT
The results of this work indicate that over a certain range of hydrogen concentrations the dissolved gas content is not an important variable in determining the porosity distribution or the mean porosity. For the Al–Cu alloy used here this range would appear to extend at least from 0·19 to 0·46 ml H_2 (STP)/100 g Al. This statement is substantiated by two observations.

First, as seen in Table 1, ingots cast from 973 K into moulds at 298 K have essentially the same mean porosity and the same type of porosity distribution at gas contents of both 0·19 and 0·30 ml H_2 (STP)/100 g Al. Secondly, ingots cast from 1 098 K and 1 223 K contained virtually the same hydrogen content and yet exhibited both very different mean porosities and different porosity distributions. Thus, within this particular range of gas concentrations, it is clear that the solidification conditions rather than the hydrogen concentration itself are of major importance in microporosity formation.

Some of the experiments with Al–8% Si alloys are illuminating in this regard. Most of these experiments were carried out with extremes of dissolved hydrogen concentration, i.e. hydrogen levels either much above or much below that found in good aluminium melting practice. At the very high gas levels (greater than about 0·7 ml H_2 (STP)/100 g Al) no clear pattern of micro-

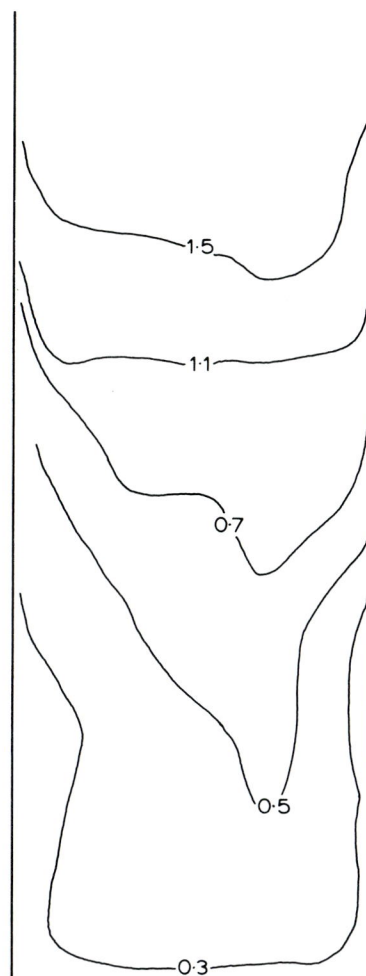

5 Porosity distribution in Al–8% Si alloy: pouring temperature 1 023 K, water-cooled mould

porosity could be discerned. Porosity was fairly uniform and high throughout the ingot. At very low gas levels (0·10 ml H_2 (STP)/100 g Al), again no clear pattern emerged and the ingot was uniformly sound with a mean porosity of 0·1%, which is approaching the level of detectability of the density technique.[19] Only at intermediate gas levels (e.g. 0·30 ml H_2 (STP)/100 g Al) were clear porosity distributions observed which could be related to the manner in which the ingot solidified (Fig. 5).

The characteristic porosity distribution is thus found only in the range of gas contents in which dissolved gas and solidification shrinkage combine to form microporosity. If the gas level is too high, shrinkage is of little importance in determining porosity formation; if the gas level is too low, the solidification shrinkage itself is ineffective in producing pores. The interesting feature of porosity is that, in the intermediate range, although both gas and shrinkage act together, it is apparently the shrinkage (the solidification mechanism) which determines where pores form and their total volume.

METALLOGRAPHIC OBSERVATIONS
Types of porosity
Metallographic observations substantiated conclusions drawn from the porosity profiles and their variation with hydrogen concentration and freezing conditions. Three types of porosity were found under different conditions.

348   *Entwistle et al.*

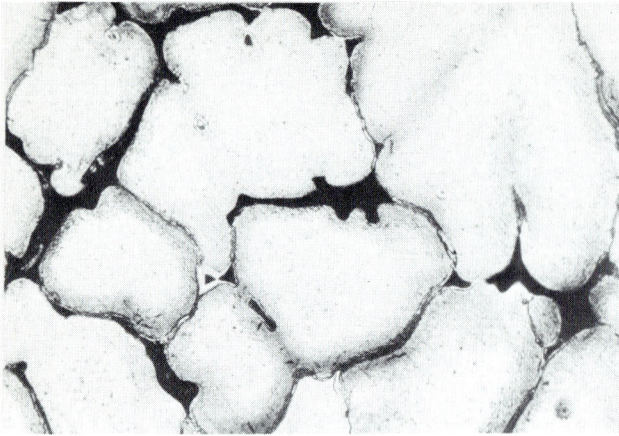

6   **Fine interdendritic porosity typically found in ingots cast from low superheats into cold moulds, Al–4·5% Cu alloy   [×160]**

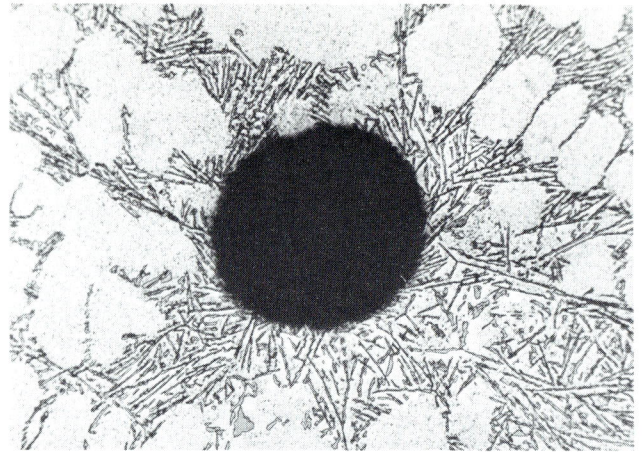

8   **Free gas pores found at high hydrogen levels, Al–8% Si alloy   [×120]**

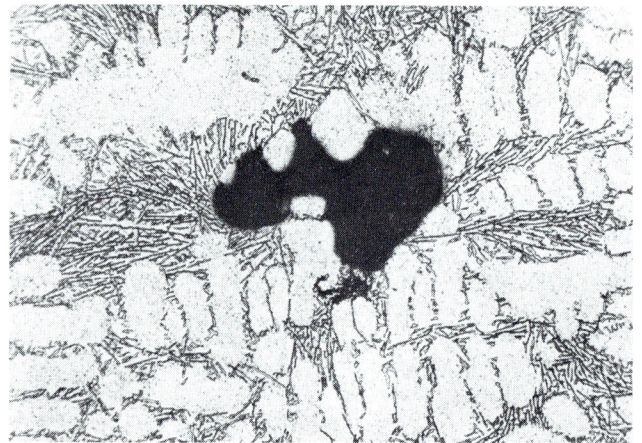

9   **Interaction of free gas pores and dendrite structure, Al–8% Si alloy   [×120]**

Ingots containing fine grains and lower amounts of porosity, such as ingots cast from 973 K or grain-refined ingots, contained fine interdendritic porosity (*see* Fig. 6). Such porosity indicates an origin in feeding difficulties on the scale of the dendrite size.

With coarser grains, produced by casting at higher temperatures, the porosity became more continuous in nature (Fig. 7) and at times was even macroscopic. This type of porosity was particularly noticeable in the areas of high central porosity found in slowly solidified ingots (Fig. 3). This is an example, no doubt, of hot tearing above the final solidification temperature. Although it is caused by thermal contraction coupled with feeding problems, it should be noted that the hot tears occur on a size scale greater than the size of a dendrite, with each pore involving many dendrites.

Both the interdendritic porosity and the hot tears are related to the solidification mechanism, and this is reflected in the porosity distributions and the relative unimportance of the hydrogen concentration. At very high hydrogen levels, however, where no characteristic distribution of porosity could be found, the pores were of the spherical gas bubble type shown in Fig. 8. Many times, these pores showed no relationship to the dendritic structure as in Fig. 8. Other pores did show some

slight interaction of pore and dendrite (Fig. 9); however, at these gas levels it was quite clear that gas, rather than interdendritic feeding, was determining the type of microporosity.

RELATIONSHIP OF POROSITY TO LOCAL FREEZING TIME

The observations recorded above indicate that microporosity is caused by local feeding difficulties only within a certain range of conditions. These conditions are those which promote relatively rapid solidification and a fine grain structure at moderate hydrogen levels. It would be expected that for these conditions the amount of microporosity would depend on the size and shape of individual dendrites.

A parameter which has been shown to influence the detailed structure of individual dendrites is the local solidification time, which is also related to the local cooling rate. The solidification of the Al–4·5% Cu alloys was computer modelled to calculate the local solidification time for the various node points at which porosity was measured. The program was verified experimentally, and in addition measured dendrite arm spacings correlated well with those predicted from the computed local solidification times.[20]

Figure 10 shows the measured amounts of porosity plotted against local solidification times for ingots cast

7   **Coarse, intergranular hot tearing found in ingots cast from high temperatures, Al–4·5% Cu alloy   [×16]**

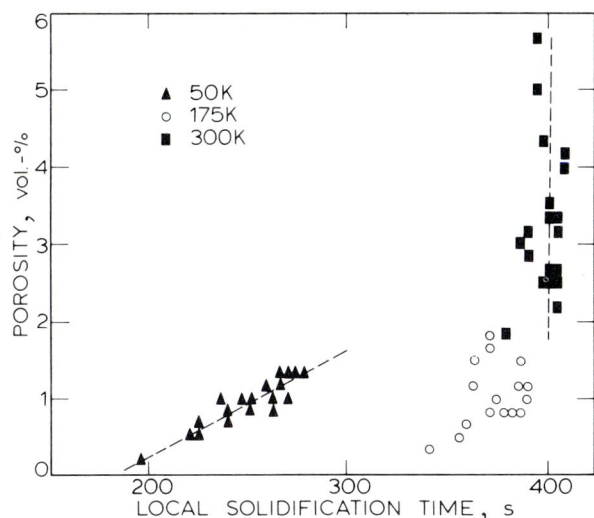

10 Relationship of vol.-% porosity to local solidification time for ingots cast from various temperatures into moulds at 298 K

from 50°, 175°, and 300° of superheat into moulds at 298 K. It may be seen that at the lower superheats, where interdendritic porosity is found, there is a marked dependence of the vol.-% of porosity on the local solidification time. At the higher superheats, where much coarser porosity is found, there is no dependence of the amount of porosity on local freezing time.

It should be remembered that the local solidification time is quite an insensitive parameter, and a large variation in it is not usual from one location to another in ingot freezing. The dependence shown in Fig. 10 indicates that the amount of microporosity increases as the freezing time increases, or alternatively as the cooling rate decreases. Greater microporosity is therefore associated with more slowly cooled and somewhat coarser structures. Thus the ease of flow in interdendritic channels may not be so important in microporosity formation as is usually assumed. Factors such as the number of flow channels, or the time available in the interdendritic liquid for pore nucleation and gas diffusion, may be significant.

CONCLUSIONS
The following main conclusions can be drawn from this work.

1 Microporosity follows a definite pattern in cast ingots of aluminium alloy over a wide range of solidification conditions and hydrogen concentrations. The pattern of microporosity can be related to the manner in which the ingot solidified.

2 The pattern of porosity disappears at either very high hydrogen levels (>0.7 ml H_2 (STP)/100 g Al) or at very low levels (<0.1 ml H_2 (STP)/100 g Al). At high levels porosity is uniformly distributed. At the low levels any porosity which was present was less than 0.1 vol.-% and not readily detectable.

3 There is a dependence of the amount of microporosity on the local solidification time for conditions which promote fine interdendritic porosity.

ACKNOWLEDGMENTS
The authors acknowledge the financial assistance of the National Research Council of Canada and the Government of Quebec. Hydrogen analyses and alloys were kindly supplied by Alcan Canada Ltd.

REFERENCES
1 V. KONDIC: *The Metallurgist*, 1976, **8**, 321
2 J. CAMPBELL: *Cast Met. Research J.*, 1969, **4**, 1
3 J. CAMPBELL: *Trans. Met. Soc., AIME*, 1967, **239**, 138
4 J. CAMPBELL: *ibid.*, 1969, **245**, 2 325
5 J. CAMPBELL: 'The solidification of metals', 18, 1967, London, The Iron and Steel Institute
6 T. S. PIWONKA AND M. C. FLEMINGS: *Trans. Met. Soc., AIME*, 1966, **236**, 1 157
7 E. J. WHITTENBERGER AND F. N. RHINES: *Trans. AIME*, 1952, **194**, 409
8 R. J. KISSLING AND J. F. WALLACE: *Foundry*, 1963, **2**, 70
9 B. CHAMBERLAIN AND J. SULZER: *Trans. AFS*, 1964, **46**, 600
10 N. S. MAHADEVAN *et al.*: *ibid.*, 1962, **76**, 77
11 S. Z. URAM *et al.*: *ibid.*, 1958, **66**, 129
12 S. NISHI AND T. KUROBUCHI: *Light Metal Industry*, 1974, **24**, 245
13 J. CAMPBELL: *Br. Foundryman*, 1969, **4**, 147
14 W. D. WALTHER *et al.*: *Trans. AFS*, 1956, **64**, 658
15 S. J. WALKER: *AFS Cast Metals Research Journal*, 1971, **7**, 13
16 A. TZAVARAS AND M. C. FLEMINGS: *Trans. Met. Soc. AIME*, 1965, **233**, 355
17 M. F. JORDAN *et al.*: *J. Inst. Met.*, 1962–63, **91**, 48
18 C. E. RANSLEY AND D. E. J. TALBOT: *ibid.*, 1955–56, **84**, 445
19 R. A. ENTWISTLE: M.Eng. thesis, McGill University, Montreal, 1976
20 K. P. YOUNG AND D. H. KIRKWOOD: *Metall. Trans. A*, 1975, **6**, 197

Numerical simulations of the solidification process

P. N. Hansen

A method is presented for the determination of the macroscopic progress of solidification, as defined by the geometry and velocity of the solidification front, from the point of view that this is one of the most important factors in the casting process. The geometry indicates whether or not the casting will be sound; the velocity of the front is an important parameter in controlling microstructure and the related mechanical properties of the casting. The present method concentrates on calculations of the unsteady temperature field in both casting and mould. The solidification front is then given at the desired time level as the domain between the liquidus isotherm of the bulk liquid and the selected isotherm at complete solidification. The present paper concerns simple casting geometries, but some ideas and initial work on real casting geometries are also presented.

The author is in the Department of Metallurgy, The Technical University of Denmark, Lyngby, Denmark

LIST OF SYMBOLS

c = coefficient of specific heat capacity, J/kg °C
c_0 = concentration in bulk liquid, wt-%
c_s = concentration at the solidus line, wt-%
F, G = functions given by Fig. 7
g = fraction solidified
H = heat content function, s °C/mm^2
i, j, k = cartesian coordinates, unity Δx_c
K = equilibrium partition coefficient
k = coefficient of thermal conductivity, W/m °C
k_0 = coefficient of thermal conductivity at 0°C, W/m °C
L = latent heat, J/kg
T = temperature, °C
t = time, s
t_p = time at beginning of hot cracking, s
Δt = incremental time step, s
u = free linear thermal expansion, mm/mm
$VSAM$ = rate, mm/s
Δx = shortest distance between net points, mm
Δx_0 = shortest distance between net points at 0°C, mm
x, y, z = cartesian coordinates, mm
α' = coefficient of thermal diffusivity, mm^2/s
β = relative linear thermal expansion, mm/mm
ρ = density, kg/m^3
ρ_c = density at 0°C, kg/m^3
θ = modified temperature, °C

Symbolism

d = total derivative
$T_{i,j,k}$ = indicating the temperature in net point (i, j, k)
∂ = partial derivative
δ = a difference approximation

In the area of numerical simulations of unsteady temperature fields, several different procedures exist.[1-5] In the case of solidification of castings, the author believes that work in the future will be concentrated upon two methods:

(i) the method of finite differences in a form as presented in this paper
(ii) the method of finite elements.

Until now the first method has been the cheapest with respect to computer time[3-5] owing to the problems in the second method when simulating the liberation of latent heat.

Among the methods of finite differences it has been shown[6] that the explicit method presented below is favourable in terms of computer time owing to the use of the heat content function H (equation (10)).

MATHEMATICAL BASIS

The mathematical model is based on the unsteady heat conduction equation:

$$\rho c \frac{\partial T}{\partial t} = \frac{\partial}{\partial x}\left(k \frac{\partial T}{\partial x}\right) + \frac{\partial}{\partial y}\left(k \frac{\partial T}{\partial y}\right) + \frac{\partial}{\partial z}\left(k \frac{\partial T}{\partial z}\right) \qquad (1)$$

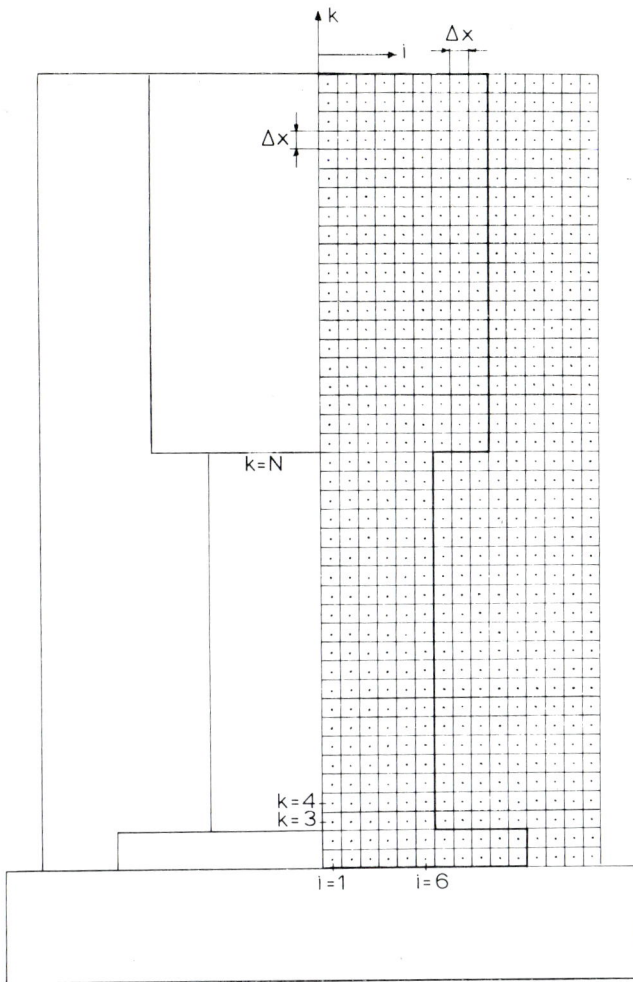

1 A normal net (the dots) which was used on the parts shown in Fig. 9

free expansion and contraction the following equations apply:

$$\Delta x = \Delta x_0 \beta \qquad (5)$$

$$\beta = 1 + u(T) \qquad (6)$$

For the specific weight:

$$\rho = \rho_0 \beta^{-3} \qquad (7)$$

Equation (4) now becomes:

$$\frac{1}{\alpha'} \frac{\partial T}{\partial t} = \frac{1}{\Delta x_0^2}(\delta_{xx} + \delta_{yy} + \delta_{zz}) \qquad (8)$$

where

$$\alpha' = \frac{k_0 \beta}{\rho_0 c} \qquad (9)$$

As mentioned above, a more economical procedure in solidification processes is the use of a heat content function, such as:

$$H(T) = \int_0^T \left(\frac{1}{\alpha'} + \frac{\rho_0}{k_0 \beta} \frac{\partial L}{\partial T} \right) dT \qquad (10)$$

where $\partial L / \partial T$ is the liberation of latent heat with temperature.

Finally, equation (8) becomes:

$$H(T_{i,j,k}^{new}) = H(T_{i,j,k}^{old}) + \frac{\Delta t}{\Delta x_0^2}(\delta_{xx} + \delta_{yy} + \delta_{zz}) \qquad (11)$$

where Δt is the incremental time step.

MATERIAL THERMAL PROPERTIES

The author has worked with castings of low-carbon steel in sand and ceramic moulds[6] and only data for these cases will be given, but the principles are general.

Data for the θ function given by equation (2) are shown in Fig. 2. $k_0^{steel}/k_0^{sand} = 43$ is used in the calculations.

To determine the heat content function given by equation (10), the first step is to find out how the latent heat of solidification is liberated. A good approximation should be to say that a given amount solidified is proportional to the liberation of a corresponding amount of latent heat.

Using the Fe/C equilibrium diagram, the amount solidified is given by the 'lever rule', which is shown in Fig. 3.

Using Scheil's equation[9]:

$$g = 1 - \left(\frac{c_s}{K \cdot c_0} \right)^{\frac{1}{K-1}} \qquad (12)$$

the curves of Fig. 4 are obtained.

The effect of these two methods of calculating the heat content function H can be seen in Fig. 5 for a carbon content of 0·1 wt-%.

When this can be considered as two extremes, the thermal calculations will not be much affected if the correct curve lying between these two curves is not used. However, it is important to use a curve which demands that most of the solidification takes place near the liquidus temperature of the bulk liquid, as in the curves of Fig. 3 or Fig. 4. The data used in the temperature range of interest are shown in Fig. 6. In this representation the liquidus depression and coring effects of other alloying elements (Mn and Si) have been ignored, since

This form is not suitable for use because it is not linear but it may be rearranged by the use of a modified temperature[7]:

$$\theta = \frac{1}{k_0} \int_0^T k \, dT \qquad (2)$$

Equation (1) can then be rewritten as:

$$\frac{\rho c}{k_0} \frac{\partial T}{\partial t} = \frac{\partial^2 \theta}{\partial x^2} + \frac{\partial^2 \theta}{\partial y^2} + \frac{\partial^2 \theta}{\partial z^2} \qquad (3)$$

The equation is now in a linear form. θ can also be set on the left-hand side of the equation if knowledge of the temperature T on each time level in the calculations is not required, thus slightly reducing computer time.

The calculations are carried out on a normal net[8] with equally spaced netpoints a distance Δx apart, as in Fig. 1. The second derivatives in equation (3) are approached by normal finite-difference approximations[8]:

$$\frac{\rho c}{k_0} \frac{\partial T}{\partial t} = \frac{\delta_{xx}}{\Delta x^2} + \frac{\delta_{yy}}{\Delta x^2} + \frac{\delta_{zz}}{\Delta x^2} \qquad (4)$$

δ_{xx}, δ_{yy}, and δ_{zz} are given below.

As the net is fixed to the casting part, the distances between the net points will vary with temperature. For

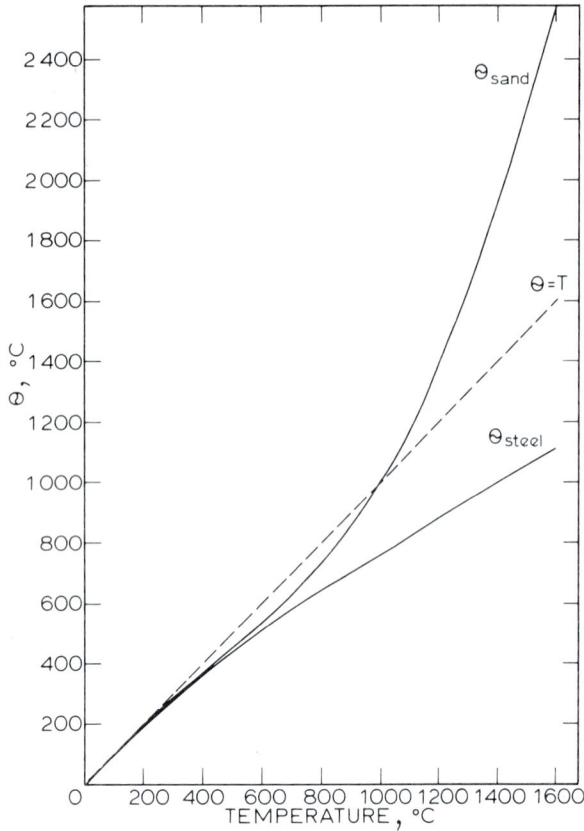

2 Modified temperature θ as a function of temperature T (data for steel from Ref. 11 and data for sand from Ref. 14)

their effect is relatively small. These effects can be included.[9]

SIMPLE CASTING GEOMETRIES
In these cases the net matches with the casting geometry in such a way that either net points are lying on the surfaces or the surfaces are lying midway between net points, as shown for example in Fig. 1.

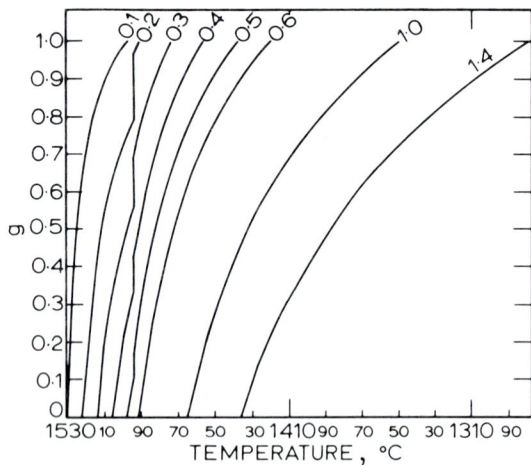

3 Fraction solidified g as a function of temperature calculated by use of the lever rule; figures on the curves are the carbon concentration in wt-% (data for calculations from Ref. 13)

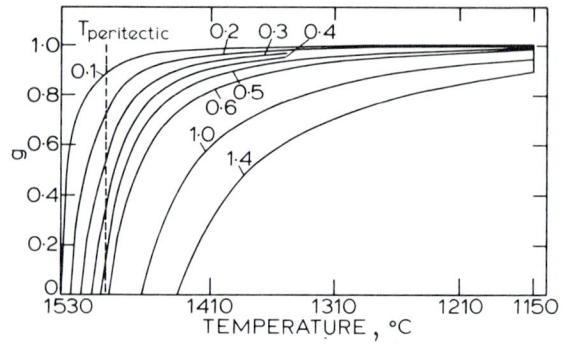

4 Fraction solidified g given by Scheil's equation; figures on the curves are the carbon concentration in wt-% (data for calculations from Ref. 13)

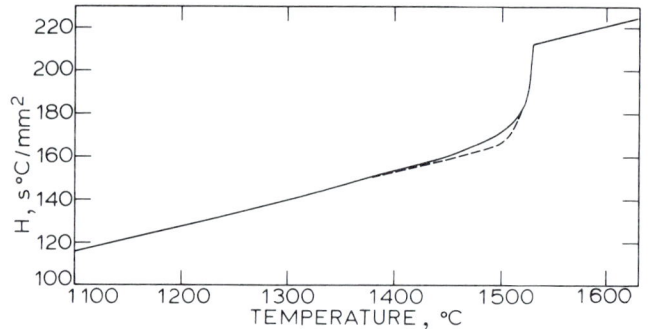

5 Heat content H for a 0·1 wt-% carbon steel; full line is fraction solidified from Fig. 4 and dashed line is fraction solidified from Fig. 3 (data for α' from Ref. 11)

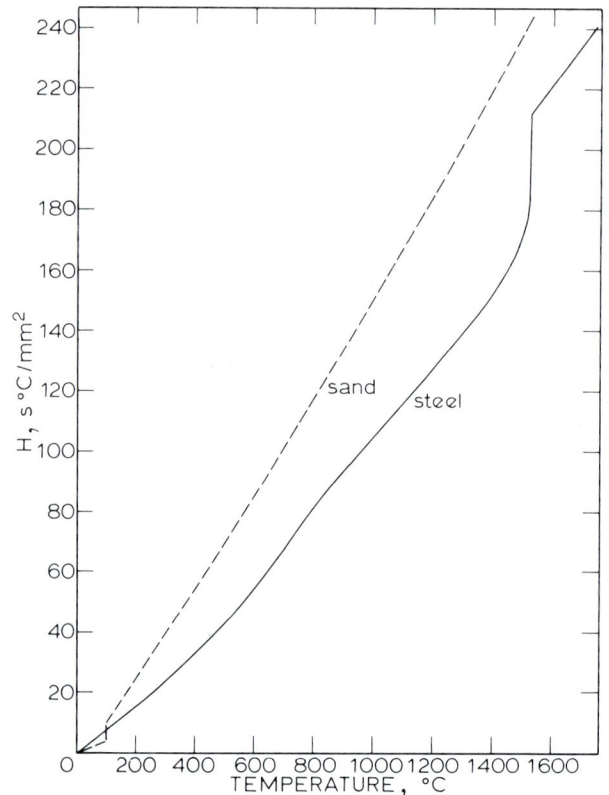

6 Heat content function H as a function of temperature T; the values for sand must be multiplied by 0·058 (data for α'_{steel} from Ref. 11 and data for α'_{sand} from Ref. 12)

Below, formulae are given for δ_{xx}, δ_{yy}, and δ_{zz} in equation (11):

Cartesian case[8]

$$\delta_{xx} = \theta_{i+1,j,k} + \theta_{i-1,j,k} - 2\theta_{i,j,k} \qquad (13)$$

$$\delta_{yy} = \theta_{i,j+1,k} + \theta_{i,j-1,k} - 2\theta_{i,j,k} \qquad (14)$$

$$\delta_{zz} = \theta_{i,j,k+1} + \theta_{i,j,k-1} - 2\theta_{i,j,k} \qquad (15)$$

Symmetry about the Z axis[6]

$$\delta_{xx} + \delta_{yy} = -G(i)(\theta_{i,k} - \theta_{i-1,k}) + F(i)(\theta_{i+1,k} - \theta_{i,k}) \qquad (16)$$

$$\delta_{zz} = \theta_{i,k+1} + \theta_{i,k-1} - 2\theta_{i,k} \qquad (17)$$

The functions $G(i)$ and $F(i)$ are shown in Fig. 7.

Spherically symmetrical case

$$\delta_{xx} + \delta_{yy} + \delta_{zz} = -G(i)(\theta_i - \theta_{i-1}) + F(i)(\theta_{i+1} - \theta_i) \qquad (18)$$

The functions $G(i)$ and $F(i)$ are shown in Fig. 7.

Boundary conditions at the metal/mould interface when casting in moulding sands and ceramics

In this case it is found to be a good approximation for the thermal coupling to have no temperature drop across the interface.[6] (The mould can transport less heat than the metal is able to transport.)

On a net such as in Fig. 1 the temperature at a point such as A on Fig. 8 the interface can be found by linear extrapolation:

$$T_A = 1 \cdot 5T_{4,j} - 0 \cdot 5T_{3,j} \qquad (19)$$

A good approximation is obtained[6] when the heat flow across the interface is determined only by the gradients in the mould.

First, the temperatures determined by equation (19) have to be transformed to θ^{mould}-values by equation (2).

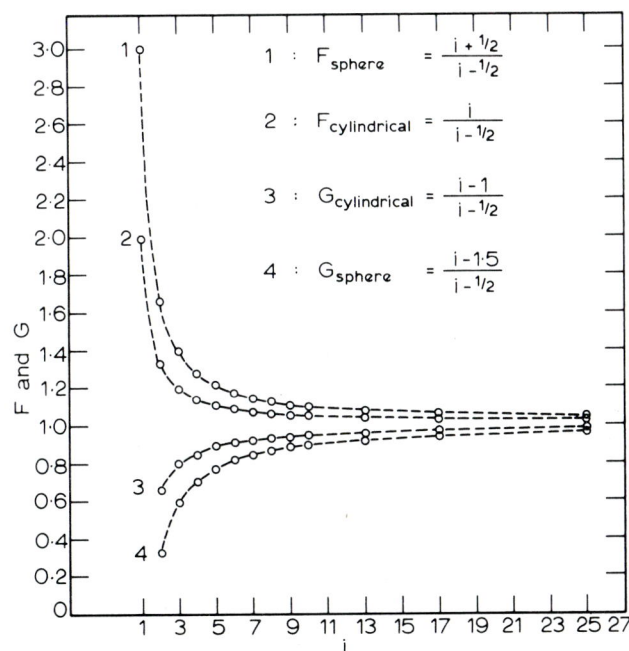

7 Functions *F* and *G*; the space parameter *i* is valid on a net as in Fig. 1 and Fig. 8

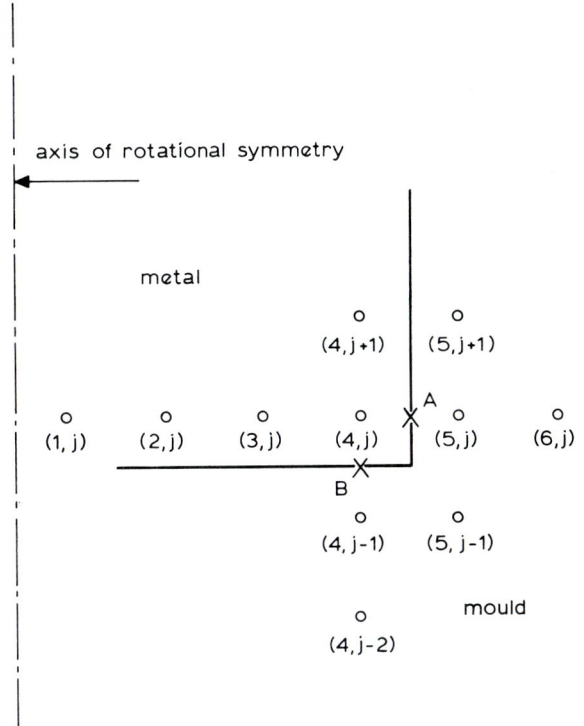

8 Net points near the metal/mould boundary in a part with axial geometry

(Note $\theta_A^{mould} \neq \theta_A^{metal}$, Fig. 2.) The heat flux at a point such as A in Fig. 8 is then given by Fourier's law:

$$Flux_A = \frac{2k_0^{mould}}{\Delta x_0}(\theta_{5,j} - \theta_A^{mould}) \qquad (20)$$

As an example, equation (11) is written in point $(4, j)$ and point $(5, j)$ of Fig. 8:

$$H(T_{4,j}^{new}) = H(T_{4,j}^{old}) + \frac{\Delta t}{\Delta x_0^2}\left(-G(4)(\theta_{4,j} - \theta_{3,j})\right.$$
$$+ \frac{2F(4)k_0^{mould}}{k_0^{metal}}(\theta_{5,j} - \theta_A^{mould})$$
$$\left. - \frac{2k_0^{mould}}{k_0^{metal}}(\theta_B^{mould} - \theta_{4,j-1}) + \theta_{4,j+1} - \theta_{4,j}\right) \qquad (21)$$

$$H(T_{5,j}^{new}) = H(T_{5,j}^{old}) + \frac{\Delta t}{\Delta x_0^2}\left(-2G(5)(\theta_{5,j} - \theta_A^{mould})\right.$$
$$\left. + F(5)(\theta_{6,j} - \theta_{5,j}) + \theta_{5,j+1} + \theta_{5,j-1} - 2\theta_{5,j}\right) \qquad (22)$$

Some results

Although the theory presented above cannot be used on real casting geometries, it can be used on testpieces, such as those in Fig. 9, in analysing and synthesizing problems concerning real castings. As an example[6] work has been carried out on hot cracking in steel castings. Some results from this work are set out below.

1 Testpieces were cast with a thin web of varying length, and hot cracking (caused by the restraint of the ceramic mould) was found to occur when a certain length was exceeded. Calculations simulating the solidification of the pieces were performed on the net of Fig. 1. Results are shown in Fig. 10 and Fig. 11. Taking the 1 490°C isotherm as indicating about 100% solidification, and

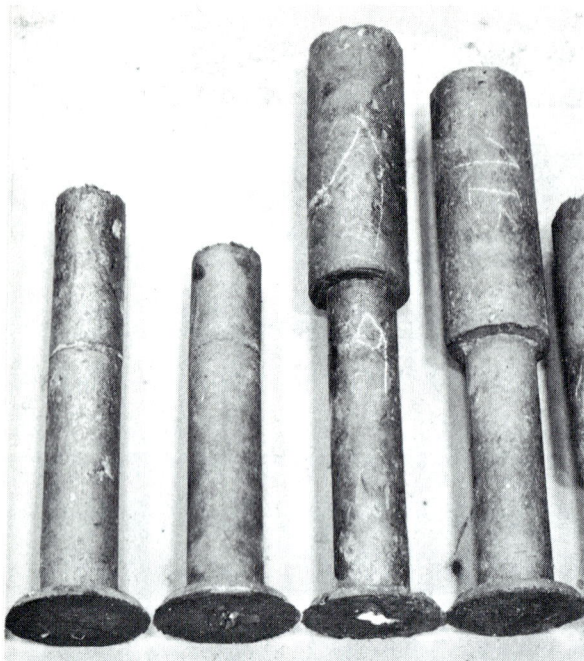

9   **Four low-carbon steel castings with axial geometry; the feeding heads of two parts are omitted**   [×$\frac{1}{7}$]

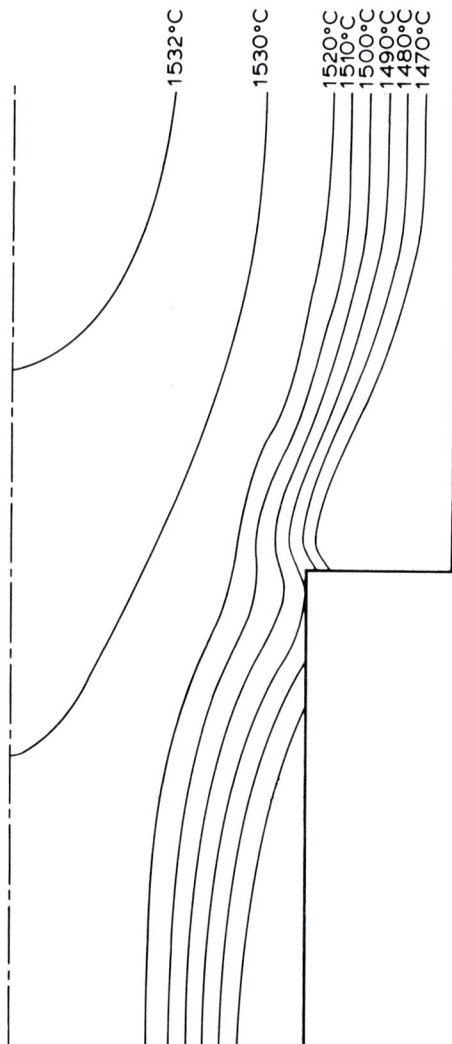

11 **The 1 490°C isotherm at different times after pouring for the parts shown in Fig. 9**

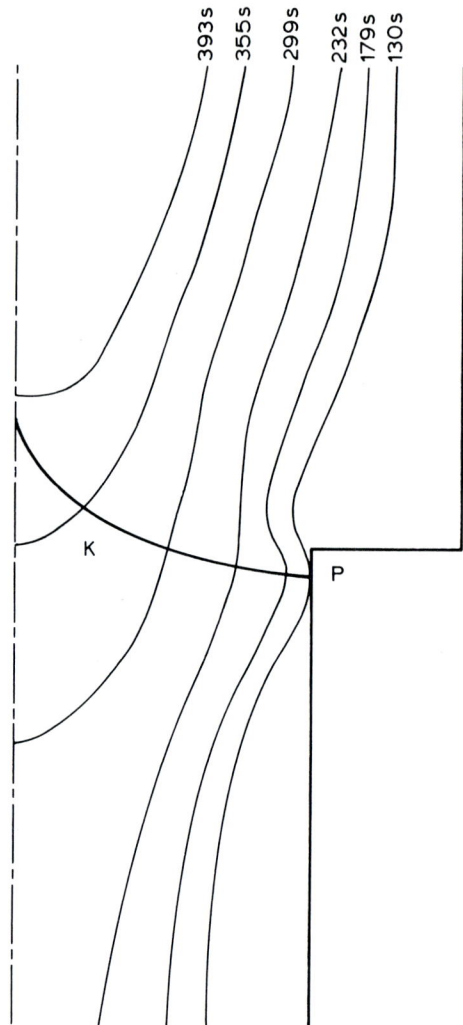

10 **Isotherms in the parts shown in Fig. 9 130 s after pouring**

12 **Geometry of the cracking profiles of two of the parts shown in Fig. 9**

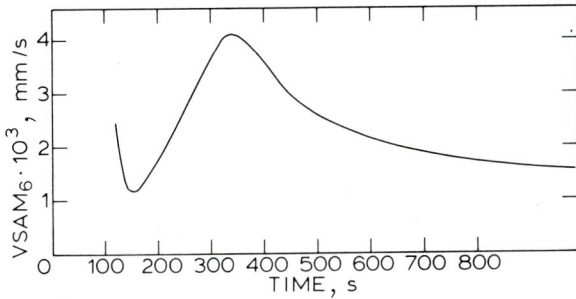

13 Rate of free contraction *VSAM* of the *i* = 6 column in Fig. 1 between *k* = 3 and *k* = *N* = 63 as a function of the time after pouring

drawing a curve *K* at right-angles to the isotherms of Fig. 11, the geometry of the cracking profile is found to be as shown in Fig. 12, indicating that cracking follows the progress of solidification. It is further seen that the temperature profiles indicate that cracking starts a little below the corner, as in fact it does on the parts of Fig. 9.

2 Calculating the speed of free concentration *VSAM* of the *i* = 6 column in Fig. 1, between *k* = 3 and *k* = *N*, using equation (5), one obtains the curve of Fig. 13. (*N* = 63, Δx = 5 mm.) To reduce the susceptibility to hot cracking it is important to minimize *VSAM* at that moment t_p when a point such as *P* on Fig. 11 solidifies completely. In Fig. 11 t_p is about 130 s and in Fig. 14 t_p is about 510 s, and *VSAM* can be reduced by increasing t_p. This can be done by increasing the "hot centre" either by increasing the radius of the thick part of the castings of Fig. 9 or by the use of exothermic padding. If $200 < t_p < 350$ s, *VSAM* is reduced when t_p is also reduced. This

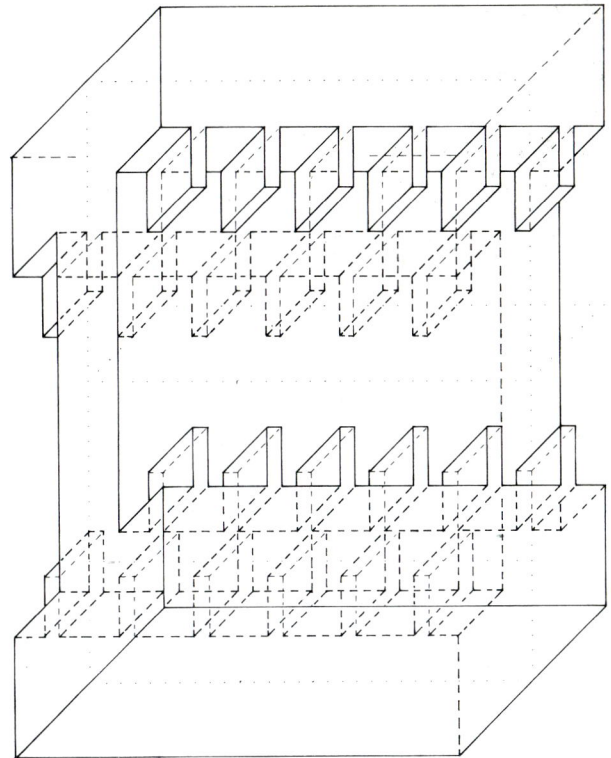

15 A casting geometry with cooling fins

can be done by reducing the radius of the thick part, by rounding the corner at point *P*, and by the use of chilling.

3. Calculations can easily be made on a thermal, 3-dimensional geometry such as that in Fig. 15. This geometry was used to examine the effect of cooling fins. The result was that fins change the value of t_p and will therefore be able either to decrease or increase the susceptibility to hot cracking.

REAL CASTING GEOMETRIES

This work is in progress but, for the moment, attempts have been made to put the real geometry into a normal net such that equation (11) will still be valid. Net points located away from the metal/mould interface can be treated as described above.

The problems to be solved concern the net points neighbouring the interface:

(i) to determine δ_{xx}, δ_{yy}, and δ_{zz} when the temperatures on the interface are known
(ii) to determine the temperature in the points of intersection between the net and the interface.

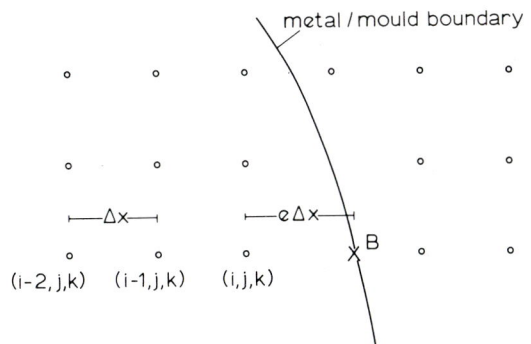

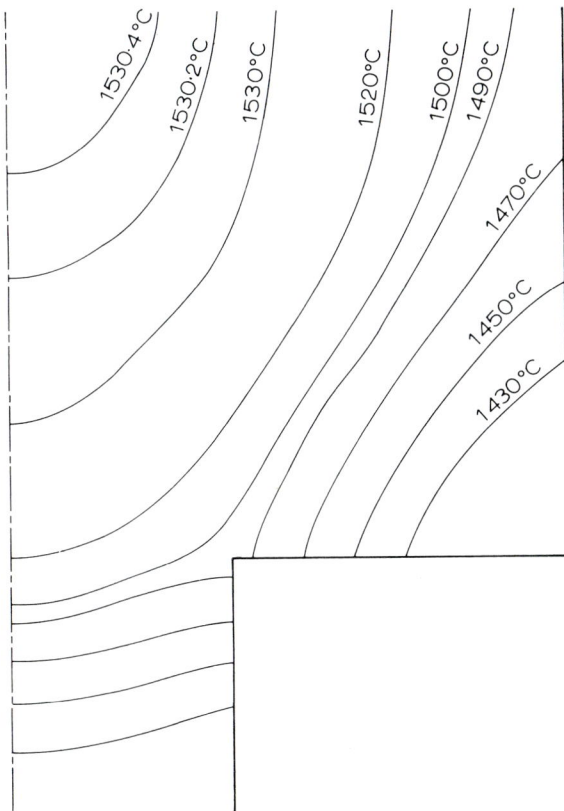

16 Net points near the metal/mould boundary

14 Isotherms 510 s after pouring in a part with a thicker feeder head than the parts shown in Fig. 9

Regarding (i), the same precision should be sought in the finite-difference approximation as in the remaining points to ensure reasonable accuracy in the calculations. This is obtained by the following formula[10]:

$$\delta_{xx} = \frac{1}{e(e+1)(e+2)}(e(e^2-1)\theta_{i-2,j,k} - 2(e^3-4e)\theta_{i-1,j,k} + (e^3-7e-6)\theta_{i,j,k} + 6\theta_B)$$ (23)

δ_{yy} and δ_{zz} similar. The symbolism is given in Fig. 16. Use of this formula has been shown to give good results if the interface temperatures are correct.

Regarding (ii), the temperatures must be determined from a suitable description of the thermal coupling of the metal and the mould domains. In the period after pouring it is a good approximation (sand mould) to say that the interface temperature is a constant, dependent on the curvature of the interface. (One can think of diffusion couples.) Later, an extrapolation technique is used, based on Fourier's law of heat conduction and the temperatures in net points on each side of the interface.

The method presented here was tested by simulating the solidification of a sphere using a computer program with three describing space parameters. It was found that the surface of the 1 490°C isotherm is a 0·1% carbon steel deviated about 4% from the sphere geometry measured in relation to the radius of the sphere.

CONCLUSIONS

Work in the areas of simple geometries is relatively easy since the computer programs are ready for use. Although lacking good material thermal properties in many cases, it is nevertheless possible to achieve results which improve the qualitative understanding of the influence of the various parameters in the casting process and which can thus be used when real casting geometries are considered. The work on real casting geometries will be continued, and it is believed that the cost of computer calculation will be reasonably low.

REFERENCES

1 'Moving boundary problems in heat flow and diffusion', (ed. J. R. Ockendon and W. R. Hodgkins), 1975, Oxford, Oxford University Press
2 B. CARNAHAN et al.: 'Applied numerical methods', 1969, John Wiley and Sons
3 A. F. EMERY AND W. W. CARLSON: J. Heat Transfer, 1971, 93, 136
4 R. V. S. YALAMANCHILI AND S. C. CHU: ibid., 1973, 95, 235
5 J. I. SOLIMAN AND E. A. FAKHROO: J. Mech. Eng. Sci., 1972, 14, (1), 19
6 P. N. HANSEN: Ph.D. thesis part 2, 1975, Department of Metallurgy, The Technical University of Denmark
7 H. S. CARSLAW AND J. C. JAEGER: 'Conduction of heat in solids', 11, 1973, Oxford, Oxford University Press
8 G. D. SMITH: 'Numerical solution of partial differential equations', 1969, Oxford University Press
9 M. C. FLEMINGS: Scand. Journal of Metall., 1976, 5, (1), 1
10 M. G. SALVADORI AND M. L. BARON: 'Numerical methods in engineering', 69, 1961, Prentice-Hall Inc
11 R. E. MARRONE et al.: AFS Cast Metals Research J., 1970, 6, (4), 184
12 V. DE L. DAVIES et al.: 'Determination of thermal data', Report 1976, Technical University of Denmark, Institute for Materials Processing, Building 425, Foundry Department
13 Y. K. CHUANG et al.: Metall. Trans., 1975, 6A, 235
14 M. I. KIRT AND R. D. PEHLKE: AFS Cast Metals Research J., 1973, 9, (3), 117

Computed feeding range for gravity die castings

V. de L. Davies and R. Moe

Two-dimensional solidification patterns of plates with feeders have been computed numerically for aluminium, copper, and some aluminium alloys. The results are presented as solidification diagrams describing the solidification along the centreline of the plate. From these diagrams the length of the centreline molten channel at which feeding problems arise can be determined. When this centreline capillary is longer than the calculated capillary feeding distance, centreline shrinkage will occur. In this way it is possible to determine the maximum feeding distance for pure metals, eutectics, and alloys with a narrow solidification range. The computed feeding distance for gravity die cast aluminium alloys, cast in dies at 300°C, is less than half the feeding distance for sand castings. The feeding distance increases with increasing die temperature and with decreasing length/thickness ratio. Very high die temperatures, giving solidification rates comparable to the solidification rate in sand castings, should give feeding distances comparable to the feeding distances of sand castings. Experimental determinations of the feeding distances for gravity die castings show scatter. Apparently the external atmospheric pressure compresses the plate, giving no visible centreline shrinkage in well degassed metal solidifying as a plate with parallel walls.

At the time of the conference, both authors were with SINTEF, Trondheim, Norway; R. Moe is now with A/S Strømmen Staal, Raufoss

LIST OF SYMBOLS

A = surface area
B = correction factor
$b' = \sqrt{k\rho c}$ = heat diffusivity for mould material
c = specific heat
H_b = heat content of elements at surface of casting
H_s^* = heat content at pouring temperature as a modified latent heat of solidification
h = heat-transfer coefficient
k = thermal conductivity
l = capillary length
l_f = capillary feeding distance
Δl = half length of element normal to surface
L = plate length
m = solidification contraction
r = capillary radius
S = plate thickness
t_S = solidification time
T_b = surface temperature of casting
T_d = die temperature before pouring
T_e = eutectic temperature
T_i = pouring temperature
T_l = liquidus temperature
T_m = mould temperature
T_0 = ambient temperature
T_s = solidus temperature
V = volume
v_s = solidus velocity along centreline
η = dynamic viscosity
ρ = density

The production of sound gravity die castings is as important as that of sound sand castings. However, the feeding of gravity die castings has received very little attention in the literature, whereas there is an abundance of papers dealing with the feeding of sand castings. The feeding range for gravity die castings is controversial, as it is variously claimed to be longer and shorter than the feeding range of sand castings.

The feeding of gravity die castings seems to be determined by the experience of the die maker, in respect of both size and position of the feeders. The size of the feeders can be calculated by the methods used for sand castings, e.g. using the modulus method.[1] There is, of course, one important limitation to feeder placement in gravity die castings in that both runners and feeders must be positioned in the parting plane of the die.

The main difference between sand and gravity die castings is the much faster solidification in dies, which influences both solidification pattern and structure. Fast solidification results in a narrow solidification band which improves feeding. On the other hand, fast solidification leaves less time available for feeding.

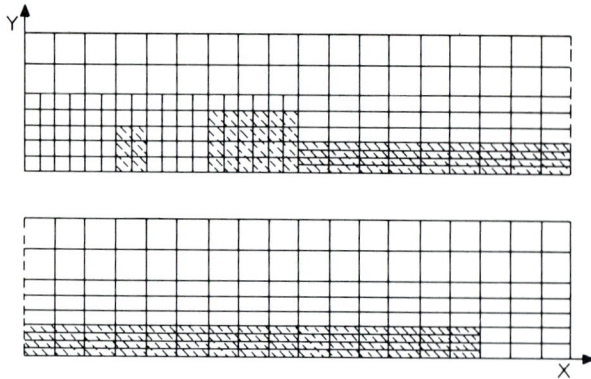

1   Half-model divided into elements; plate length $L = 24$ cm and plate thickness $s = 2$ cm

NUMERICAL COMPUTATION OF HEAT CONTENT

The method used in computing the feeding distance for gravity die castings is the same as that employed previously for determining the feeding distance for sand castings.[2] The computed results for sand cast plates agreed with experimentally determined feeding distances.

To simplify the computations a 2-dimensional model has been used, and this applies only to plates. In the computer program finite differences are used in an explicit solution of the partial differential equation for transient heat conduction. To include heat of transformation and variations in specific heat, heat content is used instead of temperature. To take care of variations in thermal conductivity with temperature a modified temperature is employed.

The geometrical half-model for computation is divided into rectangular elements as shown in Fig. 1, where the X-axis is a line of symmetry. This model is a cross-section of a die for a plate-formed casting (see Fig. 2).

When starting the computation, each element is given a heat content corresponding to the chosen temperature. When the mould is poured, all elements representing cast metal are at the casting temperature and all elements representing the mould are at the mould temperature. This temperature distribution is a simplification. If desired it is possible to start with any chosen temperature gradient in mould and cast metal.

In gravity die casting the main resistance to heat transfer from the casting takes place at the interface between casting and die. This heat transfer is complicated by the air gap which is formed during solidification. The air gap appears shortly after the

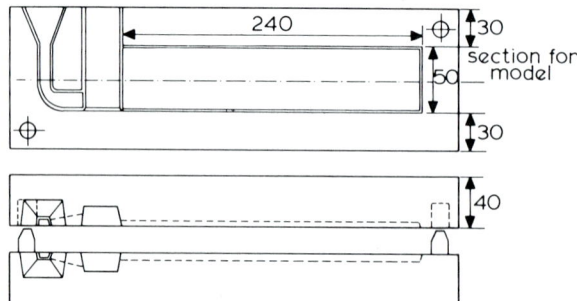

2   Cast iron die

Table 1   Conditions for air gap formation*

Alloy	Air gap formation
Aluminium	$H_b = 52400\Delta l + 628$
Al–12 Si	$H_b = 52400\,\Delta l + 628$
Al–7 Si–Mg	$T_b = 577°C$ for $T_d < 260°C$ or
	$T_b = 557 + 0.077 T_d$
Al–5 Si–3 Cu	$H_b = 21000\,\Delta l + 800$
Al–4 Cu–Ti	$H_b = 21000\,\Delta l + 880$
Copper	$H_b = 25100\,\Delta l + 400$

* Heat-transfer coefficients:
  Before air gap formation, 4.2×10^3 W/m²/°C
  After air gap formation, 1.0×10^3 W/m²/°C.

mould has been filled, as soon as the layer of metal has solidified against the mould wall. Measurements on aluminium and aluminium alloys cast in a cast iron mould at mould temperatures between 250 and 400°C have shown that the air gap appears after 3–5 s.[3] In the numerical simulation the change in the heat-transfer coefficient must be controlled by

(i) the temperature at the surface of the casting, or
(ii) the heat content of the elements in the casting against the die wall.

Table 1 gives the conditions for air-gap formation used for the different alloys. The values are determined by comparing experimentally measured and computed cooling curves. The heat-transfer coefficients refer to a commercial ceramic die coating of thickness ~ 50 μm.

The results of the numerical computations are given as solidification diagrams. These location–time diagrams give the state of solidification as a function of distance along the plate and time after pouring for elements adjacent to the centreline of the plate. The solidification diagram for aluminium is shown in Fig. 3. Lines are plotted for the liquidus (the temperature has fallen to the solidification temperature), start of solidification, this being delayed owing to release of latent heat of fusion for elements closer to the surface of the plate, and for solidus (complete solidification). As desired curves can be plotted for any fraction solidified, e.g. 70, 80, 90, and 95% solidified. For pure metals the events between 90 and 100% solidified will be of most interest with respect to feeding.

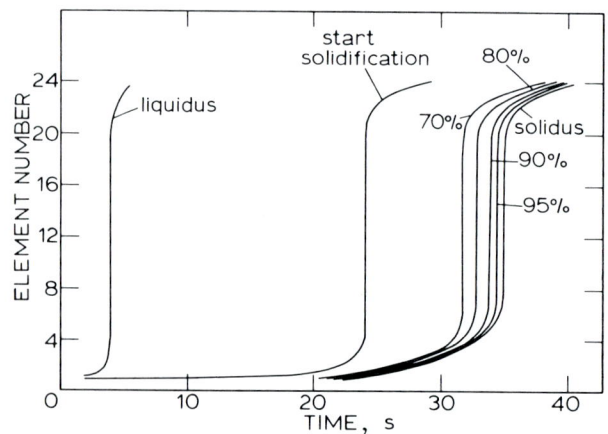

3   Solidification diagram for aluminium; plate length 24 cm, plate thickness 2 cm; pouring temperature 700°C, die temperature 250°C

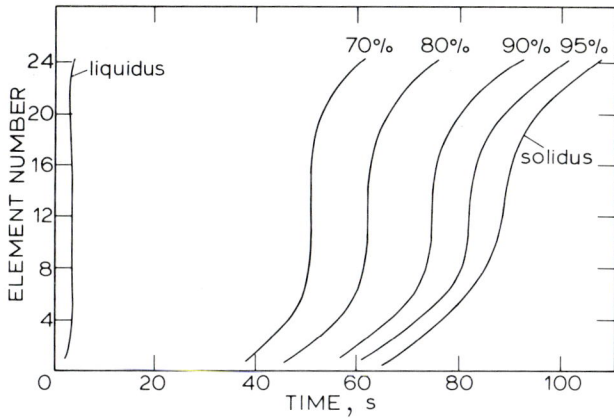

4 Solidification diagram for Al–5 Si–3 Cu; plate length 24 cm, plate thickness 2 cm; pouring temperature 700°C, die temperature 300°C

As shown in Fig. 4, the solidification diagram for an alloy with a wide solidification range is rather different from that of a pure metal. Curves for liquidus, different fractions solidified, and solidus are plotted. For these alloys the events between 70 and 100% solidified are of particular interest. It is generally considered that mass feeding ends when 70% is solidified.

FEEDING OF METALS AND ALLOYS
When feeding metals and alloys it is necessary to distinguish between several modes of solidification. The extremes are plane front solidification from the mould wall, and the 'mushy' solidification of alloys with a wide solidification range. Depending on alloy composition, nuclei present, and solidification range a number of modes of solidification may appear.

Pure metals
Very pure metals solidify by growth from the mould wall with a plane solidification front. Late in the solidification process a channel will be formed in the middle of the casting, where feeding must take place over a fairly long distance, as shown in Fig. 5. This feeding may continue as capillary flow until the cross-section of the remaining melt is very small.

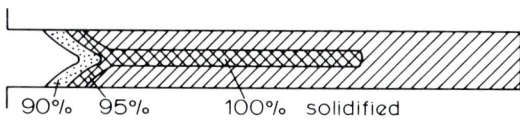

5 Centreline channel

Analyses of capillary flow have been made elsewhere,[4] but the theoretical results are applicable only to very pure metals. In this work it is intended to try to correlate the feeding ranges of commercial metals and alloys by means of the relevant physical data and solidification diagrams.

When a plate solidifies there is of course no circular capillary but a thin flat layer of melt in the centre of the casting. For the sake of simplicity the remaining melt is considered to be confined in a capillary.

From the shape of both the solidification diagram in Fig. 3 and the solidus velocity along the centreline in Fig. 6, it is obvious where centreline shrinkage will appear, but the boundary between sound material and shrinkage is not well defined. To determine the boundary more exactly the capillary feeding distance is calculated using the semi-empirical equation[2]

$$l_f = B \frac{\rho r^2}{\eta m v_s}$$

From earlier work[2] it was determined that $B = 0.375$. The data employed are given in Table 2.

A casting is considered sound when there is no visible centreline porosity. A normal sight can just discern between points 1.5×10^{-4} m apart. Thus, the transition from invisible to visible porosity takes place when $r = 0.75 \times 10^{-4}$ m.

For aluminium:

$$l_f = B \frac{2400(0.75 \times 10^{-4})^2}{1.5 \times 6 \times 10^{-2} v_s}$$
$$= \frac{1.5 \times 10^{-4} B}{v_s} = \frac{0.56 \times 10^{-4}}{v_s}$$

A similar calculation for the other metals and alloys investigated gives the following results:

$$\text{Al–7 Si–Mg } l_f = \frac{0.34 \times 10^{-4}}{v_s}$$

$$\text{Al–12 Si } \quad l_f = \frac{0.85 \times 10^{-4}}{v_s}$$

$$\text{Cu } \quad l_f = \frac{0.85 \times 10^{-4}}{v_s}$$

The solidus velocity along the centreline of the plate v_s can be calculated from the computer output and is shown in Fig. 6 for aluminium. The movement of the solidus along the centreline is, of course, not the growth rate as the crystals grow from the surface towards the centre of the plate.

Table 2 Physical data

Alloy	ρ 20°C, kg/m³	ρ_1, kg/m²	k, W/m/°C	c, kJ/kg °C	H_s, kJ/kg	m, %	η, mPa s	T_1, °C	T_s, °C	T_e, °C
Aluminium	2 700	2 400	230	1·05	394	6	1·5	660	—	—
Al–12 Si	2 650	2 300	161	1·05	460	4	1·5	—	—	577
Al–7 Si–Mg	2 670	2 300	184	1·05	433	5	3·0	610	—	577
Al–5 Si–3 Cu	2 700	2 400	126	1·05	490*	6	1·5	615	520	—
Al–4 Cu–Ti	2 750	2 400	155	1·00	494*	6	1·5	640	550	—
Copper	8 960	8 300	373	0·42	208	5	4·0	1083	—	—
Cast iron $C_E = 3·86$	7 600	—	54	0·60	251	—	3·5	1210	—	1 154

* Inclusive $c\Delta T$ for alloys with a solidification range ΔT

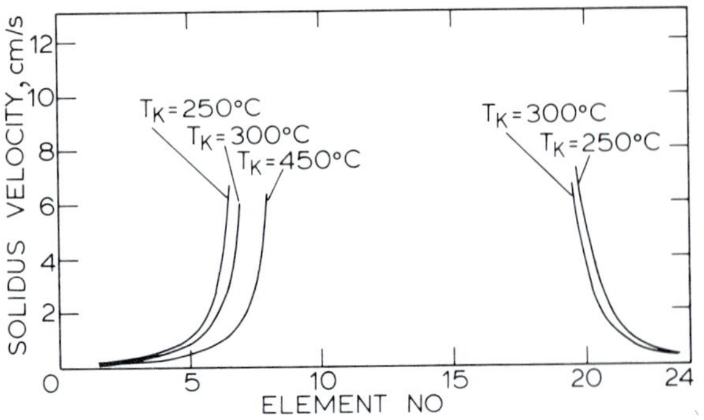

6 Solidus velocity along centreline for aluminium at different die temperatures

When the capillary feeding distance is shorter than the real capillary length, centreline shrinkage will appear. The capillary length can be found from the solidification diagram as the vertical distance between the fraction solidified curve chosen and the solidus. For aluminium of commercial purity it is assumed that feeding problems in the centreline channel start when 95% has solidified. The capillary length is then the vertical distance between the 95% solidified curve and the solidus. This length l is plotted in Fig. 7. The limits of centreline shrinkage are the points where the curves for l_f and l intersect as shown in Fig. 7. From this figure the portions of the plate without centreline shrinkage can be determined. Expressing the feeding distance as a multiple of plate thickness s gives a value of $4 \cdot 1s$.

The solidification diagram for pure copper is very similar to the diagram shown for aluminium in Fig. 3. The resulting computed feeding distance is $4 \cdot 0s$ for copper.

Aluminium alloys
The same procedure has been applied to two aluminium alloys, Al–7 Si–Mg and Al–12 Si. The computed solidification diagram for Al–7 Si–Mg is shown in Fig. 8. Diagrams have been computed for different die temperatures. The computed feeding distances for Al–7 Si–Mg and Al–12 Si are given in Table 3.

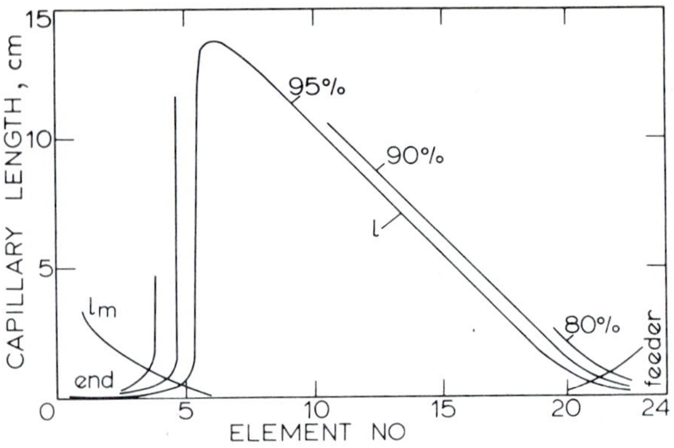

7 Capillary feeding distance and capillary length for aluminium

Alloys with a wide solidification range solidify in a mushy or pasty manner and during solidification the whole casting, or major parts of it consists of a mixture of solid and liquid. The solidification diagram for Al–5 Si–3 Cu is shown in Fig. 4.

During the first part of the solidification mass feeding will take place, i.e. movement of liquid and solidified crystals. The mass feeding will continue until a network of solidifying crystals prevents further movement. This is

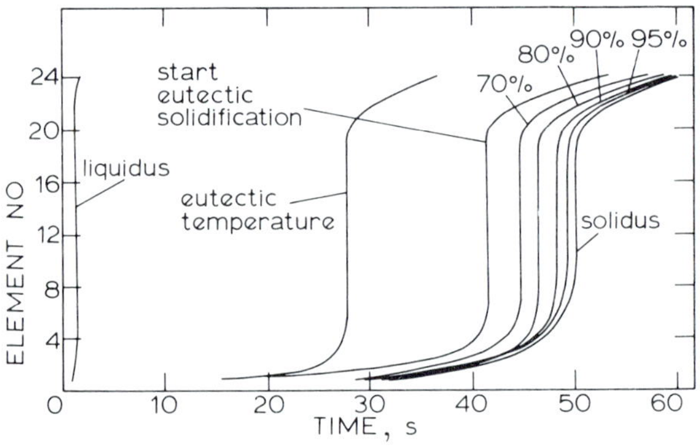

8 Solidification diagram for Al–7 Si–Mg; plate length 24 cm, plate thickness 2 cm; pouring temperature 650°C; die temperature 250°C

Table 3 Computed feeding distances

Alloy	T_i, °C	s, cm	L, cm	Length/thickness ratio	T_d, °C	Feeding distance, expressed as multiple of s End	Feeder	Total
Aluminium	700	2	24	$12s$	250	2·3	1·5	3·8
					300	2·5	1·6	4·1
					450	2·9	1·9*	4·8*
Al–7 Si–Mg	650	1	16	$16s$	250	1·7	1·4	3·1
					300	1·8	1·7	3·5
					350	2·0	1·9	3·9
					450	3·0	2·9	5·9
	650	2	24	$12s$	250	2·2	1·5	3·7
					300	2·5	1·7	4·2
					350	2·8	2·1	4·9
					450	4·1	2·0*	6·1*
	650	2	11	$5·5s$	250	2·3	1·6	3·9
Al–12 Si	650	2	24	$12s$	300	3·2	2·3	5·5
Copper	1 150	2	24	$12s$	300	2·5	1·5	4·0

* Too short because of too small a feeder.

generally believed to take place at 70% solidified. When mass feeding terminates a continuing feeding may take place interdendritically, but this will be only a short-range effect.[2] In gravity die castings, there is very little time available for interdendritic feeding. This means that gravity die castings of aluminium alloys with a wide solidification range will contain ~1·8% porosity as microshrinkage, even with favourable feeding conditions.

Casting conditions

An important variable in gravity die castings is the die temperature. By changing the temperature of the die, the solidification conditions can be changed, and in practice the temperature may vary between 250 and 450°C.

Most of the work done to correlate feeding distance with die temperature was performed with the alloy Al–7 Si–Mg. The feeding distance was computed for the die temperatures 250°, 300°, 350°, and 450°C. The casting temperature was 650°C. The resulting feeding distances are given in Table 3.

Experimental results

It was desirable to compare the computed feeding distances with measured ones. Some plate castings were made in the cast iron die shown in Fig. 2. The measurements resulted in feeding distances with a wide scatter and the feeding distance apparently increased with degassing of the melt.

Discussion of results

When the results of computations for gravity die castings and sand castings are compared (*see* Table 4), it is seen that the feeding distances in gravity die castings are much shorter than the feeding distances for sand castings.

The general opinion seems to contradict this result,[5] as do the authors' own experiments with well degassed aluminium alloys. The discrepancies may be explained as follows.[6] The thin sections employed in gravity die castings result in thin centreline shrinkage and also some finely distributed microporosity. Thin centreline shrinkage may easily collapse owing to external atmospheric pressure and disappear. If gas is present this may collect in the centreline shrinkage and prevent the compression. In well degassed melts no gas will be available to prevent the compression of centreline shrinkage and the casting will be sound.

To indicate the importance of the external atmospheric pressure one may imagine the solidification

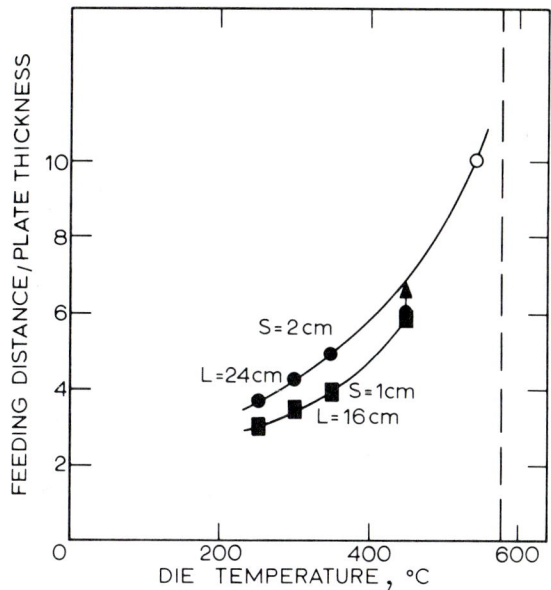

● plate thickness 2 cm, plate length 24 cm; ■ plate thickness 1 cm, plate length 16 cm; ○ calculated die temperature for $t_{S_{die}} = t_{S_{sand}}$

9 Feeding distance *v.* die temperature for Al–7 Si–Mg

shrinkage of 6% of the last 10% to solidify to be concentrated at the centreline in a plate 10 mm thick. This would result in a gap $10 \times 6/100 \times 10/100 = 0.06$ mm. To resist the compression of this gap due to the external pressure of 1 bar, the yield strength would have to be at least 22 MPa (22 N/mm²). The yield strength of aluminium just below the solidus is only a small fraction of this.[6] The resulting plate would thus be 0·06 mm thinner than expected. This compression of the shrinkage applies to plane parallel plates. Small irregularities in thickness or in mould coating will result in local hot spots and local shrinkage which will give scatter in experimental results.

The influence of die temperature on feeding distance is shown in Fig. 9. The feeding distance increases with increasing die temperature, i.e. with a decrease in solidification rate. It might be expected that at the same solidification rate the feeding distances would be comparable for gravity die and sand castings. By using the following very simple equations for the solidification times in sand moulds and gravity dies and making them equal, the die temperature T_m may be calculated as follows:

$$\text{sand:} \qquad t_S = \frac{\pi \rho^2 H_S^{*2}}{4b'^2(T_1 - T_0)^2}\left(\frac{V}{A}\right)^2$$

$$\text{gravity die:} \quad t_S = \frac{\rho H_S^*}{h(T_S - T_m)}\frac{V}{A}$$

where half the plate thickness $S/2 = V/A$, and the plate thickness is 2 cm.

For this calculation a mean heat-transfer coefficient of 2×10^3 W/m² °C has been used. The die temperature has been calculated as 543°C. If such a high die temperature were used for Al–7 Si–Mg a feeding range comparable to the feeding range for sand castings should be realized, but this is never achieved in commercial castings.

Table 4 Comparison of computed feeding distances for gravity die and sand castings

Alloy	Sand cast	Gravity die cast (die temp.) 250 °C	300 °C	350 °C	450 °C	L
Aluminium	7·5	3·8	4·1	—	>4·8	12s
Al–7 Si–Mg	10	3·1	3·5	3·9	5·9	16s
		3·7	4·2	4·9	>6·1	12s
Al–12 Si	10·5	—	5·5	—	—	12s
Copper	9	—	4·0	—	—	12s

Feeding distance, expressed as multiple of s

CONCLUSIONS

The feeding distance for gravity die castings is considerably shorter than that for sand castings when using dies at ordinary temperatures. Only at very high die temperatures will the feeding distance approach that of sand castings.

In real plate castings which are well degassed, the centreline shrinkage will be compressed owing to external atmospheric pressure and the casting will apparently exhibit a considerable feeding length. Local hot spots and high gas content may cause local shrinkage.

ACKNOWLEDGMENTS

This investigation was sponsored by the Royal Norwegian Council for Scientific and Industrial Research under contract B0512.3089. The authors wish to acknowledge the cooperation of members of the staff of SINTEF, The Foundation of Scientific and Industrial Research at the University of Trondheim. Numerical computations were made using UNIVAC 1108 at the Computation Centre, University of Trondheim, The Norwegian Institute of Technology.

REFERENCES

1 R. WLODAWER: 'Directional solidification of steel castings', 1966, Oxford, Pergamon Press
2 V. DE L. DAVIES: *AFS Cast Metals Res. J.*, 1975, Jun. 33
3 R. MOE: 'Numerically determined feeding ranges for gravity die castings', (in Norwegian), Rep. STF16 A75028 1975, Trondheim, SINTEF
4 T. S. PIWONKA AND M. C. FLEMINGS: *Trans. Met. Soc. AIME*, 1966, **236**, (8), 1 157
5 R. A. FLINN: 'Fundamentals of metal casting', 1963, Reading, Mass., Addison-Wesley
6 V. DE L. DAVIES: *Støperitidende*, 1975, **41**, (6), 129

Prediction of structure in industrial VAR, ESR, and PAR ingots using computed local solidification times

A. S. Ballantyne and A. Mitchell

Previous work has shown that computed values of the local solidification time (LST) can be used to deduce dendrite spacings in ingots on the basis of published solidification structure data. This paper reports the extension of the work to different alloy systems: AISI 52100 steel, AISI M2 steel, and Inconel 718. The results suggest that the computed LST values in remelted ingots, used in conjunction with laboratory directional solidification data, produce a good approximate picture of the ingot structure. Although phenomena involving factors other than LST, such as 'freckles' or the columnar-equiaxial transition cannot be predicted, the columnar region can be described with adequate accuracy for most industrial purposes.

The authors are in the Department of Metallurgy, University of British Columbia, Vancouver, Canada

Previous workers in the field of solidification in electroslag (ESR), vacuum arc (VAR), and plasma arc (PAR) remelted ingots have commented extensively on the relationship between dendrite spacings and ingot thermal regime. Mellberg,[1] Holzgruber,[2,3] Kroneis et al.,[4] and Eckstein[5] have attempted to derive ingot temperature gradients in the solidification region by correlating either primary or secondary dendrite arm spacings between ESR ingots and the results of laboratory directional solidification experiments. The inaccuracy of this technique had been discussed by Ballantyne et al.[6] who reversed the above procedure by computing the ingot local solidification time (LST) and used this value to deduce dendrite spacings in the ingot on the basis of published solidification structure data. It is clear that the latter procedure provides a good way of confirming the results of thermal computations within the accuracy to be expected from measurements of dendrite growth. In the present work the aim has been to extend the method used by Ballantyne to a number of alloy systems and further to demonstrate the use of the technique in aiding structure improvement and prediction.

SOLIDIFICATION STRUCTURES
Several industrial ingots were sectioned for structural analysis and are described as follows.

AISI 52100 steel
A transverse slice of a VAR ingot, 400 mm dia., was obtained from Latrobe Steel Corporation at an ingot length of 2 200 mm. The slice was etched to obtain dendrite spacings as shown in Fig. 1. The LST values were computed using the method of Ballantyne et al.[6]

AISI M2 steel
A transverse slice of an ESR ingot, 390 mm dia., was obtained from Latrobe Steel Corporation at an ingot length of 1 050 mm. Results of the subsequent structure analysis are shown in Fig. 2. A second sample, from an ESR ingot of 500 mm dia., was also examined (from Stora Kopparberg), and the results are shown in Fig. 3.

Solidification parameters of 52100 and M2
The solidification of 52100 has been extensively studied in respect of the spacings shown in Fig. 1. Flemings et al.,[7] Melford and Granger,[8,9] and Rickinson[10] have reported primary and secondary dendrite spacings for this alloy. The results are collected in Fig. 4 and are recalculated in the form of the local solidification time. It will be seen that the present results of Fig. 1 agree well with the values of Fig. 4, within the precision to be expected of the solidification measurements. These values also correlate well with the small-scale ESR ingot measurements of Mellberg[1] on 52100. The authors have used the computation method of Ballantyne to derive values of the LST in the 52100 VAR ingots studied by Shcherbakov et al.[11] who report primary dendrite arm spacings as a function of melting rate. Although the precision of their structure examination is low, the results agree well with the present values, as shown in

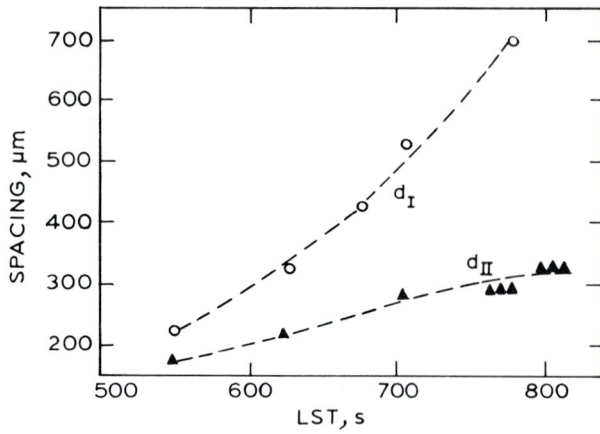

1   LST values and structure in VAR 52100 (400 mm dia.)

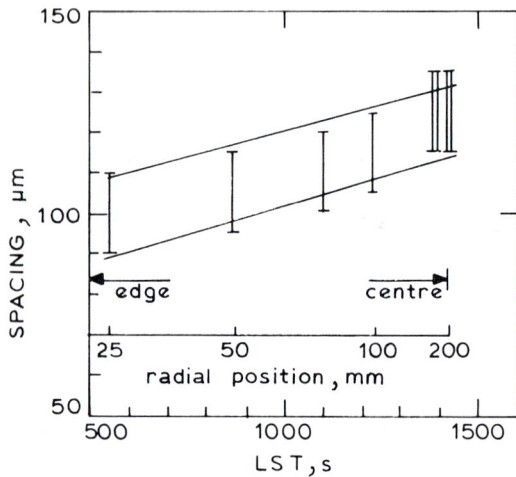

2   LST values and structure in ESR M2 (390 mm dia.)

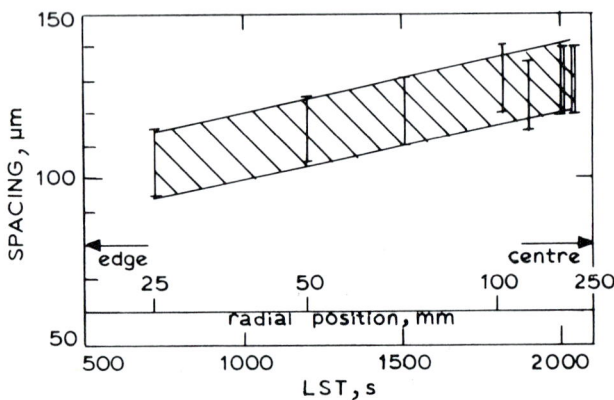

3   LST values and structure in ESR M2 (500 mm dia.)

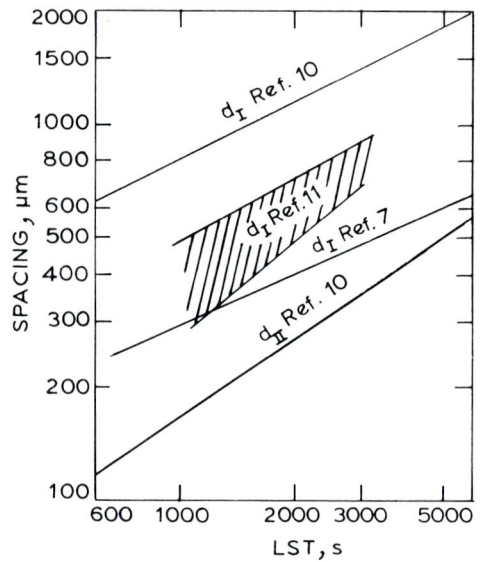

4   Collected published values for LST and structure in 52100

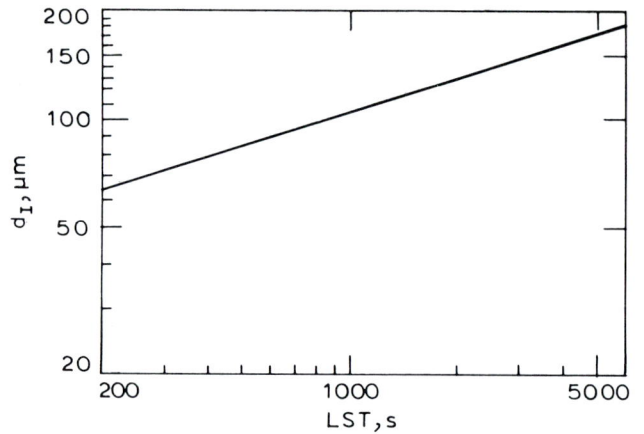

5   LST values and ledeburite cell size in M2 (after Mellberg[1])

Fig. 4. It may be concluded that the method of computing the LST in a remelting process is a valid route to the prediction of local structure under these conditions.

The solidification of M2 has been studied by Mellberg,[1] and also by Barkalow,[12] in respect of the ledeburite cell size as a function of solidification time and temperature gradients. Their results are reproduced in

Fig. 5 for the range of LST appropriate to the ingots studied. As with the 52100 alloy results, it will be seen that the present values of cell size agree well with the experimental solidification values. It is concluded that in this alloy also the computation method represents a valid route to structure prediction in respect of cell size.

The above conclusions are not unexpected in view of the extensive previous body of experimental evidence relating solidification time to structure. However, they serve to emphasize that the computation of LST is significantly more accurate than the correlations of LST and dendrite spacings. Further, once having demonstrated the relationship between computed LST and ingot structure the computed values may be used to optimize the melting rates in remelting processes. The application of this technique has been demonstrated by Ballantyne *et al.*[6] for the superalloy Udimet-700.

PRECIPITATION AND LST

The relationship between LST and dendrite spacing ought, in principle, to be extended to a relationship between LST and primary precipitation, i.e. precipitation of a second solid phase during solidification. In

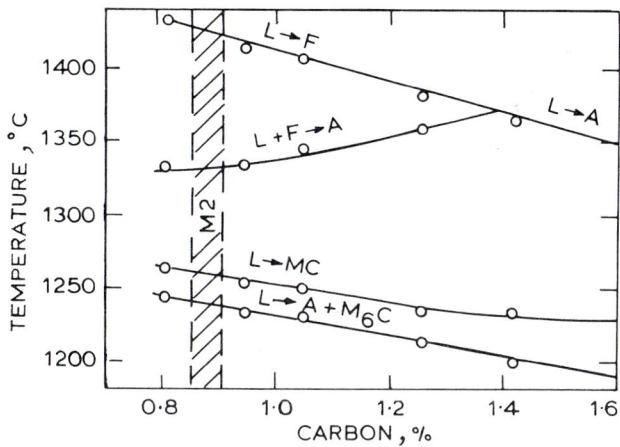

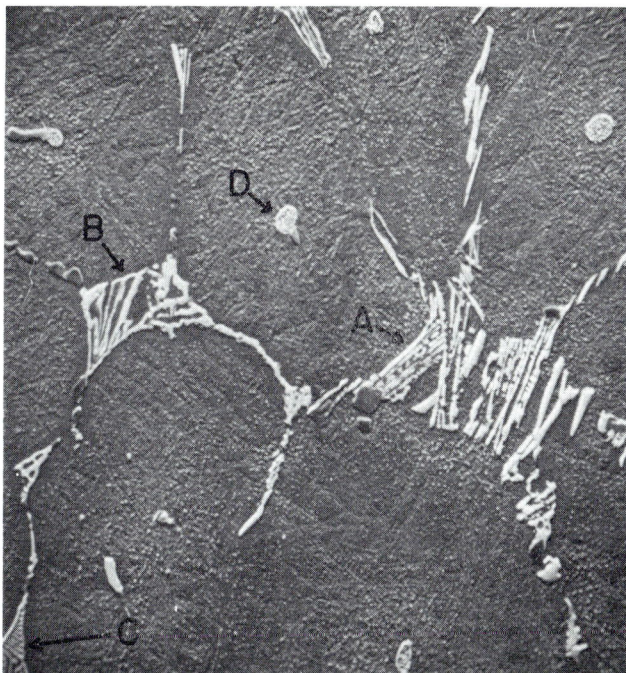

6 *a* Equilibrium solidification and structure of M2, A2, M₂C+MC precipitates, B and C: M₆C+MC precipitates, D: transformed δ-core of the dendrite [×230]; *b* M₂C+MC precipitate of area A in Fig. 6*a* [×500]; *c* M₆C+MC precipitate of area B in Fig. 6*a* [×500]

order to examine such a relationship the precipitates formed during the freezing of M2 and Inconel 718 were investigated.

Solidification of M2

The solidification sequence of M2 steel has been determined[1,11] and is summarized in Fig. 6. From this data one may make a schematic diagram of the solidifying centreline zone in the ESR ingot of Fig. 3, as shown in Fig. 7, assuming equilibrium solidification. It will be noted from Fig. 7 that the carbide structure of M2 is controlled by precipitation in a very narrow section of the ingot.

The structure which results from the solidification of the ingot of Fig. 7 is illustrated in Fig. 8 as a function of computed LST. The transformed δ-core and γ-primary dendrite structures are clearly visible, together with the three forms of carbide, namely, M₆C, M₂C, and MC. The carbides M₆C and MC are held to represent[13] the

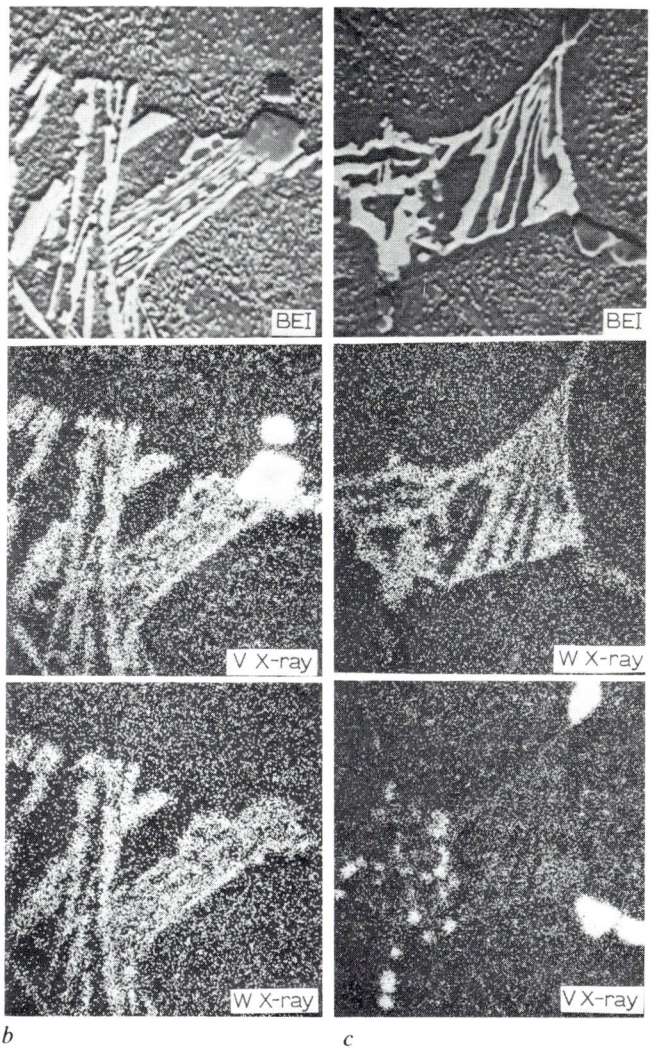

equilibrium structure, with M₂C appearing at low LST values.

It appears from the literature[14] that the extensive presence of M₆C in this alloy leads to large, undesirable average carbide size after heat treatment in either cast or wrought product. Barkalow *et al.*[12] state that the M₆C appears in alloys of low vanadium content (e.g. T1), and that it is suppressed by high vanadium, high carbon contents. One can establish that the M₆C–M₂C mixture is controlled primarily by LST in an M2 ESR ingot by examining a 'freckled' area, as shown in Fig. 9. It is clear that in this highly segregated region, the primary carbide is M₂C. This dependence on LST may be further confirmed by examining the structures of Fig. 8, where it is seen that M₆C can form in areas where the segregation is normal and the dendrites have δ-cores indicating that the alloy has frozen following the predicted δ → γ transition.

From the above discussion it can be concluded that a low LST is desirable in this alloy, not only from the point

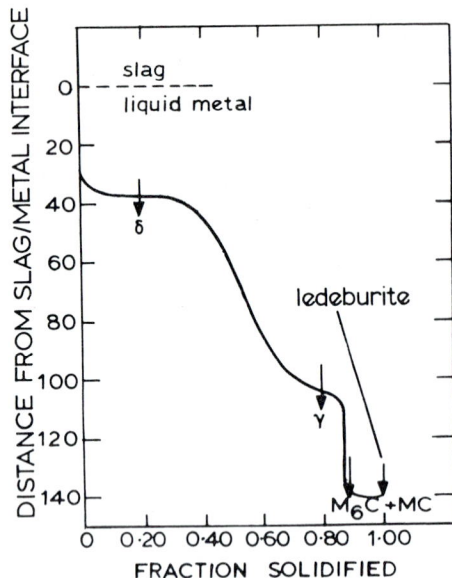

7 Centreline solidification schematic for ESR M2 ingot, 500 mm dia., 430 kg/h

of view of minimizing the dendrite primary spacing, but also in suppressing the formation of M_6C.

In order to identify the critical range of LST for the suppression of M_6C, the authors heat treated samples of a 500 mm dia. M2 ESR ingot in the as-cast condition containing solely M_2C and MC carbides. The sequence and results are shown in Figs 10–14. It is recognized that this series of tests does not reproduce the ESR freezing sequence exactly, and is hence only a qualitative indication of the LST necessary for M_6C formation. Nevertheless, the results indicate that the transformation

$$M_2C + A \rightarrow M_6C + MC$$

requires a LST approaching 2 500–3 000 s for a 1 : 1 ratio of M_2C/M_6C.

The validity of the above discussion may be illustrated by reference to Fig. 8 where it is clear that the central section of the ingot contains approximately 20–30% M_6C, with a LST of 2 100 s: the outer portion of the ingot with a LST < 1 500 s contains virtually no M_6C. The ingot in Fig. 2 has a LST < 1 500 s throughout its section and contains no M_6C.

The authors examined a further ESR ingot of M2 and of 500 mm dia. with a melting rate of 570 kg/h. At the position where the LST was computed, the values were almost identical to those of the ingot in Fig. 8 and the structures were also identical. In addition, the hot-top regions of these ingots were examined where the LST was between 3 000 and 4 500 s. In all cases, the carbides were entirely $M_6C + MC$.

It is concluded from the above results that the remelting sequence for M2 must result in a LST ≤ 2 000 s for good ingot structure and also for the desired carbide composition.

Solidification of Inconel 718

The equilibrium solidification sequence in Inconel 718 has been defined by Eisenstein[15] and others[16–21]. The alloy is aged to the desired strength condition by a number of thermal treatments designed to precipitate γ' phases (A_3B compounds: A is Ni, B is Nb, Al, and Ti). Ideally, the alloy would solidify as a single-phase solid, subsequently precipitating these compounds at a lower temperature. In practice, the alloy also contains two

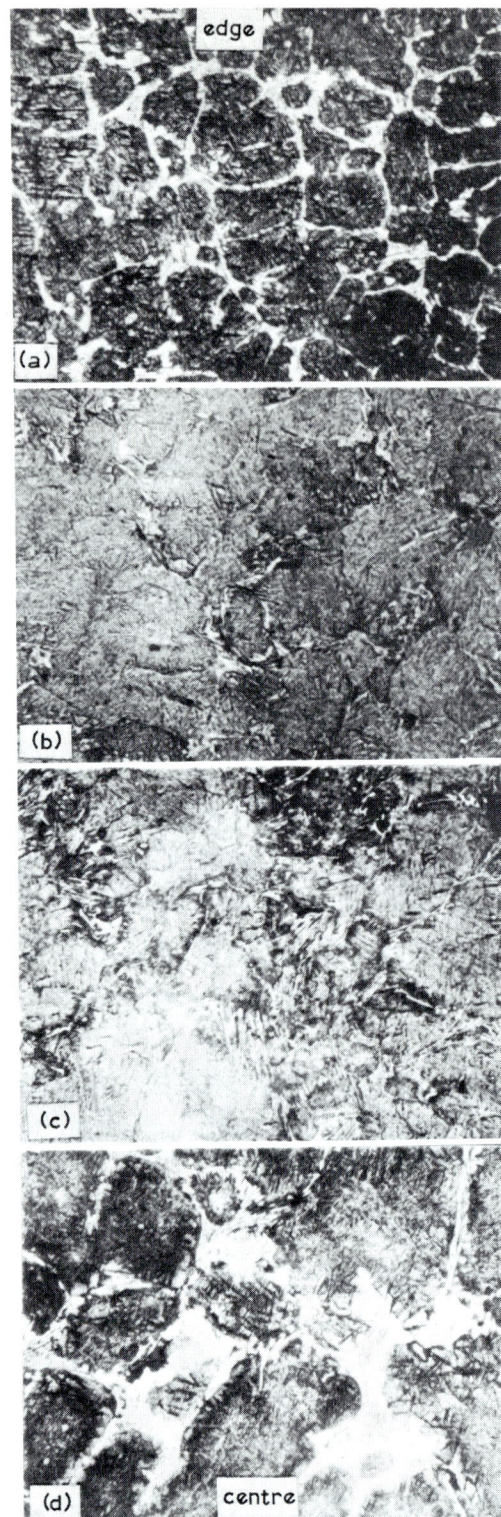

a LST = 720 s, 60 mm from edge, 100% M_2C + MC; *b* LST = 1 530 s, 100 mm from edge, 100% M_2C + MC; *c* LST = 1 900 s, 200 mm from edge, 5–10% M_6C; *d* LST = 2 100 s, centreline, 20–30% M_6C

8 Solidification structures and LST values for ESR M2, 500 mm dia., 430 kg/h [× 100]

other phases, Nb–Ti carbonitrides and an A_2B compound (A = Ni, Fe, and Cr; B = Nb, Mo, and Ti) of Laves phase structure. The carbonitrides may be minimized by compositional control, but the appearance of Laves phase causes a deterioration in ductility.

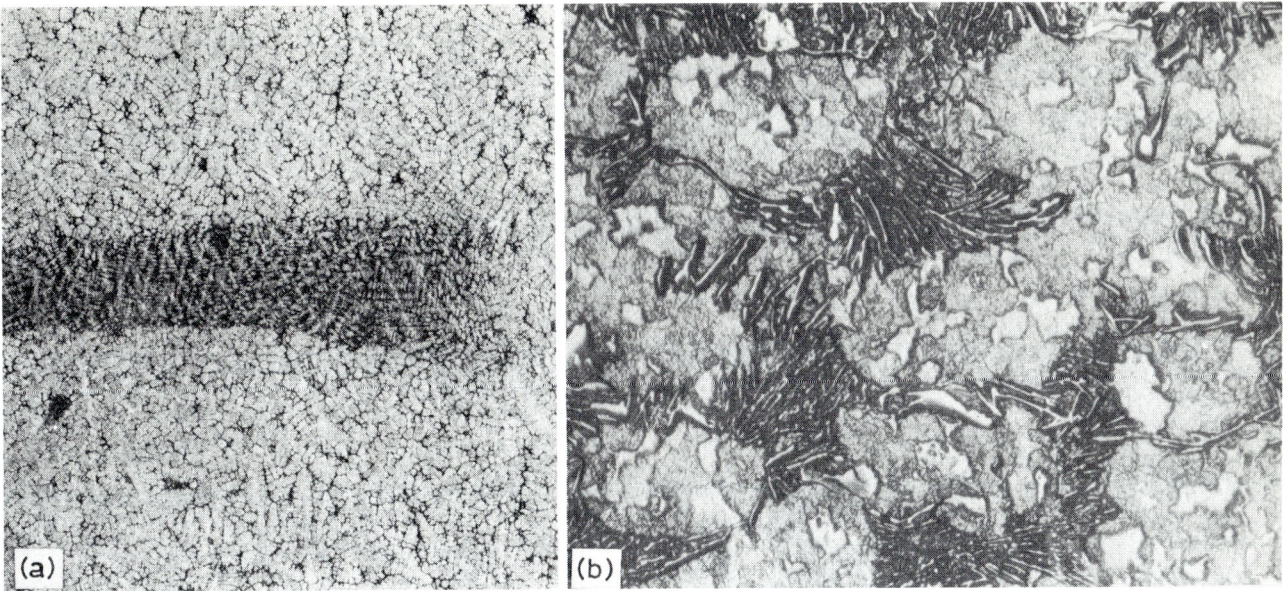

a [×5·5]; *b* [×100]

9 Freckled area in M2

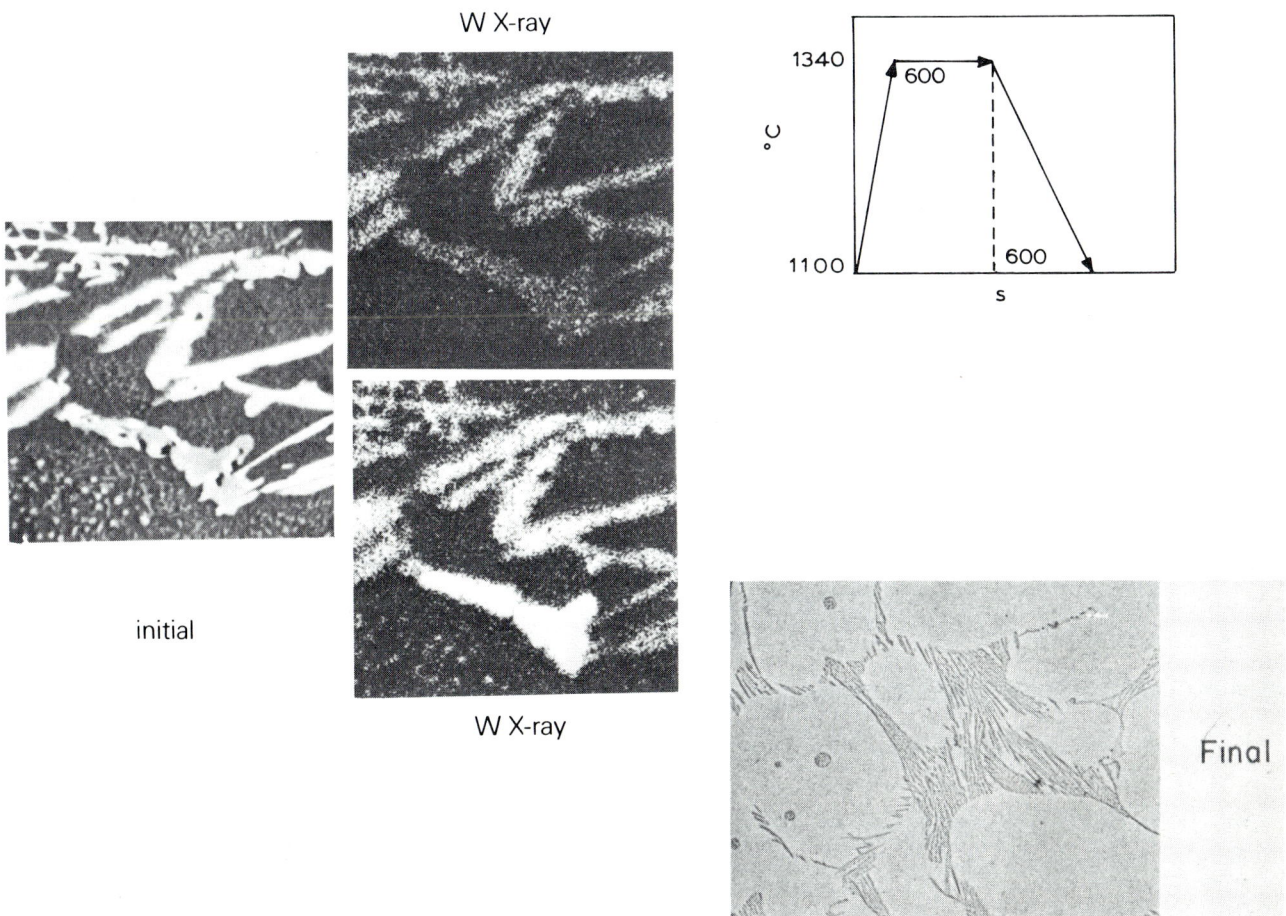

initial

W X-ray

W X-ray

Final

10 Carbide structures before and after heat treatment into the γ-range are shown in sequence in Figs 10–14, this figure showing no M_6C, M_2C pre-spheroidized [×1 000]

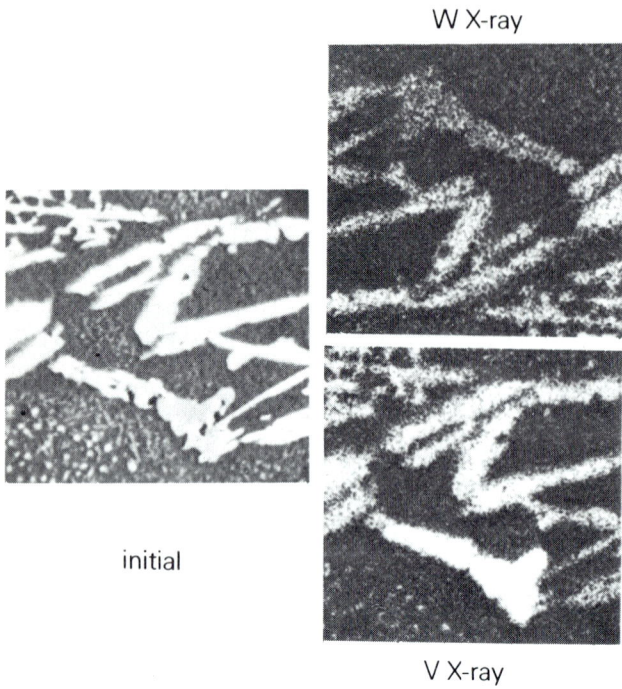

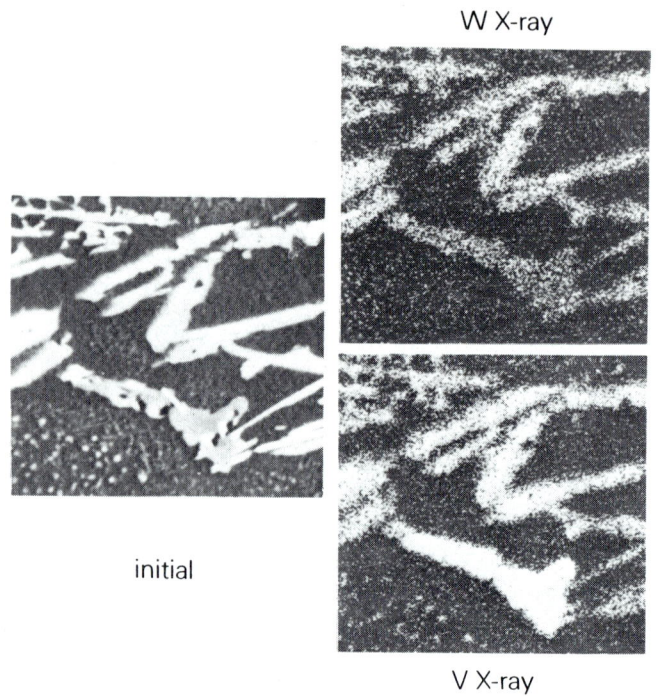

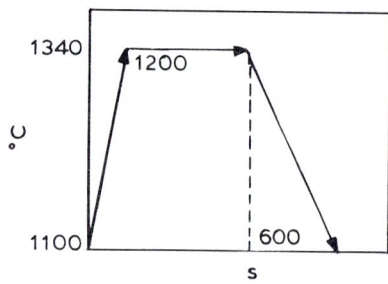

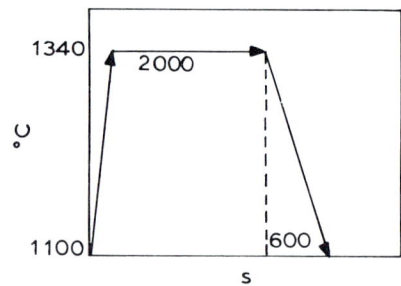

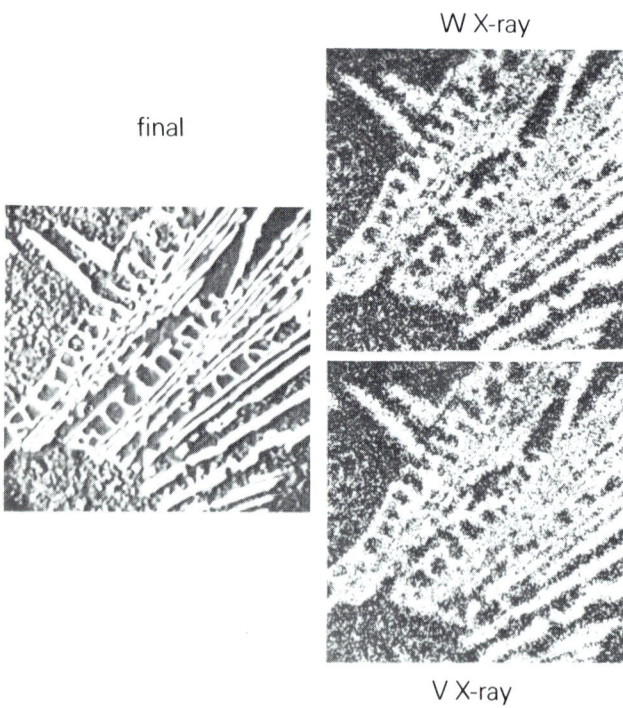

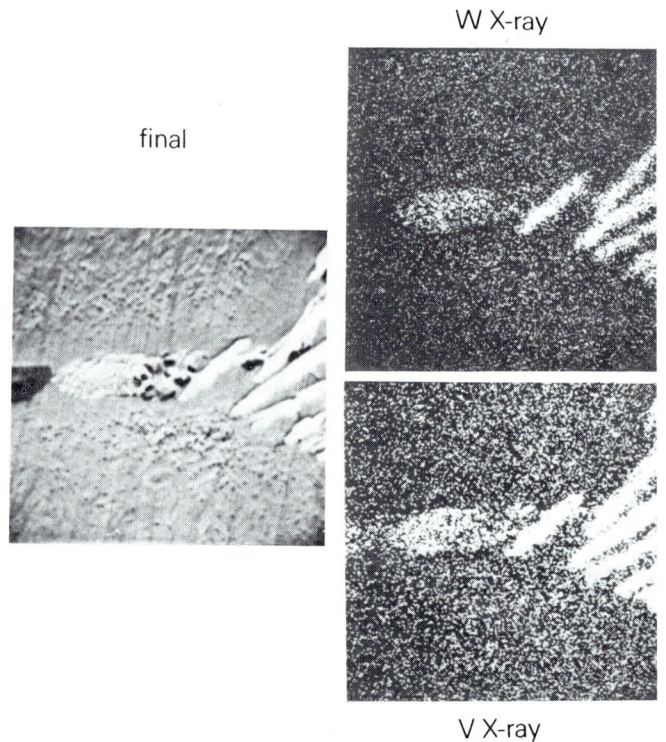

11 Carbide structures before and after heat
treatment: no M_6C, M_2C partially melted
[×1 000]

12 Carbide structure before and after heat
treatment: some appearance of M_6C
[×1 000]

W X-ray

W X-ray

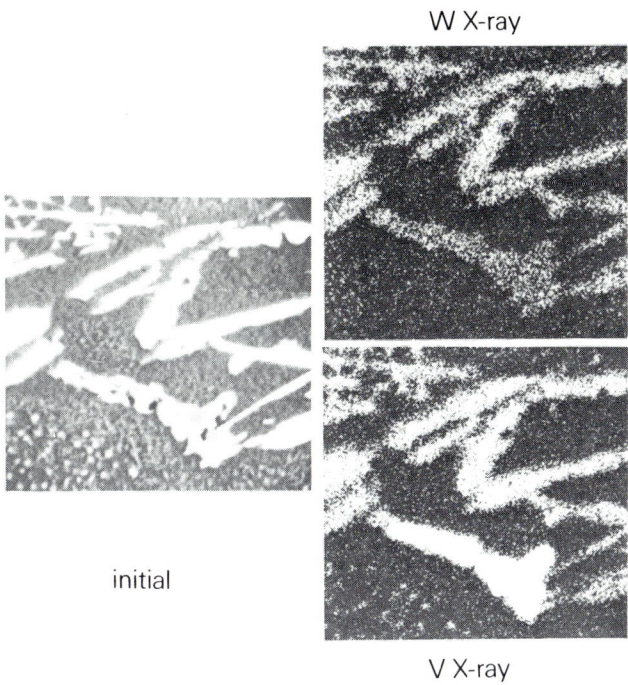

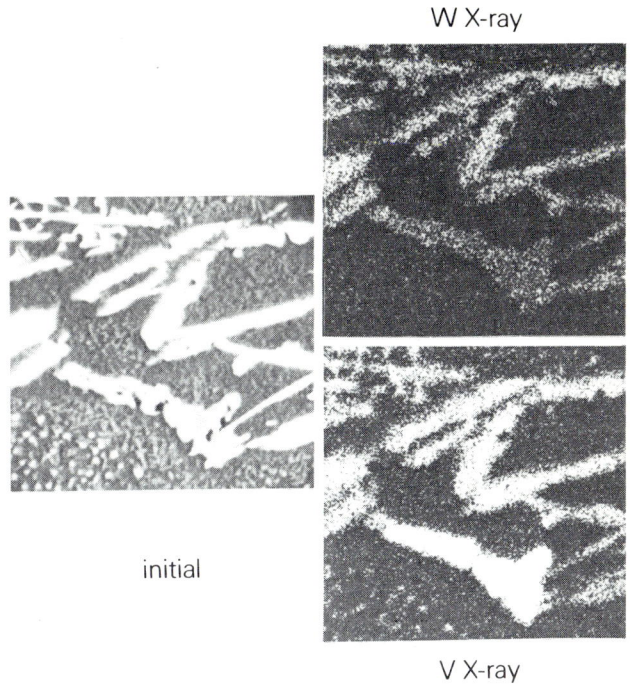

initial

V X-ray

initial

V X-ray

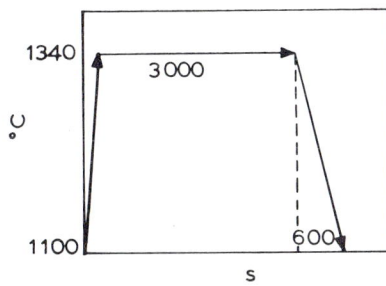

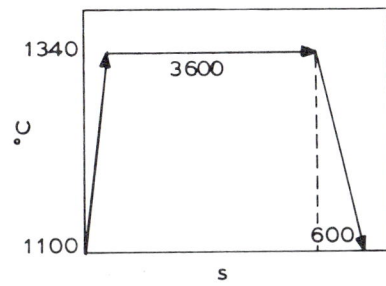

W X-ray

W X-ray

final

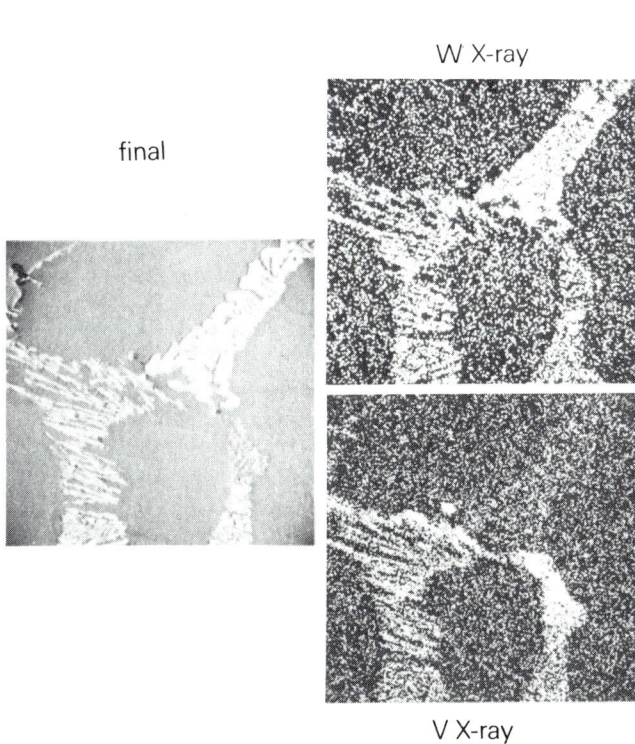

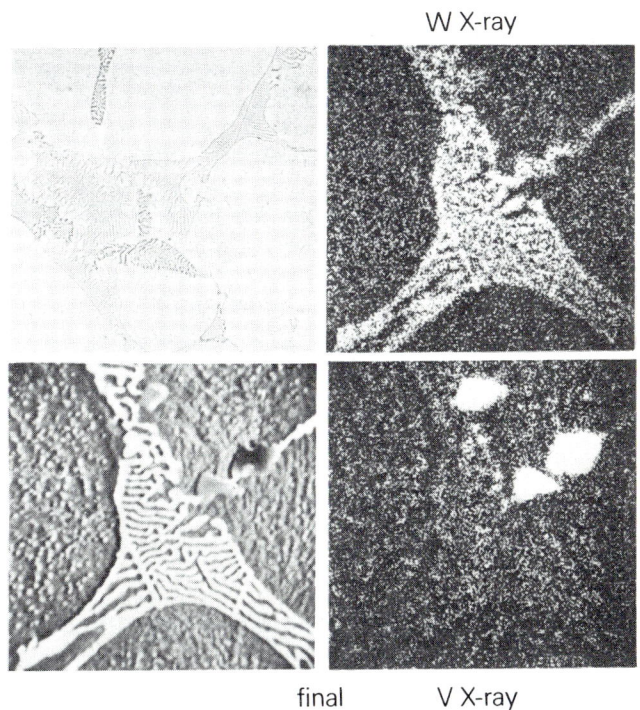

V X-ray

final   V X-ray

13   Carbide structures before and after heat treatment: 20–30% M₆C   [×1 000]

14   Carbide structures before and after heat treatment: fully transformed to M₆C [×1 000]

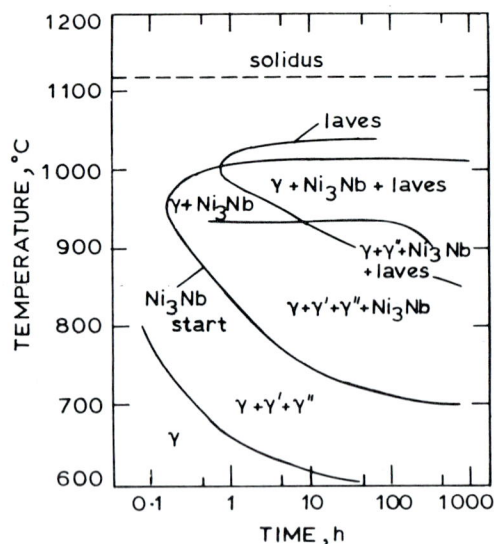

15 TTT diagram[21] for Inconel 718

Data on Laves phase formation indicate that it forms sluggishly. It is found primarily in segregated areas, and its presence is therefore a function of LST. No solidification data equivalent to that for M2 above are available, but the TTT diagram shown in Fig. 15 indicates the relative stability range of the phase.

It has been suggested that ESR ingots of 718 are superior to both PAR and VAR in the same size because they contain less Laves phase. The authors examined two samples, from radial sections of 325 mm dia. ESR and VAR ingots, and conclude that this statement is qualitatively correct. The computed LST values for these ingots have been found[6] to differ by a factor of two, those for ESR being consistently lower.

An examination of 'freckled' areas in a 718 ingot confirmed that an extensive Laves phase was present in these areas. This fact suggests that, unlike the M2 case, the occurrence of Laves phase is not the direct result of cooling rate but is the consequence of segregation arising from the freezing process. In this event, the difference in microsegregation between ESR and VAR accounts for the relative presence of the phase.

Although the LST for Laves phase precipitation cannot yet be defined in terms of laboratory solidification experiments, the experimental ingot LST values can be correlated with the proportion of Laves phase present. Following this procedure it is estimated that under VAR, ESR, and PAR solidification conditions the LST must be less than approximately 1 200 s if the Laves phase is to be restricted to ≤ 5 vol.-%. By following this restriction it is possible to compute the appropriate melting rate sequence for remelting ingots of 718.

CONCLUSIONS

The work presented above suggests that the computed LST in remelted ingots may be used in conjunction with laboratory directional solidification data to obtain a good approximate picture of the ingot structure. The authors are unable to predict phenomena involving factors other than LST, such as 'freckles' or the columnar–equiaxial transition, but can describe the columnar region with adequate accuracy for most industrial purposes.

In one of the cases where precipitation has been examined it has been found that a slow M_2C/M_6C transition is described well by considerations of LST in M2 ingots. The correlation between LST, laboratory heat treatments, and industrial ingots permits one to define a maximum LST criterion for the remelting of this alloy.

In the case of Inconel 718 it appears that the quantity of the deleterious Laves phase, A_2B, is proportional to the level of microsegregation. There are not at present sufficient data on the solidification sequence of 718 to enable the critical LST to be predicted from first principles. However a critical value may be derived by a comparison of structures in experimental VAR, PAR, and ESR ingots.

REFERENCES

1 P. O. MELLBERG: thesis, KTH Stockholm 1975
2 W. HOLZGRUBER: Proc. 5th Int. Conf. on ESR, 70, 1974, Pittsburgh, Mellon Institute
3 W. HOLZGRUBER: Proc. 3rd Int. Conf. on ESR, 1971, Pittsburgh, Mellon Institute
4 M. KRONEIS et al.: Berg. Hutten. Monat., 1968, **113**, 416
5 H. J. ECKSTEIN: ibid, 1968, **113**, 424
6 A. S. BALLANTYNE et al.: Proc. 5th Int. CVM, Munich, 1976; to be published
7 M. C. FLEMINGS et al.: J. Iron Steel Inst., 1970, **208**, 371
8 D. A. MELFORD AND D. A. GRANGER: 'Solidification of metals', 289, 1968, London, The Iron and Steel Institute
9 R. D. DOHERTY AND D. A. MELFORD: J. Iron Steel Inst., 1966, **204**, 1 131
10 B. A. RICKINSON: Ph.D. thesis, University of Sheffield, 1975
11 A. I. SHCHERBAKOV et al.: Stahl, July 1972, 535
12 R. H. BARKALOW et al.: Metall. Trans, 1972, **3**, 919
13 E. HORN: DEW Technische Berichte, 1972, **3**, 217
14 G. HOYLE AND E. INESON: BISRA Reports, MG/L/209/59, MG/L/197/58, MG/L/44/59, MG/L/50/59, MG/L/241/59
15 H. L. EISENSTEIN: ASTM STP 379, Philadelphia, Apr. 1965
16 R. C. HALL: J. Basic Eng., Sep. 1967, 504
17 H. J. WAGNER AND A. M. HALL: DMIC Report 217, Battelle Memorial Institute, 1 June, 1965
18 D. F. PAULONIS et al.: Trans. ASM, 1969, **62**, 611
19 W. J. BOESCH AND H. B. CANADA: J. Met., 1969, **21**, (10), 34
20 F. J. RIZZO AND J. D. BUZZANELL: ibid., **21**, (10), 24
21 D. D. KEISER AND H. L. BROWN: Rep. ANCR-1292, 20 pp, 1976, NTIS, Springfield Va., USA

Simulation of shaped casting solidification

R. D. Pehlke, J. T. Berry, W. Erickson, and C. H. Jacobs

The paper traces the growth of simulation studies as applied to shaped castings through the work of various contributors, including research sponsored by the American Foundrymen's Society (AFS). Difficulties pertaining to selection of thermal properties, interfacial problems, and validation experiments are discussed, together with the long range aspects of implementation in the casting industry.

R. D. Pehlke is at the University of Michigan and W. Erickson is at the Los Alamos Scientific Laboratory. J. T. Berry and C. H. Jacobs are at the Georgia Institute of Technology, USA

WORK OF AFS HEAT-TRANSFER COMMITTEE

Early work

Since its formation in the early 1940s the AFS Heat-Transfer Committee has provided considerable guidance to the metal casting industry on the application of computing techniques to solving some of the many practical heat transfer problems occurring in the foundry. From the early 1940s, the Society sponsored work under Paschkis at Columbia University, utilising a large analog computer.

The publication of no fewer than twenty papers in the Society's literature provided the foundryman with invaluable information on solidification patterns, heat losses from ladles, runner design, and heat flow in moulding materials.[1] Although all of the above problems are highly important to the foundryman, that of freezing pattern prediction has some of the most immediate economic returns. In predicting or attempting to predict such patterns, the foundryman is examining the efficiency of his feeding (or risering) system, which essentially provides reservoirs of molten metal to counteract the shrinkage seen in the cooling liquid as well as in the solidifying metal. He is also attempting to guarantee the integrity of the casting in so doing, since a correctly fed or sequentially solidified casting would be expected to be a sound casting. Also, a high yield, brought about by using the minimum amount of feed metal per pound of sound casting poured, is synonymous with efficient and productive systems.

The contributions of Paschkis and his various students to furthering our understanding of casting design, chill effectiveness, and the contribution of the thermal properties of the mould to producing sound castings are clearly seen by reference to the many papers in the transactions of the AFS between 1944 and 1961. In addition, the vigorous discussions in the Society's transactions through the years concerning this topic indicate that at least some attempt was made to diffuse this early computer-based technology into foundry units themselves.

However, in spite of Paschkis's efforts, the very nature of the analog computer and the lack of familiarity of the practising foundrymen with its operation prevented a wider use of the technique.

The digital computer-borne simulation technique, which has now largely supplanted the analog method, is based on the method of finite-difference approximation (FDA), pioneered by Dusinberre[2] and others some 20 to 30 years ago in various engineering applications.

Forsund in Denmark (1962)[3] appears to have been the first to apply the computer to solving heat-transfer problems in the foundry using FDAs and the digital computer. Three years later, Henzel and Keverian of the General Electric Company (both also associated with the AFS and its research activity) applied this new tool to predicting freezing patterns in large steel castings.[4]

When the terms of reference for the current phase of the committee's work were drawn up in 1966, it was recognized that the digital computer would be the principal tool involved throughout the various stages of the long-term investigation envisaged at that time. The proposal formulated a series of small-scale 2-dimensional solidification pattern simulations which would then be followed by more ambitious 3-dimensional tasks in which complex combinations of shape moulding materials and riser-insulating compounds would be simulated.

Table 1   Details of castings simulated in the University of Michigan Program

Shape	Casting medium	Mould
L-shaped section $10 \times 10 \times 4$in thick section 2-dimensional simulation	0·25% C steel poured at 2850–2900°F (1565–1595°C)	Silica sand (dry)
Barrel-flange shape $7\frac{5}{8}$ o.d. $\times$ 6 i.d. $\times 9\frac{3}{4}$in barrel $12\frac{3}{4}$ o.d. $\times$ 6 i.d. $\times 1\frac{3}{4}$in flange Axisymmetric simulation	85–5–5–5 Brass poured at 2050°F (1121°C)	Silica sand (dry and core) plus graphite chills

Bar shape
$3\frac{1}{2} \times 3\frac{1}{2} \times 8$in
with double end risers (not simulated)
2-dimensional simulation
	(i) Silicon brass poured at
	    2000°F· (1093°C)
	(ii) Commercial aluminium poured at
	    1350°F (732°C)
		Silica sand (dry)

Connected disc casting (Three cylindrical sections, various diameters) Axisymmetric simulation	As bar shape	As bar shape with addition of chills

Current work of the Committee at the University of Michigan

The work conducted at the University of Michigan, begun in 1968, has, to date, followed the general outline drawn up in 1966. The work was initially directed towards the calculation of solidification patterns in sand castings by matching the computed results with existing experimental data reported in the literature. The casting shapes involved were T and L sections in steel and a flanged barrel shape in 85–5–5–5 bronze. The work continued by pouring thermally monitored castings in 2-dimensional and axisymmetric shapes moulded in sands of known thermal properties.

Throughout the work, finite-difference approximations of the diffusion equation have been solved using computer-borne numerical techniques. Both explicit and implicit methods have been used. The advantages and disadvantages of the respective methods and their variants have been discussed in considerable detail in an AFS Monograph *Computer simulation of solidification*[5] by one of the present authors (RDP) and his colleagues Wilkes and Marrone. The authors also discuss the problems of handling the interfacial areas occurring between the casting and a sand mould, the selection and determination of mould thermal properties, and certain of the techniques available for handling the liberation of latent heat during the solidification of the molten metal. The specific casting shapes and the various mould and casting media examined are reported in Table 1.

It was appreciated on the completion of the *L* and flanged-barrel section simulations, which were aimed at reproducing time–temperature histories generated a number of years ago by Pellini, Flinn, and their respective co-workers[6,7] that the possession of accurate data describing the thermal properties of the moulding media was of paramount importance. Secondly, the exact location of thermocouples within the mould cavity was also felt to be an important factor in determining agreement between simulation and practice. Nonetheless, as Chapters 4 and 5 of the monograph show, reasonable agreement between predicted solidification times and those observed were obtained and subsequently the decision was made to continue with bar and disc simulations. The major difference, however, was that in these instances the castings were poured by the investigators

themselves using moulds of known thermal properties (determined in independent experiments).*

That the experiments and the simulations provided excellent practical insight into the problem of casting soundness is clear from examining results of the disc-casting experiments.

The principal objective of this study was to demonstrate the ability of computer simulation of solidification to predict shrinkage unsoundness and its location. Figure 1*a* summarizes the Wlodawer's modulus[†] of each section of the sound casting which was simulated and experimentally cast in commercial aluminium. This

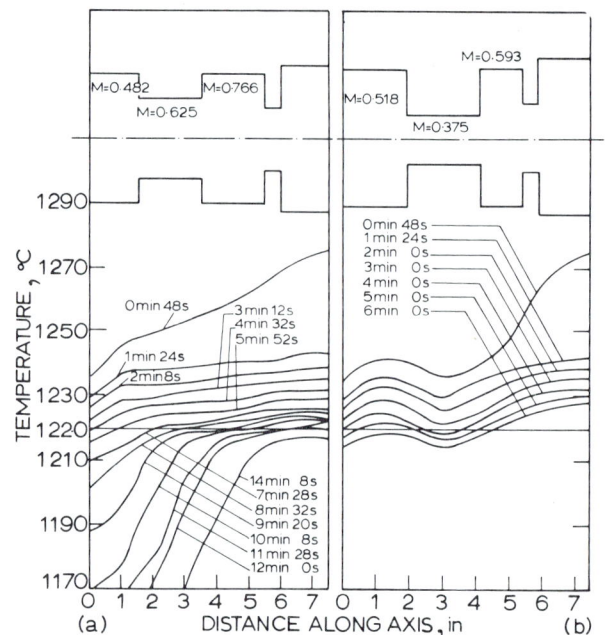

1   Simulated temperature distributions for disc castings reproduced from AFS Monograph; solidification moduli (M) of the three sections and computed axial temperature distributions in the sound aluminium casting at various times

* *See* Ref. 32

† $\dfrac{V}{A} = \dfrac{\text{volume}}{\text{area}}$

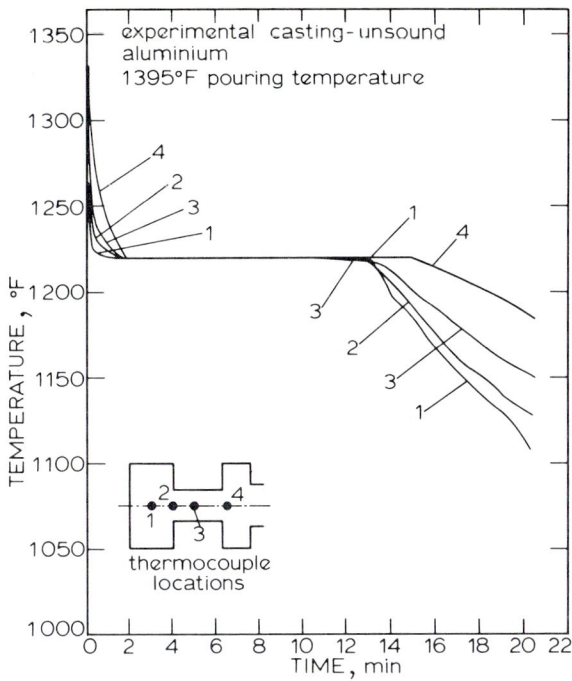

2 **Cooling curve data for unsound aluminium casting reproduced from AFS Monograph (poured at 758°C, 1 395°F with 20 s pouring time)**

figure also gives the simulated temperature distribution along the axis of the casting at various times. Cooling below the solidification temperature takes place progressively from left to right and finally in the riser, so that no shrinkage would be predicted. The casting dimensions were modified as shown in Fig. 1b so that the central section had a lower modulus than the section farthest removed from the riser. The simulated temperature distribution along the central axis is shown as a function of time. The central section of the casting is observed to cool below the solidification temperature before the section on the left, giving rise to a numerical prediction of shrinkage in the left section of the casting. This casting was then experimentally cast and the temperature monitored along the central axis as summarized in Fig. 2. The point of solidification is arrived at first at location 3 as predicted by numerical simulation. Freezing times for other points along the central axis were also correctly predicted by the computer. Figures 3a and b are photographs of the central cross-sections of the sound and unsound aluminium castings,† respectively.

From these and previous results it was seen that this experimental and computer study had demonstrated the usefulness of numerical simulation in predicting heat-transfer and solidification patterns in casting. The thermal properties of the materials involved in these studies were well known. In the case of the dry moulding

† *See* Ref. 32

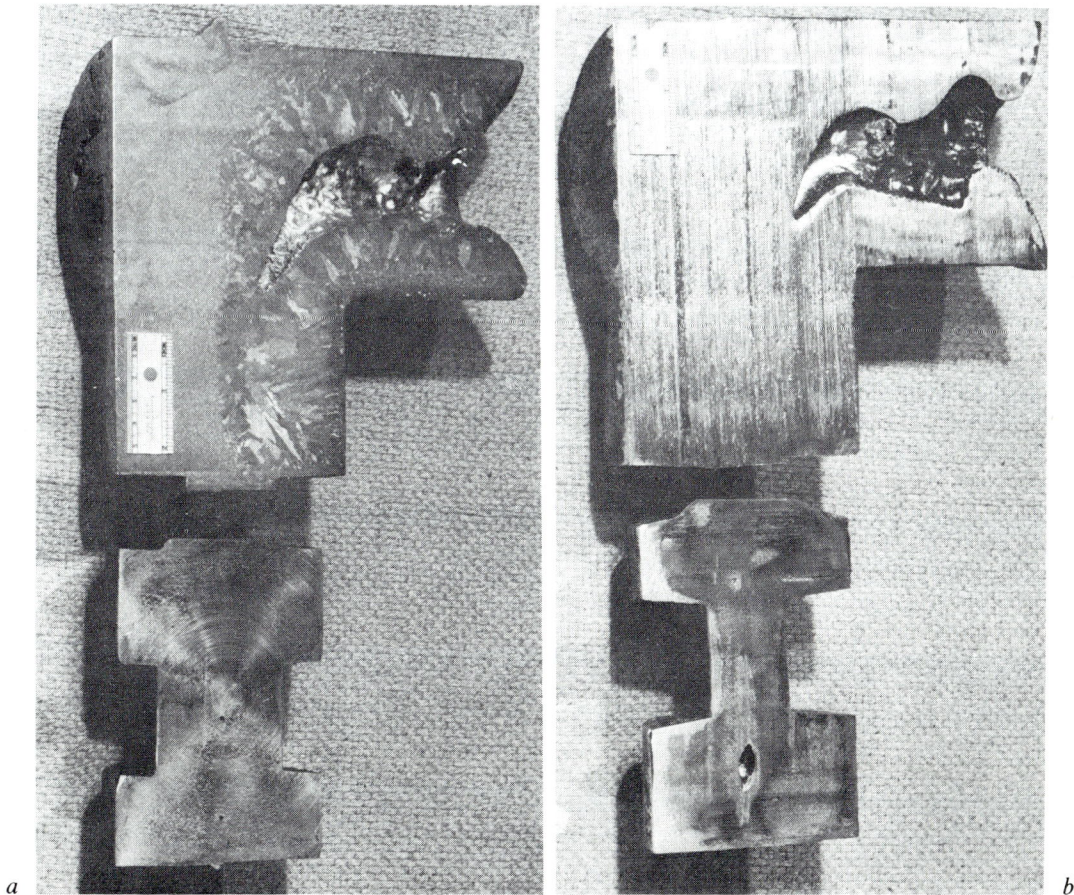

3 *a* Sections of sound aluminium castings examined in monograph (as per Fig. 1a);
   b sections of unsound aluminium castings examined in monograph (as per Figs 1b and 2)

sand used, the thermal diffusivity was independently measured experimentally and developed by computer analysis into the form of an equation as a temperature function which could be used in the computer simulation of the casting.

The experimental procedures and their reproducibility were verified in several ways. Reproducibility of measurements, symmetry of heat flow for those cases where that assumption was necessary for the numerical analysis, as well as the effects of superheat and pouring time, were evaluated.

Castings in the bar and connected disc shape were thermally monitored in silicon brass and aluminium. Computer simulations for these castings were compared with the experimental results and good agreement was demonstrated.

In the case of the disc-shaped casting in aluminium, the geometry was varied slightly to produce an unsound as well as a sound casting. Computer simulation and thermal monitoring of this casting indicated that a shrinkage cavity should appear in the modified shape: this result was indeed observed.

At present, casting design involves art, experience, and a number of empirical rules. In many cases expensive experimentation is required. The design of castings using the digital computer, based on numerical simulation of solidification, would eliminate such experimentation and improve yield, reliability, and accuracy of the casting process. This initial work in predicting temperature distribution during solidification is the first stage of the development of casting design by computer. Hence the committee is now turning its attention to the problems of:

(i) demonstrating the *utility* of this tool in a production application
(ii) demonstrating *how* computer simulation can be used by foundrymen
(iii) making computer simulation data available to foundrymen in *usable* form.

The next step in the current program is to evaluate the economic aspects of using such a program on a part which can be simulated using a 2-dimensional production casting with axial symmetry.

For computer simulation to be used in an economically sound manner, the program must not only be able to optimize the solidification sequence but also to do so at a minimum cost and in a manner which can be readily implemented. Currently, therefore, a 150 kg hub-risered, sand-cast production wheel blank is being simulated. The simulation will be iterative in nature: that is, the solidification pattern will be calculated, the results evaluated by experts in casting design and foundry procedures, design changes made, and the sequence repeated.

During the course of this work, detailed information concerning manpower and computer time will be recorded. From this information, cost formulae will be developed to enable potential users to determine cost in their own work environment. Data will also be generated relating to the economic aspects of the design changes suggested by the computer simulation. Factors which will be considered include metal savings, manpower costs, changes in production costs (i.e. insulating sleeves, chills, etc.), reduced scrap estimates, etc. Such

economic analyses will be essentially similar to those found in recent issues of *Trans. Amer. Foundrymen's Soc.*, which discuss foundry economics rather than the technological aspects of casting yield.[8,9]

Beyond this, larger-scale, non-symmetric 3-dimensional simulations are planned, utilizing existing commercially available computer programs, with validations involving the pouring of corresponding commercial castings risered in different ways. The committee has recognized the enormity of this eventual phase of the task and is seeking the cooperation of outside agencies to help perform this assignment.

RECENT WORK BY AMERICAN GROUPS INDEPENDENT OF AFS HEAT-TRANSFER COMMITTEE

In addition to the work reported above, a variety of other American groups have made important contributions to the progress of applying computer simulation techniques to practical solidification problems. These are set out below.

In industry

(i) the General Electric group under Henzel, Keverian, and others, concerned with large steel castings (*see* above)
(ii) the Tecumseh group under Riegger and colleagues, who have been concerned with the simulation of die-casting freezing.[10]

In universities and research institutes

(i) the ILZRO/Battelle group under Kaiser, which studied the die-casting of zinc-base alloys[11]
(ii) the SFSA/Case-Western Reserve Group under Wallace, concerned with steel casting solidification, found largely in the restricted literature but with contributions concerning chill contact in *Trans. Amer. Foundrymen's Soc.*[12]
(iii) the group at Penn State University under Draper and Samuels concerned with ferrous die-casting simulation[13]
(iv) the cooperating groups at the University of Vermont and the Georgia Institute of Technology, concerned with the problems of thermal contact at chills (to be discussed below)
(v) the group at the University of Pittsburgh under Brody, which has been concerned with axisymmetric (roll) castings as well as the continuous casting work represented at the present conference, using finite-difference and finite-element techniques, respectively
(vi) the group at the University of Bridgeport under Poirier who has also applied the finite-element technique to steel castings[14]
(vii) the group at Los Alamos under Erickson, which has examined special aspects of experimental parameters; for example, thermocouple/perturbation effects, mould filling effects, and the use of commercially available computer codes in casting simulation.

As examples, the work of the groups mentioned under (iv) and (vii) above will now be described.

Work of cooperating groups at the University of Vermont and the Georgia Institute of Technology

The work of these cooperating groups began in the late 1960s when Tillman and Berry[15] examined the possibility of simulating the time–temperature history during the unidirectional freezing of a high-purity (4·5%) copper–aluminium alloy, used in a mechanical property study sponsored by the AFS Light and Reactive Metals Division at the University of Bridgeport.[16–18]

In the original work, in which a variety of thermal contact possibilities at the chill/metal interface were

examined, the authors concluded that air-gap formation was particularly important and that an interfacial heat-transfer conductance of 0·03–0·07 cgsu was probably appropriate to describing the overall conditions of contact at that interface for the given experimental conditions. Another, less well publicized, finding of this paper was the apparent variation of the implied *h* value with the presence or absence of Ti–Al grain refiner. The work has continued and the following casting media have been examined:

(i) lead
(ii) tin
(iii) lead–tin alloys
(iv) Wood's metal
(v) zamak.

The variety of chill materials has included:

(i) copper
(ii) 70–30 brass
(iii) AISI 304 stainless steel
(iv) mild steel
(v) cast iron (grey).

The interfacial conditions have also been varied to include:

(i) carbon black wash
(ii) polished chill surfaces
(iii) fluxed chill surface.

The superheat of the casting media has also been studied to a limited extent. A special feature of the work to date[19–22] has been the validation of all computer runs using appropriate experiments. Furthermore, all the work to date has involved probing techniques rather than the use of thermocouple networks (*see* comments on thermal field perturbation in the following section). Figures 4 and 5 show some typical results of the validation using lead and tin. Several interesting features of the work to date are of special interest:

(i) the general order of magnitude of the *h* values observed in the spectrum of experiments thus far performed: 0·1–2·0 cgs units
(ii) the importance of the interfacial condition with respect to pouring temperature in controlling the speed of solidification (Figs 6 and 7)

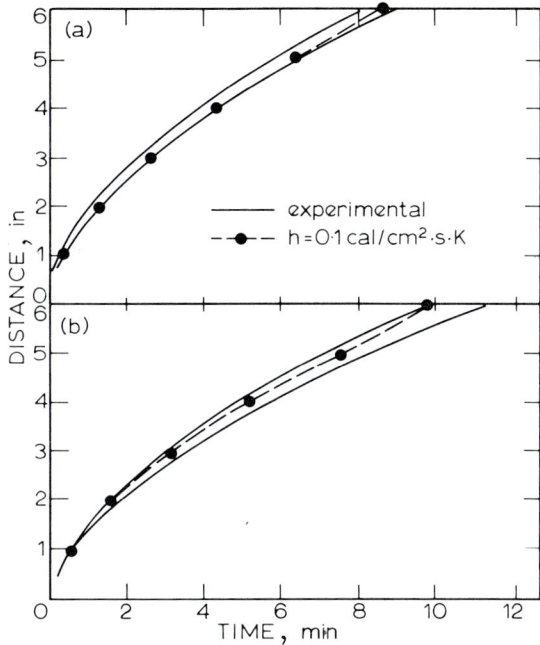

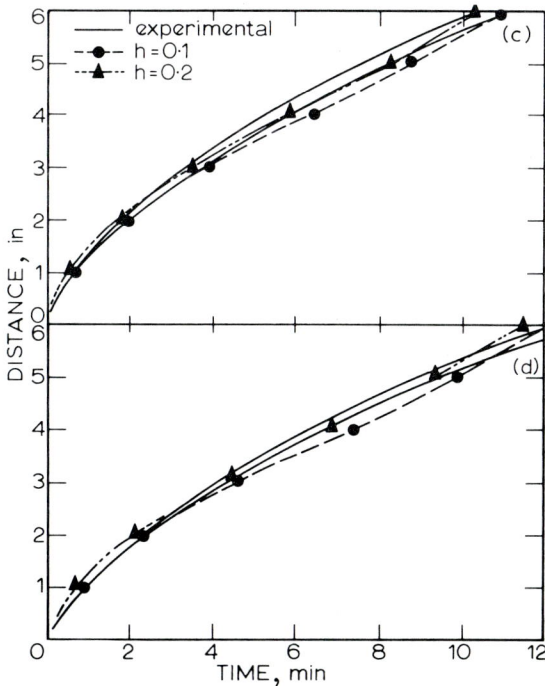

a 603 K; *b* 623 K pouring temperature; *c* 648 K; *d* 673 K

4 Distance solidified *v.* time; lead poured against brass chill

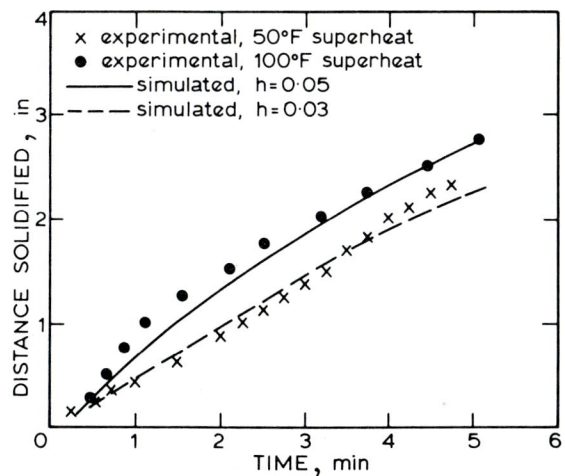

5 Distance solidified *v.* time; tin poured against brass chill, two superheats

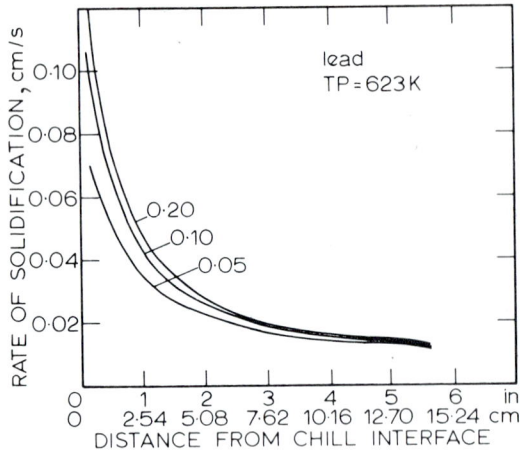

6   **Effect of interfacial conductance (cgsu) on rate of solidification (computed)**

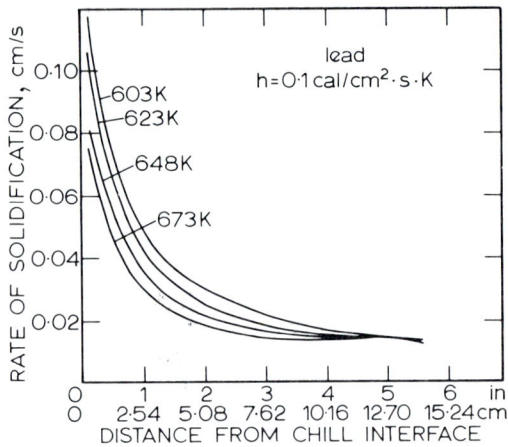

7   **Effect of superheat on rate of solidification (computed)**

(iii) the condition at, and preparation of, the chill surface in affecting solidification rates

(iv) the conditions at the upper casting surface and their effects on distance solidified as a function of time (Fig. 8).

Although the work is oriented to heavier sectioned castings with solidification times of several minutes (heavy-sectioned permanent mould, low-pressure die castings and chilled sand castings) the results should prove to be of special interest to physical metallurgists interested in unidirectional solidification techniques.

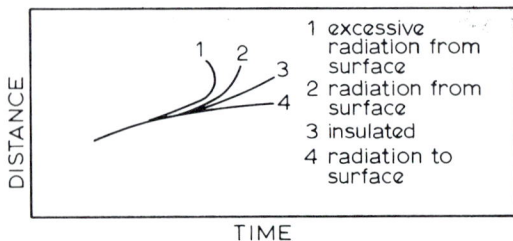

8   **Effect of condition at upper casting surface (schematic)**

Casting Simulation Work at Los Alamos

Work being conducted at the Los Alamos Scientific Laboratory (LASL) is primarily involved with the simulation of congruent melting materials. This work has addressed items relating to the effects of experimental and modelling parameters; minimizing human resource requirements through the use of general-purpose heat-transfer codes and colour microfilm output; comparing finite-element and finite-difference modelling techniques; and calculating metal flow patterns during mould filling. Work on the latter two problems is in its early stages, and results are not yet available.

The simulation work has used NASA's general-purpose heat-transfer code CINDA-3G.[23] This code was developed for applications completely unrelated to casting solidification problems, but has proved effective in solidification applications.[24,25] Using a general purpose code both reduces the manpower required to prepare a model for simulation and eliminates the need for detailed knowledge of numerical analysis techniques.

A further reduction in manpower has been realized through the use of colour microfilm output.[26] This technique consists of plotting different colours for specified temperature ranges. These colours are then plotted for the calculated temperatures at various time steps. Figure 9 shows the results for a hypothetical flanged copper barrel casting. For this particular simulation, the effect of a chill positioned at the bottom of the casting was being evaluated.

In Fig. 9, the ability to detect molten metal and potential defective material is readily demonstrated.

The results can be plotted as either 35 mm slides or 16 mm movies. The effectiveness of computer-generated movies is shown by the film evaluating design parameters in a directional solidification mould.

Work has been performed on evaluating both modelling and experimental parameters. This work has shown that the absolute accuracy of calculated data is

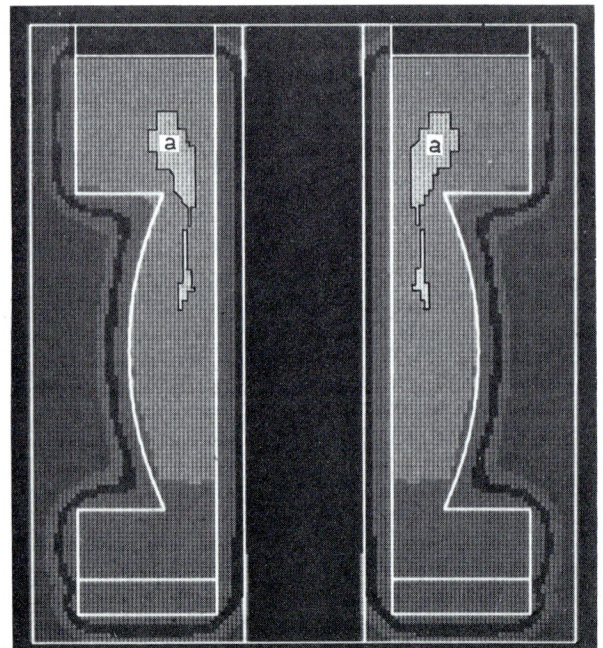

9   **Solidification pattern of flanged barrel (*a* indicates molten metal)**

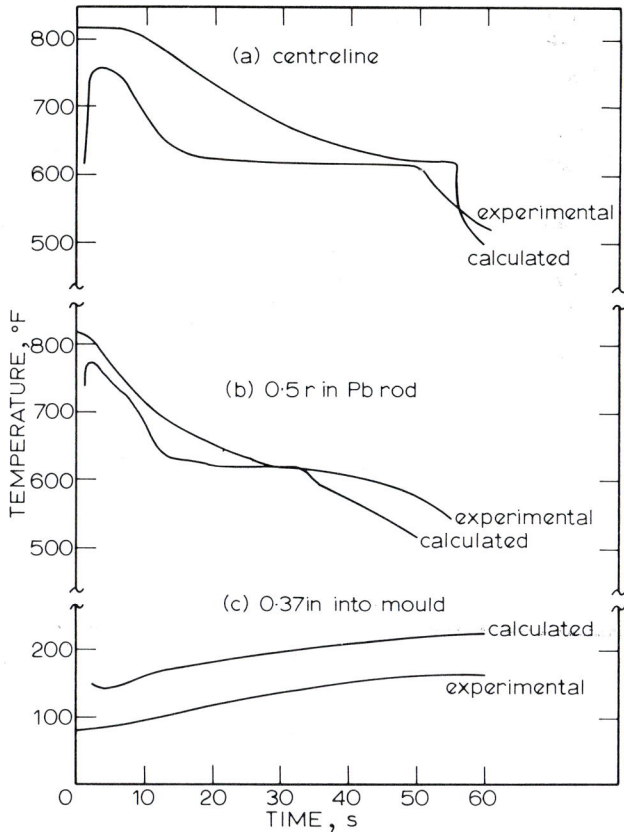

10 Calculated and experimental time–temperature data for lead rod

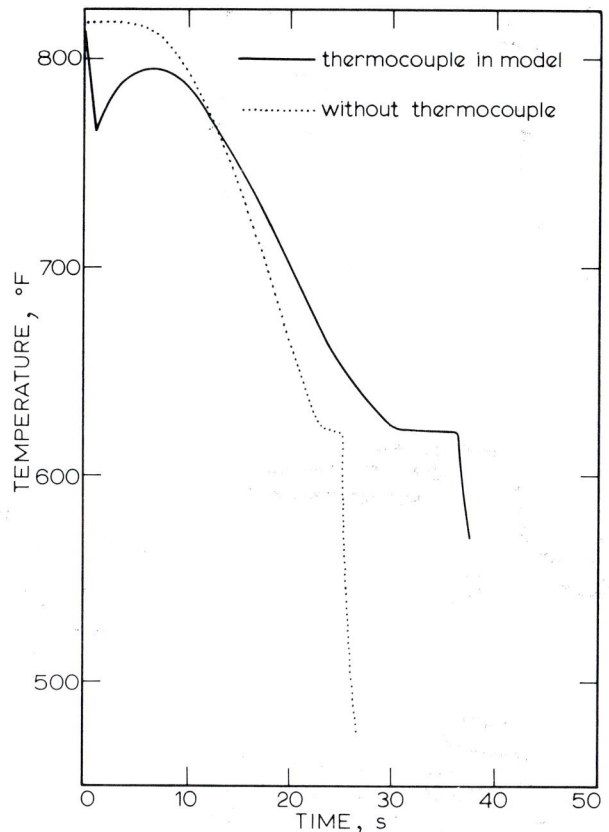

11 Effect of a thermocouple assembly on the calculated cooling curve of a lead rod

significantly affected by the model constructed for the CINDA-3G heat-transfer analysis. For example, the solidification time was calculated for a lead bar using three models containing 6, 18, and 30 nodes. From these models, solidification times of 25, 21·5, and 20 s, respectively, were calculated. The reasons for these differences are not yet fully understood, although it is obvious that the calculational model (i.e. number of nodes) has a significant effect on the results.

An area of specific engineering interest is simulating metal solidification in permanent moulds. Figure 10 shows three sets of calculated and experimental curves for lead cast in acetylene soot-coated graphite moulds. The results show two distinct trends. Above the melting point the calculated temperatures are greater, while below the melting point the calculated temperatures are less, than the experimentally determined temperatures.

Several factors contribute to the lack of agreement. Below the melting point, the calculated and experimental values can be brought into closer agreement by varying the thermal conductance values for the mould/metal interface. Above the melting point, factors such as non-instantaneous mould filling and convection currents must be taken into account. It was found that multiplying the thermal conductivity of the liquid metal by a factor of 8 will approximate the effect of convection currents and bring the calculated and experimental curves into closer agreement.[27]

Computer simulation has also been used to calculate the effect of ceramic-sheathed thermocouple assemblies on the cooling curves of metal cast in permanent moulds. Figure 11 shows that this type of assembly initially acts as a chill, but the net effect is to increase the local

solidification time. The effect is noted both radially and circumferentially. A 3-dimensional model and the CINDA-3G heat-transfer code were used to make these calculations.

The work at LASL is directed toward simulating permanent mould casting procedures for congruent melting materials. This work has shown that assumptions applicable to sand castings cannot be made in casting processes characterized by rapid cooling of the metal. While simplifying assumptions must still be made which reduce the absolute accuracy of calculated time–temperature data, the assumptions do not prevent valuable information from being obtained through parametric simulation studies.

CONCLUSIONS

It is interesting to take stock of the results of over 30 years of work associated with the AFS Heat-Transfer Committee and the various independent groups referred to in this paper in the area of solidification simulation, starting with the electrical analog work of Paschkis and continuing through to today's efforts using the digital computer. From the point of view of the engineer/scientist many, but not all, of the problems that faced Paschkis's group continue.

1 There is still a lack of accurate data concerning thermal properties of the various media, especially at elevated temperatures. In particular, data concerning mould materials remain of vital importance. There is still no convenient model for thermal transport in green sand (or for that matter any sand containing a significant volume fraction of volatile materials), although some progress has been reported very recently in Japan.[28]

2 Some increased understanding of the conditions of thermal contact between casting media and chills has become available. There seems little doubt that Sully's work in this area establishes the need for regarding the equivalent *h* value as a highly variable quantity during the solidification of thin sectioned castings in metal moulds. However, for more massive shapes the overall values encompassed by *h* do not seem to span more than about one order of magnitude.

3 The effects of convection, turbulence, and related transients are still poorly understood in commercial-scale castings. However, these effects are suggested as being sometimes significant especially in castings in metal moulds.

4 With the growing availability of large-scale 3-dimensional generalized computer programs, both of the finite-element and finite-difference varieties, computations are no longer limited to 1- and 2-dimensional geometries, and casting complexity is no longer a major problem.

From the point of view of the foundryman, where the eventual justification of this research lies:

1 A major effort must be mounted to obtain easy access to the commercially available computer codes alluded to above: otherwise the communication gap which has built up over the years is hardly likely to be bridged.

2 A similar effort would then seem to be in order beyond the above, to demonstrate the powerful nature of the computer as a problem-solving tool in the hands of foundrymen themselves.

3 The above will require validation experiments, undertaken not necessarily by thermal analysis – especially since thermocouples are now seen as presenting problems themselves in the perturbation of temperature fields – but simply by pouring and evaluating certain alternative casting–rigging design combinations which relate to continuing foundry practice.

4 At some juncture, the completion of the feedback loop, the interaction of the foundryman with the casting designer, possibly through interactive computer graphics, will be secured, facilitating optimization of design from the points of view of both manufacturability and fitness for purpose, for the benefit of the public at large.

ACKNOWLEDGMENTS

The authors would like to thank their colleagues for their many contributions in the conduct of the work described. In particular, they would like to acknowledge the many graduate and undergraduate students involved in the studies at the University of Michigan, the University of Vermont and the Georgia Institute of Technology, and especially the signal contributions made by Professors J. O. Wilkes, E. Tillman and A. Houghton in the computational areas, of Mrs Delcie Durham, who has brought some unique approaches to both experimental and simulation techniques, and finally to the members of the AFS Heat-Transfer Committee and Headquarters Staff.

REFERENCES

1 *Trans. of the Amer. Foundrymen's Soc.*, 1944–1961, Publications of V. Paschkis. (For a complete list of the papers of Paschkis and co-workers *see* pp. 0–11 of Ref. 5)

2 G. M. DUSINBERRE: 'Heat-transfer calculations by finite differences', 1961, Scranton, Pa, USA, Intl. Textbook Co

3 K. FURSUND: *Giesserei*, **14**, 1962, 51

4 J. G. HENZEL AND J. KEVERIAN: *AFS Cast Metals Res. J.*, **1**, 1965, 19

5 R. D. PEHLKE *et al.*: 'Computer simulation of solidification', monograph published by AFS, Des Plaines, Ill., 1976, 232 (*see* Ref. 1)

6 F. A. BRANDT *et al.*: 'Solidification at corner and core positions', AFS, 1953, **61**, 451

7 P. K. TROJAN *et al.*: *Trans. Amer. Foundrymen's Soc.*, 1963, **71**, 656

8 R. W. RUDDLE: *ibid.*, 1975, **83**, 575

9 D. R. POIRIER AND N. V. GHANDI: *ibid.*, 1976, **84**, 577

10 R. B. WEATHERWAX AND O. K. RIEGGER: *SAE Trans.* 1971, **80**, 2112 (paper no. 710600)

11 W. D. KAISER *et al.*: 'A computer model for studying transient heat transfer conditions during die filling', Conference Society Die Casting Engineers, 1972, Paper No. 5472

12 L. J. D. SULLY: *Trans. Amer. Foundrymen's Soc.*, 1976, **84**, 735

13 J. M. SAMUELS *et al.*: *Die Casting Engr.*, 1974, **18**, 14

14 D. R. POIRIER AND C. A. SNYDER: 'The finite element technique applied in casting solidification', paper presented at AIME Spring Meeting, Atlanta, Ga, March 1977

15 E. TILLMAN AND J. T. BERRY: *AFS Cast Metals Res. J.*, 1972, **8**, 1

16 F. ST. JOHN *et al.*: *Trans. Amer. Foundrymen's Soc.*, 1968, **76**, 645

17 J. T. BERRY: (Discussion contribution) 'The solidification of metals', 127, 408, 1968, London, The Iron and Steel Institute

18 J. T. BERRY: *Trans. Amer. Foundrymen's Soc.*, 1970, **78**, 421

19 D. R. DURHAM AND J. T. BERRY: *ibid.*, 1974. **82**, 101

20 D. R. DURHAM *et al.*: *ibid.*, 1976, **84**, 787

21 D. R. DURHAM AND J. T. BERRY: 'Effects of mould material and surface upon solidification heat transfer', paper presented at AIME Spring Meeting, Atlanta, Ga, March 1977

22 D. R. DURHAM: To be published in Weld. Res. (supplement to *Weld. J.*), 1977

23 P. R. LEWIS *et al.*: 'Chrysler improved numerical differencing analyzer for 3rd generation computers', Chrysler Corporation Space Divisions, New Orleans, TN-AP-67-287, 1967

24 W. C. ERICKSON: *Metall. Trans. B*, 1970, 8B, (1), 93

25 W. C. ERICKSON: 'The use of a general purpose heat transfer code for casting simulation', MS thesis, University of New Mexico, 1975

26 W. C. ERICKSON: *AFS Cast Metals Res. J.*, 1974, **10**, 161

27 W. C. ERICKSON AND A. V. HOUGHTON: *Trans. Amer. Foundrymen's Soc.*, 1977, **85**, 59

28 H. SAITO AND N. SEKI: *Trans. Amer. Soc. Mech. Eng.*, *J. Heat Transfer*, 1977, **100**, 105

29 R. E. MARRONE *et al.*: *Amer. Foundrymen's Soc. Cast Metals Res. J.*, 1970, **6**, 184

30 R. E. MARRONE *et al.*: *ibid.*, 1970, **6**, 188

31 R. E. MARRONE *et al.*: Copper-base alloy casting –
flanged barrel shape, AFS Research Report, 1972,
Amer. Foundrymens' Soc. Cast Metals Res. J., 1972,
8, 94

32 M. J. KIRT AND R. D. PEHLKE: *ibid.*, 1973, **9**, 117

33 R. D. PEHLKE *et al.*: *ibid.*, 1973, **9**, 49

34 A. JEYARAJAN AND R. D. PEHLKE: *Trans. Amer.
Foundrymen's Soc.*, 1975, **83**, 405

35 A. JEYARAJAN AND R. D. PEHLKE: *ibid.*, 1976, **84**,
647

APPENDIX

The complete record of the work sponsored at the
University of Michigan by the AFS Heat-Transfer
Committee is contained in a variety of publications
found in either *Trans. Amer. Foundrymen's Soc.* or
Amer. Foundrymen's Soc. Cast Metals Res. J. (now
Amer. Foundrymen's Soc. Cast Metals J.), over the past
few years. These are listed as references 29–35.

DISCUSSION

Session 5: Solidification and quality control: macrosegregation, porosity, and simulation of solidification

Chairman: J. A. Reynolds (Steel Castings Research and Trade Association)

The following papers were discussed: *Keynote Address: Solidification and aspects of cast metal quality* by P. R. Beeley; *Observations on the nature and extent of gravitational interdendritic fluid flow* by K. M. Fisher and J. D. Hunt; *Effect of fluid flow on macrosegregation in steel ingots* by T. Takahashi, K. Ichikawa, and M. Kudou; *Solidification of high-carbon steel ingots* by F. Weinberg, J. Lait, and R. Pugh; *Observation of porosity formation during solidification of aluminium alloys by acoustic emission measurements* by U. Feurer and R. Wunderlin; *Development of porosity in aluminium-base alloys* by R. A. Entwistle, J. E. Gruzleski, and P. M. Thomas; *Numerical simulations of the solidification process* by P. N. Hansen; *Computed feeding range for gravity die castings* by V. de L. Davies and R. Moe; *Prediction of structure in industrial VAR, ESR, and PAR ingots using computed local solidification times* by A. S. Ballantyne and A. Mitchell; *Simulation of shaped casting solidification; a review of recent US work* by R. D. Pehlke, J. T. Berry, W. Erickson, and C. H. Jacobs.

PROFESSOR M. C. FLEMINGS (MIT): I was very interested in Dr Feurer's excellent experimental work, and I would like to know what makes the noise?

DR FEURER: You can imagine that there is a concentration of hydrogen in the melt before the pore is formed, and that at the moment when pores are very rapidly forming there is a displacement of the melt. This rapid displacement will give a pressure wave in the melt, which travels to the mould wall and is there transferred into a shear wave, which is finally detected by the acoustic transmission transducer.

DR V. KONDIC (University of Birmingham): I would like to comment on the papers, and perhaps on Professor Flemings' question. In the experiments we have been conducting in this area in Birmingham, we have used a rather different experimental approach.

We have used an enclosed system, i.e. enclosed in the sense that as soon as solidification forms a skin this creates a vacuum in the remaining liquid, in which voids

are created by two distinct mechanisms. Either the skin is broken by the ingress of atmospheric air, shown on the right-hand side, or alternatively gas liberation occurs, as discussed in the previous paper. This could explain where the noise is coming from.

Regarding the second paper, the authors mentioned the large number of parameters influencing the formation of porosity. My comment on that would be that to investigate porosity formation, which is a constant problem, it is advantageous to reduce the number of variables, which we attempted to do by using the method I have described.

DR B. CALLMER: *A guide to the solidification of steels* (written contribution). The work presented is a systematic compilation of solidification data describing the formation of the as-cast microstructures in steels of commercial importance. Theoretical treatments of solidification processes in steels are not included. It is hoped, however, that the reaction temperatures given

Table 1 Steels investigated

Carbon and low alloy steels 0·1–1·0%C	16
5% and 13% chromium steels 0·04–1·0%C	9
Stainless and heat resistant steels, ferritic-austenitic and austenitic	15
High-speed steels	2
	—
Total	42

and the descriptions of the microstructures formed on solidification will be useful in solving or avoiding problems which may arise during casting.

The 42 compositions examined are a selection of steels in production (*see* Table 1). The experimental techniques indicated in Table 2 were used for all steels investigated.

Data from the literature show that cooling rates during the solidification of ingots and during continuous casting, with the exception of the outermost 2–3 mm, vary between 3° and 0·05°C/s. To simulate full size ingot solidification, 35 g samples were solidified in a resistance-heated furnace (*see* Fig. 1). Three furnace cooling rates were used: 0·1, 0·5, and 2·0°C/s. Samples for metallographic investigation were quenched during

Table 2 Experimental techniques

Thermal analysis using three cooling rates giving:
Temperatures of liquidus and solidus
Temperatures of formation of secondary phase and precipitates
Fraction solid phase

Metallographic investigation:
Microphotography
Identification of phases
Dendrite arm measurements
Microprobe analysis
Measurements of δ-ferrite and carbide content

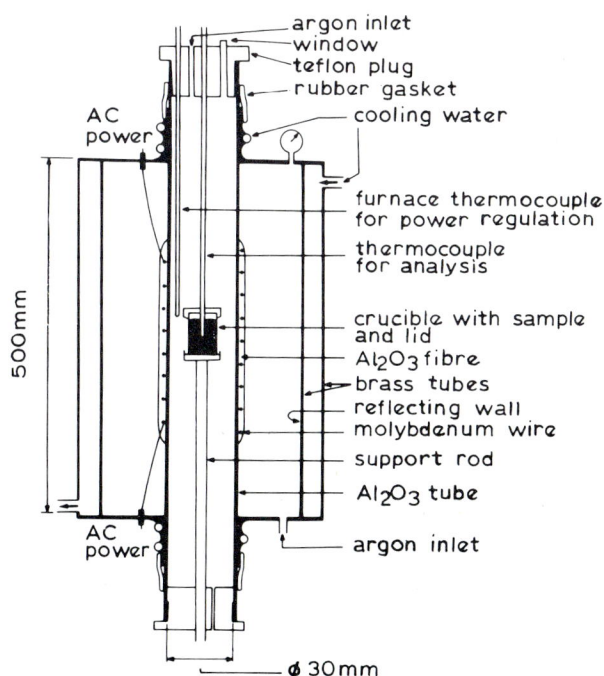

1 Experimental furnace

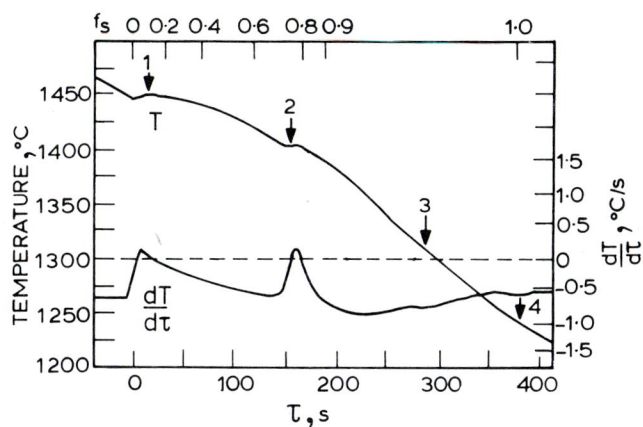

Steel 303; 0·5% C Mo V 5% Chromium steel

Designations

SIS   AISI
Compositions, wt-%

C	Si	Mn	P	S	Cr	Ni
0·50	1·00	0·48	0·025	0·010	5·1	0·18

Mo	Cu	W	V	Al	N
1·36	0·10	0·02	1·10	0·013	0·036

	Average cooling rate, R, °C/s		
	2·0	0·5	0·1
Liquidus temperature, ferritic primary phase, C 1	1 460	1 460	1 460
Temperature of austenite formation, C 2	1 410	1 410	1 412
Temperature of formation of MC austenite eutectic, C 3	1 320, 1 240	1 345	1 300–1 320
Solidus temperature, C 4	1 140	1 240	1 260
Solidification range, C	320	220	200
Solidification time, s	170	380	1 900

Precipitates
1 Interdendritic MC austenite eutectic; MC was of the VC type
2 Small amount of interdendritic $M_{23}C_6$ austenite eutectic (M was Cr, Fe and Mo) precipitated after the MC carbide

Microsegregation

Element	Cr	Mo	V	
	1·3	1·5	1·3	$R = 0·5°C/s$
				$T_q = 1\,200°C$

2 Example of a data page and diagram of thermal analysis (above)

solidification and from the solidus temperature or just below. The stainless steels were also quenched at 1 200°C.

The information from thermal analysis and from metallographic investigations was collected for each of the 42 steels and is presented on data pages of which Fig. 2 is an example. A diagram of the thermal analysis is given. The arrow marked 1 indicates the liquidus temperature, 2 indicates the formation of austenite, 3 indicates the formation of vanadium carbide–austenite eutectic, and 4 indicates the solidus temperature. The diagram also shows the fraction solid phase, f_s, as a function of time, while at the bottom of the data page the segregation ratio, I, is given.

STEEL 303   5% CHROMIUM STEEL

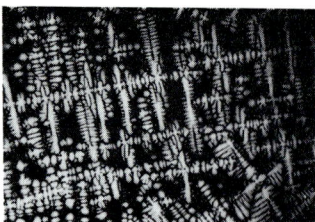

PARTLY SOLIDIFIED
1.
R = 0.5°C/s
T_Q = 1445°C
D = 60 µm
                            x25

COMPLETELY SOLIDIFIED
2.
R = 2.0°C/s
T_Q = 1140°C
D = 60 µm   x25

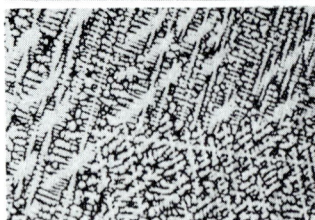

3.
R = 0.5°C/s
T_Q = 1200°C
D = 80 µm
                            x25

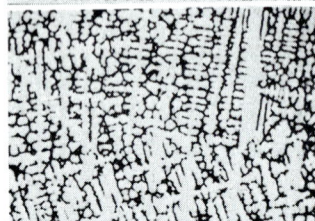

4.
R = 0.1°C/s
T_Q = 1200°C
D = 110 µm
                            x25

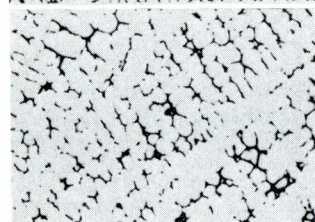

3 Example of a page with microstructures (magnification reduced on reproduction)

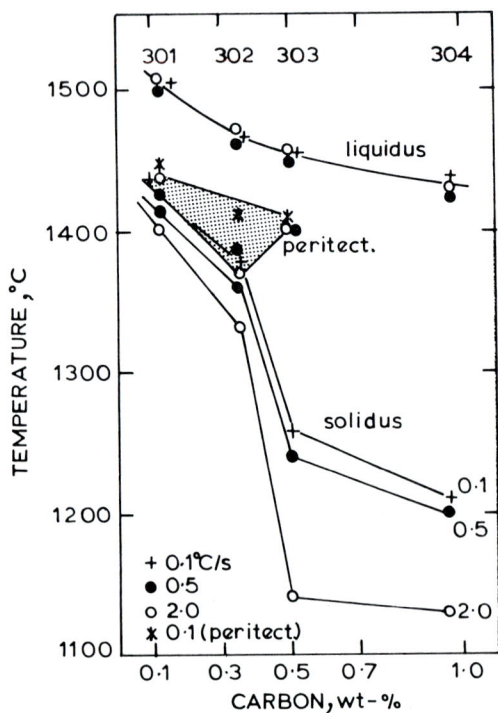

4 Liquidus, peritectic, and solidus temperatures for 5% chromium steels

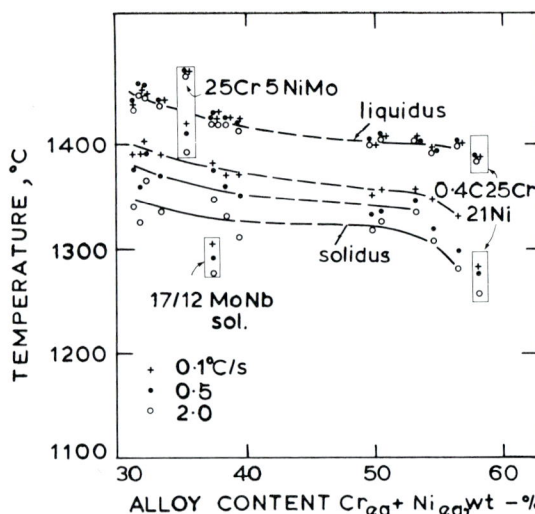

5 Liquidus and solidus temperatures for stainless and heat resistant steels

The microstructures for each steel are systematically presented as in Fig. 3, showing partly and completely solidified structures. The increase in dendrite arm spacings with decreasing cooling rate is clear. Where they occur, precipitates and residual ferrite are shown at a higher magnification.

Results from all the steels are combined in diagrams which illustrate the influence of parameters such as cooling rate and composition on the microstructure and solidification process. Examples of such results are shown below. Temperature data for five 5% chromium steels are shown in Fig. 4. The carbon content and cooling rate strongly affect the solidus temperature. Figure 5 shows the liquidus and solidus temperatures as a function of cooling rate and alloy content ($Cr_{eq} + Ni_{eq}$) for the stainless and heat resistant steels. The secondary dendrite arm spacing's dependence on cooling rate and steel type is illustrated in Fig. 6. The segregation ratios, I, for chromium as measured in low alloy, 5, and 13%

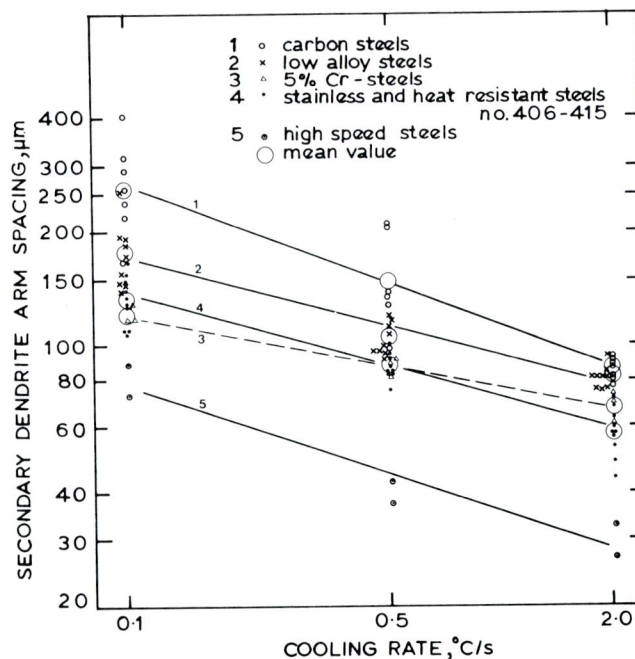

6 Secondary dendrite arm spacings

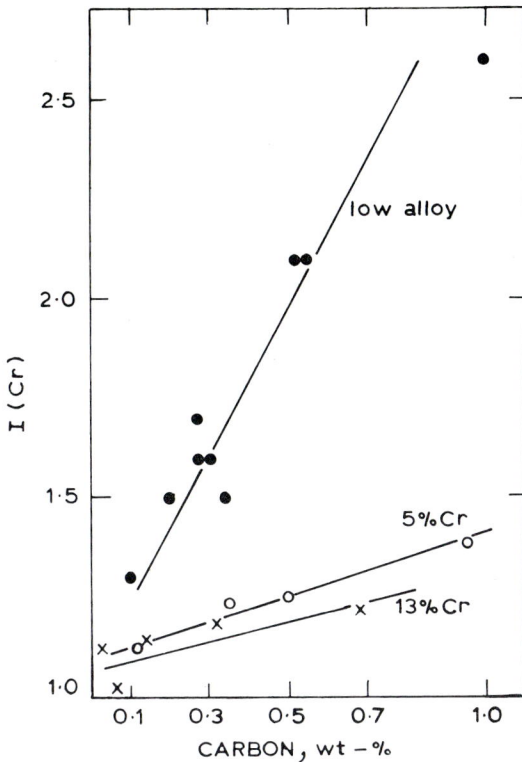

7 Microsegregation of chromium in low alloy and chromium steels

chromium steels (*see* Fig. 7) show that the segregation increases with increased carbon content, which also increases the fraction solidified as austenite.

From thermal analysis the fraction solidified as primary ferrite was calculated. Figure 8 illustrates the drastic change in mode of solidification caused by a small change in the composition of stainless steels. Measurements of ferrite content for these steels were made in samples quenched from the solidus and from 1 200°C (*see* Fig. 9). The ferrite content clearly decreases with

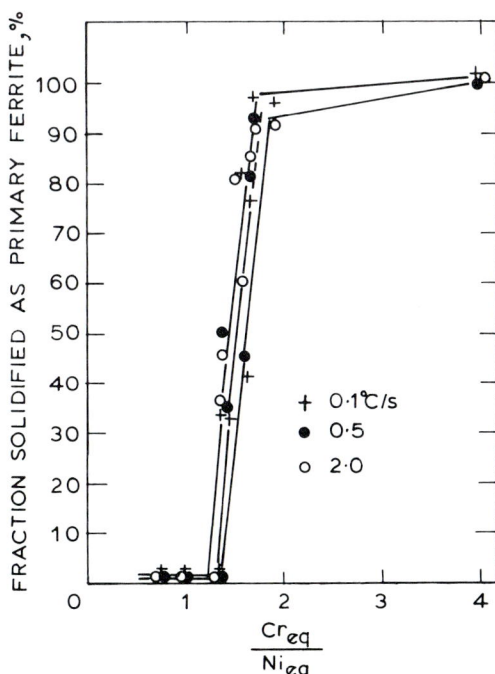

8 Fraction solidified as δ-ferrite in stainless steels

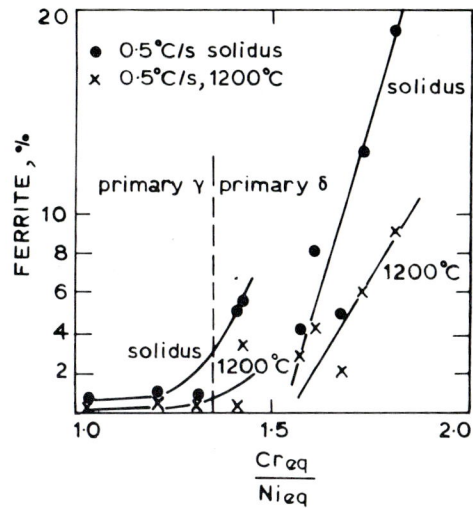

9 δ-ferrite in stainless steels at the solidus and 1 200°C

temperature, but from these data two separate populations can be distinguished. The population with a Cr–Ni equivalent ratio larger than 1·5 corresponds to primary dendritic ferrite and that with a Cr–Ni equivalent ratio of less than 1·5 corresponds to mainly interdendritic ferrite, formed as a result of segregation during freezing. A book containing this and other information will be published by Jernkontoret in Stockholm in January or February 1978.

DR I. G. DAVIES (BSC, Sheffield Laboratories): I would like to ask some questions on Professor Weinberg's paper. First, could I ask for some clarification of the tapers in the narrow-end-up or broad-end-down ingots he was employing, since he claims that variation in the taper explains the variation in the distribution of A-segregates? I am interested to know if in the narrow-end-up ingots he observed any bridging at the top of the ingot which might have led to secondary piping and possibly to associated severe axial segregation. There was no evidence of this in the paper, and I would be interested to know if he has encountered it.

As far as the carbon segregates are concerned, perhaps one feature which was significant but which was overlooked was not so much the increase in carbon level at the very top but the significant reduction in carbon level a little way below that. Could the author please comment on this?

Finally, I would like to make an observation on the appearance of the columnar zone in his ingots, and the suggestion that all these were upward pointing and originated from the very onset of solidification of the ingot. Observations I made some time ago suggested that the columnar zone is essentially horizontal until you get a distinct incidence of branching, which may be associated with the entrapment of descending dendrite fragments. I believe this effect was also seen by Professor Fredriksson, who distinguishes between the ordinary columnar zone and the branched columnar zone. I wonder if Professor Weinberg believes that there is such a distinction to be made?

PROFESSOR WEINBERG: About the tapers, I do not remember the actual numbers. We only used two—big-end-up, big-end-down—but I can find out the actual numbers for you if it is of interest. I realise that the taper itself is significant but when cutting up 3-ton ingots, the scope for variation is somewhat confined.

In terms of carbon segregation, there was a decrease in the middle. I would have liked to have talked about this more but for time and the limitations of space in the paper. The main feature in steel is the high concentration at the top and I still think that most of the carbon is associated with the pipes, not with flow through the interdendritic regions.

With regard to the columnar zone, the basis of the evidence I presented was the sulphur prints. Looking at these the columnar zone was very difficult to find, very close to the edge of the ingot and there was no indication of either secondary dendrites or anything else. They simply pointed up for a long distance, and exactly what happens right at the beginning I do not know, but I am convinced that there is no clear evidence that there is metal flow causing the upward pointing. It might be something associated with the very beginning of the solidification process in the outside zone.

DR D. A. MELFORD (Tube Investments Research Laboratories): My question is directed to Professor Weinberg. I would like to refer to the work of Kohn on determining the profiles of the liquid pool in an ingot by radioactive additions which I think was presented at the Brighton Conference. Kohn obtained a shape which resembled that shown by Professor Weinberg in his first figure, but there was an important difference. In Kohn's observation there was a ledge in the curve, and as solidification proceeded it moved inward and became a little more pronounced. We were all rather happy about that because we said that it was produced by a group of equiaxed dendrites falling down the front of the columnars. In Professor Weinberg's paper, it was one of his conclusions that there was no evidence of falling crystals accumulating at the bottom. Can he explain this difference between the behaviour of his ingots and those studied by Kohn?

PROFESSOR WEINBERG: I am of course conscious of the points raised. They were made at the Brighton Conference and they are certainly very significant. In the four ingots we looked at, we saw no evidence of the profile effect referred to and I also suggest that the lower part does not reflect the solidus/liquidus interface either. I did not see any evidence of debris falling to the bottom. If debris started at the top it would contain radioactive tracer and as it fell to the bottom we would see piles of tracer-containing material there. I saw no evidence of this at all and quite frankly, I do not understand why.

MR K. M. FISHER (University of Oxford): I agree with Professor Weinberg that the A-segregate pipes are responsible for the large amounts of carbon at the top of his ingots. I am probably explaining things in the same way that he was, but if you have a big-end-down ingot and a big-end-up ingot, presumably the isotherms in these two cases must be orientated differently with respect to the vertical (*see* Fig. A). Carbon rich interdendritic liquid flowing upwards in the big-end-up case will be moving from a colder to a hotter region, so that there is unstable flow and preferential channels form. In the narrow-end-up case, however, upward flow will not necessarily be from a cold to a hot region. It may well be from a cold to a colder region and so material will be deposited and will

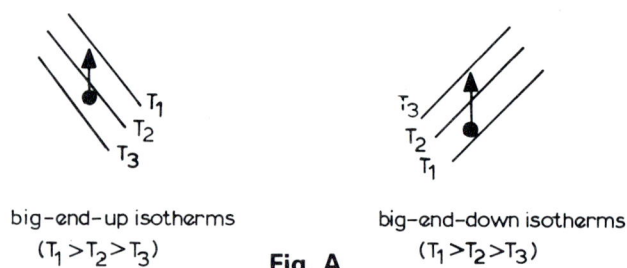

big-end-up isotherms
$(T_1 > T_2 > T_3)$

Fig. A

big-end-down isotherms
$(T_1 > T_2 > T_3)$

effectively strangle the convective cycle by restricting the channel.

THE CHAIRMAN: Can I just repeat what I said to Professor Weinberg earlier, in view of the way the discussion is going. Twenty-five years ago, Basil Gray, who was at the time Director of English Steel Corporation and did a lot of work in this area, astounded his contemporaries, who were trying very hard to get the tops of ingots open, by chilling the top with a very large mass of cast iron. In doing so, he succeeded in obtaining an ingot entirely free of columnar growth. I am talking of very large ingots that were full of fine equiaxed crystals, free from segregation, free from gross porosity and contained only finely distributed microporosity which, of course, healed up in the subsequent working. He demonstrated this on very large scale ingot production and was widely regarded either as a man before his time or an eccentric.

PROFESSOR WEINBERG: I would suggest that the isotherms in both cases are primarily controlled by heat extraction from the wall, so I suggest that there is no difference between the thickness of the shell in big-end-up and big-end-down ingots. I emphasise that the A-segregates form in this region, but that they form in the upper central region in a relatively tight area. You have to look well away from the shell and I do not think you get the kind of convection curve described here in the region where the A-segregates occur.

DR R. D. DOHERTY (University of Sussex): I would like to ask two questions, one of which is directed to Professor Weinberg. I was very interested in his observation that the radioactive tracer in the bottom part of the ingot was not penetrating up to the solidification front. It casts grave doubt on the picture one previously had, as did Kohn, of the interface shape. I wonder if he has any explanation why, in the liquid, we are not getting, in the bottom part of the ingot, the tracer going right up to the edge of the dendrites.

My other question concerns the observations of porosity in the small aluminium ingots studied at McGill University. It was observed that when the mould temperature was increased, there was no change in the distribution of the porosity; in other parts of the paper, it was said that porosity was controlled by the structure. The implication to me was that changing the mould temperature in small aluminium–copper ingots made no difference to the columnar-equiaxed transition. Is that inference correct? Recently I have published a paper (R. D. Doherty *et al.*: *Metall. Trans.*, 1977) showing that in somewhat smaller ingots than were studied at McGill, the mould temperature had a very big influence in promoting a higher incidence of columnars. Would the authors say whether their columnar-equiaxed macrostructure changed with mould temperature.

PROFESSOR WEINBERG: On the basis of the results I showed plus some extensive measurements we made on fluid flow in small cells some years ago, I feel confident in saying that the flow in the pool does not penetrate into the mushy zone. The reason there is no flow to the bottom is that after a short time the ingot loses its superheat, the hot liquid rises, the cold liquid stays down, and the thermal driving force stops. In experiments by Stewart several years ago with a small square block, as the flow went down the side wall and approached the bottom, it deviated so that the interfaces tended to be sloped. The isotherms are distorted by flow and by the geometry of the situation: flow occurs down the wall, hitting the bottom and moving sideways away from the wall. This is not surprising: I think it is quite normal. Let me emphasise that there is no reaction between the pool and the solid/liquid zone. I am not saying that there is no flow anywhere near the top.

DR GRUZLESKI: With regard to the other question, I think you may have slightly misunderstood me. I stated that for the first set of results presented—the ones which were cast at a superheat of 50°—there did not appear to be any influence of the mould temperature. The second set shows that there was an influence of the mould temperature. We have three sets of experiments: those at low and very high superheat are extreme cases and there is in fact a transition region in the centre. Thus there is an influence of mould temperature, at least in the intermediate casting conditions.

We did not, however, measure the relative amounts of columnar and equiaxed grains.

With regard to the further point, one of the major problems of determining microporosity by a density method is macrosegregation. Microsegregation is not a problem because the samples are very much larger than the scale of microsegregation. We analysed some of the ingots for macrosegregation and as expected we found it, particularly in the ingots that were very slowly solidified, i.e. the ones cast at very high superheats. It is possible, however, to make a correction for it. In the most severe case, it amounted to changes in porosity of the order of 0·2 or 0·3 vol.-% in the central region of the ingot. Looking at the magnitude of the isopores for the very slowly solidified ingots, they will be found to be of the order of several percent porosity. The macrosegregation thus makes a small modification to the numerical value of the isopore but this is not really a major change.

DR J. BEECH (University of Sheffield): I would like to comment on Mr Fisher's paper. We also did some work on macrosegregation using the *in situ* technique reported in Session 2 of this conference. We can confirm that there is some activity at a considerable distance behind the interface. In fact we found similar behaviour to that shown in Fig. 1 of his paper.

My other point is that we seem to know a lot about what goes on in macrosegregation: we have a lot of theories and yet many people still have problems. What I would like to ask all those who have spoken about macrosegregation is what we are going to do for those people who still have problems? There has been talk about dendrite arm spacings, about density differences and we know that the isotherms in relation to the fluid flow are important. I have seen work published by Professor Flemings that talked about how something can be done in relation to the latter point, but what about

doing something in other respects? There are things we can do and I would like to ask the speakers how useful they believe some of their work would be in dealing with real problems.

MR K. FISHER: I was very interested to hear of Dr Beech's work and I am pleased to see that the flow still occurs in the thin sections that he used. Our thicknesses are about 30 mm so we have quite a lot of dendrites across the section and thus a lot of flow channels. We have observed many effects, including small particles coming out of the channels at the bottom, as would be expected from dendrite fragmentation, and these are growing in the liquid later on.

With regard to the second point, I think it is very important that we now relate some of these macrosegregation data to industry. The need is to find out their specific problems and then to adapt ourselves to look at them.

PROFESSOR WEINBERG: Let me point out that very fancy casting has been done for 4 000 years; by 2 000 years ago we were doing very elegant things, and if you iterate long enough you reach a pretty high degree of perfection. I suggest that the casting people know their business quite well. They still have to live with alloys that segregate. There is no way that they can be made not to segregate unless you invent a new alloy. They always freeze dendritically and cannot be stopped from doing so in normal casting conditions, so improvements, yes, but miracles, I doubt it.

THE CHAIRMAN: That is a realistic comment but the steel casting industry needs to improve in spite of its 4 000 years of experience, and I am hoping that the body of information that we are receiving this morning will make a contribution to that in due course.

PROFESSOR M. C. FLEMINGS (MIT): I have a short comment, more on the fundamental side. I was very interested in the work of Dr Takahashi and his co-workers on bulk fluid flow and its analysis. I think one of the interesting areas for further work would be tying up the mathematics of that kind of flow with the mathematics of the flow that goes up inside the dendrites. They have certainly got to be related and, in connection with the current discussion, I think that would be a useful problem with some practical applications.

DR K. G. DAVIS (Canadian Department of Energy, Mines and Resources): This is a question on the very elegant work on acoustic emission. I have made fairly extensive observations for several years now on beer, and I always had the impression that the noise came when the bubbles emerged from the top surface. I wonder if this is a complicating factor in your observations?

DR FEURER: We have made other experiments where the bubbles were created in the liquid and it is not the noise of the bubbles when they reach the surface, but when they are formed in the liquid that accounts for the major part of the acoustic emission.

DR V. KONDIC (Birmingham University): I was asked during the interval what is the evidence for the mechanism I propose about skin rupture. The answer is that a pressure gauge is placed inside the liquid so that as soon as the skin forms, we are able to observe the creation of a vacuum within the liquid and associate it with the subsequent rupture.

I would like to make a comment on the last paper. It is great news for the foundry industry that the work has been initiated, and my point really is to make an observation for the sake of completeness.

I am sure that the people doing this work are well aware that they omitted to deal extensively with the relevance of temperature gradients as well as physical phenomena to the solution of feeding problems.

Forecasting the temperature distribution in a casting during its freezing, as shown in Fig. 1, could provide very useful information for suggesting improvements in casting and mould design and for the calculation of feeders to improve the soundness of castings. But the remaining task is still a formidable one, as suggested in the closing remarks of paper by Pehlke *et al*. I would like to add two more items to the list of future activities proposed in the paper. First, an obvious and simple one, the conversion of the data in Fig. 1 to temperature gradients and their directions, particularly around the cores and near to and in the feeders. Secondly, the establishment of relationships between the temperature data and flow phenomena of the liquid metal which occur in the feeding process during freezing. I should like to ask the authors how they propose to deal with this second problem, as ultimately the soundness of castings depends on the physical events in the freezing zone.

Session 6

Structure and property relationships and welding

Chairman: F. B. Pickering (Sheffield Polytechnic)

Editor: A. G. Fuller (British Cast Iron Research Association)

Keynote Address: Solidification phenomena and properties of cast and welded microstructures

P. R. Sahm and F. Schubert

Both casting and welding microstructural phenomena may be rationalized in terms of accepted solidification parameters. To show this, two advanced technologies, i.e. precision casting and electron beam (EB) welding, are briefly being considered from three points of view: nucleation, solidification parameters, and alloy composition. The importance of the mechanisms of dendritic crystallization and involved segregational effects, including pore formation, is stressed: a considerable tolerance band in mechanical properties of superalloys and steels results as a function of the solidification mode, both in precision cast and EB-welded products. Other similarities of consideration are to be found where N_v- or Cr-equivalence numbers are utilized or where polyphase system solidification studies yield essential insight into both precision cast and EB-welded microstructural properties.

P. R. Sahm is with Brown Boveri and Cie Mannheim, Central Research Laboratory, Heidelberg, and F. Schubert is with Thyssen Giesserei AG, Bochum, West Germany

GENERAL FEATURES

Microstructure and property relationships are relevant for casting and for welding processes of all kinds. These processes are utilized to produce machine and structural components. They have to pass certain minimal property requirements. In both types of process, therefore, the mode of solidification is to be observed. The object of this paper is to discuss existing solidification models in terms of two advanced technologies:

(i) precision casting including directional solidification primarily of high temperature materials (superalloys) and
(ii) electron beam (EB) welding of steels.

Generally applicable solidification diagrams have been proposed on numerous occasions. They normally indicate various microstructural regimes as a function of growth rate v. temperature gradient G, and sometimes compositional parameters. Figure 1[1] represents such a diagram as it shows a plot of the single parameters v versus G, simultaneously delineating a network of iso-cooling rates, important in judging any type of solidification microstructure. The ideal situation for the casting or welding engineer would be reached if he possessed recipes for moving such plots up and down at will, tailoring the properties sought for.

The regime of most practical concern is the upper diagram portion in Fig. 1, i.e. the transitions 'cellular–dendritic–equiaxed'. Strictly planar front growth is normally verified in single crystal synthesis or directionally solidified one- and more-phase (eutectic or monotectic) alloys, i.e. where non-variant reactions may be utilized. The enormous amount of knowledge obtained from such simplified considerations helps us to understand the more complex cases encountered in practice. The processes to be considered together cover a wide range of conditions, with normal (that is, non-EB) welding methods lying somewhere in between (*see* Fig. 1).

One and the same composition solidified under different experimental conditions may yield considerable differences in mechanical or physical behaviour because of both (a) the mode of crystallization; and (b) the type, number, and distribution of defects. The control of properties, therefore, requires a sound understanding of the solidification mechanisms and the application of such models for the evaluation of process parameters, along with a high level of practical experience.

PROCESSING FEATURES
Investment casting
For the production of a 'sound' casting, Wlodawer's[2] formulae, tables, and graphics may be utilized. They are based on Chvorinov's rule[3]:

$$t = KM^2 \qquad (1)$$

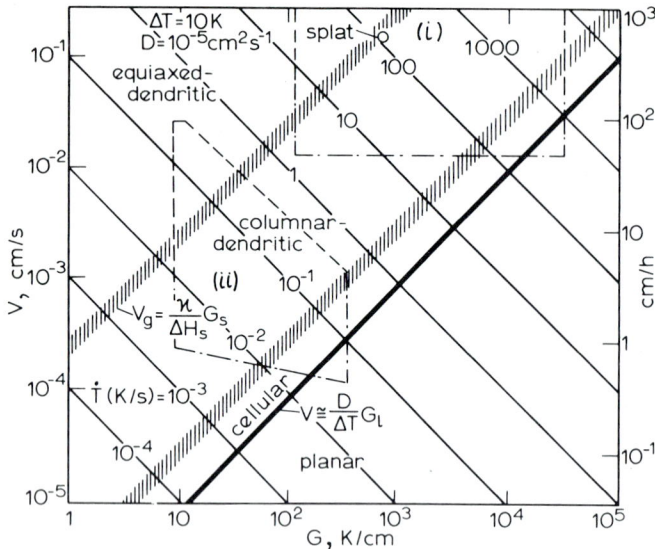

1 Simplified schematic for regimes of solidification microstructures within *G/v* network for fixed solidus–liquidus interval ($\Delta T = 10$ K) and diffusion coefficient (10^{-5} cm^2/s) and a certain (not defined) state of convection, extrapolated after Kurz 1976; approximate regimes of EB welding (upper) and of precision casting (lower) conditions are outlined

in which the solidification time *t* of a certain casting section is equal to the square of the casting modulus *M* and a constant *K*. *K* depends on the alloy composition, the pouring parameters, and the type of mould materials. Equation (1) helps greatly in determining steel castings. Very little data have been published concerning solutions for investment castings similar to equation (1), (Cook[4]).

For steel castings, Fig. 2[5] demonstrates the role of the mould material and solidification time (Antony and

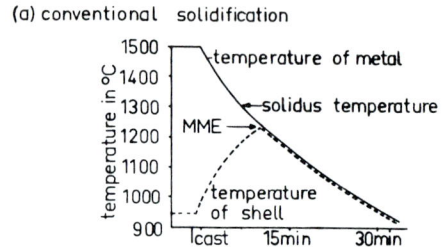

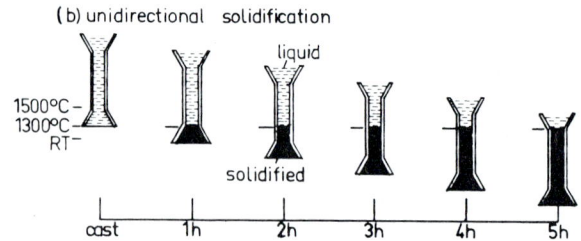

2 Solidification time of sphere with 150 mm dia. in different mould materials (after Locke, Briggs, and Ashbrook)

3 Comparison of temperature–time history between conventional and directional solidification processes

Radavick[6]). First experiments with a constant sample dia. of 150 mm, although only partly representative for investment castings, lead to the estimated regime of solidification times for investment casting shell materials. For a given shell mould and an optimized feeding system (Fig. 3), the microstructure of an investment casting depends on (*a*) pouring temperature; (*b*) mould temperature; (*c*) metal-mould-equilibrium (MME) temperature; and (*d*) cooling history down to MME-point.

In the case of non-heat treatable superalloys, the entire cooling history is important not only with respect to the solidification structure but also with respect to the distribution and morphology of γ-prime precipitates, in other words, for the verifiable level of mechanical properties. In the case of the directional solidification (DS) process, for obtaining a well aligned and defect-free microstructure, the pouring temperature and the mould temperature have to be nearly equal. The required plane front solidification across the entire component is obtained by continuous withdrawal of mould or heating furnace relative to each other. Directional solidification processes obviously demand more stable shell mould materials than conventional investment casting, particularly with respect to good chemical stability, thermomechanical resistance, and thermal properties.

EB-welding

An excellent review on 'Solidification structures and properties of fusion welds' was published by Davies and Garland in 1975.[7] They very aptly state: '... while the process of weld pool solidification is frequently compared with that of an ingot in miniature, a number of basic differences exist ...', particularly with respect to (*a*) nucleation; (*b*) macroscopic solidification rates; (*c*) interface shape; and (*d*) motion of melt; points (*a*)–(*c*) resulting from the different kind of 'mould' utilized (i.e. its own envelope), and (*d*) from the very high, and thus local, specific energy input. Solidification rates and melt motion are even more pronounced in EB-welding than in 'conventional' welding processes.

Figure 4[8,9] illustrates the present model of an EB-welding channel (see for example Eggers *et al.*[10] 1972 or Dorn[8]). The essential features for a non-oscillating

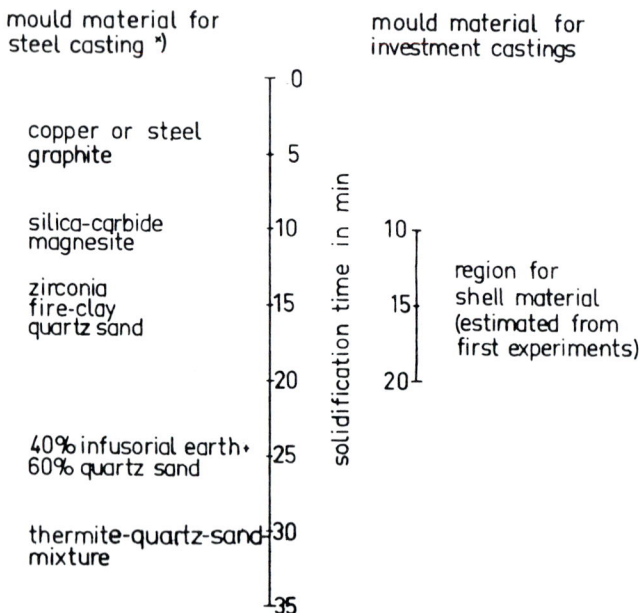

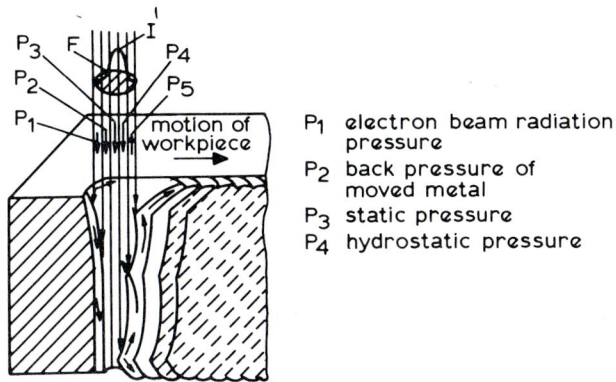

P1 electron beam radiation pressure
P2 back pressure of moved metal
P3 static pressure
P4 hydrostatic pressure

4 Model of EB-weld channel (Dorn 1975) and visualized in real EB-weld after fracturing along seam of steel X 22 CrMoV 12 1 (Raupach and Speil 1977)

beam (concerning some fundamental differences to other welding techniques, *see* Table 1[11]) are:

(i) formation of a vapour channel in the centre of a weld on account of very high local temperatures; this vapour capillary 'breathes', i.e. it opens and closes in 'statistically regular' intervals; the 'breathing process', however, can be influenced by oscillating the beam utilizing x- and y-generators for a variety of basic mathematical functions such as sinusoidal, triangular, rectangular modes, etc. Keeping the capillary open increases the ability for deep penetration welds (for example, by using an x-oscillation mode)

(ii) a liquid metal skin showing strong convection with the main flow directions indicated in Fig. 4 and stabilized essentially by interfacial energy forces

Table 1 Specific energy densities of different joining techniques (partly after Remund[11])

Process	Achievable specific energy density		Reference
	Continuous	Pulsed	
Laser welding	10^7	10^9	Bennighof[39]
Electron-beam welding	10^6	10^8	Steigerwald[40]
Arc welding	10^5	—	Steigerwald
Flame welding	10^4	—	Steigerwald

(iii) the solidification process normally being governed by extreme values of temperature gradient and cooling rate.

The latter point will be discussed in some detail, particularly in view of the various possibilities of welding channel modification by the option of inertia-less beam oscillation and/or modulation. Another important processing parameter is given by preheating the workpiece to a preselected temperature and thus affecting the thermal conductivity of the channel ('mould') walls.

NUCLEATION CONTROL
General
Both as-cast and as-welded microstructures make an important contribution to the final mechanical properties. Of the two principal nucleation mechanisms, i.e.

(i) homogeneous nucleation (crystallization centres develop in the melt due to fluctuating energy: a considerable supercooling is required) and
(ii) heterogeneous nucleation (here nucleus formation is primarily governed by the effect of mould wall, solid particles, impurities, etc)

the latter is all-important so far, both in casting and welding practice.

The energy ΔG, necessary for the formation of the nucleus is

$$\Delta G \cong \frac{\sigma^3}{\Delta G_v^2} f(\theta) \text{ with } f(\theta) \cong \frac{(2+\cos\theta)(1-\cos\theta)^2}{4} \quad (2)$$

where σ is energy per unit surface of the melt/nucleus interface, ΔG_v is the change of free enthalpy per unit volume, θ is the wetting angle. Lowering the wetting angle between melt (to be nucleated) and container wall or other heterogeneous nucleus site, the 'nucleus-forming energy' decreases. At $\theta = 0°$ non-epitaxial nucleus formation on heterogeneous nucleus substrate may take place, at $\theta = 180°$ homogeneous nucleation becomes mandatory.

For homogeneous nucleation the nucleation rate I_N, besides being indirectly proportional to the cooling rate $\dot{T}$, is determined by interaction of two supercooling dependent activation energies

$$I_N = \exp -[(\Delta Q + \Delta G_s)/kT] \quad (3)$$

where ΔG_s is critical nucleus formation energy (surface formation term) and ΔQ is diffusional activation energy. Thus it is possible to calculate, at least theoretically, necessary cooling conditions for even entirely suppressing nucleation, that is for obtaining an amorphous 'microstructure'. Modern high power, laser beam technology enables one to achieve solidification rates sufficient to pass by the nose in TTT-diagrams of this type, even for steels and superalloys (Breinan et al.,[12] 1977), getting up into the upper right corner of Fig. 1. However, before reaching the stage of amorphous solidification, there are various stages in between to be considered, especially in welding.

Nucleation in precision casting
In superalloy solidification two cases must be distinguished: aligned grain structure (DS process), and equiaxed grain structure (conventional process). In the first, nucleation is not a matter for concern (excepting

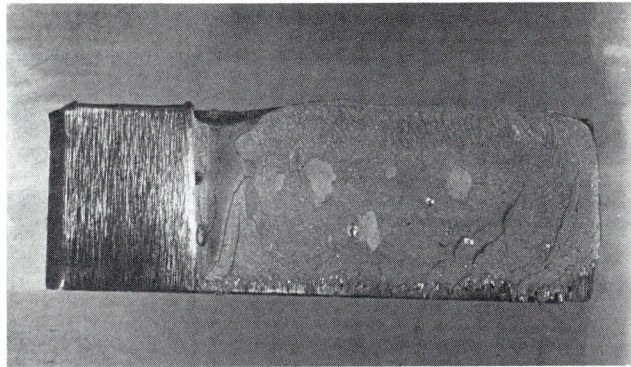

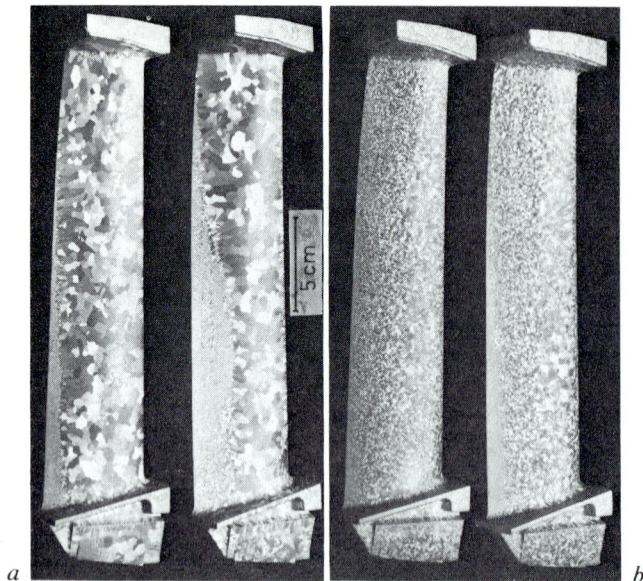

a without nucleants; *b* with 6% nucleants

5 Macrostructure of Ni-base superalloy (IN 713) turbine blades cast without and with nucleants on the inner shell layer

the very beginning of growth); in the second, nucleation is to be closely controlled.

In investment casting of Al-alloys grain refining is achieved by doping the melt with titanium borate as inoculant. Investment castings of high temperature nickel-base superalloys, which are both melted and cast in fired ceramic shells under vacuum, are being nucleated by way of the first dip coating, e.g. at the inner mould layer. For otherwise constant pouring conditions (i.e. pouring temperature, shell temperature, and shell base material), different grain sizes and distributions are achievable by altering the first shell coat, for example such as shown in Fig. 5 without and with 6 vol.-% cobalt–aluminate. The blades produced without nucleant agent in the first dip are not acceptable for turbine service because of unsatisfactory mechanical properties.

Role of nucleation in EB-welding
Essentially, there are two distinguishable nucleation modes in welding solidification:

(i) the 'epitaxial growth mode', i.e. heterogeneous nucleation from the weld channel walls
(ii) independent nucleation in the fused weld metal.

Clearly, in practice hybrid modes, more or less determined by the one or the other, are being observed. Figure 6 illustrates what may be such a 'misch-mode' with an 'epitaxial' layer at the channel walls and an apparently independently nucleated central zone.

The possibilities to influence the nucleation mode in welding are numerous (see, for example, Davies and Garland[7]). In EB-welding many of these methods (inoculation, mechanical vibration) can be replaced effectively by beam modulation and/or oscillation, and fall more into the realm of modelling the shape of the solidification front. However, for a clean separation of the mechanisms involved much work remains to be done.

6 EB weld of X 10 CrNiTi 18 9 with apparent zone on independently nucleated (non-epitaxial) microstructure ('apparent' indicates that it had not been determined how much of the central microstructure was due to curved growth of epitaxial dendrites and how much to independent nucleation)

SOLIDIFICATION PARAMETERS
Solidification behaviour of investment castings
Fulfilling all requirements of the processing conditions discussed above, microstructure and mechanical properties of a casting are determined by the solidification time and local cooling rate of each part of a casting. The freezing time for an investment casting piece may be reasonably well described by a kind of relationship similar to that of Chvorinov's[3]:

$$t_f = \left(\frac{\rho_m \Delta H}{T_s - T_{sh}}\right)^2 \left(1 + \frac{c_l T_{su}}{\Delta H}\right)^2 \left(\frac{1}{K_{sh}\rho sh^c sh}\right) M^2 \qquad (4)$$

where

t_f = solidification (freezing) time of the casting section
ρ_m = density of the metal
ΔH = heat of fusion of the metal
T_s = solidus temperature
T_{sh} = temperature of the shell mould
c_l = specific heat of the liquid alloy
T_{su} = superheating at pouring
K_{sh} = thermal conductivity of the shell mould
ρ_{sh} = density of the shell mould
c_{sh} = specific heat of the shell mould
M = casting modulus (= local casting volume/local casting area).

In investment casting most of these physical constants are not known. For the microstructure t_f can be understood as an 'internal parameter' describing both globular

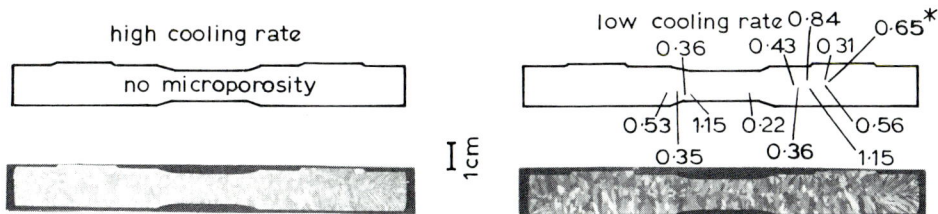

results of stress–rupture at room temperature in as–cast condition:

	φ8mm	φ6mm	φ8mm	φ6mm
$\sigma_{0.2}$ (N/mm^2)	805	815	735	725
σ_B (N/mm^2)	1020	1040	845	820
A_5 (%)	5·0	6·7	6·3	3·3
Z (%)	10·0	8·0	3·5	7·5

*% of micropores

7   Influence of cooling rate on microstructure and stress rupture properties of Ni-base superalloy (IN 738 LC) castings for two different specimen diameters

and columnar grain growth independently of the casting's cross section. This relationship, therefore, is of great qualitative value for explaining numerous day to day observations.

At constant T_{su} and T_{sh} (shell geometry was the same) the freezing time could be retarded by back-filling one of the shells with fired clay (Fig. 7). The rapidly cooled mould resulted in a pore-free microstructure with medium sized grains and little tendency for interdendritic segregation. Consequently a higher yield strength was measured. Poorer 'etchability' of the microstructure in this case was another indication for differences in morphology, including the distribution of γ' precipitates. Figure 8 shows typical as-cast microstructures for both cases: fine equiaxed dendritic without microporosity arm spacings and with microporosity in interdendritic spaces after slow cooling. Figure 8[13] shows a magnification of the microstructure in Fig. 7. Figure 9[14,15] demonstrates the possible kinds of microporosity or microshrinkage in investment castings and the toleration limits. For example, an aircraft turbine blade is only acceptable over a very restricted regime of defects (portion above the line).

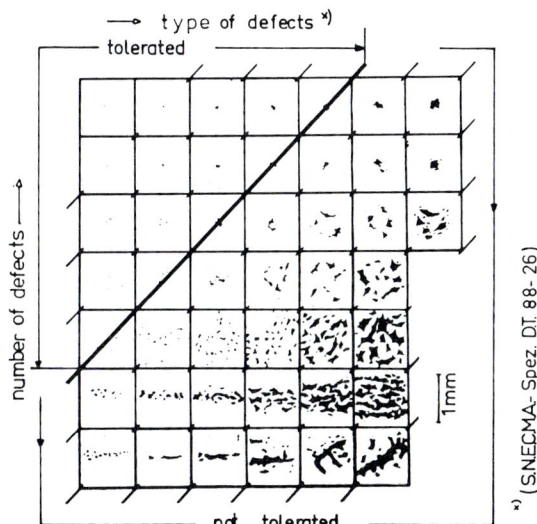

9   Example for a microporosity standard in turbine blades (SNECMA sp.DT 88-26)

Besides difficulties in feeding, microporosity in open air castings is very often caused by an interaction between gas evolution and solidification shrinkage. In vacuum cast material, according to Antony and Radavick,[6] shrinkage-induced microporosity is thought to occur when the local melt pressure becomes negative during the latter stage of freezing. For the pouring process in vacuum, i.e. with no ambient pressure, the feeding systems serve to minimize microporosity. In practice, certain 'tricks' help in reducing microporosity, e.g.:

(i) quick inert gas flushing of the vacuum chamber
(ii) utilizing differential pressure levels during the pouring process (CLV- and CLA-processes)
(iii) adjusted shell-temperature and thermal conductivity of shell material.

An essential phenomenon governing microporosity is the dendrite growth mode. Their solidification is governed by the local cooling rate. It may be equated to

$$\dot{T} = dT/dt = Gv \qquad (5)$$

Structural features, such as dendrite arm spacing λ_D may be explained in terms of these parameters (compare

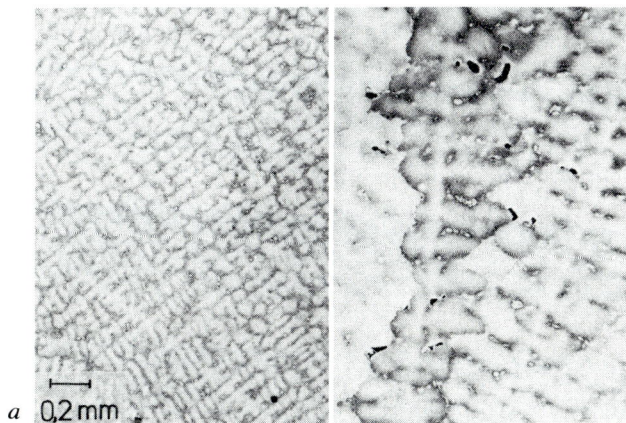

a high cooling rate; b low cooling rate

8   Influence of cooling rate on microstructure and microporosity of Ni-base superalloy (IN 738 LC); same sample as in Fig. 7

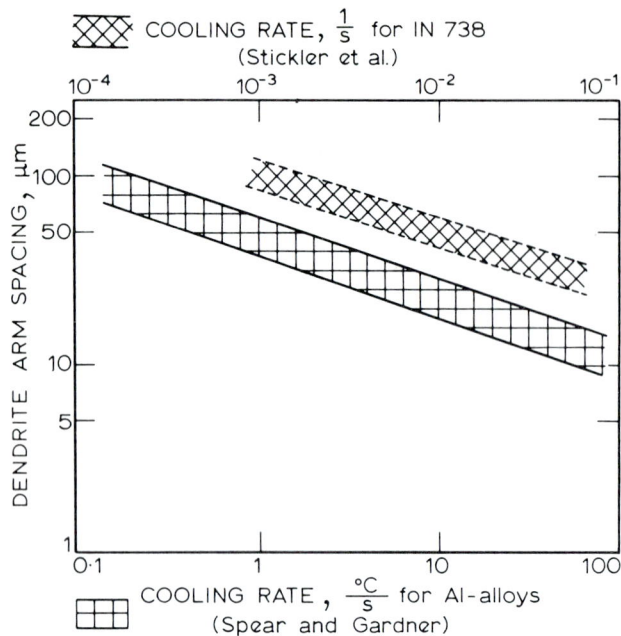

10 Cooling rate dependence of dendrite arm spacing for Ni-base superalloy (IN 738, Stickler and Kay) and for Al alloys (Spear and Gardner)

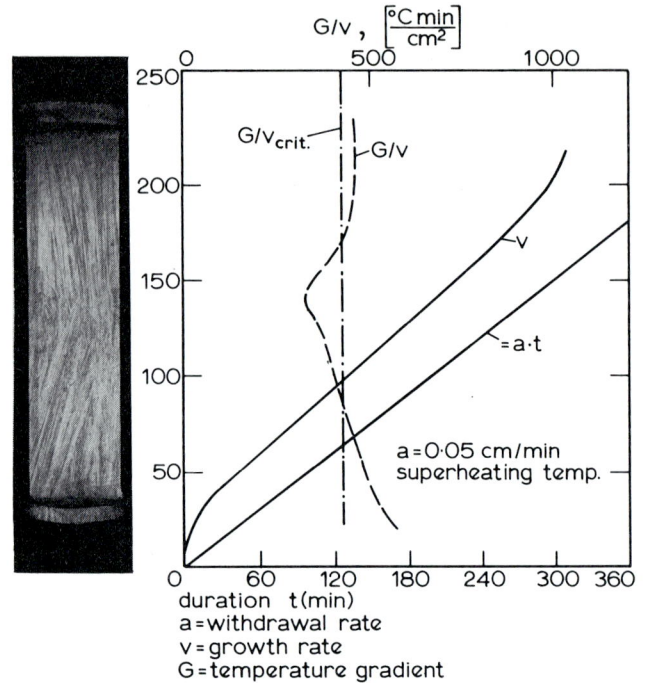

11 G/v dependence of Co–Cr₇C₃ eutectic superalloy macrostructure

Spear and Gardner,[15] Flemings,[16] Stickler and Kay,[14] and Kurz[1]).

$$\lambda_D = K \cdot \dot{T}^{-a} = K'v^{-b}G^{-c} = K''t_f^{-d} \qquad (6)$$

with a, b, c, d = constant. In conventional castings local G and v values are difficult to measure. A fairly clear relationship between dendrite arm spacing and cooling rate has been plotted in Fig. 10. Dendrite arm spacing greatly influences the mechanical properties. In the aluminium foundry industry suitable mould chilling is used to achieve a higher cooling rate, in other words, smaller dendrite arm spacings which guarantee mechanical properties of the so-called 'premium quality' with nearly twice the strength and ductility data of normal material. Hofweber and Fiore[17] have demonstrated that one and the same Ni-base alloy can be solidified in different solidification modes by varying the growth rate. Accordingly, different mechanical properties were measured. In directional solidification processing the G/v value at the solidification front is of greatest importance during the entire solidification process. The first computer controlled d.s. turbine blade production was described by Giamei and Erickson[18]. An illustrative example was verified with the d.s. processed Co–Cr₇C₃-eutectic superalloy: decreasing G/v values below an alloy dependent critical number lead to misoriented columnar grains (Fig. 11). At the beginning of the directional solidification process the solidification rate is higher than the withdrawal velocity, the difference being dependent on mould geometry.

An interesting proposal to circumvent the purely mechanical method of G/v control has been presented by Haour.[19] Highly undercooled melts for high rate (low gradient) solidification are utilized and obtain well oriented, fine microstructure with improved mechanical properties.

From the practical point of view, the G/v value alone is not sufficient to guarantee a well aligned columnar grain. Figure 12 shows how total specimen length (= starter rod length) can affect microstructure. This 'starter rod length' effect may not be equated to the local G/v differences such as is illustrated in Fig. 11, but is due to selective growth of the best oriented crystals or even monocrystal.

Directional solidification processing of turbine blades yields higher thermal shock resistance and longer creep rupture lives as compared to conventional material. Figure 13 shows creep-curve plots of conventionally cast and directionally solidified test-specimens utilizing the same master heat of IN 738 LC. At lower stress levels the gain in creep rupture life is a result of an extended steady state and ternary creep stage. In many cases, however, directionally solidified specimens of Ni-base superalloys yield higher primary creep stage elongation than equiaxed material. The primary creep velocity of Co-Cr₇C₃, directional solidification eutectic, is lower than in conventionally cast superalloys.

Creep rupture strength σ_c depends sensitively on the degree of alignment of the 'fibres' in the *in-situ* composite and is normally expected to increase with smaller mean interfibre distances. Processing requirements for good alignment have been indicated above (Fig. 11). The mean interfibre distance λ can be decreased by higher growth rates according to the relationship

$$\lambda^2 v = K \ (= \text{constant}) \qquad (7)$$

However, creep measurements indicate that λ and σ_c are not indirectly proportional in a straightforward manner (*see*, for example, Kurz and Sahm[20]) and particular emphasis will have to be paid to parameters like temperature gradient and cooling rate affecting defect type, number, and distribution.

short   long

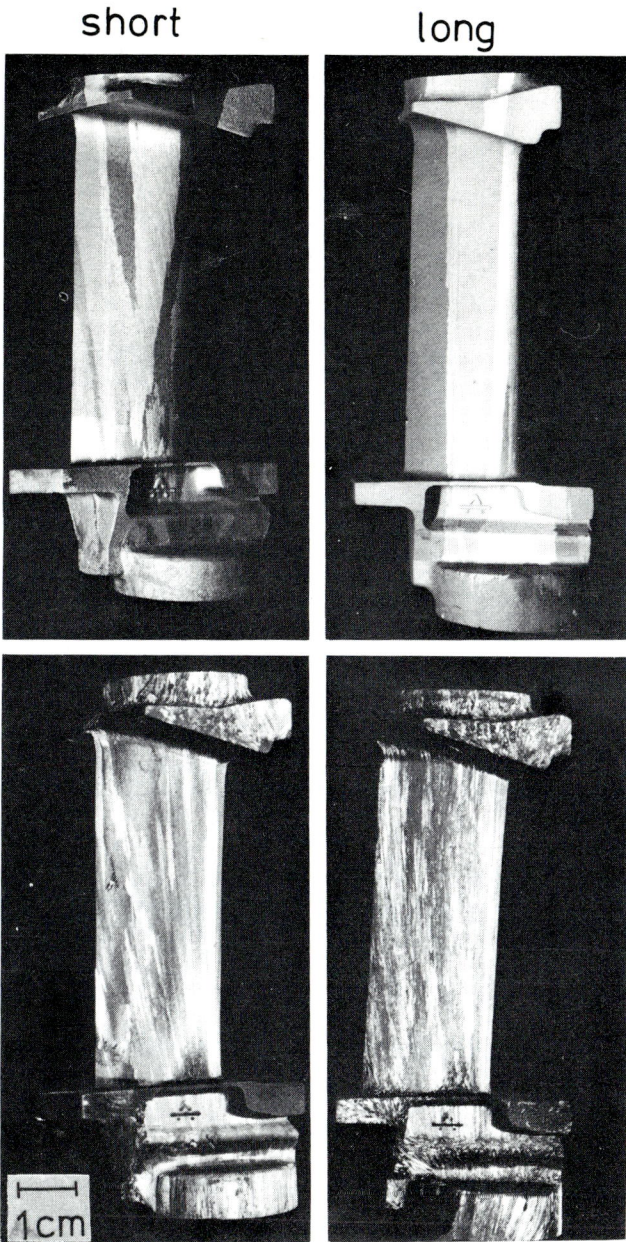

a Ni-base superalloy, FIS 145; Co-base eutectic alloy, C73

12 Influence of starter rod length on macrostructural morphology of directionally solidified turbine blades (competitive growth effect)

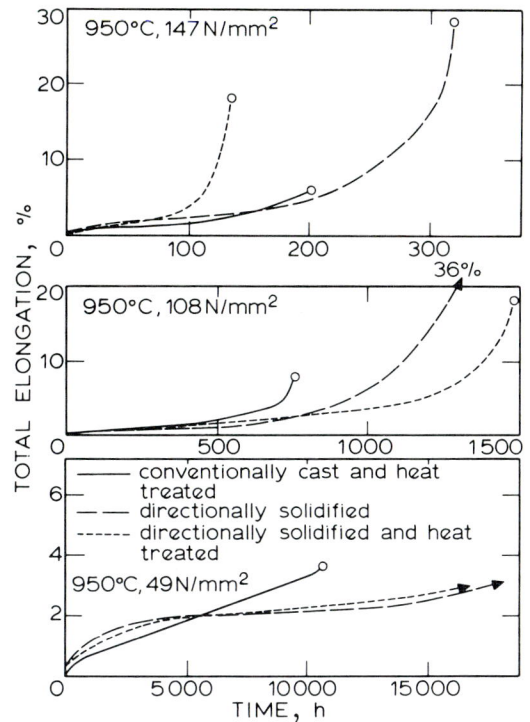

13 Creep rupture curves at 950°C of directionally solidified and conventionally processed Ni-base superalloy (IN 738 LC)

tionship for the local growth rate v_s (in terms of the welding speed v_w):

$$v_s = \frac{v_w \cdot \cos \alpha}{\cos(\alpha' - \alpha)} \qquad (8)$$

to not only account for welding channel shape with respect to the welding direction, but also for the case that

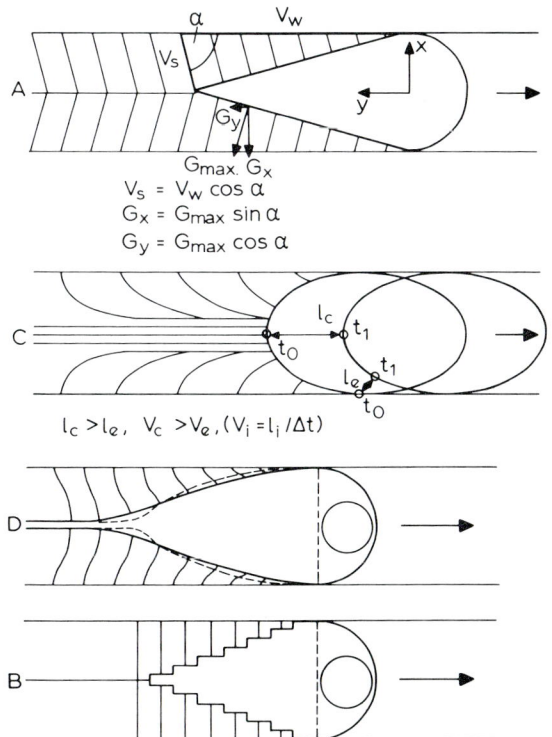

$V_s = V_w \cos \alpha$
$G_x = G_{max} \sin \alpha$
$G_y = G_{max} \cos \alpha$

$l_c > l_e, \; V_c > V_e, (V_i = l_i /\Delta t)$

14 Possible solidification front morphologies in EB-welds (Sahm and Janke) $\Delta t = t_1 - t_0 =$ solidification time

Solidification in the EB-welding channel

There have been numerous attempts to describe welding channel solidification in terms of accepted solidification models (*see*, for example, Savage *et al.*[21], Shabinian *et al.*[22], Wittke[23], and Eichhorn and Schuhmacher[24]). As Davies and Garland[7] point out, the relationship of G to v in welding channel solidification stands in contrast to normal castings: maximal growth rates may occur in the channel centre where minimal temperature gradients are encountered and vice versa at the channel edge (*see* Fig. 14). Nakagawa *et al.*[25] propose the following rela-

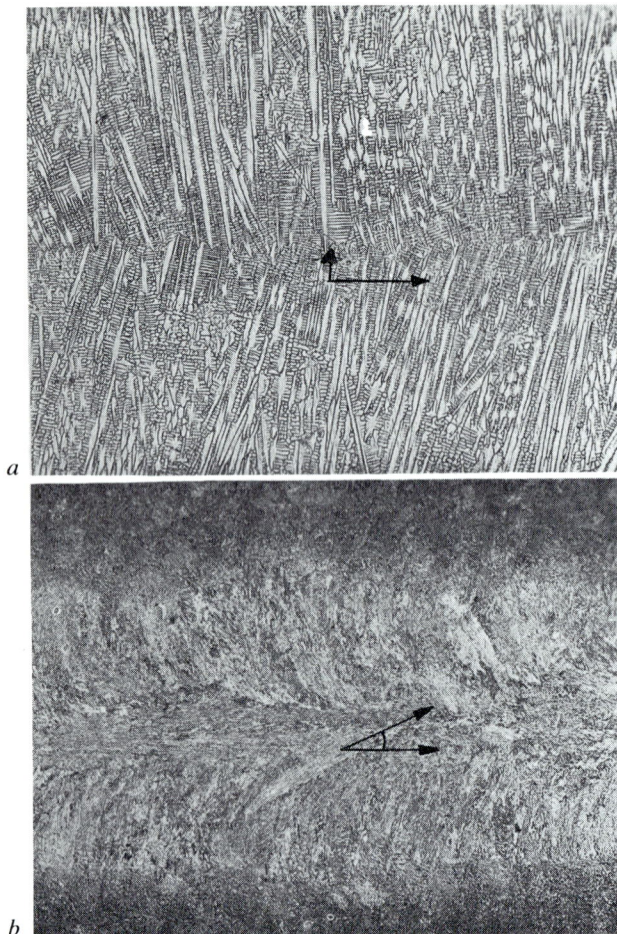

a $V_w = 20$ mm/s, $V_s = 6$ mm/s ($\times 10$ CrNiTi 189)
b $V_w = 6$ mm/s, $V_s = 5$ mm/s ($\times 22$ CrNoV 12 1)

15 Welding and solidification velocities according to microstructural features (Sahm and Janke)

Table 2 Solidification parameters in EB-welded X 22 CrMoV 12 1 steel and measured dendrite spacings compared to estimated C- and Cr-diffusion distances

Temperature gradient, G, K/cm	End of solidification (centreline) 10^3		Beginning of solidification (channel edge) 10^5	
Welding speed, V, cm/s	0·3	1·0	3	10
Cooling rate, $\dot{T}$, K/s	300	1 000	3 000	10 000
Cooling time, t, s	3	0·4	0·3	0·04
Main dendrite arm spacing, λ_1, μm	30–40	15–25	30–40	15–25
Carbon diffusion distance, d^*, μm $D = 10^{-7} - 10^{-6}$ cm^2/s (Seith[41])	5·5–17	2–6	1·7–5·5	0·6–2
Diffusion distance Cr, d, μm $D \simeq 10^{-10}$ (Fridberg[42])	0·17	0·06	0·05	0·006

* $d \sim \sqrt{D \cdot t}$; t = cooling time

crystallization is not isotropic (and thus following uncritically any change of direction) which amounts to saying that in cases of non-curved crystallization (e.g. Fig. 15a[26]) a 'faceted welding channel wall geometry' may be assumed (Fig. 14b). Estimates of growth rates from microphotographs such as those shown in Fig. 15, therefore, yield maximal local values for the growth rate.

Some growth rates in EB-welds have been estimated from experiments. The shape of the interface was assumed to have continuously alternated between a faceted type (Fig. 14a) and a curved one, similar to Fig. 14c. The estimates gave values of $v_s \simeq 0\cdot3$ to $0\cdot9v_w$ ($\alpha \simeq 20$ to $\leqslant 90°$ or else v_s: $v_w \simeq 6$:20 to 5:6, *see* Fig. 15). With the knowledge of v_s a first handle on crystallization mode criteria is possible; in this context Fig. 1 may be referred to:

$$v_s \leqslant G \cdot D/\Delta T_{l-s} \qquad (9)$$

One may deduce whether planar, cellular, or dendritic interface crystallization is to be expected. While the solid–liquid temperature interval, ΔT_{l-s} (K), and an average diffusion coefficient D(cm^2/s), have to be determined experimentally for complex alloys, G may also be calculated through the cooling rate $\dot{T}$ (K/s)

(compare equation (5)) and according to Goldak *et al.*[27] and Rosenthal[28],

$$\partial T/\partial t = \alpha[(\partial^2 T/\partial x^2)+(\partial^2 T/\partial x^2)] \qquad (10)$$

(solving the heat conduction equation with α = heat diffusivity, cm^2/s). Our own evaluation of equations (9) and (10) for a 12%Cr steel (X 22 CrMoV 12 1), (*see* Table 2) show G values of between 10^3 and 10^5K/cm and corresponding cooling rates of 300–1 000K/s. With $\Delta T_{l-s} = 10$ to 100K for a typical alloyed steel, equation (9) yields minimally

$$v_s \leqslant \frac{(10^3 \text{ to } 10^5) \cdot 10^{-5}}{(10^1 \text{ to } 10^2)} = 10^{-1} \text{ cm/s,}$$

i.e., the dendritic mode—which is normally encountered in EB-weld microstructures (*see* Fig. 15). Realistically attainable minimal welding speeds and $v_w = 10^{-1}$ cm/s, correspond with values of $v_s \simeq 5 \cdot 10^{-2}$ cm/s, still well above planar front growth for the main portion solidified.

Different dendrite arm spacings in weld microstructures have been shown to deliver similar relationships to mechanical properties as in castings (Jordan and Coleman[29] or Schmid and Hildebrandt[30]). EB-welds may not be as susceptible to defects in interdendritic spaces, for example pores, as precision castings, since much higher growth rates are involved. By directionally solidifying the X 22 CrMoV 12 1 it is possible to create interdendritic pores at will, (Fig. 16).[8,9] The same steel when EB-welded (e.g. Fig. 15b) solidifies at much larger T-gradients, higher speeds, and in smaller sections (typically 0·1 cm). Here, the critical set of pore enhancing solidification conditions has yet to be determined.

As noted above, the possibilities of inertialess modulation or oscillation of the electron beam may be effectively utilized to adjust weld channel shape, a field which is so far very little explored. Also, the accompanying melt convection has yet to be tapped for quantitative solidification control.

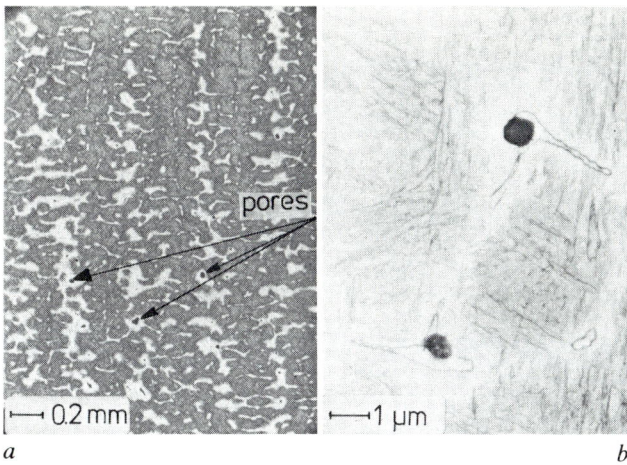

16 Longitudinal sections obtained for unidirectionally solidified 0·5 cm dia. steel × 22 CrMoV 12 1 rods; $V_s = 7·10^{-3}$ cm/s, $G = 33$ K/cm

EFFECT OF ALLOY COMPOSITION ON MICROSTRUCTURE AND PROPERTIES

The discussion which follows will only be concerned with segregational effects, i.e. local composition.

Compositional effects in castings

The chemical composition of modern high-temperature alloys is to be more and more accurately controlled. As an example of the effect of small compositional variations, the creep rupture properties of four IN 738 LC melts, cast and heat treated under the same conditions, are given in Fig. 17. The loss of creep rupture strength for the higher Cr- or (Al + Ti) level alloys (although still within tolerance limits) was due to σ-phase precipitation, which was predictable by the calculation of N_v-numbers (electron vacancy number of the γ-residual matrix).

The chemical composition of modern high-temperature nickel-base superalloys, therefore, has to be N_v

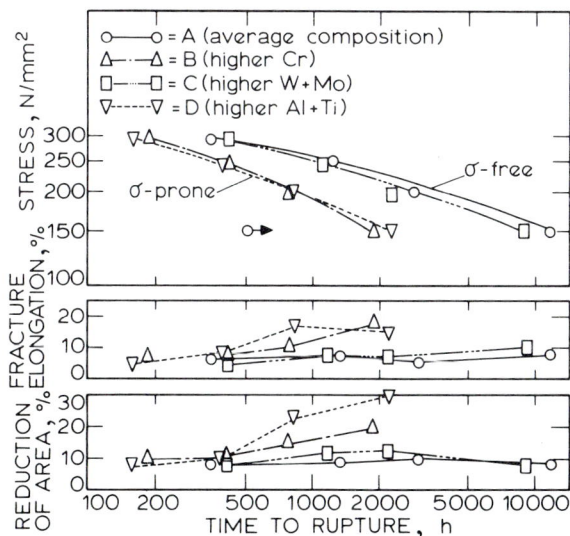

17 Differences in creep rupture data at 850°C resulting from variable location within tolerance limits of chemical composition for precision cast superalloy (IN 738 LC)

controlled. For overcritical N_v, σ-needles are precipitated after exposure in interdendritic spaces and at grain boundaries, from there growing through the entire bulk of the grain (Fig. 18).[31] By N_v control, a certain guarantee for σ-free material may be given. However, due to interdendritic microsegregation, σ-needles can be found even in cases where the average master heat alloy had had a 'safe' N_v number.

The tendency to microsegregation is determined by the size of solidus–liquidus temperature range ΔT_{l-s}. It may be expressed in terms of the distribution coefficient k ($= c_s/c_l$).

$$\Delta T_{l-s} = c_x \text{ (or } c_{\Sigma i})m(1-k)/k \qquad (11)$$

where $c_{x,\Sigma i} =$ composition of element x or the sum of alloying elements; $m =$ liquidus slope for c_x, c_s, and c_l being the concentration of x, Σi in the solid or liquid state respectively. Small compositional changes of alloying or even doping elements may have remarkable effects upon the properties of the casting. Thus, the doping of alloys may help to ameliorate the properties, e.g.

(i) Mg treatment of Ni-base superalloys to increase hot workability
(ii) alloying small amounts of rare-earth metals to increase hot corrosion resistance of Co-base alloys, etc.

On the other hand, small amounts of the so-called trace elements, (Sb, Bi, Pb, etc.) may have detrimental effects on creep-rupture strength and ductility in superalloys (Kleemann et al.[32]).

For predicting the solidification behaviour and segregation problems of modern multi-element alloys, knowledge of the phase-diagrams, especially the liquidus and solidus temperatures, is mandatory. Concerning highly alloyed steels, very important phase-relationships are to be learned from the Fe-rich corner of the ternary Fe–Ni–Cr system published by Schürmann and Brauckmann.[33] The three-phase equilibria are being determined both by the binary Fe–Cr system, with a minimum of solidus and liquidus temperatures, T_s and T_l, and an increase of T_l with increasing Cr-content, as well as by the binary Fe–Ni system with a peritectic reaction, a regime of primary γ-crystallization, and slowly decreasing liquidus and solidus temperatures. The expected solidification reactions are illustrated by a temperature-concentration cut (Fig. 19). For 75% iron, according to Bungardt et al.[26], considering a composition close to 18 Cr–8 Ni stainless steel, δ-ferrite should first precipitate from the melt, becoming Cr-enriched with progressing crystallization due the shallow solidus slope. As the temperature is lowered the melt goes through the ternary phase space (M + δ + γ) and δ-ferrite starts enriching in Ni. Upon further cooling, part of the δ-ferrite transforms to austenite (cooling rate dependent). The amount of remaining δ-ferrite then also depends on the duration and temperature of the austenitization treatment. Cr–Ni equivalence numbers may be used to a certain extent for the prediction of phases to be expected (see also below).

Due to segregation while solidifying, compositionally well adjusted 13% Cr steels (by equivalence number computations) nevertheless show residual austenite and δ-ferrite in the as-cast microstructure. Unless succeeding thermomechanical treatment satisfactorily evens out

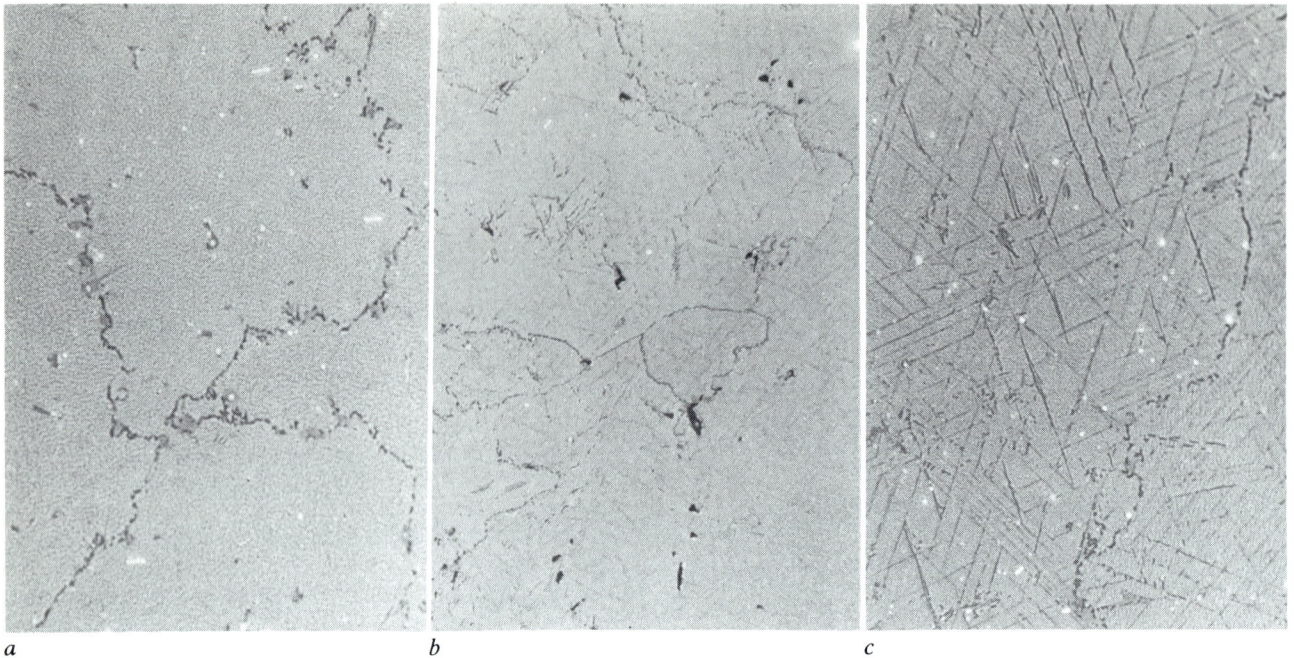

a   b   c

a σ-free (N_v = 2 186) overaged, 800°C, 10 000 h; *b* σ-prone (N_v = 2 395) overaged, 700°C, 10 000 h;
c σ-prone (N_v = 2 395) overaged, 800°C, 10 000 h

18   σ-proneness and Ni-base superalloy (IN 738 LC) as function of N_v number and aging treatment

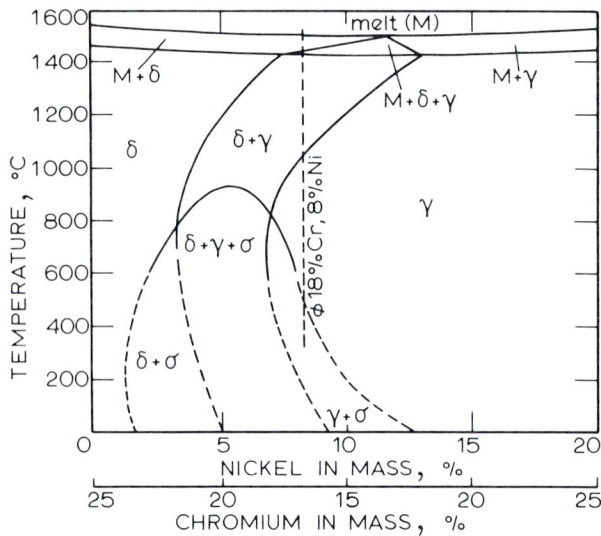

19   75% Fe corner of Fe–Cr–Ni ternary system indicating cooling path for 18% Cr–8% Ni stainless steel (after Bungardt *et al.*)

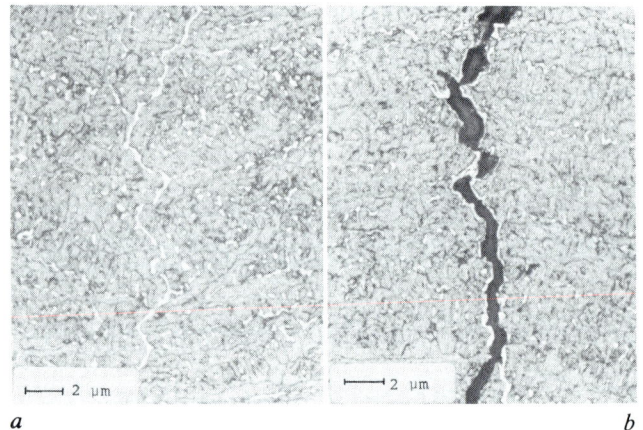

a   b

20   Centreline segregation and crack in EB welded ×22 CrMoV 12 1; *b* being the continuation of *a* (Sahm and Janke)

these segregations, insufficient impact strength is the result.

Segregational effects in EB-welds

Weld centreline forming modes (Figs 14*c* and 20) will concentrate segregational material in the centreline and thus introduce a tendency to hot tearing (Fig. 20*b*). The more probable case of a $c_{\Sigma i}$ effect (rather than c_x) (compare equation (11)) often results in the formation of interdendritic eutectic or monotectic microstructure yielding finely dispersed hard phases—a classic case being the ternary phosphide eutectic in cast iron (with a

melting point of 950°C, i.e. 200 K lower than the Fe–C eutectic with $m = 1\,152$°C).

Figure 21 conveys an idea of the metamorphosis of a dendrite from the start of its growth in the melt to the resulting final microstructure. Welding a complex steel yields particularly good examples for such metamorphosis, and directional solidification of such steels helps to understand solidification behaviour (Figs 20 and 16, respectively).

Figure 16 shows the dendritic microstructure of an EB-welded and directionally solidified X 22 CrMoV 12 1. Three areas of different coloration may be discerned: a dark central spine, the dendrite 'bulk', and the light interdendritic spaces. Considering an isocarbon concentration (0·2 wt-%) cut from the Fe-rich corner of

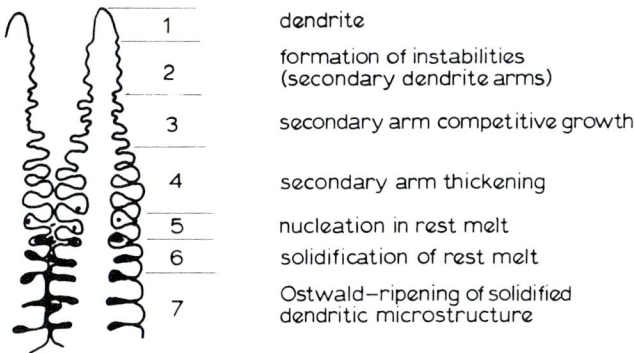

21 Metamorphosis of a dendritic microstructure, assembled after Kurz and Kattamis

1 dendrite
2 formation of instabilities (secondary dendrite arms)
3 secondary arm competitive growth
4 secondary arm thickening
5 nucleation in rest melt
6 solidification of rest melt
7 Ostwald–ripening of solidified dendritic microstructure

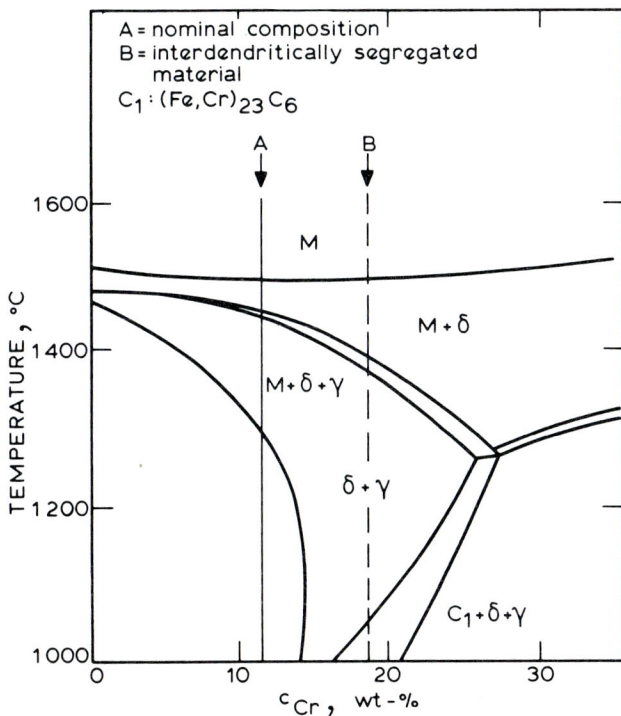

22 Effect of segregation on Cr-equivalence number (Cr) (see text) and shift in phase formation of EB welded ×22 CrMoV 12 1 implied from ternary cut corner 0·2 wt-% C cut ternary Fe–Cr–C system (Burghardt et al.)

Table 3 Microprobe analysis of interdendritic segregation in X 22 CrMoV 12 1 steel

Place of analysis	Cr	Mo	V	Ni	Mn	S
Centreline	14·0	1–1·5	0·65	0·2	0·6	<0·1
Average weld (50×50 μm^2 spot)	11·7	0·49	0·37	0·33	0·38	<0·1
Not welded matrix (50×50 μm^2 spot)	12	0·40	0·37	0·40	0·38	<0·1

the Fe–Cr–C ternary system the micrograph is explained as follows, (Fig. 22);[34,35]

(i) primary δ-ferrite is formed first (dark central spine); cooling beneath the transition temperature

$$\delta + \text{melt} \rightarrow \gamma + \delta + \text{melt}$$

occurs
(ii) the δ-ferrite is transformed and, upon further cooling, eventually primary γ-crystallization becomes possible
(iii) meanwhile, several alloying elements segregate heavily and are pushed ahead of the advancing dendritic protrusions; Table 3 lists the measured values. If one inserts these values to calculate the Cr-equivalence number [Cr], one has to assume a reversion to primary δ-crystallization (Fig. 22)

$$[\text{Cr}] = \text{wt-\% Cr} + 2(\text{wt-\% Si}) + 1·5(\text{wt-\% Mo}) + 5(\text{wt-\% V}) \tag{12}$$

(Schneider[36]).

Due to the fact that several carbide formers, i.e. V, Mo, and Cr are concentrated in the interdendritic melt, the specific carbide particle density is expected to be higher as well. However, this latter assumption has to be proved. Proof is also required for the δ-dendrite spine hypothesis pronounced above (compare Chuang et al.[37]).

The interaction between solidification microstructure and properties in weldments is as obvious as in the case of precision castings. The local compositional variations, due to segregational effects, can have a deleterious effect as may be seen from Fig. 20b. It was concluded that this material separation in Fig. 20b must have occurred at about 1 300°C, the transition from δ to γ of the nominal alloy composition being accompanied by a negative change in volume. Hot tearing tendencies, as well as simple pore or void formation (such as is illustrated in Figs 7 and 8 for precision castings and in Fig. 16 for an EB-weld), determine material properties. With regard to weldments, definitions for critical pore volumes have been attempted (Herberg et al.[38]). Such considerations could be rendered even more meaningful if standards such as those given by Fig. 9 could be agreed upon. They, of course, presuppose excellent and available non-destructive testing methods.

CONCLUSIONS

In broad outline, standard solidification models are applicable to both precision cast and EB-welded microstructures. In accordance with the line of argumentation presented in this paper the following differences and/or similarities between precision casting and EB-welding have been emphasized.

1 Nucleation. While the microstructure of precision castings is being controlled to a large extent by nucleant agents (excepting directionally solidified products) the microstructure of EB-weldings is influenced far more by epitaxial growth conditions.

2 Solidification parameters. Differences between the two technologies primarily arise due to the much more constricted solidification conditions in EB-welding as compared to precision casting. They are comparatively high solidification rates and temperature gradients, very small molten volumes and an interesting (although very

little explored) potential for weld channel (= mould) shape. Great similarities are found between the two technologies when considering compositionally induced dendritic crystallization behaviour and accompanying segregational effects.

3 Compositional effects. If quantitatively analysed, dendrite arm spacing and composition of interdendritic material may both yield a quantitative measure for mechanical properties as well as reasons for (and the loci of) micropore or void formation.

INCONEL (= IN) is a trademark of International Nickel Company

REFERENCES

1 W. KURZ: Proc. 5th International Conference on Vacuum Metallurgy, 1976, Munich, Germany

2 R. WLODAWER: 'Die gelenkte Erstarrung von Stahlguß', 1967, Giesserei-Verlag, Düsseldorf

3 N. CHVORINOV: *Giesserei*, 1940, **27**, 177, 201, 222

4 W. H. COOK: BICTA-Conference, Apr. 1977, Blackpool

5 C. LOCKE et al.: *Trans. Amer. Foundrym. Soc.* 1954, **62**, 589

6 K. C. ANTONY AND J. F. RADAVICK: 'Solute effects of boron and zirkonium on microporosity', 3rd International Symposium on Superalloys, 1976, Seven Springs, USA

7 G. J. DAVIES AND J. G. GARLAND: *Int. Metall. Rev.*, 1975, **20**, 83

8 L. DORN: *Strahltechnik VI*, 1975, (DVS-Bericht No. 38) 15

9 H. RAUPACH AND P. SPEIL: personal communication, 1977, Brown Boveri and Cie, Heidelberg

10 E. EGGERS et al.: *Schweissen Schneiden*, 1972, **24**, 200

11 R. REMUND: *Bull. Schweiz. Elektrotechn. Ver.*, 1973, **64**, 977

12 E. M. BREINAN et al.: 'Superalloys, metallurgy and manufacture', Proceedings of 3rd International Symposium, 1977, 435

13 SNECMA-Specification D.T., 88–26

14 R. STICKLER AND E. KAY: private communication (COST 50), 1976

15 R. E. SPEAR AND G. R. GARDNER: *Trans. Amer. Foundryman Soc.*, 1963, **71**, 209

16 M. C. FLEMINGS: *Metall. Trans.*, 1974, **5**, 2 121

17 J. HOFWEBER AND N. F. FIORE: *Mat. Sci. Eng.*, 1977, **27**, 157

18 A. F. GIAMEI AND J. S. ERICKSON: 'Computer application in directional solidification processing', 3rd International Symposium on Superalloys, 1976, Seven Springs, USA

19 G. HAOUR: Paper on seminar: 'Kristallisationsvorgänge beim Giessen', 1977, ETH-Lausanne, Switzerland

20 W. KURZ AND P. R. SAHM: 'Gerichtet erstarrte eutektische Werkstoffe', 1975, Springer-Verlag, Berlin

21 W. F. SAVAGE et al.: *Weld. Res. Suppl.*, 1976, 213

22 P. SHABINIAN et al.: Proc. 3rd. International Conference on Electron Ion Beam Science and Technology, 1968, Boston, 385

23 K. WITTKE: *Schweisstechnik*, 1975, **25**, 290

24 F. EICHHORN AND D. SCHUHMACHER: *Strahltechn. VI (6. Kolloq. Strahltechnik)*, 1973, 27

25 H. NAKAGAWA et al.: *J. Japan. Weld. Soc.* 1970, **39**, 94

26 K. BUNGARDT et al.: *DEW Technische Berichte*, 1970, **10**, (a), 298

27 J. A. GOLDAK et al.: *Con. Metall. Quart.*, 1970, **9**, 459

28 D. ROSENTHAL: *Trans. ASME*, 1946, **68**, 849

29 M. F. JORDAN AND M. C. COLEMAN: *Brit. Weld. J.* 1968, **15**, 552

30 H. E. SCHMID AND U. W. HILDEBRANDT: *Gefüge und Bruch, Materialkundl.-Techn. Reihe*, 1977, **3**, 136

31 P. R. SAHM AND B. JAHNKE: Paper on seminar: 'Kristallisationsvorgänge beim Giessen', 1977, ETH-Lausanne, Switzerland

32 W. KLEEMANN et al.: 'The effect of two trace elements on creep rupture properties and microstructure in a vacuum melted cobalt base superalloy' 3rd International Symposium on Superalloys, 1976, Seven Springs, USA

33 E. SCHÜRMANN AND J. BRAUCKMANN: *Arch. Eisenhüttenwesen*, 1977, **48**, 3

34 W. KURZ: Paper on seminar: 'Kristallisationsvorgänge beim Giessen', 1977, ETH-Lausanne, Switzerland

35 T. Z. KATTAMIS AND J. C. LECOMTE: Paper on seminar: 'Kristallisationsvorgänge beim Giessen', 1977, ETH-Lausanne, Switzerland

36 H. SCHNEIDER: *Foundry Trade J.*, 1960, **108**, 562

37 Y. K. CHUANG et al.: *Metall. Trans.*, 1975, **6A**, 235

38 G. HERBERG et al.: *Z. Werkstofftech.*, 1976, **7**, 107

39 H. BENNIGHOF: *Techn. Rundsch.*, 1974, **66**, 33, (9)

40 K. H. STEIGERWALD: *Neue Züricher Zeitung*, 1963

41 W. SEITH: 'Diffusion in Metallen', 56, 1955, Berlin, Springer-Verlag

42 J. FRIDBERG et al.: *Jernkontorets Ann.*, 1969, **153**, 263

43 K. BURGHARDT et al.: *Arch. Eisenhw.*, 1935–36, **29**, 607

Influence of steel composition on segregation and microstructure during solidification of austenitic stainless steels

Ö. Hammar and U. Svensson

In austenitic stainless steels primary ferritic and primary austenitic freezing can take place. The purpose of this investigation was to relate the steel composition to the mode of solidification, and to show that by choosing an appropriate composition, such properties as the distribution of δ-ferrite and the segregation can be controlled in ingot production. Thermal analysis and solidification experiments have been performed using 30 g samples and varying the elements C, N, Cr, Ni, Mn and Mo. The experiments were extended to trials with 170 kg ingots. The influence of the elements on the segregation, microstructure, and hot ductility were examined. The primary phase could be successfully predicted by calculating the chromium and nickel equivalents. Carbon, nitrogen, manganese, and nickel favours primary precipitation of austenite. An evaluation of the solute distribution coefficients during solidification and subsequent phase transformations in the 30 g samples explained the different modifications of δ-ferrite and segregations found in the ingots. With respect to the distribution of δ-ferrite in the ingots, the ferritic freezing improves the hot workability.

The authors are with the Metallurgical Research Laboratory of Sandvik AB, Sweden

Austenitic stainless steels often have a duplex as-cast structure incorporating a greater or lesser amount of δ-ferrite. The proportion δ-ferrite is primarily determined by the segregation of the alloying elements during solidification and cooling. A general rule of hot working is that rising ferrite content impairs hot ductility.[1-4] In austenitic weld deposits a certain ferrite content, between 0 and 6%, depending on grade, can help to prevent hot cracking.[5-6] Quantity apart, the shape and distribution of the ferrite also have a major bearing on the properties of the material. As a check on the solidification structure, Schaeffler charts are often used to calculate ferrite content.[7-8]

Austenitic stainless steels can freeze with primary precipitation of ferrite or austenite.[9-13] These steels can be given a full austenitic structure after a homogenization annealing in the 1 000–1 300°C range. The purpose of the present investigation was to see whether Cr and Ni equivalents, according to the Schaeffler model, could be used to predict the way in which a stainless steel would freeze. The intention was also to study the way in which different modes of solidification affected segregation, the distribution of the δ-ferrite in the ingot structure and, finally, hot ductility.

METHODS OF INVESTIGATION
Thermal analysis and quenching technique
The alloys were produced in a 50 kg induction furnace in 26 charge series with 4–7 alloys in each series (*see* Table 1). Table 2 exemplifies some of the typical compositions which were analysed. Altogether the investigation included more than 130 alloys. Thermal analysis was conducted in a vertical resistance coiled Pt–40 Rh tube-type furnace in an argon gas shield. After heating to about 50°C above the presumed liquidus, a Pt–10 Rh thermocouple was lowered into the centre of a 30 g sample in an Al_2O_3 crucible. The cooling rate inside the furnace was kept constant at 20°C/min and was governed by a programmable regulator. The melt temperature was recorded on a punch tape for further analysis in a computer to plot the cooling curve. Samples were quenched in water at selected temperatures on the cooling curve while solidification was in progress. The samples were electrolytically etched in oxalic acid for structural studies.

Microprobe investigation
The distribution of the alloying elements in the solidification specimens was analysed by stepwise spot

401

Table 1   Nominal compositions of alloys for thermal analysis (wt-%)

Alloy series	C	Si	Mn	Cr	Ni	Mo	N
1A–G	0·04	0·60	1·25	18·4	10·0	—	0·01–0·20
2A–F	0·04–0·10	0·60	1·25	18·4	10·0	—	0·10
3A–E	0·04	0·60	1·25	17·8	9·0–10·3	—	0·03
4A–E	0·04	0·60	1·25	18·4	9·0–10·3	—	0·03
5A–E	0·04	0·60	1·25	19·0	9·0–10·3	—	0·03
6A–E	0·04	0·60	1·70	17·0	12·5	2·5	0·01–0·20
7A–E	0·01–0·10	0·60	1·70	17·0	12·5	2·5	0·03
8A–F	0·04	0·60	1·70	17·0	12·2–13·7	2·5	0·03
9A–F	0·04	0·60	1·70	17·4	12·2–13·7	2·5	0·03
10A–F	0·04	0·60	1·70	17·8	12·2–13·7	2·5	0·03
11A–F	0·04	0·60	1·70	17·0	12·1–13·7	2·9	0·03
12A–F	0·04	0·60	1·70	17·4	12·2–13·7	2·9	0·03
13A–F	0·04	0·60	1·70	17·8	12·2–13·7	2·9	0·03
14A–D	0·04	0·60	0·0	17·4	11·4–14·0	2·7	0·03
15A–D	0·04	0·60	0·5	17·4	11·4–14·0	2·7	0·03
16A–D	0·04	0·60	1·0	17·4	11·4–14·0	2·7	0·03
17A–D	0·04	0·60	1·5	17·4	11·4–14·0	2·7	0·03
18A–D	0·04	0·60	2·0	17·4	11·4–14·0	2·7	0·03
19A–E	0·02	0·40	1·85	23·0–25·0	12·0	—	0·03
20A–E	0·02	0·40	1·85	23·0–25·0	12·5	—	0·03
21A–E	0·02	0·40	1·85	23·0–25·0	13·0	—	0·03
22A–E	0·02	0·40	1·85	23·0–25·0	13·5	—	0·03
23A–E	0·02	0·40	1·85	23·0–25·0	14·0	—	0·03
24A–E	0·02	0·20	1·70	22·5	10·0–12·0	—	0·03
25A–E	0·02	0·20	1·70	23·0	10·0–12·0	—	0·03
26A–E	0·02	0·20	1·70	23·5	10·0–12·0	—	0·03

Table 2   Typical analysed compositions according to Table 1 (wt-%)

Alloy		C	Si	Mn	Cr	Ni	Mo	N
6	A	0·041	0·72	1·71	16·9	12·4	2·46	0·008
	B	0·039	0·71	1·87	16·9	12·5	2·45	0·025
	C	0·039	0·63	1·84	17·0	12·5	2·56	0·074
	D	0·041	0·63	1·75	16·9	12·6	2·52	0·11
	E	0·038	0·60	1·93	16·9	12·4	2·45	0·18
7	A	0·006	0·62	1·77	17·0	12·5	2·47	0·033
	B	0·019	0·63	1·77	16·9	12·5	2·46	0·033
	C	0·042	0·66	1·75	16·9	12·4	2·45	0·032
	D	0·066	0·65	1·74	17·0	12·5	2·44	0·032
	E	0·11	0·69	1·74	17·0	12·4	2·41	0·031

analysis using a CAMECA microprobe. Homogenized austenitic material (several days at 1 200°C) of type 18 Cr 12 Ni 2·5 Mo and 18 Cr 10 Ni were used as references. All results have been corrected for the atomic number, absorption, and fluorescence. The following partition coefficients were used in the presentation of results;

$$k_e = C_D/C_0, \qquad I_e = C_{ID}/C_0, \qquad I_s = C_{ID}/C_D$$

where C_D is the concentration measured in primary dendrite, C_{ID} is the concentration in the last solidified regions, and C_0 is the charge analysis.

Trials with 170 kg ingots

The influence of the mode of solidification on ingot structure and hot ductility was studied in 170 kg ingots. Three ingots with differing analyses were produced (*see* Table 3). Alloy G1 was selected so that the ingot would freeze with primary austenite, while alloys G2 and G3 were selected for ferritic freezing. Tensile specimens were taken from the columnar zone of the ingot in varying states of heat treatment; (*a*) untreated structure; (*b*) annealed 1 150°C/4 h; (*c*) annealed 1 300°C/2 h, (*d*) annealed 1 300°C/8 h.

Material with a hot-worked structure, obtained by forging the bottoms of the ingots down to 15 mm dia., was also tensile tested. After annealing for 2 h at 1 250°C, all three alloys acquired a uniform coarse grain size of about 270 μm. Hot tensile testing was carried out on cooling in a Gleeble machine at a strain rate of about 5 s^{-1}.

RESULTS

The determination of the Cr and Ni equivalents for the transition from ferritic to primary austenitic freezing will be discussed first. There seems no point in presenting

Table 3 Compositions of materials for hot ductility tests (wt-%)

Alloy	C	Si	Mn	P	S	Cr	Ni	Mo	Al	N
G1	0·029	0·59	1·59	0·006	0·015	17·2	14·4	2·67	0·015	0·033
G2	0·027	0·57	1·54	0·006	0·013	17·3	13·4	2·65	0·015	0·036
G3	0·029	0·55	1·61	0·006	0·012	17·5	12·4	2·66	0·013	0·035

here the enormous mass of primary data. Figure 1 will serve as an example of a thermal analysis curve. The temperature difference, ΔT, and the resident time, Δt, between liquidus and incipient austenitic growth in the melt were determined for all alloys which solidified with primary precipitation of ferrite (*see* Fig. 1). The margin of error in the determination of ΔT is less than $\mp 1°C$. The effect of the analysed composition on ΔT was summarized for 70 alloys and investigated by means of regression analysis, and the Cr and Ni equivalents were calculated at $\Delta T = 0$, i.e. the critical analysis for the transition from ferritic to austenitic freezing. Table 4

shows the result of the regression analysis, which has a high degree of significance. For $\Delta T = 0$ one obtains

$$\text{wt-\% Ni} = -0·257 - 0·31\,\text{Mn} - 22\text{C} - 14·2\,\text{N}$$
$$+ 0·75\,\text{Cr} + 1·03\,\text{Mo} \qquad (1)$$

where Mn is the percentage of Mn in the alloy by weight, etc. and the Ni and Cr equivalent can be expressed by

$$\text{Ni} + 0·31\,\text{Mn} + 22\text{C} + 14·2\,\text{N}$$
$$= -0·257 + 0·75\,(\text{Cr} + 1·37\,\text{Mo}) \qquad (2)$$

$$\text{Ni}_{eq} = -0·257 + 0·75\,\text{Cr}_{eq} \qquad (3)$$

Figure 2 shows how carbon and nitrogen affect ΔT and

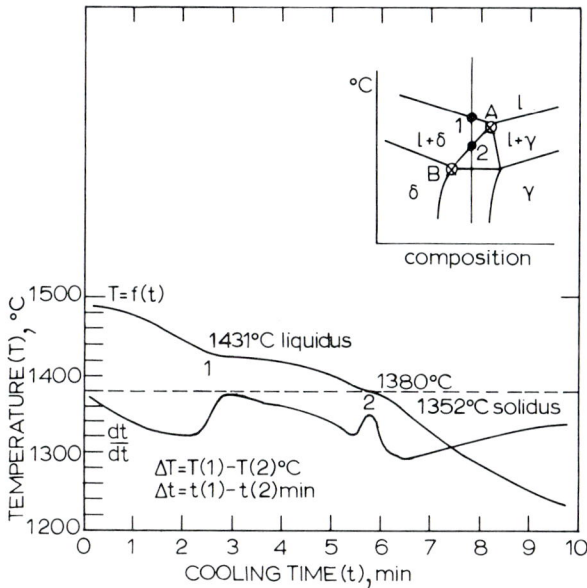

1 Cooling curve and its differential for alloy 18A characteristic for ferritic-freezing alloys according to the quasibinary phase diagram in the upper right hand corner

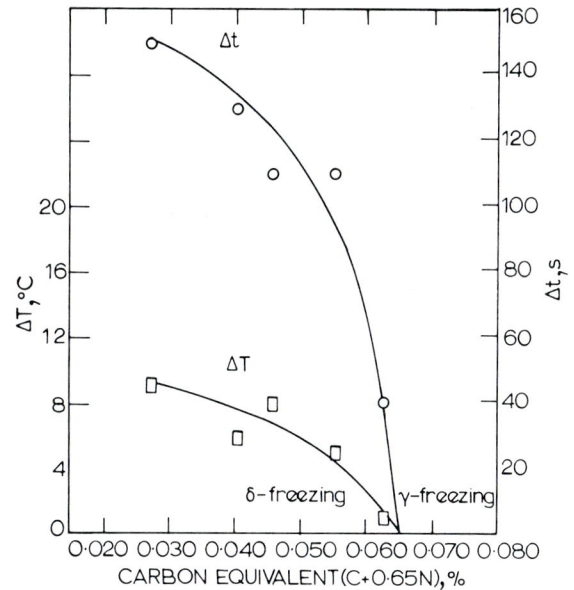

2 Resident time (Δt) and solidification range (ΔT) in the ($l + \delta$) region for alloy 6A–B and 7A–C

Table 4 Coefficients, standard errors, F values, and Cr- and Ni-equivalents of alloying elements determined by regression analysis from experimental data with respect to ΔT (Fig. 1)

Element	Regn coeff.	S error	F value	Ni-equivalent	Cr-equivalent
C	−197·4	36·1	29·9	22 C	
Si	—	—	0·605*		
Mn	−2·784	0·83	11·2	0·31 Mn	
Cr	6·768	0·37	333	−0·75 Cr	1·0 Cr
Ni	−8·990	0·57	246	1·0 Ni	
Mo	9·250	0·78	139	−1·03 Mo	1·37 Mo
N	−127·8	20·2	39·9	14·2 N	
Const.	−2·307	5·07			

* Term is not statistically significant.
F-value for regn.: 157
Multiple correlation coefficient: 0·959
Standard error: 3·06
No. of observations: 70

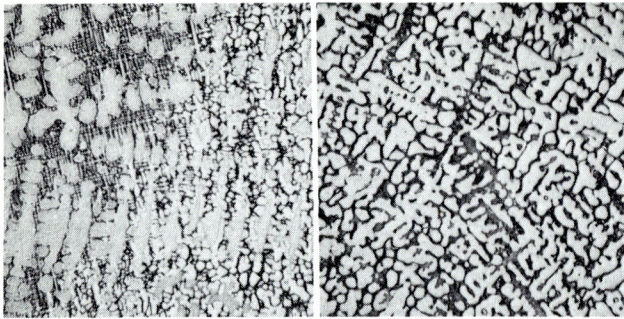

a   *b*

3   Ferritic freezing; alloy *6B* cooled at 20°C/min
and quenched from different temperatures
during solidification. Quenched from *a*
1 425°C; *b* 1 400°C   [×10]

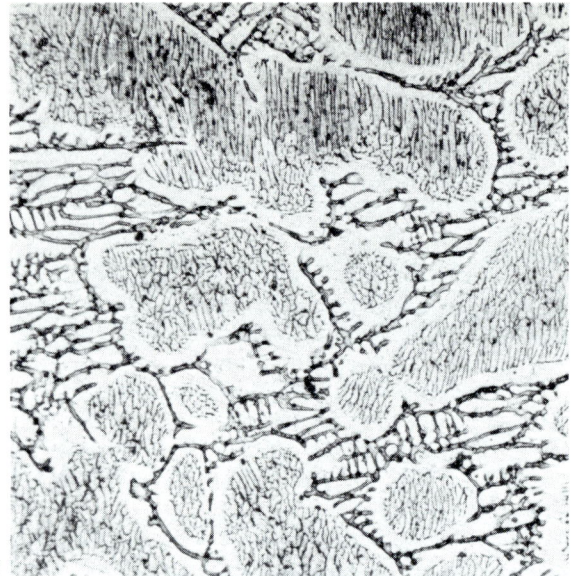

a

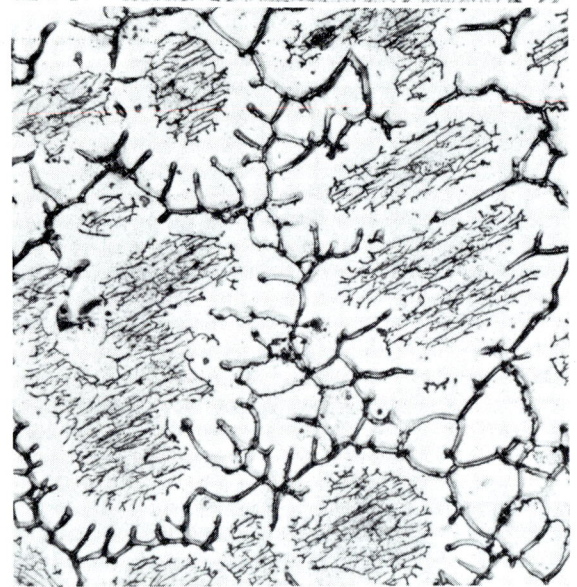

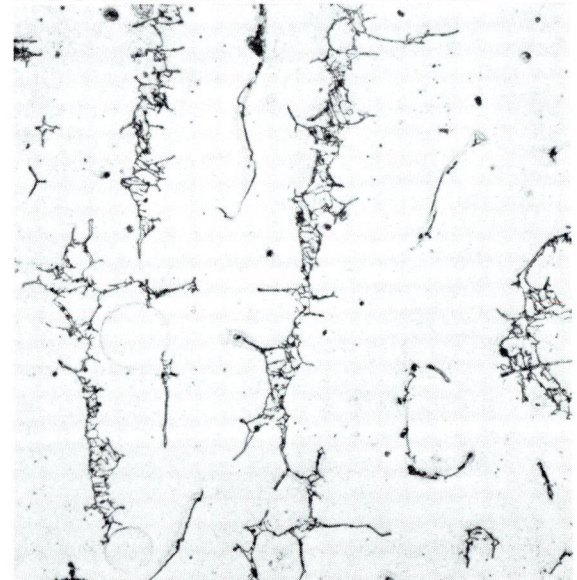

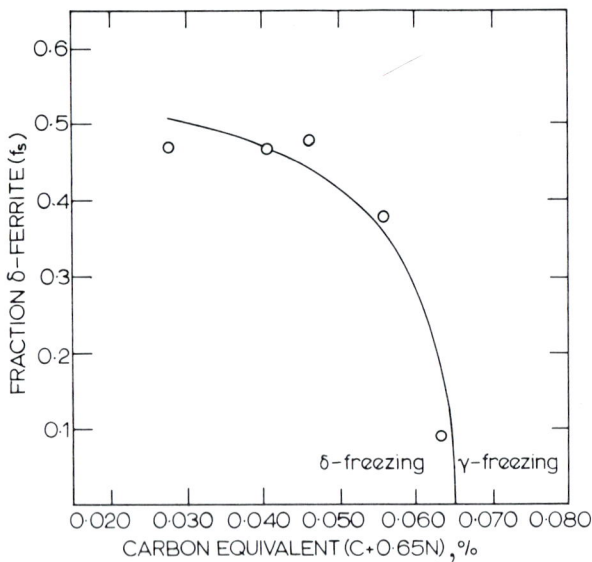

4   Fraction solidified δ-ferrite in the (*l* + δ) region
at the start of austenite growth in alloy *6A–B*
and *7A–C*

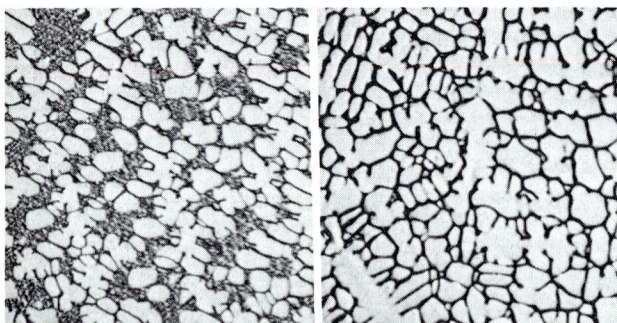

a   *b*

5   Austenitic freezing; alloy *6C* cooled at
20°C/min and quenched from different
temperatures during solidification. Quenched
from *a* 1 422°C; *b* 1 400°C   [×10]

6   Ferritic freezing; alloy *2A* (0·036% C) cooled
at 20°C/min and quenched from different
temperatures during solidification. Quenched
from *a* 1 436°C; *b* 1 434°C; *c* 1 400°C   [×130]

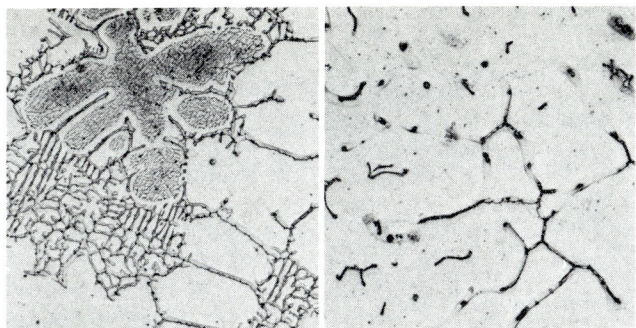

a   b

7 Simultaneous growth of austenite and ferrite; alloy 2F (0·099% C) cooled at 20°C/min and quenched from different temperatures during solidification. Quenched from a 1 439°C; b 1 400°C [×75]

Δt for alloy series 6 and 7 (*see* Table 2). The carbon equivalent $C_{eq} = (C + 0·65 N)$ is obtained from equation (1), where $C_{eq} = 22[C + (14·2/22)N]$. At $C_{eq} = 0·065\%$ the alloys solidify austenitically instead of ferritically.

Structure development in 30 g samples

To exemplify the modes of solidification occurring, we have chosen here to show the successive solidification processes occurring in a 17 Cr 13 Ni 2·5 Mo steel (Figs 3–5), and in an 18 Cr 10 Ni steel (Figs 6 and 7). In the former steel the nitrogen content has been made to vary (alloy series 6) while in the latter the carbon content has been varied (alloy series 2).

In ferritic freezing the austenite begins to grow on the ferrite dendrites as per the three-phase reaction $l + \delta \rightarrow l + \delta + \gamma$ (Figs 3a and 6a–b). The austenite growth causes the continuous ferritic dendrites to be broken down into a network of alternating ferrite and austenite. Figure 4 shows the fraction of solidified δ-ferrite

measured by the interception method at the beginning of the three-phase reaction. The delta ferrite is unstable at low temperatures because of its low content of ferrite stabilizing alloying elements, and it is converted into Widmanstätten austenite during quenching (Fig. 6a and b).

Austenitic freezing occurs in the Mo alloyed steels when $C_{eq} > 0·065\%$ (Figs 4 and 5). The transition from ferritic to austenitic freezing can result in a simultaneous growth of ferritic and austenitic dendrites. This will happen in the 18 Cr 10 Ni steels if $C_{eq} = 0·16\%$ (Fig. 7). The ferritic dendrites are dissolved, however, and the final solidification structure assumes an appearance typical of austenitic freezing, with secondary δ-ferrite in the last solidified regions (Fig. 7b).

The solidification range ΔT occurring in ferritic freezing influences the morphology and distribution of the ferrite. A large ΔT, greater than 10° in our instances, means that the primary δ-ferrite forms a network in the final solidification structure (Fig. 8), while a small ΔT, i.e. 10° and below, gives continuous films or scales of primary δ-ferrite (*see* Fig. 9).

Microsegregation during solidification

Alloys 8A, 8C, and 8E were investigated with a microprobe to determine the distribution of alloying elements during successive solidification and subsequent cooling down to 1 200°C. Table 5 shows the partition coefficients k_e, I_e, and $k^{\delta/\gamma}$. The last mentioned of these

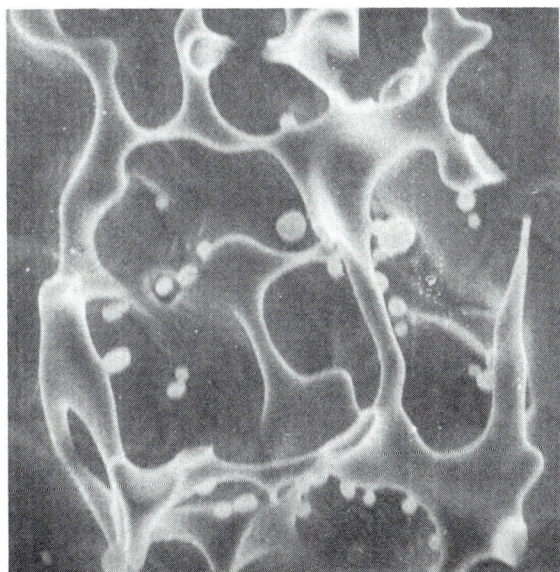

8 Primary δ-ferrite and chromium sulphides in alloy 15B, cooled at 20°C/min and quenched from 1 200°C ($\Delta T = 18$°C); composition = 0·035 C, 0·68 Si, 0·54 Mn, 0·035 S, 17·5 Cr, 12·3 Ni, 2·63 Mo, 0·03 N: deep etched in a solution of 2% bromine in methyl alcohol [×1 000]

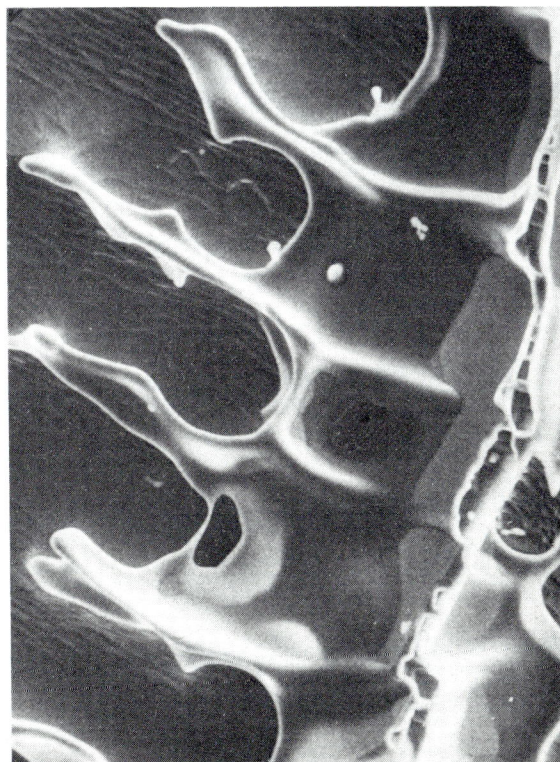

9 Primary δ-ferrite and manganese sulphides in alloy 17C, cooled at 20°C/min and quenched from 1 200°C ($\Delta T = 9$°C); composition = 0·035 C, 0·67 Si, 1·48 Mn, 0·039 S, 17·4 Cr, 13·1 Ni, 2·59 Mo, 0·030 N: deep etched in a solution of 2% bromine in methyl alcohol [×500]

Table 5   Partition coefficients in alloy 8A, 8C, and 8E cooled at 20°C/min and quenched from 1 424°C and 1 200°C

Alloy		Cr	Mo	Mn	Ni	Mode of solidification	Temp., °C
8C	k_e	1·01	0·91	0·88	0·87	δ	1 424
	I_e	1·0	1·1	1·1	1·2	δ	1 424
8E	k_e	0·98	0·81	0·90	1·03	γ	1 424
	I_e	1·1	1·1	1·1	1·0	γ	1 424
8A	$k^{\delta/\gamma}$	1·3	2·7	1·1	0·8	δ	1 200
8C	$k^{\delta/\gamma}$	1·4	3·4	1·1	0·8	δ	1 200
8E	$k^{\delta/\gamma}$	1·4	4·0	1·3	0·8	γ	1 200

indicates the relationship between the content of alloying elements in the secondary δ-ferrite in the last solidified regions and the content in the austenite.

In ferritic freezing (alloys 8A and 8C) the elements Ni, Mn, and Mo segregate to the melt, while chromium does not segregate at all. In austenitic freezing (alloy 8E) Cr, Mo, and Mn segregate to the melt, while nickel is enriched somewhat in the austenite. Figures 10 and 11 illustrate the segregation process during solidification and subsequent cooling in ferritic and austenitic freezing, respectively. During the growth of austenite in the ferritically freezing alloys (8A and 8C), which begins at 1 416°C and 1 422°C respectively, the alloying elements are reapportioned; k_e (Ni) and I_e (Ni) are reduced while k_e (Cr) and I_e (Cr) increased (*see* Fig. 10). This means that nickel diffuses into the growing austenite at the same time as chromium accumulates in the delta ferrite and in the melt. This process continues long after the sample has solidified. Austenitic growth in the melt, whether it occurs in ferritic or austenitic solidification, implies that chromium and molybdenum segregate to the melt.

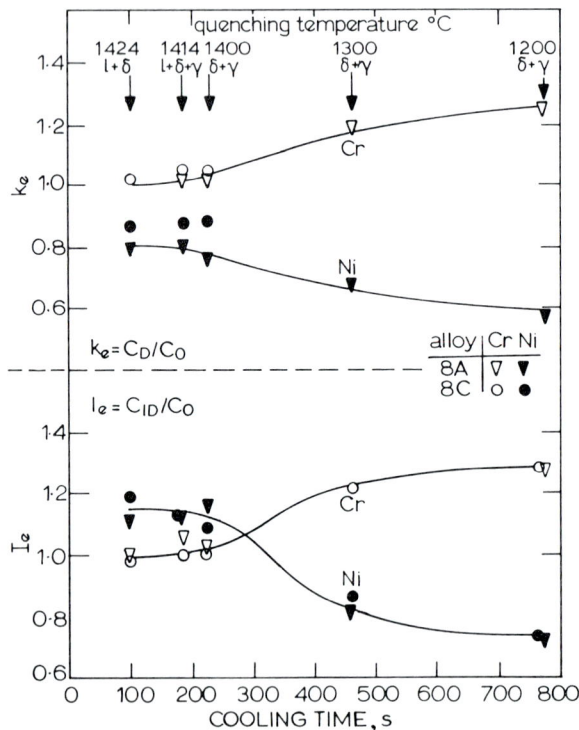

11   Austenitic freezing: variation of k_e and I_e during solidification and cooling in alloy 8E

The influence of different carbon and nitrogen contents on the partition coefficients was studied in alloy series 6 and 7 (*see* Table 2). In Fig. 12, k_e and the segregation ratio I_s are given as a function of the carbon equivalent for samples quenched at 1 400°C. With a rising carbon equivalent, i.e. an increasing amount of growing austenite in the melt, the k_e for both chromium and nickel approaches unity. Nickel segregation is very heavy in ferritic freezing but declines with an elevated carbon equivalent. In these experiments a value of k_e (Ni) < 1 was obtained with austenitic freezing, i.e. nickel accumulated in the melt.

Microstructure in 170 kg ingots after different heat treatments

The structure in those surface areas of the ingots from which hot tensile specimens were taken consists of coarse columnar crystals extending at right angles to the direction of deformation. The modes of solidification were determined by thermal analysis and by structural studies of ingots. Table 6 summarizes modes of solidification and the structural changes occurring in the

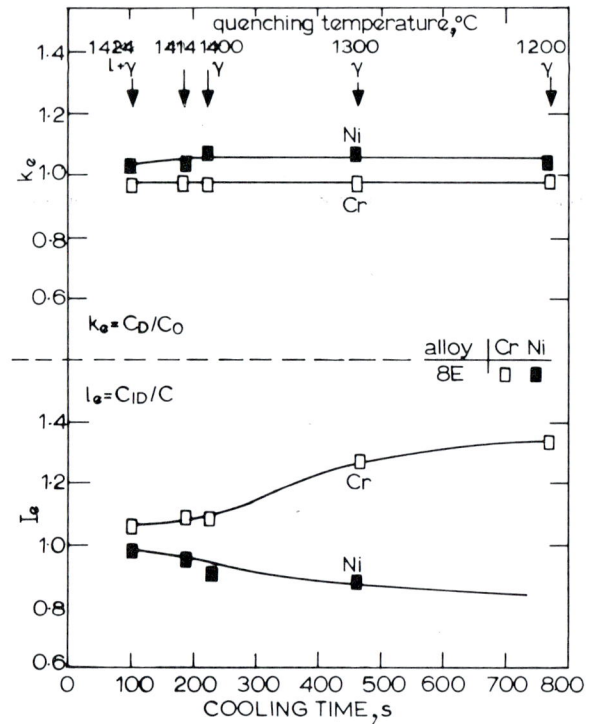

10   Ferritic freezing: variation of k_e and I_e during solidification and cooling in alloy 8A and 8C

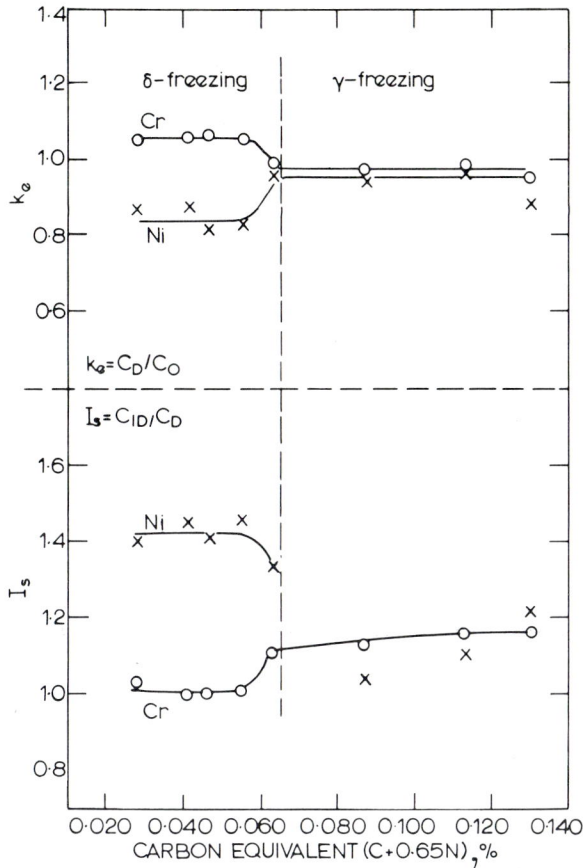

12 **Influence of different mode of solidification on k_e and segregation ratio I_s in quenched specimens at 1 400°C of alloy series 6 and 7**

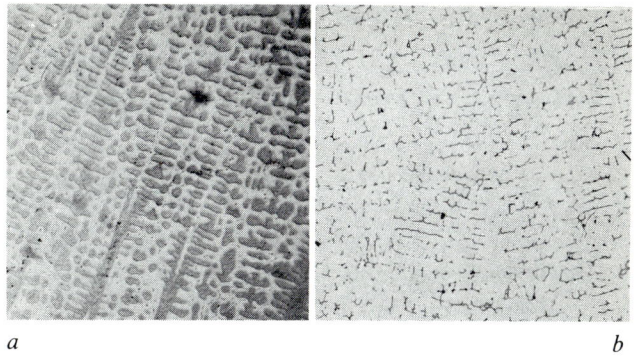

a alloy G2, interdendritic ferrite; *b* alloy G3 ferrite in dendritic areas [×25]

13 **As-cast structure at ingot surface**

the interdendritic areas (Fig. 14*a*). The structure of variant G3 (Fig. 14*b*) comprises a ferrite network in residues of δ-dendrites. Solidification here has been primarily ferritic. In variant G2 both the above mentioned structural types occur intermingled throughout the central area (Fig. 14*c*).

While ingots are being annealed a radical reapportion of alloying elements occurs, but microsegregation is fully measurable after 4 h at 1 150°C (*see* Table 7). Ingots G1 and G2 display the same segregation pattern, while ingot G3 deviates mainly as regards the distribution of chromium and molybdenum, which have accumulated in and stabilized the δ-ferrite.

surface zone of the ingots in the different states of heat treatment which were ductile tested. In the austenitically solidified variant G1, and in the ferritically solidified alloy G2, the ferrite residues are located in interdendritic areas (*see* Fig. 13*a*). In the ferritically solidified variant G3, on the other hand, this ferrite is mainly found in residues of former dendrites (*see* Fig. 13*b*).

Figures 14*a*–*c* show examples of the structure at the centre of the ingot. The structure here is consistently coarser than in the surface zone. Here too, the G1 variant freezes austenitically, with residues of ferrite in

Effects of solidification and heat treatment on hot-working properties

Figure 15 shows the hot ductility of forged material. Ductility is comparable between the three variants. Since all the material is ferrite free and since grain sizes are of the same order of magnitude in the three variants, the results imply that a change of Ni content does not influence the ductility of a homogeneous austenitic matrix.

Figures 16 and 17 compare the ductility of the three ingots in corresponding states of heat treatment. The primarily ferritically solidified variant G3 exhibits the best ductility in all states of heat treatment, regardless of whether it contains ferrite (Fig. 16) or is ferrite free (Fig. 17).

Table 6 Microstructure in materials for hot ductility tests after different heat treatments

	Mode of solidification	G1 γ	G2 δ	G3 δ
	Position of ferrite	Interdendritic	Interdendritic	Dendritic
Ferrite, %	No heat treatment	0·3	2	6
	4 h, 1 150°C	0	0·3	2
	2 h, 1 300°C	0	0	0·3
	8 h, 1 300°C	0	0	0
	Wrought material	0	0	0
Grain size,* mm	4 h, 1 150°C	0·5	0·6	0·5
	2 h, 1 300°C	0·9	1·0	1·0
	8 h, 1 300°C	1·1	1·1	1·3
	Wrought material	0·29	0·29	0·25

* 10 mm from ingot surface, measured perpendicular to growth direction of the columnar crystals.

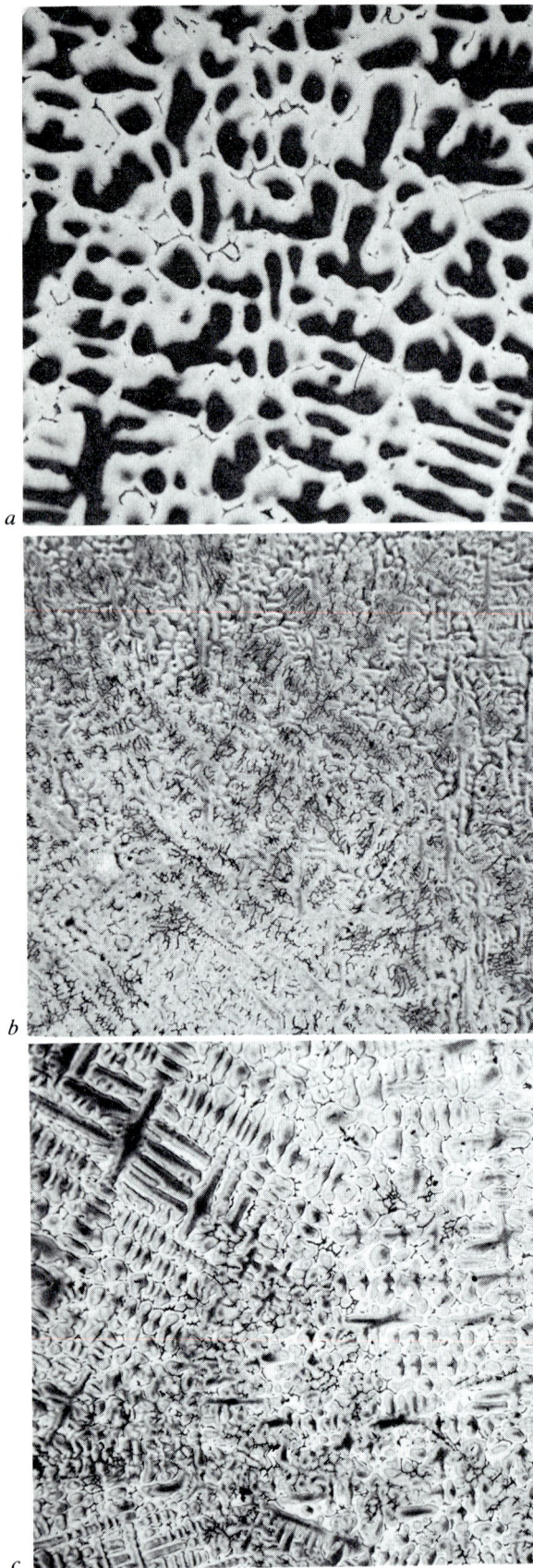

a alloy G1, interdendritic ferrite; *b* alloy G3, ferrite in dendritic areas; *c* alloy G2, both types of ferrite   [×23]

14   As-cast structure in ingot centre

Table 7   Partition coefficients in surface of ingot materials heat treated, 4 h 1 150°C

Alloy		Mn	Cr	Ni	Mo	Mode of solidification
G1	k_e	0·88	0·93	0·97	0·67	γ
	I_e	1·07	1·01	1·05	1·05	
G2	k_e	0·91	0·94	0·94	0·87	δ
	I_e	1·10	1·04	1·05	1·25	
G3	k_e	0·89	1·03	0·89	1·13	δ
	I_e	1·12	0·95	1·07	0·90	

$k_e = C_D/C_0$ where C_D = alloy content in former dendrite centre

$I_e = C_{ID}/C_0$ where C_{ID} = alloy content in former interdendritic areas

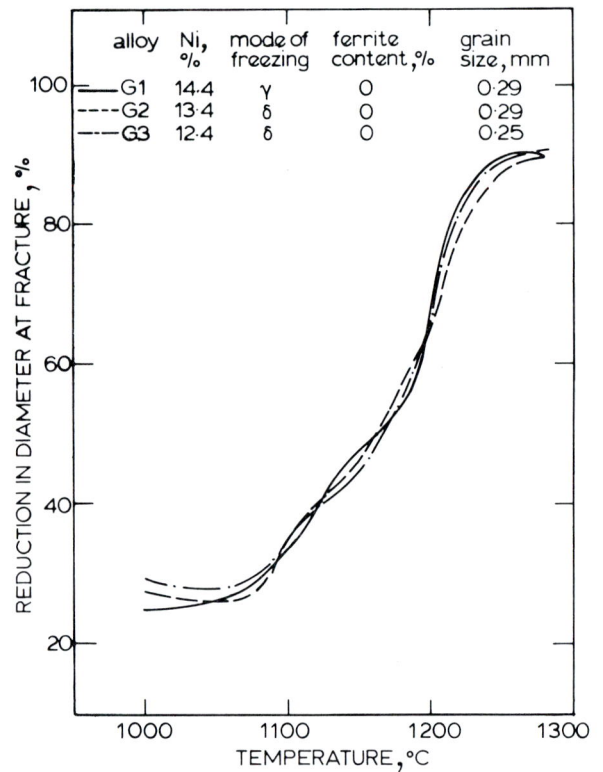

15   Influence of nickel on hot ductility in wrought materials for different mode of solidification

DISCUSSION
Mechanism of solidification

We have shown by thermal analysis and regression analysis that it is possible to estimate and predict the mode of solidification of stainless steel. The following function can be applied:

$$\Phi = Ni_{eq} - 0·75\, Cr_{eq} + 0·257 \qquad (4)$$

where Ni_{eq} and Cr_{eq} have been defined in equations (2) and (3). The various modes of solidification are as follows:

$$\Phi < 0: l \rightarrow l + \delta \rightarrow l + \delta + \gamma \rightarrow \gamma + \delta \quad \text{(Fig. 3}a, 6a, b)$$

$$\Phi = 0: l \rightarrow l + \gamma + \delta \rightarrow \gamma + \delta \qquad \text{(Fig. 7)}$$

$$\Phi > 0: l \rightarrow l + \gamma \rightarrow \gamma \qquad \text{(Fig. 5)}$$

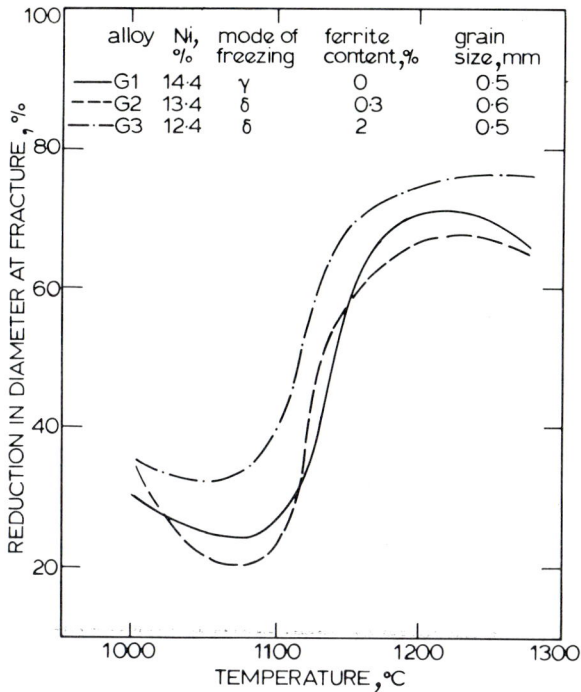

16 Influence of mode of solidification on hot ductility in ingots after heat treatment 4 h, 1 150°C

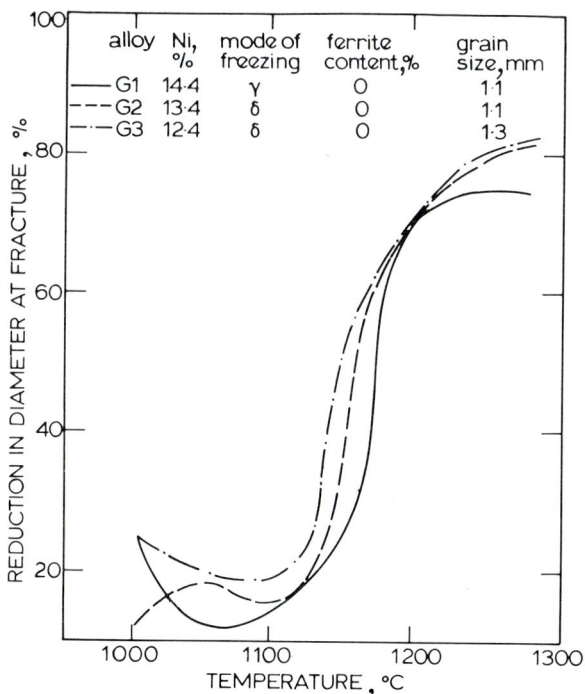

17 Influence of mode of solidification on hot ductility in ingots after heat treatment 8 h, 1 300°C

The change between ferritic and austenitic freezing occurs at Cr/Ni equivalent conditions just below or equal to 1·35 ($\Phi = 0$). The equivalents calculated in Table 4 agree well with equivalents used to determine the ferrite content of weld deposits as per the Schaeffler model. Ferrite contents determined by annealing experiments give similar equivalents.[3] It ought therefore to be possible for a solidification model to include isoferrite lines for a particular temperature as well as equation (3). In an internal project,[14] an empirical determination was made of the line of demarcation for alloys solidifying ferritically without a three phase reaction, i.e. in accordance with the reaction $l + \delta \rightarrow \delta$. Examples of a solidification model are shown in Fig. 18, which in addition to equation (3) includes the subsequent line of demarcation, line *B*, and a number of isoferrite lines for ferrite contents determined in quenched specimens after 8 h annealing at 1 200°C.[14]

During ferritic solidification a considerable amount of δ-ferrite grows even at relatively short resident times in the solidification range ($l + \delta$). The growth of the delta ferrite can be described by the functions $f_s = k \cdot t^n$, where f_s is the fraction of δ-ferrite according to Fig. 4 and t is the resident period Δt as per Fig. 2. The coefficient n is 0.5–1. With very small solidification ranges, ΔT or small Δt, the primary δ-ferrite is dissolved during the growth of the austenite, because only a small fraction of ferrite has had time to grow and branch out. The resident time is short in the surfaces of the ingots, and the primary δ-ferrite is therefore completely dissolved in ingot G2, which has a small ΔT (Fig. 13a). Secondary ferrite occurs instead in the last solidified regions, owing to heavy segregation, especially of chromium and molybdenum, and the solidification structure resembles that obtained with austenitic freezing. In the borderline case close to complete dissolution, the primary δ-ferrite occurs in the solidification structure in the form of thin 2–4 μm scales (Figs 9 and 13b). With longer Δt, the δ-ferrite has time to grow and coarsen and is stabilized by the indiffusion of ferrite stabilizing alloying elements. The δ-ferrite then forms a network in the solidification structure as shown in Fig. 8. This network appears in the samples from ingot centre in alloys G2 and G3 (Figs 14b, c).

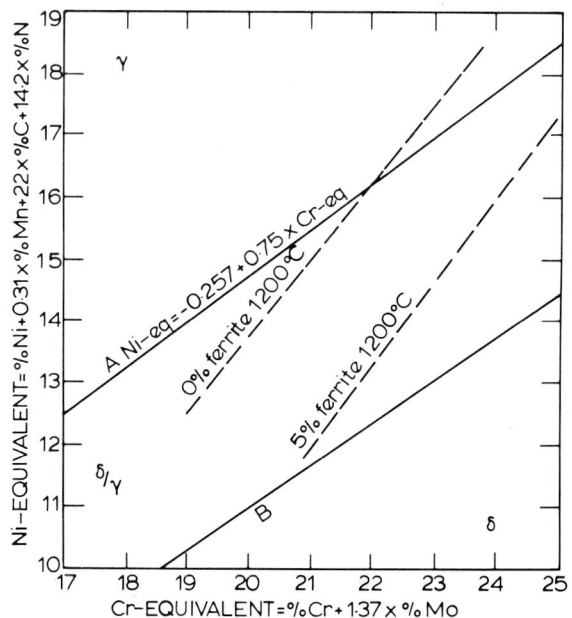

18 Influence of Cr- and Ni-equivalents on mode of solidification; for line *A* and *B*, see quasibinary phase diagram in Fig. 1: γ and δ denote austenitic and ferritic freezing, respectively; ferritic freezing with a three-phase reaction occurs in the area δ/γ

Table 8 Equilibrium partition coefficients between solid and liquid phase in austenitic stainless steels

	$k^{\delta/l}$		$k^{\gamma/l}$	
	This paper	Ref. 10	This paper	Ref. 10
Cr	1·0	1·0	0·98	0·96
Ni	0·8–0·85	0·79	0·96–1·05	0·94

Since the rediffusion of both chromium and nickel into the austenite is negligible, k_e in connection with austenitic freezing can be taken as equal to the equilibrium partition coefficient $k^{\gamma/l}$ for distribution between the solid phase and the melt. In connection with ferritic freezing, k_e can be taken to equal the equilibrium partition coefficient $k^{\delta/l}$ in rapidly quenched specimens near liquidus when no rediffusion has been possible. In Table 8 the equilibrium partition coefficients obtained have been collated and compared with the coefficients calculated by Fredriksson.[10] There is good agreement between the two.

Effects of solidification on as cast properties

Ferritic freezing of the kind giving a residual ferrite in areas situated in former dendrites is an advantage in terms of ductility (*see* Figs 16 and 17). Since all ingots exhibit the same ductility when forged (Fig. 15), the difference of ductility in the as-cast condition must be due to the changes in segregation pattern and ferrite distribution which are caused by the change in solidification. In the annealed as-cast material there is a cosegregation of Cr, Ni, Mo, and Mn to interdendritic areas in ingots G1 and G2. In G3 on the other hand, Cr and Mo accumulate in areas situated in dendrite residues. These areas are also deprived of Ni and Mn (Table 7).

The variant having the best ductility (ingot G3) also involves a lower force requirement than other variants in the hot deformation of as-cast material. Tests of forged material did not reveal any such difference in force requirements between the variants. This means that the segregation pattern influences the deformation resistance of the as-cast structure and that deformation resistance is lowest in the mode of solidification represented by variant G3. The critical tension required to initiate grain boundary fracture in alloy G3 is attained after a higher degree of deformation than for variants G1 and G2.

CONCLUSIONS

1 The mode of solidification of austenitic stainless steels can be predicted by calculating the Cr and Ni equivalents given in Table 4 and testing the function $\Phi = Ni_{eq} - 0.75\,Cr_{eq} + 0.257$. If $\Phi < 0$ the steel will freeze ferritically; if $\Phi = 0$ ferrite and austenite will be simultaneously precipitated; if $\Phi > 0$ the steel will freeze austenitically.

2 In stable ferritic freezing the δ-ferrite in the as-cast structure is mainly located in former dendrites where chromium and molybdenum accumulate. In austenitic freezing the δ-ferrite in the as-cast structure is located in interdendritic areas and there is a cosegregation of the elements Mn, Cr, Ni and Mo to these areas.

3 Ferritic freezing with a structure as per point 2 is an advantage in terms of hot ductility.

ACKNOWLEDGMENTS

We wish to thank Dr H. Widmark and G. Grünbaum for helpful discussions and Professor R. Kiessling, Director of Steel Research and Development, for permission to publish this paper.

REFERENCES

1 J. H. DECROIX *et al*.: 'Deformation under hot-working conditions', 135, 1968, London, The Iron and Steel Institute

2 M. HILDEBRAND: *Neue Hütte*, 1972, **12**, 724

3 K. SKUIN *et al*.: *Neue Hütte*, 1973, **11**, 663

4 K. MAYLAND *et al*.: *Metals Technology*, 1976, **8**, 350

5 J. C. BORLAND AND R. N. YOUNGER: *British Weld. J.*, 1960, **1**, 22

6 C. D. LUNDIN *et al*.: *Weld. J., Research Supplement*, 1975, **8**, 241s

7 A. L. SCHAEFFLER: *Met. Progr.*, 1949, 680

8 F. HULL: *Weld. J., Research Supplement*, 1973, **5**, 193s

9 G. BLANC AND R. TRICOT: *Mém. Sci. Rev. Met.*, 1971, **11**, 735

10 H. FREDRIKSSON: *Met. Trans.*, 1972, **11**, 2 989

11 U. SIEGEL AND M. GÜNZEL: *Neue Hütte*, 1973, **7**, 422 and 1973, **10**, 599

12 H. HOFFMEISTER: *Schweissen Schneiden*, 1973, **5**, 164

13 J. MASUMOTO *et al*.: *Schweissen Schneiden*, 1975, **11**, 450

14 K. WENNSKOG: Internal report Sandvik AB

Solidification cracking in 18 Cr–13 Ni–1 Nb stainless steel weld metal: the role of magnesium additions

G. Jolley and J. E. Geraghty

A series of experiments has been carried out on the solidification cracking tendency of 18 Cr–13 Ni–1 Nb stainless steel weld metal. The cracking tendency has been associated with the presence of low-melting-point niobium-rich eutectics. Increasing additions of magnesium as nickel–magnesium alloy have been shown to cause a globularization of these eutectics. Hot tensile tests have shown that this globularization is associated with increased ductility during solidification and reduced susceptibility to weld metal cracking.

G. Jolley is with the Department of Aeronautical and Mechanical Engineering, University of Salford, and J. E. Geraghty is with British Nuclear Design and Construction, Leicester

Weld metal cracking of austenitic stainless steels of composition 18 Cr–13 Ni–1 Nb is known to be associated with the presence of continuous grain-boundary eutectics. This does not normally produce problems during welding as it can be avoided by the use of electrodes of composition 18 Cr–13 Ni–1 Nb which contain around 5% delta ferrite in the room-temperature microstructure and are not susceptible to weld metal cracking. However, in instances where such welds come into contact with certain corrosive environments the juxtaposition of weld metal and parent plate of differing compositions can cause hazards owing to galvanic corrosion. Often, in such cases, the root run is carried out by tungsten inert gas welding which involves fusion of the 18 Cr–13 Ni–1 Nb material and there is a risk of cracking. It would therefore be extremely beneficial if a weld metal of composition 18 Cr–13 Ni–1 Nb could be used which had a much reduced susceptibility to solidification cracking. The authors describe in this paper an investigation into the possibilities of reducing the solidification cracking tendency of 18 Cr–13 Ni–1 Nb stainless steel.

SOLIDIFICATION CRACKING IN STAINLESS STEEL

The problem of solidification cracking in austenitic stainless steels has received a great deal of experimental and theoretical attention, and the accumulated knowledge on the subject up to 1960 is adequately presented in the excellent review by Borland and Younger.[1] Since

18 Cr–13 Ni–1 Nb stainless steels have a narrow solidification range, any susceptibility to low ductility at temperatures near the melting point might be thought to be associated with low-melting-point liquid films. Younger and Baker[2] observed liquid phases of this type in austenitic steels which were associated with interdendritic areas of high solute concentration. Fredricks and Van der Toorn[3] showed that these phases were rich in sulphur and phosphorus: this was substantiated by Honeycombe and Gooch[4] who found that these phases solidified just below the bulk solidus in fully austenitic stainless steel weld metal. While it is argued that these films can cause weld metal cracking, it must be pointed out that cracking beneath the bulk solidus has occurred without evidence of the prior presence of liquid films.[5,6] However, it has been shown that an increase in sulphur content can lower the ductility near the solidus[7,8] and hence increase the hot cracking tendency, and other authors have shown the adverse effect of phosphorus on hot cracking in weldments[9] and castings.[10,11]

The mechanism by which the introduction of delta ferrite into the microstructure of stainless steels reduces the hot cracking tendency of austenitic stainless steels is perhaps not as well understood as might be thought. Although several theories have been proposed to explain the effect it is felt there are two worthy of special attention. Several authors have pointed out that harmful elements such as sulphur and phosphorus are more soluble in delta ferrite than in austenite[3,12–14] and have concluded that delta ferrite can reduce hot cracking by

411

Table 1   Chemical composition of some of the materials described in the investigation

Material	Composition, wt-%										
	Cr	Ni	Nb	Si	Mn	Mo	Ti	P	S	C	Mg
18 Cr–13 Ni–1 Nb base material	18·2	12·5	0·88	0·77	1·29	0·2	0·01	0·02	0·022	0·08	
18 Cr–9 Ni–1 Nb base material	17·9	8·9	0·82	0·61	1·10	0·2	0·01	0·01	0·015	0·06	
18 Cr–13 Ni–1 Nb as cast	17·9	12·6	0·93	0·69	1·07	0·2	—	0·02	0·029	0·08	
18 Cr–13 Ni–1 Nb +0·4% Ni–Mg	17·5	13·0	0·79	0·74	1·14	0·2	—	0·014	0·026	0·09	0·1
18 Cr–13 Ni–1 Nb +1·5% Ni–Mg	17·5	13·3	0·76	0·68	1·07	0·2	—	0·017	0·017	0·11	0·17

absorbing more of these elements and leaving fewer available for the formation of low-melting-point films. Hull[15] criticised this explanation and argued that if absorption of sulphur and phosphorus were the only criterion then a fully ferritic alloy would be even more crack resistant, which is not always the case. Hull explained the effect of delta ferrite on the basis of interfacial energies. The work of Taylor[16] had shown that the austenite/ferrite boundary had a lower interfacial energy than either an austenite/austenite or a ferrite/ferrite boundary. Hence, it was argued, it is possible that during the last stages of solidification the remaining liquid occupies only the austenite/austenite and not the austenite/ferrite boundaries. If delta ferrite particles are sufficiently numerous and well distributed at the austenite grain boundaries they can prevent hot cracking by helping to withstand the stresses imposed by thermal contraction and also by helping to break up the continuity of the liquid films.

Whatever the mechanism of hot cracking, any investigation into the possibility of reducing hot cracking tendency must concern itself with the size and shape of low-melting-point films. Since solidification cracking takes place with little or no ductility Williams and Singer adopted a Griffith approach and quantified the phenomenon by the expression:

$$\sigma = \sqrt{\frac{8\mu\gamma}{\pi(1-\nu)l}}$$

where

σ = stress to cause failure
μ = shear modulus
ν = Poisson's ratio
l = crack (liquid film) length
γ = the effective fracture surface energy.

Obviously if l can be decreased either by decreasing the amount of the liquid films or by reducing the dihedral angle of the films, then the stress to cause failure will be increased thus reducing the hot cracking susceptibility.

EXPERIMENTAL PROCEDURE

The preliminary part of the investigation was concerned with the effects of increasing amounts of delta ferrite on the microstructure of austenitic stainless steel. The base materials for this investigation were 18 Cr–13 Ni–1 Nb bar and 18 Cr–9 Ni–1 Nb electrode wire. The chemical compositions of these materials are given in Table 1. The effect of composition on the as-cast microstructure was

investigated by producing six melts from the 18 Cr–13 Ni–1 Nb and 18 Cr–9 Ni–1 Nb material with increasing amounts of 18 Cr–9 Ni–1 Nb wire from 0–100%, thus producing a range of microstructures containing up to about 5% delta ferrite. Melting was carried out in a 20 kW high-frequency induction furnace using recrystallized alumina crucibles with an argon atmosphere. The metal was poured at 1 700°C into sand moulds to produce 1·4 cm dia. bars. The cast bars were sectioned to allow a longitudinal section to be produced from the centre of each bar. Figure 1 shows the effect of decreasing nickel content which is to produce increasing amounts of delta ferrite and to reduce the quantity and continuity of the grain-boundary eutectics which are at a maximum in the 18 Cr–13 Ni–1 Nb material. Figure 1*d* shows that some eutectic does exist even in the 18 Cr–9 Ni–1 Nb material and that it is not associated with the delta ferrite, a fact which will be discussed later. Figure 2 shows the effect of nickel content on the amounts of delta ferrite and grain-boundary eutectics. These figures were obtained by quantitative optical microscopy and it is the unmistakable trends rather than the accuracy of each individual figure which is of greater importance.

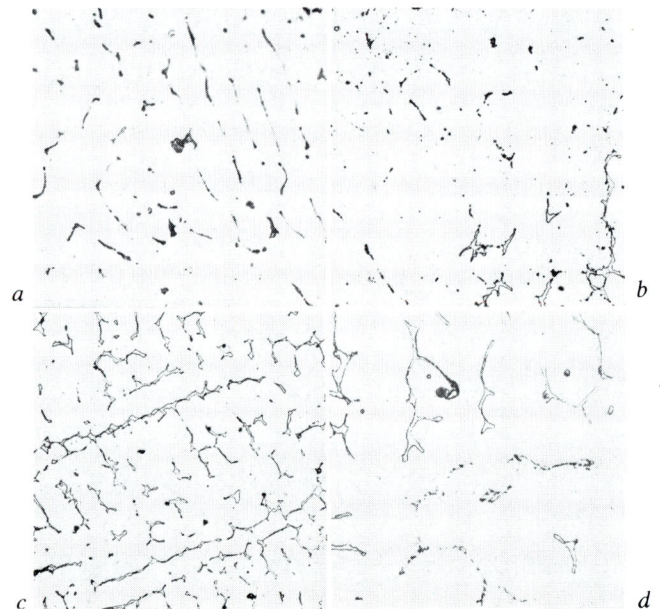

a 12·6% [×140]; *b* 10·6% [×140]; *c* 9·0% [×140]; *d* 9·0% [×230]

1   Effect of varying nickel content on microstructure

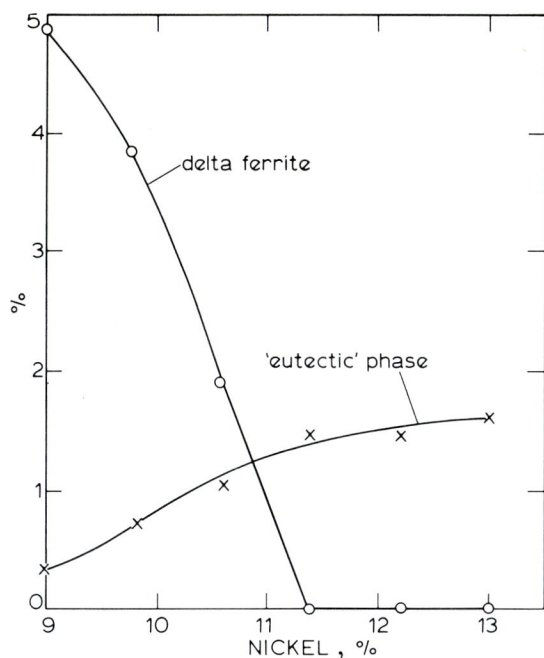

2   Effect of nickel content on amounts of delta ferrite and grain-boundary eutectics

Extensive electron-probe X-ray microanalysis was carried out on the samples with rather disappointing results. Apart from the three main constituent elements the only element found in the eutectic areas was niobium which was present in quantities varying from 10 to 12%. No sulphur, phosphorus, or silicon could be found in these areas although sulphur and manganese were found to be present in spherical inclusions which can be seen particularly in Fig. 1d.

Pt/Pt–13 Rh thermocouples were inserted in bars of 18 Cr–13 Ni–1 Nb and 18 Cr–9 Ni–1 Nb composition, respectively. In both cases the resultant cooling curve consisted of a plateau for the bulk of solidification followed by a secondary arrest about 150°C beneath this temperature corresponding, presumably, to solidification of grain-boundary eutectics. The bulk solidification arrest for the 18 Cr–13 Ni–1 Nb material was 1 430°C followed by an inflexion at 1 275°C, while the figures for

the 18 Cr–9 Ni–1 Nb material were 1 420° and 1 265°C. The inflexion on the 18 Cr–9 Ni–1 Nb curve was much less discernible, as could be expected from the fact that this alloy exhibited smaller amounts of eutectic in the microstructure.

Since niobium is intentionally added to stainless steels to prevent the formation of chromium carbide with its associated corrosion hazards, the eutectic films cannot be removed by omitting this element. The melting experiments were repeated with 18 Cr–13 Ni–1 Nb material and increasing additions of magnesium as nickel–magnesium alloy. Magnesium has a strong affinity for sulphur which would prevent its formation of low-melting-point eutectics but the element can also affect the surface energy of certain phases which might have been important in this particular instance.

A series of 1·4 cm dia. bars were cast with increasing additions of magnesium as nickel–magnesium alloy. Precautions were necessary during the additions which were accompanied by violent flaring necessitating the use of protective clothing and eye shielding. The analyses of some of the bars are given in Table 1. The effect of increasing magnesium additions was to cause a globularization of the eutectics which became very marked above about 0·15% magnesium: these effects can be seen in Fig. 3. Thermal analyses on bars containing 0·25% magnesium did not display an inflexion below the bulk arrest temperature.

The additions of magnesium had obviously changed the nature of the low-melting-point films in a manner likely to make them less harmful during solidification. The effect of magnesium additions on the hot ductility of 18 Cr–13 Ni–1 Nb stainless steel was therefore investigated using a 'Gleeble' hot ductility testing machine. Since the low-melting-point eutectic arrests had been established at less than 1 300°C tensile specimens machined from cast bars were tested at 1 320°C when these films would be molten. Testing was carried out in an argon atmosphere, the specimens were heated to 1 320°C in 18 s, held for 60 s at temperature, and then strained at 4·0 cm/s. The temperature was recorded by chromel/alumel thermocouples spot welded to the specimens which had threaded ends and were 1 cm in diameter with a gauge length of 6 cm. The 18 Cr–13 Ni–1 Nb material failed at comparatively low load in a

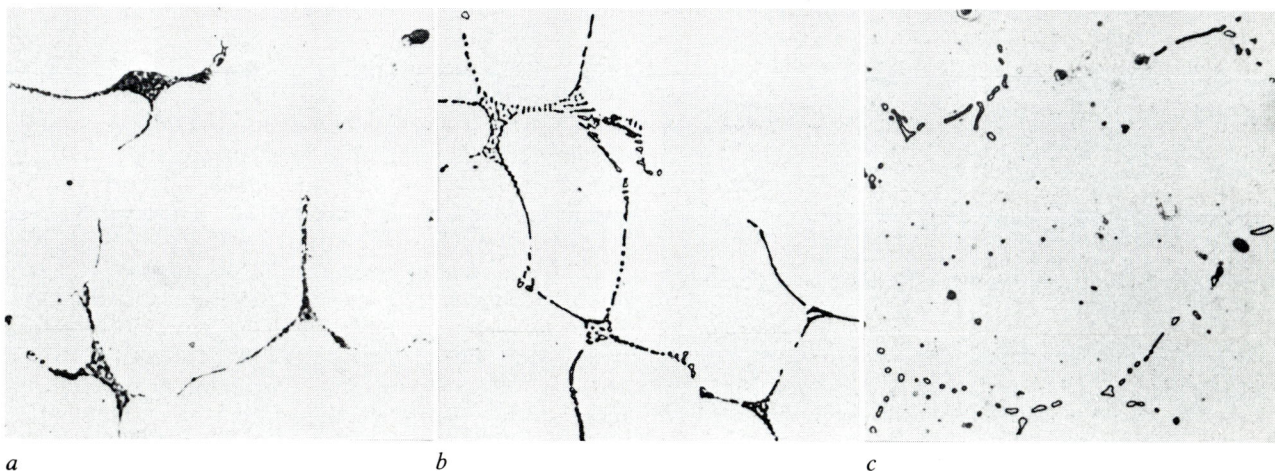

a 18 Cr–13 Ni–1 Nb; *b* 18 Cr–13 Ni–1 Nb + 0·1% Mg; *c* 18 Cr–13 Ni–1 Nb + 0·25% Mg

3   Effect of magnesium additions on the microstructure of 18 Cr–13 Ni–1 Nb material   [×315]

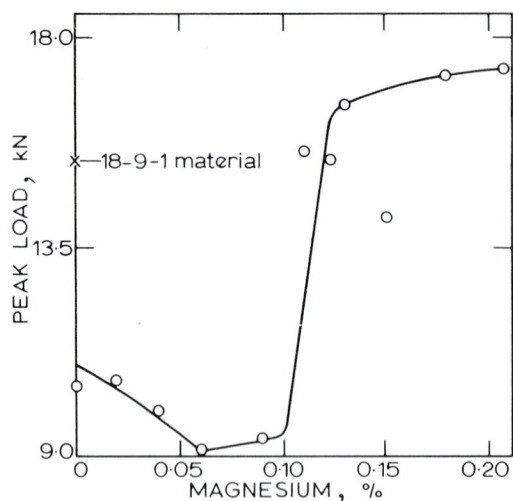

4 Effect of magnesium additions on fracture energy at 1 320°C

brittle manner while the 18 Cr–9 Ni–1 Nb material, as expected, exhibited much greater ductility and had a much higher fracture load. However, as Fig. 4 shows, additions of magnesium in excess of 0·15% raised the fracture energy of the 18 Cr–13 Ni–1 Nb material in excess of the 18 Cr–9 Ni–1 Nb specimens. The increased ductility associated with magnesium additions is shown in Fig. 5 which is a photograph of fractured 'Gleeble' specimens. The fracture surfaces were examined in greater detail by scanning electron microscopy. Figure 6

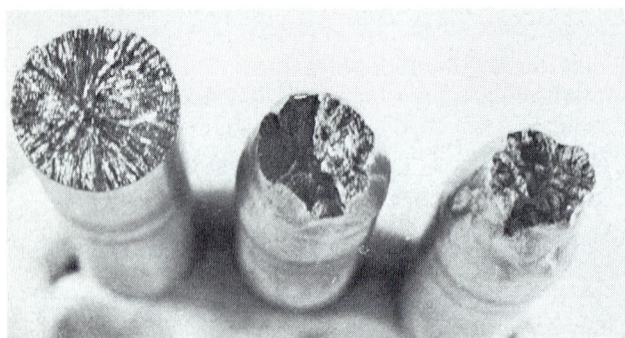

5 Fractured Gleeble specimens; left to right: 18 Cr–13 Ni–1 Nb, 18 Cr–9 Ni–1 Nb, 18 Cr–13 Ni–1 Nb + 0·22% Mg

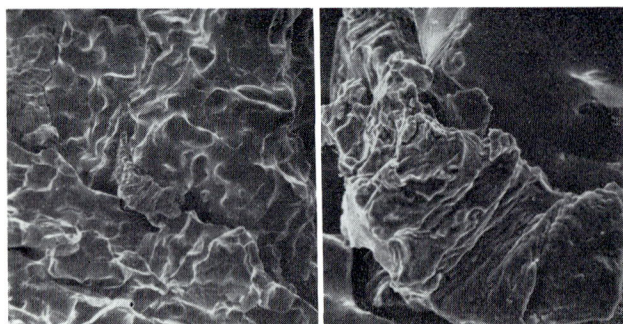

a [×115]; *b* [×575]

6 Fracture surface of Gleeble specimen of 18 Cr–13 Ni–1 Nb material

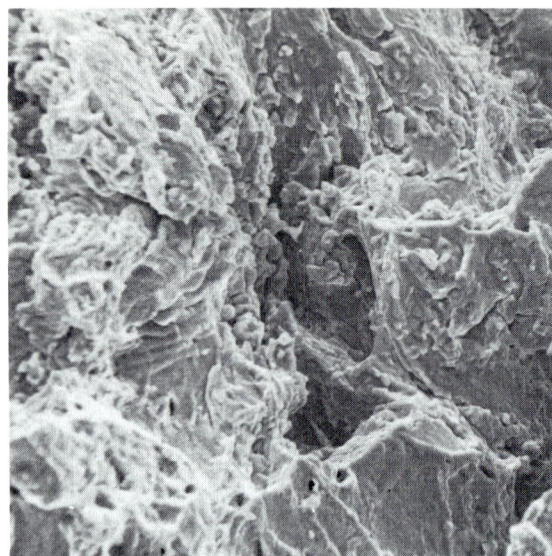

7 Fracture surface of Gleeble specimen of 18 Cr–9 Ni–1 Nb material

8 Fracture surface of Gleeble specimen of 18 Cr–13 Ni–1 Nb + 0·22% Mg

shows the fracture surface of the 18 Cr–13 Ni–1 Nb specimen which has an interdendritic appearance and reveals evidence of small amounts of resolidified eutectic phase (Fig. 6b). Figure 7 shows the fracture surface of the 18 Cr–9 Ni–1 Nb material which is largely ductile and fibrous and much more indicative of a solid-phase fracture. Figure 8 is taken from the 18 Cr–13 Ni–1 Nb + 0·22% magnesium specimen. This fracture is again interdendritic but reveals much more evidence of plastic deformation than is shown in Fig. 6. Figures 9 and 10 are details from the surfaces of the 18 Cr–13 Ni–1 Nb and 18 Cr–13 Ni–1 Nb + 0·22% magnesium fractures, respectively. Figure 9 shows that the sulphur-rich inclusions also played a part in the fracture process and Fig. 10 shows that in areas of the fracture of the more ductile magnesium-containing material, thermal steps on the surface suggest that some melting and re-solidification has taken place.

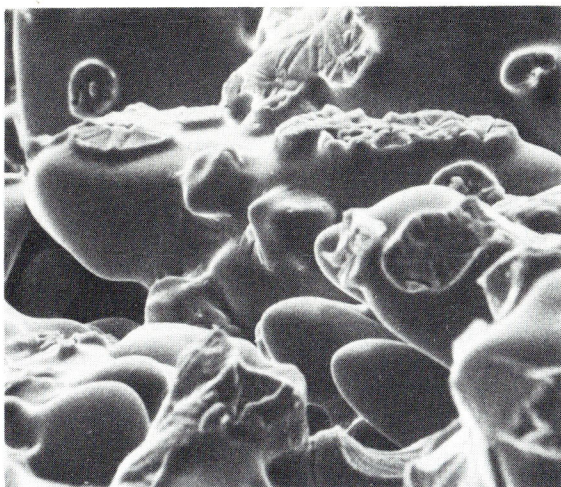

9   Detail from specimen shown in Fig. 6

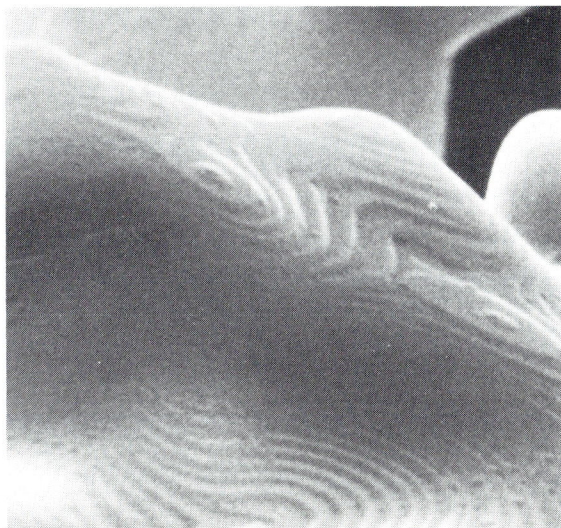

10   Detail from specimen shown in Fig. 8

DISCUSSION

The investigation has shown that increasing additions of magnesium to 18 Cr–13 Ni–1 Nb stainless steel increases the hot ductility of the material owing to its action of globularizing low-melting-point grain-boundary eutectics. The effect of delta ferrite in reducing the cracking susceptibility has also been partly related to a decrease in eutectic content (Fig. 2). The evidence proposed in this paper suggests that the eutectics are niobium rich and do not contain significant amounts of sulphur and phosphorus. It is known[1] that delta ferrite absorbs more niobium (4·5% at 1 360°C) than austenite (2·0% at 1 220°C), although both phases are capable of absorbing more than the niobium content of the base material under equilibrium conditions. It is, therefore, not surprising that when delta ferrite is present the amount of niobium-rich eutectic in the microstructure diminishes, resulting in reduced solidification cracking

susceptibility. While no direct evidence has been proposed on the nature of the eutectics it is felt that they must consist mainly of niobium carbide and austenite.

Although Table 1 shows that magnesium addition reduced the sulphur level of the melt, this is not considered to be significant to the solidification cracking problem for reasons discussed previously. The beneficial effect of magnesium must be associated with its effect on the amount and continuity of the eutectic phase. These effects are obvious in the microstructure (Fig. 4) and from the results of the hot tensile tests (Fig. 5).

SUMMARY

It is possible to increase the hot ductility of 18 Cr–13 Ni–1 Nb material with magnesium additions to a level comparable to that of 18 Cr–9 Ni–1 Nb electrode material. Extensive weld metal cracking trials are now necessary to corroborate the evidence presented in this paper. It will also be necessary to carry out extensive corrosion trials on magnesium-containing 18 Cr–13 Ni–1 Nb material to ensure that it can meet with service requirements. It is hoped that the results presented in this paper will generate further interest in this topic.

ACKNOWLEDGMENTS

The authors are grateful to the United Kingdom Atomic Energy Authority who financed the project and in particular to Mr F. S. Dickinson and Mr P. Lees, UKAEA, Culcheth, for provision of materials and helpful discussion. The authors would also like to acknowledge the invaluable assistance of Mr D. Crichton, UKAEA, Springfields, who carried out the Gleeble tests.

REFERENCES

1 J. C. BORLAND AND R. N. YOUNGER: *Brit. Weld. J.*, 1960, **7**, 667
2 R. N. YOUNGER AND R. G. BAKER: *J. Iron Steel Inst.*, 1968, **206**, 1 124
3 H. FREDRIKS AND L. J. VAN DER TOORN: *Brit. Weld. J.*, 1968, **15**, 178
4 J. HONEYCOMBE AND T. G. GOOCH: *Metal Construction*, 1970, **2**, 37
5 O. A. KALNER AND X. RUSSIYAN: *Weld. Prod.*, 1970, 17 5, 55
6 T. M. CULLEN AND J. W. FREEMAN: *Trans. ASME*, 1963, **85A**, 151
7 R. P. SOPHER: *Weld. J.*, 1958, **37**, 481s
8 N. SMITH AND B. I. BAGNALL: *Brit. Weld. J.*, 1968, **15**, 63
9 T. J. MOORE AND R. B. GUNIA: ASAEC N.Y.O. 9725, July 1961
10 J. VAN EAGHAM AND A. DESY: 'Modern castings', 1969, p. 100
11 F. C. HULL: *Proc. ASTM*, 1960, **60**, 667
12 A. BERNSTEIN: *Eng. Dig.*, 1969, **30**, 41
13 E. C. ROLLASON AND M. C. T. BYSTRAM: *J. Iron Steel Inst.*, 1951, **69**, 347
14 J. C. BORLAND: *Brit. Weld. J.*, 1960, **7**, 558
15 F. C. HULL: *Weld. J.*, 1967, **46**, 399s
16 J. W. TAYLOR: *J. Inst. of Met.*, 1958, **86**, 456

Studies of TIG weld pool solidification and weld bead microstructure in stainless steel tubes

S. De Rosa, M. H. Jacobs, D. G. Jones, and C. Sherhod

This paper is principally concerned with the physical metallurgy of autogenous TIG (tungsten–inert gas) welded stainless steel tubes. A very brief description of the tube welding process is followed by a qualitative description of the salient features of the weld pool solidification. The chief non-equilibrium characteristics of the solidified weld bead are described, and form a platform from which four specific topics are discussed. These are (a) the formation of 'weld blobs' by processes similar to those well known in the technique of zone melting; (b) the formation of metastable precipitates in the heat-affected zone of the weld; (c) an interesting, but rare, example of centreline embrittlement; and (d) non-equilibrium dislocation structures in as-welded beads and their responses to certain annealing treatments. It is concluded that to understand fully the nature and consequences of these non-equilibrium effects they must be investigated with the aid of several modern, microanalytical techniques.

The authors are with TI Research Laboratories, Saffron Walden, Essex

This paper is concerned with some interesting phenomena associated with stainless steel tubes produced by longitudinal, autogenous tungsten–inert gas[1] (TIG) welding of cylindrically formed strip. The tubes are manufactured on a mill, and the process includes the following operations. First, the steel strip, of appropriate width and gauge, is roll-formed into tubular shape. Secondly, the longitudinal gap in the formed tube is progressively fusion welded with an arc struck between the moving tube and a stationary tungsten electrode. The atmosphere of the arc is predominantly argon, although a little hydrogen is sometimes added to promote arc stability. The welded tube is water cooled and sized on the outlet side of the mill. The welding speed used depends on the dimensions of the tube—the samples referred to in this paper were a few centimetres in diameter, with wall thicknesses of a few millimetres, and these were welded at speeds between 2·3 and 4 m/min. The corresponding power input was of the order of 100 J/m of tube length.

Let us now turn to some general features of TIG welded stainless steel tubes. The most pronounced microstructural characteristic, which is common to all the ferritic and austenitic stainless steel examples discussed in this paper, is the pattern of columnar dendrites which constitutes the welded bead and which is illustrated in Figs 1a and b. These columnar dendrites initiate in the very early stages of weld pool solidification by epitaxial 'seeding' at the two liquid/solid interfaces. The steep thermal gradients between the centreline of the weld pool and the parent solid metal on either side force the columnar grains to grow approximately at right angles towards the centreline. In other words, the major directions of heat flow are circumferentially outwards from the weld pool. The solidifying dendrites (of ferrite in ferritic steels and of austenite in austenitic steels) grow most rapidly along ⟨100⟩ directions; hence, those seeds which are oriented with their ⟨100⟩ directions parallel to the heat flow predominate in the formation of columnar grains which, in the solidified weld bead, exhibit a very marked texture with ⟨100⟩ almost exclusively normal to the centreline of the weld. It is common for the columnar grains to impinge at the weld centreline, as illustrated in Figs 1a and b. Furthermore, it is not uncommon for the columnar grains to be separated by a narrow region of equiaxed grains, as exemplified by Fig. 1c, and the presence of this feature depends critically on the welding parameters employed, on the diameter and the

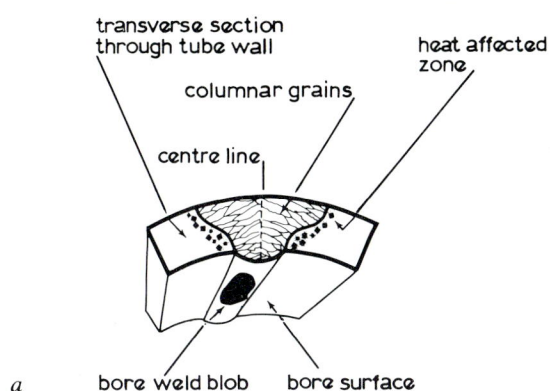

a

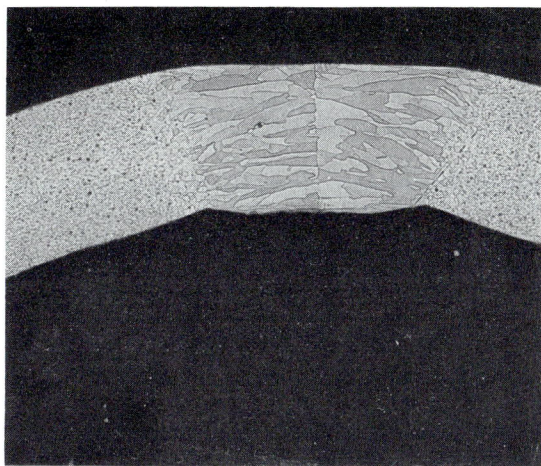

b

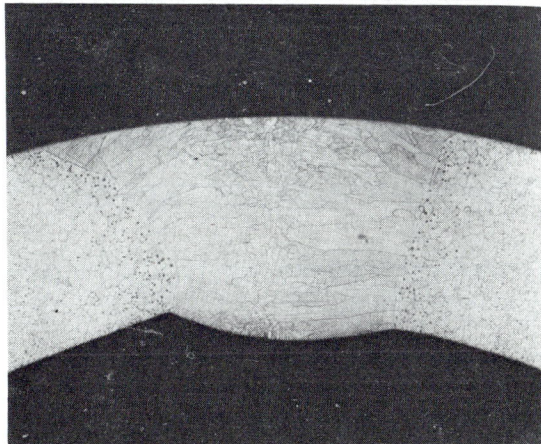

c

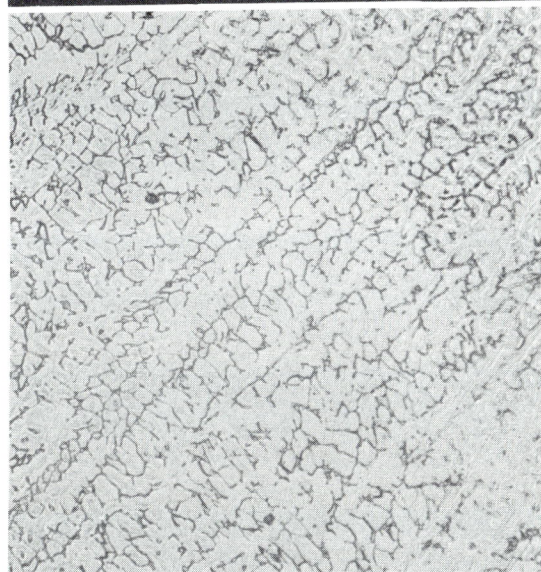

d

gauge of the tube, and on the type of alloy being welded.

Another feature characteristic of many weld beads in austenitic stainless steels is the decoration of the columnar austenite dendrites with δ-ferrite. One may use the mean spacing of the secondary arms of these dendrites to obtain an estimate of the cooling rate of the weld bead during solidification. An example is shown in Fig. 1*d*; the spacing is $\sim 5\ \mu$m, which indicates[2] a high cooling rate of the order of 1 000°C/s.

From the point of view of this paper it is convenient to distinguish between the following specific features and processes:

(i) the redistribution of particles and trace elements in the weld pool during welding

(ii) the non-equilibrium precipitates which may form in the heat-affected zone (Fig. 1*a*) of some alloys

(iii) specific characteristics which may be associated with the weld centreline (Fig. 1*a*) of some alloys

(iv) non-equilibrium dislocation structures in as-welded beads and their responses to annealing treatments.

These four topics will be illustrated by reference to selected examples, each chosen for its scientific interest as well as its practical relevance. Studies such as these are necessary if stainless steel tubes are to be manufactured to high standards of quality and suitability for service.

EXPERIMENTAL PROCEDURES

Several modern electron metallographic techniques were used in these studies. The scanning electron microscopy was performed with a Stereoscan S4 equipped with an Ortec energy dispersive X-ray detector. The analytical electron microscope,[3] EMMA-3, which combines transmission electron microscopy with electron probe microanalysis of thin foils and extracted small particles, was developed at TI Research Laboratories in conjunction with AEI and was the forerunner of AEI's EMMA-4. The Auger electron spectroscopy was carried out by the staff and with the facilities at the United Kingdom Atomic Energy Research Establishment, Harwell.

The specifications of the stainless steels referred to in this report are listed in Table 1. In the text they will be designated according to their British Standard 1449 type number.

---

a schematic diagram showing certain features associated with TIG weld beads; *b* optical cross-section of a weld in a ferritic stainless steel (niobium stabilized 430) showing fully columnar microstructure [×12]; *c* optical cross-section of a weld in a ferritic stainless steel (18 Cr–2 Mo) showing central equiaxed grain structure [×12]; *d* decoration of columnar austenite dendrites with δ-ferrite in an 18 Cr–8 Ni TIG weld bead [×420]

1   Characteristic features of weld beads in TIG welded stainless steel tubes

RESULTS AND DISCUSSION
Weld blobs

The relative movement of the tube beneath the stationary welding torch produces effects with similarities to classical zone melting.[4] In practice, the molten weld pool traverses the complete length of the welded tube, which may be many metres long. Two processes occur, each with its counterpart in the technique of zone refining; this is in spite of the relatively much higher rate of traverse of the weld pool compared with classical zone melting practice:

   (i) small particles, derived from inclusions in the parent strip, may sometimes collect on the top and bottom surfaces of the molten bead; this applies especially to particles which have restricted solubility in the weld pool

   (ii) trace elements which are soluble in liquid steel but which have low solubility in solid steel tend to partition to the weld pool: some of the elements oxidize at the melt surface and thereby form a floating dross enriched in certain trace elements.

In both cases the accumulated surface dross becomes intermittently trapped in the solidifying metal, and thus gives rise to irregularly spaced (10–40 cm) blemishes along the weld line. These blemishes are known as 'weld blobs'.

The constitution of the blobs depends strongly on the deoxidation and stabilization practice used during the manufacture of the steel. Blobbing caused by the presence of titanium in the alloy, in the form of titanium nitride (TiN) and titanium carbide (TiC) particles, is an excellent example (see Figs 2a and b). In this instance, the dark-coloured blobs consist of a single layer principally of TiN particles and a much smaller amount of Al, S, P, and Ca bearing particles. (TiN particles can be identified optically by their characteristic pink colour which also distinguishes them from grey TiC particles*). Their size and shape are similar to those of the TiN inclusions in the parent steel. Hence it may be deduced that the welding operation redistributes a small proportion of the primary TiN inclusions into surface clusters which subsequently become weld blobs. (As shown below, most of the TiC and the remainder of the TiN inclusions dissolve during welding and then reprecipitate during cooling of the weld bead.) Furthermore, a

bulk chemical analysis of regions of weld bead, which contained several blobs, revealed no overall depletion in either titanium or carbon levels compared with the matrix.

Weld blobs are also found in steels not deliberately stabilized with titanium: for example, the alloy types 430 S15, 316 S16, and 304 S16—see Table 1. In this case, the blobs are frequently aggregates of silicate particles which contain the trace elements magnesium and calcium. A particularly interesting example concerned a type 304 S16 austenitic stainless steel which had been treated with rare-earth elements as a means of modifying the shape and dispersion of sulphide inclusions. The weld blob in this case (see Figs 2c and d) consisted of spheroidal inclusions, rich in rare-earth elements (e.g. Ce, Nd, and Pm).

Finally, it should be said that weld blobs are normally very minor surface blemishes and, as such, exert no adverse influence on the quality of the tubing.

Metastable precipitates

The regions of the parent metal adjacent to either side of the weld bead are subjected to rapid rates of heating and cooling. These 'heat-affected zones' (HAZs)—see Fig. 1a—frequently display non-equilibrium microstructures. As an example of this behaviour, the authors have chosen the formation of metastable precipitates in titanium-stabilized ferritic stainless steels (e.g. 409 S17, 430 S15, and 18 Cr–2 Mo alloys).

The strip steel contains primary TiN and TiC particles—see Fig. 3a—which are usually several micrometres in diameter. Within the weld pool the TiC particles dissolve and then subsequently reprecipitate as a much finer dispersion during solidification (Fig. 3b), whereas some of the TiN particles become incorporated in weld blobs (see above). In marked contrast, a very different reaction proceeds in the HAZ, which remains solid throughout the welding cycle. Rapid heating, to a temperature in excess of 1 200°C, causes the primary TiC particles to decompose partially by solid solution during which the Ti and C diffuse away. Upon subsequent rapid cooling, reprecipitation occurs under non-equilibrium conditions because the interstitial carbon atoms have diffused more rapidly, and hence further, from the parent particle. This creates a halo of fine particles around the parent, partly decomposed, inclusion, Fig. 3c.

An analysis of the fine particles in the halo with the aid of EMMA reveals their metastable nature. Figure 4a is a micrograph of typical particles extracted on a carbon support film. Their characteristic diffraction pattern,

---

* It should be noted that Duncumb and Melford[5] used electron-probe microanalysis to demonstrate that pink-coloured titanium nitride can contain up to about 4% carbon; likewise, grey titanium carbide may contain a very small amount of nitrogen

Table 1   Specifications of stainless steels referred to in the paper

BS 1449 type no.	Element, wt-% C(max.)	Si(max.)	Mn(max.)	P(max.)	S(max.)	Cr	Ni	Mo	Ti	N
304 S16	0·06	0·2–1·0	0·5–2·0	0·045	0·03	17·5–19·0	9·0–11·0	—	—	—
304 S12	0·03	0·2–1·0	0·5–2·0	0·045	0·03	17·5–19·0	9·0–12·0	—	—	—
316 S16	0·07	0·2–1·0	0·5–2·0	0·045	0·03	16·5–18·5	10·0–13·0	2·25–3·0	—	—
409 S17	0·09	0·8	1·00	0·04	0·03	11·0–13·0	0·7	—	5×C–6·0	—
430 S15	0·10	0·8	1·00	0·04	0·03	16·0–18·0	0·5	—	—	—
430 Nb	As 430 S15 with Nb at 10×C									
430 Ti	As 430 S15 with Ti at 5–10×C									
18–2	0·025	1·0	0·5	0·03	0·02	17·0–19·0	Ni+Cu 0·5 (max.)	1·75–2·5	0·2+4(C+N) –0·6	0·025

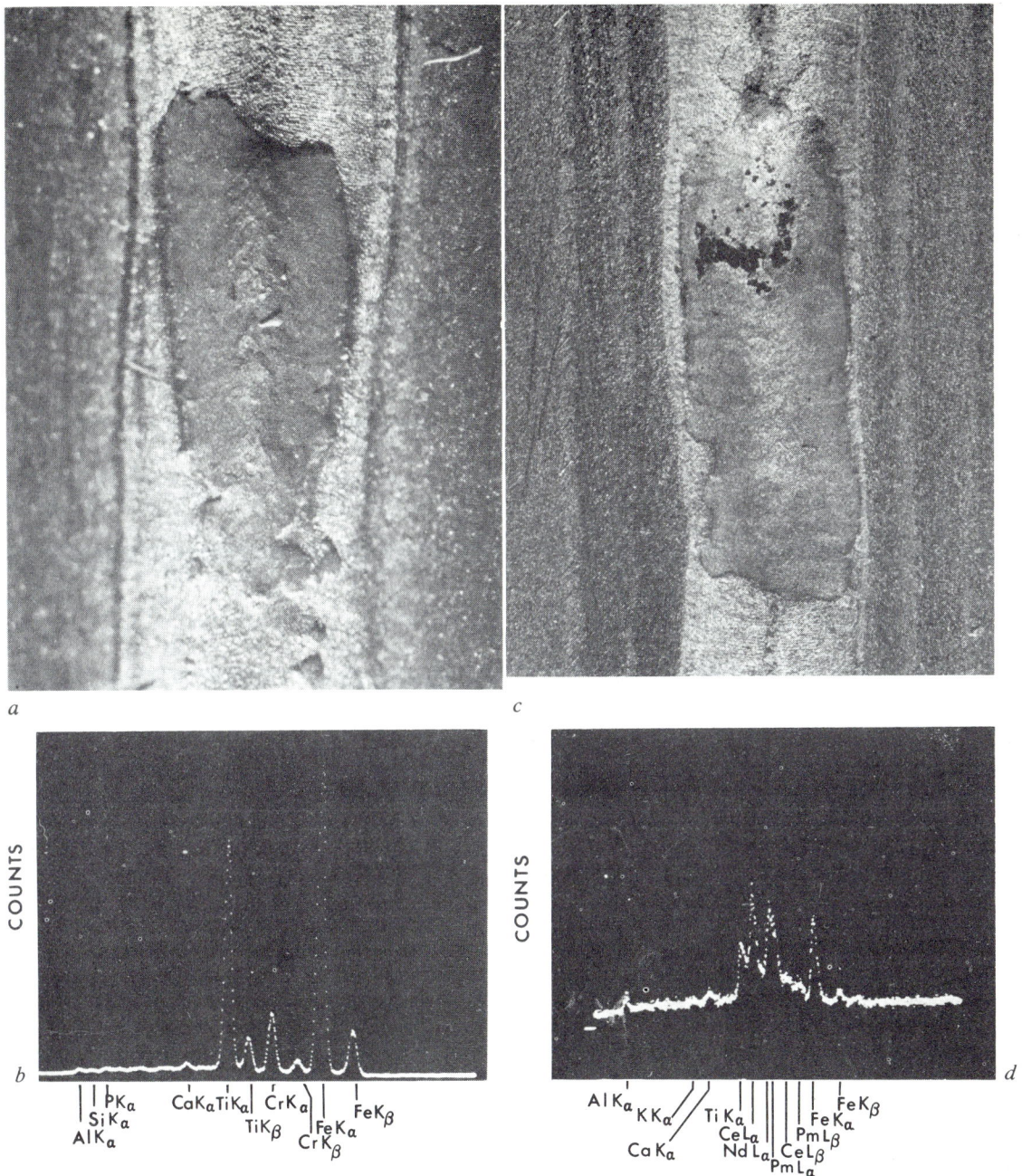

a a blob containing TiN particles on a ferritic stainless tube (409 S17) [×20]; *b* X-ray spectrum from blob shown in Fig. 2*a*; *c* a blob containing rare-earth elements on an austenitic stainless tube (304 S16)[×10]; *d* X-ray spectrum from blob shown in Fig. 2*c*

2 Characteristic weld blobs on the bore surfaces of TIG welded stainless steel tubes

Fig. 4*b*, does not allow them to be distinguished from TiC; they have closely similar, if not identical, cubic structures with $a = 0.431 \pm 0.005$ nm. However, high resolution X-ray analysis with EMMA reveals that they contain not only titanium but also an appreciable proportion of chromium (*see* Fig. 4*c*). Hence, these fine particles may be designated as metastable titanium–chromium carbides.

In ferritic steels with low levels of interstitials, the thermal decomposition effect described above has negligible influence on the properties of the steel by virtue of the low density of TiC inclusions present. However, in a 430 Ti steel (*see* Table 1) which contained 0.06% carbon, the degree of decomposition was very severe and the metastable precipitates so embrittled the HAZs that the weld beam could be wound from the tube, Fig. 4*d*. In this case, intergranular embrittlement was promoted by severe decoration of the grain boundaries by the metastable precipitates. Fortunately, these detrimental effects can be readily eliminated by conventional heat treatment.

Centreline embrittlement

Because of the columnar mode of solidification, the centreline of the weld pool is frequently the last liquid to freeze and may become a highly microsegregated

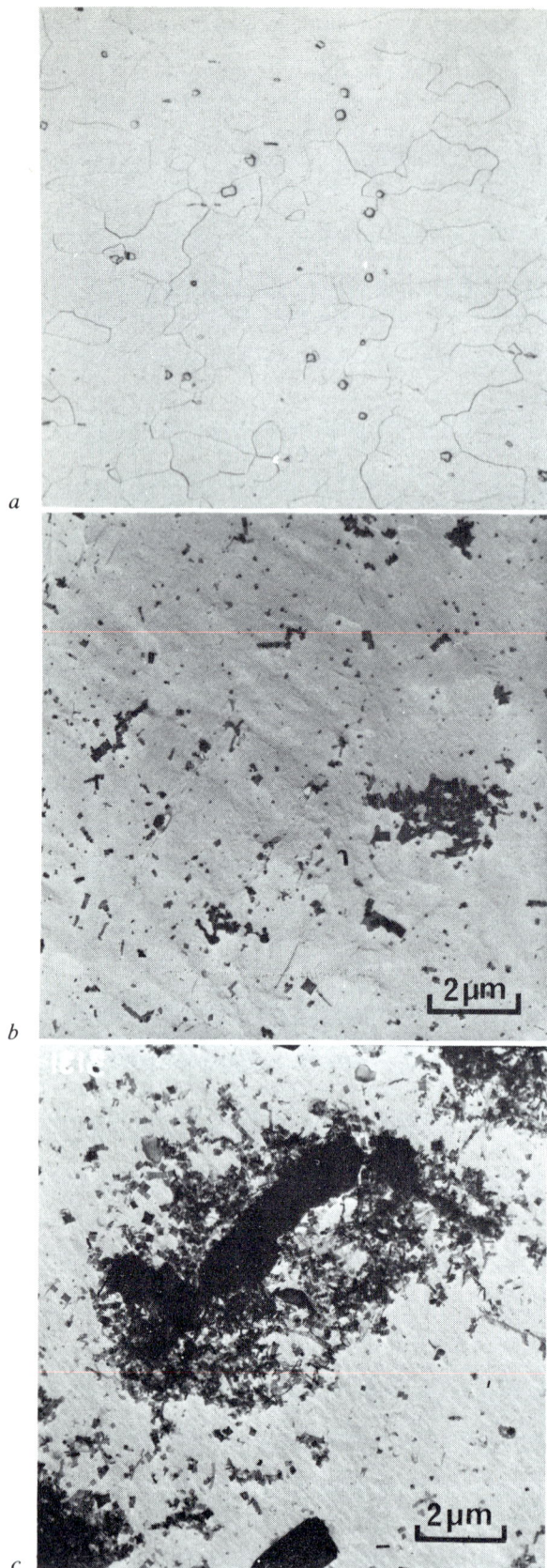

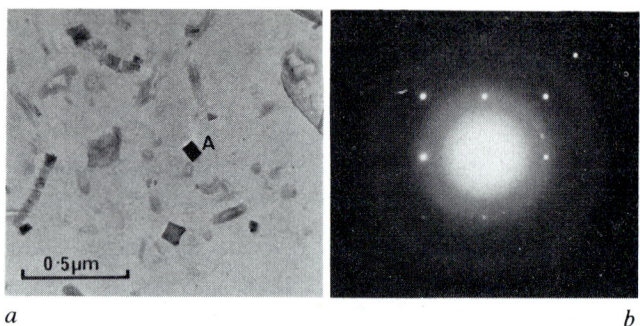

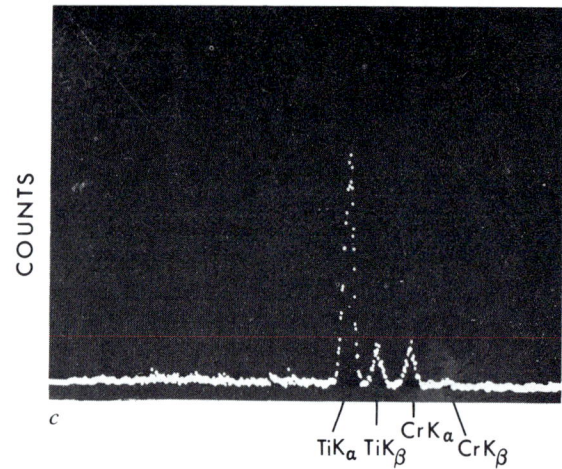

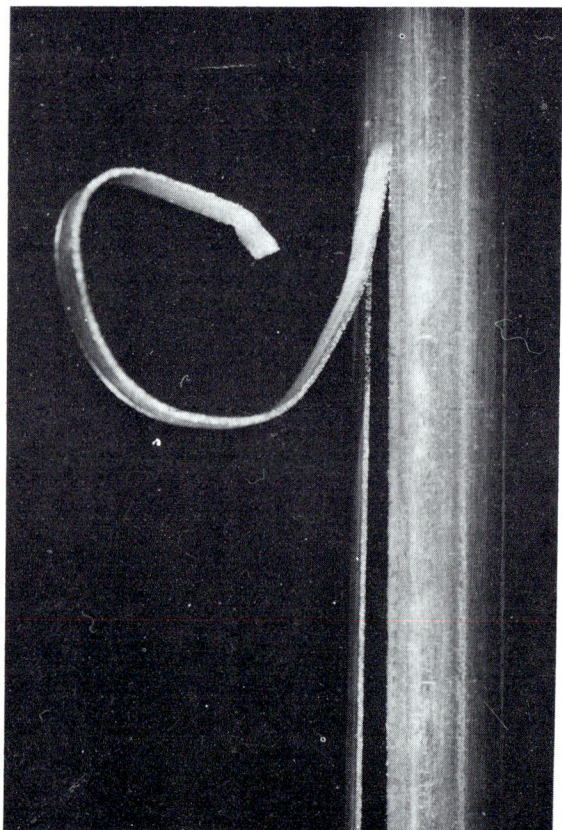

a optical micrograph of titanium carbonitride particles in the steel strip (409 S17) [×300]; *b* extraction replica of fine-scale precipitates in a fusion weld (18 Cr–2 Mo); *c* extraction replica of metastable precipitates surrounding a decomposed primary particle

3   Effect of welding on titanium carbonitride particles

a extraction replica; *b* electron diffraction pattern from particle A; *c* X-ray spectrum from particle A; *d* an illustration of the embrittled heat-affected zone in as-welded 430 Ti (NB embrittlement is easily eliminated by post heat treatment) [×1·5]

4   EMMA analysis of metastable precipitates

CLEAVAGE

INTERGRANULAR

25µm

a

0·5µm

b

nitrogen
peak x 15

(c)

bulk nitrogen level
0·1 at.-%

(d)

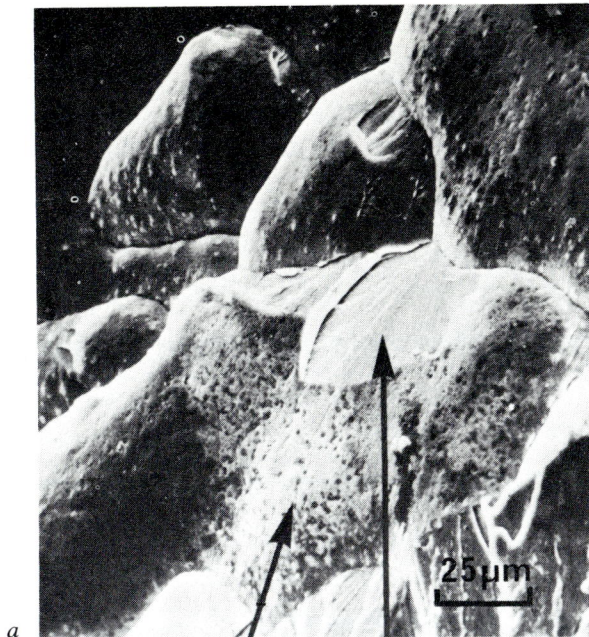

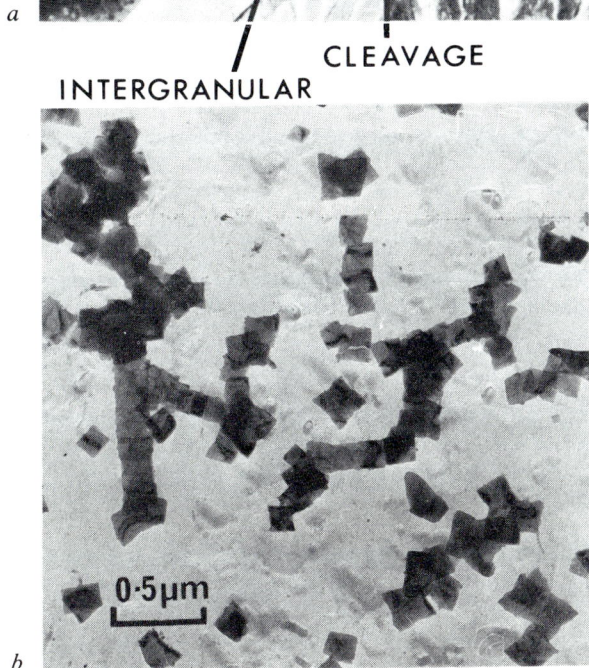

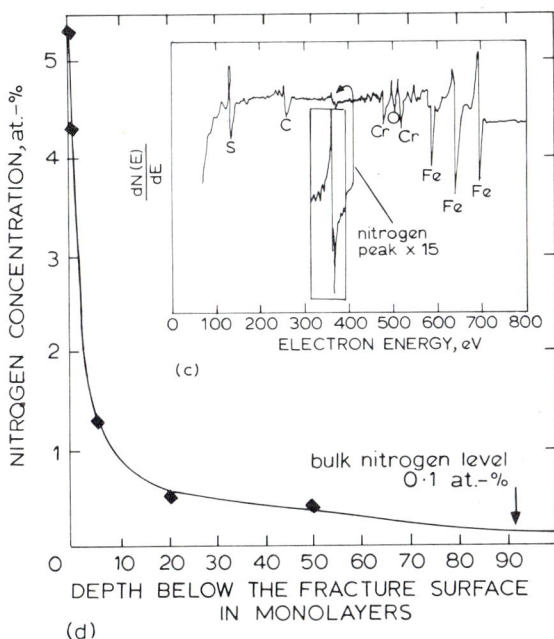

region. This very occasionally leads to centreline weakness, revealed by intergranular failure when a tube is subjected to a flare test at room temperature.

Figure 5*a* shows a predominantly intergranular, centreline fracture obtained with a niobium-stabilized ferritic stainless steel. Similar results were obtained with several specimens. From careful studies with the scanning electron microscope it was tentatively concluded that the intergranular fracture surfaces were covered with very thin surface films, although the nature of the films could not be established with the SEM. The following studies were undertaken to elucidate the cause of the intergranular embrittlement.

First, carbon extraction replicas were prepared from a fracture surface and these contained extracted particles of niobium carbonitride (Fig. 5*b* is representative of the specimens examined in EMMA). Niobium, unlike titanium, forms eutectiferous carbonitride particles,[1] and these particles were all small (~0·1 µm across) and therefore did not fully substantiate the 'intergranular film' suspected from the scanning microscopy.

Secondly, another sample was examined by Auger electron spectroscopy, by staff at Harwell, after it had been fractured *in situ* under ultra-high-vacuum conditions to avoid surface contamination. Some of their results are reproduced in Figs 5*c* and *d*. An unexpectedly high level of nitrogen was confined to the first few monolayers of the intergranular fracture surface (some neighbouring transgranular cleavage fracture faces did not display this effect). Also, the shape of the Auger peak was characteristic of a nitride, although the type of nitride was not determined. There was evidence of microsegregation of sulphur to the intergranular boundary, but niobium was not detected. Hence, although niobium carbonitride precipitates could be extracted from the vicinity of the fracture surface, the Auger results suggest a very thin layer of some as-yet-undetermined nitride at the centreline boundary.

Clearly, all the factors which led to this particular centreline embrittlement have not been resolved, but the example does illustrate the unique nature of the weld centreline in this case.

Weld bead softening

In as-welded tube, the weld is often significantly harder than the matrix. As we shall see shortly, this may be predominantly attributed to a very high dislocation density generated by the thermal stresses associated with rapid cooling after welding. Also, with the austenitic 304 S16 and 304 S12 steels (*see* Table 1, where it will be seen that the latter alloy contains much less carbon), a contribution to the hardness may be derived from the presence of the non-equilibrium network of δ-ferrite. Moreover, it is normal practice to anneal austenitic

a scanning electron micrograph of centreline fracture induced at room temperature [×500]; *b* extracted niobium carbonitride particles; *c* auger spectrum from intergranular fracture surface; *d* depth profile of nitrogen concentration as a function of monolayers below the intergranular fracture surface

5 Centreline embrittlement of 430 Nb alloy

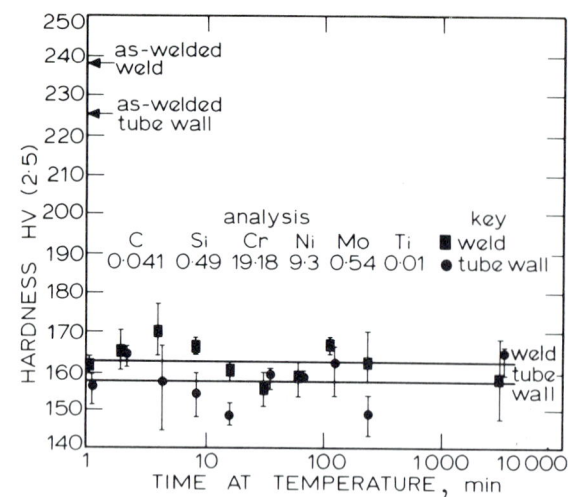

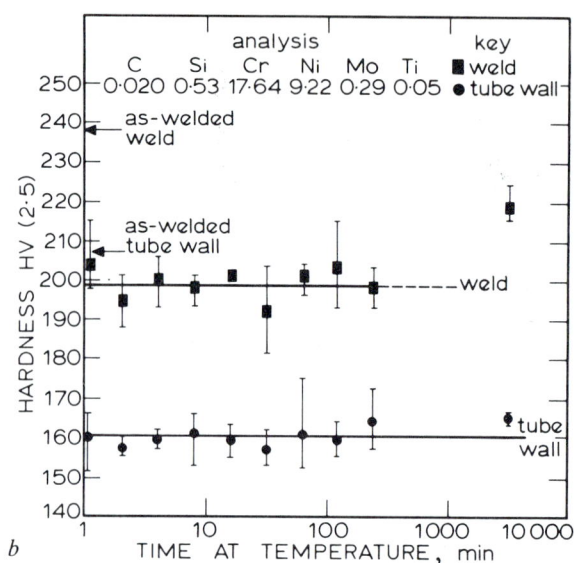

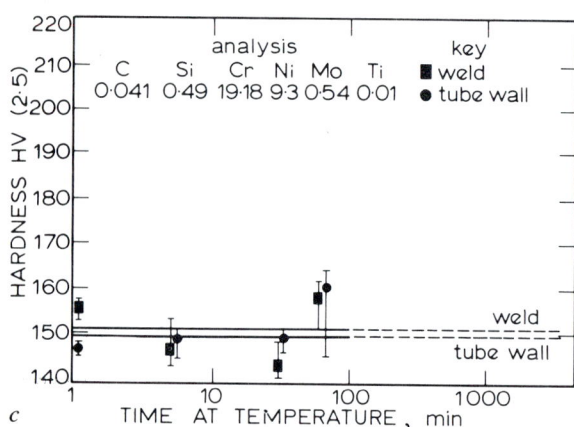

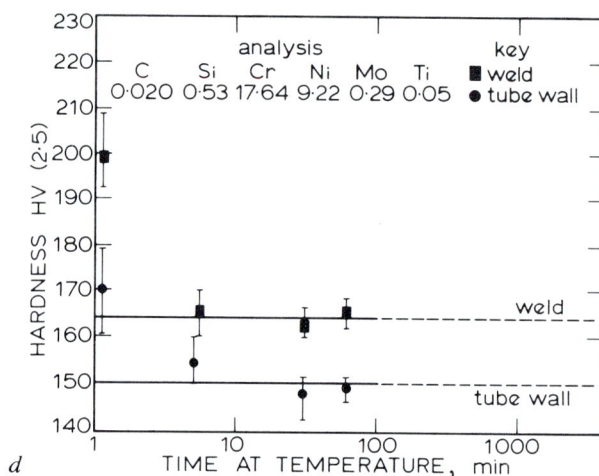

a softening of 304 S16 at 1 050°C; *b* softening of 304 S12 at 1 050°C; *c* softening of 304 S16 at 1 300°C; *d* softening of 304 S12 at 1 300°C

6 Hardness changes in annealed weld beads

stainless tubes at 1 050°C for 10 min to optimize the corrosion resistance and to soften the weld bead to a hardness comparable with the matrix. This treatment is adequate for 304 S16 (Fig. 6*a*), but treatments of much longer duration at 1 050°C fail to realize the desired softening in the exceptional case of the lower carbon alloy (Fig. 6*b*). However, satisfactory softening of both compositions of weld bead can be achieved in 10 min at 1 300°C (compare Figs 6*c* and *d*).

Some insight into this remarkable behaviour has been obtained by transmission electron microscopy of thin foils prepared from weld beads and neighbouring matrices. Compared with the remainder of the tube, the dislocation density in a weld bead is high; this applies equally to the high- and low-carbon grades of the alloy. Figure 7*a* illustrates this point for a low-carbon, 304 S12 sample, and the micrograph also shows a region of characteristic ferrite in the centre of the field of view.

Different aspects of annealing behaviour are illustrated in Figs 7*b*, *c*, and *d*. Figure 7*b* confirms that a high dislocation density persisted in the low-carbon, 304 S12 weld bead annealed at 1 050°C for 1 h; whereas Fig. 7*c* shows that a high proportion of the dislocations were

annealed out in a similar sample treated for only 10 min at 1 300°C. In contrast, the sample of higher carbon-weld bead which was annealed for 10 min at 1 050°C (*see* Fig. 7*d*) was almost completely polygonized—a mode of behaviour not observed with 304 S12 samples. Another difference observed was that the δ-ferrite was completely removed in the higher carbon weld bead after 10 min at 1 050°C, but there was a small amount of persistent, spheroidized δ-ferrite in the lower carbon weld bead even after 1 h at 1 050°C.

The above observations go some way towards understanding the non-equilibrium dislocation structures in as-solidified beads and their responses to annealing treatments, but the root causes of the differences in behaviour have yet to be resolved.

CONCLUSIONS

This paper is principally concerned with selected aspects of the physical metallurgy of TIG weld beads in stainless steel tubes. The consequences of the welding operation are strongly reflected in the characteristic microstructural features of the weld bead. If the origin and nature of these microstructures are properly understood, then

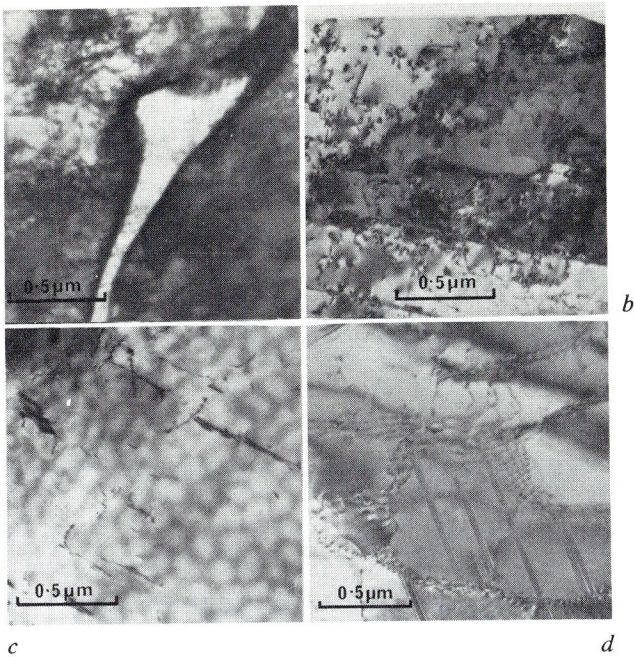

a high dislocation density in as welded, 304 S12 weld bead; *b* 304 S12 annealed 1 h at 1 050°C; *c* 304 S12 annealed 10 min at 1 300°C; *d* polygonized structure of 304 S16 annealed for 10 min at 1 050°C

7 Dislocation structures in thin foils

optimum welding conditions and remedial post-welding heat treatments may be devised to realize the full mechanical property and corrosion resistance potential of the alloy. Furthermore, it is only by deploying the microanalytical facilities made available by a range of modern electronmetallographic techniques that sufficient insight will be forthcoming to elucidate the complex processes which occur during TIG welding.

ACKNOWLEDGMENTS
The authors wish to acknowledge the support and encouragement of Dr G. Thursfield and his colleagues at TI Rollo-Hardy in the preparation of this paper. They are indebted to Dr J. Riviere and Dr B. C. Edwards of UKAERE, Harwell, for carrying out and interpreting their studies by Auger electron microscopy and for permission to include their results in this paper. We also wish to thank Dr D. A. Melford for his helpful comments on the manuscript.

REFERENCES
1 R. J. CASTRO AND J. J. DE CADENET: 'Welding metallurgy of stainless and heat-resisting steels' (Translation of second French edition); 1974, Cambridge, Cambridge University Press
2 W. E. BROWER AND M. C. FLEMINGS: 'Solidification of iron based alloys at large degrees of undercooling', AMMRC CR 69-14/1, 1969, Watertown, Massachusetts, US Army Materials and Mechanics Center
3 M. H. JACOBS: *J. Microscopy*, 1973, **99**, (2), 165
4 W. G. PFANN: 'Zone melting', 2 ed., 1966, New York, John Wiley
5 P. DUNCUMB AND D. A. MELFORD: 'X-ray optics and microanalysis', (ed. R. Castaing *et al.*), 1966, 240, Paris, Hermann

Control of the solidification process during submerged arc welding

T. Edvardsson

The influence of the process variables power input and travel speed on primary structure has been studied during the submerged arc welding of thick panels of low-carbon manganese steel; it has been shown to be possible to influence and direct the development of the primary crystals so that both the morphology of the growing interface and the growth directions of the crystals can be controlled. With a technique for temperature measurement, temperature gradients relevant to the solidification process have been calculated. The growth conditions in the weld pool have also been influenced by altering the weld geometry and by changing the thermal conditions by addition of extra cold material to the melt. If the crystals growing from the sides of the weld pool are prevented from meeting in the middle of the weld by the presence of an equiaxed crystal zone, enrichment of solute material in this region can be reduced. The development of such an equiaxed zone has been greatly enhanced by additions of grain refiners. By simultaneous addition of cold material to the weld melt, conditions have been produced which increase the effectiveness of the added grain refiner.

The author is with the Swedish Institute for Metal Research, Stockholm, Sweden

During the solidification process in a weld, stresses and segregations build up which can greatly influence weld properties[1-6] and susceptibility to solidification cracking.[7,8] The deleterious effects of the stresses generated can be reduced by control of the growth directions of the primary crystals. This control should be carried out in such a way that areas of high segregation will be affected as little as possible by the tensile stresses.

Grains with specific crystallographic directions aligned close to the temperature gradient will be favoured during the initial period of solidification in a weld pool.[9,10] These growth directions will then be maintained during the growth process so that when one is trying to influence the solidification conditions of the primary crystals it is of great interest to have information about the temperature and temperature distribution in the system.

The segregation generated during solidification is closely related to the crystal morphology of the growing solid phase. A larger amount of solute material in the liquid phase is thus obtained in front of a planar interface as compared with an interface with a dendritic morphology. It has been shown[11] that when crystals of cellular morphology meet at a seam in the weld the resulting segregation facilitates the formation of cracks down the centre of the weld, whereas when crystals have developed with a dendritic morphology and interlock in the middle of the weld the formation of cracks is less favoured. The large amounts of solute material which can build up in front of a cellular interface may no longer be found when the interface develops a dendritic morphology and a good welded joint can be obtained.

The actual heat flow in the weld as determined by the heat input and the heat removed gives the temperature gradient and the growth rate of the liquid/solid interface and hence the morphology developed. Information about temperature and temperature distribution in the pool during welding is thus of importance in the understanding and control of the solidification process. This has been studied from a theoretical point of view in a number of papers and theoretical calculations of temperature conditions during the welding process have been reported.[12,13] Direct measurements of the temperature conditions have not been studied so extensively.

Therefore, a method was used[14] which allows temperature measurements to be made in the weld pool during submerged arc welding in the liquid phase in front of the growing interface, and in the totally solid phase in the weld deposit, and in the liquid–solid region. A Pt/Pt 10 Rh thermocouple, flame sprayed with aluminium oxide, was first adjusted with a special holder to a given distance from the symmetry line (y) in the weld pool and to a given depth below the surface (z) of the steel sheet during the welding process and then frozen in at the given coordinates (*see* Fig. 1). This procedure

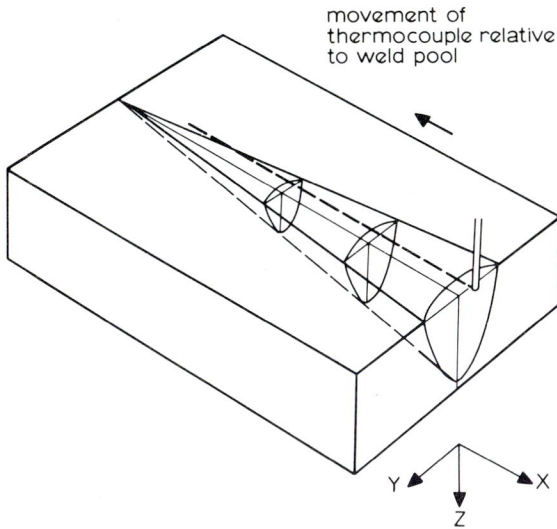

1 Geometrical conditions during the welding tests; the direction covered by the temperature measurements is also marked

made it possible to study the variation of temperature with time during the solidification process in the welding direction for different y and z coordinates.

In order to control the structural variables and thereby the solidification process in the experiments, the appropriate process variables for the system concerned were used. It should be noted that a change in a process variable will normally influence many structural variables. For example, a change in travel speed may lead to changes in the temperature gradient and in the growth rates of the crystals, which will in turn affect the morphology of the growing interface and the amount of solute material in front of the interface.

The aim of this investigation has been to study the control of the solidification process during submerged arc welding in thick panels of low-carbon manganese steel as used in the shipbuilding industry.

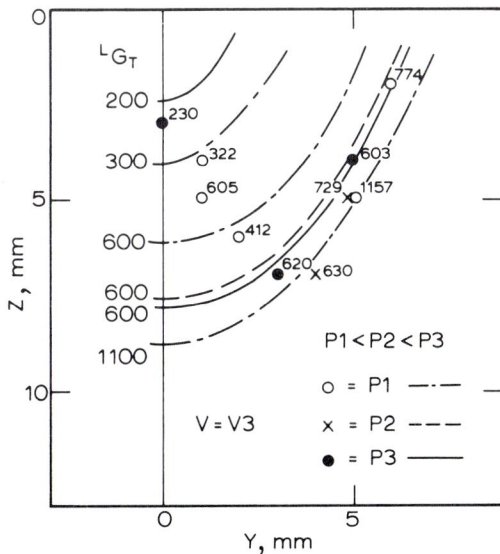

2 Temperature gradient as a function of the measuring position for different values of power input at constant travel speed

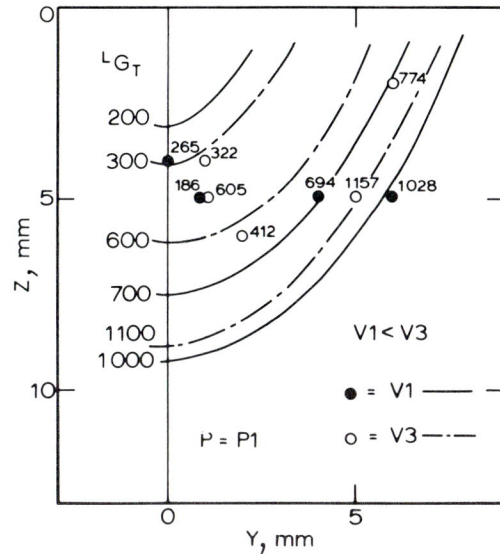

3 Temperature gradient as a function of the measuring position for different values of the travel speed at constant power input

PROCESS VARIABLES

The influence of the process variables on the primary structure was studied during welding and the growth conditions of the crystals were related to the power input and the travel speed.[15] Test welds were made using a nominal travel speed (V) of 42 m/h and voltage and currents for the two electrodes of 38 V/1 200 A and 45 V/900 A, giving a power input (P) of 86 kW, and then with combinations when travel speed and power input were changed by about 15%. With the technique for temperature measurement, temperature gradients of interest for the solidification process could be calculated.

Values obtained for the temperature gradient in the liquid in front of the interface $^{L}G_{T}$ at different coordinates are given in Fig. 2 for varying values of the power input, and isogradient curves have been indicated. Figure 3 shows the indicated isogradient curves and values of the temperature gradient at different coordinates for varying travel speed. The curves show that the temperature gradient in the upper central part of the weld is affected by the power input and travel speed: with increasing power input decreasing values for the temperature gradient are obtained, and with increasing travel speed the temperature gradient increases. A more fully developed dendritic interface morphology should thus be favoured by increasing the power input and by decreasing the travel speed.

Changes in the power input and travel speed also affect the growth direction of the crystals, so it is possible to control the crystal growth and in this way try to minimize the injurious influence of the tensile stresses generated during the welding process.

If the tensile stresses act perpendicularly to the symmetry line of the weld influencing the growth direction of the crystals, so that the angle between the growth direction and the symmetry line of the weld increases, this can lead to the enriched areas between the growing crystals developing in a direction which is more nearly parallel to the tensile stresses. In this way the susceptibility to cracking in areas with high amounts of solute material may be reduced as compared to the case when the angle is small and the tensile stresses can more easily separate the growing crystals.

Values of the power input and travel speed used in practice during submerged arc welding both with two electrodes and with three electrodes often give a vertically growing crystal zone in the weld deposit which prevents the crystals growing from the sides meeting in the middle of the weld. This zone can develop vertically growing diverging crystals as shown in Fig. 4 or vertically growing parallel crystals as shown in Fig. 5. Under these conditions the tensile stresses built up during welding act perpendicularly to the growth direction of the crystals and to the areas between the crystals with enriched solute material content.

It is shown in Fig. 6, however, that by varying the power input this zone of vertically growing crystals can be completely suppressed. Furthermore, Fig. 7 demonstrates that by control of the thermal conditions the crystals growing from the sides can be induced to grow in a dendritic manner, forming an interlocked seam in the middle of the weld. The build up of solute at the interface is simultaneously reduced and a sound join favoured.

It has thus been shown to be possible to influence the development of primary crystals by variation of the process variables in such a way that both the morphology of the growing interface and the growth directions of the crystals can be controlled.

GROOVE ANGLE

The influence of the groove angle on the primary structure and on the growth conditions has been studied.[16]

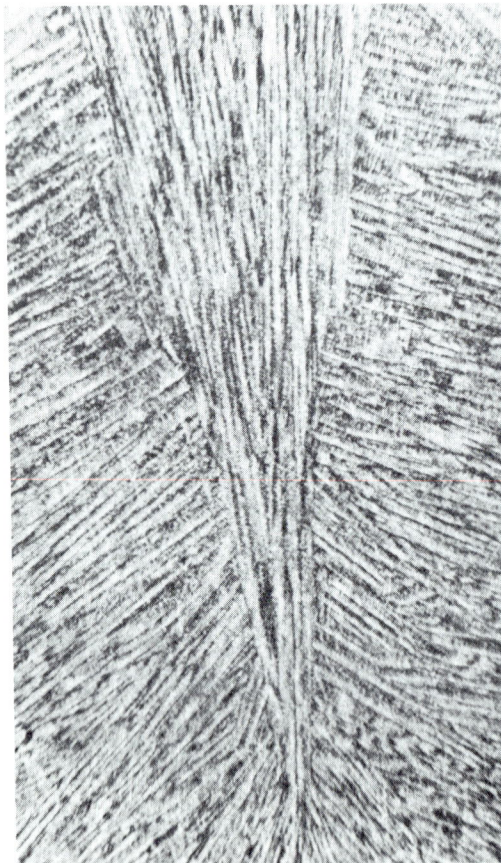

4   Primary structure of the central parts of the weld deposit with vertically growing diverging crystals   [×20]

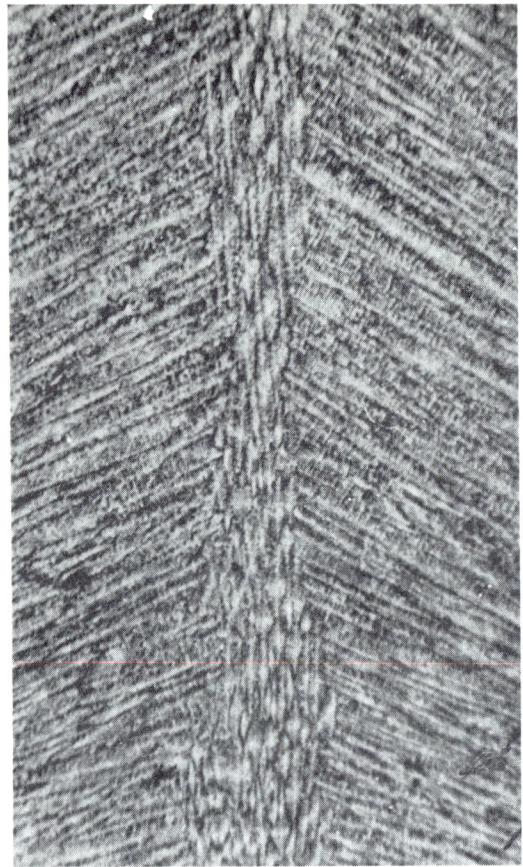

5   Primary structure of the central parts of the weld deposit with vertically growing parallel crystals   [×20]

Test welds were made with the five groove angles 0°, 30°, 60°, 90°, and 120° using the same nominal data as before. The primary structures obtained in cross-section for the groove angles 0° and 120° are given in Figs 8 and 9. It can be seen from these that the growth directions of

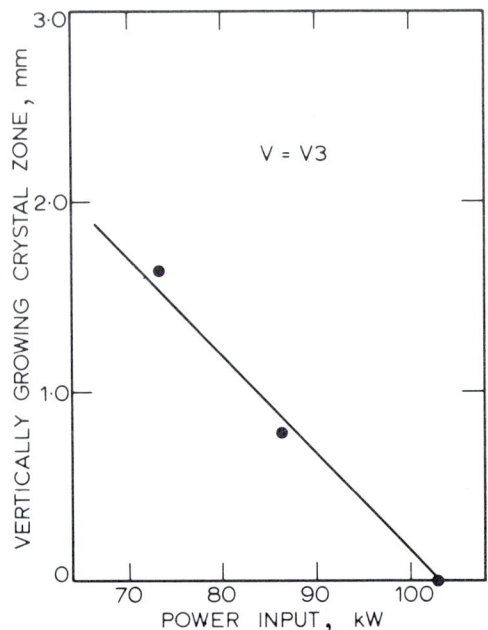

6   Width of the vertically growing crystal zone as a function of the power input

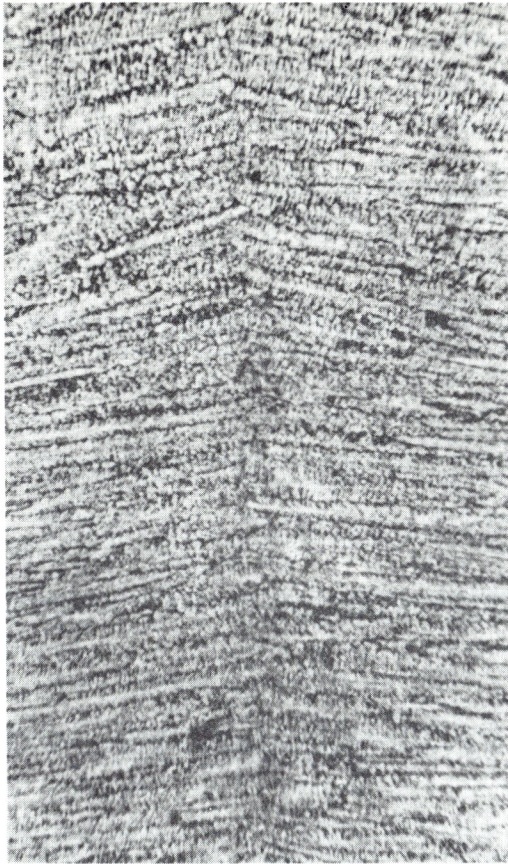

7 **Primary structure of the central parts of the weld deposit without a vertically growing crystal zone [×20]**

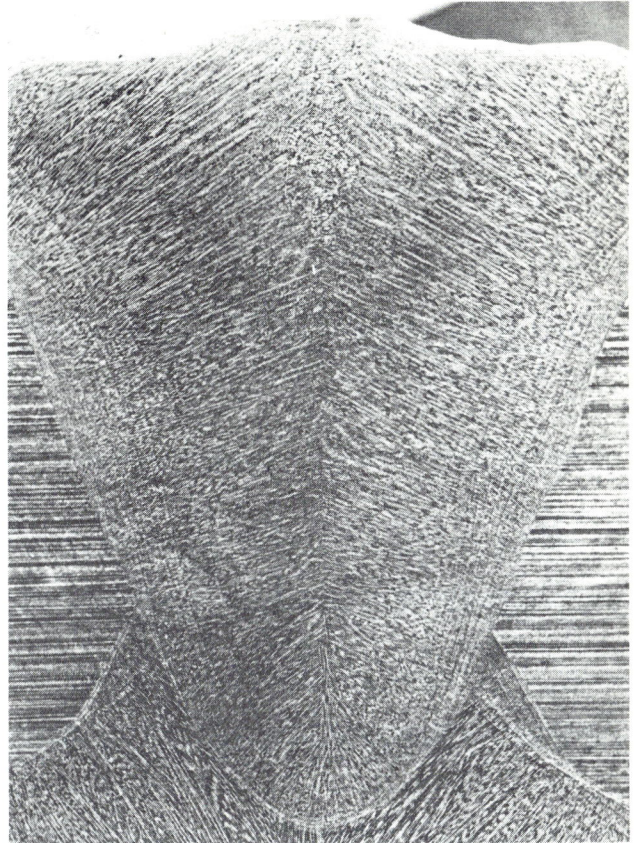

9 **Primary structure in cross-section for the groove angle 120° [×4]**

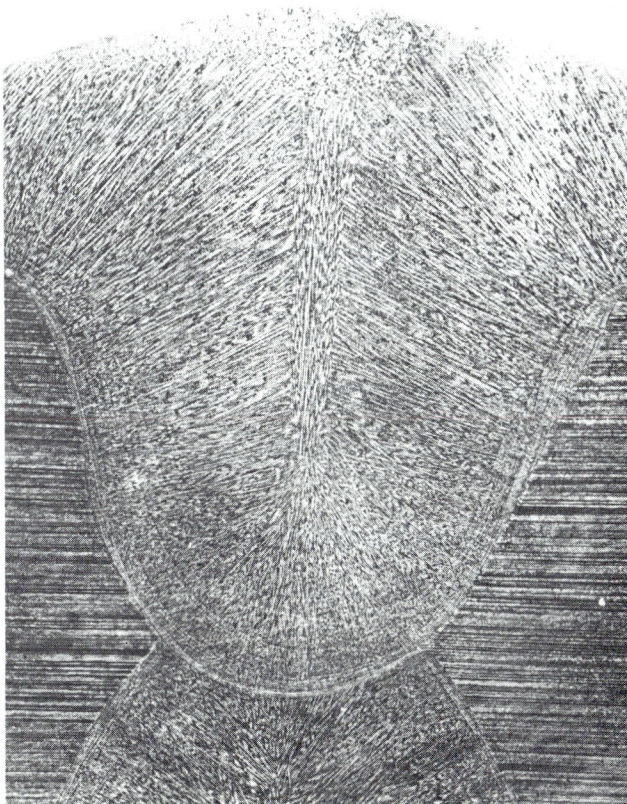

8 **Primary structure in cross-section for the groove angle 0° [×4]**

the primary crystals are affected by variations in the groove angle in that the angle formed with the symmetry line in the weld increases with increasing groove angle.

The vertically growing crystal zone can be controlled by the groove angle in that with increased groove angle the width of this crystal zone decreases. These vertically growing crystals in the middle of the weld can also be completely suppressed by having high values of the groove angle.

The temperature gradient in the liquid phase in front of the interface and the temperature gradient within the liquid–solid region can be affected by variations in the groove angle, and the temperature measurements indicate that these gradients decrease with increased groove angle. This observation can be understood by considering the changed cooling efficiency of the steel sheet surrounding the weld pool due to differences in the penetration depth and the geometry of the pool (*see* Fig. 10). With increasing groove angle the morphology of the growing solid phase is thus influenced and a more dendritic form of the interface will be favoured.

As an insulating flux layer covers the weld during submerged arc welding, the heat from the melt should in principle be transported to the surrounding cold material in the steel sheet. This transport might then be directed to the sides and downwards from the weld pool. When the ratio of the heat flow directed downwards to the heat flow directed to the side is comparatively small, as it is when the penetration depth is increased and a thinner layer of steel remains below the weld pool, the development of crystals growing from the sides may be

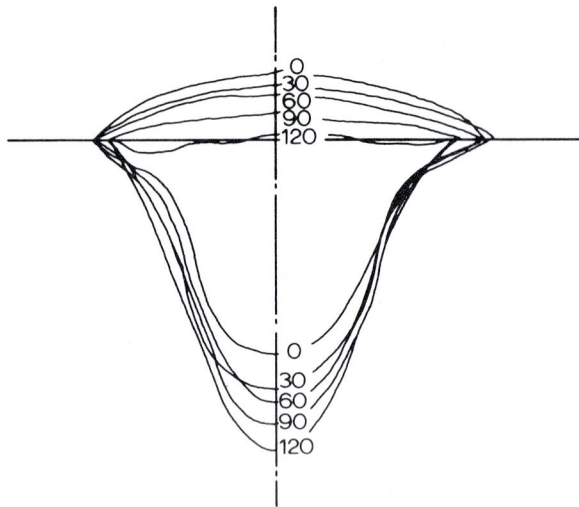

10   Weld pool geometry in cross-section for different values of the groove angle

favoured relative to that of crystals growing vertically in the weld pool.

It is shown that with varying groove angle these heat flow conditions can be affected and that an increased groove angle promotes the crystals growing from the sides while crystals growing vertically in the weld pool may be suppressed. As the depth of penetration increases with increased groove angle, the cooling efficiency of the steel below the weld pool can decrease relative to that of the steel beside the weld. Measured temperature gradients then show decreasing values with increased groove angle and a more fully dendritic interface morphology is promoted.

The growth conditions of the crystals during submerged arc welding can be controlled by the process variables. It is thus possible to generate structural variables in the welding process by variation of the groove angle and hence to obtain desired growth directions of the crystals, vertically growing crystal zone, and crystal morphology developed.

ADDITION OF A COLD WELDING WIRE

Investigations of the addition of cold material during arc welding have been reported in the literature. In some, the addition has been made in front of the electrodes into the groove during the welding process as granulated material mixed into the flux,[17] as small metal pieces in the form of cut wire fed into the groove,[18] and as a metal powder placed in the groove before the welding flux.[19] The addition of cold material has also been made directly into the weld pool via a cold welding wire in order to increase the travel speed and sometimes to add alloying elements.[20-22]

By addition of cold material to a weld, changes in the thermal conditions in the weld pool are brought about which have a direct influence on the solidification process in the weld. The crystal growth conditions during the solidification process for varying amounts of added cold solid welding wire into the weld melt were studied both for welding with constant power input and travel speed and for welding with varying values of the power input and travel speed.[23] For the test welds made at constant power input, the nominal data was used and the cold welding wire of 4 mm diameter was fed into the weld pool 5 mm behind the second electrode at feed

rates of $V' = 0.5, 1.0, 1.5, 2.0, 2.5,$ and 3.0 m/min. For the test welds made at constant energy input per unit length, the travel speed was increased with increased feed rate of the cold welding wire in order to keep the height of the weld over the steel sheet constant, so the power input was correspondingly increased to give a constant energy supply per unit length. Temperature measurements were made and the calculated temperature gradients have been related to the feed rate of the cold welding wire.

For the test welds made with constant power input and travel speed, the height of the weld over the steel sheet increases and the width of the weld decreases with increasing feed rate of the cold solid welding wire. The depth of penetration of the weld is fairly constant for different amounts of added cold material when this is added to the weld pool behind the second electrode. In this increasing upper part of the weld, equiaxed crystals are favoured giving an increasing area of equiaxed crystal zone in the cross-section of the weld with increased feed rate of the cold welding wire (Fig. 11). With an increasing amount of cold material added to the weld melt, the temperature gradients in the growth directions of the crystals tend to increase. The addition of a cold welding wire may increase the cooling efficiency of the steel sheet beside the weld and this increase could lead to increased temperature gradients in the middle of the weld as indicated.

The extent of the vertically growing crystal zone in the cross-section can be controlled by the feed rate of the cold welding wire, and with an increasing amount of

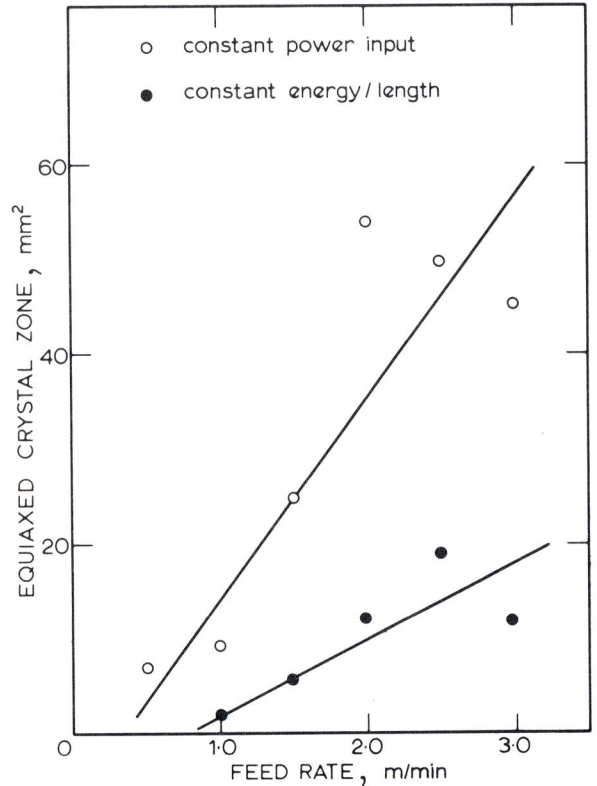

11   Area of the equiaxed crystal zone in cross-section as a function of the feed rate of the cold welding wire in welding tests with constant power input and in tests with constant energy per length

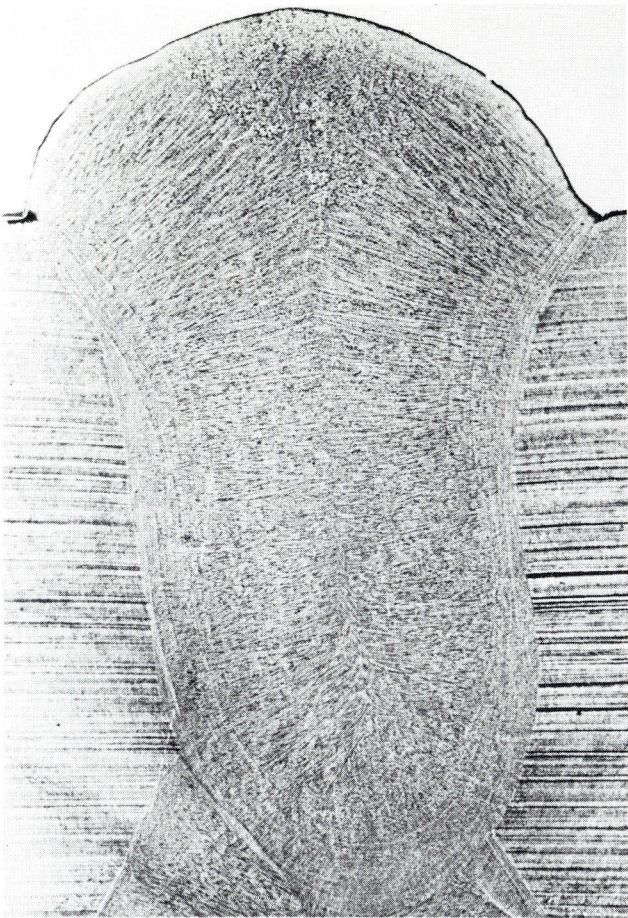

12 **Primary structure in cross-section for a feed rate of the cold welding wire of 3·0 m/min [×4]**

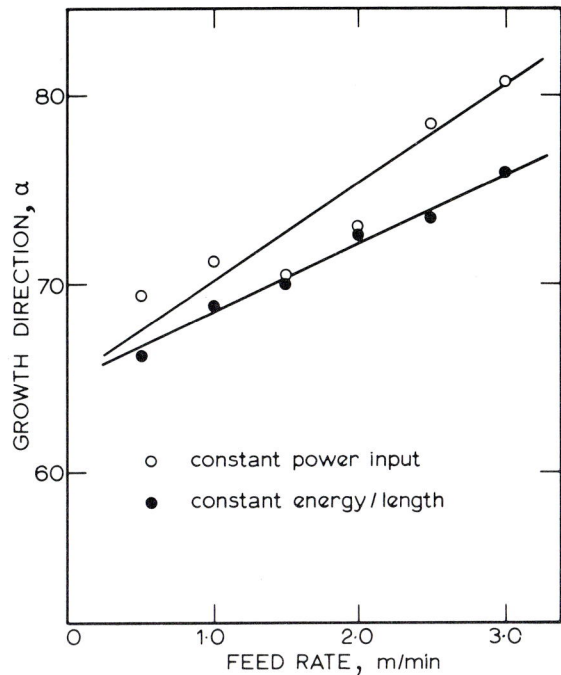

13 **Growth direction of the crystals in the upper central part of the weld as functions of the feed rate of the cold welding wire in welding tests with constant power input and for tests with constant energy per unit length**

added cold material at high feed rates the vertically growing crystal zone can be completely suppressed allowing the crystals growing from the sides to meet in the middle of the weld. The primary structure in the cross-section of the weld for a feed rate of the cold welding wire of 3·0 m/min is shown in Fig. 12.

The growth directions of the primary crystals growing from the fusion boundary into the melt are influenced by the addition of cold material to the weld pool, and the angle between the symmetry line of the weld and the growth direction of the crystals in the cross-section increases with increased feed rate of the cold welding wire (Fig. 13). This effect is related to the heat transport from the hot melt to the surrounding cold sheet material. With increased feed rate of the cold welding wire the ratio of the heat flow directed to the side to that directed downwards may increase due to the energy absorbed by the heating and melting of the extra added material. This would give more efficient cooling from the sides of the weld and thus promote the growth of crystals growing from the sides, hence giving increased values of the above-mentioned angle.

For those test welds made with constant power input and travel speed with varying amounts of cold material added to the melt, a direct relation can be observed between the growth conditions of the crystals and the feed rate of the cold welding wire. In order to keep the height of the weld over the steel sheet constant, the travel speed can be increased with increasing feed rate of

the cold welding wire while in order to give a constant energy supply per unit length the power input can be correspondingly increased. Test welds made, having a fairly constant profile in cross-section, show that the growth conditions for the primary crystals in this case are influenced in the same way by the addition of cold material as for the test welds made with constant power input and travel speed.

It is thus possible to influence the thermal conditions in the weld pool by addition of cold material during welding and thus to control the solidification process.

GRAIN REFINEMENT
Generally, it is thought that a fine equiaxed primary structure can lead to improved properties. Investigations have therefore been carried out on the grain refinement of castings by inoculation with different grain refiners and interesting results on steel castings have been reported.[24-26]

Methods for the grain refinement of welds have been reported, the techniques used including the application of mechanical vibration during solidification,[27] ultrasonic stimulation of the weld pool,[28,29] electrode weaving,[30-32] and application of an alternating magnetic field.[33-35] Investigations on grain refinement of steel welds have also been made with various methods of inoculation such as the introduction of grain refiner into the weld from the flux,[36] addition during welding via the welding electrode,[36,37] and inoculation of the weld pool via a steel tube fed into the pool.[38] However, it has been found that inoculation with grain refiner gives reduced impact resistance which is believed to be the result of intergranular precipitation caused by the additions.[37-41] It is thus of great interest to grain refine the weld structure under conditions that makes it possible to use a minimum amount of grain refiner. By simultaneous

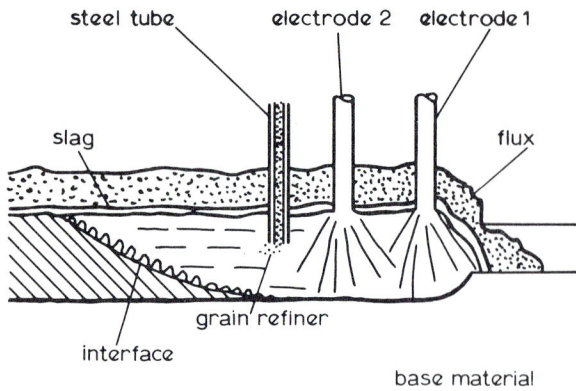

14 Experimental arrangement for addition of grain refiner to the melt via a steel tube fed into the weld pool

addition of cold material to the weld melt, conditions may be produced which increase the effectiveness of an added grain refiner. The solidification process in the weld was influenced and studied for different amounts of added grain refiner and extra addition of cold material into the weld melt.[42]

Grain refiner was added to the weld pool first by addition directly to the weld groove in front of the hot welding electrodes, and secondly by addition via a steel tube containing a mixture of iron and grain refiner powder which was fed into the weld pool (Fig. 14). For the welds with grain refiner added directly to the weld groove, the tests were made both with and without

simultaneous addition to the melt of a cold solid welding wire. The test welds were made using the nominal data which were adjusted when the steel tube was added in the same way as described for addition of a cold welding wire.

The primary structure for a given TiC inoculation is shown in Fig. 15 with no cold material added, and with simultaneous addition of cold material in Fig. 16. If the area of the equiaxed crystal zone in the cross-section for the test welds is related to the amount of added grain refiner, it is found (Fig. 17) that this area increases with increased amount of added grain refiner per unit length weld in all cases. For a given addition of grain refiner it is found that an increasing amount of cold material added to the weld pool increases the area of the equiaxed crystal zone in cross-section. The best grain refinement effects were observed for large additions of grain refiner in combination with a high rate of addition of cold material to the weld pool. By appropriate choice of combinations of welding conditions, the amount of added cold material and of added grain refiner, almost the whole melt could be induced to solidify as equiaxed crystals.

The effect of an added grain refiner is dependent on the conditions in the system and one factor of importance is the surface energy. By appropriate choice of an alloy made under different cooling conditions, the wetting conditions may be affected. The grain refiner is

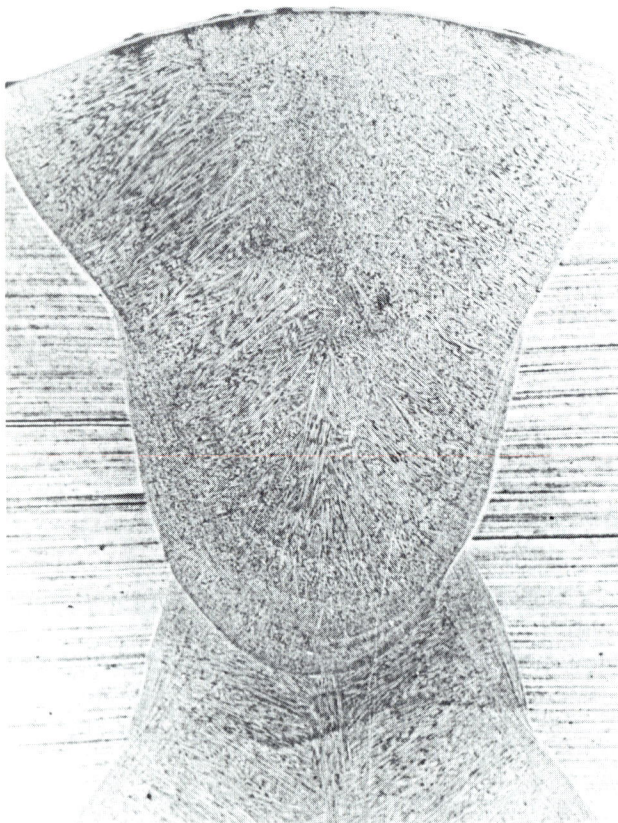

15 Primary structure in cross-section for a TiC inoculation of 1·9 g/dm in the weld groove with no cold material added   [×4]

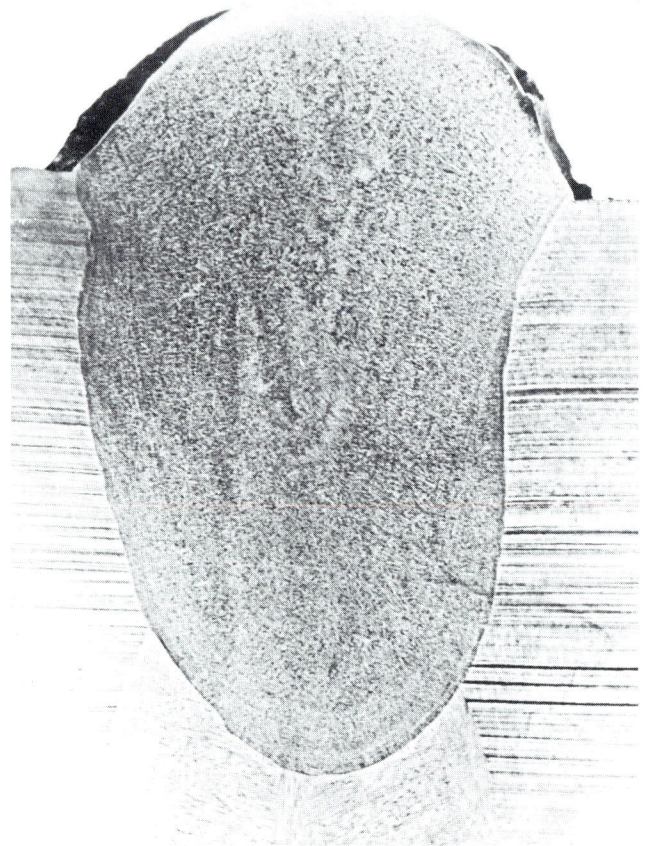

16 Primary structure in cross-section for a TiC inoculation of 1·9 g/dm via a steel tube, with a feed rate of 3 m/min for the tests at constant power input   [×4]

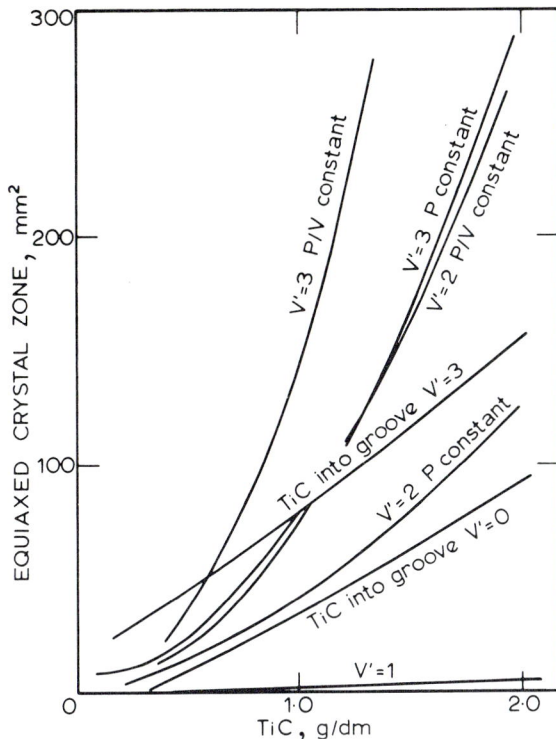

17 **Area of the equiaxed crystal zone in cross-section as a function of the amount of TiC added for the different series of test welds**

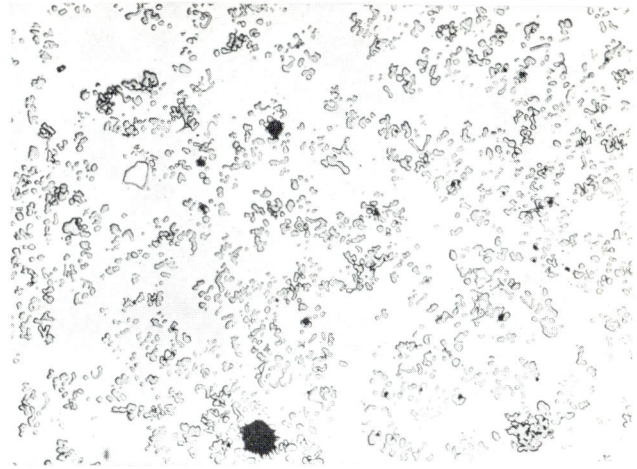

18 **Precipitated particles of titanium carbide in an iron matrix** [×120]

then precipitated at a temperature whose value can affect the stability of the precipitate. In this way good grain refinement of the primary structure might be possible even with simultaneous reduction of the amount of added grain refiner. Methods of adding these alloys to the weld pool have been studied and test welds were made where these alloys with precipitated particles were added (Fig. 18). The results indicate that the amount of added grain refiner can be reduced and still allow good grain refinement of the primary structure by a method whereby the alloy is part of cold welding wire fed into the pool behind the hot electrodes.

ACKNOWLEDGMENTS
This work has been carried out at the Swedish Institute for Metal Research in Stockholm, at the Royal Institute of Technology in Stockholm, at the Kockums laboratory in Malmö, and at the ESAB laboratory in Göteborg. The help of numerous colleagues at these places is gratefully acknowledged. The author would like to thank Dr L. Bäckerud, Swedish Institute for Metal Research, Stockholm, for valuable advice and discussions.

REFERENCES
1 T. WADA: *Trans. Natl. Res. Inst. Met. (Jpn.)*, 1966, **8**, 136
2 A. T. D'ANNESSA: *Weld. J.*, 1967, **46**, 491s
3 V. S. ARISTOV: *Autom. Weld. (U.S.S.R.)*, 1960, **9**, 43
4 H. BACH: *Oerlikon Schweissmitteilungen*, 1965, **54**, 16
5 P. COLVIN AND A. F. BUSH: *Weld. Met. Fabr.*, 1971, **39**, 40
6 J. WARING: *Aust. Weld. J.*, 1967, Nov., 15
7 W. F. SAVAGE et al.: *Weld. J.*, 1968, **47**, 420s
8 J. C. BORLAND: *Br. Weld. J.*, 1961, **8**, 526
9 W. F. SAVAGE AND A. H. ARONSSON: *Weld. J.*, 1966, **45**, 85s
10 W. F. SAVAGE et al.: *Weld. J.*, 1968, **47**, 522s
11 L. BÄCKERUD AND K. NILSSON: IM-870B, Swedish Institute for Metal Research, Stockholm, 1972
12 P. CHRISTENSEN et al.: *Br. Weld. J.*, 1965, **2**, 54
13 N. RYKALIN: *Soudage*, 1961, **15**, Jan./Feb., 5
14 T. EDVARDSSON AND K. OLSSON: IM-1096, Swedish Institute for Metal Research, Stockholm, 1975
15 L. BÄCKERUD AND T. EDVARDSSON: *Scand. J. Metall.*, 1975, **4**, 267
16 L. BÄCKERUD AND T. EDVARDSSON: *Scand. J. Metall.*, 1976, **5**, 92
17 E. L. FROST: U.S. Patent 2 511 976, 1950
18 K. OKADA et al.: H. W. Doc. No. X11-219-64, 1964
19 L. WOLFF: Swedish Patent, 12875/66, 1966
20 S. SOKAL: UK Patent Spec. 472 363, 1936
21 N. G. SMITH: Swedish Patent, 122012, 1944
22 W. M. CONN: German Patent, Auslegeschrift 1012712, 1957
23 L. BÄCKERUD AND T. EDVARDSSON: *Scand. J. Metall.*, 1976, **5**, 262
24 P. F. WIESER et al.: *J. Met.*, 1967, **19**, Jun., 44
25 B. L. BRAMFITT: *Metall. Trans.*, 1970, **1**, Jul., 1 987
26 J. CAMPBELL: 'The grain refinement of electroslag remelted iron alloys', Oct. 1973, Stoke Poges, Fulmer Research Institute
27 D. C. BROWN et al.: *Weld. J.*, 1962, **41**, Jun., 241
28 A. A. EROKHIN AND L. L. SILIN: *Weld. Prod. (U.S.S.R.)*, 1960, **5**, 8
29 V. A. LEBIGA: *Automatic Weld.*, 1960, **13**, 21
30 S. L. MANDELBERG: *Automatic Weld.*, 1965, **18**, 8
31 P. O. ANTONETS AND F. I. BUKIN: *Automatic Weld.*, 1964, **17**, 47
32 O. P. ANTONETS AND G. G. PSARAS: *Automatic Weld.*, 1966, **19**, 54
33 D. A. DUDKO AND I. N. RUBLEVSKII: *Automatic Weld.*, 1960, **13**, 14
34 I. P. TROCHUN AND V. P. CHERNYSH: *Weld. Prod. (U.S.S.R.)*, 1965, **12**, 6
35 M. N. GAPCHENKO: *Automatic Weld.*, 1970, **23**, 36

36 K. K. KHRENOV *et al.*: *Weld. Prod.*, 1959, **6**, 12

37 G. K. TURNBULL *et al.*: *Trans. AFS*, 1961, **69**, 792

38 J. G. CARLAND: M/70/72, The Welding Institute, May 1972

39 C. DA CASA AND V. B. NILESHWAR: *JISI*, 1969, **207**, 1 003

40 P. STULAR: *Schweissen Schneiden*, 1967, **6**, 270

41 F. BONOMO: BWRA, M28/68, Apr. 1968

42 L. BÄCKERUD AND T. EDVARDSSON: IM 1107, Swedish Institute for Metal Research Stockholm, Sweden, 1975

Non-epitaxial growth in weld metal

C. R. Loper, Jr. and J. T. Gregory

Autogenous, bead-on-plate GTA welds were made on 6 mm thick alloy 1100 aluminium plate and prepared for microstructural examination by sectioning on the weld longitudinal centreline plane normal to the plate surface. Welds at speeds of up to 850 mm/s were made on cold-pressed base metal. At high welding speeds substantial lengths of weld fusion boundary displayed non-epitaxial growth, i.e. crystallographic orientations of fusion zone columnar grains were distinctly independent of their partially melted base metal parent grains. Polarized light microscopy on anodized aluminium specimens revealed this loss of epitaxy phenomenon by brilliant colour differences between fusion zone and base metal grains. Investigation of this unusual growth phenomenon was conducted using X-ray diffraction techniques as well as polarized light microscopy. The probability of the existence of non-epitaxial growth was found to be a function of welding speed, base metal grain size, and weld pool size. A metallurgical grain boundary was found to exist wherever non-epitaxy occurred. This growth mode created a single crystal within the fusion zone with $\langle 100 \rangle$ oriented vertical as well as along the welding direction. A crystallographic growth mechanism is proposed.

C. R. Loper, Jr. is in the Department of Metallurgical and Mineral Engineering, University of Wisconsin, USA, and J. T. Gregory is in the Department of Chemical and Metallurgical Engineering, University of Auckland, New Zealand

A common starting point from which to consider weld metal solidification is the site of initial freezing at the fusion boundary, that is the growth of the cooling weld metal upon the partially melted base metal substrate. This stage of growth is characterized by perfect or near-perfect wetting, depending upon chemical and metallurgical aspects of filler and base metals. Investigators have found the term epitaxy[1] useful to convey the idea of: (a) perfect or near-perfect wetting of the base metal by the molten weld metal; (b) favourable substrate characteristics to preclude distinct nucleation; (c) fusion zone grains having the identical crystallographic orientation as the respective base metal grains upon which they initially grow; and (d) initial fusion zone grains having diameters identical to their parent base metal grain.

Metallurgical and structural discontinuities or other peculiarities may develop at the fusion boundary as a result of some event or feature which may affect ideal continuity, but the initial stage of weld metal growth is by essentially perfect epitaxial growth.

Since the most rapid growth direction for freezing weld metal grains is that opposite to the heat flow direction, initial growth at the fusion boundary occurs just after the molten weld pool moves away. At this instant the heat flow direction is normal to the fusion boundary and weld metal grains begin to grow inward towards the molten pool.

The relationship between columnar grains morphology in the fusion zone and weld pool shape has been reported previously.[2] Essentially, the longer the weld pool (for instance, as a result of holding the arc energy input constant while increasing the welding speed), the less the columnar grains bend towards the welding direction. Figure 1 shows the centreline section of a weld made on commercially pure aluminium at 760 mm/s. Note the straight, near-parallel, columnar grains indicative of an extremely long and gently sloping trailing weld pool interface.

Long, straight, columnar grains in ingot solidification are indicative of unidirectional solidification where growth is parallel to the maximum thermal gradient. Curved columnar grains are generally found in conventional fusion zones since the heat source is constantly moving at a speed sufficiently slow to allow heat to flow out not only to the left and right of the weld pool and possibly downward, but also to the rear. The columnar grains of Fig. 1 are indicative of extremely high welding speeds wherein the heat flow is essentially only in those directions normal to the welding direction; the rate of heat conduction opposite to the welding direction is relatively negligible, thus columnar grains do not bend.

A characteristic which makes polarized light illumination on anodized aluminium a valuable metallographic technique is evident in Fig. 1. Here the micrographs are of an identical area but the specimen has been

1 A single area of non-epitaxial growth; photomicrograph *a* differs from *b* by about 15° rotation; welded left to right at 760 mm/s [×40]

rotated about the optical axis by about 15°. The significant feature of this pair of micrographs is that what appears in Fig. 1*a* to be separate and distinct columnar grains appears in Fig. 1*b* as essentially one long single crystal. This appearance is typical of extensive non-epitaxial growth found in welds made at extremely high welding speeds.

For anodized aluminium viewed under polarized light illumination, a similarity in colour between two neighbouring grains provides evidence that those two grains have similar crystallographic orientation.[3] Since essentially the entire fusion zone of Fig. 1 is identically coloured at some particular rotational position, it may be concluded that the entire area has some strong preferred crystallographic orientation. Initial work by Gregory[4] has suggested that for a similar-appearing fusion zone area the ⟨100⟩ is aligned vertically and that the ⟨110⟩ or ⟨100⟩ is parallel to the welding direction. This would account for the similarity of colour intensity under polarized light illumination. Sufficiently rapid growth of favourably oriented crystals having strong growth components oriented in the welding direction could consume the entire fusion zone and inhibit conventional expitaxial growth.

EXPERIMENTAL PROCEDURE

To determine the frequency and extent of non-epitaxial growth regions at the fusion boundary as a function of welding conditions, numerous specimens were produced using welding speeds varying from 210 to 850 mm/s in steps of 42 mm/s. Specimens were cold compressed to 6 mm from various initial annealed thicknesses before being welded to produce controlled heat-affected zone grain sizes. Some specimens were cold compressed at liquid nitrogen temperatures to produce a fine grain size heat-affected zone. Welding parameters of travel speed and current were varied to produce weld beads having nearly identical widths. All welds were melt runs using 3 mm thoriated tungsten electrode, argon shielding, DCSP.

Specimens were sectioned on the longitudinal weld centreline plane for microstructural examination. Polarized light microscopy was used on specimens electropolished and anodized using electrochemical techniques described elsewhere.[3]

X-ray diffraction studies were conducted using the Laue back reflection technique. Specimens were slowly moved in a plane normal to the X-ray beam to produce an integrated structure pattern for non-epitaxial fusion zone regions. Crystallographic orientation analysis using stereographic projection techniques was used to construct a standard triangle representation taking into consideration the welding direction, the specimen vertical axis, and the X-ray beam axis, as described by the authors previously.[3]

RESULTS AND DISCUSSION

By the use of polarized light illumination on anodized aluminium, non-epitaxial growth at the fusion boundary is revealed clearly and distinctly by the brilliant colour differences existing between what would normally be identically coloured, hence identically oriented, epitaxial parent base metal grain/columnar grain pairs. Figure 2 shows a region of a fusion zone which displays significant non-epitaxial growth surrounding small epitaxial columnar grains. In this case it is argued that although the epitaxial columnar grains could grow to some extent, they were quickly eliminated by competitive growth mechanisms of the more favourably oriented and massive non-epitaxial grain.

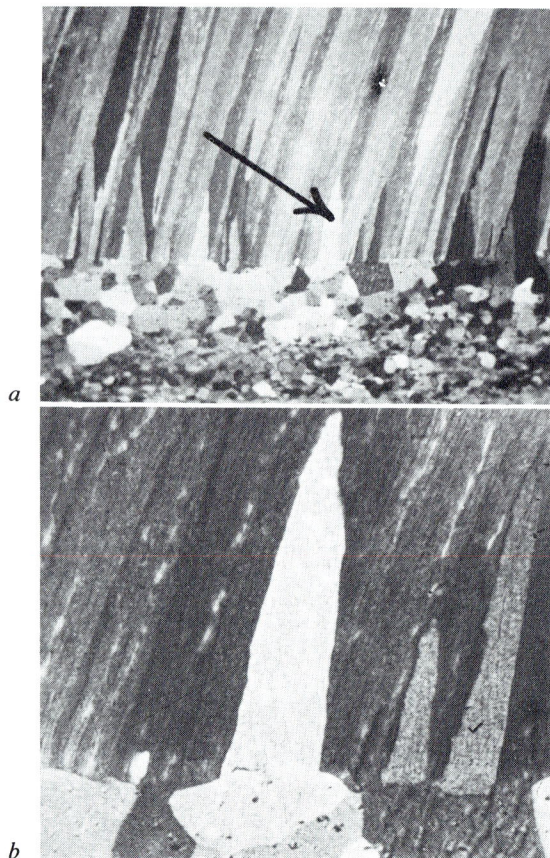

2 Unfavourably oriented epitaxial grains surrounded and eliminated by the massive non-epitaxial zone grain; photomicrograph *b* [×500] is an enlargement of central portion of *a* [×100]; welded left to right at 850 mm/s

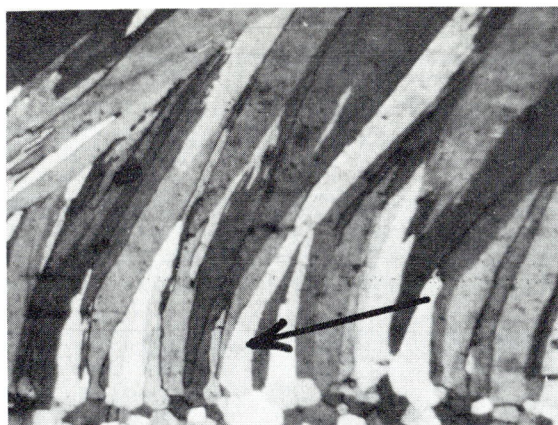

3 Successful growth of a favourably oriented columnar grain; welded left to right at 8 mm/s [×60]

Figure 2*b* illustrates this growth morphology for one particular region. Here it is seen that the short, centrally located, columnar grain grows from only a small portion of a larger heat-affected zone parent grain. Non-epitaxial growth has almost completely inhibited this partially melted grain from growing; non-epitaxial growth virtually surrounds the short columnar grain.

After initial solidification takes place at the fusion boundary, individual columnar grains grow in competition with their neighbours. Bray *et al.*[2] have shown that the direction of fusion zone columnar grain boundaries is influenced by the respective crystallographic orientation of the columnar grains on either side of the boundary. Thus, the growth characteristics of one columnar grain can markedly affect the growth morphology of its neighbours resulting in competitive growth.

This growth mechanism is seen to have occurred in Fig. 3 which shows the longitudinal centreline section of a weld made at 8 mm/s. Note the single, long, thin, columnar grain growing up through the fusion zone after initial epitaxial growth upon the partially melted base metal grain.

Although the fusion zone of Fig. 2*a* appears to be composed of numerous, straight-sided, long columnar grains, the entire fusion zone apart from the few short epitaxial grains mentioned above is essentially identical in kind to the fusion zone of Fig. 1.

Conditions favouring non-epitaxial growth
The occurrence of non-epitaxial growth at the fusion boundary was found to be a function of the welding speed, the base metal (heat-affected zone) grain size, and the size of the weld bead. Table 1 summarizes the relationships between these three factors.

For specific welding conditions and base metal grain size, results were not always reproducible. Slight variations towards larger grain size and larger weld bead dimensions significantly reduced the occurrences of non-epitaxial growth. Even seemingly minor differences in the shape of the electrode or its distance from the specimen markedly influenced the shape and size of the weld bead. Figure 2 illustrates what happens when the heat-affected zone grain size is somewhat larger than that favouring complete non-epitaxial growth: growth of

Table 1 Fraction of fusion boundary displaying non-epitaxial growth, %

Grain size	Narrow bead width (0·8 mm) Welding speed, mm/s Low (210)	High (850)	Wide bead width (1·4 mm) Welding speed, mm/s Low (210)	High (850)
Large (0·15 mm)	0	<10	0	<10
Small (0·012 mm)	<50	>90	<25	<75

less-favourably oriented partially melted grains begins to become substantial.

Characteristics of non-epitaxial growth
The following metallographic features are characteristic of non-epitaxial fusion boundaries:

(i) a metallurgical grain boundary separates the partially melted heat-affected zone grains from adjacent fusion zone columnar grains

(ii) although sufficient solute is present to form a marked cellular substructure within columnar grains of the fusion zone in autogenous welds, no solute rejection or other precipitates were found in the non-epitaxial grain boundary of (i) above

(iii) frequently, growth on partially melted heat-affected zone grains occurs to only a few microns into the fusion zone

(iv) in all cases the classical transition of growth modes from planar to open cells to well developed cellular substructure of columnar grains as described by Maltesta and Buffum[5] occurs within columnar grains

(v) fusion zones are essentially single crystals having low-angle subgrain boundaries in place of what would normally be considered columnar grain boundaries.

Metallographic evidence that the non-epitaxial fusion zone is in fact single crystal in nature is shown in Fig. 4. Here it is seen that although Fig. 4*a* resembles a conventional columnar growth morphology, rotation of the specimen under polarized light illumination reveals the uniformity of structure throughout the entire fusion zone. Figures 4*b*–*d* are the same fusion zone as Fig. 4*a*, differing only by approximately 15° rotation about the optical axis from one position to the next. The optical anisotropy of the anodized aluminium oxide layer in Fig. 4*a* is caused by the slight misorientation of columnar ·subgrains producing the maximum colour differences and thus somewhat resembling conventional columnar growth.

Crystallography of non-epitaxial growth
Rapid solidification which occurs at high welding speeds dictates that columnar grains which are poorly oriented for rapid growth be crowded out by their more favourably oriented neighbours. Such favourably oriented grains have stronger lateral components of growth which allow them to widen with distance from the fusion boundary at the expense of their poorly oriented neighbours.

An extreme manifestation of this basic rule of weld metal solidification mechanics is found in non-epitaxial

4   A single area of non-epitaxial growth; a–d differ by about 15° rotation from one photomicrograph to the next; welded left to right at 850 mm/s   [×40]

columnar growth microstructures. In Fig. 2 it is seen that little or no growth of unfavourably oriented columnar grains occurred. Those few epitaxial grains that did originate at the fusion boundary were quickly crowded out by the one single massive fusion zone grain.

X-ray diffraction studies using the Laue back reflection technique on several non-epitaxial fusion zones, similar to those shown in Figs 1 and 4, were conducted to determine the crystallographic orientation of the massive single crystal fusion zone. Stereographic projection representation of the X-ray data revealed that the direction of growth normal to the fusion boundary was in all cases ⟨100⟩.

Further analysis showed that the direction of welding corresponded within a few degrees of rotation to the ⟨100⟩. Exact values could not be obtained due to the inherent error in estimating the centres of arc segments of Debye rings on the Laue film indicative of slight crystallographic orientation variation within the fusion zone. Additional estimated errors resulted from:

sectioning the weld exactly normal to the specimen plate surface and exactly on the weld centreline; mounting the specimen on the X-ray unit; and usual measurement and plotting errors.

The non-epitaxial growth mechanism may now be explained with the aid of the schematic diagram of Fig. 5. In Fig. 5 the small arrows represent the most favourable growth direction for aluminium, i.e. ⟨100⟩. As the weld pool moves at high speed, the rear solid/liquid interface slopes gently due to the long trailing molten pool. Grains A, B, and C are sufficiently favourably oriented to grow upwards into the fusion zone after conventional epitaxial growth at the fusion boundary.

However, grain D happens by chance to be favourably oriented not only for growth normal to the fusion boundary but also parallel to it. Grain E is unfavourably oriented for normal growth. As the weld pool moves past the grain boundary between partially melted grains D and E, the strong ⟨100⟩ growth direction of grain D parallel to the fusion boundary allows grain D to grow forward faster than grain E can grow upward. The net result is that grain E is completely inhibited as grain D proceeds to widen and inhibit growth of grains further down the line.

Some relatively minor interaction occurs between advancing grain D and other suppressed, partially melted, heat-affected zone grains to cause a slight deviation in the orientation of grain D, thus creating subgrain boundaries in the single crystal grain D.

There are three-dimensional aspects also to consider when dealing with non-epitaxial growth but they are beyond the scope of the present paper. However, initial stages of work done on transverse sections of non-epitaxial growth fusion zones seem to indicate that this growth morphology can readily extend throughout the entire fusion zone. Considering the orientation of the two ⟨100⟩ growth directions, it is obvious that the third ⟨100⟩ must also be in the most favourable orientation for lateral growth from both sides of the weld pool.

CONCLUSIONS

Metallographic studies revealed that three factors favour non-epitaxial growth at the fusion boundary of welds made on commercially pure aluminium: a fine heat-affected zone grain size; a high welding speed; and a small weld bead size.

Non-epitaxial growth creates a metallurgical grain boundary at the fusion boundary which contains no rejected solute or other precipitates.

Fusion zones are frequently essentially single crystals composed of numerous parallel subgrain boundaries having ⟨100⟩ directions parallel to the vertical and welding directions.

Conventional epitaxial columnar grains having ⟨100⟩ orientations vertical as well as parallel to the welding direction can grow forward sufficiently fast to

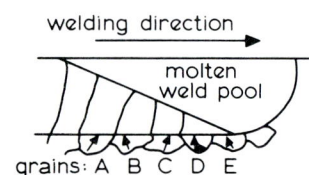

5   Schematic diagram of non-epitaxial growth mechanism

completely suppress upward growth in their less favourably oriented neighbours. This mechanism is favoured by the long, gently sloping, rear position of the rapidly moving weld pool.

REFERENCES

1 G. J. DAVIES AND J. G. GARLAND: *Int. Met. Rev.*, 1975, **20**, 83

2 R. BRAY *et al.*: *Weld. J.*, 1969, **48**, May, Res. Suppl., 181s

3 C. R. LOPER, JR AND J. T. GREGORY: *Weld. J.*, 1974, **53**, Mar., Res. Suppl., 126s

4 J. T. GREGORY: *Aust. Weld. J.*, 1974, **18**, (6), 19

5 W. C. MALTESTA AND D. C. BUFFUM: *Weld. J.*, 1967, **46**, Feb., Res. Suppl., 58s

Oriented structures and properties in type 316 stainless-steel weld metal

B. L. Baikie and D. Yapp

The development of oriented structures has been investigated for austenitic multipass manual metal arc welds. It has been shown that highly aligned grain structures are produced in welds built up from planar layers of weld beads. The welds have a strong [100] fibre texture with the [100] fibre axis parallel to the mean grain orientation. Epitaxial growth occurs from weld bead to weld bead in spite of frequent small changes in fusion boundary orientation. However, if the changes in fusion boundary orientation between beads exceed about 30° new grains may appear, and for still larger changes it is possible for epitaxial growth to take place from the dendrite secondary arms in the previous bead. Interpretation of weld microstructures is complicated by solid-state transformations from δ-ferrite to austenite on cooling. However, recrystallization does not take place in the weld metal, and the weld grain structure is equivalent to that produced on solidification. Simple models of heat flow allow prediction of grain orientation, and techniques have been developed to produce welds with a specified grain orientation. Oriented structures in austenitic welds have been shown to have a significant influence on properties. For instance, there is a strong effect of grain orientation on ultrasonic attenuation, with low attenuation for compression waves travelling at 45° to the grain orientation. Use has been made of this observation together with the grain orientation control technique to enable the ultrasonic inspection of a heavy section weld in a nuclear power plant.

B. L. Baikie is with the CEGB, North West Region Scientific Services Division, Manchester, and D. Yapp is with the CEGB, Research Division, Marchwood Engineering Laboratories, Southampton

Type 316 stainless steel[1] is widely used in power plant because of its good corrosion resistance and high-temperature strength. It is normally fully austenitic at room temperature, but the composition of type 316 weld metal is adjusted to contain about 5% δ-ferrite in order to minimize the risk of hot cracking. Welds in power plant components made using the manual metal arc process may be from 10 to 75 mm thick, and the cross-section of the weld may contain over 100 weld beads. Figure 1 shows a typical weld in 25 mm thick plate.

OBSERVATION AND CONTROL OF ORIENTED STRUCTURES IN TYPE 316 WELDS
Weld manufacture
Initial investigations were made on 75 mm cubic blocks of weld metal made by depositing manual metal arc weld beads onto a stainless steel base plate. A variety of different welding techniques were used, including the 'flat', 'horizontal–vertical', and 'vertical' welding positions. Block *A* (horizontal–vertical weld) was made by depositing horizontal beads of weld metal one above

another to build up planar layers on a vertical base plate, welding always being carried out in the same direction and with the same bead sequence (Fig. 6). The manual metal arc electrode used to make the welds had a lime–titania based coating, and deposited weld metal of approximately matching composition to the parent material. A typical analysis (wt-%) is as follows:

C	Cr	Ni	Mo	Mn	Si	S	P
0·03	17·6	11·1	2·46	0·77	0·27	0·011	0·013

Metallography
Several etchants were used to reveal different aspects of the structure. The δ-ferrite was revealed by 10% oxalic acid used electrolytically at 4 V. Etchant I, a concentrated version of acid ferric chloride, was used to show the grain structure. Etchant II was specially developed to reveal the solidification structure, and was particularly useful since it showed the outlines of the weld beads as well as the instantaneous fusion boundaries. (For details *see* Table 1.)

1 Grain structure of a butt weld (etchant I) [×2]

Table 1 Macroetchants for stainless steel

Etchant I		Etchant II	
340 g	FeCl$_3$.6 H$_2$O	132 g	FeCl$_3$.6 H$_2$O
126 ml	HCl	80 ml	HCl
42 ml	HNO$_3$	20 ml	HNO$_3$
252 ml	H$_2$O	59 ml	H$_2$O
		200 ml	C$_2$H$_5$.OH
		40 ml	CH$_3$.COOH

X-ray metallography was carried out using a Siemens X-ray diffractometer fitted with a Schulz reflection stage one and a Cu Kα source. Pole figures for the selected (hkl) reflection were plotted manually on to a Wulff net.

Macroscopic grain orientation

Figure 2 shows the grain structure of block A (horizontal–vertical weld) sectioned perpendicular to the welding direction. Long columnar grains are present in a highly aligned grain structure, with deviations of only a few degrees from the mean grain orientation. Figure 3 shows the bead structure of the same section, with the curved fusion boundaries typical of the manual metal arc process. It is apparent from Figs 2 and 3 that the grains are almost perpendicular to the fusion boundaries near the centre of the weld beads. However, the normal to the fusion boundary may deviate from the grain growth direction near the edge of the weld beads by up to 30°, the grains remaining essentially parallel. Epitaxial growth from bead to bead is the dominant mechanism in block A, but there are some regions where new grains have been nucleated and epitaxial growth has been prevented.

In order to determine the orientation of the grains in a weld in three dimensions, it is necessary to make a series of sequential sections from initially perpendicular planes and subsequently parallel to the grain directions on

2 Grain structure of block A (etchant I) [×5]

those planes. In the case of block A, this reveals that the grains have an orientation of 30° relative to the base plate normal in a plane perpendicular to the welding direction and 5° relative to the base plate normal in a plane parallel to the welding direction (*see* Fig. 8).

Microstructure

Examination of the microstructure of block A and other welds produced the following observations:

(i) dendrite orientation is usually parallel to grain orientation; however, in some cases the dendrite direction may deviate from the grain orientation by up to 30° (*see* Fig. 4)

3 Bead structure of block A (etchant II) [×5]

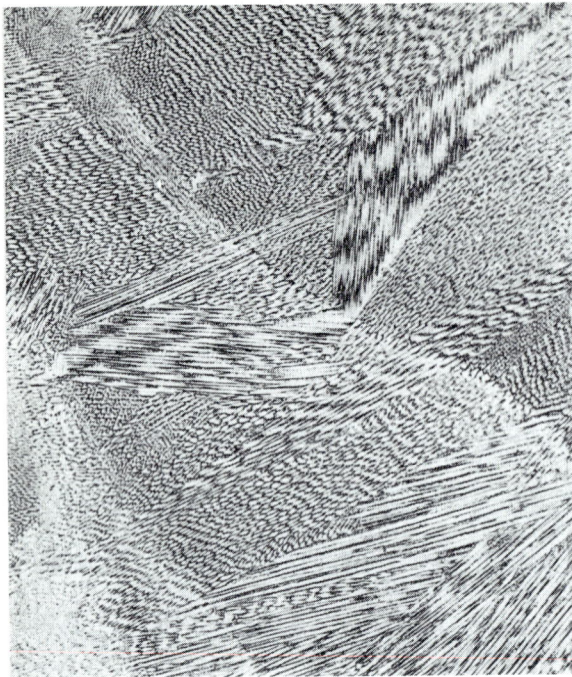

4   Microstructure of type 316 weld metal (etchant: 10% oxalic acid at 4 V) [×50]

(ii) there is usually continuity of dendrite structures across bead boundaries, although there is often a change in dendrite spacing due to changes in solidification rate

(iii) new grains appear to be nucleated and grow to prevent epitaxial growth in some areas, especially where there is a large deviation of the normal to the fusion boundary from the dendrite growth direction

(iv) occasional examples of epitaxial growth from dendrite secondary arms in the previous bead are observed.

Some of these effects can be seen in Fig. 4.

The microstructure of all the welds was examined carefully for correspondence between the grain structure indicated by the dendrites and the grains revealed by grain contrast etching. In all cases where the grains were clearly defined, there was exact correspondence between the structures.

Crystallographic orientation

Figure 5 shows that block *A* has a strong fibre texture, with the [100] fibre axis parallel to the average grain orientation. Examination of a wide range of type 316 welds has shown that this characteristic fibre texture is present in each case, and that the [100] fibre axis is parallel to the mean grain orientation determined by macroetching. The fibre texture was generally well developed, with a spread of about ±10° about the fibre axis.

Prediction and control of grain orientation

Examination of pole figures and macrostructures from a wide range of welds revealed that each welding technique has associated with it a characteristic grain orientation which is only slightly affected by welding variables such as electrode size, welding current, and speed. From the observation that grains appear to grow perpendicular to the 'average' fusion boundary orientation, it proved possible to construct a simple model of grain orientation based on heat flow arguments. Figure 6 illustrates the model for a horizontal–vertical weld. In this case, part of the energy, H_y, flows into the previous weld bead, and the remainder, H_x, flows into the previous layer of weld beads, with a resultant heat flow vector **H**. The orientation of the mean heat flow vector **H**, and hence of the average fusion boundary orientation and grain orientation, then depends on the ratio $H_y : H_x$. The effect of gravity in the horizontal–vertical welding position produces relatively thick layers of weld beads, and the mean heat flow vector makes an angle of 30° to the base plate normal (Fig. 6). However, for the flat

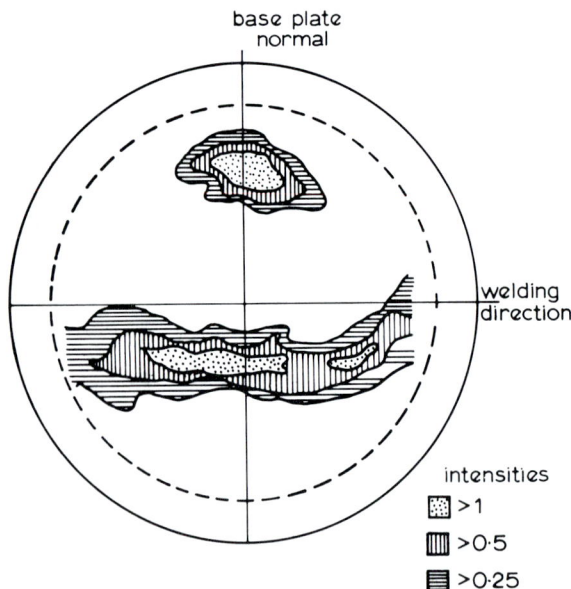

5   (100) pole figure for block *A* (horizontal–vertical weld)

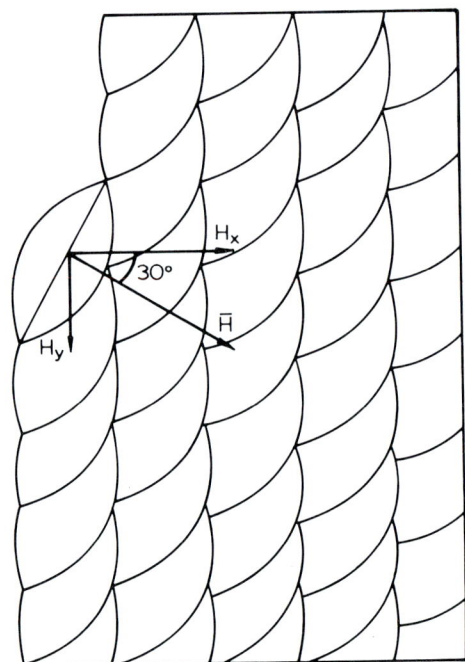

6   Schematic diagram of heat flow for a horizontal–vertical weld

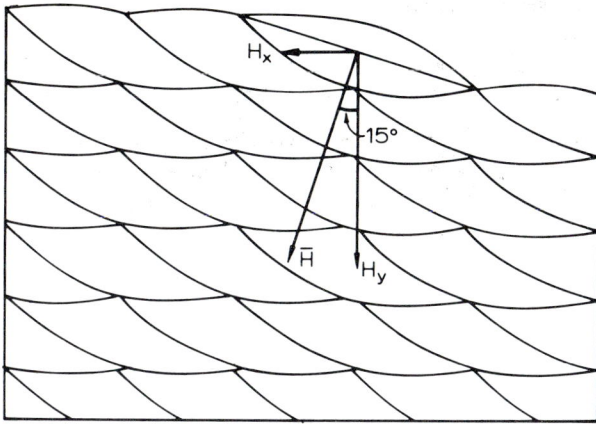

7 Schematic diagram of heat flow for a weld in the 'flat' position

9 Grain structure of controlled orientation weld (etchant I) [×0·7]

welding position (Fig. 7) thinner layers of weld beads are produced, and the mean heat flow vector makes an angle of about 15° relative to the base plate normal.

The information collected for a variety of welding techniques has been generalized using the stereographic projection. Figure 8 shows the results for three common welding positions. Using this scheme it is possible to predict grain orientations for welds built up from planar layers of weld beads. It is also possible to make welds with a specified grain orientation, subject to the limits imposed by the requirement to maintain good welding practice. Figure 9 shows a weld between two concentric pipes in which the weld beads were deposited in a controlled sequence in order to produce a grain orientation at 45° to the free surface of the weld. The desirability of producing welds with highly aligned grain structures depends on the effect of grain orientation on properties. As will be shown later, these effects can be large, and control over grain orientation can sometimes be used to practical benefit.

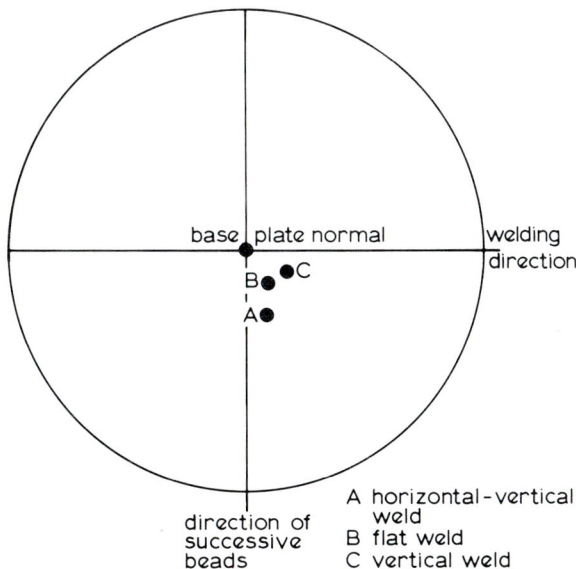

GROWTH OF ORIENTED STRUCTURES IN TYPE 316 WELD METAL
Solidification sequence and δ-ferrite transformation

The interpretation of solidification structures in type 316 welds is complicated by the presence of solid-state transformations from δ-ferrite to austenite on cooling. The possibility of these high-temperature transformations has been known for some time,[2] but detailed investigations have only been made recently.[3-7] The Fe–Cr–Ni ternary diagram can be used to indicate the course of solidification and transformation, although it does not of course relate exactly to commercial steels. Depending on composition, the following sequences are possible:

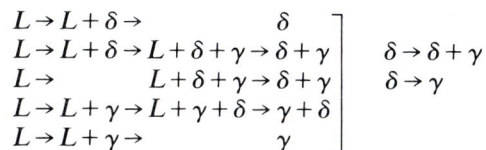

$$\left.\begin{array}{l} L \rightarrow L+\delta \rightarrow \qquad\qquad \delta \\ L \rightarrow L+\delta \rightarrow L+\delta+\gamma \rightarrow \delta+\gamma \\ L \rightarrow \qquad\quad L+\delta+\gamma \rightarrow \delta+\gamma \\ L \rightarrow L+\gamma \rightarrow L+\gamma+\delta \rightarrow \gamma+\delta \\ L \rightarrow L+\gamma \rightarrow \qquad\qquad \gamma \end{array}\right] \begin{array}{l} \delta \rightarrow \delta+\gamma \\ \\ \delta \rightarrow \gamma \end{array}$$

Blanc and Tricot,[6] and Fredriksson,[3] have explained some of the complex microstructures which may result in commercial steels. If primary δ-ferrite dendrites are precipitated, they may transform wholly or partially to austenite on cooling, depending on composition and cooling rate. Similarly, if δ-ferrite is precipitated from the last liquid to solidify, the interdendritic δ-ferrite may transform to austenite on cooling. Furthermore, the original solidification grain structure may be replaced by long columnar grains as a result of the transformation. These so-called 'pseudo-columnar' grains tend to grow perpendicularly to the isotherms in ingots, and hence they often follow a curved path.

Calculation of the Cr and Ni equivalents[6] for the type 316 weld metal used in this investigation suggests the following solidification and transformation sequence:

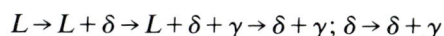

$$L \rightarrow L+\delta \rightarrow L+\delta+\gamma \rightarrow \delta+\gamma; \delta \rightarrow \delta+\gamma$$

The microstructure of the weld metal (Fig. 10) is consistent with this sequence. However, as indicated earlier,

8 Stereographic projection for the generalized scheme of grain orientations

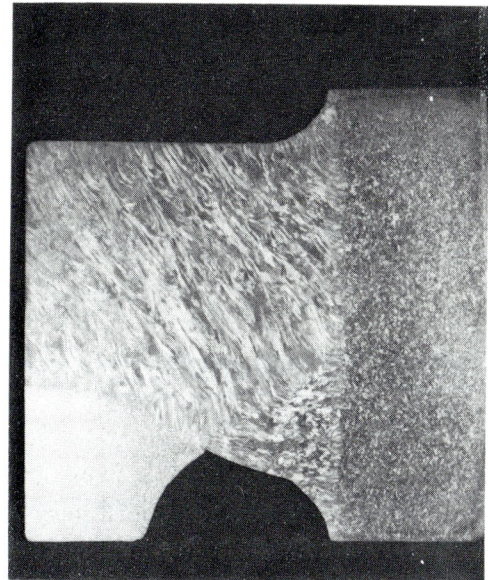

base plate normal — welding direction

B● ●C
A●

direction of successive beads

A horizontal-vertical weld
B flat weld
C vertical weld

10   δ-Ferrite distribution in type 316 weld metal   [×400]

11   Grain structure of horizontal–vertical weld reversed between layers (etchant I)   [×3]

exact correspondence was observed between the dendritic structure and the grain structure. Thus transformation did not destroy the original grain structure, even though there is little doubt that part of the δ-ferrite present after solidification transformed to austenite on cooling. This is probably due to a combination of factors: the transformation may take place at a relatively high temperature with a steep thermal gradient and rapid cooling, and austenite is probably precipitated with δ-ferrite on solidification, so that the nucleation of new austenite grains is not required.

Growth mechanisms

The microstructure of type 316 weld metal is dendritic, with at least secondary dendrite arms, but the dendritic structure is not highly developed because of the high temperature gradient near the solid/liquid boundary. The welds have a strong [100] fibre texture, produced by the preferred [100] growth direction of the dendrites. Several other effects have already been noted, including competitive growth to eliminate unfavourably oriented grains, limited curvature of grains to follow changes in direction of the heat flow vector, apparent nucleation of new grains for large changes in direction of the heat flow vector, and, under special conditions, growth of grains from side arms in an underlying weld bead. Similar effects have been reported for single-pass welds in thin sheet by several authors,[8-11] but understanding of the basic mechanisms is still limited. For instance, there is at present no convincing model for the growth of dendrites along the [100] axis.

The tendency of the curvature of grain boundaries in type 316 weld metal to follow changes in the direction of the heat flow vector is less than that found for some other systems.[9,10] This may possibly be due to the dendrite morphology—if a plate-like structure is developed, the repeated side-branching necessary to produce curvature may be difficult.

The criteria which determine whether, and at what stage, new grains may be nucleated are unclear. However, the mechanism probably involves growth of random dendrite fragments which circulate in the relatively turbulent weld pool and which can grow in the increased undercooling in front of a slow growing grain.

In order to illustrate the mechanism of dendrite secondary arms in a previous bead forming the primary arm of a succeeding bead, a weld block was produced by the horizontal–vertical technique with a rotation of 180° between layers. From Fig. 6, it can be shown that there is a change of 60° in the direction of heat flow vector between layers and hence some dendrite secondary arms from the first layer will be closer to the heat flow vector in the second layer than the primary arms. Figure 11 shows that the grain growth direction switches through about 90° at each layer boundary which is consistent with a switch from primary to secondary arms. For comparison, a weld block was made in the 'flat' position with 180° reversals between each layer. In this case the +15° to −15° change at the layer boundary is insufficient to cause switching from secondary to primary dendrite arms, and the grains grow, on average, perpendicular to the base plate, i.e. in the direction of the heat flow vector averaged over several layers (*see* Fig. 12).

EFFECT OF ORIENTED STRUCTURES ON PROPERTIES

It has been shown that orientation is present on several levels in type 316 metal: in the dendrite structure, the grain structure, and the crystallographic structure. Each of these structures may exert important effects on the properties of type 316 welds. For instance, crystallography will affect elastic properties, grain structure affects creep properties, and dendrite structure affects creep crack growth.[12]

12 Grain structure of 'flat' position weld reversed between layers (etchant I) [×3]

A brief summary of the effects of orientation on properties is given in Table 2. Clearly some of the effects are of practical importance. For example, there is an increase of 180% in minimum creep rate for a specimen with grains perpendicular to the axis compared with a specimen with grains parallel to its axis. A detailed investigation is now in progress to determine the importance of anisotropic properties on weld performance.

The effect of grain orientation on ultrasonic properties of austentic welds has been put to practical use.[13]

Ultrasonic attenuation is strongly affected by grain orientation, with a minimum in attenuation for a compression wave travelling at 45° to the grain orientation. For pulse-echo measurements on a 50 mm thick weld the material attenuation can be reduced to a total of 5 dB for the best orientation compared with 40 dB for the worst case. The grain orientation techniques discussed earlier can be used to produce welds with grains at 45° to their free surfaces in order to reduce the attenuation of compression waves launched from the surface. Figure 9 shows an example in which the grain orientation technique was used to enable the inspection of a weld between two concentric pipes in a nuclear power plant.

ACKNOWLEDGMENTS
This paper is published with the permission of the Central Electricity Generating Board.

REFERENCES
1 British Standard 1501; Part 3: 1973. Grade 316S12
2 A. HULTGREN: *JISI*, 1929, **120**, 69
3 H. FREDRIKSSON: *Metall. Trans.*, 1972, **3**, 2 989
4 R. CASTRO: *Mém. Sci. Rev. Metall.*, 1961, **58**, 881
5 J. BEECH: 'Grain structures in austenitic steels' in Seminar on Grain Control in Cast Metals, Institution of Metallurgists, Loughborough, 1973
6 G. BLANC AND R. TRICOT: *Mém. Sci. Rev. Metall.*, 1971, **68**, 735
7 H. HOFFMEISTER: *Schweissen Schneiden*, 1973, **25**, 164
8 C. R. LOPER et al.: *Weld. J.*, 1969, **48**, 171s
9 W. F. SAVAGE et al.: *Weld. J.*, 1968, **47**, 522s
10 R. S. BRAY et al.: *Weld. J.*, 1969, **48**, 181s
11 Y. ARATA et al.: *Trans. Jpn. Weld. Res. Inst.*, 1974, **3**, 89
12 J. HAIGH: private communication
13 B. L. BAIKIE et al.: *J. Br. Nucl. Energy Soc.*, 1976, **15**, 257

Table 2 Effect of grain orientation on weld properties

Orientation	0·2% Proof stress at 20°C, N/mm²	UTS at 20°C, N/mm²	0·2% Proof stress at 488°C, N/mm²	UTS at 488°C, N/mm²	Minimum creep rate at 538°C and 250 N/mm², %/h	Longitudinal wave velocity, km/s	Ultrasonic attenuation, dB/mm
Parallel to grains	—	—	—	—	7×10^{-5}	5·45	0·40
45° to grains	488	761	340	436	—	6·0	0·05
Perpendicular to grains	491	685	406	492	2×10^{-4}	5·25	0·40

Structure–property correlation and fracture studies on directionally solidified Fe–C alloys

K. V. Prabhakar, H. Nieswaag, and A. J. Zuithoff

Unidirectional solidification experiments were conducted on pure Fe–C eutectic alloys with and without silicon (up to 3%), phosphorus (up to 0·11%), and sulphur (up to 0·06%) using the Bridgman technique at a constant temperature gradient (33°C/cm) and different growth rates (1·8 mm/h to 360 mm/h). Increasing solidification rates brought about a gradual increase in graphite fineness with a corresponding improvement in tensile strength. The fine flake graphite irons with a ferritic matrix exhibited strengths comparable to those obtainable in ferritic nodular irons, while even a small amount of coarse graphite reduced the strength drastically. Alloying with silicon increased the strength further by solution hardening of the matrix, while phosphorus and sulphur coarsened the graphite structure, particularly at the eutectic cell boundaries, resulting in a decrease in strength. Sulphur caused excessive coarsening of graphite at cell boundaries with the result that an increase in solidification rate produced virtually no improvement in strength. Scanning electron microscope examination of fracture surfaces showed that fine graphite tends to decohere from the matrix rather than cleave along its own planes while coarse graphite prefers to cleave. The behaviour of stress–strain curves of irons with different graphite fineness indicates that fine graphite promotes strain hardening of the matrix, thus contributing to the strength. Coarse graphite irons show plastic deformation (creep) at constant stress soon after yielding.

The authors are in the Department of Metals Science and Technology, Delft University of Technology, Delft, The Netherlands

The shape and dimensions of graphite flakes strongly influence the mechanical properties of lamellar grey cast irons. The graphite skeleton formed during solidification is influenced by many factors, among them the most important being the solidification rate and the impurities in the alloy. Hence it is possible to obtain a wide range of mechanical properties within even a limited range of compositions. In general, finer graphite flakes produce relatively higher tensile strength. Faster cooling rates and low impurity levels help to produce fine graphite flakes. Sulphur is an important and commonly found impurity. If the sulphur content is kept below 0·001%, the mechanical strength of eutectic cast iron with 2% Si and a ferritic matrix can reach as high as 350 N/mm^2 and an elongation of more than 2% at fast solidification rates.[1] In the present investigation, the unidirectional solidification technique was used to produce a variety of graphite structures in eutectic cast irons, with and without varying amounts of sulphur and phosphorus. An attempt was made to understand the underlying influence of graphite structure on mechanical properties with the aid of normal tensile testing, intermittent tensile testing, and fracture studies using a scanning electron microscope.

MORPHOLOGY OF THE SOLIDIFYING AUSTENITE–GRAPHITE EUTECTIC

Nieswaag and Zuithoff,[2,3] and others[4,5] have shown that with an increase in the growth rate, R (solidification rate), the morphology of the solid/liquid interface in synthetic cast irons of eutectic composition undergoes the following transitions: planar → cellular → endogenous. These transitions take place at lower growth rates if impurities such as phosphorus or sulphur are present.

Figure 1 illustrates the morphology as a function of R and phosphorus content[2,3] when the temperature gradient G ahead of the interface was maintained constant (33°C/cm). Besides the influence on morphology, an increase in R will gradually diminish the size of graphite flakes since $\lambda^2 R = $ constant, where λ is the mean distance between flakes. At a distinct R, however, a sudden refinement of the graphite occurs resulting perhaps from a change in growth mechanism. This sudden transition from coarse to fine graphite is

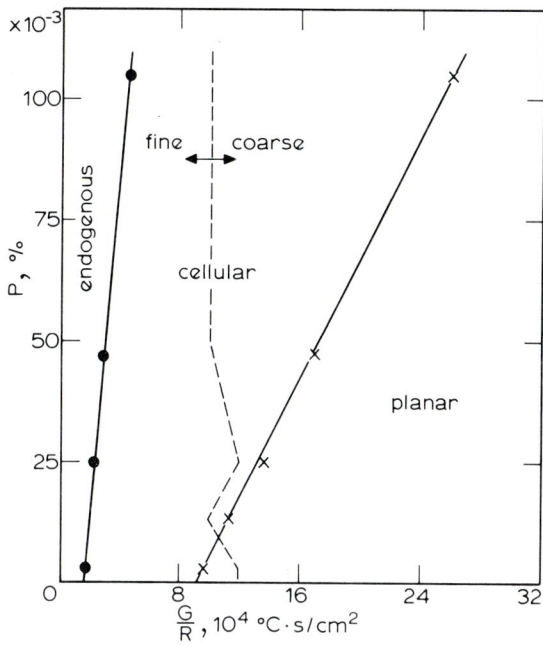

1 Influence of phosphorus on the solidification interface morphology of eutectic grey cast iron[3]

indicated in Fig. 1 by the dashed line. In the phosphorus-containing alloys the transition is independent of the phosphorus content.

The influence of sulphur on the morphology of the solid/liquid interface and on the dimensions of the graphite flakes is rather complex. In particular, its strong influence on the coarse to fine transition, Fig. 2, may be noted. Initially, an increase in sulphur content causes the transition to occur at lower R, but afterwards however at higher R. Above a sulphur level of about 0·04%, the transition does not occur at all, with relatively coarse graphite being present even at high R. If a planar

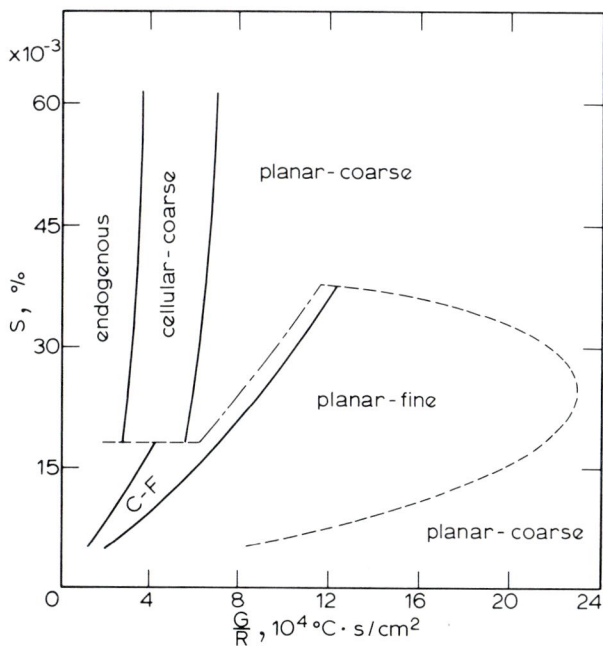

2 Influence of sulphur on the solidification interface morphology of eutectic grey cast iron[3] (C–F = cellular–fine)

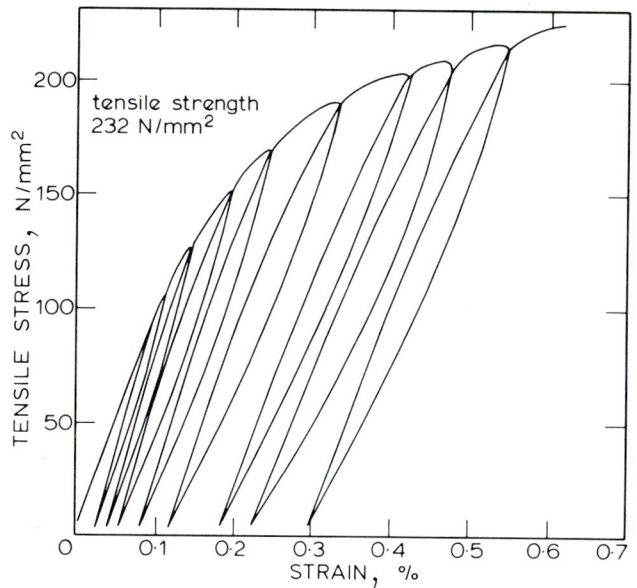

3 Tensile stress–strain curve of a flake graphite cast iron subjected to cyclical loading[7]

interface in alloys with $S > 0·02\%$ changes to a cellular one, coarse graphite reappears at cell boundaries. The amount of coarse graphite increases with sulphur content and growth rate.

Two other influences of sulphur must also be mentioned. An increase in sulphur level from 0·001 to 0·03% causes a gradual increase in the interlamellar distance of coarse graphite. From Fig. 2 it can be seen that at high sulphur contents the planar interface is relatively stable. This phenomenon seems to be due to an abnormal segregation behaviour of sulphur during the directional solidification experiments.[6]

TENSILE BEHAVIOUR AND FRACTURE

Gilbert[7] has conducted extensive investigations on mechanical properties of normal flake cast irons and their behaviour during tensile deformation. Amongst other things he concludes that a permanent plastic deformation occurs even at low stresses during tensile testing, and that the slope of the recoverable strain curve changes gradually with increasing stress (Fig. 3). Hence the elastic modulus decreases if the tensile stress is increased. This suggests that under stress the metal envelope around the graphite skeleton might elongate. The total strain at fracture is also very low.

Literature on the influence of graphite shape and size on fracture is generally inconclusive. Clough and Shank[8], and Gilbert[7] observed that the surface structure of a grey cast iron contained graphite flakes which had separated under stress, while Glover and Pollard[9] observed the presence of cleaved graphite flakes at all stress levels. The flakes had cleaved along the basal planes (0001) and short sections of prismatic fracture linked the islands of cleavage within the flake. In their opinion this caused a deviation from the linear elastic region during tensile testing. However, Kuroda and Takada[10] found from a study of tensile fracture of flake graphite cast irons that fracture surfaces were formed along the boundary between the graphite and the matrix. Maillard[11] developed a special etching technique to discriminate between cleavage and decohesion of the

Table 1   Chemical composition of carbonyl iron and graphite powder, %

	C	S	P	Ni	Cr	Mo	O_2	N_2
Carbonyl iron	0·03	<0·001	<0·001	<0·01	<0·01	<0·01	0·15	0·01
Graphite powder		0·014						

graphite. With this technique, it is shown[12] that during both fatigue and impact testing, the coarse type A and B graphite can cleave or decohere. During fatigue testing, no progressive cracking of graphite was observed, irrespective of its form.

EXPERIMENTAL PROCEDURE
Alloys used
Carbonyl iron and pure graphite powder were used to prepare the alloys. Table 1 gives the composition of the iron and graphite powder. Suitable additions were made to the melts in a vacuum furnace to obtain the desired content of phosphorus, sulphur or silicon respectively. 9·7 mm diameter and 200 mm long rods were cast in ceramic moulds. These castings were subsequently used for experiments in the unidirectional solidification furnace. Table 2 lists the chemical composition of the alloys.

Apparatus
The experiments were conducted in a unidirectional solidification furnace described elsewhere[2] at a temperature gradient of 33°C/cm according to the Bridgman technique in which the molten metal was lowered through the furnace at a fixed speed, R, ranging from 1·8 mm/h to 360 mm/h. Pure argon (with less than 5 ppm oxygen) served as a protective cover during the experiment. At the end of solidification cylindrical specimens of about 130 mm long and 10 mm diameter were obtained.

Table 2   Chemical composition of the alloys used

| | Alloy number | | | | | |
	100	101	102	109	112	114
Carbon	4·29	4·29	4·29	4·29	3·86	3·56
Silicon	0·01	0·01	0·01	0·01	1·92	2·94
Phosphorus	0·003	0·003	0·003	0·11	0·003	0·003
Sulphur	0·003	0·01	0·02	0·003	0·003	0·003

In all the alloys, Mn = 0·004%

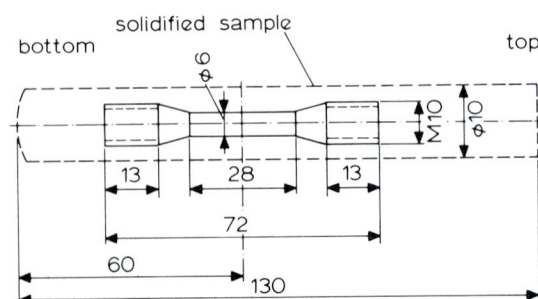

4   Details of tensile test specimen (dimensions in mm)

Heat treatment and tensile testing
The specimens were subjected to a ferritizing heat treatment to eliminate the influence of any pearlite present,[13] and were machined to obtain a tensile specimen such that its central section was at 60 mm from the bottom (start of solidification). This facilitated comparison of mechanical properties with interface morphology experiments carried out earlier.[2,3] Figure 4 shows the details of the tensile test specimen. Both normal tensile testing and intermittent tensile testing were carried out with an Instron testing machine at a cross-head pulling rate of 0·05 cm/min. The strain was measured using an electronic strain gauge extensometer with a magnification of ×400.

The aim of the intermittent tensile tests was to determine the onset of the plastic region more accurately. Specimens of material no. 100, solidified at the rates of 7·2, 36, and 144 mm/h were pulled with loads applied successively in stages of 100 to 250 N. When each load increment was reached, the specimen was held at its load for 3 min before it was unloaded to the no-load condition. However, after a certain load level—at about the elastic limit—the specimen was found to creep. The load was kept constant in that case by repeated micro-loading and when the specimen no longer showed any creep, it was held for a further 3 min before being unloaded. This operation was repeated with the next highest load.

Examination of fracture using SEM
Three different tensile specimens with an interlamellar distance of about 100 μm (coarse graphite), 50 μm (medium graphite), and 5 μm (fine graphite) were specially prepared and were studied under a JEOL scanning electron microscope and microprobe analyser, after fracture.

Graphite in Fe–C alloys can either: cleave under tension ('cleavage' fracture) or decohere from the ferrite matrix ('decohesion' fracture). Examination of exactly matched surfaces of two fracture sides is necessary to decide whether the mode of fracture is decohesion or cleavage. If cleavage is the dominating mode, both sides of matched detail would show graphite at most of the places. However, if decohesion is operating, one side will be ferrite and the corresponding part of the other side will be graphite and vice versa. In addition, graphite flakes may imprint on soft ferrite their characteristic hexagonal pattern leading to possible misinterpretation of SEM photographs. To avoid this, the two sides of the fractured specimens were extensively photographed and precise matching details were recognized. These were etched with an etchant that would attack only ferrite leaving the graphite intact. The matching details on two sides were extensively re-photographed.[14] Figure 5 shows the scheme followed for fracture study in this investigation.

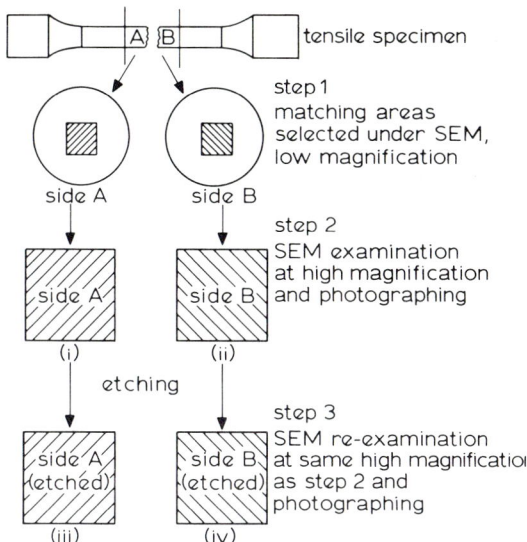

5 Scheme for fracture study using SEM[14]

RESULTS
Influence of phosphorus, sulphur, and silicon on graphite structure

The influence of phosphorus on the interface morphology and graphite structure closely resembled that of earlier work which is shown in Fig. 1. The experiments of the present investigation also showed typical coarse to fine transition in graphite structures.

In the phosphorus-containing alloys the coarse to fine transition of the graphite was not as noticeable and not as sudden as in sulphur-containing alloys. As the growth rate increased, the interlamellar spacing decreased more gradually in the P series. Figures 6 and 7 show a few examples of the graphite structure obtained. Phosphorus slightly coarsened the graphite, particularly at the eutectic cell boundaries (Fig. 7a). Eutectic cell boundaries in pure alloy seem to be dense aggregates of graphite with no noticeable coarsening, seen particularly at the highest growth rates. The fine graphite at the highest growth rates appears like the so-called 'rod-like' or 'coral' graphite, which is a rounded type of branched graphite, (Fig. 6c). At low growth rates, however, long graphite flakes were present (Figs 6a and 6b).

The graphite structure in sulphur-containing alloys is much more complex than in the phosphorus-containing alloys. In nearly all cases, areas with coarse graphite were present, as for instance in alloy no. 101 (Fig. 8). These coarse areas increase in size if the solidified length of the specimen is increased, or if the sulphur content in the alloy is higher (Fig. 7b). Compared to phosphorus, however, sulphur caused much more marked coarsening of graphite. Silicon had no significant effect on graphite structure. At any given rate of solidification, the graphite structure of silicon-containing alloy closely resembled that of pure eutectic alloy.

In this description it should be borne in mind that 'fineness' is only a qualitative term, and the graphite fineness at different sulphur contents should be compared only with actual measurement of the flake size and interflake distance. When a transition from coarse graphite to fine graphite takes place, the type of graphite itself changes to 'rod-like' graphite.

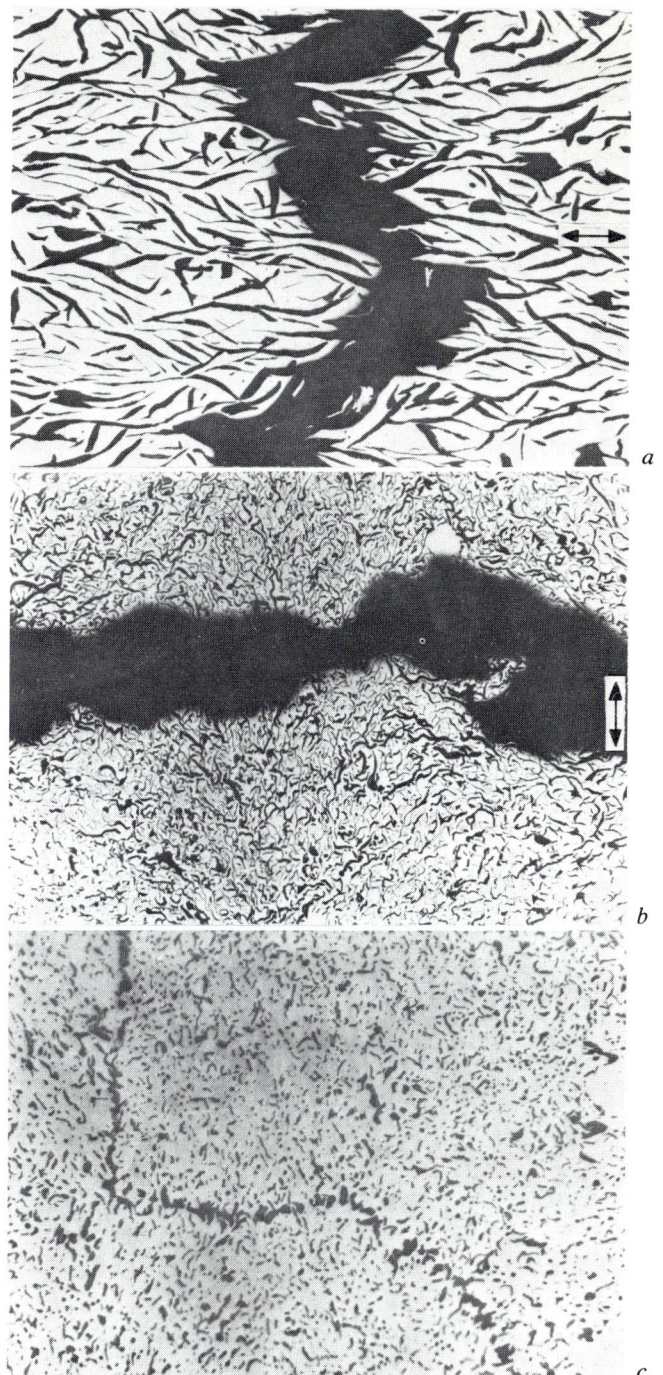

a growth rate $R = 7\cdot2$ mm/h, [×60]; b $R = 36$ mm/h, [×40]; c $R = 144$ mm/h, [×160]; the arrows show direction of loading

6 Graphite structure in pure Fe–C eutectic alloy

Influence of growth rate on mechanical properties

Figure 9 summarizes the variation of ultimate tensile strength with growth rate, R, for a number of alloys investigated. The change in interface morphology in the pure alloy no. 100 has been superimposed in the same figure. Two distinct morphologies were observed: planar solidification up to a growth rate of about 48 mm/h (133×10^{-5} cm/s) and cellular–endogenous solidification beyond this rate. There was also a smooth transition from coarse to fine graphite taking place at about

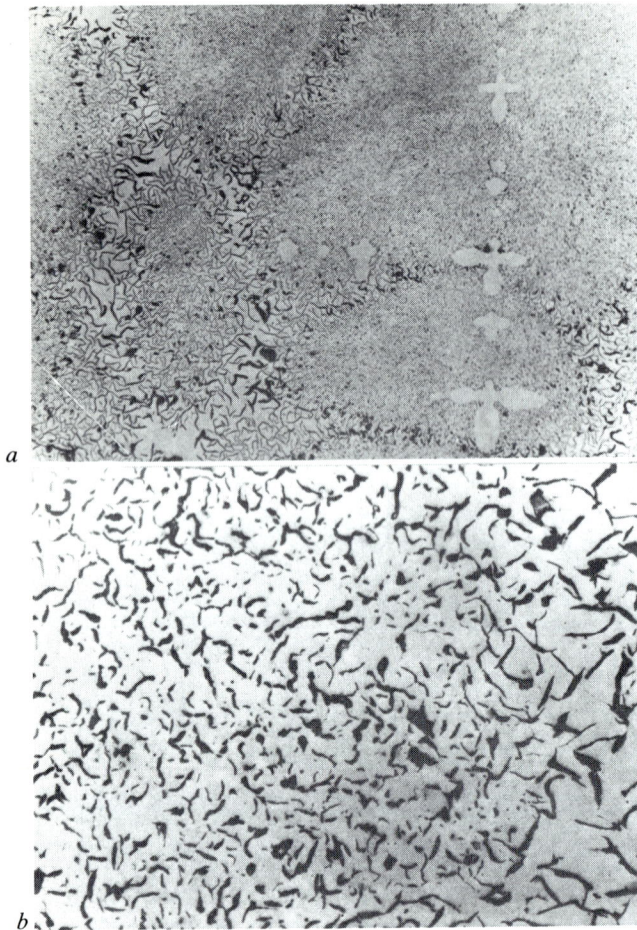

7   Graphite structure in: *a* 0·02% S-containing alloy, [×40]; *b* 0·11% P-containing alloy, [×160]; *R* = 144 mm/h

18 mm/h (50×10^{-5} cm/s). With this superimposition it may be observed that after the coarse to fine transition, there is a marked improvement in strength which is further enhanced during cellular solidification. An approximately $2\frac{1}{2}$-fold increase in strength is observed with an increase in growth rate from 7·2 mm/h (20×10^{-5} cm/s) to 144 mm/h (400×10^{-5} cm/s) in this pure alloy.

Additions of more than 0·03% phosphorus reduced the strength at all growth rates, though the reduction in strength is not so marked as can be noted for alloy no. 109 with 0·11% P. Unlike phosphorus, however, sulphur reduces the strength drastically, especially at higher growth rates. Even additions as small as 0·01% S reduce the strength by nearly 45%. With this alloy, in nearly all the specimens fine graphite was present but the fracture was invariably along regions of coarse graphite

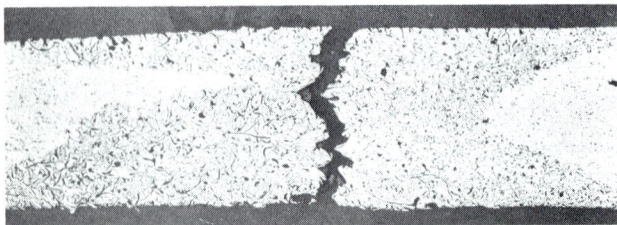

8   Fracture path and coarse–fine transition in graphite in an alloy with 0·01% S, R = 7·2 mm/h   [×4]

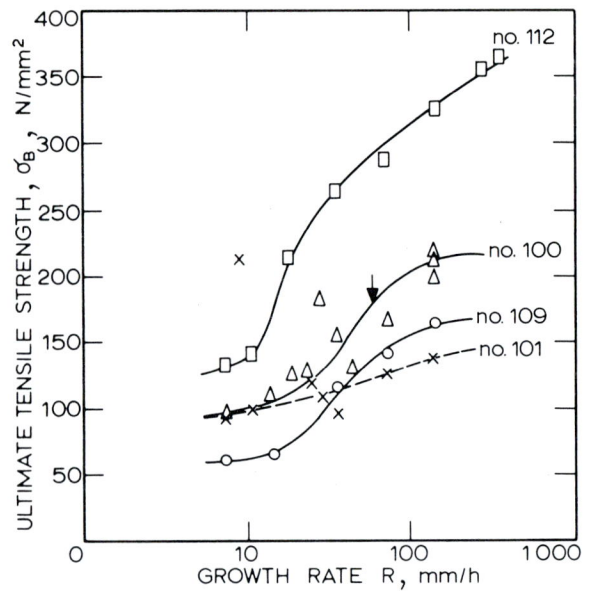

9   Tensile strength variation with growth rate for different alloys (arrow indicates the cellular-endogenous transition)

(Fig. 8) and the strength was low. Silicon improves the strength for all growth rates due to solution hardening. With silicon present, solidification at higher growth rates was possible since this element prevented the formation of ledeburite. Table 3 lists the strengths of alloys tested during this investigation.

Fracture

An example of the results obtained with the procedure described earlier is shown in Fig. 10. The fracture mode was described after examination of all the four sets of photographs (i–iv in Fig. 5) together. Figure 10, however, shows the matching surfaces after etching which correspond to iii and iv. Corresponding unetched surfaces are not shown. Two distinct types of fracture mode could be observed: cleavage and decohesion. When fracture took place by cleavage, the fracture path passed through the graphite flake along its weak planes which resulted in the appearance of the characteristic hexagonal pattern of graphite flakes on both matching surfaces. Etching did not alter this pattern, dispelling any doubts that they are graphite imprints on a ferrite matrix. On the other hand, when fracture was by decohesion of the ferrite/graphite interface, one surface showed graphite while the other showed ferrite; the latter shows graphite imprints, but could later be removed by etching. Figure 10 illustrate these phenomena in detail. The SEM photographs in Figs 10a and 10b are two exactly matched surfaces shown here after etching. At the places marked *D*, fracture is by decohesion and at the places marked *C*, fracture is by cleavage.

The type of fracture mode operating seemed to depend upon graphite fineness. Examination of fractured specimens solidified at different growth rates indicated that: (*a*) when the graphite was coarse and more or less aligned in the growth direction, the fracture was predominantly cleavage, (*b*) when the graphite was coarse, but randomly oriented, the fracture mode was partly cleavage and partly decohesion, and (*c*) when the graphite was fine and randomly oriented, the fracture was predominantly by decohesion.

Table 3 Mechanical properties

Growth rate, R, mm/h	Ultimate tensile strength of alloys, N/mm²						0.2% proof stress,* N/mm²	Elastic modulus,* N/mm² × 10⁴
	100	109	101	102	112	114		
7·2	96	62	96	64	132	188	67·5	5·8
9·0	—	—	212**	101	—	—	—	—
10·8	—	—	99	114	140	283	—	—
14·4	106	65	—	67	—	—	64	9·4
18·0	126	—	—	62	214	249	67	9·9
24·0	128	—	120	—	—	—	—	—
28·8	182	—	109	58	—	—	65	10·1
36·0	156 104	118	97	59	264	294	79	11·8
45·0	129	—	—	—	—	—	—	—
72·0	167	140	125	82	287	340	87	14·1
144·0	212 213 199	165	139	115	325	416	94	17·7
288·0	white	—	—	—	355	467	—	—
360·0	white	—	—	—	364	440	—	—

* Material no. 100; ** Structure with rod-like graphite

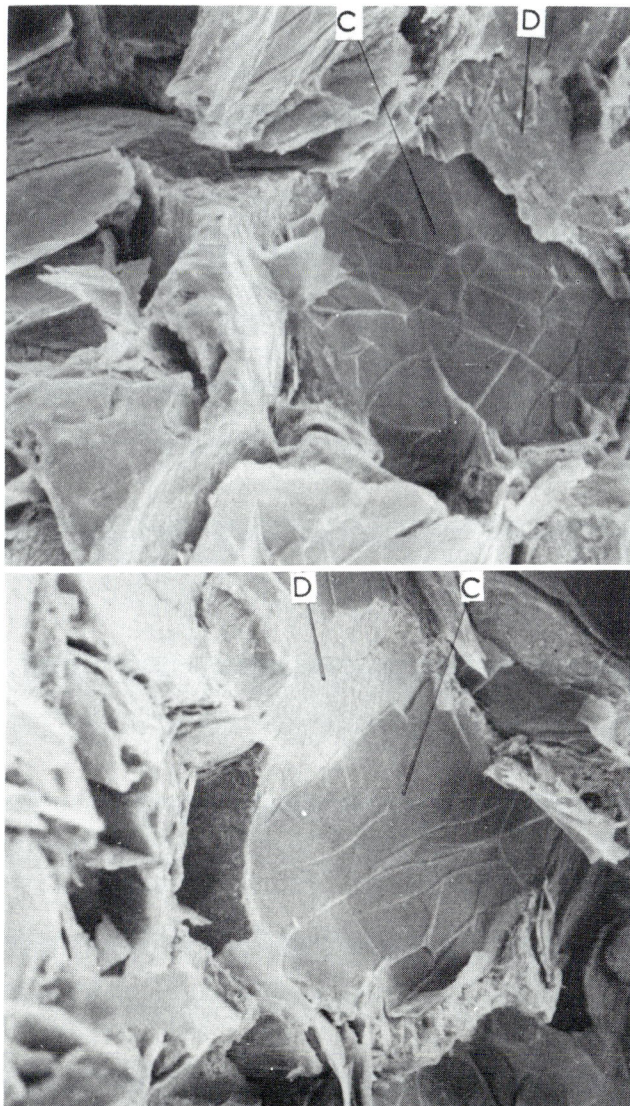

10 Scanning electron photomicrographs of matching surfaces, etched; the place marked *D* has fractured by decohesion while the place marked *C* has fractured by cleavage of graphite flakes[14] [× 200]

Tensile testing

Typical load extension diagrams from the normal tensile testing are given in Fig. 11 which shows a distinct plastic region. The total elongation was about 2–3%, which is much higher than that of normal cast iron. 0·2% proof stress increased with the growth rate but to a lower extent than the tensile strength. Table 4 summarizes the results for the pure alloy no. 100. A gradual decrease in the ratio $\sigma_{0.2}/\sigma_B$ would signify that strain hardening is influenced by the graphite size. A typical result from the intermittent tensile test is given in Fig. 12, which can be

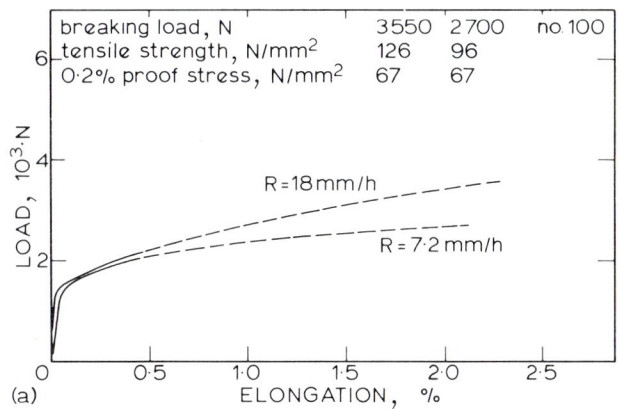

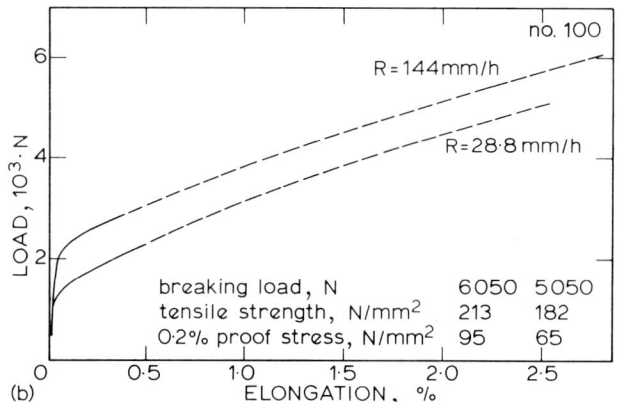

11 Load–extension diagrams; (—— measured by extensometer; – – – extrapolated)

Table 4   Tensile and 0·2% proof stress of pure Fe–C eutectic alloy

Growth rate, mm/h	$\sigma_{0.2}$, N/mm^2	σ_B, N/mm^2	$\sigma_{0.2}/\sigma_B$
7·2	67·5	96	0·70
14·4	64·0	106	0·60
18	67·0	126	0·53
28·8	65·0	182	0·36
36	79·0	156	0·50
72	87·0	167	0·52
144	94·0	212	0·44

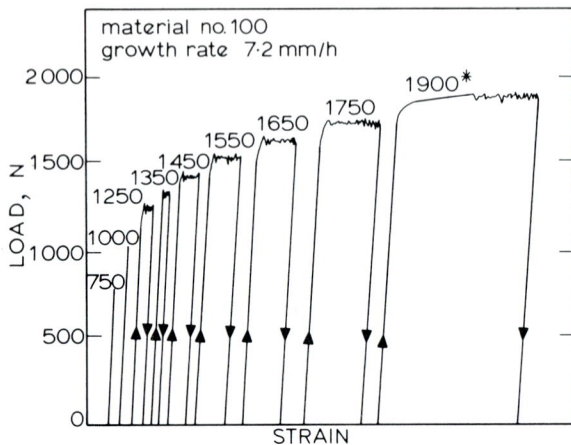

12   Typical intermittent tensile test curves (*these figures are the load in Newtons reached in the cycle)

DISCUSSION

From the results of the present investigation obtained so far, it appears that coarse graphite generally produces low strength while specimens with fine graphite exhibit relatively high strength (Fig. 9). The higher the growth rate, the finer the graphite will be with a corresponding improvement in strength. This trend is similar to that found in other metal/non-metal eutectics, such as Al–Si[16] and Zn–Ge.[17]

However, in these alloys, the increase in tensile strength was not as marked as in Fe–C alloys and also the increase required a 100-fold increase in the growth rate.

Thompson et al.[18] demonstrated for unidirecionally solidified eutectics that yield strength and ultimate tensile strength are proportional to $\lambda^{-\frac{1}{2}}$, λ representing the interlamellar distance. Since $\lambda^2 R = $ constant, the following relations are valid:

$$\sigma_{ys} = \sigma_0 + K_1 . R^{\frac{1}{4}} \tag{1}$$

$$\sigma_{UTS} = K_2 . R^{\frac{1}{4}} \tag{2}$$

in which K_1, K_2, and σ_0 are constants. Equation (1) originates from the well known Hall–Petch relation, which describes the influence of grain fineness on the yield strength. Holecek and Stolar,[19] and at the same time Justi and Bragg,[20] have proved equation (1) and (2) for the Al–Si eutectic; Garmong and Rhodes[21] investigated the ternary eutectic in the Al–Cu–Mg system.

Figure 13 illustrates the results obtained with the very pure alloy no. 100. In spite of the scatter in the measurements it seems that the strength is proportional to $R^{\frac{1}{4}}$.

compared with Gilbert's result (Fig. 3). From such curves the total and plastic strain can be computed. It was found that, even in coarse graphite irons (solidified at 7·2 mm/h), the stress–strain relationship showed a straight line relationship, up to at least 37 N/mm^2 in contrast to 8 N/mm^2 as reported by Gilbert.[7] This 'limit of proportionality' was higher when graphite became finer, up to 67 N/mm^2. This would mean that the unidirectionally solidified flake graphite cast irons do not show curvature in stress–strain relations at very low stresses, in contrast to normal cast iron.[7] It was also noted that after this limit of proportionality, when the specimens were unloaded immediately after the test stage was reached, creep appeared during unloading making the unloading curve slightly curved towards the origin. If, instead, the specimen was held at the load for some time, the creep appeared at the loaded stage and not during unloading, and the curve was a line parallel to the original loading line. The creep could be easily identified since the specimen required repeated microloading to maintain a constant load level. The amount of creep seemed to be a function of graphite size and load level. With coarse graphite, creep was considerably higher than with fine graphite. In general, creep increased with increasing stress level. Near the 'yield point', the creep was more than after the 'yield point'. Also, at certain load levels, a sudden increase in the amount of creep was observed. This phenomenon of creep was also reported by Gilbert[7] and was also observed by Plénard and Plessier[15] on commercial flake and spheroidal graphite iron.

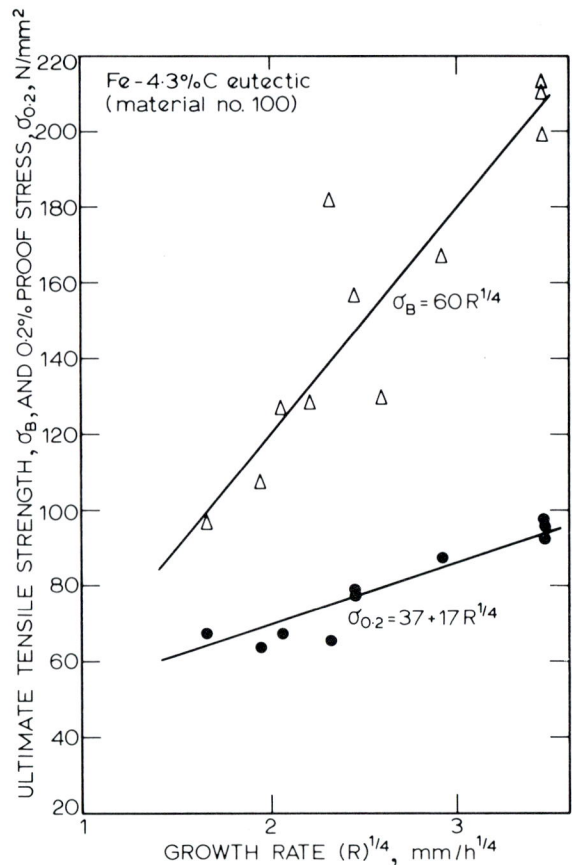

13   Ultimate tensile strength and 0·2% proof stress in pure eutectic alloy

In alloy no. 109 with 0·11% P and in the silicon-containing alloys, however, the ultimate tensile strength does not fit equation (2) very well. More measurements have to be done to investigate the exact significance of equations (1) and (2) for the iron–graphite eutectic.

A high growth rate gives fine graphite, but on the other hand it may introduce eutectic cell boundaries due to the onset of the endogenous type of solidification. Although it is not quite clear in the literature about the influence of cell boundaries on strength, it may be concluded from several investigations,[22–24] that an increasing cell number improves the tensile strength.

It was observed in the present investigation in *pure* Fe–C eutectic alloy that the fracture path in tensile test specimens followed randomly across the section of the specimens with no preference being shown to follow along the cell boundaries. However, it was often observed in *phosphorus*-containing alloys that fracture was along the cell boundaries. In *sulphur*-containing alloys, fracture was mainly along the cell boundaries which generally contained coarse graphite. It appears, therefore, that the cell boundaries are the weakest parts of the specimen, if relatively coarse graphite is present in these boundaries brought about by segregation of impurities such as phosphorus and sulphur. The above-mentioned phenomenon: more eutectic cells → higher strength in commercial cast irons, seems to be a result of finer graphite and a more uniform distribution of segregation products and not of a higher eutectic cell count in itself.

In the absence of any such impurities the non-preferential path of fracture indicates that there exists little or no difference in strength between cell boundaries and cell interiors. The advantage of a structure with fully fine graphite is illustrated in a specimen of a sulphur-containing alloy (marked no. 101) when it was solidified at 9 mm/h. The interface was planar, the structure consisting completely of fine graphite brought about by the coarse–fine transition (Fig. 14), and eutectic cell boundaries were absent.

This alloy also exhibited a high tensile strength of 215 N/mm², which is fully comparable with the best results obtained in alloy no. 100.

Not only the interlamellar distance, but also the shape of the graphite has an important influence on the ultimate tensile strength since rounded flakes diminish the stress concentrations and so the tendency to nucleate fracture. In the very pure alloy no. 100, alloys up to 0·03% P and in the silicon-containing alloys the graphite is very fine and additionally rounded at growth rates of 72 mm/h and higher. The graphite looks like the so-called rod-like or coral graphite in these cases. The high tensile strength is then understandable. If 0·01% or more sulphur is present in the alloy, the refinement of the graphite at high growth rates is far less, and at the same time the flakes are not so rounded. The result is a relatively low tensile strength.

The structure of the unidirectionally solidified cast iron of the present investigation can be considered to consist of a matrix of armco-iron with a fine branched interconnected graphite skeleton. The present results are compared with those from recent investigations[1,25,26] in Fig. 15. The ultimate tensile strength of unidirectionally solidified cast iron appears to be rather high and comparable with ductile iron. The very fine graphite present in the material gives hardly any loss in strength. The ductility, however, is much lower than in ductile iron.

In both the normal tensile testing and the intermittent tensile tests a clearly defined elastic region could be observed in all the specimens irrespective of growth rate. There was no observable plastic deformation at low loads. A straight-line relationship exists between stress and strain in the elastic region. The differences in stress–strain behaviour are obvious. If coarse graphite is present in the structure there is a sharp transition between elastic and plastic deformation. For $R = 7·2$ mm/h the curve is comparatively flat in the plastic region which means that the strain hardening is rather low. With increase in growth rate, however, the slope of this part of the curve increases, indicating progressively higher strain hardening as the graphite becomes finer. Also, if the structure consists of fine graphite the change from elastic to plastic deformation is more gradual.

From the intermittent tensile tests the Young's modulus, E, was calculated (material no. 100); E improved from 57×10^3 N/mm² at a solidification rate of 7·2 mm/h to 174×10^3 N/mm² at 144 mm/h showing that material with fine graphite is considerably stiffer. This phenomenon has been observed already by Gilbert,[7] and also reported by Plénard,[27] although their

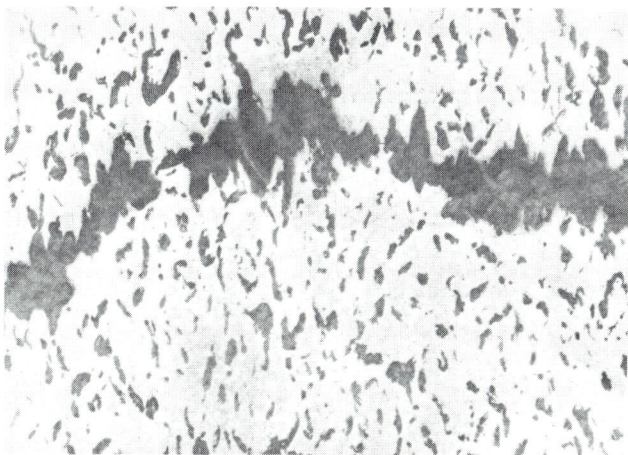

14 Rod-like eutectic and fracture path in an alloy containing 0·01% S; $R =$ 9 mm/h [× 160]

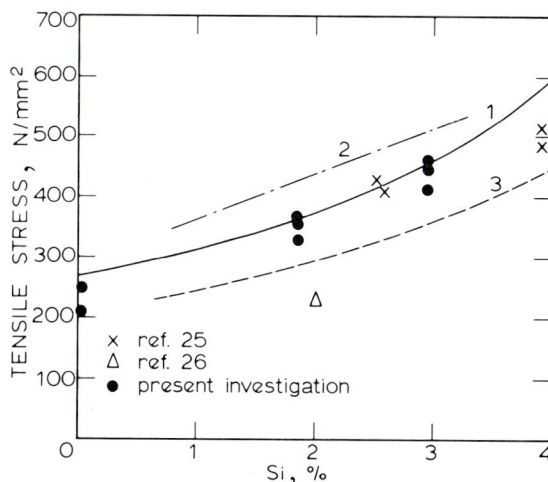

15 Tensile strengths of: *1* armco-iron alloyed with silicon,[1] *2* ferritic nodular iron, and *3* rod-like graphite grey iron

alloys are mainly hypoeutectic. Lux[1] measured Young's modulus between 137×10^3 and 157×10^3 N/mm^2 on pure cast iron with very fine type D graphite.

Using a model with a number of simplifications, Plénard[27] has derived two relations to calculate the Young's modulus of cast iron. Equation (3) concerns normal cast iron with flake graphite type A, while equation (4) is valid for irons with spheroidal graphite.

$$\frac{1}{E^*} = \frac{1}{E_m} + \frac{v_g}{v_g + v_m}\left(\frac{1}{E_g} - \frac{1}{E_m}\right) \quad (3)$$

$$\frac{E^*}{E_m} = 1 - 1\cdot93\frac{v_g}{v_g + v_m} \quad (4)$$

where

E^* = Young's modulus of cast iron
E_m = Young's modulus of the matrix (206×10^3 N/mm^2)
E_g = Young's modulus of graphite ($9\cdot8 \times 10^3$ N/mm^2)
v_g = total volume of graphite
v_m = total volume of matrix
ρ_g = density of graphite ($2\cdot2$ g/cm^3)
ρ_m = density of matrix ($7\cdot8$ g/cm^3)

Since the amount of graphite in the alloy = $4\cdot3\%$,

$$\frac{v_g}{v_m + v_g} = 0\cdot137$$

and equation (3) yields $E^* = 55 \times 10^3$ N/mm^2; equation (4) yields $E^* = 153 \times 10^3$ N/mm^2.

It appears that in the case of coarse graphite (growth rate 7·2 mm/h) the observed value of E^* is in good agreement with the calculated value according to equation (3). In the presence of very fine graphite (growth rate 144 mm/h) the observed Young's modulus is comparable with that of ductile iron, although there is some difference with the calculated value of equation (4) and the measurements of Lux.[1] Thus it is seen that unidirectionally solidified cast iron can exhibit interesting engineering properties, although more research is necessary to understand better the elastic–plastic and creep behaviour of this type of cast iron compared with commercial irons.

Examination of the matching fracture surfaces of tensile test specimens revealed that the fracture mode depends upon the fineness and the orientation of the graphite flakes. Figure 16 shows schematically the fracture paths in fine and coarse graphite alloys, with actual and alternate paths. As far as possible, the fracture path seems to prefer the graphite/ferrite interface. In fine graphite alloys, the actual path is an easier path, since the area of graphite interface to be separated is comparable to the area that needs to be separated if the alternate path is followed. This results in a decohesion type of fracture. In coarse graphite, however, the fracture perhaps initiated at some graphite tip passes along the graphite interface initially, enters the matrix, and encounters a large graphite flake at its middle. At this stage, the area that needs to be separated by cleavage (actual path) is much smaller than the large area that needs to be separated by decohesion (alternate path). Hence, cleavage will be a predominant mode in such alloys, especially if the graphite flakes are well aligned in the direction of the tensile stress. This discussion should, of course, include the decohesion force between graphite and ferrite, as well as the cleavage force between atomic layers of graphite and the influence of impurities on it. However, sufficient data are not available at present; nevertheless the points mentioned appear to be valid.

CONCLUSIONS

1 The graphite structure is strongly influenced by growth rate as well as impurity levels. Increase in growth rate decreases the interlamellar spacing between graphite flakes, the flakes becoming smaller in length and more rounded. The presence of impurities such as phosphorus, and to a larger extent sulphur, coarsens the graphite at the eutectic cell boundaries. Increase in sulphur content produces large regions of coarse graphite with sharp edges.

2 Alloys with fine graphite exhibit superior mechanical properties, a large degree of strain hardening, and relatively less creep compared to alloys with coarse graphite. Even a small quantity of coarse graphite (produced because of impurity) is sufficient to reduce the mechanical properties considerably.

3 Alloys with randomly oriented fine graphite tend to fracture by a decohesion of graphite matrix interface while those with more or less well aligned coarse graphite prefer to cleave along weak planes in graphite. Alloys with a graphite of intermediate size show partly decohesion and partly cleavage.

4 Graphite shape plays an important role in determining the mechanical strength. Well rounded rod-like graphite flakes contribute to superior mechanical properties. This type of graphite can be obtained by controlling the graphite coarsening impurity elements such as sulphur to very low levels and by resorting to high solidification rates.

5 Silicon prevents the formation of ledeburite during solidification and as such it is possible to directionally solidify at much higher growth rates than otherwise is possible without silicon. These higher growth rates promote fine rod-like graphite with a consequent improvement in tensile strength.

REFERENCES

1 B. LUX: *Giesserei Forschung*, 1967, **19**, 141
2 H. NIESWAAG AND A. J. ZUITHOFF: *Giesserei Forschung*, 1974, **26**, 19
3 H. NIESWAAG AND A. J. ZUITHOFF: 'The metallurgy of cast iron', 327, 1975, Switzerland, Georgi Publishing Co.

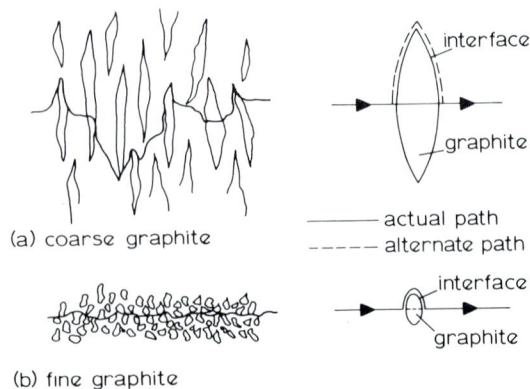

(a) coarse graphite
——— actual path
----- alternate path
(b) fine graphite

16 Paths of fracture (schematic)

4   K. D. LAKELAND: *J. Aust. Inst. Metals,* 1975, **10**, 55

5   Y. SAYAMA *et al.*: *J. Crystal Growth,* 1974, **22**, 272

6   TH. LUYENDIJK *et al.*: to be published

7   G. N. J. GILBERT: *Br. Foundryman,* 1968, **61**, 264

8   W. R. CLOUGH AND M. F. SHANK: *Trans. ASM,* 1957, **49**, 241

9   A. G. GLOVER AND G. POLLARD: *JISI,* 1971, **209**, 138

10   Y. KURODA AND H. TAKADA: Study of fracture of cast iron by using scanning type electron microscope, 36th Int. Foundry Congress, Beograd, 1969

11   A. MAILLARD: *Mém. Sci. Rev. Métall.,* 1975, **72**, 367

12   A. MAILLARD *et al.*: *Mém. Sci. Rev. Métall.,* 1975, **72**, 585

13   H. NIESWAAG *et al.*: Properties and structure of unidirectionally solidified eutectic grey cast iron, 43rd Int. Foundry Congress, Bucaresti, 1976

14   K. V. PRABHAKAR AND H. L. EWALDS: Internal rep., Lab. of Metallurgy, Delft University of Technology, 1976

15   E. PLÉNARD AND J. PLESSIER: *Fonderie,* 1967, (251), 1

16   M. SAHOO AND R. W. SMITH: *Metal Sci.,* 1975, **9**, 217

17   M. SAHOO AND R. W. SMITH: *J. Mater. Sci.,* 1976, **11**, 1 125

18   E. R. THOMPSON *et al.*: Publication NMAB—308–11, p. 71, 1973, Washington, National Academy of Sciences

19   S. HOLECEK AND P. STOLAR: *Aluminium,* 1976, **52**, 741

20   S. JUSTI AND H. BRAGG: *Metall. Trans. A,* 1976, **7**, 1 954

21   G. GARMONG AND C. G. RHODES: *Metall. Trans. A,* 1972, **3**, 533

22   G. N. J. GILBERT: *Iron Steel,* 1957, **30**, 19; 45; 103

23   G. N. J. GILBERT: *BCIRA J.,* 1959, **7**, 692

24   M. N. SRINIVASAN AND V. KONDIC: 'The metallurgy of cast iron', 753, 1975, Switzerland, Georgi Publishing Co.

25   E. CAMPOMANES AND R. GOLLER: *Trans. AFS,* 1975, **83**, 55

26   M. PALLADINO *et al.*: Fibrous graphite cast iron, 42nd Int. Foundry Congress, Lisboa, 1975

27   E. PLÉNARD: *Fonderie,* 1962, (191), 1

Effect of decreased section thickness on the formation, structure, and properties of a chill-cast aluminium-silicon alloy

G. R. Armstrong and H. Jones

Mechanical properties, microstructure, and conditions of cooling and freezing have been studied as a function of section size, z, in the range 5–0·05 mm for Al–10·5% Si made by gravity, burst-injection, and counterpressure casting and by twin-piston and roller quenching. Thermal analysis as a function of z indicated that conditions were in the intermediate or upper Newtonian régimes of heat transfer. Both αAl dendrite arm spacing, d, and fraction of residual eutectic showed power law decreases with increasing cooling rate within the range covered (400 to $1·2 \times 10^6$ K/s) and corresponding strength increases (e.g. of UTS from 240 to 370 MN/m^2) were related to d by Hall–Petch relationships.

G. R. Armstrong is with Tube Investments Research Laboratories, Saffron Walden, Essex, and H. Jones is in the Department of Metallurgy, University of Sheffield where the work was carried out

It is well established that increased cooling rate during solidification results in increased tensile strength and hardness of grey cast irons at decreased cast section sizes (e.g. ref. 1) and in increased tensile strength and elongation to fracture in aluminium alloys,[2–9] normally related to decreased dendrite spacing. These results are however limited to dendrite spacings in the range ~ 10–100 μm and cooling rates < 100 K/s, various governing relationships for properties having been suggested or implied,[4–8,10] and there is evidence for[5,7,9] and against[3,4] any dependence for yield strength.

The present study explored these effects for cooling rates during solidification in the range 400 to 10^6 K/s and dendrite spacings in the range ~5 down to ~0·5 μm. The work reported here was carried out on the Al–10·5% Si alloy as a compromise between the Al–7% Si base alloy studied earlier[3,4] for more conventional cooling rates and susceptible to hot-tearing under our conditions, and the Al–13% Si alloy for which microstructures were frequently complicated by primary silicon. The range of cooling rates was obtained by chill-casting sections of thicknesses 5 down to 0·05 mm, involving a number of methods ranging from gravity-drop casting to twin-roller quenching. A feature was that, except for twin-rolled tape, cooling rates and durations of freezing were measured directly by thermal analysis, rather than relying on predictions based on heat-transfer coefficients reported for related but not necessarily identical conditions.

PROCEDURES AND RESULTS

Figures 1a–d illustrate the basic methods employed to produce specimens suitable for tensile testing for the range of thicknesses studied. Gravity-drop casting (Fig. 1a) involved gravity discharge of ~ 30 g of melt via a stoppered crucible from a standard height of 300 mm into a vertical split chill mould with 45° tapered entry into a parallel-sided cavity of section thickness set by feeler gauges.[11] Burst-injection casting (Fig. 1b) is a form of injection chill-casting (e.g. ref. 12) in which a pressure differential between a pressurized melting chamber and an evacuated mould drives liquid metal into the mould when its entry stalk is immersed below the surface of the melt. Metal transfer is triggered by melting through the thinned tip of an aluminium sheath enclosing the entry stalk. Counterpressure casting[13] used the same apparatus with the aluminium sheath omitted, the mould and melting chamber being equally pressurized. Metal transfer occurred by immersion of the stalk followed by venting of the mould. Twin-piston quenching (Fig. 1c) involved the spreading of the melt between two copper or dural pistons driven together by pressurized gas. Samples were RF-induction melted on an alumina plaque at a level below the pistons and then

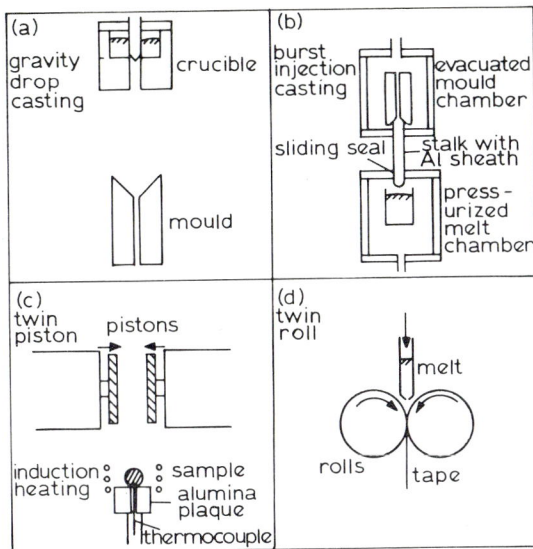

a gravity-drop casting; *b* burst-injection and counterpressure casting; *c* twin-piston quenching; *d* twin-roller quenching

1 Schematic diagrams of methods of producing samples

moved into the gap between them just before closure.[14] Twin-roller quenching (Fig. 1*d*) produced tape or ribbon 2 to 5 mm wide and 50 to 150 μm thick by argon pressure discharge of the melt through a nozzle into the nip of a special rolling mill[15] with rolls of 100 mm diameter held in contact and rotating at 1 000–4 000 rev/min. A sufficient area of sample to allow a standard tensile specimen of 20 mm gauge length and 3 mm width to be stamped out from relatively sound material was obtained by gravity casting at thicknesses above 0·6 mm, counterpressure casting above 0·5 mm, injection casting above 0·2 mm, and twin-piston quenching above 0·1 mm.

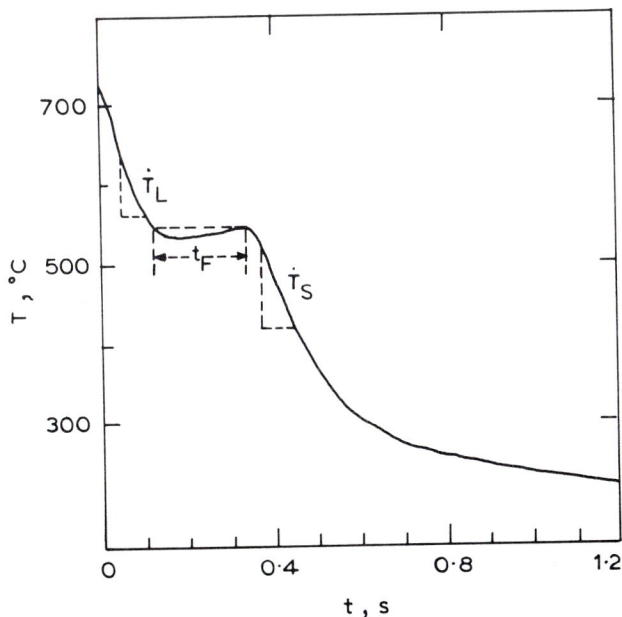

2 Typical cooling curve obtained by thermal analysis, defining cooling rates $\dot{T}_L$ and $\dot{T}_S$ through the liquid and solid phases, and duration of freezing t_F

Thermal analysis of cooling and freezing employed chromel–alumel thermocouples of 0·1 mm wire diameter inserted through the mould wall for gravity and counterpressure methods and through the alumina plaque into the melt for the twin-piston method. Figure 2 shows a typical cooling curve and indicates the significance of the thermal parameters derived from it. Standard procedures were employed for metallographic measurement of dendrite spacings and fraction eutectic (typical microstructures are shown in Fig. 3*a–c*) and for X-ray determination of lattice parameters of the α Al solid solution. Tensile testing was carried out at a cross-head speed of 0·5 mm/min. Results are presented as a function of section thickness *z* in Table 1. Many specimens failing in tension well below the maximum UTS values recorded in Table 1 exhibited porosity on fracture faces (Fig. 3*d*) indicating premature failure, consistent with observations of decreasing elongation to fracture with decreasing UTS at a given section thickness (Fig. 4). Higher UTS and proof stress values were invariably obtained with twin-roller compared with twin-piston quenching and for gravity or counterpressure casting compared to burst-injection casting. UTS, proof stress, and hardness decreased and elongation to fracture increased with increased temperature of isochronal ($\frac{3}{4}$ h) treatment as shown in Table 2.

HEAT FLOW AND SECTION THICKNESS

Heat flow analysis predicts that cooling rate $\dot{T}$ and duration of freezing t_F are related to section thickness *z* by power relationships of the form:

$$\dot{T} = A_1 z^{-a} \qquad (1)$$

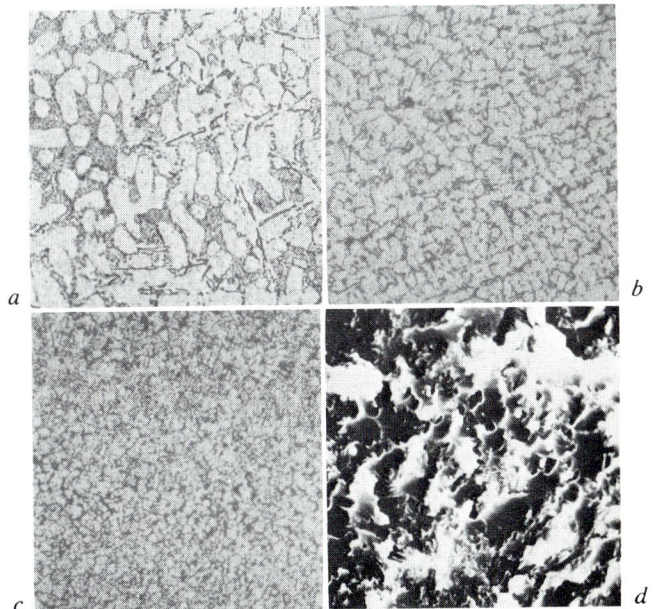

a z = 1 mm, counterpressure cast; *b z = 0·35* mm, twin-piston quenched; *c z = 0·12* mm, twin-roller quenched; *d* fracture surface, *z = 1* mm, injection cast

3 Typical microstructures as a function of section thickness *z* and a fracture surface [× 600]

Table 1 Cooling rates $\dot{T}_L$ and $\dot{T}_S$ through the liquid and solid states and duration of freezing t_F; αAl dendrite arm spacing d_a, dendrite cell size d_c, fraction eutectic f_{EU} and αAl lattice parameter a; microhardness H_v (50 g load), 0.1% and 0.2% proof stresses, $\sigma_{0.1}$ and $\sigma_{0.2}$, and ultimate tensile strength, σ_u (maximum values); all as a function of section thickness z for Al–10.5% Si

z, mm	$\dot{T}_L$, K/s	$\dot{T}_S$, K/s	t_F, s	d_a, μm	d_c, μm	f_{EU}	a, nm	H_v, kg/mm^2	$\sigma_{0.1}$, MN/m^2	$\sigma_{0.2}$, MN/m^2	σ_u, MN/m^2
0.05*	1.2×10^6¶	—	—	0.6	2.0	0.22	—	—	196	215	334
0.07*	—	—	—	—	—	—	0.404 57	125	166	200	370
0.10*	—	—	—	—	—	—	0.404 65	106	208	229	339
0.14†	1.2×10^5	3.0×10^3	0.02	1.0	2.9	0.29	0.404 81	99*	180*	196*	351*
0.17*	—	—	—	—	—	—	—	100	196	213	312
0.35†	4.0×10^4	2.0×10^3	0.04	1.4	4.3	—	—	86*	154*	168*	305*
0.5†	1.8×10^4	1.1×10^3	0.09	2.0	5.4	0.36	0.404 83	88‡	147‡	171‡	320‡
0.7‡	—	—	—	—	—	—	—	93	118	151	305
1.0‡	5.5×10^3	1.2×10^3	0.14	3.2	6.5	0.40	0.404 84	91	142	154	288
2.0‡	1.8×10^3	1.5×10^3	0.24	4.1	8.5	0.52	—	90§	117§	144§	277§
2.4§	1.0×10^3	8.4×10^2	0.39	4.6	—	—	—	—	—	—	—
3.0§	7.5×10^2	1.8×10^2	0.66	5.0	—	—	—	—	—	—	—
5.0§	4.0×10^2	3.6×10^2	—	7.0	—	—	—	—	120	138	240

* twin-roll; † twin-piston; ‡ counterpressure; § gravity; ¶ estimate from eutectic spacing

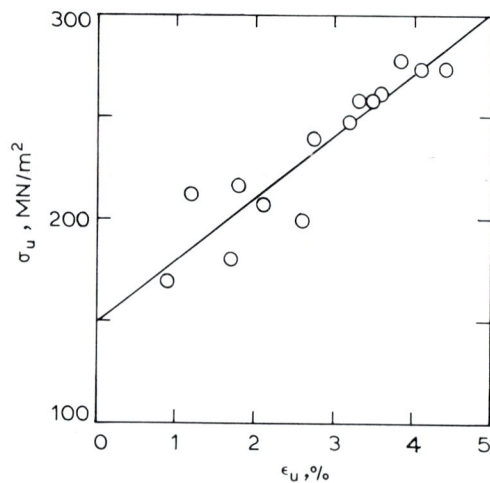

4 Ultimate tensile strength, σ_U, as a function of elongation to fracture, ϵ_U, for gravity-cast Al–10.5% Si of section thickness 2 mm

Table 2 Effect of treatment at 473–773 K on the mechanical properties of Al–10.5% Si gravity-cast at 2 mm section thickness

Treatment	H_v, kg/mm^2	$\sigma_{0.1}$, MN/m^2	$\sigma_{0.2}$, MN/m^2	σ_u, MN/m^2	ϵ_u, %
As-cast	90	117*	144*	277*	0.9–4.4
473 K for 0.75 h	93	120	138	229	2.3
573 K for 0.75 h	68	98	108	190	6.2
673 K for 0.75 h	59	79	95	160	9.0
773 K for 0.75 h	52	77	85	151	13.6
773 K for 18.5 h	43	69	77	141	12.5
773 K for 48 h	48	68	74	129	14.0

H_v, $\sigma_{0.1}$, $\sigma_{0.2}$, σ_u as Table 1; ϵ_u is elongation to fracture; * maximum values

and

$$t_F = A_2 z^b \qquad (2)$$

where for Newtonian heat transfer a and b are both unity while for ideal contact conditions they are equal to 2. Figure 5 shows that $\log \dot{T}_L$ and $\log t_F$ from Table 1 vary linearly with $\log z$ as predicted with $A_1 = 5\,500\pm400$, $a = 1.74\pm0.04$, $A_2 = 0.15\pm0.01$, and $b = 1.08\pm0.08$ with $\dot{T}$ in K/s, z in mm, and t_F in s. The implication of a and b less than 2 that contact was less than perfect is supported by the observation that cooling rates were factors between 2 and 10 times less than predicted for ideal cooling of aluminium against copper.[16] Our values of $\dot{T}_L$ are within a factor of 2 of those reported for 0.13 mm thick twin-piston quenched Al–5% Cr[14] and for wedge-cast Al–2% Cr between 3 and 12 mm thick.[17] Values nearer to the predicted ideal have also been reported.[12,18] Apparent heat transfer coefficients h (kW/m^{2}K) calculated from our results assuming purely Newtonian cooling[18] range from 6 to 40 for $\dot{T}_L$, from 4 to 8 for t_F, and from 1.4 to 8 for $\dot{T}_s$ as z decreases from 3 to 0.14 mm. Nusselt numbers Nu range from 0.1 to 0.04 from $\dot{T}_L$, 0.007 to 0.03 from t_F and 0.000 4 to 0.04 from $\dot{T}_s$, and confirm that Newtonian conditions ($Nu < 0.015$)[16] prevail for thinner sections. The decline in h and Nu in proceeding from liquid cooling to freezing and solid cooling is in agreement with the results of Harbur et al.[18] and probably stems from the fact that shrinkage reduces thermal contact. Our values of h for z in the range 1 to 5 mm are 2 to 9 times larger than those used by Bardes and Flemings[19] to calculate solidification times of chill-cast plates in this thickness range while our higher values are an order of magnitude less than obtained by Harbur et al. for 0.13 mm thick twin-piston aluminium.

MICROSTRUCTURE AND COOLING RATE

Dendrite secondary arm spacing d_a and dendrite cell size d_c have been related experimentally to cooling rate $\dot{T}$ by power functions of the form:

$$d = A_3 \dot{T}^{-c} \qquad (3)$$

where typically $0.3 < c < 0.5$.[3,10] Figure 6a shows that

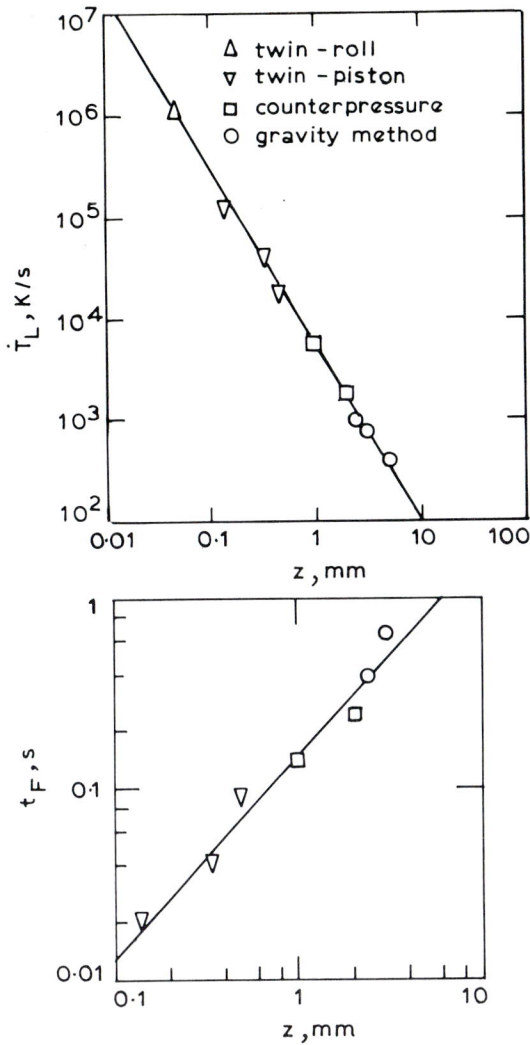

5 Cooling rate, $\dot{T}_L$, before freezing and duration of freezing, t_F, as a function of section thickness, z, for Al–10·5% Si

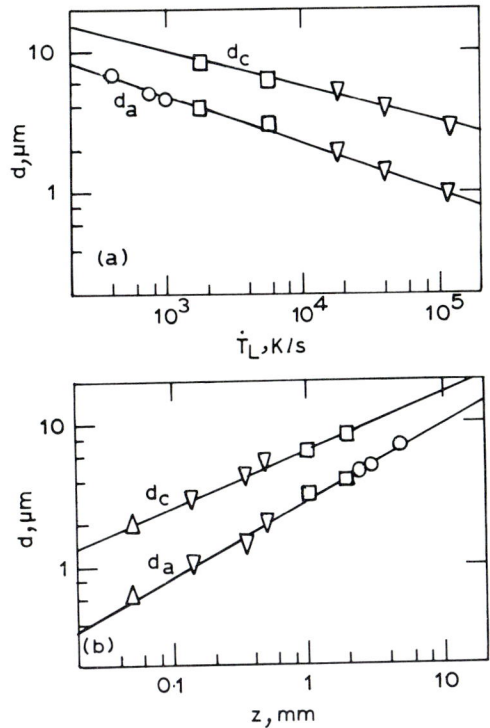

6 Dendrite arm spacing, d_a, and dendrite cell size, d_c, as function of cooling rate, $\dot{T}_L$, and of section thickness, z, for Al–10·5% Si

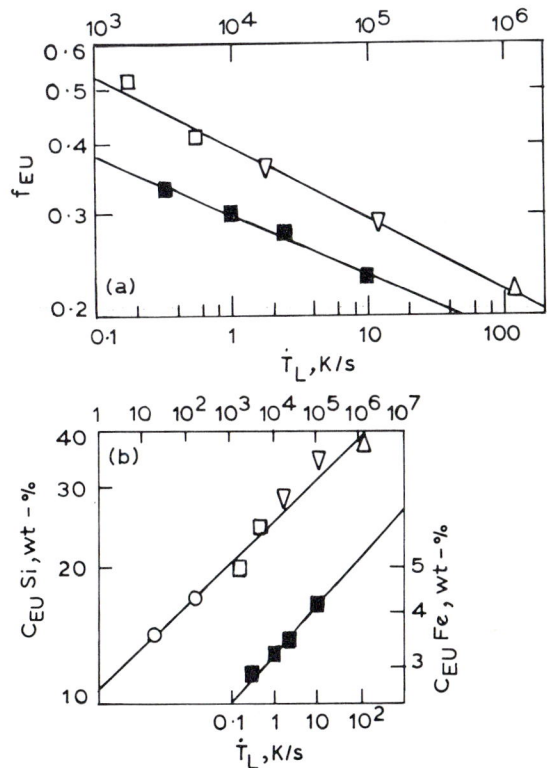

upper scales and open points for Al–10·5% Si from Table 1; lower scales and filled points for Al–10·5% Si from Bäckerud[20]; O for Al–Si from Steen and Hellawell[21]

7 Fraction eutectic, f_{EU}, and equivalent eutectic composition, C_{EU}, as a function of cooling rate, $\dot{T}_L$

log d_a and log d_c from Table 1 vary linearly with log $\dot{T}_L$ in conformity with $A_3 = 47 \pm 6$ and $c = 0.33 \pm 0.01$ for d_a and $A_3 = 55 \pm 12$ and $c = 0.25 \pm 0.02$ for d_c (d in μm and $\dot{T}$ in K/s). These give values of d_a in good agreement with results collected by Dean and Spear[10] for several aluminium alloys containing copper and silicon, values of d_c being a factor of 2 to 3 higher. Equations (1) and (3) combine to give

$$d = A_4 z^m \qquad (4)$$

where m equals ac and A_4 is A_3/A_1. This relationship shown in Fig. 6b gives $A_4 = 2.9 \pm 0.1$ and $m = 0.54 \pm 0.02$ for d_a and $A_4 = 6.6 \pm 0.2$ and $m = 0.40 \pm 0.02$ for d_c (d in μm, z in mm). Figure 7a shows that volume fraction of residual eutectic f_{EU} also conforms to a power relationship with cooling rate, i.e.

$$f_{EU} = A_5 \dot{T}^{-n} \qquad (5)$$

where A_5 is 1.25 ± 0.13 and n is 0.13 ± 0.01 for Al–10·5% Si ($\dot{T}_L$ in K/s). The results of Bäckerud[20] for Al–1 wt-% Fe also shown give $A_5 = 3.35 \pm 0.05$ and $n = 0.10 \pm 0.01$. Taking the level of silicon in solid solution in aluminium indicated by the lattice parameter (Table 1) and applying conservation of solute gives the composition of the residual eutectic C_{EU} as increasing from 19·5 to 37% Si as cooling rate $\dot{T}_L$ increases from 1 800 to $1·2 \times 10^6$ K/s. These values as a function of $\dot{T}_L$ are consistent with those given for coupled eutectic growth by Steen and Hellawell[21] for lower C_{EU}, together fitting the relationship

$$C_{EU} = A_6 \dot{T}_L^r \qquad (6)$$

with A_6 as $10·9 \pm 0·7$ and r as $0·091 \pm 0·007$ (C_{EU} in wt-%, $\dot{T}_L$ in K/s), as shown in Fig. 7b. The implication that the interdendritic eutectic in our specimens has the composition appropriate to coupled eutectic growth for the prevailing growth conditions extends to the regime of high growth velocity, the identical conclusion of Sharp and Flemings[22] for Al–Cu alloys grown slowly at high temperature gradients. An effective distribution coefficient k of 0·36 is obtained from the slope $(k - 1)$ of $\log C_{EU}$ plotted against $\log f_{EU}$ for our results in conformity with the Scheil equation, compared with the equilibrium value of 0·13.

MECHANICAL PROPERTIES AND MICROSTRUCTURE

Previous workers have related increases in UTS and yield stress with increasing cooling rate during solidification to grain size L,[5] dendrite cell size d_c[3,5–8,10] or to spacing λ between second-phase particles.[5] Figure 8 shows 0·1 and 0·2% proof stress, $\sigma_{0·1}$ and $\sigma_{0·2}$, and ultimate tensile strength, σ_u, plotted against $d_a^{-\frac{1}{2}}$ according to the Hall–Petch relation

$$\sigma = \sigma_0 + KD^{-\frac{1}{2}} \qquad (7)$$

where in general D is L, d or λ. Erginer and Gurland[5] found that $\sigma_{0·2}$ for aluminium of grain size 0·57 mm, Al–1·6 to 2·5% Si with d_c 1·3 to 5·0 μm and Al–5·3 to 25% Si with λ 1·7 to 22 μm fitted equation (7) with σ_0 of 7·8 MN/m^2 and K of 0·105 MN/m$^{\frac{3}{2}}$. This value of K accords with our value of $0·101 \pm 0·015$ MN/m$^{\frac{3}{2}}$ for Al–10·5% Si for d_a between 0·6 and 7 μm, somewhat higher than the value of 0·07 MN/m$^{\frac{3}{2}}$ for polycrystalline aluminium.[23] The two cases[3,4] where no effect of d_a on flow stress was detectable both involved significant solution and age-hardening treatments following solidification masking any remaining effect of d_a, which nevertheless continues to limit fracture properties. Gangulee and Gurland[24] have shown that deformation in Al–2·8 to 13·2% Si proceeds by slip in the matrix accommodated by progressive fracture in increasing numbers of silicon particles, leading eventually to complete fracture of the specimen. For cracks to form, the strain energy in the particle must be sufficient to provide the surface energy of the newly formed crack surface. Assuming a stress field with the same order of dimensions as the particle radius ρ, then the critical stress for fracture $\sigma_{CRIT} \sim (\bar{E}\gamma/\rho)^{\frac{1}{2}}/q$ where $\bar{E}$ is the weighted mean elastic modulus of matrix and particle, γ is specific surface energy, and q is a stress concentration factor.[25] Taking typical values $E \sim 10^{11}$ N/m^2, $\gamma \sim$ 1J/m^2, and $q = 2$ gives σ_{CRIT} of the same order as our proof stresses if ρ increases from $\sim 0·5$ to ~ 2 μm with decreasing cooling rate. These values are not unrealistic if the applicable r is taken to be the width of the interdendritic areas. Our highest levels of UTS are in good agreement with those of Lux and Hiller[26] for Al–11% Si solidified at comparably high cooling rates. These values represent an increase of some 80% over

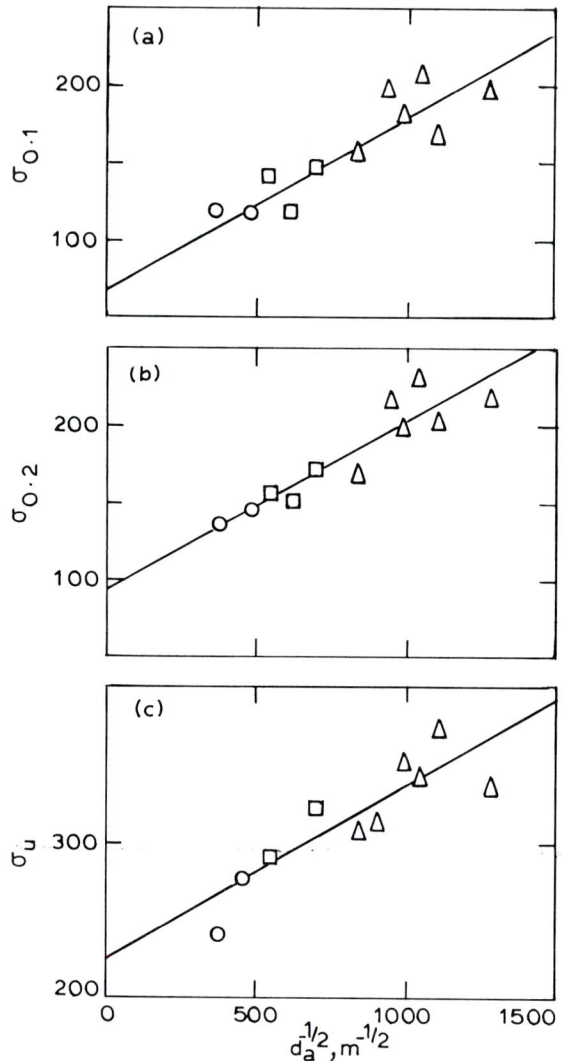

8   **0·1 and 0·2% proof stresses, $\sigma_{0·1}$ and $\sigma_{0·2}$, and ultimate tensile strength, σ_U, in MN/m^2 as a function of dendrite arm spacing for Al–10·5% Si**

values for the normal chill-cast alloy entirely attributable to increased cooling rate during solidification. Thus, equation (7) can be combined with equations (3) or (4) to give

$$\sigma - \sigma_0 = K'\dot{T}^u \quad \text{or} \quad K''z^{-v} \qquad (8)$$

with $K' = K/A_3^{\frac{1}{2}}$, $K'' = K/A_4^{\frac{1}{2}}$, $u = c/2$, and $v = m/2$. Our value of 0·27 for v contrasts with 0·17 reported by Suzuki and Furumoto.[7] Their work did not include cooling rate measurements but their result is consistent with cooling conditions more completely in the Newtonian regime than in our studies.

CONCLUSIONS

1   Cooling rate $\dot{T}_L$ in K/s before freezing and duration of freezing t_F in seconds for chill cast Al–10·5% Si are related to section thickness z in the range 0·14 to 3 mm by:

$$\dot{T}_L = 5\,500(\pm 400)z^{-1·74(\pm 0·04)}$$

and

$$t_F = 0·15(\pm 0·01)z^{1·07(\pm 0·08)}$$

with cooling conditions in the intermediate or upper Newtonian regimes.

2 Dendrite secondary arm spacing d_a of α Al in μm and fraction f_{EU} of eutectic over the ranges 0·6 to 7 μm and 0·22 to 0·52 respectively are related to cooling rate $\dot{T}_L$ over the range 400 to 10^6 K/s by

$$d_a = 47(\pm 6)\dot{T}_L^{-0.33(\pm 0.01)}$$

and

$$f_{EU} = 1.25(\pm 0.13)\dot{T}_L^{-0.13(\pm 0.01)}$$

associated with interdendritic eutectic containing as much as 37 wt-% Si at 10^6 K/s compared to the equilibrium eutectic composition of 12·7 wt-% Si.

3 0·1 and 0·2% proof stresses, $\sigma_{0.1}$ and $\sigma_{0.2}$, and ultimate tensile strength, σ_U, in MN/m^2 were related to dendrite spacing d_a (in m) over the same range by:

$$\sigma_{0.1} = 76(\pm 16) + 0.101(\pm 0.019)d_a^{-\frac{1}{2}}$$

$$\sigma_{0.2} = 97(\pm 13) + 0.101(\pm 0.015)d_a^{-\frac{1}{2}}$$

$$\sigma_u = 223(\pm 18) + 0.109(\pm 0.021)d_a^{-\frac{1}{2}}$$

Strength levels at 10^6 K/s approach twice those of normal chill-cast material.

ACKNOWLEDGMENTS

The authors would like to thank Professor B. B. Argent for providing laboratory facilities, the Science Research Council for supporting one of us (G. R. A.), and Dr B. G. Lewis for use of the twin-roller quenching apparatus.

REFERENCES

1 G. N. J. GILBERT: *Iron Steel,* 1957, **30**, 19, 45, 103

2 W. D. WALTHER *et al.: Trans. Am. Foundrymen's Soc.,* 1954, **62**, 219

3 R. E. SPEAR AND G. R. GARDNER: *Trans. Am. Foundrymen's Soc.,* 1963, **71**, 209

4 S. F. FREDERICK AND W. A. BAILEY: *Trans. Met. Soc. AIME,* 1968, **242**, 2 063

5 E. ERGINER AND J. GURLAND: *Z. Metallkd,* 1970, **61**, 606

6 V. S. ZOLOTEREVSKIY AND V. V. TELESHOV: *Russ. Metall.,* 1971, (5), 135

7 M. SUZUKI AND K. FURUMOTO: *J. Jpn. Inst. Met.,* 1971, **21**, 36–45 and 350–357

8 T. FUKUI AND K. NAMBA: *Trans. Jpn. Inst. Met.,* 1971, **12**, 355

9 J. N. LANZAFAME AND T. Z. KATTAMIS: *Weld. J.,* 1973, **52**, suppl., 226

10 W. A. DEAN AND R. E. SPEAR: 'Strengthening mechanisms: metals and ceramics', (ed. J. J. Burke *et al.*), 268, 1966, Syracuse, New York, Syracuse University Press

11 Y. SERITA *et al.: J. Jpn. Inst. Light Met.,* 1970, **20**, (1), 1

12 G. FALKENHAGEN AND W. HOFMANN: *Z. Metallk.,* 1952, **42**, 69

13 A. T. BALEVSKY AND T. DIMOV: *Br. Foundryman,* 1965, **7**, 280

14 I. S. MIROSHNICHENKO AND V. O. ZAKHAROV: *Ind. Lab. (U.S.S.R.)* 1969, **35**, 362

15 E. BABIĆ *et al.: J. Phys. E. (Sci. Instrum.),* 1970, **3**, 1 014; *Fizika,* 1970, **2**, suppl. 2, 1

16 R. C. RUHL: *Mater. Sci. Eng.,* 1967, **1**, 313

17 R. ICHIKAWA AND T. OHASHI: *Trans. Jpn. Inst. Met.,* 1971, **12**, 179

18 D. R. HARBUR *et al.: Trans. Met. Soc. AIME,* 1969, **245**, 1 055

19 B. P. BARDES AND M. C. FLEMINGS: *Trans. Am. Foundrymen's Soc.,* 1966, **74**, 406

20 L. BÄCKERUD: *Jernkontorets. Ann.,* 1968, **152**, 109

21 H. A. H. STEEN AND A. HELLAWELL: *Acta. Metall.,* 1972, **20**, 363

22 R. SHARP AND M. C. FLEMINGS: *Metall. Trans.,* 1972, **4**, 997

23 E. O. HALL: 'Yield point phenomena in metals and alloys', 1970, Chap. 1, New York, Plenum Press

24 A. GANGULEE AND J. GURLAND: *Trans. Met. Soc. AIME,* 1967, **239**, 269

25 J. GURLAND AND J. PLATEAU: *Trans. Am. Soc. Met.,* 1963, **55**, 442

26 B. LUX AND W. HILLER: *Prakt. Metallogr.,* 1971, **8**, 218

Effect of porosity on tensile properties of two Al–Si alloys

A. Herrera and V. Kondic

The effect of shrinkage porosity volume on the tensile properties of two aluminium alloys containing 12·1% Si (LM6), tested as cast, and 10·6% Si–0·9% Mg (LM13), tested as fully heat treated, was investigated using standard sand cast test bar design. The volume of porosity (0–0·45%) was varied using cast iron chills located along the gauge length of the test bar in such a way as to concentrate shrinkage porosity in the geometrical centre of the bar. The effects of porosity on the tensile properties were related to: (a) percentage of porosity, and (b) projected pore area in the fracture as measured in the macrographs. In addition, an attempt was made to relate the pore damage to the properties using the method of fracture mechanics. For this purpose, K_{IC} values of LM13 alloy were obtained using specimens cut from cast ingots ($2·5 \times 15 \times 25$ cm) and applying standard crack displacement bend tests. The location of the major pore and its geometry were assessed from the radiographs obtained in two directions. Finally, an attempt was made to correlate the measured effect of the pore with that predicted from K_{IC} values and assuming various types of flaw geometry. The results relating tensile properties to the percentage of porosity show that strength and ductility properties decrease more rapidly with increasing porosity volume for LM13 than for LM6 alloy. The yield strength remains unaffected with porosity in LM6 but it decreases rapidly in LM13 alloy. Using regression analyses, quantitative relationships are well defined, but the implicit cause of scatter using porosity % method still remains, i.e. it takes into account the total porosity rather than the porosity in the fracture only. The changes in properties related to the pore area in the fracture show less scatter than those obtained using total porosity values. The UTS values for both LM6 and LM13 alloys fall below the theoretical line based on the reduction in the load-bearing area. The average K_{IC} value for LM13 alloy was found to be $492·53 \ N/mm^{-3/2}$. Assuming that through the thickness flaws of various geometry occur the calculated and experimental critical flaw sizes were not in good agreement. Theoretical, experimental, and practical problems in assessing porosity damage in castings using the methods of fracture mechanics are discussed.

The authors are in the Department of Industrial Metallurgy, University of Birmingham

Studies of the effects of porosity on the mechanical properties of cast alloys have been making progress along two different lines. On the one hand, the causes and the mechanisms involved in the formation of porosity, mainly of gaseous or shrinkage origins, have been investigated from the point of view of their control or elimination (i.e. degassing and feeding theories and technologies).[1-5] At the same time, the effects of various morphologies, distributions, and volumes of porosity on the mechanical properties have been studied using a variety of approaches and techniques. One of the earliest methods was that of obtaining the volume of porosity by density measurements and relating various mechanical properties to the percentage of porosity.[6-8] The main limitation of this approach is that the changes in properties obtained in this way are, as a rule, subject to a very wide scatter. Consequently, no quantitative relations are obtained which could be used either to explain the effect of porosity on the mechanism of fracture or to predict the behaviour of partially porous castings in service. By the use of the porosity volume method some improvement in reduction of scatter was obtained by relating the area of porosity in the fracture to a given property measured.[9,10] More recently, the methods of fracture mechanics have been extended to cast metals also, in the development of design criteria for porous castings as well as in the study of the effects of various types of porosity on the mechanism of fracture.[11-13] In the present work a comparative investigation has been carried out with two Al–Si casting alloys using the density or porosity techniques as well as the fracture mechanics method of interpretation of shrinkage porosity effects.

EXPERIMENTAL

The alloys used were of commercial purity: LM6 (Si 12·09, Fe 0·06, Mn 0·5, Ti 0·2, Zn, Pb, Ni, Cu < 0·1 and Sn < 0·05%) and LM13 (Si 10·6, Fe 0·1, Mg 0·9,

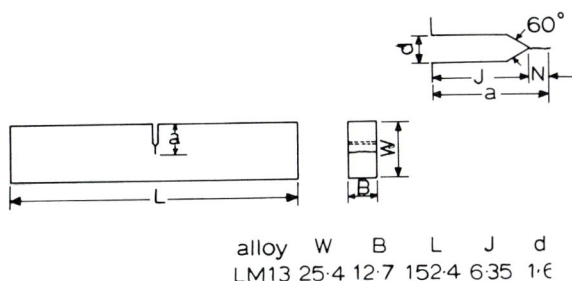

1 Testpiece design for fracture toughness measurement

alloy W B L J d
LM13 25·4 12·7 152·4 6·35 1·6

Mn 0·5, Ni 1·5, Zn 0·5, Ti 0·2 and Cu, Pb, Sn < 0·1%). The melts, fully degassed and 'modified' using salt treatment, were poured at 720°C (LM6) or 700°C (LM13). LM6 was tested as cast and LM13 after solution treatment (8 h at 520°C and water quenched) and aging (6 h at 180°C).

Two types of test bar design were used: standard gravity die test bars for obtaining sound bars, and the design shown in Fig. 2 for controlled shrinkage porosity bars. By varying the location, thickness, and design of cast iron chills in sand moulds (Fig. 2), it was possible to obtain various volumes of shrinkage porosity located in the centre of the bars. All the test bars were radiographed in two directions normal to the gauge length, their density measured and the dimensions of porosity in the tensile fracture determined optically. Sizes and areas of pores were also measured from the radiographs.

K_{IC} values for LM13 alloy were obtained on 12·7 × 25·4 × 152·4 mm bend testpieces designed to meet the standard requirements of plane strain tests (Fig. 1). These testpieces were cut from 25·4 × 146·1 × 298·5 mm fully sound ingots which were produced in cast iron moulds. After heat treatment the testpieces were ground to 0·05 mm tolerances, the slot was cut using a 60° cutter, and the notch radius was kept to < 0·1 mm. Fatigue cracks were obtained using a 2t Amsler Vibrophore, applying standard testing procedure except that a lower than normally recommended stress was found to be more favourable in the initiation of a smooth crack. The fracture toughness tests were carried out on an Instron machine using a tapered beam displacement gauge output connected to Y-axes of the recorder while the load applied at the crosshead speed of 3·1 mm/min was connected to the X-axes.

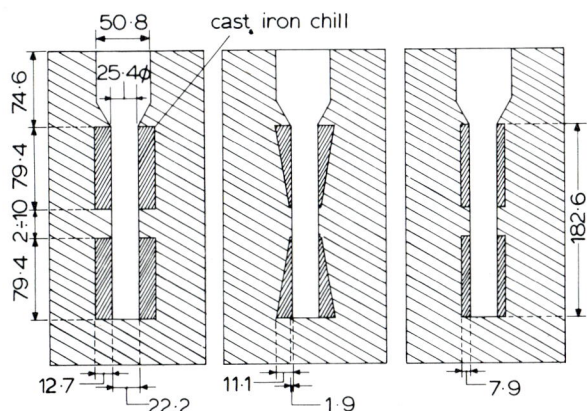

2 Test bar and mould design (dimensions in mm)

3 Typical central shrinkage porosity, LM6 alloy [×32]

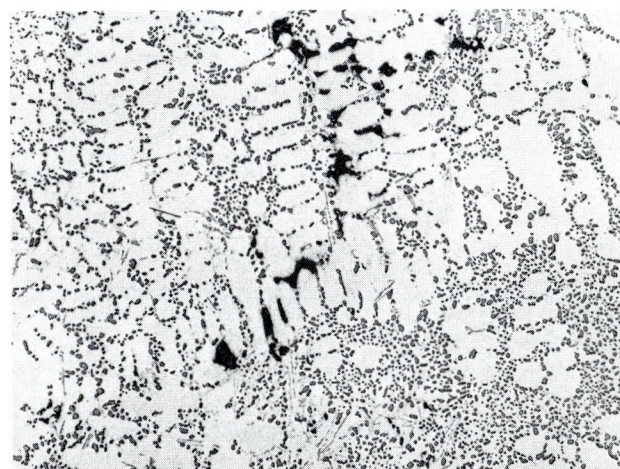

4 An example of interdendritic porosity, LM13 alloy [×115]

RESULTS

The values of porosity in controlled shrinkage bars were calculated from density measurements using the densities of sound bars, 2·665 and 2·684 g/cm³ for LM6 and LM13 alloys respectively, as reference points. The porosity observed was generally of the concentrated shrinkage type (Fig. 3) but occasionally it was interdendritically dispersed (Fig. 4). In either case a single pore located centrally in the test bar was seldom present, and several pores located in the heat centres of the bars could be identified as shown in the radiographs (Fig. 5). On the other hand, one macro type cavity of irregular morphology was invariably observed in the fracture (Figs 6 and 7).

Tensile properties were plotted both against total volume porosity of test bars and against pore area in the fracture. Detailed results have been given elsewhere,[14] while some typical relationships obtained are shown in Figs 8–14. The scatter obtained with total porosity volume in LM6 alloy (Fig. 8) was such that it was necessary to apply regression analyses to evaluate the trends numerically (Fig. 9). The scatter in the pore length or pore area in the fracture results (Figs 10 and 11), was less in comparison. The tensile strength falls off

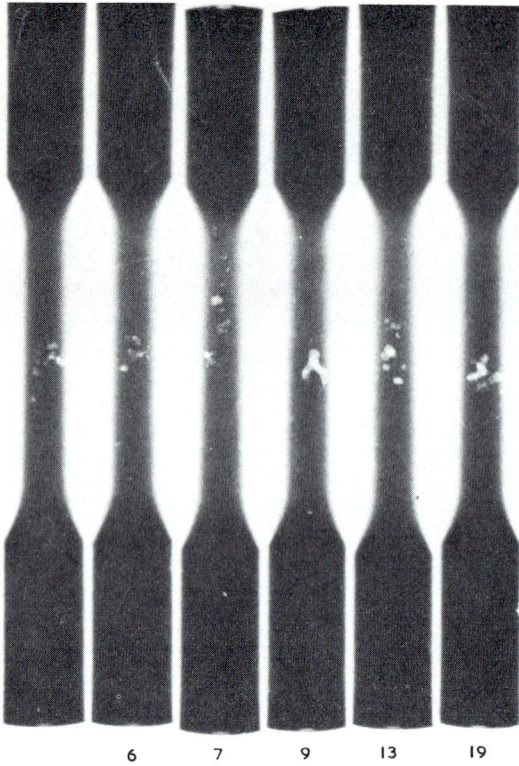

5   **Typical radiographs, LM6 alloy**

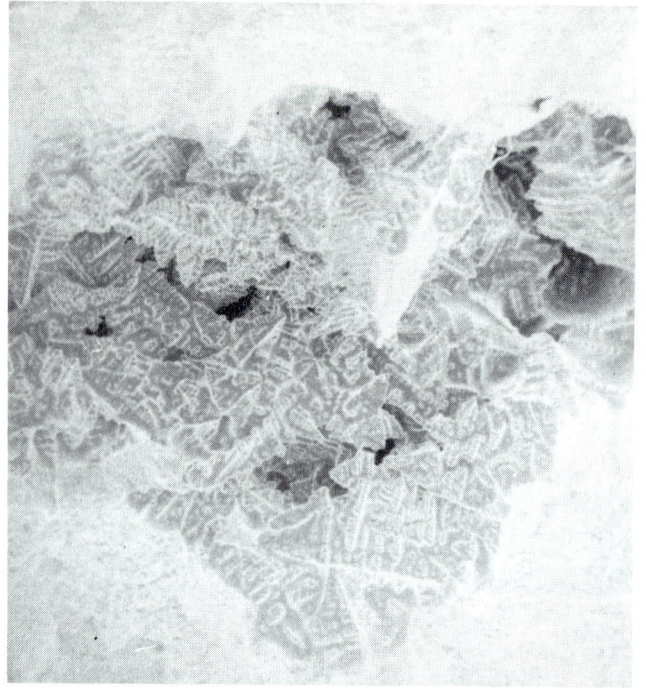

7   **Scanning micrograph of typical porosity in the fracture   [×24]**

more rapidly than that predicted by calculating the strength based on the true load-bearing area (Fig. 11). Similar results were obtained with LM13 alloy (Figs 12–14).

An increase in porosity volume does not affect the yield strength of LM6 alloy, as shown in Fig. 15, but the yield strength decreases with the pore area in the same alloy (Fig. 16). With LM13 alloy, on the other hand, even a small volume of porosity was sufficient for test-pieces to fracture at a stress w·ll below the yield strength

of sound bars. Unlike the strength properties, which fell off gradually with the porosity present, both ductility properties, elongation, and reduction of area decreased much more rapidly with the percentage of porosity (e.g.

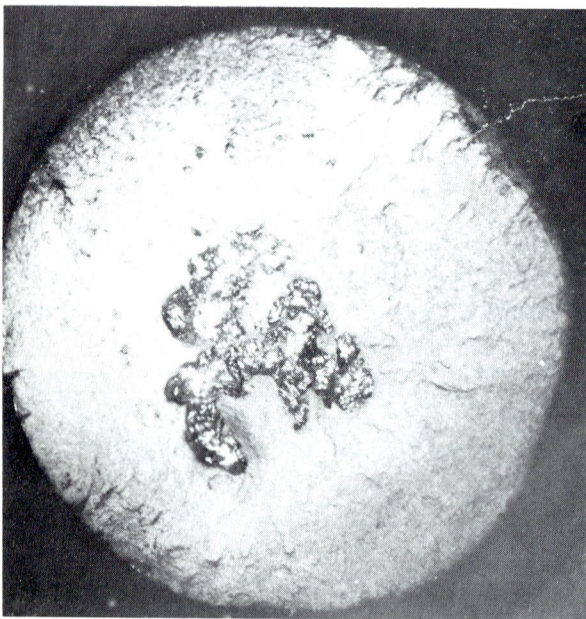

6   **Typical porosity in the fracture, LM13 alloy [×5]**

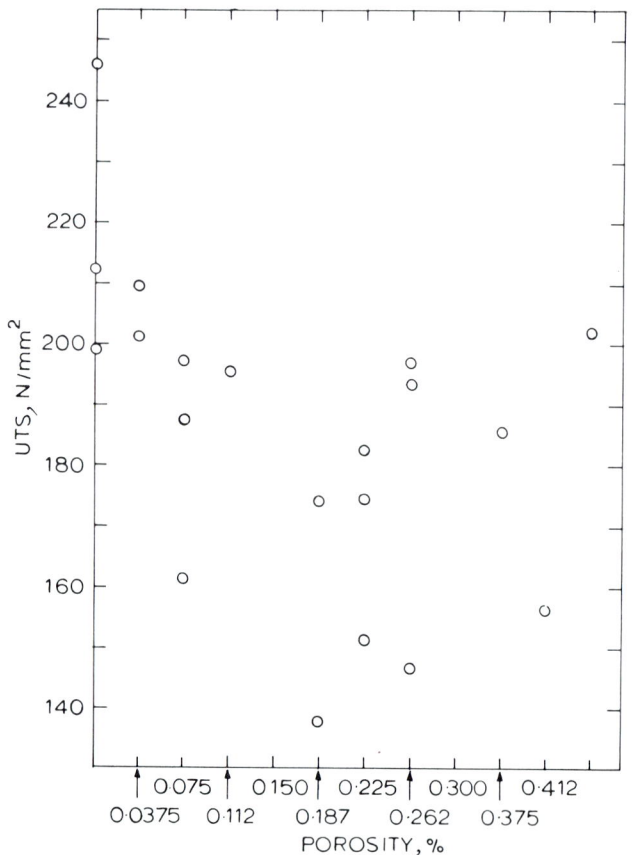

8   **Relationship of tensile strength to porosity for LM6 alloy**

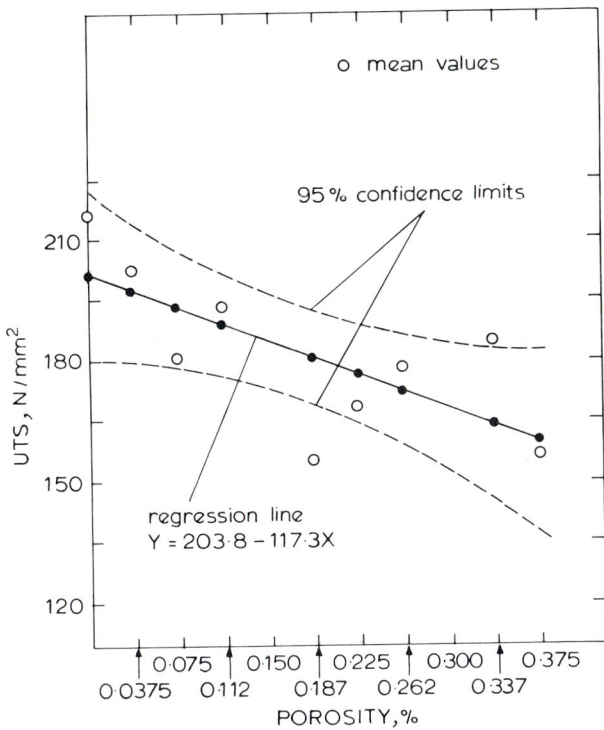

9 Regression line of tensile strength with total porosity of the test bars for LM6 alloy

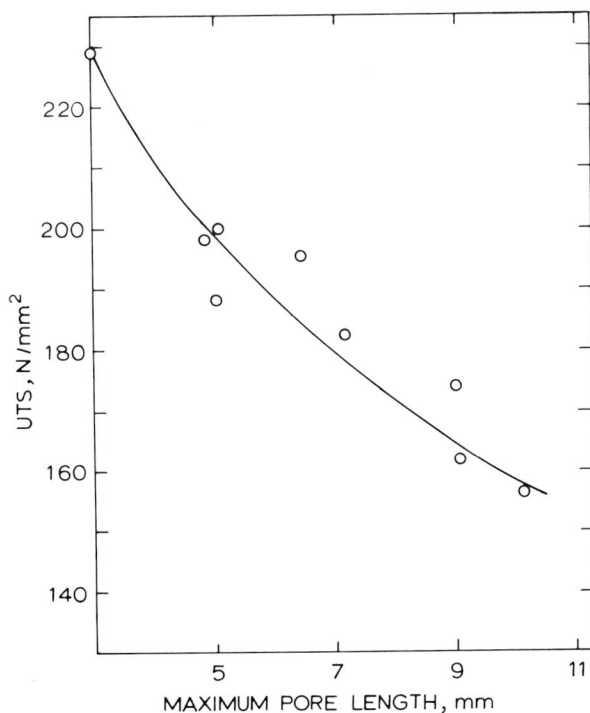

11 Effect of pore area in the fracture on tensile strength of LM6 alloy

with LM6 alloy from 11 to 1%). With LM13 alloy, however, the elongation decreased from 1 to 0·2%.

The results of K_{IC} evaluation for LM13 alloy are given in Table 1. Using the mean value $K_{IC} = 492.53 \, \text{N/mm}^{-3/2}$, the relationship between the critical flaw size, a_c, and the breaking stress, σ_w, can be obtained (for the conditions under which K_{IC} was measured) from

$$a_c = \frac{K_{IC}^2}{\pi \sigma_w^2}$$

The calculated values of flaw sizes (i.e. crack length as indicated in Fig. 1) and those obtained in the present experiments are shown in Fig. 17. The experimental values were taken from Fig. 14. For an embedded flaw, the critical flaw length 'a' was assumed to be equal to one half of the cavity's largest linear dimension as measured in the testpiece fracture. The dimension of cavities in the fracture measured in the radiographs followed the same trend but always gave slightly smaller values than those measured directly in the fracture.

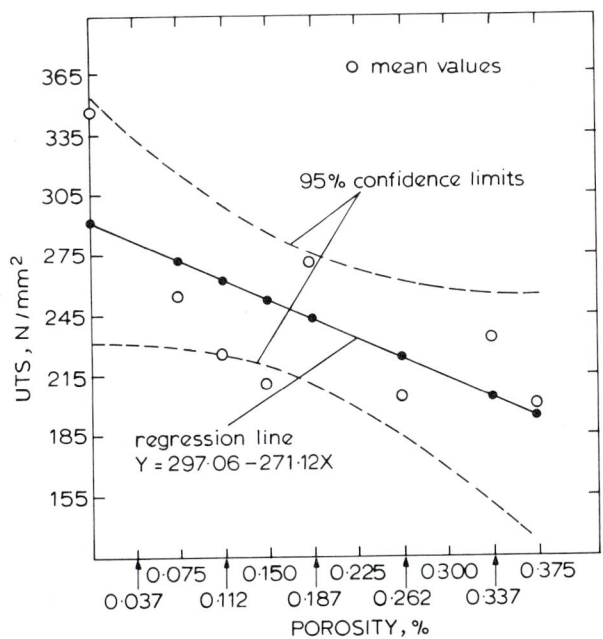

10 Variation of tensile strength with pore length for LM6 alloy

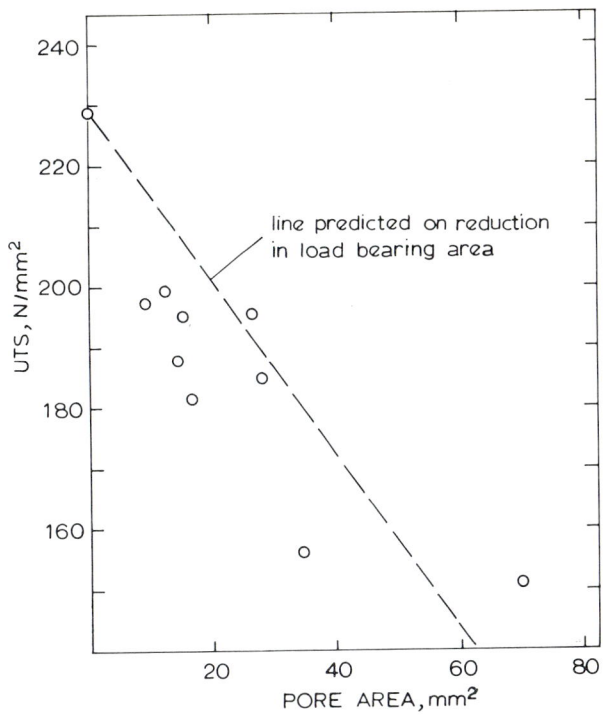

12 Regression line of tensile strength with total porosity of test bar for LM13 alloy

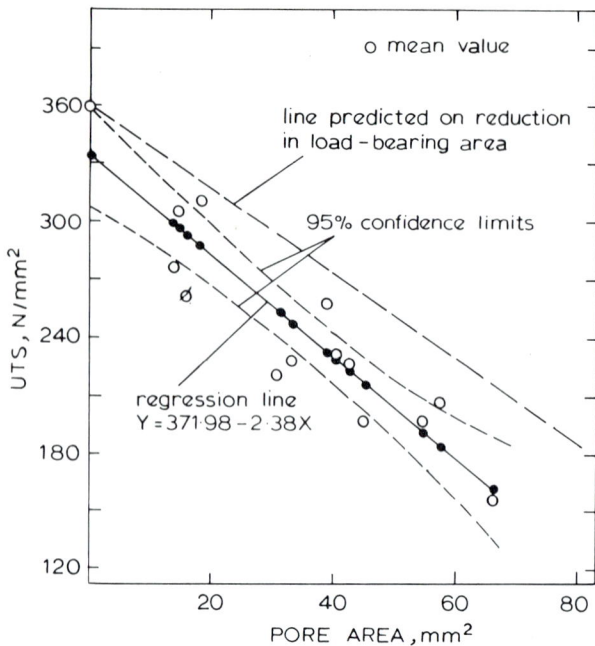

13 Regression line of tensile strength with pore area in the fracture for LM13 alloy

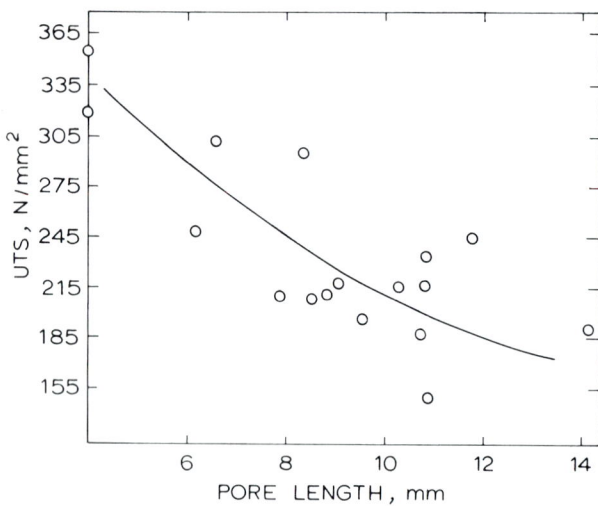

14 Variation of tensile strength with pore length in the fracture for LM13 alloy

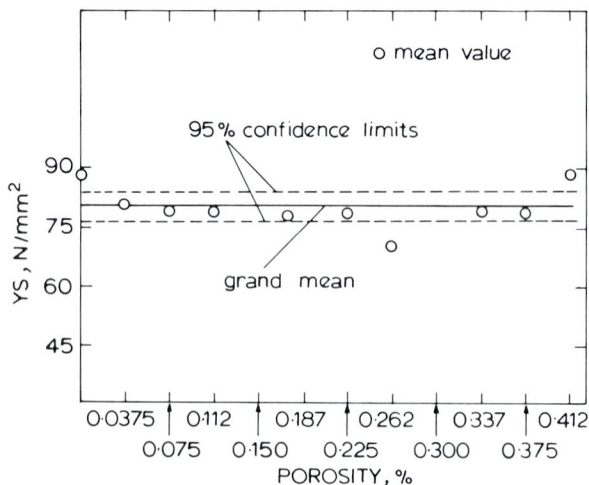

15 Variation of yield strength with total porosity in the test bar for LM6 alloy

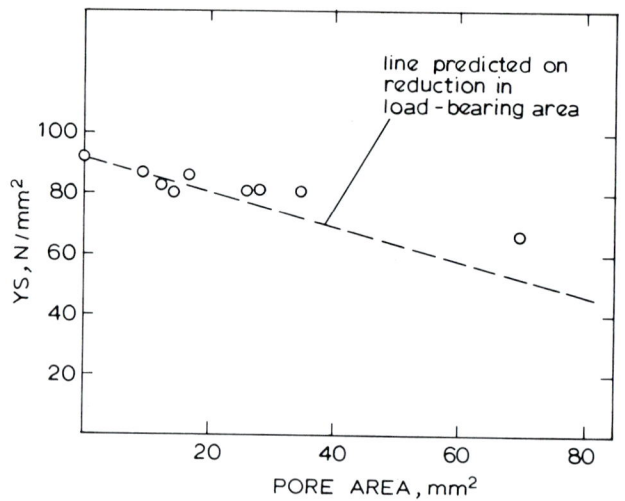

16 Effect of pore area in the fracture on yield strength of LM6 alloy

DISCUSSION

The two alloys selected for the present work are compositionally and structurally similar apart from the higher Mg content in LM13 alloy which allows subsequent age hardening heat treatment, accompanied by higher strength but lower ductility compared with LM6 alloy. Freezing behaviour and shrinkage cavities of both alloys are similar, i.e. both solidify in a skin growth manner resulting largely in centrally located intercellular shrinkage cavities of irregular morphology. The results obtained allow therefore a direct comparison of the effects of similar shrinkage cavities on strength and fracture of a relatively strong and ductile alloy (LM6) and a stronger but more brittle alloy (LM13).

The results obtained confirm earlier observations that density is the least promising parameter for studying quantitative effects of cavities on the strength property changes of a cast alloy. A possible exception would be a uniformly distributed porosity which seldom occurs either in typical casting alloys or in castings. Both the area or linear dimensions of cavities in the fracture show a general reduction in the scatter of results.

In general, the trend in the changes in properties with cavities is similar with both alloys but the magnitude of effects is different, the strength of LM13 being reduced about twice as much as for LM6 alloy and for a comparative condition of cavities. On the other hand, while the yield strength of LM6 alloy is lowered only slightly with an increase in cavities in the fracture, the fall in the yield strength of LM13 alloy is as rapid as that of the strength. Changes in the ductility properties of both alloys show exactly the opposite behaviour to that of the yield strength, with LM6 these changes being large and with LM13 being small.

The observed changes in properties in both alloys could be qualitatively explained in the light of current theories of ductile and brittle fractures. On the other hand, greater progress has been made in the quantitative interpretation of brittle fractures, and in particular that of the effects of defects on the fracture, using the methods of fracture mechanics.

The application of fracture mechanics in predicting the effects of flaws on the fracture stress of a brittle alloy

Table 1 K_{IC} values for LM13 alloy

Specimen	W, mm	B, mm	a, mm	$\dfrac{a}{W}$	Y	P_Q, N	K_Q, $\left(\dfrac{N}{mm^{-3/2}}\right)$	$\dfrac{K_Q}{\sigma_Y}$	$\dfrac{K_Q^2}{\sigma_Y}$	$2 \cdot 5\dfrac{K_Q^2}{\sigma_Y}$	K_{IC}
1	25·4	12·7	12·115 6	0·477	9·87	3 292	508·64	1·670	2·78	6·95	508·64
2	25·4	12·7	11·43	0·45	9·10	3 465	493·60	1·626	2·64	6·60	493·60
3	25·4	12·7	12·49	0·492	10·34	3 078	498·20	1·640	2·68	6·70	498·20
4	25·4	12·7	12·446	0·49	10·28	2 919	469·70	1·540	2·37	5·92	469·70

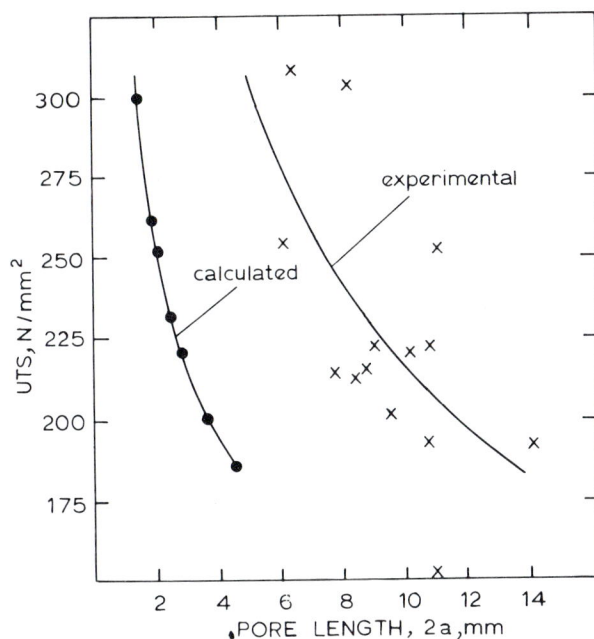

17 Calculated and experimental values of critical crack length in alloy LM13

rests in the establishment of a relationship between: fracture toughness, K_C of the alloy, defect dimension a of the testpiece, and of the stress at fracture, σ_w. For the experimental conditions used in the present work such a relationship is given by[15]

$$K_{IC} = \sigma_w \phi^{-1} \gamma \sqrt{\pi(a + r_y)}$$

where ϕ is a constant representing specific effects of defect morphology, and γ that of the test specimen, and r_y is the radius of the plastic zone at the defect tip.

Amongst these parameters, K_{IC} which represents fracture toughness for tensile loading can be readily measured for different structural conditions of an alloy. Fracture stress σ_w can be obtained from direct experiments. The value of 'a' represents the length of a through edge crack in a plate satisfying plane strain conditions. The main practical difficulties in applying the above relation to resolve such problems as studied in the present work are to be found in evaluating the specific effects of morphologies of cavities (ϕ), of the testpiece (γ), and of plastic yielding at the crack tip (r_y).

The results obtained (Fig. 17) show that only one of the linear dimensions of the cavity (the length) cannot be taken as the single critical parameter and that further

experimental work and refinements in defining shrinkage cavity parameters are necessary. An attempt was also made to utilize alternative cavity measurements, but the major problem in interpretation of the present work appears to be that of scatter of results. This is believed to be largely due to the small size of the testpiece used in relation to the dimensions of the cavity in the fracture.

CONCLUSIONS

By concentrating shrinkage cavities in the central region of the test bar it was possible to obtain quantitative relationships of the effects of shrinkage cavities on the tensile properties of two aluminium base alloys. The same technique could be developed further for applying the methods of fracture mechanics for assessing cavity effects in engineering applications of castings.

ACKNOWLEDGMENTS
The authors wish to acknowledge Centro de Investigacion de Materiales, Mexico, for a scholarship and the Department of Industrial Metallurgy, University of Birmingham, for laboratory facilities.

REFERENCES
1 M. C. FLEMINGS: 'Solidification processing', 1973, New York, McGraw Hill
2 D. E. J. TALBOT: *Inst. Metall. Rev.*, 1975, **20**, 166
3 F. A. BRANDT *et al.*: *Trans. Am. Foundrymen Soc.*, 1954, **62**, 646
4 R. W. RUDDLE: *Foundry*, 1970, **98**, (11), (12), 45
5 J. CAMPBELL: *The Cast Metals Res. J.*, 1969, **5**, (1), 1
6 R. W. RUDDLE: *J. Inst. Met.*, 1950, **77**, 37
7 G. V. KATUMBA RAO *et al.*: *Br. Foundryman*, 1973, **66**, (5), 135
8 D. IRANI *et al.*: *Trans. AFS*, 1969, **77**, 208
9 S. J. WALKER: thesis, University of Birmingham, 1968
10 C. H. CARMAN: 'Fracture toughness of high-strength materials: theory and practice', 116, 1970, London, The Iron and Steel Institute
11 S. R. HOLDSWORTH *et al.*: *Br. Foundryman*, 1975, **68**, (6), 169
12 J. C. WRIGHT: *Br. Foundryman*, 1976, **69**, 161
13 H. D. GREENBURG *et al.*: *Metals Eng. Quart.*, 1969, **9**, 30
14 A. HERRERA: thesis, University of Birmingham, 1976
15 J. T. BARNBY: *Non-Dest. Test.*, 1972, **5**, 32

Effect of variations in growth rate and temperature gradient on microstructure and creep properties of directionally solidified eutectic alloys

P. N. Quested, E. Bullock, and M. McLean

The temperature gradient G and withdrawal rate R have been varied in the directional solidification of five eutectic alloys and aspects of the creep behaviour have been related to the resulting microstructure. Superior creep properties are obtained in all systems when plane front solidification is maintained and this regime is extended by the use of higher temperature gradients. In Ni_3Nb reinforced systems, higher G can give substantial improvements in the rupture lives and reductions in the notch sensitivity. Reduced microstructural scale gives increased rupture lives, reduced creep rates, and greater creep ductilities in alloys which remain stable during testing. The opposite trend is observed in systems exhibiting a phase transformation. Electroslag processing gives an aligned but variable microstructure with rupture lives at 700°C intermediate between those of the directionally solidified and cast and extruded materials.

The authors are with the National Physical Laboratory, Teddington, Middlesex

The potential advantage of *in situ* composites, particularly for blading in gas turbines, has led to considerable effort aimed at developing practical directionally solidified eutectic alloys.[1] A wide range of materials has now been studied and two important relationships between the solidification microstructures and the important processing variables, temperature gradient G and solidification rate R, are now quite well established:

(i) G/R must exceed a critical value $(G/R)_{crit}$, which is a material parameter, in order to maintain a planar solidification front which leads to a well aligned composite microstructure; otherwise a cellular microstructure is produced

(ii) the microstructural scale is related to R by the expression $\lambda^2 R = $ constant A, where λ is the mean fibre radius or lamellar thickness of the composite; as discussed by McLean,[10] A is not always independent of G.

Hellawell[3] has recently reviewed the techniques used to produce these materials, the most common being variations of the Bridgman–Stockbarger crystal growing method although, in principle, other approaches could have certain advantages.

Both the quality and scale of microstructure has been shown to influence the tensile and yield strengths of several eutectic composites.[4] The relationships between the creep behaviour and the solidification microstructures, although of more practical relevance, have not been studied so fully[5] although there are indications that finer microstructures can be beneficial.

In this paper the results of creep tests on five eutectic composites prepared under a variety of solidification conditions will be presented and related to the types and scales of microstructures produced. On the basis of the results on the alloys Ni, Ni_3Al–Ni_3Nb,[6] Ni_3Al–Ni_3Nb,[7] (CoCr)–Cr_7C_3,[8] Ni, Ni_3Al–Cr_3C_2[9] and (Fe, Cr, Al, Y)–$Cr_7C_3^*$ the most beneficial microstructures and the most appropriate processing conditions will be suggested. In addition, preliminary results on the last-mentioned system prepared by an electroslag refining process will be described.

EXPERIMENTAL PROCEDURES

Precast ingots of eutectic composition, 13 mm in diameter, were solidified at various rates in an atmosphere of flowing argon in two modifications of a

* An unpublished experimental alloy developed by D. E. Miles (NPL)

Table 1   Nominal compositions of eutectic alloys in present study (wt-%)

Alloy	Ni	Al	Nb	Cr	C	Co	Fe	Y
Ni, Ni$_3$Al–Ni$_3$Nb (γ–γ'–δ)	71·8	2·5	19·7	6·0	—	—	—	—
Ni$_3$Al–Ni$_3$Nb (γ'–δ)	72·5	4·4	23·1	—	—	—	—	—
(Co, Cr)–Cr$_7$C$_3$ (73C)	—	—	—	41·0	2·4	56·6	—	—
Ni,Ni$_3$Al–Cr$_3$C$_2$ (γ–γ'–Cr$_3$C$_2$)	78·6	6·9	—	12·7	1·8	—	—	—
(Fe,Cr,Al,Y)–Cr$_7$C$_3$	—	3·2	—	33·2	2·6	—	60·2	0·8

Bridgman crystal growing apparatus described in more detail by McLean.[10] The two rigs designated CH and LMC gave temperature gradients of 5 and 13 K/mm respectively. The compositions of the alloys used in this study are shown in Table 1.

A single ingot, 50 mm in diameter, of (Fe, Cr, Al, Y)–Cr$_7$C$_3$ was directionally solidified in a d.c. electroslag refining unit. The electrode was made by welding six 30 mm diameter × 75 mm long ingots of the material. The apparatus consisted of a water-cooled copper jacket and base with a graphite spacer. A flux of 70 wt-% CaF$_2$; 15 wt-% Al$_2$O$_3$, and 15 wt-% CaO with a melting point of 1 300°C was used. The ingot was maintained at a positive voltage of 18 V and a current of 500 amps passed. The melt rate was 36 g/min and an atmosphere of argon was maintained.

Creep testing was performed in air under constant load. The extensions of the specimens were monitored by capacitance transducers via extensometers attached to the shoulders of the specimens. The extensometer heads and the specimen holders were made from Nimonic 80A and Nimonic 105 or 115 respectively. With these materials, the assembly was capable of testing to 1 000°C although the specimen holders had limited lives at these temperatures when highly stressed. The sample temperature was maintained within ±3°C of test temperature with typical temperature gradients of 1°C along the specimen length.

Notch creep testing was carried out on specimens with a circumferential notch conforming to the standard BS1 notch. Miniature rupture specimens were used to test the esr processed Fe, Cr, Al, Y–Cr$_7$C$_3$.

RESULTS
Ni, Ni$_3$Al–Cr$_3$C$_2$ (γ–γ'–Cr$_3$C$_2$)
Since this alloy has a very low $(G/R)_{crit}$ ratio a wide range of solidification rates can be used to produce well aligned composites with a substantial difference in fibre sizes. Consequently this material is particularly suitable for demonstrating any structure–property relationships.

The CH material apparently maintained a well aligned composite microstructure at all growth rates below 500 mm/h. However, an undesirable cellular microstructure developed at higher rates of solidification. At the lower solidification rates the carbide fibres were relatively coarse and irregular; at higher rates ($\bar{r}$) the mean fibre radius (and spacing) was reduced and the fibre morphologies became smoother. The γ' size is also a function of solidification rate, the mean cube side increasing with decreasing R. When the

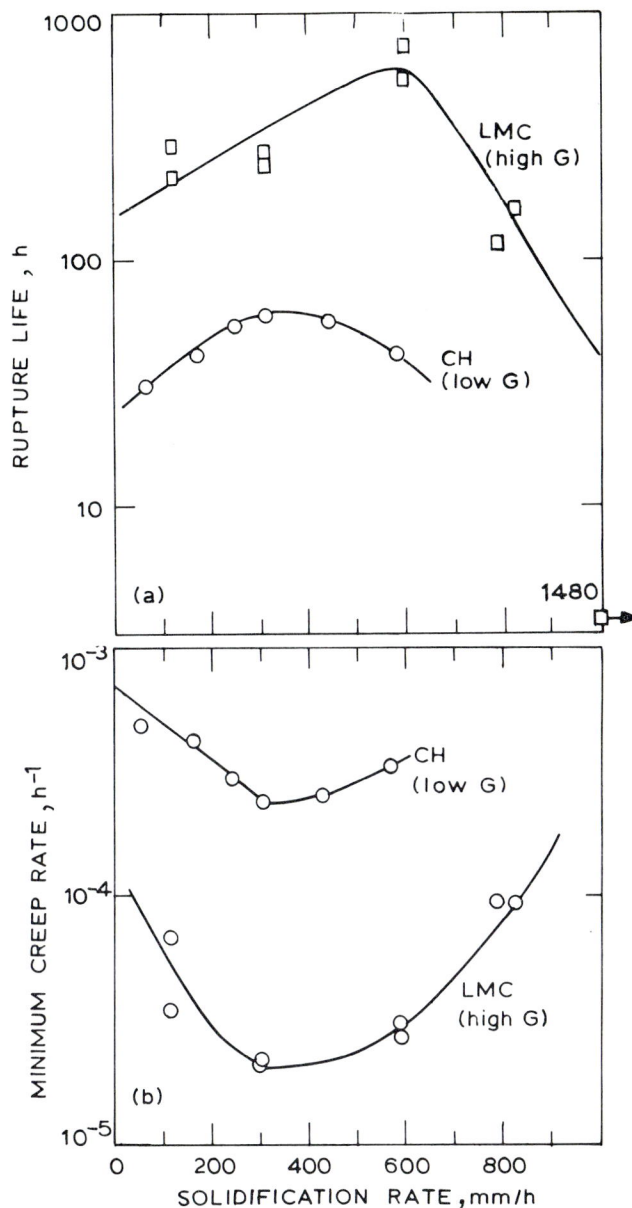

1 a Log (rupture life) v. solidification rate; b log (minimum creep rate) v. solidification rate for γ/γ'–Cr$_3$C$_2$ in two different temperature gradients (testing conditions 980°C, 122 MPa)

temperature gradient G was increased using the LMC apparatus a well aligned composite microstructure was maintained at all solidification rates considered to date in this study. The maximum rate attempted has been 800 mm/h. $\bar{r}$ was significantly reduced relative to the CH material. Again the γ' size was also reduced due to the increased temperature gradient.

The variations of rupture life and minimum creep rate with rate of solidification are shown in Figs 1a and b respectively. The data for material produced in CH and LMC show self-consistent variations but the two curves are displaced. The initial systematic improvements in creep behaviour, i.e. increasing rupture life and decreasing creep rate, with increasing R and G are associated with a progressive reduction in the dimensions of the reinforcing fibres. The reduction in rupture life and increase in creep rate at higher values of R are due to the onset of cellular microstructure.

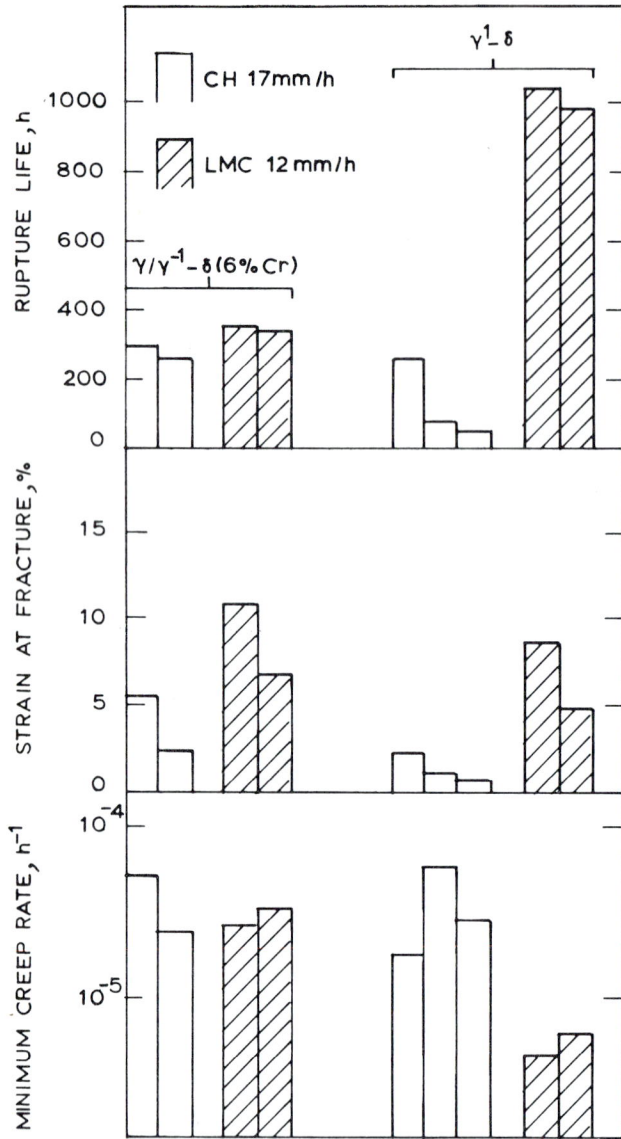

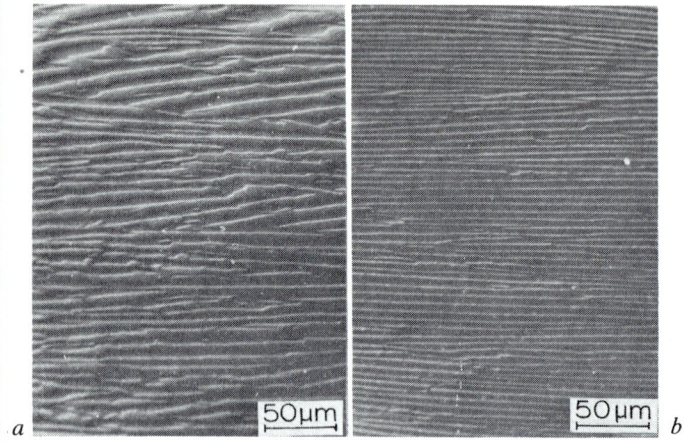

a grown at 17 mm/h in CH; *b* grown at 12 mm/h in LMC

3 Micrographs showing the longitudinal sections of $\gamma'-\delta$ creep tested at 825°C with a stress of 555 MPa

The effect of varying R on the creep behaviour of $\gamma'-\delta$ solidified in CH is shown in Fig. 4. As with $\gamma-\gamma'-Cr_3C_2$ there is an improvement in creep behaviour with increasing R, or decreasing interlamellar spacing, as long as plane front solidification takes place ($R <$ 30 mm/h). However at higher growth rates there is a sudden drop in rupture life and increase in creep rate corresponding to the onset of cellular microstructure.

A few tests with notched rupture specimens were performed on $\gamma/\gamma'-\delta$ with 510 MPa/825°C. This gives a 100 h rupture life wth a plane parallel specimen. The CH material solidified at 17 mm/h showed the creep rupture lives to be independent of the specimen geometry but for LMC material solidified at 12 mm/h substantial notch strengthening was observed wth increasing lives of between 6 and 12 times.

(Co, Cr)–Cr_7C_3 (73C)
In Figs 5*a* and *b*, minimum creep rates and rupture lives are shown as a function of rate of growth for CH material (low temperature gradient) using conditions 250 MPa/825°C and 159 MPa/1 000°C. It was established that a planar interface could be retained for solidification rates

2 Histograms showing the minimum creep rates, strains at fracture, and rupture lives of $\gamma'-\delta$ and $\gamma-\gamma'-\delta$ solidified at 17 mm/h in CH and 12 mm/h in LMB and tested at 825°C; the stresses were 555 MPa on $\gamma'-\delta$ and 500 MPa on $\gamma-\gamma'-\delta$.

Ni_3Al–Ni_3Nb ($\gamma'-\delta$) and Ni, Ni_3Al–NiNb ($\gamma-\gamma'-\delta$)
Both alloys have a high $(G/R)_{crit}$ ratio and this limits the withdrawal rates that may be used to maintain lamellar structure. In Fig. 2 a histogram shows the creep behaviour of $\gamma/\gamma'-\delta$ (6% Cr) at 500 MPa/825°C and $\gamma'-\delta$ at 555 MPa/825°C grown at 12 mm/h and 17 mm/h with LMC and CH respectively. These two conditions lead to similar interlamellar spacings so that we are observing the effect of quality of microstructure and not a scale effect. With $\gamma'-\delta$ the rupture life is increased while the minimum creep rate is lowered whereas with $\gamma/\gamma'-\delta$ the total elongation rises when using high G. There is little effect on the rupture life or minimum creep rate. Post-test examination showed that the quality of alignment of the CH material was poorer than the LMC material. Figure 3 shows examples from $\gamma'-\delta$ solidified in CH and LMC.

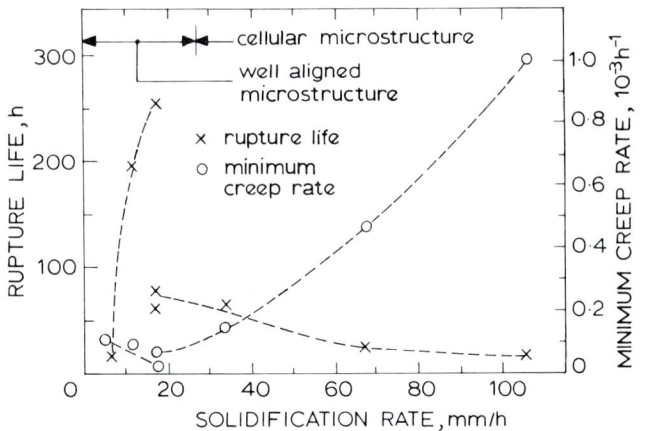

4 Creep rupture life and minimum creep rate of $\gamma'-\delta$ as a function of growth rate; all tests carried out at 825°C with a stress of 555 MPa

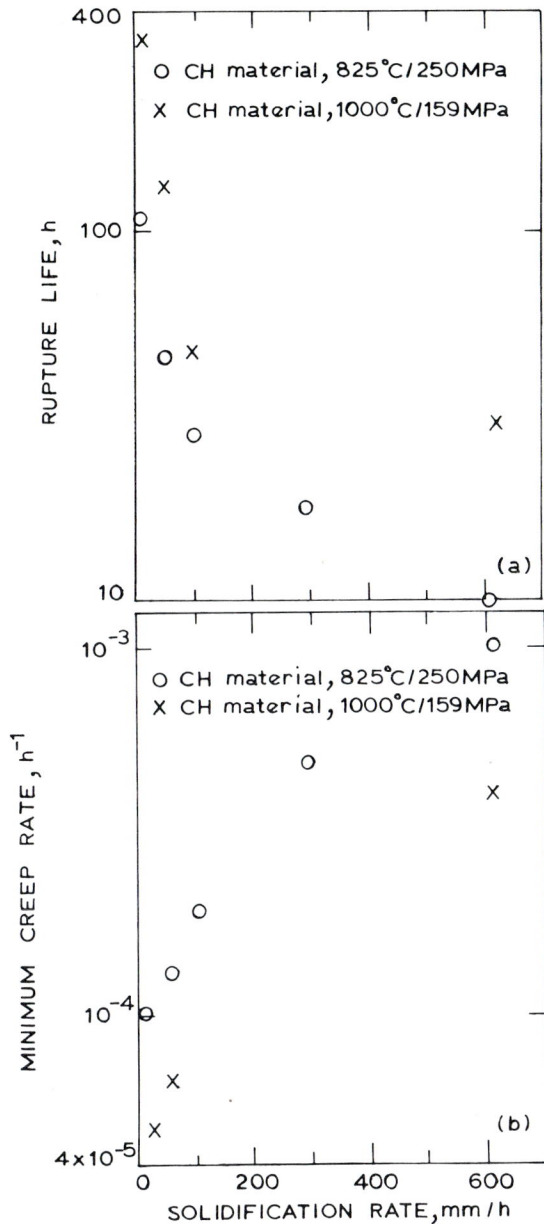

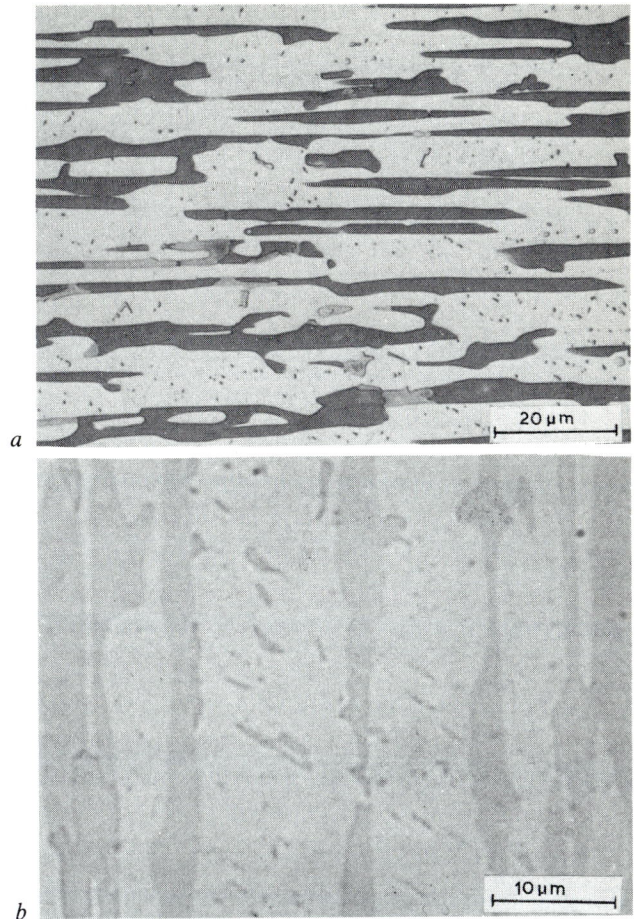

a transformation of the fibre; *b* precipitate in the matrix

6   Micrographs showing degradation of the carbide in 73C solidified in CH after creep testings at 825°C

5   *a* Log (rupture life) *v.* solidification rate; *b* log (minimum creep) rate *v.* solidification rate for 73C; tested at 825°C/250 MPa for the CH material and 1 000°C/159 MPa for CH material

below about 300 mm/h with CH, and within the planar interface regime the rupture life decreases and the minimum creep rate increases as the rate of withdrawal increases. This is the opposite trend from that found for γ–γ'–Cr_3C_2 and γ'–δ.

Post-fracture examination showed transformation of the carbide, probably M_7C_3 going to $M_{23}C_6$ together with small precipitate particles deposited within the matrix (Fig. 6).

Fe, Cr, Al, Y–Cr₇C₃

This alloy was prepared in the two temperature gradients at differing rates of growth. In addition a single ingot was prepared by electroslag remelting. Again the higher temperature gradients of LMC lead to smaller fibre radii because of the suppression of post-solidification coarsening.

The results of creep rupture tests on material grown under various conditions are summarized in Fig. 7 which plots the creep rupture life as a function of rate of growth for constant temperature and stress conditions. Two sets of data are illustrated: (700°C, 172 MPa) and (950°C, 40 MPa). At 700°C the rupture life increases as the average fibre radius increases from 1·6 μm to 2·0 μm although coarser microstructure ($\sim$3·6 μm) gives a substantially lower rupture life. At 950°C the coarsest microstructure ($\sim$3·6 μm) leads to slightly longer rupture lives. This behaviour is similar to that found for the alloy 73C.

A comparison of the microstructures of the cast and esr processed alloys is shown in Fig. 8. These show the phases in the lower 60% of the ingot to be reasonably well aligned but discontinuities run across the ingot probably due to thermal fluctuations during solidification. Miniature rupture specimens tested at 700°C enabled the stress to produce a 100 h rupture life to be established. Figure 9 shows a comparison of this value together with other iron-based superalloys.

DISCUSSION

These results clearly confirm that the creep behaviour of eutectic composites is sensitive to both the quality and

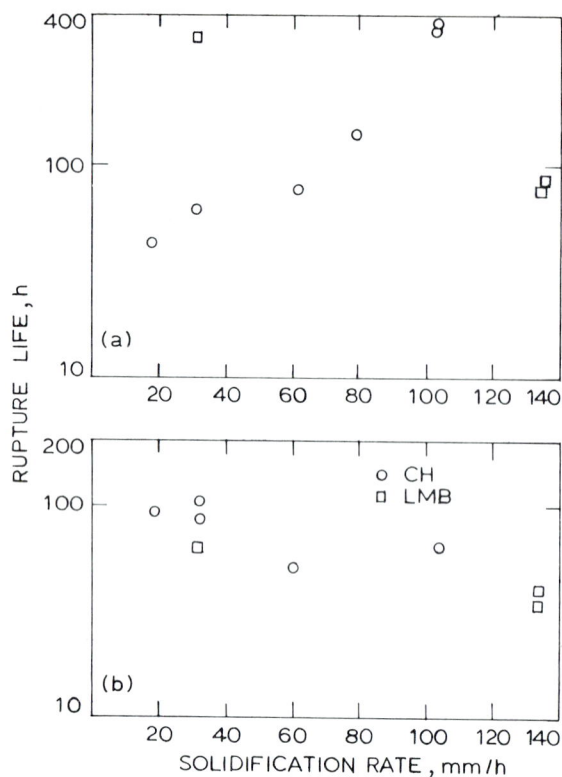

a 700°C, 172 MPa; *b* 950°C, 40 MPa

7 Log (rupture) life *v.* rate of solidification for the (Fe, Cr, Al, Y)–Cr₇C₃ alloy

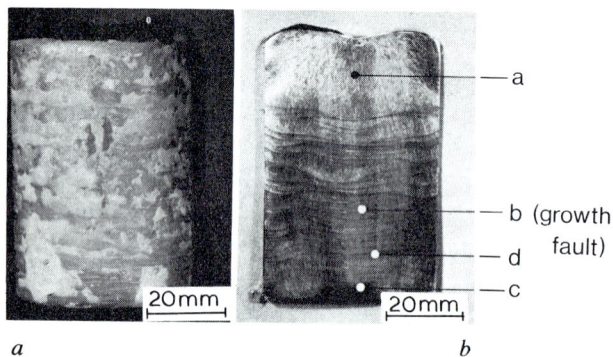

a external surface after removal of residual slag; *b* longitudinal section through the axis of the ingot

8 Macrographs of the electroslag-refined ingot of (Fe, Cr, Al, Y)–Cr₇C₃

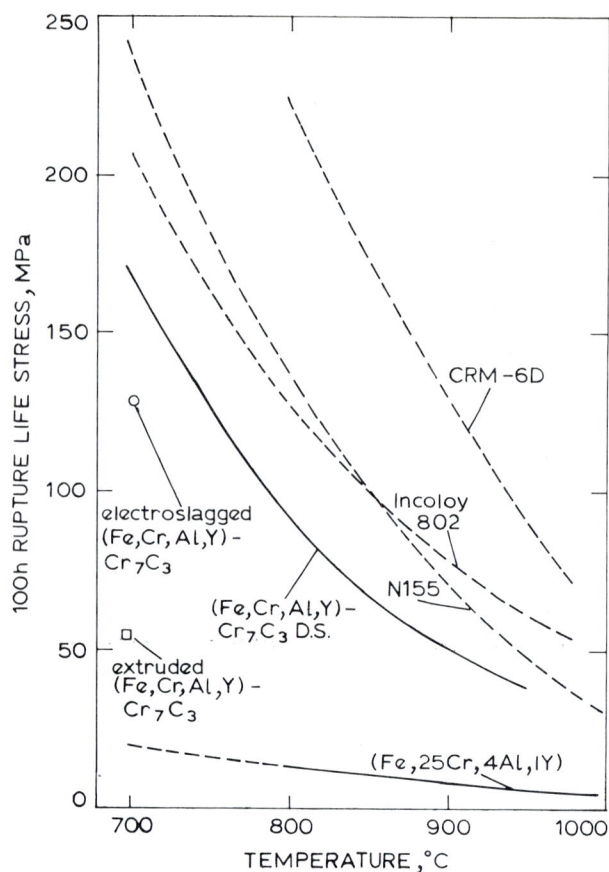

9 Plot of 100 h rupture stress *v.* temperature for (Fe–Cr–Al–Y)–(Fe, Cr)₇C₃ alloy, prepared by extrusion, electroslag refining, and directional solidification; the basic Fe, Cr, 4Al–Y[14] and CRM 6D Incoloy 802 and N155[15] are included for comparison

scale of the composite microstructure which are controlled by the G and R conditions. However, the effect of those processing variables on the response of creep properties is not the same for all the systems investigated.

The occurrence of a cellular microstructure when $G/R < (G/R)_{crit}$ is detrimental to the creep behaviour of all alloys examined although there are differences in the extent of property degradation. Thus the carbide reinforced systems γ–γ'–Cr₃C₂, 73C, and (Fe, Cr, Al, Y)–Cr₇C₃ show a gradual deterioration in properties when R exceeds a critical value whereas there is a discontinuous drop in rupture life for the intermetallic

γ'–δ system. A similar result has been observed with γ–γ'–δ.[11] Since γ–γ'–δ and γ–γ'–δ have rather higher values of $(G/R)_{crit}$, considerable advantage is derived from use of high temperature gradients which reduce the density of growth imperfections and lead to substantial improvements in creep behaviour. In particular, the creep ductility is greatly increased and this can lead to a reduction in creep notch sensitivity. This result differs from the recent work of Barkalow *et al.*[12]

The effect of decreasing fibre or lamellar dimensions within the planar solidification regime on the creep behaviour is less clear. In γ–γ'–Cr₃C₂, γ'–δ and γ–γ'–δ[11] reducing microstructural scale leads to longer rupture life, reduced creep rates, and increased ductility. These results are in accord with the predictions of a recent theory developed by Bullock *et al.*[13] In the other carbide reinforced systems examined, 73C and (Fe, Cr, Al, Y)–Cr₇C₃, the opposite trend of deteriorating creep behaviour with reducing microstructural scale is observed. In 73C there is metallographic evidence of the degradation of the carbide during creep testing leading to both a transformation of the fibres and the deposition of a fine precipitate within the matrix. This is probably associated with the transformation M₇C₃ → M₂₃C₆ during testing. Thus the finer carbide fibres, although having higher intrinsic strength, would degrade more rapidly. A similar effect may occur in (Fe, Cr, Al, Y)–Cr₇C₃ where the maximum in the rupture life versus fibre size is a consequence of competition between fibre degradation, dominating at finer microstructures, and intrinsic strengthening at coarser microstructures.

Electroslag refining of (Fe, Cr, Al, Y)–C_7C_3, in the present study, has led to a very variable microstructure although this could probably be improved substantially. Although the 100 h rupture life is less than that of the directionally solidified material it offers a significant improvement over typical (Fe, Cr, Al, Y) matrix alloys. Because of the lower costs of electroslag processing this could potentially lead to a wider use of eutectic composites.

CONCLUSIONS

1 The creep behaviour of cellular composites is worse than that of those produced by plane front solidification. The use of higher temperature gradients extends the regime of plane front solidification and leads to substantial improvements in creep properties particularly in creep ductility.

2 The finer microstructures produced by high solidification rates and high temperature gradients lead to improved creep properties in systems that are stable during testing (e.g. γ–γ'–Cr_3C_2, γ'–δ and γ–γ'–δ) but to worse creep properties in systems subject to phase transformations (e.g. 73C and (Fe, Cr, Al, Y)–Cr_7C_3).

3 Electroslag refining of (Fe, Cr, Al, Y)–Cr_7C_3 gives an aligned but variable microstructure with creep properties inferior to material directionally solidified by the modified Bridgman process but superior to the cast alloy.

ACKNOWLEDGMENTS
Part of this work was funded by the Ministry of Defence, Procurement Executive. We would like to thank D. E. Miles and S. Powell for valuable assistance during the course of the programme.

REFERENCES

1 Proc. 2nd Conf. on *In Situ* Composites, 2–5 Sept. 1975, Bolton Landing
2 G. A. CHADWICK: Proc. 1st Conf. on *In Situ* Composites, Lakeville, 5–8 Sept. 1972, I, 25
3 A. HELLAWELL: AGARD Conf. Proc. No. 156, Washington, DC, 23–24 Apr. 1974, 57
4 A. LAWLEY: Proc. 2nd Conf. on *In Situ* Composites, 2–5 Sept. 1975, 451, Bolton Landing
5 P. HENRIQUES AND W. KURZ: Proc. Conf. Verbundwerkstoffe, Deutsche Gesellschaft fur Metallkunde, Konstantz, 1974
6 F. D. LEMKEY AND E. R. THOMPSON: Proc. 1st Conf. on *In Situ* Composites, 5–8 Sept. 1972, 105, Lakeville
7 E. R. THOMPSON AND F. D. LEMKEY: *Trans. Quart. ASM*, 1969, **62**, 140
8 E. R. THOMPSON AND F. D. LEMKEY: *Metall. Trans.*, 1970, **1**, 2 799
9 D. C. TIDY: Ph.D. thesis, University of Cambridge, 1970
10 M. MCLEAN: this volume
11 K. D. SHEFFLER *et al.*: Proc. 2nd Conf. on *In Situ* Composites, Bolton Landing, 2–5 Sept. 1975, 353
12 R. H. BARKALOW *et al.*: Proc. 2nd Conf. on *In Situ* Composites, Bolton Landing, 2–5 Sept. 1975, 549
13 E. BULLOCK *et al.*: *Acta Metall.*, in press
14 C. S. WUKUSICK AND J. F. COLLINS: *Mater. Res. Stand.*, 1964, **4**, 637
15 W. F. SIMMONS: ASTM Data Series Publication No. DS9E

DISCUSSION

Session 6: Structure and property relationships and welding

Chairman: F. B. Pickering
(Sheffield Polytechnic)

The following papers were discussed: *Keynote Address: Solidification phenomena and properties of cast and welded microstructures* by P. R. Sahm; *Influence of steel composition on segregation and microstructure during solidification of austenitic stainless steels* by O. Hammar and U. Svensson; *Solidification cracking in 18 Cr–13 Ni–1 Nb stainless steel weld metal: the role of magnesium additions* by G. Jolley and J. E. Geraghty; *Studies of TIG weld-pool solidification and weld bead microstructures in stainless steel tubes* by S. de Rosa, M. H. Jacobs, and D. G. Jones; *Control of the solidification process during submerged arc welding* by T. Edvardsson; *Non-epitaxial growth in weld metal* by C. R. Loper and J. T. Gregory; *Oriented structures and properties in type 316 stainless-steel weld metal* by B. L. Baike and D. Yapp; *Structure–property correlation fracture studies on directionally solidified Fe–C alloys* by K. V. Prabhakar, H. Nieswag, and A. J. Zuithoff; *Effect of decreased section thickness on the formation, structure, and properties of a chill-cast aluminium–silicon alloy* by G. R. Armstrong and H. Jones; *Effect of porosity on tensile properties of Al–Si alloys* by A. Herrera and V. Kondic; *Effect of variations in growth rate and temperature gradient on the microstructure and creep properties of directionally solidified eutectic alloys* by P. N. Quested, E. Bullock, and M. McLean.

THE CHAIRMAN: This is where we come to the nitty-gritty of the process we are dealing with, namely, structure and property relationships, because in the end we have, in fact, to be able to sell the weld metals or castings that are produced.

DR E. SELCUK (Middle East Technical University, Ankara, Turkey): I would like to comment on Dr Geraghty's paper. As we all know Mg has a very high vapour pressure, nearly atmospheric, at its melting point. Therefore, it is unlikely to keep Mg within the melt if there is no sulphur. My first question is 'have they any quantitative evidence that Mg is present in the vicinity of their modified eutectic'. My second comment is that I think the proposed electrode material 18/13/1 with Mg will hardly help to introduce Mg into the weld pool even under an argon shield.

DR JOLLEY: In answer to the first part of the question by electron probe microanalysis, some magnesium was found to be associated with the eutectic grain boundary areas and also with the inclusions present in the material in the eutectic and other areas. With regard to the

second part of the question, the melts which were carried out in the laboratory took longer to solidify than would occur with a weld bead under argon. Provided that the magnesium was incorporated in the electrode wire no problem of the type suggested by Dr Selcuk was foreseen.

MR M. McLEAN (NPL): I would like to comment on the paper by Jolley and Geraghty. The enhanced spheroidization observed after the addition of magnesium suggests that the interfacial energy which provides the driving force for the process has been increased. However, in metallic systems interfacial segregants almost invariably lead to a reduction in interfacial energy. It is more likely that magnesium scavenges a minor impurity present in the original alloy in much the same manner as described by Johnson and Smartt earlier in this Conference. The reason that rogue elements were not identified is probably due to the lack of resolution of the electron microprobe microanalyser. Hondros at NPL has shown that interfacial energy can be very significantly reduced by segregation at monolayer levels to interfaces. Such impurities are only likely to be detected by a surface analytical technique, such as Auger Electron Spectroscopy.

DR JOLLEY: We did not find any sulphur or phosphorus in the grain-boundary eutectics and the micro-analyser was sensitive to 0·5% of the elements. If a scavenging reaction was associated with smaller amounts of these elements, our work neither proves nor disproves Mr McLean's interesting speculation.

PROFESSOR R. SMITH (Queen's, Canada): In connection with the paper by Dr Jolley on the effect of magnesium in modifying the 'precipitate' form in 18/13/1 stainless steel, we have a general interest in the microstructure that a eutectic may adopt in the presence of surface active additions. We have looked at many non-faceted/faceted systems and found that an increase in the degree of modification goes hand in hand with an increase in absorption energy on the faceting phase.

DR S. A. BALOGUN (Sheffield University): In Yapp and Baike's work on austenitic stainless-steel welds, they suggested that oblique ultrasound penetration makes flaw detection easier. I would like to ask whether they observed a difference in grain structure and hence ultrasonic wave attenuation in the heat-affected zone and in the weld bead area. It has been widely suggested that the delta ferrite/austenite ratio, the grain orientation, and secondary grain size all influence wave attenuation in stainless steels. Have the authors tried modifying the weld grain size and, if so, are their results consistent with the suggestions in the literature.

DR YAPP: We believe that the grain size in austenitic multi-pass welds is a result of the balance between competitive growth and nucleation of new grains which are both eventually controlled by frequent misorientation of the fusion boundary in multi-pass welds. Thus the angle θ between fusion boundary normal and the dendrite growth direction will determine to what extent competitive elimination of grains will occur, and at the same time if the angle θ exceeds about 30° new grains may be nucleated and grow. Hence, a possibility to refine grain size is to arrange for frequent renucleation by producing a large percentage of misorientations between 30° and 60°. This can be achieved in practice by using small electrode gauge sizes in manual metal arc welding or by using welding processes with a deep penetration profile on normal metal arc welding. In the grain oriented welds produced by the methods described in our paper, epitaxial growth from the grain boundary of the parent plate occurs in the normal way with competitive elimination of unfavourably orientated grains, and epitaxial growth of favourably orientated grains.

DR BEECH (University of Sheffield): If you consider the steels that Mr Hammar studied and lower the nickel content you produce another structure in which the solidification is totally as δ with a solid state change to $\delta + \gamma$. The transformation and the resulting structure are similar to the $\alpha\beta$ brasses.

It is even possible to produce both this type of structure and the one that occurs via a peritectic reaction in the same ingot. You find the former structure towards the ingot walls and the latter nearer the centre. This produces an ingot which is equiaxed on the outside and has a secondary columnar structure in the interior.

In response to Dr Yapp's contribution I agree with his findings but unfortunately the steel caster cannot make use of the orientation effect that he describes as can the welder. In our experiments we have found that the alloys that solidify as δ and subsequently transform to $\delta + \gamma$ have a lower ultrasonic attenuation than the others when a random orientation is considered. It may be, therefore, that this type of structure is useful to the steel castings producer.

Finally, the amount of δ ferrite is considered to be critical in many applications but evidence on the effect of cooling rate on the amount present is conflicting. Does anyone have any comments on the effect of cooling rate in this respect?

MR HAMMAR: I will only verify the statement by Dr Beech. With respect to the distribution of the δ ferrite and its morphology at the surface and ingot centre you can often see both types of mode of solidification occurring in full scale ingot casting. For stainless steels there are two points in steel composition which are of real interest. First, you have the critical composition for transition from austenitic to ferritic freezing, secondly you have the critical composition when you get a completely ferritic freezing without a peritectic reaction. For these two types of stainless steels we can always see the influence of the cooling rate on the final as-cast structure since you have different rates of solidification at the surface and at the ingot centre.

PROFESSOR H. W. KERR (University of Waterloo, Canada): With regard to Dr Beech's question on the effect of cooling rate on ferrite in stainless steels we have been doing some work on the effect of cooling rates on various austenitic and ferritic type steels. One can in some circumstances either inhibit the peritectic transformation because of fast cooling rates or in fact miss primary ferrite nucleation altogether. The latter can then subsequently give interdendritic ferrite rather than intradendritic ferrite. It remains to be seen whether the type of ferrite affects the hot cracking susceptibilities. The nucleation of one or both of austenite and ferrite, as separate grains also appears to affect the columnar or equiaxed transition during solidification.

DR YAPP: I am not sure that I understood Dr Beech's remarks. Some work was done about 15 years ago showing that, in welds, delta ferrite does not play a particularly significant part in the compositions we are considering.

DR BEECH: What I was saying was that the structure which results, starting as delta, and then changing over by solid-state reaction, behaves better in that you can get an ultrasonic beam through it more easily than you can through an alloy which undergoes a peritectic reaction because the latter has a secondary austenite grain structure which is generally coarser.

DR YAPP: As I understand it, you are saying that there is a difference in the final effective grain size. Perhaps I could talk about some theoretical work on stainless steel which is relevant here. The ultrasonic attenuation has been theoretically described by the Merkvlov relationship which says the attenuation of alpha is equal to a constant times the elastic anisotropy squared, times grain size over velocity to the sixth power. This shows that we expect grain size to have a large effect on attenuation and with larger grain sizes one expects to see increased ultrasonic attenuation. However, the details of this sort of analysis are very complicated and this particular analysis applies only for a grain size range which is greater than $\lambda/3$ and less than 3λ. It works for our welds and we can in fact predict a minimum of 45° ultrasonic attenuation. If, however, we are talking about castings having a much larger grain size compared to the testing wavelength, we have to do a different sort of analysis. Then it may be unreasonable to expect the same sort or indeed any sort of attenuation relationship with grain size.

MR J. E. NORTHWOOD (National Gas Turbine Establishment): I would like to comment on Dr Sahm's paper, with regard to his Fig. 12. It is unfortunate that one of the few properties of unidirectionally solidified material reported at this Conference should show a reduction in creep life compared with the conventionally cast material, particularly at the highest stress condition. In most nickel alloys I have cast by unidirectional solidification the opposite is the case. There is an improvement in the creep/rupture life, although the extent is often greatest at the intermediate temperature range of 700°–800°C. Another point I would like to raise in this connection, which I think has some bearing on a number of the other papers, is that Dr Sahm mentions that thermal shock resistance is improved in directionally solidified alloys. One of the points of importance here is that in a directionally solidified material the Young's Modulus is lower in the preferred ⟨100⟩ growth direction. Therefore, for a given amount of thermal strain the stress is very much reduced, and this of course helps to reduce the effects of thermal shock. In his Fig. 11, Dr Sahm shows some examples of castings with different starter lengths. This is the length of growth to produce the preferred orientation before actually starting to grow the casting. He shows that there is grain coarsening and a somewhat inferior product with some short starter lengths. Casting geometry will affect the grain structure. If you have a sudden widening of section and the grains grow sideways, you do not get any fresh nucleation, so that you tend to get a coarser structure above this particular wide section. It is difficult to tell from the figure but I wonder whether casting the blade the opposite way up would have given better results. The shape looks as though it would be amenable. The extent of this problem will be partly governed by the quality of the unidirectional structure.

DR F. SCHUBERT (Thyssen Precision Foundry, West Germany): In answer to the first question by Mr Northwood regarding the creep behaviour of directionally solidified material in comparison with equiaxed material, it has been shown that the directionally solidified material has a higher primary creep rate than the conventional equiaxed material. This can be for two reasons. One is that the cooling rate with the directionally solidified process after solidification is much lower than the cooling rate of equiaxed material. Due to the slower cooling rate below the solidification point the gamma morphology is different from the morphology and distribution resulting from conventional investment casting. The other fact is that the directionally solidified material can have fewer dislocations. However, the directionally solidified materials have longer duration in the secondary creep stage and gain in tertiary creep at stress levels which are realistic in stationary gas turbine engines. The thermal fatigue properties of turbine blades are mainly influenced by crack initiation along the grain boundaries transverse to the trailing edge. Therefore, a well aligned grain structure gives no chance for this kind of crack initiation with the result that directionally solidified material has a superior thermal fatigue resistance. Figure 11 is shown by us to indicate how important the construction of the whole shell mould is to the obtaining of a geometrical way of solidification, in order to get a well aligned blade.

DR P. R. SAHM (Brown Boveri and Co.): In Dr Quested's presentation a figure was projected showing Cr_7C_3 transformed into $Cr_{23}C_6$ after creep. Did you make a conscious effort to shift the solvus line between the 2-phase field Co/Cr_7C_3 and the 3-phase field $Co/Cr_2C_3/Cr_{23}C_6$ by alloying towards a Cr-richer composition to ameliorate this tendency to transform on heat treatment?

DR QUESTED: We have not tried alloy development. We saw the system we examined as a model to gain experience and not as a gas turbine material.

DR SCHUBERT: Could Dr Quested show a typical creep curve for directionally solidified materials defining: primary creep rate, maximum creep rate, and 1% creep deformation limit?

DR QUESTED: The diagram referred to as Fig. A shows a typical creep curve obtained for well aligned Ni_3Al–Ni_3Nb tested at a stress of 555 MPa and a temperature of 825°C. The curve shows well defined primary, secondary, and tertiary stages. In this example the primary creep elongation is small whereas with poorly aligned structures the elongation during primary creep is larger with a less well defined secondary creep stage. The other systems considered in this work broadly show the same behaviour.

PROFESSOR R. SMITH (Queen's, Canada): I was pleased to see Dr Kondic's attempt to quantify the deleterious effects of porosity in cast metals. Most commercial alloys contain inclusions and when pores form they frequently do so in association with them. As a result the pore geometry may be dependent on that of the inclusion. Because of this, it might be expected that the resultant strength of a casting for a given pore size may be dependent on pore shape. To what extent did Dr Kondic consider this influence when measuring pore size?

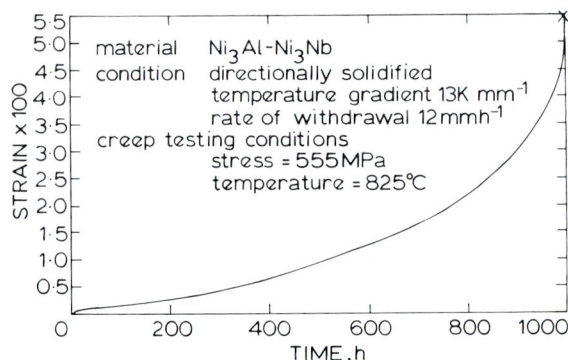

Fig. A   **Typical creep curve for well aligned Ni$_3$Al-Ni$_3$Nb**

DR V. KONDIC: Relating the strength of flake graphite cast iron to their microstructure has provided a challenge to metallurgists for many years and is likely to do so for a long time. Dr Prabhakar's contribution to this knowledge is very welcome. Another possible field worth exploring would be the application of fracture mechanics methods. The assumption could be made that the cast irons provide matrix structures of various kinds in which graphite flakes represent 'cracks' or notch effects. In this interpretation there should be a relationship between graphite flake length and cast iron strength. The necessary data to apply fracture mechanics methods to cast iron are still lacking, but collation of data from a recent paper by Duff and Wallace (*Tr. AFS*, 1976) suggests that perhaps this line of attack on the problem would be worth pursuing further.

PROFESSOR M. HILLERT (RIT, Stockholm): In his presentation Dr Kondic suggested that we should pay more attention to the eutectic reaction $L \rightarrow \alpha + P$, where α is the solid metal and P is porosity. I completely agree with Dr Kondic and am sure he will be happy to hear that such work is already under way in Stockholm. In fact, Dr Fredriksson has already published a directional-solidification study of this reaction showing that it gives rise to a microstructure of the rod-like eutectic type. (H. Fredriksson and I. Svensson: *Metall. Trans.*, 1976).

PROFESSOR I. MINKOFF (Technion, Haifa, Israel): The problem of estimating the strength of graphite iron composites has been tackled in my laboratory and the analysis suggested by Dr Prabhakar cannot be simply applied, because the strength of the composite *decreases* as a function of the volume of the second phase. Therefore, the theory, which is based fundamentally on Griffith analysis, is not completely applicable.

DR K. V. PRABHAKAR: First, in answer to Professor Minkoff. The amount of graphite in the alloys of our investigation was kept constant, in all our experiments, at 4·3%. The effect of growth rate is to bring about an increase in fineness of graphite with increase in growth rate. Correspondingly, the specimens showed an improvement in ultimate tensile strength and yield strength. We are currently very busy varying the graphite content but still retaining the eutectic composition by adding silicon. The model proposed by me has not yet been applied to these ternary alloys.

Secondly, in answer to Dr Kondic, we have attempted to analyse the mechanical behaviour and predict the mechanical strength of our cast iron assuming graphite particles as ellipsoidal holes (K. V. Prabhakar and H. L. Ewalds: Internal report, Delft University of Technology, Metallurgy Laboratory, 1977, to be published). We measured the principal dimensions of the ellipsoidal graphite particles in specimens solidified at various growth rates. The measurements ranged from 200 at low growth rates to 800 at high growth rates. From these principal dimensions and assuming the matrix strength, we could estimate the ultimate strength of cast irons solidified at different rates R. Very good agreement between the measured values of tensile strength and those estimated from fundamental principles was observed. However, the agreement was obtained only after assuming a matrix strength which was three times more than the actual value. The complete implications

of our model are not clear yet. There are also certain limitations in our model, e.g., the interconnected nature of graphite (holes) is not taken into account. However, the agreement between the measured results and the estimated results is rather encouraging. We are currently busy refining our model.

DR YAPP: This concerns Dr Gregory's paper. I would like to congratulate him on his excellent micrographs and rather remarkable results on non-epitaxial growth in high-speed welding of aluminium alloys. However, I found it difficult to visualize how his proposed mechanism can operate. If we assume that growth of a dendrite side arm blocks epitaxial growth in a neighbouring grain, the dendrite side arm would necessarily be in a region of higher temperature weld in the strong thermal gradient within the weld pool, and there would appear to be no driving force for extended sideways growth. The maximum undercooling would occur at the fusion boundary of the unfavourably oriented grain, and hence if this grain were to be prevented from growing, one might expect the nucleation of new grains at the fusion boundary. However, this would not explain the nature of the grain orientations found in Dr Gregory's work.

DR GREGORY: In reply to Dr Yapp, the initial weld metal freezing at the fusion boundary is by planar growth, usually of the order of several microns thick. Subsequently, open-end cells develop, followed by full cellular growth as described by Malatesta and Buffum (Ref. 5 in our paper). It is proposed that the suppression of growth of neighbouring grains to produce non-epitaxial growth occurs primarily by strong horizontal growth components within this initial planar region, and to a much lesser extent by branching within the subsequently cellular growth zone. In response to the second part of the question, earlier work on extremely large-grain base metal, in the order of 5 mm dia., revealed that welding at 60 mm/s produced no nucleation whatsoever in the fusion zone when the ⟨100⟩ of the partially-melted base metal grain was oriented normal to the fusion boundary. Since these high welding speed, non-epitaxially-grown fusion zones may be viewed to a first approximation, as single crystals with ⟨100⟩ oriented normal and parallel to the fusion boundary, it is concluded that nucleation is not expected in these fusion zones and indeed no discrete nucleation events have been observed.

DR YAPP: In answer to Dr Gregory's reply (that dendrite growth would not have time to develop near to the fusion boundary), it is difficult to visualize sideways growth of a cellular structure and I am not sure that preferred growth directions have been observed for planar fronts.

DR H. SOLARI (Comision Nacional de Energia, Argentina): With regard to the paper by Drs Loper and Gregory, usually, the perfect wetting of the base metal by the molten weld metal is followed by columnar grain growth controlled by a competitive mechanism. In the model proposed for non-epitaxial growth, the columnar grains that appear do so not by independent nucleation events but by epitaxial growth of a small number of favourably orientated base metal grains. The only differences with the conventional growth during the solidification of a molten metal arise from the fact that the competitive growth becomes more important under some particular conditions. So for high welding speed, some of the grains are inhibited from growing during the first step of the solidification. The principal parameter affecting the competitive growth is the shape of the liquid metal pool which is strongly dependent upon the welding velocity.

Session 7

New processes and products

Chairman: J. Nutting (University of Leeds)

Editor: H. A. Davies (University of Sheffield)

Keynote Address: New solidification processes and products

M. C. Flemings

This paper attempts to summarize some recent important advances in solidification theory and practice. As solidification technology is advancing so rapidly it is possible to discuss a representative sample of topics only, including: ingot and shape casting processes, grain size and grain size control, dendrite structure, and inclusions and microporosity.

The author is with the Massachusetts Institute of Technology, Cambridge, Massachusetts, USA

The problems for solidification science and engineering range from the macroscopic to the submicroscopic. On the science side we try to understand, quantify, and predict phenomena, and on the engineering side the aim is to control solidification so as to produce materials more economically, or with improved properties and performance characteristics.

On the macroscopic scale we are concerned with ingot or casting design, and with problems such as the macrosegregation sketched in Fig. 1. On the microscopic side we are concerned with grain size and orientation (Fig. 2). On a finer scale we are concerned with the dendritic structure and microsegregation within the grains (when dendrites exist) or with various 'non-dendritic' structures possible when dendrites are not present (Fig. 3).

On a still finer scale, we study and attempt to control inclusions and microporosity (Fig. 4), and on yet a finer scale we finally reach the atomic level and are concerned with the nature of the liquid/solid interface and the kinetics of attachment of new atoms during solidification (Fig. 5).

At all these levels, excellent science has been underway over the last decade, and the engineering achievements have been significant. Selected engineering examples are described below.

INGOT AND SHAPE CASTING PROCESSES

Continuous casting throughout the world has become increasingly important over the last ten years. Continuous (or semi-continuous) processes now account for essentially all aluminium and magnesium production, and for some 15% of the production of carbon steels in the USA and 30% in Japan. Aluminium is typically cast in semi-continuous ('D.C.') ingots. Steel is cast in billets from about 4in square to slabs 104in × 8in. The bulk of production is in conventional stationary water-cooled moulds, but new processes involving movable moulds are appearing: so far, the most successful of these are the rotating wheel processes for producing copper (or aluminium) rod for electrical applications. Here, the metal is solidified in a 'D' shaped groove in the rim of a rotating water-cooled wheel. A steel band, at the rim of the wheel, completes the mould cavity. In the process of one US company (Southwire), rolling mills are placed in line with the casting unit, producing $\frac{1}{4}$in–$\frac{5}{8}$in dia. copper rod at rates of up to 40 t/h.[1]

Another, startlingly innovative, semi-continuous casting process is the Watts–Technicon process, wherein the casting is stationary, the water-cooled mould moves, liquid/solid metal travels through the lengthening ingot before it reaches the mould, where it solidifies (Fig. 6).[2]

Electroslag remelting is another type of ingot casting process that has grown rapidly in the last decade. One advantage is that the use of a desulphurizing slag permits the reduction of sulphur to extremely low levels, but the major advantage is the improved solidification control that can be obtained by the process.[3]

New shape casting processes have been developed over the last ten years or have seen greatly increased applications over this period. The major emphasis in US foundries is on cost reduction through greater mechanization, automation, and computer control. One important new process is the 'CLA' process developed at the Hitchiner Corporation, New Hampshire, by G. D. Chandley and J. Lamb. This process permits substantial savings in the manufacture of lost wax investment castings by eliminating the conventional pouring and cut-off operations. The process is shown schematically in Fig. 7. Liquid metal is 'drawn' into the cavity under partial vacuum and allowed to remain there only until the castings are solidified. The remaining liquid in the 'riser' then drains back to the metal pot to be used again,

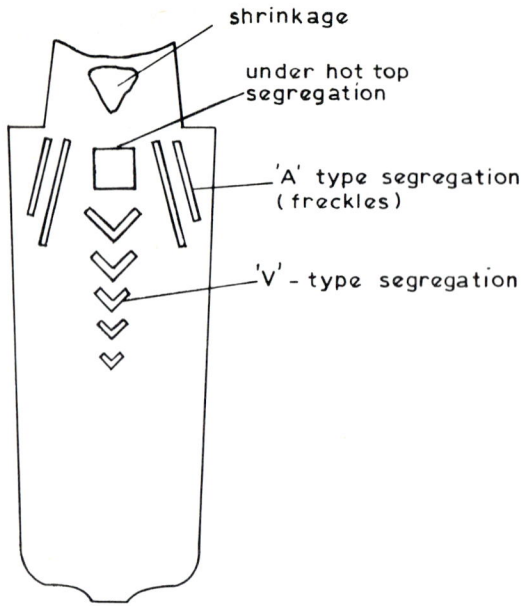

1   **Some macroscopic features of solidification structure of a large ingot**

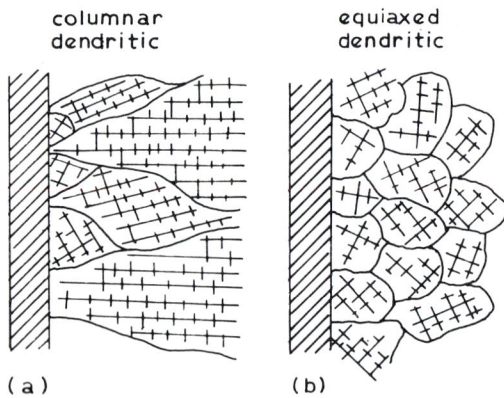

2   **Grain structure in a solidified casting or ingot**

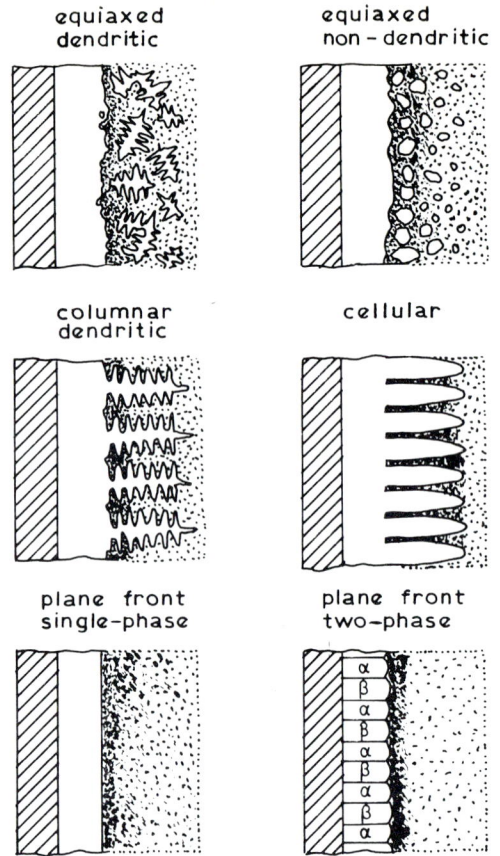

3   **Types of solidification structures that can be obtained in directionally solidified binary alloys (of the non-faceting type)**

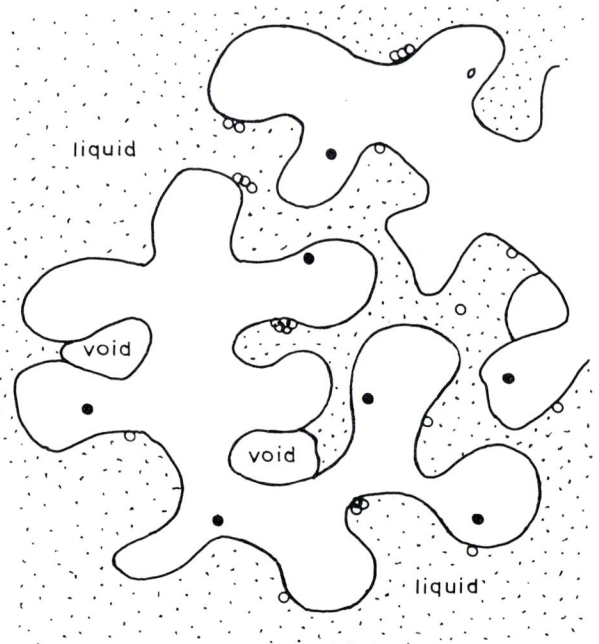

4   **Some microscopic features of solidification structures (sketched during solidification): black circles represent trapped inclusions; open circles represent 'pushed' inclusions; microvoids also shown**

and the castings come out of the mould individually with only small 'gate stubs' to be ground down.

The fundamental work on the macroscopic behaviour of large castings and ingots has taken place primarily in two areas, on computer simulation of heat flow and solidification, and on fundamental analyses of macrosegregation.[4,5] In this latter area, we now know the basic mechanism of formation of the important types of macrosegregation and are in a position to apply this knowledge to improve the homogeneity of large ingots. As one example, Fig. 8a is a sketch of isotherms during solidification in a small Sn–Pb ingot being solidified in a manner to simulate solidification of a large ESR ingot. Resulting radial segregation in this ingot is pronounced, as seen in Fig. 8b. It is now understood that this segregation results from gravity-driven convection within the liquid/solid 'mushy' zone. When the ingot is rotated at a proper angular velocity during solidification, radial acceleration effectively counterbalances the gravitational acceleration and segregation is essentially eliminated (Fig. 8c). This work is the current doctoral thesis work of Mr Sindo Kou in our laboratory.

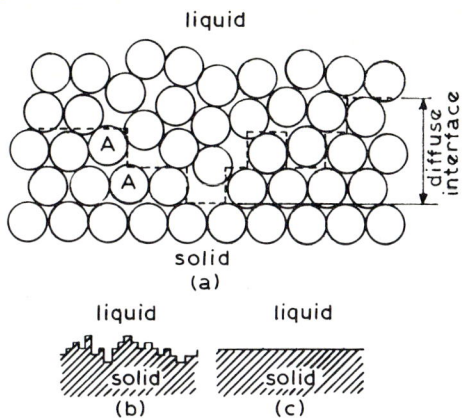

a atom packing in a diffuse interface;
b schematic representation of a diffuse
interface; *c* atomically flat interface

5 Two types of interface

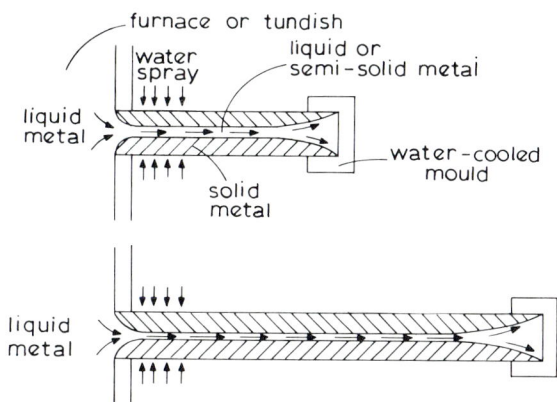

**6 The Watts–Technicon process: top, early in
solidification, water-cooled mould moves,
ingot is stationary; bottom, later in
solidification**

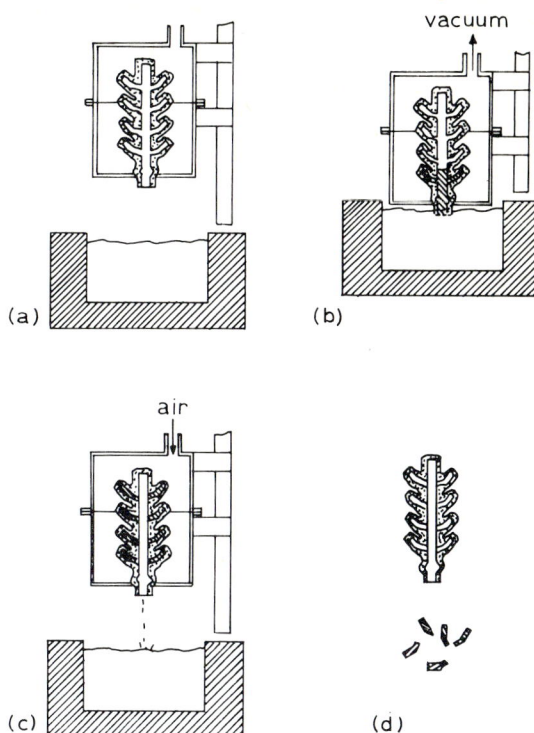

a hot mould in chamber, pour cup down; *b* pour
cup immersed in melt, metal vacuumed into
mould; *c* castings solidify, vacuum released,
castings stay in mould, most gating returns to
melt; *d* when ceramic is removed, castings
within short gates ready for final cut and grind

**7 Sequence of CLA casting of investment
moulds**

GRAIN SIZE AND GRAIN SIZE CONTROL

In the 1950s and early 1960s considerable research was
conducted on grain refinement of metals by chemical
additions. Improved refining agents and procedures
were developed during this period for aluminium and
magnesium alloys. Surface grain refiners, now commercially important for super alloy investment castings
were also developed at this time.[6] During this period
great effort was also devoted to finding effective grain
refining agents for steel, especially stainless steel, but
without success.

During the last decade, we have learned that there is
another factor working to promote fine grain size, in
addition to heterogeneous nucleation, i.e. dendrite
'remelting' (Fig. 9). The remelting is enhanced by agitation during the initial stages of solidification, and it is
now clear that this mechanism is of great importance in
determining the final grain size of many cast metals.

In the laboratory, we have exploited this phenomenon
by the vigorous agitation of metal during the entire first
50% or so of solidification, so that we completely break
up the dendritic structure into fine, more or less
spheroidal, solid particles. The resulting liquid–solid
mixture comprises a 'slurry' that can be formed directly

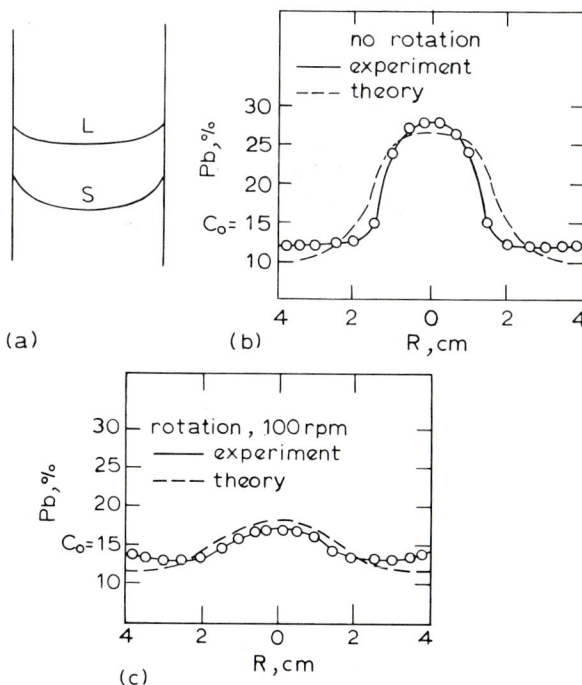

a isotherms; *b* segregation in a non-rotated
ingot, *c* segregation in a rotated ingot

**8 Segregation in a stationary and rotated Sn–Pb
ingot**

before flow   after flow

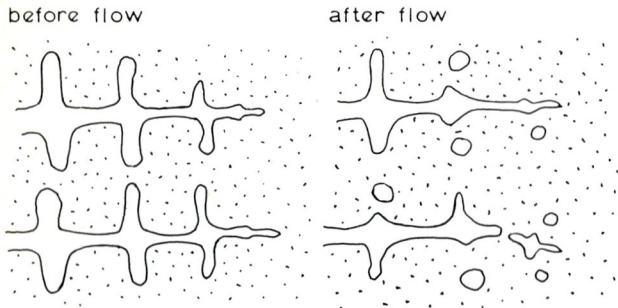

9   'Dendrite multiplication' resulting from convection and associated thermal pulses

(we term this process 'Rheocasting', Fig. 10) or can be solidified, then reheated to its partially liquid state and cast (we term this process 'Thixocasting', Fig. 11).[7] This very new process, which has great commercial potential, was the direct result of a fundamental study on deformation behaviour of semi-solid alloys.[8]

This process is now in pilot production of parts of a number of different metals, including bronze and steel. Primarily, the work at MIT is currently on steel and we are finding that steel die castings can be made, using this process, with excellent quality and with what we believe is a more than adequate die life to make the process economic. Some results of recent die life studies are shown in Fig. 12. Die determination occurs by surface checking and an arbitrary crack rating has been developed to compare die materials. About 500 castings were made in H-13 die steel before reaching a crack rating of 5. After 500 castings, the die of H-21 steel reached a crack rating of only 3 and a water-cooled

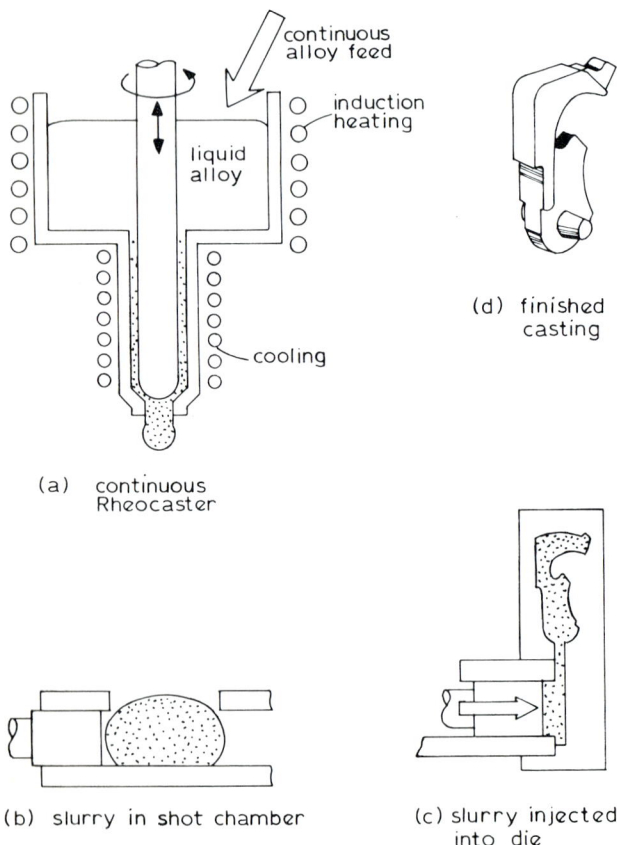

11   Schematic of the Thixocast process

copper mould showed no cracking whatsoever. Figure 13 shows the five hundredth casting in each of the three moulds. This study is continuing in the laboratory under the direct supervision of Dr Kenneth P. Young with the aim of optimizing process variables. The results to date have made us very optimistic about the commercial potential of ferrous Rheocasting and Thixocasting.

Of course, fine-grained structures are not always desired in cast metals. In some cases (notably for high-strength magnets and for high-temperature alloys) we sometimes prefer columnar grains. Columnar 'directionally solidified' turbine blades have now become of great commercial importance in the USA. Large mechanized vacuum furnaces are used to melt, cast, and directionally solidify these blades over periods of several

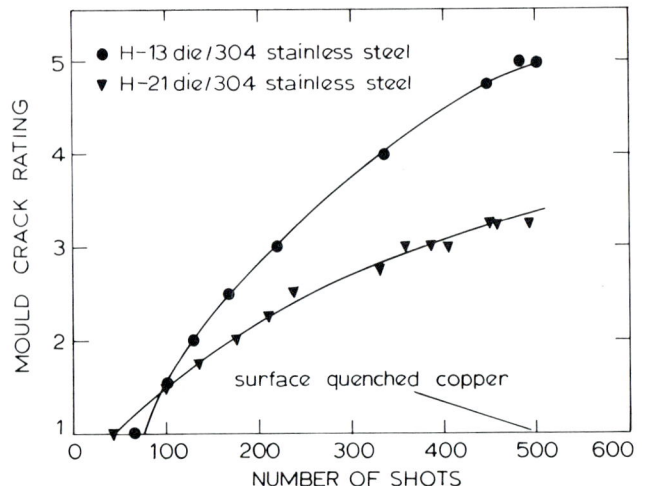

10   Schematic of the Rheocast process

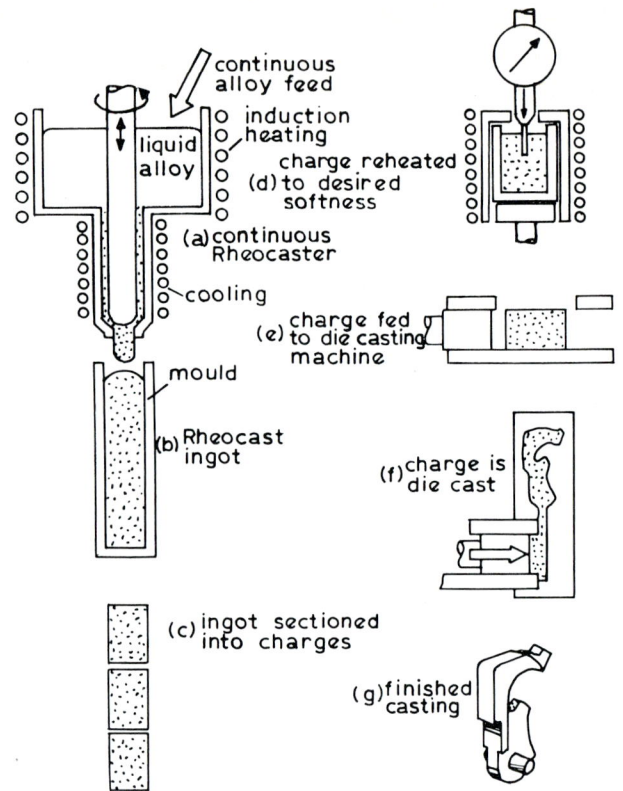

12   Crack rating for Thixocasting trials on three different die materials

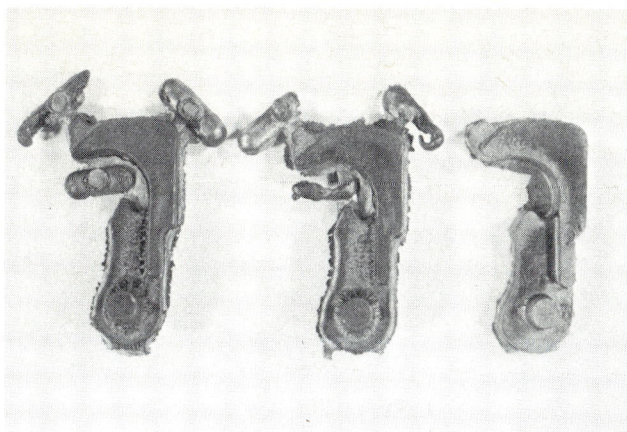

left to right: H-13 steel; H-21 steel; surface quenched copper

13 Five hundredth castings in three different mould materials

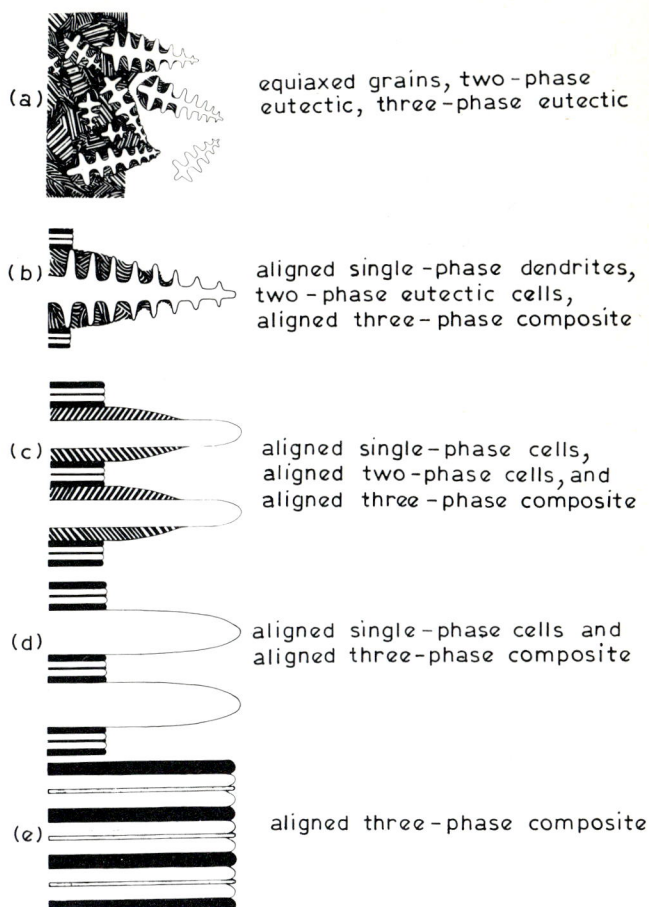

(a) equiaxed grains, two-phase eutectic, three-phase eutectic

(b) aligned single-phase dendrites, two-phase eutectic cells, aligned three-phase composite

(c) aligned single-phase cells, aligned two-phase cells, and aligned three-phase composite

(d) aligned single-phase cells and aligned three-phase composite

(e) aligned three-phase composite

14 Various possible microstructures obtained from three-phase alloys, in order of increasing *G/R*

hours. The directional solidification is achieved by employing a water-cooled chill at the base of the blades, while heating the investment casting mould up to, or above, the melting point of the metal cast. Special procedures, which are now the subject of renewed research interest, enable us to obtain an entire turbine blade with a single oriented columnar grain. Whether single crystal, or polycrystalline, these directionally solidified ('d.s.') turbine blades possess properties and serviceability substantially above those of conventionally cast blades.

The directionally solidified dendritic studies described above are achieved at modest values of G/R where G is the thermal gradient and R is the growth rate. At still lower values, structures tend to be equiaxed, while at higher values, cellular structures are obtained (Fig. 14). At still higher temperature gradients, in alloys that are two-phase on equilibrium solidification, eutectic-like *in situ* composites are obtained which have properties at high temperatures that are significantly superior to the best dendritic directionally solidified alloys now produced. These '*in situ* composites' have not yet reached the production stage, but there are a number of research and development activities throughout the world aimed at bringing them to commercial reality. A recent conference has summarized much on-going work in the area.[9]

DENDRITE STRUCTURE

Whether dendrites are aligned, as in the columnar turbine blades described above, or equiaxed, as in many castings and ingots, the important parameter describing dendrite structure that has correlated best with engineering properties is dendrite arm spacing, usually secondary dendrite arm spacing. Finer dendrite arm spacings have been found to produce stronger, more ductile castings of a variety of different metals. Ingots of finer dendrite arm spacing have been shown to result in wrought material of improved properties, and this, for example, appears to be a significant factor in the improved properties found in ESR-melted metals.

Dendrite arm spacing depends almost exclusively on local cooling rate in a casting or ingot. Typically, the dendrite arm spacing is linearly proportional (on a log–log scale) to cooling rate, with slope being about $-\frac{1}{3}$. The major factor governing secondary dendrite arm spacing is now known to be 'ripening' or 'coarsening' which proceeds during the solidification process. The less time we allow for this ripening (by increasing the cooling rate), the finer will be the resulting structure.[10]

Another example from a very different field is the manufacture of alumina–zirconia abrasives. Here, a molten alumina–zirconia alloy is poured into thin section metal moulds (e.g. into the interstices between small diameter iron balls) and rapidly solidified. The alumina–zirconia is then removed from the mould, crushed, and fabricated into grinding wheels. The resulting abrasive is vastly superior, for many operations, to conventionally produced abrasives. Important advances have been and are being made by research and development engineers at Norton Company in this industrially important area.

It is possible to obtain very rapid solidification rates (greater than, say, about 10^4 °C/s) only in very thin sections. We cannot hope to achieve them in bulk specimens.* One option we have is to go to atomized melts and cool the liquid metal droplets very quickly. A major research activity is now going on at Pratt and Whitney

---

* Unless we can some day learn to supercool bulk melts greater than the critical undercooling given by H/C_p, where H is heat of fusion and C_p is specific heat of solid/liquid

Aircraft, Florida, in which superalloy droplets are cooled, in some cases probably at rates $> 10^6\,°C/s$. The powders are then compacted into billets for subsequent processing into high-temperature engineering components. The results of this programme, as they continue to develop, will be of considerable scientific and engineering interest.

A still different example of obtaining very rapidly solidified structures is that of surface melting a thin layer on a metal part with a high-energy laser and then allowing it to resolidify. The cold metal beneath the layer acts as an ideal chill and solidification rates can be obtained that are well in excess of $10^6\,°C/s$. The special structures obtained may be found to have special engineering advantages in a number of applications.[11]

When solidification rates approach or exceed $10^6\,°C/s$, solidification of some metals no longer takes place in the usual dendritic or cellular fashion. The resulting structure can be an essentially homogeneous crystalline structure or a non-crystalline 'glassy' structure, and material of this latter type is being studied extensively throughout the world, and even being produced for sale by one US company. Much room remains for further exploitation of rapid solidification to produce either very fine dendritic structures, homogeneous equilibrium structures, or non-equilibrium structures such as 'glasses'. A recent conference summarizes much work in this area.[11]

On the theoretical side of dendrite formation and growth, significant progress has been made in the last decade and work in several laboratories has produced some particularly exciting results recently. Following the work of Ivantsov in 1947, a number of workers in the 1960s, including notably Horvay and Cahn[13] and Trivedi,[14] developed analyses for dendritic growth, considering only the dendrite tip, and its growth into an isothermal or isoconcentrated liquid. These studies led to a detailed understanding of certain controlled solidification phenomena but were too idealized to be of use in understanding solidification behaviour of real castings and ingots.

Also during the 1960s, Flemings and co-workers took the different approach of neglecting dendrite tip related phenomena and concentrating on processes occurring behind the tip during the solidification of alloy melts. This led to a quantitative understanding of a variety of solidification phenomena, including microsegregation, macrosegregation, the mechanism of establishment of secondary dendrite arm spacing, etc.[15]

Now, Burden and Hunt have shown that the basic ideas emanating from the two different approaches can be unified, and that the result is a fresh basis on which to view the dendrite growth process.[16-18] Figure 15 shows data from Burden and Hunt on tip undercooling versus growth rate for an Al–2% Cu alloy and compares these data with the unified theory. This unified view of the dendritic growth process is a sound approach that will, I believe, lead to solution of other important remaining dendrite growth problems, such as the mechanism of establishment of primary arm spacing.

INCLUSIONS AND MICROPOROSITY

Only in the last decade have we begun to apply modern solidification theory to any great extent to the formation of inclusions in metal during solidification. Inclusions, like primary metal phases, often grow dendritically and

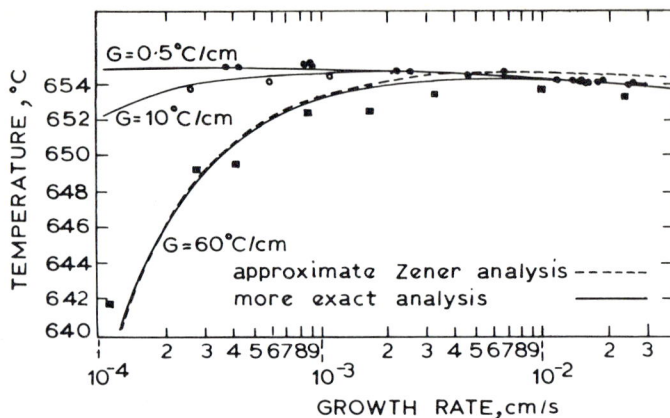

15 A comparison between the experimental points in the Al–2 wt-% Cu alloy, with a theoretical line using the Zener approximation for a gradient of 60°C/cm and the theoretical line for a more accurate analysis for 0·5, 10, and 60°C/cm[16]

'remelt' to produce new inclusions. They also collide with each other, often sticking to build up massive inclusions; they also coarsen and coalesce.

An important engineering development in the field of inclusions in recent years is the tracing of nozzle blockage in steel to the build-up of alumina inclusions in the nozzle, not by a solution precipitation and growth mechanism, but by a collision and sticking mechanism.[19] Another has been the discovery that residual calcium in steel melts eliminates results of sulphide inclusion stringers, indigenous alumina 'galaxies', and results in fine oxide inclusions with a CaS 'halo' about them. The steels produced with calcium treatment have significantly improved machinability.[20] Calcium-treated steels can be produced with improved transverse and bending properties.

The filtering out of inclusions is an important step in the production of high-quality wrought aluminium, and of a variety of cast materials including aluminium, magnesium, and sometimes copper-base alloys. Perhaps we will see the time when high-quality ferrous alloys are filtered as well.

A small amount of microporosity is always present in cast structures and we can strive to reduce this by good metal degassing and by adequate feeding. The porosity can also be removed by 'hot isostatic processing' at up to 20 or 30 atm of argon at high temperature.[21] It has been found that improved properties of cast metals result when the last remaining microporosity is removed in this way, and certain 'shrink-prone' superalloy parts are now being routinely hot isostatically pressed in the USA. Much smaller pressures should be effective in eliminating this porosity if the pressures were applied during solidification, as shown by Uram *et al.*[22] on an aluminium alloy, and by G. D. Chandley[23] on a superalloy.

CONCLUSIONS

This paper has attempted to summarize some recent important advances in solidification theory and practice. On the practical side, solidification technology is advancing so rapidly that it is possible in a paper of this type to discuss at best a representative sample. This paper attempts to do that, with emphasis on new ingot and shaped casting processes. Equally, there is now much activity throughout the world on advancing solidification theory, and in this paper reference is made to a small fraction of that work, particularly to aspects that have special relevance to understanding of engineering problems, or development of engineering processes.

REFERENCES

1 D. B. COFER: 'Southwire continuous rod system', paper presented at Institute of Metals Meeting, Sept. 20, 1971 (Southwire Company, Carrollton, Georgia)

2 L. WATTS: US Patent No. 3 517 725 (Jun. 30, 1970)

3 Electroslag Remelting and Plasma Arc Remelting, National Materials Advisory Board Report, National Academy of Sciences, Washington DC, 1976

4 M. C. FLEMINGS: 'Principles of control of soundness and homogeneity of large ingots', Axel Hultgren Memorial Lecture, May 28, 1975, Stockholm, Sweden (also *Scand. J. Met.*, 1976, **5**, 1)

5 R. MEHRABIAN *et al.*: *Met. Trans.*, 1970, **1**, 1 209

6 J. A. REYNOLDS AND C. R. TOTTLE: *J. Inst. Metals*, 1951, **80**, 93

7 M. C. FLEMINGS *et al.*: *Mater. Sci. Eng.*, 1976, **25**, 103

8 D. B. SPENCER *et al.*: *Metall. Trans.*, 1972, **3**, 1 925

9 Conference on *In Situ* Composites — II, (eds Jackson *et al.*) 1976, Lexington, Mass. Xerox Publishing

10 T. Z. KATTAMIS *et al.*: *Trans. Met. Soc. AIME*, 1967, **239**, 1 504

11 E. M. BREINAN *et al.*: 'Surface treatment of superalloys by laser skin melting', 1976, United Aircraft Co., to be published

12 Proceedings of the Second International Conference on Rapidly Quenched Metals, *Mat. Sci. Eng.*, 1976, **23**, (eds N. J. Grant and B. C. Giessen)

13 G. HORVAY AND J. W. CAHN: *Acta Metall.*, 1961, **9**, 695

14 R. TRIVEDI: *Acta Metall.*, 1970, **18**, 287

15 M. C. FLEMINGS: 'Solidification processing', 1974, New York, N.Y., McGraw–Hill

16 M. H. BURDEN AND J. D. HUNT: *J. Cryst. Growth*, 1974, **22**, 99

17 *Ibid.*: 109

18 *Ibid.*: 328

19 S. N. SINGH: *Metall. Trans.*, 1974, **5**, 21

20 D. C. HILTY AND J. W. FARRELL: *Iron Steelmaker*, May and Jun., 1975

21 M. BASARAN *et al.*: *Metall. Trans.*, 1973, **4**, 2 429

22 S. Z. URAM *et al.*: Transactions AFS, 1958, **66**, 129

23 G. D. CHANDLEY: US Patent No. 3 420 291

Sheet casting process for aluminium and aluminium alloys

A. Handasyde Dick, M. C. Simms, D. D. Double, and A. Hellawell

A historic method for the casting of low melting point metals in sheet form has been successfully adopted for the semi-continuous casting of aluminium and aluminium-base alloys. The design and operation of a laboratory scale plant are outlined and its use described for aluminium, aluminium–copper, and aluminium–silicon alloys, each with a variety of minor additions.

The first three authors are with the Department of Metallurgy and Science of Materials, University of Oxford. A. Hellawell is now with the Department of Materials, University of Wisconsin, USA

This paper describes a method of metal casting whereby the melt is allowed to run across a single cool surface in such a way that it solidifies as sheet. This is not a new process and has been used for low melting point materials—lead and lead–tin alloys—for at least two centuries,[1] originally being the only obvious way to produce sheet such as that used in the construction of organ pipes. The process is still in practice and there are numerous examples of lead–tin sheet so produced,[2] while lead sheet for roofing purposes is also produced in a similar manner. The essential requirements are some means of moving a bath of metal and horizontal bed relatively, with a suitable gating system and the provision of a porous upper covering across the casting bed so that air or moisture are not trapped under the metal sheet. For organ pipe material, lead–tin alloy (typically on the lead rich side of the eutectic composition) is run across a stone bench which is covered with linen; for roofing lead the metal may be cast upon a sand bed. Figure 1 illustrates the arrangement used in the former case, and the product solidifies as sheet of but a few millimetres thick, having a eutectic grain size of 5–10 mm wide which nucleates on the lower cloth substrate. This grain structure is delineated on the upper surface by intergranular contraction grooves and has an attractive mottled appearance which is termed 'spotty metal'.[2] The simplicity of the procedure makes it attractive for wide application—to more refractory metals—and it has been extended in this laboratory to aluminium and its alloys.

EXPERIMENTAL PROCEDURE

A variety of arrangements have been used, including casting beds made of asbestos-base sheet and metal, with or without water cooling, using in each case a substrate cloth of woven silica—'refrasil'.[3] The width of the operation was limited to between 80–100 mm and some 3 m in length.

In earlier experiments the procedure was almost exactly that depicted in Fig. 1 (the original practice), molten metal being poured into a preheated asbestolite box which was then pulled along a stationary bed at rates from 50 to 300 mm/s. With a poor conducting asbestolite bed the sheet metal froze relatively slowly and did not begin to do so until ~1 m behind the moving melt box. The width of the sheet was predetermined by the gate width (100 mm), and with a gate having a height of 1–2 mm the thickness of the sheet varied from ~5 mm at slower rates to ~3 mm at the faster rates, i.e. it was a relatively insensitive function of the traction rate and depended primarily on spreading out the liquid sheet before it began to solidify. The sheet thickness was also a relatively insensitive function of gate height and length, being apparently determined more by the liquid surface tension than its viscosity (see later section).

Subsequently, to make the arrangement more continuous, apparatus was constructed (Fig. 2) in which the melt container and feeding system were stationary and continuously heated, while refrasil cloth was drawn beneath and across a horizontal water-cooled metal plate ~2 m long. In this arrangement the melt was contained in a cylindrical graphite crucible surrounded by an HF induction coil, the base being sealed by a graphite slab through which channels were cut for liquid metal to run under a milled slot or gate ~2 mm high, 80 mm wide, and 20 mm long. The feeding system was blocked by a retractable graphite plug until the melt reached a desirable superheat—typically ~40°K. About 1 litre of liquid produced some 3 m of sheet. The underlying bed was cooled from a point ~50 mm from the

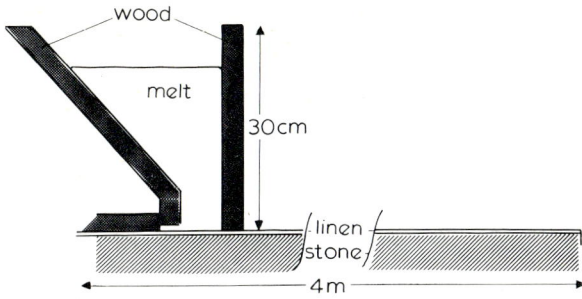

1 Method of casting lead–tin sheets

gate and metal began to solidify ~250 mm from it when the traction rate was ~50 mm/s. A sheet thickness between 2–5 mm was then obtained by varying the feed rate (width of channels in the base of the crucible) and the belt traction rate. The thickness was again determined by liquid surface tension and the volume of liquid which was allowed to run out between gate and solidification front.

RESULTS

Sheet metal was produced as above with (*a*) aluminium of commercial purity without/with sodium additions and with Al_3Ti grain refinement; (*b*) Al–15 wt-% Cu; and (*c*) aluminium–silicon eutectic alloy, LM6, with and without sodium and strontium additions.

(*a*) Commercial purity aluminium of ~98·5% was used in this process. With no additives the lower surface pattern was indistinct in this material, but the upper surface (Fig. 3*a*) showed an attractive dendritic pattern which was produced by intergranular and interdendritic shrinkage, the surface relief corresponding to fluctuations up to ~0.5 mm in amplitude. Macroscopical examination of the cross section (Fig. 3*b*) revealed a grain size of a few millimetres so that the sheet appeared to be effectively equiaxed with a coarse cored microstructure. Tensile tests and rolling behaviour compared well with as-cast aluminium and 3 mm cast sheet could be cold rolled to 0·1 mm foil without cracking, although some traces of surface irregularity persisted as smears. Minor additions of sodium or Al_3Ti grain refiner produced notable changes in the product.

About 0·1 wt-% sodium, enclosed in a pellet of aluminium foil, was immersed in the melt immediately before casting—as for the modification of Al–Si eutectic alloys. The upper surface of the sheet was then quite smooth and matt (Fig. 4*a*), the lower cloth pattern more distinct, and the thickness reduced by ~two-thirds. Despite these outward changes, however, examination of the grain size in section (Fig. 4*b*) showed remarkably little change from the untreated metal, and it was concluded that the difference was caused almost entirely by the liquid surface tension and the way in which this influences the spreading of the liquid layer. As is known, one effect of sodium is to reduce the surface tension of aluminium-base alloys[4] by ~25%, and this is relatively

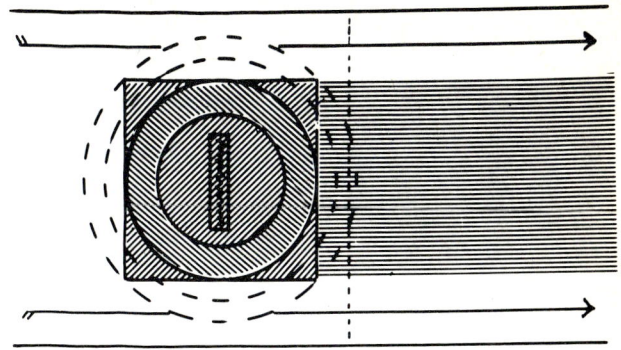

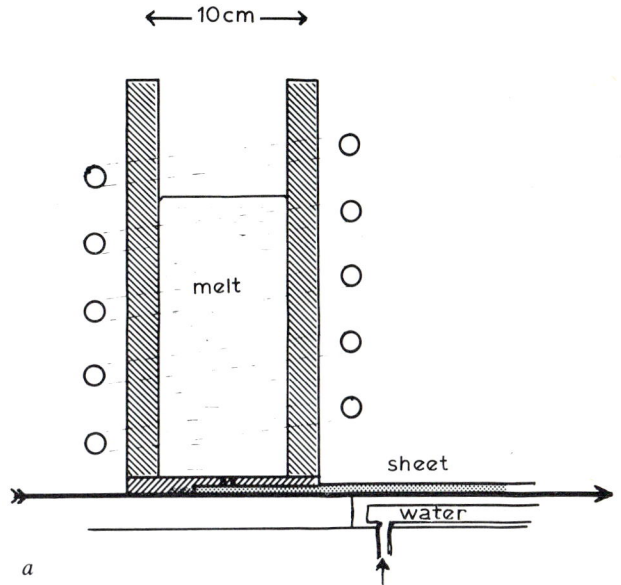

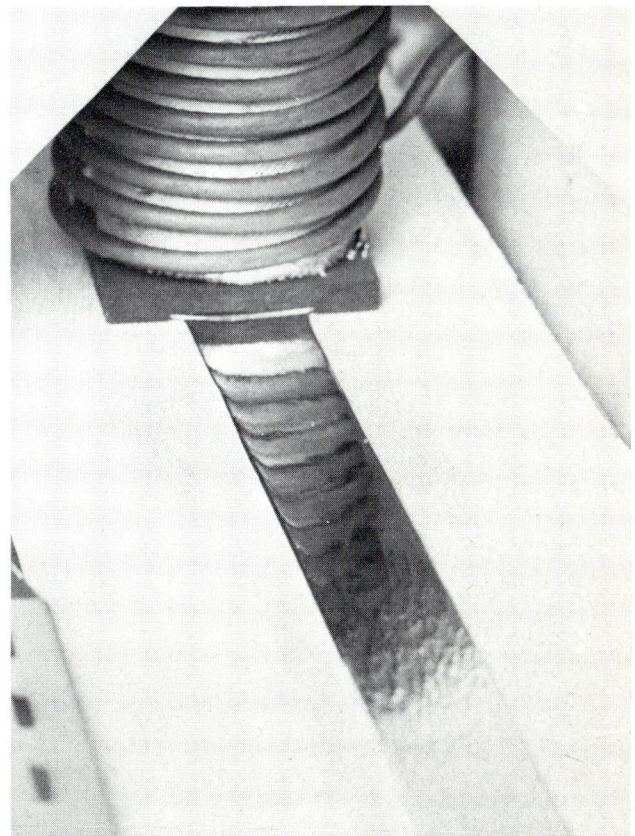

a

b

2 *a* Plan and elevation of apparatus used for semi-continuous casting of aluminium-base alloys; *b* in action

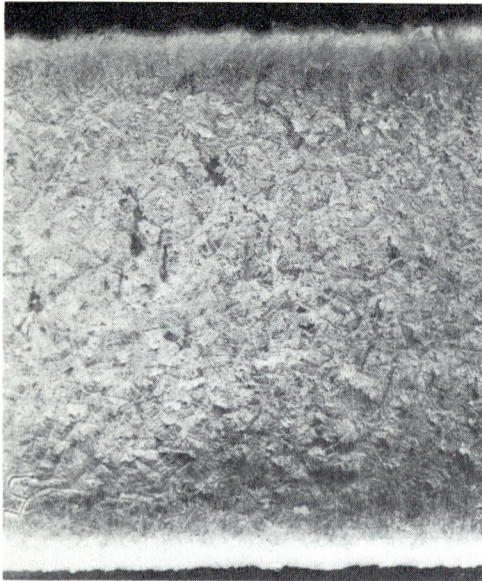

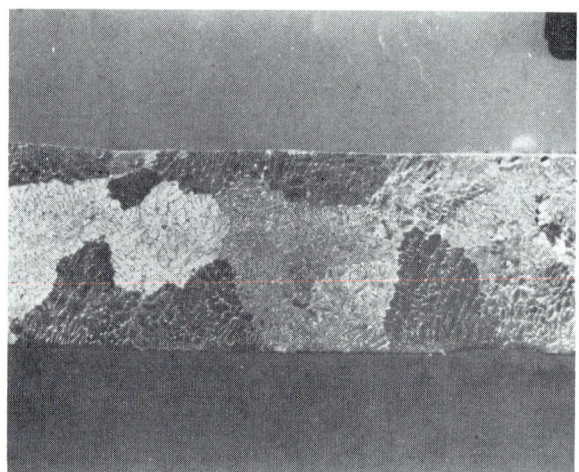

3 *a* **Upper surface of commercially pure Al sheet** [×⅔]; **and** *b* **heavily etched cross-section of the sheet showing grain size and coring** [×6]

4 *a* **Upper surface of commercially pure Al sheet after 0·1 wt-% addition of sodium to the melt,** [×⅔]; **and** *b* **corresponding cross-section of the sheet for comparison with Fig. 3*b*** [×6]

insensitive to other major solute additions such as silicon. This conclusion is consistent with the improved lower surface replication of the cloth pattern although we do not understand exactly how the contraction relief on the upper surface is removed.

In contrast to the above, an addition of ~0·2% Ti to the melt produced a fine grain size (<1 mm), a correspondingly fine upper surface pattern, and a well defined lower surface relief. The enhanced nucleation effect led to more rapid solidification in the early stages so that the liquid did not spread as uniformly and was somewhat thicker than the untreated or sodium 'modified' material.

(*b*) Experiments with Al–Cu alloys having freezing ranges of ~50–60 K were not significantly different from those with aluminium of commercial purity. The upper surface relief pattern was somewhat finer, as one might expect from a finer dendritic structure, but the wider freezing range did not appear to introduce extra

problems, and the distribution of copper within the width and thickness of the sheet was uniform, i.e. there was the usual microsegregation.

(*c*) Aluminium–silicon alloy of near-eutectic composition (LM6) was cast both with and without modifying additions. Without additions the sheet was relatively thick for given casting conditions (e.g. 5 mm) and exhibited a eutectic cell pattern about 5–10 mm across with relatively deep intercellular contraction cavities. The sheet was brittle and cracked easily on rolling. The addition of sodium, as for the purer metal, then produced thinner sheet with a very smooth upper surface and clear lower surface relief. The eutectic microstructure was modified as it is in conventionally cast alloy and the mechanical properties correspondingly improved. By comparison with aluminium, this alloy combined the advantages of surface tension reduction with those of a modified microstructure—the latter being related to eutectic silicon shape and scale rather than grain size.⁵ A strontium addition of ~1 wt-% produced an intermediate effect between the previous two examples, with a finer eutectic cell pattern apparent on the upper surface, but much less smooth and sound than the sodium modified sheet.

CONCLUSIONS

This historical method of sheet casting has proved to be applicable to aluminium and aluminium alloys in a small semi-continuous operation. There should be no great difficulty in increasing the scale of the procedure, and in principle it should be also applicable to other more refractory metals and alloys, e.g. those based on copper or iron. Apart from its inherent simplicity, the advantages which this casting method might possess by comparison with existing continuous casting plants[6] (based on contrarotatory drums or belts) are:

(i) it can be used to produce metal sheet of 2–5 mm thickness, which is considerably less than that typically possible in existing sheet or slab casting machines

(ii) the method is applicable to alloys having both narrow and wide freezing ranges and the product does not suffer from macroscopic inhomogeneity

(iii) without certain minor additions there is considerable surface texture and for subsequent rolling this would probably require a skimming treatment; the surfaces do, however, have some potential for decorative purposes and can be rendered smooth by minor additions which reduce the liquid surface tension: relief patterns can be imprinted on the lower surface from the substrate material

(iv) the as-cast products have mechanical properties comparable with those found in more conventional castings

(v) the necessary plant is simple, relatively cheap, and requires a minimum of maintenance.

ACKNOWLEDGMENTS

The authors would like to acknowledge Professor C. S. Smith, who originally drew attention to this method of casting, and Dr L. R. Morris whose enthusiasm for historical practice led to a closer acquaintance with existing foundries. The present research was supported in part by the Science Research Council.

REFERENCES

1 C. S. SMITH: 'A History of Metallography', 62, 1960, Chicago University Press
2 North Transept, St. John's Church, Ranmoor, Sheffield
3 The Chemical and Insulating Co. Ltd, Darlington, Co. Durham
4 V. DE L. DAVIES AND J. M. WEST: *J. Inst. Metals*, 1963–64, **92**, 208
5 A. HELLAWELL: *Prog. Mater. Sci.*, 1973, **15**, 1
6 E. F. EMLEY: *Int. Met. Rev.*, 1976, **21**, (206), 75

Formation of filamentary metallic glasses by ultra-rapid quenching from the melt

B. G. Lewis, I. W. Donald, and H. A. Davies

Over the last decade ultra-rapid quenching from the melt has become a well established technique for radically altering the structure and hence the properties of many metallic alloys. Thus, depending on the alloy composition and the applied cooling rate, structural modifications are feasible, including grain refinement, significant increase in solute supersaturation, formation of new metastable crystalline phases, and ultimately, if the cooling rate is high enough, the complete suppression of crystallization and hence the formation of non-crystalline metallic phases. In particular, the formation of metallic glasses has stimulated interest, owing to the exceptional combination of properties that many of them possess, including very high strength, approaching the theoretical limit, and high corrosion resistance and/or soft ferromagnetic properties. In this paper the conditions required for the formation of metallic glasses in general are discussed, and some of the process mechanisms involved in the production of continuous filament or tape from the melt by the roller-quenching and melt-spinning techniques are compared. It is concluded that the melt-spinning technique offers the more satisfactory means of producing good quality glassy alloy tape, and is also applicable to a wider range of alloys in view of the longer contact time between ribbon and substrate.

I. W. Donald and H. A. Davies are with the Department of Metallurgy, University of Sheffield, and B. G. Lewis is with the Department of Metallurgy, University of Connecticut, USA

The advent of splat quenching, i.e. ultra-rapid solidification, has led to many interesting possibilities for structural modifications in alloys and thus for marked improvements in their properties compared with those obtainable by conventional routes. The structural changes include substantial refinement of dendrite, grain, and multiphase structures, enhanced solute supersaturation in terminal and intermediate solid solutions, and the formation of new metastable crystalline and amorphous phases. In parallel with the studies of structures and properties of rapidly quenched alloys, a number of techniques have been developed and refined to produce continuous filament, either tape or wire, by direct quenching from the melt. Most of these can give high cooling rates and thus offer the possibility of practical utilization of the enhanced properties associated with many splat-quenched alloys. Of particular interest are the glassy alloys, which have unique combinations of mechanical and magnetic properties and/or corrosion resistance. The glassy structure represents the ultimate in quenched-in metastability, and the ability of a technique to vitrify an alloy is a stringent test of its quenching efficacy.

In this paper we compare the process physics of the melt-spinning[1] and roller-quenching techniques[2,3] and

their effectiveness for rapid continuous casting of metallic glasses in the form of thin ribbon.

THE CRITICAL COOLING RATE FOR GLASS FORMATION

It is appropriate first to summarize the melt cooling conditions required to avoid crystallization in alloys in general. It has been shown previously by the authors,[4–7] that since metallic melts have relatively uniform viscosities, η, at their equilibrium melting or liquidus temperature, T_m, the critical cooling rate for glass formation, R_c, is largely dependent on the reduced glass temperature $T_{rg} = T_g/T_m$, where T_g is the glass transition temperature, at which $\eta \sim 10^{13}$ poise. This approach is based on the Johnson–Mehl–Avrami theory of transformation kinetics and accepted theories of homogeneous nucleation[8] and of crystal growth from the melt.[9] Time–temperature–transformation curves are thereby constructed expressing the time to a small and barely detectable fraction of crystal, X, taken here to be 10^{-6}, as a function of temperature. R_c is then given as the cooling rate to just avoid this fraction X, i.e. to just bypass the nose of the T–T–T curve (see Fig. 1). The construction of these T–T–T curves requires an assumption concerning the temperature-dependence of

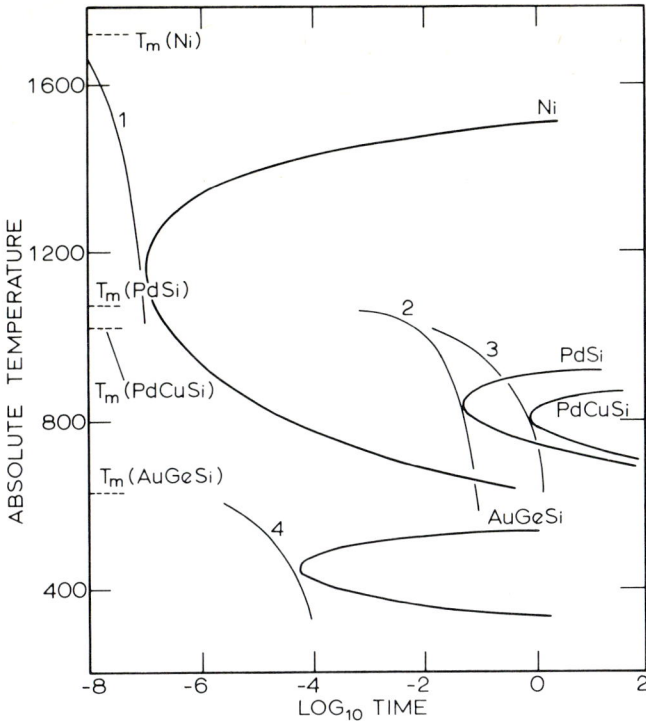

1 ~5×10^9 K/s; 2 ~3×10^3; 3 ~2×10^2;
4 ~8×10^6

1 T–T–T curves corresponding to a fraction crystal of 10^{-6} for several metallic systems: also indicated are the critical cooling rates required to avoid crystallization

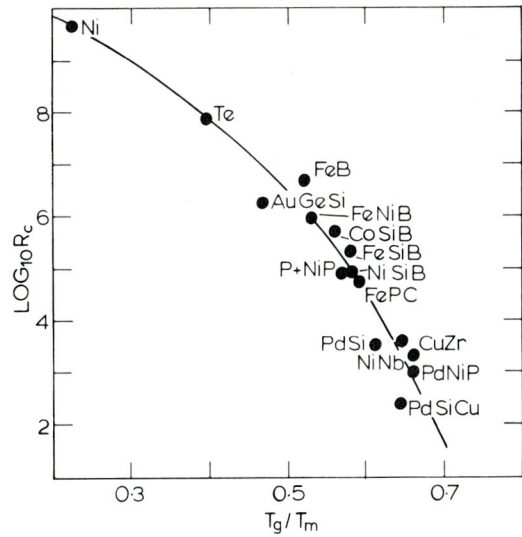

2 Theoretically calculated values of the critical cooling rate for glass formation (R_c) as a function of the reduced glass temperature (T_{rg}) (actual compositions of the alloys are: $Fe_{83}B_{17}$, $Au_{77.8}Ge_{13.8}Si_{8.4}$, $Fe_{41.5}N_{41.5}B_{17}$, $Co_{75}Si_{15}B_{10}$, $Fe_{79}Si_{10}B_{11}$, $Ni_{75}Si_8B_{17}$, $Pt_{60}Ni_{15}P_{25}$, $Fe_{80}P_{13}C_7$, $Cu_{60}Zr_{40}$, $Ni_{62.4}Nb_{37.6}$, $Pd_{40}Ni_{40}P_{20}$, $Pd_{77.5}Si_{16.5}Cu_6$, $Pd_{82}Si_{18}$)

Table 1   Thermal properties for some metallic glasses

Alloy	T_m, K	T_g (or T_c), (at 80 K/min)	$T_m - T_g$	T_g/T_m	R_c, K/s
$Pd_{40}Ni_{40}P_{20}$	916	602	314	0·66	1×10^3
$Pt_{60}Ni_{15}P_{25}$	875	500	375	0·57	8×10^4
$Ni_{62.4}Nb_{37.6}$	1 442	945	497	0·66	2×10^3
$Fe_{80}P_{13}C_7$	1 258	736	522	0·59	5×10^4
$Ni_{75}Si_8B_{17}$	1 340	782	558	0·58	8×10^4
$Fe_{79}Si_{10}B_{11}$	1 419	818	601	0·58	2×10^5
$Co_{75}Si_{15}B_{10}$	1 393	785	608	0·56	5×10^5
$Fe_{41.5}Ni_{41.5}B_{17}$	1 352	720	632	0·53	9×10^5
$Fe_{83}B_{17}$	1 448	760	688	0·52	5×10^6

η in the range T_m to T_g; the higher is T_g relative to T_m, i.e. the higher is T_{rg}, the more steeply η increases with decreasing T, and hence the longer the time to a given fraction of crystal and the lower is R_c.

The theoretically predicted value of R_c plotted as a function of T_{rg} for several metallic glasses are shown in Fig. 2 (*see also* Table 1). The predicted R_c varies from ~10^{10}K/s for a nominally pure metal Ni[10], down to ~100 K/s for the easy glass forming alloy $Pd_{77.5}Si_{16.5}Cu_6$. These predicted values are in satisfactory agreement with experimentally estimated cooling rates for the systems investigated.

The practical implications of this are that for unidirectional cooling of a rectangular section the maximum thickness, x_m, of glassy phase obtainable[7] can be related approximately to the time, t_n, at the nose of the T–T–T curve, by:

$$x_m = (D_\lambda t_n)^{\frac{1}{2}}$$

where D_λ is the average thermal diffusivity of the liquid alloys. Hence, the lower is R_c, the greater is x_m. For instance, x_m is ~200 μm for $Fe_{80}P_{13}C_7$ and ~2 mm for $Pd_{77.5}Si_{16.5}Cu_6$, which is consistent with experimental observation.

MELT SPINNING AND ROLLER QUENCHING

Both of these rapid quenching techniques require the formation of a stable jet of molten alloy which was obtained in the present case by ejecting, using argon gas

pressure, the rf melted alloys from fused silica crucibles drawn down to fine (i.d. 0·3–1·0 mm) outlet nozzles. In the case of melt spinning, the jet is directed on to the outer surface of a rapidly rotating wheel, and in the case of roller quenching it is quenched between contra-rotating rollers. General views of the melt-spinning and roller-quenching units used in the present work are given in Fig. 3.

In detail, the quenching efficacy of these techniques depends on several factors: jet diameter and velocity, roll speed, the thermal conductivities of the melt and of the substrate, the heat transfer coefficient at the melt/substrate interface(s), and the time that the melt is in contact with the substrate. The jet diameter and roll speed together determine the final ribbon dimensions, high speeds and fine nozzles giving thin tapes which, for a given melt ejection pressure (jet velocity), increase in

a

4 **Still photograph of melt-spinning process showing the molten jet, melt pool, and metallic tape departing from the quench wheel**

b

3 *a* **Melt-spinning apparatus;** *b* **roller-quenching apparatus**

width with increasing roll speed owing to greater spreading of the melt.[11] The contact time is also important since the very high cooling rate is only available while the melt/ribbon is in contact with the roll(s), after which cooling is by radiation and convection only. We will now consider aspects of the two techniques separately.

During melt spinning the single roll experiences relatively little wear so that it is possible to use a high conductivity copper substrate with the surface polished with fine (600 grade) emery cloth to achieve good thermal contact.

In our melt spinning unit where the roll radius was 75 mm, the ribbon was observed (by high-speed still photography) to remain in circumferential contact with the copper substrate for between about 22 and 31 mm (Fig. 4), depending on the dimensions of the tape and the apparently random oscillations of the tape on leaving the wheel. This is in fair agreement with the observations of Walter and Luborsky.[12] Assuming that there is negligible dwell of the jet stream on the roll surface, for a roll speed of 6 000 rev/min, these contact lengths represent quench times in the range 0·47–0·66 ms. Whether or not this is a sufficient contact time for complete vitrification depends on the cooling rate during the contact with the wheel and on the interval between the liquidus temperature of the alloy, T_m, and its glass transition temperature T_g (see next section).

In roller quenching the ribbon dimensions are determined also by the roll pressure, or, where they are not in contact, the roll gap. Hence to obtain tape of uniform dimensions high tolerances are required when machining roll surfaces and their axles, and also high grade roller bearings are needed. To obtain thin, rapidly quenched tape the unit should be run with the rolls initially in contact, but with solid rollers this procedure tended to result in non-uniform and badly deformed tape.[13,14] This problem was minimized by incorporating an accurately cast thin annulus of silicon rubber into each roll to limit the effects of any slight eccentricity and vibration.

Deterioration of the roll surfaces is a significant problem in roller quenching and to minimize this the rollers were chromium plated and then cylindrically ground to give a final thickness of about 1 mm of chromium. Although the thermal conductivity of chromium is substantially lower than that of copper, the thermal contact was expected to be better than for melt spinning because of the mutual pressure of the rollers, and with quenching by *two* substrates the overall heat transfer rate was thereby expected to be comparable with melt spinning.

For the ideal geometry of roller quenching of a well defined stream of liquid it can easily be shown that when the roll radius $r \gg$ the jet diameter d the contact distance is approximately $\sqrt{rd}$. For the 40 mm rolls used in our apparatus and for $d = 1$ mm, the contact length is ~6·3 mm. At an operating speed of 6 000 rev/min, and assuming that the mean speed of the tape equals the circumferential speed of the rolls, this represents a contact time of ~0·25 ms. This is subject to uncertainty since there may be appreciable build-up of melt in the nip of the rolls, which would increase the contact time, and also the mean tape speed may be higher than that of the roll surface, which may shorten the contact time.

COOLING RATES IN MELT SPINNING AND ROLLER QUENCHING

The cooling rates, $\dot{T}$, produced by the two techniques were estimated from measurements of the interlamellar spacings in quenched ribbons of eutectic Al–17·3% Cu[15] and of the dendrite cell size in ribbons of Al–10·5% Si. The interlamellar spacings were measured in electron transparent areas of foils, that were ion-beam thinned from sections of ribbon, using a Philips EM301 electron microscope. Although a large proportion of the eutectic was degenerate, the mean spacing for the lamellar regions of the melt-spun tape was 0·055 μm which, assuming a section thickness equal to the final tape thickness of 20 μm, corresponds to a cooling rate of ~9·6 × 10^5 K/s. For roller-quenched samples mean spacings of 0·03 μm were observed in 40 μm thick tape, which represents a cooling rate of 2·9 × 10^6 K/s.

Optical determination of the dendrite cell size, d_c, in quenched ribbons of Al–10·5% Si indicated cooling rates of ≥ 10^7 K/s ($d_c = 0·6–0·8$ μm at the limit of optical resolution) for 30 μm thick melt-spun ribbon, and ~1·2 × 10^6 K/s ($d_c \simeq 2$ μm) for roller-quenched ribbon 50 μm thick. Correcting the roller-quenched cooling rate for a copper substrate would increase it to ~3·5 × 10^6 K/s which, although closer to the melt-spun value, is still significantly smaller.

Although there is some uncertainty in the absolute values of the cooling rates estimated by these techniques, the relative values for the two quenching techniques should in principle be quite reliable. The discrepancy between the cooling rates derived from lamellar spacings and those derived from dendrite cell sizes for the roller quenched specimens may arise partly from assuming an incorrect section thickness in the derivation of the former. The section thickness in roller quenching varies substantially with the distance traversed between the rolls, and it is not clear at which point solidification is complete and to what extent deformation of the finally solid tape determines its final dimensions. In the case of melt spinning also, part of the cooling occurs in a relatively thick pool of melt immediately below the orifice, before the ribbon is extracted from it by the roll surface. Since the calculation of cooling rates from the microstructures of quenched Al–10·5% Si is independent of thickness these will be used in preference to the values derived from Al–Cu lamellar spacings. Assuming that, for the tape thicknesses being considered here (<50 μm), Newtonian cooling conditions apply, the above cooling rates will apply also to transition-metal based alloys considered in this work.

If we combine the contact times estimated above with the experimentally estimated cooling rates, and assume that the cooling rates remain constant over the whole of the contact times, the resulting predicted temperature drops, while the tape is in contact with the roll(s), are approximately 300 K for roller quenching and > 10^3 K for melt spinning. The very high cooling increment

predicted for melt spinning is obviously too large. It is implicit however from Fig. 4, and also from high-speed cine photography of Walter and Luborsky,[12] that quenching at the highest rate does not occur over the whole contact length. The tape apparently loses its red heat, a drop of typically ~400 K, within about 3 mm of the centre of the melt pool, which would indicate a cooling rate of ~10^7 K/s for iron-based glass forming alloys over that section of the contact length. The predicted cooling increment of 300 K for roller quenching is in agreement with the experimental observation that tapes of $Fe_{80}P_{13}C_7$ emerge from the rollers red hot. After parting from the rolls the ribbon is cooled only by radiation and convection, and we have estimated this latter cooling rate to be ~400 K/s.

The problem associated with complete vitrification of an alloy by these techniques is therefore determined by two factors. First, the cooling rate must be sufficient to bypass the 'nose' of the relevant $T–T–T$ curve (see 2nd section and Table 1). Secondly, the temperature of the ribbon on leaving the roll surface should ideally be less than a temperature T'_c, below the nose of the $T–T–T$ curve, in order that the cooling curve resulting from the subsequent radiation and convection does not intersect the undercooled liquid → crystal $T–T–T$ curve and cause partial crystallization (Fig. 5). The critical cooling rate for the formation of an amorphous phase can be estimated from the reduced glass temperature T_g/T_m of the alloy (1st section), and a measure of the required cooling increment is given by $(T_m–T_g)$. These parameters are given in Table 1 for a number of standard glass forming alloys. It is clearly more difficult to retain a fully amorphous structure, by the two quenching techniques described here, in alloys where the increment $(T_m–T_g)$ is large and/or the value of T_g/T_m predicts a critical cooling rate near the limit of the apparatus.

All the alloys shown in Table 1 were satisfactorily quenched to a fully glassy state by the melt spinning unit. However, the roller-quenching unit was less flexible and produced fully vitreous tape only for the alloys $Pd_{82}Si_{18}$, $Pd_{40}Ni_{40}P_{20}$ and $Pt_{60}Ni_{15}P_{25}$. This would lend support to the proposition that the cooling increment in the roller-quenching unit used here is considerably smaller than for the melt-spinning unit, and also that the available cooling rate using chromium-plated rolls is less than that obtainable for the melt-spinning unit. The contact time, and thus the cooling increment, could be enhanced by

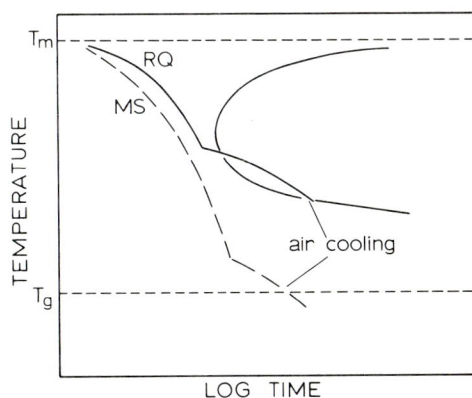

5 Diagram comparing the quench mechanisms in roller quenching (RQ) and melt spinning (MS)

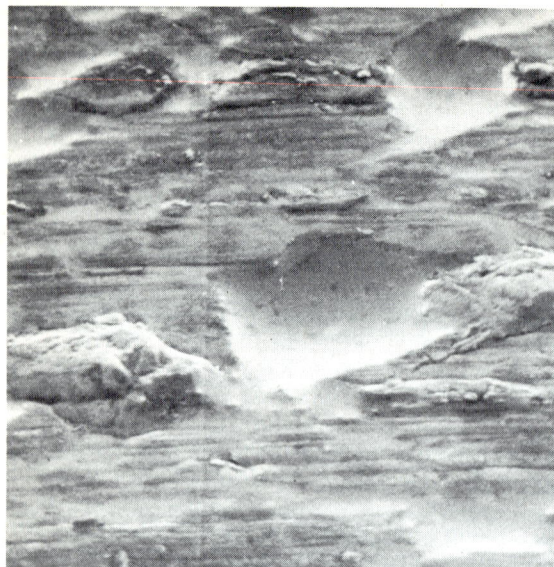

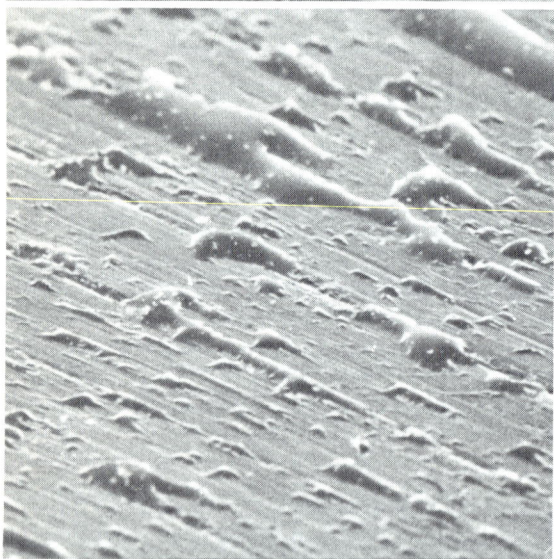

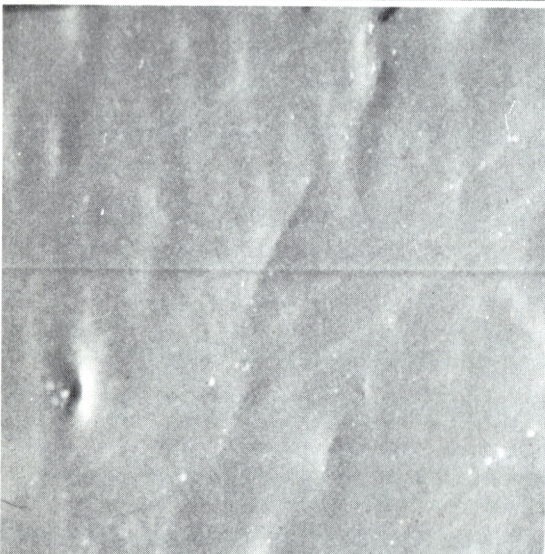

a bottom surface, [×450]; *b* bottom surface, [×100]; *c* top surface, [×100]

6 Scanning electron micrographs of a melt-spun Fe₈₀P₁₃C₇ ribbon

increasing the roll diameter, but the practical difficulties associated with machining rollers to high tolerances is also increased. The contact time could also be increased by decreasing the roll speed, but this may be compensated by a decrease in the cooling rate. The cooling rate should in theory be increased by a factor of $\sim 1\cdot 8$[16], while still maintaining reasonable wear resistance, by employing hardened steel rolls.

It is also possible to extend the quench time for melt spinning, typically by means of a gas manifold or a flexible metal belt, and these devices have been described by Bedell.[17] Preliminary tests in this laboratory indicate that a gas manifold does lead to an improved performance, probably owing to enhanced cooling due to the gas flow since, as discussed above, the contact distance appears to be sufficient without a manifold, even for considerably thicker tape than is considered here. The experiments are continuing.

THE SURFACE STRUCTURE OF MELT SPUN AND ROLLER QUENCHED RIBBON

Apart from the differences in the quenching efficiency of melt spinning and roller quenching, there are significant differences between the tapes produced by the two techniques. For instance, Figs 6*a*, *b*, and *c* show scanning electron micrographs of the upper and lower surfaces of a melt-spun $Fe_{80}P_{13}C_7$ glass ribbon; the side of the tape in contact with the roller exhibits surface defects in the form of voids, elongated in a direction parallel to the tape edge, and associated with these voids are protrusions situated towards the rear of the tape, relative to the spinning direction. The mechanism responsible for the formation of these defects is most probably gas entrapment during solidification,[11] before the tape leaves the roller, although microdefects in the surface of the wheel may also influence their incidence and distribution. The overall effect of these defects is to give the lower side of the tape a matt-grey appearance. The shiny 'free-melt' side of the tape is usually relatively featureless, as may be expected, consisting of a gently undulating and relatively defect-free surface.

The surfaces of roller-quenched tapes, Figs 7*a* and *b* are markedly different from those of melt-spun tapes, although this is perhaps not surprising in view of the fact that superimposed on the quenching and solidification process is a state of compressive-tensile deformation, i.e. solidification and deformation are occurring simultaneously. Thus, the surface of roller-quenched $Pd_{40}Ni_{40}P_{20}$ for instance, exhibited localized areas of highly deformed and locally ruptured material which gives rise to a series of distorted parabolic cracks and fissures, aligned in the rolling direction. Consequently, the mechanical properties of roller-quenched glassy alloy tapes are generally inferior to similar melt-spun tapes, and this has been noted by Chen and Polk.[18] Similarly, roller-quenched crystalline Al–Si alloy exhibits a highly distorted surface consisting of numerous cracks and fissures, although in this instance the defects are more extensive and not of a well defined profile.

The precise mechanisms responsible for the formation of the defects observed in roller-quenched tapes and their variation for the different alloys studied are open to speculation at the present time, but are presumably related to the complex solidification/deformation interaction that occurs during the quench process.

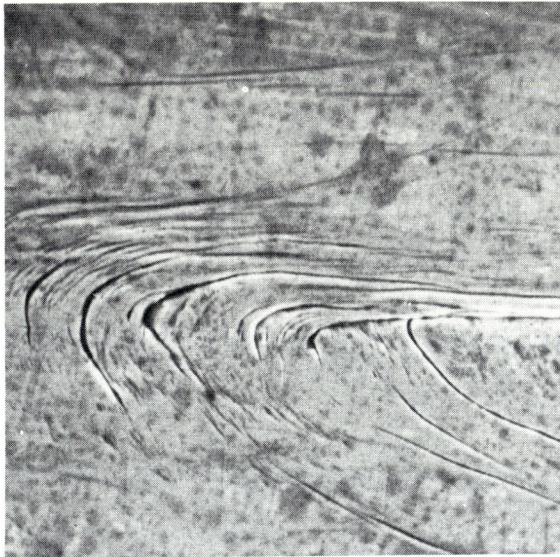

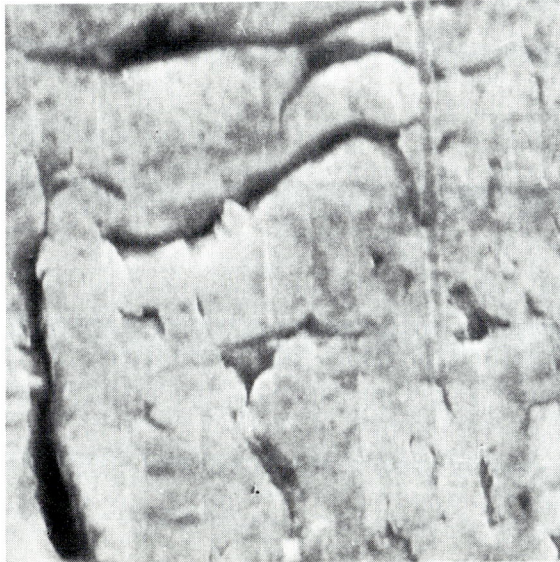

a Pd$_{40}$Ni$_{40}$P$_{20}$, glassy, [×950]; *b* Al$_{89.5}$Si$_{10.5}$, crystalline, [×1 900]

7 Scanning electron micrographs of roller-quenched ribbon

ACKNOWLEDGMENTS
The authors are grateful to the Science Research Council and the Royal Society for financial support, and to Professors G. W. Greenwood and B. B. Argent for the provision of laboratory facilities. They also wish to thank Dr Emil Babič for helpful discussion regarding the design of a roller quenching unit.

REFERENCES
1 R. B. POND: US Patent 2 825 108, 1958 (*see also* ref. 11)
2 H. S. CHEN AND C. E. MILLER: *Rev. Sci. Instrum.*, 1970, **41**, 1 237
3 E. BABIČ et al.: *J. Phys. E (Sci. Instrum.)*, 1970, **3**, 1 014
4 H. A. DAVIES et al.: *Scr. Metall.*, 1974, **8**, 1 179
5 H. A. DAVIES AND B. G. LEWIS: *Scr. Metall.*, 1975, **9**, 1 107
6 B. G. LEWIS AND H. A. DAVIES: 'Liquid metals', 274, 1976, London, Institute of Physics
7 H. A. DAVIES: *Phys. Chem. Glasses*, 1976, **17**, 159
8 D. R. UHLMANN et al.: *Phys. Chem. Glasses*, 1966, **7**, 159
9 D. TURNBULL: *J. Phys. Chem.*, 1962, **62**, 609
10 H. A. DAVIES et al.: *Nature Phys. Sci.*, 1973, **246**, 13
11 H. H. LIEBERMANN AND C. D. GRAHAM: A.I.P. Conference Proceedings No. 34, Magnetism and Magnetic Materials, 1976, paper 6 D1
12 J. L. WALTER AND F. E. LUBORSKY: G. E. Research and Development Lab., Schenectady, personal communication
13 B. G. LEWIS: unpublished data
14 P. M. THOMAS: University College, Swansea, personal communication
15 M. H. BURDEN AND H. JONES: JIM, 1970, **98**, 249
16 S. KAVESH: 'Metallic glasses', (eds J. J. Gilman and H. J. Learny), 36, 1978, ASM, Metals Park, OHIO
17 J. R. BEDELL: US Patent 3 862 658, 1975
18 H. S. CHEN AND D. E. POLK: *J. Non-Cryst. Solids*, 1974, **15**, 174

Rapid casting of metallic wires

J. V. Wood

The techniques for the rapid quenching of metallic wires and ribbons from the melt are described and compared. The 'pendant drop' melt extraction method, which has been developed by the Battelle Laboratories in the USA, is described in detail and a number of experimental results are given to elucidate the more important production variables, and to show areas where this technique might provide alloys with significantly different properties.

The author is with the Department of Materials Science, The Open University

For many years the techniques involved in the production of metals and alloys that have been rapidly quenched from the molten state have limited the amount of material produced to a few milligrams. Predominant among these processes is that termed 'gun' splat-quenching, which was originally devised by Duwez and co-workers.[1] The technique involved the melting of a small charge of material in a crucible and then ejecting the melt through a small orifice in the base of the crucible. The atomized droplets were propelled on to a solid heat-conducting substrate by the shock wave and solidified immediately upon impact. The cooling rates claimed for this technique were of the order of 10^6 K/s. Three basic results could be obtained from this technique on certain alloy systems, as follows;

(i) extension of equilibrium terminal solid solubility of solute and defects (normally vacancies)
(ii) production of new metastable phases including a large range of amorphous structures
(iii) if the product remains crystalline, then a significant reduction in grain size and segregation can be achieved (grain size is typically less than 1 μm).

Early classifications made the division between crystalline and amorphous phases, but it has become clear that there is a gradual progression of structure, through the completely random structure, via clusters and short-range order, to loose crystalline materials and finally to close packed crystal structures. While there still is a case for making this division on a property criterion, it is not justified when considering the processing side. It appeared that such results would remain of academic interest only until the advent of large scale processing techniques about five years ago.

BULK TECHNIQUES

Two predominant techniques for producing fine wire or ribbon at cooling rates between 10^4–10^6 K/s are available. These are melt spinning[2] and melt extraction,[3] which have been developed in the USA by the Battelle Laboratories at Columbus, Ohio. With the large commercial interest in amorphous wires these techniques have been rapidly developed. There is little distinction between the two techniques and, indeed, both techniques can be performed on the same apparatus with very little disruption.[4]

In the present work a variation on conventional melt extraction from a crucible was used, namely the pendant drop technique.

Pendant drop melt extraction

Instead of melting and maintaining a constant head of metal, as in the crucible process, this variation uses the concept of a levitated drop held above the rotating disc to supply the molten feed. Heating techniques such as oxy-acetylene, electron beam, and laser beam have been employed, but in the present case a radio frequency induction coil was used with a concentrator to focus power into the molten drop. A sharp-edged wheel contour with an angle of about 90° pulls a fine wire filament from the drop. The drop is maintained at the end of a rod which is lowered into the hot area continuously by a constant lowering device. In this way a continuous wire can be made of almost indefinite length.

Two wheels were used for the present research. The first was an air-cooled copper disc of 18 cm dia. attached to a variable speed motor. This allowed a number of system variables to be examined and the results are outlined below. The second design involved the use of a water-cooled brass wheel (220 mm dia. and 30 mm thick) which was balanced on a water-cooled stainless steel shaft between two high tolerance bearings. This latter design allowed no movement in the wheel at speeds up to 2 500 rev/min. Unlike the air-cooled wheel, this wheel did not heat up during long production runs. Also the sharp edge on the copper disc soon became rounded and brass was used instead. There is no evidence to show that this reduces the heat-removal efficiency. Results on the specific systems below were obtained on this apparatus using an argon flow round the molten droplet. In aluminium-based systems the wire often adhered to the spinning disc and was removed by a pad or scraper before the disc returned to the droplet. However, the largest proportion of wire came directly off the wheel and was collected in a bucket filled with water placed about 350 mm from the droplet.

GENERAL RESULTS

As with 'gun' splat quenching it is necessary to define a number of parameters and their effect on the resultant structure before the specific results become meaningful. Such results will be given later.

Alloys

Both low and high melting point alloys can be melted easily by a radio frequency induction coil using a concentrator to focus the beam on to the droplet above the wheel. In the case of Pb–Sn alloys a small gas flame was used instead to give greater control of the droplet temperature. In all the cases presented in this paper, alloys were prealloyed and cast into moulds 10 mm in diameter before they were used as feedstock in the pendant drop apparatus.

Disc condition

The condition of the wheel appears to have a marked effect on the resulting wire. In the present case, the wheel was always rubbed with emery paper and washed with acetone before its final finish. The final condition of the wheel was varied between roughening with 400 grade emery cloth to polishing with 1 μm diamond paste. In general, reasonable results for all alloy systems can be achieved using 600 grade emery cloth. The initial wheel condition is of greater importance for short runs as it is necessary to maximize heat extraction and yet ensure that the wire comes away cleanly. On long production runs the wheel is gradually eroded, thus removing the finish unless steps are taken to apply a surface treatment continuously.

Using this technique wires of the order 30–50 μm in diameter are cast. A general view of such a product is shown in Fig. 1*a*. A detail of a portion of Fig. 1*a* is shown in Fig. 1*b*. This is the free surface (i.e. the last surface to solidify and furthest from the wheel), and is characterized by a dimple structure on the surface. The dimensional variations are reasonable (± 5 μm) in view of the production speed. If the molten drop and the wheel edge are incorrectly aligned then a tape is produced of unpredictable section. If a reliable *tape* is required then the sharp edge can be replaced by a flat plateau. The width of this is limited by the drop size. Using a sharp edge the product is not exactly round, as can be seen in Fig. 2; there is a flute running the length of the wire corresponding to the contact with the disc. This makes for a slightly kidney-shaped cross section. The sample shown in Fig. 2 has been pulled in tension and hence there is necking at the free end.

Figure 3*a* is a scanning electron micrograph from an area which has been in direct contact with the wheel. The striations are a replica of the wheel condition and indicate where thermal contact has occurred. In between, a grain structure is observed which has solidified without touching the disc. These regions are

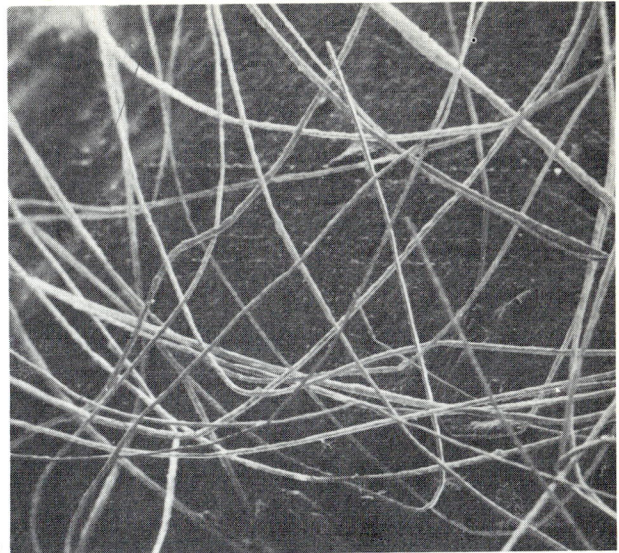

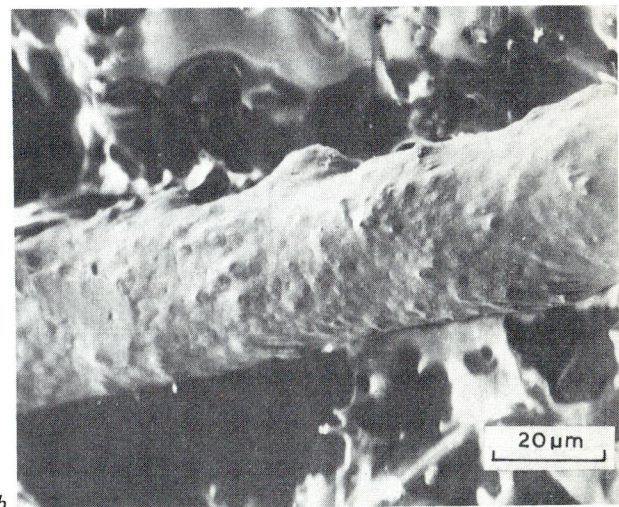

a

b

1 Scanning electron micrographs of melt extracted wire

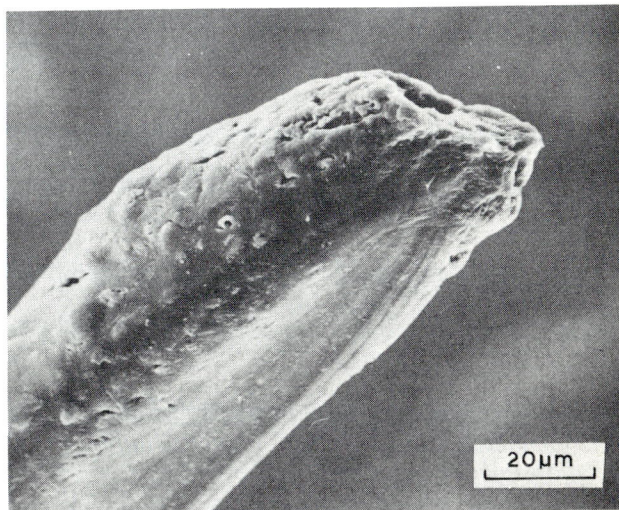

2 Dimple in wire formed by a sharp edged mould (sem)

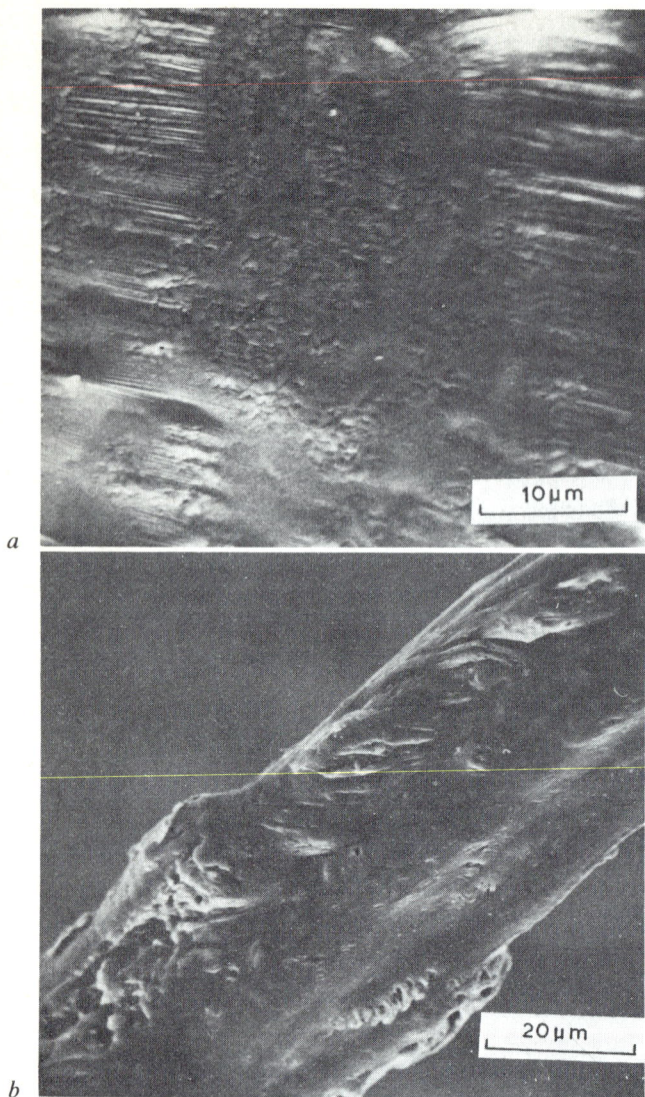

3 a Underside of wire showing 'lift off' regions; b oxide inclusions incorporated during the solidification process (sem)

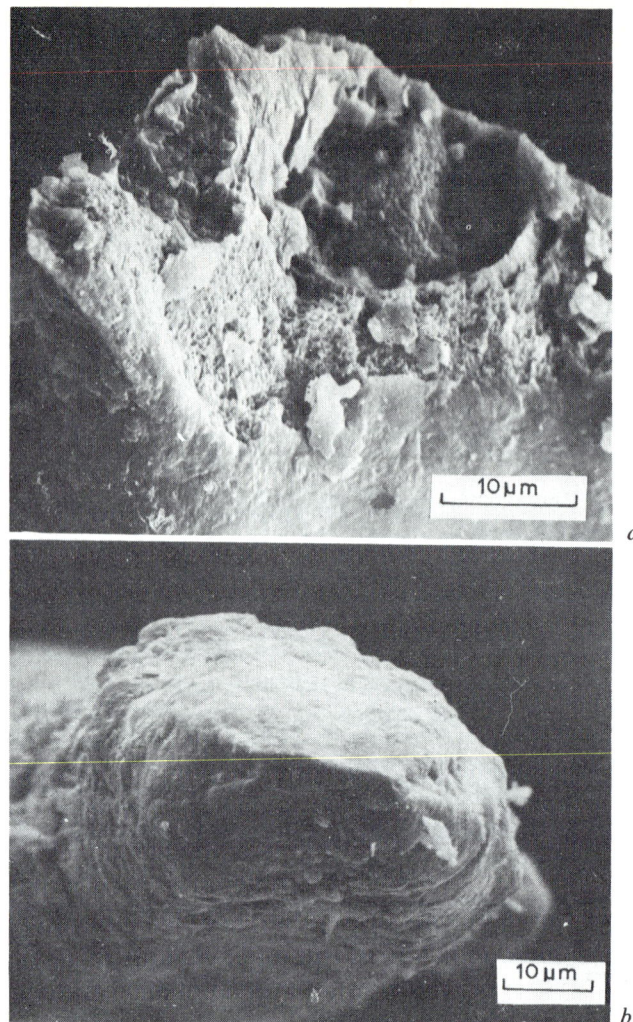

4 a Tensile specimen after fracture showing effect of large gas pore; b same material without gas entrapment which has necked down after testing (sem)

analogous to 'lift off' areas that have been well documented in 'splat-quenching' studies[8] and relate to the slowest cooled regions. There is often a change in microstructure in these areas which would be significant in wires of such small dimensions.

Atmosphere

The large surface area to volume ratio inherent in this technique makes the problem of oxidation and gas pick-up all too apparent. Figure 3b shows the effect of air pick-up on an aluminium-base alloy which has not been protected. Not only are large oxide inclusions incorporated in the wire, but it can be seen that the uniformity of cross section of the wire has been destroyed. At present, a study of the mechanical properties of these wires is being undertaken on a miniload tensile machine. In the case of a number of aluminium wires that were tested, there was an extremely large variation in values for the ultimate tensile strength. Figures 4a and 4b show the fracture surfaces for two of these wires. It can be seen in Fig. 4a that over a third of the surface is accounted for by a gas pore. Grains are

observed in the cavity which indicate that these have grown into 'free space' rather than a particle falling out after testing. Figure 4b shows the typical necking-down expected in a ductile material. This sample was produced with an argon blanket covering the molten drop.

Fe–0·4% C steels have been processed in this way, both under atmospheric conditions and in vacuum. In air, most of the filament oxidizes and that which is unoxidized is extremely brittle because of the large density of oxide inclusions. The product made in vacuum is, by contrast, extremely strong and tough without any inclusions. However, on X-ray analysis it is found that the product made in air is fcc while that made in vacuum is bcc. There is much evidence from splat-quenching that at high cooling rates the bcc product transforms to fcc.[9] This is backed up by the observation that in air the filament remains red hot only about 5 cm from the wheel, whereas in vacuum the distance is about 30 cm. Hence it can be seen that apart from providing protection, the gaseous environment of argon contributes significantly to the cooling rate.

Cooling rate

Fundamental measurements of cooling rate can be undertaken by measuring the interlamellar spacing of the Al–Cu eutectic after quenching in the apparatus. In the present case this gives a value between 10^5–10^6 under the best available conditions.

SPECIFIC RESULTS

While this paper is mainly involved with the description of the melt extraction technique, three studies are outlined which show how the early results of 'gun' splat quenching are now being exploited for technological systems.

Aluminium–transition metal alloys

Much work has been published on the large increases in supersaturation that can be achieved for Fe, Cr, and Zr in aluminium after splat quenching.[7] Al–3 wt-% Fe and Al–3 wt-% Cr wires have been produced on the above apparatus and were used as control alloys for the general investigations into the pendant drop technique (Figs 1–4). In both cases all the transition metal was found to have remained in solution when wires were examined by X-ray diffraction. The change in aluminium lattice parameter with solute content confirmed that there had been no significant solute loss during the quenching process. As mentioned above, the wires were pulled in tension on a microtensile machine initially designed for whisker testing (manufactured by Techne Instruments, Cambridge). The load is applied via a torsion bar of known calibration, and a 'frictionless' mirror system is attached to the end of the specimen to measure elongation. Hence a small load is applied and the mirror system is then adjusted to give a reading. This stepwise loading does not enable strain rate sensitivity to be considered. In general, gauge lengths between 1 and 4 mm were employed with the sample being glued between two quartz holders. For samples that did not undergo premature failure because of gas pickup (see 'Atmosphere' above) the yield stress was about $150\ \text{MN m}^{-2}$ and the ultimate tensile strength was $230\ \text{MN m}^{-2}$. These results are significantly better than for chill-cast aluminium and fall in the lower regions of hard-drawn wire. The greatest error in making these measurements is in estimating the area of cross section. The wires in this case were between 30 and 35 μm dia. and were assumed to be of circular cross section. However, because of the kidney-shaped cross section (Fig. 2), the stress values are underestimated. The ductility of these alloys can be observed in Fig. 4b, showing a considerable reduction of area in the necked region.

The increase in stress loads is directly associated with the reduction in grain size ($\sim 1\ \mu$m) and solid solution strengthening by the transition metal. It is known[10] that the solute atoms are associated with excess vacancies which have been quenched-in, thus complicating the solid solution strengthening effect. Work on aged wires has still to be carried out.

Ag₃Sn

Ag_3Sn is the basis of most dental alloys which are amalgamated with mercury to give the familiar cavity filling amalgam. The reaction with Hg proceeds along grain boundaries and diffuses into the solid γ (Ag_3Sn) particles forming two distinct intermetallic phases γ_1 and γ_2. Work on 'gun' splat quenching of this basic alloy showed a number of exciting features in respect to the deformation processes in the alloy and in its amalgamation properties.[5] Melt extraction provided a route for producing large quantities of this alloy by a technique that was considerably easier to use and as economical as present practice. Once again the wires are considerably stronger than chill cast alloys, having a UTS of $340\ \text{MN m}^{-2}$ (nearly twice the chill cast value). Of more significance in this normally very brittle compound is that an elongation of almost 2% was observed. It is assumed that this is associated with the excess vacancy concentration which diffuses to dissociated dislocations in order for the complicated slip–climb process found in this system to occur. As would be expected the wires react rapidly with mercury, giving a very fine dispersion of γ, γ_1, and γ_2. The increase in ductility of the γ particles makes accomodation within the tooth cavity easier without the risk of chipping.

Cu–Zn–Al

Cu–25 Zn–4 Al has been quenched on the above apparatus in an attempt to ascertain what effect grain size had on the martensitic transformation. These alloys produce a memory effect and there is considerable interest in the temperature and mode of transformation. The as-received material has a transformation temperature (in this case the austenite start temperature found by heating in a differential scanning calorimeter) of 40°C whereas the melt extracted material transforms at 80°C. Zinc loss can account for a large increase in A_s temperature but no zinc loss was detected by chemical analysis. The original grain size of the as-received material is between 1 and 3 mm whereas the wire has a grain size of $\sim 10\ \mu$m.

The wire was heat treated in the β region in order to allow grain growth to occur (~ 850°C). Unfortunately, this did not happen and the grain size remained unchanged, yet the transformation temperature was reduced to 20°C. The crystallography of these reactions is extremely complicated and is described by the author elsewhere.[6] However, apart from the metastable phases produced by quenching, the actual results can be explained in terms of a grain size and vacancy mechanisms. The as-quenched wire is softer than the as-received material, which is indicative of more transformable martensite containing the premartensitic omega phase.

CONCLUDING REMARKS

Results have been presented on the type of product that can be made by using the 'pendant drop melt extraction' technique. Fine wires of almost any alloy can be produced in this way, providing adequate attention is paid to the main variables outlined. Specific data have been presented for the Al–Fe, Al–Cr, Ag₃Sn, Cu–Zn–Al, and Fe–C alloy systems to demonstrate how this technique can be applied to technologically useful alloy systems.

ACKNOWLEDGMENTS

I should like to thank Dr Mobley and Dr Collings for introducing me to this technique, and Professor R. W. K. Honeycombe for encouraging this interest and for supplying laboratory facilities. Maintenance was provided by a Foseco post-doctoral research fellowship.

REFERENCES

1 P. DUWEZ: *Trans. ASM*, 1967, **60**, 607
2 R. B. POND: US Patent No. 2 825 108, 1958
3 R. E. MARINGER *et al.*: Proc. 2nd Internat. Conf. on rapidly quenched metals, 29, 1976, MIT Press
4 R. B. POND *et al.*: 'New trends in materials processing', 128, 1974, American Society of Metals
5 J. V. WOOD AND V. JACOMBS: *J. Mater. Sci.* 1976, **11**, 865
6 J. V. WOOD AND W. M. STOBBS: *Acta Metall.*, to be published
7 M. H. JACOBS *et al.*: *Fizika*, 1970, **2**, (2), 18.1
8 J. V. WOOD AND I. R. SARE: *Metall. Trans.*, 1975, **6A**, 2 153
9 I. R. SARE: 1975 Ph.D. dissertation, University of Cambridge
10 P. FURRER AND M. WARLIMONT: *Z. Metall.*, 1973, **64**, 236

Laser glazing — a new process for production and control of rapidly chilled metallurgical microstructures

B. H. Kear, E. M. Breinan, and L. E. Greenwald

A new process for producing rapidly chilled metallurgical microstructures under reproducible cooling rate conditions has been developed. Processing is accomplished by rapidly traversing a continuous, high energy–density laser beam over the material to produce a very thin molten layer at the material surface. The high specific rate of energy delivery facilitates surface-localized melting at very high melting efficiencies, i.e. the major portion of the absorbed energy is used for melting, with only a very small fraction going into heating of the solid, subsurface material. This ability to maintain a cold substrate while melting a thin surface layer of material results in rapid quenching of the molten layer once the energy source is removed. Calculations made for pure nickel indicate that cooling rates of $5 \times 10^8 \,°C/s$ are attainable in layers that are $2 \cdot 5 \times 10^{-3}$ mm thick, while $2 \cdot 5 \times 10^{-2}$ mm thick layers cool at rates of up to $5 \times 10^6 \,°C/s$. The importance of rapid quenching is based on the fact that the unsurpassed homogeneity of the liquid can be preserved (or nearly preserved) in the solid, which can then be utilized in that form, or these structures can be further modified in order to produce other desired properties. The subject process, termed 'laser glazing', has been shown to be capable of producing a wide variety of novel and interesting metallurgical microstructures on the surfaces of bulk metals. Amorphous surface layers have been produced on bulk substrates by laser glazing compositions at or near deep eutectic troughs. In other compositions, extended solid solution phases have been obtained, which have the potential for being decomposed into finely divided multiphase microstructures in the solid state. Such compositions are candidates for new dispersion strengthened alloys. At higher second phase contents, extremely fine filamentary eutectic microstructures have been obtained. A strong tendency for oriented overgrowth (epitaxial growth), in various orientations, has been demonstrated for a variety of monocrystalline substrate materials. Laser glazing (in conjunction with surface alloying, post-glaze heat treatment, and/or deformation) has great flexibility as a means to generate a broad range of novel surface microstructures and properties. Furthermore, although currently being employed primarily for the surface treatment of materials, it also has the potential for processing bulk materials by the sequential build-up of one glazed layer upon another.

The authors are with the United Technologies Research Center, Connecticut, USA

The interaction between a laser beam and a metal or alloy surface is controlled by a number of variables including the wavelength of the radiation, the incident power density, and the available interaction time. For the case of a continuous, convectively-cooled higher power CO_2 laser with a wavelength of 106 000 Å the laser–material interaction as a function of incident power density and available interaction time is as depicted in Fig. 1. A particular combination of power density and interaction time defines a specific operational regime within the 'interaction spectrum', and each operational regime results in the occurrence of a unique materials processing effect. In order of increasing power density, the various processing effects include transformation hardening,[1,2] bulk surface alloying and cladding,[3] deep penetration welding[4-6] and the LASER-GLAZE™ effect,[7,8] drilling of holes and metal removal effects,[9] and laser shock hardening.[10,11] In Fig. 1 it may be noted that the materials processing effects, or operational regimes, are clustered along a diagonal running from 'high power density/short interaction time' to 'low power density/long interaction time'. This is owing to

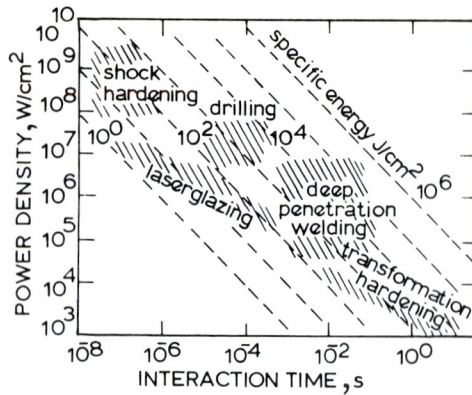

1 Operational regimes for laser materials processing techniques

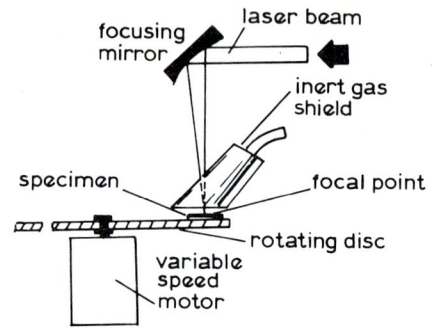

2 Schematic of laser glazing apparatus

several factors. First, the very high power densities necessary for explosive material vaporization effects, such as those producing shock hardening and drilling, are obtained with pulsed lasers and thus are available only for short times. Repeated pulses generally are not additive. Also, the quantities of energy necessary to generate the various thermal effects ranging from heating to 'red heat' up to vaporization do not differ in energy consumed by more than two orders of magnitude. Thus, it is not primarily the quantity of energy applied, but the rate and power density at which it is applied, which gives rise to the specific materials processing effect desired. Since both power density and interaction time span six or seven orders of magnitude, the diagonal is the primary area of interest. A study of Fig. 1 should make this clear, along with the fact that, since the range of power densities extends far above that produced by most common thermal sources, even the long interaction time laser processes are in fact quite rapid, with interaction times rarely approaching 0·1 s.

The LASERGLAZE[TM] process was conceived to utilize the high power densities available from focused laser beams in conjunction with short interaction times to effectively confine thermal effects to a shallow surface layer,[7,8] with specific energy inputs ranging from 10^{-2} to 1 J/mm². Rapid surface melting thus occurs in a time in which an almost negligible amount of thermal energy can be conducted into the base metal, producing extremely sharp temperature gradients between the solid and the liquid. Since there is always intimate solid/liquid contact, very rapid quenching of the melt to the cold, solid substrate material results from these high thermal gradients. As will be indicated below, the quench rate is ultimately dependent on melt layer thickness, with cooling rates of $10^4 - 10^8$ °C/s attainable in thin melt layers. As a result of this rapid chilling, a variety of interesting metallurgical structures are produced.

EXPERIMENTAL

A continuous convectively-cooled, high-power carbon dioxide laser was utilized for the laser glazing studies. Both a Gaussian energy profile and a hollow beam output were made available by interchanging the cavity optics. The output beam, which had a 75 mm initial dia., was directed at the workpiece and simultaneously focused using reflective optics, as indicated in Fig. 2. The spherical focusing mirror employed had an effective

focal length of 460 mm, which produced a focal spot of 0·5 mm dia. at the workpiece surface. When 3 kW of laser power were applied, for example, an incident power density of $1·5 \times 10^6$ W/cm² was obtained. Such a power density is equivalent to that provided by a black body radiative source at 22 800°C. This high power density promotes effective coupling of the laser energy into the workpiece despite the initially high reflectivity of metallic surfaces to the 106 000 Å wavelength of CO_2 laser radiation. The high power density is also essential for localizing the energy input into the surface. Although substantially higher values were attainable, the conditions above were selected in order to promote laser glazing effects through efficient coupling and yet not create significant plasma generation problems. The inert gas shield indicated in Fig. 2 was used to suppress plasma generation while simultaneously shielding the melt from atmospheric contamination. Although the possibility of enhanced cooling from this gas cover was considered initially, heat flow calculations showed that its contribution, even under the highest assumed heat transfer coefficients, would be negligible.

A range of melt depths was obtained by varying the translational speed of the workpiece under the focused beam over a linear speed range of from 1·5 to > 60 m/min. This speed range resulted in a variation in the specific incident laser energy from 120 down to 3 J/mm. The high specific energy inputs were used as homogenizing passes, whereas passes at the lower end of the specific energy range yielded rapidly cooled glazed surfaces. The homogenizing passes were used in order to provide additional dissolution time in the molten state, since the extremely short liquid-state residence time characteristic of laser glazing may not always allow full homogenization of all the phases present in some alloys.

By arranging the samples to be laser glazed around the circumference of a rotating disc, it is possible to subject a large number of specimens to identical processing parameters. Samples were arranged side by side with their top surfaces located at the focal plane of the focused laser beam. Surfaces were ground and sanded to a uniform 240 grit finish, and no absorptive coatings were added to the surface. Coatings are often used in surface hardening to promote absorption of the low power densities normally utilized in this process. Such coatings are not effective at the higher power densities employed for laser glazing, and could also lead to contamination of the melt zone. Specimen thicknesses of 5 mm were employed to ensure sufficient thermal capacity in the substrate to minimize bulk temperature increase. This minimum thickness was well in excess of the analytically predicted requirement that the specimen

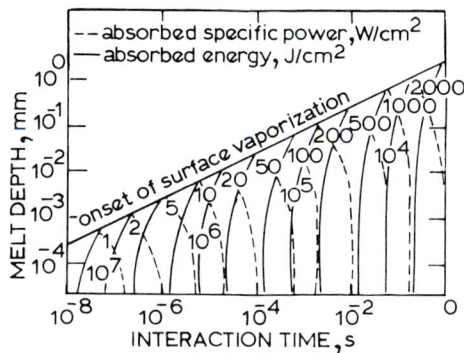

3 Transient surface melting characteristics of nickel

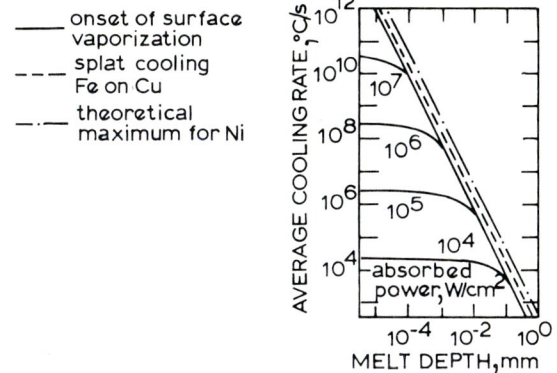

4 Effect of melt depth and power density on average cooling rate

be at least four times the glaze depth so as to facilitate rapid self-quenching without substantially increasing the subsurface temperature.

THEORETICAL

In order to determine the effects of different process parameters on the melting and rapid cooling processes resulting from high power density laser glazing, a one-dimensional, finite element heat transfer analysis was applied to the transient cooling processes.[12] Results of an analog computation[13] were used for the surface melting process occurring on heating. Using this method, if the laser power density and interaction time are specified, the melt depth can be calculated (Fig. 3). Temperature profiles in the solid and molten regions were also obtained, and were used as initial conditions in the finite element program in order to compute the transient cooling of the material for the specified power and interaction time. The average cooling rate reported was calculated over the time from the cessation of power input to the time at which 800°C was reached.

Transient surface melting characteristics for nickel are shown in Fig. 3. The solid lines represent melt depth as a function of interaction time for a variety of power densities. Interaction time is defined as the time required for the laser beam to traverse one spot diameter, and is computed by dividing the spot diameter by the surface traverse velocity (D/V). The period initiates from a uniform initial temperature of 21°C and thus includes the time that is required to heat the material in the solid state to its melting temperature. After the onset of melting, the advance of the melt interface was found to be linear with time. The dashed lines in Fig. 3 represent constant specific energy inputs and are computed as the product of the absorbed specific power and the inter-action time. It is obvious from Fig. 3 that as the power density increases the specific energy necessary to obtain a given melt depth decreases, and the melting is also accomplished in a shorter time. This is the result of the fact that higher melting efficiencies are obtained at higher power densities since less time is available for conduction into the substrate material. In addition to promoting higher melting efficiencies, the reduction in conduction to the substrate also leads to increased temperature gradients in the alloys following melting and thereby increases cooling rates in the glazed samples.

In Fig. 4, the effect of melt depth on cooling rate is shown for selected power densities. The maximum cool-

ing rate which can be attained at each power density occurs for the limiting case in which the melt depth approaches zero. This maximum cooling rate increases with the square of the power density. For a given melt depth, the increase in cooling rate with absorbed specific power is a consequence of the increased temperature gradients; doubling the power density approximately doubles the cooling rate for a given melt depth. Figure 4 also shows that there is a maximum melt depth obtainable for a given power density after which surface vaporization begins, and that the lowest cooling rates are realized under these conditions. A theoretical maximum cooling rate for nickel is also included in Fig. 4 as well as typical results for splat cooling of iron on a copper substrate.[14] The theoretical maximum cooling rate is obtained by considering that all of the energy is initially concentrated in the melt and that a temperature discontinuity equal to the melting temperatures exists at the solid/liquid interface. It is noteworthy that the splat cooling rate, the average cooling rate at the threshold of surface vaporization, and the theoretical maximum cooling rate are all inversely proportional to the square of the melt depth. The possibility of contributing effects from convection and radiation at the surface on cooling rates was investigated for a wide range of parameters. Even assuming a radiative emissivity of 1·0 and a high convective heat transfer film coefficient (0·57 W/cm² °C), the average cooling rate was not changed by more than a few per cent for absorbed power densities in excess of 10^5 W/cm². It was concluded, therefore, that at the high power densities employed for laser glazing, rapid cooling occurs primarily through conductive heat transfer to the substrate.

An analytical study has also been made of the influence of power density on the transient behaviour of temperature gradient G, freezing rate R, and cooling rate $\dot{T}$ for various critical melt depths. The main conclusions derived from this study may be summarized as follows:

1 For a given initial depth of melt, the temperature at any point beneath the surface rises initially when power is removed. The effect increases with the power density for a given depth of melt. This phenomenon means that after the cessation of energy input, melting continues, however briefly, before solidification commences (Fig. 5). Thus, cooling rate and freezing rate are both initially zero.

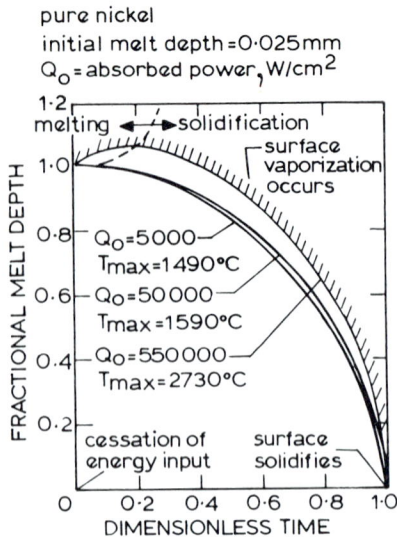

pure nickel
initial melt depth =0·025mm
Q_O = absorbed power, W/cm^2

5   **Transient behaviour of melt surface**

2 The temperature gradient G at the melt interface decreases from a maximum, at the cessation of energy input, to zero as the surface solidifies, and has a strong dependence on absorbed power (Fig. 6).

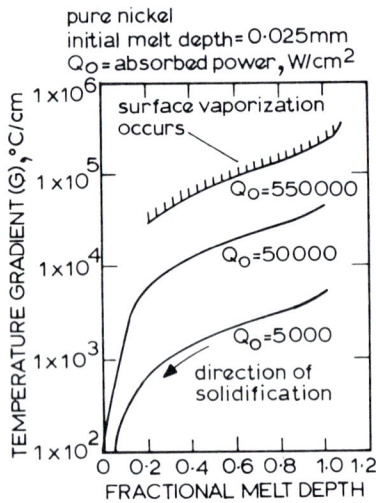

pure nickel
initial melt depth= 0·025mm
Q_O = absorbed power, W/cm^2

6   **Transient behaviour of temperature gradient**

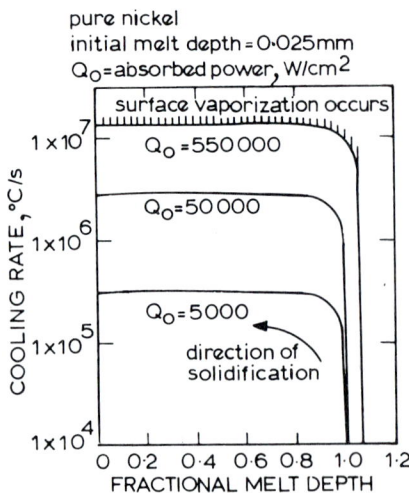

pure nickel
initial melt depth= 0·025mm
Q_O = absorbed power, W/cm^2

7   **Transient behaviour of cooling rate**

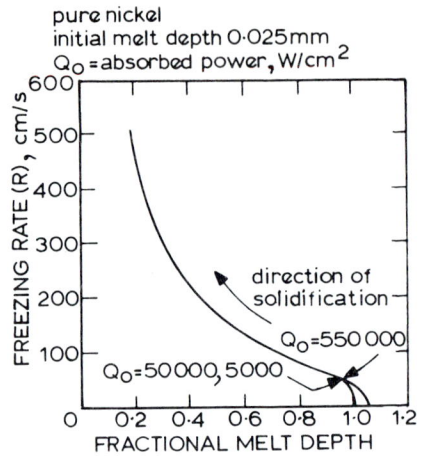

pure nickel
initial melt depth 0·025mm
Q_O = absorbed power, W/cm^2

8   **Transient behaviour of freezing rate**

3 The cooling rate $\dot{T}$ at the melt interface starts at zero and tends to a constant limit as solidification proceeds to the surface. The cooling rate has a strong dependence on power density (Fig. 7).

4 The freezing rate R at the melt interface starts at zero as solidification begins and goes to infinity as the surface solidifies. For a given initial depth of melt, the freezing rate is essentially independent of power density after solidification has commenced (Fig. 8).

5 The ratio G/R is infinite at the start of solidification (because the freezing rate is zero) and goes to zero as the surface solidifies (because R becomes infinite). G/R increases with power density for a given depth of melt (Fig. 9).

6 The parameters G/R, R, G, T, etc. follow definite scaling laws. G/R will have the same dependence on relative depth of solidification if, when the initial melt depth is increased by a factor M, power density is decreased by M and time is increased by M^2.

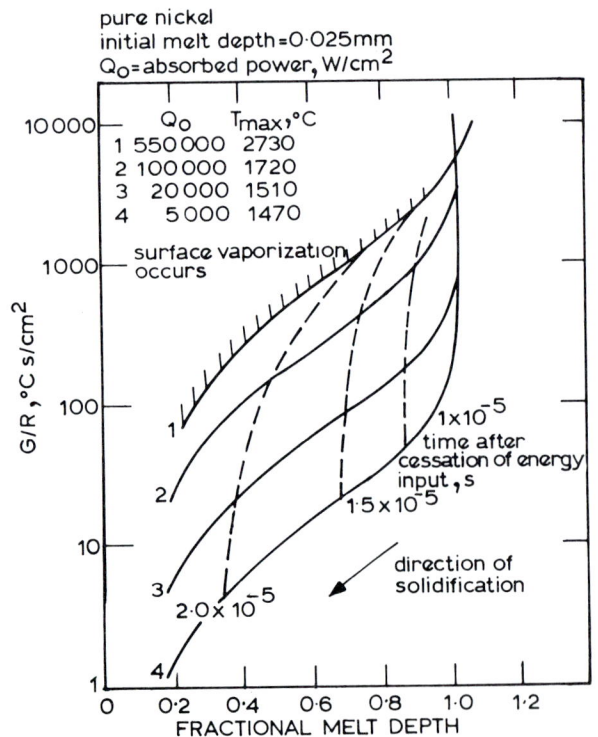

pure nickel
initial melt depth =0·025mm
Q_O = absorbed power, W/cm^2

	Q_O	T_{max}, °C
1	550 000	2730
2	100 000	1720
3	20 000	1510
4	5 000	1470

9   **Variation of *G/R* at solidification interface**

RESULTS AND DISCUSSION

The initial question usually asked about laser glazing is whether or not the technique can be used to generate a non-crystalline or amorphous (glassy) surface layer on a bulk material. This is a reasonable question since the presence of a crystalline substrate in intimate contact with the melt should inhibit the undercooling necessary to form an amorphous solid. As shown elsewhere,[7] the results obtained for the low melting point Pd–4·2 Cu–5·1 Si eutectic alloy indicate that at least in this alloy an amorphous solid can be formed. However, all other alloys examined to date have yielded extended solid solution phases, with varying degrees of stable or metastable phase decomposition, depending sensitively on cooling rate and composition. Typical examples will now be considered.

Conventional superalloys

A superalloy is essentially a nickel-base solid solution phase (γ-phase) that has been strengthened by a fine precipitate of $Ni_3(Al, Ti)$ phase (γ'-phase). A typical microstructure consists of dendrites of cooling γ' in γ, interspersed with interdendritic γ/γ' eutectic and MC carbides, the latter being the product of gross dendritic segregation.

For the glazing experiments on the superalloys, monocrystalline materials were selected. This choice was made because it was of interest to determine if epitaxial growth could be induced under rapid solidification conditions, where substantial modifications in microstructure inevitably occur. The monocrystalline specimens were glazed using a wide variety of processing conditions, corresponding with a broad range of cooling rates. The preferred procedure was to employ an initial deep-penetration, relatively low cooling-rate glazing pass to homogenize the material properly, before the application of one or more superimposed high cooling-rate passes. Without the application of an initial homogenizing pass, glazed layers in superalloys exhibit incomplete dissolution of interdendritic MC carbide particles, apparently because of insufficient residence time in the molten state of the alloy.

Figure 10 shows photomicrographs of areas containing superimposed glazing passes in Mar-M200 and B-1900, both with {100} surface orientations. In both cases, it is clear that glazing has produced a marked refinement in the scale of the dendritic structure, with a corresponding reduction in the wavelength of dendritic segregation. Moreover, there is a strong tendency for epitaxial growth on the substrate, as well as between successive glazed layers. The few misoriented grains visible in these micrographs have their origin primarily, if not exclusively, in γ/γ' eutectic regions residing at the initial melt/substrate interface. Misoriented grains that have orientations not far from the $\langle 100 \rangle$ orientation of the adjacent epitaxial regions tend to propagate alongside the epitaxial layer, irrespective of the number of superimposed glazed layers. Grossly misoriented grains, however, are usually eliminated in the competition in growth with the more favourably oriented epitaxial layer.

Figure 11 shows higher magnification photomicrographs of the central region (Fig. 10b) of $\langle 100 \rangle$ epitaxial growth in B-1900. As shown, the scale of the dendritic structure is markedly reduced at both the interface between the substrate and the first homogenizing pass, as well as between the first and second superimposed glazing passes. In the region of epitaxial growth between the first pass and the substrate, the structure has a distinctly cellular character. With increasing distance from this interface, the structure becomes cellular–dendritic and finally fully dendritic. It is proposed that these structural changes merely reflect a diminishing ratio of temperature gradient to solidification rate G/R with increasing distance from the initial substrate/melt interface (Fig. 9). The general appearance of the second pass indicates cellular, or one-dimensional dendritic growth throughout, which is indicative of persistently high values of G/R. The fine scale of all the structures in the glazed regions, whether cellular or dendritic, is a consequence of the high cooling rates $\dot{T}$ (Fig. 7).

As shown in Fig. 11, the region encompassing the second pass, including a thin heat affected zone (HAZ) in the underlying first pass, does not etch up as well as the rest of the material. The reason for this difference in etching behaviour is that the cooling rate is fast enough to prevent solid state precipitation of γ'-phase in the γ-matrix only in pass 2 and its associated HAZ. Presumably, in the HAZ the elimination of the γ'-particles is a consequence of much faster kinetics for solution of the γ'-phase than for its precipitation. The region including the HAZ and pass 2 also shows striking differences in size, distribution, and morphology of the interdendritic MC carbide particles (Fig. 11). In pass 2, small platelet carbides are unidirectionally aligned in the interdendritic regions of the cellular structure. On the other hand, in pass 1, the carbide platelets are distributed in an irregular manner following the tortuous path defined by the interstices of a fully dendritic structure. The finer scale of the carbide particles in pass 2 is consistent with the higher average cooling rate in this region, compared with pass 1. The nature of epitaxial dendritic growth in superalloys has also been observed for substrate orientations other than {100}. In all cases, it has been found that $\langle 100 \rangle$ dendritic growth is preferred, even when this fast-growing dendrite growth direction is sharply inclined to the direction of solidification or normal to the free surface. An interesting case is the {110} surface orientation, where, owing to the fact that two equivalent $\langle 100 \rangle$ growth directions are favoured, the

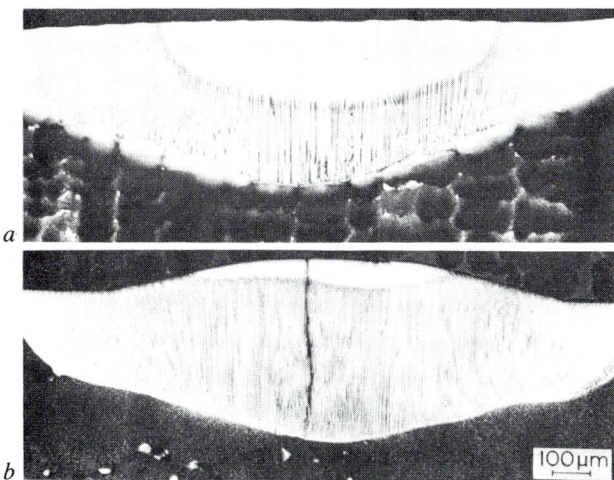

a Mar-M200; *b* B-1900

10 Superimposed glazing passes in monocrystals with {100} surface orientations

12 Dendrite boundary in {110} surface orientation

dendritic colonies impinge upon one another along the centre line of solidification to form a 'dendritic boundary', with no change in crystallographic orientation. Such dendrite boundaries tend to become sharply delineated as the interface between melt zone and substrate assumes some curvature. Thus, in a deep-penetration homogenizing pass, the dendritic boundary is sharply defined (Fig. 12).

Eutectic superalloys

A recent development in the superalloy field has been the introduction of the eutectic superalloys. Typical of this new class of superalloys are the filamentary reinforced Ni- and Co-base alloys, such as CoTaC, NiTaC and $\gamma/\gamma'-\delta$. In the CoTaC and NiTaC series of alloys, matrix reinforcement is achieved with directionally aligned fibres of TaC phase, whereas in $\gamma/\gamma'-\delta$ this is accomplished with thin platelets of δ-phase (Ni$_3$Nb).

Essentially two types of microstructure are observed in the glazed carbide eutectic superalloys irrespective of minor deviations in alloy composition from the ideal eutectic, i.e. whether or not the alloy contains moderate amounts of primary γ or TaC carbide. Under low-moderate cooling rates, the microstructure is distinctly dendritic. Moreover, the dendritic structure typically develops by epitaxial growth on the underlying eutectic substrate (Fig. 13). As the cooling rate increases, the structure becomes less obviously dendritic, at least under the optical microscope, and eventually all semblance of a dendritic structure is lost (Fig. 14). In the dendritic regime of growth, the dendrites are composed of a γ-solid solution phase, whereas the interdendritic regions are made up of a fine filamentary eutectic of γ+TaC (Fig. 15a). In the 'dendrite-free' regime of growth (high cooling rate condition), the TaC particles assume a more discrete particulate morphology, rather

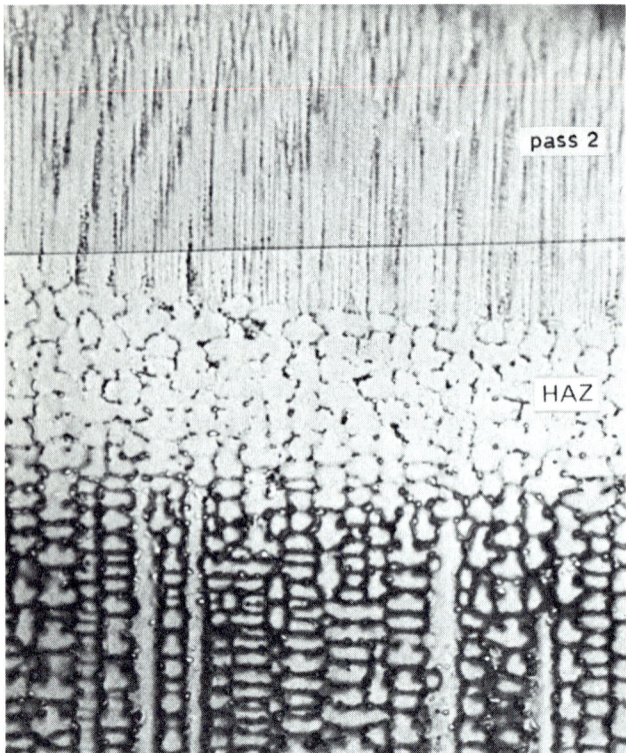

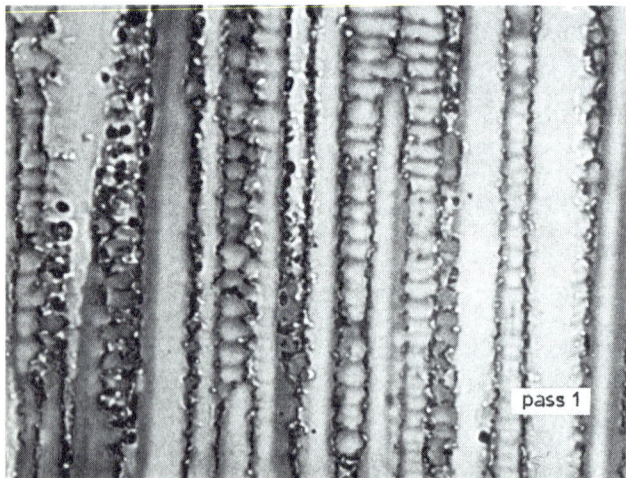

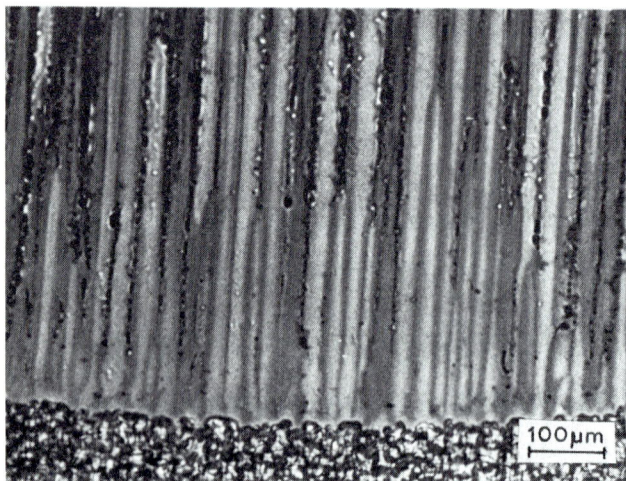

11 Enlargement of Fig. 10b showing evidence of epitaxial growth

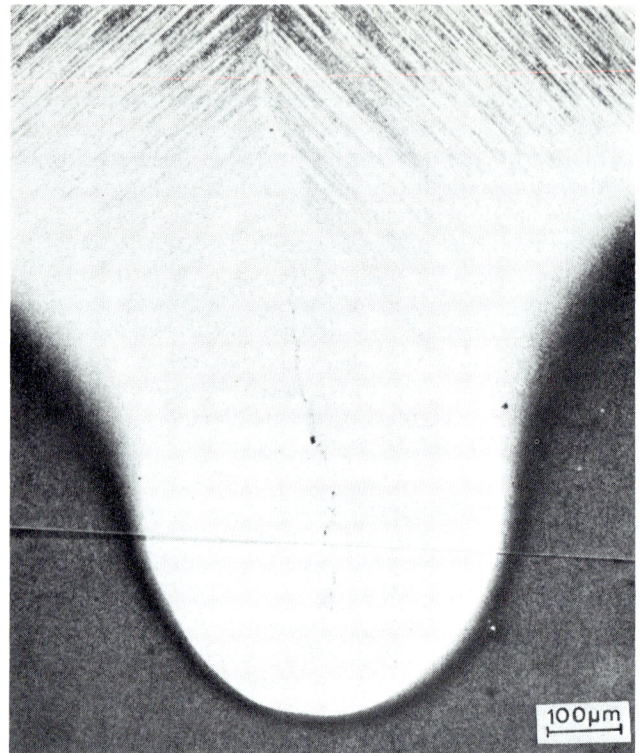

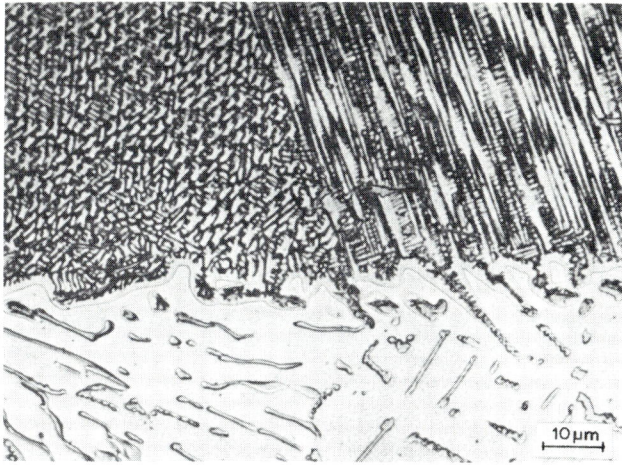

13 Epitaxial growth in coarse-grained CoTaC–3 alloy

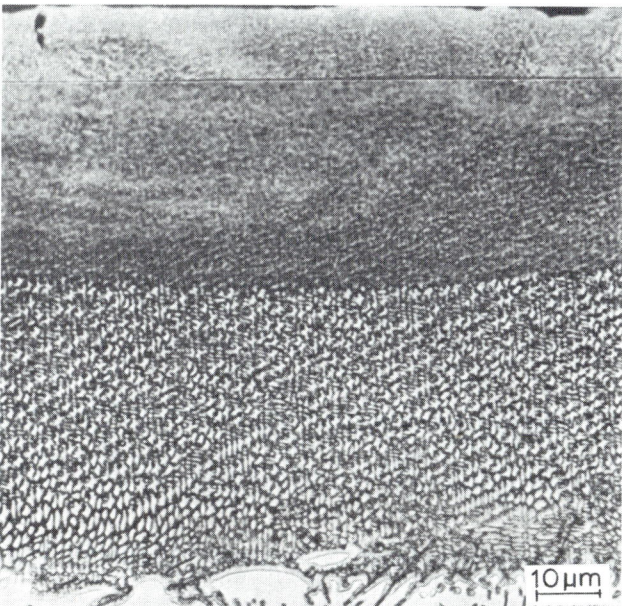

14 Dendritic and 'non-dendritic' (perhaps unresolved) growth in CoTaC-3 alloy (double pass)

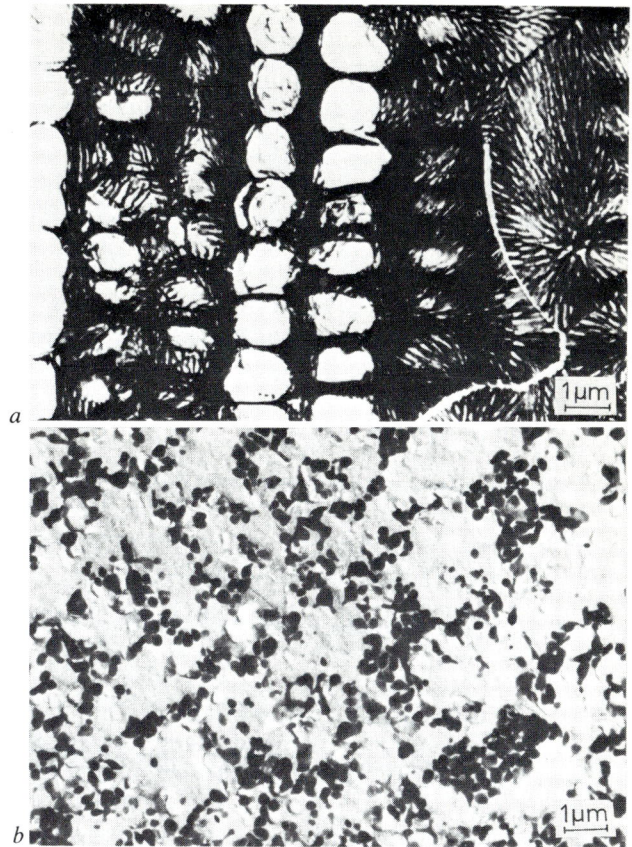

a dendritic region; *b* non-dendritic region

15 Extraction replicas of MC carbide phase shown in Fig. 16

than a filamentary morphology (Fig. 15*b*). Dendritic growth is also a characteristic of glazed $\gamma/\gamma'-\delta$, at least under moderate cooling rates (Fig. 16). It is not yet known, however, to what extent the precipitation of γ', δ, or γ'' is influenced by cooling rate. A notable feature of all these dendritic structures is their high hardness e.g. HV = 600–700 kg/mm^2 for CoTaC-3 and HV = 750–850 kg/mm^2 for $\gamma/\gamma'-\delta$.

An interesting feature of the glazed eutectic superalloys is the prevalence of the dendritic structure, at least under moderate cooling rates, despite the fact that the underlying substrate is fully eutectic. This is interpreted to mean that in the undercooled condition, not only is there a shift (extension) of the solid solution phase boundary, but also a shift in the position of the eutectic point. A simple line of reasoning suggests that the shift in the eutectic with undercooling is towards the phase boundary for single-phase TaC. This being the case, it seems reasonable that the laser glazing of these alloys

will yield a fully eutectic structure only if the substrate composition is on the primary TaC side of the monovariant eutectic trough. Moreover, since the phase decomposition takes place in the undercooled state, the eutectic structure should be on a much finer scale than is usual for conventional solidification processing. Evidence for such behaviour in boron-rich nickel-base alloys (TLP-21) has been reported previously.[7] In that case, a lamellar eutectic was produced, with interlamellar spacing of 300–500 Å. Further work is now under way to determine if a similar ultrafine eutectic can be generated in the glazed eutectic superalloys.

16 Dendritic growth in $\gamma/\gamma'-\delta$ alloy

Alloy steels

Most alloy steels in the normal heat-treated condition contain a fine dispersion of one or more carbides. The high-speed steels have a relatively large volume fraction of strengthening carbide phases. In vanadium steels, the predominant carbide is MC. In steels with high concentrations of the refractory elements W and Mo, the principal carbide phase is M_6C. Nearly all the materials contain Cr-rich $M_{23}C_6$ carbide. Alloys selected for the laser glazing experiments were 440C, M-50, M-2, and 4350. The primary interest in the study was to determine the influence of cooling rate on measured Vickers hardness in the glazed material, including both melt and heat affected zones.

In the glazed condition, alloy 440C exhibited a dendritic structure of Cr-stabilized ferrite, interspersed with thin sheets of carbide in the interdendritic region (Fig. 17). The melt zone was associated with a broad heat-affected zone (HAZ) in the substrate. The dark-etching band in the substrate appeared to define the lower limit of microstructural changes in the HAZ. The measured hardness in the melt zone (dendritic structure) was HV ~ 470 kg/mm^2. This is to be compared with HV ~ 825 kg/mm^2 for the carbide-strengthened martensitic substrate. So in this case the effect of glazing was to produce softening of the material. On the other hand, the region of the HAZ just below the melt interface was somewhat harder (HV ~ 900 kg/mm^2) than the substrate. It appears, therefore, that laser heat treatment, rather than melting, is the approach to take to further harden this particular material.

Alloy M-50 also showed evidence of dendritic growth in the melt zone (Fig. 18), albeit in a form more difficult

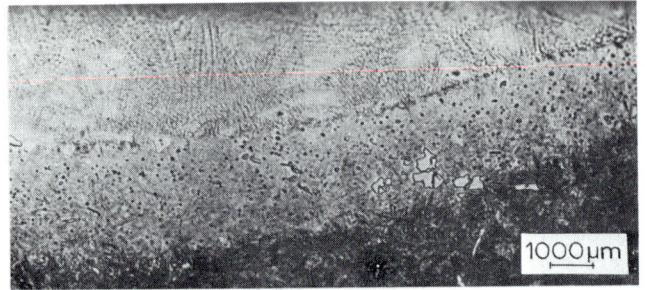

18   Glazed alloy M-50 showing undissolved carbides in heat affected zone

to detect than in alloy 440C. Moreover, in marked contrast to the behaviour of alloy 440C, M-2 did show some improvement in hardness in the melt zone compared with the substrate: HV ~ 850 v. 800 kg/mm^2, respectively. The heat affected zone also showed a small increase in hardness. Since the original carbide particles in the substrate are still visible in the HAZ (Fig. 18), it seems likely that the high hardness of this zone is caused by the formation of fine martensite, not resolvable under the optical microscope. A similar situation exists in the melt zone, except that the fine martensite forms within the dendritic structure, probably interspersed with fine carbides. It is noteworthy that the original coarse carbide particles readily dissolve in the melt during glazing. In double or multiple passes, a clear indication was obtained of the formation of martensite (or bainite) on a coarse scale, accompanied by some reduction in hardness.

In complete contrast to 440C and M-50, alloy M-2 showed no indications of a dendritic structure. On the contrary, the melt zone appeared to be composed of a fine-grained, two-phase structure (Fig. 19). Hardness measurements showed this two-phase structure to be softer (HV ~ 600 kg/mm^2) than the substrate (HV ~ 1000 kg/mm^2), at least in the region of a single pass. In the HAZ associated with a double pass, hardness values as high as HV ~ 900 kg/mm^2 were recorded, demonstrating that the strength in the melt zone can be recovered by a post-glazing heat treatment. Possibly, this occurs as a result of the formation of tempered martensite. Recently[15] it has been shown that the as-glazed, two-phase structure is composed of ferrite and austenite, but no martensite. The absence of martensite explains the low hardness of the glazed material.

The preceding results show the varied response of optimally hardened alloy steels to glazing. It is of interest now to consider what happens to a material that is initially in the softened condition, and in fact normally used as such. Illustrative of the significant changes that can be induced by glazing of such materials is the behaviour of spheroidized annealed alloy 4350 (HV ~ 450 kg/mm^2). Laser glazing of this material produces a dramatic hardening, with the maximum hardness corresponding to the thinnest glazing passes or, in other words, the fastest cooling rates. Typical hardness values range from 800 to 850 kg/mm^2, depending on cooling rate. In the HAZs associated with double or multiple passes (Fig. 20) the hardness is much reduced compared with that in the as-glazed material, but still higher than that of the annealed substrate.

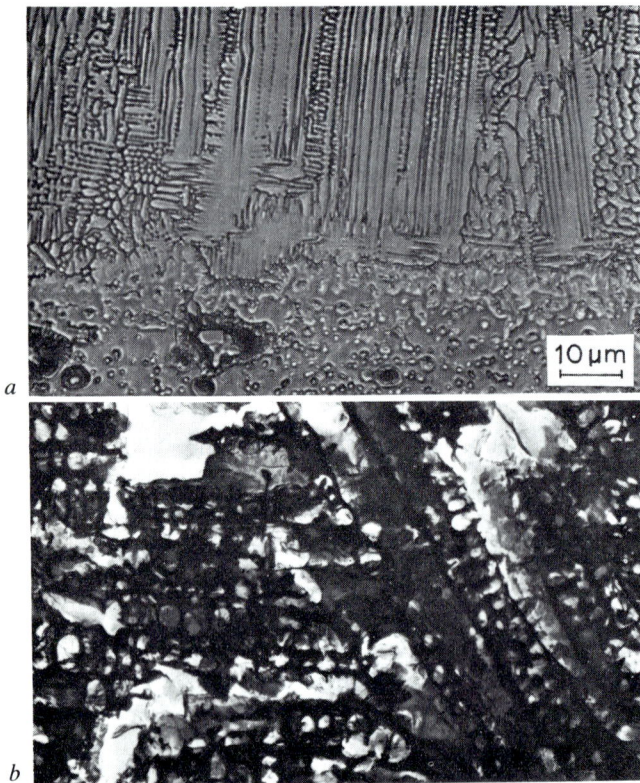

a dendrites; *b* extracted carbides in interdendritic regions

17   Structure of glazed alloy 440C in interdendritic regions

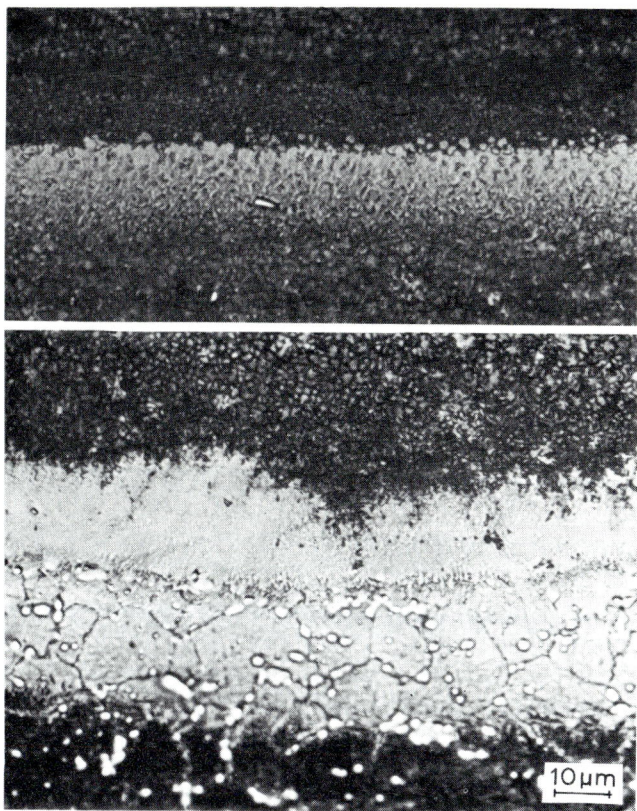

19 Double glazing pass in alloy M-2 showing duplex $\alpha + \gamma$ structure

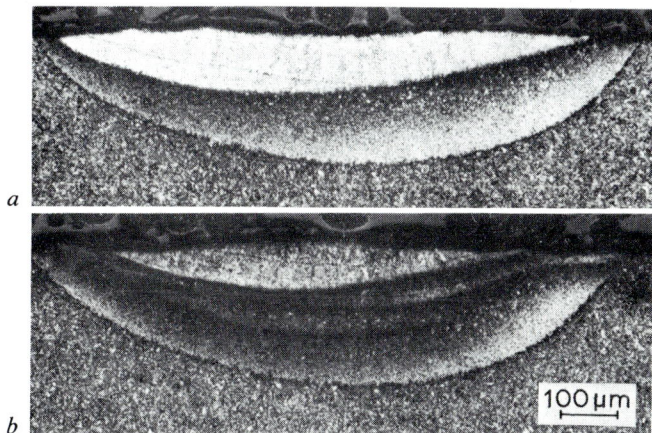

20 Superimposed glazing passes in alloy 4350

CONCLUSIONS

It has been shown that rapid chilling of alloys by laser glazing gives rise to a variety of interesting and potentially useful metallurgical microstructures. In addition, consideration of these structures and the means by which they were produced has highlighted the laser as a potentially important tool for future materials processing applications. The incentive for applying lasers to additional new areas of materials processing stems from the unique capabilities of lasers as means for energy delivery. These capabilities include precise control of energy delivery at high power densities, precise control and limitation of process interaction times, accurate control of the location of energy deposition because of the ability to focus finely, the ability to operate out-of-vacuum in a variety of gaseous atmospheres and at various pressures, the cleanliness and remote nature of the laser as a heat source, and the adaptability of lasers to automation. These characteristics of the laser have led to its being considered as the energy source for a variety of new materials processing techniques. Apart from its ability to produce rapidly chilled structures and its potential for precise aging of metastable structures so produced, the laser also shows potential for melting of previously difficult-to-melt materials at high melt superheats, machining and atomization processes, alloying and shape fabrication from the melt, directional solidification and gradient annealing, epitaxial solidification, coating, and weld bonding. These concepts, when integrated, project the laser as an important energy source for future materials processing. The laser glazing technique, which is finding its first applications as a technique for alloy screening and as a research tool for work involving the production of novel and potentially useful metallurgical microstructures, has opened the door for lasers in special materials processing applications. As a result, the unique capabilities of lasers as energy sources now promise to provide a new method for control of structure and properties through more sophisticated materials processing than has previously been possible.

REFERENCES

1 R. A. HELLA AND D. S. GNANAMUTHU: Annual Meeting AWS, St Louis, 10–14 May 1976
2 E. M. BREINAN et al.: Proc. ASM Conf. on Laser Surface Treatment for Automotive Applications, Detroit, Mich., 17 February 1976
3 D. S. GNANAMUTHU AND E. V. LOCKE: US Patent No. 4, 015, 100
4 C. O. BROWN AND C. M. BANAS: 52nd Annual Meeting, AWS, San Francisco, 26–29 April 1971
5 E. V. LOCKE et al.: *IEEE J. Quantum Electronics*, 1972, **8**, 132–135
6 E. M. BREINAN et al.: IIW Document IV–181–75, June 1975
7 E. M. BREINAN et al.: Proc. 3rd Int. Symp. on Superalloys, Seven Springs, 435–450; Baton Rouge, 1976, Claitor's Publishing Division
8 E. M. BREINAN et al.: *Phys. Today*, 1976, (1), 44–50
9 W. G. VOORHIS: Proc. Ion and Electron Beam Symposium, 1964
10 B. P. FAIRAND et al.: *J. Appl. Phys.*, 1972, **43**, 3 875–3 893
11 B. P. FAIRAND et al.: *Appl. Phys. Lett.*, 1974, **25**, 431–433
12 L. E. GREENWALD: United Technologies Research Center Report R75–111321–1, October 1975
13 M. I. COHEN: *J. Franklin Inst.*, 1967, **283**, (4), 271–285
14 R. C. RUHL: *Mater. Sci. Eng.*, 1967, **1**, 313–320
15 P. R. STRUTT: personal communication

Structures and properties of Thixocast steels

K. P. Young, R. G. Riek, and M. C. Flemings

A die-casting system is described which utilizes the thixotropic nature of semi-solid Rheocast slurries. This system has been developed to a pilot-plant scale at Massachusetts Institute of Technology (MIT). It comprises a continuous Rheocaster which is used to produce ingot stock, a reheating furnace to reheat individual charges to the semi-solid state, a 'softness indicator' to determine when the reheated charges are suitable for casting, and a commercial cold chamber die-casting machine. This paper deals primarily with the structure and properties of Thixocast ferrous alloys. Discussion is given of control of primary particle size, phases, oxide inclusions and microsegregation in AISI 304, 440C, and other steels. Structure–property relationships are also presented and comparisons made with properties of conventionally cast and wrought alloys.

The authors are at the Massachusetts Institute of Technology, Cambridge, Massachusetts, USA

Fundamental investigations[1,2] of the mechanical and rheological properties of semi-solid alloys have led to discoveries upon which a radically new family of metal-working concepts are based. These Rheocasting processes[3] are based upon the production, by the application of vigorous agitation during solidification, of a unique, highly fluid, non-dendritic alloy slurry which can exhibit apparent viscosities of less than 10 poise at volume fractions solid in excess of 0·60.[4] The availability of such Rheocast slurries exposes a variety of new shape-forming processes and can offer substantial advantages to existing processes, both those currently applied to fully solid metal such as forging and extrusion, and conventional casting processes.

The engineering developments which have, over the course of a four year programme at MIT, led to equipment capable of producing high temperature alloy slurries (such as bronze, cast irons, steels, superalloys) have been described elsewhere,[5–7] as have the developments which have led to the pilot-plant stage operation of a viable ferrous alloy die-casting process.[8] This paper briefly reviews those developments and, subsequently, describes the structures and properties that characterize these processes.

EQUIPMENT

Major emphasis at MIT to date has been on the development of the Thixocasting process, shown schematically in Fig. 1, for ferrous alloy die casting. In the Thixocasting process, a Continuous Rheocaster (Fig. 1a) is used to produce slurry containing a high volume fraction primary solids (~0·50) which is teemed to produce ingots. After final solidification and cooling to room temperature, these ingots are then sectioned into weights or 'charges' appropriate for one die-casting shot (Fig. 1c). When desired, each individual charge is reheated to the semi-solid range and at the appropriate softness (or consistency) (Fig. 1d) it is die cast (Fig. 1e, f).

Continuous Rheocaster

The design for a Continuous Rheocaster is dictated to a large extent by the intended application of the product. Thus, to exploit the grain-refining advantages of Rheocasting in the manufacture of large ingots or castings a relatively high throughput rate of slurry containing numerous small primary solid particles may be desired. High cooling rates, coupled with relatively low volume fractions of primary solid particles to act as nuclei, may be a suitable approach. In the production of ingot stock for Thixocasting high-temperature alloys, however, high volume fractions of primary solid particles are desired to maximize the advantages with respect to both die filling and machine component life.[9,10]

The design of the high-temperature Continuous Rheocaster at MIT is shown schematically in Fig. 2. It consists essentially of two vertically connected, concentric cylinders. An upper reservoir chamber holds about 20 kg of molten alloy at about 30°C superheat. Beneath this, a smaller diameter chamber, lined with recrystallized alumina, functions essentially as a heat exchanger and agitation zone.

Shearing is provided by a hollow recrystallized alumina rotor which is concentric with and passes through

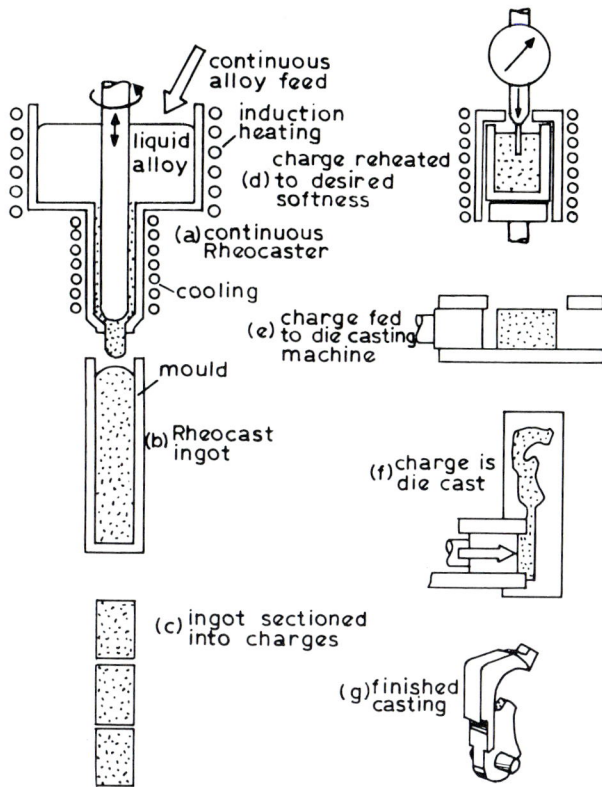

1 The Thixocasting process

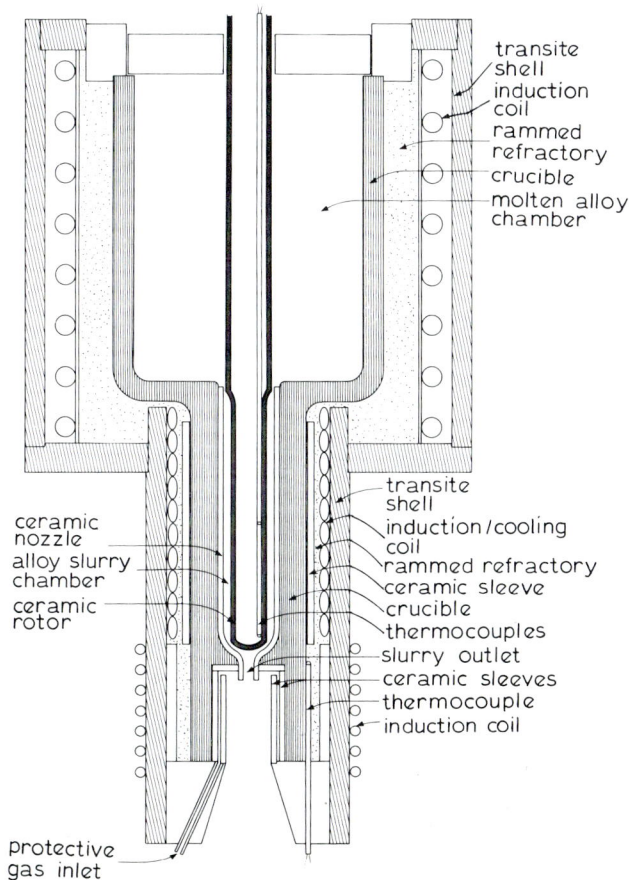

2 Schematic of the Continuous Rheocaster

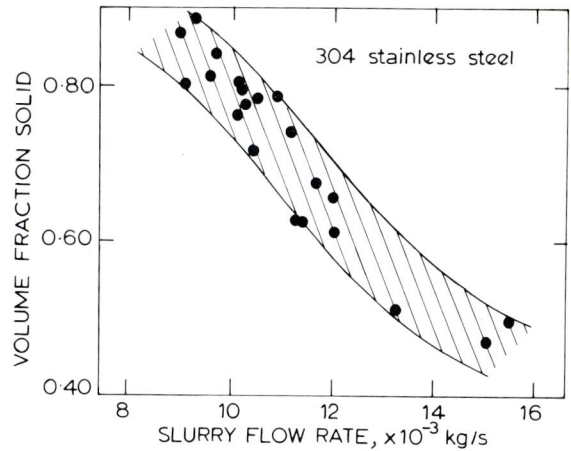

3 Volume fraction solid *v.* slurry flow rate for Continuously Rheocast AISI 304 stainless steel

the reservoir and mixing chambers. This rotor also functions as a valve on a small exit port at the base of the mixing chamber. Rheocast slurry is produced continuously by allowing alloy to flow through the mixing tube walls by the surrounding water-cooled copper coils. These coils also function as induction heating coils for start-up. The rate of heat extraction through the crucible walls is a function of crucible conductivity (sensitive to operating temperature) cooling coil/crucible separation, and other factors such as cooling coil density.

For a given furnace configuration, the rate of heat extraction is primarily dependent upon operating temperature. Thus, for relatively narrow freezing range alloys such as most steels, the rate of heat extraction remains fairly constant regardless of fraction solid in the mixing chamber. Under these conditions, slurry flow rate and slurry volume fraction solid are roughly inversely proportional, as shown in Fig. 3 for 304 stainless steel.

To date, more than 3 000 kg of ferrous and other high-temperature alloys including 304 and 440C stainless steel, M2 tool steel, X-40 cobalt based superalloy, and 4340 steel, have been Continuously Rheocast at MIT at rates up to 5 lb/min and in quantities ranging from 10 to 200 kg. These have in most part been teemed into ingot moulds 15 cm long × 3 cm dia. contained in carousels (*see* Fig. 4).

Casting system

The casting system, represented schematically by Figs 1*c–e*, consists of a small induction furnace powered by a 60 kW, 3·8 kc supply, a 'softness indicator' and a 125 t cold chamber horizontal die-casting machine. The automatic reheating furnace which represents the heart of the system is shown in Fig. 5. It comprises the induction furnace which is controlled via the 'softness indicator'.

The 'softness indicator', or penetrometer, has been developed during the course of this work to eliminate the dependence upon thermocouples to control the reheating stage. It consists of a rod (alumina in the case of ferrous alloys) which impinges upon the flat surface of a 'charge' while it is reheated. For a given depth of penetration, typically 3 mm, the load on the rod correlates well with fraction solid, as shown in Fig. 6. In

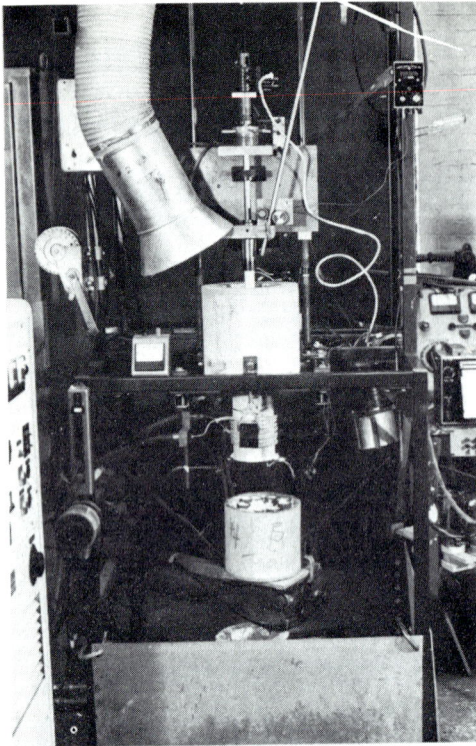

4   The Continuous Rheocaster

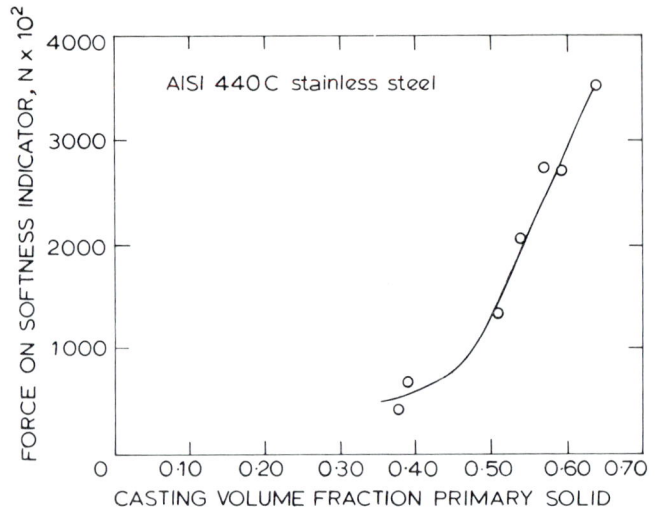

5   The Thixocast reheat station

6   Operating curve for the softness indicator, AISI 440C stainless steel

essence therefore it measures the softness or consistency of charges as they are reheated.

At the appropriate softness or consistency, the charges are transferred to the die-casting machine and cast. At the time of transfer the unsheared charges have an apparent viscosity of 10^7 P (estimated from a simple Stokes' law analysis of the penetrometer and associated experiments in our laboratory[11]) and behave as 'soft solids', (Fig. 7). Upon casting, the shearing which occurs lowers the material viscosity to a level at which it flows smoothly to fill the die cavity.

The influence of the higher viscosity possessed by the Thixocast alloys can be seen in Fig. 8 which plots the regimes of ingate velocity and pressures to obtain Thixo-castings of acceptable internal and surface quality. The optimum regime is displaced markedly to lower velocities and pressures compared to conventional liquid practice.[12] Additional experimental evidence of the improved die-filling characteristics achieved by die-casting semi-solid Rheocast alloy slurries have been obtained by direct observation of die-filling behaviour.[13] These and other benefits, including improved machine component life, are discussed in detail elsewhere.[3,8,9]

7   Reheated charge of AISI 440C stainless steel about to be cast

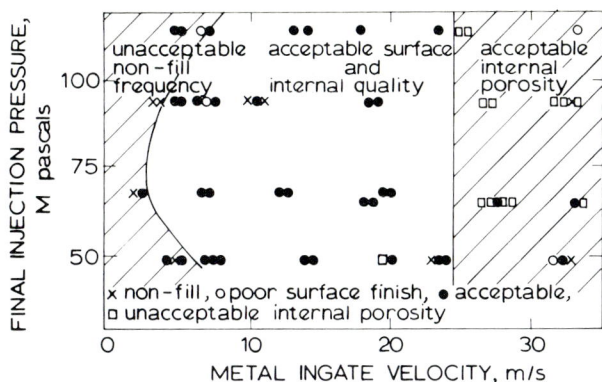

8 Die-casting parameter optimization for Thixocast AISI 440C stainless steel cast at 0·5–0·6 volume fraction solid; cast part weighs approximately 50 gm with ingate cross-section of 0·22 cm^2

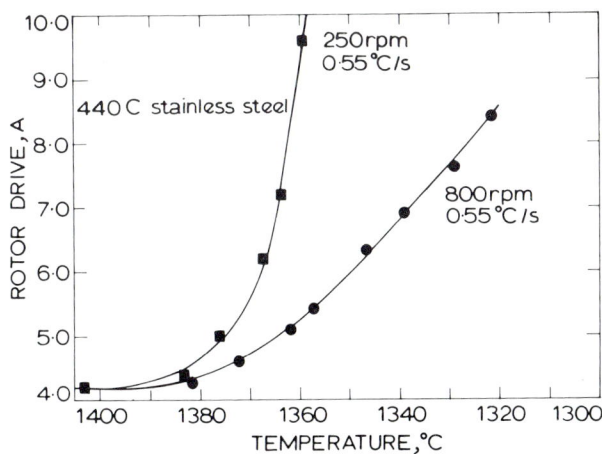

10 Approximate apparent viscosity v. volume fraction solid for AISI 440C stainless steel at different shear and cooling rates

RHEOLOGY OF HIGH–TEMPERATURE ALLOYS

With the exit valve closed off, the mixing chamber of the Continuous Rheocaster can be operated as a simple Stormer viscometer. The mixing chamber can be maintained isothermal by balancing the upper and lower induction coils and power supplied to the central mixing chamber coil. The temperature within the mixing chamber can be monitored via the emf's of the three Pt–Pt10Rh thermocouples located within the rotor which are transmitted via a rotating contact system. Alternatively, continuous cooling can also be obtained by reducing the power supplied to the middle mixing chamber coil. Operated in this way, since resistance of the fully liquid alloy in the reservoir chamber is negligible, the power required to drive the rotor at constant speed is an indication of the apparent viscosity of the material within the mixing chamber. Thus, data such as are given in Fig. 9 may be generated for high-temperature alloys. This is a plot of current required to drive the rotor v. temperature of 440C stainless steel. As can be seen, this is qualitatively similar to data generated for Sn–Pb alloys in the more sophisticated Couette viscometer.[2,4]

Substitution of various standard viscosity fluids into the Rheocaster also permits the construction of approximate apparent viscosity v. volume fraction solid curves such as Fig. 10. These curves bear even stronger qualitative similarity to the earlier Sn–Pb data. Increasing the cooling rate at a given shear rate is seen to create slurries of increasing apparent viscosity. In this way comparisons can also be made from alloy to alloy. Figure 11 shows such a set of data for M2 tool steel and 440C stainless steel. Clearly M2 tool steel exhibits lower apparent viscosities at a given fraction solid. This is consistent with experimental observations during actual slurry production.

Data such as those of Figs 10 and 11 are essential for the full understanding and exploitation of the potential of Rheocast slurries. In our own laboratory it has provided a reliable means, independent of temperature, by which the Continuous Rheocaster may be controlled. This is particularly advantageous when narrow freezing range alloys (the majority of engineering alloys) are Rheocast.

At a given shear and cooling rate (more precisely rate of solidification), rotor power consumption is a sensitive indicator of slurry volume fraction solid, as shown in Fig. 9. Maintenance of constant rotor power consumption during continuous slurry production therefore ensures consistent slurry production. This becomes more

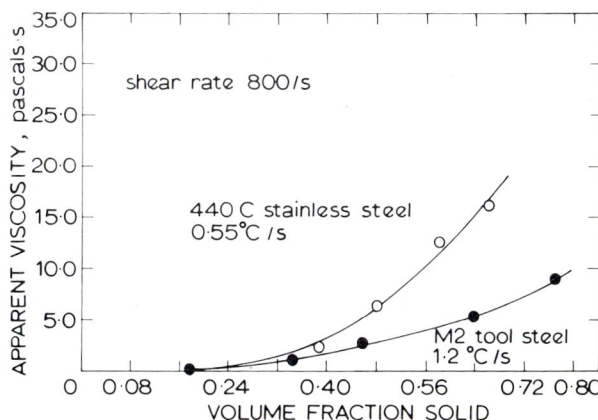

9 Rotor drive amps v. temperature for AISI 440C stainless steel at different rotor speeds and alloy cooling rates

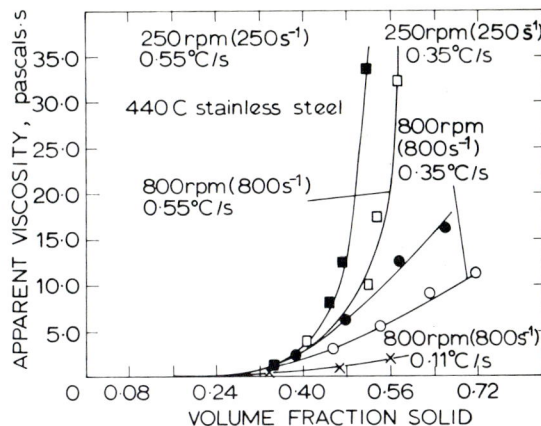

11 Comparison of apparent viscosity of AISI 440C and M2 steels v. volume fraction solid

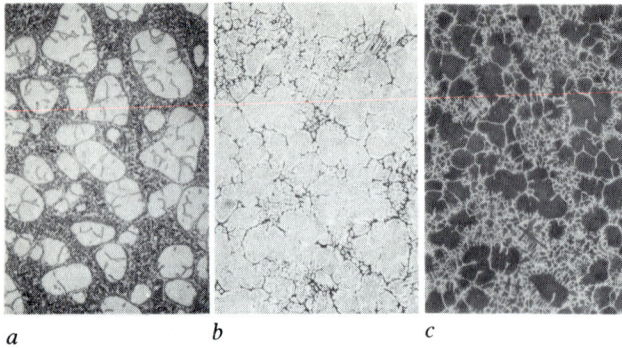

a copper alloy CDA 905, (88 Cu–10 Sn–2 Zn);
b AISI 304 stainless steel, (18 Cr–8 Ni stainless
steel); *c* HS 31 (X-40) cobalt-base superalloy,
(Co–25·5 Cr–11 Ni–7·5 W–0·5 C)

12 Micrographs of Continuously Rheocast alloys [×38]

sensitive as a control device the higher the fraction solid (Fig. 9) and can be automated easily.

MICROSTRUCTURES

Typical microstructures of several alloys, obtained by direct quenching of small droplets leaving the Continuous Rheocaster, are shown in Fig. 12. In these micrographs the large rounded particles are those formed in the Rheocaster, clearly delineated from the fine dendritic structure formed by the matrix liquid at the time of the quench.

As can be seen from Figs 9–11, Continuously Rheocast slurries can exhibit diverse rheological properties. From earlier work on model Sn–Pb alloys[4,13] a strong correlation exists between these rheological properties and slurry microstructure. Thus, for example, Fig. 13 is a plot of primary particle diameter *v.* cooling rate for Sn–Pb alloys. The Continuously Rheocast data (open circles) are compared in this figure with data for primary dendrite arm spacings from earlier work.[14] Also plotted are data taken from two Continuously Rheocast steels.

Clearly the mechanisms responsible for primary dendrite arm spacings and primary particle diameters will be different but this figure provides a good indication of the order of magnitude involved. It is also apparent that primary particle diameter is strongly dependent upon cooling rate. This is consistent over a

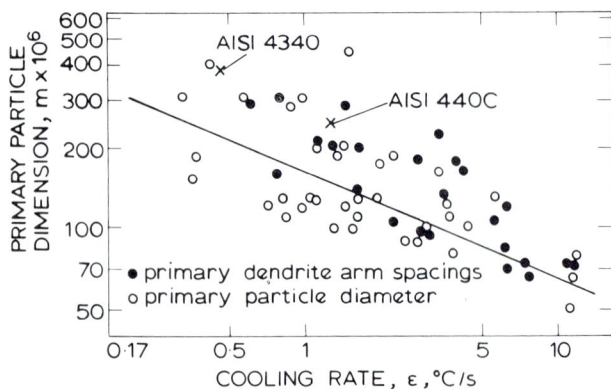

13 Primary dendrite arm spacings[14] and primary Rheocast particle diameters[13] *v.* cooling rate for Sn–Pb alloys

wide range of shear rates except for very low cooling rates (rates of solidification), at which increasing shear rates tend to produce smaller primary particles.[4]

In the production of Continuously Rheocast slurries there are three main independent variables which need to be considered. These are volume throughput, cooling rate (rate of solidification), and shear rate. For a given throughput, shear rate and cooling rate must be balanced to produce microstructures that are acceptable both from a rheological standpoint and from other metallurgical considerations such as homogenization times.

As can be seen in Fig. 13, primary particle size is dictated essentially by cooling rate. However, while shear rate does not substantially influence particle diameter, it strongly influences particle shape. At low shear rates and moderate cooling rates, typically 0·67°C/s, particles tend to revert to a dendritic appearance. At the same time, as can be seen in Figs 9–11, apparent viscosity rises and the alloy slurry becomes more difficult to shape or cast.

A typical composition profile, obtained by electron probe microanalysis on a Continuously Rheocast AISI 304 stainless-steel slurry,[15] is shown in Fig. 14. The primary particle exhibits an essentially flat concentration profile, typical of all Continuously Rheocast alloys examined to date. This particular alloy composition freezes initially in the Rheocaster as δ-ferrite which subsequently transforms to austenite in a cellular manner described by McTighe.[16] This decomposition leaves the black veins of δ-ferrite which can be seen in the micrograph, surrounded by the halo of γ-austenite which froze directly from the melt just before leaving the Rheocaster.

The typical oxide inclusion distribution of this alloy, when Rheocast, is not significantly different from the charging rod stock, and we conclude that the shielding gas of argon–4% hydrogen above and below the melt effectively eliminates oxygen pickup in the Continuous Rheocaster.

In the Thixocasting process, the slurry is teemed directly into ingot moulds. Here, the slower cooling rate of the final solidification stage produces large dendrites which are often structurally indistinguishable from the primary particles. The secondary solid so formed, however, is of a different composition to that of the primary solid formed in the Rheocaster and will therefore have a lower melting point. Thus by reheating the ingots to the semi-solid range the slurry microstructure may be reclaimed. A typical ingot microstructure of AISI 440C is shown in Fig. 15.

Thixocast microstructures display many of the characteristics of the Rheocast slurries, particularly when charges are reheated and cast at approximately the same volume fraction solid as they were Rheocast. However, in the case of aluminium alloys the mode of secondary solidification is such that it can occur in a cellular fashion radiating outward from primary particles. This structure coarsens rapidly during cooling and reheating such that an initially 0·50 volume fraction solid slurry may be Thixocast to yield 0·72 volume fraction solid of apparent primary solid (Fig. 16). For aluminium alloys, therefore, substantial grain refinement may be possible while producing only relatively low-volume fractions primary solids in the Continuous Rheocaster.

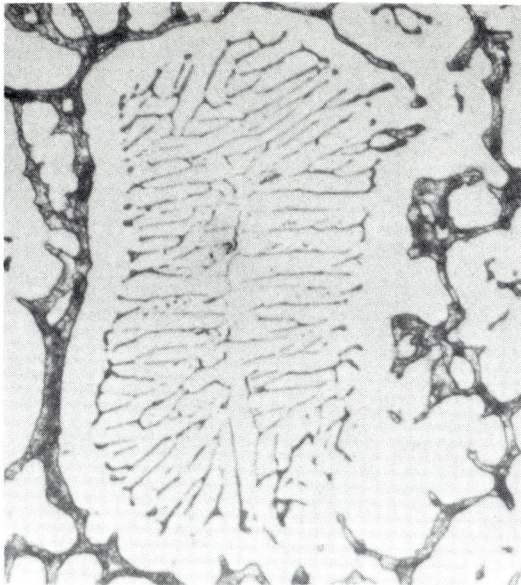

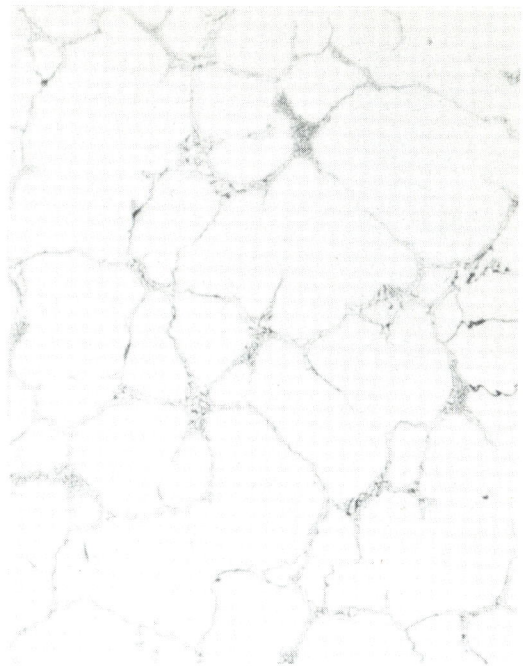

15 Micrograph taken from a Continuously Rheocast ingot of AISI 440C stainless steel [×100]

MECHANICAL PROPERTIES

At the moment only preliminary data are available on the mechanical properties of Rheocast and Thixocast alloys. Recognizing the need for comprehensive dynamic as well as static properties, and the multiplicity of various thermomechanical treatments available, our data has at the present time been restricted to characterizing the Thixocast product. The basic tensile data for

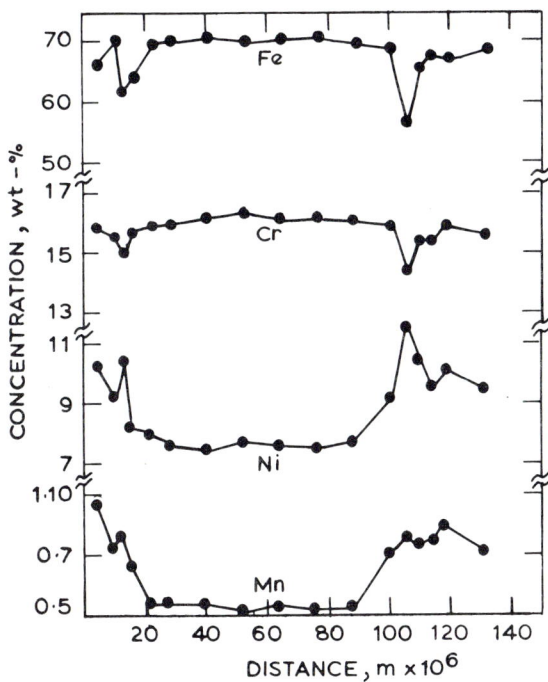

14 Electron microprobe analysis data showing composition variations in a primary particle of AISI 304 stainless steel

CASTING QUALITY

The major emphasis at MIT has been directed at the development of the Thixocasting process for ferrous alloys. The part produced in this work is the hammer of an M16 rifle weighing about 50 g. The results of a radiographic examination of a sample of about 300 of these castings can be seen in Fig. 17. Most parts exhibit good surface quality and fall within the first rating category corresponding to the absence of detectable internal flaws. This is consistent with the predictions of earlier model Sn–Pb and bronze[7,9,13] studies, and is the result of the improved die filling characteristics obtained by Thixocasting.

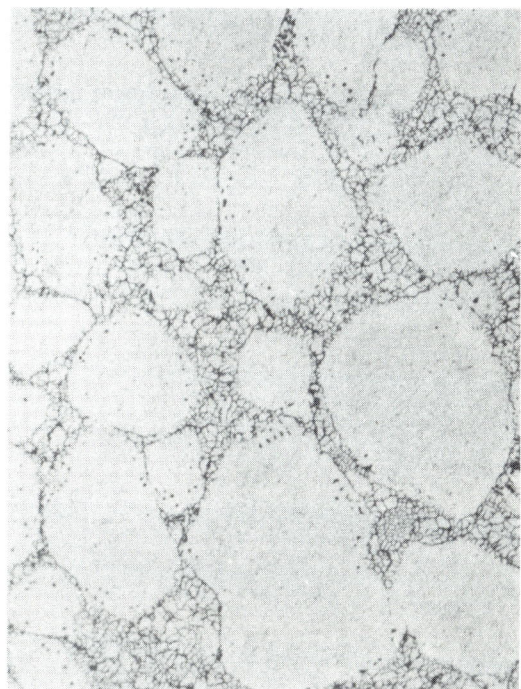

16 Micrograph of 7075 aluminium Thixocast at 0·72 volume fraction solid from an ingot originally Rheocast at 0·50 volume fraction solid [×100]

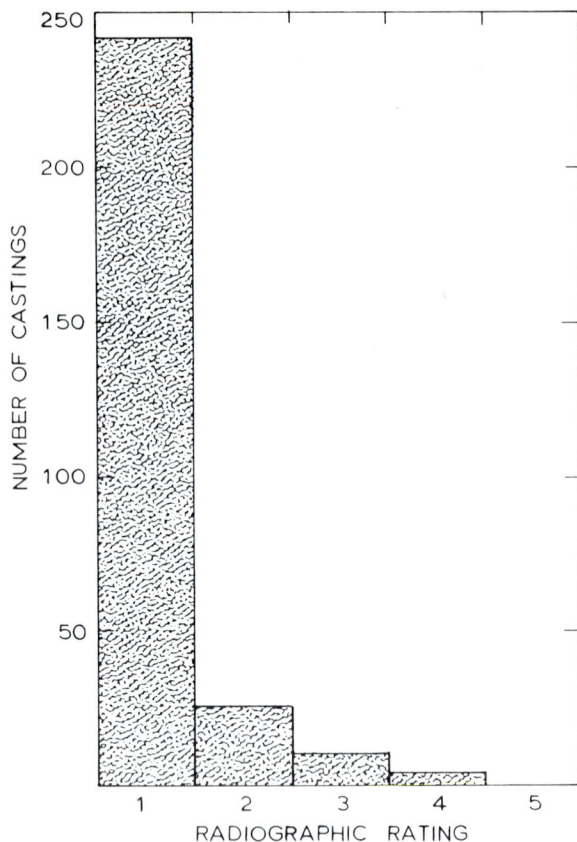

17 Results of radiographic inspection of about 300 M16 rifle hammers Thixocast in AISI 304 stainless steel at 0·30 volume fraction solid; rating 1 corresponds to absence of detectable flaws[9]

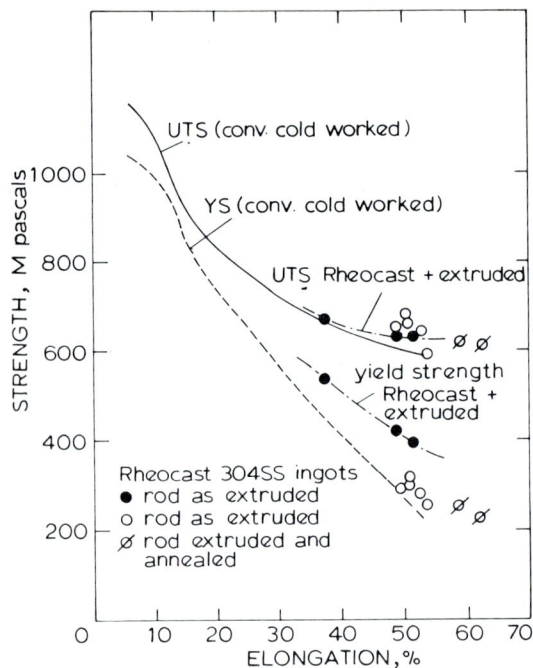

18 Comparison of post hydrostatically extruded properties of *a* as-Rheocast ingot and *b* wrought material (from Ref. 18)

two alloys, copper alloy CDA 905 (used extensively as a model system in our development work) and AISI 304 stainless steel, is presented in Table 1.[17] Tensile strengths are equivalent or slightly superior to conventionally cast alloy with somewhat lower ductility.

In addition to these data some limited exploratory investigations are in progress to determine the suitability of Rheocast microstructures to alternative processing techniques. Preliminary results from one such investigation[18] are presented in Fig. 18. This figure compares the tensile properties of hydrostatically extruded 304 stainless steel, (*a*) from as-Rheocast ingot; and (*b*) from wrought stock. The as-Rheocast material is readily extrudable and results in equivalent or slightly better post-extrusion properties than the wrought material. Significant potential savings therefore seem possible, by directly extruding Rheocast preforms, eliminating the current steps of 'working a large ingot to obtain suitable size and structure for extruding'.

ACKNOWLEDGMENTS

This review paper describes primarily an ongoing major university–industry programme on machine casting, supported by the Defense Advanced Research Projects Agency, monitored by the Army Materials and Mechanics Research Centre, Watertown, Massachusetts, under contract number DAAG46-73-C-0110. The basic work leading to this program was sponsored by the Army Research Office. Work reported herein was carried out at MIT by a large number of graduate students and staff members, most of whom are listed in the references as co-authors of various papers and publications.

REFERENCES

1   S. A. METZ AND M. C. FLEMINGS: *Trans. AFS*, 1970, **78**, 453
2   D. B. SPENCER *et al.*: *Metall. Trans.* 1972, **3**, 1 925
3   M. C. FLEMINGS *et al.*: *Int. Cast Met. J.*, 1976, **1**, 11
4   P. A. JOLY AND R. MEHRABIAN: *J. Mater. Sci.*, 1976, **11**, 1 393
5   E. F. FASCETTA *et al.*: *Casting Engineer*, Sept.–Oct. 1973, 44 (*also AFS Cast Met. Res. J.*, 1973, **9**, (4), 167; *Trans. AFS*, 1973, **81**, 95)
6   R. G. RIEK *et al.*: Eighth SDCE International Die-Casting Exposition and Congress, Detroit, Michigan, 1975, paper No. G-T-75-153
7   R. G. RIEK *et al.*: *Trans. AFS*, 1975, **83**, 25
8   M. C. FLEMINGS *et al.*: Ninth SDCE International Die-Casting Exposition and Congress, Milwaukee, 1977, paper No. G-T77-092
9   K. P. YOUNG *et al.*: *Trans. AFS*, 1976, **80**, 169

Table 1   Mechanical properties of two Thixocast alloys

	Ultimate tensile strength, Pascals	0·2% offset, Pascals	Elongation at fracture, %
AISI 304			
Thixocast	$6·6 \times 10^8$	$2·76 \times 10^8$	19
Typical investment cast	$5·16 \times 10^8$	$2·74 \times 10^8$	30
Copper alloy CDA 905			
Thixocast	$2·81 \times 10^8$	$1·55 \times 10^8$	7
Sand cast	$3·90 \times 10^8$	$1·52 \times 10^8$	30

10 B. E. BOND: Sc.D. Thesis, Department of Materials Science and Engineering, Massachusetts Institute of Technology (thesis in progress)

11 V. LAXMANAN: Sc.D. Thesis, Department of Materials Science and Engineering, Massachusetts Institute of Technology (thesis in progress)

12 A. A. MACHONIS *et al.*: 'Copper alloy pressure die castings', 1975, New York, International Copper Research Association, Inc.

13 F. J. SCHOTTMAN: Sc.D. Thesis, Department of Materials Science and Engineering, Massachusetts Institute of Technology (thesis in progress)

14 J. A. E. BELL AND W. C. WINEGARD: *J. Inst. Met.*, 1973, **92**, 357

15 A. SHIBUTANI: S.M. Thesis, Department of Materials Science and Engineering, Massachusetts Institute of Technology, Jun. 1977

16 J. MCTIGHE: Ph.D. Thesis, Faculty of Materials Technology, University of Sheffield, Apr. 1975

17 F. E. GOODWIN: S.M. Thesis, Department of Materials Science and Engineering, Massachusetts Institute of Technology, Sept. 1976

18 H. E. ANDREWS: ITT Corporation, unpublished work

Stir-cast microstructure and slow crack growth

A. Vogel, R. D. Doherty, and B. Cantor

Al–Cu alloys were stirred at various speeds while slowly solidifying. The microstructures produced were examined both by sampling the melt during stirring and by quenching the alloys at a temperature just above the eutectic temperature. The effect of stirring was to transform the normal dendritic structure to one consisting of spherical or rosette shaped particles of primary aluminium owing to fragmentation of the dendrites. A model is proposed for this fragmentation based on the formation of high-angle grain boundaries in the dendrites following bending of the dendrite arms. The grain boundaries are then fully wetted by a liquid film, leading to dendrite arm separation. Evidence for this mechanism is described. Other observations made include the apparently accelerated Ostwald ripening of the primary aluminium particles due to faster solute transport with stirring. The fracture toughness of stir cast Al–Cu was briefly studied to examine if a dispersion of ductile particles in a brittle eutectic matrix could give a worthwhile increase of toughness. The results were inconclusive, due apparently to the formation of a brittle $CuAl_2$ layer around all the ductile particles.

A. Vogel is at the Fulmer Research Institute, Stoke Poges, and R. D. Doherty and B. Cantor are in the School of Engineering and Applied Sciences, University of Sussex, Brighton

Turbulent fluid flow, induced by stirring a solidifying alloy, can produce a novel microstructure in the fully solidified material.[1] In the normal form of this process, the alloy is slowly cooled through the freezing range, while the two-phase mixture is stirred. At a temperature between liquidus and solidus the stirring is stopped and the remaining liquid rapidly frozen by quenching the two-phase mixture. The stir-cast* microstructure is of considerable interest for a number of reasons. The ones which attracted our attention were the following:

(i) the physical metallurgy of how such a structure forms
(ii) whether improved mechanical properties might be achieved with such a microstructure.

During stir casting the first solid to form will be depleted in solute, while the surrounding fine dendritic liquid will be solute rich (Fig. 1). If the alloy is one in which the solute additions give significant strength increases, then after appropriate heat treatment the stir-cast structure

should give rise to a composite material with extremely good bonding between the strong, solute rich, matrix and soft primary solid. It was considered possible that this composite material might offer a new way of increasing the toughness of high-strength alloys. Kelly[3] has pointed out that current high-strength precipitation hardening alloys 'have strengths which it appears pointless to exceed while relying on the ductility of the material to provide adequate notch insensitivity'. The stir-cast microstructure might provide the means to overcome this limitation.

The present work describes two parts of a current investigation into aspects of the development of the stir-cast microstructure ·and the assessment of its fracture toughness. While high-strength aluminium–zinc–magnesium alloys are the alloys of long-term interest, the difficulties in handling small quantities of these materials on a laboratory scale led us to investigate the stir-casting process mainly with aluminium copper alloys, in which the solute-enriched liquid would be the eutectic composition (33 wt-% copper). Since this composition is normally brittle the same alloy system was thought suitable for initial fracture toughness testing. For the toughness testing the double torsion test developed originally for studies of slow crack growth in glass was used. As described recently by Outwater[4] this test has advantages for the assessment of toughness in metals.

---

* Despite the popularity of the term 'Rheocasting' for the process[1] the authors prefer the name of stir-casting since both the verbs to stir and to cast come from the same source—Old and Middle English—while the noun rheology comes from the Greek word meaning a stream.[2] Stream casting does not seem to make much sense as a description

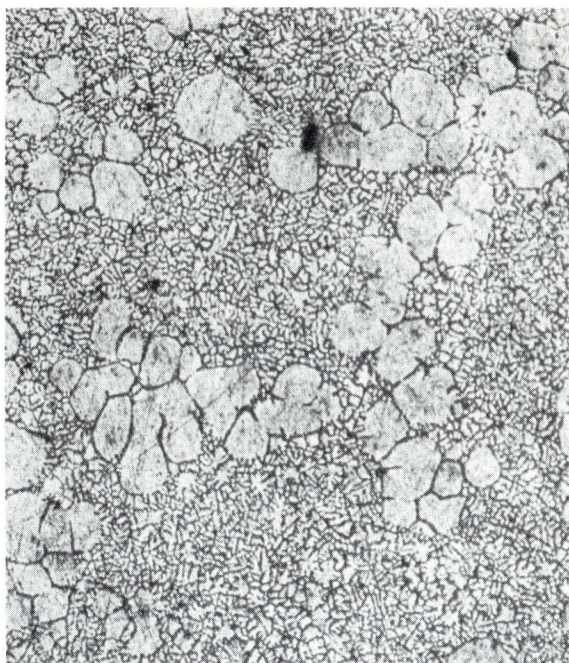

1 Al–Si–Mg alloy stirred at 1 000 rev/min and quenched 5°C below liquidus temperature [×40]

EXPERIMENTAL METHODS

Al–Cu alloys with 20–30% Cu were melted in graphite crucibles in a large resistance furnace, heated to about 100°C above the liquidus and slowly cooled, at the natural furnace cooling rate of a few degrees a minute. Stirring was carried out with a titanium propellor rotating in the lower part of the melt. Stirring speeds were restricted to ≤ 1 000 rev/min, to avoid the formation of a vortex with air entrapment that occurred at higher stirring speeds. The temperature of the metal was continuously monitored with a thermocouple in an alumina tube. Normally the stir casting was terminated a few degrees above the eutectic temperature by removing the propellor, and then quenching the two-phase mixture with a large water-cooled copper chill inserted into the melt. Full details have been given by Vogel.[12]

RESULTS

Gravity induced segregation

In Al–Cu alloys, the primary solid is depleted in Cu, and is therefore lighter than the remaining liquid, and readily floats to the top of lightly stirred ingots. Figure 2 shows the top, middle and bottom regions of an ingot stirred at 100 rev/min. The top of the ingot contains a high density of large 'rosette-like' particles while the central region contains only small rounded particles of a size similar to that of the individual arms of rosettes in the top of the ingot. Material from the bottom of the ingot contained no primary solid. This form of gravity segregation persisted for all stirring speeds below the two highest ones, 750 and 1 000 rev/min. Figure 3a shows examples of the fully developed stir-cast particles found uniformly distributed throughout the volume of rapidly stirred ingots.

Morphology of primary aluminium

Figure 3 shows the structures produced in the same alloy stir cast at 1 000 and 750 rpm, and solidified without stirring. In the absence of stirring the material shows a conventional dendritic structure; at 750 rev/min the dendritic structure has been replaced by spherical and rosette-like particles and at the highest speed the change to spherical particles is more complete. The irregular particles seen in Fig. 3a and b are more distorted than the rosettes seen at slower stirring speeds (Fig. 2a). In particular the occasional apparently *bent* rosettes, (Fig. 3a and b) may be of significance (*see* section on 'Observations on the stir-cast microstructure').

The smallest structural unit of the primary aluminium particles, the secondary dendrite arms in Fig. 3c, the isolated spheres and rosette arms of Fig. 3a and b, show clearly that more complete particle coarsening[5,6] (Ostwald ripening) has occurred with stirring. The mean radius of curvature of the solid/liquid interface is smallest in the unstirred alloy even though all three alloys had similar solidification times.

Figure 4 shows that the maximum particle size, measured as the longest distance across a particle, seen in the plane of section to be connected, fell steadily from over 600 μm to less than 200 μm as the stirring increased from 100 to 1 000 rev/min. It is likely that only the results for the highest stirring speeds are appropriate; for the lower speeds the flotation noted earlier may have ensured that the large particles, seen at the top of the more slowly stirred alloys, did not pass regularly through the region of the ingot containing the propellor. The smaller particles seen at higher stirring speeds were, however, fully suspended in the liquid and subjected to the full turbulence of the propellor.

Other observations on the stir-casting process

In order to gain some further insights into the process, three other types of investigation were carried out.

1 During furnace cooling through the solidification range the stirred alloys were regularly sampled by suction into a Pyrex glass tube. This gave a somewhat slower quench than that obtained with a water-cooled chill, but sampling allowed the *time evolution* of the primary aluminium particles to be studied. Figure 5 shows examples of this for an Al–20% Cu alloy stirred at 1 000 rev/min and sampled at various times after the onset of solidification, as determined by thermal analysis. Figs 5a–c show that the primary solid particles start life as small rosettes and grow to their maximum size after approximately 4 min; subsequent solidification appeared to occur by the formation of *more* particles rather than the continued growth of the existing particles. The coarsening of the rosette arms during continued solidification under stirring is seen in Fig. 5.

2 A feature of stir casting is the apparent increase in the *numbers* of primary particles between the unstirred and stirred material and the increased density of particles during stir casting. The higher density of solid particles could be due either to increased primary nucleation or to increased 'secondary nucleation'—the fragmentation or remelting[7] of the highly branched dendrites that usually grow from primary nucleation. To try to gain some insight as to whether or not stirring might increase the frequency of primary nucleation an ingot of Al–20% Cu was furnace cooled without stirring and quenched 1 min after the start of solidification (Fig. 6a). After examination the same material was remelted and again solidified by furnace cooling while stirred at

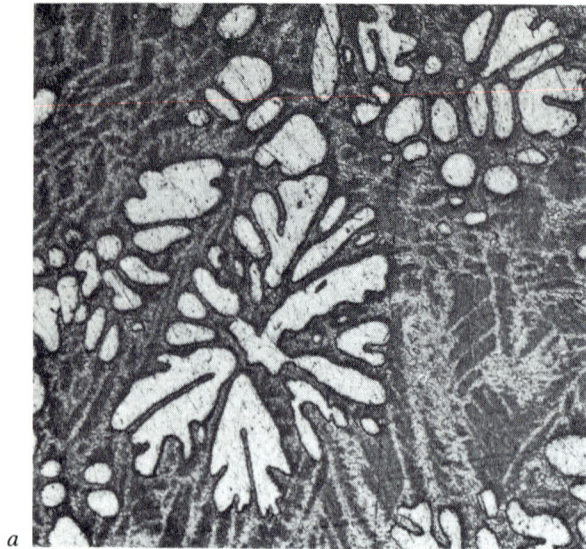

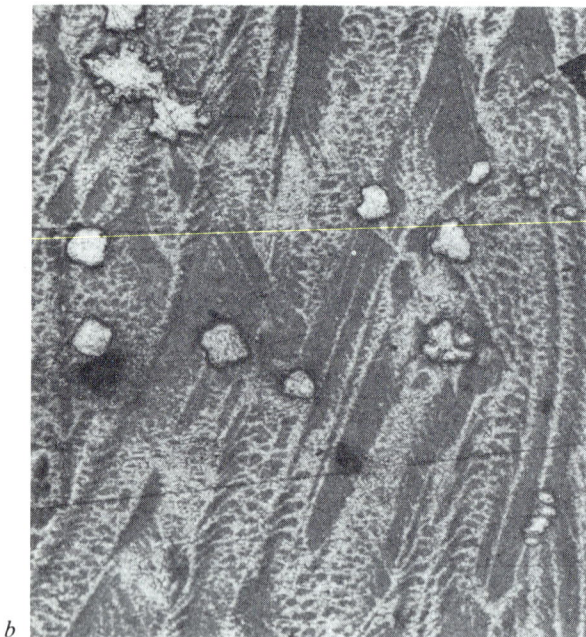

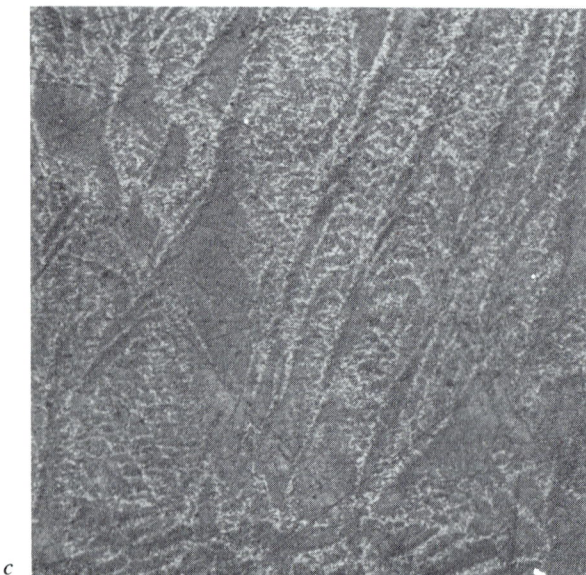

a top; *b* middle; *c* bottom of the ingot

2 Al–24 wt-% Cu alloy stirred at 100 rev/min and quenched a few degrees above the eutectic temperature [×64]

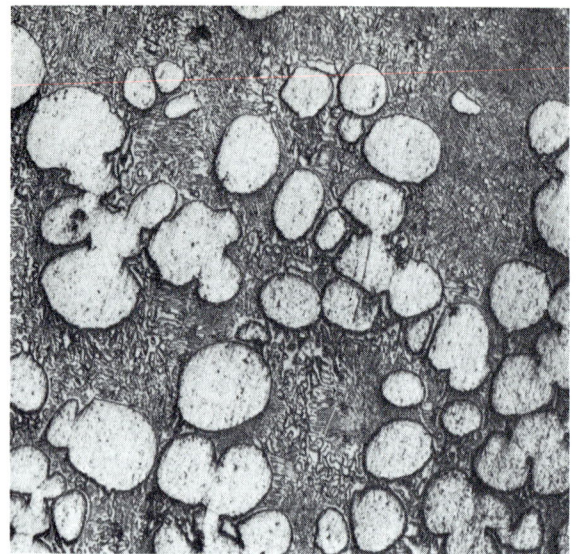

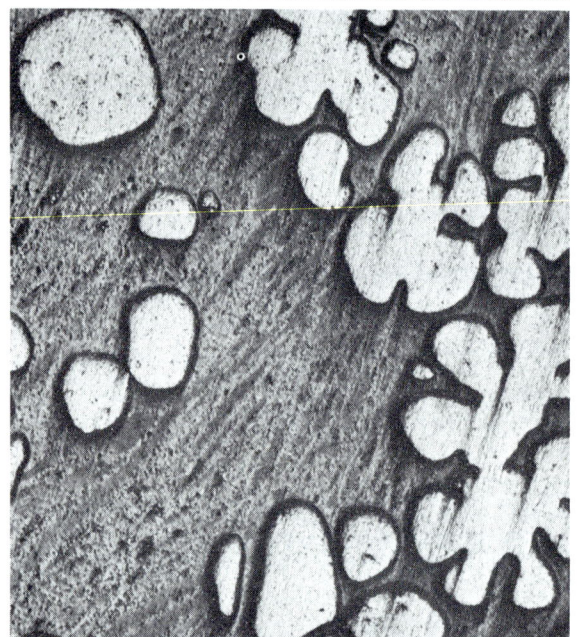

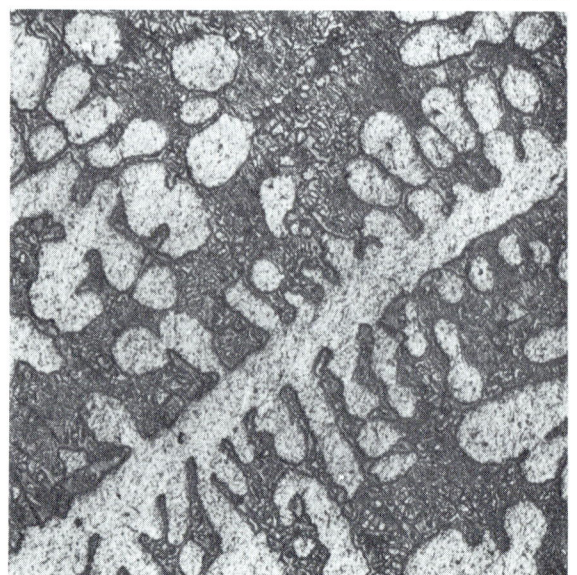

3 Al–20 wt-% Cu stir cast at the following speeds: *a* 1 000 rev/min; *b* 750 rev/min; and *c* without stirring

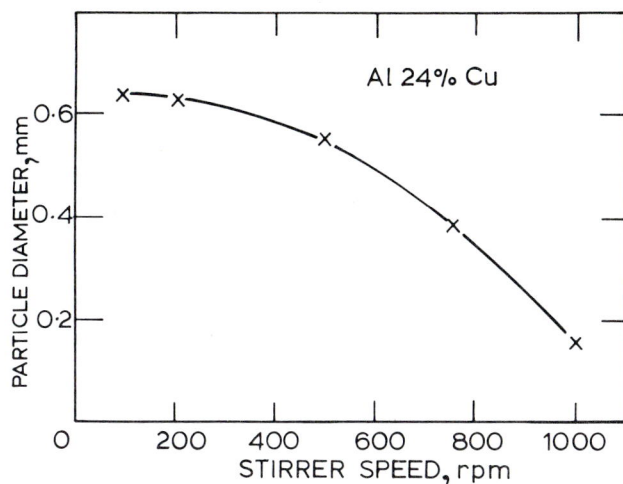

4 Particle size as a function of stirrer speed for stir-cast Al–24 wt-% Cu

1 000 rev/min and quenched 1 min after solidification began (Fig. 6*b*). Though the unstirred alloy *appears* to have a higher density of larger particles than the stirred alloy, this is due to the gravity segregation in the unstirred alloy. A study throughout the height of the two ingots showed a similar total number of growth centres in both samples. This result suggests that stirring caused no significant increase in primary nucleation during the first minute of solidification and so lends support to the alternative model of dendrite fragmentation. It is interesting that recent evidence for dendrite fragmentation in NH_4Cl–H_2O ingots has been obtained by comparison of solidification in near zero gravity (free-fall) conditions with normal solidification.[8]

3 Spencer *et al.*[9] have shown that the stir-cast microstructure can be developed in material in which stirring is applied after some initial solidification. A similar experiment was carried out in Al–20% Cu with stirring at 1 000 rpm applied 4 min after solidification began. Figure 7*a* and *b* show the structure of a sample from this alloy just before stirring was applied, and Fig. 7*c* shows the final as-quenched structure, at a higher magnification. Figure 7*c* has many of the features of the normal stir-cast structure (compare with Figs 3*a* and 5*c*). The main distinctions are the apparently longer particles, and greater evidence of dendrite, or rosette, *bending* in Fig. 7*c*.

Many features of these structures were seen in the other systems studied—aluminium–silicon, aluminium–zinc–magnesium and NH_4Cl–H_2O. However, limitations of space prevent any discussion of these results in this paper.

DISCUSSION OF THE OBSERVATIONS ON THE STIR-CAST MICROSTRUCTURE

The results obtained clearly imply that some form of dendrite fragmentation process is occurring during stir

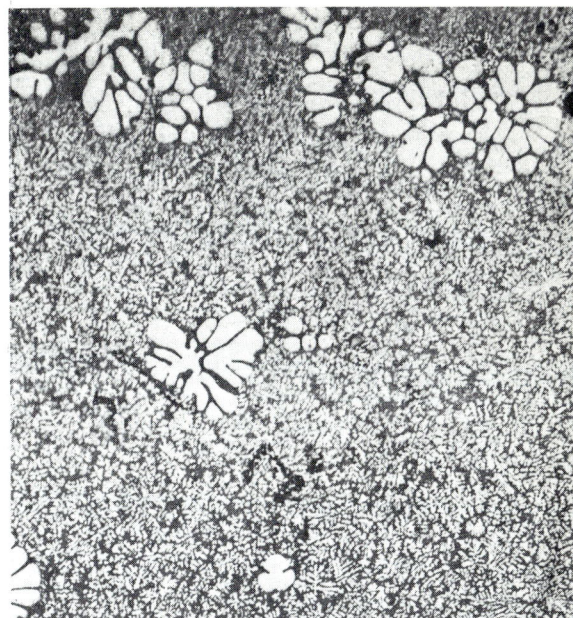

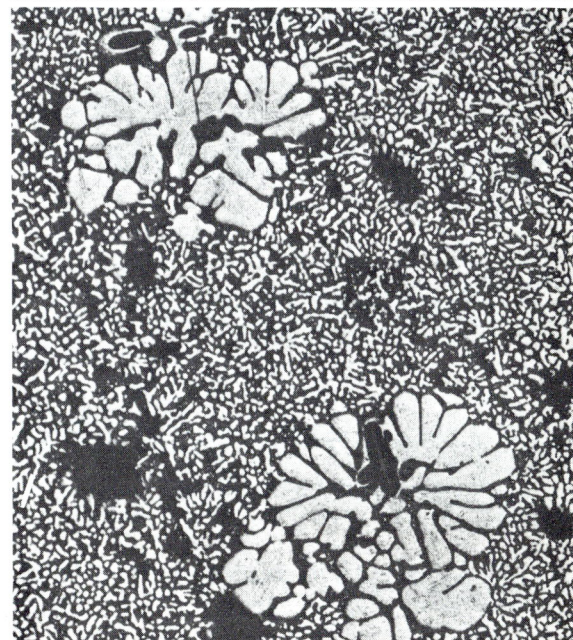

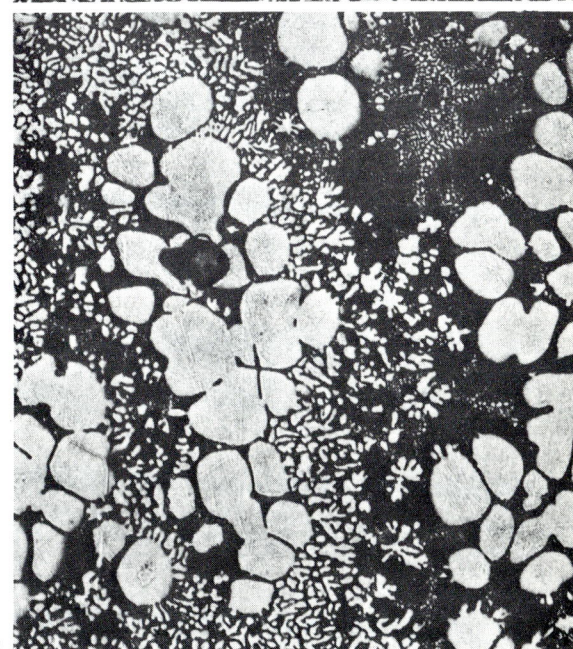

5 Al–20 wt-% Cu stir cast at 1 000 rev/min; samples taken at the following times after the start of solidification: *a* 1·5 min; *b* 4 min; and *c* 17·5 min [×60]

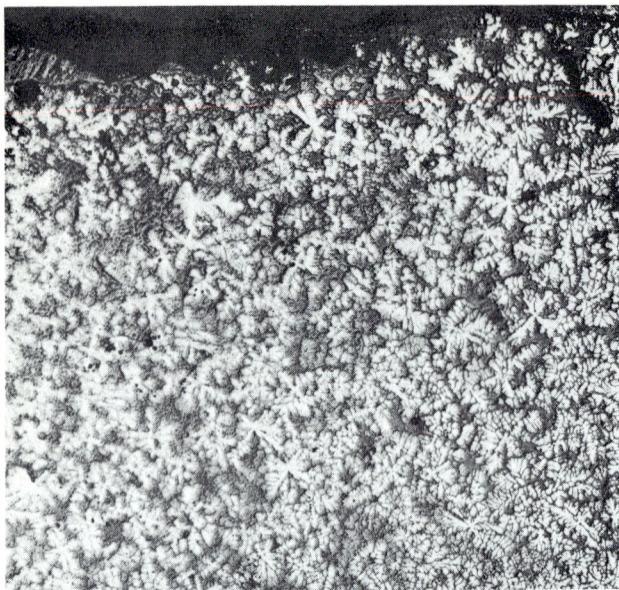

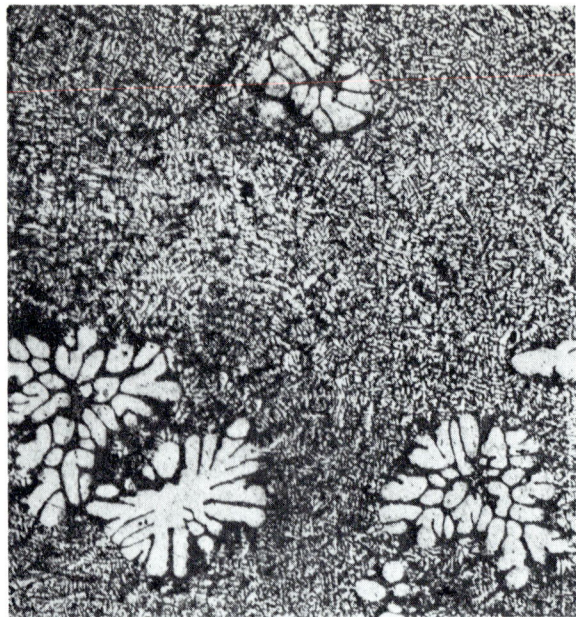

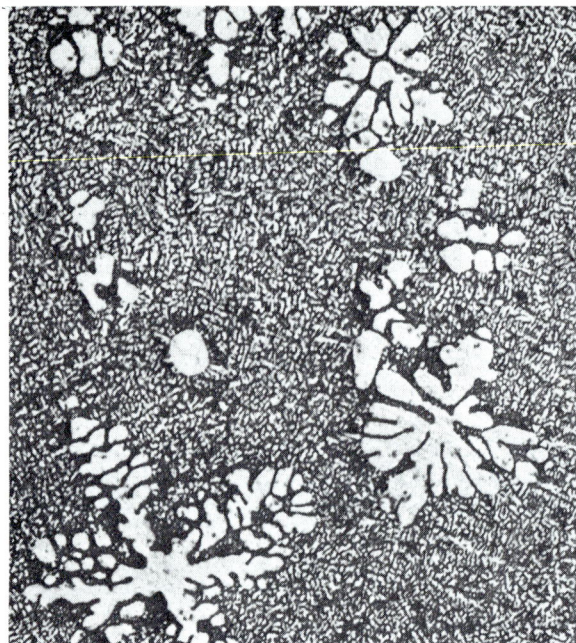

**6   Al–20 wt-% Cu slow cooled and quenched
1 min after the start of solidification;
a without stirring showing the top of the
ingot with a concentration of floating
dendrites; *b* same material stirred at
1 000 rev/min before quenching but with all
the particles uniformly suspended   [×12]**

a sampled by suction into a glass tube
1·6 min after the start of solidification,   [×60];
b sampled after 4 min just before stirring was
started,   [×60]; and *c* as-quenched after
15·3 min from the start of solidification   [×120]

**7   Al–20 wt-% Cu furnace cooled without stirring
until 4 min after the onset of solidification and
then stirred at 1 000 rev/min until quenched in
the normal way at a temperature just above
the eutectic temperature**

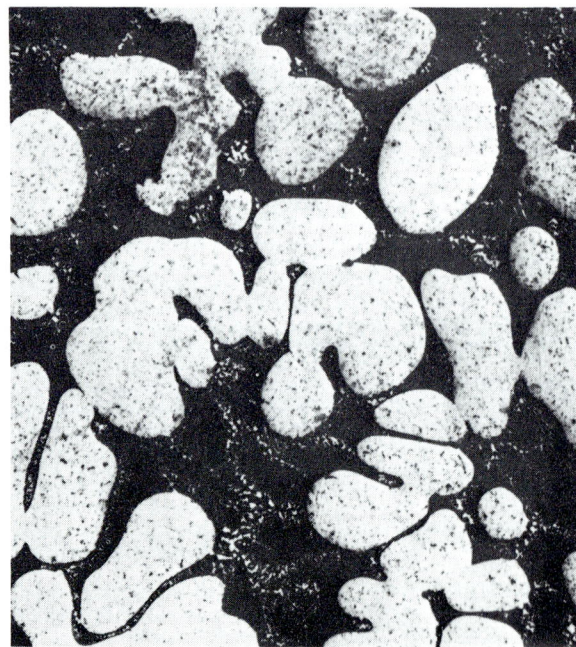

casting. Theoretical analysis[10] has shown first that stirring would be expected to *increase* the shape instability of spherical particles solidifying with a given undercooling, and secondly that fully established dendritic growth should be faster, and with a finer tip radius, in a stirred melt.[11] However the observations made here, as those reported elsewhere[1] are of non-dendritic small spherical particles. Since stirring should promote dendritic growth the observation of spherical particles after stir casting can only be explained either by a high rate of primary nucleation or dendrite fragmentation. A high *density* of growing particles has been shown[12] to lead to shape *stability* by virtue of a reduction of thermal and solute gradients because of 'soft impingement' of the diffusion fields, between adjacent growth centres. The observations on samples quenched with and without stirring after 1 min of solidification appear to rule out increased nucleation and therefore the evidence for dendrite fragmentation is strongly supported. Additional evidence for dendrite fragmentation is given by the observations here, of the development of the stir-cast microstructure by stirring introduced *after* solidification had occurred. Spencer *et al.*[9] also showed clear evidence for dendrite fragmentation and dendrite *bending*.

On the basis of this experimental evidence, the following model is proposed for the mechanism of dendrite fragmentation. The starting point for the model is the expected *toughness* of primary aluminium dendrites at temperatures close to the melting point. This expected toughness should make it likely that the albeit fragile, dendrite structure would be expected to *bend* rather than fracture when exposed to turbulent fluid flow.

If a dendrite arm is bent through an angle θ with respect to the dendrite stem, the extra dislocations required to achieve this arm rotation should collapse by recovery and recrystallization processes to a *grain boundary* with a misorientation of angle θ. The energy of grain boundaries increases with misorientation,[13] and for a high-angle grain boundary the energy of the boundary σ_{gb} is normally greater than *twice* the solid liquid interfacial energy σ_{sl}. Hondros[14] quotes a value for σ_{gb} of $0.6\,J\,m^{-2}$ for aluminium and the value of σ_{sl} of aluminium is $0.09\,J\,m^{-2}$.[15] Such a relationship between σ_{gb} and σ_{sl} will lead to the complete 'wetting' of the grain boundary by a liquid film, so that the formation of a highly bent dendrite arm should lead, after a short time delay, first to the formation of a grain boundary, followed by liquid attacking the grain boundary, and finally the separation of the dendrite arm removed from the dendrite by *grain-boundary induced melting*. Evidence in support of this model is the following:

(i) the frequent observation of bent dendrites— particularly for material stirred after solidification had commenced (Fig. 7*b*)

(ii) the observation in many cases (Fig. 8) of grain boundaries in stir-cast primary aluminium

(iii) the observation in aluminium alloys[12] of the frequent complete wetting of grain boundaries by liquid films in fine-grained samples annealed at temperatures just above the solidus.

A final point on this model should be noted—if the rotation angle between two grains is small, the value of σ_{gb} will be reduced[13] and complete wetting will not occur, only the formation of a grain-boundary groove,

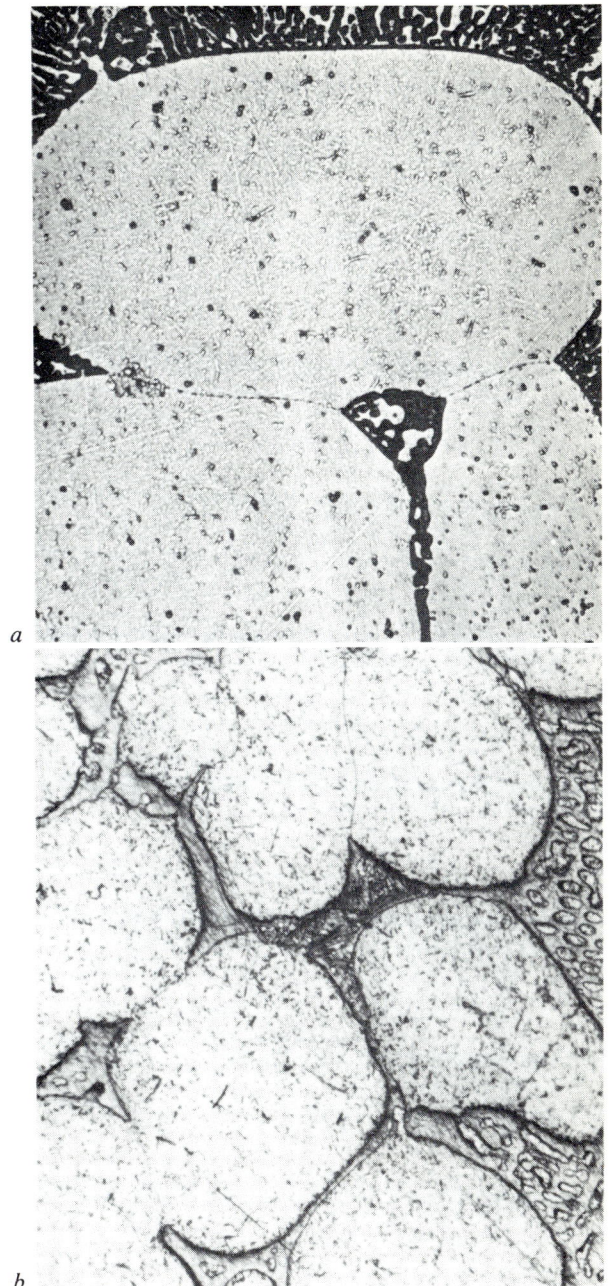

8 Grain boundaries seen in stir-cast Al–Cu alloys: *a* part of the bent dendrite seen in Fig. 7*c*, [×600]; *b* Al–24 wt-% Cu fully stir-cast sample, 1 000 rev/min [×256]; in both these structures the layer of CuAl$_2$ surrounding the primary aluminium can be seen, this being believed to be responsible for the poor toughness of these structures

the situation in Fig. 8. The converse of the fragmentation process is of course also possible, that of the welding together of particles, should they collide with a low crystal misorientation.

The apparently accelerated dendrite arm coarsening under stirring, (Fig. 3*a–c*) is readily understood using the model developed for theoretical studies of stirring-promoted shape instability and dendritic growth.[10,11] With stirring, solute transport by diffusion need only occur across a *thin boundary layer* around a growing or dissolving particle. Transport across the remaining interparticle distances can be achieved by fluid flow,

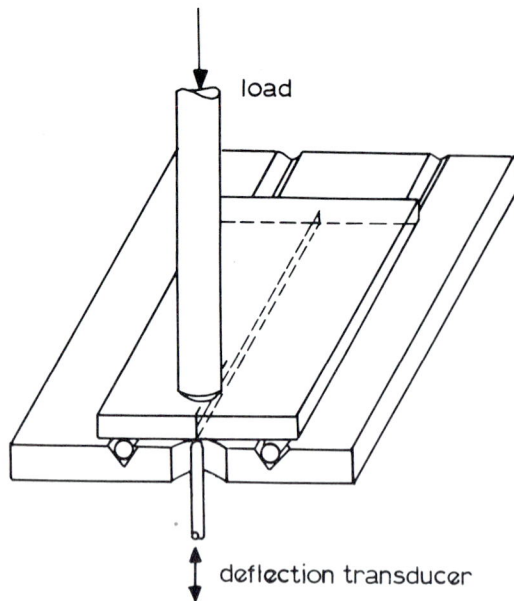

9 **Schematic diagram of the double torsion test apparatus; the test plate is loaded above a 4 mm starter notch and the crack is held central by a shallow crack guiding groove in the base of the plate; crack growth away from the starter notch increases the specimen compliance but not the stress intensity at the tip of the crack**

giving a higher rate of solute transport and thereby faster coarsening. The surface energy-driven changes inside an individual particle, i.e. the change of a rosette to a sphere, should not however be accelerated by fluid flow, since fluid flow will be unlikely to penetrate between the rosette arms.

TOUGHNESS STUDIES

Samples of aluminium–copper alloys with a eutectic matrix and various volume fractions of primary aluminium particles were obtained by stir casting at 1 000 rpm a series of alloys with copper contents in the range 22–33 wt-% Cu. Each of these was quenched at a temperature just above the eutectic temperature. Rectangular plates were cut out of the quenched ingot with a thickness of about 5 mm, and a 4 mm edge notch and a crack guiding groove of 2 mm depth were cut by spark machining (Fig. 9).

The crack propagation energy G_c for this geometry is given[12] as

$$G_c = 3P_{max}^2 W_M^2 / 2\mu t_1 t^3 W$$

where P_{max} is the maximum load, $2W_M$ is the full specimen width, $2W$ the width between the supports, t is the full specimen thickness, t_1 the thickness in the crack plane, and μ is the shear modulus. As described by Outwater,[4] and also found in this study, the load-deflection curve produced by material undergoing this test gives a series of peaks in load—each of these peaks gives a value of P that can be used to evaluate G_c. Table 1 lists the results with the average value of P, together with the standard deviation of the P values and the range of resultant toughness values.

The results indicate two effects, first that small volume fraction of the primary aluminium appears to *reduce* the

Table 1 Results of toughness studies

Composition, wt-% Cu	Wt. fraction of primary Al, %	P_{max}, N	Crack propagation energy G_c^*, J m^{-2}
33	0	883	$3\cdot1\times10^3$
33	0	849 ± 13	$3\cdot7\pm0\cdot1\times10^3$
32	4	579 ± 69	$0\cdot33\pm0\cdot05\times10^3$
30	11	$1\ 196\pm35$	$2\cdot1\pm0\cdot2\times10^3$
28	18	806 ± 20	$3\cdot7\pm0\cdot2\times10^3$
28	18	747 ± 25	$2\cdot4\pm0\cdot2\times10^3$
22	39	999 ± 37	$5\cdot6\pm0\cdot4\times10^3$
22	39	$1\ 079\pm69$	$4\cdot9\pm0\cdot6\times10^3$

* The estimated errors in G_c come only from the variation in load maxima P_{max}

fracture toughness, and secondly that higher volume fractions give a small increase of toughness compared to that of 100% eutectic, but a larger increase of toughness when compared to the alloys with small volume fraction of primary solid. Detailed observations of the fracture surfaces showed that in most cases the fracture had passed *around* and not through the primary particles (Fig. 10). The origin of this behaviour may have been in the layer of brittle intermetallic compound surrounding each primary aluminium particle. Such layers can be seen in the high magnification micrographs of Fig. 8. The brittle behaviour found here might be expected whenever the brittle intermetallic component of the eutectic has a low entropy of melting—giving non-faceted growth[16,17] and thereby a *coating of second phase* around the primary solid. Aluminium–silicon alloys did not show such behaviour.

These initial results are clearly rather inconclusive as to whether the proposed toughening model will give significant toughness increments in high strength precipitation hardening alloys. Further work on such material will be reported in due course.

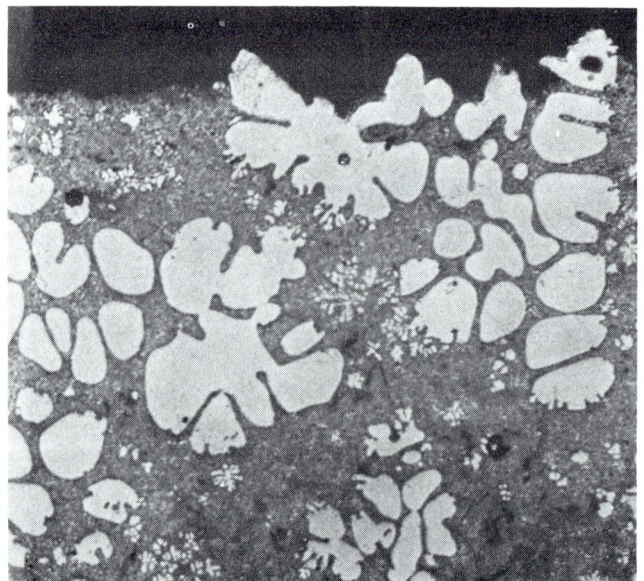

10 **Section normal to fracture surface of stir-cast Al–28 wt-% Cu; the upper fracture surface has been protected by mounting resin; the crack has apparently run around most of the primary aluminium particles [×80]**

ACKNOWLEDGMENTS
The authors would like to thank Mr P. Enright for valuable discussions and Alcan Laboratories for partial financial support for this work and financial support for A. Vogel.

REFERENCES
1 M. C. FLEMINGS *et al.*: *Mater. Sci. Eng.*, 1976, **25**, 103
2 Shorter Oxford English Dictionary, Oxford University Press
3 A. KELLY: 'Strong solids', 2nd ed., 134, 1973, Oxford University Press
4 J. O. OUTWATER *et al.*: 'Fracture toughness and slow stable cracking', ASTM STP 559, 1973, 127
5 T. Z. KATTAMIS *et al.*: *Trans. Met. Soc. AIME*, 1967, **239**, 1 504
6 M. KAHLWEIT: *Scripta Metall.*, 1968, **2**, 251
7 K. A. JACKSON *et al.*: *Trans. Met. Soc. AIME*, 1966, **236**, 149
8 M. H. JOHNSTON AND C. S. GRIMER: *Metall. Trans.*, 1977, **8A**, 77
9 D. B. SPENCER *et al.*: *Metall. Trans.*, 1972, **3**, 1 925
10 A. VOGEL AND B. CANTOR: *J. Cryst. Growth*, 1977, **37**, 309
11 B. CANTOR AND A. VOGEL: *J. Cryst. Growth*, 1977, **41**, 109
12 A. VOGEL: D.Phil. Thesis, University of Sussex, 1977
13 N. GJOSTEIN AND F. N. RHINES: *Acta Metall.*, 1959, **7**, 319
14 E. D. HONDROS: Interfaces Conference (ed. R. C. Gifkins), 77, 1969, London, Butterworth
15 D. TURNBULL: *J. App. Phys.*, 1950, **21**, 1 022
16 J. D. HUNT AND K. A. JACKSON: *Acta Metall.*, 1965, **13**, 1 212
17 J. D. HUNT AND K. A. JACKSON: *Trans. Met. Soc. AIME*, 1966, **236**, 843

Solidification and properties of $\gamma/\gamma'-\alpha$ Mo ductile/ductile eutectic superalloy

D. D. Pearson and F. D. Lemkey

The existence of a monovariant eutectic trough within the Ni–Mo–Al ternary system permitted the generation of in situ composites consisting of a ductile matrix of gamma/gamma prime phases reinforced by a ductile fibrous molybdenum phase. These alloys have been further characterized as having exceptional tensile and creep properties in the direction of phase alignment, high ductilities, combined with good oxidation resistance at elevated temperatures. The solidification reaction and the development of γ' within γ and the α Mo fibre morphology were examined by quenching the solid/liquid interface during steady-state directional solidification. A monovariant eutectic reaction was observed between γ and α as aluminium was varied between 2·5 and 7·5wt-% Al. As aluminium was increased to amounts greater than 7·5wt-% a ternary reaction involving β (NiAl) was observed with the γ and α Mo phases. The β was present only at temperatures near the solidus of the alloys and dissolved during gradient cooling with concurrent precipitation of γ' from β and γ; this led to a matrix which was nearly 100% γ' at room temperature. The effect of β and γ' on fibre morphology is discussed as well as evidence of the existence of a $\gamma + \alpha + \beta$ ternary eutectic.

The authors are with United Technologies Research Center, Connecticut, USA

The nickel-rich corner of the Ni–Mo–Al ternary system has been the subject of several recent investigations[1-3] because it contains a monovariant eutectic trough which permits the growth of aligned composite structures by directional solidification. The aligned structure can be described as a rod eutectic consisting of Mo (α) fibres in a matrix of Ni solid solution (γ) with precipitated Ni$_3$Al (γ'). The relative amounts of γ and γ' depend on the Al content of the alloys with the volume fraction γ' increasing with Al. Fully aligned $\gamma/\gamma' + \alpha$ eutectics have been observed through a range of compositions between Ni–39·5 wt-% Mo–3·5 wt-% Al and Ni–25·4 wt-% Mo–8·6 wt-% Al.[1] The α(Mo) fibres are highly faceted with a square cross-section (Fig. 1). The crystallographic orientation relationship between the fibres and the matrix has been determined by transmission electron microscopy to be:

$[001]^\gamma \| [001]^\alpha \|$ growth direction
$(100)^\gamma \| (110)^\alpha \|$ interfacial planes

These ternary monovariant eutectics are composed of three ductile phases which exhibit an improved balance in the anisotropy of longitudinal and off-axis properties. In addition, the alloys exhibit good tensile and creep strength together with excellent ductility when compared with first generation eutectic superalloys such

as $\gamma/\gamma'-\delta^4$ and $\gamma/\gamma'-MC^{5,6}$ currently undergoing evaluation in P & WA, SNECMA, and General Electric gas turbine test programmes.

Current eutectic alloy development programmes have as their objective the optimization of key mechanical and physical properties, such as creep shear parallel to the growth direction and oxidation and sulphidation resistance. In order to best obtain these properties, the nature of the solidification reaction and the extent to which further ternary and quaternary additions could be made were investigated. In a previous paper,[1] it was reported that the phase equilibria of these eutectics appeared to consist of two monovariant troughs, one between γ' and α, and another between γ and α. The two troughs were believed to end in a ternary eutectic, $\gamma + \gamma' + \alpha$, based on the observed microstructures and the published phase equilibria of the system.[7-9] However, DTA analysis of alloys lying along the proposed troughs could not distinguish the exact class of the reactions because of small differences between the liquidus and solidus of the alloys and the shallow slopes of the troughs. Sprenger *et al.*[2] also reported difficulty in identifying the eutectic reactions but implied the existence of a monovariant reaction between γ' and α from observations on alloys slightly on the Ni-rich side of a stoichiometric tie between Ni$_3$Al and Mo. Henry[3] has

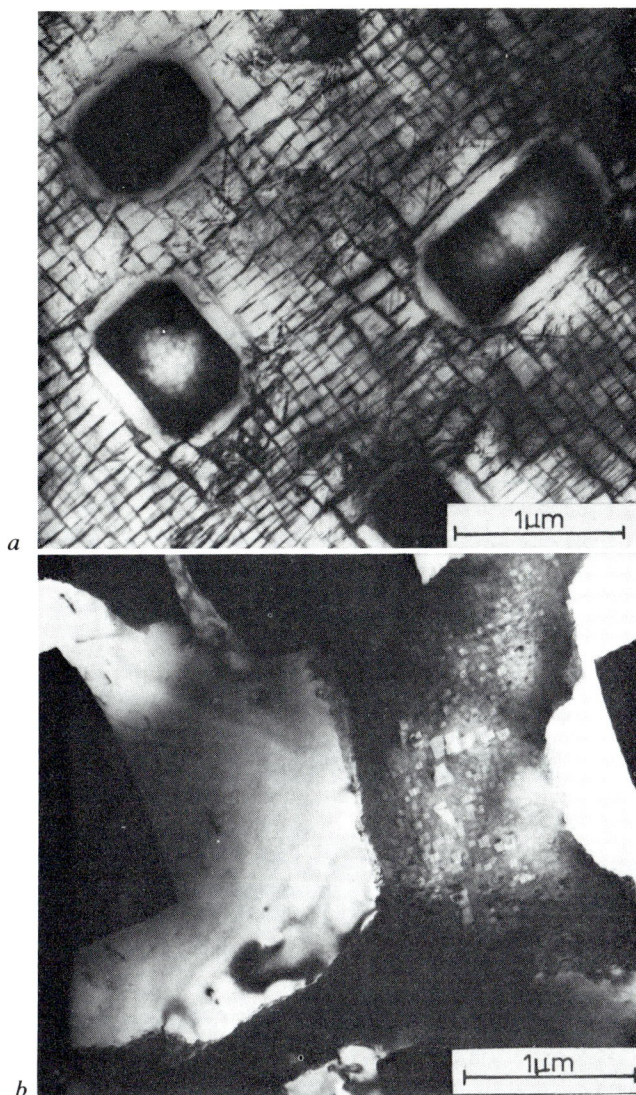

a at quenched solid/liquid interface, Ni–
31·5 Mo–6·2 Al; *b* equilibrium distribution, Ni–
31·5 Mo–6·2 Al

**1 Transmission electron micrographs of
transverse sections of γ/γ' + α DS eutectic**

shown that alloys which have essentially γ' matrices at
room temperature gradually decompose on heating to
form increasing amounts of γ at temperatures near the
solidus and as a result inferred that the γ + α trough
extends to at least 8 wt-% Al. The purpose of this paper
is to describe the solidification reactions and the
development of the γ, γ' and α phase morphologies as
examined by rapidly quenching the solid/liquid interface during steady-state directional solidification and by
application of more sensitive differential thermal analysis techniques. In addition, improvements in the high-
temperature properties of the ternary eutectics through
additional alloying are discussed.

EXPERIMENTAL PROCEDURE

Trial alloys designed to locate the monovariant troughs
were prepared by melting high-purity (>99·9%) Ni,
Mo, and Al under argon in Al₂O₃ crucibles and casting
rods in copper chill moulds. The rods were remelted in
99·7% recrystallized Al₂O₃ cylindrical crucibles using
an inductively heated graphite core, high thermal

gradient apparatus described elsewhere.[10] The crucibles
had an approximate internal diameter of 1·3 cm with a
wall thickness of 0·14 cm. Directional solidification of
the alloys was accomplished by withdrawing the crucibles through a direct impinging water spray. High thermal gradients of the order of 300°C/cm at the solidifying
interface were achieved by maintaining a 400°C superheat in the melt. The steady-state solidification interface
was quenched by rapidly traversing the crucibles into
water.

Specimens for differential thermal analysis of the
liquidus and solidus of the alloys were cut from the
directionally solidified ingots. The specimens were
melted in Al₂O₃ crucibles into which a Pt *v*. Pt/13% Rh
thermocouple extended. The thermocouple tip was
covered with an Al₂O₃ protection tube which was
ground thin at the tip to provide extra sensitivity. A
similar arrangement using Ni or Mo was used for the
standard. DTA specimens for γ' solvus temperature
determinations were not melted but instead had a small
hole drilled in them into which an exposed thermocouple was inserted. Heating and cooling occurred in
an argon atmosphere furnace at a rate of 4°C/min. The
differential output between the thermocouples was
expanded using a dc mV amplifier. The entire arrangement was calibrated using Cu and Ni as standards. The
accuracy of the temperature measurements is probably
within ± 2°C. The solidus was determined from the point
where melting first occurred and the liquidus was
obtained from the cooling curves. The γ' solvus was
recorded both on heating and cooling.

Stress rupture tests on samples machined parallel to
the solidification direction were performed between
760–1 115°C in air. Tests were performed in accordance
to ASTM specification E139–6 with specimens of gauge
diameter 0·28 cm and nominal gauge length 1·27 cm.
Cross-head extension was measured during testing and
elongation and reduction in area measurements were
made from caliper measurements of the specimen after
mating the fractured halves.

RESULTS AND DISCUSSION
Phase equilibria: ternary Ni–Mo–Al

The range of composition studied extended into the
ternary system from the binary Ni–NiMo(δ) eutectic.
The eutectic between γ and δ persisted until an Al level
of ~2·5wt-% was reached at which point a monovariant
eutectic reaction between γ and α began. The exact
transition between these two reactions was not studied.
The quenched solid/liquid interfaces of the γ + α eutectics were sectioned longitudinally and the phase
development at various positions in the temperature
gradient was observed. In general, a definite line corresponding to the precipitation of γ' in γ could be easily
seen as shown in Fig. 2. As the Al level was increased
from 3·5 to 7·5 wt-% Al, the position of the γ' precipitation line moved closer to the solid/liquid interface,
indicating an increase in the γ' solvus temperature.
Quantitative metallography of transverse sections cut
from the ingots at the solid/liquid interface, and from
various positions in the gradient, showed that the
volume fraction of α increases during gradient cooling
from the eutectic reaction. The volume fraction of α
becomes approximately stabilized at the γ' solvus at
which point the γ' forms preferentially at the
fibre/matrix interface. The diffusion of Mo to the

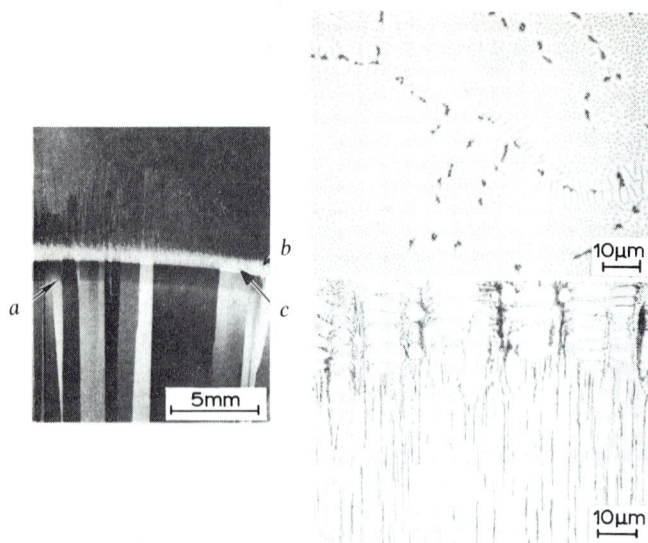

a arrow on left indicates γ' solvus; *b* transverse section $\gamma + \beta + \alpha$; *c* breakdown of plane front to γ dendrites at quench, longitudinal section

$$\alpha + \beta + \gamma \xrightarrow{\Delta T} \gamma/\gamma' + \alpha$$

2   Microstructure of quenched Ni–29·3 Mo–7·5 Al solid/liquid interfaces

'whiskers' together with the γ' precipitation leads to the sharp faceting and square cross-section of the fibres shown in Fig. 1.

An increase in the Al content above 7·5 wt-% Al led to a ternary reaction involving NiAl (β). However, the β was present only at high temperatures near the solidus of the alloys and decomposed on cooling with the concurrent precipitation of γ' (Fig. 2). The β solidified in the form of an elongated phase within clusters of α rods which appeared as rows within the eutectic (Fig. 2). Fully coupled growth between $\gamma + \beta + \alpha$ was observed between 7·5 and 8·6 wt-% Al. At higher Al contents, plane front growth could no longer be maintained and the solidification interface degenerated to β dendrites or complex $\beta + \alpha$ eutectic dendrites. The formation of complex dendrites in ternary systems has been described by others[11,12] and occurs generally in monovariant eutectics which have a wide freezing range between the trough and the solidus. Therefore, the occurrence of the β and $\beta + \alpha$ dendrites is probably the result of a steeply rising $\gamma + \beta$ eutectic trough from the $\gamma + \beta + \alpha$ composition plane.

The range of microstructures observed and the projection of the monovariant eutectic troughs on the Ni–Mo–Al composition triangle is shown in Fig. 3. The solubility limit of Mo in γ was computed from an analysis of the volume fraction of α Mo rods at the quenched solid/liquid interface. The fibres were assumed to contain less than 3 at.-% Ni and negligible Al. This is in accordance with a chemical analysis of extracted Mo fibres and the results of a computer calculation of the phase diagram using the techniques of Kaufman *et al.*[13,14] The solubility curve for Mo in γ shown in Fig. 3 agrees fairly well with the phase diagrams of Markiv *et al.*[9] but extends to higher Al contents. Figure 3 also shows what we believe to be good evidence of a ternary $\gamma + \beta + \alpha$ eutectic at 1 300°C. Some of the characteristic DTA traces observed during

melting and solidifying are shown in Fig. 4. Surprisingly, the γ' solvus can be clearly seen as an additional exotherm or endotherm on the cooling and heating curves. The γ' solvus temperatures determined from these traces agree very well with the temperatures determined from traces taken from unmelted samples containing an imbedded thermocouple. The γ' solvus temperature reached a maximum of 1 288°C at 7·5 wt-% Al and was constant at all compositions greater than 7·5 wt-% Al. This observation together with the metallographic observations of the decomposition of β on cooling supports the existence of the solid-state reaction $\gamma + \beta \rightleftarrows \gamma' + \alpha$ at 1 265°C as reported by Pryakhina *et al.*,[7] although we observe this at a somewhat higher temperature of 1 288°C. The results of the DTA analysis of the liquidus, solidus, and γ' solvus temperatures of the alloys lying along the monovariant eutectic $\gamma + \alpha$ and $\gamma + \beta$ troughs are shown in Fig. 5. The projected eutectic troughs do not represent a true isopleth through the Ni–Mo–Al ternary diagram because the eutectic troughs are not compositional straight paths but exhibit some curvature (Fig. 3). Figure 5 does, however, show the small slope of the $\gamma + \alpha$ eutectic trough and the intersection of the steep $\gamma + \beta$ trough at the proposed $\gamma + \beta + \alpha$ ternary eutectic. In addition to the DTA analysis of the eutectic troughs, the intersection of γ and α liquidus surfaces perpendicular to the eutectic trough at a nominal composition of Ni–33·8 wt-% Mo–6 wt-% Al were analysed and a section of the isopleth is shown in Fig. 6. The γ liquidus is only approximate because of the small volume fraction of γ dendrites in the high Mo side of the eutectic which produced an insufficient change in the DTA traces to determine accurately the onset of freezing. The analysis of the γ and α liquidus surfaces was made in order to determine the extent to which Mo could be varied and still maintain coupled growth. Generally, plane front off-eutectic growth can be achieved if the freezing range does not exceed an amount approximated by the relation

$$\Delta T \leqslant \frac{GD}{R}$$

If it is assumed that a freezing range of 10°C can be tolerated, then the variation in Mo content can be about 2–3 wt-% and still maintain coupled growth.

Phase equilibria: Ni–Mo–Al + Cr and Ni–Mo–Al + Ta

In order to improve the oxidation and hot corrosion behaviour of the ternary $\gamma/\gamma' + \alpha$ eutectics, chromium was added in 1 wt-% intervals in alloys with Al contents of 5·5, 6·2, and 7·5 wt-%. The relative amounts of Ni and Mo were adjusted so that plane front growth was achieved. The low Al content alloys exhibited good tolerance for Cr additions with as much as 9 wt-% added without cellular or dendritic interface break down. Alloying Cr at 7·5 wt-% Al was somewhat more difficult with the additions tending to stabilize the formation of β. The effect of Cr on the γ' solvus temperature in alloys with an aluminium content of 7·5 wt-% is shown in Fig. 7.

Chromium was expected to partition to both the γ- and α-phase as it is extensively soluble in Ni and completely miscible in Mo. In order to determine the distribution of Cr between γ and α, the α whiskers were extracted by dissolving the γ/γ' matrix in hot HCl

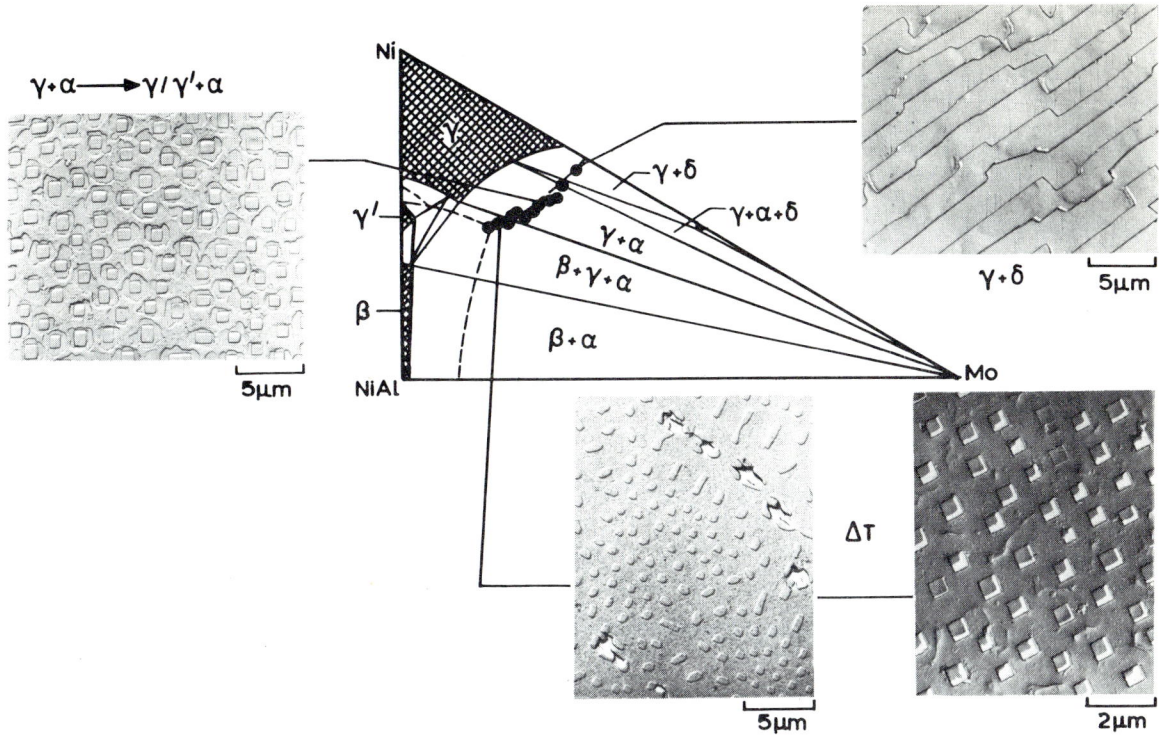

3   Projected monovariant liquidus troughs on Ni–Al–Mo isothermal section at 1 300 °C

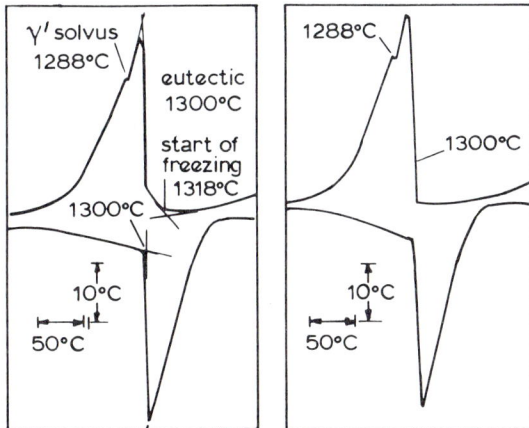

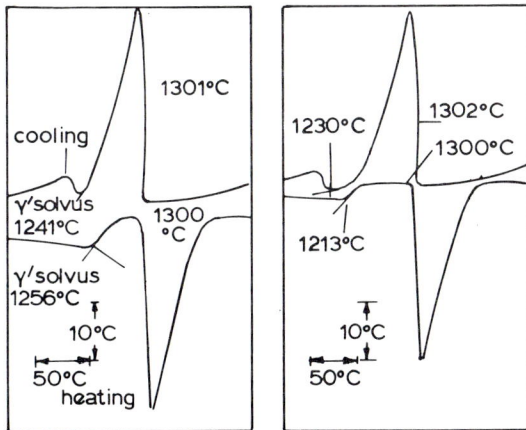

4   Differential thermal analysis traces of various Ni–Al–Mo eutectic alloys; heating rate 4°C/min

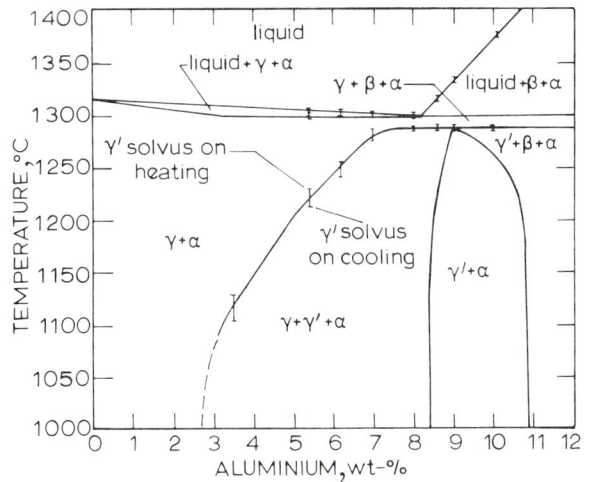

5   DTA analysis of vertical section through γ + α eutectic trough

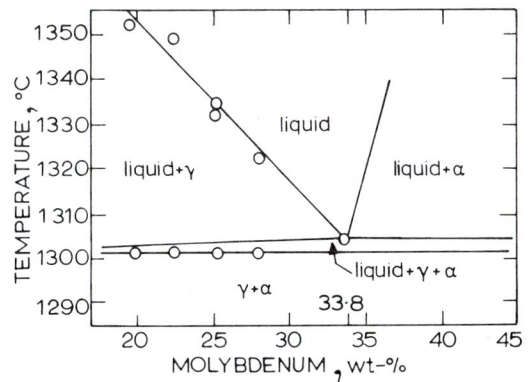

6   DTA analysis of vertical section perpendicular to γ + α eutectic trough (C_E = Ni–33·8 Mo–6·0 Al)

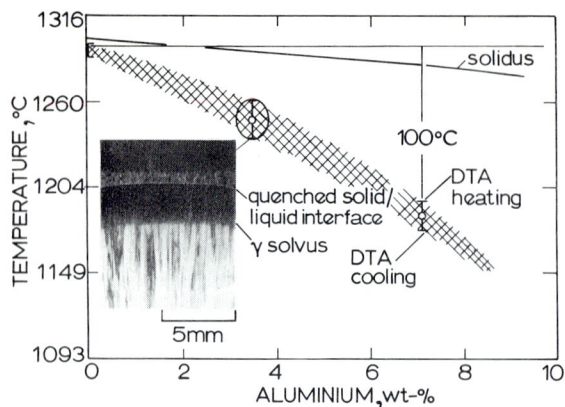

7 Effect of chromium on γ' solvus temperature at 7·5 wt-% Al

activated with hydrogen peroxide.[15] Wet chemical analysis of the extracted α whiskers is presented in Fig. 8. Figure 8 shows that Cr partitions to the α-phase on an approximately equal atomic basis with the nominal Cr content of the alloy. From this it can be inferred that the Cr distributes on nearly an equal atomic basis between the matrix and the fibres. As was discussed in a previous paper[1] the partitioning of Cr to the matrix phase promotes desirable oxidation resistance.

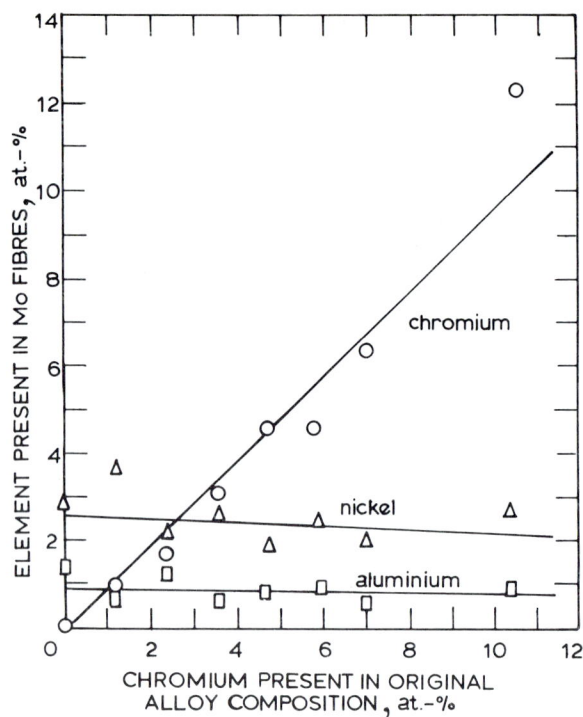

8 Solubility of Cr, Ni, and Al in extracted Mo fibres from DS γ/γ'–α alloys

Table 1 Results of DTA of γ/γ' + α alloys

	Ni, wt-%	Mo, wt-%	Al, wt-%	Cr, wt-%	Liquidus, °C	Solidus, °C	γ' solvus (cooling), °C	γ' solvus (heating), °C
					Eutectic γ + α alloys			
1	55·0	39·5	3·5	—	1 302	1 305	1 104	1 129
2	62·1	32·5	5·4	—	1 301	1 304	1 213	1 230
3	62·3	31·5	6·2	—	1 301	1 302	1 241	1 257
4	62	31·0	7·0	—	1 300	1 301	1 276	1 287
					Eutectic γ + β + α alloys			
5	63·2	29·3	7·5	—	1 301	1 300	1 288	
6	65·0	27·2	7·8	—	1 301	1 302	1 286	
7	65·0	27·0	8·0	—	1 300	1 300	1 288	does not
8	66·0	25·4	8·6	—	1 316	1 300	1 286	appear
9	65·5	25·5	9·0	—	1 335	1 301	1 287	
10	65·9	24·0	10·1	—	1 370	1 301	1 286	
					Off Eutectic γ + α			
	Primary γ alloys							
13	72·8	19·8	7·4	—	1 301	1 352	—	—
14	70·1	22·8	7·1	—	1 302	1 349	—	—
15	67·6	25·6	6·6	—	1 301	1 330	—	—
16	65·0	28·4	6·6	—	1 301	1 321	—	—
17	60·2	32·8	6·0	—	1 301	1 302	—	—
					Effects of chromium			
18	62·3	31·5	6·2	0	—	—	1 241	1 257
19	59·8	31·0	6·2	3·0	—	—	1 218	1 230
20	57·3	31·5	6·2	5·0	—	—	1 202	1 214
21	57·3	27·5	6·2	3·0	—	—	1 151	1 169
22	63·2	29·3	7·5	0	—	—	1 288	—
23	62·3	26·6	7·6	3·5	—	—	1 236	1 259
24	60·4	25·1	7·4	7·1	—	—	1 177	1 195

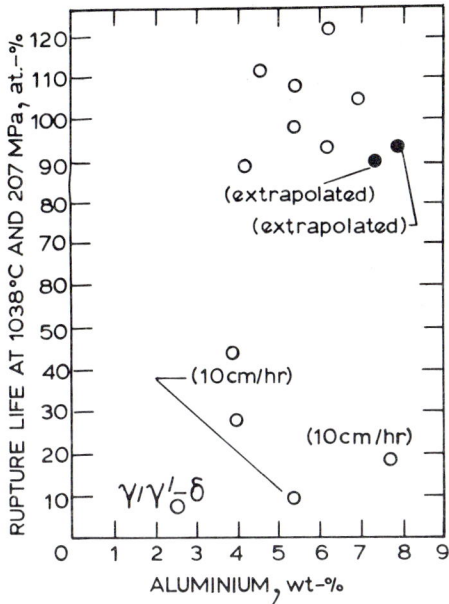

9 Effect of aluminium content on rupture life at 1 038°C and 207 MPa of α–Mo reinforced Ni–Al–Mo ternary alloys ($R = 3$ cm/h, G_L 300°C/cm)

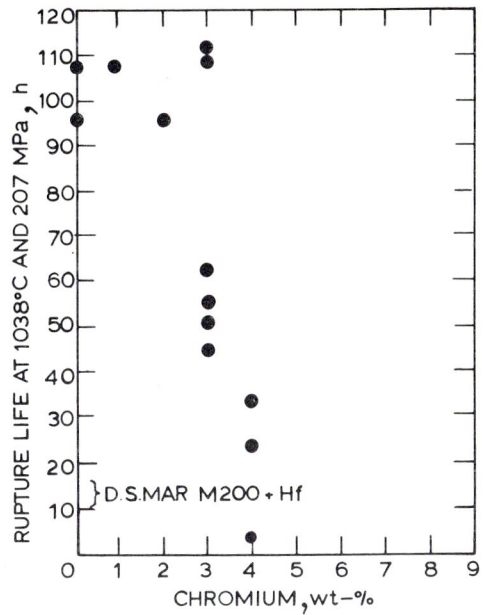

10 Effect of chromium content on rupture life (1 038 °C, 207 MPa) of $\gamma/\gamma'-\alpha$ alloys containing 5·5 wt-% Al

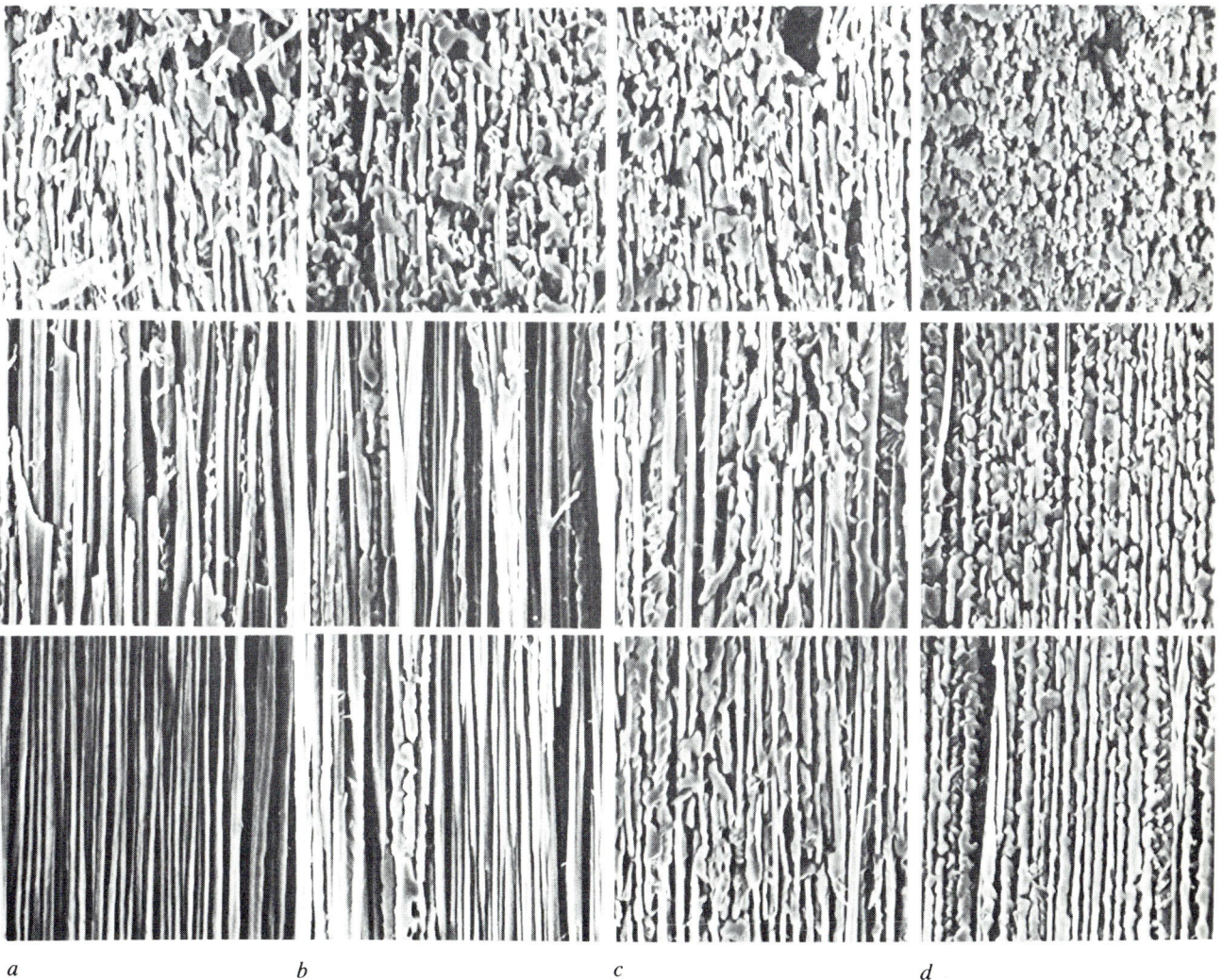

top row at fracture; *middle row* 3 mm away from fracture; *bottom row* unstressed region at 1 900°F
a No Cr (93 h); *b* 3 wt-% Cr (62 h); *c* 6 wt-% Cr (20 h); *d* 9 wt-% Cr (0·3 h)

11 Breakup of α–Mo rods containing varying amounts of Cr at 1038°C and 207 MPa

Ta was observed to have an opposite effect to Cr on the phase equilibria of the $\gamma/\gamma' + \alpha$ alloys. At low Al-contents (5·5 wt-%), Ta could only be added to approximately 3–4 wt-% before the solid/liquid interface developed $\gamma' + \alpha$ cells. At high Al-contents, a pseudobinary reaction between γ' and α and a ternary $\gamma' + \beta + \alpha$ reaction was observed, and Ta could be extensively added from 6 to 15 wt-%. These alloys also had the advantage of having reduced Mo contents which may improve the hot corrosion behaviour of the alloy. Work is presently in progress to determine the partitioning of Ta between γ' and α in a similar manner as was for Cr.

Mechanical behaviour

The rupture strengths of various ternary Ni–Mo–Al monovariant eutectic alloys tested in air are presented in Fig. 9. The optimum aluminium content can be observed to vary between 4–8 wt-% for maximum rupture life at 1 038°C and 207 MPa. Two data points were extrapolated from the Larson–Miller parameter-stress plot for 7·0 and 7·8 wt-% Al alloys, from tests performed at higher temperatures and stress, to complete the range of aluminium contents. The deleterious effect that cellular microstructure has on the high temperature stress rupture lives of this class of alloy is noted for alloys directionally solidified at 10 cm/h. For comparison purposes the rupture life of DS MarM 200 + Hf at these conditions of stress and temperature would be 10–15 h.

The effect of chromium on the rupture life at 1 038°C and 207 MPa was studied most extensively at aluminium levels of 5·5 and 6·2 wt-%.[1] A significant scatter in the rupture lives of the 3 wt-% Cr specimens in particular was observed, as shown in Fig. 10. On examination of the fracture surface of each specimen it was determined that cellular and dendritic α structural defects existed within the gauge section of two of the four specimens containing 3 wt-% Cr which exhibited rupture lives below 100 h. It appeared, however, that specimens which contained totally microduplex eutectic structures and possessed bimodal α-rod distributions within grains exhibited a slight but significant reduction in strength owing to the presence of 3 wt-% Cr.

The previous observation of a marked dependence of rupture life on chromium content at an aluminium level of 6·2 wt-%[1] was examined metallographically. Although some of the scatter in data could again be attributed to the presence of an occasional α-dendrite existing within the gauge volume, a marked degree of accelerated coarsening of the α-rods was observed both in the stressed and more surprisingly the unstressed regions of the rupture specimen. A montage depicting this breakup of the highly faceted α, Mo rods is shown in Fig. 11. In the unstressed grip regions of the test specimens in the lower portion of the figure, an accelerated rate of α Mo coarsening is observed to occur with increasing Cr content. The rate of coarsening is expected to be determined by either a bulk diffusion or an interface controlled reaction.[16,17] Chromium appears to be catalyzing coarsening by reducing the interfacial energy of the system when present in the bulk alloy above 3 wt-%. Experience with equivalent amounts of larger substitutional atoms, i.e. W and Ta indicate that a slower diffusing addition is not as catalytic in coarsening the α-phase and experiments on combinations of solute atoms are to be reported elsewhere.[18]

ACKNOWLEDGMENT

Part of the research presented herein was performed under a contract, N62269-76-C-0107, with the Naval Air Development Centre, with Mr I. Machlin, Naval Air Systems Command, as technical consultant.

REFERENCES

1 F. D. LEMKEY: 'Superalloys: metallurgy and manufacture', 3rd International Symposium, Seven Springs, Pa., 321, 1976, Claitor's Publication Division, Baton Rouge, La
2 H. SPRENGER et al.: *J. Mats. Sci.*, 1976, **11**, 2 075
3 M. HENRY: *Scripta Metall.*, 1976, **10**, 955
4 F. D. LEMKEY AND E. R. THOMPSON: Proceedings of Conference on *In Situ* Composites, Vol. II, NMAB 308, 105, 1973, Washington, DC, National Academy of Sciences
5 H. BIBRING: *ibid.*, 1; 1973
6 L. P. JAHNKE AND C. A. BRUCH; Proceedings of Conference on Directionally Solidified *In Situ* Composites, 3, 1974, AGARD, CP No. 156
7 L. I. PRYAKHINA et al.: *Diagrammy Sostoyaniya Metall. Sistem*, 1971, Nanka, Moscow
8 YU. A. BAGARYATSKII AND L. E. IVANOVSKAYA: *Proceedings of Academy of Sciences USSR*, May–June 1969, 491
9 V. YA. MARKIV et al.: *Izv, AN SSR Metally*, 1969, **5**, 180
10 E. R. THOMPSON et al.: Proceedings of Conference on Directionally Solidified *In Situ* Composites, Vol. II, NMAB-308, 71, 1973, National Academy of Sciences, Washington, DC
11 M. D. RINALDI et al.: *Metall. Trans.*, 1972, **3**, 3 139
12 J. D. HOLDER AND B. F. OLIVER: *Metall. Trans.*, 1974, **5**, 2 423
13 L. KAUFMAN AND H. BERNSTEIN: 'Computer calculation of phase diagrams', 1970, New York, Academic Press
14 L. KAUFMAN AND H. NESOR: *Metall. Trans.*, 1974, **5**, 1 623
15 J. L. WALTER: GE Research & Development Centre, Schenectady, NY, private communication
16 C. WAGNER: *Z. Electrochem.*, 1961, **65**, 581
17 J. M. LIFSHITZ AND U. V. SLYOSOV: *J. Phys. Chem. Solids*, 1961, **19**, 35
18 E. P. WHELAN et al.: 'Hot corrosion oxidation and stress-rupture properties of modified Ni–Mo–Al–Cr eutectic alloys', Fall AIME Meeting, Chicago, Ill., Oct. 1977

Development of composite structures in cast steels by liquid metal treatment

P. A. Blackmore, A. Segal, A. J. Baker, and P. R. Beeley

Coarse carbide dispersions have been developed in cast steels by the addition of substantial proportions of carbide-forming elements to the molten bath shortly before casting. Volume fractions of up to 14% of massive primary carbides have been achieved with matrix compositions mainly of maraging steel types. Niobium and vanadium carbides were employed, the former having a similar density to that of the matrix alloy. Heat treatments were applied for development of matrix structures and the materials have been examined by optical and electron microscopy and mechanical testing. Analytical assessments were made by X-ray fluorescence techniques. Evidence is presented for the formation of the homogeneous dispersions of coarse carbides as primary phases in the melt; finer carbides of lamellar morphology were formed interdendritically at some later stage. The coarse dispersions, with particular sizes in the range 3–10 μm, were unaffected by normal maraging solution treatments, while the treated matrices showed the rapid age hardening response and other normal characteristics of maraging steels. Fracture behaviour of the material is discussed. The results, with analogous findings from other steels, indicate that the duplex structures produced may be regarded as in situ *composites with potential for die and tool applications.*

The authors are in the Department of Metallurgy, University of Leeds

In situ composite materials of the aligned multiphase structure type have attracted much interest recently for their mechanical properties, especially at high temperatures, while in applications where surface as well as bulk properties are of major concern, the special role of duplex microstructures has long been exemplified in the typical bearing alloy requirement of hard particles in a soft matrix.

The present work arose from industrial interest in the development of special duplex cast steels which might offer a combination of surface and bulk properties suitable for use as cast dies and tooling for materials-forming operations. Early industrial experiments[1] had indicated that unusually coarse vanadium carbide dispersions could be developed in a steel of maraging type by the addition of substantial proportions of the carbide-forming elements to molten baths. The present paper describes the first findings from an experimental programme designed to investigate the production and characteristics of such structures and factors influencing properties of the materials. Air induction melts were produced under various conditions to develop dispersions, initially of vanadium and subsequently of nio-bium carbide, in a number of maraging steels. Simple block castings produced from the melts were metallurgically examined in as-cast and heat-treated conditions, while other melts were used for quench sampling and examination. Initial experiments were also performed on other types of steel to establish whether analogous dispersions could be formed under similar conditions.

MELTING PRACTICE AND LIQUID METAL TREATMENT

Melts were produced in induction furnaces of 15 kg and 5 kg capacity for the block casting and quench experiments respectively and the base compositions of the maraging steels are given in Table 1.

For each melt a base charge of cobalt, nickel, ferro-molybdenum, and EN3 steel bar was melted down and deoxidized with aluminium. Additions of ferro-aluminium, and where appropriate ferrotitanium, were made at a bath temperature of about 1 550°C. The temperature was then raised to 1 600°C–1 620°C for the

Table 1 Summary of nominal base alloy compositions

Alloy series	Composition, wt-%				
	Ni	Co	Mo	Al	Ti
MS1	16	6·5	4·3	0·4	0·5
MS2	17	—	5·0	0·4	0·5*
MS3	18	6·0	5·0	0·4	0·5

* Only where indicated in the text and in Table 2.

addition of the carbide-forming elements. This addition was made using two different techniques:

(i) an intimate mixture of finely divided graphite powder and either ferrovanadium or ferroniobium powder was added to the molten bath
(ii) finely divided graphite powder was added to a bath already containing niobium in solution.

The proportions of the graphite and ferroalloys were calculated on the basis of the stoichiometry of the expected carbide products. In both cases the powder addition was wrapped in aluminium foil to reduce oxidation loss of carbon. The bath was then mechanically stirred to facilitate homogeneous mixing and reaction of the vanadium or niobium with carbon to form the carbide dispersion.

A thick dross was observed on melts containing vanadium carbide dispersions. X-ray diffraction analysis showed this to be rich in vanadium carbide. This was attributed to primary vanadium carbide particles floating out of the melt due to their low density with respect to that of the liquid steel ($5\,400\;kg/m^3$ for VC, $7\,200\;kg/m^3$ approx. for the liquid steel[2]). This problem was not encountered with niobium carbide due to its higher density ($7\,850\;kg/m^3$). The maraging steel–vanadium carbide system, at temperatures in excess of 1 700°C, showed a marked tendency for the slag to be re-absorbed by the bath, while on progressive cooling the slag reformed, suggesting a marked change of solubility of this carbide in the liquid with temperature. Melts based on the maraging steel–niobium carbide system, by contrast, gave much clearer molten metal surfaces at the concentrations investigated, with no evidence of separation of the carbide dispersion.

The larger melts were cast into blocks $300 \times 80 \times 80$ mm in Shaw Process ceramic moulds to produce surface cooling conditions approximating to those which would be obtained in casting shaped dies by the same process.[3–5] The blocks were sectioned for the removal of metallographic specimens and testpieces. Some specimens were examined in the as-cast condition while others were aged in evacuated sealed silica tubes to establish aging response.

In the experiments designed to determine the reactions taking place in the melt, molten samples were withdrawn in closed ended vitreous silica tubes, 12 mm in diameter, and quenched in iced water to provide metallographic evidence for the existence and form of the melt dispersions. These experiments, on both vanadium and niobium carbide dispersions, were carried out in the following manner:

(i) base charge melted down as previously described
(ii) carbide former (V or Nb) and graphite added, time = 0

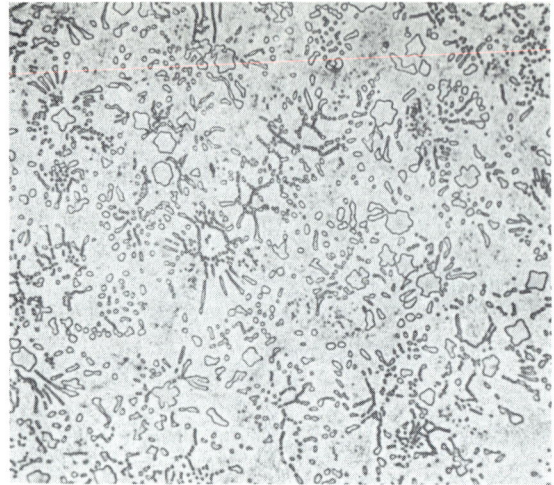

1 Optical micrograph of alloy MS1/10VT as cast; etched in acidified $CuCl_2$ [×540]

(iii) bath temperature raised to predetermined level, 1 600°, 1 650°, 1 700 or 1 750°C, and metal held for 15 min.

Liquid metal samples were taken every 3 min. After the bath had been held at temperature for the requisite time the furnace was switched off and the bath was allowed to cool. Sampling continued at 3 min intervals until solidification of the bath occurred.

METALLOGRAPHIC EXAMINATION

Representative micrographs from the block castings are given in Figs 1–3. A summary of carbide volume fractions, measured by the normal point counting technique,[6] and characteristics of the dispersions, with hardness values in the solution treated state, for each of the alloys is given in Table 2. In all the alloys the coarse dispersoid had a particle size range of 3–10 μm. The alloys are identified by the base maraging steel number (MS1–MS3) with the weight percentage of either vanadium or niobium carbide: 'T' indicates a titanium addition.

In each alloy the matrix is typical of that expected in a maraging steel.[7] In the case of high volume fraction

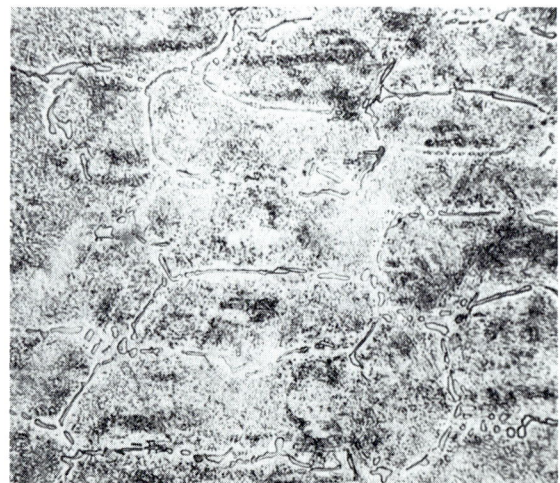

2 Optical micrograph of alloy MS2/5VT as cast [×420]

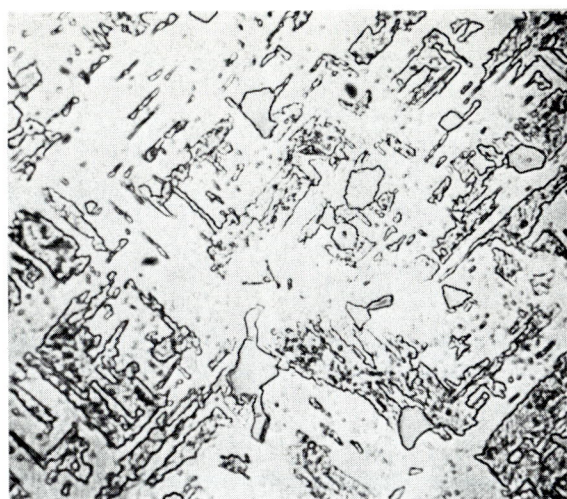

3 Optical micrograph of alloy MS3/3NT as cast; etched in saturated picral [×420]

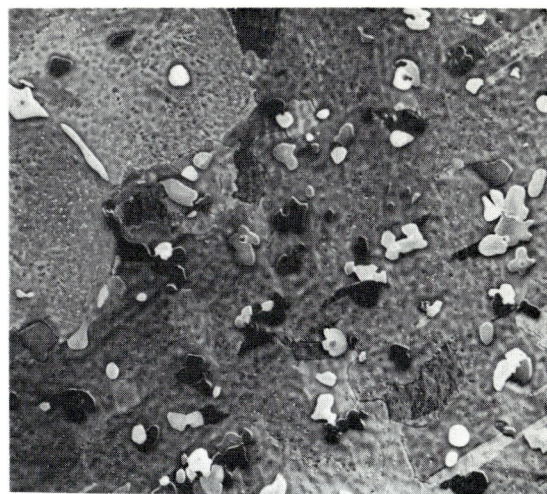

4 Photoemission electron micrograph of alloy MS1/10VT at 800°C [×500]

Table 2 Summary of carbide volume fractions, carbide morphologies, and hardness values for the alloys in the solution treated condition (treatment 900°C for 1 h followed by water quench)

Alloy	Volume fraction of dispersoid, %	Carbide morphology	Solution treated hardness, H_v
MS1/10VT	14·7	Angular and interdendritic	404
MS2/7V	9·0	Interdendritic	373
MS2/5V	8·3	Interdendritic	370
MS2/5VT	8·4	Interdendritic	—
MS2/5N	6·9	Angular homogeneous	345
MS2/5NT	7·8	Angular homogeneous	356
MS3/5NT	7·7	Angular homogeneous	383
MS3/3NT	4·5	Angular homogeneous	307

Fig. 4). To investigate the nature of these particles, X-ray fluorescence studies were carried out using a modified SEM. A set of X-ray distribution micrographs for a carbide particle in alloy MS3/5NT are shown in Fig. 5. This carbide shows a nucleus, although the X-ray micrographs do not reveal gross segregation of titanium in the nucleus. However, spot analyses taken from the nucleus and from the edge of the particle clearly exhibit a concentration of titanium in the nucleus (Fig. 6). Thus it is concluded that certain of the primary carbides are nucleated by Ti-rich particles present in the melt.

Detailed physical examination has only been carried out on alloy MS3/5NT. The aging response is characteristic of the base maraging steel,[8] with a very rapid increase in hardness on aging at 500°C, peak hardness

vanadium carbide materials (MS1/10VT) a large number of coarse angular carbides, believed to have been present in the liquid, and finer interdendritic carbides, the latter having the appearance of a eutectic structure, were observed (Fig. 1). When the volume fraction of vanadium carbide was reduced to 9% (MS2/5VT) no coarse carbides were observed and the microstructure showed only strings of interdendritic carbides (Fig. 2). In the case of niobium carbide, even volume fractions as low as 4·5% (MS3/3NT) showed homogeneous dispersions of coarse angular carbides (Fig. 3).

Using X-ray diffraction analysis the carbides were identified as vanadium monocarbide [VC(14 wt-% Cu)] and niobium monocarbide [NbC(11·45 wt-% C)].

The size and morphology of the coarser carbides in both systems suggested strongly that they were primary particles formed in the melts. Optical microscopy showed that many of the coarse carbides, in the presence of titanium, possessed what appeared to be nucleating particles in their centres. This was also clearly shown on photoemission electronmicrographs in which the carbide dispersion itself appears in good contrast (*see*

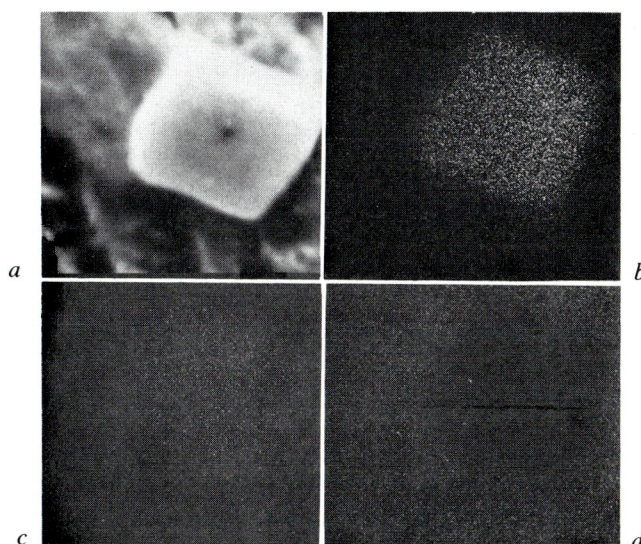

5 *a* Scanning electron micrograph of alloy MS3/5NT, deeply etched in acidified CuCl₂, showing a NbC particle with central nucleus [×5 500]. *b*, *c*, and *d* X-ray fluorescence micrographs of the carbide particle shown in Fig. 5*a*: *b* niobium distribution; *c* titanium distribution; *d* iron distribution

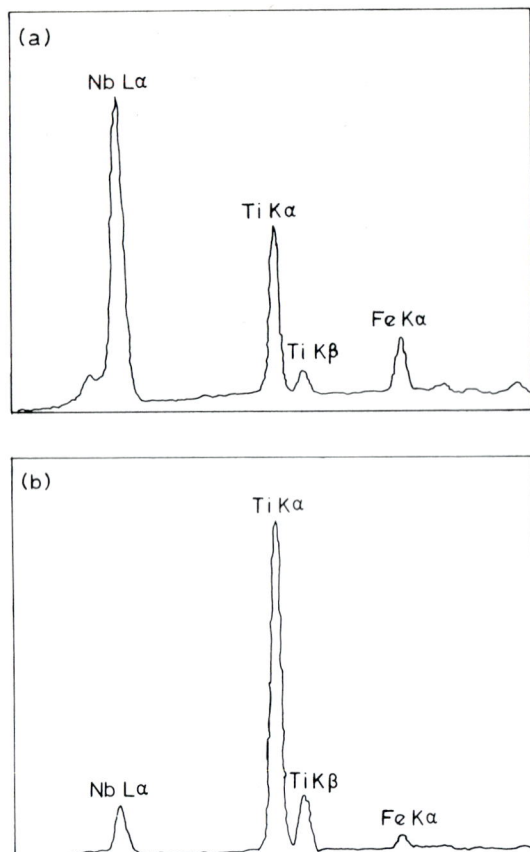

6   X-ray fluorescence spectra of *a* the edge, and *b* the centre of the carbide particle shown in Fig. 5*a*

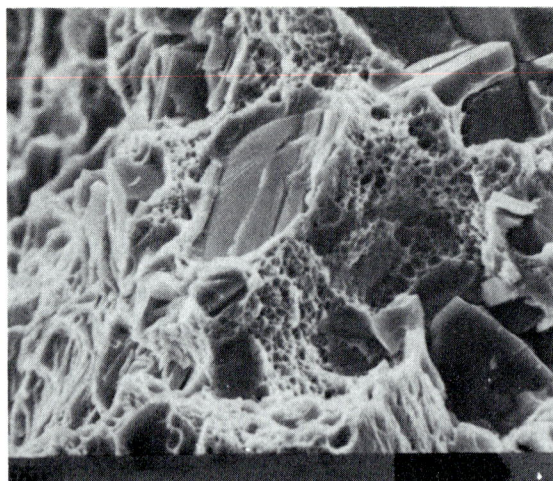

8   Scanning electron micrograph of impact fracture surface in alloy MS1/10VT   [×1 400]

at 300°C. No plastic deformation occurred, in tension, at either temperature but the elastic elongation to fracture was greater at 300°C (4% and 10·5% respectively). Notched bar impact tests were carried out on the material in the solution treated and peak hardness conditions. Some specimens were also tested after aging to develop about 30% reverted austenite, which has previously been shown to improve toughness in conventional maraging steels.[9,10] However, in all cases examined in the present work the impact energy was very low and no differences could be detected in the fracture mechanism. The fracture surface showed cracking and cleavage of the carbides with some ductile failure occurring in the matrix. A typical fractograph is shown in Fig. 8.

In the quenched samples taken from various melts of alloys MS3/5NT and MS3/10VT it was observed that primary carbide particles, showing Ti-rich nuclei, had been present in the liquid. In the case of NbC the dispersion and particle morphologies were not significantly affected by increasing bath temperature. However, primary VC particles were observed to have serrated surfaces, suggesting that re-solution of these particles was occurring with increasing bath temperature. Carbide remaining in solution in the liquid steel was precipitated, on quenching, as an interdendritic network. In the case of the NbC dispersions this was in the form of interdendritic particles, but in the case of the VC it showed a eutectic-like rod morphology similar to that observed in the slowly cooled material. The differences in primary carbide morphologies between the NbC and VC dispersions are shown in Figs 9*a* and 9*b* while Fig. 9*c* shows the VC eutectic structure together with some primary carbides.

being retained even on extended heat treatment (e.g. 10 h at 500°C). Aging curves for the alloys MS3 and MS3/5NT are shown in Fig. 7. It will be noted that the aging characteristics of the two alloys are similar and that the greater hardness of MS3/5NT, evidently due to the presence of the NbC dispersion, is maintained throughout.

The tensile strength of the alloy was found to be about 900 N/mm^2 at room temperature, falling to 600 N/mm^2

OTHER ALLOY SYSTEMS
Preliminary work has been carried out[11] to investigate the possibility of forming similar carbide dispersions in the conventional die steels BS 224 No. 5 and BH13.[12] Melts were made using identical procedures to those previously described for the maraging steel matrix.

Photomicrographs from sections of cast blocks are shown in Fig. 10 and it is observed that similar dispersions of primary carbides were obtained.

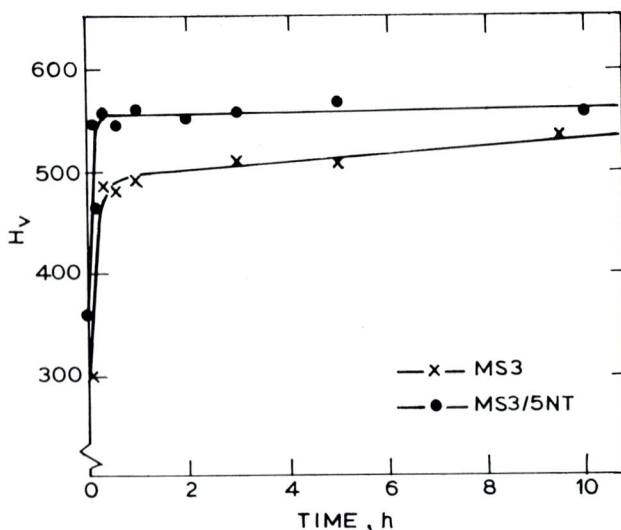

7   Aging curves for alloys MS3 and MS3/5NT at 500°C

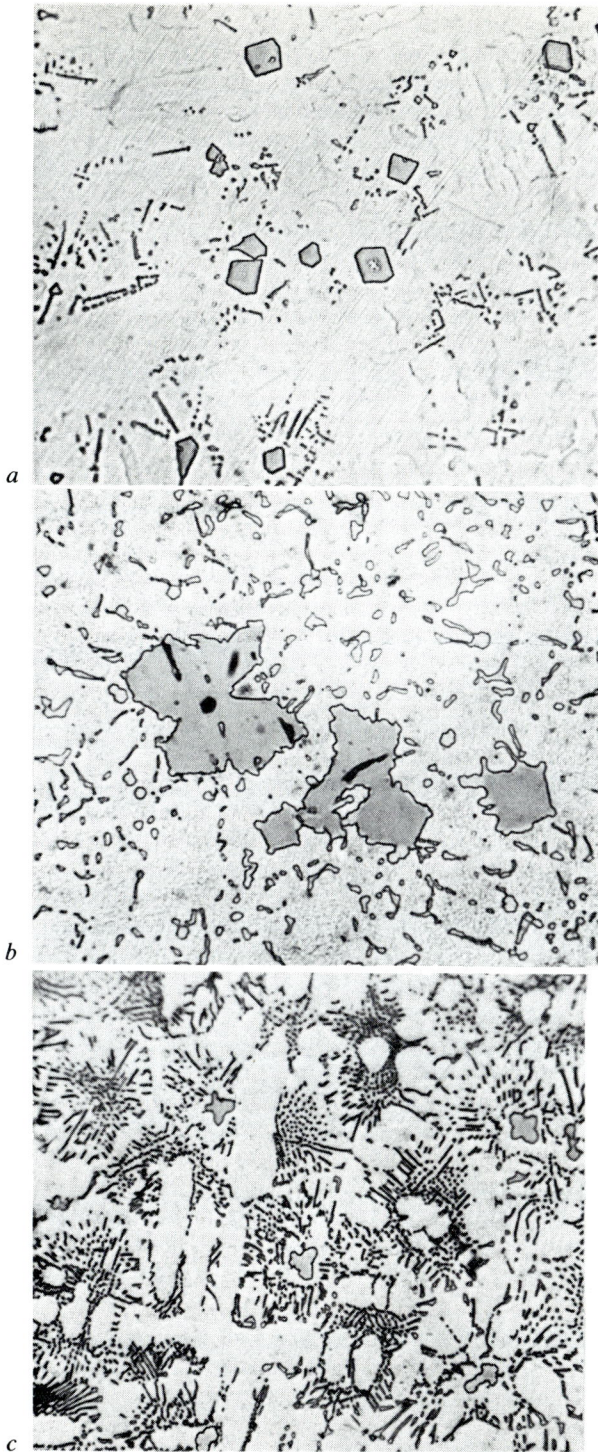

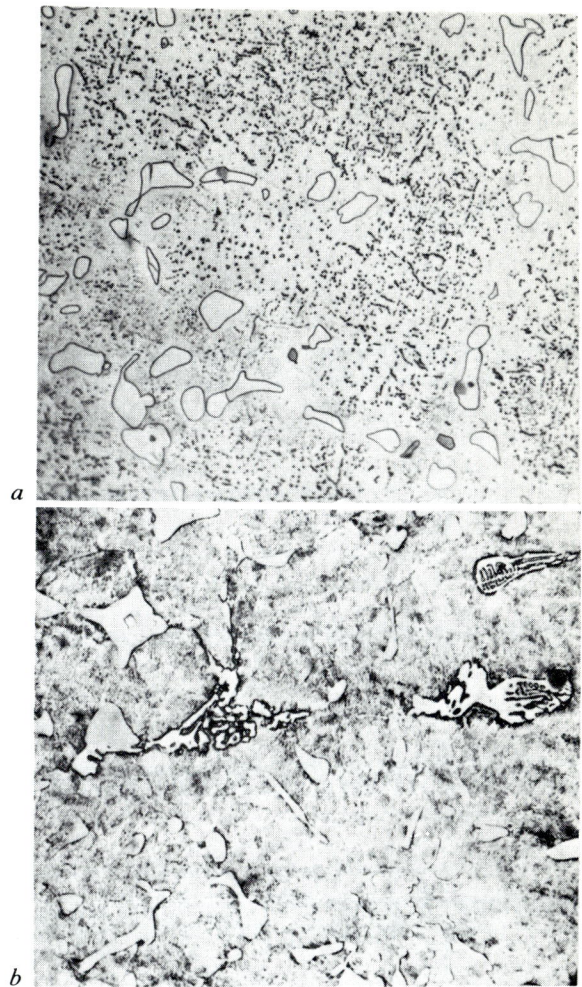

9 *a* Optical micrograph of quenched liquid metal sample of alloy MS3/5NT etched in saturated picral, [×540]; *b* optical micrograph of quenched liquid metal sample of alloy MS3/10VT etched in saturated picral, [×675]; *c* optical micrograph of quenched liquid metal sample of alloy MS3/10VT etched in saturated picral [×240]

10 *a* Optical micrograph of VC dispersion in as cast BS224 No. 5 die steel etched in saturated picral, [×540]; *b* optical micrograph of VC dispersion in as cast BH13 hot work tool steel etched in saturated picral, [×540]

DISCUSSION AND CONCLUSIONS

The experimental evidence demonstrates that coarse solid carbide dispersions can be produced in liquid steel by substantial additions of carbide-forming elements to molten baths and that these dispersions can be largely retained as a feature of the final solid state structures.

The characteristics of an *in situ* composite are thus achieved: volume fractions of at least 14% of the dispersed phase have proved feasible in the cases of niobium and vanadium carbides, with particle sizes mainly in the range 3–10 μm.

That the coarse carbides originate in the melt is shown both by the faceted shape of the niobium carbide and by the presence of the coarser carbides in the quenched as well as the slowly cooled microstructures. The fact that they occur irrespective of whether the carbide formers were added simultaneously or sequentially indicates formation by homogeneous reactions within the melt. The carbides were evidently nucleated heterogeneously on pre-existing particles. Some of the latter were directly observed and identified as titanium rich but other substrates, effective at equivalent undercoolings, may also be involved, since the carbide particles themselves were of similar size in the presence and absence of titanium as a major alloy constituent.

Evidence at present available suggests a difference between the relationships of the niobium and vanadium carbide dispersions with the melt. The former remained stable with respect to size and shape over a 100°C range of bath temperature, indicating little change in solubility and a steep liquidus surface. The vanadium carbide showed an apparent tendency for the particle size to diminish with increasing bath temperature over a 150°C range: together with the observed serration of particle/melt interfaces and the drossing behaviour on the melt surface, this suggested some re-solution due to an increase in solubility with temperature.

The further carbides visible in the microstructures are quite distinct with respect both to size and shape from the melt dispersoid. Their interdendritic distribution pattern can be attributed to dendritic segregation of the carbide formers during the freezing of the bulk of the alloy. The eutectic or eutectoid morphology of these carbides also stands in marked contrast to that of the primary particles. It can thus be inferred that the finer lamellar-type carbides were formed much later in the cooling sequence, probably in the last stages of freezing or possibly in the solid state.

The fact that the aging responses of the steels are similar in the presence and absence of the primary carbide dispersion suggests that the matrix retains its essential maraging characteristics. The exact nature of the precipitation reactions during aging in these particular alloys is not yet known. The extremely rapid aging response of the alloys is typical of maraging materials, in which intermetallic compounds such as Ni_3Ti and Ni_3Mo predominate. However, at the high residual carbon contents likely in the present materials, the possibility of alloy carbide precipitation also exists.

The low impact energy of the material is attributable to the presence of the massive carbides, since cleavage in these carbides and failure of the carbide/matrix interface have both been observed and reduction in the volume fraction of the carbides produced a slight increase in impact resistance. However, the observation of a ductile fracture mode in the maraging matrix between the massive carbide, implies that this matrix is still relatively tough. The ductile behaviour of the matrix under impact fracture conditions is somewhat unexpected since the carbon content of the matrix is likely to be relatively high and the alloys are produced by air melting. Air melting is particularly disadvantageous to the toughness of the ultra-high strength maraging steels and these materials are also known to suffer embrittlement at higher levels of residual carbon content. Since these forms of embrittlement are associated with the formation of intergranular Ti compounds, the apparent ductility of the present material may be due to removal of any Ti addition during the initial carbide creation reactions in the melt. Once formed, the Ti carbide may be incorporated as a nucleant in either NbC or VC, and could then no longer give rise to solid state reactions during the subsequent heat treatments needed for age-hardening.

That the formation of coarse carbide dispersions in the melt could be applied more widely is shown by the preliminary work with other steels: such *in situ* composites should offer potential for tool, die, and other applications requiring specialized forms of wear resistance.

ACKNOWLEDGMENTS

The authors are indebted to the Science Research Council for financial support, to Professor J. Nutting and the University of Leeds for facilities, and to numerous industrial friends for helpful discussions.

REFERENCES

1 J. REYNOLDS AND A. K. MCDONALD: Associated Engineering Developments Ltd., 1975, private communication
2 C. J. SMITHELLS: 'Metals reference book, 5th Edn, 1976, London and Boston, Butterworths
3 H. G. EMBLEM: *Foundry Trade J.*, 1972, **132**, 2 884, 379
4 O. MADONO: Proc. S.A.E. Conf., Detroit, Michigan, Jan. 1963
5 T. MURAO: *Foundry Trade J.*, 1961, **110**, 2 304, 149
6 R. T. DE HOFF AND F. R. RHINES: 'Quantitative microscopy', 1st Edn, 45, 1969, New York, McGraw–Hill
7 S. FLOREEN: *Int. Met. Rev.*, 1968, **13**, 115
8 R. F. DECKER *et al.*: *Trans. ASM*, 1962, **55**, 58
9 S. D. ANTOLOVICH *et al.*: *Metall. Trans.*, 1974, **5**, 623
10 B. FRANCIS: *Metall. Trans.*, 1976, **7A**, 465
11 Department of Metallurgy, Leeds University, 1976, unpublished project work
12 British Standards Institution, BS 4659: 1971

Structure of cellular composites

K. P. Young, B. A. Rickinson, and M. C. Flemings

This paper summarizes and compares criteria for cellular growth and for solid/liquid interface stability in binary and ternary alloys which solidify with two or three phases. At sufficiently high values of G/R (thermal gradient divided by growth rate) the alloys solidify with a fully plane front. At somewhat lower values of G/R both the binary and ternary alloys exhibit a cellular composite structure with single-phase cells and intercellular root material which grows with a 'plane front'. For ternary alloys that are three-phase on equilibrium solidification, the intercellular material is three-phase. For binary alloys, and for ternary alloys that are two-phase on equilibrium solidification, the intercellular material is two-phase. At still lower values of G/R, the cells in both binary and ternary alloys lengthen. In addition, in ternary alloys, two-phase cells appear behind the single-phase cells. Some examples of application of the theories are given for representative alloys from the Al-rich corner of the Al–Cu–Ni phase diagram. These give good quantitative agreement with experiment. Good qualitative agreement is also shown for the nickel-base superalloy system designated γ/γ' − δ which is in an advanced stage of development as an in situ *composite turbine blade material.*

The authors are with the Massachusetts Institute of Technology, Cambridge, Massachusetts, USA

This paper reviews theoretical and experimental work conducted at Massachusetts Institute of Technology on the structure of 'cellular composites'.[1–8] Cellular composites are those obtained when the ratio of thermal gradient to growth rate (G/R) lies between that required for dendritic and for fully plane front *in situ* composite type growth. Such structures are virtually unexplored as engineering materials. Recent work, however, has suggested that they may offer substantial mechanical property advantages over columnar dendritic structures.[5,9]

The analyses reviewed here are all based on the simplified model employed in the earlier works. This model includes the assumption of equilibrium interface kinetics, no effect of radius of curvature on melting point, planar isotherms, negligible lateral constitutional supercooling in intercellular region, and no convection. More exact analyses are available in the literature[10,11] but results of these appear not to differ greatly from the results based on the simplified model presented here.

Primary phases shown schematically in this paper are drawn as 'cells' and referred to as 'cells'. It should be recognized, however, that at the higher values of R, lower values of G, and higher values of C_0, the cells tend to form cross branches, i.e. to become dendritic. The analyses presented here are equally valid whether the structure is cellular or dendritic.

The analyses have been developed with the aid of model Al–Cu–Ni alloys and some calculated examples are given. Good qualitative agreement between these analyses and results on a nickel-base superalloy system designated $\gamma/\gamma' - \delta$ has also been obtained and is discussed.[12]

BINARY ALLOYS

Consider first nominally† single-phase alloys (such as alloy 1 of Fig. 1) whose composition, C_0, is less than the maximum solid solubility, $C_{\alpha m}$. The required conditions for plane front growth are given by the well known constitutional supercooling criterion:[13]

$$\left(\frac{G}{R}\right)^* = -\frac{m(1-k)C_0}{kD} \quad (C_0 < C_{\alpha m}) \tag{1}$$

where $(G/R)^*$ is the critical (i.e. minimum) G/R for plane front growth, m is the liquidus slope, k is the equilibrium partition coefficient, and D is the liquid diffusion coefficient of solute. When G/R is less than that given by equation (1), cells (or dendrites) form (Fig. 1b).

The temperature, and hence location, of the cell tips depends on G/R. As G/R decreases, the temperature of the cell tips increases progressively reaching a maximum temperature which is the liquidus temperature of the alloy, T_L, at $G/R = 0$, Fig. 2a. From the work of Bower et al.[1] and Sharp et al.,[4] the tip temperature, T_t, is given by:

$$T_t - T_L = -\frac{DG}{R} \tag{2}$$

† 'Nominally' as used in this paper in connection with structure is that structure that would be obtained on equilibrium solidification

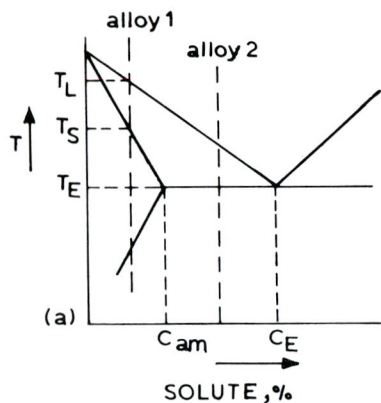

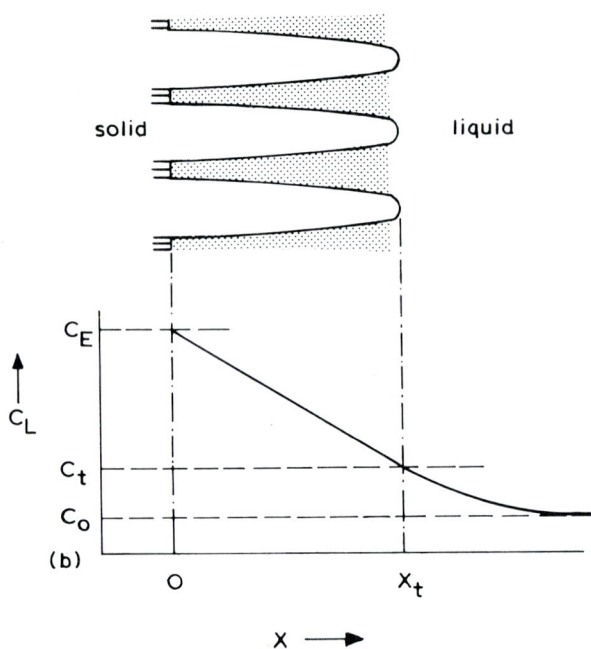

a phase diagram: *b* concentration of solute as a function of distance

1 Schematic illustration of the unidirectional solidification of cells of a binary alloy

Since the cell root is always at the eutectic temperature, T_E, the length the cells protrude in front of the fully solid front, ΔL, is

$$\Delta L = \frac{T_L - T_E}{G} - \frac{D}{R} \qquad (3)$$

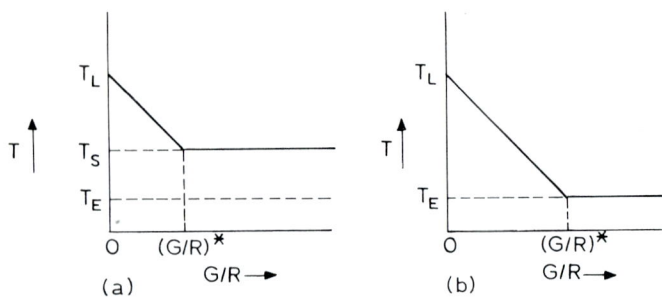

2 Schematic variation of single-phase cell tip temperature (T_t) for *a* nominally single-phase; and *b* nominally two-phase binary eutectic alloys such as alloy 1 and 2 of Fig. 1

The composition of the intercellular liquid is as shown schematically in Fig. 1*b*, dropping from eutectic composition, C_E, at the cell roots to C_t at the tips, and then exponentially to the bulk alloy composition, C_0, in front of the tips. Negligible concentration variation is assumed in the transverse direction. This solute redistribution has been discussed in detail. Quantitative treatment of the problem has been given[4] and can be employed to calculate cell shape (although not spacing) and final microsegregation in the solidified ingot.

In the case of binary alloys that are nominally two-phase (i.e. alloys whose composition lies between $C_{\alpha m}$ and C_E, alloy 2 of Fig. 1) the stability criterion for plane front solidification is:[2]

$$C_E - C_0 = -\frac{D}{m}\left(\frac{G}{R}\right)^* \quad (C_{\alpha m} < C_0 < C_E) \qquad (4)$$

and cell length, ΔL, is given by equation (3).

It is of interest to note here, that for alloys such as alloy 1, which are nominally single-phase, decreasing G/R below $(G/R)^*$ produces a sudden discontinuous change of morphology as cells of the full length $(T_t - T_E)/G$ form, where T_t is given by equation (2). In contrast, nominally two-phase binary alloys break down to form cells of essentially zero length which progressively lengthen as G/R decreases below $(G/R)^*$. This is shown schematically in Fig. 2.

TERNARY ALLOYS, NOMINALLY THREE-PHASE

Plane front stability of three-phase ternary alloys such as groups A and B in Fig. 3 has been described by Rinaldi *et al.*[3] who arrived at the following simple condition for plane front stability:

$$\left(\frac{G}{R}\right)^* = -m_m\frac{(C_{Em} - C_{0m})}{D_m} = m_n\frac{(C_{En} - C_{0n})}{D_n} \qquad (5)$$

where m_m, m_n are the liquidus slopes with respect to elements m and n, C_{0m}, C_{0n} are the bulk alloy concentrations and D_m, D_n are the respective liquid interdiffusion coefficients, and C_{Em} and C_{En} are the interfacial liquid concentrations, which for nominally three-phase ternary alloys are the ternary eutectic concentrations. Results from that work are shown in Figs 4 and 5 showing reasonable agreement with experiment.

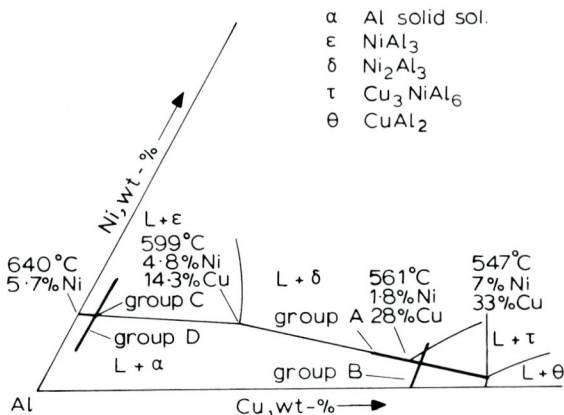

3 Composition ranges of specimens examined by Flemings *et al.*[3–5,6–8]

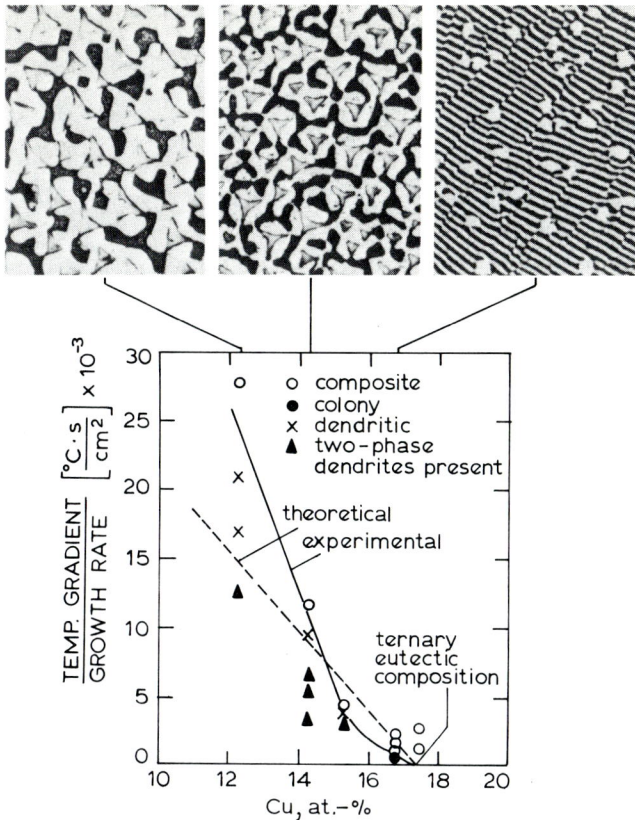

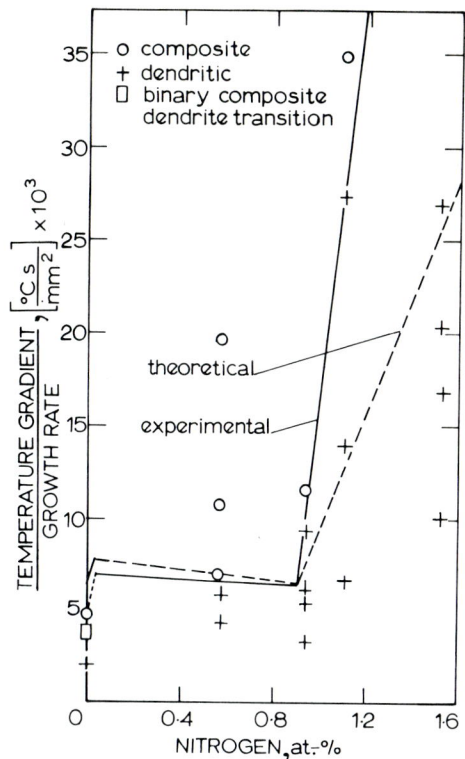

4 Comparison between theoretical predictions and experiment for overall plane front stability of alloys of group A, Fig. 3; micrographs show structures obtained with plane front growth of different alloy compositions (from Rinaldi et al.[3]) [×60]

5 Comparison between theoretical predictions and experiment for overall plane front stability of alloys of group B, Fig. 3 (from Rinaldi et al.[3]; the binary composite–dendrite transition is from Jordan and Hunt[14])

The various possible microstructures obtained by directional growth of nominally three-phase ternary eutectic alloys are summarized schematically in Fig. 6. As can be seen, at all G/R values the structures terminate with three-phase intercellular material.[7] Sharp et al.[4] have analyzed these structures and shown that, analogous to the binary alloys, both single- and two-phase cell tip temperatures, T_t and T_{2t} respectively, increase smoothly from T_E, the ternary eutectic temperature, with decreasing G/R as shown schematically in Fig. 7. A typical microstructure from that work showing both a single-phase and a two-phase dendrite in an alloy from group A, Fig. 3, is shown in Fig. 8. Theoretical results from Sharp et al.[4] compared well with experiment as shown in Fig. 9 taken from that work.

TERNARY ALLOYS, NOMINALLY TWO-PHASE

There are similarities but also significant differences in results when the same theories are applied to ternary alloys that are 'nominally' two-phase such as group D in Fig. 3.

The condition for plane front stability in these alloys is similar to equation (5) except that the liquid composition at the interface is no longer the ternary eutectic composition:

$$\left(\frac{G}{R}\right)^* = -\frac{m_m(C_{Lm}^* - C_{0m})}{D_m} - \frac{m_n(C_{Ln}^* - C_{0n})}{D_n} \quad (6)$$

where C_{Lm}^* and C_{Ln}^* are the liquid composition at the interface. Thus, evaluation of this equation, unlike equation (5), requires that the tie line at the temperature corresponding to the equilibrium solids of C_{0m}, C_{0n}, be known. Typical results of calculations from Dunn et al.[6] are plotted in Fig. 10 which gives $(G/R)^*$ v. weight percent nickel for the series of nominally two-phase Al–Cu–Ni alloys shown in Fig. 3 (group D). Reasonable agreement with experiment is apparent.

The various possible structures obtained by directional growth of nominally two-phase ternary alloys are shown schematically in Fig. 11. At low values of G/R the structure is as shown schematically in Fig. 11a. This is similar to that obtained with nominally three-phase alloys at low G/R and consists in the case of a simple ternary eutectic phase diagram of single-phase cells, two-phase cells, and three-phase eutectic-like root material, each growing at successively lower interface temperature. These structures terminate with three-phase intercellular root material which is in equilibrium with liquid of the ternary eutectic composition and which grows at the ternary eutectic temperature. The solute redistribution in these structures can therefore be treated essentially as Sharp and Flemings have described.[4]

At intermediate values of G/R, however, the structure can be as shown in Fig. 11b. Single-phase cells protrude ahead of two-phase root material which grows with plane front between the single-phase cells. Dunn et al.[6] examined the solute balance at the freezing, barely stable, intercellular interface and they arrived at the

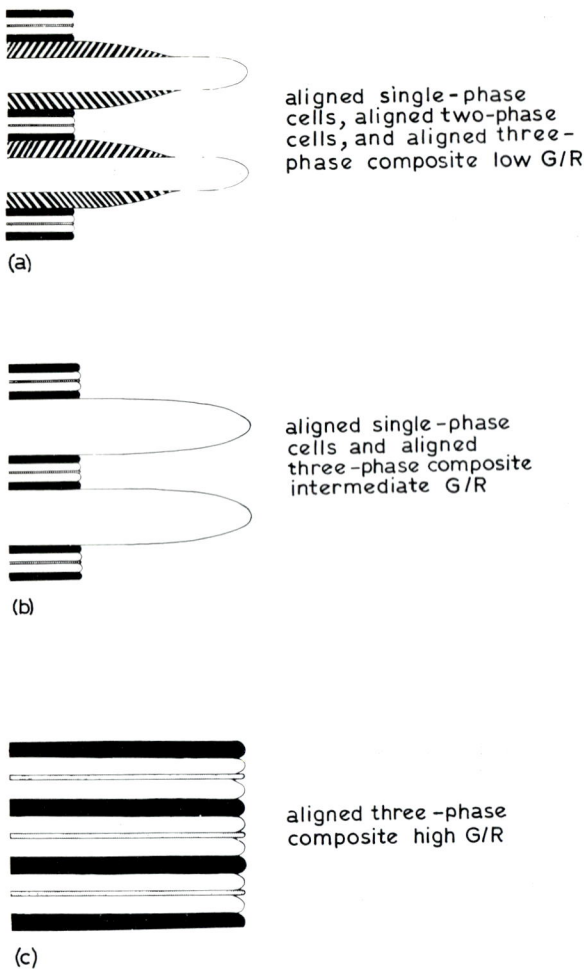

a aligned single-phase cells, aligned two-phase cells, and aligned three-phase composite low G/R

aligned single-phase cells and aligned three-phase composite intermediate G/R

aligned three-phase composite high G/R

a aligned single-phase cells, aligned two-phase cells, and aligned three-phase composite, low G/R; b aligned single-phase cells and aligned three-phase composite, intermediate G/R; c aligned three-phase composite, high G/R

6   Schematic of the various microstructural possibilities obtained by directional growth of nominally three-phase ternary eutectic alloys

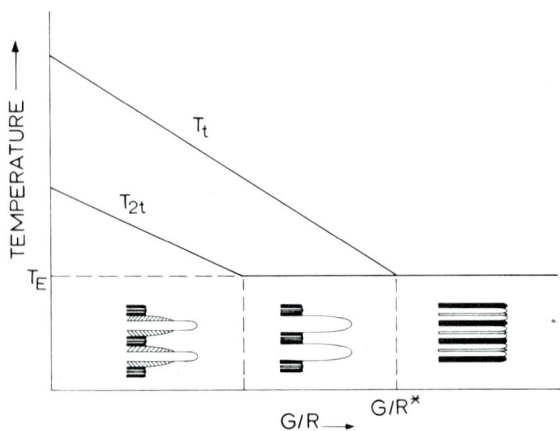

7   Schematic showing structure types and variation of one-phase (Tₜ) and two-phase (T₂ₜ) cell tip temperature v. G/R for nominally three-phase ternary eutectic alloys (from Sharp and Flemings⁴)

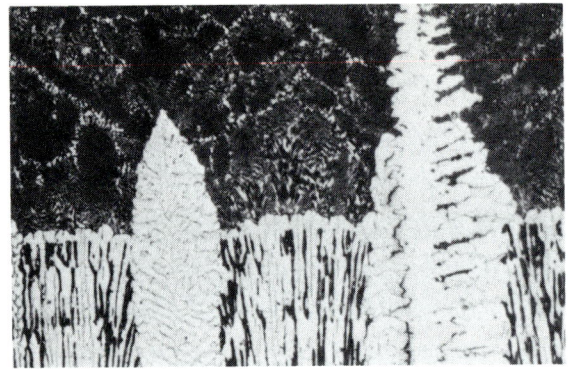

8   Micrograph of the quenched interface of an Al–24·5 Cu–3·2 Ni alloy from group A, Fig. 3, grown at $G/R = 12\cdot7 \times 10^3$ °C s mm^{-2} showing one- and two-phase dendrites (from Rinaldi et al.³)   [× 128]

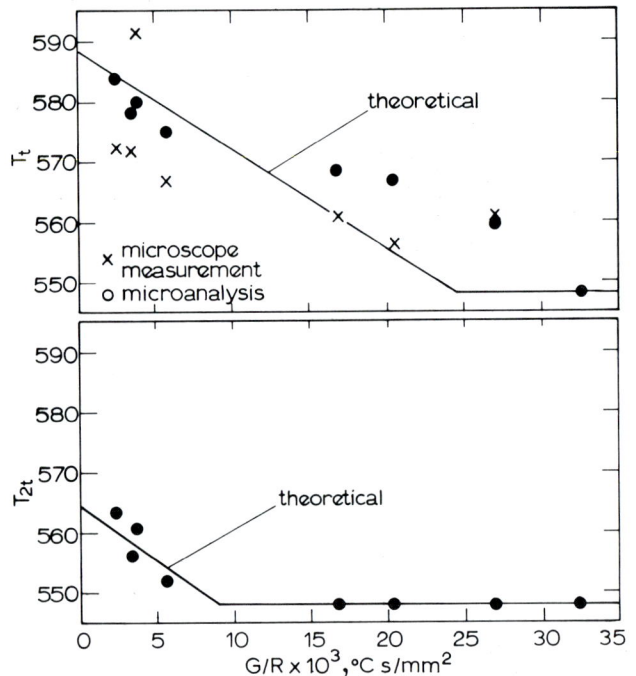

9   Comparison between theoretical predictions and experimental measurements of single-phase cell tip temperatures (T_t) and two-phase cell tip temperatures (T_{2t}) v. G/R for a three-phase ternary Al–Cu–Ni alloy (from Sharp and Flemings⁴)

following condition for the intercellular material to remain plane front:

$$\left(\frac{G}{R}\right)^* = -\,m_m \frac{(C'_{2tm} - \bar{C}'_{sm})}{D_m} - m_n \frac{(C'_{2tn} - \bar{C}'_{sn})}{D_n} \quad (7)$$

where $\bar{C}'_{sm}$, $\bar{C}'_{sn}$ are average concentrations of m and n respectively in the barely stable (denoted by the prime) intercellular solid and C'_{2tm}, C'_{2tn} are concentrations of liquid in contact with this two-phase solid. Now, by comparing equations (6) and (7), it can be seen that depending on the numerical values of C'_{2tm}, C'_{2tn}, $\bar{C}'_{sm}$, $\bar{C}'_{sn}$, the equality (7) can be less than that of equation (6). Under these conditions structures such as Fig. 11b are

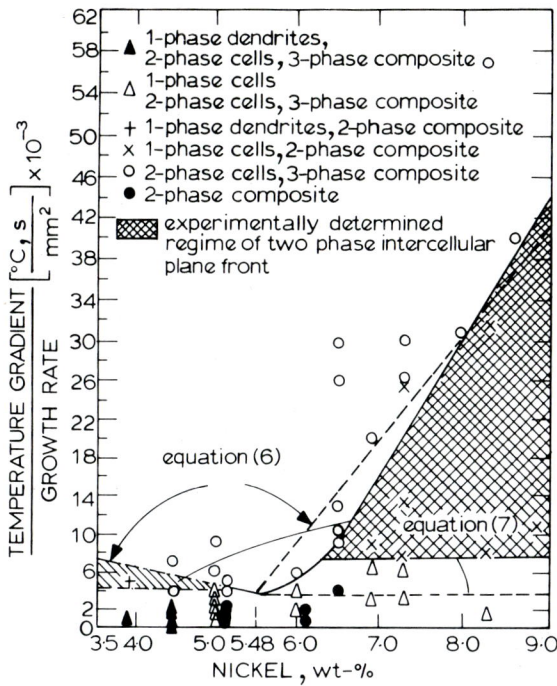

10 Comparison between experimentally determined and theoretical predictions of intercellular and overall stability of the 1 wt-% Cu, Al–Cu–Ni alloys shown in Fig. 3 (group D) (from Dunn *et al.*[6])

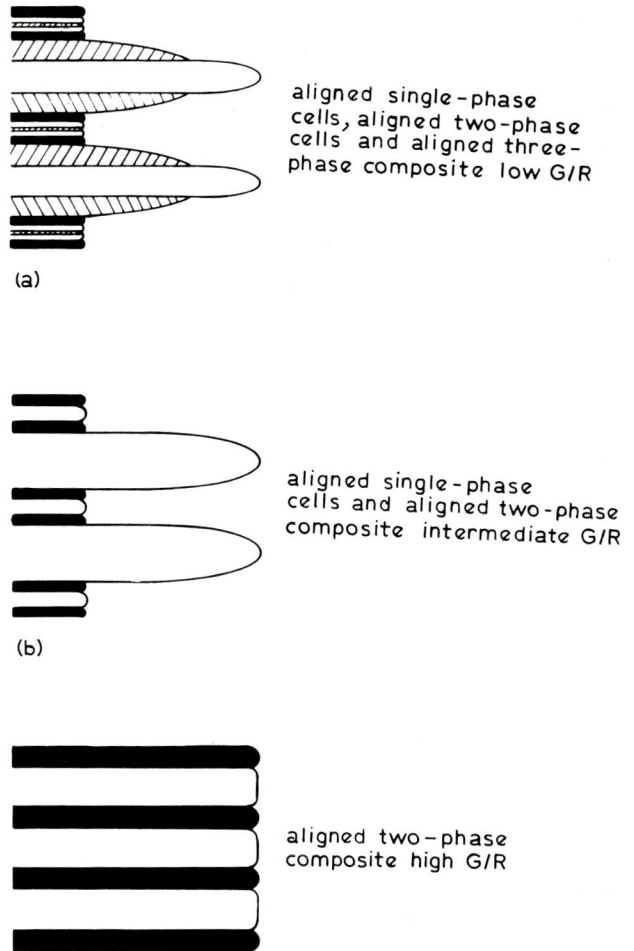

a aligned single-phase cells, aligned two-phase cells, and aligned three-phase composite, low G/R; *b* aligned single-phase cells and aligned two-phase composite, intermediate G/R; *c* aligned two-phase composite, high G/R

11 Schematic of the microstructural possibilities obtained by directional growth of nominally two-phase ternary eutectic alloys

predicted to occur at all G/R values within the range bounded by equations (6) and (7).

However, unlike equation (6), evaluation of equation (7) requires solution of a complex solute redistribution equation to determine interfacial compositions. In the general case this requires numerical solution. Since the intercellular interface is two-phase it is not fixed either in temperature or liquid composition but free to vary along the line of two-fold saturation with G/R.

The results obtained by Dunn *et al.*[6] for alloys of group D are also plotted in Fig. 10. Representative microstructures from that work obtained at moderate and low G/R values are shown in Figs 12 and 13. These transverse sections reveal the absence (Fig. 12) or presence (Fig. 13) of two-phase cell boundaries and it was these observations that are reported in Fig. 10. While the observed regimes for structures such as Fig. 12 were found to be somewhat more constricted than predicted, Fig. 10 again shows reasonable agreement between theory and experiment for the stability of the two-phase intercellular plane front.

Dunn *et al.*[6] have also examined cell tip temperatures for nominally two-phase ternary alloys and results from that work are shown schematically in Fig. 14. Single-phase cell tip temperature T_t increases smoothly from T_s the bulk alloy solidus temperature (at which plane front growth occurs) with decreasing G/R in exactly the same way as in the other alloys considered. However, T_{2t}, the two-phase cell tip temperature, varies with decreasing G/R as $\bar{C}_{sm}$, $\bar{C}_{sn}$, the average composition of the two-phase intercellular solid migrates away from C_0 the bulk alloy composition towards the composition $\bar{C}'_{sm}$, $\bar{C}'_{sn}$ (equation 7). This is analogous to the way in which the composition of the two-phase intercellular material of nominally two-phase binary alloys migrates from C_0, the bulk alloy composition, toward C_E, the eutectic composition, with decreasing G/R.

Thus, in general, T_{2t} first decreases with decreasing G/R to T'_s at which point G/R is that given by equation (7). As G/R decreases further, two-phase cells form and T_{2t} behaves exactly as discussed for nominally three-phase ternary alloys increasing smoothly with further decreases in G/R.

COMPARISON WITH NICKEL-BASED SUPERALLOYS

Similar microstructural variations have also been observed[12,15] in the nominally two-phase nickel-base superalloy system $\gamma/\gamma' - \delta$, which is an advanced stage of development for high-temperature gas turbine blade applications. Thus Figs 15 and 16 show typical microstructures obtained at moderate G/R with both nickel-rich and niobium-rich alloys of that family.[12] Figure 15 shows a faceted $\delta(Ni_3Nb)$ cell protruding ahead of a two-phase plane front intercellular interface. Figure 16,

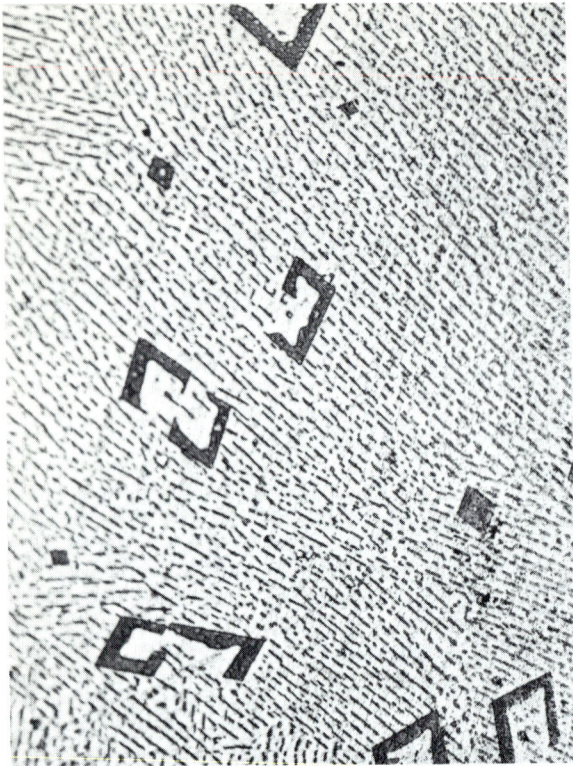

12 Transverse section taken from an Al–1 Cu–8·3 Ni alloy from group D, Fig. 3, grown at a $G/R = 8·4 \times 10^3 °C$ s mm^{-2} such that two-phase intercellular plane front solid was obtained (from Dunn *et al.*[6])   [×256]

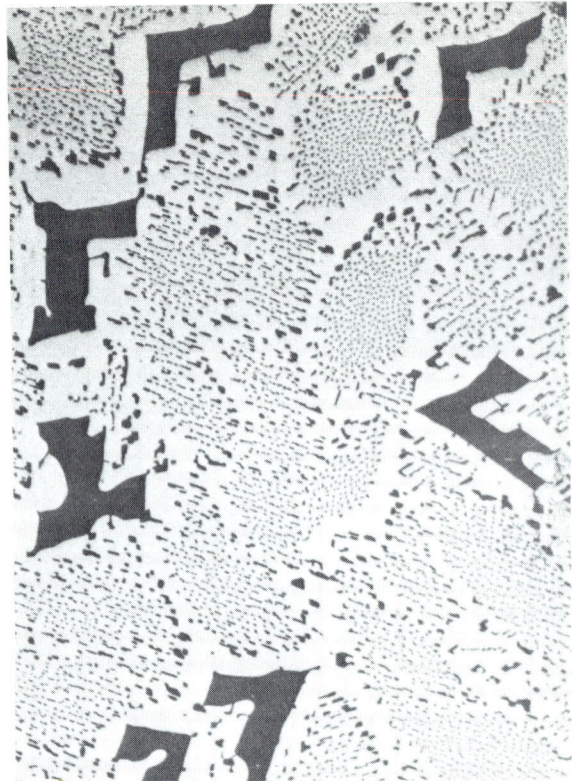

13 Transverse section taken from the same alloy as Fig. 12 but grown at a $G/R = 3 \times 10^3 °C$ s mm^{-2} to produce both single- and two-phase cells (from Dunn *et al.*[6]) [×256]

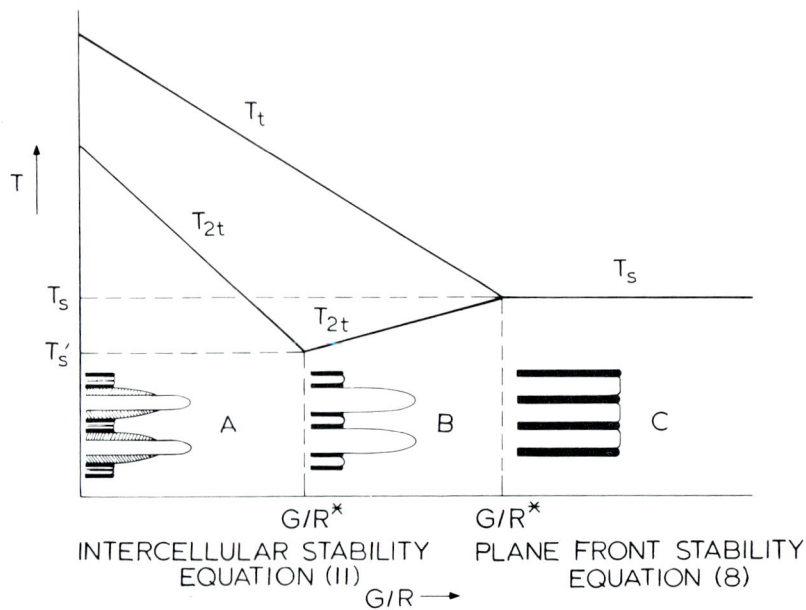

14 Schematic showing structure types and variation of one-phase (T_t) and two-phase (T_{2t}) cell tip temperatures *v.* G/R for nominally two-phase ternary eutectic alloys (from Dunn *et al.*[6])

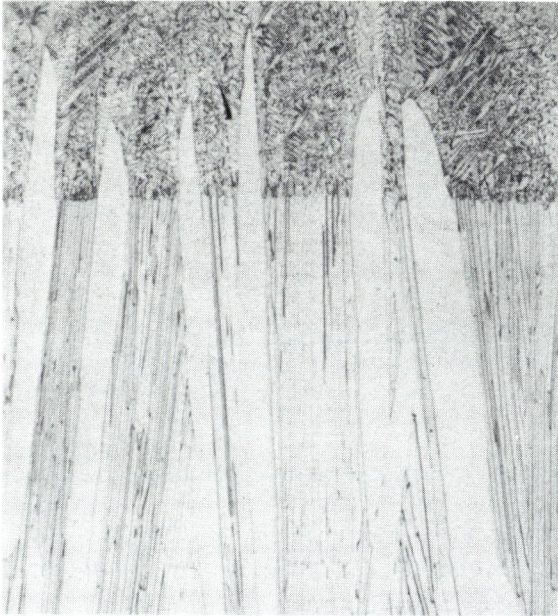

15 Longitudinal section taken from a Ni–23·2 Nb–3·1 Al–2·5 Cr alloy grown at G/R = 30×10^3°C s mm^{-2} (from Neff *et al.*[12])   [× 128]

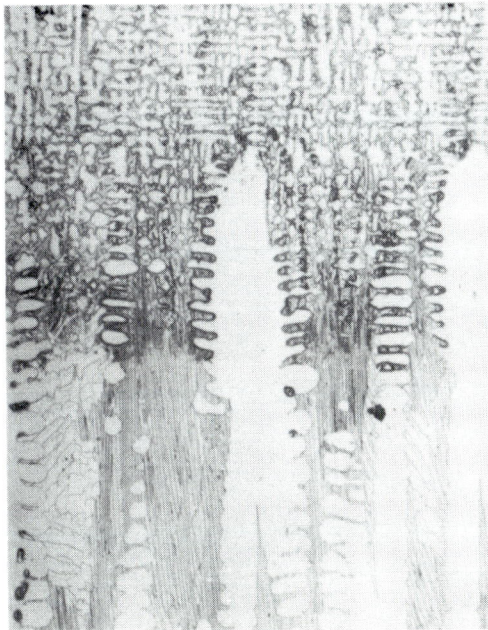

16 Longitudinal section taken from a Ni–19·3 Nb–2·83 Al alloy grown at G/R = $3·2 \times 10^3$°C s mm^3 (from Neff *et al.*[12])   [× 128]

Table 1   Composition of $\gamma/\gamma'–\delta$ alloys examined by Neff *et al.*[12]

	Ni, wt-%	Nb, wt-%	Al, wt-%	Cr, wt-%
Alloy 1	Balance	23·2	3·1	2·46
Alloy 2	Balance	21·89	2·48	7·04
Alloy 3	Balance	19·3	2·83	—

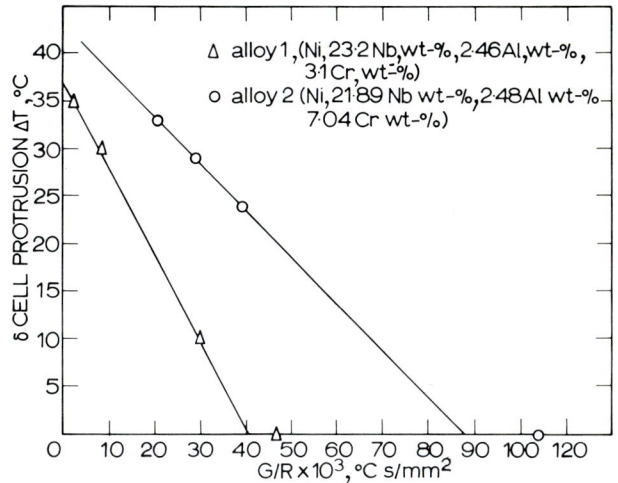

17 δ(Ni$_3$Nb) cell protrusion ahead of two-phase plane front intercellular interface *v. G/R* for two $\gamma/\gamma'–\delta$ alloys (from Neff *et al.*[12])

the cells have zero length. This has advantages in many alloys of engineering importance in which the high alloy melting point makes steep thermal gradients (to obtain high G/Rs at acceptable growth rates) difficult to achieve experimentally.

As an example Neff *et al.*[12] examined the nickel-base $\gamma/\gamma' - \delta$ alloys listed in Table 1. While two of these

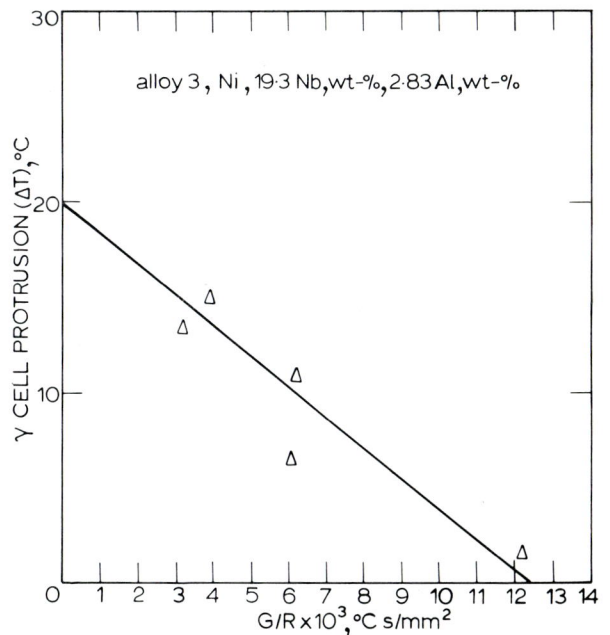

18 γ (Ni solid solution) dendrite protrusion ahead of two-phase plane front intercellular interface *v. G/R* for a Ni–19·3 Nb–2·83 Al alloy (from Neff *et al.*[12])

taken from the ternary alloy 3 of Table 1 shows a γ (Ni solid solution) dendrite again protruding ahead of a two-phase intercellular interface.

The theoretical basis established during the course of this work and the qualitative agreement with the $\gamma/\gamma' - \delta$ alloys also suggests a refined experimental technique by which the G/R required for fully plane front growth of nominally two-phase ternary alloys may be determined from a minimum of experiments. Single-phase and two-phase cell tip temperatures are a function of G/R as we have discussed. Thus, by plotting cell protrusion *v.* G/R we may accurately define the required G/R for fully plane front growth by extrapolation to the value at which alloys contain four components, the structures each exhibit with varying G/R are exactly similar to the model Al–Cu–Ni alloys discussed above.

Figure 17 shows data for alloys 1 and 2, both of which exhibit primary δ cells at moderate G/R as shown in Fig. 15. These data are plotted in terms of δ cell protrusion in degrees Centigrade (calculated with the aid of the experimentally measured thermal gradient) $v.$ G/R. For one alloy, (alloy 2) rich in chromium it can be seen that three points up to a maximum G/R ratio of $40\times 10^3\,^\circ\mathrm{C\,s\,cm}^{-2}$ are sufficient to determine $(G/R)^*$ of $82\times 10^3\,^\circ\mathrm{C\,s\,cm}^{-2}$. Similar data are shown in Fig. 18 for the ternary alloy 3 and this provides particularly good qualitative agreement with the theory presented here.

ACKNOWLEDGMENTS

Portions of this work were supported by the National Science Foundation, the National Aeronautics and Space Administration, the Office of Naval Research, and the U.S.A.F. Air Force Systems Command.

REFERENCES

1 T. F. BOWER *et al.*: *Trans. TMS-AIME*, 1966, **236**, 624

2 F. R. MOLLARD AND M. C. FLEMINGS: *Trans. TMS-AIME*, 1967, **239**, 1 526

3 M. D. RINALDI *et al.*: *Metall. Trans.*, 1972, **3**, 3 133–3 148

4 R. M. SHARP AND M. C. FLEMINGS: *Metall. Trans.*, 1974, **5**, 823

5 E. M. DUNN *et al.*: CISC-II, New York, 1975, 365

6 E. M. DUNN *et al.*: 'Structure of directionally solidified two phase ternary alloys', to be published

7 R. M. SHARP AND M. C. FLEMINGS: *Metall. Trans.*, 1973, **4**, 997

8 E. M. DUNN: 'Morphology and microsegregation of two phase *in-situ* and cellular *in-situ* composites', Sc.D. Thesis, MIT, 1976

9 M. M. FARAG AND M. C. FLEMINGS: Proc. 2nd Int. Conf. on Mechanical Behaviour of Materials, 1 857, 1976, Boston, Mass.

10 M. H. BURDEN AND J. D. HUNT: *J. Cryst. Growth*, 1974, **22**, 109

11 M. H. BURDEN AND J. D. HUNT: *J. Cryst. Growth*, 1974, **22**, 328

12 M. A. NEFF *et al.*: 'The growth and morphology of directionally solidified nickel based $\gamma/\gamma' - \delta$ superalloys', to be published

13 W. A. TILLER *et al.*: *Acta Metall.*, 1953, **1**, 428

14 R. M. JORDAN AND J. D. HUNT: *Metall. Trans.*, 1971, **2**, 3 401

15 E. H. KRAFT AND E. R. THOMPSON: Proc. Conf. on *In Situ* Composites II, 297, 1976, Lexington, Mass., Xerox Corporation

Metallurgical aspects of electromagnetic stirring during the solidification of low-carbon steels

R. Widdowson and H. S. Marr

The application of electromagnetic stirring to the continuous casting process is reviewed. Within-mould stirring has been shown to offer potentially improved surface quality through reduction in surface and subsurface oxide and sulphide inclusion contents. Work has progressed to the pilot plant stage with the aim of applying the technique to the production of aluminium-killed low-carbon strip steels. Stirring below the mould is practised on a limited number of continuous casting plants and improved internal soundness, structure, and increased tolerance to less than optimum casting conditions are claimed.

The authors are with BSC Teesside Laboratories, Middlesbrough

This paper describes the results of casting trials carried out to assess the feasibility of applying electromagnetic stirring forces within a continuous casting mould in order to modify the initial stages of product solidification. The work is based on a patent[1] which claims the creation of electromagnetically induced flow along the solidification front to redistribute solute elements having high liquid solid partition coefficients and second-phase inclusions. Figure 1 shows the preferred stirring pattern generated within the mould, i.e. an upward movement of the liquid at the liquid/solid interface.

Although there is an appreciable history of the principle of electromagnetic stirring being applied to the continuous casting process, such applications have been largely limited to influencing the solidification process at an advanced stage, i.e. at a point within the secondary cooling zone. The fluid flow patterns created have generally rotated around the longitudinal axis of the strand with the object of dendrite fragmentation in order to transform the columnar structures characteristic of short freezing range alloys to equiaxed. A finding which is possibly more important is that the induced flow can also be instrumental in preventing dendrite bridging, thus ensuring effective liquid feeding and hence reducing central unsoundness and axial segregation in the product. Figure 2 illustrates the principle of below-mould billet stirring developed by IRSID originally on 120 mm square billets but now applied to the continuous casting of 240 mm square blooms at Hagondange. Development of electrical hardware and software was

carried out by CEM (a subsidiary of Brown Boveri Co.) who are now marketing made-to-measure inductors with power supplies and control equipment.

It can be seen that the principle of electromagnetic stirring may in theory be applied to influence the total solidification process of the continuously cast product, that is by a combination of in-mould and below-mould stirrers extending to the point of complete solidification. In practice it is very doubtful whether such elaboration is necessary to obtain the product benefits which have been shown to result from stirring either within the mould or for a limited strand length several metres below the mould. It should also be stressed at this point that below-mould stirring is commercially established for billet and bloom products as noted above and is also under development for slab plants (primarily in Japan). In contrast the application of in-mould stirring, which is the subject of this paper, has first to overcome the problem of transmitting sufficient energy through the copper mould and is still in the experimental stage.

Before describing the results of initial solidification studies within an electromagnetically induced field, existing applications and claimed benefits are reviewed briefly in order to put the present work in context.

APPLICATION OF ELECTROMAGNETIC STIRRING IN CONTINUOUS CASTING
In-mould stirring
Although in-mould stirring is less advanced compared with below-mould applications the former technique was used at an early stage (1966) on the Kapfenberg

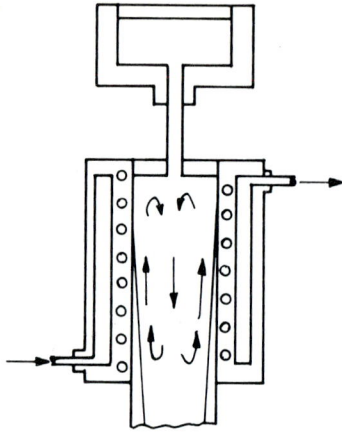

1   Diagrammatic representation of electromagnetic stirring within a continuous casting mould

plant of Bohler Bros. and elsewhere. The stirring pattern used was rotary, i.e. around the product longitudinal axis, and structural refinement together with reduced internal pipe and improved workability were reported[2] for 150 mm diameter low- and high-alloy steel rounds. However, the problem of adequate penetration of stirring energy at mains frequency had been approached by the use of lower electrical conductivity materials such as aluminium bronzes and molybdenum for the mould tubes. Such materials consequently had lower thermal conductivities and were found to be impracticable on grounds of heat transfer. It is worth noting also, in view of the later move away from rotary stirring patterns, that rotary stirring tended to concentrate slag within the centre of the paraboloid top surface of the metal in the

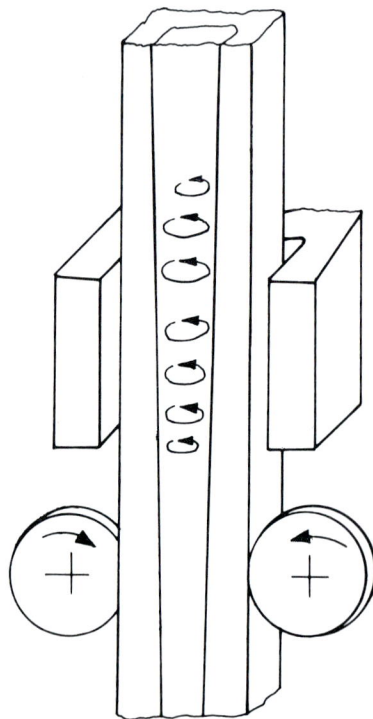

2   Principle of IRSID–CEM technique for electromagnetic stirring within the secondary cooling zone

mould with the possibility of entrainment in the solidified product with a free teeming stream. Similarly, certain non-optimum conditions could result in deterioration of product surface quality by amplification of the normal transverse reciprocation marks. More recently IRSID have commenced work on a technique for billets based on vertical stirring at low frequencies (5–10 Hz) in thin-walled moulds. Improved surface and subsurface quality based on step tests was reported, although this benefit was reduced when a submerged pouring practice was used.

In summary, provided the problems of adequate electrical penetration of the casting mould can be overcome and that conditions can be maintained where slag and inclusions accumulating at the meniscus are not stirred into the product, electromagnetic stirring in the mould offers the potentiality of improved surface and subsurface quality and structure.

Below-mould stirring

Below-mould stirring has been established for three years on the SAFE caster at Hagondange.[3] All four strands are equipped on a plant which originally produced 120 mm and 200 mm strands and now casts 240 mm blooms. The 50 Hz stirring coils are situated 4 metres below meniscus level corresponding to about 50% solidification and produce a rotary stirring pattern. Use of the technique is claimed to result in an increased fraction of equiaxed structure with reduced axial porosity and central segregation. The practical view appears to be that similar results could be achieved when the casting procedure was optimum, but the technique compensated for less than optimum conditions such as excessive superheat.

Similar work has been reported by Republic Steel Corporation who have applied the technique on a trial basis to 254 mm × 222 mm low-alloy blooms.[4] It was shown that high-intensity stirring could result in pronounced depletion of solute elements in the fraction solidifying within the length of the stirring coils with an associated positively segregated band. This segregation pattern was absent at lower stirring intensities while still effective in eliminating columnar growth. Standard deviations of carbon, sulphur, and phosphorus levels at the axis were about half those of unstirred casts. Similar benefits have been reported for 115 mm square high-carbon wire billets produced by Sumitomo Denki at Itami.[5,6] Reduction in billet centreline segregation was confirmed by reduced frequency of cup fractures during wire drawing, a defect normally resulting from excessive axial segregation.

Of greater interest to tonnage steel producers is the application in Japan of electromagnetic stirring to slab casting. Nippon Steel Corporation have reported[7] below-mould rotary stirring of 210 × 1 280–2 050 mm wide carbon–manganese slab for plate with consequent reduction or elimination of centreline segregation. Nippon Kokan have also reported results[8] of vertical stirring of 250 mm × 1 900 mm slab with similar benefits. It should be noted that termination of growth of the columnar crystals at the initiation of stirring was accompanied by negative segregation bands as wide as 20 mm, and the two plants have opposing views on their effect on product quality.

In summary, electromagnetic stirring may be used to modify solidification patterns and influence segregation of both solute elements and inclusions, but the system cannot affect the layers which solidify before reaching the zone of influence of the stirring coils. Hence only in-mould stirring can influence the surface and subsurface region: below-mould stirring is confined in influence to the central regions of the product. Having illustrated existing and possible applications of electromagnetic stirring in continuous casting, the results of feasibility studies aimed at applying in-mould stirring to a particular continuously cast product are presented in the following.

ELECTROMAGNETIC STIRRING OF LOW-CARBON STEELS

Low-carbon sheet and strip for applications such as vehicle body panels and deep-drawn components have stringent surface quality requirements. Rimming steel and aluminium-killed qualities have traditionally been supplied by the conventional ingot route for such products. The induced stirring resulting from the gas evolution which accompanies solidification of the rimming steel ingot resulted in scavenging both a proportion of solute elements and inclusions from the element solidifying under these conditions—the 'rim zone'. The rim zone was associated with high drawability and good surface finish in the strip or sheet product.

The increased adoption of continuous casting has stimulated the development of casting techniques for low-carbon slab after early work had shown the impracticability of casting the traditional rimmed steel because of more stringent requirements for oxidation control and frequent gas-induced surging in the mould. The trend has been towards the adoption of the aluminium-killed steel thus avoiding the problems of control of gas evolution but introducing the problems of characteristically poor surface quality of the concast slab through alumina defects. The approach of BSC Corporate Laboratories has been to create the stirring pattern characteristic of the rimming steel ingot by applying electromagnetic stirring to the continuous casting mould.

Figure 3 shows the apparatus used in the pilot plant to cast 300 mm square ingots. While work was proceeding to overcome the problem of providing sufficient energy to penetrate the copper continuous casting mould, the

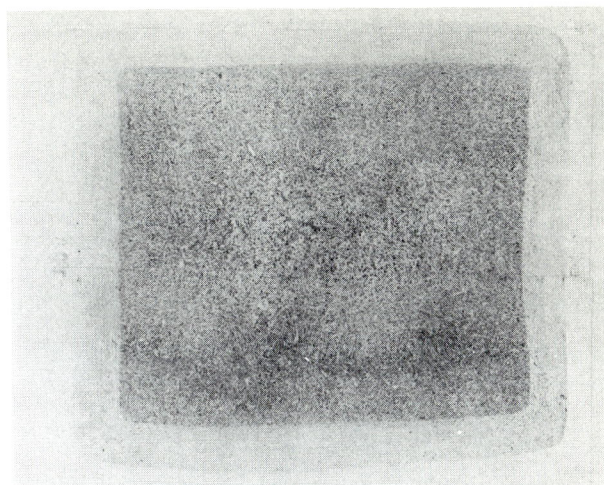

4 Transverse sulphur print of electromagnetically stirred ingot

work described was restricted to the use of a water-cooled 6·4 mm thick non-magnetic stainless steel mould. This was surrounded by a mains frequency stirring coil which could traverse the mould vertically providing a 3-phase travelling field and inducing a liquid metal velocity of about 400 mm/s. The continuous casting situation was simulated by casting from a ladle and traversing the coil to follow the rising meniscus at a rate of 0·6 m/min.

0·2% CARBON SILICON-KILLED STEELS

Two preliminary casting trials were carried out using this quality to emphasize the measurable effects on solute segregation. In the second trial the electromagnetic stirring forces were reversed to produce a stirring pattern direction downwards over the solidifying front and up the centre. Sulphur prints and etch tests showed that the originally defined upward stirring direction produced the more extensive rim zone, about 25 mm thick, under these conditions (Fig. 4).

Table 1 illustrates the negative rim zone segregation in both ingots. Table 2 details Quantimet inclusion counts carried out on ingot samples illustrating the effectiveness of the scavenging action set up in reducing particularly oxide concentrations in the rim zone.

3 Pilot plant casting rig

Table 1 Extent of negative segregation in silicon-killed steel

Ingot	Height, %	Distance from ingot surface, mm	C	S	P	Si	Mn
1	95	12	0·21	0·028	0·063	0·15	0·94
		50	0·25	0·039	>0·070	0·16	1·01
	50	12	0·22	0·032	>0·070	0·16	0·96
		50	0·24	0·043	>0·070	0·16	1·01
2	95	12	0·25	0·033	0·036	0·22	1·06
		50	0·29	0·041	0·041	0·22	1·10
	50	12	0·23	0·031	0·035	0·21	1·05
		50	0·30	0·040	0·044	0·21	1·12

Table 2   Inclusion contents of electromagnetically stirred silicon-killed steels

Ingot	Height, %	Distance from ingot surface, mm	Area of oxides, %	Area of sulphides, %	No. of oxides/100 mm² >3 μm	No. of oxides/100 mm² >10 μm	No. of sulphides/100 mm² >3 μm	No. of sulphides/100 mm² >10 μm
1	95	10	0·010	0·133	94	12	1 776	41
		20	0·033	0·163	490	35	3 387	123
		70	0·078	0·248	1 210	153	3 015	573
		80	0·053	0·198	1 103	24	2 631	472
	50	10	0·012	0·122	118	0	1 876	53
		20	0·019	0·160	177	18	2 720	106
		70	0·048	0·173	478	112	3 133	413
		80	0·036	0·180	519	53	3 027	425
2	95	10	0·012	0·090	30	0	1 280	53
		20	0·016	0·151	201	30	3 446	147
		70	0·040	0·285	625	112	4 738	690
		80	0·032	0·184	442	65	3 457	413
	50	10	0·012	0·075	177	12	909	88
		20	0·009	0·099	130	6	1 652	53
		70	0·023	0·133	283	65	1 947	319
		80	0·105	0·297	1 032	301	4 077	549

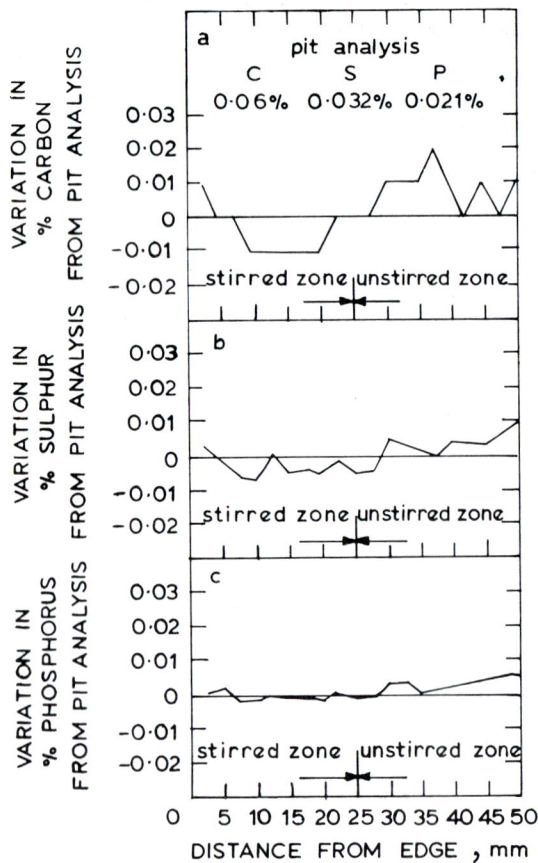

5   Variation in carbon, sulphur, and phosphorus for aluminium-killed low-carbon steel

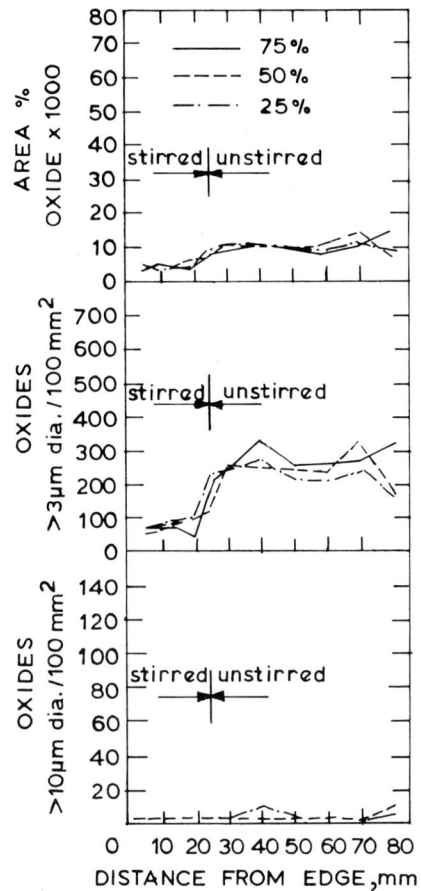

6   Oxide inclusion distribution for aluminium-killed low-carbon steel

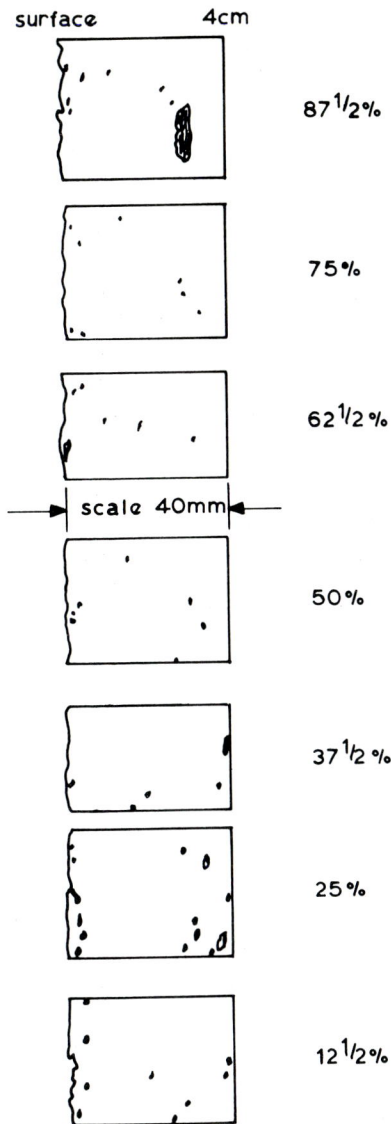

7 Alumina cluster severity distribution in subsurface layers of aluminium-killed low-carbon steel

Aluminium-killed low-carbon steels

The steel specification used for these casting trials was:

	max., %	aim, %
C	0·06	0·045
S	0·02	0·018
P	0·02	0·012
Mn	0·32	0·28
Al	0·06	0·03–0·06
Cu	0·06	—
Ni	0·06	—
Cr	0·06	—
Sn	0·015	—
N_2	0·007	0·006

Teeming rate and rate of coil traverse were again fixed at 0·6 m/min. Sulphur printing of ingot sections revealed a 25 mm negatively segregated rim corresponding to the element solidifying under electromagnetic stirring conditions. Figure 5 illustrates carbon, sulphur, and phosphorus segregation indicating maximum negative segregation in the rim zone of 17% for carbon, 15% for sulphur, and negligible phosphorus segregation.

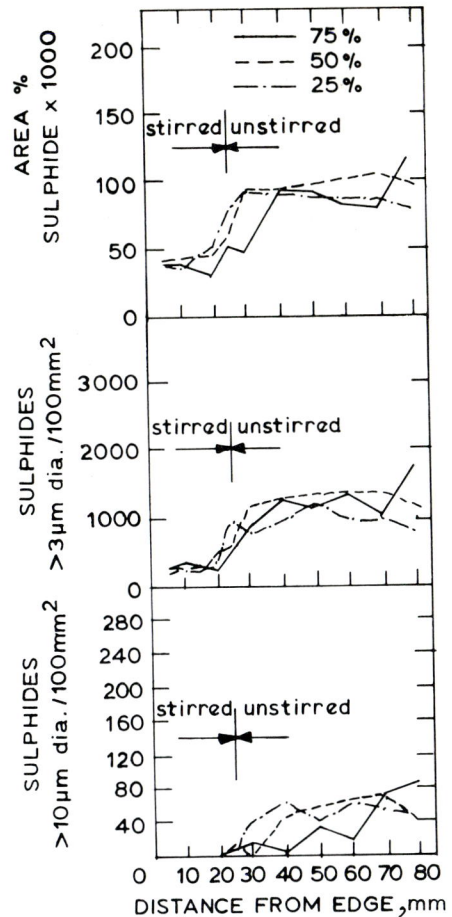

8 Sulphide inclusion distribution for aluminium-killed low-carbon steel

In relation to the final product, inclusion content and distribution were considered to be of greater significance and this is illustrated by Fig. 6. A pronounced reduction in alumina inclusions $> 3 \mu m$ is shown. In practical terms individual alumina inclusions are of little consequence, but become significant where they form agglomerates. In order to record agglomerates, inclusion counts were carried out for all groups of 10 particles or more visible at ×50. The results are shown in Fig. 7 which shows relatively low cluster contents within the stirred zone, although there is evidence of some entrapment at the end of stirring.

9 Transverse sulphur print of ingot stirred after initial period of normal solidification

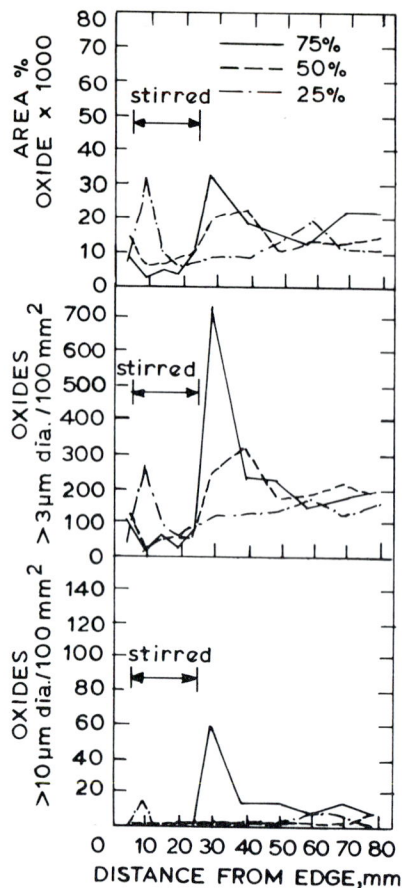

10 Oxide inclusion distribution in ingot stirred after initial period of normal solidification

Figure 8 illustrates the effect on sulphide inclusion distributions, again showing a marked reduction in both volume fraction and sulphide inclusions >3 μm in the stirred rim. This observation may offer an alternative means of reducing surface defectives where these have been correlated with high overall sulphur contents.

The above observations were positively confirmed by examination of a 0·1% C Al–Si killed ingot in which 5 mm of skin was allowed to solidify normally before electromagnetic stirring was initiated. The sulphur print in Fig. 9 shows an internal rim corresponding to the stirring period. The efficiency of electromagnetic stirring is further shown by oxide inclusion distributions (Fig. 10) which reflect the delayed stirring effect.

CONCLUSIONS

1 The results reported indicate the metallurgical feasibility of applying electromagnetic stirring within the continuous casting mould as a means of improving surface and subsurface cleanness in aluminium-killed strip steels.

2 Such negative segregation of non-metallics may be accompanied by heavy positive segregates corresponding to the termination of the stirred zone. Other workers have also pointed to the possibility of re-entrainment of non-metallics, and both aspects require further development involving the use of mould flux additions.

3 Similar negative segregation of sulphur may relax the necessity to desulphurize to low levels in order to reduce surface defectives.

4 Complementary benefits may be obtained by electromagnetically stirring during the later stages of solidification in terms of reduced central porosity and segregation and improved internal structure. Again conditions must be adequately controlled to avoid the creation of highly negatively segregated bands.

REFERENCES

1 B.P. 1 326 728
2 W. POPPMEIER *et al.*: *J. Met.*, 1966, **18**, Oct., 1 109
3 R. ALBERNY *et al.*: Proc. 31st AIME Elec. Furn. Conf., 1973, 245
4 D. J. HURTAK AND A. A. TZAVARAS: Spring TMS AIME Conf., Feb. 1976, Las Vegas, Nevada
5 H. IWATA *et al.*: *Trans. Iron Steel Inst. Jpn*, 1976, **16**, (7), 374
6 H. IWATA *et al.*: *Tetsu-to-Hagané*, 1973, **59**, (11), S373
7 M. NISHIWAKI *et al.*: *Tetsu-to-Hagané*, 1973, **59**, (11), S374

DISCUSSION

Session 7: New processes and products

Chairman: J. Nutting (University of Leeds)

The following papers were discussed: *Keynote Address*: *New solidification processes and products* by M. C. Flemings; *Sheet casting process for aluminium and aluminium alloys* by A. Handasyde Dick, M. C. Simms, D. D. Double, and A. Hellawell; *Formation of filamentary metallic glasses by ultra rapid quenching from the melt* by B. G. Lewis, I. W. Donald and H. A. Davies; *Rapid casting of metallic wires* by J. V. Wood; *Laser glazing: a new process for production and control of rapidly chilled metallurgical microstructures* by B. H. Kear, E. M. Breinan, and L. E. Greenwald; *Structures and properties of Thixocast steels* by K. P. Young, R. G. Riek, and M. C. Flemings; *Stir-cast microstructure and slow crack growth* by A. Vogel, R. D. Doherty, and B. Cantor; *Solidification and properties of* $\gamma/\gamma'-\alpha Mo$ *ductile/ductile eutectic superalloy* by D. D. Pearson and F. D. Lemkey; *Development of composite structures in cast steels by liquid metal treatment* by P. A. Blackmore, A. Segal, A. J. Baker and P. R. Beeley; *Structure of cellular composites* by K. P. Young, B. A. Rickinson, and M. C. Flemings; *Metallurgical aspects of electromagnetic stirring during the solidification of low-carbon steels* by R. Widdowson and H. S. Marr.

PROFESSOR R. W. SMITH (Queens University, Canada): I have a question for Dr Wood. You commented that your rapidly cooled material had a useful enhanced vacancy concentration. How long is this retained?

DR J. V. WOOD (University of Cambridge): The exact vacancy concentration depends on the quench rate and the number of immediate regions where vacancies can diffuse to, e.g. grain boundaries, low angle boundaries, and the particular alloy system. If solute atoms are present then there is definite evidence of solute-vacancy interaction.

DR D. H. KIRKWOOD (University of Sheffield): I have a question for Dr Vogel. How did the copper enriched film around the soft particles arise? During the slurry formation period, or in the final quench?

DR A. VOGEL (Fulmer Research): The copper-rich layer is formed as a thin layer of θ, $CuAl_2$ phase, as the first part of the eutectic reaction. It is perhaps significant that the $CuAl_2$ phase, though a brittle intermetallic, is non-faceted and so can easily grow around the primary aluminium-rich phase.

PROFESSOR V. DeL DAVIES (Norwegian Institute of Technology): I have a very short question for Professor Flemings. As is well known, a ductile alloy does not

crack as easily as a stronger one in thermal fatigue. The copper alloys Professor Flemings has used are probably rather ductile. He showed that they did not crack.

Did the moulds deform so that deformation may be the limit of their life?

PROFESSOR FLEMINGS: I think you are quite right. I think that the limit on copper will be the deformation. We began to get some dishing on the mould cavity but within the 500 castings that we made there was no dimensional change of the cavity itself. I think that would come eventually.

DR R. WIDDOWSON (BSC, Teesside Labs.): I would like to ask Professor Flemings what the life of the remelted Rheocast material is? Does it have to be cast fairly quickly? Do the dendrites rapidly coarsen during holding, preparatory to die casting, or is the slurry stable over a long period of time?

PROFESSOR FLEMINGS: The spheres are usually somewhat larger than comparable dendrite arm spacings, and of course they will coarsen eventually but in some metals, for example, we can hold them as long as half an hour without a major change in the size of the particles.

DR L. R. MORRIS (Aluminium Company of Canada Ltd): As regards the papers by Handasyde Dick *et al.*,

Young *et al.*, and Vogel and Cantor, I would like to point out the connection between the historic strip-casting technique for Pb–Sn organ-pipe sheet and the more recent work on Rheocasting, Thixocasting, and Stir casting.

For casting the Pb–Sn sheet the metal (before it was poured into the 'well aged mahogany' casting box) was stirred in an iron pot until it was 'porridgy', i.e. the Pb dendrites had formed. The reasons for this Rheocasting or stir-casting of the organ-pipe sheet are the same as for the more recent examples: *a* structural uniformity, *b* control of shape; and *c* lower working temperatures. Also the structure (i.e. rounded primary Pb-phases) in the organ-pipe sheet is quite similar to that shown by Professor Flemings and Dr Vogel.

DR P. R. SAHM (Brown Boveri and Co., West Germany): What about Thixocast superalloys?

PROFESSOR FLEMINGS: We have done quite a bit of that work in connection also with United Aircraft. The superalloys Thixocast and Rheocast marvellously. They melt at relatively low temperature and have long freezing ranges. They also die-cast very easily. The high-temperature properties apparently look very good. Dr Kear tells me he is worried about the size of the carbides, but I think we will have to see about that as time goes on.

DR M. McLEAN (NPL): I have a comment and a question on the paper by Dr Lemkey. The equation describing diffusion-controlled fibre coarsening is of the form

$$r^3 = r_0^3 + \frac{4}{9}\frac{CD\lambda\Omega^2}{kT}t$$

where r, r_0 are final and initial fibre radius, C and D are the concentration and diffusivity of the fibrous phase in the matrix, γ is the fibre matrix interfacial energy, Ω is atomic volume, kT is the thermal energy, and t is time.

If γ were reduced by the addition of chromium, as suggested, then the microstructure would become *more* and not *less* stable. The reduction in stability is probably due to an increased solubility of the fibre in the matrix when chromium is present.

It might be expected that the significant solubility of the fibrous phase, and the temperature variation of this solubility, would lead to further instability problems in complex thermal environments such as cyclic temperatures and spatial temperature gradients. For example, Dunbery and Wallace have shown COTAC3 to be subject to large morphological changes to the fibres because of continual dissolution and non-fibrous reprecipitation as the temperature is raised and lowered. Has γ–γ'–α been examined after temperature cycling tests?

The reduced rupture lines of material solidified at 10 cm/h might also suggest problems associated with stability of the reinforcement phase. In the NPL work, reported earlier, reduced rupture lines at high growth rates, providing plane front solidification has been maintained, were only observed when phase instabilities occurred.

DR LEMKEY: Dr McLean's points are well taken. Because of the extensive solubility of Cr in γ we measured the γ' solvus temperature as a function of Cr content. One observes γ' precipitation around the molybdenum fibres where the solubility of the molybdenum phase is very small in γ' but very extensive in γ.

So, as we bring up the temperature we have more γ close in contact with the molybdenum phase where the solubility is much greater in γ than it is in γ'. Therefore, we expected the γ/γ'-α alloys with the larger fraction of γ' to be much more stable thermally than those which have more γ in contact with the molybdenum fibres or which have the lower γ' solvus temperatures. That was the main point of measuring the solvus temperature as a function of aluminium and chromium content. It was a mistake to say Cr lowered the interfacial surface energy term. Obviously, it might raise it. I think that the surface energy term is secondary to the solubility term. Also, chromium may raise the diffusion coefficient of Mo in nickel and we have looked at slower diffusing elemental modifications like rhenium and tungsten to 3 wt-% and they do not exhibit accelerated coarsening. We had to add chromium to these alloys to improve their oxidation resistance. We would not have chosen Cr to enhance microstructural stability from the knowledge that it is quite soluble in the γ-phase and that it significantly reduced the γ' solvus temperature.

The last point was in respect to thermal cyclic stability. We have just begun a programme to evaluate transverse and longitudinal cyclic stability. The first results on longitudinal specimens under 10 ksi axial stress after 3 000 cycles between 750° and 2 050°F indicate no fibre breakage and very little axial extension. The γ/γ'-α alloy is a very thermally stable material but its response can vary with aluminium content, and I believe when I start the chromium-containing alloys their behaviour will vary also with Cr content. All of these elemental modifications will be sorted out empirically for thermal fatigue response; however, we would like to have a predictive model for anticipating changes. As I said, we find that chromium is necessary for enhancing oxidation resistance and have been fortunate that nature has not been too perverse since such alloys do not significantly drop in cyclic thermal stability nor creep resistance at the 3 wt-% level.

PROFESSOR F. WEINBERG (University of British Columbia): I have a question for Dr Widdowson. Assuming you are able to use electromagnetic stirring in the mould in the manner described, how would the billet quality compare with the standard practice of immersed bifurcated nozzles?

DR WIDDOWSON: We believe that electromagnetically induced flow within the mould is a more controllable and, therefore, reproducible technique than hydrodynamically induced flow using the shaped pouring tubes to which he refers. This is particularly so on very wide slab moulds where it would be difficult using the latter technique to create uniform fluid flow at the solidification front. Additionally, the immersed nozzle approach suffers from inconsistencies due to partial nozzle blocking and deliberately reduced flow during ladle changeover in sequence casting.

THE CHAIRMAN: Might not rotary stirring be better?

DR WIDDOWSON: This is the stirring pattern developed for the commercially applied below-mould stirring process. Our original conception was to develop the stirring pattern known to exist in traditional rimmed steel ingots, known to be effective in reducing inclusion content of the (gas) stirred zone. We believe this is the more efficient pattern for reducing inclusion content of the solidifying 'rim' and think that rotary stirring about the longitudinal axis may give less efficient bulk transfer of our metallics from the solidification interface to the bulk melt.

PROFESSOR D. W. KURZ (Ecole Polytechnique Federale de Lausanne, Switzerland): Dr Rickinson, you said that you can substantially increase the growth rate compared to plane-front growth. As far as metals are concerned, the range of growth conditions within which one can obtain cellular growth is very narrow and is close to that for plane-front growth. How do you achieve the high rate in this case?

DR RICKINSON: Certainly the zone of stability for cellular composite structures varies markedly with alloy composition. However, the opportunity to produce a cellular morphology at high production rates can be enlarged by a technique developed at MIT, based upon isothermal coarsening of a very rapidly solidified fully dendritic structure. Such a treatment transforms the dendrites to rod-like cells and, therefore, generates a pseudocellular composite structure in a minimum of time.

DR L. R. MORRIS (Aluminium Company of Canada Ltd): Concerning the paper by Young and Rickinson, I would question whether a eutectic with a cellular or colony structure can be useful for engineering applications. My experience, which I believe is in agreement with others, is that the yield strength of these cellular structures tends to be low because it tends to be that of the cell boundary phase. The strain is also concentrated in localized regions at the cell boundaries and therefore the fatigue and fracture characteristics are poor. I feel that for a cellular material this is inherently so.

DR RICKINSON: Stress rupture and creep properties of cellular composite structures are not always substantially inferior to a fully composite morphology. A case in point, to illustrate such a situation, is in the work of Sheffler of Pratt and Whitney Laboratories who optimized mechanical properties of some $\gamma/\gamma'-\delta$ type alloys with 5% δ cells.

DR B. CANTOR (University of Sussex): In Fig. 5 of Dr Lemkey's paper, he shows a vertical section through the $\gamma + \alpha$ eutectic trough in Ni–Mo–Al with a $\gamma + \alpha$ region extending in 0 wt-% Al. If my memory serves me correctly, there is no $\gamma + \alpha$ region in binary Ni–Mo alloys. Could Dr Lemkey say if this is correct, and if so, that there should be something else on the left hand side of this figure?

DR LEMKEY: You are correct; there should be something else there. There is a δ-reaction, of course, which is pinched off at relatively small amounts of aluminium which we did not study.

DR M. NEFF (MIT): Dr Lemkey, since the $\gamma/\gamma'-\alpha$ alloy is not resistant to Na_2SO_4 corrosion, why is it necessary to make a deleterious chromium addition, when both oxidation and corrosion could be prevented by a protective coating?

DR LEMKEY: That is a very practical engineering question. When we first performed this study, we thought we could put enough chromium in to improve both the hot sulphidation resistance as well as the oxida-

tion resistance, but we were not successful in keeping the creep rupture properties acceptable to gas turbine requirements. Therefore, our alloy activity is directed toward reducing the molybdenum content to below 20% by weight and looking at the resulting resistance to sulphidation.

The other more engineering path is a coating programme. This alloy presents a particular challenge to those people who apply gas-turbine coatings in that its coefficient of expansion is relatively low compared with current alloys, and a coating programme is indeed under way in the US.

DR CANTOR: I have a simple question of clarification for Dr Kear. When he passes a laser beam across the surface, is he able to control the melt depth and power density independently and if so, how is this done?

DR KEAR: The simplest procedure is to fix the power density and vary the interaction time. In the apparatus I showed, that is done merely by varying the tangential velocity and the interaction time determines the depth of melt that is achieved. So, if one is interested in obtaining a range of cooling rates, one simply makes a series of passes for a fixed power density and varies the interaction time. This gives varying melt depths and the computations give the cooling rates corresponding to those melt depths.

PROFESSOR F. WEINBERG (University of British Columbia): Dr Vogel, what evidence do you have for the generation of large-angle grain boundaries resulting from the bending of dendrite arms?

DR R. D. DOHERTY (University of Sussex): We have no experimental information on the dislocation behaviour during the dendrite arm bending process that we discussed. The argument for the existence of grain boundaries comes from the geometric necessity to take up in some way the misorientation introduced, in the originally single crystal dendrite, by bending. If the misorientation is $\theta°$ then there has to be, after recovery, either one boundary with a misorientation of $\theta°$, or a series of boundaries where the net total misorientation is $\theta°$.

DR P. R. BEELEY (University of Leeds): Would Dr Davies please give a brief outline of the current position with respect to the possible industrial application of these materials, especially in heavy products where their high strength could be fully exploited?

We already know of the Swedish development in which rapidly cooled droplets of tool steel are consolidated into large ingots by partial remelting and pressing, producing zero porosity yet retaining the very fine dendrite structure of the original droplets, with fine carbide distribution.

Has any analogous work yet been done to consolidate the various tapes, wires or droplets to form material of sufficiently heavy section to make engineering components?

DR DAVIES: I will attempt to summarize the current position briefly. The work on tool steels that Dr Beeley referred to is, of course, well known. Tube Investments Research Laboratories had a research programme several years ago where rapidly solidified powder, of aluminium–iron alloys, was produced and subsequently consolidated and processed. Substantially greater supersaturation of the iron was achieved than would be possible by solid-state quenching, which, on aging, gave higher strengths than conventional alloys together with enhanced thermal stability. This was not adopted commercially at the time but there is a renewed interest in this type of process route in several quarters. The possible additional advantages to be gained from rapid solidification, such as reduced grain size leading to superplastic behaviour, in the production of superalloys by the new powder route are also, I believe, being investigated.

With regard to fibres, there is now an interest in the direct production, i.e. high-speed continuous casting, of wire by the melt extraction process which in common with all the direct methods consumes much less energy (about 80% less, in fact) than the conventional routes for producing wire, via ingot solidification and several stages of reduction. The rapid solidification leads to highly refined grain, dendrite and precipitate structures and in many alloys to structures and properties that are not attainable by conventional routes.

Similarly, there is much interest in the possibility of producing sheet and foil directly from the melt, again having enhanced mechanical and physical properties. One of the present constraints of some techniques such as melt spinning is the limited width of the strip (up to about 50 mm) but attempts are being made to increase this and to generally increase the versatility of various techniques. Twin-roll quenching is, in principle, rather attractive since it can produce a greater range of thicknesses than other techniques and, indeed, a variant of this is presently being employed for continuous casting of large sections though, of course, not at high cooling rates. It appears that several techniques hold considerable promise for commercial exploitation for rapid, continuous quenching.

Turning now to amorphous alloys produced by rapid quenching from the melt, Allied Chemical Corporation in the US are producing high quality, narrow glassy strip known as Metglas in various alloys based on iron, nickel or titanium. These alloys have remarkable combinations of properties. Since there are no mobile dislocations present in the amorphous structure their mechanical strengths approach the theoretical limit, combined with significant ductility and fracture toughness. Those based on iron, nickel, and cobalt have particularly attractive magnetic properties, comparable with Permalloys, but produced by a substantially cheaper process route. Several applications are envisaged and flexible sheet, woven from Metglas ribbon, is available for magnetic screening. There are some of the opinion that if the width of iron–boron glassy alloy strip could be increased to about 150 mm, which, in principle, is possible, it would become a competitive material for distribution transformers, for which there is, of course, a substantial market. Application in larger transformers is, I believe, rather more speculative.

Some of the glassy alloys that contain both chromium and phosphorus have substantially higher corrosion resistance than conventional stainless steels in a wide range of environments. Full utilization of this property may be difficult since there are limitations on the processing possibilities because of the inherent instability of the glassy phase. One application that might be envisaged, for instance, is for stainless razor blades. The laser glazing process described by Dr Kear might be utilised for rendering a corrosion resistant glassy surface layer on large components, perhaps after surface impregnation with an appropriate 'glass-forming' element.

In contrast to non-metallic fibres, metallic glasses have isotropic mechanical properties and applications as reinforcing of resin matrices are currently being experimentally studied.

In summary, there is a wide spectrum of possible technological uses for rapidly quenched alloys and for the processes used to produce them and many organisations are either actively involved in research and development in this area or are considering becoming so.

PROFESSOR D. W. KURZ (Ecole Polytechnique Federale de Lausanne): Professor Hellawell, the surface quality of your Na-treated alloy strip was impressive. As the surface quality in thin, cast products—cooled from one side—is always poor, this observation might be a practically important one. Could you comment a little more about this?

PROFESSOR HELLAWELL: We do not really understand why sodium is effective in smoothing the upper surfaces of the solidified sheet. The microstructures—grain sizes—of modified and unmodified sheets are the same and we suppose that sodium must be exerting an influence because it reduces the surface tension of the liquid.

I suppose that surface relief occurs from interdendritic and intergranular shrinkage as solid grows through the liquid surface, but I can't understand how or why this should differ in the two cases. Incidentally, in Al–Si alloys, there is the same smoothing of the surface but in this case the eutectic microstructure is also modified.

MR J. HERBERTSON (BSC, Sheffield Laboratories): Professor Hellawell and co-authors attributed the strip thickness in their process mainly to surface tension/wettability considerations. The thickness below which liquid will not spread because of surface tension constraints is given by $2\gamma_L(1 - \cos\theta)/\rho g$ where γ_L is the melt surface tension, θ is the contact angle, ρ is the melt density, and g is the gravitational constant. The samples of aluminium alloy strips mainly showed contact angles greater than 90°. For a surface tension of say 900 dynes/cm, this suggests a spreading thickness of about 1 cm. Thus the 3–5 mm strip thicknesses suggest the metal is smeared onto the substrate rather than spread across it. This explains the absence of sideways spreading in the process. Also, breaks in the liquid sheet are not likely to recoalesce and an example of this type of defect is evident on one of the samples. I would imagine that the characteristics of the metal delivery system control the product thickness.

PROFESSOR HELLAWELL: Certainly, surface tension is not the only variable. The liquid running into the gate is effectively sheared between the upper and lower surfaces and subject to the belt speed, superheat and bed temperature, begins to freeze some 10 cm to 1 m beyond it. Thus, the liquid sheet is effectively stretched between the gate and freezing front. If the belt speed is too slow or the feed rate too high then the liquid just wells out on to the substrate cloth and gives a layer as thick as that which you suggest and which we observe for accidentally isolated globules.

DR V. KONDIC (University of Birmingham): With regard to the paper by Dr Lemkey, the stress rupture life of normally cast, high temperature alloys is frequently subject to side scatter, the causes of which cannot be readily eliminated, particularly under production conditions. Would Dr Lemkey be good enough to comment on the scatter in stress rupture life of directionally solidified cast alloys, produced and tested as given in his paper. How many test pieces are tested for each test condition?

DR LEMKEY: To those who fly in aircraft, the number of data for a research programme is of very great concern. The points on the slide were for individual tests. A maximum of only five tests at different stresses and temperatures were performed for each composition, because the composition was varied greatly. The test specimens are cylindrical and are made in a crystal growth apparatus; they are quite separate from specimens that we would use in engineering tests and for screening tests on a research programme. However, for a given alloy composition, as for $\gamma–\gamma'–\alpha$, we do a full 'Larsen–Miller parameter v. stress' plot. In the paper (Ref. 1 of our paper) there is such a plot which shows the degree of scatter, which is very small in the ductile–ductile system. We get very good reproducibility.

The biggest problem influencing reproducibility is structure. If one develops cellular structure in these alloys, in the test gauge length, this will give scatter in the properties. So, when we are making engine sets of eutectics, this has to be done in perfect growth conditions so that there are no bands in the sections. It is costly and we have to face the problem. The French, I believe, go to the extent of electrochemically machining turbine blades.

DR N. P. FITZPATRICK (Aluminium Company of Canada Ltd): I was interested in the paper on the method of continuous casting and thought it might be worth making a comment about continuous casting. The ratios of capital invested in the processes of making aluminium sheet indicate that the capital cost of the caster is of the order of 1% of the capital cost of making the aluminium from bauxite. We are not looking for more economic methods of casting but rather for superior and more flexible ones. The casting method proposed by Hellawell's group is an excellent means of investigating the process variables of other strip casters.

DR E. SELCUK (Middle East Technical University, Turkey): I have a question on the paper by Widdowson. I would like to know whether Mr Widdowson has observed any morphological changes of inclusions as a result of stirring. I assume one would expect some emulsification of the sulphide inclusions and some other morphological changes for oxide inclusions.

MR WIDDOWSON: We have noted changes in both oxides and sulphides. In the former case there appears to be greater agglomeration of individual alumina inclusions, although the total density of agglomerates is reduced by the process. We have also noted a reduction in sulphide inclusion density in the stirred zone with a reduction in size and reduction in the properties of type II sulphides.

PROFESSOR J. NUTTING (Leeds University) *Concluding remarks and closing of conference*: I gather from comments made earlier that I was to be asked to review almost in a technical manner the contents of this meeting. To say that I know little about the topics that we have been discussing these last few days would be wrong, because for the past 17 years I have almost had to live daily with Peter Beeley extolling the virtues of cast products so that if I have gained any knowledge of casting and foundry processes, it has not been possibly the way most of you have gained it, by reading learned papers or actually doing experiments. I have gained it by osmosis. It has just had to sink in through the skin, a continuous nagging by Peter telling me what a miserable person I was to confine my activities to wrought material.

In making inquiries and listening to some of the papers and comments that various people have made to me, on one side there has been a feeling perhaps of disappointment in that there were so many academics here and that much of the material was of an academic nature.

After hearing these comments, I then tried to analyse what might be the problem. It immediately set me thinking of the changes that could well have taken place, say over the last 30 years. If a meeting of this type had been held in Sheffield 30 years ago, it would have been a very different type of meeting from the one we have had. It would have taken us back to 1947 and at that time there would still have been the remnants of the group that worked in Sheffield on the heterogeneity of steel ingots and produced the rather amazing reports which formed one of the really serious parts of the activities of the Iron and Steel Institute in the immediate pre-war period. The gentlemen who would have gathered here would have been rather elderly and would have been joined by a few other elderly men puffing pipes and talking in simple terms of our problems with feeder heads. One thing that would have been absolutely certain is that there would have been no academics and even more certain, there would have been no women.

If we move on and come to 20 years ago, one would perhaps have become aware of the first stirrings or reawakening interest in the fundamentals of solidification and casting; the work of the Canadian group of Chalmers, Weingarten and Tiller would just about have been beginning to be felt, and if a meeting on this topic had been held, then again there would have been very few academics.

We could move on perhaps to 10 years ago, and this was probably the reason for working in decades, in that 10 years ago there was the Brighton Conference, and I think the Brighton Conference could be very adequately summarized by the statement that Dr Kondic made at the Conference when, with a certain irony, I will not say bitterness, he said that the physical metallurgists had now discovered metals are formed by the process of solidification from a melt. He has been hammering that theme in the subsequent 10 years.

Now we can perhaps summarize the present situation by saying that the academics have discovered that there are practical problems associated with the production of ingots and the casting of metals to shape. Again if we look at the composition of this meeting, what one finds is that if we took the average age it really would be very low—in the sense that I am a young man and most of you are younger! This I find is a rather welcome change. Points on youth, as it were, were made earlier and the interesting thing that comes about from this is that if one looks at the criticism that the Conference is full of academics, they are young academics. However, there is one thing that is quite certain in the present financial climate in this country, and I think in most other countries, that although they may be young academics, they will not be young academics for much longer. They will very quickly be having to move out into industry because, certainly, there is not the money available to keep them working in an academic environment.

If one then takes that point into account, one cannot help but be pleased because, in the industrial hierarchy, we are going to find young men and, very pleasingly from this meeting, we are obviously going to find young women, who are working in various aspects of the metal-founding industry and who are acquainted with the developments that have gone on—the very extensive developments—during the last 10 years. I feel that this will be of very great benefit.

One has perhaps lived through a somewhat similar situation where, in the late forties and early fifties, one then got the development of dislocation theory, the whole new approach to studying phase transformations in the solid state. Now one finds that the people who were involved in that sort of development are very largely working in industry and the concepts that were developed then, which the older and greyer beards said were a load of nonsense, are now a part of everyday thinking. And I am quite sure, 10 years from now, the things that we have been discussing at this meeting will be part of everyday life from the foundry and ingot production point of view.

On that basis, I think this meeting is rather significant in the sense that it is a pointer to the way in which I feel that the foundry industry will develop. To some extent, the arguments that Dr Kondic has been making over the years will, I feel, all be met but not quite in the way that he anticipated.

Author index

*Refers to Discussion

559

Subject index

*Refers to discussion